PROCESS HEAT TRANSFER

DEDICATION

This book is dedicated to C.C.S.

The authors would also like to dedicate this edition to all the students and engineering professionals who provided thoughtful feedback on the first edition, identifying many corrections and improvements. We would especially like to thank members of HTRI who have shared their expertise and experience to further the development of thermal analysis methods for process heat exchangers.

PROCESS HEAT TRANSFER

PRINCIPLES, APPLICATIONS AND RULES OF THUMB

Robert W. Serth

Department of Chemical and Natural Gas Engineering,
Texas A&M University-Kingsville, Kingsville, Texas, USA

Thomas G. Lestina

Vice President, Research & Engineering Services,
Heat Transfer Research, Inc.,
College Station, Texas, USA

AMSTERDAM • BOSTON • HEIDELBERG • LONDON • NEW YORK • OXFORD
PARIS • SAN DIEGO • SAN FRANCISCO • SINGAPORE • SYDNEY • TOKYO
Academic Press is an imprint of Elsevier

Academic Press is an imprint of Elsevier
The Boulevard, Langford Lane, Kidlington, Oxford OX5 1GB, UK
225 Wyman Street, Waltham, MA 02451, USA
Radarweg 29, PO Box 211, 1000 AE Amsterdam, The Netherlands
525 B Street, Suite 1800, San Diego, CA 92101-4495, USA

First edition 2007
Second edition 2014

British Library Cataloguing in Publication Data
A catalogue record for this book is available from the British Library

Library of Congress Catalog Number
A catalog record for this book is available from the Library of Congress

ISBN 978-0-12-397195-1

For information on all Academic Press publications
visit our website at http://store.elsevier.com/

Typeset by TNQ Books and Journals Pvt Ltd.
www.tnq.co.in

Printed and bound in the United States of America

14 15 16 17 18 10 9 8 7 6 5 4 3 2 1

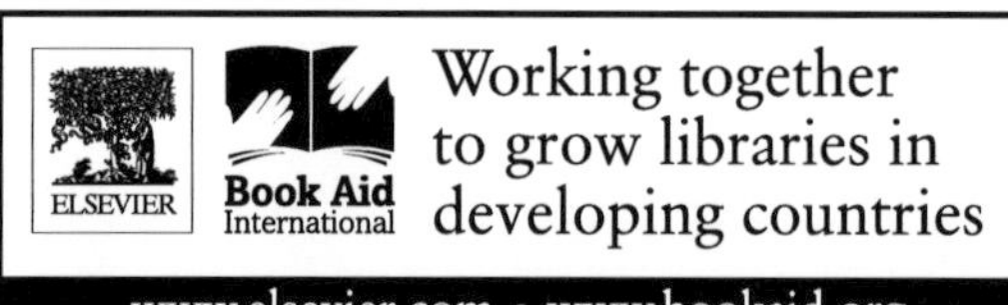

CONTENTS

PREFACE TO FIRST EDITION

This book is based on a course in process heat transfer that I have taught for many years. The course has been taken by seniors and first-year graduate students who have completed an introductory course in engineering heat transfer. Although this background is assumed, nearly all students need some review before proceeding to more advanced material. For this reason, and also to make the book self-contained, the first three chapters provide a review of essential material normally covered in an introductory heat transfer course. Furthermore, the book is intended for use by practicing engineers as well as university students, and it has been written with the aim of facilitating self-study.

Unlike some books in this field, no attempt is made herein to cover the entire panoply of heat transfer equipment. Instead, the book focuses on the types of equipment most widely used in the chemical process industries, namely, shell-and-tube heat exchangers (including condensers and reboilers), air-cooled heat exchangers and double-pipe (hairpin) heat exchangers. Within the confines of a single volume, this approach allows an in-depth treatment of the material that is most relevant from an industrial perspective, and provides students with the detailed knowledge needed for engineering practice. This approach is also consistent with the time available in a one-semester course.

Design of double-pipe exchangers is presented in Chapter 4. Chapters 5 to 7 comprise a unit dealing with shell-and-tube exchangers in operations involving single-phase fluids. Design of shell-and-tube exchangers is covered in Chapter 5 using the Simplified Delaware method for shell-side calculations. For pedagogical reasons, more sophisticated methods for performing shell-side heat-transfer and pressure-drop calculations are presented separately in Chapter 6 (full Delaware method) and Chapter 7 (Stream Analysis method). Heat exchanger networks are covered in Chapter 8. I normally present this topic at this point in the course to provide a change of pace. However, Chapter 8 is essentially self-contained and can, therefore, be covered at any time. Phase-change operations are covered in Chapters 9 to 11. Chapter 9 presents the basics of boiling heat transfer and two-phase flow. The latter is encountered in both Chapter 10, which deals with the design of reboilers, and Chapter 11, which covers condensation and condenser design. Design of air-cooled heat exchangers is presented in Chapter 12. The material in this chapter is essentially self-contained and, hence, it can be covered at any time.

Since the primary goal of both the book and the course is to provide students with the knowledge and skills needed for modern industrial practice, computer applications play an integral role, and the book is intended for use with one or more commercial software packages. HEXTRAN (SimSci-Esscor), HTRI *Xchanger Suite* (Heat Transfer Research, Inc.) and the HTFS Suite (Aspen Technology, Inc.) are used in the book, along with HX-Net (Aspen Technology, Inc.) for pinch calculations. HEXTRAN affords the most complete coverage of topics, as it handles all types of heat exchangers and also performs pinch calculations for design of heat exchanger networks. It does not perform mechanical design calculations for shell-and-tube exchangers, however, nor does it generate detailed tube layouts or setting plans. Furthermore, the methodology used by HEXTRAN is based on publicly available technology and is generally less refined than that of the other software packages. The HTRI and HTFS packages use proprietary methods developed by their respective research organizations, and are similar in their level of refinement. HTFS Suite handles all types of heat exchangers; it also performs mechanical design calculations and develops detailed tube layouts and setting plans for shell-and-tube exchangers. HTRI Xchanger Suite lacks a mechanical design feature, and the module for hairpin exchangers is not included with an academic license. Neither HTRI nor HTFS has the capability to perform pinch calculations.

As of this writing, Aspen Technology is not providing the TASC and ACOL modules of the HTFS Suite under its university program. Instead, it is offering the HTFS-plus design package. This package basically consists of the TASC and ACOL computational engines combined with slightly modified GUI's from the corresponding BJAC programs (HETRAN and AEROTRAN), and packaged with the BJAC TEAMS mechanical design program. This package differs greatly in appearance and to some extent in available features from HTFS Suite. However, most of the results presented in the text using TASC and ACOL can be generated using the HTFS-plus package.

Software companies are continually modifying their products, making differences between the text and current versions of the software packages unavoidable. However, many modifications involve only superficial changes in format that have little, if any, effect on results. More substantive changes occur less frequently, and even then the effects tend to be relatively minor. Nevertheless, readers should expect some divergence of the software from the versions used herein, and they should not be unduly concerned if their results differ somewhat from those presented in the text. Indeed, even the same version of a code, when run on different machines, can produce slightly different results due to differences in round-off errors. With these caveats, it is hoped that the detailed computer examples will prove helpful in learning to use the software packages, as well as in understanding their idiosyncrasies and limitations.

I have made a concerted effort to introduce the complexities of the subject matter gradually throughout the book in order to avoid overwhelming the reader with a massive amount of detail at any one time. As a result, information on shell-and-tube exchangers is spread over a number of chapters, and some of the finer details are introduced in the context of example problems, including computer examples. Although there is an obvious downside to this strategy, I nevertheless believe that it represents good pedagogy.

Both English units, which are still widely used by American industry, and SI units are used in this book. Students in the United States need to be proficient in both sets of units, and the same is true of students in countries that do a large amount of business with U.S. firms. In order to minimize the need for unit conversion, however, working equations are either given in dimensionless form or, when this is not practical, they are given in both sets of units.

I would like to take this opportunity to thank the many students who have contributed to this effort over the years, both directly and indirectly through their participation in my course. I would also like to express my deep appreciation to my colleagues in the Department of Chemical and Natural Gas Engineering at TAMUK, Dr. Ali Pilehvari and Mrs. Wanda Pounds. Without their help, encouragement, and friendship, this book would not have been written.

PREFACE TO SECOND EDITION

This edition provides an improved discussion of practical industry considerations in the design and operation of process heat exchangers. Many of the revisions are based on recent advances of HTRI research, plus the ongoing feedback of process engineers and heat-exchanger designers via the HTRI Technical Support Group. A number of new examples have also been included that illustrate additional aspects of heat-exchanger design. The following is a summary of the most important additions:

- A section on radiation has been added in Chapter 2 to provide more complete coverage of heat transfer fundamentals. Material on mixed convection and non-ideal heat transfer from fins has also been added in this chapter.
- A section on plate heat exchangers has been added in Chapter 3 to provide more complete coverage of the most widely used types of industrial heat exchangers. New material on baffles and shell selection criteria for shell-and-tube exchangers has also been added.
- Two examples involving the design of a multi-tube hairpin exchanger have been added in Chapter 4.
- An example illustrating the significance of temperature profiles in the design of shell-and-tube heat exchangers has been added in Chapter 5.
- A section covering shell-side design guidelines based on stream analysis has been added in Chapter 7, and Example 7.4 has been re-worked on this basis. Two new computer examples have also been added.
- In Chapter 8, a case study has been added that illustrates the application of Pinch Analysis to a real-world process, the production of gasoline from bio-ethanol.
- A new example has been added in Chapter 10 that illustrates some of the unintended consequences that can arise from the use of fouling factors in reboiler design.
- In Chapter 11, a section on condenser venting, draining, and subcooling has been added. A new example comparing different baffle configurations and different types of tubing for condensing applications has also been added, and Example 11.10 has been completely re-worked.
- An example involving the design of an air-cooled condenser has been added in Chapter 12.

All material pertaining to HTRI *Xchanger Suite* has been updated to Version 7.0, the most recent release at the time of writing. We have also corrected numerous errors that unfortunately escaped the review process for the first edition. It is our sincere hope that these revisions will enhance the utility of the book for both students and practitioners of the subject.

CONVERSION FACTORS

Acceleration	$1\ m/s^2 = 4.2520 \times 10^7\ ft/h^2$
Area	$1\ m^2 = 10.764\ ft^2$
Density	$1\ kg/m^3 = 0.06243\ lbm/ft^3$
Energy	$1\ J = 0.239\ cal$ $= 9.4787 \times 10^{-4}\ Btu$
Force	$1\ N = 0.22481\ lbf$
Fouling factor	$1\ m^2 \cdot K/W = 5.6779\ h \cdot ft^2 \cdot °F/Btu$
Heat capacity flow rate	$1\ kW/K = 1\ kW/°C$ $= 1895.6\ Btu/h \cdot °F$
Heat flux	$1\ W/m^2 = 0.3171\ Btu/h \cdot ft^2$
Heat generation rate	$1\ W/m^3 = 0.09665\ Btu/h \cdot ft^3$
Heat transfer coefficient	$1\ W/m^2 \cdot K = 0.17612\ Btu/h \cdot ft^2 \cdot °F$
Heat transfer rate	$1\ W = 3.4123\ Btu/h$
Kinematic viscosity and thermal diffusivity	$1\ m^2/s = 3.875 \times 10^4\ ft^2/h$
Latent heat and specific enthalpy	$1\ kJ/kg = 0.42995\ Btu/lbm$
Length	$1\ m = 3.2808\ ft$
Mass	$1\ kg = 2.2046\ lbm$
Mass flow rate	$1\ kg/s = 7936.6\ lbm/h$
Mass flux	$1\ kg/s \cdot m^2 = 737.35\ lbm/h \cdot ft^2$
Power	$1\ kW = 3412\ Btu/h$ $= 1.341\ hp$
Pressure (stress)	$1\ Pa\ (1\ N/m^2) = 0.020886\ lbf/ft^2$ $= 1.4504 \times 10^{-4}\ psi$ $= 4.015 \times 10^{-3}\ in.\ H_2O$
Pressure	$1.01325 \times 10^5\ Pa = 1\ atm$ $= 14.696\ psi$ $= 760\ torr$ $= 406.8\ in.\ H_2O$
Specific heat	$1\ kJ/kg \cdot K = 0.2389\ Btu/lbm \cdot °F$
Surface tension	$1\ N/m = 1000\ dyne/cm$ $= 0.068523\ lbf/ft$
Temperature	$K = °C + 273.15$ $= (5/9)\ (°F + 459.67)$ $= (5/9)\ (°R)$
Temperature difference	$1\ K = 1°C$ $= 1.8°F$ $= 1.8°R$
Thermal conductivity	$1\ W/m \cdot K = 0.57782\ Btu/h \cdot ft \cdot °F$

Thermal resistance	$1\ \text{K/W} = 0.52750°\text{F} \cdot \text{h/Btu}$
Viscosity	$1\ \text{kg/m} \cdot \text{s} = 1000\ \text{cp}$
	$= 2419\ \text{lbm/ft} \cdot \text{h}$
Volume	$1\ \text{m}^3 = 35.314\ \text{ft}^3$
	$= 264.17\ \text{gal}$
Volumetric flow rate	$1\ \text{m}^3/\text{s} = 2118.9\ \text{ft}^3/\text{min(cfm)}$
	$= 1.5850 \times 10^4\ \text{gal/min (gpm)}$

lbf: pound force and lbm: pound mass.

PHYSICAL CONSTANTS

Quantity	Symbol	Value
Universal gas constant	$\tilde{R}$	0.08205 atm · m^3/kmol · K 0.08314 bar · m^3/kmol · K 8314 J/kmol · K 1.986 cal/mol · K 1.986 Btu/lb mole · °R 10.73 psia· ft^3/lb mole · °R 1545 ft · lbf/lb mole · °R
Standard gravitational acceleration	g	9.8067 m/s^2 32.174 ft/s^2 4.1698×10^8 ft/h^2
Stefan–Boltzman constant	σ_{SB}	5.670×10^{-8} $W/m^2 \cdot K^4$ 1.714×10^{-9} Btu/h· ft^2 · $°R^4$

ACKNOWLEDGMENTS

Item	Special Credit Line
Figure 3.1	Reprinted, with permission, from *Extended Surface Heat Transfer* by D. Q. Kern and A. D. Kraus. Copyright © 1972 by The McGraw-Hill Companies, Inc.
Table 3.1	Reprinted, with permission, from *Perry's Chemical Engineers' Handbook*, 7th edn., R. H. Perry and D. W. Green, eds. Copyright © 1997 by The McGraw-Hill Companies, Inc.
Figure 3.6	Reprinted, with permission, from *Extended Surface Heat Transfer* by D. Q. Kern and A. D. Kraus. Copyright © 1972 by The McGraw-Hill Companies, Inc.
Table 3.2	Reproduced, with permission, from J. W. Palen and J. Taborek, Solution of shell side flow pressure drop and heat transfer by stream analysis method, *Chem. Eng. Prog. Symposium Series*, 65, No. 92, 53–63, 1969. Copyright © 1969 by AIChE.
Table 3.5	Reprinted, with permission, from *Perry's Chemical Engineers' Handbook*, 7th edn., R. H. Perry and D. W. Green, eds. Copyright © 1997 by The McGraw-Hill Companies, Inc.
Figure 4.1	Copyright © 1998 from *Heat Exchangers: Selection, Rating and Thermal Design* by S. Kakac and H. Liu. Reproduced by permission of Taylor & Francis, a division of Informa plc.
Figure 4.2	Copyright © 1998 from *Heat Exchangers: Selection, Rating and Thermal Design* by S. Kakac and H. Liu. Reproduced by permission of Taylor & Francis, a division of Informa plc.
Figure 4.4	Reprinted, with permission, from *Extended Surface Heat Transfer* by D. Q. Kern and A. D. Kraus. Copyright © 1972 by The McGraw-Hill Companies, Inc.
Figure 4.5	Reprinted, with permission, from *Extended Surface Heat Transfer* by D. Q. Kern and A. D. Kraus. Copyright © 1972 by The McGraw-Hill Companies, Inc.
Figure 5.3	Copyright © 1988 from *Heat Exchanger Design Handbook* by E. U. Schlünder, Editor-in-Chief. Reproduced by permission of Taylor & Francis, a division of Informa plc.
Figures 6.1–6.5	Copyright © 1988 from *Heat Exchanger Design Handbook* by E. U. Schlunder, Editor-in-Chief. Reproduced by permission of Taylor & Francis, a division of Informa pic.
Table 6.1	Copyright © 1988 from *Heat Exchanger Design Handbook* by E. U. Schlünder, Editor-in-Chief. Reproduced by permission of Taylor & Francis, a division of Informa plc.
Figure 6.10	Copyright © 1988 from *Heat Exchanger Design Handbook* by E. U. Schlünder, Editor-in-Chief. Reproduced by permission of Taylor & Francis, a division of Informa plc.
Figure 7.1	Reproduced, with permission, from J. W. Palen and J. Taborek, Solution of shell side flow pressure drop and heat transfer by stream analysis method, *Chem. Eng. Prog. Symposium Series*, 65, No. 92, 53–63, 1969. Copyright © 1969 by AIChE.
Table, p. 283	Reproduced, with permission, from R. Mukherjee, Effectively design shell-and-tube heat exchangers, *Chem. Eng. Prog.*, 94, No. 2, 21–37, 1998. Copyright © 1998 by AIChE.
Figure 8.20	Reprinted from *Computers and Chemical Engineering*, Vol. 26, X. X. Zhu and X. R. Nie, Pressure Drop Considerations for Heat Exchanger Network Grassroots Design, pp. 1661–1676, Copyright © 2002, with permission from Elsevier.
Figure 9.2	Copyright © 1997 from *Boiling Heat Transfer and Two-Phase Flow*, 2nd edn., by L. S. Tong and Y. S. Tang. Reproduced by permission of Taylor & Francis, a division of Informa plc.

(*Continued*)

—cont'd

Item	Special Credit Line
Figures 10.1–10.5	Copyright © 1988 from *Heat Exchanger Design Handbook by* E. U. Schliinder, Editor-in-Chief. Reproduced by permission of Taylor & Francis, a division of Informa plc.
Figure 10.6	Reproduced, with permission, from A. W. Sloley, Properly design thermosyphon reboilers, *Chem. Eng Prog.* 93, No. 3, 52–64, 1997. Copyright © 1997 by AIChE.
Table 10.1	Copyright © 1988 from *Heat Exchanger Design Handbook* by E. U. Schlünder, Editor-in-Chief. Reproduced by permission of Taylor & Francis, a division of Informa plc.
Appendix 10.A	Reprinted, with permission, from *Chemical Engineers' Handbook*, 5th edn., R. H. Perry and C. H. Chilton, eds. Copyright © 1973 by The McGraw-Hill Companies, Inc.
Figure 11.1	Copyright © 1988 from *Heat Exchanger Design Handbook* by E. U. Schlünder, Editor-in-Chief. Reproduced by permission of Taylor & Francis, a division of Informa plc.
Figure 11.3	Copyright © 1998 from *Heat Exchangers: Selection, Rating and Thermal Design* by S. Kakac and H. Liu. Reproduced by permission of Taylor & Francis, a division of Informa plc.
Figure 11.6	Copyright © 1988 from *Heat Exchanger Design Handbook* by E. U. Schlünder, Editor-in-Chief. Reproduced by permission of Taylor & Francis, a division of Informa plc.
Figure 11.7	Copyright © 1988 from *Heat Exchanger Design Handbook* by E. U. Schlünder, Editor-in-Chief. Reproduced by permission of Taylor & Francis, a division of Informa plc.
Figure 11.8	Reprinted, with permission, from *Distillation Operation* by H. Z. Kister. Copyright © 1990 by The McGraw-Hill Companies, Inc.
Figure 11.14	Reprinted, with permission, from G. Breber, J. W. Palen and J. Taborek, Prediction of tubeside condensation of pure components using flow regime criteria, *J. Heat Transfer*, 102, 471–476, 1980. Originally published by ASME.
Figure 11.15	Copyright © 1998 from *Heat Exchangers: Selection, Rating and Thermal Design* by S. Kakac and H. Liu. Reproduced by permission of Taylor & Francis, a division of Informa plc.
Figures 11.A1–11.A3	Copyright © 1988 from *Heat Exchanger Design Handbook* by E. U. Schlünder, Editor-in-Chief. Reproduced by permission of Taylor & Francis, a division of Informa plc.
Figure 12.5	Copyright © 1991 from *Heat Transfer Design Methods* by J. J. McKetta, Editor. Reproduced by permission of Taylor & Francis, a division of Informa plc.
Figures 12.A1–12.A5	Copyright © 1988 from *Heat Exchanger Design Handbook* by E. U. Schlünder, Editor-in-Chief. Reproduced by permission of Taylor & Francis, a division of Informa plc.
Table A. 1	Copyright © 1972 from *Handbook of Thermodynamic Tables and Charts* by K. Raznjevič. Reproduced by permission of Taylor & Francis, a division of Informa plc.
Table A.3	Reprinted, with permission, from *Heat Transfer*, 7th edn., by J. P. Holman. Copyright © 1990 by The McGraw-Hill Companies, Inc.
Table A.4	Copyright © 1972 from *Handbook of Thermodynamic Tables and Charts* by K. Raznjevič. Reproduced by permission of Taylor & Francis, a division of Informa plc.
Table A.7	Copyright © 1972 from *Handbook of Thermodynamic Tables and Charts* by K. Raznjevič. Reproduced by permission of Taylor & Francis, a division of Informa plc.
Table A.8	Reprinted, with permission, from *ASME Steam Tables*, American Society of Mechanical Engineers, New York, 1967. Originally published by ASME.
Table A.9	Reprinted, with permission, from *Flow of Fluids Through Valves, Fittings and Pipe*, Technical Paper 410, 1988, Crane Company. All rights reserved.
Table A.11	Copyright © 1975 from *Tables of Thermophysical Properties of Liquids and Gases*, 2nd edn., by N. B. Vargaftik. Reproduced by permission of Taylor & Francis, a division of Informa plc.
Table A.13	Copyright © 1972 from *Handbook of Thermodynamic Tables and Charts* by K. Raznjevič. Reproduced by permission of Taylor & Francis, a division of Informa plc.
Table A.15	Reprinted, with permission, from *Chemical Engineers' Handbook*, 5th edn., R. H. Perry and C. H. Chilton, eds. Copyright © 1973 by The McGraw-Hill Companies, Inc.

(*Continued*)

Item	Special Credit Line
Table A.17	Reprinted, with permission, from *Chemical Engineers' Handbook,* 5th edn., R. H. Perry and C. H. Chilton, eds. Copyright © 1973 by The McGraw-Hill Companies, Inc.
Figure A.1	Reprinted, with permission, from *Chemical Engineers' Handbook,* 5th edn., R. H. Perry and C. H. Chilton, eds. Copyright © 1973 by The McGraw-Hill Companies, Inc.
Table A.18	Reprinted, with permission, from *Chemical Engineers' Handbook,* 5th edn., R. H. Perry and C. H. Chilton, eds. Copyright © 1973 by The McGraw-Hill Companies, Inc.
Figure A.2	Reprinted, with permission, from *Chemical Engineers' Handbook,* 5th edn., R. H. Perry and C. H. Chilton, eds. Copyright © 1973 by The McGraw-Hill Companies, Inc.

1 Heat Conduction

1.1 Introduction

Heat conduction is one of the three basic modes of thermal energy transport (convection and radiation being the other two) and is involved in virtually all process heat-transfer operations. In commercial heat exchange equipment, for example, heat is conducted through a solid wall (often a tube wall) that separates two fluids having different temperatures. Furthermore, the concept of thermal resistance, which follows from the fundamental equations of heat conduction, is widely used in the analysis of problems arising in the design and operation of industrial equipment. In addition, many routine process engineering problems can be solved with acceptable accuracy using simple solutions of the heat conduction equation for rectangular, cylindrical, and spherical geometries.

This chapter provides an introduction to the macroscopic theory of heat conduction and its engineering applications. The key concept of thermal resistance, used throughout the text, is developed here, and its utility in analyzing and solving problems of practical interest is illustrated.

1.2 Fourier's Law of Heat Conduction

The mathematical theory of heat conduction was developed early in the nineteenth century by Joseph Fourier [1]. The theory was based on the results of experiments similar to that illustrated in Figure 1.1 in which one side of a rectangular solid is held at temperature T_1, while the opposite side is held at a lower temperature, T_2. The other four sides are insulated so that heat can flow only in the x-direction. For a given material, it is found that the rate, q_x, at which heat (thermal energy) is transferred from the hot side to the cold side is proportional to the cross-sectional area, A, across which the heat flows; the temperature difference, $T_1 - T_2$, and inversely proportional to the thickness, B, of the material. That is:

$$q_x \propto \frac{A(T_1 - T_2)}{B}$$

Writing this relationship as an equality, we have:

$$q_x = \frac{k\,A(T_1 - T_2)}{B} \tag{1.1}$$

The constant of proportionality, k, is called the thermal conductivity. Equation (1.1) is also applicable to heat conduction in liquids and gases. However, when temperature differences exist in fluids, convection currents tend to be set up, so that heat is generally not transferred solely by the mechanism of conduction.

The thermal conductivity is a property of the material and, as such, it is not really a constant, but rather it depends on the thermodynamic state of the material, i.e., on the temperature and pressure of the material. However, for solids, liquids, and low-pressure gases, the pressure dependence is usually negligible. The temperature dependence also tends to be fairly weak, so that it is

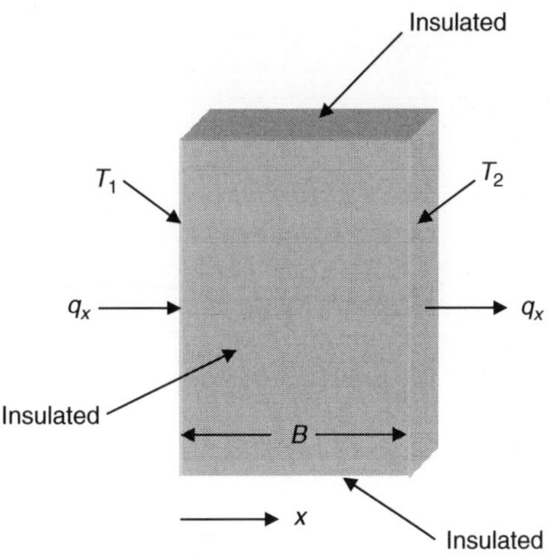

FIGURE 1.1 One-dimensional heat conduction in a solid.

often acceptable to treat k as a constant, particularly if the temperature difference is moderate. When the temperature dependence must be taken into account, a linear function is often adequate, particularly for solids. In this case,

$$k = a + bT \tag{1.2}$$

where a and b are constants.

Thermal conductivities of a number of materials are given in Appendix A. Many other values may be found in various handbooks and compendiums of physical property data. Process simulation software is also an excellent source of physical property data (see, e.g., Appendix E). Methods for estimating thermal conductivities of fluids when data are unavailable can be found in the authoritative book by Poling et al. [2].

The form of Fourier's law given by Equation (1.1) is valid only when the thermal conductivity can be assumed constant. A more general result can be obtained by writing the equation for an element of differential thickness. Thus, let the thickness be Δx and let $\Delta T = T_2 - T_1$. Substituting in Equation (1.1) gives:

$$q_x = -k\,A\frac{\Delta T}{\Delta x} \tag{1.3}$$

Now in the limit as Δx approaches zero,

$$\frac{\Delta T}{\Delta x} \rightarrow \frac{dT}{dx}$$

and Equation (1.3) becomes:

$$q_x = -k\,A\frac{dT}{dx} \tag{1.4}$$

Equation (1.4) is not subject to the restriction of constant k. Furthermore, when k is constant, it can be integrated to yield Equation (1.1). Hence, Equation (1.4) is the general one-dimensional form of Fourier's law. The negative sign is necessary because heat flows in the positive x-direction when the temperature decreases in the x-direction. Thus, according to the standard sign convention that q_x is positive when the heat flow is in the positive x-direction, q_x must be positive when dT/dx is negative.

It is often convenient to divide Equation (1.4) by the area to give:

$$\hat{q}_x \equiv q_x/A = -k\frac{dT}{dx} \tag{1.5}$$

where $\hat{q}_x$ is the heat flux. It has units of $\text{J/s} \cdot \text{m}^2 = \text{W/m}^2$ or $\text{Btu/h} \cdot \text{ft}^2$. Thus, the units of k are $\text{W/m} \cdot \text{K}$ or $\text{Btu/h} \cdot \text{ft} \cdot {}^\circ\text{F}$.

Equations (1.1), (1.4), and (1.5) are restricted to the situation in which heat flows in the x-direction only. In the general case in which heat flows in all three coordinate directions, the total heat flux is obtained by adding vectorially the fluxes in the coordinate directions. Thus,

$$\vec{\hat{q}} = \hat{q}_x\vec{i} + \hat{q}_y\vec{j} + \hat{q}_z\vec{k} \tag{1.6}$$

where $\vec{\hat{q}}$ is the heat flux vector and $\vec{i}, \vec{j}, \vec{k}$ are unit vectors in the x-, y-, z-directions, respectively. Each of the component fluxes is given by a one-dimensional Fourier expression as follows:

$$\hat{q}_x = -k\frac{\partial T}{\partial x} \quad \hat{q}_y = -k\frac{\partial T}{\partial y} \quad \hat{q}_z = -k\frac{\partial T}{\partial z} \tag{1.7}$$

Partial derivatives are used here since the temperature now varies in all three directions. Substituting the above expressions for the fluxes into Equation (1.6) gives:

$$\vec{\hat{q}} = -k\left(\frac{\partial T}{\partial x}\vec{i} + \frac{\partial T}{\partial y}\vec{j} + \frac{\partial T}{\partial z}\vec{k}\right) \tag{1.8}$$

The vector in parentheses is the temperature gradient vector, and is denoted by $\vec{\nabla}T$. Hence,

$$\vec{\hat{q}} = -k\vec{\nabla}T \tag{1.9}$$

Equation (1.9) is the three-dimensional form of Fourier's law. It is valid for homogeneous, isotropic materials for which the thermal conductivity is the same in all directions.

Equation (1.9) states that the heat flux vector is proportional to the negative of the temperature gradient vector. Since the gradient direction is the direction of (locally) greatest temperature increase, the negative gradient direction is the direction of greatest temperature decrease. Hence, Fourier's law states that heat flows in the direction of greatest temperature decrease.

Example 1.1

The block of 304 stainless steel shown below is well insulated on the front and back surfaces, and the temperature in the block varies linearly in both the x- and y-directions. Find:

(a) The heat fluxes and heat flows in the x- and y-directions.
(b) The magnitude and direction of the heat flux vector.

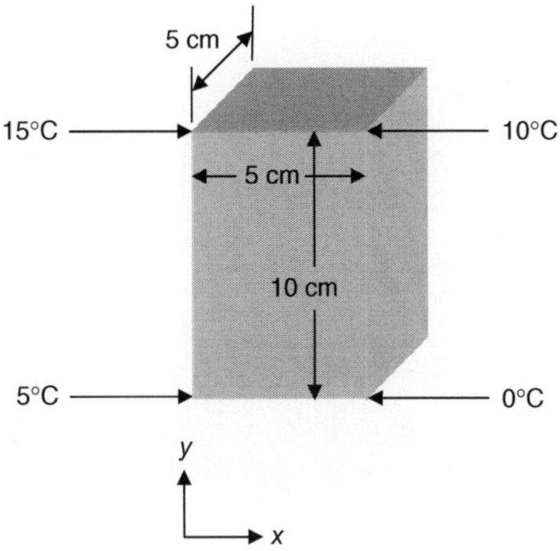

Solution

(a) From Table A.1, the thermal conductivity of 304 stainless steel is 14.4 W/m · K. The cross-sectional areas are:

$$A_x = 10 \times 5 = 50 \text{ cm}^2 = 0.0050 \text{ m}^2$$

$$A_y = 5 \times 5 = 25 \text{ cm}^2 = 0.0025 \text{ m}^2$$

Using Equation (1.7) and replacing the partial derivatives with finite differences (since the temperature variation is linear), the heat fluxes are:

$$\hat{q}_x = -k\frac{\partial T}{\partial x} = -k\frac{\Delta T}{\Delta x} = -14.4\left(\frac{-5}{0.05}\right) = 1440 \text{ W/m}^2$$

$$\hat{q}_y = -k\frac{\partial T}{\partial y} = -k\frac{\Delta T}{\Delta y} = -14.4\left(\frac{10}{0.1}\right) = -1440 \text{ W/m}^2$$

The heat flows are obtained by multiplying the fluxes by the corresponding cross-sectional areas:

$$q_x = \hat{q}_x A_x = 1440 \times 0.005 = 7.2 \text{ W}$$

$$q_y = \hat{q}_y A_y = -1440 \times 0.0025 = -3.6 \text{ W}$$

(b) From Equation (1.6):

$$\vec{q} = \hat{q}_x \vec{i} + \hat{q}_y \vec{j}$$

$$\vec{q} = 1440\,\vec{i} - 1440\,\vec{j}$$

$$|\vec{q}| = \left[(1440)^2 + (-1440)^2\right]^{0.5} = 2036.5 \text{ W/m}^2$$

The angle, θ, between the heat flux vector and the x-axis is calculated as follows:

$$\tan\theta = \hat{q}_y/\hat{q}_x = -1440/1440 = -1.0$$

$$\theta = -45°$$

The direction of the heat flux vector, which is the direction in which heat flows, is indicated in the sketch below.

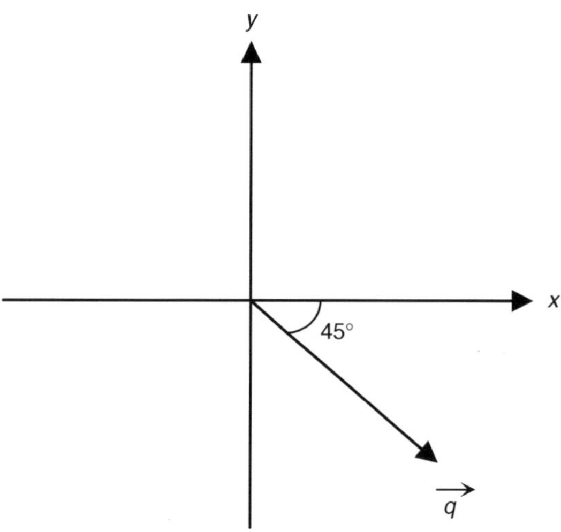

1.3 The Heat Conduction Equation

The solution of problems involving heat conduction in solids can, in principle, be reduced to the solution of a single differential equation, the heat conduction equation. The equation can be derived by making a thermal energy balance on a differential volume element in the solid. For the case of conduction only in the x-direction, such a volume element is illustrated in Figure 1.2. The balance equation for the volume element is:

$$\{\text{rate of thermal energy in}\} - \{\text{rate of thermal energy out}\} + \{\text{net rate of thermal energy generation}\}$$
$$= \{\text{rate of accumulation of thermal energy}\} \quad (1.10)$$

The generation term appears in the equation because the balance is made on thermal energy, not total energy. For example, thermal energy may be generated within a solid by an electric current or by decay of a radioactive material.

The rate at which thermal energy enters the volume element across the face at x is given by the product of the heat flux and the cross-sectional area, $\hat{q}_x|_x A$. Similarly, the rate at which thermal energy leaves the element across the face at $x + \Delta x$ is $\hat{q}_x|_{x+\Delta x} A$. For a homogeneous heat source of strength $\dot{q}$ per unit volume, the net rate of generation is $\dot{q}A\Delta x$. Finally, the rate of accumulation is given by the time derivative of the thermal energy content of the volume element, which is $\rho c(T - T_{ref})\, A\Delta x$, where T_{ref} is an arbitrary reference temperature. Thus, the balance equation becomes:

$$\left(\hat{q}_x|_x - \hat{q}_x|_{x+\Delta x}\right) A + \dot{q}A\Delta x = \rho c \frac{\partial T}{\partial t} A\Delta x$$

It has been assumed here that the density, ρ, and heat capacity, c, are constant. Dividing by $A\Delta x$ and taking the limit as $\Delta x \to 0$ yields:

$$\rho c \frac{\partial T}{\partial t} = -\frac{\partial \hat{q}_x}{\partial x} + \dot{q}$$

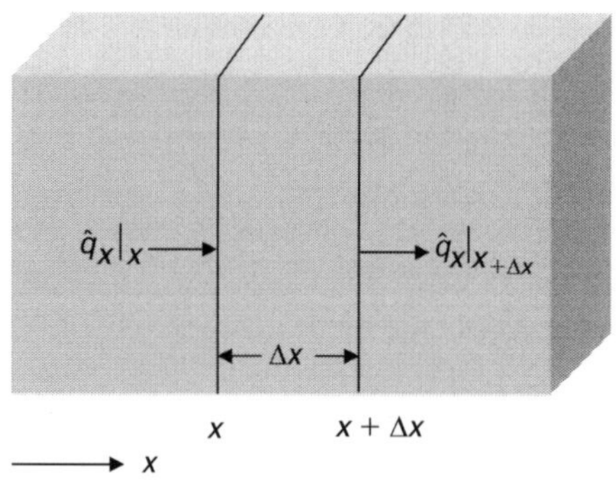

FIGURE 1.2 Differential volume element used in derivation of conduction equation.

Using Fourier's law as given by Equation (1.5), the balance equation becomes:

$$\rho c \frac{\partial T}{\partial t} = \frac{\partial}{\partial x}\left(\frac{k\partial T}{\partial x}\right) + \dot{q}$$

When conduction occurs in all three coordinate directions, the balance equation contains y- and z-derivatives analogous to the x-derivative. The balance equation then becomes:

$$\rho c \frac{\partial T}{\partial t} = \frac{\partial}{\partial x}\left(\frac{k\partial T}{\partial x}\right) + \frac{\partial}{\partial y}\left(\frac{k\partial T}{\partial y}\right) + \frac{\partial}{\partial z}\left(\frac{k\partial T}{\partial z}\right) + \dot{q} \quad (1.11)$$

Equation (1.11) is listed in Table 1.1 along with the corresponding forms that the equation takes in cylindrical and spherical coordinates. Also listed in Table 1.1 are the components of the heat flux vector in the three coordinate systems.

When k is constant, it can be taken outside the derivatives and Equation (1.11) can be written as:

$$\frac{\rho c}{k}\frac{\partial T}{\partial t} = \frac{\partial^2 T}{\partial x^2} + \frac{\partial^2 T}{\partial y^2} + \frac{\partial^2 T}{\partial z^2} + \frac{\dot{q}}{k} \quad (1.12)$$

or

$$\frac{1}{\alpha}\frac{\partial T}{\partial t} = \nabla^2 T + \frac{\dot{q}}{k} \quad (1.13)$$

where $\alpha \equiv k/\rho c$ is the thermal diffusivity and ∇^2 is the Laplacian operator. The thermal diffusivity has units of m^2/s or ft^2/h.

The use of the conduction equation is illustrated in the following examples.

Example 1.2

Apply the conduction equation to the situation illustrated in Figure 1.1.

Solution

In order to make the mathematics conform to the physical situation, the following conditions are imposed:

(1) Conduction only in x-direction $\Rightarrow T = T(x)$, so $\frac{\partial T}{\partial y} = \frac{\partial T}{\partial z} = 0$

(2) No heat source $\Rightarrow \dot{q} = 0$

(3) Steady state $\Rightarrow \frac{\partial T}{\partial t} = 0$

(4) Constant k

The conduction equation in Cartesian coordinates then becomes:

$$0 = k\frac{\partial^2 T}{\partial x^2} \quad \text{or} \quad \frac{d^2 T}{dx^2} = 0$$

(The partial derivative is replaced by a total derivative because x is the only independent variable in the equation.) Integrating on both sides of the equation gives:

$$\frac{dT}{dx} = C_1$$

A second integration gives:

$$T = C_1 x + C_2$$

Thus, it is seen that the temperature varies linearly across the solid. The constants of integration can be found by applying the boundary conditions:

(1) At $x = 0$, $T = T_1$

(2) At $x = B$, $T = T_2$

The first boundary condition gives $T_1 = C_2$ and the second then gives:

$$T_2 = C_1 B + T_1$$

Solving for C_1 we find:

$$C_1 = \frac{T_2 - T_1}{B}$$

The heat flux is obtained from Fourier's law:

$$\hat{q}_x = -k\frac{dT}{dx} = -kC_1 = -k\frac{(T_2 - T_1)}{B} = k\frac{(T_1 - T_2)}{B}$$

TABLE 1.1 The Heat Conduction Equation

A. Cartesian coordinates (x,y,z)

$$\rho c \frac{\partial T}{\partial t} = \frac{\partial}{\partial x}\left(k\frac{\partial T}{\partial x}\right) + \frac{\partial}{\partial y}\left(k\frac{\partial T}{\partial y}\right) + \frac{\partial}{\partial z}\left(k\frac{\partial T}{\partial z}\right) + \dot{q}$$

The components of the heat flux vector, $\vec{\hat{q}}$, are:

$$\hat{q}_x = -k\frac{\partial T}{\partial x};\ \hat{q}_y = -k\frac{\partial T}{\partial y};\ \hat{q}_z = -k\frac{\partial T}{\partial z}$$

B. Cylindrical coordinates (r, ϕ, z)

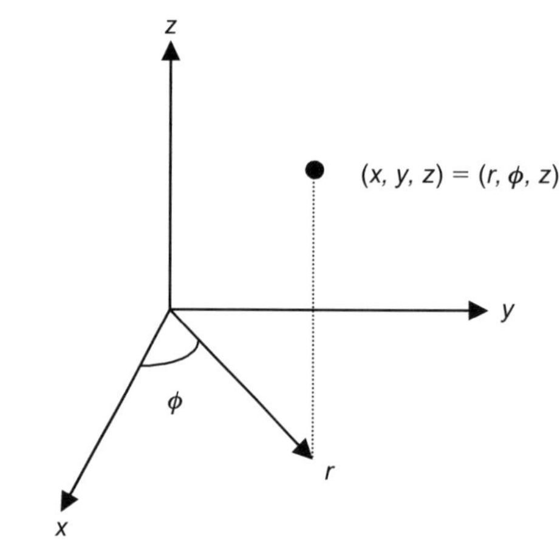

$$\rho c \frac{\partial T}{\partial t} = \frac{1}{r}\frac{\partial}{\partial r}\left(kr\frac{\partial T}{\partial r}\right) + \frac{1}{r^2}\frac{\partial}{\partial \phi}\left(k\frac{\partial T}{\partial \phi}\right) + \frac{\partial}{\partial z}\left(k\frac{\partial T}{\partial z}\right) + \dot{q}$$

The components of $\vec{q}$ are:

$$\hat{q}_r = -k\frac{\partial T}{\partial r};\ \hat{q}_\phi = \frac{-k}{r}\frac{\partial T}{\partial \phi};\ \hat{q}_z = -k\frac{\partial T}{\partial z}$$

C. Spherical coordinates (r, θ, ϕ)

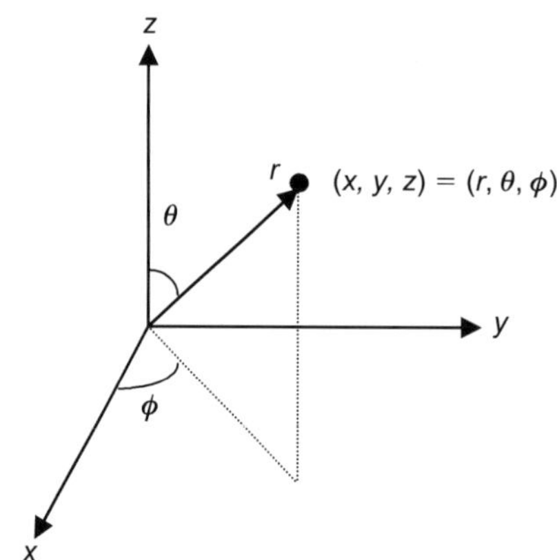

$$\rho c \frac{\partial T}{\partial t} = \frac{1}{r^2}\frac{\partial}{\partial r}\left(kr^2\frac{\partial T}{\partial r}\right) + \frac{1}{r^2 \sin\theta}\frac{\partial}{\partial \theta}\left(k\sin\theta\frac{\partial T}{\partial \theta}\right) + \frac{1}{r^2\sin^2\theta}\frac{\partial}{\partial \phi}\left(k\frac{\partial T}{\partial \phi}\right) + \dot{q}$$

The components of $\vec{q}$ are:

$$\hat{q}_r = -k\frac{\partial T}{\partial r};\ \hat{q}_\theta = -\frac{k}{r}\frac{\partial T}{\partial \theta};\ \hat{q}_\phi = -\frac{k}{r\sin\theta}\frac{\partial T}{\partial \phi}$$

Multiplying by the area gives the heat flow:

$$q_x = \hat{q}_x A = \frac{kA(T_1 - T_2)}{B}$$

Since this is the same as Equation (1.1), we conclude that the mathematics are consistent with the experimental results.

Example 1.3

Apply the conduction equation to the situation illustrated in Figure 1.1, but let $k = a + bT$, where a and b are constants.

Solution

Conditions 1–3 of the previous example are imposed. The conduction equation then becomes:

$$0 = \frac{d}{dx}\left(k\frac{dT}{dx}\right)$$

Integrating once gives:

$$k\frac{dT}{dx} = C_1$$

The variables can now be separated and a second integration performed. Substituting for k, we have:

$$(a + bT)dT = C_1 dx$$

$$aT + \frac{bT^2}{2} = C_1 x + C_2$$

It is seen that in this case of variable k, the temperature profile is not linear across the solid.

The constants of integration can be evaluated by applying the same boundary conditions as in the previous example, although the algebra is a little more tedious. The results are:

$$C_2 = aT_1 + \frac{bT_1^2}{2}$$

$$C_1 = a\frac{(T_2 - T_1)}{B} + \frac{b}{2B}(T_2^2 - T_1^2)$$

As before, the heat flow is found using Fourier's law:

$$q_x = -kA\frac{dT}{dx} = -AC_1$$

$$q_x = \frac{A}{B}\left[a(T_1 - T_2) + \frac{b}{2}(T_1^2 - T_2^2)\right]$$

This equation is seldom used in practice. Instead, when k cannot be assumed constant, Equation (1.1) is used with an average value of k. Thus, taking the arithmetic average of the conductivities at the two sides of the block:

$$k_{ave} = \frac{k(T_1) + k(T_2)}{2}$$

$$= \frac{(a + bT_1) + (a + bT_2)}{2}$$

$$k_{ave} = a + \frac{b}{2}(T_1 + T_2)$$

Using this value of k in Equation (1.1) yields:

$$q_x = \frac{k_{ave}A(T_1 - T_2)}{B}$$

$$= \left[a + \frac{b(T_1 + T_2)}{2}\right]\frac{A}{B}(T_1 - T_2)$$

$$q_x = \frac{A}{B}\left[a(T_1 - T_2) + \frac{b}{2}(T_1^2 - T_2^2)\right]$$

This equation is exactly the same as the one obtained above by solving the conduction equation. Hence, using Equation (1.1) with an average value of k gives the correct result. This is a consequence of the assumed linear relationship between k and T.

Example 1.4

Use the conduction equation to find an expression for the rate of heat transfer for the cylindrical analog of the situation depicted in Figure 1.1.

Solution

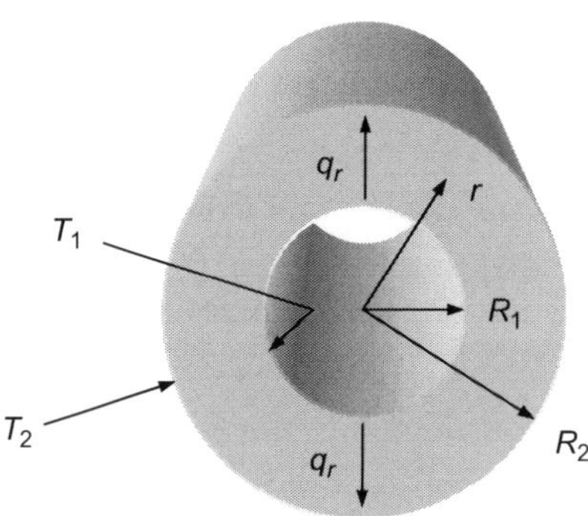

As shown in the sketch, the solid is in the form of a hollow cylinder and the outer and inner surfaces are maintained at temperatures T_1 and T_2, respectively. The ends of the cylinder are insulated so that heat can flow only in the radial direction. There is no heat flow in the angular (ϕ) direction because the temperature is the same all the way around the circumference of the cylinder. The following conditions apply:

(1) No heat flow in z-direction $\Rightarrow \dfrac{\partial T}{\partial z} = 0$

(2) Uniform temperature in ϕ-direction $\Rightarrow \dfrac{\partial T}{\partial \varphi} = 0$

(3) No heat generation $\Rightarrow \dot{q} = 0$

(4) Steady state $\Rightarrow \dfrac{\partial T}{\partial t} = 0$

(5) Constant k

With these conditions, the conduction equation in cylindrical coordinates becomes:

$$\frac{1}{r}\frac{\partial}{\partial r}\left(kr\frac{\partial T}{\partial r}\right) = 0$$

or

$$\frac{d}{dr}\left(r\frac{dT}{dr}\right) = 0$$

Integrating once gives:

$$r\frac{dT}{dr} = C_1$$

Separating variables and integrating again gives:

$$T = C_1 \ln r + C_2$$

It is seen that, even with constant k, the temperature profile in curvilinear systems is nonlinear.

The boundary conditions for this case are:

(1) At $r = R_1, T = T_1 \Rightarrow T_1 = C_1 \ln\, R_1 + C_2$
(2) At $r = R_2, T = T_2 \Rightarrow T_2 = C_1 \ln\, R_2 + C_2$

Solving for C_1 by subtracting the second equation from the first gives:

$$C_1 = \frac{T_1 - T_2}{\ln R_1 - \ln R_2} = -\frac{T_1 - T_2}{\ln(R_2/R_1)}$$

From Table 1.1, the appropriate form of Fourier's law is:

$$\hat{q}_r = -k\frac{dT}{dr} = -k\frac{C_1}{r} = \frac{k(T_1 - T_2)}{r\,\ln(R_2/R_1)}$$

The area across which the heat flows is:

$$A_r = 2\pi rL$$

where L is the length of the cylinder. Thus,

$$q_r = \hat{q}_r A_r = \frac{2\pi kL\,(T_1 - T_2)}{\ln(R_2/R_1)}$$

Notice that the heat-transfer rate is independent of radial position. The heat flux, however, depends on r because the cross-sectional area changes with radial position.

Example 1.5

The block shown in the diagram below is insulated on the top, bottom, front, back, and the side at $x = B$. The side at $x = 0$ is maintained at a fixed temperature, T_1. Heat is generated within the block at a rate per unit volume given by:

$$\dot{q} = \Gamma e^{-\gamma x}$$

where $\Gamma, \gamma > 0$ are constants. Find the maximum steady-state temperature in the block. Data are as follows:

$\Gamma = 10\ \text{W/m}^3$ $\quad B = 1.0\ \text{m}$ $\quad \text{k} = 0.5\ \text{W/m}\cdot\text{K}$ = block thermal conductivity
$\gamma = 0.1\ \text{m}^{-1}$ $\quad T_1 = 20°\text{C}$

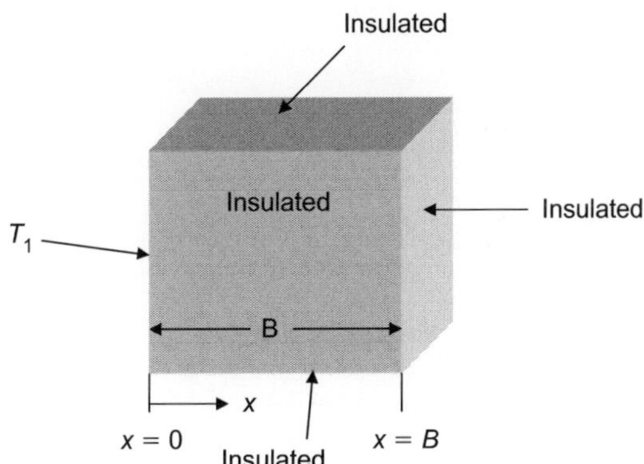

Solution

The first step is to find the temperature profile in the block by solving the heat conduction equation. The applicable conditions are:

- Steady state
- Conduction only in x-direction
- Constant thermal conductivity

The appropriate form of the heat conduction equation is then:

$$\frac{d(kdT/dx)}{dx} + \dot{q} = 0$$

$$k\frac{d^2T}{dx^2} + \Gamma\, e^{-\gamma x} = 0$$

$$\frac{d^2T}{dx^2} = -\frac{\Gamma e^{-\gamma x}}{k}$$

Integrating once gives:

$$\frac{dT}{dx} = \frac{\Gamma e^{-\gamma x}}{k\gamma} + C_1$$

A second integration yields:

$$T = -\frac{\Gamma e^{-\gamma x}}{k\gamma^2} + C_1 x + C_2$$

The boundary conditions are:

(1) At $x = 0$, $T = T_1$

(2) At $x = \mathrm{B}$, $\frac{dT}{dx} = 0$

The second boundary condition results from assuming zero heat flow through the insulated boundary (perfect insulation). Thus, at $x = \mathrm{B}$:

$$q_x = -kA\frac{dT}{dx} = 0 \Rightarrow \frac{dT}{dx} = 0$$

This condition is applied using the equation for dT/dx resulting from the first integration:

$$0 = \frac{\Gamma e^{-\gamma B}}{k\gamma} + C_1$$

Hence,

$$C_1 = -\frac{\Gamma e^{-\gamma B}}{k\gamma}$$

Applying the first boundary condition to the equation for T:

$$T_1 = -\frac{\Gamma e^{(0)}}{k\gamma^2} + C_1(0) + C_2$$

Hence,

$$C_2 = T_1 + \frac{\Gamma}{k\gamma^2}$$

With the above values for C_1 and C_2, the temperature profile becomes:

$$T = T_1 + \frac{\Gamma}{k\gamma^2}\left(1 - e^{-\gamma x}\right) - \frac{\Gamma e^{-\gamma B}}{k\gamma}x$$

Now at steady state, all the heat generated in the block must flow out through the un-insulated side at $x = 0$. Hence, the maximum temperature must occur at the insulated boundary, i.e., at $x = B$. (This intuitive result can be confirmed by setting the first derivative of T equal to zero and solving for x.) Thus, setting $x = B$ in the last equation gives:

$$T_{max} = T_1 + \frac{\Gamma}{k\gamma^2}\left(1 - e^{-\gamma B}\right) - \frac{\Gamma B e^{-\gamma B}}{k\gamma}$$

Finally, the solution is obtained by substituting the numerical values of the parameters:

$$T_{max} = 20 + \frac{10}{0.5(0.1)^2}\left(1 - e^{-0.1}\right) - \frac{10 \times 1.0\ e^{-0.1}}{0.5 \times 0.1}$$

$$T_{max} \cong 29.4°\mathrm{C}$$

The procedure illustrated in the above examples can be summarized as follows:

(1) Write down the conduction equation in the appropriate coordinate system.
(2) Impose any restrictions dictated by the physical situation to eliminate terms that are zero or negligible.
(3) Integrate the resulting differential equation to obtain the temperature profile.
(4) Use the boundary conditions to evaluate the constants of integration.
(5) Use the appropriate form of Fourier's law to obtain the heat flux.
(6) Multiply the heat flux by the cross-sectional area to obtain the rate of heat transfer.

In each of the above examples there is only one independent variable so that an ordinary differential equation results. In unsteady-state problems and problems in which heat flows in more than one direction, a partial differential equation must be solved. Analytical solutions are often possible if the geometry is sufficiently simple. Otherwise, numerical solutions are obtained with the aid of a computer.

1.4 Thermal Resistance

The concept of thermal resistance is based on the observation that many diverse physical phenomena can be described by a general rate equation that may be stated as follows:

$$\text{Flow rate} = \frac{\text{Driving force}}{\text{Resistance}} \tag{1.14}$$

Ohm's Law of Electricity is one example:

$$I = \frac{E}{R} \tag{1.15}$$

In this case, the quantity that flows is electric charge, the driving force is the electrical potential difference, E, and the resistance is the electrical resistance, R, of the conductor.

In the case of heat transfer, the quantity that flows is heat (thermal energy) and the driving force is the temperature difference. The resistance to heat transfer is termed the thermal resistance, and is denoted by R_{th}. Thus, the general rate equation may be written as:

$$q = \frac{\Delta T}{R_{th}} \tag{1.16}$$

In this equation, all quantities take on positive values only, so that q and ΔT represent the absolute values of the heat-transfer rate and temperature difference, respectively.

An expression for the thermal resistance in a rectangular system can be obtained by comparing Equations (1.1) and (1.16):

$$q_x = \frac{kA(T_1 - T_2)}{B} = \frac{\Delta T}{R_{th}} = \frac{T_1 - T_2}{R_{th}} \tag{1.17}$$

$$R_{th} = \frac{B}{kA} \tag{1.18}$$

Similarly, using the equation derived in Example 1.4 for a cylindrical system gives:

$$q_r = \frac{2\pi kL(T_1 - T_2)}{\ln(R_2/R_1)} = \frac{T_1 - T_2}{R_{th}} \tag{1.19}$$

$$R_{th} = \frac{\ln(R_2/R_1)}{2\pi kL} \tag{1.20}$$

These results, along with a number of others that will be considered subsequently, are summarized in Table 1.2. When k cannot be assumed constant, the average thermal conductivity, as defined in the previous section, should be used in the expressions for thermal resistance.

The thermal resistance concept permits some relatively complex heat-transfer problems to be solved in a very simple manner. The reason is that thermal resistances can be combined in the same way as electrical resistances. Thus, for resistances in series, the total resistance is the sum of the individual resistances:

$$R_{Tot} = \sum_i R_i \tag{1.21}$$

TABLE 1.2 Expressions for Thermal Resistance

Configuration	R_{th}
Conduction, Cartesian coordinates	B/kA
Conduction, radial direction, cylindrical coordinates	$\dfrac{\ln(R_2/R_1)}{2\pi\, k\, L}$
Conduction, radial direction, spherical coordinates	$\dfrac{R_2 - R_1}{4\pi\, k\, R_1 R_2}$
Conduction, shape factor	$1/kS$
Convection, un-finned surface	$1/\text{hA}$
Convection, finned surface	$\dfrac{1}{h\eta_w A}$

S = shape factor
h = heat-transfer coefficient
η_w = weighted efficiency of finned surface $= \dfrac{A_p + \eta_f A_f}{A_p + A_f}$
A_p = prime surface area
A_f = fin surface area
η_f = fin efficiency

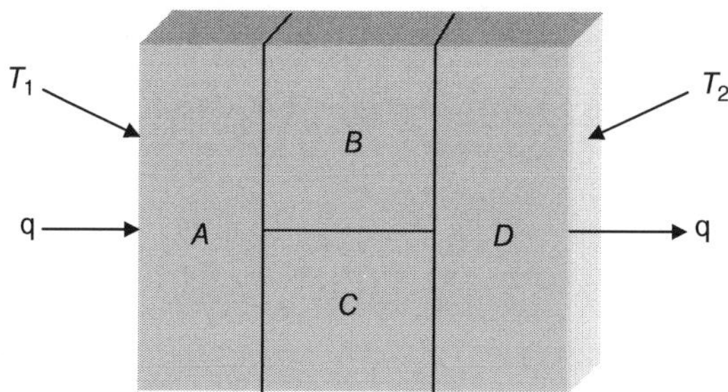

FIGURE 1.3 Heat transfer through a composite material.

Likewise, for resistances in parallel:

$$R_{Tot} = \left(\sum_i 1/R_i\right)^{-1} \tag{1.22}$$

Thus, for the composite solid shown in Figure 1.3, the thermal resistance is given by:

$$R_{th} = R_A + R_{BC} + R_D \tag{1.23}$$

where R_{BC}, the resistance of materials B and C in parallel, is:

$$R_{BC} = (1/R_B + 1/R_C)^{-1} = \frac{R_B R_C}{R_B + R_C} \tag{1.24}$$

In general, when thermal resistances occur in parallel, heat will flow in more than one direction. In Figure 1.3, for example, heat will tend to flow between materials B and C, and this flow will be normal to the primary direction of heat transfer. In this case, the one-dimensional calculation of q using Equations (1.16) and (1.22) represents an approximation, albeit one that is generally quite acceptable for process engineering purposes.

Example 1.6

A 5-cm (2-in.) schedule 40 steel pipe carries a heat-transfer fluid and is covered with a 2-cm layer of calcium silicate insulation ($k = 0.06$ W/m · K) to reduce the heat loss. The inside and outside pipe diameters are 5.25 cm and 6.03 cm, respectively. If the inner pipe surface is at 150°C and the exterior surface of the insulation is at 25°C, calculate:

(a) The rate of heat loss per unit length of pipe.
(b) The temperature of the outer pipe surface.

Solution

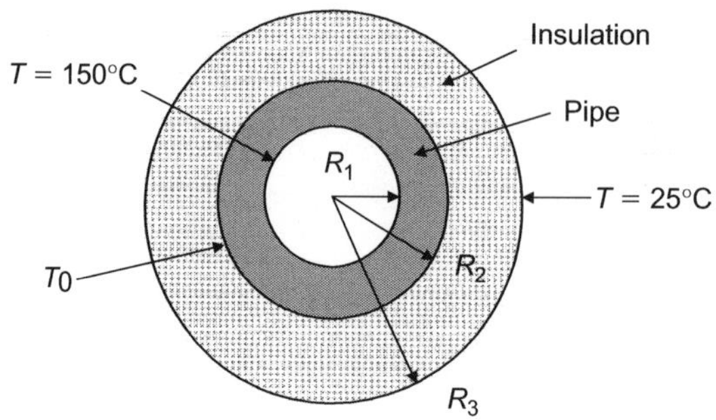

(a)

$$q_r = \frac{\Delta T}{R_{th}} = \frac{150 - 25}{R_{th}}$$

$$R_{th} = R_{pipe} + R_{insulation}$$

$$R_{th} = \frac{\ln(R_2/R_1)}{2\pi k_{steel} L} + \frac{\ln(R_3/R_2)}{2\pi k_{ins} L}$$

$$R_1 = 5.25/2 = 2.625 \text{ cm}$$

$$R_2 = 6.03/2 = 3.015 \text{ cm}$$

$$R_3 = 3.015 + 2 = 5.015 \text{ cm}$$

$$k_{steel} = 43 \text{ W/m·K (Table A.1)}$$

$$k_{ins} = 0.06 \text{ W/m·K (given)}$$

$$L = 1 \text{ m}$$

$$R_{th} = \frac{\ln\left(\frac{3.015}{2.625}\right)}{2\pi \times 43} + \frac{\ln\left(\frac{5.015}{3.015}\right)}{2\pi \times 0.06} = 0.000513 + 1.349723$$

$$= 1.350236 \text{ K/W}$$

$$q_r = \frac{125}{1.350236} \cong 92.6 \text{ W/m of pipe}$$

(b) Writing Equation (1.16) for the pipe wall only:

$$q_r = \frac{150 - T_0}{R_{pipe}}$$

$$92.6 = \frac{150 - T_0}{0.000513}$$

$$T_0 = 150 - 0.0475 \cong 149.95°\text{C}$$

Clearly, the resistance of the pipe wall is negligible compared with that of the insulation, and the temperature difference across the pipe wall is a correspondingly small fraction of the total temperature difference in the system.

It should be pointed out that the calculation in Example 1.6 tends to overestimate the rate of heat transfer because it assumes that the insulation is in perfect thermal contact with the pipe wall. Since solid surfaces are not perfectly smooth, there will generally be air gaps between two adjacent solid materials. Since air is a very poor conductor of heat, even a thin layer of air can result in a substantial thermal resistance. This additional resistance at the interface between two materials is called the contact resistance. Thus, the thermal resistance in Example 1.5 should really be written as:

$$R_{th} = R_{pipe} + R_{insulation} + R_{contact} \tag{1.25}$$

The effect of the additional resistance is to decrease the rate of heat transfer according to Equation (1.16). Since the contact resistance is difficult to determine, it is often neglected or a rough approximation is used. For example, a value equivalent to an additional 5 mm of material thickness is sometimes used for the contact resistance between two pieces of the same material [3]. A more rigorous method for estimating contact resistance can be found in Ref. [4].

A slightly modified form of the thermal resistance, the *R*-value, is commonly used for insulations and other building materials. The *R*-value is defined as:

$$R\text{-value} = \frac{B(\text{ft})}{k(\text{Btu/h·ft·°F})} \tag{1.26}$$

where B is the thickness of the material and k is its thermal conductivity. Comparison with Equation (1.18) shows that the *R*-value is the thermal resistance, in English units, of a slab of material having a cross-sectional area of 1 ft^2. Since the *R*-value is always given for a specified thickness, the thermal conductivity of a material can be obtained from its *R*-value using Equation (1.26). Also, since *R*-values are essentially thermal resistances, they are additive for materials arranged in series.

Example 1.7

Triple-glazed windows like the one shown in the sketch below are often used in very cold climates. Calculate the *R*-value for the window shown. The thermal conductivity of air at normal room temperature is approximately 0.015 Btu/h · ft · °F

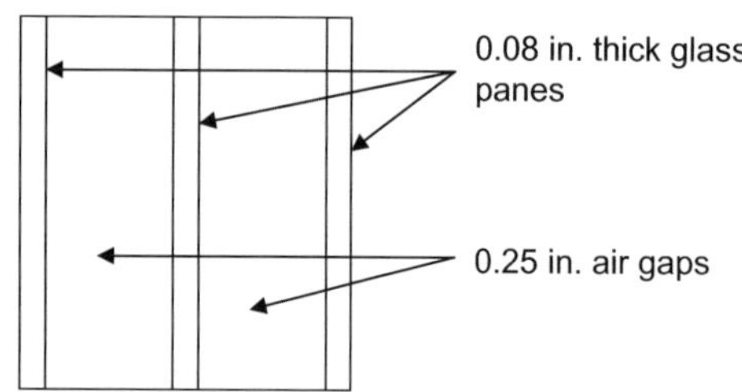

Triple-pane window

Solution

From Table A.3, the thermal conductivity of window glass is 0.78 W/m · K. Converting to English units gives:

$$k_{glass} = 0.78 \times 0.57782 = 0.45 \text{ Btu/h·ft·°F}$$

The R-values for one pane of glass and one air gap are calculated from Equation (1.26):

$$R_{glass} = \frac{0.08/12}{0.45} \cong 0.0148$$

$$R_{air} = \frac{0.25/12}{0.015} \cong 1.3889$$

The R-value for the window is obtained using the additive property for materials in series:

$$\begin{aligned} R\text{-value} &= 3R_{glass} + 2R_{air} \\ &= 3 \times 0.0148 + 2 \times 1.3889 \\ R\text{-value} &\cong 2.8 \end{aligned}$$

1.5 The Conduction Shape Factor

The conduction shape factor is a device whereby analytical solutions to multi-dimensional heat conduction problems are cast into the form of one-dimensional solutions. Although quite restricted in scope, the shape factor method permits rapid and easy solution of multi-dimensional heat-transfer problems when it is applicable. The conduction shape factor, S, is defined by the relation:

$$q = kS\Delta T \tag{1.27}$$

where ΔT is a specified temperature difference. Notice that S has the dimension of length. Shape factors for a number of geometrical configurations are given in Table 1.3. The solution of a problem involving one of these configurations is thus reduced to the calculation of S by the appropriate formula listed in the table.

TABLE 1.3 Conduction Shape Factors

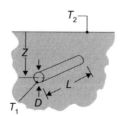

Configuration	Diagram	Restriction	Shape Factor
Case 1 Isothermal sphere buried in a semi-infinite medium	T_2, z, T_1, D	$z > D/2$	$S = \dfrac{2\pi D}{1 - D/4z}$
Case 2 Horizontal isothermal cylinder of length L buried in a semi-infinite medium	T_2, z, L, T_1, D	$L >> D$	$S = \dfrac{2\pi L}{\cosh^{-1}(2z/D)}$

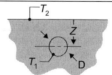

TABLE 1.3 Conduction Shape Factors—cont'd

Case 3 Vertical cylinder in a semi-infinite medium	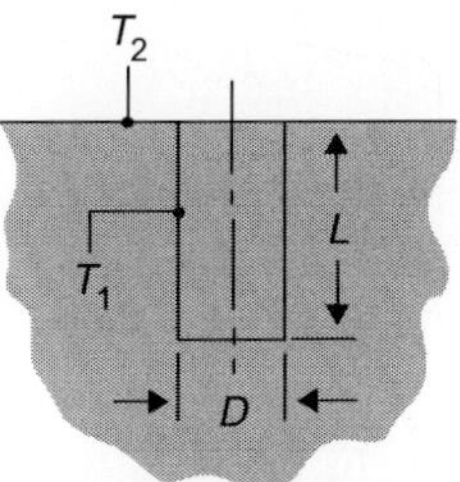	$L >> D$	$S = \dfrac{2\pi L}{\ln(4L/D)}$
Case 4 Conduction between two cylinders of length L in infinite medium	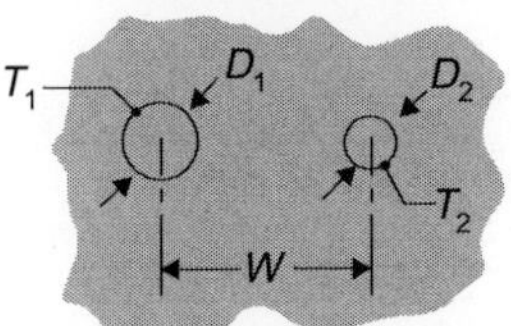	$L >> D_1, D_2$ $L >> w$	$S = \dfrac{2\pi L}{\cosh^{-1}\left(\dfrac{4w^2 - D_1^2 - D_2^2}{2D_1 D_2}\right)}$
Case 5 Horizontal circular cylinder of length L midway between parallel planes of equal length and infinite width	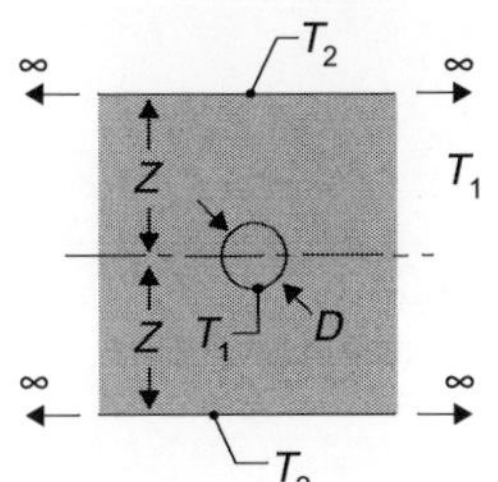	$z >> D/2$ $L >> z$	$S = \dfrac{2\pi L}{\ln(8z/\pi D)}$
Case 6 Circular cylinder of length L centered in a square solid of equal length	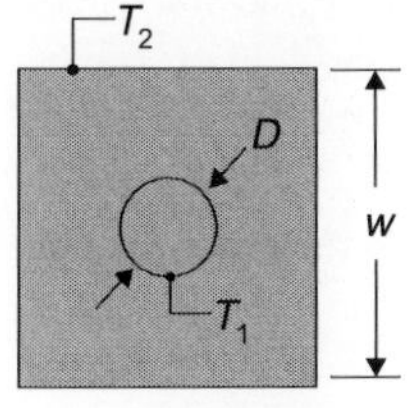	$w > D$ $L >> w$	$S = \dfrac{2\pi L}{\ln(1.08w/D)}$
Case 7 Eccentric circular cylinder of length L in a cylinder of equal length	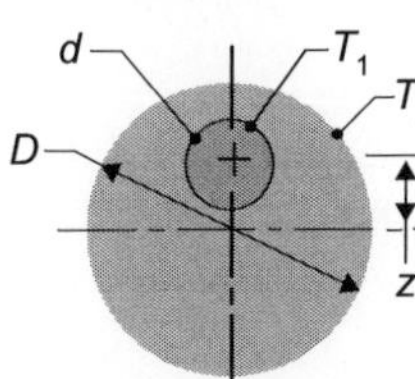	$D > d$ $L >> D$	$S = \dfrac{2\pi L}{\cosh^{-1}\left(\dfrac{D^2 + d^2 - 4z^2}{2Dd}\right)}$
Case 8 Conduction through the edge of adjoining walls	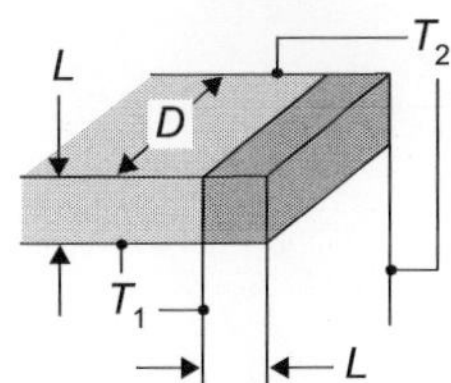	$D > L/5$	$S = 0.54\ D$
Case 9 Conduction through corner of three walls with a temperature difference T_1–T_2 across the walls	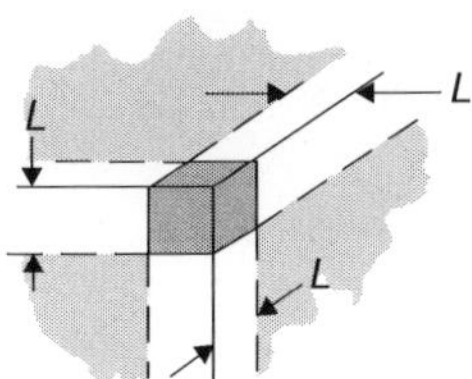	$L <<$ length and width of wall	$S = 0.15\ L$

(*Continued*)

TABLE 1.3 Conduction Shape Factors—cont'd

Case 10 Disk of diameter D and T_1 on a semi-finite medium of thermal conductivity k and T_2	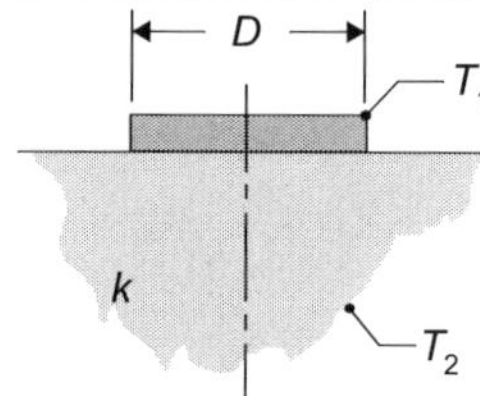		$S = 2\,D$
Case 11 Square channel of length L	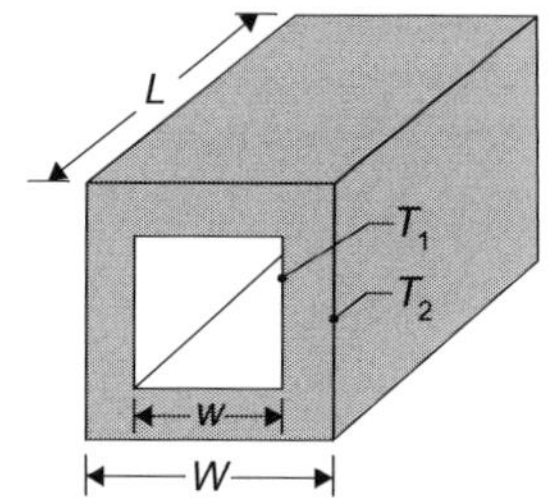	$\frac{W}{w} < 1.4$ $\frac{W}{w} > 1.4$	$S = \frac{2\pi L}{0.785\ \ln(W/w)}$ $S = \frac{2\pi L}{0.930\ \ln(W/w) - 0.050}$

(Source: Ref. [5])

The thermal resistance corresponding to the shape factor can be found by comparing Equation (1.16) with Equation (1.27). The result is:

$$R_{th} = 1/kS \tag{1.28}$$

This is one of the thermal resistance formulas listed in Table 1.2. Since shape-factor problems are inherently multi-dimensional, however, use of the thermal resistance concept in such cases will, in general, yield only approximate solutions. Nevertheless, these solutions are usually entirely adequate for process engineering calculations.

Example 1.8

An underground pipeline transporting hot oil has an outside diameter of 1 ft and its centerline is 2 ft below the surface of the earth. If the pipe wall is at 200°F and the earth's surface is at –50°F, what is the rate of heat loss per foot of pipe? Assume $k_{earth} = 0.5$ Btu/h · ft · °F.

Solution

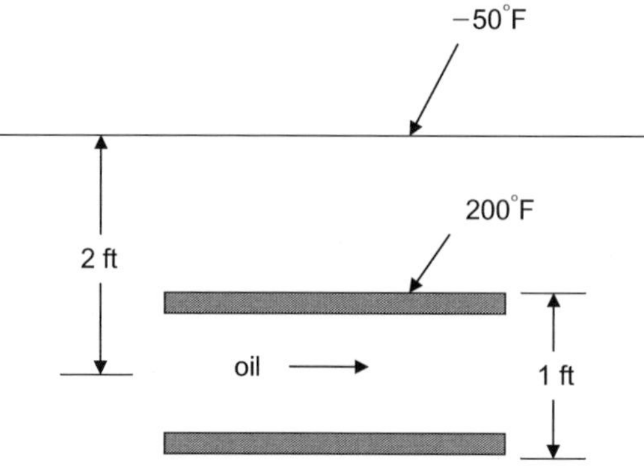

From Table 1.3, the shape factor for a buried horizontal cylinder is:

$$S = \frac{2\pi\ L}{\cosh^{-1}(2z/D)}$$

In this case, $z = 2$ ft and $D = 1.0$ ft. Taking $L = 1$ ft we have:

$$S = \frac{2\pi L}{\cosh^{-1}(4)} = 3.045 \text{ ft}$$

$$q = k_{earth} S \Delta T$$

$$= 0.5 \times 3.045 \times [200 - (-50)]$$

$$q \cong 380 \text{ Btu/h} \cdot \text{ft of pipe}$$

Note: If necessary, the following mathematical identity can be used to evaluate $\cosh^{-1}(x)$:

$$\cosh^{-1}(x) = \ln\left(x + \sqrt{x^2 - 1}\right)$$

Example 1.9

Suppose the pipeline of the previous example is covered with 1 in. of magnesia insulation ($k = 0.07$ W/m · K). What is the rate of heat loss per foot of pipe?

Solution

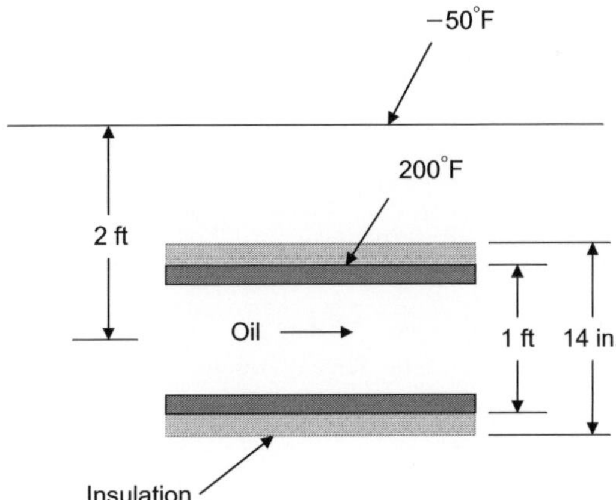

This problem can be solved by treating the earth and the insulation as two resistances in series. Thus,

$$q = \frac{\Delta T}{R_{th}} = \frac{200 - (-50)}{R_{earth} + R_{insulation}}$$

The resistance of the earth is obtained by means of the shape factor for a buried horizontal cylinder. In this case, however, the diameter of the cylinder is the diameter of the exterior surface of the insulation. Thus,

$$z = 2 \text{ ft} = 24 \text{ in.}$$

$$D = 12 + 2 = 14 \text{ in.}$$

$$2z/D = \frac{48}{14} = 3.4286$$

Therefore,

$$S = \frac{2\pi L}{\cosh^{-1}(2z/D)} = \frac{2\pi \times 1}{\cosh^{-1}(3.4286)} = 3.3012 \text{ ft}$$

$$R_{earth} = \frac{1}{k_{earth} S} = \frac{1}{0.5 \times 3.3012} = 0.6058 \text{ h} \cdot {}^\circ\text{F/Btu}$$

Converting the thermal conductivity of the insulation to English units gives:

$$k_{ins} = 0.07 \times 0.57782 = 0.0404 \text{ Btu/h} \cdot \text{ft} \cdot {}^\circ\text{F}$$

Hence,

$$R_{insulation} = \frac{\ln(R_2/R_1)}{2\pi k_{ins}L} = \frac{\ln(7/6)}{2\pi \times 0.0404 \times 1} = 0.6073 \text{ h} \cdot {}^\circ\text{F/Btu}$$

Then

$$q = \frac{250}{0.6058 + 0.6073} = 206 \text{ Btu/h} \cdot \text{ft of pipe}$$

1.6 Unsteady-State Conduction

The heat conduction problems considered thus far have all been steady state, i.e., time-independent, problems. In this section, solutions of a few unsteady-state problems are presented. Solutions to many other unsteady-state problems can be found in heat-transfer textbooks and monographs, e.g., Refs. [5–10].

We consider first the case of a semi-infinite solid illustrated in Figure 1.4. The rectangular solid occupies the region from $x = 0$ to $x = \infty$. The solid is initially at a uniform temperature, T_0. At time $t = 0$, the temperature of the surface at $x = 0$ is changed to T_s and held at that value. The temperature within the solid is assumed to be uniform in the y-and z-directions at all times, so that heat flows only in the x-direction. This condition can be achieved mathematically by allowing the solid to extend to infinity in the $\pm y$- and $\pm z$-directions. If T_s is greater that T_0, heat will begin to penetrate into the solid, so that the temperature at any point within the solid will gradually increase with time. That is, $T = T(x,t)$, and the problem is to determine the temperature as a function of position and time.

Assuming no internal heat generation and constant thermal conductivity, the conduction equation for this situation is:

$$\frac{1}{\alpha}\frac{\partial T}{\partial t} = \frac{\partial^2 T}{\partial x^2} \tag{1.29}$$

The boundary conditions are:

(1) At $t = 0$, $T = T_0$ for all $x \geq 0$
(2) At $x = 0$, $T = T_s$ for all $t > 0$
(3) As $x \rightarrow \infty$, $T \rightarrow T_0$ for all $t \geq 0$

The last condition follows because it takes an infinite time for heat to penetrate an infinite distance into the solid.

The solution of Equation (1.29) subject to these boundary conditions can be obtained by the method of combination of variables [11]. The result is:

$$\frac{T(x,t) - T_s}{T_0 - T_s} = \text{erf}\left(\frac{x}{2\sqrt{\alpha t}}\right) \tag{1.30}$$

The error function, erf, is defined by:

$$\text{erf}\left(\frac{x}{2\sqrt{\alpha t}}\right) = \frac{2}{\sqrt{\pi}} \int_0^{\frac{x}{2\sqrt{\alpha t}}} e^{-z^2} dz \tag{1.31}$$

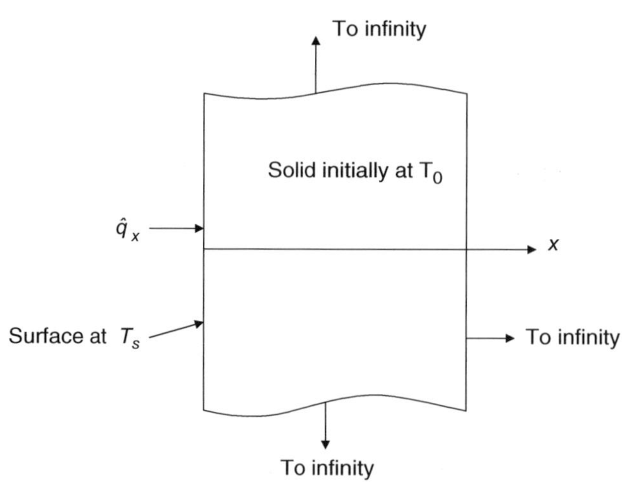

FIGURE 1.4 Semi-infinite solid.

TABLE 1.4 The Error Function

x	erf x	x	erf x	X	erf x
0.00	0.00000	0.76	0.71754	1.52	0.96841
0.02	0.02256	0.78	0.73001	1.54	0.97059
0.04	0.04511	0.80	0.74210	1.56	0.97263
0.06	0.06762	0.82	0.75381	1.58	0.97455
0.08	0.09008	0.84	0.76514	1.60	0.97635
0.10	0.11246	0.86	0.77610	1.62	0.97804
0.12	0.13476	0.88	0.78669	1.64	0.97962
0.14	0.15695	0.90	0.79691	1.66	0.98110
0.16	0.17901	0.92	0.80677	1.68	0.98249
0.18	0.20094	0.94	0.81627	1.70	0.98379
0.20	0.22270	0.96	0.82542	1.72	0.98500
0.22	0.24430	0.98	0.83423	1.74	0.98613
0.24	0.26570	1.00	0.84270	1.76	0.98719
0.26	0.28690	1.02	0.85084	1.78	0.98817
0.28	0.30788	1.04	0.85865	1.80	0.98909
0.30	0.32863	1.06	0.86614	1.82	0.98994
0.32	0.34913	1.08	0.87333	1.84	0.99074
0.34	0.36936	1.10	0.88020	1.86	0.99147
0.36	0.38933	1.12	0.88079	1.88	0.99216
0.38	0.40901	1.14	0.89308	1.90	0.99279
0.40	0.42839	1.16	0.89910	1.92	0.99338
0.42	0.44749	1.18	0.90484	1.94	0.99392
0.44	0.46622	1.20	0.91031	1.96	0.99443
0.46	0.48466	1.22	0.91553	1.98	0.99489
0.48	0.50275	1.24	0.92050	2.00	0.995322
0.50	0.52050	1.26	0.92524	2.10	0.997020
0.52	0.53790	1.28	0.92973	2.20	0.998137
0.54	0.55494	1.30	0.93401	2.30	0.998857
0.56	0.57162	1.32	0.93806	2.40	0.999311
0.58	0.58792	1.34	0.94191	2.50	0.999593
0.60	0.60386	1.36	0.94556	2.60	0.999764
0.62	0.61941	1.38	0.94902	2.70	0.999866
0.64	0.63459	1.40	0.95228	2.80	0.999925
0.66	0.64938	1.42	0.95538	2.90	0.999959
0.68	0.66278	1.44	0.95830	3.00	0.999978
0.70	0.67780	1.46	0.96105	3.20	0.999994
0.72	0.69143	1.48	0.96365	3.40	0.999998
0.74	0.70468	1.50	0.96610	3.60	1.000000

This function, which occurs in many diverse applications in engineering and applied science, can be evaluated by numerical integration. Values are listed in Table 1.4.

The heat flux is given by:

$$\hat{q}_x = \frac{k(T_s - T_0)}{\sqrt{\pi \alpha\, t}} \exp\left(-x^2/4\alpha t\right) \tag{1.32}$$

The total amount of heat transferred per unit area across the surface at $x = 0$ in time t is given by:

$$\frac{Q}{A} = 2k(T_s - T_0)\sqrt{\frac{t}{\pi\alpha}} \tag{1.33}$$

Although the semi-infinite solid may appear to be a purely academic construct, it has a number of practical applications. For example, the earth behaves essentially as a semi-infinite solid. A solid of any finite thickness can be considered a semi-infinite solid if the time interval of interest is sufficiently short that heat penetrates only a small distance into the solid. The approximation is generally acceptable if the following inequality is satisfied:

$$\frac{\alpha t}{B^2} < 0.1 \tag{1.34}$$

where B is the thickness of the solid. The dimensionless group $\alpha t/B^2$ is called the Fourier number and is designated *Fo*.

Example 1.10

The steel panel of a firewall is 5-cm thick and is initially at 25°C. The exterior surface of the panel is suddenly exposed to a temperature of 250°C. Estimate the temperature at the center and at the interior surface of the panel after 20 s of exposure to this temperature. The thermal diffusivity of the panel is $0.97 \times 10^{-5}\ \text{m}^2/\text{s}$.

Solution

To determine if the panel can be approximated by a semi-infinite solid, we calculate the Fourier number:

$$Fo = \frac{\alpha t}{B^2} = \frac{0.97 \times 10^{-5} \times 20}{(0.05)^2} \cong 0.0776$$

Since $Fo < 0.1$, the approximation should be acceptable. Thus, using Equation (1.30) with $x = 0.025$ for the temperature at the center,

$$\frac{T - T_s}{T_0 - T_s} = \text{erf}\left(\frac{x}{2\sqrt{\alpha\, t}}\right)$$

$$\frac{T - 250}{25 - 250} = \text{erf}\left(\frac{0.025}{2\sqrt{0.97 \times 10^{-5} \times 20}}\right) = \text{erf}\ (0.8974)$$

$$\frac{T - 250}{-225} = 0.7969 \ \text{(from Table 1.4)}$$

$$T \cong 70.7°\text{C}$$

For the interior surface, $x = 0.05$ and Equation (1.30) gives:

$$\frac{T - 250}{-225} = \text{erf}\left(\frac{0.05}{2\sqrt{0.97 \times 10^{-5} \times 20}}\right) = \text{erf}\ (1.795)$$

$$= 0.9891$$

$$\text{T} \cong 27.5°\text{C}$$

Thus, the temperature of the interior surface has not changed greatly from its initial value of 25°C, and treating the panel as a semi-infinite solid is therefore a reasonable approximation.

Consider now the rectangular solid of finite thickness illustrated in Figure 1.5. The configuration is the same as that for the semi-infinite solid except that the solid now occupies the region from $x = 0$ to $x = 2s$. The solid is initially at uniform temperature

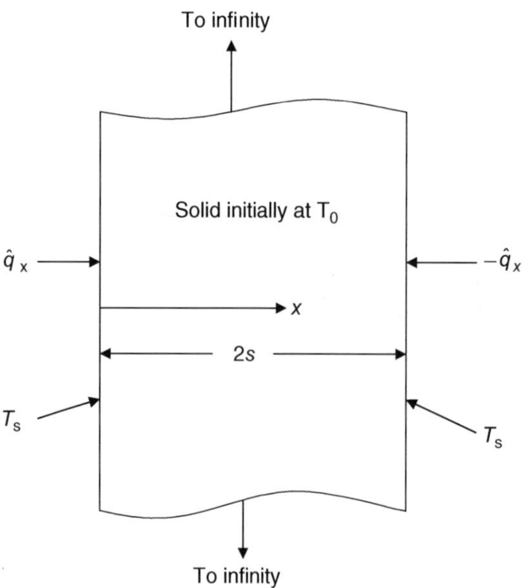

FIGURE 1.5 Infinite solid of finite thickness.

T_0 and at time $t = 0$ the temperature of the surfaces at $x = 0$ and $x = 2s$ are changed to T_s. If $T_s > T_0$, then heat will flow into the solid from both sides. It is assumed that heat flows only in the x-direction, which again can be achieved mathematically by making the solid of infinite extent in the $\pm y$- and $\pm z$-directions. This condition will be approximated in practice when the areas of the surfaces normal to the y- and z-directions are much smaller than the area of the surface normal to the x-direction, or when the former surfaces are insulated.

The mathematical statement of this problem is the same as that of the semi-infinite solid except that the third boundary condition is replaced by:

$$(3')\text{At } x = 2s, T = T_s$$

The solution for $T\ (x,\ t)$ can be found in the textbooks cited at the beginning of this section. Frequently, however, one is interested in determining the average temperature, $\overline{T}$, of the solid as a function of time, where:

$$\overline{T}(t) = \frac{1}{2s}\int_0^{2s} T(x,t)dx \tag{1.35}$$

That is, $\overline{T}$ is the temperature averaged over the thickness of the solid at a given instant of time. The solution for $\overline{T}$ is in the form of an infinite series [12]:

$$\frac{T_s - \overline{T}}{T_s - T_0} = \frac{8}{\pi^2}\left(e^{-aFo} + \frac{1}{9}e^{-9aFo} + \frac{1}{25}e^{-25aFo} + \cdots\right) \tag{1.36}$$

where $a = (\pi/2)^2 \cong 2.4674$ and $Fo = \alpha t/s^2$.

The solution given by Equation (1.36) is shown graphically in Figure 1.6. When the Fourier number, Fo, is greater than about 0.1, the series converges very rapidly so that only the first term is significant. Under these conditions, Equation (1.36) can be solved for the time to give:

$$t = \frac{1}{\alpha}\left(\frac{2s}{\pi}\right)^2 \ln\left[\frac{8\ (T_s - T_0)}{\pi^2 (T_s - \overline{T})}\right] \tag{1.37}$$

The total amount of heat, Q, transferred to the solid per unit area, A, in time t is:

$$\frac{Q(t)}{A} = \frac{mc}{A}\left[\overline{T}(t) - T_0\right] \tag{1.38}$$

where m, the mass of solid, is equal to $2\rho sA$. Thus,

$$\frac{Q(t)}{A} = 2\rho cs\left[\overline{T}(t) - T_0\right] \tag{1.39}$$

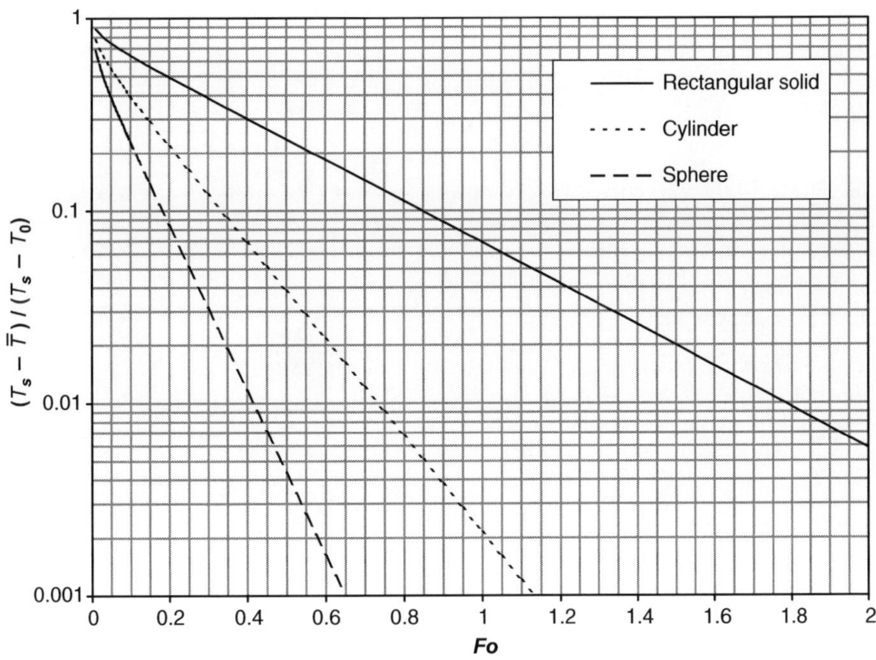

FIGURE 1.6 Average temperatures during unsteady-state heating or cooling of a rectangular solid, an infinitely long cylinder, and a sphere.

The analogous problem in cylindrical geometry is that of an infinitely long solid cylinder of radius, R, initially at uniform temperature, T_0. At time $t = 0$ the temperature of the surface is changed to T_s. This situation will be approximated in practice by a finite cylinder whose length is much greater than its diameter, or whose ends are insulated. The solutions corresponding to Equations (1.36), (1.37), and (1.39) are [12]:

$$\frac{T_s - \overline{T}}{T_s - T_0} = 0.692e^{-5.78Fo} + 0.131e^{-30.5Fo} + 0.0534e^{-74.9Fo} + \cdots \tag{1.40}$$

$$t = \frac{R^2}{5.78\alpha} \ln\left[\frac{0.692(T_s - T_0)}{T_s - \overline{T}}\right] \tag{1.41}$$

$$\frac{Q(t)}{A} = \frac{\rho c R}{2}\left[\overline{T}(t) - T_0\right] \tag{1.42}$$

where

$$Fo = \frac{\alpha t}{R^2} \tag{1.43}$$

Here A is the circumferential area, which is equal to $2\pi R$ times the length of the cylinder. Equation (1.40) is shown graphically in Figure 1.6.

The corresponding equations for a solid sphere of radius R are [12]:

$$\frac{T_s - \overline{T}}{T_s - T_0} = 0.608\, e^{-9.87Fo} + 0.152e^{-39.5Fo} + 0.067e^{-88.8Fo} + \cdots \tag{1.44}$$

$$t = \frac{R^2}{9.87\alpha} \ln\left[\frac{0.608\,(T_s - T_0)}{T_s - \overline{T}}\right] \tag{1.45}$$

$$Q(t) = \frac{4}{3}\pi R^3 \rho c\left[\overline{T}(t) - T_0\right] \tag{1.46}$$

The Fourier number for this case is also given by Equation (1.43). Equation (1.44) is shown graphically in Figure 1.6.

Example 1.11

A 12-ounce can of beer initially at 80°F is placed in a refrigerator, which is at 36°F. Estimate the time required for the beer to reach 40°F.

Solution

Application to this problem of the equations presented in this section requires a considerable amount of approximation, a situation that is not uncommon in practice. Since a 12-ounce beer can has a diameter of 2.5 in. and a length of 4.75 in., we have:

$$L/D = \frac{4.75}{2.5} = 1.9$$

Hence, the assumption of an infinite cylinder will not be a particularly good one. In effect, we will be neglecting the heat transfer through the ends of the can. The effect of this approximation will be to overestimate the required time.

Next, we must assume that the temperature of the surface of the can suddenly drops to 36°F when it is placed in the refrigerator. That is, we neglect the resistance to heat transfer between the air in the refrigerator and the surface of the can. The effect of this approximation will be to underestimate the required time. Hence, there will be at least a partial cancellation of errors.

We must also neglect the heat transfer due to convection currents set up in the liquid inside the can by the cooling process. The effect of this approximation will be to overestimate the required time.

Finally, we will neglect the resistance of the aluminum can and will approximate the physical properties of beer by those of water. We thus take:

$$\begin{aligned} &k = 0.341 \text{ Btu/h}\cdot\text{ft}\cdot{}^\circ\text{F} && T_s = 36^\circ\text{F} \\ &\rho = 62.4 \text{ lbm/ft}^3 && T_0 = 80^\circ\text{F} \\ &c = 1.0 \text{ Btu/lbm}\cdot{}^\circ\text{F} && \overline{T} = 40^\circ\text{F} \end{aligned}$$

With these values we have:

$$\alpha = \frac{k}{\rho c} = 0.0055\text{ft}^2/\text{h}$$

$$\frac{T_s - \overline{T}}{T_s - T_0} = \frac{36 - 40}{36 - 90} = 0.0909$$

From Figure 1.6, we find a Fourier number of about 0.35. Thus,

$$Fo = \frac{\alpha t}{R^2} = 0.35$$

$$t = \frac{0.35\, R^2}{\alpha} = \frac{0.35(1.25/12)^2}{0.0055} \cong 0.69 \text{ h}$$

Alternatively, since $Fo > 0.1$, we can use Equation (1.41):

$$t = \frac{R^2}{5.78\alpha} \ln\left[\frac{0.692(T_s - T_0)}{T_s - \overline{T}}\right]$$

$$= \frac{(1.25/12)^2}{5.78 \times 0.0055} \ln\left[\frac{0.629}{0.0909}\right]$$

$$t = 0.66 \text{ h}$$

This agrees with the previous calculation to within the accuracy with which one can read the graph of Figure 1.6. Experience suggests that this estimate is somewhat optimistic and, hence, that the error introduced by neglecting the thermal resistance between the air and the can is predominant. Nevertheless, if the answer is rounded to the nearest hour (a reasonable thing to do considering the many approximations that were made), the result is a cooling time of 1 h, which is essentially correct. In any event, the calculations show that the time required is more than a few minutes but less than a day, and in many practical situations this level of detail is all that is needed.

1.7 Mechanisms of Heat Conduction

This chapter has dealt with the computational aspects of heat conduction. In this concluding section we briefly discuss the mechanisms of heat conduction in solids and fluids. Although Fourier's law accurately describes heat conduction in both solids and fluids, the underlying mechanisms differ. In all media, however, the processes responsible for conduction take place at the molecular or atomic level.

Heat conduction in fluids is the result of random molecular motion. Thermal energy is the energy associated with translational, vibrational, and rotational motions of the molecules comprising a substance. When a high-energy molecule moves from a high-temperature region of a fluid toward a region of lower temperature (and, hence, lower thermal energy), it carries its thermal energy along with it. Likewise, when a high-energy molecule collides with one of lower energy, there is a partial transfer of energy to the lower-energy molecule. The result of these molecular motions and interactions is a net transfer of thermal energy from regions of higher temperature to regions of lower temperature.

Heat conduction in solids is the result of vibrations of the solid lattice and of the motion of free electrons in the material. In metals, where free electrons are plentiful, thermal energy transport by electrons predominates. Thus, good electrical conductors, such as copper and aluminum, are also good conductors of heat. Metal alloys, however, generally have lower (often much lower) thermal and electrical conductivities than the corresponding pure metals due to disruption of free electron movement by the alloying atoms, which act as impurities.

Thermal energy transport in non-metallic solids occurs primarily by lattice vibrations. In general, the more regular the lattice structure of a material is, the higher its thermal conductivity. For example, quartz, which is a crystalline solid, is a better heat conductor than glass, which is an amorphous solid. Also, materials that are poor electrical conductors may nevertheless be good heat conductors. Diamond, for instance, is an excellent conductor of heat due to transport by lattice vibrations.

Most common insulating materials, both natural and man-made, owe their effectiveness to air or other gases trapped in small compartments formed by fibers, feathers, hairs, pores, or rigid foam. Isolation of the air in these small spaces prevents convection currents from forming within the material, and the relatively low thermal conductivity of air (and other gases) thereby imparts a low effective thermal conductivity to the material as a whole. Insulating materials with effective thermal conductivities much less than that of air are available; they are made by incorporating evacuated layers within the material.

References

[1] Fourier JB. The Analytical Theory of Heat, translated by A Freeman. New York: Dover Publications, Inc.; 1955 (originally published in 1822).
[2] Poling BE, Prausnitz JM, O'Connell JP. The Properties of Gases and Liquids. 5th ed. New York: McGraw-Hill; 2000.

[3] White FM. , Heat Transfer. Reading, MA: Addison-Wesley; 1984.
[4] Irvine Jr TF. Thermal contact resistance. In: Heat Exchanger Design Handbook, vol. 2. New York: Hemisphere Publishing Corp; 1988.
[5] Incropera FP, DeWitt DP. Introduction to Heat Transfer. 4th ed. New York: John Wiley & Sons; 2002.
[6] Kreith F, Black WZ. Basic Heat Transfer. New York: Harper & Row; 1980.
[7] Holman JP. Heat Transfer. 7th ed. New York: McGraw-Hill; 1990.
[8] Kreith F, Bohn MS. Principles of Heat Transfer. 6th ed. Pacific Grove, CA: Brooks/Cole; 2001.
[9] Schneider PJ. Conduction Heat Transfer. Reading, MA: Addison-Wesley; 1955.
[10] Carslaw HS, Jaeger JC. Conduction of Heat in Solids. 2nd ed. New York: Oxford University Press; 1959.
[11] Jensen VG, Jeffreys GV. Mathematical Methods in Chemical Engineering. 2nd ed. New York: Academic Press; 1977.
[12] McCabe WL, Smith JC. Unit Operations of Chemical Engineering. 3rd ed. New York: McGraw-Hill; 1976.

Notations

A	Area
A_f	Fin surface area (Table 1.2)
A_p	Prime surface area (Table 1.2)
A_r	$2\pi rL$
A_x, A_y	Cross-sectional area perpendicular to x-or y-direction
a	Constant in Equation (1.2); constant equal to $(\pi/2)^2$ in Equation (1.36)
B	Thickness of solid in direction of heat flow
b	Constant in Equation (1.2)
c	specific heat of solid
C_1, C_2	Constants of integration
D	Diameter; distance between adjoining walls (Table 1.3)
d	diameter of eccentric cylinder (Table 1.3)
E	Voltage difference in Ohm's law
erf	Gaussian error function defined by Equation (1.31)
Fo	Fourier number
h	Heat-transfer coefficient (Table 1.2)
I	Electrical current in Ohm's law
$\vec{i}$	Unit vector in x-direction
$\vec{j}$	Unit vector in y-direction
k	Thermal conductivity
$\vec{k}$	Unit vector in z-direction
L	Length; thickness of edge or corner of wall (Table 1.3)
Q	Total amount of heat transferred
q	Rate of heat transfer
q_x, q_y, q_r	Rate of heat transfer in x-, y-, or r-direction
$\hat{q}$	q/A = Heat flux
$\dot{q}$	Rate of heat generation per unit volume
$\vec{q}$	Heat flow vector
$\vec{\hat{q}}$	Heat flux vector
R	Resistance; radius of cylinder or sphere
R_{th}	Thermal resistance
R-value	Ratio of a material's thickness to its thermal conductivity, in English units
r	Radial coordinate in cylindrical or spherical coordinate system
S	Conduction shape factor defined by Equation (1.27)
s	Half-width of solid in Figure 1.5
T	Temperature
$\overline{T}$	Average temperature
t	Time
W	Width
w	Width or displacement (Table 1.3)
x	Coordinate in Cartesian system
y	Coordinate in Cartesian system
z	Coordinate in Cartesian or cylindrical system; depth or displacement (Table 1.3)

Greek Letters

α	$k/\rho c$ = Thermal diffusivity
Γ	Constant in Example 1.5
γ	Constant in Example 1.5
ΔT, Δx, *etc.*	Difference in T, x, etc.
η	Efficiency
η_f	Fin efficiency (Table 1.2)
η_w	Weighted efficiency of a finned surface (Table 1.2)
θ	Angular coordinate in spherical system; angle between heat flux vector and x-axis (Example 1.1)
ρ	Density
ϕ	Angular coordinate in cylindrical or spherical system

Other Symbols

$\vec{\nabla} T$	Temperature gradient vector
∇^2	Laplacian operator $= \dfrac{\partial^2}{\partial x^2} + \dfrac{\partial^2}{\partial y^2} + \dfrac{\partial^2}{\partial z^2}$ in Cartesian coordinates
$\rightarrow$	Overstrike to denote a vector
$\vert_x$	Evaluated at x

Problems

(1.1) The temperature distribution in a bakelite block ($k = 0.233$ W/m · K) is given by:

$$T(x, y, z) = x^2 - 2y^2 + z^2 - xy + 2yz$$

where $T \propto$ °C and $x,y,z \propto$ m. Find the magnitude of the heat flux vector at the point $(x,y,z) = (0.5, 0, 0.2)$.

Ans. 0.252 W/m^2.

(1.2) The temperature distribution in a Teflon rod ($k = 0.35$ W/m · K) is:

$$T(r, \phi, z) = r \sin \phi + 2z$$

where
$T \propto$ °C
r = radial position (m)
ϕ = circumferential position (rad)
z = axial position (m)
Find the magnitude of the heat flux vector at the position $(r,\phi, z) = (0.1, 0, 0.5)$.

Ans. 0.78 W/m^2.

(1.3) The rectangular block shown below has a thermal conductivity of 1.4 W/m · K. The block is well insulated on the front and back surfaces, and the temperature in the block varies linearly from left to right and from top to bottom. Determine the magnitude and direction of the heat flux vector. What are the heat flows in the horizontal and vertical directions?

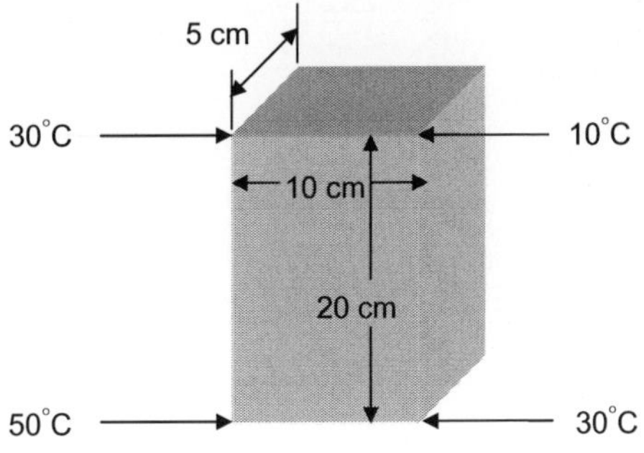

Ans. 313 W/m^2 at an angle of 26.6° with the horizontal; 2.8 W and 0.7 W.

(1.4) The temperature on one side of a 6-in. thick solid wall is 200°F and the temperature on the other side is 100°F. The thermal conductivity of the wall can be represented by:

$$k(\text{Btu/h}\cdot\text{ft}\cdot{}^\circ\text{F}) = 0.1 + 0.001\ T({}^\circ\text{F})$$

(a) Calculate the heat flux through the wall under steady-state conditions.
(b) Calculate the thermal resistance for a 1 ft^2 cross-section of the wall.

Ans. (a) 50 Btu/h · ft^2. (b) 2 h · ft^2 · °F/Btu.

(1.5) A long hollow cylinder has an inner radius of 1.5 in. and an outer radius of 2.5 in. The temperature of the inner surface is 150°F and the outer surface is at 110°F. The thermal conductivity of the material can be represented by:

$$k(\text{Btu/h}\cdot\text{ft}\cdot{}^\circ\text{F}) = 0.1 + 0.001\ T({}^\circ\text{F})$$

(a) Find the steady-state heat flux in the radial direction:
 (i) At the inner surface
 (ii) At the outer surface
(b) Calculate the thermal resistance for a 1 ft length of the cylinder.

Ans: (a) 144.1 Btu/h · ft^2, 86.4 Btu/h · ft^2, (b) 0.3535 h · ft^2 · °F/Btu.

(1.6) A rectangular block has thickness B in the x-direction. The side at $x = 0$ is held at temperature T_1 while the side at $x = B$ is held at T_2. The other four sides are well insulated. Heat is generated in the block at a uniform rate per unit volume of Γ.
(a) Use the conduction equation to derive an expression for the steady-state temperature profile, $T(x)$. Assume constant thermal conductivity.
(b) Use the result of part (a) to calculate the maximum temperature in the block for the following values of the parameters:

$$T_1 = 100^\circ\text{C} \quad k = 0.2\ \text{W/m}\cdot\text{K} \quad \text{B} = 1.0\ \text{m}$$

$$T_2 = 0^\circ\text{C} \quad \Gamma = 100\ \text{W/m}^3$$

Ans. (a) $T(x) = T_1 + \left(\frac{T_2 - T_1}{B} + \frac{\Gamma B}{2k}\right)x - \frac{\Gamma x^2}{2k}$. (b) $T_{max} = 122.5^\circ\text{C}$ at $x = 0.3$ m.

(1.7) Repeat Problem 1.6 for the situation in which the side of the block at $x = 0$ is well insulated.

Ans. (a) $T(x) = T_2 + \frac{\Gamma}{2k}(B^2 - x^2)$. (b) $T_{max} = 250^\circ\text{C}$.

(1.8) Repeat Problem 1.6 for the situation in which the side of the block at $x = 0$ is exposed to an external heat flux, $\hat{q}_0$, of 20 W/m^2. Note that the boundary condition at $x = 0$ for this case becomes

$$\frac{dT}{dx} = -\frac{\hat{q}_o}{k}.$$

Ans. (a) $T(x) = T_2 + \frac{\hat{q}_o}{k}(B - x) + \frac{\Gamma}{2k}(B^2 - x^2)$. (b) $T_{max} = 350^\circ\text{C}$.

(1.9) A long hollow cylinder has inner and outer radii R_1 and R_2, respectively. The temperature of the inner surface at radius R_1 is held at a constant value, T_1, while that of the outer surface at radius R_2 is held constant at a value of T_2. Heat is generated in the wall of the cylinder at a rate per unit volume given by $\dot{q} = \Gamma r$, where r is radial position and Γ is a constant. Assuming constant thermal conductivity and heat flow only in the radial direction, derive expressions for:
(a) The steady-state temperature profile, $T(r)$, in the cylinder wall.
(b) The heat flux at the outer surface of the cylinder.

Ans.

$$(a)T(r) = T_1 + (\Gamma/9k)(R_1^3 - r^3) + \frac{\left\{T_2 - T_1 + \frac{\Gamma}{9k}(R_2^3 - R_1^3)\right\}\ \ln(r/R_1)}{\ln(R_2/R_1)}.$$

$$(b)\ \hat{q}_r|_{r=R_2} = \frac{\Gamma R_2^3}{3} + \frac{k\{T_1 - T_2 - (\Gamma/9k)\ (R_2^3 - R_1^3)\}}{R_2\ \ln(R_2/R_1)}.$$

(1.10) Repeat Problem 1.9 for the situation in which the inner surface of the cylinder at R_1 is well insulated.

Ans. (a) $T(r) = T_2 - (\Gamma/9k)(R_2^3 - r^3) + \frac{\Gamma R_1^3 \ln(r/R_2)}{3k}$. (b) $\hat{q}_r\big|_{r=R_2} = \frac{\Gamma(R_2^3 - R_1^3)}{3R_2}$.

(1.11) A hollow sphere has inner and outer radii R_1 and R_2, respectively. The inner surface at radius R_1 is held at a uniform temperature T_1, while the outer surface at radius R_2 is held at temperature T_2. Assuming constant thermal conductivity, no heat generation and steady-state conditions, use the conduction equation to derive expressions for:

(a) The temperature profile, $T(r)$.

(b) The rate of heat transfer, q_r, in the radial direction.

(c) The thermal resistance.

Ans.

(a) $T(r) = T_1 + \dfrac{R_1R_2(T_1 - T_2)\left(\dfrac{1}{r} - \dfrac{1}{R_1}\right)}{R_2 - R_1}.$

(b) $q_r = \dfrac{4\pi\, kR_1R_2(T_1 - T_2)}{R_2 - R_1}.$

(c) See Table 1.2.

(1.12) A hollow sphere with inner and outer radii R_1 and R_2 has fixed uniform temperatures of T_1 on the inner surface at radius R_1 and T_2 on the outer surface at radius R_2. Heat is generated in the wall of the sphere at a rate per unit volume given by $\dot{q} = \Gamma r$, where r is radial position and Γ is a constant. Assuming constant thermal conductivity, use the conduction equation to derive expressions for:

(a) The steady-state temperature profile, $T(r)$, in the wall.

(b) The heat flux at the outer surface of the sphere.

Ans.

(a) $T(r) = T_1 + (\Gamma/12k)\,(R_1^3 - r^3) + \dfrac{R_1R_2\{T_1 - T_2 - (\Gamma/12k)\,(R_2^3 - R_1^3)\}\left(\dfrac{1}{r} - \dfrac{1}{R_1}\right)}{R_2 - R_1}.$

(b) $\hat{q}_r\big|_{r=R_2} = \dfrac{\Gamma R_2^3}{4} + \{kR_1/R_2(R_2 - R_1)\}\{T_1 - T_2 - (\Gamma/12k)(R_2^3 - R_1^3)\}$

(1.13) Repeat Problem 1.12 for the situation in which the inner surface at radial position R_1 is well insulated.

Ans.

(a) $T(r) = T_2 + (\Gamma/12k)(R_2^3 - r^3) + \left(\dfrac{\Gamma R_1^4}{4k}\right)\left(\dfrac{1}{R_2} - \dfrac{1}{r}\right)$

(b) $\hat{q}_r\big|_{r=R_2} = \dfrac{\Gamma(R_2^4 - R_1^4)}{4R_2^2}$

(1.14) When conduction occurs in the radial direction in a solid rod or sphere, the heat flux must be zero at the center ($r = 0$) in order for a finite temperature to exist there. Hence, an appropriate boundary condition is:

$$\frac{dT}{dr} = 0 \text{ at } r = 0$$

Consider a solid sphere of radius R with a fixed surface temperature, T_R. Heat is generated within the solid at a rate per unit volume given by $\dot{q} = \Gamma_1 + \Gamma_2 r$, where Γ_1 and Γ_2 are constants.

(a) Assuming constant thermal conductivity, use the conduction equation to derive an expression for the steady-state temperature profile, $T(r)$, in the sphere.

(b) Calculate the temperature at the center of the sphere for the following parameter values:

$$R = 1.5 \text{ m} \quad \Gamma_1 = 20\ \text{W/m}^3 \quad T_R = 20°\text{C}$$

$$k = 0.5\text{W/m·K} \quad \Gamma_2 = 10\text{W/m}^4$$

Ans. (a) $T(r) = T_R + (\Gamma_1/6k)(R^2 - r^2) + (\Gamma_2/12k)(R^3 - r^3)$. (b) 40.625°C.

(1.15) A solid cylinder of radius R is well insulated at both ends, and its exterior surface at $r = R$ is held at a fixed temperature, T_R. Heat is generated in the solid at a rate per unit volume given by $\dot{q} = \Gamma(1 - r/R)$, where Γ = constant. The thermal conductivity of the solid may be assumed constant. Use the conduction equation together with an appropriate set of boundary conditions to derive an expression for the steady-state temperature profile, $T(r)$, in the solid.

Ans. $T(r) = T_R + (\Gamma/36k)(5\,R^2 + 4r^3/R - 9r^2)$.

(1.16) A rectangular wall has thickness B in the x-direction and is insulated on all sides except the one at $x = B$, which is held at a constant temperature, T_w. Heat is generated in the wall at a rate per unit volume given by $\dot{q} = \Gamma(B - x)$, where Γ is a constant.

(a) Assuming constant thermal conductivity, derive an expression for the steady-state temperature profile, $T(x)$, in the wall.
(b) Calculate the temperature of the block at the side $x = 0$ for the following parameter values:

$$\Gamma = 0.3 \times 10^6 \text{ W/m}^4 \quad B = 0.1 \text{ m}$$

$$T_w = 90°\text{C} \quad k = 25 \text{ W/m·K}$$

Ans. (a) $T(x) = T_w + (\Gamma/k)\left(\frac{x^3}{6} - \frac{x^2 B}{2} + \frac{B^3}{3}\right)$ (b) 94°C.

(1.17) The exterior wall of an industrial furnace is to be covered with a 2-in. thick layer of high-temperature insulation having an *R*-value of 2.8, followed by a layer of magnesia (85%) insulation. The furnace wall may reach 1200°F, and for safety reasons, the exterior of the magnesia insulation should not exceed 120°F. At this temperature, the heat flux from the insulation to the surrounding air has been estimated for design purposes to be 200 Btu/h · ft^2.
(a) What is the thermal conductivity of the high-temperature insulation?
(b) What thickness of magnesia insulation should be used?
(c) Estimate the temperature at the interface between the high-temperature insulation and the magnesia insulation.

Ans. (a) $k = 0.0595$ Btu/h · ft · °F. (c) 640°F.

(1.18) A storage tank to be used in a chemical process is spherical in shape and is covered with a 3-in. thick layer of insulation having an *R*-value of 12. The tank diameter is 15 ft. The tank will hold a chemical intermediate that must be maintained at 150°F. A heating unit is required to maintain this temperature in the tank.
(a) What is the thermal conductivity of the insulation?
(b) Determine the duty for the heating unit assuming as a worst-case scenario that the exterior surface of the insulation reaches a temperature of 20°F.
(c) What thermal resistances were neglected in your calculation?

Ans. (a) $k = 0.02083$ Btu/h · ft · °F. (b) $q \cong 7900$ Btu/h.

(1.19) A 4-in. schedule 80 steel pipe (ID = 3.826 in., OD = 4.5 in.) carries a heat-transfer fluid at 600°F and is covered with a $^1/_2$-in. thick layer of pipe insulation. The pipe is surrounded by air at 80°F. The vendor's literature states that a 1-in. thick layer of the pipe insulation has an *R*-value of 3. Neglecting convective resistances, the resistance of the pipe wall, and thermal radiation, estimate the rate of heat loss from the pipe per foot of length.

Ans. 453 Btu/h · ft of pipe.

(1.20) A pipe with an OD of 6.03 cm and an ID of 4.93 cm carries steam at 250°C. The pipe is covered with 2.5 cm of magnesia (85%) insulation followed by 2.5 cm of polystyrene insulation ($k = 0.025$ W/m · K). The temperature of the exterior surface of the polystyrene is 25°C. The thermal resistance of the pipe wall may be neglected in this problem. Also neglect the convective and contact resistances.
(a) Calculate the rate of heat loss per meter of pipe length.
(b) Calculate the temperature at the interface between the two types of insulation.

Ans. (a) 63 W/m of pipe. (b) 174.5°C.

(1.21) It is desired to reduce the heat loss from the storage tank of Problem 1.18 by 90%. What additional thickness of insulation will be required?

(1.22) A steel pipe with an OD of 2.375 in. is covered with a 0.5-in. thick layer of asbestos insulation ($k = 0.048$ Btu/h · ft · °F) followed by a 1-in. thick layer of fiberglass insulation ($k = 0.022$ Btu/h · ft · °F). The temperature of the pipe wall is 600°F and the exterior surface of the fiberglass insulation is at 100°F. Calculate:
(a) The rate of heat loss per foot of pipe length.
(b) The temperature at the interface between the asbestos and fiberglass insulations.

Ans. (a) 110 Btu/h · ft of pipe. (b) 471°F.

(1.23) A building contains 6000 ft^2 of wall surface area constructed of panels shown in the sketch below. The interior sheathing is gypsum wallboard and the wood is yellow pine. Calculate the rate of heat loss through the walls if the interior wall surface is at 70°F and the exterior surface is at 30°F.

Ans. 22,300 Btu/h.

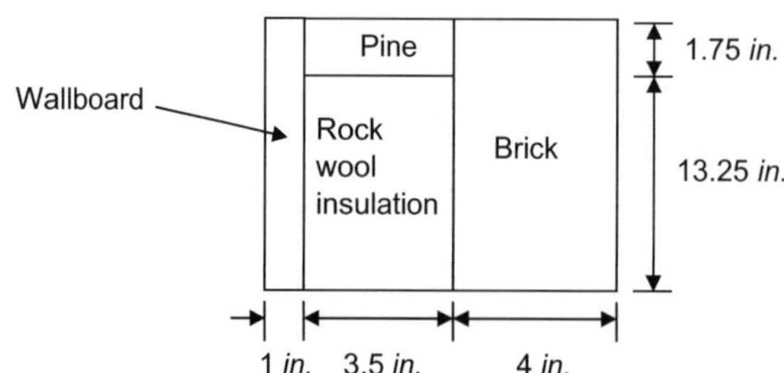

(1.24) A 6-in. schedule 80 steel pipe (OD = 6.62 in.) will be used to transport 450°F steam from a boilerhouse to a new process unit. The pipe will be buried at a depth of 3 ft (to the pipe centerline). The soil at the plant site has an average thermal conductivity of 0.4 Btu/h · ft · °F and the minimum expected ground surface temperature is 20°F. Estimate the rate of heat loss per foot of pipe length for the following cases:

(a) The pipe is not insulated.

(b) The pipe is covered with a 2-in. thick layer of magnesia insulation.

Neglect the thermal resistance of the pipe wall and the contact resistance between the insulation and pipe wall.

Ans. (a) 350 Btu/h · ft of pipe. (b) 160 Btu/h · ft of pipe.

(1.25) The cross-section of an industrial chimney is shown in the sketch below. The flue has a diameter of 2 ft and the process waste gas flowing through it is at 400°F. If the exterior surface of the brick is at 120°F, calculate the rate of heat loss from the waste gas per foot of chimney height. Neglect the convective resistance between the waste gas and interior surface of the flue for this calculation.

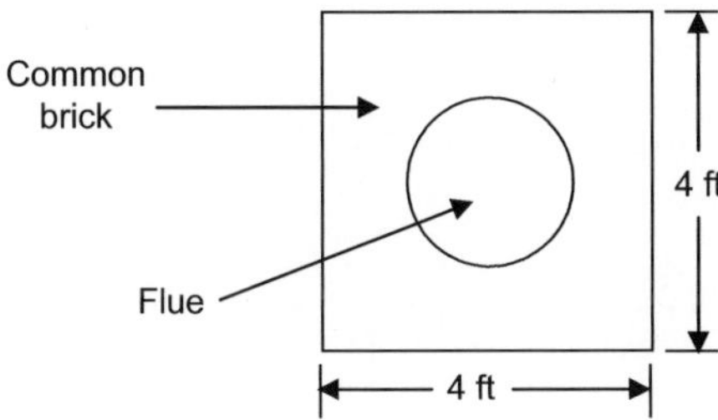

Ans. 910 Btu/h · ft of chimney height.

(1.26) A new underground pipeline at a chemical complex is to be placed parallel to an existing underground pipeline. The existing line has an OD of 8.9 cm, carries a fluid at 283 K and is not insulated. The new line will have an OD of 11.4 cm and will carry a fluid at 335 K. The center-to-center distance between the two pipelines will be 0.76 m. The ground at the plant site has an average thermal conductivity of 0.7 W/m · K. In order to determine whether the new line will need to be insulated, calculate the rate of heat transfer between the two pipelines per meter of pipe length if the new line is not insulated. For the purpose of this calculation, neglect the resistances of the pipe walls and the convective resistances between the fluids and pipe walls.

Ans. 42 W/m of pipe length.

(1.27) Hot waste gas at 350°F will be transported from a new process unit to a pollution control device via an underground duct. The duct will be rectangular in cross-section with a height of 3 ft and a width of 5 ft. The top of the duct will be 1.25 ft below the ground surface, which for design purposes has been assigned a temperature of 40°F. The average thermal conductivity of the ground at the plant site is 0.4 Btu/h · ft · °F. Calculate the rate of heat loss from the waste gas per foot of duct length. What thermal resistances are neglected in your calculation?

The following shape factor for a buried rectangular solid is available in the literature:

$$S = 2.756\,L\left[\ln\left(1+\frac{h}{a}\right)\right]^{-0.59}\left(\frac{h}{b}\right)^{-0.078}$$

$$L >> h, a, b$$

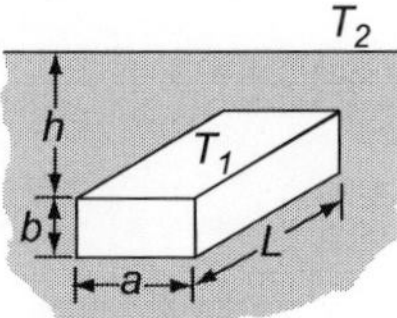

(1.28) An industrial furnace wall will be made of diatomaceous refractory brick ($\alpha = 1.3228 \times 10^{-7}$ m^2/s) and is to be designed so that the exterior surface will remain cool enough for safety purposes. The design criterion is that the mid-plane temperature in the wall will not exceed 400 K after 8 h of operation with an interior wall surface temperature of 1100 K.

(a) Assume that the furnace wall can be approximated as a semi-infinite solid. Calculate the wall thickness required to meet the design specification assuming that the wall is initially at a uniform temperature of 300 K.

(b) Using the wall thickness obtained in part (a), calculate the exterior wall surface temperature after 8 h of operation.

(c) Based on the above results, is the assumption that the furnace wall can be approximated as a semi-infinite solid justified, i.e., is the wall thickness calculated in part (a) acceptable for design purposes? Explain why or why not.

Ans. (a) 26.8 cm. (b) 301.7 K.

(1.29) The steel panel ($\alpha = 0.97 \times 10^{-5}\ \text{m}^2/\text{s}$) of a firewall is 5 cm thick and its interior surface is insulated. The panel is initially at 25°C when its exterior surface is suddenly exposed to a temperature of 250°C. Calculate the average temperature of the panel after 2 min of exposure to this temperature.

Note: A wall of width s with the temperature of one side suddenly raised to T_s and the opposite side insulated is mathematically equivalent to a wall of width 2s with the temperature of both sides suddenly raised to T_s. In the latter case, $dT/dx = 0$ at the mid-plane due to symmetry, which is the same condition that exists at a perfectly insulated boundary.

Ans. 192°C.

(1.30) An un-insulated metal storage tank at a chemical plant is cylindrical in shape with a diameter of 4 ft and a length of 25 ft. The liquid in the tank, which has properties similar to those of water, is at a temperature of 70°F when a frontal passage rapidly drops the ambient temperature to 40°F. Assuming that ambient conditions remain constant for an extended period of time, estimate:

(a) The average temperature of the liquid in the tank 12 h after the frontal passage.
(b) The time required for the average temperature of the liquid to reach 50°F

Ans. (a) 59°F. (b) 92 h.

(1.31) Repeat Problem 1.30 for the situation in which the fluid in the tank is

(a) Methyl alcohol.
(b) Aniline.

(1.32) Repeat Problem 1.30 for the situation in which the tank is spherical in shape with a diameter of 4.2 ft.

Ans. (a) 63°F. (b) 207 h (From Equation (1.44). Note that $Fo < 0.1$.).

2 Convective and Radiative Heat Transfer

2.1 Introduction

Convective heat transfer occurs when a gas or liquid flows past a solid surface whose temperature is different from that of the fluid. Examples include an organic heat-transfer fluid flowing inside a pipe whose wall is heated by electrical heating tape, and air flowing over the outside of a tube whose wall is chilled by evaporation of a refrigerant inside the tube. Two broad categories of convective heat transfer are distinguished; namely, forced convection and natural (or free) convection. In forced convection, the fluid motion is caused by an external agent such as a pump or blower. In natural convection, the fluid motion is the result of buoyancy forces created by temperature differences within the fluid.

In contrast to conductive heat transfer, convective heat-transfer problems are usually solved by means of empirical correlations derived from experimental data and dimensional analysis. The reason is that in order to solve a convection problem from first principles, one must solve the equations of fluid motion along with the energy balance equation. Although many important results have been obtained by solving the fundamental equations for convection problems in which the flow is laminar, no method has yet been devised to solve the turbulent flow equations entirely from first principles.

The empirical correlations are usually expressed in terms of a heat-transfer coefficient, h, which is defined by the relation:

$$q = hA\,\Delta T \tag{2.1}$$

In this equation, q is the rate of heat transfer between the solid surface and the fluid, A is the area over which the heat transfer occurs, and ΔT is a characteristic temperature difference between the solid and the fluid. Equation (2.1) is often referred to as Newton's Law of Cooling, even though Newton had little to do with its development, and it is not really a physical law. It is simply a definition of the quantity, h. Note that the units of h are $W/m^2 \cdot K$ or $Btu/h \cdot ft^2 \cdot °F$.

Equation (2.1) may appear to be similar to Fourier's Law of heat conduction. However, the coefficient, h, is an entirely different kind of entity from the thermal conductivity, k, which appears as the constant of proportionality in Fourier's Law. In particular, h is not a material property. It depends not only on temperature and pressure, but also on such factors as geometry, the hydrodynamic regime (laminar or turbulent), and in turbulent flow, the intensity of the turbulence and the roughness of the solid surface. Hence, the heat-transfer coefficient should not be regarded as a fundamental quantity, but simply as a vehicle through which the empirical methods are implemented.

From the standpoint of transferring heat, turbulent flow is highly desirable. In general, heat-transfer rates can be ordered according to the mechanism of heat transfer as follows:

conduction < natural convection < laminar forced convection < turbulent forced convection

The reason that turbulent flow is so effective at transferring heat is that the turbulent eddies can rapidly transport fluid from one area to another. When this occurs, the thermal energy of the fluid is transported along with the fluid itself. This eddy transport mechanism is much faster (typically, about two orders of magnitude) than the molecular transport mechanism of heat conduction.

Radiative heat transfer is important in high-temperature operations such as those involving boilers, furnaces, fired heaters and some chemical reactors. It can also be significant at moderate temperatures when the rate of heat transport by convection is relatively low, as is often the case with natural convection.

The heat-transfer correlations presented in this chapter are valid for most common Newtonian fluids. They are not valid for liquid metals or for non-Newtonian fluids. Special correlations are required for these types of fluids.

2.2 Combined Conduction and Convection

Heat-transfer problems involving both conduction and convection can be conveniently handled by means of the thermal resistance concept. Equation (2.1) can be put into the form of a thermal resistance as follows:

$$q = hA\,\Delta T = \Delta T / R_{th}$$

Thus,

$$R_{th} = \frac{1}{hA} \tag{2.2}$$

This expression for convective thermal resistance was given previously in Table 1.2. Convective resistances can be combined with other thermal resistances according to the rules for adding resistances in series and in parallel.

Example 2.1

A 5 cm (2 in.) schedule 40 steel pipe carries a heat-transfer fluid and is covered with a 2 cm layer of calcium silicate insulation to reduce the heat loss. The inside and outside pipe diameters are 5.25 cm and 6.03 cm, respectively. The fluid temperature is 150°C, and the temperature of the exterior surface of the insulation is 25°C. The coefficient of heat transfer between the fluid and the inner pipe wall is 700 W/m^2 · K. Calculate the rate of heat loss per unit length of pipe.

Solution

$$q_r = \frac{\Delta T}{R_{th}} = \frac{150 - 25}{R_{th}}$$

$$R_{th} = R_{convection} + R_{pipe} + R_{insulation}$$

$$R_{convection} = \frac{1}{hA} = \frac{1}{h\pi DL} = \frac{1}{700 \times \pi \times 0.0525 \times 1} = 0.008661 \text{ K/W}$$

$$R_{pipe} = 0.000513 \text{ K/W} \text{ (from Example 1.6)}$$

$$R_{insulation} = 1.349723 \text{ K/W} \text{ (from Example 1.6)}$$

$$R_{th} = 0.008661 + 0.000513 + 1.349723 = 1.358897 \text{ K/W}$$

$$q_r = \frac{125}{1.358897} \cong 92.0 \text{ W/m of pipe}$$

Another situation involving conduction and convection is the transient heating or cooling of a solid in contact with a fluid as depicted in Figure 2.1. For the special case in which the thermal resistance within the solid is small compared with the convective resistance between the fluid and solid, temperature variations in the solid are small and can be neglected. The temperature of the solid can then be approximated as a function of time only.

A thermal energy balance on the solid gives:

$$\{\text{rate of accumulation of thermal energy}\} = \{\text{rate of heat transfer from fluid}\} \tag{2.3}$$

$$\rho v c \frac{dT}{dt} = hA_s(T_\infty - T)$$

$$\frac{dT}{dt} = \frac{hA_s}{\rho v c}(T_\infty - T) \tag{2.4}$$

where ρ, c, v, and A_s are, respectively, the density, heat capacity, volume, and surface area of the solid.

The solution to this linear first-order differential equation subject to the initial condition that $T = T_0$ when $t = 0$ is:

$$T(t) = T_\infty + (T_0 - T_\infty)\exp\left(-\frac{hA_s t}{\rho v c}\right) \tag{2.5}$$

The approximation $T = T(t)$ is generally acceptable when the following condition is satisfied [1]:

$$\frac{h(v/A_s)}{k_{solid}} < 0.1 \tag{2.6}$$

where k_{solid} is the thermal conductivity of the solid. Equation (2.5) can also be applied to the transient heating or cooling of stirred process vessels. In this case, the fluid inside the vessel is maintained at a nearly uniform temperature by the mixing. The physical properties (ρ and c) of the fluid in the vessel are used in Equation (2.5).

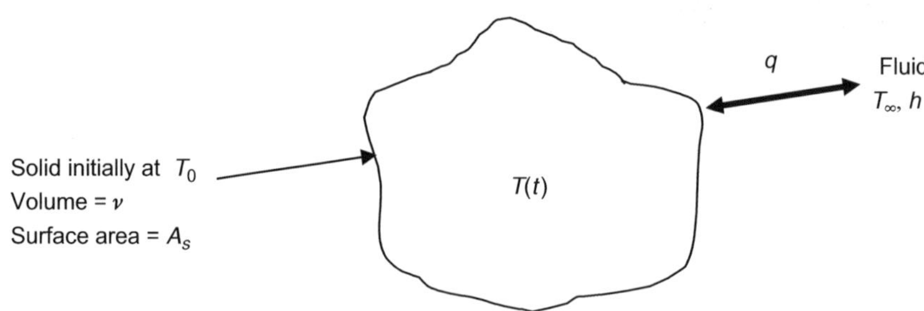

FIGURE 2.1 Transient heat transfer between a solid and a fluid.

Example 2.2

A batch chemical reactor operates at 400 K and the contents are well mixed. The reactor volume is 0.8 m^3 with a surface area of 4.7 m^2. After the reaction is complete, the contents are cooled to 320 K before the reactor is emptied. The cooling is accomplished with ambient air at 300 K and $h = 75$ W/$m^2 \cdot$ K. The agitator continues to operate during cool down. The reactor contents have a density of 840 kg/m^3 and a heat capacity of 2200 J/kg $\cdot$ K. Determine the cooling time required.

Solution

Neglect the thermal capacity of the reactor vessel and heat loss by radiation. Also, neglect the thermal resistance of the vessel wall and the convective resistance between the wall and the fluid in the reactor. Equation (2.5) is then applicable since the reactor contents remain well mixed during cooling.

$$T(t) = T_\infty + (T_0 - T_\infty)\exp\left(-\frac{hA_s t}{\rho v c}\right)$$

$$320 = 300 + (400 - 300)\ \exp\left(-\frac{75 \times 4.7t}{840 \times 0.8 \times 2200}\right)$$

$$t = 6,750\ \text{s} \cong 1.9\ \text{h}$$

2.3 Extended Surfaces

An important application involving both conduction and convection is that of heat-transfer fins, also referred to as extended surfaces. Although fins come in a variety of shapes, the two most frequently used in process heat exchangers are rectangular fins (Figure 2.2) and annular (radial or transverse) fins (Figure 2.3). The basic idea behind fins is to compensate for a low heat-transfer coefficient, h, by increasing the surface area, A. Thus, fins are almost always used when heat is transferred to or from air or other gases, because heat-transfer coefficients for gases are generally quite low compared with those for liquids. Since fins usually consist of very thin pieces of metal attached to the primary heat-transfer surface (the prime surface), a relatively large amount of additional surface area is achieved with a small amount of material.

All the heat transferred by convection from the fin surface must first be transferred by conduction through the fin from its base, which is generally taken to be at the same temperature as the prime surface. Therefore, a temperature gradient must exist along the length of the fin, and as the distance from the base of the fin increases, the temperature of the fin becomes closer to the temperature of the surrounding fluid. As a result, the ΔT for convective heat transfer decreases along the length of the fin, and hence the extended surface is less effective in transferring heat than the prime surface. In order to calculate the rate of heat transfer from a fin, it is

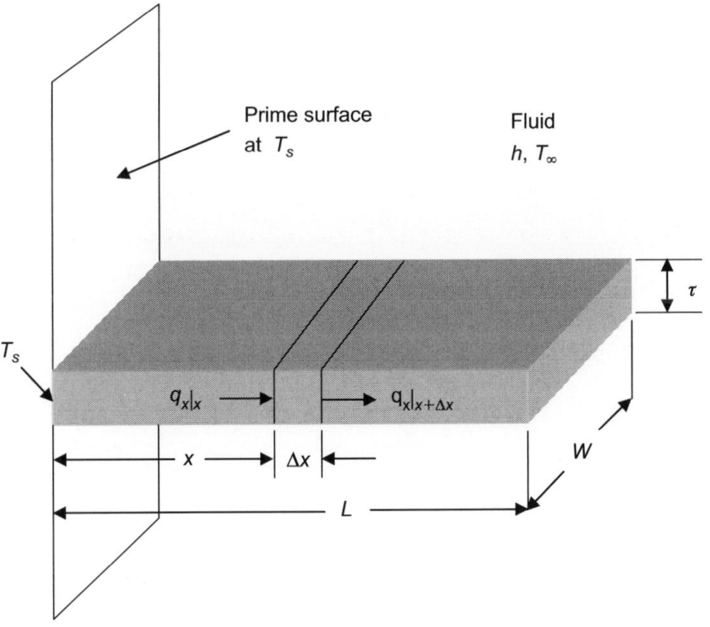

FIGURE 2.2 Rectangular fin attached to a flat surface.

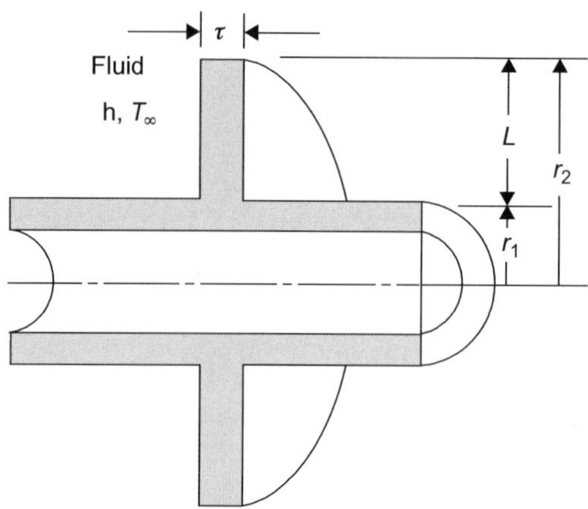

FIGURE 2.3 Annular fin attached to a tube wall.

necessary to determine the temperature profile along the length of the fin. We consider a rectangular fin attached to a flat surface as depicted in Figure 2.2, and make a thermal energy balance on a differential volume element. The following assumptions are made:

- Steady state
- Heat transfer from fin surface by convection only (no radiation)
- No heat generation in the fin
- Constant thermal conductivity in the fin
- Constant heat-transfer coefficient
- Constant fluid temperature surrounding the fin
- Negligible temperature difference across the thickness of the fin
- Negligible edge effects

The last two assumptions are justified because the fin is assumed to be very thin (like a knife blade), and made of metal (a good heat conductor). Therefore, there is very little thermal resistance between the top and bottom of the fin, and hence little temperature difference. Furthermore, the fin width, W, is much greater than the thickness, τ, and so nearly all the surface area resides on the top and bottom of the fin. Hence, nearly all the convective heat transfer occurs at the top and bottom surfaces, and very little occurs at the sides and tip of the fin. The upshot of all this is that the temperature in the fin can be considered a function only of x, i.e., $T = T(x)$.

With steady state and no heat generation, the thermal energy balance on the control volume is simply:

$$\{\text{rate of thermal energy in}\} - \{\text{rate of thermal energy out}\} = 0 \tag{2.7}$$

Heat enters the control volume by conduction at position x, and leaves by conduction at position $x + \Delta x$ and by convection at the surface of the control volume. Thus, the balance equation can be written:

$$\hat{q}_x\big|_x A - \hat{q}_x\big|_{x+\Delta x} A - \hat{q}_{cv} P\ \Delta x = 0 \tag{2.8}$$

where

$A = W\tau$ = fin cross-sectional area
$P = 2(W + \tau)$ = fin perimeter

Note that $P\Delta x$ is the total surface area of the differential element. Dividing by Δx and taking the limit as $\Delta x \to 0$ yields:

$$-A\frac{d\hat{q}_x}{dx} - \hat{q}_{cv}P = 0 \tag{2.9}$$

Now from Fourier's Law,

$$\hat{q}_x = -k\frac{dT}{dx}$$

and from Equation (2.1), the convective heat flux is:

$$\hat{q}_{cv} = h\ \Delta T = h(T - T_\infty) \tag{2.10}$$

Substituting and utilizing the fact that k is constant gives:

$$kA\frac{d^2T}{dx^2} - hP(T - T_\infty) = 0 \tag{2.11}$$

or

$$\frac{d^2T}{dx^2} - \frac{hP}{kA}(T - T_\infty) = 0 \tag{2.12}$$

Equation (2.12) can be simplified by making the following definitions:

$$\theta \equiv T - T_\infty \tag{2.13}$$

$$m^2 \equiv \frac{hP}{kA} = \frac{2h(W+\tau)}{kW\tau} \cong \frac{2h}{k\tau} \quad (\text{since } \tau << W) \tag{2.14}$$

(*Note:* The above approximation for m^2 is frequently used in practice.)

Making these substitutions and assuming that the fluid temperature, T_∞, is a constant yields:

$$\frac{d^2\theta}{dx^2} - m^2\theta = 0 \tag{2.15}$$

This equation is subject to the following boundary conditions:

$$\begin{aligned} &1.\ \text{At } x = 0,\ \theta = T_s - T_\infty \\ &2.\ \text{At } x = L,\ \frac{d\theta}{dx} = 0 \end{aligned} \tag{2.16}$$

The first condition results from the assumption that the temperature at the base of the fin is equal to the prime surface temperature, T_s. The second condition follows from the fact that there is very little heat transfer from the tip of the fin. Setting the heat flux at the end of the fin to zero gives $dT/dx = d\theta/dx = 0$.

The general solution to Equation (2.15) can be expressed either as a sum of exponential functions or as a sum of hyperbolic functions, but the latter is more convenient in this context:

$$\theta = C_1 \sinh(mx) + C_2 \cosh(mx) \tag{2.17}$$

Applying the boundary conditions (2.16) gives:

$$C_1 = -(T_s - T_\infty)\tanh(mL)$$

$$C_2 = T_s - T_\infty$$

With these results, the temperature profile along the fin becomes:

$$T = T_\infty - (T_s - T_\infty)\,[\tanh(mL)\sinh(mx) - \cosh(mx)] \tag{2.18}$$

To find the rate of heat transfer from the fin, note that all the heat must enter the fin by conduction at the base, where $x = 0$. Therefore,

$$q_{fin} = -kA\left.\frac{dT}{dx}\right|_{x=0} \tag{2.19}$$

Differentiating Equation (2.18) and evaluating at $x = 0$ gives:

$$\left.\frac{dT}{dx}\right|_{x=0} = -m(T_s - T_\infty)\ \tanh(mL) \tag{2.20}$$

Substituting into Equation (2.19) gives:

$$q_{fin} = kAm(T_s - T_\infty)\ \tanh(mL)$$

or

$$q_{fin} = (hPkA)^{1/2}(T_s - T_\infty)\ \tanh(mL) \tag{2.21}$$

Now the maximum possible rate of heat transfer occurs when the entire fin is at the prime surface temperature, T_s:

$$q_{max} = hPL(T_s - T_\infty) \tag{2.22}$$

The fin efficiency, η_f, is defined as the ratio of q_{fin} to q_{max}:

$$\eta_f = \frac{q_{fin}}{q_{max}} = \frac{(hPk\ A)^{1/2}(T_s - T_\infty)\ \tanh(mL)}{hPL\ (T_s - T_\infty)}$$

or

$$\eta_f = \frac{\tanh(mL)}{mL} \tag{2.23}$$

The rate of heat transfer from the fin can now be expressed in terms of the fin efficiency as:

$$q_{fin} = \eta_f q_{\max}$$

or

$$q_{fin} = \eta_f h A_{fin} \left| T_s - T_\infty \right| \tag{2.24}$$

where the surface area, A_{fin}, is equal to PL for the rectangular fin, and absolute value signs have been added to make q_{fin} positive regardless of the direction of heat flow.

One further modification is made to the fin equations to account (approximately) for the heat transfer from the fin tip, which was neglected in the above analysis. This is accomplished by, in effect, adding the tip area to the periphery of the fin. To this end, the fin length is increased by an amount, ΔL, determined as follows:

$$P\Delta L = \text{tip area} = W\tau$$

$$\Delta L = \frac{W\tau}{P} = \frac{W\tau}{2(W+\tau)} \cong \frac{\tau}{2} \tag{2.25}$$

The corrected length L_c is defined by:

$$L_c = L + \Delta L = L + \frac{\tau}{2} \tag{2.26}$$

The corrected length is then used in place of L in Equations (2.23) and (2.24).

The analysis of an annular fin is somewhat more complicated and the equation for fin efficiency involves Bessel functions [1,2], making it inconvenient to use. However, the following simple modification of Equation (2.23) provides a very good approximation for annular fins [3]:

$$\eta_f \cong \frac{\tanh(m\psi)}{m\psi} \tag{2.27}$$

where

$\psi = (r_{2c} - r_1)\,[1 + 0.35\,\ln(r_{2c}/r_1)]$ = effective fin height
r_1 = inner radius of fin = external radius of tube (prime surface)
r_2 = outer radius of fin
$r_{2c} = r_2 + \tau/2$ = corrected fin radius

The rate of heat transfer from the fin is given by:

$$q_{fin} = 2\eta_f h\pi\left(r_{2c}^2 - r_1^2\right)\,|T_s - T_\infty| \tag{2.28}$$

The factor of two in Equation (2.28) results from the fact that heat is transferred from both sides of the fin.

The total rate of heat transfer from a finned surface is the sum of the heat-transfer rates from the prime surface and from all of the fins. Thus,

$$\begin{aligned} q &= q_{prime} + q_{fins} \\ &= hA_{prime}|T_s - T_\infty| + \eta_f h A_{fins}|T_s - T_\infty| \\ &= h\left[\frac{A_{prime} + \eta_f A_{fins}}{A_{Tot}}\right] A_{Tot}|T_s - T_\infty| \end{aligned} \tag{2.29}$$

$$q = h\eta_w A_{Tot}|T_s - T_\infty| \tag{2.30}$$

where

$$A_{Tot} = A_{prime} + A_{fins} = \text{total heat-transfer area}$$

$$\eta_w = \textit{weighted efficiency} = \frac{A_{prime} + \eta_f A_{fins}}{A_{Tot}} \tag{2.31}$$

It follows from Equation (2.30) that the thermal resistance of a finned surface is:

$$R_{th} = \frac{1}{h\eta_w A_{Tot}} \tag{2.32}$$

This result was previously listed in Table 1.2.

While Equations (2.23) and (2.27) are widely applied for fin efficiency calculation, they are only as accurate as the simplifying assumptions used in the above derivation. Variation in fluid temperature from base to tip and fin bond resistance both can change the fin efficiency, but a non-uniform heat-transfer coefficient has the most significant effect on the efficiency. The effect of non-uniform heat-transfer coefficient has been studied extensively and some useful results are discussed by Owen [4], Lymar and Ridal [5] and Zhukauskas et al. [6]. However, there is little consensus on a method to account for this non-ideal variation. For annular fins, the following formula has been proposed to estimate the actual fin efficiency, η_{af}, which includes the effect of non-uniform heat-transfer coefficient [6]:

$$\eta_{af} = [0.97 - 0.056m\psi]\eta_f \tag{2.33}$$

The equation is valid for $0.3 < m\psi < 3.0$.

It has been confirmed that at high fin efficiencies ($\eta_f > 0.9$), the actual and theoretical fin efficiencies deviate by less than 5%, whereas at lower fin efficiencies ($\eta_f < 0.6$) the deviation is more than 10%. While the deviation is large in some conditions, overall heat-transfer predictions are generally accurate provided that the analyst uses the same method of calculating fin efficiency as the researcher who experimentally derived the heat-transfer correlation. Nevertheless, caution is needed when extrapolating finned heat-transfer methods to conditions outside of experimental and industry experience.

Example 2.3

A rectangular aluminum alloy (Duralumin) fin is 2 in. long, 0.1 in. thick, and 40 in. wide. It is attached to a prime surface at 150°F and is surrounded by a fluid at 100°F with a heat-transfer coefficient of 75 Btu/h · ft^2 · °F. Calculate the fin efficiency and the rate of heat transfer from the fin.

Solution

$$k = 164 \times 0.57782 = 94.76 \text{ Btu/h·ft·°F (Table A.2)}$$

$$m^2 \cong \frac{2h}{k\tau} = \frac{2 \times 75}{94.76(0.1/12)} = 189.95$$

$$m = \sqrt{189.95} = 13.78 \text{ ft}^{-1}$$

$$L_c = L + \tau/2 = \frac{2 + (0.1/2)}{12} = 0.17083 \text{ ft}$$

$$mL_c = 13.78 \times 0.17083 = 2.354$$

Using Equation (2.23) with L replaced by L_c

$$\eta_f = \frac{\tanh(mL_c)}{mL_c} = \frac{\tanh(2.354)}{2.354} \cong 0.417$$

The rate of heat transfer from the fin is computed from Equation (2.24).

$$q_{fin} = \eta_f hPL_c|T_s - T_\infty|$$

$$P = 2\ (W + \tau) = 2(40 + 0.1)/12 = 6.6833 \text{ ft}$$

$$q_{fin} = 0.417 \times 75 \times 6.6833 \times 0.17083\ (150 - 100)$$

$$q_{fin} \cong 1,785 \text{ Btu/h}$$

Example 2.4

A 0.1-in. thick annular Duralumin fin with an external diameter of 3 in. is attached to a tube having an external diameter of 2 in. The tube wall temperature is 150°F and the tube is surrounded by a fluid at 100°F with a heat-transfer coefficient of 75 Btu/h · ft^2 · °F.

(a) Calculate the fin efficiency and the rate of heat transfer from the fin.
(b) If the total heat-transfer surface consists of 75% fin surface and 25% prime surface, calculate the weighted efficiency of the surface.
(c) Estimate the effect on fin efficiency due to a non-uniform heat-transfer coefficient.

Solution

(a)

$$\text{Tube } OD = 2 \text{ in.} \Rightarrow r_1 = 1 \text{ in.}$$

$$\text{Fin } OD = 3 \text{ in.} \Rightarrow r_2 = 1.5 \text{ in.}$$

$$r_{2c} = r_2 + \tau/2 = 1.5 + 0.1/2 = 1.55 \text{ in.}$$

$$k = 94.76 \text{ Btu/h·ft·°F (from Example 2.3)}$$

$$m = \left(\frac{2h}{k\tau}\right)^{0.5} = \left(\frac{2 \times 75}{94.76(0.1/12)}\right)^{0.5} = 13.782 \text{ ft}^{-1}$$

$$\psi = (r_{2c} - r_1)\ [1 + 0.35 \ln (r_{2c}/r_1)] = (1.55 - 1.0)\ [1 + 0.35 \ln(1.55/1.0)]$$

$$\psi = 0.63436 \text{ in.} = 0.05286 \text{ ft}$$

$$m\psi = 13.782 \times 0.05286 = 0.7285$$

From Equation (2.27), the fin efficiency is:

$$\eta_f = \frac{\tanh(m\psi)}{m\psi} = \frac{\tanh(0.7285)}{0.7285} = 0.854$$

The rate of heat transfer from the fin is given by Equation (2.28):

$$q_{fin} = 2\eta_f h\pi\left(r_{2c}^2 - r_1^2\right)\ |T_s - T_\infty|$$
$$= 2 \times 0.854 \times 75\pi\left[(1.55/12)^2 - (1.0/12)^2\right]\ (150 - 100)$$
$$q_{fin} \cong 196 \text{ Btu/h}$$

(b) The weighted efficiency is computed from Equation (2.31):

$$\eta_w = \left(A_{prime}/A_{Tot}\right) + \eta_f\left(A_{fins}/A_{Tot}\right)$$
$$= 0.25 + 0.854 \times 0.75$$
$$\eta_w = 0.89$$

(c) Equation (2.33) is used to estimate the actual fin efficiency:

$$\eta_{af} = [0.97 - 0.056 m\psi]\eta_f$$
$$= [0.97 - 0.056 \times 0.7285] \times 0.854$$
$$\eta_{af} \cong 0.794$$

Using this value in Equation (2.31), the weighted efficiency becomes:

$$\eta_w = 0.25 + 0.794 \times 0.75 \cong 0.85$$

Thus, the Zhukauskas equation predicts a 7% reduction in the fin efficiency, with a corresponding reduction of about 4.5% in the weighted efficiency of the finned surface.

2.4 Forced Convection in Pipes and Ducts

Heat transfer to fluids flowing inside pipes and ducts is a subject of great practical importance. Consequently, many correlations for calculating heat-transfer coefficients in this type of flow have been proposed. However, the correlations presented below are adequate to handle the vast majority of process heat-transfer applications. Correlations are given for each of the three flow regimes: turbulent, transition, and laminar. The physical situation described by the correlations is depicted in Figure 2.4. Fluid enters the pipe at an average temperature T_{b1} and leaves at an average temperature T_{b2}. (T_b stands for bulk temperature. It is the fluid temperature averaged over the flow area.) The pipe wall temperature is T_{w1} at the entrance and T_{w2} at the exit. The fluid is either heated or cooled as it flows through the pipe, depending on whether the wall temperature is greater or less than the fluid temperature.

FIGURE 2.4 Heat transfer in pipe flow.

With respect to heat transfer in circular pipes, fully developed turbulent flow is achieved at a Reynolds number of approximately 10^4. For this flow regime ($Re \geq 10^4$), the following correlation is widely used:

$$Nu = 0.027\, Re^{0.8}\, Pr^{1/3}(\mu/\mu_w)^{0.14} \tag{2.34}$$

where

$Nu \equiv$ Nusselt Number $\equiv hD/k$
$Re \equiv$ Reynolds Number $\equiv DV\rho/\mu$
$Pr \equiv$ Prandtl Number $= C_P\mu/k$
$D =$ inside pipe diameter
$V =$ average fluid velocity
$C_P, \mu, \rho, k =$ fluid properties evaluated at the average bulk fluid temperature
$\mu_w =$ fluid viscosity evaluated at average wall temperature

The average bulk fluid temperature and average wall temperature are given by:

$$T_{b,ave} = \frac{T_{b1} + T_{b2}}{2} \tag{2.35}$$

$$T_{w,ave} = \frac{T_{w1} + T_{w2}}{2} \tag{2.36}$$

Equation (2.34) is generally attributed to Seider and Tate [7], although they did not explicitly formulate it. The equation is frequently written with the coefficient 0.027 replaced by 0.023. The latter value gives a somewhat more conservative estimate for the heat-transfer coefficient, which is often desirable for design purposes. Both coefficients will be used in this book. Although this may cause some consternation, it does reflect current practice, as both values are in common use.

Equation (2.34) is valid for fluids with Prandtl numbers between 0.5 and 17,000, and for pipes with $L/D > 10$. However, for short pipes with $10 < L/D < 60$, the right-hand side of the equation is often multiplied by the factor $[1 + (D/L)^{2/3}]$ to correct for entrance and exit effects. The correlation is generally accurate to within ±20% to ±40%. It is most accurate for fluids with low to moderate Prandtl numbers ($0.5 \leq Pr \leq 100$), which includes all gases and low-viscosity process liquids such as water, organic solvents, light hydrocarbons, etc. It is less accurate for highly viscous liquids, which have correspondingly large Prandtl numbers.

For laminar flow in circular pipes ($Re < 2100$), the Seider-Tate correlation takes the form:

$$Nu = 1.86[RePrD/L]^{1/3}\ (\mu/\mu_w)^{0.14} \tag{2.37}$$

This equation is valid for $0.5 < Pr < 17{,}000$ and $(RePrD/L)^{1/3}\ (\mu/\mu_w)^{0.14} > 2$, and is generally accurate to within ±25%. Fluid properties are evaluated at $T_{b,ave}$ except for μ_w, which is evaluated at $T_{w,ave}$. For $(RePrD/L)^{1/3}\ (\mu/\mu_w)^{0.14} < 2$, the Nusselt number should be set to 3.66, which is the theoretical value for laminar flow in an infinitely long pipe with constant wall temperature. Also, at low Reynolds numbers heat transfer by natural convection can be significant (see Section 2.6 below), and this effect is not accounted for in the Seider–Tate correlation.

Note: A critical Reynolds number of 2100 is assumed herein for the attainment of sustained turbulence in pipe flow. A value of 2040± 10 has recently been determined based on both experimental and computer simulation studies [8].

For flow in the transition region ($2100 < Re < 10^4$), the Hausen correlation is recommended [2,9]:

$$Nu = 0.116\left[Re^{2/3} - 125\right]Pr^{1/3}(\mu/\mu_w)^{0.14}\left[1 + (D/L)^{2/3}\right] \tag{2.38}$$

Heat-transfer calculations in the transition region are subject to a higher degree of uncertainty than those in the laminar and fully developed turbulent regimes. Although industrial equipment is sometimes designed to operate in the transition region, it is generally recommended to avoid working in this flow regime if possible. Fluid properties in Equation (2.38) are evaluated in the same manner as with the Seider–Tate correlations.

An alternative equation for the transition and turbulent regimes has been proposed by Gnielinski [10]:

$$Nu = \frac{(f/8)\ (Re - 1{,}000)\ Pr}{1 + 12.7\sqrt{f/8}\left(Pr^{2/3} - 1\right)}\left[1 + (D/L)^{2/3}\right] \tag{2.39}$$

Here, f is the Darcy friction factor, which can be computed from the following explicit approximation of the Colebrook equation [11]:

$$f = (0.782\ \ln Re - 1.51)^{-2} \tag{2.40}$$

Equation (2.39) is valid for $2100 < Re < 10^6$ and $0.6 < Pr < 2000$. It is generally accurate to within $\pm$ 20%.

For flow in ducts and conduits with non-circular cross-sections, Equations (2.34) and (2.38)–(2.40), can be used if the diameter is replaced everywhere by the equivalent diameter, D_e, where

$$D_e = 4 \times \text{hydraulic radius} = 4 \times \text{flow area/wetted perimeter} \tag{2.41}$$

This approximation generally gives reliable results for turbulent flow. However, it is not recommended for laminar flow.

The most frequently encountered case of laminar flow in non-circular ducts is flow in the annulus of a double-pipe heat exchanger. For laminar annular flow, the following equation given by Gnielinski [10] is recommended:

$$Nu = 3.66 + 1.2(D_2/D_1)^{0.8} + \frac{0.19\left[1 + 0.14\ (D_2/D_1)^{0.5}\right]\ [Re\,Pr\,D_e/L]^{0.8}}{1 + 0.117[Re\,Pr\,D_e/L]^{0.467}} \tag{2.42}$$

where

D_1 = outside diameter of inner pipe
D_2 = inside diameter of outer pipe
D_e = equivalent diameter = $D_2 - D_1$

The Nusselt number in Equation (2.42) is based on the equivalent diameter, D_e.

All of the above correlations give average values of the heat-transfer coefficient over the entire length, L, of the pipe. Hence, the total rate of heat transfer between the fluid and the pipe wall can be calculated from Equation (2.1), which takes the following form:

$$q = hA\Delta T_{\ln} \tag{2.43}$$

In this equation, A is the total heat-transfer surface area (πDL for a circular pipe) and $\Delta T_{\ln}$ is an average temperature difference between the fluid and the pipe wall. A logarithmic average is used; it is termed the logarithmic mean temperature difference (LMTD) and is defined by:

$$\Delta T_{\ln} = \frac{\Delta T_1 - \Delta T_2}{\ln(\Delta T_1/\Delta T_2)} \tag{2.44}$$

where

$$\Delta T_1 = |T_{w1} - T_{b1}|$$
$$\Delta T_2 = |T_{w2} - T_{b2}|$$

Like any mean value, the LMTD lies between the extreme values, ΔT_1 and ΔT_2. Hence, when ΔT_1 and ΔT_2 are not greatly different, the LMTD will be approximately equal to the arithmetic mean temperature difference, by virtue of the fact that they both lie between ΔT_1 and ΔT_2. The arithmetic mean temperature difference, ΔT_{ave}, is given by:

$$\Delta T_{ave} = \frac{\Delta T_1 + \Delta T_2}{2} \tag{2.45}$$

The difference between $\Delta T_{\ln}$ and ΔT_{ave} is small when the flow is laminar and $RePrD/L > 10$ [12].

Example 2.5

Carbon dioxide at 300 K and 1 atm is to be pumped through a 5 cm ID pipe at a rate of 50 kg/h. The pipe wall will be maintained at a temperature of 450 K in order to raise the carbon dioxide temperature to 400 K. What length of pipe will be required?

Solution

Equation (2.43) may be written as:

$$q = hA\Delta T_{\ln} = h\pi DL\Delta T_{\ln}$$

In order to solve the equation for the length, L, we must first determine q, h, and ΔT_{ln}. The required physical properties of CO_2 are obtained from Table A.10.

Property	At $T_{b,ave} = 350$ K	At $T_{w,ave} = 450$ K
μ (N · s/m²)	17.21×10^{-6}	21.34×10^{-6}
C_P (J/kg · K)	900	
k (W/m · K)	0.02047	
Pr	0.755	

(i)

$$Re = \frac{DV\rho}{\mu} = \frac{4\dot{m}}{\pi D\mu} = \frac{4 \times (50/3600)}{\pi \times 0.05 \times 17.21 \times 10^{-6}} = 2.06 \times 10^4$$

Since $Re > 10^4$, the flow is turbulent, and Equation (2.34) is applicable.

(ii)

$$Nu = 0.027Re^{0.8}Pr^{1/3}(\mu/\mu_w)^{0.14}$$

$$Nu = 0.027\left(2.06 \times 10^4\right)^{0.8}(0.755)^{1/3}\left(\frac{17.21}{21.34}\right)^{0.14} = 67.41$$

$$h = (k/D)Nu = (0.02047/0.05) \times 67.41$$

$$h = 27.6\ \text{W/m}^2\cdot\text{K}$$

(iii)

$$\Delta T_{ln} = \frac{\Delta T_1 - \Delta T_2}{\ln(\Delta T_1/\Delta T_2)}$$

$$= \frac{(450-300)-(450-400)}{\ln\left(\frac{450-300}{450-400}\right)} = \frac{100}{\ln(3)}$$

$$\Delta T_{ln} = 91.02\ \text{K}$$

(iv) The rate of heat transfer, q, is obtained from an energy balance on the CO_2.

$$q = \dot{m}C_P(\Delta T)_{CO2}$$

$$= \left(\frac{50}{3600}\right) \times 900 \times (400-300)$$

$$q = 1250\text{W}$$

(v)

$$q = h\pi DL\Delta T_{ln}$$

$$L = \frac{q}{\pi Dh\Delta T_{ln}} = \frac{1250}{\pi \times 0.05 \times 27.6 \times 91.02} = 3.17\ \text{m}$$

Checking the length to diameter ratio,

$$L/D = \frac{3.17}{0.05} = 63.4 > 60$$

Hence, the correction factor for short pipes is not required. Therefore, a length of about 3.2 m is required. However, if it is important that the outlet temperature be no less than 400 K, then it is advisable to provide additional heat-transfer area to compensate for the uncertainty in the Seider–Tate correlation. Thus, if the actual heat-transfer coefficient turns out to be 20% lower than the calculated value (about the worst error to be expected with the given values of Re and Pr), then a length of

$$L = \frac{3.17}{0.8} = 3.96 \cong 4.0\ \text{m}$$

will be required to achieve the specified outlet temperature.

When the length of the pipe is known and the outlet temperature is required, an iterative solution is usually necessary. A general calculation sequence is as follows:

(1) Estimate or assume a value of T_{b2}
(2) Calculate $T_{b,ave} = \dfrac{T_{b1}+T_{b2}}{2}$ and $T_{w,ave} = \dfrac{T_{w1}+T_{w2}}{2}$
(3) Obtain fluid properties at $T_{b,ave}$ and μ_w
(4) Calculate the Reynolds number ($Re = 4\dot{m}/\pi D\mu$ for a circular duct)
(5) Calculate h from appropriate correlation
(6) Calculate $\Delta T_{\ln}$
(7) Calculate $q = hA\Delta T_{\ln}$
(8) Calculate new value of T_{b2} from $q = \dot{m}C_P(T_{b2} - T_{b1})$
(9) Go to step 2 and iterate until two successive values of T_{b2} agree to within the desired accuracy.

Example 2.6

1000 lbm/h of oil at 100°F enters a 1-in. ID heated copper tube. The tube is 12 ft long and its inner surface is maintained at 215°F. Determine the outlet temperature of the oil. The following physical property data are available for the oil:

$$C_P = 0.5 \text{ Btu/lbm·°F}$$
$$\rho = 55\text{lbm/ft}^3$$
$$\mu = 1.5\text{lbm/ft·h}$$
$$k = 0.10 \text{ Btu/h·ft·°F}$$

Solution

Since the temperature dependencies of the fluid properties are not given, the properties will be assumed constant. In this case, the heat-transfer coefficient is independent of the outlet temperature, and can be computed at the outset.

$$Re = \frac{4\dot{m}}{\pi D\mu} = \frac{4 \times 1000}{\pi \times 1/12 \times 1.5} = 10,186 \Rightarrow \text{turbulent flow}$$

$$h = \frac{k}{D} \times 0.027\ Re^{0.8} Pr^{1/3} (\mu/\mu_w)^{0.14}$$

$$h = \frac{0.1}{(1/12)} \times 0.027\ (10,186)^{0.8} \left(\frac{0.5 \times 1.5}{0.1}\right)^{1/3} (1.0)$$

$$h = 102 \text{ Btu/h·ft}^2\text{·°F}$$

In order to obtain a first approximation for the outlet temperature, T_{b2}, we assume $\Delta T_{\ln} \cong \Delta T_{ave}$. Then,

$$q = \dot{m}C_P\Delta T_{oil} = h\pi DL\Delta T_{ave}$$

$$\dot{m}C_P(T_{b2} - 100) = h\pi DL\left[\frac{115 + (215 - T_{b2})}{2}\right]$$

Solving for T_{b2},

$$1000 \times 0.05(T_{b2} - 100) = 102 \times \pi \times \frac{1}{12} \times 12 \times \left[\frac{330 - T_{b2}}{2}\right]$$

$$500\ T_{b2} - 50,000 = 52,873 - 160.2\ T_{b2}$$

$$660.2\ T_{b2} = 102,873$$

$$T_{b2} = 155.8°\text{F}$$

To obtain a second approximation for T_{b2}, we next calculate the LMTD:

$$\Delta T_{\ln} = \frac{115 - (215 - 155.8)}{\ln\left(\dfrac{115}{215 - 155.8}\right)} = 84°\text{F}$$

Then,

$$q = h\pi DL\Delta T_{\ln}$$

$$= 102 \times \pi \times \frac{1}{12} \times 12 \times 84$$

$$q = 26,917 \text{ Btu/h}$$

But,

$$
\begin{aligned}
q &= \dot{m}C_P(T_{b2} - 100)\\
26,917 &= 1000 \times 0.5(T_{b2} - 100)\\
T_{b2} &= 154°\text{F}
\end{aligned}
$$

This value is sufficiently close to the previous approximation that no further iterations are necessary.

Convergence of the iterative procedure, as given above, is often very slow, and it is then necessary to use a convergence accelerator. Wegstein's method for solving a nonlinear equation of the form $x = f(x)$ is well suited for this purpose. The new value of T_{b2} calculated in step 8 of the iterative procedure depends on the old value of T_{b2} used in step 2, i.e.,

$$T_{b2,new} = F(T_{b2,old})$$

where the function, F, consists of steps 2 through 8. When the solution is reached, $T_{b2,new} = T_{b2,old}$, so that:

$$T_{b2,new} = F(T_{b2,new})$$

or

$$T = F(T)$$

which is the form to which Wegstein's method applies. The Wegstein iteration formula is:

$$T_{i+1} = T_i + \frac{(T_i - T_{i-1})}{\left(\dfrac{T_{i-1} - F(T_{i-1})}{T_i - F(T_i)} - 1\right)} \tag{2.46}$$

where T_{i-1}, T_i, and T_{i+1} are three successive approximations to the outlet temperature, T_{b2}.

Example 2.7

Calculate the outlet temperature of the oil stream of Example 2.6 using Wegstein's method and an initial guess of $T_{b2} = 120°$F.

Solution

Designate the initial approximation to the solution as T_0 and set $T_0 = 120$. A second approximation is obtained by following the iterative procedure to get a new value of T_{b2}.

$$
\begin{aligned}
h &= 102\ \text{Btu/h·ft}^2\text{·°F (from Example 2.6)}\\
\Delta T_{\ln} &= \frac{115 - (215 - 120)}{\ln[115/(215 - 120)]} = 104.68°\text{F}\\
q &= h\pi DL\Delta T_{\ln} = 102 \times \pi \times 1/12 \times 12 \times 104.68 = 33,544\ \text{Btu/h}\\
33,544 = q &= \dot{m}C_P(T_{b2} - T_{b1}) = 500(T_{b2} - 100)\\
T_{b2} &= 167.09°\text{F}
\end{aligned}
$$

Now this value becomes both T_1 and $F(T_0)$, i.e.,

$$T_1 = 167.09 \text{ and } F(T_0) = 167.09$$

To calculate $F(T_1)$, repeat the above calculations with T_1 as the initial guess for the outlet temperature:

$$
\begin{aligned}
\Delta T_{\ln} &= \frac{115 - (215 - 167.09)}{\ln[115/215 - 167.09]} = 76.62°\text{F}\\
q &= h\pi DL\Delta T_{\ln} = 102 \times \pi \times 1/12 \times 12 \times 76.62 = 24,552\ \text{Btu/h}\\
24,552 = q &= 500(T_{b2} - 100)\\
T_{b2} &= 149.10°F = F(T_1)
\end{aligned}
$$

We now have the following values:

$$
\begin{aligned}
T_0 &= 120 \qquad & F(T_0) &= 167.09\\
T_1 &= 167.09 \qquad & F(T_1) &= 149.10
\end{aligned}
$$

The next approximation, T_2, is obtained from Wegstein's formula:

$$T_2 = T_1 + \frac{(T_1 - T_0)}{\left[\dfrac{T_0 - F(T_0)}{T_1 - F(T_1)} - 1\right]}$$

$$T_2 = 167.09 + \frac{(167.09 - 120)}{\left[\dfrac{120 - 167.09}{167.09 - 149.10} - 1\right]} = 154.07°\text{F}$$

Next, $F(T_2)$ is calculated in the same manner as $F(T_1)$:

$$\Delta T_{\text{ln}} = \frac{115 - (215 - 154.07)}{\ln[115/(215 - 154.07)]} = 85.12°\text{F}$$

$$q = 102 \times \pi \times 1/12 \times 12 \times 85.12 = 27{,}276 \text{ Btu/h}$$

$$27{,}276 = 500(T_{b2} - 100)$$

$$T_{b2} = 154.55°\text{F} = F(T_2)$$

Applying Wegstein's formula again gives:

$$T_3 = T_2 + \frac{(T_2 - T_1)}{\left[\dfrac{T_1 - F(T_1)}{T_2 - F(T_2)} - 1\right]}$$

$$T_3 = 154.07 + \frac{(154.07 - 167.09)}{\left[\dfrac{167.09 - 149.10}{154.07 - 154.55} - 1\right]} = 154.41°\text{F}$$

Calculating $F(T_3)$ in the same manner as above:

$$\Delta T_{\text{ln}} = \frac{115 - (215 - 154.41)}{\ln[115/(215 - 154.41)]} = 84.91°\text{F}$$

$$q = 102 \times \pi \times 1/12 \times 12 \times 84.91 = 27{,}209 \text{ Btu/h}$$

$$27{,}209 = 500(T_{b2} - 100)$$

$$T_{b2} = 154.42°\text{F} = F(T_3)$$

The results are summarized below.

i	T_i	$F(T_i)$
0	120.00	167.09
1	167.09	149.10
2	154.07	154.55
3	154.41	154.42

Since $T_3 = F(T_3)$ to four significant figures, the procedure has converged and the final result is $T_{b2} = 154.4°\text{F}$.

Example 2.8

Carbon dioxide at 300 K and 1 atm is to be pumped through a duct with a 10 cm x 10 cm square cross-section at a rate of 250 kg/h. The walls of the duct will be at a temperature of 450 K. What distance will the CO_2 travel through the duct before its temperature reaches 400 K?

Solution

The required physical properties are tabulated in Example 2.5. The equivalent diameter is calculated from Equation (2.41):

$$D_e = 4 \times \frac{100}{40} = 10 \text{ cm} = 0.1 \text{ m}$$

(i)

$$Re = \frac{D_e V \rho}{\mu} = \frac{D_e \dot{m}}{\mu A_f} = \frac{0.1 \times (250/3600)}{17.21 \times 10^{-6} \times 0.01}$$

where A_f = flow area = 0.01m^2

$$Re = 4.04 \times 10^4 \Rightarrow \text{turbulent flow}$$

Note that the Reynolds number is not equal to $4\dot{m}/\pi D_e \mu$ for non-circular ducts. Rather the continuity equation ($\dot{m} = \rho V A_f$) is used to express the Reynolds number in terms of mass flow rate.

(ii)

$$Nu = 0.027 Re^{0.8} Pr^{1/3} (\mu/\mu_w)^{0.14}$$

$$= 0.027 \left(4.04 \times 10^4\right)^{0.8} (0.755)^{1/3} \left(\frac{17.21}{21.34}\right)^{0.14}$$

$$Nu = 115.5$$

$$\frac{hD_e}{k} = 115.5$$

$$h = \frac{115.5 \times 0.02}{0.1} = 23.1 \text{ W/m}^2 \cdot \text{K}$$

(iii)

$$q = \dot{m} C_P \Delta T_{CO_2}$$

$$= (250/3600) \times 900 \times 100$$

$$q = 6250 \text{ W}$$

(iv)

$$q = hA\Delta T_{\ln} = hPL\Delta T_{\ln} \quad \text{where } P = \text{perimeter of duct cross-section}$$

$$6250 = 23.1 \times 0.4 \times L \times 91.02$$

$$L = 7.4 \text{ m}$$

Since $L/D_e = 7.4/0.1 = 74 > 60$, no correction for entrance effects is required.

2.5 Forced Convection in External Flow

Many problems of engineering interest involve heat transfer to fluids flowing over objects such as pipes, tanks, buildings or other structures. In this section, correlations for several such systems are presented in order to illustrate the method of calculation. Many other correlations can be found in heat-transfer textbooks, e.g., [1,11,13–16], and in engineering handbooks.

Flow over a flat plate is illustrated in Figure 2.5. The undisturbed fluid velocity and temperature upstream of the plate are V_∞ and T_∞, respectively. The surface temperature of the plate is T_s and L is the length of the plate in the direction of flow. The fluid may flow over one or both sides of the plate. The heat-transfer coefficient is obtained from the following correlations [11]:

$$Nu = 0.664\ Re^{1/2} Pr^{1/3} \qquad \text{for } Re < 5 \times 10^5 \tag{2.47}$$

$$Nu = \left(0.037\ Re^{0.8} - 870\right) Pr^{1/3} \text{ for } Re > 5 \times 10^5 \tag{2.48}$$

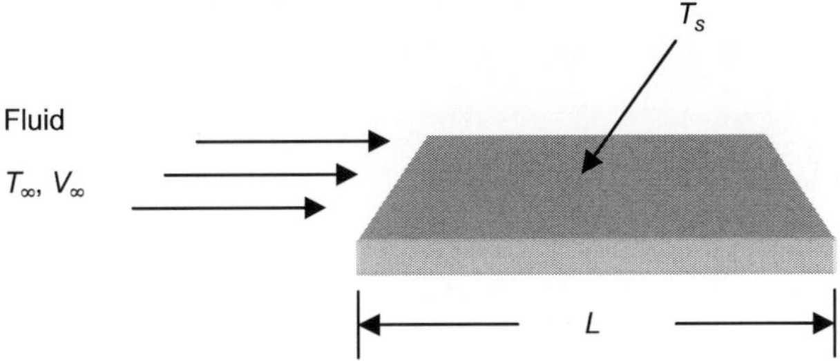

FIGURE 2.5 Forced convection in flow over a flat plate.

where

$$Nu = \frac{hL}{k} \text{ and } Re = \frac{LV_\infty \rho}{\mu}$$

Equation (2.47) is valid for Prandtl numbers greater than about 0.6. Equation (2.48) is applicable for Prandtl numbers between 0.6 and 60, and Reynolds numbers up to 10^8. In these equations all fluid properties are evaluated at the film temperature, T_f, defined by:

$$T_f = \frac{T_\infty + T_s}{2} \tag{2.49}$$

The heat-transfer coefficients computed from Equations (2.47) and (2.48) are average values for the entire plate. Hence, the rate of heat transfer between the plate and the fluid is given by:

$$q = hA|T_s - T_\infty| \tag{2.50}$$

where A is the total surface area contacted by the fluid.

For flow perpendicular to a circular cylinder of diameter D, the average heat-transfer coefficient can be obtained from the correlation [17]:

$$Nu = 0.3 + \frac{0.62\ Re^{1/2}Pr^{1/3}}{\left[1 + (0.4/Pr)^{2/3}\right]^{1/4}} \left[1 + \left(\frac{Re}{282,000}\right)^{5/8}\right]^{4/5} \tag{2.51}$$

where

$$Nu = \frac{hD}{k} \text{ and } Re = \frac{DV_\infty \rho}{\mu}$$

The correlation is valid for $Re\,Pr > 0.2$, and all fluid properties are evaluated at the film temperature, T_f. The rate of heat transfer is given by Equation (2.50) with $A = \pi DL$, where L is the length of the cylinder.

For flow over a sphere of diameter D, the following correlation is recommended [18]:

$$Nu = 2 + \left(0.4\ Re^{1/2} + 0.06\ Re^{2/3}\right)Pr^{0.4}\left(\frac{\mu_\infty}{\mu_s}\right)^{0.25} \tag{2.52}$$

where

$$Nu = \frac{hD}{k} \text{ and } Re = \frac{DV_\infty \rho}{\mu}$$

All fluid properties, except μ_s, are evaluated at the free-stream temperature, T_∞. The viscosity, μ_s, is evaluated at the surface temperature, T_s. Equation (2.52) is valid for Reynolds numbers between 3.5 and 80,000, and Prandtl numbers between 0.7 and 380.

For a gas flowing perpendicular to a cylinder having a square cross-section (such as an air duct), the following equation is applicable [1]:

$$Nu = 0.102\ Re^{0.675}Pr^{1/3} \tag{2.53}$$

The Nusselt and Reynolds numbers are calculated as for a circular cylinder but with the diameter, D, replaced by the length of a side of the square cross-section. All fluid properties are evaluated at the film temperature. Equation (2.53) is valid for Reynolds numbers in the range 5000 to 10^5.

The equations presented in this (and the following) section, as well as similar correlations which appear in the literature, are not highly accurate. In general, one should not expect the accuracy of computed values to be better than ±25% to ±30% when using these equations. This limitation should be taken into consideration when interpreting the results of heat-transfer calculations.

Example 2.9

Air at 20°C is blown over a 6 cm OD pipe that has a surface temperature of 140°C. The free-stream air velocity is 10 m/s. What is the rate of heat transfer per meter of pipe?

Solution

For forced convection from a circular cylinder, Equation (2.51) is applicable. The film temperature is:

$$T_f = \frac{T_s + T_\infty}{2} = \frac{140 + 20}{2} = 80°\text{C}$$

From Table A.4, for air at 80°C:

$$\begin{aligned}
v &= 21.5 \times 10^{-6} \text{m}^2/\text{s} = \text{kinematic viscosity} \\
k &= 0.0293 \ \text{W/m·K} \\
Pr &= 0.71 \\
Re &= DV_\infty/v = 0.06 \times 10/(21.5 \times 10^{-6}) = 2.79 \times 10^4
\end{aligned}$$

Substituting in equation (2.51) gives:

$$Nu = 0.3 + \frac{0.62(27,900)^{1/2}(0.71)^{1/3}}{\left[1 + \left(\frac{0.4}{0.71}\right)^{2/3}\right]^{1/4}}\left[1 + \left(\frac{27,900}{282,000}\right)^{5/8}\right]^{4/5}$$

$$Nu = 96.38$$

$$h = \frac{Nu \times k}{D} = \frac{96.38 \times 0.0293}{0.06}$$

$$h \cong 47 \ \text{W/m}^2\text{·K}$$

The rate of heat transfer is given by Equation (2.50). Taking a length of 1 m gives:

$$\begin{aligned}
q &= h\pi DL(T_s - T_\infty) \\
q &= 47 \times \pi \times 0.06 \times 1.0(140 - 20) \\
q &= 1065 \cong 1100 \ \text{W/m of pipe}
\end{aligned}$$

Example 2.10

A duct carries hot waste gas from a process unit to a pollution control device. The duct cross-section is 4 ft × 4 ft and it has a surface temperature of 140°C. Ambient air at 20° C blows across the duct with a wind speed of 10 m/s. Estimate the rate of heat loss per meter of duct length.

Solution

The film temperature is the same as in the previous example:

$$T_f = \frac{140 + 20}{2} = 80°\text{C}$$

Therefore, the same values are used for the properties of air. The Reynolds number is based on the length of the side of the duct cross-section. Hence, we set:

$$D = 4 \text{ ft} \cong 1.22 \text{ m}$$

$$Re = \frac{DV_\infty}{v} = 1.22 \times 10/(21.5 \times 10^{-6}) = 5.674 \times 10^5$$

The Reynolds number is outside the range of Equation (2.53). However, the equation will be used anyway since no alternative is available. Thus,

$$\begin{aligned}
Nu &= 0.102Re^{0.675}Pr^{1/3} \\
&= 0.102 \ (5.674 \times 10^5)^{0.675}(0.71)^{1/3} \\
Nu &= 696.5 \\
h &= \frac{k}{D} \times 696.5 \\
h &= \frac{0.0293}{1.22} \times 696.5 \cong 16.7 \ \text{W/m}^2 \text{·K}
\end{aligned}$$

The surface area of the duct per meter of length is:

$$A = PL = 4 \times 1.22 \times 1.0 = 4.88 \text{ m}^2$$

Finally, the rate of heat loss is:

$$\begin{aligned}
q &= hA\Delta T \\
&= 16.7 \times 4.88 \times (140 - 20) \\
q &\cong 9800 \ \text{W/m of duct length}
\end{aligned}$$

2.6 Free Convection

Engineering correlations for free-convection heat transfer are similar in nature to those presented in the previous section for forced convection in external flows. However, in free-convection problems there is no free-stream velocity, V_∞, upon which to base a Reynolds number. The dimensionless group that takes the place of the Reynolds number in characterizing free convection is called the Grashof number, *Gr*, and is defined by:

$$Gr = \frac{g\beta\,|T_s - T_\infty|\,L^3}{\nu^2} \tag{2.54}$$

where

g = gravitational acceleration
T_s = surface temperature
T_∞ = fluid temperature far from solid surface
L = characteristic length
$\nu = \mu/\rho$ = kinematic viscosity
$\beta = \frac{1}{\hat{v}}\left(\frac{\partial \hat{v}}{\partial T}\right)_P$ = coefficient of volume expansion
$\hat{v}$ = specific volume

For gases at low density, the ideal gas law may be used to evaluate $(\partial\hat{v}/\partial T)_P$ and show that:

$$\beta = 1/T \quad \text{(ideal gas)} \tag{2.55}$$

where T is absolute temperature.

For dense gases and liquids, β can be approximated by:

$$\beta \cong \frac{1}{\hat{v}}\frac{\Delta\hat{v}}{\Delta T} \tag{2.56}$$

if the specific volume (or density) is known at two different temperatures.

Free-convection heat transfer from a heated vertical plate is illustrated in Figure 2.6. Fluid adjacent to the surface is heated and rises by virtue of its lower density relative to the bulk of the fluid. This rising layer of fluid entrains cooler fluid from the nearly quiescent region further from the heated surface, so that the mass flow rate of the rising fluid increases with distance along the plate. Near the bottom of the plate, the flow is laminar, but at some point, transition to turbulent flow may occur if the plate is long enough. The free-convection circulation is completed by a region of cold, sinking fluid far removed from the heated surface.

Heat-transfer coefficients for free convection from a heated or cooled vertical plate of height L can be obtained from the correlation [19]:

$$Nu = \left\{0.825 + \frac{0.387(Gr\,Pr)^{1/6}}{\left[1 + (0.492/Pr)^{9/16}\right]^{8/27}}\right\}^2 \tag{2.57}$$

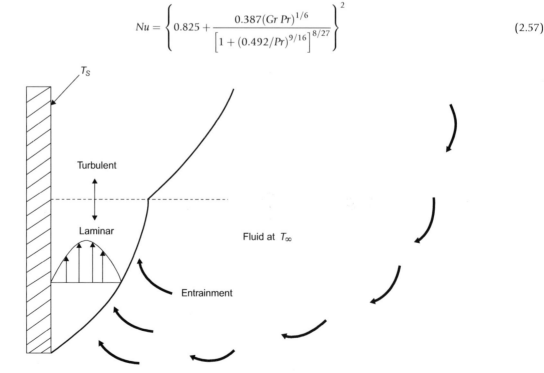

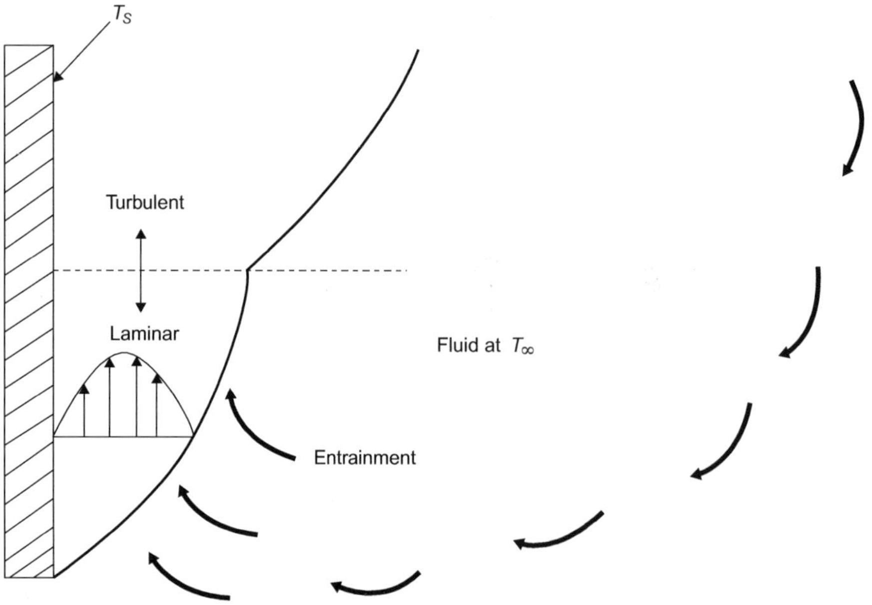

FIGURE 2.6 Free-convection circulation pattern near a heated vertical plate.

The characteristic length used in the Nusselt and Grashof numbers is the plate height, L. The fluid properties, including β, are evaluated at the film temperature, T_f, defined by Equation (2.49). Equation (2.57) is valid for $0.1 \le Gr\,Pr \le 10^{12}$, and is accurate to within about $\pm30\%$.

Equation (2.57) may also be used for free convection from vertical cylinders if the diameter-to-height ratio is greater than $(35/Gr^{0.25})$. For smaller values of D/L, the effects of surface curvature become significant. However, a correction factor can be applied as follows [11]:

$$Nu_{cylinder} = Nu_{plate}\left[1 + 1.43\left(\frac{L}{D\,Gr^{0.25}}\right)^{0.9}\right] \tag{2.58}$$

where Nu_{plate} is the Nusselt number calculated from Equation (2.57).

For free convection from a horizontal cylinder, a correlation similar to Equation (2.57) can be used [20]:

$$Nu = \left\{0.60 + \frac{0.387\,(Gr\,Pr)^{1/6}}{\left[1 + (0.559/Pr)^{9/16}\right]^{8/27}}\right\}^2 \tag{2.59}$$

In this case, however, the characteristic length used in the Nusselt and Grashof numbers is the diameter of the cylinder. Equation (2.59) is valid over the range $10^{-5} \le Gr\,Pr \le 10^{12}$. Fluid properties are evaluated at the film temperature.

For free convection from horizontal plates, the following correlations are available [21,22]:

Upper surface hot or lower surface cool:

$$Nu = 0.54\,(Gr\,Pr)^{1/4} \quad \text{for } 10^4 \le Gr\,Pr \le 10^7 \tag{2.60}$$

$$Nu = 0.15\,(Gr\,Pr)^{1/3} \quad \text{for } 10^7 \le Gr\,Pr \le 10^{11} \tag{2.61}$$

Upper surface cool or lower surface hot:

$$Nu = 0.27\,(Gr\,Pr)^{1/4} \quad \text{for } 10^5 \le Gr\,Pr \le 10^{11} \tag{2.62}$$

In these equations, fluid properties are again evaluated at the film temperature. The characteristic length used in both the Nusselt and Grashof numbers is the surface area of the plate divided by its perimeter.

For free convection from spheres, the following correlation is recommended [23]:

$$Nu = 2 + \frac{0.598\,(Gr\,Pr)^{1/4}}{\left[1 + (0.469\,/Pr)^{9/16}\right]^{4/9}}\left\{1 + \frac{7.44\times10^{-8}Gr\,Pr}{\left[1 + (0.469\,/Pr)^{9/16}\right]^{16/9}}\right\}^{1/12} \tag{2.63}$$

The characteristic length is the sphere diameter and the correlation is valid for $Gr\,Pr \le 10^{13}$ and $0.6 \le Pr \le 100$. For larger values of the Prandtl number, better agreement with experimental data is obtained by dropping the term in brackets raised to the l/12th power. Fluid properties are evaluated at the film temperature.

Correlations for free convection heat transfer in other geometrical configurations can be found in the references cited at the end of the chapter. It should be noted that free convection effects can sometimes be significant in forced convection problems, especially in the case of laminar flow at low Reynolds numbers. The relative importance of forced and free convection effects can be determined by comparison of the Reynolds and Grashof numbers as follows:

$Re^2/Gr >> 1$: Forced convection predominates

$Re^2/Gr \cong 1$: Both modes of heat transfer are significant

$Re^2/Gr << 1$: Free convection predominates

A combination of free and forced convection is often referred to as mixed convection. Mixed convection can have a significant effect on the heat-transfer coefficient and the predictions can be complicated. For vertical tubes, the laminar flow coefficient is enhanced for aiding flow (up-flow heating and down-flow cooling). For opposing flow (down-flow heating and up-flow cooling), heat transfer is generally reduced. However, there are exceptions, such as when the flow remains laminar with Reynolds numbers greater than 2000 for up-flow heating in some cases. For horizontal tubes, mixed convection enhances heat transfer and the increase is often substantial. While there is not a universally accepted method of predicting heat transfer for cases with significant mixed convection, a method proposed by Ghajar and Tam [24] for flow in horizontal tubes has been shown to be accurate over the range $280 < Re < 3800$:

$$Nu = 1.24\left[RePr\frac{D}{L} + 0.025(GrPr)^{0.75}\right]^{1/3}\left(\frac{\mu}{\mu_w}\right)^{0.14} \tag{2.64}$$

In this equation all fluid properties, with the exception of μ_w, are evaluated at the average bulk fluid temperature. The characteristic length is the tube ID, and the Grashof number is computed using the difference between the average tube-wall temperature and the average bulk fluid temperature. Prediction of thermal performance in mixed convection remains an open topic of research and an area of uncertainty in process heat transfer.

Example 2.11

A 12-ounce can of beer at 80°F is placed on end in a refrigerator which is at 36°F. Estimate the value of the external heat-transfer coefficient for this situation. A 12-ounce beer can has dimensions, $D = 2.5$ in. and $L = 4.75$ in.

Solution

It will be assumed that the refrigerator blower is not running and that the can is not in contact with any other item in the refrigerator. Under these conditions, the heat transfer will be by free convection from a vertical cylinder. We have:

$$T_s = 80°\text{F}$$

$$T_\infty = 36°\text{F}$$

$$T_f = \frac{80 + 36}{2} = 58°\text{F} = 14.44°\text{C}$$

From Table A.4, we find by interpolation for air at 14.44°C:

$$\frac{g\beta}{\nu^2} = 1.496 \times 10^8\ \text{K}^{-1}\ \text{m}^{-3}$$

$$k = 0.0247\ \text{W/m}\cdot\text{K}$$

$$Pr = 0.71$$

The characteristic length is the height of the cylinder, i.e.,

$$L = 4.75\ \text{in.} = 0.1207\ \text{m}$$

Thus,

$$Gr = \frac{g\beta}{\nu^2}|T_s - T_\infty|L^3$$

$$= 1.496 \times 10^8 \left(\frac{80 - 36}{1.8}\right)(0.1207)^3$$

$$Gr = 6.430 \times 10^6$$

$$Gr\,Pr = 6.430 \times 10^6 \times 0.71 = 4.566 \times 10^6$$

Substituting in Equation (2.57) gives:

$$Nu = \left\{0.825 + \frac{0.387\left(4.566 \times 10^6\right)^{1/6}}{\left[1 + (0.492/0.71)^{9/16}\right]^{8/27}}\right\}^2 \quad Nu = 25$$

Now,

$$\frac{D}{L} = \frac{2.5}{4.75} = 0.5263$$

and

$$\frac{35}{Gr^{0.25}} = \frac{35}{\left(6.430 \times 10^6\right)^{0.25}} = 0.695$$

Therefore, Equation (2.58) should be used to correct for the effects of surface curvature:

$$Nu_{cylinder} = Nu_{plate}\left[1 + 1.43\left(\frac{L}{DGr^{0.25}}\right)^{0.9}\right]$$

$$= 25\left[1 + 1.43\left(\frac{4.75}{2.5\left(6.43 \times 10^6\right)^{0.25}}\right)^{0.9}\right] = 25\,[1.0749]$$

$$Nu_{cylinder} = 26.9$$

$$\frac{hL}{k} = 26.9$$

$$h = \frac{26.9 \times k}{L} = \frac{26.9 \times 0.0247}{0.1207} \cong 5.5\ \text{W/m}^2\cdot\text{K}$$

$$h = 5.5 \times 0.17612 \cong 0.97\ \text{Btu/h}\cdot\text{ft}^2\cdot°\text{F}$$

Example 2.12

A small holding tank in a chemical plant contains a corrosive liquid that is maintained at a temperature of 120°F by means of an electrical heater. The heating element consists of a refractory disk 2 ft in diameter situated at the bottom of the tank. Estimate the power required to maintain the surface of the heating element at 160°F. The liquid properties are as follows:

Property	At 140°F	At 60°F
ν (ft^2/h)	0.023	
k (Btu/h · ft · °F)	0.4	
ρ (lbm/ft^3)	69.3	70.0
Pr	4.8	

Solution

The heat transfer is by free convection from a horizontal plate with the upper surface hot. We first use the density data to estimate β:

$$\hat{v} = 1/\rho = 1/70 = 0.014286\ \text{ft}^3/\text{lbm at } 60°\text{F}$$
$$\hat{v} = 1/\rho = 1/69.3 = 0.014430\ \text{ft}^3/\text{lbm at } 140°\text{F}$$

$$\hat{v}_{ave} = 0.5\,(0.014286 + 0.014430) = 0.014358\ \text{ft}^3/\text{lbm}$$

$$\beta \cong \frac{1}{\hat{v}_{ave}}\frac{\Delta\hat{v}}{\Delta T} = \frac{1}{0.014358} \times \frac{(0.014430 - 0.014286)}{(140 - 60)}$$

$$\beta = 0.000125\ (°R)^{-1}$$

Next, the characteristic length is calculated as the ratio of surface area to perimeter of the heating element:

$$L = A_s/P = \frac{\pi D^2/4}{\pi D} = \frac{D}{4} = \frac{2}{4} = 0.5\ \text{ft}$$

The Grashof number is calculated using the fluid properties at the film temperature:

$$T_f = \frac{T_s + T_\infty}{2} = \frac{160 + 120}{2} = 140°\text{F}$$

$$Gr = \frac{g\beta\,|T_s - T_\infty|\,L^3}{\nu^2} = \frac{4.17 \times 10^8 \times 0.000125\,|160 - 120|\,(0.5)^3}{(0.023)^2}$$

$$Gr = 4.9267 \times 10^8$$

$$GrPr = 4.9267 \times 10^8 \times 4.8 = 2.3648 \times 10^9$$

For this value of $Gr\,Pr$, Equation (2.61) is applicable:

$$Nu = 0.15\,(Gr\,Pr)^{1/3} = 0.15\,\left(2.3648 \times 10^9\right)^{1/3} = 199.84$$

$$h = \frac{199.84 \times k}{L} = \frac{199.84 \times 0.4}{0.5} \cong 160\ \text{Btu/h}\cdot\text{ft}^2\cdot°\text{F}$$

$$q = hA\Delta T = h\left(\pi D^2/4\right)|T_s - T_\infty| = 160 \times \frac{\pi\,(2)^2}{4}\,(160 - 120)$$

$$q = 20{,}090\ \text{Btu/h}$$

$$q = 20{,}090/3412 \cong 5.9\ \text{kW}$$

2.7 Radiation

All material objects, including living organisms, emit electromagnetic energy at a rate that depends on their surface temperature. The radiated energy comes at the expense of an object's internal energy, and is referred to as thermal radiation. The equations used to calculate radiant energy transfer are based on the concept of an ideal radiator, called a black body, which absorbs all of the electromagnetic radiation that strikes its surface; none of the radiation is reflected or transmitted. Such an object would appear to be completely black provided the radiation that it emitted was not in the visible part of the spectrum. In fact, at normal ambient temperatures, thermal radiation is almost entirely in the infrared. It becomes visible only when the temperature of the emitting surface reaches about 800 K, the point at which objects become "red hot."

Although an ideal radiator does not exist in nature, a very close approximation consists of a closed box, or cavity, with a small pinhole in one side. A photon entering through the pinhole has a negligible chance of re-emerging, since this would generally require that the photon undergo numerous reflections from the interior walls of the box without ever being absorbed. If the walls of the box are held at a uniform fixed temperature, the radiation emanating from the pinhole will closely approximate that from a black body at the given temperature. Some familiar materials, such as carbon black (soot) and graphite, also approximate black body behavior reasonably well, though not as closely as a cavity.

A convex black body, i.e., one that does not intercept any of its own radiation, emits electromagnetic energy at a rate, q_{rad}, given by the Stefan–Boltzman equation:

$$q_{rad} = \sigma_{SB} A T^4 \tag{2.65}$$

where

T = absolute temperature of the black body
A = surface area of the black body
σ_{SB} = Stefan–Boltzman constant
$= 5.67 \times 10^{-8} \mathrm{W/m^2 \cdot K^4} = 1.714 \times 10^{-9} \mathrm{Btu/h \cdot ft^2 \cdot {}^\circ R^4}$

All real objects emit thermal radiation at a smaller rate than a black body at the same temperature. Thus, Equation (2.65) provides an upper bound on the radiation emitted by any object.

A (convex) semi-ideal radiator, called a gray body, emits electromagnetic energy at a rate given by:

$$q_{rad} = \epsilon \sigma_{SB} A T^4 \tag{2.66}$$

where the emissivity, ϵ, is a constant having a value between 0 and 1.0, with a value of 1.0 corresponding to a black body. Obviously, Equation (2.65) is a special case of Equation (2.66). In addition to composition, the emissivity depends strongly on the surface characteristics of an object, e.g., whether it is clean or dirty, polished or corroded, smooth or rough, and if painted, the color and condition of the paint. Values for common materials range from less than 0.05 for some polished metals to greater than 0.95 for graphite and carbon black. The emissivity of real objects is not constant, but depends on temperature, the wavelength of the emitted radiation, and the angle at which it is emitted. Experimental values of emissivity are usually given for a specified temperature or temperature range. They may also be given as a function of wavelength, but more often they are averaged over all wavelengths, in which case they are referred to as total emissivities. Both spectral and total emissivities are usually measured either for radiation emitted normal to the surface of an object or for the radiation emitted in all directions encompassing a hemispherical region about the object. The corresponding values are called normal emissivity and hemispherical emissivity, respectively. For many materials, the values of normal and hemispherical emissivity do not differ greatly.

When radiant energy is transferred between two or more objects at different temperatures, the net rate of energy transfer depends on the geometrical configuration of the objects, which determines the amount of radiation emitted from one object that impinges on another. The net rate of radiant energy transfer from object 1 to object 2, denoted by $q_{1,2}$, can be represented by:

$$q_{1,2} = \sigma_{SB} A_1 F_{1,2} \left(T_1^4 - T_2^4\right) \tag{2.67}$$

where $F_{1,2}$ is a radiation shape and emissivity function, which includes the radiation shape factor and the emissivities of objects 1 and 2. The radiation shape factor quantifies the geometrical configuration of the two objects. For industrial heat transfer equipment, there are a number of different surfaces receiving and emitting radiation. A more general expression for net radiation heat transfer from object 1 to a group of objects 2, 3,..., N can be used in this case:

$$q_1 = \sigma_{SB} A_1 \sum_{i=2}^{N} F_{1,i} \left(T_1^4 - T_i^4\right) \tag{2.68}$$

Shape factors used to calculate $F_{1,i}$ have been tabulated in a number of heat transfer texts, but most have been originally cited from work by Hottel and have been assembled in [25].

One simple but useful configuration is that of a convex gray body (1) that is completely surrounded, or enclosed, by a second gray body (2). In this case, Equation (2.67) takes the following form:

$$q_{1,2} = \frac{\sigma_{SB}\left(T_1^4 - T_2^4\right)}{\dfrac{1}{\epsilon_1 A_1} + \dfrac{1 - \epsilon_2}{\epsilon_2 A_2}} \tag{2.69}$$

When the enclosure is much larger than the enclosed object, so that $A_2 >> A_1$, Equation (2.69) can be simplified by taking the limit as A_1/A_2 approaches zero to yield:

$$q_{1,2} = \epsilon_1 \sigma_{SB} A_1 \left(T_1^4 - T_2^4\right) \tag{2.70}$$

For problems involving both radiation and convection, it is sometimes convenient to define a radiative heat-transfer coefficient, h_r, by analogy with convective heat transfer:

$$q_{rad} = h_r A \Delta T \tag{2.71}$$

Now Equation (2.67), for example, can be expressed in terms of h_r by equating q_{rad} with $q_{1,2}$, A with A_1, and setting $\Delta T = T_1 - T_2$. The result is:

$$h_r = F_{1,2}\left[\frac{\sigma_{SB}\left(T_1^4 - T_2^4\right)}{T_1 - T_2}\right] \tag{2.72}$$

Similarly, Equation (2.70) is equivalent to the following expression for h_r:

$$h_r = \frac{\epsilon_1 \sigma_{SB}\left(T_1^4 - T_2^4\right)}{T_1 - T_2} \tag{2.73}$$

We will have occasion to use this result in connection with film boiling in Chapter 9.

The radiative and convective coefficients can be combined to obtain a total heat-transfer coefficient, h_t, encompassing both radiation and convection:

$$h_t = h_r + h_c \tag{2.74}$$

This equation implies that the rates of heat transfer by radiation and convection are additive, which is correct in most situations. However, the case of film boiling mentioned above is an exception, as explained in Chapter 9.

The term in brackets in Equation (2.72) is sometimes called the temperature factor and denoted by F_T. It represents the maximum possible radiative heat-transfer coefficient for an object at temperature T_1 radiating to a surface at T_2. Figure 2.7 shows the temperature factor for a range of temperatures. These results show that for typical unfired heat exchangers, the radiative coefficient is much less than the convective coefficient and can be neglected at low temperatures.

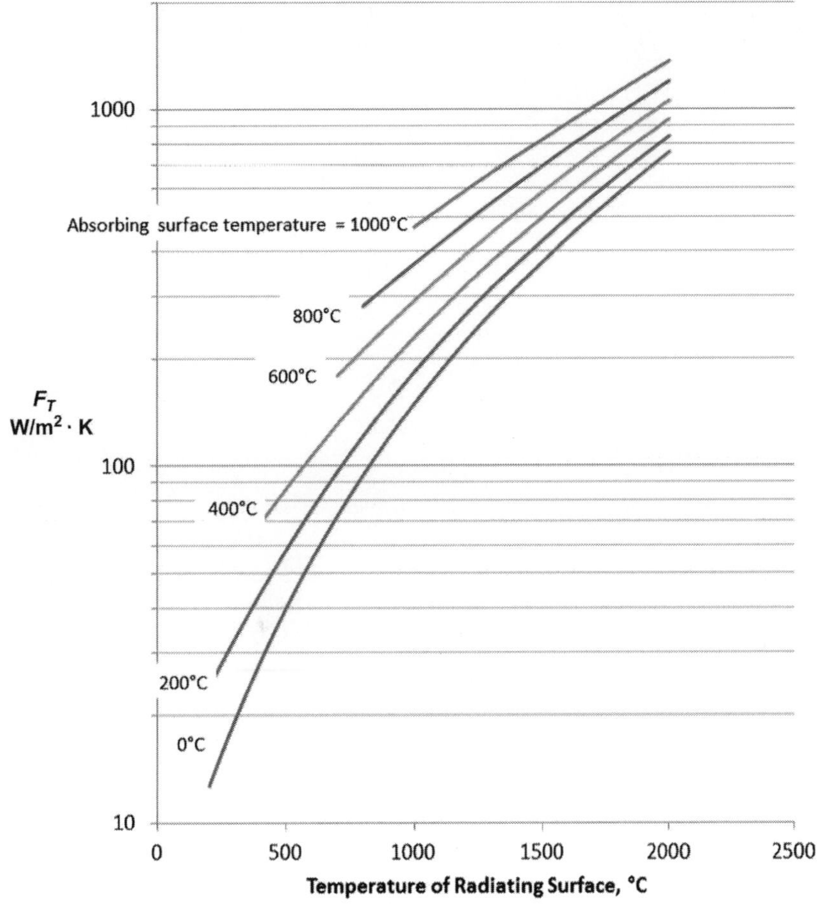

FIGURE 2.7 Radiation temperature factor versus temperature of radiating surface.

Example 2.13

A horizontal steam pipe with an outside diameter of 6.0 cm passes through a large room whose walls are at 298 K. The outer surface of the pipe wall has a temperature of 443 K and an emissivity of 0.8. The air temperature in the room is 303 K. Calculate the rate of heat loss per unit length from the pipe by radiation and natural convection.

Solution

The radiative heat loss is calculated using Equation (2.70). Taking a unit length of 1 m, the surface area of the pipe is:

$$A_1 = A = \pi DL = \pi \times 0.06 \times 1 = 0.1885\text{m}^2$$

Substituting in Equation (2.70) with $T_1 = 443$ K and $T_2 = 298$ K gives:

$$q_{rad} = q_{1,2} = \epsilon_1 \ \sigma_{SB} A_1 \left(T_1^4 - T_2^4\right)$$
$$= 0.8 \times 5.67 \times 10^{-8} \times 0.1885 \left[(443)^4 - (298)^4\right]$$
$$q_{rad} = 262 \text{ W/m of pipe}$$

The convective heat-transfer coefficient is calculated using Equation (2.59). The film temperature is:

$$T_f = \frac{443 + 303}{2} = 373\text{K}$$

The properties of air at 373 K are obtained from Table A.4:

$$\frac{g\beta}{\nu^2} = 0.472 \times 10^8 \text{ K}^{-1}\text{m}^{-3}$$

$$k = 0.0307 \text{ W/m} \cdot \text{K}$$

$$Pr = 0.71$$

The appropriate characteristic length for a horizontal cylinder is the diameter of the cylinder. Hence,

$$Gr = \frac{g\beta}{\nu^2}|T_s - T_\infty|D^3$$

$$= 0.472 \times 10^8 (443 - 303)(0.06)^3$$

$$Gr = 1.4273 \times 10^6$$

$$GrPr = 1.4273 \times 10^6 \times 0.71 = 1.0134 \times 10^6$$

Substituting in Equation (2.59) gives:

$$Nu = \left\{0.6 + \frac{0.387(GrPr)^{1/6}}{\left[1 + (0.559/Pr)^{9/16}\right]^{8/27}}\right\}^2$$

$$= \left\{0.6 + \frac{0.387\left(1.0134 \times 10^6\right)^{1/6}}{\left[1 + (0.559/0.71)^{9/16}\right]^{8/27}}\right\}^2$$

$$Nu = 14.59$$

$$\frac{hD}{k} = 14.59$$

$$h = \frac{14.59 \times k}{D} = \frac{14.59 \times 0.0307}{0.06} = 7.47 \text{ W/m}^2 \cdot \text{K}$$

The convective heat loss is:

$$q_{cv} = hA(T_s - T_\infty) = 7.47 \times 0.1885(443 - 303)$$
$$q_{cv} = 197 \text{ W/m of pipe}$$

The total rate of heat loss is:

$$q = q_{rad} + q_{cv} = 262 + 197 = 459 \text{ W/m of pipe}$$

References

[1] Incropera FP, DeWitt DP. Introduction to Heat Transfer. 4th ed. New York: John Wiley & Sons; 2002.
[2] Kern DQ, Krause AD. Extended Surface Heat Transfer. New York: McGraw-Hill; 1972.
[3] McQuiston FC, Tree DR. Optimum space envelopes of the finned tube heat transfer surface. Trans ASHRAE 1972;78:144–52.
[4] Owen FK. Heat transfer from plain and finned cylinders in crossflow. J IHVE 1967:213–26.
[5] Lymer A, Ridal BF. Finned tubes in a crossflow of gas. J Brit Nucl Energy Conf 1961:307–13.
[6] Zhukauskas A, Stasiuleyichius J, Skriniska A. Experimental investigation of heat transfer of a tube with spiral fins in crossflow. Proc Third Int Heat Transfer Conf 1966;3:299–305.
[7] Seider EN, Tate CE. Heat transfer and pressure drop of liquids in tubes. Ind Eng Chem 1936;28:1429–35.
[8] Avila K, Moxey D, de Lozar A, Avila M, Barkley D, Hof B. The onset of turbulence in pipe flow,. Science 2011;333:192–6.
[9] Hausen H. Darstellung des Warmeuberganges in Rohren durch Verallgemeinert Potenzbeziehungen. Z VDI Beih Verfahrenstech 1943;4:91.
[10] Gnielinski V. Forced convection in ducts. In: Heat Exchanger Design Handbook, vol. 2. New York: Hemisphere Publishing Corp.; 1988.
[11] White FM. Heat Transfer. Reading, MA: Addison-Wesley; 1984.
[12] Taborek J. Design methods for heat transfer equipment-A critical survey of the state-of the art. In: Afgan N, Schlunder EU, editors. Heat Exchangers: Design and Theory Sourcebook. New York: McGraw-Hill; 1974.
[13] Holman JP. Heat Transfer. 7th ed. New York: McGraw-Hill; 1990.
[14] Kreith F, Black WZ. Basic Heat Transfer. New York: Harper & Row; 1980.
[15] Kreith F, Bohn MS. Principles of Heat Transfer. Pacific Grove, CA: Brooks/Cole; 2001.
[16] Mills AF. Heat Transfer. 2nd ed. Upper Saddle River, NJ: Prentice-Hall; 1999.
[17] Churchill SW, Bernstein M. A correlating equation for forced convection from gases and liquids to a circular cylinder in crossflow. J Heat Transfer 1977;99:300–6.
[18] Whitaker S. Forced convection heat transfer correlations for flow in pipes, past flat plates, single cylinders, single spheres, and for flow in packed beds and tube bundles. AIChE J 1972;18:361–71.
[19] Churchill SW, Chu HHS. Correlating equations for laminar and turbulent free convection from a vertical plate. Int J Heat Mass Transfer 1975;18:1323–9.
[20] Churchill SW, Chu HHS. Correlating equations for laminar and turbulent free convection from a horizontal cylinder. Int J Heat Mass Transfer 1975;18:1049–53.
[21] Fujii T, Imura H. Natural convection heat transfer from a plate with arbitrary inclination. Int J Heat Mass Transfer 1972;15:755–64.
[22] Goldstein RJ, Sparrow EM, Jones DC. Natural convection mass transfer adjacent to horizontal plates. Int J Heat Mass Transfer 1973;16:1025–35.
[23] Churchill SW. Free Convection around immersed bodies. In: Heat Exchanger Design Handbook, vol. 2. New York: Hemisphere Publishing Corp.; 1988.
[24] Ghajar AJ, Tam LM. Heat transfer measurements and correlations in transition region for a circular tube in three different inlet configurations. Exp Thermal Fluid Sci 1994;8:79–90.
[25] Hottel HC, Sarofin AF. Radiative Transfer. New York: McGraw-Hill; 1967.

Notations

A	Area
A_{fin}	Surface area of fin
A_{prime}	Prime surface area
A_{Tot}	Total area of finned surface
A_s	Surface area of solid
C_1, C_2	Constants of integration
C_P	Constant pressure heat capacity of fluid
c	Specific heat of solid
D	Diameter
D_e	Equivalent diameter
D_1	Outer diameter of inner pipe of an annulus
D_2	Inner diameter of outer pipe of an annulus
F_T	$\sigma_{SB}(T_1^4 - T_2^4)/(T_1 - T_2)$ = Radiation temperature factor
$F_{1,2}$, $F_{1,i}$	Radiation shape and emissivity functions
f	Darcy friction factor
Gr	Grashof number
g	Gravitational acceleration
h	Heat-transfer coefficient
h_c	Convective heat-transfer coefficient
h_r	Radiative heat-transfer coefficient
h_t	Total heat-transfer coefficient for radiation and convection
k	Thermal conductivity
L	Length
L_c	Corrected fin length
m	$(hP/kA)^{1/2} \cong (2h/k\tau)^{1/2}$ = Fin parameter
Nu	Nusselt number
P	Perimeter
Pr	Prandtl number

q	Rate of heat transfer
q_{cv}	Rate of heat transfer by convection
q_{fin}	Rate of heat transfer from a single fin
q_{fins}	Rate of heat transfer from all fins on a finned surface
q_{max}	Maximum possible rate of heat transfer from a fin
q_{prime}	Rate of heat transfer from prime surface
q_r, q_x	Rate of heat transfer in r- or x-direction
q_{rad}	Rate of heat transfer by radiation
q_1	Net rate of radiant heat transfer from object 1
$q_{1,2}$	Net rate of radiant heat transfer from object 1 to object 2
$\hat{q}$	Heat flux
$\hat{q}_{cv}$	Convective heat flux
Re	Reynolds number
R_{th}	Thermal resistance
r_1,r_2	Inner and outer radii of annular fin
r_{2c}	$r_2 + \tau/2 =$ Corrected fin radius
T	Temperature
T_b	Bulk fluid temperature
T_f	Film temperature
T_0	Initial temperature
T_s	Surface temperature
T_w	Temperature of pipe wall
T_∞	Fluid temperature far from solid surface
t	Time
V	Fluid velocity
V_∞	Fluid velocity far from solid surface
v	Volume
$\hat{v}$	$v/\rho =$ Specific volume
W	Width of rectangular fin
x	Coordinate in Cartesian coordinate system

Greek Letters

β	Coefficient of volume expansion
$\Delta T, \Delta x$	Difference in T or x
ΔT_{ave}	Arithmetic mean temperature difference
$\Delta T_{\ln}$	Logarithmic mean temperature difference (LMTD)
ϵ	Emissivity
η_{af}	Actual fin efficiency
η_f	Theoretical fin efficiency
η_w	Weighted efficiency of finned surface
θ	$T - T_\infty$
μ	Fluid viscosity
μ_w	Fluid viscosity at average temperature of pipe wall
μ_∞	Fluid viscosity at temperature T_∞
v	$\mu/\rho =$ Kinematic viscosity
ρ	Density
σ_{SB}	Stefan–Boltzman constant
τ	Fin thickness
ψ	Effective height of annular fin

Problems

(2.1) A pipe with an OD of 6.03 cm and an ID of 4.93 cm carries steam at 250°C with a heat-transfer coefficient of 200 W/m^2 · K. The pipe is covered with 2.5 cm of magnesia (85%) insulation followed by 2.5 cm of polystyrene insulation ($k = 0.025$ W/m · K).

The outer surface of the polystyrene is exposed to air at a temperature of 20°C with a heat-transfer coefficient of 10 W/m^2 · K. What is the rate of heat loss per meter of pipe length? The resistance of the pipe wall, contact resistances, and thermal radiation effects may be neglected in this problem.

Ans. 60 W/m of pipe.

(2.2) A plastic ($k = 0.5$ W/m · K) pipe carries a coolant at –35°C with a heat-transfer coefficient of 300 W/m^2 · K. The pipe ID is 3 cm and the OD is 4 cm. The exterior pipe surface is exposed to air at 25°C with a heat-transfer coefficient of 20 W/m^2 · K. Radiative heat transfer may be neglected in this problem.

(a) Calculate the rate of heat transfer to the coolant per meter of pipe length.

(b) Calculate the temperature of the exterior pipe surface.

(c) It is desired to reduce the rate of heat transfer by 90%. Will a 2 cm layer of magnesia insulation be sufficient for this purpose?

Ans. (a) 114 W/m of pipe. (b) –20.5°C.

(2.3) A 2-in. schedule 40 steel pipe (ID = 2.067 in., OD = 2.375 in.) carries steam at 350°F with a heat-transfer coefficient of 50 Btu/h · ft^2 · °F. The pipe is exposed to air at 70°F with a heat-transfer coefficient of 2.0 Btu/h · ft^2 · °F. For safety reasons, as well as to conserve energy, it is desired to insulate the pipe so that the temperature of the surface exposed to air will be no greater than 120°F. A commercial insulation vendor claims that using just 0.5 in. of his "high-efficiency" insulation will do the job. Is his claim correct?

Notes:

1. The vendor's literature states that a 6-in. thick layer of "high-efficiency" insulation has an *R*-value of 23.7.
2. The thermal resistance of the pipe wall may be neglected in this calculation. Also neglect contact resistance and thermal radiation effects.

(2.4) Suppose the steam pipe of Problem 2.3 is a 4-in. schedule 40 steel pipe (ID = 4.026 in., OD = 4.5 in.). What thickness of "high-efficiency" insulation will be required in this case?

(2.5) A 4-in. schedule 80 steel pipe (ID = 3.826 in., OD = 4.5 in.) carries a heat-transfer fluid ($T = 80$°F, $h = 200$ Btu/h · ft^2 · °F) and is covered with a 0.5-in. thick layer of pipe insulation. The pipe is exposed to ambient air ($T = 80$°F, $h = 10$ Btu/h · ft^2 · °F). The insulation vendor's literature states that a 1-in. layer of the pipe insulation has an *R*-value of 3. Neglecting the resistance of the pipe wall and thermal radiation effects, calculate the rate of heat loss from the pipe per foot of length. Compare your answer with the result from Problem 1.19 in which convective resistances were neglected.

Ans. 423 Btu/h · ft of pipe.

(2.6) A steel pipe (ID = 2.067 in., OD = 2.375 in.) is covered with a 0.5-in. thick layer of asbestos insulation ($k = 0.048$ Btu/h · ft · °F) followed by a 1-in. thick layer of fiberglass insulation ($k = 0.022$ Btu/h · ft · °F). The fluid inside the pipe is at 600°F with $h =$ 100 Btu/h · ft^2 · °F, and the exterior surface of the fiberglass insulation is exposed to air at 100°F with $h = 5$ Btu/h · ft^2 · °F. Calculate the rate of heat loss per foot of pipe length. How does your answer compare with the result from Problem 1.22 in which the convective resistances were neglected?

(2.7) Consider again the building wall of Problem 1.23. Suppose the interior surface is exposed to air at 70°F with $h = 10$ Btu/h · ft^2 · °F while the exterior surface is exposed to air at 30°F with $h = 25$ Btu/h · ft^2 · F. Calculate the rate of heat loss for this situation and compare with the result from Problem 1.23 in which the convective resistances were neglected.

Ans. 21,300 Btu/h.

(2.8) A consulting firm has designed a steam system for a new process unit at your plant. Superheated steam is delivered to the unit via a 3-in. steel pipe that has an ID of 3.068 in. and an OD of 3.5 in. The steam is at 800°F with a heat-transfer coefficient of 100 Btu/h · ft^2 · °F. The pipe is covered with a 0.5-in. thick layer of Type I insulation having an *R*-value of 1, followed by a 1-in. thick layer of Type II insulation having an *R*-value of 4. The steam line will be exposed to ambient air at 80°F with a heat-transfer coefficient of 5 Btu/h · ft^2 · °F. Design specifications call for the external surface temperature of the insulation to be no greater than 110°F and the manufacturer's literature states that Type I insulation is suitable for temperatures up to 1200°F while Type II insulation is suitable for temperatures up to 500°F. Your job is to evaluate the consulting firm's design. Proceed as follows:

(a) Calculate the rate of heat loss from the steam line per foot of pipe length that will actually occur if the consulting firm's design is implemented.

(b) Calculate the appropriate temperatures needed to determine whether or not all design and material specifications have been met.

(c) State whether or not the consulting firm's design is acceptable and the reason(s) for your conclusion.In these calculations, neglect the thermal resistance of the pipe wall, contact resistances, and radiation effects.

(2.9) Consider again the underground waste gas duct of Problem 1.27. If the heat-transfer coefficient for the waste gas is 25 Btu/h · ft · °F, calculate the rate of heat loss from the waste gas per foot of duct length.

Ans. 880 Btu/h · ft of length.

(2.10) Consider again the industrial chimney of Problem 1.25. The heat-transfer coefficient for the flue gas is 25 Btu/h · ft^2 · °F and the ambient air surrounding the chimney is at 60°F with $h = 10$ Btu/h · ft^2 · °F. Calculate the rate of heat loss from the flue gas per foot of chimney height.

Ans. 1100 Btu/h · ft of height.

(2.11) A graphite heat exchanger is constructed by drilling circular channels in a solid block of graphite. Consider a large graphite block ($k = 800$ W/m · K) containing two 1-in. diameter channels spaced 2 in. apart (center-to-center). A highly corrosive liquid ($T = 350$°F, $h = 120$ Btu/h · ft^2 · °F) flows through one channel and a coolant ($T = 170$°F, $h = 200$ Btu/h · ft^2 · °F) flows in the other channel. Calculate the rate of heat transfer from the corrosive liquid to the coolant per foot of exchanger length.

Ans. 3470 Btu/h · ft of length.

(2.12) A highly corrosive liquid is cooled by passing it through a circular channel with a diameter of 5 cm drilled in a rectangular block of graphite ($k = 800$ W/m · K). Ambient air is blown over the exterior of the block as shown in the sketch below. The heat-transfer coefficient for the air is 100 W/m^2 · K, based on the total exterior surface area of the block. Calculate the rate of heat loss from the liquid per meter of channel length.

Ans. 10,500 W/m of length.

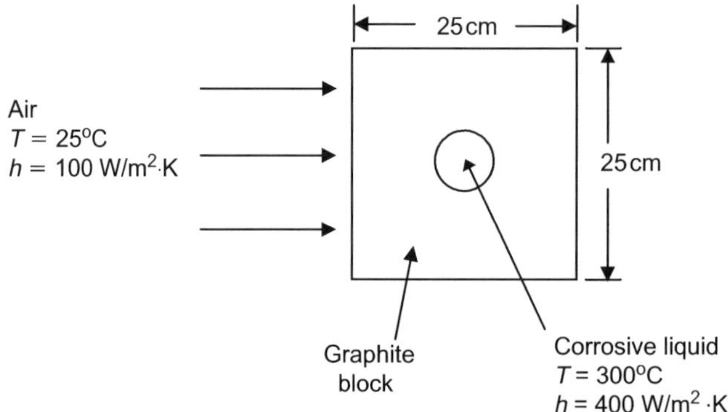

(2.13) A pipeline to transport chilled water from the utility building to a process unit is being designed at a chemical plant. The steel pipe will have an ID of 7.8 cm, an OD of 8.9 cm, and will be buried with its centerline at a depth of 0.92 m. Chilled water will enter the pipe at 285 K with a flow rate of 1.2 kg/s. The rate of heat transfer to the water is to be calculated to determine if the pipe will require insulation. For design purposes, the ground surface temperature is assumed to be 335 K and the pipe is assumed to be uninsulated. The process unit is located 40 m from the utility building.

(a) Make a reasonable guess for the pipe wall temperature and use it to calculate the heat-transfer coefficient for the water.

(b) Use the heat-transfer coefficient found in part (a), together with the design value for ground surface temperature, to calculate the rate of heat transfer to the water per meter of pipe length. (Do not use the wall temperature assumed in part (a) for this calculation because it was just a guess. The correct value for T_w is unknown.)

(c) Use the result of part (b) to find the temperature of the water delivered to the process unit.

(d) Check the value assumed for T_w in part (a) by calculating the pipe wall temperature. Is the difference between assumed and calculated values large enough to warrant further iteration?

(2.14) Chromium steel ($\rho = 7836$ kg/m^3, $c = 520$ J/kg · K, $k = 45$ W/m · K) ball bearings are to be heat treated by heating to 300°C followed by immersion in a quenching bath which is maintained at 40°C. The heat-transfer coefficient between the bearings and the quenching fluid is estimated to be 300 W/m^2 · K, and the bearings are 4.0 cm in diameter.

(a) How long should the bearings be quenched to cool them to 80°C?

(b) The quenching bath consists of unused engine oil. Is the value given for the heat-transfer coefficient in the problem statement reasonable? Give calculations to support your answer.

Ans. (a) 170 s.

(2.15) At 7:15 PM a process upset at a chemical plant resulted in the accidental release of a hazardous material to the atmosphere. The process was shut down, and an inspection revealed a faulty heater on a holding tank in the process. The tank holds a chemical intermediate which is normally maintained at 300°F by the heater, and a mechanical agitator keeps the temperature uniform throughout the tank. It is known that if the temperature of the intermediate drops below 100°F, the process will not operate properly. Hence, it is postulated that a failed heater is the cause of the accidental release. Plant records show that the heater was operating normally at the start of the second shift (4:00 PM). The ambient temperature at the time of the accident was 50°F, and the heat-transfer coefficient between the ambient air and the tank is estimated to have been 10 Btu/h · ft^2 · °F. Properties of the chemical intermediate and tank dimensions are given below. Are the data consistent with the hypothesis that the failed heater caused the process upset?

Chemical Intermediate Properties	Cylindrical Tank Dimensions
$\rho = 50$ lbm/ft^3	$D = 10$ ft
$C_P = 0.5$ Btu/lbm · °F	Height = 15 ft
$k = 0.1$ Btu/h · ft · °F	

Notes:
(1) The agitator did not fail, so the temperature in the tank remained uniform at all times.
(2) The bottom of the tank sits on a concrete pad, so assume heat transfer through the side and top of the tank only.

(2.16) A prilling tower is a device which converts a liquid into a granular solid. The liquid is formed into droplets by forcing it through nozzles at the top of the tower. As the droplets fall under gravity, they are contacted by air blown upward through the tower. The droplets cool and solidify as they fall, and the solid particles are collected at the bottom of the tower. A key design parameter for a prilling tower is the time required for a droplet to cool and solidify.
In a new chemical process, the liquid organic chemical product will be fed to a prilling tower at 130°C. Preliminary design parameters include a droplet diameter of 6 mm, an air temperature of 30°C, and a heat-transfer coefficient of 40 W/m^2 · K between air and droplet. The organic chemical has the following properties:
Melting point = 122°C
Liquid density = 1.3 g/cm^3
Heat capacity = 1460 J/kg · K
Thermal conductivity = 0.42 W/m · K
Latent heat of melting = 180 J/g
(a) Calculate $h(v/A_s)/k$ for a droplet. Is $T = T(t)$ a valid approximation for a droplet?
(b) Determine the time required for a droplet to reach the melting point.
(c) After the droplet has reached the melting point, estimate the additional time required for complete solidification of the droplet. Neglect any effect of changes in physical properties as solidification occurs.
(d) Is the specified value of 40 W/m^2 · K for the heat-transfer coefficient reasonable? Assume a relative velocity of 10 m/s between air and droplets, and give calculations to support your answer.

Ans. (a) 0.095. (b) 4 s. (c) 64 s.

(2.17) A rectangular steel ($k = 58.7$ W/m · K) fin 25 mm thick and 150 mm long is attached to a wall that is at 250°C. The ambient air around the fin is at 20°C and the heat-transfer coefficient is 26 W/m^2 · K. Calculate:
(a) The fin efficiency.
(b) The rate of heat transfer from the fin per meter of fin width.

Ans. (a) 77.9%. (b) 1456 W/m of width.

(2.18) A rectangular fin 1.0 ft long, 1.0 ft wide, and 2 in. thick loses heat to the atmosphere at 70°F. The base of the fin is at 270°F. If $h = 2$ Btu/h · ft^2 · °F and the thermal conductivity of the fin is 24 Btu/h · ft · °F, find:
(a) The fin efficiency.
(b) The total rate of heat loss from the fin.

Ans. (a) 70.9%. (b) 709 Btu/h.

(2.19) An annular stainless steel ($k = 16.6$ W/m · K) fin of thickness 5 mm and outside diameter 3 cm is attached to a 1-cm OD tube. The fin is surrounded by a fluid at 50°C with a heat-transfer coefficient of 40 W/m^2 · K. The tube wall temperature is 150°C. Calculate:
(a) The fin efficiency.
(b) The rate of heat transfer from the fin.

Ans. (a) 91%. (b) 6.4 W.

(2.20) The inner pipe of a double-pipe heat exchanger has an OD of 1.9 in. and contains 36 rectangular fins of the type shown in the sketch below. The fins are made of steel ($k = 34.9$ Btu/h · ft · °F) and are 0.5 in. high and 0.035 in. thick. The pipe wall temperature is 250°F and the fluid in the annulus surrounding the fins is at 150°F with a heat-transfer coefficient of 30 Btu/h · ft^2 · °F. Calculate:
(a) The fin efficiency.
(b) The rate of heat transfer from one fin per foot of pipe length.
(c) The prime surface area per foot of pipe length.
(d) The rate of heat transfer from the prime surface per foot of pipe length.
(e) The weighted efficiency of the finned surface.

(f) The total rate of heat transfer (from fins and prime surface) per foot of pipe length.
(g) The thermal duty for the exchanger is 390,000 Btu/h. What length of pipe is required to satisfy this duty?

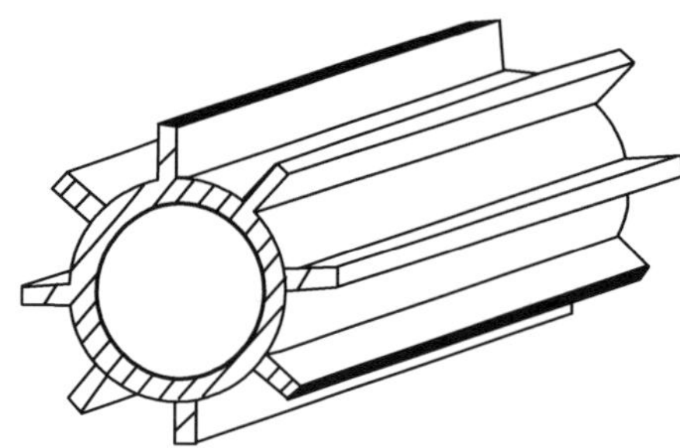

Ans. (a) 75%. (b) 194 Btu/h. (c) 0.3924 ft^2. (d) 1177 Btu/h.
(e) 77.8%. (f) 8160 Btu/h. (g) 48 ft.

(2.21) A carbon steel ($k = 45$ W/m · K) heat exchanger tube has an outside diameter of 2.4 cm and contains 630 annular fins per meter of length. Each fin extends 0.16 cm beyond the exterior tube wall and has a thickness of 0.04 cm. The fluid surrounding the tube is at 25°C and the heat-transfer coefficient is 1200 W/m^2 · K. The tube wall temperature is 100°C. Determine:
(a) The fin efficiency.
(b) The surface area of the fins per meter of tube length.
(c) The prime surface area per meter of tube length.
(d) The weighted efficiency of the finned surface.
(e) The rate of heat transfer from the fins per meter of tube length.
(f) The rate of heat transfer from the prime surface per meter of tube length.
(g) The total rate of heat transfer per meter of tube length.

Ans. (a) 86.7%. (b) 0.183828 m^2. (c) 0.056398 m^2. (d) 89.8%.
(e) 14,344 W. (f) 5076 W. (g) 19,420 W.

(2.22) Annular steel fins ($k = 56.7$ W/m · K) are attached to a steel tube that is 30 mm in external diameter. The fins are 2 mm thick and 15 mm long. The tube wall temperature is 350 K and the surrounding fluid temperature is 450 K with a heat-transfer coefficient of 75 W/m^2 · K. There are 200 fins per meter of tube length. Calculate:
(a) The fin efficiency.
(b) The fin surface area per meter of tube length.
(c) The prime surface area per meter of the tube length.
(d) The weighted efficiency of the finned surface.
(e) The rate of heat transfer per meter of tube length.

Ans. (a) 85%. (b) 0.9248 m^2. (c) 0.0565 m^2. (d) 86%. (e) 6330 W.

(2.23) A finned heat exchanger tube is made of aluminum alloy ($k = 186$ W/m · K) and contains 125 annular fins per meter of tube length. The bare tube between fins has an OD of 50 mm. The fins are 4 mm thick and extend 15 mm beyond the external surface of the tube. The outer surface of the tube will be at 200°C and the tube will be exposed to a fluid at 20°C with a heat-transfer coefficient of 40 W/m^2 · K. Calculate:
(a) The rate of heat transfer per meter of tube length for a plain (un-finned) tube.
(b) The fin efficiency.
(c) The fin and prime surface areas per meter of tube length.
(d) The weighted efficiency of the finned surface.
(e) The rate of heat transfer per meter of tube length for a finned tube.
(f) If the cost per unit length of finned tubing is 25% greater than for plain tubing, determine whether plain or finned tubing is more economical for this service.

Ans. (a) 1130 W. (b) 98.6%. (c) 0.8946 m^2 and 0.07854 m^2.
(d) 98.7%. (e) 6920 W.

(2.24) A stream of ethylene glycol having a flow rate of 1.6 kg/s is to be cooled from 350 to 310 K by pumping it through a 3-cm ID tube, the wall of which will be maintained at a temperature of 300 K. What length of tubing will be required?

Ans. 49 m.

(2.25) A small steam superheater will be made of 0.75-in. schedule 80 stainless steel pipe which will be exposed to hot flue gas from a boiler. 50 lbm/h of saturated steam at 320°F will enter the pipe and be heated to 380°F. Assuming that the pipe wall temperature will vary from 375°F at the steam inlet to 415°F at the outlet, calculate the length of pipe required for the superheater.

Ans. 6.1 ft.

(2.26) Suppose that the steam superheater of Problem 2.25 has a length of 10 ft and the wall temperature is constant at 415°F over the entire length of the pipe. Using the value of the heat-transfer coefficient calculated in Problem 2.25, estimate the temperature of the steam leaving the superheater.

Ans. 405°F.

(2.27) A gas heater shown below in cross-section consists of a square sheet metal duct insulated on the outside. A 5-cm OD steel pipe passes through the center of the duct and the pipe wall is maintained at 250°C by condensing steam flowing through the pipe. Air at 20°C will be fed to the heater at a rate of 0.35 kg/s.
(a) Calculate the equivalent diameter for the heater.
(b) Calculate the length of the heater required to heat the air stream to 60°C.

Ans. (a) $D_e = 0.1083$ m. (b) 8.1 m.

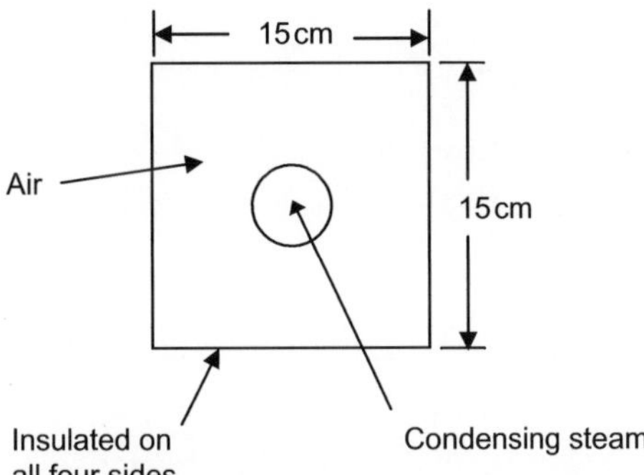

(2.28) Water at 20°C with a mass flow rate of 1.0 kg/s enters an annulus formed by an inner pipe having an OD of 2.5 cm and an outer pipe having an ID of 10 cm. The wall temperature of the inner pipe varies from 80°C at the inlet to 100°C at the outlet. The outer surface of the outer pipe is well insulated. Calculate the length of the annulus required to heat the water to 75°C.

Ans. 75.3 m.

(2.29) A stream of ethylene glycol having a flow rate of 4000 lb/h is to be heated from 17°C to 37°C by passing it through a circular annular heater. The outer pipe of the annulus will have an ID of 5 cm and the inner pipe will have an OD of 4 cm. The exterior surface of the outer pipe will be well insulated, while the inner pipe wall temperature will vary from 67°C at the entrance to 87°C at the exit. The required length of the heater is desired. Assuming as a first approximation that $L = 20$ m, calculate a second approximation for the length.

Ans. 20.8 m.

(2.30) 100 lbm/h of oil at 130°F is needed for a process modification. The oil will be available at 70°F, so a plant engineer has designed a heater consisting of a 1-in. ID tube, 10 ft long, the wall of which will be maintained at 215°F by condensing steam. Will the heater work as required? Properties of the oil may be assumed constant at the following values:

$C_P = 0.49$ Btu/lbm · °F	$\mu = 1.42$ lbm/ft · h
$k = 0.0825$ Btu/h · ft · °F	$\rho = 55$ lbm/ft^3

(2.31) A stream of Freon-12 with a flow rate of 4000 lb/h is to be heated from −20°C to 30°C for use in a chemical processing operation. Available for this service is an annular heater 3.05 m in length. The outer pipe of the annulus has an ID of 5 cm and an O.D. of 5.5 cm. The inner pipe has an ID of 3.75 cm and an OD of 4 cm. When in operation, the walls of both pipes are maintained at a constant temperature that can be set to any value between 30°C and 70°C. Will the heater be suitable for this service?

(2.32) A feed stream to a chemical reactor consists of 0.2 kg/s of ammonia vapor at 300 K. Prior to entering the reactor, the ammonia is preheated by passing it through a rectangular duct whose walls are heated to 500 K by hot process waste gas. The duct cross-section is 9 cm by 20 cm, and it is 2.5-m long. At what temperature does the ammonia enter the reactor?

(2.33) An air preheater is required to heat 0.2 kg/s of process air from 15°C to 115°C. The preheater will be constructed from rectangular ducting having a cross-section of 7.5 cm by 15 cm. The air will flow inside the duct and the duct walls will be maintained at 250°C by hot flue gas. What length of ducting will be required?

(2.34) A cylindrical storage tank with a diameter of 4 m and a length of 10 m will hold a fluid whose temperature must be maintained at 347 K. In order to size the heater required for the tank, the following worst-case scenario is considered:
Ambient air temperature = -20°C
Wind speed = 20 m/s
Tank wall temperature = 347 K
What size (kW) heater will be needed?

Ans. 400 kW

(2.35) A viscous liquid is to be pumped between two buildings at a chemical plant in an above-ground pipe that has an OD of 22 cm and is 110 m long. To facilitate pumping, the liquid will be heated to 40° C in order to reduce its viscosity. The liquid flow rate will be 20 kg/s and the specific heat of the liquid is 1300 J/kg · K. Determine the temperature drop that the liquid will experience over the length of the pipe under the following worst-case conditions:
Ambient air temperature = −10°C
Wind speed = 14 m/s
Pipe surface temperature = 37°C

Ans. 5.4°C.

(2.36) A surge tank to be used in a chemical process is spherical in shape with a diameter of 10 ft. The tank will hold a liquid that must be maintained at 180°F by means of a heating unit. The following parameters have been established for design purposes:
Ambient air temperature = 20°F
Wind speed = 20 miles/h
Tank wall temperature = 180° F
(a) Estimate the duty (Btu/h) that the heater must supply.
(b) Comment on the probable accuracy of your estimate.

Ans. (a) 88,500 Btu/h.

(2.37) A duct is being designed to transport waste gas from a processing unit to a pollution control device. The duct will be 5 ft high, 6 ft wide, and 100 ft long. In order to determine whether the duct will need to be insulated, the rate of heat loss to the environment must be estimated. Based on the following design conditions, compute the rate of heat loss from the duct.
Average duct surface temperature = 250° F
Ambient air temperature = 10°F
Wind speed = 20 miles/h

Ans. 1.4×10^6 Btu/h.

(2.38) A pipeline at a chemical complex has an OD of 10 cm, an ID of 9.4 cm, and is covered with a layer of insulation ($k = 0.055$ W/m · K) that is 4 cm thick. A process liquid at 50°C flows in the pipeline with a heat-transfer coefficient of 400 W/m^2 · K. The pipeline is exposed to the environment on a day when the air temperature is -20°C and the wind speed is 15 m/s. What is the rate of heat loss from the pipeline per meter of length?

Ans. 40 W/m of length.

(2.39) Consider again the graphite heat exchanger of Problem 2.12. What air velocity is required to achieve the stated heat-transfer coefficient of 100 W/m^2 · K?

(2.40) An energy recovery system is being considered to preheat process air using hot flue gas. The process requires air at 80°C. In the proposed energy recovery system, air at 25°C will enter a rectangular duct with a flow rate of 2 kg/s. The duct dimensions are 1 m × 2 m × 6 m long. The duct walls will be maintained at a temperature of 250°C by hot flue gas flowing over the outside of the duct. Estimate the air temperature that will be achieved with this system and thereby determine whether or not the energy recovery system will satisfy the process requirement.

(2.41) The surface temperature of the electronic chip shown below is 75°C and it is surrounded by ambient air at 25°C. Calculate the rate of heat loss by natural convection from the upper surface of the chip.

Ans. 0.178 W.

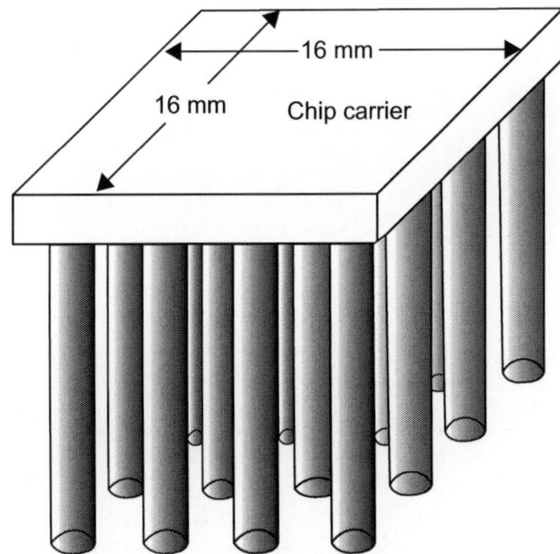

(2.42) A petrochemical storage tank is cylindrical in shape with a diameter of 6 m and a height of 10 m. The surface temperature of the tank is 10°C when, on a calm clear night, the air temperature drops rapidly to –10°C. Estimate the rate of convective heat loss from the tank under these conditions.

(2.43) A horizontal elevated pipeline in a chemical plant has an OD of 10 cm, an ID of 9.4 cm, and is covered with a layer of magnesia insulation that is 4 cm thick. A process liquid at 50°C flows in the pipeline with a heat-transfer coefficient of 400 $W/m^2 \cdot K$. The pipeline is exposed to the environment on a calm night when the air temperature is –5°C.

(a) Make a reasonable guess for the temperature of the exterior surface of the insulation and use it to calculate the heat-transfer coefficient between the insulation and the ambient air.
(b) Use the result of part (a) together with the other information given in the problem to calculate the rate of heat loss per meter of pipe length.
(c) Use the result of part (b) to calculate the temperature of the exterior surface of the insulation and compare it with the value that you assumed in part (a).

(2.44) A spherical storage tank has a diameter of 5 m. The temperature of the exterior surface of the tank is 10°C when, on a calm clear night, the air temperature drops rapidly to –10°C. Estimate the rate of convective heat loss from the tank under these conditions.

Ans. 5550 W.

(2.45) Estimate the maximum possible rate of heat loss due to radiation for:

(a) The cylindrical storage tank of problem 2.42.
(b) The spherical storage tank of Problem 2.44.

Ans. (a) 78,837 W. (b) 28,564 W.

(2.46) Calculate the rate of heat loss by radiation from the upper surface of the electronic chip of Problem 2.41. Assume that the chip is contained in an enclosure that is large compared with the size of the chip and whose walls are at 25°C. Also assume an emissivity of 0.5 for the surface of the chip.

Ans. 0.049 W.

(2.47) A horizontal steam pipe has an outer diameter of 8.9 cm and passes through a large room whose walls are at 293 K. The pipe wall has a temperature of 410 K and an emissivity of 0.85. The air temperature in the room is 296 K. Calculate the rate of heat loss per unit length from the pipe by radiation and natural convection.

Ans. 281 W/m of pipe by radiation; 217 W/m of pipe by convection.

(2.48) At a convective boundary such as the one at $x = B$ in the sketch below, heat is transferred between a solid and a fluid. In order for a finite temperature to exist at the solid-fluid interface, the rate at which heat is transferred to the interface by conduction through the solid must equal the rate at which heat is transferred away from the interface by convection in the fluid. The reason is that the interface has no volume and no mass, and so has zero heat capacity. The boundary condition at $x = B$ is thus:

$$\hat{q}_{conduction} = \hat{q}_{convection}$$

$$-k\frac{dT}{dx} = h\,(T - T_\infty)$$

$$h\,(T - T_\infty) + \frac{dT}{dx} = 0$$

Expressions for both T and dT/dx (obtained by integration of the conduction equation) are substituted into this equation to obtain a relationship between the constants of integration.

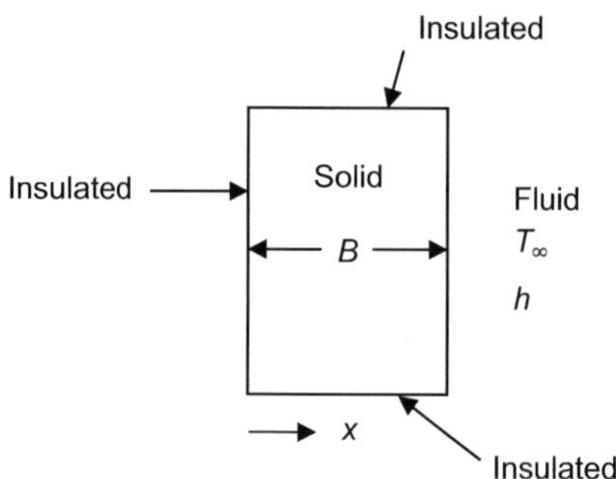

The rectangular solid shown above is insulated on all sides except the one at $x = B$, which is exposed to a fluid at temperature T_∞ with heat-transfer coefficient, h. Heat is generated within the solid at a rate per unit volume given by $\dot{q} = \Gamma x$, where Γ is a constant.
(a) Assuming constant thermal conductivity, derive an expression for the steady-state temperature distribution, $T(x)$, in the solid.
(b) Calculate the temperature of the insulated boundary at $x = 0$ for the following parameter values:

$T_\infty = 20°C$	$k = 1.5\ W/m \cdot K$	$\Gamma = 50\ W/m^4$
$B = 2\ m$	$h = 30\ W/m^2 \cdot K$	

Ans. (a) $T(x) = T_\infty + \dfrac{\Gamma B^2}{2h} + \dfrac{\Gamma}{6k}(B^3 - x^3)$. (b) 67.8°C.

(2.49) Consider the rectangular solid shown below that is insulated on four sides. The side at $x = B$ is held at a fixed temperature, T_w, while the side at $x = 0$ is exposed to a fluid at temperature T_∞. Heat is generated within the wall at a rate per unit volume given by $\dot{q} = \Gamma x$, where Γ is a constant. The thermal conductivity of the solid may be assumed constant.
(a) Formulate an appropriate set of boundary conditions for this configuration.
(b) Use the conduction equation to derive an expression for the steady-state heat flux, $\hat{q}_x$, in the solid.

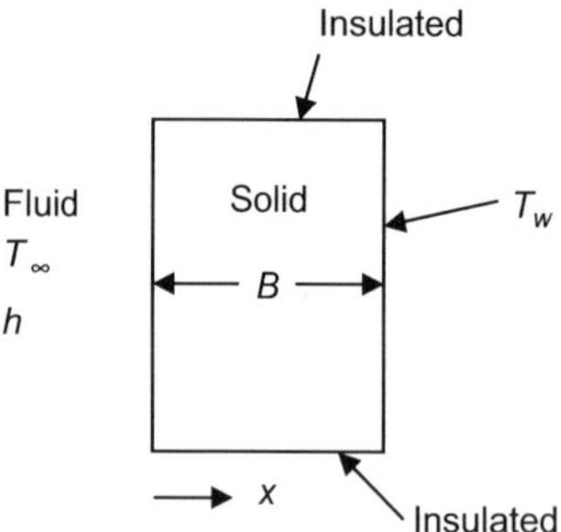

Ans. (b) $\hat{q}_x = \dfrac{\Gamma x^2}{2} + \dfrac{h\{k(T_\infty - T_w) - \Gamma B^3/6\}}{(k + Bh)}$.

(2.50) A long solid cylindrical rod of radius R contains a heat source that generates heat per unit volume at a rate $\dot{q} = \Gamma r$, where Γ is a constant and r is radial position measured from the centerline. The rod is completely surrounded by a fluid at temperature T_∞ with heat-transfer coefficient, h. Assuming heat flow only in the radial direction:
(a) Formulate an appropriate set of boundary conditions to be used with the heat conduction equation.
(b) Derive an expression for the steady-state temperature profile, $T(r)$, in the rod. State any assumptions that you make in your derivation.
(c) Obtain an expression for the maximum temperature in the rod at steady state.

Ans. (b) $T(r) = T_\infty + \dfrac{\Gamma R^2}{3h} + \dfrac{\Gamma(R^3 - r^3)}{9k}$.

(2.51) A long hollow cylinder has inner and outer radii R_1 and R_2, respectively. Heat is generated in the cylinder wall at a uniform rate, Γ, per unit volume. The outer surface of the cylinder is well insulated. A fluid flows through the inside of the cylinder to provide cooling. The fluid temperature is T_∞ and the heat-transfer coefficient is h. Assuming constant thermal conductivity, k, derive an expression for the steady-state temperature profile, $T(r)$, in the cylinder wall.

Ans. $T(r) = T_\infty + \frac{\Gamma\,(R_2^2 - R_1^2)}{2\,hR_1} + \frac{\Gamma\,(R_1^2 - r^2)}{4\,k} + \frac{\Gamma\,R_2^2\ln(r/R_1)}{2\,k}.$

(2.52) A solid sphere of radius R is immersed in a fluid with temperature T_∞ and heat-transfer coefficient h. Heat is generated within the sphere at a rate per unit volume given by $\dot{q} = \Gamma r$, where Γ is a constant and r is radial position measured from the center of the sphere. Assuming constant thermal conductivity, k:

(a) Derive an expression for the steady-state temperature profile, $T(r)$, in the sphere.

(b) Calculate the maximum temperature in the sphere under steady-state conditions for the following parameter values:

$R = 1.5$ m	$\Gamma = 100$ W/m^4	$h = 100$ W/m$^2\cdot$K
$k = 0.5$ W/m $\cdot$ K	$T_\infty = 20°$C	

Ans. (a) $T(r) = T_\infty + \frac{\Gamma R^2}{4h} + \frac{\Gamma(R^3 - r^3)}{12\,k}$. (b) 76.8°C.

3 Heat Exchangers

3.1 Introduction

In this chapter we describe the most common types of heat exchangers used in the chemical process industries; namely, double-pipe, shell-and-tube, and plate heat exchangers. Air-cooled heat exchangers, which are widely used in petroleum and petrochemical processing, are covered separately in Chapter 12. Each of these exchanger types has been successfully employed for decades in a wide variety of industrial applications. The descriptions given here are intended as a general introduction. Detailed analyses of specific applications are usually required to fully understand the advantages and limitations of any type of heat exchanger. It should also be noted that improvements in heat exchanger design and manufacturing are ongoing, and future developments will no doubt result in equipment whose thermal capability and reliability exceed those upon which many current practices are based.

This chapter also presents some of the basic techniques that are used in the analysis of industrial heat-exchange equipment. The primary focus is on double-pipe and shell-and-tube exchangers. The double-pipe is the simplest type of heat exchanger and the most straightforward to analyze, while the shell-and-tube is the most widely used type of exchanger in the chemical process industries. For information regarding methods used in analyzing other types of heat-transfer equipment, the reader is referred to Refs. [1–3], as well as Chapter 12.

Heat-exchanger calculations can be divided into two distinct categories; namely, thermal and hydraulic calculations on the one hand and mechanical design calculations on the other. Thermal and hydraulic calculations are made to determine heat-transfer rates and pressure drops needed for equipment sizing. Mechanical design calculations are concerned with detailed equipment specifications, and include considerations such as stress and tube vibration analyses. In this chapter we will be concerned with thermal calculations only. Hydraulic calculations are considered in subsequent chapters. Mechanical design calculations are not covered in detail in this book; however, some aspects of mechanical design are discussed in the chapters that deal specifically with equipment design.

Heat-exchanger problems may also be categorized as rating problems or design problems. In a rating problem, one must determine whether a given, fully specified exchanger will perform a given heat-transfer duty satisfactorily. It is immaterial whether the exchanger physically exists or whether it is specified only on paper. In a design problem, one must determine the specifications for a heat exchanger that will handle a given heat-transfer duty. A rating calculation is generally an integral part of a design calculation. However, a rating problem also arises when it is desired to use an existing exchanger in a new or modified application. In this chapter we will be primarily concerned with rating calculations, although some aspects of the design problem will also be considered. Design procedures are covered in greater detail in the following chapters.

3.2 Double-Pipe Equipment

A simple double-pipe exchanger consists of two pairs of concentric pipes arranged as shown in Figure 3.1. Such a configuration is called a hairpin, for obvious reasons. Batteries of hairpins connected in series or in series-parallel arrangements are commonly employed to provide adequate surface area for heat transfer. The two fluids that are transferring heat flow in the inner and outer pipes, respectively. The fluids usually flow through the exchanger in opposite directions as shown in Figure 3.1. Such a flow pattern is called counter flow or counter-current flow. In some special-purpose applications, parallel (or co-current) flow is employed in which the two streams flow in the same direction. The outer pipe may be insulated to minimize heat transfer to or from the environment. Also, nozzles may be provided on the inner-pipe (tube) side as well as on the annulus side to facilitate connection to process piping. The maximum nozzle size is generally equal to the size of the outer heat-exchanger pipe.

Multi-tube exchangers are also available in which the inner pipe is replaced by a bundle of U-tubes, as shown in Figure 3.2. The tubes may be either plain or equipped with longitudinal fins.

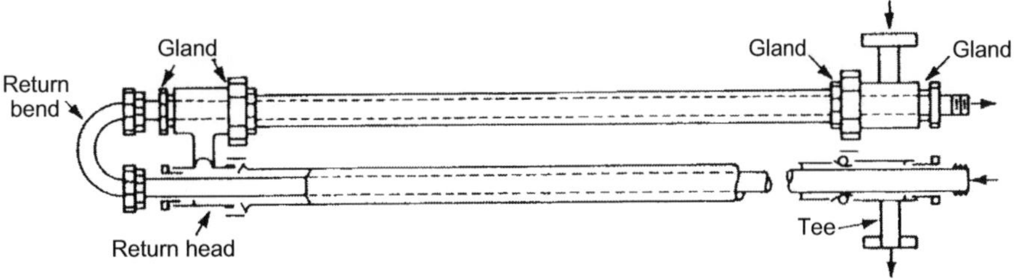

FIGURE 3.1 Simple double-pipe exchanger (Source: Ref. [4]).

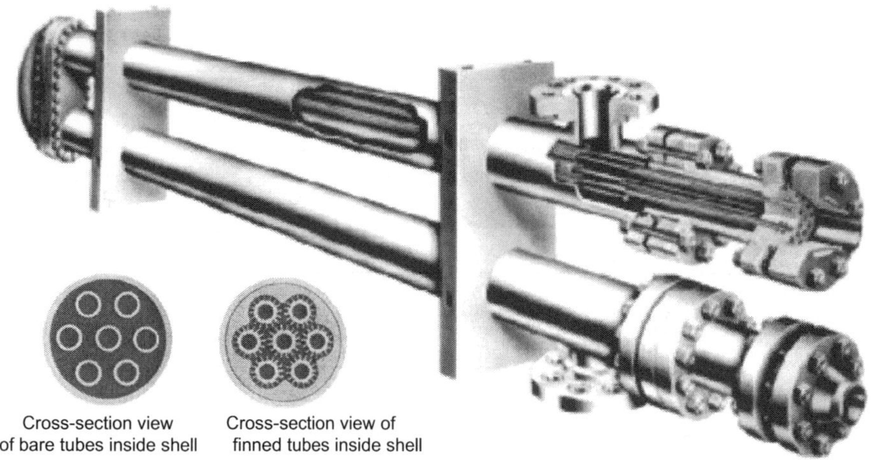

FIGURE 3.2 Multi-tube hairpin exchanger (Source: Koch Heat Transfer Company, LP).

Simple double-pipe exchangers are commercially available with outer-pipe sizes ranging from 2 to 8 in. and inner pipes from 3/4 to 6 in. Multi-tube exchangers typically have outer pipes ranging in size from 3 to 16 in. with tubing of various sizes. However, multi-tube units with outer-pipe sizes as large as 36 in. are commercially available.

Double-pipe exchangers are commonly used in applications involving relatively low flow rates and high temperatures or pressures, for which they are well suited. Other advantages include low installation cost, ease of maintenance, and flexibility. Hairpins can easily be added to or removed from an existing battery, or arranged in different series-parallel combinations to accommodate changes in process conditions. Simple un-finned double-pipe exchangers tend to become unwieldy, however, when heat-transfer areas greater than about 1000 ft^2 are required. They are also relatively expensive per unit of heat-transfer surface. Much larger heat-transfer areas are practical using multi-tube units with finned tubes. In fact, multi-tube hairpin exchangers with heat-transfer areas in excess of 12,000 ft^2 have been built and are commercially available.

3.3 Shell-and-Tube Equipment

A shell-and-tube exchanger (Figure 3.3) consists of a bundle of tubes contained in a cylindrical shell. Most shell-and-tube exchangers are designed in accordance with standards [5] defined by the Tubular Exchanger Manufacturers Association (TEMA). These standards have been used for more than 70 years and provide manufacturing practices and design guidelines to ensure that equipment designed and fabricated by different organizations will perform similarly and predictably. While the mechanical design is usually dictated by the ASME Boiler and Pressure Vessel code, the TEMA requirements also serve to standardize exchanger thermal performance.

The tubes may be permanently positioned inside the shell (fixed tubesheet exchanger) or may be removable for ease of cleaning and replacement (floating-head or U-tube exchanger). In addition, a number of different head and shell designs are commercially available as shown in Figure 3.4.

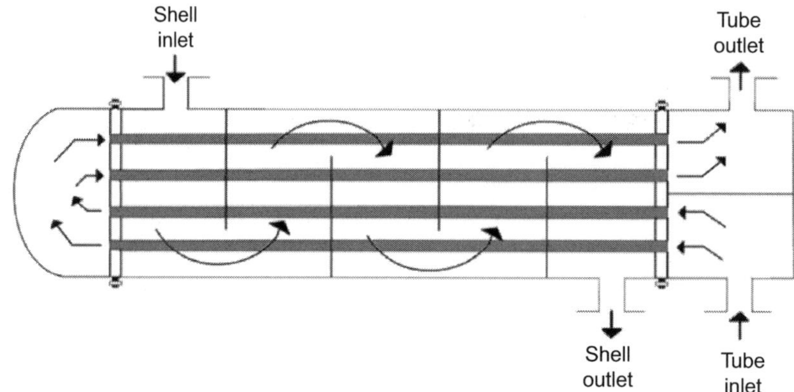

FIGURE 3.3 Schematic illustration of a shell-and-tube heat exchanger with one shell pass and two tube passes.

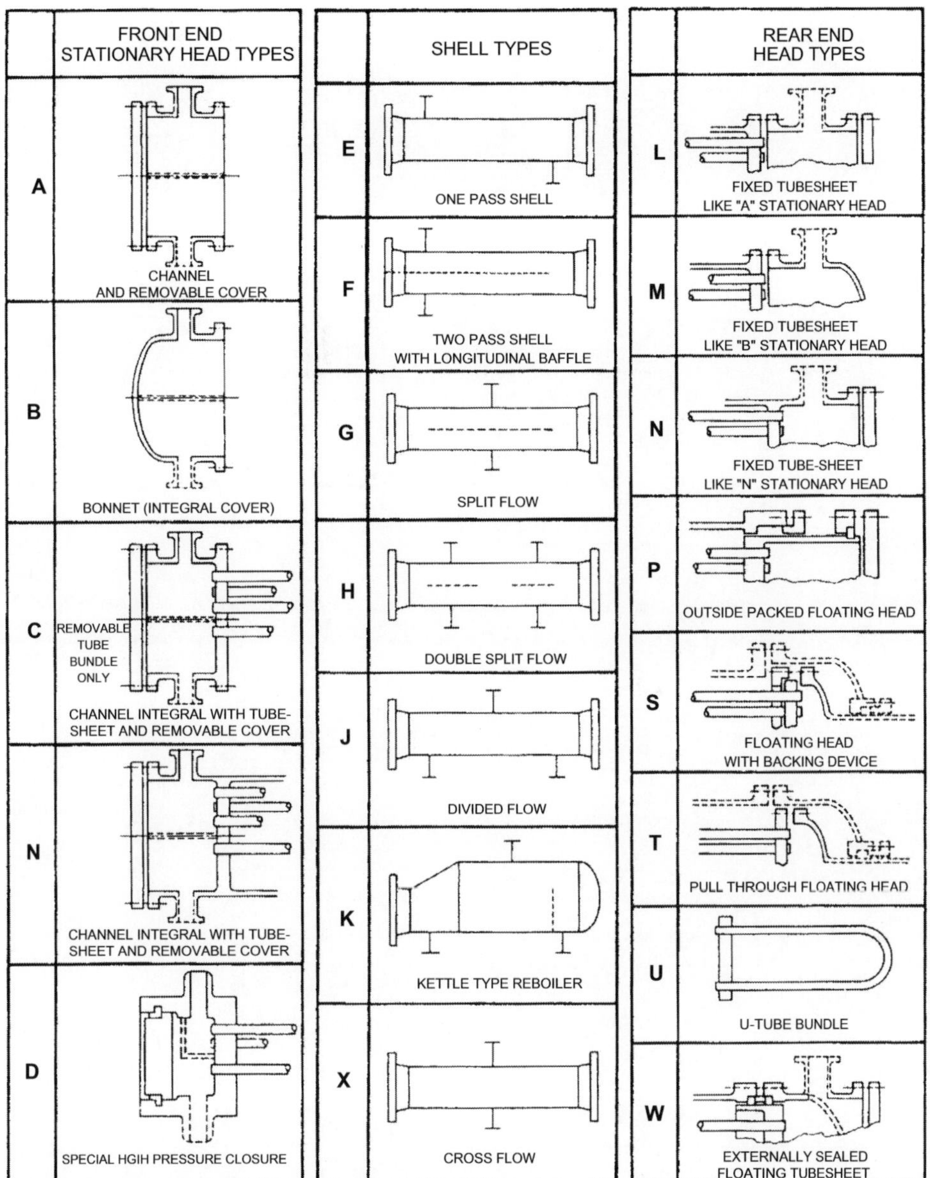

FIGURE 3.4 TEMA designations for shell-and-tube exchangers (Source: Ref. [5])

TEMA employs a three-letter code to specify the front-end, shell, and rear-end types. For example, a fixed tubesheet type BEM exchanger is shown in Figure 3.5(b). An internal-floating-head type AES exchanger is shown in Figure 3.5(a). Shells are available with inside diameters in discrete sizes up to 120 in. Shells up to 24 in. in diameter are generally made from steel pipes, while larger sizes are made from rolled steel plate. Some key features of shell-and-tube exchangers are summarized in Table 3.1.

Successful TEMA designs depend upon proper shell selection. For most general-purpose applications, E shells are appropriate since they simplify the design and fabrication processes. Depending on the allowable pressure drop, required process temperature change, temperature approach[1], and fluid phase, other shell types should be selected. Some general guidelines for shell selection are summarized in Table 3.2.

Heat exchanger and condenser tubing is available in a wide range of metals in sizes from 1/4 to 2 in. outside diameter. Finned tubing is also available. The dimensions of plain tubing are listed in Table B.l. Notice that the size designations for tubing are different from those of pipe (Table B.2). The nominal outside diameter of a heat-exchanger tube is the actual diameter. Also, the wall thickness is specified by the Birmingham Wire Gage (BWG) rather than by schedule number.

[1] Temperature approach is the minimum difference between hot and cold stream temperatures in the exchanger. For shell-and-tube exchangers, temperature approaches less than 3°C can be a design challenge.

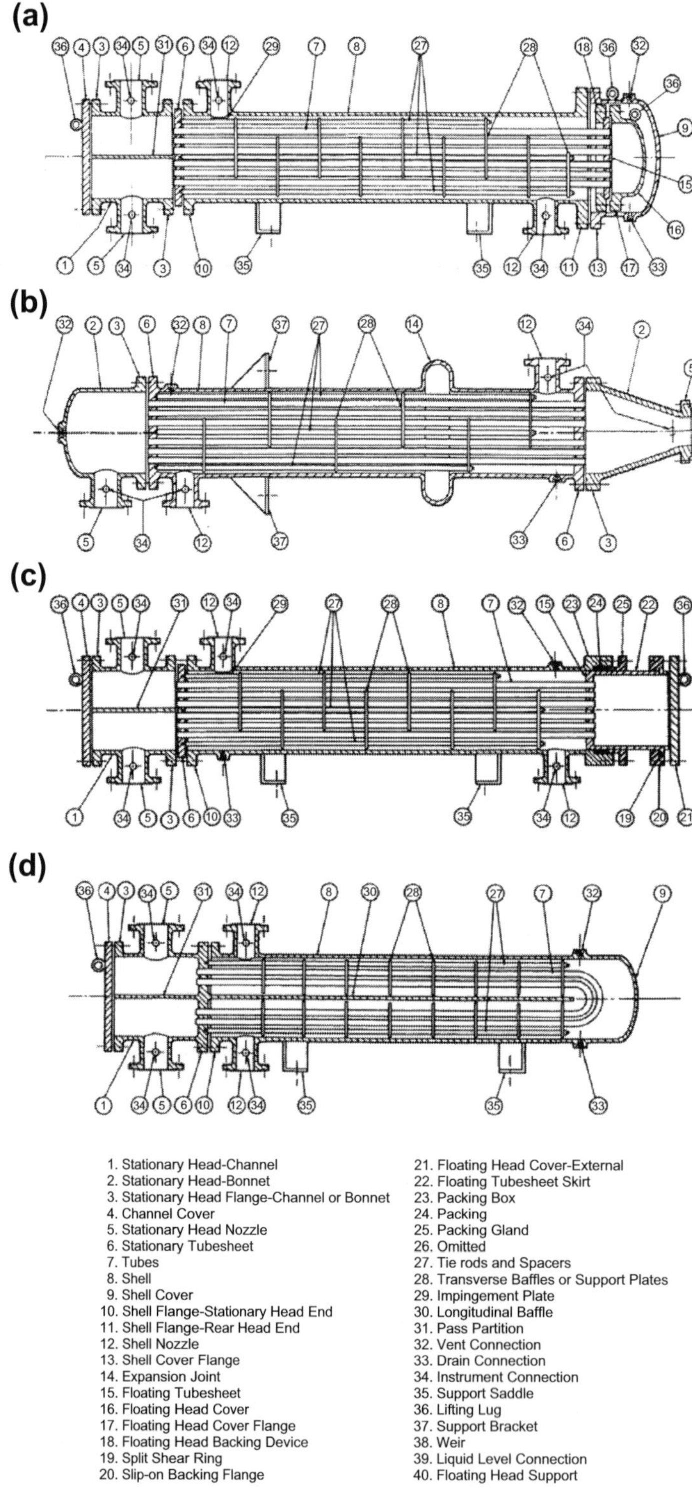

FIGURE 3.5 (a) Type AES floating-head exchanger; (b) type BEM fixed tubesheet exchanger with conical rear head; (c) Type AEP floating-head exchanger; (d) type CFU U-tube exchanger with two-pass shell (Source: Ref. [5]).

The tubes are arranged in the bundle according to specific patterns. The most common are square (90°), rotated square (45°), and equilateral triangle (30°), as shown in Figure 3.6. The square and rotated square layouts permit mechanical cleaning of the outsides of the tubes. The center-to-center distance between tubes is called the *tube pitch*. The minimum distance between adjacent tubes is the *gap*. From Figure 3.6, it can be seen that the pitch is equal to the gap plus the outside diameter.

The tube bundle is supported by baffles of various types. Figure 3.7 shows the most common baffle types covered by the TEMA standards.

TABLE 3.1 Features of Shell-and-Tube Heat Exchangers[a]

Type of Design	Fixed Tubesheet	U-Tube	Packed Lantern-Ring Floating Head	Internal Floating Head (Split Backing Ring)	Outside-Packed Floating Head	Pull-Through Floating Head
TEMA rear-head type	L, M or N	U	W	S	P	T
Relative cost increases from A (least expensive) through E (most expensive)	B	A	C	E	D	E
Provision for differential expansion	Expansion joint in shell	Individual tubes free to expand	Floating head	Floating head	Floating head	Floating head
Removal bundle	No	Yes	Yes	Yes	Yes	Yes
Replacement bundle possible	No	Yes	Yes	Yes	Yes	Yes
Individual tubes replaceable	Yes	Only those in outside row[b]	Yes	Yes	Yes	Yes
Tube cleaning by chemicals inside and outside	Yes	Yes	Yes	Yes	Yes	Yes
Interior tube cleaning mechanically	Yes	Special tools required	Yes	Yes	Yes	Yes
Exterior tube cleaning mechanically:						
Triangular pitch	No	No[c]	No[c]	No[c]	No[c]	No[c]
Square pitch	No	Yes	Yes	Yes	Yes	Yes
Hydraulic-jet cleaning:						
Tube interior	Yes	Special tools required	Yes	Yes	Yes	Yes
Tube exterior	No	Yes	Yes	Yes	Yes	Yes
Double tube sheet feasible	Yes	Yes	No	No	Yes	No
Number of tube passes	No practical limitations	Any even number possible	Limited to one or two passes	No practical limitations[d]	No practical limitations	No practical limitations[d]
Internal gaskets eliminated	Yes	Yes	Yes	No	Yes	No

[a]Modified from page a-8 of the Patterson-Kelley Co. Manual No. 700 A, Heat Exchangers.

[b]U-tube bundles have been built with tube supports that permit the U-bends to be spread apart and tubes inside of the bundle replaced.

[c]Normal triangular pitch does not permit mechanical cleaning. With a wide triangular pitch, which is equal to 2 (tube diameter plus cleaning lane)/$\sqrt{3}$, mechanical cleaning is possible on removable bundles. This wide spacing is infrequently used.

[d]For odd number of tube-side passes floating head requires packed joint or expansion joint.

Note: Relative costs A and B are not significantly different and interchange for long lengths of tubing.

Source: Ref. [1].

TABLE 3.2 Advantages and Disadvantages of TEMA Shell Types

Shell Type	Advantages	Disadvantages
E	Wide application for single-phase, boiling, and condensing services Temperature cross[a] possible without reverse heat transfer[b] with a single tube pass	Reverse heat transfer possible with even number of tube passes and no fouling
F	Temperature change for shell-side flow streams can be higher than E-shell Fewer shells-in-series needed to achieve multiple shell passes	Longitudinal baffle can leak if not welded Thermal conduction across longitudinal baffle Removable bundles more costly to maintain
G	Suitable for horizontal shell-side reboilers Split flow lowers entrance/exit velocities Lowers vibration potential Improved tube support under nozzle	Fewer tube-pass options with removable bundle Thermal conduction across longitudinal baffle Temperature profile not ideal as compared with counter- or co-current flow in E-shell
H	Suitable for horizontal shell-side reboilers Double split flow lowers entrance/exit velocities and provides additional tube support compared to G-shell	More nozzles than G-shell Thermal conduction across longitudinal baffle Temperature profile not ideal as compared with counter- or co-current flow
J	Split flow lowers velocities Lowers vibration potential Shorter flow path reduces pressure drop	More nozzles than E-shell Temperature profile not ideal as compared with counter- or co-current flow
K	Low-pressure drop Circulation promotes wet-wall boiling	Larger shell Entrainment calculations needed Circulation complicated Buildup of heavy components possible
X	Low-pressure drop due to single cross-pass Wide application for single-phase, boiling and condensing services Temperature cross possible without reverse heat transfer	Maldistribution possible Distribution plate often used Multiple nozzles common Non-condensable removal is complicated for condensers

[a]A temperature cross exists when the hot fluid outlet temperature is less than the cold fluid outlet temperature.

[b]Reverse heat transfer occurs where the hot fluid temperature is less than the cold fluid temperature. Reverse heat transfer is possible, but not desirable, in local regions of a heat exchanger with some shell types and tube-pass arrangements when a temperature cross is specified.

Source: Ref. [6].

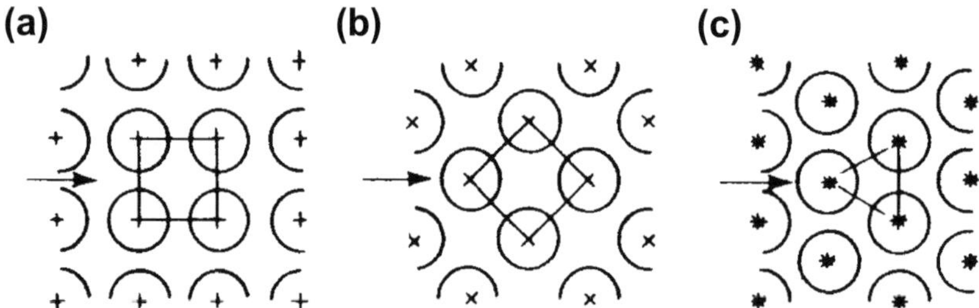

FIGURE 3.6 Commonly used tube layouts. (a) Square pitch (90°); (b) rotated square pitch (45°); and (c) triangular pitch (30°) (Source: Ref. [4]).

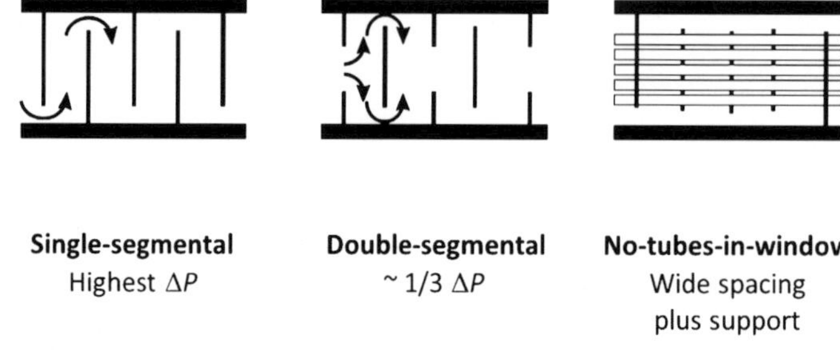

FIGURE 3.7 Common TEMA baffle types (Source: HTRI).

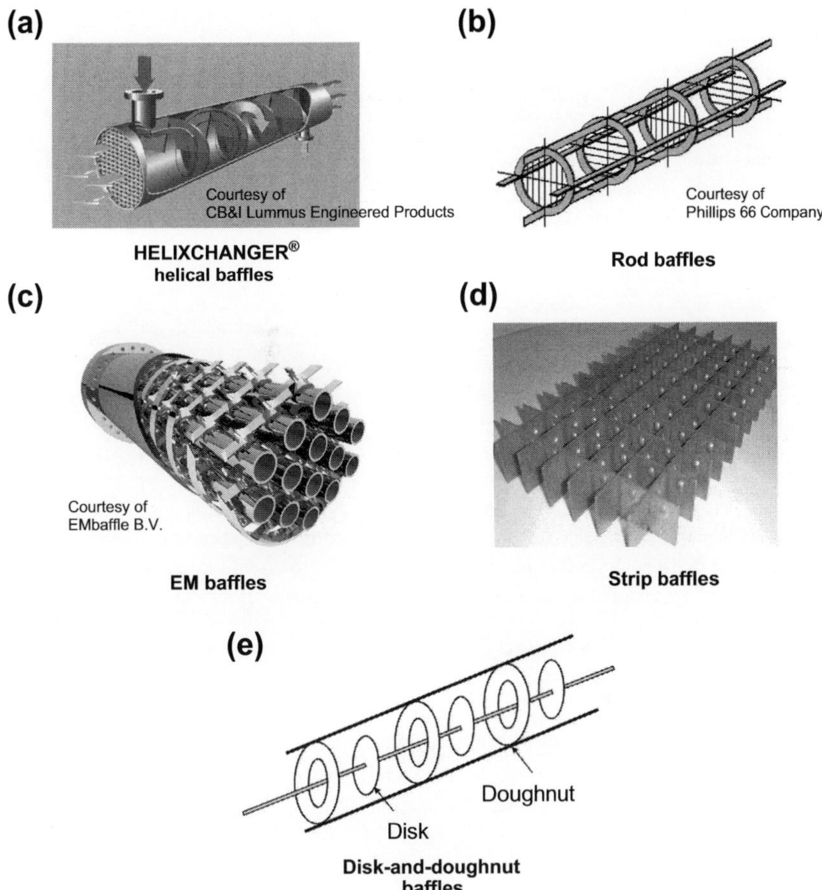

FIGURE 3.8 Non-TEMA baffles.

Baffle selection is important to control shell-side flow distribution and pressure drop. Single segmental baffles have the highest pressure drop and heat transfer since they maximize the number of tubes in cross flow. The typical baffle cut[2] for single segmental baffles varies from 20 to 35%, with 25% being the most common. Double segmental baffles lower the pressure drop (and hence the heat transfer) compared to single segmental baffles, as indicated in Figure 3.7. These baffles minimize the tubes in cross flow, with typically two rows of overlap.[3] The heat transfer area is dominated by window[4] flow, with the flow area of the central window being equal to the total flow area of the wing windows to provide proper flow distribution. Another commonly used baffle type is the no-tubes-in-window (NTIW) configuration. As the name implies, no tubes are located in the baffle windows, an arrangement which facilitates large baffle spacing. The larger baffle spacing lowers the pressure drop, while intermediate support plates provide the required tube support. Since the bundle is not full of tubes, designers need to ensure that NTIW designs are within proven operating limits. Some suggested rules-of-thumb include using a 15% baffle cut and ensuring that the cross-flow area is less than three times the window flow area.

In recent years, non-TEMA baffles have become more popular. These alternative baffles provide for improved flow distribution, which can also reduce pressure drop (for the same heat-transfer capacity). They are particularly advantageous for tube-bundle replacements where flow rate and duty are increased. See Bouhairie [7] for more details. Figure 3.8 shows some examples of non-TEMA baffles used in the process industries. Figure 3.8(a) shows helical baffles, which consist of quadrant-shaped baffles installed at an inclined helix angle as measured from the perpendicular to the axis of the exchanger. Helix angles typically vary from 10 to 40°. Rod baffles (Figure 3.8(b)) consist of rods installed in a baffle ring which contact the tube at one point. Four rod baffles are needed to fully support a tube on all four sides. Typical spacing for these baffles is 6 inches. EM baffles (Figure 3.8(c)) are formed from expanded metal inserted in a baffle ring and installed with spacing varying from 10 to 30 cm. Strip baffles (Figure 3.8(d)) are formed from interlocking metal strips and can be fabricated to permit typical TEMA pitch ratios and baffle spacing. Disk-and-doughnut baffles (Figure 3.8(e)) create a radial flow pattern which distributes the flow more effectively than TEMA baffles.

[2] Baffle cut is the ratio of the height of the baffle cutout to inside shell diameter, usually expressed as a percentage.
[3] Overlap is the distance between the baffle tips for tubes supported by adjacent baffles. Overlap facilitates tie-rod placement.
[4] The baffle window is the cut-out portion of the baffle.

Either the tube-side fluid, the shell-side fluid, or both may make more than one pass through the heat exchanger. On the tube side, multiple passes are achieved by means of U-tubes or by partitioning the headers. The number of tube-side passes is usually one, two, four, or six, but may be as high as 16. Multiple passes on the shell side are achieved by partitioning the shell with a longitudinal baffle (F-shell) or by connecting two or more single-pass shells together. The number of shell-side passes is usually between one and six.

The prevalence of shell-and-tube exchangers in the process industries is probably due to their relative ease of manufacture and extensive operating experience. They are also extremely versatile and can be designed to meet almost any heat-transfer service. As a result, they are common in a wide variety of applications.

3.4 Plate Heat Exchangers

3.4.1 Gasketed Plate-Frame Heat Exchangers

Gasketed plate-frame heat exchangers consist of a plate pack compressed between two thicker pressure plates and all mounted on a frame for easy assembly and disassembly as shown in Figure 3.9. These exchangers were originally developed for the food industry where frequent cleanings are required, and they have subsequently found their way into other process industries due to their compact size. The plates are corrugated with a chevron pattern to provide for flow distribution and to promote turbulence. Plate thickness is typically from 0.5 mm to 1.2 mm, and the gap between the plates is typically from 2 to 5 mm. Gaskets seal the plate edges to prevent the cold and hot fluids from mixing, while also preventing leakage to the environment. Plates are made from malleable corrosion-resistant materials, such as stainless steel and titanium, in a wide range of sizes.

Figure 3.10 shows the corrugations and gasket arrangement for adjacent plates in a pack. For the hot fluid channel, gaskets seal the cold fluid port; for the cold fluid channel, gaskets seal the hot fluid port. Corrugations near the ports facilitate fluid distribution. In general, design methods require proprietary heat-transfer and pressure-drop correlations, which depend upon the details of the corrugations in addition to the Reynolds number.

The flow pattern is typically pure counter flow, and heat-transfer coefficients tend to be greater than for shell-and-tube exchangers due to the small gap between the plates and turbulence induced by the corrugations. As a result, gasketed plate-frame designs tend to be more compact than shell-and-tube designs. However, these units are limited to pressures below 25 bar and temperatures not exceeding 250°C. API Standard 662 [8] provides guidelines for the design and fabrication of this type of plate heat exchanger. Kakac and Liu [3] provide a useful summary of heat-transfer and pressure-drop correlations available in the open literature.

3.4.2 Shell-and-Plate Heat Exchangers

Shell-and-plate exchangers (Figure 3.11) consist of a welded pack of circular plates that is, in turn, welded inside a cylindrical shell. They combine the advantages of compact and thermally efficient plate technology with ease of mechanical design and a fully welded cylindrical pressure vessel. This robust construction can accommodate design pressures as high as 150 bar and temperatures up to 900°C. A number of different plate materials are supported, including 304L and 316L stainless steels, duplex stainless steel, titanium and Hastelloy®. Specialized welding techniques make this technology possible, and companies capable of performing this work are

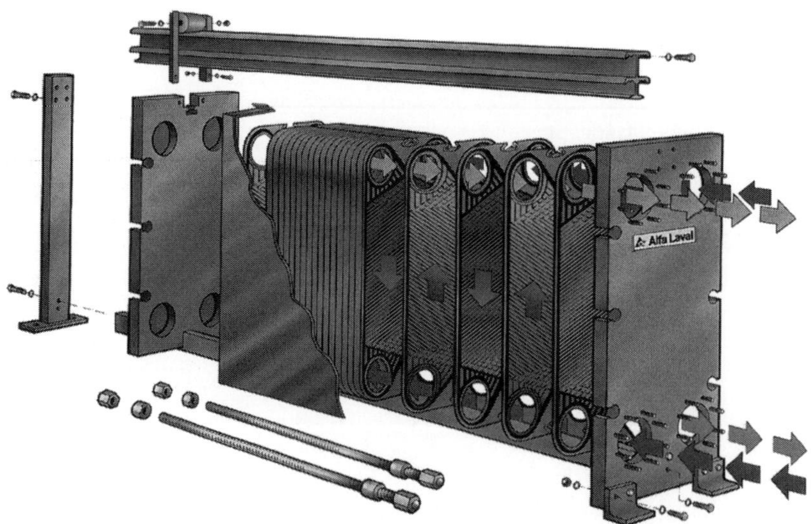

FIGURE 3.9 Exploded view of gasketed plate-frame heat exchanger (Source: Alfa Laval).

FIGURE 3.10 Plate corrugations and gasket pattern (Source: Alfa Laval).

FIGURE 3.11 A typical shell-and-plate heat exchanger (Source: Tranter, Inc.).

limited. However, due to the mechanical advantages of these exchangers compared to gasketed plate-frame units, their use in the process industries is expected to increase.

3.4.3 Plate-fin Heat Exchangers

Plate-fin heat exchangers consist of fins sandwiched between pressure plates called parting sheets, which separate the hot and cold fluids. Plate-fin geometry has been applied to a variety of applications including gas production, petrochemical and hydrocarbon processing, refrigeration, gas turbine recuperation, fuel cell technology and electronics cooling. The common element in all of these

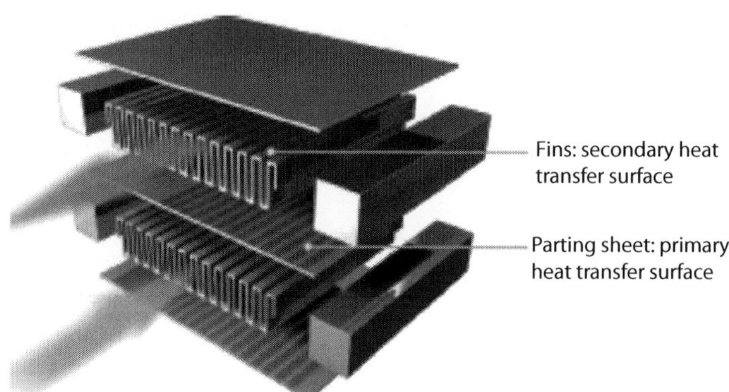

FIGURE 3.12 Plate-fin parting sheet configuration (Source: HTRI).

applications is that the fluids are clean and do not foul appreciably. An expanded view of a typical plate-fin core is shown in Figure 3.12. Flow passages tend to be small, with widths varying from 1 to 5 mm and heights from 2 to 10 mm.

Plate-fin cores are often fabricated by brazing the plates, fins, and sealing bars in a furnace. Aluminum is the most common plate-fin material of construction because of the ease of brazing, and ALPEMA standards [9] provide requirements for the design and fabrication of aluminum plate-fin exchangers. For aluminum construction, temperatures need to remain below 204°C. At higher temperatures, specialized brazing alloys are often used, and in some high- temperature applications, the cores are welded.

The main advantage of plate-fin exchangers is their compact volume, which is an important consideration where size and weight constraints dominate the design. The disadvantages of this technology are the difficulty in cleaning and in handling thermal cycling without developing leaks.

3.5 The Overall Heat-Transfer Coefficient

Consider the section of a double-pipe exchanger shown in Figure 3.13, in which the hot fluid is arbitrarily assumed to flow through the inner pipe.

Heat is transferred by convection from the hot fluid to the wall of the inner pipe, by conduction through the pipe wall, and then by convection from the pipe wall to the cold fluid. The driving force for the heat transfer is the difference in temperature between the hot and the cold streams. To describe this overall process, an overall heat-transfer coefficient, U, is defined by:

$$q = UA\Delta T_m \tag{3.1}$$

The heat-transfer area, A, is the surface area of the inner pipe, and may be based on either the inside or outside diameter. In practice, however, the outside diameter is virtually always used, so that:

$$A = A_o = \pi D_o L \tag{3.2}$$

The temperature difference, ΔT_m, is the mean temperature difference between the two fluid streams. It can be shown (see Appendix 3.A) that when U is independent of position along the exchanger, ΔT_m is the logarithmic mean temperature difference, i.e.,

$$\Delta T_m = \Delta T_{\ln} = \frac{\Delta T_2 - \Delta T_1}{\ln(\Delta T_2/\Delta T_1)} \tag{3.3}$$

where ΔT_1 and ΔT_2 are the temperature differences at the two ends of the exchanger. Equation (3.3) is valid regardless of whether counter flow or parallel flow is employed. Now, according to Equation (3.1), the thermal resistance is:

$$R_{th} = \frac{1}{UA} \tag{3.4}$$

FIGURE 3.13 Section of a double-pipe exchanger.

But this resistance is made up of three resistances in series, namely, the convective resistance between the hot fluid and the pipe wall, the conductive resistance of the pipe wall, and the convective resistance between the pipe wall and the cold fluid. Hence,

$$\frac{1}{UA_o} = \frac{1}{h_iA_i} + \frac{\ln(D_o/D_i)}{2\pi\, kL} + \frac{1}{h_oA_o} \tag{3.5}$$

where h_i and h_o are the heat-transfer coefficients for flow in the inner and outer pipes, respectively, and A_i is the surface area of the inner pipe based on the inside diameter, i.e.,

$$A_i = \pi D_i L \tag{3.6}$$

Multiplying Equation (3.5) by A_o and inverting yields:

$$U = \left[\frac{D_o}{h_iD_i} + \frac{D_o \ln(D_o/D_i)}{2k} + \frac{1}{h_o}\right]^{-1} \tag{3.7}$$

Equation (3.7) is correct when the heat exchanger is new and the heat-transfer surfaces are clean. With most fluids, however, a film of dirt or scale will build up on the heat-transfer surfaces over a period of time. This process is called fouling and results in decreased performance of the heat exchanger due to the added thermal resistances of the dirt films. Fouling is taken into account by means of empirically determined fouling factors, R_{Di} and R_{Do}, which represent the thermal resistances of the dirt films on the inside and outside of the inner pipe multiplied by the respective surface areas. Thus, for the inner dirt film:

$$R_{Dth} = \frac{R_{Di}}{A_i} \tag{3.8}$$

Adding these two additional resistances to Equation (3.5) and proceeding as before yields:

$$U_D = \left[\frac{D_o}{h_iD_i} + \frac{D_o \ln(D_o/D_i)}{2k} + \frac{1}{h_o} + \frac{R_{Di}D_o}{D_i} + R_{Do}\right]^{-1} \tag{3.9}$$

where U_D is the overall coefficient after fouling has occurred. Design calculations are generally made on the basis of U_D since it is necessary that the exchanger be operable after fouling has occurred. More precisely, the fouling factors should be chosen so that the exchanger will have a reasonable operating period before requiring cleaning. The operating period must be sufficient to ensure that exchanger cleanings coincide with scheduled process shutdowns. It should be noted that the effect of the fouling factors in Equation (3.9) is to decrease the value of the overall heat-transfer coefficient, which increases the heat-transfer area calculated from Equation (3.1). Hence, fouling factors can be viewed as safety factors in the design procedure. In any case, it is necessary to provide more heat-transfer area than is actually required when the exchanger is clean. As a result, outlet temperatures will exceed design specifications when the exchanger is clean, unless bypass streams are provided.

Fouling factors are best determined from experience with similar units in the same or similar service. When such information is not available, recourse may be had to published data. The most comprehensive tabulation of fouling factors is the one developed by TEMA, which is available in Refs. [3,5]. The values given below in Table 3.3 are representative of the data available in the public domain. It should be realized that there are very significant uncertainties associated with all such data. Fouling is a complex process that can be influenced by many variables that are not specifically accounted for in these tabulations.

Fouling can occur by a number of mechanisms operating either alone or in combination. These include:

- **Corrosion**
  Corrosion products such as rust can gradually build up on tube walls, resulting in reduced heat transmission and eventual tube failure. This type of fouling can be minimized or eliminated by the proper choice of corrosion-resistant materials of construction in the design process.
- **Crystallization**
  Crystallization typically occurs with cooling water streams containing dissolved sulfates and carbonates. Since the solubility of these salts decreases with increasing temperature, they tend to precipitate on heat-transfer surfaces when the water is heated, forming scale. This type of fouling can be minimized by restricting the outlet water temperature to a maximum of 110°F to 125°F.
- **Decomposition**
  Some organic compounds may decompose when they are heated or come in contact with a hot surface, forming carbonaceous deposits such as coke and tar. In cracking furnaces, partial decomposition of the hydrocarbon feedstock is the objective and coke formation is an undesired but unavoidable result.
- **Polymerization**
  Polymerization reactions can be initiated when certain unsaturated organic compounds are heated or come in contact with a hot metal tube wall. The resulting reaction products can form a very tough plastic-like layer that can be extremely difficult to remove from heat-transfer surfaces.
- **Sedimentation**
  Sedimentation fouling results from the deposition of suspended solids entrained in many process streams such as cooling water and flue gases. High fluid velocities tend to minimize the accumulation of deposits on heat-transfer surfaces.

TABLE 3.3 Typical Values of Fouling Factors (h · ft^2 · °F/Btu)

Cooling water streams[a]		
●	Seawater	0.0005–0.001
●	Brackish water	0.001–0.002
●	Treated cooling tower water	0.001–0.002
●	Municipal water supply	0.001–0.002
●	River water	0.001–0.003
●	Engine jacket water	0.001
●	Distilled or demineralized water	0.0005
●	Treated boiler feedwater	0.0005–0.001
●	Boiler blowdown	0.002
Service gas streams		
●	Ambient air (in air-cooled units)	0–0.0005
●	Compressed air	0.001–0.002
●	Steam (clean)	0–0.0005
●	Steam (with oil traces)	0.001–0.002
●	Refrigerants (with oil traces)	0.002
●	Ammonia	0.001
●	Carbon dioxide	0.002
●	Flue gases	0.005–0.01
Service liquid streams		
●	Fuel oil	0.002–0.005
●	Lubrication oil	0.001
●	Transformer oil	0.001
●	Hydraulic fluid	0.001
●	Organic heat-transfer fluids	0.001–0.002
●	Refrigerants	0.001
●	Brine	0.003
Process gas streams		
●	Hydrogen	0.001
●	Organic solvent vapors	0.001
●	Acid gases	0.002–0.003
●	Stable distillation overhead products	0.001
Process liquid streams		
●	Amine solutions	0.002
●	Glycol solutions	0.002
●	Caustic solutions	0.002
●	Alcohol solutions	0.002
●	Ammonia	0.001
●	Vegetable oils	0.003
●	Stable distillation side-draw and bottom products	0.001–0.002
Natural gas processing streams		
●	Natural gas	0.001
●	Overhead vapor products	0.001–0.002
●	C_3 or C_4 vapor (condensing)	0.001
●	Lean oil	0.002
●	Rich oil	0.001
●	LNG and LPG	0.001
Oil refinery streams		
●	Crude oil[b]	
–	Temperature less than 250°F	0.002–0.003
–	Temperature between 250°F and 350°F	0.003–0.004
–	Temperature between 350°F and 450°F	0.004–0.005
–	Temperature greater than 450°F	0.005–0.006
●	Liquid product streams	
–	Gasoline	0.001–0.002
–	Naphtha and light distillates	0.001–0.003
–	Kerosene	0.001–0.003
–	Light gas oil	0.002–0.003
–	Heavy gas oil	0.003–0.005
–	Heavy fuel oils	0.003–0.007
–	Asphalt and residuum	0.007–0.01

TABLE 3.3 Typical Values of Fouling Factors (h · ft² · °F/Btu)—cont'd

●	Other oil streams	
–	Refined lube oil	0.001
–	Cycle oil	0.002–0.004
–	Coker gas oil	0.003–0.005
–	Absorption oils	0.002

[a]Assumes water velocity greater than 3 ft/s. Lower values of ranges correspond to water temperature below about 120°F and hot stream temperature below about 250°F.

[b]Assumes desalting at approximately 250°F and a minimum oil velocity of 2 ft/s.

Source: Refs. [10,11] and www.engineeringpage.com.

- **Biological activity**

  Biological fouling is most commonly caused by micro-organisms, although macroscopic marine organisms can sometimes cause problems as well. Cooling water and some other process streams may contain algae or bacteria that can attach and grow on heat-transfer surfaces, forming slimes that are very poor heat conductors. Metabolic products of these organisms can also cause corrosion of metal surfaces. Biocides and copper-nickel alloy tubing can be used to inhibit the growth of micro-organisms and mitigate this type of fouling.

It can be seen from Table 3.3 that the range of values of fouling factors spans more than an order of magnitude. For very clean streams, values of 0.0005 h · ft² · °F/Btu or less are appropriate, whereas very dirty streams require values of 0.005–0.01 h · ft² · °F/Btu. However, values in the range 0.001–0.003 h · ft² · °F/Btu are appropriate for the majority of cases.

Example 3.1

A double-pipe heat exchanger will be used to cool a hot stream from 350°F to 250°F by heating a cold stream from 80°F to 120°F. The hot stream will flow in the inner pipe, which is 2-in. schedule 40 carbon steel with a thermal conductivity of 26 Btu/h · ft² · °F. Fouling factors of 0.001 h · ft² · °F/Btu should be provided for each stream. The heat-transfer coefficients are estimated to be $h_i = 200$ and $h_o = 350$ Btu/h · ft² · °F, and the heat load is 3.5×10^6 Btu/h.

(a) For counter-current operation, what surface area is required?
(b) For co-current operation, what surface area is required?

Solution

(a) *Counter-current operation*
From Equation (3.1) we have,

$$A = \frac{q}{U_D \Delta T_{ln}}$$

$$A = \frac{3.5 \times 10^6}{U_D \Delta T_{ln}}$$

Clearly, we must compute values of U_D and ΔT_{ln} in order to solve the problem. Counter flow can be represented schematically as follows:

$$\Delta T = 230°F \left\{ \begin{matrix} 350°F & \rightarrow & 250°F \\ 120°F & \leftarrow & 80°F \end{matrix} \right\} \Delta T = 170°F$$

$$\Delta T_{ln} = \frac{\Delta T_2 - \Delta T_1}{\ln(\Delta T_2/\Delta T_1)} = \frac{230 - 170}{\ln(230/170)} = 198.5°F$$

In order to compute U_D from Equation (3.9), we first find D_o and D_i for 2-in. schedule 40 pipe. From Table B.2, $D_i = 2.067$ in. and $D_o = 2.375$ in. Then,

$$U_D = \left[\frac{D_o}{h_i D_i} + \frac{D_o \ln(D_o/D_i)}{2k} + \frac{1}{h_o} + \frac{R_{Di} D_o}{D_i} + R_{Do}\right]^{-1}$$

$$U_D = \left[\frac{2.375}{200 \times 2.067} + \frac{(2.375/12) \times \ln(2.375/2.067)}{2 \times 26} + \frac{1}{350} + \frac{0.001 \times 2.375}{2.067} + 0.001\right]^{-1}$$

$$U_D = [0.0057 + 0.0005 + 0.0029 + 0.0011 + 0.001]^{-1}$$

$$U_D = 88.65 \text{ Btu/h}\cdot\text{ft}^2\cdot°\text{F}$$

Therefore,

$$A = \frac{3.5 \times 10^6}{88.65 \times 198.5} = 198.9 \text{ ft}^2 \cong 199 \text{ ft}^2$$

(b) *Co-current operation*
For parallel flow, the value of ΔT_{ln} will be different.

$$\Delta T = 270°\text{F} \left\{ \begin{array}{ccc} 350°\text{F} & \rightarrow & 250°\text{F} \\ 80°\text{F} & \rightarrow & 120°\text{F} \end{array} \right\} \Delta T = 130°\text{F}$$

$$\Delta T_{ln} = \frac{\Delta T_2 - \Delta T_1}{\ln(\Delta T_2/\Delta T_1)} = \frac{270 - 130}{\ln(270/130)} = 191.5°\text{F}$$

$$A = \frac{3.5 \times 10^6}{88.65 \times 191.5} = 206.2 \cong 206 \text{ ft}^2$$

This example illustrates the advantage of counter-current operation as opposed to co-current operation. Although the difference between the modes of operation is small in this case, it is not always so. In fact, if the outlet temperature of the cold stream were required to be 260°F instead of 120°F, parallel flow would be impossible.

3.6 The LMTD Correction Factor

In multi-pass shell-and-tube exchangers, the flow pattern is a mixture of co-current and counter-current flow, as the two streams flow through the exchanger in the same direction on some passes and in the opposite direction on others. For this reason, the mean temperature difference is not equal to the logarithmic mean. However, it is convenient to retain the LMTD by introducing a correction factor, *F*, which is appropriately termed the LMTD correction factor:

$$\Delta T_m = F(\Delta T_{ln})_{cf} \tag{3.10}$$

Note that in this relationship, the LMTD is computed as if the flow were counter current. The LMTD correction factor can be computed analytically for any number of shell-side passes and any even number of tube-side passes as follows [12]:

Let

N = number of shell-side passes

$$R = \frac{T_a - T_b}{t_b - t_a} \tag{3.11}$$

$$P = \frac{t_b - t_a}{T_a - t_a} \tag{3.12}$$

where

T_a = inlet temperature of shell-side fluid
T_b = outlet temperature of shell-side fluid
t_a = inlet temperature of tube-side fluid
t_b, = outlet temperature of tube-side fluid

For $R \neq 1$, compute:

$$\alpha = \left(\frac{1 - RP}{1 - P}\right)^{1/N} \tag{3.13}$$

$$S = \frac{\alpha - 1}{\alpha - R} \tag{3.14}$$

$$F = \frac{\sqrt{R^2 + 1}\, \ln\left(\frac{1 - S}{1 - RS}\right)}{(R - 1)\ln\left[\frac{2 - S\left(R + 1 - \sqrt{R^2 + 1}\right)}{2 - S\left(R + 1 + \sqrt{R^2 + 1}\right)}\right]} \tag{3.15}$$

For $R = 1$, compute:

$$S = \frac{P}{N - (N - 1)P} \tag{3.16}$$

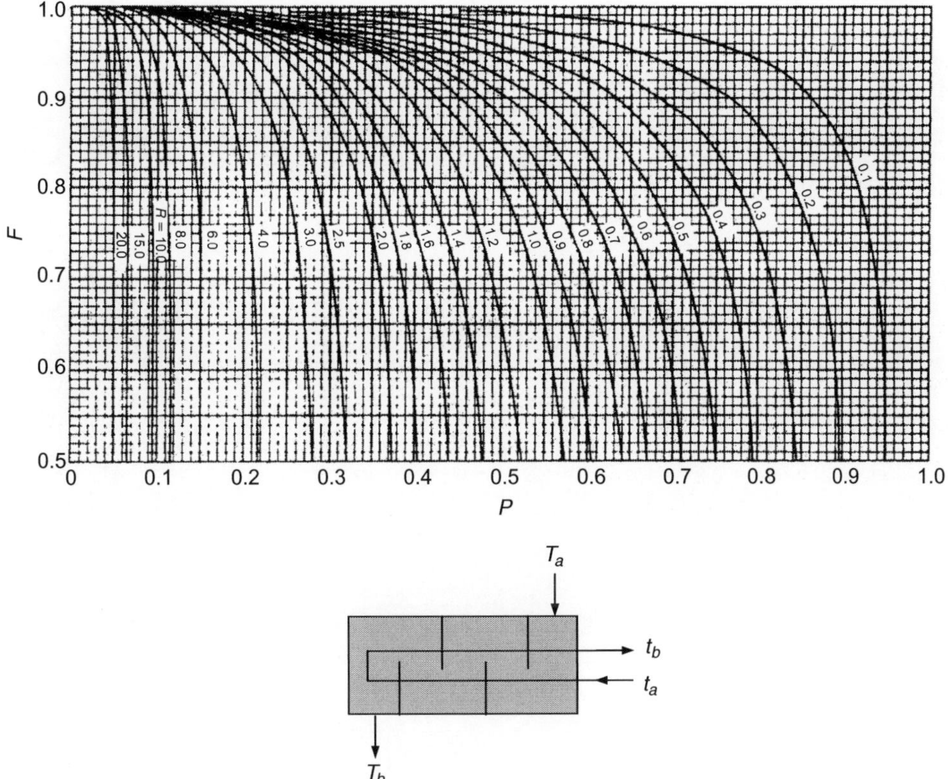

FIGURE 3.14 LMTD correction factor for 1–2 exchangers (Source: Ref [11]).

$$F = \frac{S\sqrt{2}}{(1-S)\ln\left[\dfrac{2-S(2-\sqrt{2})}{2-S(2+\sqrt{2})}\right]} \tag{3.17}$$

Although not readily apparent, these equations are symmetric with respect to fluid placement, i.e., switching the two fluids between tubes and shell changes the values of R and P, but leaves F unchanged. Also, $F = 1.0$ if either fluid is isothermal. The above relations are available in the form of graphs that are convenient for making quick estimates [2,5,11,13]. The graphs corresponding to one and two shell passes are given in **Figures 3.14 and 3.15**, respectively.

Configurations for which the calculated value of F is less than about 0.8 should not be used in practice for several reasons. First, the heat transfer in such a case is relatively inefficient, being less than 80% of that for a counter-flow exchanger. More importantly, the restriction on F is intended to avoid the steeply sloping parts of the curves in **Figures 3.14 and 3.15**. There, small deviations from simplifying assumptions (such as no heat loss and constant U) upon which the equations for F are based, can result in an actual F-value much lower than the calculated value. Also, in these regions large changes in F can result from small changes in P, making exchanger performance very sensitive to any deviations from design temperatures. Examination of **Figures 3.14 and 3.15** shows that the value of 0.8 is approximate; higher values of F are required at very large and small R-values, for example. The value of F can be increased by increasing the number of shell passes, up to a practical limit of about six. (Note, however, that the curves become steeper with increasing number of shell passes, requiring still higher values of F.) In extreme cases, true counter flow, for which $F = 1.0$, may be the only practical configuration.

Example 3.2

A fluid is to be heated from 100°F to 160°F by heat exchange with a hot fluid that will be cooled from 230°F to 150°F. The heat-transfer rate will be 540,000 Btu/h and the hot fluid will flow in the tubes. Will a 1-2 exchanger (i.e., an exchanger with one shell pass and a multiple of two tube passes) be suitable for this service? Find the mean temperature difference in the exchanger.

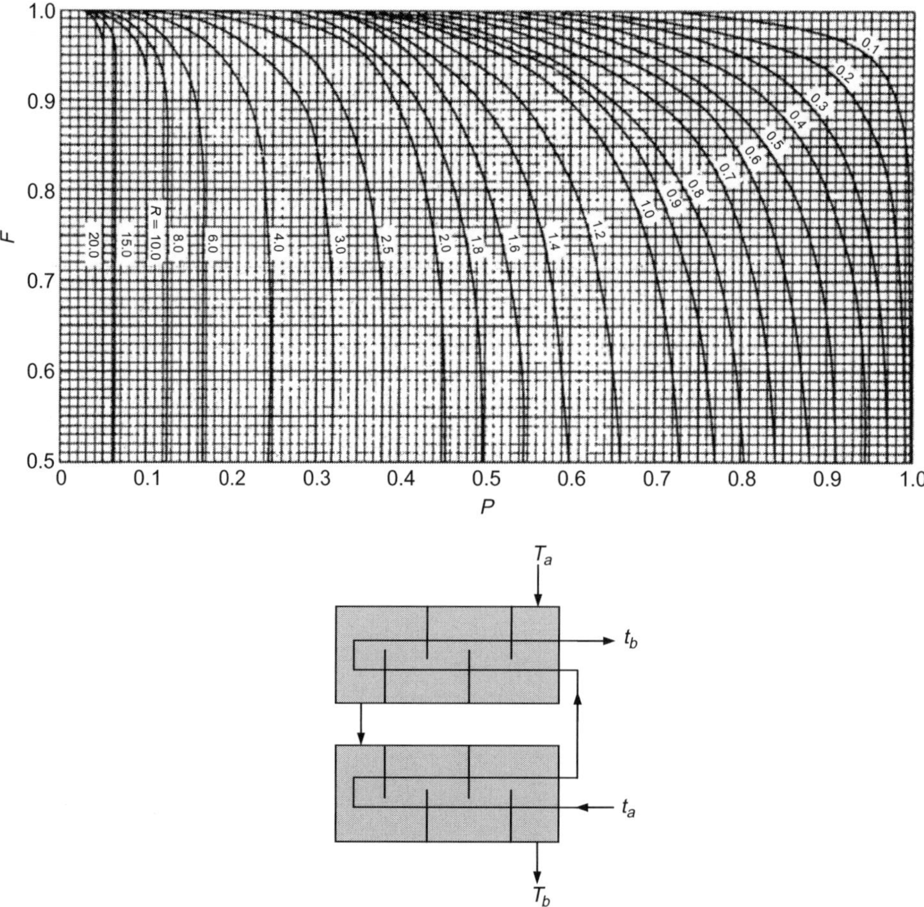

FIGURE 3.15 LMTD correction factor for 2-4 exchangers (Source: Ref. [11]).

Solution

$$R = \frac{T_a - T_b}{t_b - t_a} = \frac{100 - 160}{150 - 230} = 0.75$$

$$P = \frac{t_b - t_a}{T_a - t_a} = \frac{150 - 230}{100 - 230} = 0.615$$

From Figure 3.14 (or Equation (3.15)), $F \cong 0.72$. Since F is less than 0.8, a 1-2 exchanger should not be used. However, with two shell passes, it is found from Figure 3.15 that $F \cong 0.94$. Hence, an exchanger with two shell passes and a multiple of four tube passes will be suitable. Such an exchanger is called a 2-4 exchanger. The mean temperature difference in the exchanger is easily computed as follows:

$$\Delta T = 50°\text{F} \left\{ \begin{array}{ccc} 100°\text{F} & \rightarrow & 160°\text{F} \\ & & \\ 150°\text{F} & \leftarrow & 230°\text{F} \end{array} \right\} \Delta T = 70°\text{F}$$

$$(\Delta T_{\ln})_{cf} = \frac{\Delta T_2 - \Delta T_1}{\ln(\Delta T_2/\Delta T_1)} = \frac{70 - 50}{\ln(70/50)} = 59.6°\text{F}$$

$$\Delta T_m = F(\Delta T_{\ln})_{cf} = 0.94 \times 59.6 = 56°\text{F}$$

3.7 Analysis of Double-Pipe Exchangers

The detailed thermal analysis of a double-pipe heat exchanger involves the calculation of the overall heat-transfer coefficient according to Equation (3.9). The inside coefficient, h_i, can be computed using the equations for flow in circular pipes presented in

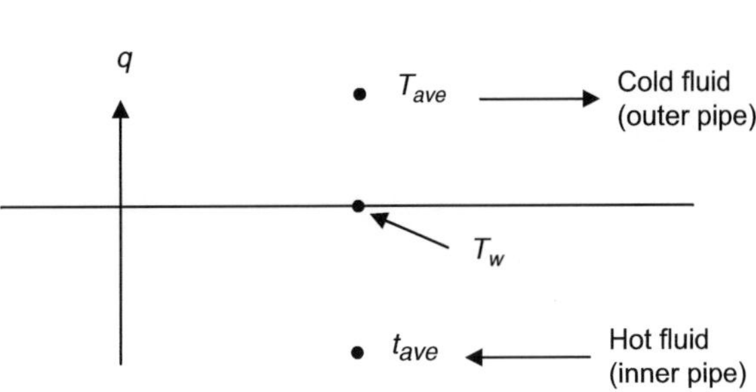

FIGURE 3.16 Configuration for derivation of equation to estimate pipe-wall temperature.

Chapter 2, i.e., Equation (2.34), (2.37), (2.38), or (2.39), depending on the flow regime. The outside coefficient, h_o, can be computed by using the equivalent diameter in Equation (2.34), (2.38), or (2.39) if the flow is turbulent or in the transition region. If the flow in the outer pipe is laminar, then Equation (2.42) should be used.

Use of the Seider–Tate or Hausen equations requires the calculation of the viscosity correction factor, $\phi = (\mu/\mu_w)^{0.14}$, which, in turn, requires that the average wall temperature of the inner pipe be determined. An equation for the wall temperature can be derived by assuming that all the heat is transferred between the streams at their average temperatures. Since h_i and h_o are calculated on the basis of a clean exchanger, the temperature drop across the pipe wall is normally very small and can be neglected. In the derivation, it will be assumed that the hot fluid is in the inner pipe, but the final equation is independent of this assumption. The configuration is indicated schematically in Figure 3.16, in which T_{ave} and t_{ave} are the average temperatures of the fluids in the outer and inner pipes, respectively, and T_w is the average wall temperature.

An energy balance gives:

$$q = h_i A_i (t_{ave} - T_w) = h_o A_o (T_w - T_{ave})$$

Solving for T_w yields:

$$T_w = \frac{h_i A_i t_{ave} + h_o A_o T_{ave}}{h_i A_i + h_o A_o}$$

Substituting $A = \pi DL$ for the areas and rearranging, we obtain:

$$T_w = \frac{h_i t_{ave} + h_o (D_o/D_i) T_{ave}}{h_i + h_o (D_o/D_i)} \tag{3.18}$$

Use of Equation (3.18) involves an iterative procedure since T_w is required for the calculation of h_i and h_o, and vice versa. However, a single iteration is usually sufficient. The values of h_i and h_o are first computed by assuming that ϕ_i and ϕ_o are both unity. These preliminary values are used in Equation (3.18) to compute T_w. The values of μ_w for the two fluids are then obtained, and the viscosity correction factors are calculated. These factors are then multiplied by the preliminary values of h_i and h_o to obtain the final values of the film coefficients.

The thermal analysis of a double-pipe exchanger is illustrated in the following example.

Example 3.3

10,000 lb/h of benzene is to be heated from 60°F to 120°F by heat exchange with an aniline stream that will be cooled from 150°F to 100°F. A number of 16ft hairpins consisting of 2-in. by 1.25-in. schedule 40 stainless steel pipe (type 316, $k = 9.4$ Btu/h · ft · °F) are available and will be used for this service. Assuming that the benzene flows in the inner pipe, how many hairpins will be required?

Solution

(a) Fluid properties at the average stream temperatures are obtained from Figures A.l and A.2, and Table A.15.

Fluid property	Benzene ($t_{ave} = 90$°F)	Aniline ($T_{ave} = 125$°F)
μ (cp)	0.55	2.0
C_P (Btu/lbm · °F)	0.42	0.52
k (Btu/h · ft · °F)	0.092	0.100

(b) Determine the heat load and aniline flow rate by energy balances on the two streams.

$$q = (\dot{m}\ C_P\Delta T)_B = 10,000 \times 0.42 \times 60 = 252,000\ \text{Btu/h}$$
$$252,000 = (\dot{m}\ C_P\Delta T)_A = \dot{m}_A \times 0.52 \times 50$$
$$\dot{m}_A = 9692\ \text{lb/h}$$

(c) Calculate the LMTD. (Counter-current flow is assumed.)

$$\Delta T = 40°\text{F}\left\{\begin{matrix} 60°\text{F} & \rightarrow & 120°\text{F} \\ 100°\text{F} & \leftarrow & 150°\text{F}\end{matrix}\right\}\Delta T = 30°\text{F}$$

$$\Delta T_{\ln} = \frac{40 - 30}{\ln(40/30)} = 34.76°\text{F}$$

(d) Calculate h$_i$ assuming $\phi_i = 1.0$.

$$D_i = \frac{1.38}{12} = 0.115\text{ft}\ \ (\text{From Table B.2})$$

$$Re = \frac{4\dot{m}}{\pi\ D_i\mu} = \frac{4 \times 10,000}{\pi \times 0.115 \times 0.55 \times 2.419} = 83,217 \Rightarrow \text{turbulent flow}$$

$$Nu = \frac{h_i D_i}{k} = 0.027 Re^{0.8} Pr^{1/3}$$

$$h_i = \frac{k}{D_i} \times 0.027 Re^{0.8} Pr^{1/3}$$

$$h_i = \frac{0.092}{0.115} \times 0.027\ (83,217)^{0.8}\left(\frac{0.42 \times 0.55 \times 2.419}{0.092}\right)^{1/3}$$

$$h_i = 340\ \text{Btu/h}\cdot\text{ft}^2\cdot°\text{F}$$

Note: The correction factor for entrance effects can usually be omitted from the Seider–Tate and Hausen equations when working with industrial heat-transfer equipment. For the inner pipe, the effect of the return bends on the heat transfer is minor, so it is reasonable (and conservative) to use the entire length of the flow path in the correction term. Although this length is unknown in the present instance, the exchanger will have at least one hairpin containing 32 ft of pipe. Hence,

$$\frac{L}{D_i} \geq \frac{32}{0.115} = 278 > 60$$

Therefore, entrance effects are negligible in this case.

(e) Calculate h_o assuming $\phi_o = 1.0$.

$$\left.\begin{matrix} D_2 = 2.067\text{ in.} \\ D_1 = 1.660\text{in.}\end{matrix}\right\}(\text{From Table B.2})$$

$$D_e = D_2 - D_1 = \frac{2.067 - 1.660}{12} = 0.0339\ \text{ft}$$

Flow area $\equiv A_f = \frac{\pi}{4}(D_2^2 - D_1^2) = 0.00826\ \text{ft}^2$

$$Re = \frac{D_e(\dot{m}/A_f)}{\mu} = \frac{0.0339 \times (9692/0.00826)}{2.0 \times 2.419} = 8222 \Rightarrow \text{transition flow}$$

$$Nu = \frac{h_o D_e}{k} = 0.116\left[Re^{2/3} - 125\right]Pr^{1/3}\left[1 + (D_e/L)^{2/3}\right]$$

Neglecting entrance effects,

$$h_o = \frac{k}{D_e} \times 0.116\left[Re^{2/3} - 125\right]Pr^{1/3}$$

$$= \frac{0.1}{0.0339} \times 0.116 \times \left[(8222)^{2/3} - 125\right]\left(\frac{0.52 \times 2.0 \times 2.419}{0.1}\right)^{1/3}$$

$$h_o = 283\ \text{Btu/h}\cdot\text{ft}^2\cdot°\text{F}$$

Note: In the annulus the flow is disrupted at the return bends, so it is appropriate to use the length of pipe in one leg of a hairpin to estimate entrance effects. Thus,

$$L/D_e = 16/0.0339 = 472 > 60$$

Hence, entrance effects are negligible. (Including the correction term in the Hausen equation gives $h_o = 288$ Btu/h · ft^2 · °F, and this does not alter the solution.)

(f) Calculate the pipe-wall temperature.

$$T_w = \frac{h_i t_{ave} + h_o\ (D_o/D_i)T_{ave}}{h_i + h_o(D_o/D_i)}$$

$$T_w = \frac{340 \times 90 + 283 \times (1.66/1.38) \times 125}{340 + 283 \times (1.66/1.38)}$$

$$T_w = 108°\text{F}$$

Calculate ϕ_i and ϕ_o, and corrected values of h_i and h_o.

At 108°F,

$$\mu_B = 0.47\text{cp and } \mu_A = 2.4\text{ cp (From Figure A.1)}$$

$$\varphi_i = \left(\frac{0.55}{0.47}\right)^{0.14} = 1.0222$$

$$\varphi_o = \left(\frac{2.0}{2.4}\right)^{0.14} = 0.9748$$

$$h_i = 340(1.0222) = 348\ \text{Btu/h·ft}^2\text{·°F}$$

$$h_o = 283(0.9748) = 276\ \text{Btu/h·ft}^2\text{·°F}$$

(h) Obtain fouling factors. A check of Table 3.3 shows no appropriate listings. However, these organic process chemicals are expected to exhibit low fouling tendencies. Hence, we take

$$R_{Di} = R_{Do} = 0.001\ \text{h·ft}^2\text{·°F/Btu}$$

(i) Compute the overall heat-transfer coefficient.

$$U_D = \left[\frac{D_o}{h_i D_i} + \frac{D_o \ln(D_o/D_i)}{2k} + \frac{1}{h_o} + \frac{R_{Di} D_o}{D_i} + R_{Do}\right]^{-1}$$

$$U_D = \left[\frac{1.66}{348 \times 1.38} + \frac{(1.66/12)\ln(1.66/1.38)}{2 \times 9.4} + \frac{1}{276} + \frac{0.001 \times 1.66}{1.38} + 0.001\right]^{-1}$$

$$U_D = 94\ \text{Btu/h·ft}^2\text{·°F}$$

(j) Calculate the required surface area and number of hairpins.

$$q = U_D A \Delta T_{ln}$$

$$A = \frac{q}{U_D \Delta T_{ln}}$$

$$A = \frac{252{,}000}{94 \times 34.76} = 77.12\ \text{ft}^2$$

From Table B.2, the external surface area per foot of 1.25-in. schedule 40 pipe is 0.435 ft^2. Therefore,

$$L = \frac{77.12}{0.435} = 177.3\ \text{ft}$$

Since each 16-ft hairpin contains 32 ft of pipe,

$$\text{Number of hairpins} = \frac{177.3}{32} = 5.5 \Rightarrow 6$$

Thus, six hairpins will be required.

The preceding example constitutes a design problem in which all of the design parameters are specified except one, namely, the total length of the heat exchanger. Design problems frequently include specifications of maximum allowable pressure drops on the two streams. In that case, pressure drops for both streams would have to be calculated in order to determine the suitability of the exchanger. A more complete analysis of the above problem will be given in the next chapter.

In the above example, benzene was specified as the tube-side fluid. Some guidelines for positioning the fluids are given in Table 3.4. It should be understood that these general guidelines, while often valid, are not ironclad rules, and optimal fluid placement depends on many factors that are service specific.

TABLE 3.4 Criteria for Fluid Placement, in Order of Priority

Tube-Side Fluid	Shell-Side Fluid
Corrosive fluid	Condensing vapor (unless corrosive)
Cooling water	Fluid with large ΔT (>100°F)
Fouling fluid	
Less viscous fluid	
Higher-pressure stream	
Hotter fluid	

For the example problem, neither fluid is corrosive to stainless steel or highly fouling, so the first applicable criterion on the list for tube-side fluid is viscosity. Benzene has a significantly lower viscosity than aniline, indicating that it should be placed in the inner pipe. However, this criterion has relatively low priority and is actually relevant to finned-pipe or shell-and-tube exchangers where the fins or tubes act as turbulence generators for the shell-side fluid. The next criterion is stream pressure, which was not specified in the example. The last criterion for tube-side fluid is the hotter fluid, which is aniline. Neither criterion for the shell-side fluid is applicable. The upshot is that either fluid could be placed in the inner pipe, and both options should be investigated to determine which results in the better design.

3.8 Preliminary Design of Shell-and-Tube Exchangers

The complete thermal design of a shell-and-tube heat exchanger is a complex and lengthy process, and is usually performed with the aid of a computer program [2]. Frequently, however, one requires only a rough approximation of the heat-transfer area for the purpose of making a preliminary cost estimate of the exchanger. Tabulations of overall heat-transfer coefficients such as the one presented in Table 3.5 are used for this purpose. One simply estimates a value for the overall coefficient based on the tabulated values and then computes the required heat-transfer area from Equation (3.1). A somewhat better procedure is to estimate the individual film coefficients, h_i and h_o, and use them to compute the overall coefficient by Equation (3.9). A table of film coefficients can be found in Ref. [14].

Example 3.4

In a petroleum refinery, it is required to cool 30,000 lb/h of kerosene from 400°F to 250°F by heat exchange with 75,000 lb/h of gas oil, which is at 110°F. A shell-and-tube exchanger will be used, and the following data are available:

Fluid Property	Kerosene	Gas Oil
C_P (Btu/lbm · °F)	0.6	0.5
μ (cp)	0.45	3.5
k Btu/h · ft · °F)	0.077	0.08

For the purpose of making a preliminary cost estimate, determine the required heat-transfer area of the exchanger.

Solution

(a) Calculate the heat load and outlet oil temperature by energy balances on the two streams.

$$q = (\dot{m}C_P\Delta T)_K = 30,000 \times 0.6 \times (400 - 250)$$
$$q = 2.7 \times 10^6 \text{ Btu/h}$$
$$q = 2.7 \times 10^6 = (\dot{m}C_P\Delta T)_{oil} = 75,000 \times 0.05 \times (T - 110)$$
$$T = 182°\text{F}$$

(b) Calculate the LMTD.

$$\Delta T = 140°\text{F} \left\{ \begin{array}{ccc} 110°\text{F} & \rightarrow & 182°\text{F} \\ 250°\text{F} & \leftarrow & 400°\text{F} \end{array} \right\} \Delta T = 218°\text{F}$$

$$(\Delta T_{\ln})_{cf} = \frac{218 - 140}{\ln(218/140)} = 176°\text{F}$$

(c) Calculate the LMTD correction factor.
For the purpose of this calculation, assume that the kerosene will flow in the shell. This assumption will not affect the result since the value of F is the same, regardless of which fluid is in the shell.

TABLE 3.5 Typical Values of Overall Heat-Transfer Coefficients in Tubular Heat Exchangers. U = Btu/h · ft^2 · °F

Shell Side	Tube Side	Design U	Includes Total Dirt
Liquid–liquid media			
Aroclor 1248	Jet fuels	100–150	0.0015
Cutback asphalt	Water	10–20	0.01
Demineralized water	Water	300–500	0.001
Ethanol amine (MEA or DEA) 10–25% solutions	Water or DEA, or MEA solutions	140–200	0.003
Fuel oil	Water	15–25	0.007
Fuel oil	Oil	10–15	0.008
Gasoline	Water	60–100	0.003
Heavy oils	Heavy oils	10–40	0.004
Heavy oils	Water	15–50	0.005
Hydrogen-rich reformer stream	Hydrogen-rich reformer stream	90–120	0.002
Kerosene or gas oil	Water	25–50	0.005
Kerosene or gas oil	Oil	20–35	0.005
Kerosene or jet fuels	Trichloroethylene	40–50	0.0015
Jacket water	Water	230–300	0.002
Lube oil (low viscosity)	Water	25–50	0.002
Lube oil (high viscosity)	Water	40–80	0.003
Lube oil	Oil	11–20	0.006
Naphtha	Water	50–70	0.005
Naphtha	Oil	25–35	0.005
Organic solvents	Water	50–150	0.003
Organic solvents	Brine	35–90	0.003
Organic solvents	Organic solvents	20–60	0.002
Tall oil derivatives, vegetable oil, etc.	Water	20–50	0.004
Water	Caustic soda solutions (10–30%)	100–250	0.003
Water	Water	200–250	0.003
Wax distillate	Water	15–25	0.005
Wax distillate	Oil	13–23	0.005
Condensing vapor-liquid media			
Alcohol vapor	Water	100–200	0.002
Asphalt (450°F)	Dowtherm vapor	40–60	0.006
Dowtherm vapor	Tall oil and derivatives	60–80	0.004
Dowtherm vapor	Dowtherm liquid	80–120	0.0015
Gas-plant tar	Steam	40–50	0.0055
High-boiling hydrocarbons V	Water	20–50	0.003
Low-boiling hydrocarbons A	Water	80–200	0.003
Hydrocarbon vapors (partial condenser)	Oil	25–40	0.004
Organic solvents A	Water	100–200	0.003
Organic solvents high NC, A	Water or brine	20–60	0.003
Organic solvents low NC, V	Water or brine	50–120	0.003
Kerosene	Water	30–65	0.004
Kerosene	Oil	20–30	0.005
Naphtha	Water	50–75	0.005
Naphtha	Oil	20–30	0.005
Stabilizer reflux vapors	Water	80–120	0.003
Steam	Feed water	400–1000	0.0005
Steam	No. 6 fuel oil	15–25	0.0055
Steam	No. 2 fuel oil	60–90	0.0025
Sulfur dioxide	Water	150–200	0.003
Tall-oil derivatives, vegetable oils (vapor)	Water	20–50	0.004
Water	Aromatic vapor-stream azeotrope	40–80	0.005
Gas-liquid media			
Air, N_2, etc. (compressed)	Water or brine	40–80	0.005
Air, N_2, etc., A	Water or brine	10–50	0.005
Water or brine	Air, N_2 (compressed)	20–40	0.005
Water or brine	Air, N_2, etc., A	5–20	0.005
Water	Hydrogen containing natural-gas mixtures	80–125	0.003

(Continued)

TABLE 3.5 Typical Values of Overall Heat-Transfer Coefficients in Tubular Heat Exchangers. U = Btu/h · ft^2 · °F—cont'd

Shell Side	Tube Side	Design U	Includes Total Dirt
Vaporizers			
Anhydrous ammonia	Steam condensing	150–300	0.0015
Chlorine	Steam condensing	150–300	0.0015
Chlorine	Light heat-transfer oil	40–60	0.0015
Propane, butane, etc.	Steam condensing	200–300	0.0015
Water	Steam condensing	250–400	0.0015

NC: non-condensable gas present; V: vacuum; A: atmospheric pressure.
Dirt (or fouling factor) units are (h) (ft^2) (°F)/Btu
Source: Ref. [1].

$$R = \frac{T_a - T_b}{t_b - t_a} = \frac{400 - 250}{182 - 110} = 2.08$$

$$P = \frac{t_b - t_a}{T_a - t_a} = \frac{182 - 110}{400 - 110} = 0.25$$

From Figure 3.14, for a 1-2 exchanger, $F \cong 0.93$.

(d) Estimate U_D.

From Table 3.5, a kerosene-oil exchanger should have an overall coefficient in the range 20 to 35 Btu/h · ft^2 · °F. Therefore, take

$$U_D = 25 \text{ Btu/h·ft}^2\text{·°F}$$

(e) Calculate the required area.

$$q = U_D AF(\Delta T_{\ln})_{cf}$$

$$A = \frac{q}{U_D F(\Delta T_{\ln})_{cf}}$$

$$A = \frac{2.7 \times 10^6}{25 \times 0.93 \times 176} \cong 660 \text{ ft}^2$$

3.9 Rating a Shell-and-Tube Exchanger

The thermal analysis of a shell-and-tube heat exchanger is similar to the analysis of a double-pipe exchanger in that an overall heat-transfer coefficient is computed from individual film coefficients, h_i and h_o. However, since the flow patterns in a shell-and-tube exchanger differ from those in a double-pipe exchanger, the procedures for calculating the film coefficients also differ. The shell-side coefficient presents the greatest difficulty due to the very complex nature of the flow in the shell. In addition, if the exchanger employs multiple tube passes, then the LMTD correction factor must be used in calculating the mean temperature difference in the exchanger.

In computing the tube-side coefficient, h_i, it is assumed that all tubes in the exchanger are exposed to the same thermal and hydraulic conditions. The value of h_i is then the same for all tubes, and the calculation can be made for a single tube. Equation (2.34), (2.37), (2.38), or (2.39) is used, depending on the flow regime. In computing the Reynolds number, however, the mass flow rate per tube must be used, where

$$\dot{m}_{per\ tube} = \frac{\dot{m}_t n_p}{n_t} \tag{3.19}$$

and

$\dot{m}_t$ = total mass flow rate of tube-side fluid
n_p = number of tube-side passes
n_t = number of tubes

Equation (3.19) simply states that the flow rate in a single tube is the total flow rate divided by the number of fluid circuits, which is n_t/n_p. With the exception of this minor modification, the calculation of h_i is the same as for a double-pipe exchanger.

As indicated in Figure 3.3, the shell-side fluid flows both across and along the tubes. Cross flow predominates between the baffle tips, whereas in the window regions there is a greater longitudinal component. The situation is further complicated by the presence of various leakages and secondary flows, so that only part of the fluid follows the primary flow path. The most rigorous method available for computing the shell-side coefficient, h_o, is the Stream Analysis method developed by *Heat Transfer Research, Inc.* (*HTRI*). However, the method is not suitable for hand calculations, and although it has been described in the open literature [15], some of the data needed for implementation remain proprietary. The most accurate method available in the public domain is the Delaware method [16,17]. Although the method is conceptually simple, it is quite lengthy and involved, and will be presented in a later chapter.

The method that will be used here is a simplified version of the Delaware method presented in Ref. [4]. Although not highly accurate, the method is very straightforward and allows us to present the overall rating procedure without being bogged down in a mass of details. Also, the method has a built-in safety factor, and therefore errors are generally on the safe side. The method utilizes the graph of modified Colburn factor, j_H, versus shell-side Reynolds number shown in Figure 3.17. All symbols used on the graph are defined in the key. The graph is valid for single segmental baffles with a 20% cut, which is within the range typically used in practice. It is also based on TEMA standards for tube-to-baffle and baffle-to-shell clearances [5]. To use Figure 3.17, one simply reads j_H from the graph and then computes h_o from:

$$h_o = j_H \left(\frac{k}{D_e}\right) Pr^{1/3} \left(\frac{\mu}{\mu_w}\right)^{0.14} \tag{3.20}$$

An approximate curve fit to Figure 3.17, which is convenient for implementation on a programmable calculator or computer, is:

$$j_H = 0.5(1 + B/d_s)\left(0.08Re^{0.6821} + 0.7Re^{0.1772}\right) \tag{3.21}$$

A general algorithm for thermal rating of a shell-and-tube heat exchanger is presented in Figure 3.18. The decision as to whether or not the exchanger is thermally suitable for a given service is based on a comparison of calculated versus required overall heat-transfer coefficients. The exchanger is suitable if the calculated value of the design coefficient, U_D, is greater than or equal to the value, U_{req}, that is needed to provide the required rate of heat transfer. If the converse is true, the exchanger is "not suitable." The quotation marks are to indicate that the final decision to reject the exchanger should be tempered by engineering judgment. For example, it may be more economical to utilize an existing exchanger that is slightly undersized, and therefore may require more frequent cleaning, than to purchase a larger exchanger. In principle, the rating decision can be based on a comparison of heat-transfer areas, heat-transfer rates, or mean temperature differences as well as heat-transfer coefficients. In fact, all of these parameters are used as decision criteria in various applications. The algorithm presented here has the advantage that the fouling factors, which are usually the parameters with the greatest associated uncertainty, do not enter the calculation until the final step. Frequently, one can unambiguously reject an exchanger in step 3 before the fouling factors enter the calculation.

It should again be noted that the rating procedure given in Figure 3.18 includes only the thermal analysis of the exchanger. A complete rating procedure must include a hydraulic analysis as well. That is, the pressure drops of both streams must be computed and compared with the specified maximum allowable pressure drops. The tube-side pressure drop is readily computed by the friction factor method for pipe flow with appropriate allowances for the additional losses in the headers and nozzles. The

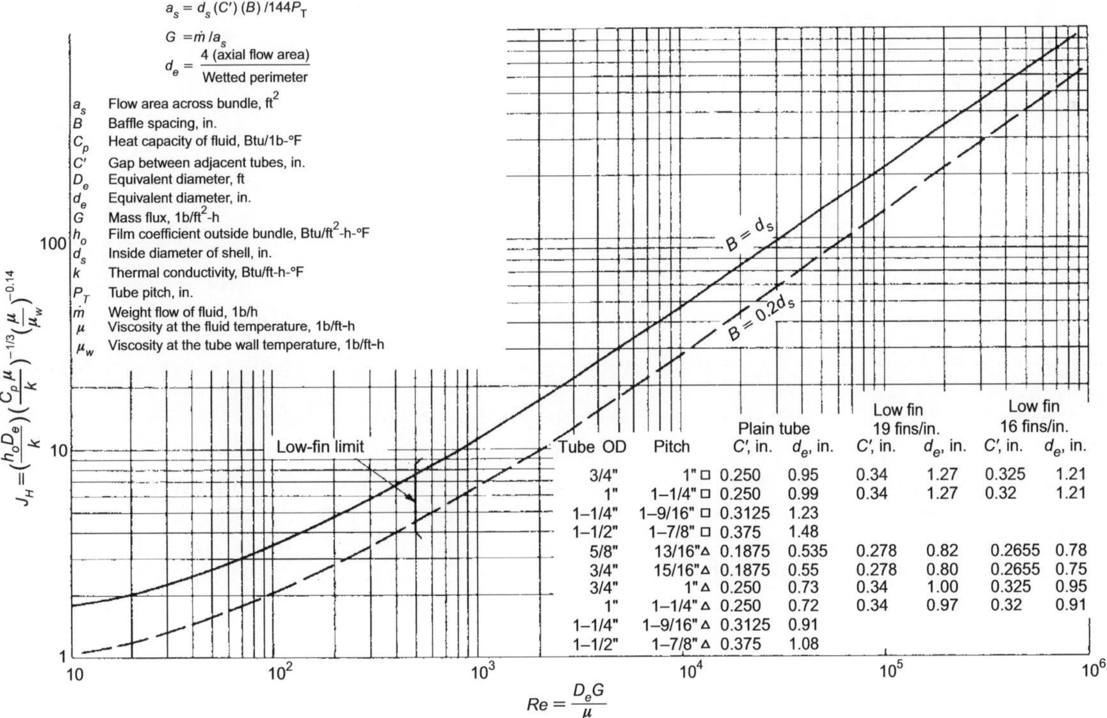

FIGURE 3.17 Correlation for shell-side heat-transfer coefficient. For rotated square tube layouts, use the parameter values for square pitch and replace P_T with $P_T/\sqrt{2}$ in the equation for a_s (Source: Wolverine Tube, Inc. Originally published in Ref. [4]).

(1) Calculate the required overall coefficient.

$$U_{req} = \frac{q}{AF(\Delta T_{\ln})_{cf}}$$

(2) Calculate the clean overall coefficient.

$$U_C = \left[\frac{D_o}{h_i D_i} + \frac{D_o \ln(D_o/D_i)}{2k} + \frac{1}{h_o}\right]^{-1}$$

(3)

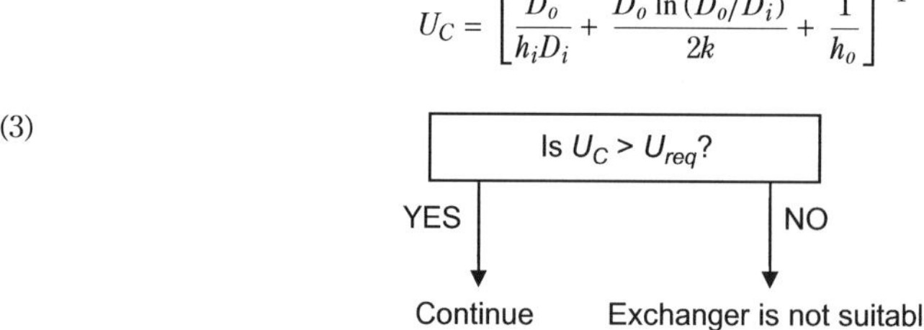

(4) Obtain required fouling factors, R_{Di} and R_{Do}, from past experience or from Table 3.2. Then compute:

$$R_D = R_{Di}(D_o/D_i) + R_{Do}$$

(5) Calculate the design overall coefficient.

$$U_D = (1/U_C + R_D)^{-1}$$

(6)

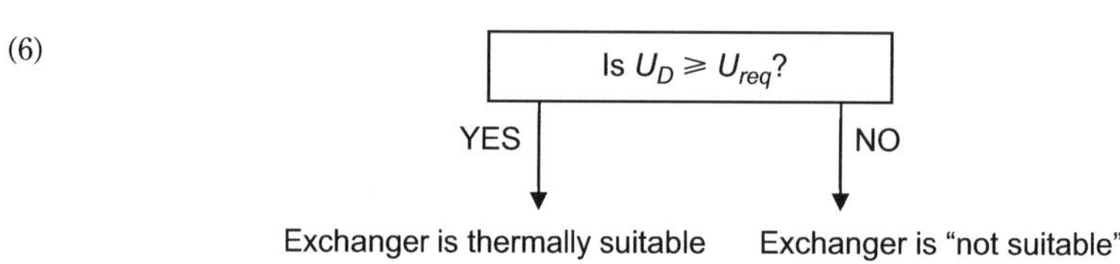

FIGURE 3.18 Thermal rating procedure for a shell-and-tube heat exchanger.

calculation of shell-side pressure drop is beset with the same difficulties as the heat-transfer coefficient. Each of the methods (Stream Analysis, Delaware, Simplified Delaware) discussed above includes an algorithm for calculation of the shell-side pressure drop. These procedures are presented in subsequent chapters.

The method outlined in Figure 3.18 is the standard approach used in the process industries, and as previously noted, rating calculations comprise an integral part of the design process. However, the standard approach often leads to poor designs due to the inherent uncertainty associated with fouling factors. In fact, the use of fouling factors can actually promote fouling under certain conditions, as described in Example 10.10. Alternate methods of specifying heat-exchanger margins exist and are becoming more popular. For applications where fouling can be avoided, 'no foul' design practices can succeed [18,19]. In these cases, flow velocities are increased to reduce fouling potential, and improved design methods are relied upon to avoid specifying large margins. For clean services, fouling factors need not be specified; instead, excess areas up to 20% can be specified to provide thermal margin. In some cases, thermal margins can be applied using heat-transfer multipliers instead of fouling factors [20]. These multipliers compensate for uncertainty in the calculation of heat-transfer coefficients while avoiding over-sizing of the heat exchanger due to unnecessarily large fouling factors.

The thermal rating procedure for a shell-and-tube heat exchanger is illustrated in the following example.

Example 3.5

30,000 lb/h of kerosene are to be cooled from 400°F to 250°F by heat exchange with 75,000 lb/h of gas oil which is at 110°F. Available for this duty is a shell-and-tube exchanger having 156 tubes in a $21^1/_4$-in. ID shell. The tubes are 1-in. OD, 14 BWG, 16 ft long on a $1^1/_4$-in. square pitch. There is one pass on the shell side and six passes on the tube side. The baffles are 20% cut segmental type and are spaced at 5-in. intervals. Both the shell and tubes are carbon steel having $k = 26$ Btu/h · ft · °F. Fluid properties are given in Example 3.4. Will the exchanger be thermally suitable for this service?

Solution

Neither fluid is corrosive, but the oil stream may cause fouling problems so it should be placed in the tubes for ease of cleaning. Also, the kerosene should be placed in the shell due to its large ΔT.

Step 1: Calculate U_{req}.

From Example 3.4, we have:

$$q = 2.7 \times 10^6 \text{ Btu/h}$$
$$F = 0.93$$
$$(\Delta T_{\ln})_{cf} = 176°\text{F}$$

The surface area is obtained from the dimensions of the exchanger:

$$A = 156 \text{ tubes} \times 16 \text{ ft} \times 0.2618 \frac{\text{ft}^2}{\text{ft of tube}} = 653 \text{ ft}^2$$

Thus,

$$U_{req} = \frac{q}{AF(\Delta T_{\ln})_{cf}} = \frac{2.7 \times 10^6}{653 \times 0.93 \times 176}$$

$$U_{req} = 25.3 \text{ Btu/h} \cdot \text{ft}^2 \cdot {}^\circ\text{F}$$

Step 2: Calculate the clean overall coefficient, U_C.

(a) Calculate the tube-side Reynolds number.

$$\dot{m}_{\text{per tube}} = \frac{75{,}000 \times 6}{156} = 2885 \text{ lb/h}$$

$$Re = \frac{4\dot{m}}{\pi D_i \mu} = \frac{4 \times 2885}{\pi \times \frac{0.834}{12} \times 3.5 \times 2.419} = 6243 \Rightarrow \text{transition region}$$

(b) Calculate

$$Nu = \frac{h_i D_i}{k} = 0.116 \left[Re^{2/3} - 125\right] Pr^{1/3} (\mu/\mu w)^{0.14} \left[1 + (D_i/L)^{2/3}\right]$$

$$h_i = \frac{0.08}{(0.834/12)} \times 0.116 \left[(6243)^{2/3} - 125\right] \left(\frac{0.5 \times 3.5 \times 2.419}{0.08}\right)^{1/3} (1) \left[1 + \left(\frac{0.834}{12 \times 16}\right)^{2/3}\right]$$

$$h_i = 110 \text{ Btu/h} \cdot \text{ft}^2 \cdot {}^\circ\text{F}$$

In this calculation, the viscosity correction factor was assumed to be unity since no data were given for the temperature dependence of the oil viscosity.

(c) Calculate the shell-side Reynolds number.
From Figure 3.17, $d_e = 0.99$ in. and $C' = 0.250$ in.

$$D_e = d_e/12 = 0.99/12 = 0.0825 \text{ ft}$$

$$a_s = \frac{d_s C' B}{144 P_T} = \frac{21.25 \times 0.250 \times 5}{144 \times 1.25} = 0.1476 \text{ ft}^2$$

$$G = \frac{\dot{m}}{a_s} = \frac{30{,}000}{0.1476} = 203{,}294 \text{ lbm/h} \cdot \text{ft}^2$$

$$Re = \frac{D_e G}{\mu} = \frac{0.0825 \times 203{,}294}{0.45 \times 2.419} = 15{,}407$$

(d) Calculate h_o.

$$B/d_s = 5/21.25 = 0.235$$

From Figure 3.17, $j_H \cong 40$. [Equation (3.21) gives $j_H = 37.9$.]

$$h_o = j_H \frac{k}{D_e} Pr^{1/3} (\mu/\mu_w)^{0.14} = \frac{40 \times 0.077}{0.0825} \left(\frac{0.60 \times 0.45 \times 2.419}{0.077}\right)^{1/3} (1)$$

$$h_o = 76 \text{ Btu/h} \cdot \text{ft}^2 \cdot {}^\circ\text{F}$$

(e) Calculate U_C.

$$U_C = \left[\frac{D_o}{h_i D_i} + \frac{D_o \ln(D_o/D_i)}{2k} + \frac{1}{h_o}\right]^{-1} = \left[\frac{1.0}{110 \times 0.834} + \frac{(1.0/12)\ln(1.0/0.834)}{2 \times 26} + \frac{1}{76}\right]^{-1}$$

$$U_C = 41.1 \text{ Btu/h} \cdot \text{ft}^2 \cdot \text{ F}$$

Since $U_C > U_{req}$, proceed to Step 3.

Step 3: Obtain the required fouling factors.

In the absence of other information, Table 3.3 indicates fouling factors of 0.002–0.003 h · ft² · °F/Btu for kerosene and 0.002–0.005 h · ft² · °F/Btu for gas oil. Taking 0.0025 for kerosene and 0.0035 for gas oil gives:

$$R_D = \frac{R_{Di} D_o}{D_i} + R_{Do} = \frac{0.0035 \times 1.0}{0.834} + 0.0025 = 0.0067 \text{ h} \cdot \text{ft}^2 \cdot {}^\circ\text{F/Btu}$$

Step 4: Calculate U_D.

$$U_D = (1/U_C + R_D)^{-1} = (1/41.1 + 0.0067)^{-1} \cong 32 \text{ Btu/h}\cdot\text{ft}^2\cdot{}^\circ\text{F}$$

Since this value is greater than the required value of 25.3 Btu/h · ft^2 · °F, the exchanger is thermally suitable. In fact, a smaller exchanger would be adequate.

In the above example, the average tube-wall temperature was not required because the variation of fluid properties with temperature was ignored. In general, however, the wall temperature is required and is calculated using Equation (3.18) in the same manner as illustrated for a double-pipe exchanger in Example 3.3.

3.10 Heat-Exchanger Effectiveness

When the area of a heat exchanger is known and the outlet temperatures of both streams are to be determined, an iterative calculation using Equation (3.1) and the energy balance equations for the two streams is generally required. Since the calculations required for each iteration are quite lengthy, the procedure is best implemented on a computer or programmable calculator. However, in those situations in which the overall coefficient is known or can be estimated *a priori*, the iterative procedure can be avoided by means of a quantity called the heat-exchanger effectiveness.

The effectiveness is defined as the ratio of the actual rate of heat transfer in a given exchanger to the maximum possible rate of heat transfer. The latter is the rate of heat transfer that would occur in a counter-flow exchanger having infinite heat-transfer area. In such an exchanger, one of the fluid streams will gain or lose heat until its outlet temperature equals the inlet temperature of the other stream. The fluid that experiences this maximum temperature change is the one having the smaller value of $C \equiv \dot{m}C_P$, as can be seen from the energy balance equations for the two streams. Thus, if the hot fluid has the lower value of C, we will have $T_{h,out} = T_{c,in}$, and:

$$q_{max} = \dot{m}_h C_{Ph}(T_{h,in} - T_{c,in}) = C_{min}(T_{h,in} - T_{c,in})$$

On the other hand, if the cold fluid has the lower value of C, then $T_{c,out} = T_{h,in}$, and:

$$q_{max} = \dot{m}_c C_{Pc}(T_{h,in} - T_{c,in}) = C_{min}(T_{h,in} - T_{c,in})$$

Thus, in either case we can write:

$$q_{max} = C_{min}(T_{h,in} - T_{c,in}) = C_{min}\Delta T_{\max} \tag{3.22}$$

where $\Delta T_{max} = T_{h,in} - T_{c,in}$ is the maximum temperature difference that can be formed from the terminal stream temperatures.

Now, by definition the effectiveness, ϵ, is given by:

$$\epsilon = \frac{q}{q_{max}} = \frac{q}{C_{min}\Delta T_{max}} \tag{3.23}$$

Thus, the actual heat-transfer rate can be expressed as:

$$q = \epsilon\ C_{min}\Delta T_{max} \tag{3.24}$$

It can be shown (see, e.g., Ref. [21]) that for a given type of exchanger the effectiveness depends on only two parameters, r and NTU, where:

$$r = C_{min}/C_{max} \tag{3.25}$$

$$NTU = UA/C_{min} \tag{3.26}$$

Here, NTU stands for number of transfer units, a terminology derived by analogy with continuous-contacting mass-transfer equipment. Equations for the effectiveness are available in the literature (e.g., Ref. [21]) for various types of heat exchangers. Equations for equipment configurations involving double-pipe and shell-and-tube exchangers are given in Table 3.6. Their use is illustrated in the following example.

Example 3.6

Determine the outlet temperatures of the kerosene and gas oil streams when the exchanger of Example 3.5 is first placed in service.

Solution

We have:

$$C_{ker} = (\dot{m}C_P)_{ker} = 30{,}000 \times 0.6 = 18{,}000 \text{ Btu/h}\cdot{}^\circ\text{F}$$
$$C_{oil} = (\dot{m}C_P)_{oil} = 75{,}000 \times 0.5 = 37{,}500 \text{ Btu/h}\cdot{}^\circ\text{F}$$

TABLE 3.6 Effectiveness Relations for Various Heat Exchangers

$r \equiv C_{min}/C_{max}$ $NTU \equiv UA/C_{min}$

Exchanger Type	Effectiveness Equation
Counter flow	$\epsilon = \dfrac{1-\exp[-NTU(1-r)]}{1-r\exp[-NTU(1-r)]}$ $(r<1)$
	$\epsilon = \dfrac{NTU}{1+NTU}$ $(r=1)$
Parallel flow	$\epsilon = \dfrac{1-\exp[-NTU(1+r)]}{1+r}$
1-2	$\epsilon = 2\left[1+r+\beta\left(\dfrac{1+\exp[-\beta\times NTU]}{1-\exp[-\beta\times NTU]}\right)\right]^{-1}$ where $\beta=\sqrt{1+r^2}$
N-2N	$\epsilon = \left[\left(\dfrac{1-\epsilon^*r}{1-\epsilon^*}\right)^N-1\right]\times\left[\left(\dfrac{1-\epsilon^*r}{1-\epsilon^*}\right)^N-r\right]^{-1}$ $(r<1)$
(For a 2-4 exchanger, $N=2$; etc.)	$\epsilon = \dfrac{N\epsilon^*}{1+(N-1)\epsilon^*}$ $(r=1)$

where ϵ^* is the effectiveness for a 1-2 exchanger with the same value of r but with $(1/N)$ times the NTU value

Therefore,

$$C_{min} = 18{,}000$$
$$C_{max} = 37{,}500$$
$$r = C_{min}/C_{max} = 18{,}000/37{,}500 = 0.48$$

When the exchanger is first placed in service, the overall coefficient will be the clean coefficient computed in Example 3.5 (neglecting the effect of different average stream temperatures on the overall coefficient). Thus,

$$NTU = UA/C_{min} = \frac{41.1\times 653}{18{,}000} = 1.4910$$

From Table 3.6, for a 1-2 exchanger:

$$\epsilon = 2\left[1+r+\beta\left(\frac{1+\exp[-\beta\times NTU]}{1-\exp[-\beta\times NTU]}\right)\right]^{-1}$$

where

$$\beta = \sqrt{1+r^2} = \sqrt{1+(0.48)^2} = 1.1092$$

Thus,

$$\epsilon = 2\times\left[1+0.48+1.1092\left(\frac{1+\exp[-1.1092\times 1.4910]}{1-\exp[-1.1092\times 1.4910]}\right)\right]^{-1}$$
$$\epsilon = 0.6423$$

From Equation (3.24) we have:

$$q = \epsilon\, C_{min}\Delta T_{max}$$
$$q = 0.6423\times 18{,}000\times(400-110)$$
$$q = 3.35\times 10^6\,\text{Btu/h}$$

The outlet temperatures can now be computed from the energy balances on the two streams.

$$q = 3.35\times 10^6 = (\dot{m}\,C_P\Delta T)_{ker} = 18{,}000\times\Delta T_{ker}$$
$$\Delta T_{ker} = 186°\text{F}$$
$$T_{ker,out} = 400-186 = 214°\text{F}$$
$$q = 3.35\times 10^6 = (\dot{m}C_P\Delta T)_{oil} = 37{,}500\times\Delta T_{oil}$$
$$\Delta T_{oil} = 89°\text{F}$$
$$T_{oil,out} = 110+89 = 199°\text{F}$$

Thus, as expected, the exchanger will initially far exceed the design specification of 250°F for the kerosene outlet temperature.

References

[1] Perry RH, Green DW, editors. Perry's Chemical Engineers' Handbook. 7th ed. New York: McGraw-Hill; 1997.
[2] Heat Exchanger Design Handbook. New York: Hemisphere Publishing Corp.; 1988.
[3] Kakac S, Liu H. Heat Exchangers: Selection, Rating and Thermal Design. Boca Raton, FL: CRC Press; 1997.
[4] Kern DQ, Kraus AD. Extended Surface Heat Transfer. New York: McGraw-Hill; 1972.
[5] Standards of the Tubular Exchanger Manufacturers Association. 8th ed. Tarrytown, NY: Tubular Exchanger Manufacturers Association, Inc.; 1999.
[6] Lestina TG. Selecting a heat exchanger shell. Chem Eng Prog 2011;107(6):34–8.
[7] Bouhairie S. Selecting baffles for shell-and-tube heat exchangers. Chem Eng Prog 2012;108(2):27–33.
[8] API Standard 662. Plate heat exchangers for general refinery services. 2nd ed. Washington, DC: American Petroleum Institute; 2002.
[9] The Standards of the Brazed Aluminum Plate-fin Heat Exchanger Manufacturer's Association. 3rd ed. ALPEMA; 2010 www.alpema.org.
[10] Chenoweth J. Final report of HTRI/TEMA joint committee to review the fouling section of TEMA Standards. Alhambra, CA: Heat Transfer Research, Inc.; 1988.
[11] Anonymous. Engineering Data Book. 11th ed. Tulsa, OK: Gas Processors Suppliers Association; 1998.
[12] Bowman RA, Mueller AC, Nagle WM. Mean temperature difference in design. Trans ASME 1940;62:283–94.
[13] Kern DQ. Process Heat Transfer. New York: McGraw-Hill; 1950.
[14] Bell KJ. Approximate sizing of shell-and-tube heat exchangers. In: Heat Exchanger Design Handbook, vol. 3. New York: Hemisphere Publishing Corp.; 1988.
[15] Palen JW, Taborek J. Solution of shell side flow pressure drop and heat transfer by stream analysis method. Chem Eng Prog Symp Series 1969;65(92):53–63.
[16] Bell KJ. Exchanger design based on the Delaware research program. Pet Eng 1960;32(11):C26.
[17] Taborek J. Shell-and-tube heat exchangers: single phase flow. In: Heat Exchanger Design Handbook, vol. 3. New York: Hemisphere Publishing Corp.; 1988.
[18] Gilmour CH. No fooling – no fouling. Chem Eng Prog 1965;61(7):49–54.
[19] Nesta J, Bennett CA. Reduce fouling in shell-and-tube heat exchangers. Hydrocarbon Proc July 2004;83:77–82.
[20] Shilling R. Fouling and uncertainty margins in tubular heat exchanger design: an alternative. Heat Transfer Engineering 2012;33(13):1094–104.
[21] Holman JP. Heat Transfer. 7th ed. New York: McGraw-Hill; 1990.

Appendix 3.A Derivation of the Logarithmic Mean Temperature Difference

Use of the logarithmic mean temperature difference and the LMTD correction factor for the thermal analysis of heat exchangers under steady-state conditions involves the following simplifying assumptions:

- The overall heat-transfer coefficient is constant throughout the heat exchanger.
- The specific enthalpy of each stream is a linear function of temperature. For a single-phase fluid, this implies that the heat capacity is constant.
- There is no heat transfer between the heat exchanger and its surroundings, i.e., no heat losses.
- For co-current and counter-current flow, the temperature of each fluid is uniform over any cross-section in the exchanger. For multi-tube exchangers, this implies that the conditions in every tube are identical.
- For multi-tube, multi-pass exchangers, the conditions in every tube within a given pass are identical.
- There is an equal amount of heat-transfer surface area in each pass of a multi-pass heat exchanger. Although not strictly necessary, the equations and graphs commonly used to calculate the LMTD correction factor for shell-and-tube exchangers are based on this assumption.
- There is no heat generation or consumption within the exchanger due to chemical reaction or other causes.
- There is no heat transfer in the axial direction along the length of the heat exchanger.

Now consider a counter-flow heat exchanger for which the temperatures of both streams increase as one moves along the length of the exchanger from the cold end (1) to the hot end (2). The rate of heat transfer between the fluids in a differential section of the exchanger is:

$$dq = U\,dA(T_h - T_c) = U\,dA\Delta T \tag{3.A.1}$$

where $\Delta T = T_h - T_c$ is the local driving force for heat transfer and subscripts h and c denote the hot and cold streams, respectively. For single-phase operation, the differential energy balances on the two streams are:

$$dq = C_h dT_h = C_c dT_c \tag{3.A.2}$$

where $C \equiv \dot{m}C_p$. The stream energy balances over the entire exchanger are:

$$q = C_h\left(T_{h,a} - T_{h,b}\right) = C_c\left(T_{c,b} - T_{c,a}\right) \tag{3.A.3}$$

where subscripts a and b denote inlet and outlet, respectively. Using Equation (3.A.2), the differential of the driving force can be written as follows:

$$d(\Delta T) = dT_h - dT_c = \frac{dq}{C_h} - \frac{dq}{C_c} = dq\left(\frac{1}{C_h} - \frac{1}{C_c}\right) \tag{3.A.4}$$

Using this relationship along with Equation (3.A.1) gives:

$$\frac{d(\Delta T)}{\Delta T} = \frac{dq\left(\frac{1}{C_h} - \frac{1}{C_c}\right)}{dq/UdA} = UdA\left(\frac{1}{C_h} - \frac{1}{C_c}\right) \tag{3.A.5}$$

Integrating this equation over the length of the heat exchanger we obtain:

$$\int_{\Delta T_1}^{\Delta T_2} \frac{d(\Delta T)}{\Delta T} = U\left(\frac{1}{C_h} - \frac{1}{C_c}\right)\int_0^A dA$$

$$\ln(\Delta T_2/\Delta T_1) = U\left(\frac{1}{C_h} - \frac{1}{C_c}\right) A \qquad (3.A.6)$$

where

$\Delta T_1 = T_{h,b} - T_{c,a}$ = driving force at cold end of exchanger

$\Delta T_2 = T_{h,a} - T_{c,b}$ = driving force at hot end of exchanger

Substituting for C_h and C_c from Equation (3.A.3) gives:

$$\begin{aligned}\ln(\Delta T_2/\Delta T_1) &= UA\left\{\frac{T_{h,a} - T_{h,b}}{q} - \frac{T_{c,b} - T_{c,a}}{q}\right\}\\ &= \frac{UA}{q}\left\{\left(T_{h,a} - T_{c,b}\right) - \left(T_{h,b} - T_{c,a}\right)\right\}\\ &= \frac{UA}{q}\left\{\Delta T_2 - \Delta T_1\right\}\end{aligned}$$

It follows from this result that:

$$q = UA\left\{\frac{\Delta T_2 - \Delta T_1}{\ln(\Delta T_2/\Delta T_1)}\right\} = UA\ \Delta T_{\ln} \qquad (3.A.7)$$

Equation (3.A.7) is the desired result, demonstrating that the mean temperature difference in the exchanger is the LMTD. The derivation for co-current flow is similar and requires only minor modifications of the foregoing analysis.

Notations

A	Surface area
A_i	$\pi D_i L$ = Interior surface area of pipe or tube
A_o	$\pi D_o L$ = Exterior surface area of pipe or tube
a_s	Flow area across tube bundle
B	Baffle spacing
C	$\dot{m}\ C_P$
C_c, C_h	Value of C for cold and hot stream, respectively
C_{max}	Larger of the values of $\dot{m}\ C_P$ for hot and cold streams
C_{min}	Smaller of the values of $\dot{m}\ C_P$ for hot and cold streams
C'	$P_T - D_o$ = Gap between tubes in tube bundle
C_P	Heat capacity at constant pressure
C_{Pc}, C_{Ph}	Heat capacity of cold and hot streams, respectively
D	Diameter
D_e	Equivalent diameter
D_i	Inside diameter of pipe or tube
D_o	Outside diameter of pipe or tube
D_1	Outside diameter of inner pipe of double-pipe heat exchanger
D_2	Inside diameter of outer pipe of double-pipe heat exchanger
d_e	Equivalent diameter expressed in inches
d_s	Inside diameter of shell
F	LMTD correction factor
G	Mass flux
h	Heat-transfer coefficient
h_i	Heat-transfer coefficient for inner (tube-side) fluid
h_o	Heat-transfer coefficient for outer (shell-side) fluid
j_H	$(h_o D_e/k)\ Pr^{-1/3}\ (\mu/\mu_w)^{-014}$ = Modified Colburn factor for shell-side heat transfer
k	Thermal conductivity

L	Length of pipe or tube
$\dot{m}$	Mass flow rate
$\dot{m}_t$	Total mass flow rate of tube-side fluid
N	Number of shell-side passes
n_p	Number of tube-side passes
n_t	Number of tubes in tube bundle
NTU	UA/C_{min} = Number of transfer units
Nu	Nusselt number
P	Parameter used to calculate LMTD correction factor; defined by Equation (3.12)
P_T	Tube pitch
Pr	Prandtl number
q	Rate of heat transfer
q_{max}	Maximum possible rate of heat transfer in a counter-flow heat exchanger
R	Parameter used to calculate LMTD correction factor; defined by Equation (3.11)
R_D	Fouling factor
R_{Di}	Fouling factor for inner (tube-side) fluid
R_{Do}	Fouling factor for outer (shell-side) fluid
R_{th}	Thermal resistance
Re	Reynolds number
r	C_{min}/C_{max} = Parameter used to calculate heat-exchanger effectiveness
S	Parameter used in calculating LMTD correction factor; defined by Equation (3.16)
T	Temperature
T_a	Inlet temperature of shell-side fluid
T_b	Outlet temperature of shell-side fluid
T_{ave}	Average temperature of shell-side fluid
T_c	Cold fluid temperature
$T_{c,a}$, $T_{c,b}$	Inlet and outlet temperatures of cold stream
T_h	Hot fluid temperature
$T_{h,a}$, $T_{h,b}$	Inlet and outlet temperatures of hot stream
T_w	Average temperature of tube wall
t_a	Inlet temperature of tube-side fluid
t_b	Outlet temperature of tube-side fluid
t_{ave}	Average temperature of tube-side fluid
U	Overall heat-transfer coefficient
U_C	Clean overall heat-transfer coefficient
U_D	Design overall heat-transfer coefficient
U_{req}	Required overall heat-transfer coefficient

Greek Letters

α	Parameter used to calculate LMTD correction factor; defined by Equation (3.13)
$\beta =$	$\sqrt{1 - r^2}$ = Parameter used in calculating heat-exchanger effectiveness
ΔT	Temperature difference
ΔT_m	Mean temperature difference in a heat exchanger
ΔT_{ln}	Logarithmic mean temperature difference
$(\Delta T_{\ln})_{cf}$	Logarithmic mean temperature difference for counter-current flow
ΔT_1, ΔT_2	Temperature differences between the fluids at the two ends of a heat exchanger
ϵ	Heat-exchanger effectiveness
ϵ^*	Parameter defined in Table 3.6 and used to calculate effectiveness of shell-and-tube heat exchanger having multiple shell passes
μ	Viscosity
μ_w	Fluid viscosity at average tube-wall temperature
ϕ	Viscosity correction factor
ϕ_i	Viscosity correction factor for inner (tube-side) fluid
ϕ_o	Viscosity correction factor for outer (shell-side) fluid

Problems

(3.1) In a double-pipe heat exchanger, the cold fluid will enter at 30°C and leave at 60°C, while the hot fluid will enter at 100°C and leave at 70°C. Find the mean temperature difference in the heat exchanger for:

(a) Co-current flow.

(b) Counter-current flow.

Ans. (a) 30.83°C.

(3.2) In Problem 3.1 the inner pipe is made of 3-in. schedule 40 carbon steel ($k = 45$ W/m · K). The cold fluid flows through the inner pipe with a heat-transfer coefficient of 600 W/m^2 · K, while the hot fluid flows in the annulus with a heat-transfer coefficient of 1000 W/m^2 · K. The exchanger duty is 140 kW. Calculate:

(a) The average wall temperature of the inner pipe.

(b) The clean overall heat-transfer coefficient.

(c) The design overall heat-transfer coefficient using a fouling factor of 0.002 h · ft^2 · °F/Btu for each stream. (Note the units.)

(d) The total length of pipe required in the heat exchanger for counter-current flow.

(e) The total length of pipe required in the heat exchanger for co-current flow.

Ans. (a) 71.2°C. (b) 330 W/m^2 · K. (c) 264 W/m^2 · K. (d) 47.5 m.

(3.3) A brine cooler in a chemical plant consists of a 1-2 shell-and-tube heat exchanger containing 80 tubes, each of which is 1-in. OD, 16 BWG and 15 ft long. 90,000 lb/h of brine flows through the tubes and is cooled from 16°F to 12°F. Ammonia enters the shell as a saturated liquid at 5°F and leaves as a saturated vapor at 5°F. The heat capacity of the brine is 0.7 Btu/lbm · °F. Refrigeration equipment is commonly rated in tons of refrigeration, where by definition, 1 ton of refrigeration is equal to 12,000 Btu/h.

(a) How many tons of refrigeration does the brine cooler deliver?

(b) How many square feet of heat-transfer surface does the cooler contain?

(c) Use Equations (3.11) to (3.17) to show that $F = 1.0$ for the brine cooler.

(d) Determine the value of the overall heat-transfer coefficient that is achieved in the brine cooler.

Ans. (a) 21. (b) 314. (d) 91 Btu/h · ft^2 · °F.

(3.4) The gas heater shown below in cross-section consists of a square sheet-metal duct insulated on the outside and a steel pipe (5 cm OD, 4.85 cm ID, $k = 56$ W/m · K) that passes through the center of the duct. A heat-transfer fluid will flow in the pipe, entering at 500 K and leaving at 450 K, with a heat-transfer coefficient of 200 W/m^2 · K. Process air will enter the heater at 300 K with a flow rate of 0.35 kg/s, leave at 350 K, and flow counter currently to the heat-transfer fluid. The heat-transfer coefficient for the air has been calculated and its value is 55 W/m^2 · K. Fouling factors of 0.00018 m^2 · K/W for the air stream and 0.00035 m^2 · K/W for the heat-transfer fluid are specified.

(a) Calculate the value of the overall heat-transfer coefficient to be used for designing the heater.

(b) What is the mean temperature difference in the heater?

(c) Calculate the required length of the heater.

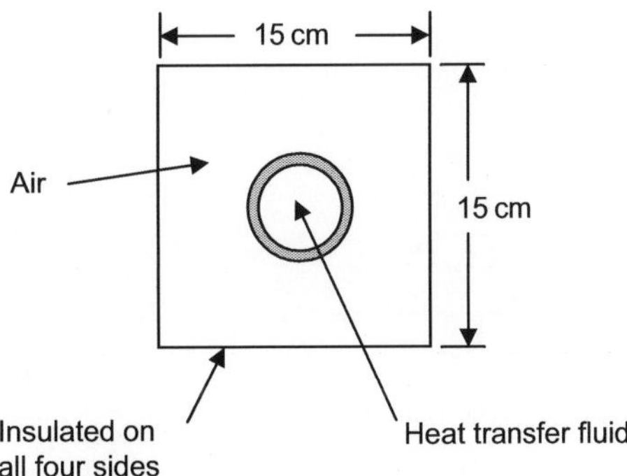

Ans. (a) 42 W/m^2 · K. (c) 18 m.

(3.5) In a refinery 100,000 lb/h of fuel oil is to be cooled from 200°F to 110°F by heat exchange with 70,000 lb/h of cooling water that is available at 60°F. A shell-and-tube exchanger will be used. The heat capacities are 0.6 Btu/lbm · °F for the oil and 1.0 Btu/lbm · °F for the water.

(a) Use stream energy balances to calculate the duty for the exchanger and the outlet temperature of the cooling water.

(b) Determine whether or not a 1-2 exchanger can be used for this service. If not, how many shell passes will be needed?

(c) Refer to Table 3.5 and obtain an approximate value for the overall heat-transfer coefficient. Use this value to calculate the heat-transfer surface area for the exchanger.

(d) If 0.75-in., 16 BWG tubes with a length of 20 ft are used in the heat exchanger, how many tubes will be required to supply the surface area calculated in part (c)?

Ans. (a) 5.4×10^6 Btu/h; 137.1°F.

(3.6) A shell-and-tube heat exchanger will be used to cool 200,000 lb/h of gasoline from 152°F to 80°F. Cooling water with a range of 60°F to 100°F will be used. Heat capacities are 0.6 Btu/lbm·°F for gasoline and 1.0 Btu/lbm·°F for water.
(a) What flow rate of cooling water will be needed?
(b) How many E-shells will be required for the heat exchanger?
(c) Make a preliminary design calculation to estimate the heat-transfer surface area required in the heat exchanger.
(d) If 1-in., 14 BWG tubes 25 ft long are used, how many tubes will be needed to supply the heat-transfer area computed in part (c)?

Ans. (a) 216,000 lb/h. (b) 2.

(3.7) In a petrochemical complex 150,000 lb/h of trichloroethylene is to be heated from 60°F to 140°F using 90,000 lb/h of kerosene that is available at 200°F. Heat capacities are 0.35 Btu/lbm · °F for trichloroethylene and 0.6 Btu/lbm · °F for kerosene. A shell-and-tube heat exchanger will be used for this service.
(a) Will a 1-2 heat exchanger be satisfactory? If not, how many shell-side passes should be used?
(b) Make a preliminary design calculation to estimate the heat-transfer surface area required in the heat exchanger.
(c) If the tube bundle is to consist of 1.25-in., 16 BWG tubes with a length of 16 ft, how many tubes will the bundle contain?

(3.8) A counter-flow heat exchanger was designed to cool 13,000 lb/h of 100% acetic acid from 250°F to 150°F by heating 19,000 lb/h of butyl alcohol from 100°F to 157°F. An overall heat-transfer coefficient of 85 Btu/h·ft^2·°F was used for the design. When first placed in service, the acetic acid outlet temperature was found to be 117°F. It gradually rose to 135°F over a period of several months and then remained essentially constant, indicating that the exchanger was over-sized.
(a) Use the design data to calculate the amount of heat-transfer surface area in the heat exchanger.
(b) Use the initial operating data to calculate the value of the clean overall heat-transfer coefficient.
(c) Use the final operating data to calculate the value of the overall heat-transfer coefficient after fouling has occurred.
(d) Use the values of U_C and U_D found in parts (b) and (c) to obtain the correct (experimental) value of the total fouling factor, $R_D \equiv R_{Di}(D_o/D_i) + R_{Do}$, for the system.

Ans. (a) 121 ft^2. (b) 196 Btu/h · ft^2 · °F. (c) 121 Btu/h · ft^2 · °F. (d) 0.0032 h · ft^2 · °F/Btu.

(3.9) A heat exchanger consists of two E-shells connected in series and contains a total of 488 tubes arranged for four passes (two per shell). The tubes are 1-in. OD, 16 BWG with a length of 10 ft. The heat exchanger was designed to cool a hot fluid from 250°F to 150°F using a cold fluid entering at 100°F. When the exchanger was first placed in service, the hot fluid outlet temperature was 130°F, but it gradually increased to 160°F over several months and then remained essentially constant. Fluid data are as follows:

Property	Hot Fluid (Tubes)	Cold Fluid (Shell)
$\dot{m}$ (lbm/h)	130,000	190,000
C_P (Btu/lbm · °F)	0.55	0.66
Inlet temperature (°F)	250	100

(a) To the nearest square foot, how much heat-transfer surface does the heat exchanger contain?
(b) Use the initial operating data to calculate the clean overall heat-transfer coefficient, U_C.
(c) Use the final operating data to calculate the fouled overall heat-transfer coefficient, U_D.
(d) Use the results of parts (b) and (c) to determine the experimental value of the total fouling factor, $R_D \equiv R_{Di}(D_o/D_i) + R_{Do}$.

Ans. (a) 1,278 ft^2. (b) 153 Btu/h · ft^2 · °F. (c) 66 Btu/h · ft^2 · °F. (d) 0.0086 h · ft^2 · °F/Btu.

(3.10) A hydrocarbon stream is to be cooled from 200°F to 130°F using 10,800 lb/h of water with a range of 75°F to 125°F. A double-pipe heat exchanger comprised of 25 ft long carbon steel hairpins will be used. The inner and outer pipes are 1.5- and 3.5-in. schedule 40, respectively. The hydrocarbon will flow through the inner pipe and the heat-transfer coefficient for this stream has been determined: $h_i = 200$ Btu/h · ft^2 · °F. Fouling factors of 0.001 h · ft^2 · °F/Btu for water and 0.002 h · ft^2 · °F/Btu for the hydrocarbon are specified. How many hairpins will be required?

(3.11) An oil stream is to be cooled from 200°F to 100°F by heating a kerosene stream from 80°F to 160°F. A fouling factor of 0.002 h · ft^2 · °F/Btu is required for each stream, and fluid properties may be assumed constant at the values given below. Ten carbon steel hairpins are available on the company's used equipment lot. Each hairpin is 20 ft long with inner and outer pipes being 0.75- and 1.5-in. schedule 40, respectively. Will these hairpins be adequate for this service?

Property	Oil	Kerosene
$\dot{m}$ (lb/h)	10,000	12,500
C_P (Btu/lbm · °F)	0.6	0.6
k (Btu/h · ft · °F)	0.08	0.077
μ (cp)	3.2	0.45
Pr	58.1	8.49

(3.12) For the double-pipe heat exchanger of Problem 3.10, calculate the outlet temperatures of the two streams when the unit is first placed in service.

(3.13) Available on the used equipment lot at a petrochemical complex is a shell-and-tube heat exchanger configured for one shell pass and one tube pass, and containing 1500 ft^2 of heat-transfer surface. Suppose this unit is used for the trichloroethylene/kerosene service of Problem 3.7, and that an overall heat-transfer coefficient of 100 Btu/h · ft^2 · °F is attained in the exchanger. Compute the outlet temperatures of the two streams if they flow:
(a) Counter-currently
(b) Co-currently

Ans. (a) 165°F (trichloroethylene), 98°F (kerosene).

(3.14) An organic liquid at 185°F is to be cooled with water that is available at 80°F. A double-pipe heat exchanger consisting of six hairpins connected in series will be used. The heat exchanger pipes are schedule 40. Each hairpin is 20 ft long and is made with 2- and 1-in. stainless steel ($k = 9.4$ Btu/h · ft · °F) pipes. Flow rates and fluid properties, which may be assumed constant, are given below. The organic liquid will flow through the inner pipe, and its heat-transfer coefficient has been determined: $h_i = 250$ Btu/h · ft^2 · °F. The streams will flow counter-currently through the exchanger.

Property	Organic liquid	Water
$\dot{m}$ (lb/h)	11,765	7,000
C_P (Btu/lbm · °F)	0.51	1.0
k (Btu/h · ft · °F)	–	0.38
μ (cp)	–	0.65
Pr	–	4.14

(a) Calculate the heat-transfer coefficient, h_o, for the water.
(b) Calculate the clean overall heat-transfer coefficient.
(c) Determine the outlet temperatures of the two streams that will be achieved when the heat exchanger is first placed in service.
(d) Calculate the average wall temperature of the inner pipe when the heat exchanger is clean.

Ans. (a) 727 Btu/h · ft^2 · °F. (b) 130 Btu/h · ft^2 · °F. (c) 115°F (organic), 140°F (water). (d) 119°F.

(3.15) A shell-and-tube heat exchanger consists of an E-shell that contains 554 tubes. The tubes are 1-in. OD, 16 BWG with a length of 20 ft, and the tube bundle is arranged for two passes. Suppose that this heat exchanger is used for the gasoline/water service of Problem 3.6, and that an overall heat-transfer coefficient of 100 Btu/h · ft^2 · °F is realized in the exchanger. Calculate the outlet temperatures of the two streams and the exchanger duty under these conditions.

Ans. 87.5°F (gasoline), 96°F (water), $q = 7.738 \times 10^6$ Btu/h.

(3.16) For the gasoline/water service of Problem 3.6, it is proposed to use a heat exchanger consisting of two E-shells connected in series. Each shell contains 302 tubes arranged for two passes. The tubes are 1-in. OD, 16 BWG with a length of 20 ft. Assuming that an overall heat-transfer coefficient of 100 Btu/h · ft^2 · °F is realized in the exchanger, what outlet temperature and exchanger duty will be attained with this unit?

Ans. 79°F (gasoline), 101°F (water), $q = 8.803 \times 10^6$ Btu/h.

(3.17) 18,000 lb/h of a petroleum fraction are to be cooled from 250°F to 150°F using cooling water with a range of 85–120°F. Properties of the petroleum fraction may be assumed constant at the following values:

Property	Value
C_P (Btu/lbm · °F)	0.52
k (Btu/h · ft · °F)	0.074
μ (cp)	2.75
ρ (lbm/ft^3)	51.2

A shell-and-tube heat exchanger with the following configuration is available:

Type: AES	Tubes: $^3/_4$ in. OD, 16 BWG, 20 ft long
Shell ID: 15.25 in.	Number of tubes: 128
Baffle type: segmental	Number of tube passes: 4
Baffle cut: 20%	Tube pitch: 1.0 in. (square)
Number of baffles: 50	Tube material: Admirality brass ($k =$ 64 Btu/h · ft · °F)
Shell material: carbon steel	

Fouling factors of 0.001 and 0.002 h · ft^2 · °F/Btu are required for the cooling water and petroleum fraction, respectively. Is the heat exchanger thermally suitable for this service?

(3.18) For the heat exchanger of Problem 3.17, calculate the outlet temperatures of the two streams that will be attained when the exchanger is clean.

(3.19) 72,500 lb/h of crude oil are to be heated from 250°F to 320°F by cooling a lube oil from 450°F to 340°F. Physical properties of the two streams may be assumed constant at the following values:

Property	Crude oil	Lube oil
C_P (Btu/lbm · °F)	0.59	0.62
k (Btu/h · ft · °F)	0.073	0.067
μ (cp)	0.83	1.5
ρ (lbm/ft^3)	47.4	48.7

It is proposed to use a carbon steel heat exchanger having the following configuration:

Type: AES	Tubes: 1.0 in. OD, 14 BWG, 24 ft long
Shell ID: 29 in.	Number of tubes: 314
Baffle type: segmental	Number of tube passes: 6
Baffle cut: 20%	Tube pitch: 1.25 in. (square)
Number of baffles: 30	

The crude oil will flow in the tubes, and published fouling factors for oil refinery streams should be used. Is the proposed unit thermally suitable for this service?

(3.20) For the heat exchanger of Problem 3.19, calculate the outlet temperatures of the two streams that will be attained when the exchanger is clean.

(3.21) Repeat Example 3.3 with the fluids switched, i.e., with aniline in the inner pipe and benzene in the annulus.

4 Design of Double-Pipe Heat Exchangers

4.1 Introduction

The basic calculations involved in the thermal analysis of double-pipe heat exchangers were presented in the previous chapter. In the present chapter we consider the design of double-pipe units in more detail. In particular, the hydraulic analysis alluded to in Chapter 3 is presented, and procedures for handling finned-tube and multi-tube exchangers are given. Series-parallel configurations of hairpins are also considered.

Design of heat-transfer equipment involves a trade-off between the two conflicting goals of low capital cost (high overall heat-transfer coefficient, small heat-transfer area) and low operating cost (small stream pressure drops). Optimal design thus involves capital and energy costs, which are constantly changing. A simpler, albeit sub-optimal, procedure is to specify a reasonable pressure-drop allowance for each stream and design the exchanger within these constraints. The pressure-drop allowances determine (approximately) the trade-off between capital and operating costs. For low-viscosity liquids such as water, organic solvents, light hydrocarbons, etc., an allowance in the range of 7 to 20 psi is commonly used. For gases, a value in the range of 1 to 5 psi is often specified.

4.2 Heat-Transfer Coefficients for Exchangers without Fins

Heat-transfer correlations for pipes and annuli were presented in Section 2.4 and their application to the analysis of double-pipe exchangers was illustrated in Section 3.7. The correlations that are used in this chapter are repeated here for convenience.

For turbulent flow ($Re \geq 10^4$), the Seider–Tate equation is used in the form:

$$Nu = 0.023\, Re^{0.8} Pr^{1/3} (\mu/\mu_w)^{0.14} \tag{4.1}$$

This is the same as Equation (2.34) except that a coefficient of 0.023 is used here rather than 0.027. As noted in Chapter 2, a coefficient of 0.023 is often preferred for design work. For the transition region ($2100 < Re < 10^4$), the Hausen equation is used:

$$Nu = 0.116 \left[Re^{2/3} - 125\right] Pr^{1/3} (\mu/\mu_w)^{0.14} \left[1 + (D_i/L)^{2/3}\right] \tag{2.38}$$

Equations (4.1) and (2.38) are used for both pipes and annuli, with the equivalent diameter replacing D_i, the pipe ID, in the case of an annulus. For laminar flow ($Re \leq 2100$) in pipes, the Seider–Tate equation is used:

$$Nu = 1.86\, [Re\ Pr\ D_i/L]^{1/3} (\mu/\mu_w)^{0.14} \tag{2.37}$$

For laminar flow in annuli, Equation (2.42) should be used. Also, Equation (2.39) may be used for the turbulent and transition regions in place of the Seider–Tate and Hausen correlations. However, these equations are not used in this chapter.

4.3 Hydraulic Calculations for Exchangers without Fins

The main contribution to pressure drop in double-pipe exchangers is from fluid friction in the straight sections of pipe. For isothermal flow, this pressure drop can be expressed in terms of the Darcy friction factor, f, as:

$$\Delta P_f = \frac{f\, L\rho\, V^2}{2 g_c D_i} \tag{4.2}$$

where

L = pipe length
D_i = pipe ID
ρ = fluid density
V = average fluid velocity

This equation can be written in terms of the mass flux, G, and the specific gravity, s, of the fluid by making the substitutions $V = G/\rho$ and $\rho = s\rho_{water}$ to give:

$$\Delta P_f = \frac{f L G^2}{2 g_c \rho_{water} D_i s} \tag{4.3}$$

The specific gravity of liquids is usually referenced to water at 4°C, which has a density of 62.43 lbm/ft^3. (The petroleum industry uses a reference temperature of 60°F, at which $\rho_{water} = 62.37$ lbm/ft^3; the difference in reference densities is insignificant in the present context.) Also, in English units,

$$g_c = 32.174 \frac{\text{lbm}\cdot\text{ft/s}^2}{\text{lbf}} = 4.16975 \times 10^8 \frac{\text{lbm}\cdot\text{ft/h}^2}{\text{lbf}}$$

With these numerical values, Equation (4.3) becomes:

$$\Delta P_f = \frac{f\,LG^2}{5.206 \times 10^{10} D_i s} \tag{4.4}$$

When L and D_i are expressed in ft and G in lbm/h · ft^2, the units of ΔP_f are lbf/ft^2.

A minor modification is made to this basic hydraulic equation by dividing the right side by 144ϕ, where

$$\phi = (\mu/\mu_w)^{0.14} \text{ for turbulent flow}$$

$$\phi = (\mu/\mu_w)^{0.25} \text{ for laminar flow}$$

The viscosity correction factor accounts for the effect of variable fluid properties on the friction factor in non-isothermal flow, while the factor of 144 converts the pressure drop from lbf/ft^2 to psi. With this modification, Equation (4.4) becomes:

$$\Delta P_f = \frac{f\,LG^2}{7.50 \times 10^{12} D_i s\phi} \tag{4.5}$$

where

$\Delta P_f \propto$ psi
$L,\ D_i \propto$ ft
$G \propto$ lbm/h · ft^2
$s,\ f,\ \phi$ are dimensionless

See Appendix 4.A for the corresponding equation in terms of SI units. Equation (4.5) is also applicable to flow in the annulus of a double-pipe exchanger if the pipe diameter, D_i, is replaced by the equivalent diameter, D_e.

The friction factor to be used with Equation (4.5) can be computed as follows. For laminar flow in the inner pipe,

$$f = 64/Re \tag{4.6}$$

For laminar flow in the annulus,

$$f = \left(\frac{64}{Re}\right)\left[\frac{(1-\kappa)^2}{1+\kappa^2+(1-\kappa^2)/\ln\kappa}\right] \tag{4.7}$$

where

$\kappa = D_1/D_2$
D_1 = OD of inner pipe
D_2 = ID of outer pipe

For practical configurations, the term in square brackets in Equation (4.7) is approximately 1.5.

For commercial pipe and turbulent flow with $Re \geq 3000$, the following equation can be used for both the pipe and the annulus:

$$f = 0.3673\,Re^{-0.2314} \tag{4.8}$$

Minor pressure losses due to entrance and exit effects and return bends are usually expressed in terms of velocity heads. The grouping $(V^2/2g)$ has dimensions of length and is called a velocity head. The change in pressure equivalent to one velocity head is:

$$\Delta P = \rho(g/g_c)(V^2/2g) = \rho V^2/2g_c \tag{4.9}$$

Expressing ρ and V in terms of s and G as above and dividing by 144 to convert to psi gives:

$$\Delta P = \left(\frac{1}{2g_c\rho_{water} \times 144}\right)\left(\frac{G^2}{s}\right) \tag{4.10}$$

$$\Delta P = 1.334 \times 10^{-13} G^2/s \tag{4.11}$$

where ΔP has units of psi and G has units of lbm/h · ft^2. These units apply in the remainder of this section. See Appendix 4.A for the corresponding results in terms of SI units.

In the return bend of a hairpin, the fluid experiences a 180° change of direction. The pressure loss for turbulent flow in a long-radius 180° bend is given as 1.2 velocity heads in Ref. [1]. In laminar flow, the number of velocity heads depends on the Reynolds number [2], but for Re between 500 and 2100, a reasonable approximation is 1.5 velocity heads. Kern and Kraus [3]

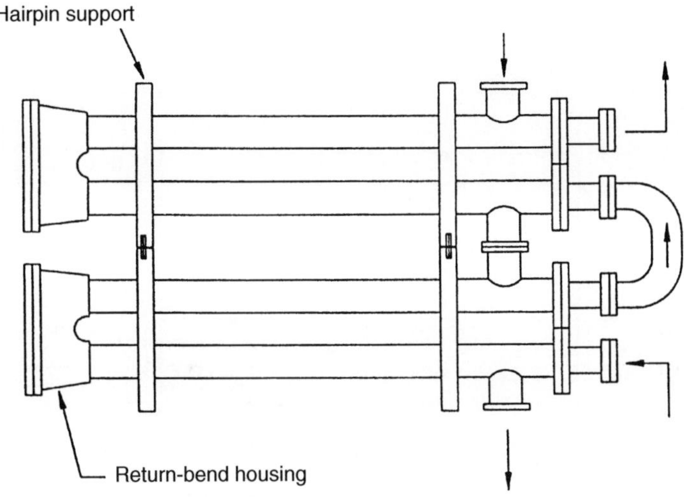

FIGURE 4.1 Two hairpins connected in series (Source: Ref. [1]).

recommend an allowance of one velocity head per hairpin for the return-bend losses, which agrees fairly well with the above value for turbulent flow.

When hairpins are connected in series as shown in Figure 4.1, both fluids make an additional 180° change of direction between the outlet of one hairpin and the inlet of the next. It follows that for N_{HP} hairpins connected in series, the total number of direction changes experienced by each fluid is $(2N_{HP} - 1)$. The total pressure drop resulting from these direction changes is thus:

$$\Delta P_r = 1.6 \times 10^{-13}(2N_{HP} - 1)G^2/s \text{ (turbulent flow)} \tag{4.12}$$

$$\Delta P_r = 2.0 \times 10^{-13}(2N_{HP} - 1)G^2/s \text{ (laminar flow, } Re \geq 500) \tag{4.13}$$

Entrance and exit losses for flow through the inner pipes of double-pipe exchangers are generally negligible because connections to process piping are inline and losses are mainly due to size mismatches, which are usually minor, between process and heat-exchanger pipes. An exception occurs in multi-tube exchangers, where the fluid experiences a significant contraction as it enters the individual tubes and a corresponding expansion as it exits the tubes at the outlet tubesheet. In turbulent flow, the sum of these losses can be approximated by one tube velocity head per hairpin. In laminar flow, the number of velocity heads varies with the Reynolds number [4].

On the annulus side, the fluid enters and leaves through standard nozzles where it experiences a sudden expansion (entering) and contraction (leaving). These losses can be approximated using the standard formulas for a sudden expansion and sudden contraction [4]. However, for turbulent flow in the nozzles, the sum of the entrance and exit losses can be estimated as 1.5 nozzle velocity heads [4], apportioned roughly as one velocity head for the entrance nozzle and 0.5 velocity head for the exit nozzle. In laminar flow, the number of velocity heads depends on the Reynolds number, but for $Re \geq 100$, a reasonable approximation is three nozzle velocity heads for the sum of the entrance and exit losses. Thus, for exchangers with internal return bends, we have the following relations for nozzle losses:

$$\Delta P_n = 2.0 \times 10^{-13} N_{HP} G_n^2/s \text{ (turbulant flow)} \tag{4.14}$$

$$\Delta P_n = 4.0 \times 10^{-13} N_{HP} G_n^2/s \text{ (laminar flow, } Re_n \geq 100) \tag{4.15}$$

where G_n and Re_n are the mass flux and Reynolds number, respectively, for the nozzle, and N_{HP} is the number of hairpins connected in series. For exchangers with external return bends, the fluid in the annulus experiences another sudden contraction and expansion at the return bends. Hence, in this case the above pressure drop should be doubled.

4.4 Series/Parallel Configurations of Hairpins

Double-pipe exchangers are extremely flexible with respect to configuration of hairpins, since both the inner pipes and annuli can be connected either in series or in parallel. In order to meet pressure-drop constraints, it is sometimes convenient to divide one stream into two or more parallel branches while leaving the other stream intact. Such a case is shown in Figure 4.2, where the inner pipes are connected in parallel, while the annuli are connected in series. Although the flow is counter-current in each hairpin of Figure 4.2, the overall flow pattern is not true counter flow. This fact can be appreciated by comparing Figure 4.1 (true counter flow) and Figure 4.2. Note in particular that the temperature of the fluid entering the inner pipe of the top hairpin is different in the two cases.

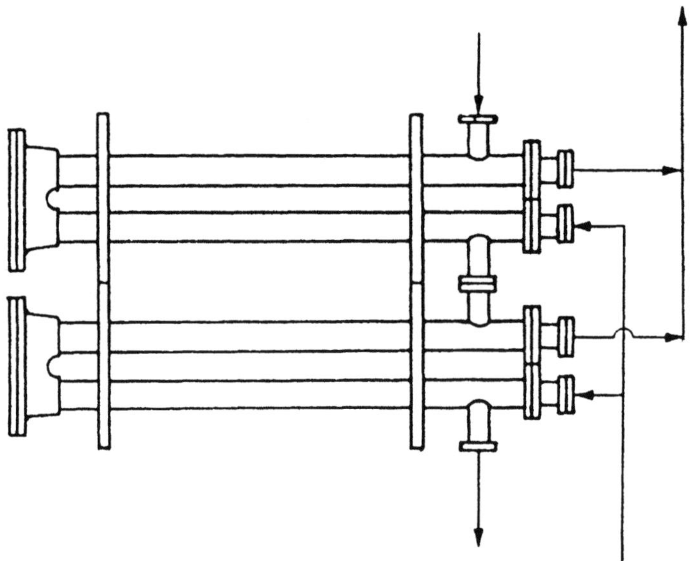

FIGURE 4.2 Hairpins with annuli connected in series and inner pipes connected in parallel (Source: Ref. [1]).

To account for the departure from true counter flow in series–parallel configurations, the counter-flow logarithmic mean temperature difference (LMTD) is multiplied by a correction factor, F, given by the following equations:

$$F = \left[\frac{(R-x)}{x(R-1)}\right] \frac{\ln\left[(1-P)/(1-PR)\right]}{\ln\left[\frac{(R-x)}{R(1-PR)^{1/x}} + \frac{x}{R}\right]} \quad (R \neq 1) \tag{4.16}$$

$$F = \frac{P(1-x)}{x(1-P)\ln\left[\frac{(1-x)}{(1-P)^{1/x}} + x\right]} \quad (R = 1) \tag{4.17}$$

$$P = (t_b - t_a)/(T_a - t_a) \tag{4.18}$$

$$R = (T_a - T_b)/(t_b - t_a) \tag{4.19}$$

where

x = number of parallel branches
T_a, T_b = inlet and outlet temperatures of series stream
t_a, t_b = inlet and outlet temperatures of parallel stream

4.5 Multi-Tube Exchangers

Un-finned multi-tube hairpin exchangers can be handled using the methods presented above with an appropriate value of the equivalent diameter. For an outer pipe with an ID of D_2 containing n_t tubes, each with OD of D_1, the flow area and wetted perimeter are:

$$A_f = (\pi/4)\left(D_2^2 - n_t D_1^2\right) \tag{4.20}$$

$$\text{Wetted perimeter} = \pi(D_2 + n_t D_1) \tag{4.21}$$

Therefore, the equivalent diameter is given by:

$$D_e = 4A_f/\text{wetted perimeter}$$

$$D_e = \left(D_2^2 - n_t D_1^2\right)/(D_2 + n_t D_1) \tag{4.22}$$

Note that for $n_t = 1$, this reduces to $D_e = D_2 - D_1$.

4.6 Over-Surface and Over-Design

Over-surface is a measure of the safety factor incorporated in the design of a heat exchanger through fouling factors and the use of standard equipment sizes. Since it deals directly with exchanger surface area, it is easier to visualize than fouling factors and calculated versus required heat-transfer coefficients. The percentage over-surface is defined as follows:

$$\%\ \text{over-surface} = \frac{A - A_C}{A_C} \times 100 \tag{4.23}$$

where

A = actual heat-transfer surface area in the exchanger
A_C = calculated heat-transfer surface area based on U_C

Over-surface depends on the relative magnitudes of the total fouling allowance and the film and wall resistances. While values of 20% to 40% may be considered typical, higher values are not unusual. Equation (4.23) is often applied with the surface area calculated using the design coefficient, U_D, rather than the clean coefficient. In this case the computed quantity is referred to as the over-design; it represents the extra surface area beyond that required to compensate for fouling. Some over-design (typically about 5–20%) is considered acceptable and (often) desirable, since it provides an additional safety margin in the final design.

The design procedure for an un-finned double-pipe exchanger is illustrated by reconsidering the benzene–aniline exchanger of Example 3.3. For completeness, the problem is restated here and worked in its entirety. However, the reader may wish to refer to Example 3.3 for additional details. (Note that a coefficient of 0.027 was used in the Seider–Tate equation in Example 3.3, whereas a value of 0.023 is used here.)

Example 4.1

10,000 lb/h of benzene will be heated from 60°F to 120°F by heat exchange with an aniline stream that will be cooled from 150°F to 100°F. A number of 16-ft hairpins consisting of 2-in. by 1.25-in. schedule 40 stainless steel pipe (type 316, $k = 9.4$ Btu/h · ft · °F) are available and will be used for this service. The hairpins have internal return bends and 1-in. schedule 40 nozzles. A maximum pressure drop of 20 psi is specified for each stream. The specific gravity of benzene is 0.879 and that of aniline is 1.022. Determine the number and configuration of hairpins that are required.

Solution

We begin by assuming that the hairpins are connected in series on both sides, since this is the simplest configuration, and that the flow pattern is counter-current. We also place the benzene in the inner pipe for the sake of continuity with Example 3.3. As discussed at the end of that example, however, either fluid could be placed in the inner pipe.

First Trial

(a) Fluid properties at the average stream temperatures are obtained from Figures A.l and A.2, and Table A.15.

Fluid property	Benzene ($t_{ave} = 90$ °F)	Aniline ($T_{ave} = 125$ °F)
μ (cp)	0.55	2.0
C_P (Btu/lbm · °F)	0.42	0.52
k (Btu/h · ft · °F)	0.092	0.100

(b) Determine the heat load and aniline flow rate by energy balances on the two streams.

$$q = (\dot{m}C_P\Delta T)_B = 10,000 \times 60 = 252,000 \text{ Btu/h}$$

$$252,000 = (\dot{m}C_P\Delta T)_A = \dot{m}_A \times 0.52 \times 50$$

$$\dot{m}_A = 9692 \text{ lb/h}$$

(c) Calculate the LMTD.

$$\Delta T_{\ln} = \frac{40 - 30}{\ln(40/30)} = 34.76°\text{F}$$

(d) Calculate h_i assuming $\phi_i = 1.0$.

$$D_i = 1.38/12 = 0.115 \text{ ft (from Table B.2)}$$

$$Re = \frac{4\dot{m}}{\pi D_i \mu} = \frac{4 \times 10{,}000}{\pi \times 0.115 \times 0.55 \times 2.419} = 83{,}217 \quad \Rightarrow \text{turbulent flow}$$

$$h_i = \frac{k}{D_i} \times 0.023\ Re^{0.8} Pr^{1/3}$$

$$h_i = \frac{0.092}{0.115} \times 0.023\ (83{,}217)^{0.8} \left(\frac{0.42 \times 0.55 \times 2.419}{0.092}\right)^{1/3}$$

$$h_i = 290 \text{ Btu/h·ft}^2\text{·°F}$$

(e) Calculate h_o assuming $\phi_o = 1.0$.

$$\left.\begin{aligned} D_2 &= 2.067 \text{ in.} \\ D_1 &= 1.660 \text{ in.} \end{aligned}\right\} \quad \text{(from Table B.2)}$$

$$D_e = D_2 - D_1 = \frac{2.067 - 1.660}{12} = 0.0339 \text{ ft}$$

$$\text{flow area} \equiv A_f = \frac{\pi}{4}(D_2^2 - D_1^2) = 0.00827 \text{ ft}^2$$

$$Re = \frac{D_e(\dot{m}/A_f)}{\mu} = \frac{0.0339 \times (9692/0.00827)}{2.0 \times 2.419} = 8212 \Rightarrow \text{transition flow}$$

Using the Hausen equation with $[1 + (D_e/L)^{2/3}] = 1.0$ gives:

$$h_o = \frac{k}{D_e} \times 0.116 \left[Re^{2/3} - 125\right] Pr^{1/3}$$

$$= \frac{0.1}{0.0339} \times 0.116 \times \left[(8212)^{2/3} - 125\right] \left(\frac{0.52 \times 2.0 \times 2.419}{0.1}\right)^{1/3}$$

$$h_o = 283 \text{ Btu/h·ft}^2\text{·°F}$$

(f) Calculate the pipe wall temperature.

$$T_w = \frac{h_i t_{ave} + h_o(D_o/D_i)T_{ave}}{h_i + h_o(D_o/D_i)}$$

$$T_w = \frac{290 \times 90 + 283(1.66/1.38) \times 125}{290 + 283 \times (1.66/1.38)}$$

$$T_w = 108.9\text{°F}$$

(g) Calculate ϕ_i and ϕ_o, and corrected values of h_i and h_o.
From Figure A.1, at 108.9°F, $\mu_B = 0.47$ cp and $\mu_A = 2.4$ cp. Therefore,

$$\phi_i = (0.55/0.47)^{0.14} = 1.0222$$

$$\phi_o = (2.0/2.4)^{0.14} = 0.9748$$

$$h_i = 290(1.0222) = 296 \text{ Btu/h·ft}^2\text{·°F}$$

$$h_o = 283(0.9748) = 276 \text{ Btu/h·ft}^2\text{·°F}$$

(h) Obtain fouling factors. For liquid organic process chemicals such as benzene and aniline, a value of 0.001 h·ft^2 ·°F/Btu is appropriate.

(i) Compute the overall heat-transfer coefficient.

$$U_D = \left[\frac{D_o}{h_i D_i} + \frac{D_o \ln(D_o/D_i)}{2k} + \frac{1}{h_o} + \frac{R_{Di} D_o}{D_i} + R_{Do}\right]^{-1}$$

$$U_D = \left[\frac{1.66}{296 \times 1.38} + \frac{(1.66/12)\ln(1.66/1.38)}{2 \times 9.4} + \frac{1}{276} + \frac{0.001 \times 1.66}{1.38} + 0.001\right]^{-1}$$

$$U_D = 89 \text{ Btu/h}\cdot\text{ft}^2\cdot{}^\circ\text{F}$$

(j) Calculate the required surface area and number of hairpins.

$$q = U_D A \Delta T_{\ln}$$

$$A = \frac{q}{U_D \Delta T_{\ln}}$$

$$A = \frac{252{,}000}{89 \times 37.76} = 81.5 \text{ ft}^2$$

From Table B.2, the external surface area per foot of 1.25-in. schedule 40 pipe is 0.435 ft^2. Therefore,

$$L = \frac{81.5}{0.435} = 187.4 \text{ ft}$$

Since each 16-ft hairpin contains 32 ft of pipe,

$$\text{Number of hairpins} = \frac{187.4}{32} = 5.9 \Rightarrow 6$$

Thus, six hairpins are required.

(k) Calculate the pressure drop for the benzene stream (inner pipe).
The friction factor is calculated from Equation (4.8):

$$f = 0.3673\, Re^{-0.2314} = 0.3673(83{,}217)^{-0.2314}$$

$$f = 0.0267$$

$$A_f = 0.0104 \text{ ft}^2 (\text{Table B.2})$$

$$G = \dot{m}/A_f = 10{,}000/0.0104 = 961{,}538 \text{ lbm/h}\cdot\text{ft}^2$$

The pressure drop in the straight sections of pipe is calculated using Equation (4.5):

$$\Delta P_f = \frac{f L G^2}{7.50 \times 10^{12} D_i s \phi}$$

$$= \frac{0.0267\,(6 \times 32)\,(961{,}538)^2}{7.50 \times 10^{12}(1.38/12) \times 0.879 \times 1.0222}$$

$$\Delta P_f = 6.1 \text{ psi}$$

The pressure drop in the return bends is obtained from Equation (4.12):

$$\Delta P_r = 1.6 \times 10^{-13}(2N_{HP} - 1)G^2/s$$

$$= 1.6 \times 10^{-13}(2 \times 6 - 1)(961{,}538)^2/0.879$$

$$\Delta P_r = 1.85 \text{ psi}$$

Since the nozzle losses associated with the inner pipes are negligible, the total pressure drop, ΔP_i is:

$$\Delta P_i = \Delta P_f + \Delta P_r = 6.1 + 1.85 = 7.95 \cong 8.0 \text{ psi}$$

(1) Calculate the pressure drop for the aniline stream (annulus).

The friction factor is calculated from Equation (4.8):

$$f = 0.3673\, Re^{-0.2314} = 0.3673(8212)^{-0.2314}$$

$$f = 0.0456$$

$$G = \dot{m}/A_f = 9692/0.00827 = 1{,}171{,}947 \text{ lbm/h·ft}^2$$

The pressure drop in the straight sections of pipe is again calculated using Equation (4.5) with the pipe diameter replaced by the equivalent diameter:

$$\Delta P_f = \frac{f\, LG^2}{7.50 \times 10^{12} D_e s \phi}$$

$$= \frac{0.0456(6 \times 32)(1{,}171{,}947)^2}{7.50 \times 10^{12} \times 0.0339 \times 1.022 \times 0.9748}$$

$$\Delta P_f = 47.5 \text{ psi}$$

Since this value greatly exceeds the allowed pressure drop, the minor losses will not be calculated. This completes the first trial. In summary, there are two problems with the initial configuration of the heat exchanger:

(1) The pressure drop on the annulus side is too large.
(2) The Reynolds number in the annulus is less than 10,000.

Since the dimensions of the hairpins are fixed in this problem, there are relatively few options for modifying the design. Two possibilities are:

(1) Switch the fluids, i.e., put the aniline in the inner pipe and the benzene in the annulus.
(2) Connect the annuli in parallel.

The effects of these changes on the Reynolds numbers and pressure drops can be estimated as follows:

(1) Switch the fluids. Since the flow rates of the two streams are approximately the same, the Reynolds numbers are essentially inversely proportional to the viscosity. Thus,

$$Re_i \to 83{,}217(0.55/2.0) \cong 23{,}000$$

$$Re_o \to 8212(2.0/0.55) \cong 30.000$$

Hence, switching the fluids will result in fully turbulent flow on both sides of the exchanger.
To estimate the effect on pressure drops, assume that the number of hairpins does not change. Then the main factors affecting ΔP are f and s. Hence,

$$\Delta P \sim f/s \sim Re^{-0.2314} s^{-1}$$

$$\Delta P_{f,i} \to 6.1\ (23{,}000/83{,}000)^{-0.2314}(1.022/0.879)^{-1} \cong 7 \text{ psi}$$

$$\Delta P_{f,o} \to 47.5\ (30{,}000/82{,}000)^{-0.2314}(0.879/1.022)^{-1} \cong 41 \text{ psi}$$

Clearly, switching the fluids does not reduce the annulus-side pressure drop nearly enough to meet the design specification (unless the number of hairpins is reduced by a factor of at least two, which is very unlikely).

(2) Connect the annuli in two parallel banks. This change will have no effect on the fluid flowing in the inner pipe. For the fluid in the annulus, however, both the flow rate and the length of the flow path will be halved. Therefore,

$$\text{Re}_o \to 8212 \times 1/2 \cong 4100$$

Assuming that the number of hairpins does not change,

$$\Delta P_{f,o} \sim f\, G^2 L$$

$$\Delta P_{f,o} \to 47.5(4100/8200)^{-0.2314}(1/2)^2(1/2) \cong 7 \text{ psi}$$

Apparently this modification will take care of the pressure-drop problem, but will push the Reynolds number further into the transition region.

Although neither modification by itself will correct the problems with the initial design, in combination they might. Hence, we consider a third alternative.

(3) Switch the fluids and connect the annuli in two parallel banks. The Reynolds numbers will become:

$$Re_i \cong 23,000$$

$$Re_o \cong 15,000$$

The pressure drops will become (assuming no change in the number of hairpins):

$$\Delta P_{f,i} \cong 7 \text{ psi}$$

$$\Delta P_{f,o} \rightarrow 47.5(15,000/8200)^{-0.2314}(1/2)^2(1/2)(0.879/1.022)^{-1} \cong 6 \text{ psi}$$

It appears that this alternative will meet all design requirements. However, it is necessary to perform the detailed calculations because h_i, h_o and the mean temperature difference will all change, and hence the number of hairpins can be expected to change as well.

Second Trial

(a) Calculate the LMTD correction factor for the series/parallel configuration. Aniline in the inner pipe is the series stream and benzene in the annulus is the parallel stream. Therefore,

$$T_a = 150°\text{F} \quad t_a = 60°\text{F}$$

$$T_b = 100°\text{F} \quad t_b = 120°\text{F}$$

$$P = (t_b - t_a)/(T_a - t_a) = (120 - 60)/(150 - 60) = 0.6667$$

$$R = (T_a - T_b)/(t_b - t_a) = (150 - 100)/(120 - 60) = 0.8333$$

$$x = 2 \text{ (number of parallel branches)}$$

Substituting into Equation (4.16) gives:

$$F = \left[\frac{(0.8333 - 2)}{2(0.8333 - 1)}\right] \frac{\ln\left[(1 - 0.6667)/(1 - 0.6667 \times 0.8333)\right]}{\ln\left[\dfrac{(0.833 - 2)}{0.8333\,(1 - 0.6667 \times 0.8333)^{1/2}} + \left(\dfrac{2}{0.8333}\right)\right]}$$

$$F = 0.836$$

(b) Calculate h_i assuming $\phi_i = 1.0$.

$$Re = \frac{4\dot{m}}{\pi D_i \mu} = \frac{4 \times 9692}{\pi \times 0.115 \times 2.0 \times 2.419} = 22,180 \Rightarrow \text{turbulent flow}$$

$$h_i = \frac{k}{D_i} \times 0.023\, Re^{0.8} Pr^{1/3}$$

$$= \frac{0.1}{0.115} \times 0.023(22,180)^{0.8}(25.158)^{1/3}$$

$$h_i = 176 \text{ Btu/h·ft}^2\text{·°F}$$

(c) Calculate h_o assuming $\phi_o = 1.0$.

$$Re = \frac{D_e(\dot{m}/A_f)}{\mu} = \frac{0.0339(5000/0.00827)}{0.55 \times 2.419} = 15,405 \Rightarrow \text{turbulent flow}$$

$$h_o = \frac{k}{D_e} \times 0.023 Re^{0.8} Pr^{1/3}$$

$$= \frac{0.092}{0.0339} \times 0.023(15,405)^{0.8}(6.074)^{1/3}$$

$$h_o = 255 \text{ Btu/h·ft}^2\text{·°F}$$

(d) Calculate the pipe wall temperature.

$$T_w = \frac{176 \times 125 + 255(1.66/1.38) \times 90}{176 + 255\ (1.66/1.38)} \cong 103°\text{F}$$

(e) Calculate ϕ_i and ϕ_o, and corrected values of h_i and h_o.
At 103°F, $\mu_A = 2.6$ cp and $\mu_B = 0.5$ cp (Figure A.1)

$$\phi_i = (2.0/2.6)^{0.14} = 0.9639$$

$$\phi_o = (0.55/0.5)^{0.14} = 1.0134$$

$$h_i = 176 \times 0.9639 = 170\ \text{Btu/h}\cdot\text{ft}^2\cdot°\text{F}$$

$$h_o = 225 \times 1.0134 = 258\ \text{Btu/h}\cdot\text{ft}^2\cdot°\text{F}$$

(f) Calculate U_D.

$$U_D = \left[\frac{D_o}{h_i D_i} + \frac{D_o \ln(D_o/D_i)}{2k} + \frac{1}{h_o} + \frac{R_{Di} D_o}{D_i} + R_{Do}\right]^{-1}$$

$$= \left[\frac{1.66}{170 \times 1.38} + \frac{(1.66/12)\ln(1.66/1.38)}{2 \times 9.4} + \frac{1}{258} + \frac{0.001 \times 1.66}{1.38} + 0.001\right]^{-1}$$

$$U_D = 69\ \text{Btu/h}\cdot\text{ft}^2\cdot°\text{F}$$

(g) Calculate the required surface area and number of hairpins.

$$q = U_D A F \Delta T_{\ln}$$

$$A = \frac{q}{U_D F \Delta T_{\ln}} = \frac{252,000}{69 \times 0.836 \times 34.76} = 125.7\ \text{ft}^2$$

$$L = \frac{125.7}{0.435} = 289\ \text{ft}$$

$$\text{Number of hairpins} = \frac{289}{32} = 9.0 \Rightarrow 9$$

Thus, nine hairpins are required. However, the equation for the LMTD correction factor is based on the assumption that both parallel branches are identical. Therefore, use two banks of five hairpins, for a total of ten hairpins.

(h) Calculate the pressure drop for the aniline stream (inner pipe).

$$f = 0.3673\ Re^{-0.2314} = 0.3673(22,180)^{-0.2314} = 0.03625$$

$$G = \dot{m}/A_f = 9692/0.0104 = 931,923\ \text{lbm/h}\cdot\text{ft}^2$$

$$\Delta P_f = \frac{0.03625\ (10 \times 32)(931,923)^2}{7.50 \times 10^{12}(1.38/12) \times 1.022 \times 0.9639} = 11.9\ \text{psi}$$

$$\Delta P_r = 1.6 \times 10^{-13}(2N_{HP} - 1)G^2/s$$
$$= 1.6 \times 10^{-13}(2 \times 10 - 1)(931,923)^2/1.022$$
$$\Delta P_r = 2.6\ \text{psi}$$

$$\Delta P_i = \Delta P_f + \Delta P_r = 11.9 + 2.6 = 14.5\ \text{psi}$$

(i) Calculate the pressure drop for the benzene stream (annulus).

$$f = 0.3673\ Re^{-0.2314} = 0.3673(15,405)^{-0.2314} = 0.03945$$

$$G = \dot{m}/A_f = 5000/0.00827 = 604,595\ \text{lbm/h}\cdot\text{ft}^2$$

$$\Delta P_f = \frac{f\,LG^2}{7.50 \times 10^{12} D_e s \phi}$$

$$= \frac{0.03945(5 \times 32)(604,595)^2}{7.50 \times 10^{12} \times 0.0339 \times 0.879 \times 1.0134}$$

$$\Delta P_f = 10.2 \text{ psi}$$

$$\Delta P_r = 1.6 \times 10^{-13}(2N_{HP} - 1)G^2/s$$
$$= 1.6 \times 10^{-13}(2 \times 5 - 1)(604,595)^2/0.879$$
$$\Delta P_r = 0.6 \text{ psi}$$

The nozzles are made from 1-in. schedule 40 pipe having a flow area of 0.006 ft^2 (Table B.2). Thus,

$$G_n = \dot{m}/A_f = 5000/0.006 = 833,333 \text{ lbm/h·ft}^2$$

For internal return bends, Equation (4.14) gives:

$$\Delta P_n = 2.0 \times 10^{-13} N_{HP} G_n^2/s = 2.0 \times 10^{-13} \times 5(833,333)^2/0.879$$

$$\Delta P_n = 0.79 \text{ psi}$$

The total pressure drop for the benzene is:

$$\Delta P_o = \Delta P_f + \Delta P_r + \Delta P_n$$
$$= 10.2 + 0.6 + 0.79$$
$$\Delta P_o \cong 11.6 \text{ psi}$$

(j) Calculate the over-surface and over-design.

$$U_C = \left[\frac{1}{U_D} - R_{D,tot}\right]^{-1} = \left[\frac{1}{69} - 0.001\ (1 + 1.66/1.38)\right]^{-1} = 81.4 \text{ Btu/h·ft}^2\text{·°F}$$

$$A_C = \frac{q}{U_C F \Delta T_{ln}} = \frac{252,000}{81.4 \times 0.836 \times 34.76} \cong 107 \text{ ft}^2$$

$$A = \pi D_o\ L = 0.435 \times (10 \times 32) = 139 \text{ ft}^2$$

$$\text{over-surface} = (A - A_C)/A_C = (139 - 107)/107 \cong 30\%$$

The required surface area is 125.7 ft^2 from Step (g). Therefore, the over-design is:

$$\text{over-design} = (139 - 125.7)/125.7 = 10.6\%$$

All design criteria are satisfied and the over-surface and over-design are reasonable; therefore, the exchanger is acceptable. The final design consists of ten hairpins with inner pipes connected in series and annuli connected in two parallel banks of five hairpins each. Aniline flows in the inner pipe and benzene flows in the annulus.

The following example illustrates the computational procedure for multi-tube hairpin exchangers.

Example 4.2

A number of stainless steel (type 316, $k = 9.4$ Btu/h · ft · °F) multi-tube hairpins are available for the benzene-aniline service of Example 4.1. The hairpins are 20 ft long with 3.5-in. schedule 40 outer pipes, internal return bends and 2-in. schedule 40 nozzles. Each hairpin contains a bundle of eight U-tubes, each having OD of 0.75 in. and ID of 0.584 in. Determine the number and configuration of hairpins required for the service.

Solution

(a) Data

Fluid properties are obtained from Example 4.1, along with the following information:

Benzene flow rate: 10,000 lb/h	Duty: 252,000 Btu/h
Aniline flow rate: 9692 lb/h	LMTD: 34.76°F (counter flow is assumed)

The ID of the outer pipe is obtained from Table B.2: $D_2 = 3.548$ in.

(b) Calculate flow areas.

The flow area in the tubes is calculated as follows:

$$(A_f)_{tubes} = n_t(\pi/4)D_i^2 = 8(\pi/4)(0.584/12)^2 = 0.01488 \text{ ft}^2$$

The flow area in the annulus is given by Equation (4.20):

$$(A_f)_{annulus} = (\pi/4)(D_2^2 - n_t D_1^2) = (\pi/4)\left[\left(\frac{3.548}{12}\right)^2 - 8\left(\frac{0.75}{12}\right)^2\right] = 0.04411 \text{ ft}^2$$

(c) Calculate equivalent diameter.

The equivalent diameter for the annulus is computed using Equation (4.22):

$$D_e = (D_2^2 - n_t D_1^2)/(D_2 + n_t D_1) = \left[(3.548)^2 - 8(0.75)^2\right]/(3.548 + 8 \times 0.75) = 0.84712 \text{ in.}$$

$$D_e \cong 0.070593 \text{ ft}$$

(d) Fluid placement

Reynolds numbers are calculated next to aid in determining the optimal fluid placement. Hairpins are assumed to be connected in series on both sides, and aniline is first assumed to flow in the tubes.

$$Re_i = \frac{D_i\left[\dot{m}_i/(A_f)_{tubes}\right]}{\mu} = \frac{(0.584/12)(9692/0.01488)}{2.0 \times 2.419} = 6552$$

$$Re_o = \frac{D_e\left[\dot{m}_o/(A_f)_{annulus}\right]}{\mu} = \frac{0.070593\,(10,000/0.04411)}{0.55 \times 2.419} = 12,029$$

Switching the fluids gives $Re_i = 24{,}583$ and $Re_o = 3206$, which is a less satisfactory result. Therefore, aniline is placed in the tubes and benzene in the annulus.

(e) Calculate h_i assuming $\varphi_i = 1.0$.

Since the flow is in the transition region, the Hausen correlation is used here.

$$h_i = (k/D_i) \times 0.116\,(Re^{2/3} - 125)Pr^{1/3}$$

$$= \frac{0.1}{(0.584/12)} \times 0.116\left[(6552)^{2/3} - 125\right](25.158)^{1/3}$$

$$h_i = 157.2 \text{ Btu/h}\cdot\text{ft}^2\cdot{}^\circ\text{F}$$

(f) Calculate h_o assuming $\varphi_o = 1.0$.

$$h_o = (k/D_e) \times 0.023 Re^{0.8} Pr^{1/3}$$

$$= (0.092/0.070593) \times 0.023\,(12,029)^{0.8}(6.074)^{1/3}$$

$$h_o = 100.5 \text{ Btu/h}\cdot\text{ft}^2\cdot{}^\circ\text{F}$$

(g) Calculate the tube wall temperature.

$$T_w = \frac{h_i t_{ave} + h_o(D_o/D_i)T_{ave}}{h_i + h_o(D_o/D_i)} = \frac{157.2 \times 125 + 100.5\,(0.75/0.584) \times 90}{157.2 + 100.5\,(0.75/0.584)} = 109.2^\circ\text{F}$$

(h) Calculate φ_i and φ_o, and corrected values of h_i and h_o.

From Figure A.1, at 109.2°F, $\mu_B \cong 0.47$ cp and $\mu_A \cong 2.4$ cp. Therefore,

$$\varphi_i = (2.0/2.4)^{0.14} = 0.9748$$

$$\varphi_o = (0.55/0.47)^{0.14} = 1.0222$$

$$h_i = 157.2 \times 0.9748 = 153.2 \cong 153 \text{ Btu/h}\cdot\text{ft}^2\cdot{}^\circ\text{F}$$

$$h_o = 100.5 \times 1.0222 = 102.7 \cong 103 \text{ Btu/h}\cdot\text{ft}^2\cdot{}^\circ\text{F}$$

(i) Compute the overall heat-transfer coefficient.

$$U_D = \left[\frac{D_o}{h_i D_i} + \frac{D_o \ln(D_o/D_i)}{2\ k_{tube}} + \frac{1}{h_o} + \frac{R_{Di} D_o}{D_i} + R_{Do}\right]^{-1}$$

$$= \left[\frac{0.75}{153 \times 0.584} + \frac{(0.75/12)\ln(0.75/0.584)}{2 \times 9.4} + \frac{1}{103} + \frac{0.001 \times 0.75}{0.584} + 0.001\right]^{-1}$$

$$U_D = 47.1\ \text{Btu/h}\cdot\text{ft}^2\cdot{}^\circ\text{F}$$

(j) Calculate required surface area and number of hairpins.

$$A_{req} = \frac{q}{U_D \Delta T_{ln}} = \frac{252,000}{47.1 \times 34.76} = 153.9\ \text{ft}^2$$

$$\textit{Area per hairpin} = n_t \pi D_o L = 8\pi \times (0.75/12) \times 40 = 62.8\ \text{ft}^2$$

$$N_{HP} = \frac{153.9}{62.8} = 2.45 \Rightarrow 3$$

Thus, three hairpins are required.

(k) Calculate the tube-side pressure drop.

$$f = 0.3673 Re^{-0.2314} = 0.3673\ (6552)^{-0.2314} = 0.04807$$

$$G = \dot{m}/A_f = 9692/0.01488 = 651,344\ \text{lbm/h}\cdot\text{ft}^2$$

$$\Delta P_f = \frac{f\,L\,G^2}{7.50 \times 10^{12} D_i s \phi} = \frac{0.04807 \times (3 \times 40)(651,344)^2}{7.50 \times 10^{12} \times (0.584/12) \times 1.022 \times 0.9748} = 6.730\ \text{psi}$$

$$\Delta P_r = 1.6 \times 10^{-13}(2N_{HP} - 1)\ G^2/s = 1.6 \times 10^{-13}(2 \times 3 - 1)(651,344)^2/1.022$$

$$\Delta P_r = 0.332\ \text{psi}$$

For the multi-tube unit, one velocity head per hairpin is used to estimate tube-side entrance and exit losses. Multiplying Equation (4.11) by N_{HP} gives:

$$\Delta P_n = 1.334 \times 10^{-13} N_{HP}\ G^2/s = 1.334 \times 10^{-13} \times 3 \times (651,344)^2/1.022$$

$$\Delta P_n \cong 0.166\ \text{psi}$$

$$\Delta P_i = \Delta P_f + \Delta P_r + \Delta P_n = 6.730 + 0.332 + 0.166 \cong 7.2\ \text{psi}$$

(l) Calculate the annulus-side pressure drop.

$$f = 0.3673 Re^{-0.2314} = 0.3673\ (12,029)^{-0.2314} = 0.04177$$

$$G = \dot{m}/A_f = 10,000/0.04411 = 226,706\ \text{lbm/h}\cdot\text{ft}^2$$

$$\Delta P_f = \frac{f\,L\,G^2}{7.50 \times 10^{12} D_e s \phi} = \frac{0.04177 \times (3 \times 40)(226,706)^2}{7.50 \times 10^{12} \times 0.070593 \times 0.879 \times 1.0222} = 0.542\ \text{psi}$$

$$\Delta P_r = 1.6 \times 10^{-13}(2N_{HP} - 1)G^2/s = 1.6 \times 10^{-13} \times (2 \times 3 - 1)(226,706)^2/0.879$$

$$\Delta P_r = 0.047\ \text{psi}$$

For 2-in. schedule 40 nozzles, the flow area is 0.0233 ft^2 from Table B.2. Hence,

$$G_n = \frac{10,000}{0.0233} = 429,185\ \text{lbm/h}\cdot\text{ft}^2$$

$$\Delta P_n = 2.0 \times 10^{-13} N_{HP} G_n^2/s = 2.0 \times 10^{-13} \times 3 \times (429,185)^2/0.879 = 0.126\ \text{psi}$$

$$\Delta P_o = \Delta P_f + \Delta P_r + \Delta P_n = 0.542 + 0.047 + 0.126 \cong 0.71\ \text{psi}$$

(m) Calculate over-surface and over-design.

$$U_C = \left[\frac{1}{U_D} - R_{D,tot}\right]^{-1} = \left[\frac{1}{47.1} - 0.001(1 + 0.75/0.584)\right]^{-1} = 52.8 \text{ Btu/h·ft}^2\text{·°F}$$

$$A_C = \frac{q}{U_C \, \Delta T_{ln}} = \frac{252,000}{52.8 \times 34.76} = 137.3 \text{ ft}^2$$

From Step (j), the heat-transfer area per hairpin is 62.8 ft^2. Hence, the total area in the exchanger is:

$$A = 3 \times 62.8 = 188.4 \text{ ft}^2$$

$$\text{over-surface} = (A - A_C)/A_C = (188.4 - 137.3)/137.3 \cong 37\%$$

The required surface area is 153.9 ft^2 from Step (j). Therefore,

$$\text{over-design} = \left(A - A_{req}\right)/A_{req} = (188.4 - 153.9)/153.9 \cong 22\%$$

All design criteria are satisfied and the over-surface and over-design are reasonable. Therefore, the exchanger is acceptable as configured. The tube-side flow is in the transition region, but as shown in Step (d), transition flow is unavoidable in this case. Furthermore, the relatively high over-design should be sufficient to compensate for any potential error in the value of h_i. The final design consists of three hairpins connected in series on both the tube side and annulus side. Aniline flows in the tubes and benzene flows in the annulus.

4.7 Finned-Pipe Exchangers

4.7.1 Finned-Pipe Characteristics

The inner pipe of a double-pipe exchanger can be equipped with rectangular fins as shown in Figure 4.3. Although fins can be attached to both the internal and external pipe surfaces, external fins are most frequently used. Pairs of fins are formed from U-shaped channels that are welded or soldered onto the pipe, depending on the materials involved. An alternative method of attaching the fins consists of cutting grooves in the pipe surface, inserting the fin material, and then peening the pipe metal back to secure the fins. The fin material need not be the same as the pipe material; for example, carbon steel fins can be attached to stainless steel pipes. Combinations of this type are used when a corrosion resistant alloy is needed for the inner fluid but not for the fluid in the annulus.

Dimensions of standard finned exchangers are given in Tables 4.1 and 4.2. The data are for units employing schedule 40 pipe. Data for exchangers intended for high-pressure service and employing schedule 80 pipe can be found in Ref. [6]. For units with a single inner pipe, the number of fins varies from 20 to 48, with fin heights, which are dictated by the clearance between the inner and outer pipes, from 0.375 to 1.0 in. The fin thickness is 0.035 in. (0.889 mm) for weldable metals. For soldered fins, the thickness is 0.0197 in. (0.5 mm) for heights of 0.5 in. or less, and 0.0315 in. (0.8 mm) for heights greater than 0.5 in. (7 mm). For multi-tube units, the number of fins per tube is either 16 or 20, and the fin height is generally less than in units with a single inner pipe. The fin thickness is the same in both single and multi-tube units.

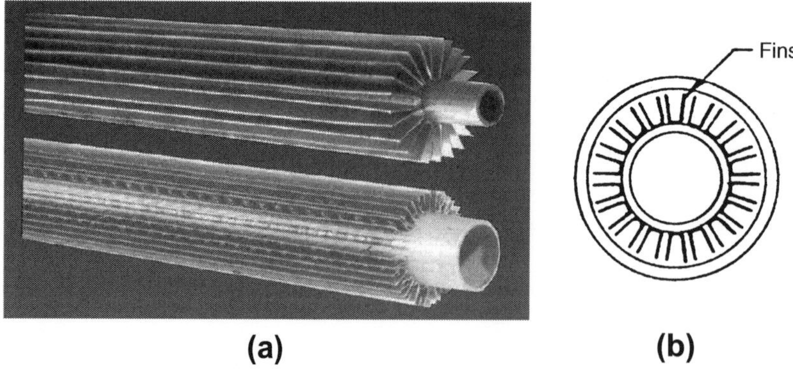

FIGURE 4.3 (a) Rectangular fins on heat-exchanger pipes, (b) Cross-section of a finned-pipe exchanger (Source: (a) Koch Heat Transfer Company, LP and (b) Ref. [5]).

TABLE 4.1 Standard Configurations for Exchangers with a Single Finned Inner Pipe; Standard Pressure (Schedule 40) Units

Nominal Diameter (in.)	Outer Pipe Thickness (mm)	Outer Pipe OD (mm)	Maximum Number of Fins	Inner Pipe OD (mm)	Inner Pipe Thickness (mm)	Fin Height (mm)
2	3.91	60.3	20	25.4	2.77	11.1
3	5.49	88.9	20	25.4	2.77	23.8
3	5.49	88.9	36	48.3	3.68	12.7
$3^1/_2$	5.74	101.6	36	48.3	3.68	19.05
$3^1/_2$	5.74	101.6	40	60.3	3.91	12.7
4	6.02	114.3	36	48.3	3.68	25.4
4	6.02	114.3	40	60.3	3.91	19.05
4	6.02	114.3	48	73.0	5.16	12.7

Source: Ref. [7].

TABLE 4.2 Standard Configurations for Multi-tube Exchangers; Standard Pressure (Schedule 40) Units

Nominal Diameter (in.)	Pipe Thickness (mm)	Pipe OD (mm)	Number of Tubes	Number of Fins	Tube OD (mm)	Tube Thickness (mm)	Fin Height (mm)
4	6.02	114.3	7	16	19.02	2.11	5.33
4	6.02	114.3	7	20	22.2	2.11	5.33
6	7.11	168.3	19	16	19.02	2.11	5.33
6	7.11	168.3	14	16	19.02	2.11	5.33
6	7.11	168.3	7	20	20.04	2.77	12.7
8	8.18	219.1	19	16	19.02	2.11	8.64
8	8.18	219.1	19	20	22.2	2.11	7.11
8	8.18	219.1	19	20	25.4	2.77	5.33
8	8.18	219.1	19	16	19.02	2.11	7.11
8	8.18	219.1	19	20	22.2	2.11	5.33

Source: Ref. [7].

4.7.2 Fin Efficiency

Heat-transfer fins were discussed in Chapter 2, where the fin height was referred to as the fin length, denoted by the symbol, L. Since L is used in the present chapter to denote pipe length, the symbol, b, will be used for fin height. Equation (2.23) for the efficiency of a rectangular fin becomes:

$$\eta_f = \frac{\tanh(mb_c)}{mb_c} \tag{4.24}$$

where

$b_c = b + \tau/2 =$ corrected fin height
$m = (2h_o/k\tau)^{1/2}$
$\tau =$ fin thickness
$k =$ fin thermal conductivity
$h_o =$ heat-transfer coefficient in annulus

The weighted efficiency of the entire finned surface is given by:

$$\eta_w = \frac{A_{prime} + \eta_f A_{fins}}{A_{Tot}} \tag{2.31}$$

where

$A_{fins} = 2n_t N_f b_c L$
$A_{prime} = (\pi D_o - N_f \tau) n_t L$
$A_{Tot} = A_{prime} + A_{fins}$
$n_t =$ number of finned pipes
$N_f =$ number of fins on each pipe
$D_o =$ pipe OD
$L =$ pipe length

4.7.3 Overall Heat-Transfer Coefficient

The overall coefficient is based on the total external surface area, A_{Tot}, of the inner pipe. The overall resistance to heat transfer is thus $1/UA_{Tot}$, and is the sum of the inner and outer convective resistances, the conductive resistance of the pipe wall, and the resistances of the fouling layers (if present). The outer convective resistance in a finned annulus is:

$$R_{th} = \frac{1}{h_o \eta_w A_{Tot}} \tag{2.32}$$

Therefore, viewing R_{Do} as a reciprocal heat-transfer coefficient, we have:

$$\frac{1}{U_D A_{Tot}} = \frac{1}{h_i A_i} + \frac{R_{Di}}{A_i} + \frac{\ln(D_o/D_i)}{2\pi k_{pipe} L} + \frac{1}{h_o \eta_w A_{Tot}} + \frac{R_{Do}}{\eta_w A_{Tot}} \tag{4.25}$$

$$U_D = \left[\frac{A_{Tot}}{h_i A_i} + \frac{R_{Di} A_{Tot}}{A_i} + \frac{A_{Tot}\ \ln(D_o/D_i)}{2\pi k_{pipe} L} + \frac{1}{h_o \eta_w} + \frac{R_{Do}}{\eta_w}\right]^{-1} \tag{4.26}$$

where $A_i = \pi D_i L$.

The equation for the clean overall coefficient is obtained by dropping the fouling terms:

$$U_C = \left[\frac{A_{Tot}}{h_i A_i} + \frac{A_{Tot}\ \ln(D_o/D_i)}{2\pi k_{pipe} L} + \frac{1}{h_o \eta_w}\right]^{-1} \tag{4.27}$$

4.7.4 Flow Area and Equivalent Diameter

Consider a finned annulus comprised of an outer pipe with ID of D_2 and n_t inner pipes, each with OD of D_1 and containing N_f rectangular fins of height b and thickness τ. The flow area and wetted perimeter are:

$$A_f = \frac{\pi}{4}\left(D_2^2 - n_t D_1^2\right) - n_t N_f b\tau \tag{4.28}$$

$$\text{wetted perimeter} = \pi(D_2 + n_t D_1) + 2n_t N_f b \tag{4.29}$$

In the case of welded or soldered fins, these equations neglect the thickness of the channels where they overlay the pipe surface.

The equivalent diameter is obtained in the usual way as four times the flow area divided by the wetted perimeter:

$$D_e = \frac{\pi\left(D_2^2 - n_t D_1^2\right) - 4n_t N_f b\tau}{\pi(D_2 + n_t D_1) + 2n_t N_f b} \tag{4.30}$$

4.8 Heat-Transfer Coefficients and Friction Factors for Finned Annuli

Heat-transfer coefficients and friction factors were determined experimentally in commercial double-pipe fin-tube exchangers by DeLorenzo and Anderson [5]. Their data were re-plotted by Kern and Kraus [3], whose graphs are reproduced in Figures 4.4 and 4.5. Note that the quantity plotted in Figure 4.5 is a Fanning friction factor that has been modified by dividing by the conversion factor, 144 in.^{2}/ft^2. Therefore, the value from Figure 4.5 must be multiplied by a factor of $4 \times 144 = 576$ to convert to a dimensionless Darcy friction factor for use in Equation (4.5). This factor of 576 is included in the curve fits to the graphs given below:

$$j_H = \left(0.0263 Re^{0.9145} + 4.9 \times 10^{-7} Re^{2.618}\right)^{1/3} \quad \text{(for 24 fins)} \tag{4.31}$$

$$j_H = \left(0.0116 Re^{1.032} + 4.9 \times 10^{-7} Re^{2.618}\right)^{1/3} \quad \text{(for 36 fins)} \tag{4.32}$$

$$f = \frac{64}{Re} \quad (Re \le 400) \tag{4.33}$$

$$f = 576 \exp\left[0.08172(\ln Re)^2 - 1.7434 \ln Re - 0.6806\right] \quad (Re > 400) \tag{4.34}$$

It should also be noted that the fins act to destabilize the laminar flow field, and as a result, the critical Reynolds number is approximately 400 in the finned annulus. Therefore, if the annulus is treated as an equivalent pipe with a critical Reynolds number of approximately 2100, both the heat-transfer coefficient and friction factor will be underestimated for $400 < Re < 2100$. It is also remarkable to note that the friction factor in the laminar region follows the Fanning equation for pipe flow ($f = 16/Re$) rather than annular flow ($f \cong 24/Re$).

The exchangers used by DeLorenzo and Anderson contained 24, 28, or 36 fins per tube, and at low Reynolds numbers the correlation for j_H depends on the fin number. The effect is most likely due to the fact that L/D_e varied from 574 for the exchangers

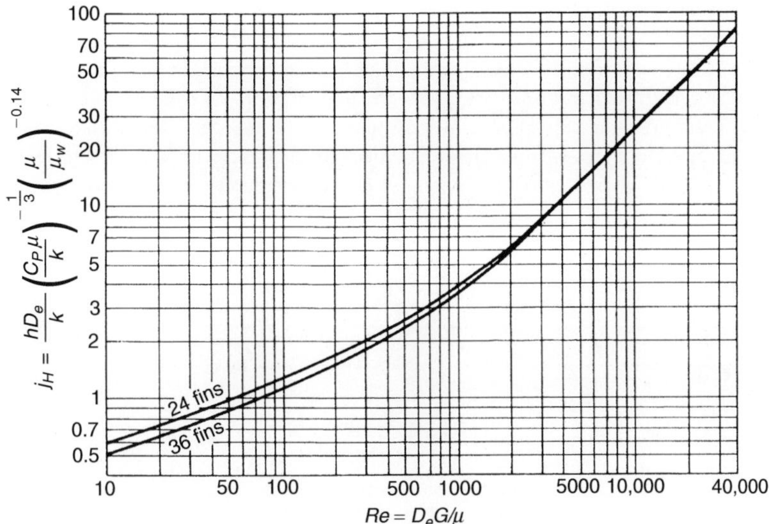

FIGURE 4.4 Heat-transfer coefficients for finned annuli (Source: Ref. [3]).

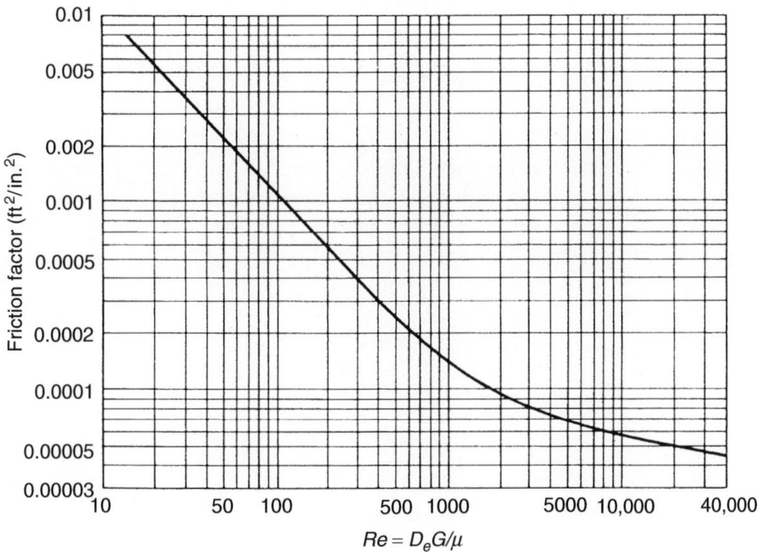

FIGURE 4.5 Friction factors for finned annuli (Source: Ref. [3]).

with 24 fins to 788 for the exchangers with 36 fins. For $Re < 1000$, the difference between the two curves in Figure 4.4 is well accounted for by a factor of $(L/D_e)^{-1/3}$. Therefore, it is suggested that Figure 4.4 (or Equations (4.31) and (4.32)) be used in the following way. Use the curve (or equation) for 24 or 36 fins, whichever has the L/D_e value closest to the exchanger being calculated. Then, for $Re \leq 1000$ scale the computed value *of* j_H by a factor of $(L/D_e)^{-1/3}$, i.e.,

$$j_H = (j_H)_{\text{Fig.4.4}}\left[\frac{(L/D_e)_{\text{Fig.4.4}}}{(L/D_e)}\right]^{1/3} \tag{4.35}$$

where

$(L/D_e)_{\text{Fig.4.4}} = 788$ (for 36 fins)
$= 574$ (for 24 fins)

For Reynolds numbers above 1000, the effect of fin number diminishes and the correction factor in Equation (4.35) can be omitted. Note that the length used in this calculation is the length of a hairpin, i.e., the length of pipe in one leg of one hairpin.

4.9 Wall Temperature for Finned Pipes

The heat-transfer correlation of DeLorenzo and Anderson contains a viscosity correction factor that is calculated using a weighted average temperature, T_{wtd}, of the extended and prime surfaces. The temperature, T_p, of the prime surface is used to calculate the viscosity correction factor for the fluid in the inner pipe. The derivation of the equations for the two pipe wall temperatures is similar to the derivation of the wall temperature presented in Chapter 3. All of the heat is assumed to be transferred between the streams at their average temperatures, t_{ave} for the fluid in the inner pipe and T_{ave} for the fluid in the annulus. An energy balance gives:

$$q = h_i A_i (t_{ave} - T_p) = h_o \eta_w A_{Tot} (T_p - T_{ave}) \tag{4.36}$$

The weighted average temperature, T_{wtd}, is defined by:

$$q = h_o A_{Tot} (T_{wtd} - T_{ave}) \tag{4.37}$$

Solving these equations for T_p and T_{wtd} yields:

$$T_p = \frac{h_i t_{ave} + h_o\, \eta_w (A_{Tot}/A_i) T_{ave}}{h_i + h_o\, \eta_w (A_{Tot}/A_i)} \tag{4.38}$$

$$T_{wtd} = \frac{h_i \eta_w t_{ave} + [h_i(1 - \eta_w) + h_o \eta_w (A_{Tot}/A_i)]\, T_{ave}}{h_i + h_o \eta_w (A_{Tot}/A_i)} \tag{4.39}$$

To reiterate, T_p is used to find ϕ_i and T_{wtd} is used for ϕ_o.

The calculation of a finned-pipe exchanger is illustrated by the following example that involves the design of an oil cooler [3]. Services, such as the one in this example, that transfer heat between an organic stream and water (usually cooling water), are good candidates for finned exchangers. The reason is that heat-transfer coefficients for water streams are often substantially higher than those for organics.

Example 4.3

Design a heat exchanger to cool 18,000 lb/h of a petroleum distillate oil from 250°F to 150°F using water with a temperature range of 85°F to 120°F. A maximum pressure drop of 20 psi for each stream is specified and a fouling factor of 0.002 h· ft^2 ·°F/Btu is required for each stream. The exchanger will use 25-ft long carbon steel hairpins with 3-in. schedule 40 outer pipes, 1.5-in. schedule 40 inner pipes, 2-in. schedule 40 nozzles, and internal return bends. Each inner pipe contains 24 carbon steel fins 0.5 in. high and 0.035 in. thick. Physical properties at the average stream temperatures are given in the following table. Note that the oil viscosity is calculated from the equation:

$$\mu_{oil}(\text{cp}) = 0.003024\ \exp\left[\frac{4495.5}{T(^\circ\text{R})}\right]$$

This result is obtained by fitting the relation:

$$\mu = \alpha\, e^{\beta/T}$$

using the two data points $\mu = 4.8$ cp at 150°F and $\mu = 1.7$ cp at 250°F.

Fluid property	Oil at 200°F	Water at 102.5°F
C_P (Btu/lb · °F)	0.52	1.0
k (Btu/h · ft · °F)	0.074	0.37
μ (cp)	2.75	0.72
Specific gravity	0.82	0.99
Pr	46.75	4.707

Solution

For the first trial, assume that the hairpins will be connected in series on both sides. Since the oil stream is expected to have the lower heat-transfer coefficient, it must flow in the annulus where the fins are located. Counter-flow is assumed.

(a) Energy balances.

$$q = (\dot{m}\, C_P \Delta T)_{oil} = 18{,}000 \times 0.52 \times 100 = 936{,}000\ \text{Btu/h}$$

$$936{,}000 = (\dot{m}\, C_P \Delta T)_{water} = \dot{m}_{water} \times 1.0 \times 35$$

$$\dot{m}_{water} = 26{,}743 \text{ lb/h}$$

(b) LMTD.

$$\Delta T_{\ln} = \frac{130 - 65}{\ln(130/65)} = 93.8°\text{F}$$

(c) Calculate h_i assuming $\phi_i = 1.0$.
For 1.5-in. schedule 40 pipe,

$$D_i = 1.61 \text{ in} = 0.1342 \text{ ft} \quad \text{(Table B.2)}$$

$$Re = \frac{4\dot{m}}{\pi D_i \mu} = \frac{4 \times 26{,}743}{\pi \times 0.1342 \times 0.72 \times 2.419} = 145{,}680 \Rightarrow \text{turbulent flow}$$

$$h_i = (k/D_i) \times 0.023 \; Re^{0.8} Pr^{1/3}$$

$$= (0.37/0.1342) \times 0.023(145{,}680)^{0.8}(4.707)^{1/3}$$

$$h_i = 1436 \text{ Btu/h·ft}^2 \cdot °\text{F}$$

(d) Calculate h_o assuming $\phi_o = 1.0$.
From Table B.2,

$$D_2 = 3.068 \text{ in. (ID of 3-in. schedule 40 pipe)}$$
$$D_1 = 1.9 \text{ in. (OD of 1.5-in. schedule 40 pipe)}$$

For the finned annulus, the flow area and wetted perimeter are calculated from Equations (4.28) and (4.29), respectively, with $n_t = 1$:

$$A_f = (\pi/4)(D_2^2 - D_1^2) - n_t N_f b\tau$$
$$= (\pi/4)\left[(3.068)^2 - (1.9)^2\right] - 1 \times 24 \times 0.5 \times 0.035$$
$$A_f = 4.137 \text{ in}^2 = 0.0287 \text{ ft}^2$$

$$\text{wetted perimeter} = \pi(D_2 + n_t D_1) + 2n_t N_f b$$
$$= \pi(3.068 + 1 \times 1.9) + 2 \times 1 \times 24 \times 0.5$$
$$\text{wetted perimeter} = 39.6074 \text{ in.} = 3.3006 \text{ ft}$$

$$D_e = 4 \times A_f/\text{wetted perimeter}$$

$$D_e = 4 \times 0.0287/3.3006 = 0.03478 \text{ ft}$$

$$G = \dot{m}/A_f = 18{,}000/0.0287 = 627{,}178 \text{ lbm/h·ft}^2$$

$$Re = D_e G/\mu = 0.03478 \times 627{,}178/(2.75 \times 2.419) = 3279$$

From Figure 4.4, $j_H \cong 9.2$. (Equations (4.31) and (4.32) both give $j_H = 9.4$.) Therefore,

$$h_o = j_H(k/D_e)Pr^{1/3} = 9.2(0.074/0.03478)(46.75)^{1/3}$$

$$h_o = 70.5 \text{ Btu/h·ft}^2 \cdot °\text{F}$$

(e) Fin efficiency.
The fin efficiency is calculated from Equation (4.24) with $k = 26$ Btu/h·ft·°F for carbon steel fins:

$$m = \left(\frac{2h_o}{k\tau}\right)^{1/2} = \left(\frac{2 \times 70.5}{26 \times 0.035/12}\right)^{1/2} = 43.12 \text{ ft}$$

$$b_c = b + \tau/2 = (0.5 + 0.035/2)/12 = 0.04313 \text{ ft}$$

$$mb_c = 43.12 \times 0.04313 = 1.8598$$

$$\eta_f = \frac{\tanh(mb_c)}{mb_c} = \frac{\tanh(1.8598)}{1.8598} = 0.5122$$

The weighted efficiency of the finned surface is given by Equation (2.31):

$$\eta_w = \frac{A_{prime} + \eta_f A_{fins}}{A_{Tot}}$$

$$A_{fins} = 2n_t N_f b_c L = 2 \times 1 \times 24 \times 0.04313 \times L = 2.07L \ \text{ft}^2$$

$$A_{prime} = (\pi D_o - N_f \tau) n_t L$$

$$= \left(\frac{\pi \times 1.9 - 24 \times 0.035}{12}\right) \times 1 \times L = 0.43L \ \text{ft}^2$$

$$A_{Tot} = A_{fins} + A_{prime} = 2.50L \ \text{ft}^2$$

The total length of pipe in the heat exchanger is unknown, but since L cancels in Equation (2.31), the areas per unit length can be used. Thus,

$$\eta_w = \frac{0.43 + 0.5122 \times 2.07}{2.50} \cong 0.60$$

(f) Wall temperatures.
The wall temperatures used to obtain viscosity correction factors are given by Equations (4.38) and (4.39). The inside surface area of the inner pipe is:

$$A_i = \pi D_i L = \pi \times 0.1342 \times L = 0.4216L \ \text{ft}^2$$

Hence,

$$A_{Tot}/A_i = 2.50/0.4216 = 5.93$$

$$T_p = \frac{h_i t_{ave} + h_o \eta_w (A_{Tot}/A_i) T_{ave}}{h_i + h_o \eta_w (A_{Tot}/A_i)}$$

$$= \frac{1436 \times 102.5 + 70.5 \times 0.6 \times 5.93 \times 200}{1436 + 70.5 \times 0.6 \times 5.93}$$

$$T_p = 117°\text{F}$$

$$T_{wtd} = \frac{h_i \eta_w t_{ave} + [h_i(1 - \eta_w) + h_o \eta_w (A_{Tot}/A_i)] T_{ave}}{h_i + h_o \eta_w (A_{Tot}/A_i)}$$

$$= \frac{1436 \times 0.6 \times 102.5 + [1436 \times 0.4 + 70.5 \times 0.6 \times 5.93] \times 200}{1436 + 70.5 \times 0.6 \times 5.93}$$

$$T_{wtd} = 150°\text{F}$$

(g) Viscosity correction factors and corrected heat-transfer coefficients.
From Figure A.1, the viscosity of water at 117°F is approximately 0.62 cp. Hence,

$$\phi_i = (0.72/0.62)^{0.14} = 1.021$$

The oil viscosity at 150°F = 610°R is:

$$\mu_{oil} = 0.003024 \exp(4495.5/610) = 4.8 \text{ cp}$$

Therefore,

$$\phi_o = (2.75/4.8)^{0.14} = 0.925$$

The corrected heat-transfer coefficients are:

$$h_i = 1436 \times 1.021 = 1466 \text{ Btu/h}\cdot\text{ft}^2\cdot°\text{F}$$

$$h_o = 70.5 \times 0.925 = 65 \text{ Btu/h}\cdot\text{ft}^2\cdot°\text{F}$$

Steps (e) to (g) could be repeated using the new values of h_i and h_o, but iteration is usually unnecessary. In the present case, recalculation yields $\eta_w = 0.61$, $T_p = 116°F$, and $T_{wtd} = 149°F$, which leaves h_i and h_o essentially unchanged.

(h) Overall coefficient.

The overall coefficient for design is calculated from Equation (4.26):

$$U_D = \left[\frac{A_{Tot}}{h_i A_i} + \frac{R_{Di} A_{Tot}}{A_i} + \frac{A_{Tot} \ln(D_o/D_i)}{2\pi\, k_{pipe} L} + \frac{1}{h_o \eta_w} + \frac{R_{Do}}{\eta_w}\right]^{-1}$$

$$= \left[\frac{5.93}{1466} + 0.002 \times 5.93 + \frac{2.50 \times L \ \ln(1.9/1.61)}{2 \times \pi \times 26 \times L} + \frac{1}{65 \times 0.6} + \frac{0.002}{0.6}\right]^{-1}$$

$$U_D = 21.1 \ \text{Btu/h}\cdot\text{ft}^2\cdot°\text{F}$$

(i) Required heat-transfer surface and number of hairpins.

$$A = \frac{q}{U_D \Delta T_{ln}} = \frac{936{,}000}{21.1 \times 93.8} = 473 \ \text{ft}^2$$

$$\text{Area per hairpin} = 2.50L = 2.50(25 \times 2) = 125 \ \text{ft}^2$$

$$\text{Number of hairpins} = 473/125 = 3.78 \Rightarrow 4$$

(j) Pressure drop for water (inner pipe).

$$A_f = 0.01414 \ \text{ft}^2 \text{ for 1.5-in. schedule 40 pipe (Table B.2)}$$

$$G = \dot{m}/A_f = 26{,}743/0.01414 = 1{,}891{,}301 \ \text{lbm/h}\cdot\text{ft}^2$$

$$Re = 145{,}680 \text{ from Step(c)}$$

$$f = 0.3673 Re^{-0.2314} = 0.3673(145{,}680)^{-0.2314}$$

$$f = 0.02345$$

$$\Delta P_f = \frac{f G^2 L}{7.50 \times 10^{12} D_i s \phi} = \frac{0.02345 \ (1{,}891{,}301)^2 \times 4 \times 50}{7.50 \times 10^{12} \times 0.1342 \times 0.99 \times 1.021}$$

$$\Delta P_f = 16.50 \ \text{psi}$$

$$\Delta P_r = 1.6 \times 10^{-13}(2N_{HP} - 1)G^2/s$$
$$= 1.6 \times 10^{-13}(2 \times 4 - 1)(1{,}891{,}301)^2/0.99$$
$$\Delta P_r = 4.05 \ \text{psi}$$

$$\Delta P_i = \Delta P_f + \Delta P_r = 16.50 + 4.05 \cong 20.6 \ \text{psi}$$

(k) Pressure drop for oil (annulus). From Step (d) we have:

$$Re = 3279$$

$$G = 627{,}178 \ \text{lbm/h}\cdot\text{ft}^2$$

$$D_e = 0.03478 \ \text{ft}$$

The friction factor is calculated using Equation (4.34):

$$f = 576 \exp\left[0.08172(\ln Re)^2 - 1.7434 \ln Re - 0.6806\right]$$
$$= 576 \exp\left[0.08172(\ln 3279)^2 - 1.7434 \ln(3279) - 0.6806\right] = 0.04585$$

$$\Delta P_f = \frac{f\, G^2 L}{7.50 \times 10^{12} D_e s \phi} = \frac{0.04585(627{,}178)^2 \times 4 \times 50}{7.50 \times 10^{12} \times 0.03478 \times 0.82 \times 0.925}$$

$$\Delta P_f = 18.24 \ \text{psi}$$

$$\Delta P_r = 1.6 \times 10^{-13}(2N_{HP} - 1)\,G^2/s$$
$$= 1.6 \times 10^{-13}(2 \times 4 - 1)(627,178)^2/0.82$$
$$\Delta P_r = 0.54 \text{ psi}$$

For 2-in. schedule 40 nozzles having a cross-sectional area of 0.0233 ft^2,

$$G_n = 18,000/0.0233 = 772,532 \text{ lbm/h·ft}^2$$

At the outlet nozzle, where the viscosity is highest, $\mu = 4.8$ cp. Hence,

$$Re_n = DG_n/\mu = (2.067/12) \times 772,532/(4.8 \times 2.419) = 11,460$$

Thus, the flow will be turbulent in both nozzles and Equation (4.14) is applicable.

$$\Delta P_n = 2.0 \times 10^{-13} N_{HP} G_n^2/s = 2.0 \times 10^{-13} \times 4(772,532)^2/0.82$$

$$\Delta P_n = 0.58 \text{ psi}$$

Hence, the total pressure drop for the annulus is:

$$\Delta P_o = 18.24 + 0.54 + 0.58 = 19.4 \text{ psi}$$

All design criteria are met, with the exception of $\Delta P_i = 20.6$ psi, which is slightly above the specified maximum pressure drop of 20 psi. This discrepancy is not large enough to be a concern in most situations. If necessary, however, it could be eliminated by using hairpins of length 24 ft rather than 25 ft.

(l) Over-surface and over-design.

$$U_C = \left[\frac{A_{Tot}}{h_0 A_i} + \frac{A_{Tot}\,\ln(D_o/D_i)}{2\pi\, k_{pipe} L} + \frac{1}{h_o \eta_w}\right]^{-1}$$

$$= \left[\frac{5.93}{1466} + \frac{2.50\,\ln(1.9/1.61)}{2 \times \pi \times 26} + \frac{1}{65 \times 0.6}\right]^{-1}$$

$$U_C = 31.0 \text{ Btu/h·ft}^2\text{·°F}$$

$$A_C = \frac{q}{U_C \Delta T_{\ln}} = \frac{936,000}{31.0 \times 93.8} = 322 \text{ ft}^2$$

$$A = 2.50\,L = 2.50 \times 200 = 500 \text{ ft}^2$$

$$\text{over-surface} = (A - A_C)/A_C = (500 - 322)/322 \cong 55\%$$

The required surface area is 473 ft^2 from Step (i). Therefore, the over-design is:

$$\text{over-design} = (500 - 473)/473 = 5.7\%$$

The over-surface is relatively high and suggests that the exchanger may be over-sized. The reason lies in the large value of the total fouling allowance:

$$R_D = R_{Di} A_{Tot}/A_i + R_{Do}/n_w = 0.002 \times 5.93 + 0.002/0.6$$

$$R_D = 0.01519 \text{ h·ft}^2\text{·°F/Btu}$$

The high over-surface is simply a reflection of the large effect that internal fouling has on the design overall heat-transfer coefficient. The effect is much greater in finned exchangers because the ratio of external to internal surface area is so high (5.93 in this case versus 1.18 for an un-finned 1.5-in. schedule 40 pipe). With the fouling factors specified in this example, the two fouling resistances account for 30% of the total thermal resistance. Clearly, in the design of finned exchangers, care must be exercised to ensure that an appropriate value is selected for the inner fouling factor.

To summarize, for the problem as given, the final design consists of four hairpins connected in series on both sides with water in the inner pipe and oil in the annulus.

In the preceding example it will be noticed that the fin efficiency was calculated based on a clean surface. The effect of fouling on η_f can be accounted for by using an effective heat-transfer coefficient, h'_o [3, 6–8]:

$$h'_o = (1/h_o + R_{Do})^{-1} \tag{4.40}$$

Since $h'_o < h_o$, the effect of fouling is to increase the fin efficiency. Therefore, neglecting the effect of fouling on η_f gives a somewhat conservative estimate of U_D. Furthermore, using h'_o requires a separate calculation using h_o in order to obtain U_C. In addition, if h'_o is used in calculating U_D, consistency then dictates that the temperatures used to obtain the viscosity correction factors should be those at the exterior surfaces of the fouling layers. Finding those temperatures involves a rather lengthy iterative calculation [3,7]. Thus, for hand calculations, it is justifiable to neglect these complications and use the fin efficiency based on clean conditions.

4.10 Computer Software

Essentially all heat-transfer equipment is designed using commercial and/or in-house computer software packages. In this section we consider one such package, HEXTRAN by SimSci-Esscor, a division of Invensys Operations Management. This program is available as part of the Process Engineering Suite that includes the PRO/II chemical process simulator. Unique among software packages devoted exclusively to heat-transfer operations, HEXTRAN is a flowsheet simulator that employs the same extensive physical property data banks and thermodynamic routines developed for PRO/II.

The double-pipe heat-exchanger module (DPE) in HEXTRAN handles exchangers with a single inner pipe, either plain or finned. A separate module (MTE) is used for multi-tube hairpin exchangers. Both of these modules operate only in rating mode. Therefore, design must be done wholly by a trial-and-error procedure. Other HEXTRAN modules can operate in either design or rating mode. The following examples explore some of the attributes of the double-pipe and multi-tube exchanger modules in HEXTRAN, version 9.2.

Example 4.4

Use HEXTRAN to rate the initial configuration (six hairpins in series) for the benzene–aniline exchanger of Example 4.1, and compare the results with those obtained previously by hand.

Solution

After program startup, login is achieved by entering *simsci* for both the User Name and Password. A new problem is then opened by entering flowsheet and database names. The flowsheet (shown below) is easily constructed using the mouse to drag and drop icons from the palette at the right of the screen to the drawing area. The items required for this problem are one double-pipe heat exchanger and four process streams (two feeds and two products). The streams are connected to the inlet and outlet ports of the heat exchanger by clicking and scrolling with the mouse. Right-clicking on any object (a stream or unit) on the flowsheet brings up the edit menu with options that include deleting or renaming the object, changing its configuration, and editing the data for the object. After renaming the streams, the flowsheet for this problem appears as shown on the next page.

In order to facilitate comparison with the hand calculations, the physical property values from Example 4.1 are used. Since the viscosity correction factors are close to unity in this problem, the temperature variation of viscosity is neglected. The two feed streams are first defined as bulk property streams by right-clicking on each stream in turn and selecting *Change Configuration* from the pop-up edit menu. There are five types of streams in HEXTRAN:

- Compositional
- Assay
- Bulk Property
- Water/Steam
- Utility

A compositional stream (the default type) is one having a defined composition. For this type of stream, methods must be chosen for generating thermodynamic and transport properties. An assay stream is a petroleum stream for which a complete assay is available. A bulk property stream is either one for which the average properties are known or a hydrocarbon stream for which the properties can be estimated from known values of specific gravity and Watson characterization factor (see Appendix E). A water/steam stream consists of pure water in the liquid and/or vapor state, for which the thermodynamic properties are obtained from steam tables. Utility streams in HEXTRAN are used only in pinch calculations (see Chapter 8). For this problem, the *Bulk Property* option is selected.

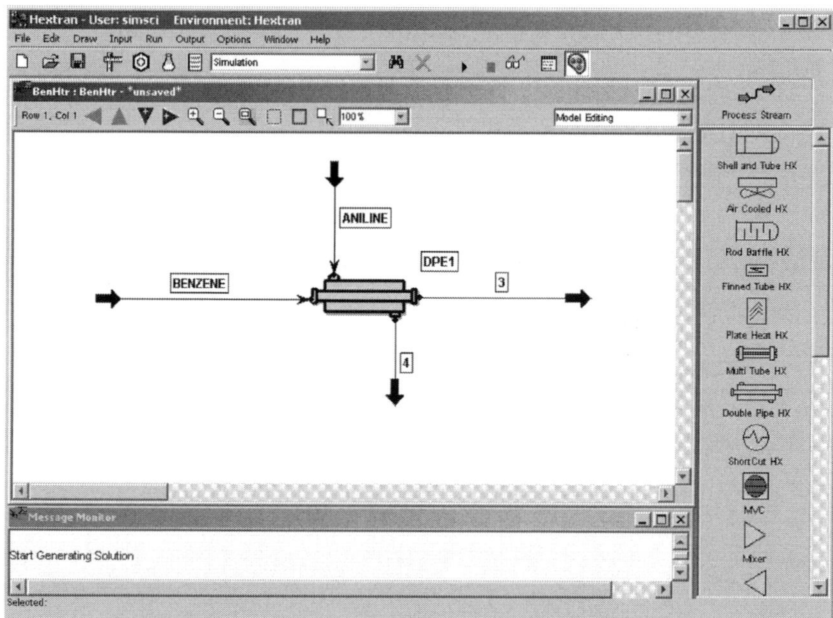

The properties of the two feed streams are entered by selecting *Edit Properties* from the pop-up edit menu or by double-clicking on the stream. The flow rate, feed temperature, and feed pressure are entered on the Specifications form as shown below for benzene:

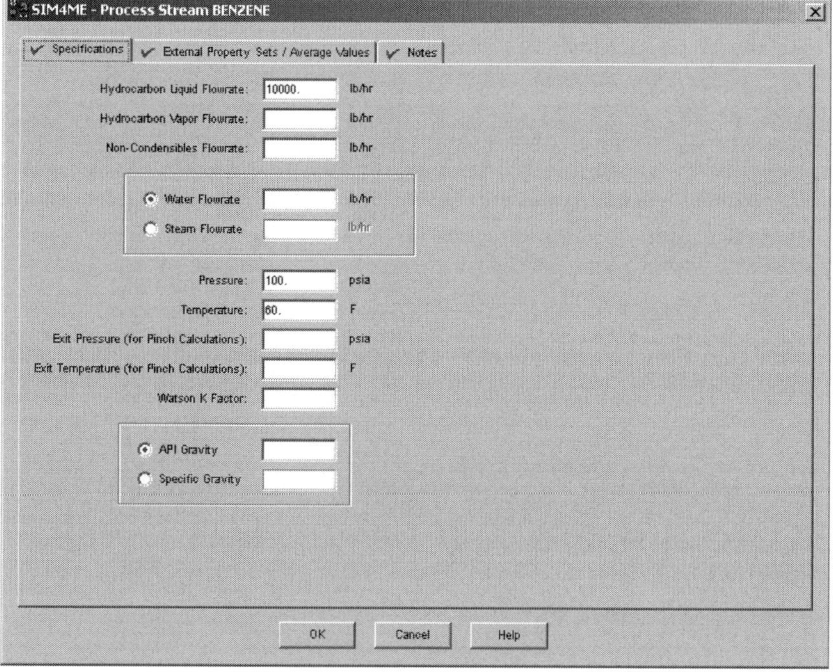

HEXTRAN requires pressures as well as temperatures for the two feed streams. Since these pressures were not specified in Example 4.1 (they are not needed for the calculations), a convenient (albeit rather high for this service) value of 100 psia is arbitrarily assigned to each stream. On the External Property Sets/Average Values form, the average stream values of specific heat, thermal conductivity, viscosity, and density are entered by selecting each property name, in turn, from the list box as shown below. Values of the first three properties for each stream are entered exactly as given in Example 4.1. The densities (54.876 lbm/ft^3 for benzene and 63.803 lbm/ft^3 for aniline) are obtained from the specific gravities given in Example 4.1. The densities, not the specific gravities, must be entered here. Otherwise, HEXTRAN will assume that stream properties are to be estimated using correlations for hydrocarbons as discussed in Appendix E.

SIM4ME - Process Stream BENZENE

Specifications | External Property Sets / Average Values | Notes

Property: Liquid Specific Heat

None

Average Value 0.42 BTU/F-lb

External Property Set

The parameters of the heat exchanger are specified by right-clicking on the unit and selecting *Edit Properties* from the pop-up menu, or by double-clicking the unit. This brings up the required forms and data are entered for the tube side (inner pipe), shell side (annulus), and nozzles as shown below. A hairpin is modeled as a single HEXTRAN shell with a length of 32 ft, the total length of pipe in both legs of the hairpin. Thus, six shells in series are specified. One-inch schedule 40 nozzles having an ID of 1.049 in. are specified for the annulus, and no nozzles are specified for the inner pipe by un-checking the box labeled *Perform Nozzle Sizing and Pressure Drop Calculations.*

On the Material form, type 316 stainless steel is selected from the list boxes for shell material and tube material. Finally, on the Film Options form the tube-side and shell-side fouling factors (both 0.001 $h \cdot ft^2 \cdot °F/Btu$) are entered under *Fouling Resistances.*

When all required data have been entered for a stream or unit, the color of the corresponding label on the flowsheet changes from red to black. The program is executed by clicking either Run or the right-pointing arrowhead on the top toolbar.

The input (keyword) file generated by the graphical user interface (GUI) is given below. Many of the items in this file are superfluous for this simulation, but are automatically included by the GUI in every case. The main items are the STREAM DATA section, where the properties of the two inlet streams are specified, and the UNIT OPERATIONS section, where the characteristics of the heat exchanger are specified. The online help file contains a detailed explanation of the keyword code used by HEXTRAN, much of which is self-evident. Note that dollar signs are used for comment statements (which are ignored by the program) while an asterisk indicates that a statement is continued on the next line. The input file is echo printed at the top of the output file along with any warnings or error messages. The output file is accessed by selecting *View Report* from the Output menu. (To access the files directly, go to C:\ProgramFiles\SIMSCI\SIM4ME12\Server\ModelApps, open the folder bearing the name of the database followed by the folder bearing the flowsheet name. The input and output files have the same name as the flowsheet with extensions .inp and .out, respectively.)

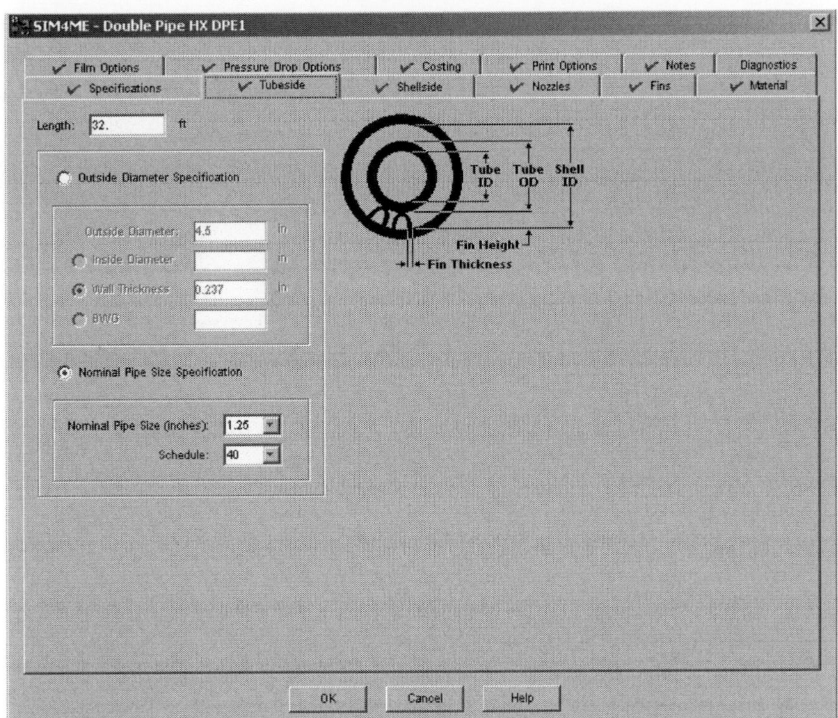

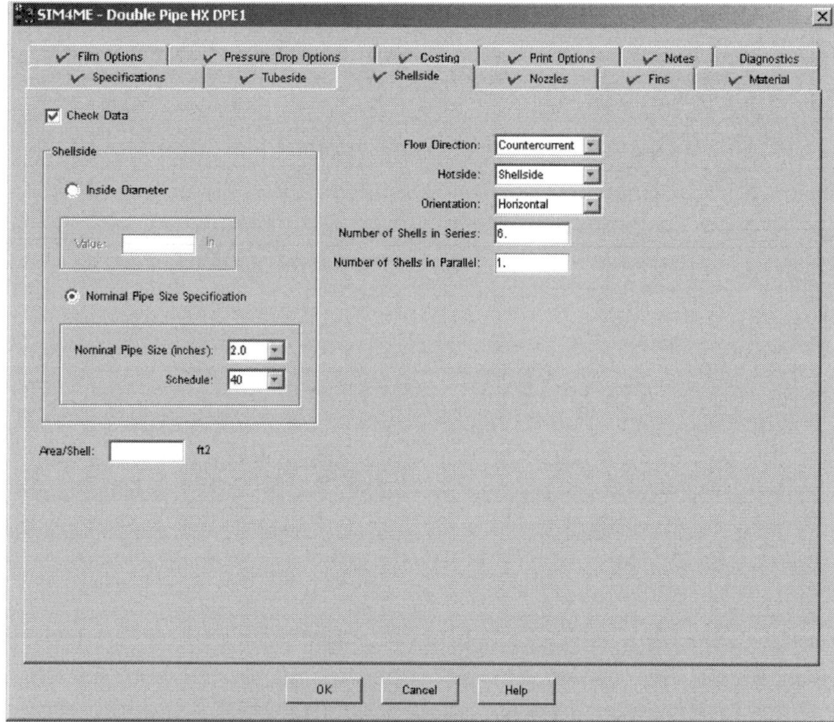

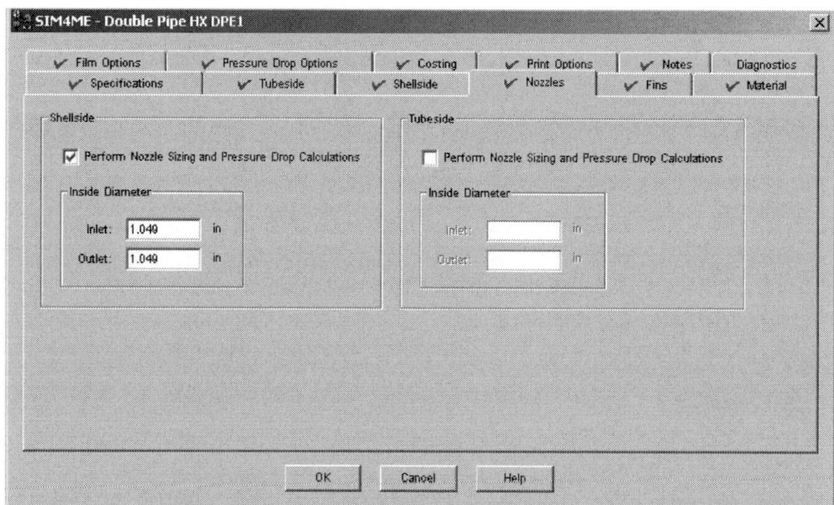

Results of the calculations are summarized in the Double Pipe Exchanger Data Sheet and Extended Data Sheet, which are given below following the input file. This information was extracted from the HEXTRAN output file and used to prepare the following comparison between computer and hand calculations:

Item	Hand calculation	HEXTRAN
Re_i	83,217	83,202
Re_o	8212	8211
h_i (Btu/h · ft^2 · °F)	290*	289.8
h_o (Btu/h · ft^2 · °F)	283*	248.7
U_D (Btu/h · ft^2 · °F)	89*	85.2
ΔP_i (psi)	8.09*	8.02
ΔP_o (excluding nozzles, psi)	48.7**	43.7

*For $\phi = 1.0$.

**For $\phi = 1.0$ and including return losses.

It can be seen that the HEXTRAN results for the inner pipe are in close agreement with the hand calculations, but significant differences exist for the annulus, where the flow is in the transition region. For Reynolds numbers between 2000 and 10,000 HEXTRAN uses linear interpolation to calculate heat-transfer coefficients and friction factors. The heat-transfer coefficient (or friction factor) is calculated for $Re = 2000$ and $Re = 10{,}000$ using the appropriate correlations for laminar and turbulent flow, respectively. Linear interpolation between these two values is then used to obtain h (or f) at the actual Reynolds number for the flow. In the present instance this procedure gives lower values of h and f compared with the hand calculations. As a result, the target temperatures for the two streams were not quite achieved in the simulation. For example, it can be seen from the output data that the benzene outlet temperature was 119.6°F versus 120°F as required by the design specifications.

Actually, the close agreement between the pressure drops for the inner pipe is fortuitous. Details of the methods used for pressure-drop calculations for double-pipe exchangers are not specified in the HEXTRAN documentation. However, by running additional simulations in which a hairpin was modeled as two or more HEXTRAN shells connected in series, it was possible to separate the pressure drop in the straight sections of pipe from that in the return bends. It was found that for shells connected in series, HEXTRAN uses 2 velocity heads per shell for the return-bend losses in the annulus and 4 × (number of shells – 1) velocity heads for the inner pipe return losses. For the present problem, this results in the following values for the inner pipe:

Item	Hand	HEXTRAN
ΔP_f (psi)	6.24	5.21
ΔP_r (psi)	1.85	2.81
ΔP_i (psi)	8.09	8.02

Clearly, the differences in ΔP_f and ΔP_r will result in greater differences in ΔP_i for different lengths of pipe. In the present case with six hairpins in series, the ΔP_i values calculated by HEXTRAN were within 10% of the values calculated by hand for hairpins ranging in length from 8 to 32 ft.

HEXTRAN Input File for Example 4.4

```
$ GENERATED FROM HEXTRAN KEYWORD EXPORTER
$
$     General Data Section
$
TITLE PROJECT=EX4-3, PROBLEM=Benzene Heater, SITE=
$
DIME English, AREA=FT2, CONDUCTIVITY=BTUH, DENSITY=LB/FT3, *
     ENERGY=BTU, FILM=BTUH, LIQVOLUME=FT3, POWER=HP, *
     PRESSURE=PSIA, SURFACE=DYNE, TIME=HR, TEMPERATURE=F, *
     UVALUE=BTUH, VAPVOLUME=FT3, VISCOSITY=CP, WT=LB, *
     XDENSITY=API, STDVAPOR=379.490
$

PRINT ALL, *
      RATE=M
$
CALC PGEN=New, WATER=Saturated
$
$     Component Data Section
$



$
$     Thermodynamic Data Section
$



$
$Stream Data Section
$
STREAM DATA


$
 PROP STRM=BENZENE, NAME=BENZENE, TEMP=60.00, PRES=100.000, *
      LIQUID(W)=10000.000, LCP(AVG)=0.42, Lcond(AVG)=0.092, *
      Lvis(AVG)=0.55, Lden(AVG)=54.876
```

```
$
 PROP STRM=3, NAME=3
$
 PROP STRM=4, NAME=4
$
 PROP STRM=ANILINE, NAME=ANILINE, TEMP=150.00, PRES=100.000, *
         LIQUID(W)=9692.000, LCP(AVG)=0.52, Lcond(AVG)=0.1, *
         Lvis(AVG)=2, Lden(AVG)=63.803
$
$ Calculation Type Section
$
SIMULATION
$
 TOLERANCE TTRIAL=0.01
$
 LIMITS AREA=200.00, 6000.00, SERIES=1, 10, PDAMP=0.00, *
        TTRIAL=30
$
 CALC TWOPHASE=New, DPSMETHOD=Stream, MINFT=0.80
$
PRINT UNITS, ECONOMICS, STREAM, STANDARD, *
       EXTENDED, ZONES
$
ECONOMICS DAYS=350, EXCHANGERATE=1.00, CURRENCY=USDOLLAR
$
 UTCOST OIL=3.50, GAS=3.50, ELECTRICITY=0.10, *
       WATER=0.03, HPSTEAM=4.10, MPSTEAM=3.90, *
       LPSTEAM=3.60, REFRIGERANT=0.00, HEATINGMEDIUM=0.00


   $
    HXCOST BSIZE=1000.00, BCOST=0.00, LINEAR=50.00, *
           EXPONENT=0.60, CONSTANT=0.00, UNIT
   $
   $      Unit Operations Data
   $
   UNIT OPERATIONS
   $
   DPE UID=DPE1
    TYPE   Old, HOTSIDE=Shellside, ORIENTATION=Horizontal, *
           FLOW=Countercurrent, *
           UESTIMATE=50.00, USCALER=1.00
    TUBE   FEED=BENZENE, PRODUCT=3, *
           LENGTH=32.00, *
           NPS=1.25, SCHEDULE=40, *
           MATERIAL=9, *
           FOUL=0.001, LAYER=0, *
           DPSCALER=1.00
   $
    SHELL   FEED=ANILINE, PRODUCT=4, *
           NPS=2.0, SCHEDULE=40, *
           SERIES=6, PARALLEL=1, *
           MATERIAL=9, *
           FOUL=0.001, LAYER=0, *
           DPSCALER=1.00
   $
    TNOZZ NONE
   $
    SNOZZ   ID=1.049, 1.049
   $
    CALC   TWOPHASE=New, *
           MINFT=0.80
   $
    PRINT STANDARD, *
           EXTENDED, *
           ZONES
   $
    COST BSIZE=1000.00, BCOST=0.00, LINEAR=50.00, *
         CONSTANT=0.00, EXPONENT=0.60, Unit
   $

   $ End of keyword file. . .
```

HEXTRAN Output Data for Example 4.4

```
===============================================================================
                    DOUBLE PIPE EXCHANGER DATA SHEET

I-----------------------------------------------------------------------------I
I EXCHANGER  NAME                              UNIT ID DPE1                   I
I SIZE    2x  384          ,  HORIZONTAL     CONNECTED 1 PARALLEL    6 SERIES I
I AREA/UNIT    83. FT2 (    83. FT2 REQUIRED)  AREA/SHELL    14. FT2          I
I-----------------------------------------------------------------------------I
I PERFORMANCE OF  ONE  UNIT          SHELL-SIDE                TUBE-SIDE      I
I-----------------------------------------------------------------------------I
I FEED STREAM NUMBER                  ANILINE                  BENZENE        I
I FEED STREAM NAME                    ANILINE                  BENZENE        I
I TOTAL FLUID          LB /HR             9692.                   10000.      I
I      VAPOR (IN/OUT) LB /HR          0./          0.          0./          0.I
I      LIQUID          LB /HR       9692./       9692.      10000./      10000.I
I      STEAM           LB /HR          0./          0.          0./          0.I
I      WATER           LB /HR          0./          0.          0./          0.I
I     NON CONDENSIBLE LB /HR               0.                          0.     I
I TEMPERATURE (IN/OUT) DEG F      150.0 /    100.4          60.0 /   119.6    I
I PRESSURE     (IN/OUT) PSIA     100.00 /     53.08        100.00 /    91.98   I
I-----------------------------------------------------------------------------I
I SP. GR., LIQ  (60F / 60F H2O)  1.023 /   1.023          0.880 /    0.880    I
I          VAP  (60F / 60F AIR)  0.000 /   0.000          0.000 /    0.000    I
I DENSITY,    LIQUID     LB/FT3  63.803 /  63.803        54.876 /   54.876    I
I             VAPOR      LB/FT3   0.000 /   0.000         0.000 /    0.000    I
I VISCOSITY, LIQUID      CP       2.000 /   2.000         0.550 /    0.550    I
I             VAPOR      CP       0.000 /   0.000         0.000 /    0.000    I
I THRML COND,LIQ  BTU/HR-FT-F   0.1000 /  0.1000        0.0920 /   0.0920    I
I             VAP  BTU/HR-FT-F   0.0000 /  0.0000        0.0000 /   0.0000    I
I SPEC.HEAT,LIQUID BTU /LB F    0.5200 /  0.5200        0.4200 /   0.4200    I
I             VAPOR  BTU /LB F   0.0000 /  0.0000        0.0000 /   0.0000    I
I LATENT HEAT          BTU /LB        0.00                    0.00            I
I VELOCITY               FT/SEC        5.10                    4.87            I
I DP/SHELL(DES/CALC)    PSI        0.00 /   7.82           0.00 /   1.34      I
I FOULING RESIST FT2-HR-F/BTU 0.00100 (0.00100 REQD)          0.00100         I
I-----------------------------------------------------------------------------I
I TRANSFER RATE  BTU/HR-FT2-F SERVICE   85.22 (  85.21 REQD), CLEAN   104.92 I
I HEAT EXCHANGED MMBTU /HR    0.250,     MTD(CORRECTED)  35.2,      FT  1.000 I
I-----------------------------------------------------------------------------I
I CONSTRUCTION OF ONE SHELL          SHELL-SIDE                TUBE-SIDE      I
I-----------------------------------------------------------------------------I
I DESIGN PRESSURE/TEMP PSIA  /F    175./  200.            175./  200.         I
I NO OF PASSES:COUNTERCURRENT          1                       1              I
I MATERIAL                          316 S.S.                316 S.S.          I
I INLET  NOZZLE ID/NO       IN        1.0/ 1                0.0/ 1            I
I OUTLET NOZZLE ID/NO       IN        1.0/ 1                0.0/ 1            I
I-----------------------------------------------------------------------------I
I TUBE: OD(IN)   1.660  ID(IN)   1.380 THK(IN) 0.140 NPS  1.250 SCHED      40 I
I TUBE: TYPE BARE,   CONDUCTIVITY   9.40 BTU/HR-FT-F                          I
I SHELL:  ID    2.07 IN,NPS  2.000 SCHEDULE  40                               I
I RHO-V2: INLET NOZZLE  3153.7 LB/FT-SEC2                                     I
I-----------------------------------------------------------------------------I
```

```
==============================================================================
                    DOUBLE PIPE EXTENDED DATA SHEET
I----------------------------------------------------------------------------I
I EXCHANGER  NAME                              UNIT ID DPE1                  I
I SIZE    2x 384                               CONNECTED 1 PARALLEL  6 SERIES I
I AREA/UNIT    83. FT2 (     83. FT2 REQUIRED)                               I
I----------------------------------------------------------------------------I
I PERFORMANCE OF ONE UNIT          SHELL-SIDE                TUBE-SIDE       I
I----------------------------------------------------------------------------I
I FEED STREAM NUMBER                 ANILINE                 BENZENE         I
I FEED STREAM NAME                   ANILINE                 BENZENE         I
I WT FRACTION LIQUID (IN/OUT)      1.00 / 1.00              1.00 / 1.00      I
I REYNOLDS NUMBER                     8211.                  83202.          I
I PRANDTL NUMBER                     25.163                  6.075           I
I UOPK,LIQUID                     0.000 /   0.000        0.000 /   0.000     I
I VAPOR                           0.000 /   0.000        0.000 /   0.000     I
I SURFACE TENSION     DYNES/CM    0.000 /   0.000        0.000 /   0.000     I
I FILM COEF(SCL) BTU/HR-FT2-F       248.7 (1.000)          289.8 (1.000)     I
I FOULING LAYER THICKNESS  IN         0.000                   0.000          I
I----------------------------------------------------------------------------I
I THERMAL RESISTANCE                                                         I
I UNITS:   (FT2-HR-F/BTU)      (PERCENT)    (ABSOLUTE)                       I
I SHELL FILM                      34.27       0.00402                        I
I TUBE  FILM                      35.37       0.00415                        I
I TUBE  METAL                     11.58       0.00136                        I
I TOTAL FOULING                   18.77       0.00220                        I
I ADJUSTMENT                       0.01       0.00000                        I
I----------------------------------------------------------------------------I
I PRESSURE DROP                    SHELL-SIDE                TUBE-SIDE       I
I UNITS: (PSIA )               (PERCENT)    (ABSOLUTE)   (PERCENT)   (ABSOLUTE)I
I WITHOUT NOZZLES                 93.04          7.28     100.00          1.34 I
I INLET   NOZZLES                  4.35          0.34       0.00          0.00 I
I OUTLET  NOZZLES                  2.61          0.20       0.00          0.00 I
I TOTAL   /SHELL                                 7.82                     1.34 I
I TOTAL   /UNIT                                 46.92                     8.02 I
I DP SCALER                                      1.00                     1.00 I
I----------------------------------------------------------------------------I
I CONSTRUCTION OF ONE SHELL                                                  I
I----------------------------------------------------------------------------I
I TUBE:OVERALL LENGTH       32.0        FT   ANNULAR HYD. DIA.    0.41    IN I
I      NET FREE FLOW AREA    0.008      FT2  AREA RATIO (OUT/IN)  1.203      I
I      THERMAL COND.         9.4BTU/HR-FT-F  DENSITY           501.10 LB/FT3I
I----------------------------------------------------------------------------I
```

Example 4.5

Use HEXTRAN to rate the final configuration (10 hairpins with inner pipes connected in series and annuli connected in two parallel banks) for the benzene-aniline exchanger of Example 4.1, and compare the results with those obtained previously by hand.

Solution

The modeling is done as in the previous example using bulk stream properties and one HEXTRAN shell per hairpin. In order to accommodate the series-parallel configuration, however, each parallel bank of hairpins is represented as a separate double-pipe heat exchanger, as shown in the diagram below. The flowsheet contains nine streams and four units (two heat exchangers, a stream divider and a mixer). Each heat exchanger consists of five hairpins connected in series. Aniline flows through the inner pipes and benzene flows through the annuli.

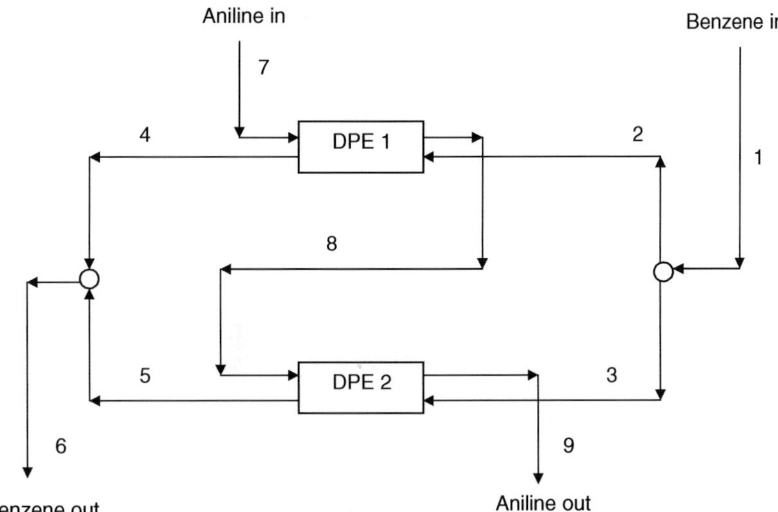

The input file generated by the HEXTRAN GUI is given below, followed by the data sheets for the two exchangers from the HEXTRAN output file. The data sheets were used to prepare the following comparison between computer and hand calculations:

Item	Hand calculation	HEXTRAN
Re_i	22,180	22,176
Re_o	15,405	15,403
h_i (Btu/h · ft^2 · °F)	176*	175.6
h_o (Btu/h · ft^2 · °F)	255*	254.9
U_D (Btu/h · ft^2 · °F)	69.8*	69.8
ΔP_i (psi)	14.0*	13.6
ΔP_o (psi)	11.7*	10.5

*For $\phi = 1.0$.

It can be seen that the heat-transfer coefficients computed by HEXTRAN agree exactly with the hand calculations, and the two sets of pressure drops are also in reasonable agreement, although the difference on the annulus side is about 10%. Note that since the two exchangers are connected in series on the tube side, the total pressure drop for the aniline stream is the sum of the pressure drops in the two units.

Finally, note that the benzene outlet temperature (after mixing) is the arithmetic average of the outlet temperatures from the two exchangers. Thus,

$$T_{Benzene\ out} = (134.7 + 108.9)/2 = 121.8°F$$

Since this temperature exceeds the design specification of 120°F, the rating procedure indicates that the series-parallel configuration is thermally and hydraulically suitable, in agreement with the hand calculations.

HEXTRAN Input File for Example 4.5

```
$ GENERATED FROM HEXTRAN KEYWORD EXPORTER
$
$      General Data Section
$
TITLE PROJECT=EX4-4, PROBLEM=Benzene Heater, SITE=
$
DIME  English, AREA=FT2, CONDUCTIVITY=BTUH, DENSITY=LB/FT3, *
      ENERGY=BTU, FILM=BTUH, LIQVOLUME=FT3, POWER=HP, *
      PRESSURE=PSIA, SURFACE=DYNE, TIME=HR, TEMPERATURE=F, *
      UVALUE=BTUH, VAPVOLUME=FT3, VISCOSITY=CP, WT=LB, *
      XDENSITY=API, STDVAPOR=379.490
$
PRINT ALL, *
      RATE=M
```

```
$
CALC  PGEN=New, WATER=Saturated
$
$     Component Data Section
$

$
$     Thermodynamic Data Section
$

$
$Stream Data Section
$
STREAM DATA

$
 PROP STRM=1, NAME=1, TEMP=60.00, PRES=100.000, *
        LIQUID(W)=10000.000, LCP(AVG)=0.42, Lcond(AVG)=0.092, *
        Lvis(AVG)=0.55, Lden(AVG)=54.876
$
 PROP STRM=7, NAME=7, TEMP=150.00, PRES=100.000, *
        LIQUID(W)=9692.000, LCP(AVG)=0.52, Lcond(AVG)=0.1, *
        Lvis(AVG)=2, Lden(AVG)=63.803
$
 PROP STRM=9, NAME=9
$
 PROP STRM=2, NAME=2
$
 PROP STRM=3, NAME=3
$
 PROP STRM=6, NAME=6
$
 PROP STRM=4, NAME=4
$
 PROP STRM=5, NAME=5
$
 PROP STRM=8, NAME=8
$
$ Calculation Type Section
$

SIMULATION
$
 TOLERANCE TTRIAL=0.01
$
 LIMITS AREA=200.00, 6000.00, SERIES=1, 10, PDAMP=0.00, *
       TTRIAL=30
$
 CALC   TWOPHASE=New, DPSMETHOD=Stream, MINFT=0.80
$
 PRINT  UNITS, ECONOMICS, STREAM, STANDARD, *
       EXTENDED, ZONES
$
ECONOMICS DAYS=350, EXCHANGERATE=1.00, CURRENCY=USDOLLAR
$
 UTCOST OIL=3.50, GAS=3.50, ELECTRICITY=0.10, *
       WATER=0.03, HPSTEAM=4.10, MPSTEAM=3.90, *
       LPSTEAM=3.60, REFRIGERANT=0.00, HEATINGMEDIUM=0.00
$
 HXCOST BSIZE=1000.00, BCOST=0.00, LINEAR=50.00, *
       EXPONENT=0.60, CONSTANT=0.00, UNIT
$
$     Unit Operations Data
$
UNIT OPERATIONS
$
SPLITTER UID=SP1
 STRMS FEED=1, PROD=2,  3
 OPERATION FRAC=0.5, 0.5
```

```
$
DPE UID=DPE1
 TYPE  Old, HOTSIDE=Tubeside, ORIENTATION=Horizontal, *
       FLOW=Countercurrent, *
       UESTIMATE=50.00, USCALER=1.00
 TUBE  FEED=7, PRODUCT=8, *
       LENGTH=32.00, *
       NPS=1.25, SCHEDULE=40, *
       MATERIAL=9, *
       FOUL=0.001, LAYER=0, *
       DPSCALER=1.00
$
 SHELL  FEED=2, PRODUCT=4, *
       NPS=2.0, SCHEDULE=40, *
       SERIES=5, PARALLEL=1, *
       MATERIAL=9, *
       FOUL=0.001, LAYER=0, *
       DPSCALER=1.00
$
 TNOZZ NONE
$
 SNOZZ  ID=1.049, 1.049
$
CALC TWOPHASE=New, *
      MINFT=0.80
$
 PRINT STANDARD, *
       EXTENDED, *
       ZONES
$
 COST  BSIZE=1000.00, BCOST=0.00, LINEAR=50.00, *
       CONSTANT=0.00, EXPONENT=0.60, Unit
$
DPE UID=DPE2
 TYPE  Old, HOTSIDE=Tubeside, ORIENTATION=Horizontal, *
       FLOW=Countercurrent, *
       UESTIMATE=50.00, USCALER=1.00
 TUBE  FEED=8, PRODUCT=9, *
       LENGTH=32.00, *
       NPS=1.25, SCHEDULE=40, *
       MATERIAL=9, *
       FOUL=0.001, LAYER=0, *
       DPSCALER=1.00
$
 SHELL  FEED=3, PRODUCT=5, *
       NPS=2.0, SCHEDULE=40, *
       SERIES=5, PARALLEL=1, *
       MATERIAL=9, *
       FOUL=0.001, LAYER=0, *
       DPSCALER=1.00
$

 TNOZZ NONE
$
 SNOZZ  ID=1.049, 1.049
$
 CALC  TWOPHASE=New, *
       MINFT=0.80
$
 PRINT STANDARD, *
       EXTENDED, *
       ZONES
$
 COST  BSIZE=1000.00, BCOST=0.00, LINEAR=50.00, *
       CONSTANT=0.00, EXPONENT=0.60, Unit
$
MIXER UID=M1
 STRMS FEED=4, 5,    PROD=6
$

$ End of keyword file...
```

HEXTRAN Output Data for Example 4.5

```
==============================================================================
                    DOUBLE PIPE EXCHANGER DATA SHEET
I----------------------------------------------------------------------------I
I EXCHANGER  NAME                                UNIT ID DPE1                I
I SIZE     2x 384            ,  HORIZONTAL      CONNECTED 1 PARALLEL  5 SERIES I
I AREA/UNIT    70. FT2 (    70. FT2 REQUIRED) AREA/SHELL    14. FT2          I
I----------------------------------------------------------------------------I
I PERFORMANCE OF ONE UNIT           SHELL-SIDE               TUBE-SIDE       I
I----------------------------------------------------------------------------I
I FEED STREAM NUMBER                   2                         7           I
I FEED STREAM NAME                     2                         7           I
I TOTAL FLUID          LB /HR              5000.                   9692.     I
I     VAPOR  (IN/OUT) LB /HR          0./         0.            0./       0. I
I     LIQUID          LB /HR      5000./      5000.         9692./     9692. I
I     STEAM           LB /HR          0./         0.            0./       0. I
I     WATER           LB /HR          0./         0.            0./       0. I
I     NON CONDENSIBLE LB /HR               0.                        0.      I
I TEMPERATURE (IN/OUT) DEG F      60.0 /   134.7          150.0 /  118.9     I
I PRESSURE    (IN/OUT) PSIA     100.00 /    89.54        100.00 /   93.21    I
I----------------------------------------------------------------------------I
I SP. GR., LIQ  (60F / 60F H2O)   0.880 /   0.880          1.023 /   1.023   I
I          VAP  (60F / 60F AIR)   0.000 /   0.000          0.000 /   0.000   I
I DENSITY,   LIQUID    LB/FT3    54.876 /  54.876         63.803 /  63.803   I
I            VAPOR     LB/FT3     0.000 /   0.000          0.000 /   0.000   I
I VISCOSITY, LIQUID    CP         0.550 /   0.550          2.000 /   2.000   I
I            VAPOR     CP         0.000 /   0.000          0.000 /   0.000   I
I THRML COND,LIQ  BTU/HR-FT-F    0.0920 /  0.0920         0.1000 /  0.1000   I
I            VAP  BTU/HR-FT-F    0.0000 /  0.0000         0.0000 /  0.0000   I
I SPEC.HEAT,LIQUID BTU /LB F     0.4200 /  0.4200         0.5200 /  0.5200   I
I            VAPOR  BTU /LB F    0.0000 /  0.0000         0.0000 /  0.0000   I
I LATENT HEAT         BTU /LB          0.00                    0.00          I
I VELOCITY             FT/SEC          3.06                    4.06          I
I DP/SHELL(DES/CALC)   PSI        0.00 /  2.09             0.00 /  1.36      I
I FOULING RESIST FT2-HR-F/BTU  0.00100 (0.00100 REQD)          0.00100       I
I----------------------------------------------------------------------------I
I TRANSFER RATE  BTU/HR-FT2-F  SERVICE    69.76 (  69.77 REQD), CLEAN   82.43 I
I HEAT EXCHANGED MMBTU /HR      0.157,      MTD(CORRECTED)  32.3,      FT ,1.000 I
I----------------------------------------------------------------------------I
I CONSTRUCTION OF ONE SHELL         SHELL-SIDE               TUBE-SIDE       I
I----------------------------------------------------------------------------I
I DESIGN PRESSURE/TEMP PSIA  /F     175./  200.            175./  200.       I
I NO OF PASSES:COUNTERCURRENT           1                       1            I
I MATERIAL                          316 S.S.                316 S.S.         I
I INLET  NOZZLE ID/NO      IN         1.0/ 1                0.0/ 1           I
I OUTLET NOZZLE ID/NO      IN         1.0/ 1                0.0/ 1           I
I----------------------------------------------------------------------------I
I TUBE: OD(IN)   1.660  ID(IN)   1.380  THK(IN)  0.140  NPS  1.250 SCHED  40 I
I TUBE: TYPE BARE,   CONDUCTIVITY   9.40 BTU/HR-FT-F                         I
I SHELL:  ID    2.07 IN,NPS  2.000 SCHEDULE  40                              I
I RHO-V2: INLET NOZZLE   975.9 LB/FT-SEC2                                    I
I----------------------------------------------------------------------------I
```

```
=============================================================================
                   DOUBLE PIPE EXTENDED DATA SHEET
I---------------------------------------------------------------------------I
I EXCHANGER  NAME                            UNIT ID DPE1                   I
I SIZE    2x 384                             CONNECTED 1 PARALLEL  5 SERIES I
I AREA/UNIT    70. FT2 (    70. FT2 REQUIRED)                               I
I---------------------------------------------------------------------------I
I PERFORMANCE OF ONE UNIT           SHELL-SIDE                TUBE-SIDE     I
I---------------------------------------------------------------------------I
I FEED STREAM NUMBER                  2                        7            I
I FEED STREAM NAME                   2                        7             I
I WT FRACTION LIQUID (IN/OUT)       1.00 / 1.00               1.00 / 1.00   I
I REYNOLDS NUMBER                     15403.                   22176.       I
I PRANDTL NUMBER                      6.075                   25.163        I
I UOPK,LIQUID                      0.000 /   0.000           0.000 /   0.000 I
I      VAPOR                       0.000 /   0.000           0.000 /   0.000 I
I SURFACE TENSION   DYNES/CM       0.000 /   0.000           0.000 /   0.000 I
I FILM COEF(SCL) BTU/HR-FT2-F         254.9 (1.000)           175.6 (1.000) I
I FOULING LAYER THICKNESS  IN           0.000                    0.000      I
I---------------------------------------------------------------------------I
I THERMAL RESISTANCE                                                        I
I UNITS:  (FT2-HR-F/BTU)       (PERCENT)   (ABSOLUTE)                       I
I SHELL FILM                       27.37      0.00392                       I
I TUBE  FILM                       47.78      0.00685                       I
I TUBE  METAL                       9.48      0.00136                       I
I TOTAL FOULING                    15.37      0.00220                       I
I ADJUSTMENT                       -0.02      0.00000                       I
I---------------------------------------------------------------------------I
I PRESSURE DROP                     SHELL-SIDE               TUBE-SIDE      I
I UNITS: (PSIA  )             (PERCENT)   (ABSOLUTE)  (PERCENT)   (ABSOLUTE) I
I WITHOUT NOZZLES                  91.95        1.92    100.00         1.36  I
I INLET   NOZZLES                   5.03        0.11      0.00         0.00  I
I OUTLET  NOZZLES                   3.02        0.06      0.00         0.00  I
I TOTAL   /SHELL                                2.09                   1.36  I
I TOTAL   /UNIT                                10.46                   6.79  I
I DP SCALER                                     1.00                   1.00  I
I---------------------------------------------------------------------------I
I CONSTRUCTION OF ONE SHELL                                                 I
I---------------------------------------------------------------------------I
I TUBE:OVERALL LENGTH        32.0         FT  ANNULAR HYD. DIA.    0.41    IN I
I      NET FREE FLOW AREA     0.008       FT2  AREA RATIO (OUT/IN)  1.203     I
I      THERMAL COND.          9.4BTU/HR-FT-F  DENSITY            501.10 LB/FT3I
I---------------------------------------------------------------------------I
```

```
==============================================================================
                    DOUBLE PIPE EXCHANGER DATA SHEET
I----------------------------------------------------------------------------I
I EXCHANGER  NAME                              UNIT ID DPE2                  I
I SIZE    2x 384           ,  HORIZONTAL      CONNECTED 1 PARALLEL  5 SERIES I
I AREA/UNIT    70. FT2 (    70. FT2 REQUIRED) AREA/SHELL    14. FT2          I
I----------------------------------------------------------------------------I
I PERFORMANCE OF ONE UNIT           SHELL-SIDE               TUBE-SIDE       I
I----------------------------------------------------------------------------I
I FEED STREAM NUMBER                 3                        8              I
I FEED STREAM NAME                   3                        8              I
I TOTAL FLUID          LB /HR           5000.                   9692.        I
I    VAPOR  (IN/OUT) LB /HR         0./         0.            0./       0.   I
I    LIQUID          LB /HR      5000./      5000.         9692./    9692.   I
I    STEAM           LB /HR         0./         0.            0./       0.   I
I    WATER           LB /HR         0./         0.            0./       0.   I
I    NON CONDENSIBLE LB /HR            0.                          0         I
I TEMPERATURE (IN/OUT) DEG F     60.0 /   108.9         118.9 /   98.5       I
I PRESSURE    (IN/OUT) PSIA    100.00 /    89.54         93.21 /   86.41     I
I----------------------------------------------------------------------------I
I SP. GR., LIQ  (60F / 60F H2O)   0.880 /   0.880        1.023 /   1.023     I
I         VAP  (60F / 60F AIR)   0.000 /  0.000         0.000 /   0.000      I
I DENSITY,   LIQUID    LB/FT3    54.876 /  54.876       63.803 /  63.803     I
I            VAPOR     LB/FT3     0.000 /   0.000        0.000 /   0.000     I
I VISCOSITY, LIQUID    CP         0.550 /   0.550        2.000 /   2.000     I
I            VAPOR     CP         0.000 /   0.000        0.000 /   0.000     I
I THRML COND,LIQ  BTU/HR-FT-F    0.0920 /  0.0920       0.1000 /  0.1000     I
I            VAP  BTU/HR-FT-F    0.0000 /  0.0000       0.0000 /  0.0000     I
I SPEC.HEAT,LIQUID BTU /LB F     0.4200 /  0.4200       0.5200 /  0.5200     I
I           VAPOR   BTU /LB F    0.0000 /  0.0000       0.0000 /  0.0000     I
I LATENT HEAT         BTU /LB         0.00                   0.00            I
I VELOCITY              FT/SEC        3.06                   4.06            I
I DP/SHELL(DES/CALC)    PSI        0.00 /  2.09          0.00 /  1.36        I
I FOULING RESIST FT2-HR-F/BTU  0.00100 (0.00100 REQD)        0.00100         I
I----------------------------------------------------------------------------I
I TRANSFER RATE  BTU/HR-FT2-F  SERVICE    69.76 (  69.78 REQD),  CLEAN  82.43 I
I HEAT EXCHANGED MMBTU /HR     0.103,      MTD(CORRECTED)  21.1,      FT 1.000 I
I----------------------------------------------------------------------------I
I CONSTRUCTION OF ONE SHELL          SHELL-SIDE               TUBE-SIDE      I
I----------------------------------------------------------------------------I
I DESIGN PRESSURE/TEMP PSIA  /F      175./  200.            150./  200.      I
I NO OF PASSES:COUNTERCURRENT            1                       1           I
I MATERIAL                           316 S.S.                316 S.S.        I
I INLET NOZZLE ID/NO       IN          1.0/ 1                0.0/ 1          I
I OUTLET NOZZLE ID/NO      IN          1.0/ 1                0.0/ 1          I
I----------------------------------------------------------------------------I
I TUBE: OD(IN)    1.660  ID(IN)   1.380  THK(IN)  0.140  NPS  1.250 SCHED  40 I
I TUBE: TYPE BARE,    CONDUCTIVITY   9.40 BTU/HR-FT-F                        I
I SHELL:  ID     2.07 IN,NPS  2.000 SCHEDULE  40                             I
I RHO-V2: INLET NOZZLE    975.9 LB/FT-SEC2                                   I
I----------------------------------------------------------------------------I
```

```
==============================================================================
                    DOUBLE PIPE EXTENDED DATA SHEET
I----------------------------------------------------------------------------I
I EXCHANGER  NAME                            UNIT ID DPE2                     I
I SIZE    2x 384                             CONNECTED 1 PARALLEL  5 SERIES I
I AREA/UNIT    70. FT2 (     70. FT2 REQUIRED)                               I
I----------------------------------------------------------------------------I
I PERFORMANCE OF ONE UNIT            SHELL-SIDE               TUBE-SIDE       I
I----------------------------------------------------------------------------I
I FEED STREAM NUMBER                   3                        8             I
I FEED STREAM NAME                     3                        8             I
I WT FRACTION LIQUID (IN/OUT)        1.00 / 1.00              1.00 / 1.00     I
I REYNOLDS NUMBER                      15403.                   22176.        I
I PRANDTL NUMBER                       6.075                   25.163         I
I UOPK,LIQUID                       0.000 /   0.000          0.000 /   0.000 I
I      VAPOR                        0.000 /   0.000          0.000 /   0.000 I
I SURFACE TENSION    DYNES/CM       0.000 /   0.000          0.000 /   0.000 I
I FILM COEF(SCL) BTU/HR-FT2-F          254.9 (1.000)           175.6 (1.000)  I
I FOULING LAYER THICKNESS  IN            0.000                    0.000       I
I----------------------------------------------------------------------------I
I THERMAL RESISTANCE                                                          I
I UNITS:   (FT2-HR-F/BTU)        (PERCENT)   (ABSOLUTE)                       I
I SHELL FILM                        27.37      0.00392                        I
I TUBE  FILM                        47.78      0.00685                        I
I TUBE  METAL                        9.48      0.00136                        I
I TOTAL FOULING                     15.37      0.00220                        I
I ADJUSTMENT                        -0.03      0.00000                        I
I----------------------------------------------------------------------------I
I PRESSURE DROP                      SHELL-SIDE               TUBE-SIDE       I
I UNITS: (PSIA  )                (PERCENT)   (ABSOLUTE)   (PERCENT)   (ABSOLUTE)I
I WITHOUT NOZZLES                   91.95          1.92      100.00        1.36 I
I INLET   NOZZLES                    5.03          0.11        0.00        0.00 I
I OUTLET  NOZZLES                    3.02          0.06        0.00        0.00 I
I TOTAL   /SHELL                                   2.09                    1.36 I
I TOTAL   /UNIT                                   10.46                    6.79 I
I DP SCALER                                        1.00                    1.00 I
I----------------------------------------------------------------------------I
I CONSTRUCTION OF ONE SHELL                                                   I
I----------------------------------------------------------------------------I
I TUBE:OVERALL LENGTH         32.0         FT  ANNULAR HYD. DIA.    0.41    IN I
I      NET FREE FLOW AREA     0.008        FT2  AREA RATIO (OUT/IN)  1.203      I
I      THERMAL COND.          9.4BTU/HR-FT-F  DENSITY            501.10 LB/FT3I
I----------------------------------------------------------------------------I
```

Example 4.6

Use HEXTRAN to rate the multi-tube hairpin exchanger of Example 4.2 and compare the results with those obtained previously by hand.

Solution

The HEXTRAN flowsheet for this case (shown below) is similar to that of Example 4.4 except that the multi-tube exchanger module (MTE) replaces the DPE module and the fluid locations are switched. The feed streams are defined as bulk property streams and the stream data are entered exactly as in Example 4.4.

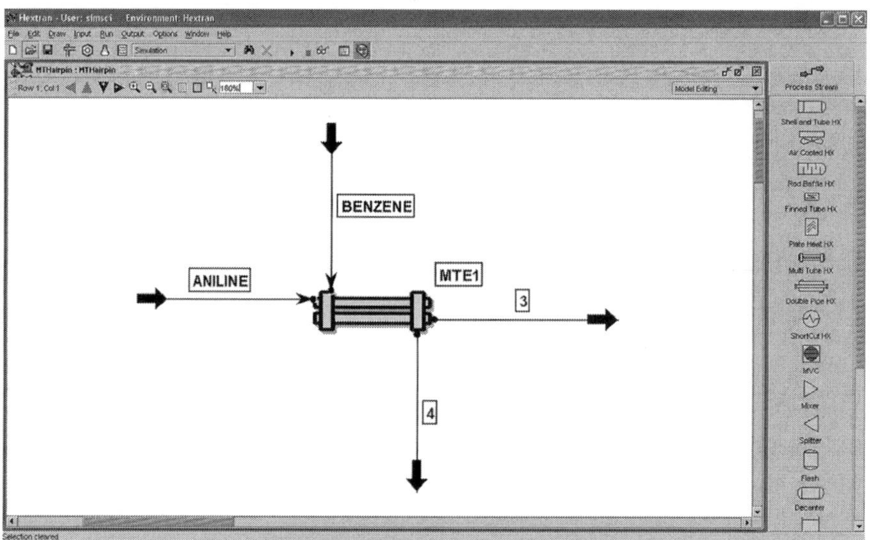

Data for the heat exchanger are entered on the appropriate forms as follows:

(a) Tube side

Data entries on the tube-side panel are as shown in the figure below.

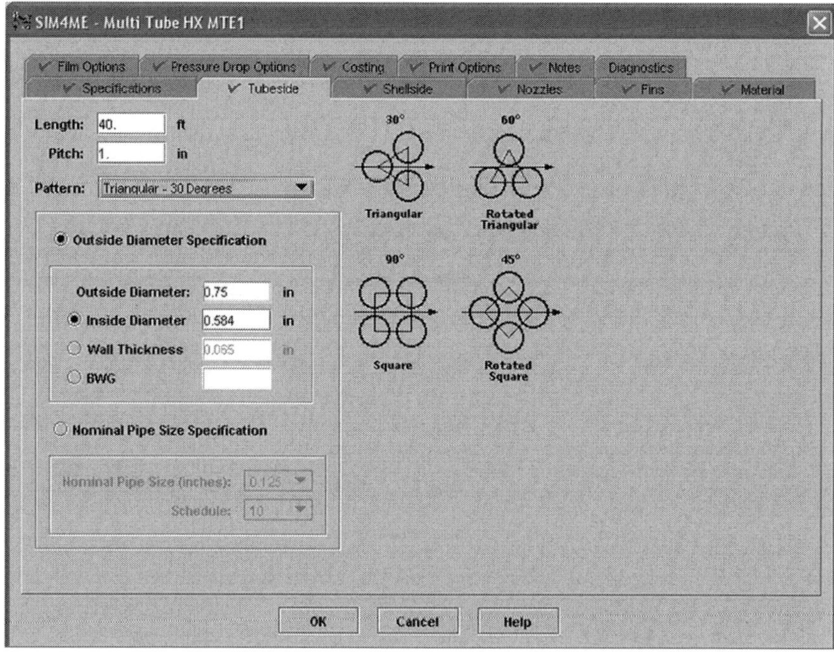

A hairpin is modeled as one HEXTRAN shell (the best option for hairpins with internal return bends), so the total straight-section length of one tube in both branches of one hairpin is entered here. The tube pitch and pattern are assumed since no information about the tube layout was given in Example 4.2. However, any of the four available patterns and any reasonable tube pitch can be specified; the results are nearly independent of these parameters.

(b) Shell side

Number of tubes /shell: 8	Hot side: Tube side
Inside diameter: 3.548 in.	Number of shells in series: 3

(c) Nozzles

Shell-side inlet ID: 2.067 in.

Shell-side outlet ID: 2.067 in.

No tube-side nozzles are specified by un-checking the box labeled *Perform Nozzle Sizing and Pressure Drop Calculations.*

(d) Material
Shell material: 316 Stainless Steel
Tube material: 316 Stainless Steel
(e) Film options
Shell-side fouling resistance: 0.001 h· ft^2· °F
Tube-side fouling resistance: 0.001 h· ft^2· °F
The input file generated by the HEXTRAN GUI is shown below, followed by the output summary tables extracted from the output file. The results are compared with the hand calculations in the following table.

Item	Hand calculation	HEXTRAN
Re_i	6552	6550
Re_o	12,029	12,023
h_i (Btu/h · ft^2 · °F)	157.2*	130.7
h_o (Btu/h · ft^2 · °F)	100.5*	100.3
U_D (Btu/h · ft^2 · °F)	47.1*	43.7
ΔP_i (psi)	7.05*	6.11
ΔP_o (psi)	0.73*	0.67

* For $\phi = 1.0$.

The hand-calculated values are listed for a viscosity correction factor of unity since the HEXTRAN values are based on constant fluid properties, including viscosity. For the annulus, where the flow is fully turbulent, the heat-transfer coefficients and pressure drops are in excellent agreement. There are greater differences in the tube-side values due to the linear interpolation procedure used by HEXTRAN for flow in the transition region. Overall, however, the HEXTRAN results confirm the hand calculations. In particular, the benzene outlet temperature of 122.8°F exceeds the design specification of 120°F, and the corresponding duty of 264,000 Btu/h provides a margin of about 5% over the required duty of 252,000 Btu/h.

HEXTRAN Input File for Example 4.6

```
$ GENERATED FROM HEXTRAN KEYWORD EXPORTER
$
$       General Data Section
$
TITLE PROJECT=, PROBLEM=, SITE=
$
DIME  English, AREA=FT2, CONDUCTIVITY=BTUH, DENSITY=LB/FT3, *
      ENERGY=BTU, FILM=BTUH, LIQVOLUME=FT3, POWER=HP, *
      PRESSURE=PSIA, SURFACE=DYNE, TIME=HR, TEMPERATURE=F, *
      UVALUE=BTUH, VAPVOLUME=FT3, VISCOSITY=CP, WT=LB, *
      XDENSITY=API, STDVAPOR=379.490
$
PRINT ALL, *
      RATE=M
$
CALC  PGEN=New, WATER=Saturated
$
$       Component Data Section
$

$
$       Thermodynamic Data Section
$

$
$Stream Data Section
$
STREAM DATA
```

```
$
 PROP STRM=ANILINE, NAME=ANILINE, TEMP=150.00, PRES=50.000, *
         LIQUID(W)=9692.000, LCP(AVG)=0.52, Lcond(AVG)=0.1, *
         Lvis(AVG)=2, Lden(AVG)=63.803
$
 PROP STRM=3, NAME=3
$
 PROP STRM=4, NAME=4
$
 PROP STRM=BENZENE, NAME=BENZENE, TEMP=60.00, PRES=50.000, *
         LIQUID(W)=10000.000, LCP(AVG)=0.42, Lcond(AVG)=0.092, *
         Lvis(AVG)=0.55, Lden(AVG)=54.876
$
$ Calculation Type Section
$
SIMULATION
$
 TOLERANCE TTRIAL=0.01
$
 LIMITS AREA=200.00, 6000.00, SERIES=1, 10, PDAMP=0.00, *
        TTRIAL=30
$
 CALC   TWOPHASE=New, DPSMETHOD=Stream, MINFT=0.80
$
 PRINT  UNITS, ECONOMICS, STREAM, STANDARD, *
        EXTENDED, ZONES
$
ECONOMICS DAYS=350, EXCHANGERATE=1.00, CURRENCY=USDOLLAR
$
 UTCOST OIL=3.50, GAS=3.50, ELECTRICITY=0.10, *
        WATER=0.03, HPSTEAM=4.10, MPSTEAM=3.90, *
        LPSTEAM=3.60, REFRIGERANT=0.00, HEATINGMEDIUM=0.00
$
 HXCOST BSIZE=1000.00, BCOST=0.00, LINEAR=50.00, *
        EXPONENT=0.60, CONSTANT=0.00, UNIT
$
$       Unit Operations Data
$
UNIT OPERATIONS
$
MTE UID=MTE1
 TYPE  Old, HOTSIDE=Tubeside, ORIENTATION=Horizontal, *
       FLOW=Countercurrent, *
       UESTIMATE=50.00, USCALER=1.00
 TUBE  FEED=ANILINE, PRODUCT=3, *
       LENGTH=40.00, *
       OD=0.750, ID=0.584, *
       NUMBER=8, PATTERN=30, *
       PITCH=1.000, MATERIAL=9, *
       FOUL=0.001, LAYER=0, *
       DPSCALER=1.00
$
 SHELL  FEED=BENZENE, PRODUCT=4, *
       ID=3.55, *
       SERIES=3, PARALLEL=1, *
       MATERIAL=9, *
       FOUL=0.001, LAYER=0, *
       DPSCALER=1.00
$

 TNOZZ NONE
$
 SNOZZ  ID=2.067, 2.067
```

```
$
 CALC  TWOPHASE=New, *
       MINFT=0.80
$
 PRINT STANDARD, *
       EXTENDED, *
       ZONES
$
 COST  BSIZE=1000.00, BCOST=0.00, LINEAR=50.00, *
       CONSTANT=0.00, EXPONENT=0.60, Unit
$

$ End of keyword file...
```

HEXTRAN Output Data for Example 4.6

```
==============================================================================
                   MULTI-TUBE EXCHANGER DATA SHEET
I----------------------------------------------------------------------------I
I EXCHANGER  NAME                              UNIT ID MTE1                  I
I SIZE    4x 480            ,  HORIZONTAL      CONNECTED 1 PARALLEL  3 SERIES I
I AREA/UNIT   188. FT2 (   188. FT2 REQUIRED) AREA/SHELL     63. FT2         I
I----------------------------------------------------------------------------I
I PERFORMANCE OF ONE UNIT            SHELL-SIDE                TUBE-SIDE     I
I----------------------------------------------------------------------------I
I FEED STREAM NUMBER                  BENZENE                  ANILINE       I
I FEED STREAM NAME                    BENZENE                  ANILINE       I
I TOTAL FLUID          LB /HR            10000.                   9692.      I
I     VAPOR  (IN/OUT) LB /HR          0./          0.          0./        0. I
I     LIQUID          LB /HR      10000./      10000.       9692./     9692. I
I     STEAM           LB /HR          0./          0.          0./        0. I
I     WATER           LB /HR          0./          0.          0./        0. I
I     NON CONDENSIBLE LB /HR               0.                      0.        I
I TEMPERATURE (IN/OUT) DEG F       60.0 /   122.8          150.0 /   97.7    I
I PRESSURE    (IN/OUT) PSIA       50.00 /    49.33          50.00 /   43.89   I
I----------------------------------------------------------------------------I
I SP. GR., LIQ  (60F / 60F H2O)   0.880 /  0.880           1.023 /  1.023    I
I          VAP  (60F / 60F AIR)   0.000 /  0.000           0.000 /  0.000    I
I DENSITY,   LIQUID    LB/FT3    54.876 / 54.876          63.803 / 63.803    I
I            VAPOR     LB/FT3     0.000 /  0.000           0.000 /  0.000    I
I VISCOSITY, LIQUID    CP         0.550 /  0.550           2.000 /  2.000    I
I            VAPOR     CP         0.000 /  0.000           0.000 /  0.000    I
I THRML COND,LIQ  BTU/HR-FT-F    0.0920 / 0.0920          0.1000 / 0.1000    I
I            VAP  BTU/HR-FT-F    0.0000 / 0.0000          0.0000 / 0.0000    I
I SPEC.HEAT,LIQUID BTU /LB F     0.4200 / 0.4200          0.5200 / 0.5200    I
I            VAPOR  BTU /LB F    0.0000 / 0.0000          0.0000 / 0.0000    I
I LATENT HEAT          BTU /LB        0.00                     0.00          I
I VELOCITY              FT/SEC        1.15                     2.84          I
I DP/SHELL(DES/CALC)    PSI       0.00 /  0.22              0.00 /  2.04     I
I FOULING RESIST FT2-HR-F/BTU  0.00100 (0.00100 REQD)        0.00100         I
I----------------------------------------------------------------------------I
I TRANSFER RATE  BTU/HR-FT2-F  SERVICE   43.65 (  43.65 REQD), CLEAN   48.48 I
I HEAT EXCHANGED MMBTU /HR     0.264,     MTD(CORRECTED)  32.2,      FT 1.000 I
I----------------------------------------------------------------------------I
I CONSTRUCTION OF ONE SHELL          SHELL-SIDE                TUBE-SIDE     I
I----------------------------------------------------------------------------I
I DESIGN PRESSURE/TEMP PSIA  /F      125./  200.           125./  200.       I
I NUMBER OF PASSES                       1                      1            I
I MATERIAL                           316 S.S.               316 S.S.         I
I INLET  NOZZLE ID/NO       IN        2.1/ 1                 0.0/ 1          I
I OUTLET NOZZLE ID/NO       IN        2.1/ 1                 0.0/ 1          I
I----------------------------------------------------------------------------I
I TUBE: NUMBER      8, OD  0.750  IN,  THICK 0.083  IN,   LENGTH 40.0 FT     I
I        TYPE BARE,                PITCH    1.0000 IN,   PATTERN 30 DEGREES  I
I SHELL:  ID    3.55 IN                                                      I
I RHO-V2: INLET NOZZLE   258.9 LB/FT-SEC2                                    I
I TOTAL WEIGHT/SHELL,LB     825.5 FULL OF WATER   0.141E+04 BUNDLE     433.2 I
I----------------------------------------------------------------------------I
```

```
==============================================================================
                    MULTI-TUBE EXTENDED DATA SHEET
I----------------------------------------------------------------------------I
I EXCHANGER  NAME                              UNIT ID MTE1                   I
I SIZE     4x  480          ,  HORIZONTAL      CONNECTED 1 PARALLEL   3 SERIES I
I AREA/UNIT    188. FT2 (    188. FT2 REQUIRED)                              I
I----------------------------------------------------------------------------I
I PERFORMANCE OF ONE UNIT           SHELL-SIDE                TUBE-SIDE      I
I----------------------------------------------------------------------------I
I FEED STREAM NUMBER                  BENZENE                  ANILINE       I
I FEED STREAM NAME                    BENZENE                  ANILINE       I
I WT FRACTION LIQUID (IN/OUT)       1.00 / 1.00              1.00 / 1.00     I
I REYNOLDS NUMBER                     12023.                    6550.        I
I PRANDTL NUMBER                      6.075                   25.163         I
I UOPK,LIQUID                      0.000 /    0.000          0.000 /    0.000 I
I      VAPOR                       0.000 /    0.000          0.000 /    0.000 I
I SURFACE TENSION     DYNES/CM     0.000 /    0.000          0.000 /    0.000 I
I FILM COEF(SCL) BTU/HR-FT2-F         100.3 (1.000)            130.7 (1.000)  I
I FOULING LAYER THICKNESS   IN          0.000                    0.000        I
I----------------------------------------------------------------------------I
I THERMAL RESISTANCE                                                         I
I UNITS:   (FT2-HR-F/BTU)        (PERCENT)    (ABSOLUTE)                     I
I SHELL FILM                        43.52       0.00997                      I
I TUBE  FILM                        42.88       0.00982                      I
I TUBE  METAL                        3.63       0.00083                      I
I TOTAL FOULING                      9.97       0.00228                      I
I ADJUSTMENT                        -0.02       0.00000                      I
I----------------------------------------------------------------------------I
I PRESSURE DROP                     SHELL-SIDE                TUBE-SIDE      I
I UNITS: (PSIA   )               (PERCENT)    (ABSOLUTE)   (PERCENT)   (ABSOLUTE)I
I WITHOUT NOZZLES                   79.96          0.18      100.00         2.04 I
I INLET   NOZZLES                   12.53          0.03        0.00         0.00 I
I OUTLET  NOZZLES                    7.52          0.02        0.00         0.00 I
I TOTAL    /SHELL                                  0.22                     2.04 I
I TOTAL    /UNIT                                   0.67                     6.11 I
I DP SCALER                                        1.00                     1.00 I
I----------------------------------------------------------------------------I
I CONSTRUCTION OF ONE SHELL                                                  I
I----------------------------------------------------------------------------I
I TUBE:OVERALL LENGTH      40.0        FT  EFFECTIVE LENGTH    39.88     FT I
I      TOTAL TUBESHEET THK  1.5        IN  AREA RATIO (OUT/IN)  1.284       I
I      THERMAL COND.        9.4BTU/HR-FT-F DENSITY            501.10 LB/FT3I
I----------------------------------------------------------------------------I
I BUNDLE: DIAMETER           3.2        IN  NET FREE FLOW AREA   0.044 FT2  I
I----------------------------------------------------------------------------I
```

References

[1] Kakac S, Liu H. Heat Exchangers: Selection, Rating and Thermal Design. Boca Raton, FL: CRC Press; 1998.

[2] Perry RH, Green DW, editors. Perry's Chemical Engineers' Handbook. 7th ed. New York: McGraw-Hill; 1997.

[3] Kern DQ, Kraus AD. Extended Surface Heat Transfer. New York: McGraw-Hill; 1972.

[4] Henry JAR. Headers, nozzles and turnarounds. In: Heat Exchanger Design Handbook, Vol. 2. New York: Hemisphere Publishing Corp.; 1988.

[5] DeLorenzo B, Anderson ED. Heat transfer and pressure drop of liquids in double-pipe fin-tube exchangers. Trans ASME 1945;67:697–702.

[6] Guy AR. Applications of double-pipe heat exchangers. In: Heat Exchanger Design Handbook, Vol. 3. New York: Hemisphere Publishing Corp.; 1983.

[7] Kraus AD, Aziz A, Welty J. Extended Surface Heat Transfer. New York: Wiley; 2001.

[8] Hewitt GF, Shires GL, Bott TR. Process Heat Transfer. Boca Raton, FL: CRC Press; 1994.

Appendix 4.A. Hydraulic Equations in SI Units

For convenience in solving problems formulated in metric units, the working equations for hydraulic calculations are given below in terms of SI units. The starting point is Equation (4.3):

$$\Delta P_f = \frac{f\, LG^2}{2 g_c \rho_{water} D_i s}$$

In the SI unit system, $g_c = 1.0$ and the density of water at 4°C is 1000 kg/m³. With these numerical values, Equation (4.3) becomes:

$$\Delta P_f = \frac{fLG^2}{2000 D_i s} \tag{4.A.1}$$

This equation is modified by inclusion of the viscosity correction factor to obtain the following equation for the friction loss in straight sections of pipe:

$$\Delta P_f = \frac{fLG^2}{2000 D_i s \phi} \tag{4.A.2}$$

where

$$\Delta P_f \propto \text{Pa}$$

$$L, D_i \propto \text{m}$$

$$G \propto \text{kg/s}\cdot\text{m}^2$$

s, f, ϕ are dimensionless

The above units apply in all of the following equations. The pressure drop associated with one velocity head is given by:

$$\Delta P = \left(\frac{1}{2 g_c \rho_{water}}\right)\left(\frac{G^2}{s}\right) \tag{4.A.3}$$

Substituting $g_c = 1.0$ and $\rho_{water} = 1000$ kg/m³ yields:

$$\Delta P = 5.0 \times 10^{-4} G^2/s \tag{4.A.4}$$

The pressure drop in return bends is estimated using the following equations:

$$\Delta P_r = 6.0 \times 10^{-4}(2N_{HP} - 1)G^2/s \quad \text{(turbulent flow)} \tag{4.A.5}$$

$$\Delta P_r = 7.5 \times 10^{-4}(2N_{HP} - 1)G^2/s \quad (\text{laminar flow}, Re \geq 500) \tag{4.A.6}$$

The following equations are used to estimate nozzle losses:

$$\Delta P_n = 7.5 \times 10^{-4} N_{HP} G_n^2/s \quad \text{(turbulant flow)} \tag{4.A.7}$$

$$\Delta P_n = 1.5 \times 10^{-3} N_{HP} G_n^2/s \quad (\text{laminar flow}, Re \geq 100) \tag{4.A.8}$$

In Equations (4.A.5–4.A.8), N_{HP} represents the number of hairpins connected in series.

Appendix 4.B. Incremental Analysis

It was noted in Appendix 3.A that the standard method for heat-exchanger analysis based on the LMTD depends on a number of simplifying assumptions, two of which are:

(1) The overall heat-transfer coefficient is constant throughout the heat exchanger.
(2) The stream enthalpies vary linearly with temperature. For single-phase operation, this implies constant heat capacities.

Although these conditions are never satisfied exactly in practice, analysis based on the LMTD is often sufficiently accurate for exchangers involving only single-phase flow. However, the above two assumptions are not generally applicable to operations involving a phase change, as in condensers, vaporizers, and reboilers. Even with single-phase operation, the overall coefficient sometimes varies greatly along the length of the exchanger. This may happen, for example, when the variation of fluid viscosity is large enough to cause a change in flow regime from turbulent at one end of the exchanger to laminar at the other end. In these situations an incremental (or zone) analysis is required, in which the governing equations are numerically integrated over the heat-transfer surface. This methodology is employed in all commercial software packages used for heat-exchanger design; however, implementation details vary considerably among computer programs.

The rate of heat transfer between the streams in a differential section of area dA in the heat exchanger is:

$$dq = U\, dA(T_h - T_c) \tag{4.B.1}$$

where T_h and T_c are the local temperatures of the hot and cold streams, respectively. This equation can be formally integrated over the heat-transfer area to obtain the total duty as follows:

$$q_{Tot} = \int_0^A U(T_h - T_c)dA \tag{4.B.2}$$

Alternatively, the heat-transfer area required to achieve a given duty, q_{Tot}, is obtained by solving Equation (4.B.1) for dA and integrating:

$$A = \int_0^{q_{Tot}} \frac{dq}{U(T_h - T_c)} \tag{4.B.3}$$

In practice the integrals are evaluated numerically by discretizing the exchanger in terms of either area or duty. In the latter case, for example, Equation (4.B.3) is replaced by:

$$A \cong \sum_i \Delta A_i = \sum_i \frac{\Delta q_i}{U(T_{h,i} - T_{c,i})} \tag{4.B.4}$$

where

$$q_{Tot} = \sum_i \Delta q_i \tag{4.B.5}$$

The stream energy balances are used to calculate the local stream temperatures. For a single-phase counter-flow exchanger in which the calculations are started at the cold end, the equations are:

$$\Delta q_i = C_{h,i}(T_{h,i+1} - T_{h,i}) = C_{c,i}(T_{c,i+1} - T_{c,i}) \tag{4.B.6}$$

where $C \equiv \dot{m}\, C_P$. Similarly, the hydraulic equations are used to obtain local values of the stream pressures. Fluid properties are evaluated at the local stream conditions and used to calculate the local film and overall heat-transfer coefficients. For operations involving a phase change, additional thermodynamic calculations are performed to determine the local flow rate and composition of each phase. An example that illustrates the computational procedure for a condenser is given in Section 11.10.

Computer programs use the results of the incremental analysis to calculate mean values of the heat-transfer coefficients and temperature difference that are reported in the output summaries. Details of the incremental calculations are generally given in the full output files. It should be noted that for rating calculations involving counter-flow or multi-pass heat exchangers, the incremental analysis involves iteration. Since the output stream pressures (and possibly the outlet temperatures as well, depending on the input specifications) are unknown, at least one stream pressure (and possibly a stream temperature) is unknown at each end of the exchanger. In order to start the incremental calculations, however, the temperatures and pressures of both streams must be known at one end of the exchanger. Hence, the missing values must be assumed and subsequently updated iteratively until convergence is attained.

Many computer programs (HEXTRAN being an exception) use the incremental method on all problems regardless of whether it is really needed. Although this method undoubtedly provides the most accurate representation of the heat-transfer process, the overall accuracy is still limited by the engineering correlations used in the calculations. Furthermore, the incremental calculations sometimes fail to converge. This can lead to the seemingly paradoxical situation in which a very sophisticated computer program fails to solve a rather simple problem.

Notations

A	Heat-transfer surface area
A_C	Calculated heat-transfer surface area based on clean overall heat-transfer coefficient
A_f	Flow area
A_{fins}	$2n_t N_f b_c L$ = Surface area of all fins in a finned-pipe heat exchanger
A_i	$\pi D_i L$ = Internal surface area of inner pipe in a double-pipe heat exchanger
A_{prime}	Prime surface area in a finned-pipe heat exchanger
A_{req}	Required heat-transfer area in a heat exchanger
A_{Tot}	$A_{prime} + A_{fins}$ = Total heat-transfer surface area in a finned-pipe heat exchanger
b	Fin height
b_c	$b + \tau/2$ = Corrected fin height
C_c, C_h	Value of $\dot{m}C_P$ for cold and hot streams, respectively
C_P	Heat capacity at constant pressure
D	Diameter
D_e	Equivalent diameter
D_i	Internal diameter of inner pipe in a double-pipe heat exchanger
D_o	External diameter of inner pipe in a double-pipe heat exchanger
D_1	External diameter of inner pipe in an annulus
D_2	Internal diameter of outer pipe in an annulus
F	LMTD correction factor for series-parallel arrangement of hairpins
f	Darcy friction factor

G	$\dot{m}/A_f$ = Mass flux
G_n	Mass flux in a nozzle
g	Gravitational acceleration
g_c	Unit conversion factor $= 32.174 \dfrac{\text{lbm}\cdot\text{ft/s}^2}{\text{lbf}} = 1.0 \dfrac{\text{kg}\cdot\text{m/s}^2}{\text{N}}$
h	Heat-transfer coefficient
h_i	Heat-transfer coefficient for inner pipe in a double-pipe heat exchanger
h_o	Heat-transfer coefficient for annulus in a double-pipe heat exchanger
h'_o	$(h_o + R_{Do})^{-1}$
j_H	$(h_o D_e/k) \times Pr^{-1/3}(\mu/\mu_w)^{-0.14}$ = Modified Colburn factor for heat transfer in the annulus of a double-pipe heat exchanger
k	Thermal conductivity
L	Pipe length
m	$(2h_o/k\,\tau)^{1/2}$ = Fin parameter
$\dot{m}$	Mass flow rate
N_f	Number of fins on a finned pipe
N_{HP}	Number of hairpins connected in series
Nu	Nusselt number
n_t	Number of tubes in a multi-tube hairpin heat exchanger
P	Pressure; parameter used to calculate LMTD correction factor
Pr	Prandtl number
q	Rate of heat transfer
q_{Tot}	Total duty for a heat exchanger
R	Parameter used to calculate LMTD correction factor
R_{Di}	Fouling factor for fluid in the inner pipe of a double-pipe heat exchanger
R_{Do}	Fouling factor for fluid in the annulus of a double-pipe heat exchanger
Re	Reynolds number
Re_i	Reynolds number for fluid in the inner pipe of a double-pipe heat exchanger
Re_o	Reynolds number for fluid in the annulus of a double-pipe heat exchanger
R_{th}	Thermal resistance
s	Specific gravity
T	Temperature
T_a	Inlet temperature of series stream for series-parallel configuration of hairpins
T_{ave}	Average temperature of fluid in the annulus of a double-pipe heat exchanger
T_b	Outlet temperature of series stream for series-parallel configuration of hairpins
T_c	Temperature of cold stream
T_h	Temperature of hot stream
T_p	Average temperature of prime surface in a finned-pipe heat exchanger
T_w	Average pipe wall temperature in an un-finned heat exchanger
T_{wtd}	Weighted average temperature of finned surface in a finned-pipe heat exchanger
t_a	Inlet temperature of parallel stream in a series-parallel configuration of hairpins
t_{ave}	Average temperature of fluid in the inner pipe of a double-pipe heat exchanger
t_b	Outlet temperature of parallel stream in a series-parallel configuration of hairpins
U_C	Clean overall heat-transfer coefficient
U_D	Design overall heat-transfer coefficient
V	Average fluid velocity
x	Number of parallel branches in a series-parallel configuration of hairpins

Greek Letters

α	Parameter in exponential relationship for temperature dependence of liquid viscosity
β	Parameter in exponential relationship for temperature dependence of liquid viscosity
ΔP	Pressure drop
ΔP_f	Pressure drop due to fluid friction in straight sections of pipe
$\Delta P_{f,i}$	Frictional pressure drop in straight sections of inner pipe
$\Delta P_{f,o}$	Frictional pressure drop in straight sections of annulus

ΔP_i	Total pressure drop for fluid flowing through inner pipe of double-pipe heat exchanger
ΔP_n	Pressure loss in nozzles
ΔP_o	Total pressure drop for fluid flowing through annulus of double-pipe heat exchanger
ΔP_r	Pressure loss in return bends of double-pipe heat exchanger
ΔT_{ln}	Logarithmic mean temperature difference
η_f	Fin efficiency
η_w	Weighted efficiency of finned surface
κ	D_2/D_1= Diameter ratio for annulus
μ	Viscosity
μ_w	Fluid viscosity evaluated at average temperature of pipe wall
ρ	Fluid density
τ	Fin thickness
ϕ	Viscosity correction factor
ϕ_i	Viscosity correction factor for fluid in the inner pipe of a double-pipe heat exchanger
ϕ_o	Viscosity correction factor for fluid in the annulus of a double-pipe heat exchanger

Problems

(4.1) In Example 4.1, repeat Steps (d)–(j) of the first trial using Equation (2.39) to calculate both h_i and h_o. The correction for short pipes can be neglected, i.e., $1 + (D/L)^{2/3} \cong 1.0$. Compared with Example 4.1, what are the percentage differences in the values of h_i and h_o? Which procedure is more conservative for design purposes?

(4.2) A double-pipe heat exchanger consists of five hairpins connected in series. The hairpins are each 20 ft long with 4-in. schedule 40 outer pipes, 2-in. schedule 40 inner pipes and internal return bends. Nozzles are made of 2-in. schedule 40 pipe. 40,000 lb/h of a petroleum fraction having the properties given below will flow in the annulus.

Petroleum fraction property	Value
k (Btu/h · ft · °F)	0.15
μ (cp)	2.0
s	0.80
Pr	17.75

(a) Calculate the heat-transfer coefficient, h_o, for the annulus.
(b) Calculate the total pressure drop, ΔP_o, for the annulus.
(c) What percentage of the total pressure loss occurs in the nozzles?
(d) If schedule 40 pipe is used, what size nozzles would be required to keep the pressure drop below 7 psi?

Ans. (a) 179 Btu/h · ft^2 · °F. (b) 8.9 psi. (c) 41%. (d) 3-in. nozzles.

(4.3) A fluid with $k = 0.173$ W/m · K and $Pr = 6$ flows in a finned annulus having the following configuration:

Length = 6.1 m	Number of fins = 48
Outer pipe ID = 102.26 mm	Fin height = 12.7 mm
Inner pipe OD = 73 mm	Fin thickness = 0.889 mm

Calculate the heat-transfer coefficient and (Darcy) friction factor for a Reynolds number of:
(a) 100.
(b) 1000.

Ans. (a) $h = 45$ W/m^2 · K; $f = 0.64$. (b) $h = 146$ W/m^2 · K; $f = 0.085$.

(4.4) The petroleum fraction of Example 4.2 flows at a rate of 1200 lb/h in the annulus of a 4-in. multi-tube hairpin exchanger having seven tubes, each with 20 fins. The fin thickness is 0.035 in. and the hairpin length is 15 ft. Other pipe and fin dimensions are given in Table 4.2. Calculate the following:
(a) The flow area for the annulus.
(b) The equivalent diameter for the annulus.
(c) The Reynolds number for the oil.

(d) The heat-transfer coefficient for the oil.
(e) The friction factor for the oil.
(f) The friction loss per unit length in the annulus.

Ans. (a) 0.0521 ft^2. (b) 0.0276 ft. (c) 131.4. (d) 19.1 Btu/h · ft^2 · °F. (e) 0.487. (f) 0.00156 psi/ft.

(4.5) A double-pipe heat exchanger consists of two hairpins connected in series. Each hairpin is 20 ft long with the following configuration:

Outer pipe: 4-in. schedule 40	Fin thickness: 0.035 in.
Inner pipe: 2-in. schedule 40 with 36 fins	Nozzles: 2-in. schedule 40
Fin height: 0.75 in.	Return bends: Internal

The petroleum fraction of Problem 4.2 will flow in the annulus at a rate of 20,000 lb/h.
(a) Calculate the heat-transfer coefficient, h_o, for the annulus.
(b) Calculate the total pressure drop, ΔP_o, for the annulus.

Ans. (a) 94 Btu/h · ft^2 · °F. (b) 3.6 psi.

(4.6) An oil cooler will consist of three finned-pipe hairpins connected in series. The hairpins have the following dimensions:

Length of one hairpin: 20 ft	Fin thickness: 0.035 in.
Outer pipe: 2-in. schedule 10	Nozzles: 1-in. schedule 40
Inner pipe: $^3/_4$ -in. schedule 40 with 30 fins	Return bends: Internal
Fin height: 0.5 in.	

12,000 lb/h of oil ($\rho = 55.2$1bm/ft^3, $\mu = 7.5$cp) will flow in the annulus. The maximum allowable pressure drop for the oil is 20 psi. Will the actual pressure drop be acceptable?

(4.7) An alternate design for the oil cooler of Problem 4.6 consists of two hairpins connected in series, each hairpin being 15 ft long. The outer pipes are 4-in. schedule 10, and each outer pipe contains two $^3/_4$ -in. schedule 40 pipes with fins as described in Problem 4.6. Nozzles and return bends are the same as in Problem 4.6. Will the oil pressure drop be acceptable for this configuration?

(4.8) A hot stream is to be cooled from 300°F to 275°F in a double-pipe heat exchanger by heating a cold stream from 100°F to 290°F. In order to control pressure drops, the heat exchanger will be comprised of three parallel banks of hairpins with the hot stream flowing in series and the cold stream divided into three parallel branches. Calculate the mean temperature difference for the heat exchanger.

Ans. 37.6°F

(4.9) 72,500 lb/h of oil (s = 0.75, $\mu = 0.83$ cp) will be cooled using a double-pipe heat exchanger consisting of eight hairpins. Each hairpin is 20 ft long with 3.5-in. schedule 40 outer pipe, 2-in. schedule 40 un-finned inner pipe, 2-in. schedule 40 nozzles and internal return bend. The maximum allowable pressure drop for the oil is 8 psi. Determine whether the pressure drop will be acceptable if:
(a) The oil flows through the inner pipes, which are connected in two parallel banks of four hairpins each.
(b) The oil flows through the annuli, which are connected in two parallel banks of four hairpins each.

(4.10) 50,000 lb/h of a fluid having a density of 53 lbm/ft^3 and a viscosity of 1.5 cp will flow in the annulus of a double-pipe heat exchanger. The hairpins are 30 ft long with 3-in. schedule 40 outer pipe, 1.5-in. schedule 40 inner pipe, internal return bends and 1.5-in. schedule 40 nozzles. A total of six hairpins will be used, and the maximum allowable pressure drop for the stream is 12 psi. Determine which of the following configurations should be used in connecting the annuli of the hairpins:

- Six hairpins in series
- Two parallel banks of three hairpins each
- Three parallel banks of two hairpins each

(4.11) In designing a double-pipe heat exchanger for a particular service it is found that the pressure drop for the stream flowing through the inner pipe is excessively high (35 psi) compared with the allowable pressure drop (7 psi). It is decided to remedy the problem by using a series-parallel configuration of hairpins. Assuming the same size and number of hairpins are used, how many parallel branches should the stream be divided into if:
(a) The flow is turbulent?
(b) The flow is laminar?

(4.12) 15,000 lb/h of benzene will be heated from 80°F to 120°F by cooling a stream of ortho-toluidine (C_7H_9N) from 160°F to 100°F. Inlet pressures are 40 psia for ortho-toluidine and 50 psia for benzene. A maximum pressure drop of 15 psia is specified for each stream. A number of used carbon steel hairpins are available for this service. The hairpins are each 20 ft long with 3-in. schedule 40 outer pipe, 1.5-in. un-finned schedule 40 inner pipe, internal return bends and 2-in. schedule 40 nozzles. Properties of ortho-toluidine at 130°F are given below. Determine the number and configuration of hairpins required for this service.

Ortho-toluidine property	Value at 130°F
C_P (Btu/lbm · °F)	0.48
k (Btu/h · ft · °F)	0.081
μ (cp)	1.85*
ρ (lbm/ft^3)	60.5
Pr	26.5

* $\log_{10}\mu(\text{cp}) = 1085.1\left(\frac{1}{T(K)} - \frac{1}{356.46}\right)$

(4.13) 92,500 lb/h of oil will be cooled from 230°F to 200°F by heating a process water stream from 60°F to 100°F in a stainless steel ($k = 9.4$ Btu/h · ft.°F) double-pipe heat exchanger. The hairpins to be used are each 20 ft long with 3.5-in. schedule 40 outer pipe, 2-in. schedule 40 un-finned inner pipe, 2-in. schedule 40 nozzles and internal return bend. Maximum allowable pressure drops are 8 psi for the oil and 15 psi for the water. Inlet pressures are 50 psia for the oil and 60 psia for the water. Fouling factors of 0.002 and 0.001 h · ft^2 · °F/Btu are required for the oil and water streams, respectively. Properties of the oil may be assumed constant at the values given below. Determine the number and configuration of hairpins to be used for this service.

Oil property	Value
C_P (Btu/lbm · °F)	0.59
k (Btu/h · ft · °F)	0.073
μ (cp)	0.83
s	0.76
Pr	16.2

(4.14) An alternative design for the heat exchanger of Problem 4.13 is to be developed that utilizes finned hairpins. The hairpin dimensions are the same as in Problem 4.13 except that each inner pipe contains 24 fins of height 0.5 in. and thickness 0.035 in. Determine the number and configuration of hairpins to be used in this case.

(4.15) Use HEXTRAN or other suitable software to solve Problem 4.12.

(4.16) Use HEXTRAN or other suitable software to solve Problem 4.13.

(4.17) Use HEXTRAN or other suitable software to solve Problem 4.14.

(4.18) 30,000 lb/h of kerosene will be heated from 75°F to 110°F by cooling a gasoline stream from 160°F to 135°F. For each stream the inlet pressure is 50 psia and a maximum pressure drop of l0 psi is specified. Published fouling factors for oil refinery streams should be used for this application. Use HEXTRAN or other suitable software to design a double-pipe heat exchanger for this service.

Note: To begin, try using 3-in. by 1.5-in. schedule 40 hairpins. The dimensions of the hairpins may be changed as necessary during the design process. However, only standard sizes should be used.

Property	Gasoline	Kerosene
C_P (Btu/lbm · °F)	0.55	0.48
k (Btu/h · ft · °F)	0.087	0.081
μ (cp)	0.42	1.60
s	0.74	0.82
Pr	6.42	22.9

(4.19) 40,000 lb/h of a hydrocarbon liquid (30°API, K_w, = 11.5) are to be cooled from 225°F to 150°F using water available from a municipal water system at 70°F and 65 psia. The inlet pressure of the hydrocarbon is 50 psia and a fouling factor of 0.0025 h·ft^2·°F/Btu is required for this stream. The maximum pressure drop for each stream is 10 psi. Carbon steel is acceptable for this application. Use HEXTRAN or other suitable software to design an un-finned hairpin heat exchanger for this service.

(4.20) Use HEXTRAN or other suitable software to design a finned-pipe hairpin heat exchanger for the service of Problem 4.19.

(4.21) 20,000 lb/h of lube oil (26°API, $K_w = 12.2$) will be cooled from 430°F to 350°F by heating a light gas oil (33°API, $K_w = 11.8$) from 310°F to 370°F. The inlet pressure of each stream will be 50 psia and maximum pressure drops of 15 psi for the lube oil and 10 psi for the gas oil are specified. Design a double pipe heat exchanger for this service using HEXTRAN or other suitable software.

(4.22) 13,000 lb/h of amyl acetate (*n*-pentyl acetate, $C_7H_{14}O_2$) will be cooled from 150°F to 100°F using 12,000 lb/h of amyl alcohol (1-pentanol, $C_5H_{12}O$) which is available at 75°F. Inlet pressures are 45 psia for the acetate and 55 psia for the alcohol. A maximum pressure drop of 10 psi is specified for each stream. Use HEXTRAN or other suitable software to design a double-pipe heat exchanger for this service.

(4.23) Use HEXTRAN and/or other available software to rate the finned-pipe heat exchanger designed by hand in Example 4.3. Compare the results from the computer solution(s) with those from the hand calculations of Example 4.3.

(4.24) Repeat Example 4.2 if each hairpin consists of a 4-in. schedule 40 outer pipe with a tube bundle containing 7 U-tubes, each having OD of 1.0 in. and ID of 0.81 in. The hairpins are 16 ft long with internal return bends and 2-in. schedule 40 nozzles on the annulus side. Material of construction is 316 stainless steel.

(4.25) As noted in Example 4.4, HEXTRAN uses linear interpolation to compute heat-transfer coefficients in the transition region. This method yields reasonable results and has the additional advantage that the calculated coefficient is a continuous function of Reynolds number over all flow regimes. For the initial configuration in Example 4.1 (hairpins connected in series, aniline in annulus), use this method to calculate h_o by proceeding as follows:

(a) Use Equation (2.42) to calculate the heat-transfer coefficient for the annulus at $Re = 2000$. As noted in Example 3.3, the length, L, for the annulus should be taken as the length of pipe in one leg of one hairpin, or 16 ft in this case.

(b) Use the Seider–Tate correlation for turbulent flow to calculate the heat-transfer coefficient for the annulus at $Re =$ 10,000.

(c) Use linear interpolation between the values calculated in parts (a) and (b) to obtain the heat-transfer coefficient at the actual Reynolds number of 8218. Compare your result with the value of 248.7 Btu/h·ft^2·°F obtained from HEXTRAN in Example 4.4.

5 Design of Shell-and-Tube Heat Exchangers

5.1 Introduction

Preliminary design- and thermal-rating calculations for shell-and-tube exchangers were presented in Chapter 3. Both are used, along with hydraulic calculations, in the complete design of shell-and-tube exchangers. As with double-pipe units, these exchangers are typically designed with specified pressure-drop constraints on the two streams. Due to the greater complexity and variety of shell-and-tube configurations, however, there are more design variables that must be taken into consideration. This complexity is also the source of computational difficulties on the shell side, which were alluded to in Chapter 3. The upshot is that design of shell-and-tube equipment is considerably more complicated than design of double-pipe exchangers.

In this chapter, the key aspects of shell-and-tube exchanger design are presented and illustrated with examples involving both plain and finned tubing. The Simplified Delaware method is used for shell-side calculations in order to keep the computations manageable. More sophisticated computational methods are considered separately in subsequent chapters.

5.2 Heat-Transfer Coefficients

The heat-transfer correlations used in this chapter were presented in Section 2.4, Section 3.9, and Section 4.2. They are repeated here for convenience.

For the tube-side heat-transfer coefficient, h_i, the Seider–Tate and Hausen equations are used as follows:

For $Re \geq 10^4$,

$$Nu = 0.023\, Re^{0.8}\, Pr^{1/3} (\mu/\mu_w)^{0.14} \tag{4.1}$$

For $2100 < Re < 10^4$,

$$Nu = 0.116\left[Re^{2/3} - 125\right] Pr^{1/3} (\mu/\mu_w)^{0.14}\left[1 + (D/L)^{2/3}\right] \tag{2.38}$$

For $Re \leq 2100$,

$$Nu = 1.86[RePrD/L]^{1/3} (\mu/\mu_w)^{0.14} \tag{2.37}$$

Alternatively, Equation (2.39) may be used for the turbulent and transition regions. The Reynolds number in these correlations is calculated using the mass flow rate per tube:

$$\dot{m}_{per\ tube} = \frac{\dot{m}_t n_p}{n_t} \tag{3.19}$$

where

$\dot{m}_t$ = total mass flow rate of tube-side fluid
n_p = number of tube-side passes
n_t = number of tubes

The shell-side heat-transfer coefficient, h_o, is computed using the following curve fit to Figure 3.17 (note that j_H is dimensionless):

$$j_H = 0.5(1 + B/d_s)\left(0.08\, Re^{0.6821} + 0.7\, Re^{0.1772}\right) \tag{3.21}$$

where

$j_H \equiv (h_o D_e/k)(C_P\mu/k)^{-1/3}(\mu/\mu_w)^{-0.14}$
B = baffle spacing
d_s = shell ID
D_e = equivalent diameter, in proper units, from Figure 3.17.

The Reynolds number in Equation (3.21) is calculated using the equivalent diameter and flow area given in Figure 3.17. The correlation is valid for tube bundles employing plain or finned heat-exchanger tubing, 20% cut segmental baffles, and one pair of sealing strips (refer to Section 5.7.8) per 10 tube rows. It is also based on TEMA standards for tube-to-baffle and baffle-to-shell clearances. (See Section 6.7 for details regarding clearances.) Although the correlation was derived from results of the Delaware study, it contains a built-in safety factor of approximately 25% [1]. Therefore, it should generally yield lower values of the heat-transfer coefficient than the Delaware method itself.

5.3 Hydraulic Calculations

5.3.1 Tube-Side Pressure Drop

The pressure drop due to fluid friction in the tubes is given by Equation (4.5) with the length of the flow path set to the tube length times the number of tube passes. Thus,

$$\Delta P_f = \frac{f\, n_p L G^2}{7.50 \times 10^{12} D_i s \phi} \tag{5.1}$$

where

f = Darcy friction factor (dimensionless)
L = tube length (ft)
G = mass flux ($\text{lbm/h} \cdot \text{ft}^2$)
D_i = tube ID (ft)
s = fluid specific gravity (dimensionless)
ϕ = viscosity correction factor (dimensionless)
= $(\mu/\mu_w)^{0.14}$ for turbulent flow
= $(\mu/\mu_w)^{0.25}$ for laminar flow

Here, $\Delta P_f \propto$ psi. (English units are used throughout this section; see Appendix 5.A for the hydraulic equations in terms of SI units.) For laminar flow, the friction factor is given by:

$$f = \frac{64}{Re} \tag{4.6}$$

For turbulent flow in commercial heat-exchanger tubes, the following equation can be used for $Re \geq 3000$:

$$f = 0.4137\, Re^{-0.2585} \tag{5.2}$$

(Note the similarity between this equation and Equation (4.8) for commercial pipe. The surface roughness tends to be lower for heat-exchanger tubes than for pipes, resulting in somewhat smaller friction factors for tubing.)

The tube-side fluid experiences a sudden contraction when it enters the tubes from the header and a sudden expansion when it exits the tubes at the opposite header. The associated pressure losses can be calculated using standard hydraulic formulas [2], and depend on the tube diameter, pitch and layout. However, for commonly used tube configurations, the sum of the entrance and exit losses can be approximated by 0.5 tube velocity heads in turbulent flow [2]. For laminar flow the number of velocity heads depends on the Reynolds number, but for $Re \geq 500$ a reasonable approximation is one velocity head for the entrance loss and 0.75 velocity head for the exit loss, giving a total of 1.75 velocity heads.

In the return header, the fluid experiences a 180° change of direction. Henry [2] has suggested an allowance of 1.5 velocity heads for the associated pressure loss, similar to that for a pipe bend. Since the flow pattern in the header is complex regardless of whether the flow in the tubes is laminar or turbulent, this allowance is probably reasonable for both flow regimes.

Combining the tube entrance, tube exit, and return losses gives an allowance for minor losses of $(2\, n_p - 1.5)$ velocity heads for turbulent flow and $(3.25\, n_p - 1.5)$ velocity heads for laminar flow and $Re \geq 500$. (Note that the first pass involves only tube entrance and exit losses, and not a return loss.)

For exchangers employing U-tubes, the situation is somewhat different because the tube entrance and exit losses occur on alternate passes rather than on each pass. Furthermore, on even-numbered passes, the return header is replaced by a 180° bend in the tubing. The pressure loss associated with the latter can be treated the same as in a double-pipe exchanger by allowing 1.2 velocity heads for turbulent flow and 1.5 velocity heads for laminar flow with $Re \geq 500$. This results in a minor loss allowance of $(1.6\, n_p - 1.5)$ velocity heads for turbulent flow and $(2.38\, n_p - 1.5)$ velocity heads for laminar flow with $Re \geq 500$.

The rather untidy situation regarding minor losses is summarized in Table 5.1. These values can be compared with allowances reported elsewhere. The shell-and-tube module in the HEXTRAN software package uses two velocity heads per pass for the tube side minor losses, which agrees (except for the first pass) with the tabled value for turbulent flow and regular tubes. Bell and Mueller [3] recommend an allowance of three velocity heads per pass. Kern [4] stated without documentation that an allowance of four velocity heads per pass was appropriate for these losses, and this value has often been cited by others, e.g., Ref. [1,5]. Although Table 5.1 indicates that Kern's allowance is conservative enough to cover nearly all situations, it appears to be excessive for turbulent flow.

TABLE 5.1 Number of Velocity Heads Allocated for Minor Losses on Tube Side

Flow Regime	Regular Tubes	U-Tubes
Turbulent	$2\, n_p - 1.5$	$1.6\, n_p - 1.5$
Laminar, $Re \geq 500$	$3.25\, n_p - 1.5$	$2.38\, n_p - 1.5$

Equation (4.11) gives the pressure drop associated with one velocity head. If we denote by α_r the number of velocity heads allocated for minor losses, then we obtain the following equation:

$$\Delta P_r = 1.334 \times 10^{-13} \alpha_r G^2/s \tag{5.3}$$

where α_r is obtained from Table 5.1. Alternatively, α_r may be selected in accordance with one of the other recommendations (Kern, Bell and Mueller, or HEXTRAN) cited above.

Each shell in a shell-and-tube exchanger contains an inlet and an outlet nozzle for the tube-side fluid. The nozzle pressure drop can be estimated in the same manner as was done for double-pipe exchangers in Chapter 4. In particular, Equations (4.14) and (4.15) are applicable if the number of hairpins is replaced by the number of shells. Thus,

$$\Delta P_n = 2.0 \times 10^{-13} N_s G_n^2/s \quad \text{(turbulent flow)} \tag{5.4}$$

$$\Delta P_n = 4.0 \times 10^{-13} N_s G_n^2/s \quad (\text{laminar flow, } Re_n \geq 100) \tag{5.5}$$

where G_n and Re_n are the mass flux and Reynolds number, respectively, for the nozzles, and N_s is the number of shells connected in series. These equations give the total pressure drop, in psi, for all the nozzles, assuming they are all the same. When the inlet and outlet nozzles differ significantly in size and/or flow regime, each nozzle can be treated individually as discussed in Section 4.3.

5.3.2 Shell-Side Pressure Drop

In the Simplified Delaware method the shell-side pressure drop per shell, excluding nozzle losses, is computed using the following equation [1], which is similar to Equation (5.1):

$$\Delta P_f = \frac{f\, G^2 d_s (n_b + 1)}{7.50 \times 10^{12}\, d_e s \phi} \tag{5.6}$$

where

f = friction factor (dimensionless)
G = mass flux = $\dot{m}/a_s$ (lbm/h · ft^2)
a_s = flow area across tube bundle (ft^2)
$= d_s C' B/(144 P_T)$
d_s = shell ID (in.)
C' = gap (in.)
B = baffle spacing (in.)
P_T = tube pitch (in.); replaced by $P_T/\sqrt{2}$ for 45° tube layout
n_b = number of baffles
d_e = equivalent diameter from Figure 3.17 (in.)
s = fluid specific gravity
ϕ = viscosity correction factor = $(\mu/\mu_w)^{0.14}$
ΔP_f = pressure drop (psi)

The shell-side friction factor is given in Figure 5.1 in dimensional form. For use in Equation (5.6), values from this figure must be multiplied by the factor 144 in.^{2}/ft^2, which is included in the following linear interpolation formula:

$$f = 144\{f_1 - 1.25(1 - B/d_s)(f_1 - f_2)\} \tag{5.7}$$

where

f_1 = friction factor from graph for $B/d_s = 1.0$
f_2 = friction factor from graph for $B/d_s = 0.2$

Approximate curve fits for f_1 and f_2 are as follows, with d_s in inches.

For $Re \geq 1000$,

$$f_1 = (0.0076 + 0.000166\, d_s) Re^{-0.125} \quad (8 \leq d_s \leq 42) \tag{5.8}$$

$$f_2 = \left(0.0016 + 5.8 \times 10^{-5} d_s\right) Re^{-0.157} \quad (8 \leq d_s \leq 23.25) \tag{5.9}$$

For $Re < 1000$

$$f_1 = \exp\left[0.092(\ln Re)^2 - 1.48 \ln Re - 0.000526\, d_s^2 + 0.0478\, d_s - 0.338\right] \quad (8 \leq d_s \leq 42) \tag{5.10}$$

$$f_2 = \exp\left[0.123\, (\ln Re)^2 - 1.78 \ln Re - 0.00132\, d_s^2 + 0.0678\, d_s - 1.34\right] \quad (8 \leq d_s \leq 23.25) \tag{5.11}$$

Note: In Equations (5.9) and (5.11), set $d_s = 23.25$ for shell diameters greater than 23.25 in.

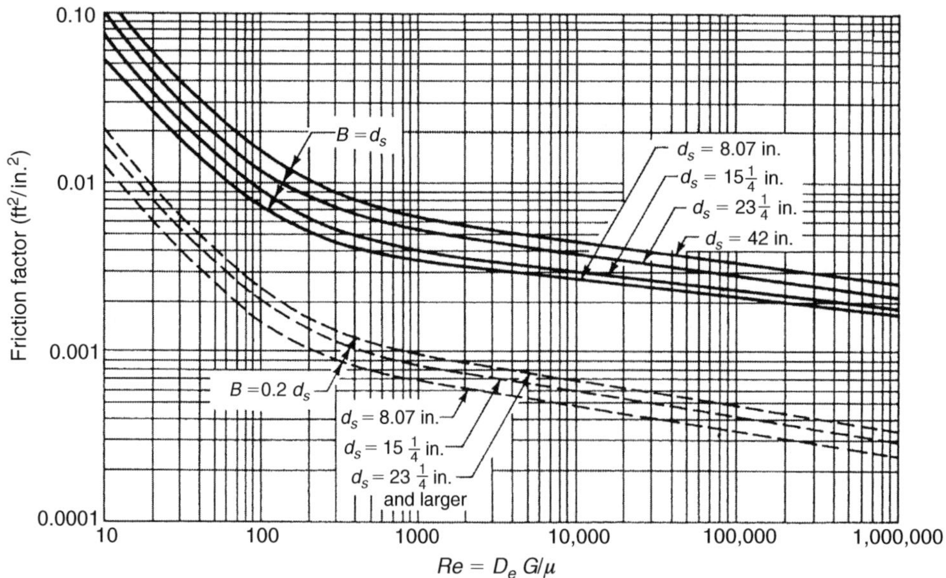

FIGURE 5.1 Shell-side friction factors (Source: Wolverine Tube, Inc., originally published in Ref. [1]).

The pressure drop due to the shell-side nozzles can be estimated in the same manner as for the tube-side nozzles using Equations (5.4) and (5.5).

5.4 Finned Tubing

The finned tubing used in shell-and-tube heat exchangers is often referred to as radial low-fin tubing. It is made by an extrusion process in which the wall of a plain tube is depressed by stacks of disks. The undisturbed portion of the tube wall between the disks forms the fins. The fins are actually helical rather that radial, but they can be approximated as radial (annular) fins of rectangular profile for computational purposes. Low-fin tubing is made in a variety of alloys and configurations. The number of fins per inch of tube length ranges from 11 to 43, fin height varies from about 0.02 to 0.125 in., and fin thickness ranges from 0.01 in. to about 0.015 in.

A diagram of a section of low-fin tubing is given in Figure 5.2. The part of the tube underneath the fins is called the root tube. The OD of the finned section is somewhat less than the diameter of the plain tube from which it was formed, so finned tubes are interchangeable with plain tubes in tube bundles. Both the ID and wall thickness of the finned section are less than those of the original tube. The ends of the tubes are left un-finned so that they can be rolled into the tubesheets. Dimensions of selected low-fin tubing from two manufacturers are given in Appendix B (Tables B.3–B.5).

Equation (2.27) can be used to calculate the fin efficiency. The equation is repeated here for convenience.

$$\eta_f = \frac{\tanh(m\psi)}{m\psi} \tag{2.27}$$

$$\psi = (r_{2c} - r_1[1 + 0.35\ln(r_{2c}/r_1)]) \tag{5.12}$$

where

$m = (2h_o/k\tau)^{1/2}$
r_1 = external radius of root tube
r_2 = outer radius of fin
$r_{2c} = r_2 + \tau/2$ = corrected fin radius
k = thermal conductivity of fin
τ = fin thickness
h_o = shell-side heat-transfer coefficient

The wall temperature of a finned tube can be estimated in the same manner as that of a finned pipe in a double-pipe exchanger. Two temperatures, T_p and T_{wtd}, are calculated using Equations (4.38) and (4.39), which are repeated here for convenience.

$$T_p = \frac{h_i t_{ave} + h_o \eta_w (A_{Tot}/A_i) T_{ave}}{h_i + h_o \eta_w (A_{Tot}/A_i)} \tag{4.38}$$

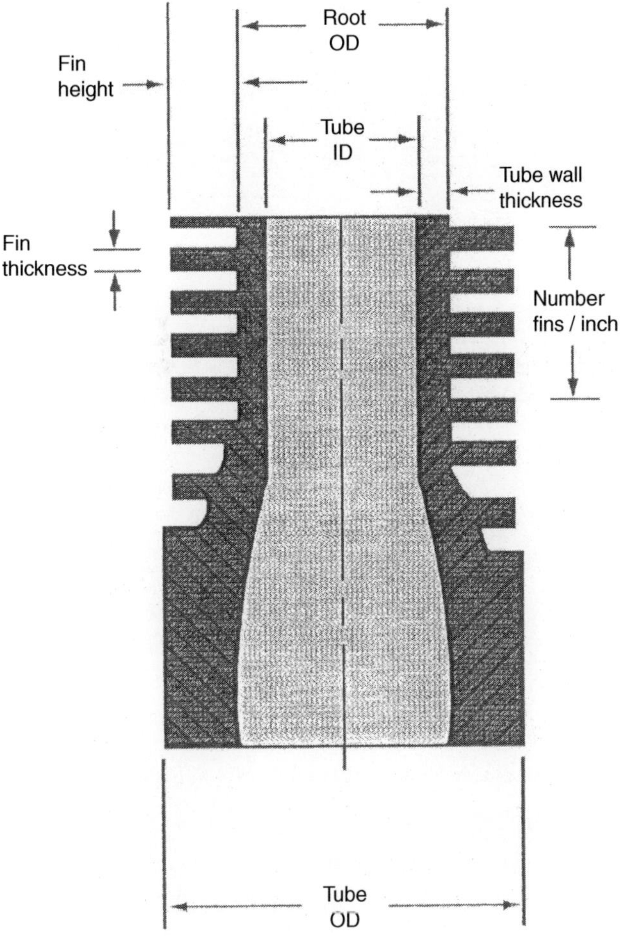

FIGURE 5.2 Schematic illustration of a radial low-fin tube (Source: Ref. [6]).

$$T_{wtd} = \frac{h_i \eta_w t_{ave} + [h_i(1 - \eta_w) + h_o \eta_w (A_{Tot}/A_i)] T_{ave}}{h_i + h_o \eta_w (A_{Tot}/A_i)} \tag{4.39}$$

In these equations,

η_w = weighted fin efficiency
t_{ave} = average temperature of tube-side fluid
T_{ave} = average temperature of shell-side fluid
A_i = interior surface area of tube
A_{Tot} = exterior surface (prime + fins) area of tube

T_p is an estimate of the average temperature of the prime surface, i.e., the root tube, and is used to calculate the viscosity correction factor, ϕ_i, for the tube-side fluid. T_{wtd} represents a weighted average of the extended and prime surface temperatures, and is used to calculate the viscosity correction factor, ϕ_o, for the shell-side fluid.

For tubing with 16 or 19 fins per inch, the gap, C', and the equivalent diameter, D_e (or d_e), that are used to calculate the shell-side heat-transfer coefficient and pressure drop are given in Figure 3.17. For other configurations these parameters must be calculated. They are based on an effective root-tube diameter obtained by adding to the root tube a cylinder whose volume is equal to the total volume of the fins. The relevant equations are as follows:

$$\hat{D}_r = \left[D_r^2 + 4n_f b\tau(D_r + b)\right]^{1/2} \tag{5.13}$$

where

$\hat{D}_r$ = effective root tube diameter
D_r = diameter of root tube
n_f = number of fins per unit length of tube
$b = r_2 - r_1$ = fin height

$$C' = P_T - \hat{D}_r \tag{5.14}$$

$$D_e = \frac{4\beta P_T^2 - \pi \hat{D}_r^2}{\pi \hat{D}_r} \tag{5.15}$$

where
$\beta = 1.0$ for square and rotated square pitch
$= 0.86$ for triangular pitch

A final comment concerns the "low-fin limit" indicated on Figure 3.17 at a Reynolds number of 500. The explanation of this notation is as follows [1]:

"When a low shell-side Reynolds number is the result of a high mass velocity and high viscosity as opposed to a low mass velocity and low viscosity some caution is suggested. Lube-oil coolers with oil on the shell side and water on the tube side illustrate the point. Often there is no temperature control on the cooling water so that in the wintertime water much below design temperature may be circulated. It might reduce the temperature of the fins below the cloud point or pour point of the oil. This may cause a viscous, insulating mass to set up at the tube circumference. The temperature of the wall, T_{fw}, will always be higher than the temperature of a plain tube for the same process conditions and equivalent coefficients. This is inherent in the difference in outside-to-inside-surface ratios between low-fin and plain tubes. Nevertheless, the low-fin tube is slightly more disposed toward insulating itself with liquids having high viscosities. The caution is intended to make mandatory a check of the viscosity μ_w against operating water temperatures as opposed to design temperatures which may be considerably higher."

5.5 Tube-Count Tables

The number of tubes of a given size that can be accommodated in a given shell depends on many factors, including number of tube passes, tube pattern and pitch, baffle cut, number of sealing strips, head type, pressure rating, and tie-rod location [7]. The actual number of tubes contained in a given shell also varies somewhat from one manufacturer to another. Although it is impractical to account for all of these factors, tables of approximate tube counts are useful in the design process for estimating the required shell size.

A set of tube-count tables is given in Appendix C. The values listed in the tables represent the number of tube holes in each tubesheet, which is equal to the number of tubes in the case of regular tubes. For U-tubes, however, the number of tube holes is twice the actual number of tubes. The tables are based on the following criteria [7]:

- Tubes have been eliminated to provide entrance area for a nozzle with a diameter equal to 0.2 d_s.
- Tube layouts are symmetrical about both horizontal and vertical axes.
- The distance from tube exterior surface to the centerline of pass partitions is 5/16 in. for shells with ID less that 22 in. and 3/8 in. for larger shells.

5.6 Factors Affecting Pressure Drop

An important aspect of heat-exchanger design is the control of pressure losses to stay within the specifications on the two streams while achieving good heat transfer. In this section we consider the design parameters that can be adjusted to control tube-side and shell-side pressure drop.

5.6.1 Tube-Side Pressure Drop

The friction loss on the tube side is given by Equation (5.1), which shows that:

$$\Delta P_f \sim \frac{f\, n_p L G^2}{D_i} \tag{5.16}$$

Furthermore, the mass flux, G, is the mass flow rate per tube $(\dot{m} n_p / n_t)$ divided by the flow area per tube $(\pi D_i^2/4)$. Substituting yields:

$$\Delta P_f \sim \frac{f\, L n_p^3}{n_t^2 D_i^5} \tag{5.17}$$

The friction factor is inversely proportional to Reynolds number in laminar flow and proportional to the –0.2585 power of Reynolds number in turbulent flow according to Equations (4.6) and (5.2). Since:

$$Re = \frac{4\dot{m}_{per\ tube}}{\pi D_i \mu} \sim \frac{n_p}{n_t D_i} \tag{5.18}$$

Proportionality (5.17) becomes:

$$\Delta P_f \sim \frac{L n_p^{2.74}}{n_t^{1.74} D_i^{4.74}} \quad \text{(turbulent flow)} \tag{5.19}$$

$$\Delta P_f \sim \frac{L n_p^2}{n_t D_i^4} \quad \text{(laminar flow)} \tag{5.20}$$

Now for a given amount of heat-transfer surface and a specified tube BWG,

$$A_i = n_t \pi D_i L = \text{constant} \tag{5.21}$$

If the tube diameter is also specified, then n_t is inversely proportional to the tube length, and proportionalities (5.19) and (5.20) become:

$$\Delta P_f \sim n_p^{2.74} L^{2.74} \quad \text{(turbulent flow)} \tag{5.22}$$

$$\Delta P_f \sim n_p^2 L^2 \quad \text{(laminar flow)} \tag{5.23}$$

Thus, regardless of flow regime, the pressure drop is a strong function of both the tube length and the number of tube passes.

5.6.2 Shell-Side Pressure Drop

According to Equation (5.6), for the shell-side pressure drop:

$$\Delta P_f \sim \frac{f\, G^2 d_s (n_b + 1)}{d_e} \tag{5.24}$$

The number of baffle spaces is approximately equal to the tube length divided by the baffle spacing in consistent units, i.e.,

$$n_b + 1 \sim L/B \tag{5.25}$$

Also,

$$G = \frac{\dot{m}}{a_s} = \frac{\dot{m}}{\left[\frac{d_s C' B}{144 P_T}\right]} \sim \frac{P_T}{d_s C' B} = \frac{P_T}{d_s B (P_T - D_o)} \tag{5.26}$$

The equivalent diameter, d_e, depends on the tube size and pitch. For plain tubes, the relationship is given by Equation (5.15) with $\hat{D}_r$ replaced by the tube OD. For square pitch,

$$d_e = 4\left(P_T^2 - \pi D_o^2/4\right)/(-\pi D_o) \tag{5.27}$$

Combining Equations (5.24) to (5.27) yields:

$$\Delta P_f \sim \frac{f\, L P_T^2 D_o^2}{B^3 d_s^2 (P_T - D_o)^2 \left(P_T^2 - \pi D_o^2/4\right)} \tag{5.28}$$

For given tube and shell diameters, this relation simplifies to:

$$\Delta P_f \sim \frac{f\, L P_T^2}{B^3 (P_T - D_o)^2 \left(P_T^2 - \pi D_o^2/4\right)} \tag{5.29}$$

Shell-side pressure drop is thus a strong function of the baffle spacing. Increasing B increases the flow area across the tube bundle, which lowers the pressure drop. The dependence is not as strong as might be inferred from the above relationship because the friction factor increases with baffle spacing. The dependence of f on B is complex since f increases directly with the ratio B/d_s and indirectly through the Reynolds number. Nevertheless, the baffle spacing is the main design parameter for controlling shell-side pressure drop.

Increasing the tube pitch also increases the flow area through the tube bundle and thereby lowers the pressure drop. However, this tactic has the disadvantage of increasing the required shell size and, hence, the cost of the heat exchanger. Therefore, the tube pitch is generally not used to control pressure drop except in situations where no alternative is available.

Shell-side pressure drop also varies directly with the tube length. However, for a specified tube diameter and a given amount of heat-transfer surface, reducing the tube length increases the number of tubes in the bundle, which may require a larger shell. It also has a major impact on the tube-side pressure drop.

There are additional ways to control shell-side pressure drop that are not accounted for in the Simplified Delaware method. Increasing the baffle cut reduces the length of the cross-flow path through the bundle, which reduces the pressure drop. In practice, however, this parameter is highly correlated with the baffle spacing because an appropriate ratio of these parameters is necessary for good flow distribution in the shell. (With the Simplified Delaware method, the baffle cut is fixed at 20% and cannot be varied.)

As discussed in Section 3.3, alternative baffle designs (double segmental, NTIW, helical, rod, etc.) can also be used to achieve lower shell-side pressure drop.

5.7 Design Guidelines

5.7.1 Fluid Placement

Guidelines for fluid placement were given in Table 3.4. The most important considerations here are corrosion and fouling. Corrosive fluids should be placed on the tube side so that only the tubes, tubesheets, and (possibly) headers need to be made of corrosion-resistant alloy. Fluids that are heavy foulers should be placed on the tube side because it is (usually) easier to clean deposits from the interior surfaces of the tubes than from the exterior surfaces. Cooling water is usually placed in the tubes due to its tendency to corrode carbon steel and to form scale, which is difficult to remove from the exterior tube surfaces. Also, in services involving cooling water and an organic stream, finned tubes are frequently used to offset the low heat-transfer coefficient of the organic stream relative to the water. This requires the organic stream on the shell side in contact with the fins.

Stream pressure is another factor that may influence the fluid placement. The reason is that it is generally less expensive to confine a high-pressure stream in the tubes rather than in the shell. Due to their small diameters, tubes of standard wall thickness can withstand quite high pressures, and only tube-side headers and nozzles normally require more robust construction.

5.7.2 Tubing Selection

The most frequently used tube sizes are $^3/_4$ and 1 in. For water service, $^3/_4$ in., 16 BWG tubes are recommended. For oil (liquid hydrocarbon) service, $^3/_4$ in., 14 BWG tubes are recommended if the fluid is non-fouling, while 1 in., 14 BWG tubes should be used for fouling fluids. Tube lengths typically range from 8 to 30 ft, and sometimes longer depending on the type of construction and the tubing material. A good value to start with is 16 or 20 ft.

5.7.3 Tube Layout

Triangular and square layouts are the most common, but rotated triangular and rotated square configurations are also used. With triangular pitch the tubes are more closely packed in the bundle, which translates to more heat-transfer surface in a given shell and somewhat higher pressure drop and heat-transfer coefficient. However, the gap between tubes is typically the larger of 0.25 in. and 0.25 D_o, and with triangular pitch this is not sufficient to allow cleaning lanes between the tube rows. Although chemical cleaning may be possible, triangular pitch is usually restricted to services with clean shell-side fluids. Rotated square pitch provides some enhancement in the heat-transfer coefficient (along with higher pressure drop) compared with square pitch, while still providing cleaning lanes between the tubes. This configuration is especially useful when the shell-side Reynolds number is relatively low (less than about 2000).

To summarize, the most commonly used tube layouts are either triangular or square, with a pitch of 1.0 in. (for $^3/_4$ -in. tubes) or 1.25 in. (for 1-in. tubes).

5.7.4 Tube Passes

For typical low-viscosity process streams, it is highly desirable to maintain fully developed turbulent flow in the tubes. Although this may not be practical with high viscosity liquids, turbulent flow provides the most effective heat transfer. Once the tube size and number of tubes have been determined, the number of tube passes can be chosen to give an appropriate Reynolds number, i.e.,

$$Re = \frac{4\dot{m}\, n_p}{\pi D_i n_t \mu} \geq 10^4 \tag{5.27}$$

Except for single-pass exchangers, an even number of tube passes is almost always used so that the tube-side fluid enters and exits at the same header. With U-tubes, this is the only feasible arrangement, and accommodating nozzles on internal (type S or T) floating heads in order to provide an odd number of passes is very cumbersome.

Fluid velocity can also be used as a criterion for setting the number of tube-side passes. It is desirable to maintain the liquid velocity in the tubes in the range of about 3–8 ft/s. Too low a velocity can cause excessive fouling, while a very high velocity can cause erosion of the tube wall. Some material-specific maximum velocities are given in Appendix 5.B. Harder tubing materials, such as steel, can withstand somewhat higher velocities than softer metals such as copper or aluminum, for example. Maximum vapor velocities are also given in Appendix 5.B.

5.7.5 Shell and Head Types

Shell and head types were presented in Figure 3.3 and attributes of each shell type were summarized in Table 3.2. As noted in Section 3.3, the single pass type E shell is standard. If multiple shell passes are required, as indicated by the logarithmic mean temperature difference (LMTD) correction factor, E-shells can be connected in series. Alternatively, a two-pass type F shell can be used, although the longitudinal baffle may be subject to leakage. An F-shell can also be used with two tube passes to obtain the equivalent of a 1-1 (true counter flow) exchanger. The other shell types listed in Figure 3.3 are used for more specialized applications, including reboilers (G, H, K, X) and units, such as condensers and gas coolers, that require low shell-side pressure drops (J, X).

The most important consideration with respect to head type is fixed tubesheet versus floating head. A fixed-tubesheet design is cheaper and less prone to leakage. However, the tube bundle cannot be removed to clean the exterior surfaces of the tubes. Therefore, fixed-tubesheet exchangers are usually restricted to services with clean shell-side fluids. Also, if the temperature difference between the two inlet streams is greater than about 100°F, an expansion joint is required in the shell of a fixed-tubesheet exchanger to accommodate the differential thermal expansion between the tubes and shell. The added cost of the expansion joint largely offsets the cost advantage of the fixed-tubesheet design.

With floating-head and U-tube exchangers, the entire tube bundle can be pulled out of the shell from the front (stationary head) end. This allows mechanical cleaning of the exterior tube surfaces, usually by high-pressure jets of water, steam, or supercritical carbon dioxide. (Of course, square or rotated square pitch must be used to provide cleaning lanes.) Also, since only one tubesheet is attached to the shell (at the front end), the tubes are free to expand or contract relative to the shell due to temperature differences.

Another consideration with respect to head type is bonnet versus channel. Bonnets are cheaper and less prone to leakage. However, to gain access to the tubesheet for inspecting or cleaning the tubes, the stationary head must be disconnected from the process piping and removed from the shell. With a channel-type head, access to the tubesheet is obtained simply by unbolting and removing the channel cover. Thus, a channel-type stationary head is preferable if the tubes will require frequent cleaning.

Other factors that may affect the choice of head type are given in Appendix 5.D. The most widely used floating-head design in the process industries is the AES exchanger. U-tube exchangers are less expensive and are also widely used. The advent of equipment for cleaning the interiors of U-tubes using high-pressure water jets has mitigated the problem of cleaning the return bends, leading to increased use of these exchangers [8].

5.7.6 Baffles and Tubesheets

Single segmental baffles are standard and by far the most widely used. In order to provide good flow distribution on the shell side, the spacing between baffles should be between 0.2 and 1.0 shell diameters (but not less than 2 in.). However, the maximum baffle spacing may be limited by tube support and vibration considerations to less than one shell diameter [9,10] (see also Appendix 5.C). The no-tubes-in-window option can be used to extend the baffle spacing in these situations. For good flow distribution, the baffle cut should be between 15% and 45%. For single-phase flow, however, a range of 20% to 35% is recommended [11]. With the Simplified Delaware method, the baffle cut is fixed at 20%.

Although baffle spacing and baffle cut are ostensibly independent parameters, in practice they are highly correlated. Figure 5.3 shows the recommendations given by Taborek [9], who states: "A sound design should not deviate substantially from the recommended values, which are based on a vast amount of practical experience as well as on studies of the shell-side flow patterns." If the baffle cut is set at 20% for the Simplified Delaware method, then Figure 5.3 indicates that the baffle spacing should be in the range of 0.2 to about 0.4 shell diameters for single-phase flow.

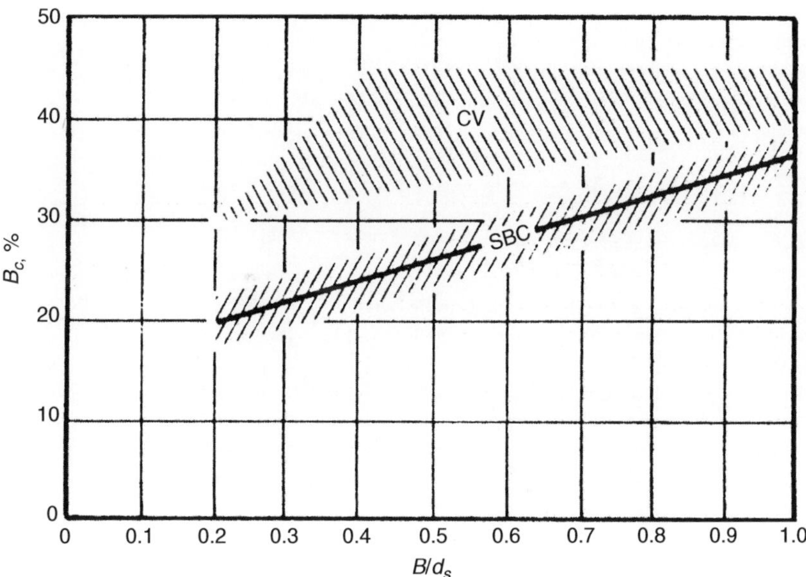

FIGURE 5.3 Recommended baffle cut, B_c, as a function of baffle spacing. SBC, for single-phase flow; CV, for condensing vapors (Source: Ref. [9]).

TABLE 5.2 Standard Values (Inches) for Baffle Thickness in Class R Heat Exchangers*

	Baffle Spacing, Inches				
Shell ID, Inches	≤24	24–36	36–48	48–60	>60
8–14	0.125	0.1875	0.250	0.375	0.375
15–28	0.1875	0.250	0.375	0.375	0.500
29–38	0.250	0.3125	0.375	0.500	0.625
39–60	0.250	0.375	0.500	0.625	0.625
61–100	0.375	0.500	0.625	0.750	0.750

*Class R exchangers are for the generally severe requirements of petroleum and related processing applications.
Source: HEXTRAN and TEAMS computer programs.

Baffle thickness is not required for the Simplified Delaware method, but it is used in the Stream Analysis method and in commercial computer programs. It varies from 1/16 to 3/4 in., and generally increases with shell size and baffle spacing. If this parameter is not specified as input, most computer programs will default to the standard values given in Table 5.2.

Tubesheet thickness is another parameter used in commercial computer programs. It varies from about 1 to 6 in., and generally increases with shell size and operating pressure. As a rough approximation, a value equal to the larger of 1 in. and 0.1 times the shell ID can be used [9]. Detailed methods for calculating required tubesheet thickness based on bending and shear stresses are given in the ASME Boiler and Pressure Vessel Code (Section VII, Part UHX) and in Ref. [10].

5.7.7 Nozzles

Nozzles can be sized to meet pressure drop limitations and/or to match process piping. The guidelines given by Kern [4] and reproduced in Table 5.3 are useful as a starting point.

Other considerations in sizing nozzles are tube vibration and erosion. The fluid entering the shell through the inlet nozzle impinges directly on the tube bundle. If the inlet velocity is too high, excessive tube vibration and/or erosion may result. TEMA specifications to prevent tube erosion are given in terms of the product of density (lbm/ft^3) and nozzle velocity (ft/s) squared [10]:

$\rho V_n^2 \leq 1500\ \text{lbm/ft} \cdot \text{s}^2$	for non-abrasive single-phase fluids
$\leq 500\ \text{lbm/ft} \cdot \text{s}^2$	for all other liquids, including bubble-point liquids

Beyond these limits (and for all other gases, including saturated vapors and vapor-liquid mixtures regardless of the ρV_n^2 value) an impingement plate is required to protect the tubes. This is a metal plate, usually about 1/4-in. thick, placed beneath the nozzle to deflect the fluid and keep it from impinging directly on the tubes. With impingement protection, values of ρV_n^2 up to twice the above values are acceptable [9]. For still higher nozzle velocities (or in lieu of an impingement plate) an annular distributor can be used to distribute the fluid more evenly around the shell periphery and thereby reduce the impingement velocity [9]. An impingement plate may not be adequate to prevent tube vibration problems, and a larger nozzle or a distributor may be needed for this purpose as well. Furthermore, impingement plates actually reduce the bundle entrance area, and as a result, tubes near plate edges may be exposed to very high velocities that can cause them to fail. Thus, impingement plates can sometimes be counterproductive. An alternative that can be used to avoid this problem is to replace the first two rows of tubes with solid rods of diameter equal to the tube OD. The rods serve to protect the tubes without reducing the bundle entrance area.

5.7.8 Sealing Strips

The purpose of sealing strips is to reduce the effect of the bundle bypass stream that flows around the outside of the tube bundle. They are usually thin strips of metal that fit into slots in the baffles and extend outward toward the shell wall to block the bypass

TABLE 5.3 Guidelines for Sizing Nozzles

Shell Size, Inches	Nominal Nozzle Diameter, Inches
4–10	2
12–17.25	3
19.25–21.25	4
23–29	6
31–37	8
39–42	10

flow and force it back into the tube bundle. They are placed in pairs on opposite sides of the baffles running lengthwise along the bundle. Sealing strips are mainly used in floating-head exchangers, where the clearance between the shell and tube bundle is relatively large. Typically, one pair is used for every four to ten rows of tubes between the baffle tips. Increasing the number of sealing strips tends to increase the shell-side heat-transfer coefficient at the expense of a somewhat larger pressure drop. In the Simplified Delaware method, the number of sealing strips is set at one pair per ten tube rows.

5.8 Design Strategy

Shell-and-tube design is an inherently iterative process, the main steps of which can be summarized as follows:

(a) Obtain an initial configuration for the heat exchanger. This can be accomplished by using the preliminary design procedure given in Section 3.8 to estimate the required heat-transfer surface area, along with the design guidelines and tube-count tables discussed above to completely specify the configuration.
(b) Rate the design to determine if it is thermally and hydraulically suitable.
(c) Modify the design, if necessary, based on the results of the rating calculations.
(d) Go to step (b) and iterate until an acceptable design is obtained.

The design procedure is illustrated in the following examples.

Example 5.1

A kerosene stream with a flow rate of 45,000 lb/h is to be cooled from 390°F to 250°F by heat exchange with 150,000 lb/h of crude oil at 100°F. A maximum pressure drop of 15 psi has been specified for each stream. Prior experience with this particular oil indicates that it exhibits significant fouling tendencies, and a fouling factor of 0.003 h · ft^2 · °F/Btu is recommended. Physical properties of the two streams are given in the table below. Design a shell-and-tube heat exchanger for this service.

Fluid Property	Kerosene	Crude Oil
C_P (Btu/lbm · °F)	0.59	0.49
k (Btu/h · ft · °F)	0.079	0.077
μ (lbm/ft · h)	0.97	8.7
Specific gravity	0.785	0.85
Pr	7.24	55.36

Solution

(a) Make initial specifications.

(i) Fluid placement

Kerosene is not corrosive, but crude oil may be, depending on salt and sulfur contents and temperature. At the low temperature of the oil stream in this application, however, corrosion should not be a problem provided the oil has been desalted (if necessary). Nevertheless, the crude oil should be placed in the tubes due to its relatively high fouling tendency. Also, the kerosene should be placed in the shell due to its large ΔT of 140°F according to the guidelines given in Table 3.4.

(ii) Shell and head types

The recommended fouling factor for kerosene is 0.001–0.003 h · ft^2 · °F/Btu (Table 3.3), indicating a significant fouling potential. Therefore, a floating-head exchanger is selected to permit mechanical cleaning of the exterior tube surfaces. Also, the floating tubesheet will allow for differential thermal expansion due to the large temperature difference between the two streams. Hence, a type AES exchanger is specified.

(iii) Tubing

Following the design guidelines for a fouling oil service, 1 in., 14 BWG tubes are selected with a length of 20 ft.

(iv) Tube layout

Since cleaning of the tube exterior surfaces will be required, square pitch is specified to provide cleaning lanes through the tube bundle. Following the design guidelines, for 1-in. tubes a tube pitch of 1.25 in. is specified.

(v) Baffles

Segmental baffles with a 20% cut are required by the Simplified Delaware method, but this is a reasonable starting point in any case. In consideration of Figure 5.3, a baffle spacing of 0.3 shell diameters is chosen, i.e., $B/d_s = 0.3$.

(vi) Sealing strips
One pair of sealing strips per ten tube rows is specified in accordance with the requirements of the Simplified Delaware method and the design guidelines.
(vii) Construction materials
Since neither fluid is corrosive, plain carbon steel is specified for tubes, shell, and other components.

(b) Energy balances.

$$q = (\dot{m}C_P\Delta T)_{ker} = 45,000 \times 0.59 \times 140 = 3,717,000 \text{ Btu/h}$$

$$3,717,000 = (\dot{m}C_P\Delta T)_{oil} = 150,000 \times 0.49 \times \Delta T_{oil}$$

$$\Delta T_{oil} = 50.6°F$$

$$\text{outlet oil temperature} = 150.6°F$$

(c) LMTD.

$$(\Delta T_{\ln})_{cf} = \frac{239.4 - 150}{\ln(239.4/150)} = 191.2°F$$

(d) LMTD correction factor.

$$R = \frac{T_a - T_b}{t_b - t_a} = \frac{390 - 250}{150.6 - 100} = 2.77$$

$$P = \frac{t_b - t_a}{T_a - t_a} = \frac{150.6 - 100}{390 - 100} = 0.174$$

From Figure 3.14 or Equation (3.15), for a 1-2 exchanger $F \cong 0.97$. Therefore, one shell pass is required.

(e) Estimate U_D.
In order to obtain an initial estimate for the size of the exchanger, an approximate value for the overall heat-transfer coefficient is used. From Table 3.5, for a kerosene/oil exchanger, it is found that $20 \le U_D \le 35$ Btu/h · ft^2 · °F. A value near the middle of the range is selected: $U_D = 25$ Btu/h · ft^2 · °F

(f) Calculate heat-transfer area and number of tubes.

$$A = \frac{q}{U_D F(\Delta T_{\ln})_{cf}} = \frac{3,717,000}{25 \times 0.97 \times 191.2}$$

$$A = 801.7 \text{ ft}^2$$

$$n_t = \frac{A}{\pi D_o L} = \frac{801.7}{\pi \times (1/12) \times 20}$$

$$n_t = 153$$

(g) Number of tube passes.
The number of tube passes is chosen to give fully developed turbulent flow in the tubes and a reasonable fluid velocity.

$$Re = \frac{4\dot{m}(n_p/n_t)}{\pi D_i \mu}$$

$$D_i = 0.834 \text{ in.} = 0.0695 \text{ ft} \quad \text{(Table B.1)}$$

$$Re = \frac{4 \times 150,000(n_p/153)}{\pi \times 0.0695 \times 8.7} = 2064.5\ n_p$$

We want $Re \ge 10^4$ and an even number of passes. Therefore, take $n_p = 6$. Checking the fluid velocity,

$$V = \frac{\dot{m}(n_p/n_t)}{\rho\pi D_i^2/4} = \frac{(150,000/3600)\ (6/153)}{0.85 \times 62.43\pi\ (0.0695)^2/4} = 8.1 \text{ ft/s}$$

The velocity is at the high end of the recommended range, but still acceptable. Therefore, six tube passes will be used.

(h) Determine shell size and actual tube count.
From the tube-count table for l in. tubes on $1^1/_4$-in. square pitch (Table C.5), with six tube passes and a type S head, the listing closest to 153 is 156 tubes in a $21^1/_4$-in. shell. Hence, the number of tubes is adjusted to $n_t = 156$ and the shell ID is taken as $d_s = 21.25$ in.

This completes the initial design of the heat exchanger. The initial design must now be rated to determine whether it is adequate for the service. Since the temperature dependence of the fluid properties is not available, they will be assumed constant; the viscosity correction factors will be set to unity and the tube wall temperature will not be calculated.

(i) Calculate the required overall coefficient.

$$U_{req} = \frac{q}{n_t \pi D_o L F(\Delta T_{ln})_{cf}} = \frac{3,717,000}{156 \times \pi \times (1.0/12) \times 20 \times 0.97 \times 191.2}$$

$$U_{req} = 24.5 \ \text{Btu/h} \cdot \text{ft}^2 \cdot {}^\circ\text{F}$$

(j) Calculate h_i.

$$Re = \frac{4\dot{m}(n_p/n_t)}{\pi D_i \mu} = \frac{4 \times 150,000(6/156)}{\pi \times 0.0695 \times 8.7} = 12,149$$

$$h_i = (k/D_i) \times 0.023 \ Re^{0.8} \ Pr^{1/3}(\mu/\mu_w)^{0.14}$$

$$= (0.077/0.0695) \times 0.023(12,149)^{0.8}(55.36)^{1/3}(1.0)$$

$$h_i = 180 \ \text{Btu/h} \cdot \text{ft}^2 \cdot {}^\circ\text{F}$$

(k) Calculate h_o.

$$B = 0.3 \ d_s = 0.3 \times 21.25 = 6.375 \ \text{in.}$$

$$a_s = \frac{d_s C' B}{144 P_T} = \frac{21.25 \times 0.25 \times 6.375}{144 \times 1.25} = 0.188 \ \text{ft}^2$$

$$G = \dot{m}/a_s = 45,000/0.188 = 239,362 \ \text{lbm/h} \cdot \text{ft}^2$$

$$D_e = 0.99/12 = 0.0825 \ \text{ft} \ (\text{from Figure 3.17})$$

$$Re = D_e G/\mu = 0.0825 \times 239,362/0.97 = 20,358$$

Equation (3.21) is used to calculate the Colburn factor, j_H.

$$j_H = 0.5(1 + B/d_s)(0.08Re^{0.6821} + 0.7Re^{0.1772})$$

$$= 0.5(1 + 0.3)[0.08(20,358)^{0.6821} + 0.7(20,358)^{0.1772}]$$

$$j_H = 47.8$$

$$h_o = j_H(k/D_e)Pr^{1/3}(\mu/\mu_w)^{0.14}$$

$$= 47.8(0.079/0.0825)(7.24)^{1/3}(1.0)$$

$$h_o = 88.5 \ \text{Btu/h} \cdot \text{ft}^2 \cdot {}^\circ\text{F}$$

(l) Calculate the clean overall coefficient.

$$U_C = \left[\frac{D_o}{h_i D_i} + \frac{D_o \ln(D_o/D_i)}{2k_{tube}} + \frac{1}{h_o}\right]^{-1}$$

$$= \left[\frac{1.0}{180 \times 0.834} + \frac{(1.0/12) \ln(1.0/0.834)}{2 \times 26} + \frac{1}{88.5}\right]^{-1}$$

$$U_C = 54.8 \ \text{Btu/h} \cdot \text{ft}^2 \cdot {}^\circ\text{F}$$

Since $U_C > U_{req}$, continue.

(m) Fouling factors

The fouling factor for the crude oil is specified as 0.003 h · ft² · °F/Btu, and from Table 3.3, a value of 0.002 h · ft² · °F/Btu is taken for kerosene. Hence, the total fouling allowance is:

$$R_D = \frac{R_{Di} D_o}{D_i} + R_{Do} = \frac{0.003 \times 1.0}{0.834} + 0.002 = 0.0056 \ \text{h} \cdot \text{ft}^2 \cdot {}^\circ\text{F/Btu}$$

(n) Calculate the design overall coefficient.

$$U_D = (1/U_C + R_D)^{-1} = (1/54.8 + 0.0056)^{-1} = 41.9 \ \text{Btu/h} \cdot \text{ft}^2 \cdot {}^\circ\text{F}$$

Since U_D is much greater than U_{req}, the exchanger is thermally workable, but over-sized.

(o) Over-surface and over-design

It is convenient to perform the calculations using overall coefficients rather than surface areas. The appropriate relationships are as follows:

$$\text{over-surface} = U_C/U_{req} - 1 = 54.8/24.5 - 1 = 124\%$$

$$\text{over-design} = U_D/U_{req} - 1 = 41.9/24.5 - 1 = 71\%$$

Clearly, the exchanger is much larger than necessary.

(p) Tube-side pressure drop

The friction factor is computed using Equation (5.2).

$$f = 0.4137\ Re^{-0.2585} = 0.4137(12,149)^{-0.2585} = 0.03638$$

$$G = \frac{\dot{m}(n_p/n_t)}{\pi D_i^2/4} = \frac{150,000(6/156)}{\left[\pi(0.0695)^2/4\right]} = 1,520,752\ \text{lbm/h}\cdot\text{ft}^2$$

The friction loss is given by Equation (5.1):

$$\Delta P_f = \frac{f\ n_p L G^2}{7.50\times 10^{12} D_i s\phi} = \frac{0.03638\times 6\times 20\ (1,520,752)^2}{7.50\times 10^{12}\times 0.0695\times 0.85\times 1.0} = 22.8\ \text{psi}$$

The tube entrance, exit, and return losses are estimated using Equation (5.3) with α_r equal to $(2n_p - 1.5)$ from Table 5.1.

$$\Delta P_r = 1.334\times 10^{-13}\left(2n_p - 1.5\right)G^2/s = 1.334\times 10^{-13}\ (10.5)(1,520,752)^2/0.85$$

$$\Delta P_r = 3.81\ \text{psi}$$

The sum of the two pressure drops is much greater than the allowed pressure drop. Therefore, the nozzle losses will not be calculated.

(q) Shell-side pressure drop

The friction factor is calculated using Equations (5.7) to (5.9).

$$f_1 = (0.0076 + 0.000166\ d_s)Re^{-0.125} = (0.0076 + 0.000166\times 21.25)(20,358)^{-0.125}$$

$$f_1 = 0.00322\ \text{ft}^2/\text{in}^2$$

$$f_2 = \left(0.0016 + 5.8\times 10^{-5} d_s\right)Re^{-0.157} = \left(0.0016 + 5.8\ \times 10^{-5}\times 21.25\right)(20,358)^{-0.157}$$

$$f_2 = 0.000597\ \text{ft}^2/\text{in}^2$$

$$f = 144\{f_1 - 1.25(1 - B/d_s)(f_1 - f_2)\}$$

$$= 144\{0.00322 - 1.25(1 - 0.3)(0.00322 - 0.000597)\}$$

$$f = 0.1332$$

The number of baffle spaces, $n_b + 1$, is estimated by neglecting the thickness of the tubesheets. The baffle spacing is commonly interpreted as the center-to-center distance between baffles, which is technically the baffle pitch. In effect, the baffle thickness is accounted for in the baffle spacing. (This interpretation is inconsistent with the equation for the flow area, a_s, where the face-to-face baffle spacing should be used rather than the center-to-center spacing. However, the difference is usually of little practical consequence and is neglected herein.) The result is:

$$n_b + 1 \cong L/B = (20\times 12)/6.375 = 37.65 \Rightarrow 38$$

The friction loss is given by Equation (5.6):

$$\Delta P_f = \frac{f\,G^2 d_s(n_b\ +\ 1)}{7.50\times 10^{12} d_e s\phi} = \frac{0.1332\ (239,362)^2\times 21.25\times 38}{7.50\times 10^{12}\times 0.99\times 0.785\times 1.0}$$

$$\Delta P_f = 1.06\ \text{psi}$$

The nozzle losses will not be calculated because the initial design requires significant modification. This completes the rating of the initial design.

In summary, there are two problems with the initial configuration of the heat exchanger:

(1) The tube-side pressure drop is too large.
(2) The exchanger is over-sized.

In addition, the pressure drop on the shell-side is quite low, suggesting a poor trade-off between pressure drop and heat transfer.

To remedy these problems, both the number of tubes and the number of tube passes can be reduced. We first calculate the number of tubes required, assuming the overall heat-transfer coefficient remains constant.

$$A_{req} = \frac{q}{U_D F(\Delta T_{\ln})_{cf}} = \frac{3,717,000}{41.9 \times 0.97 \times 191.2} = 478 \text{ ft}^2$$

$$(n_t)_{req} = \frac{A_{req}}{\pi D_o L} = \frac{478}{\pi(1.0/12) \times 20} = 91.3 \Rightarrow 92$$

Taking four tube passes, the tube-count table shows that the closest count is 104 tubes in a 17.25-in. shell. The effect of these changes on the tube-side flow and pressure drop are estimated as follows.

$$Re \rightarrow 12,149(4/6)(156/104) = 12,149$$

From Equation (5.19),

$$\Delta P_f \sim n_p^{2.74} n_t^{-1.74}$$

$$\Delta P_f \rightarrow 22.8(4/6)^{2.74}(104/156)^{-1.74} = 15.2 \text{ psi}$$

Since these changes leave Re (and hence G) unchanged, the minor losses are easily calculated using Equation (5.3).

$$\Delta P_r = 1.334 \times 10^{-13}(6.5)(1,520,752)^2/0.85 = 2.36 \text{ psi}$$

Thus, in order to meet the pressure-drop constraint, the tube length will have to be reduced significantly, to about 15 ft when allowance for nozzle losses is included. The resulting under-surfacing will have to be compensated by a corresponding increase in h_o, which is problematic. (Since Re_i remains unchanged, so does h_i.) It is left as an exercise for the reader to check the viability of this configuration.

In order to further reduce the tube-side pressure drop, we next consider an exchanger with more tubes. Referring again to the tube-count table, the next largest unit is a 19.25-in. shell containing a maximum of 130 tubes (for four passes). The tube-side Reynolds number for this configuration is:

$$Re \rightarrow 12,149(4/6)(156/130) = 9719$$

Reducing the number of tubes to 124 (31 per pass) gives $Re_i = 10{,}189$. The friction loss then becomes:

$$\Delta P_f \rightarrow 22.8(4/6)^{2.74}(124/156)^{-1.74} = 11.2 \text{ psi}$$

The minor losses will also be lower, and shorter tubes can be used in this unit, further assuring that the pressure-drop constraint will be met. (The final tube length will be determined after the overall heat-transfer coefficient has been recalculated.)

Thus, for the second trial a 19.25-in. shell containing 124 tubes arranged for four passes is specified. Due to the low shell-side pressure drop, the baffle spacing is also reduced to the minimum of 0.2 d_s. (Using a baffle spacing at or near the minimum can cause excessive leakage and bypass flows on the shell-side, resulting in reduced performance of the unit [11]. The low shell-side pressure drop in the present application should mitigate this problem.)

Second Trial

Based on the foregoing analysis, we anticipate that the exchanger will have sufficient heat-transfer area to satisfy the duty. Therefore, we will calculate the overall coefficient, U_D, and use it to determine the tube length that is needed. The pressure drops will then be checked.

(a) Calculate h_i.

$$Re = \frac{4\dot{m}(n_p/n_t)}{\pi D_i \mu} = \frac{4 \times 150,000(4/124)}{\pi \times 0.0695 \times 8.7} = 10,189$$

$$h_i = (k/D_i) \times 0.023 \; Re^{0.8} \; Pr^{1/3}(\mu/\mu_w)^{0.14}$$

$$= (0.077/0.0695) \times 0.023\,(10,189)^{0.8}\,(55.36)^{1/3}(1.0)$$

$$h_i = 156 \text{ Btu/h} \cdot \text{ft}^2 \cdot {}^\circ\text{F}$$

(b) Calculate h_o.

$$B = 0.2\ d_s = 0.2 \times 19.25 = 3.85\ \text{in.}$$

$$a_s = \frac{d_s C' B}{144 P_T} = \frac{19.25 \times 0.25 \times 3.85}{144 \times 1.25} = 0.103\ \text{ft}^2$$

$$G = \dot{m}/a_s = 45{,}000/0.103 = 436{,}893\ \text{lbm/h}\cdot\text{ft}^2$$

$$Re = D_e G/\mu = 0.0825 \times 436{,}893/0.97 = 37{,}158$$

$$j_H = 0.5(1 + B/d_s)\left(0.08 Re^{0.6821} + 0.7\ Re^{0.1772}\right)$$

$$= 0.5(1 + 0.2)\ \left[0.08(37{,}158)^{0.6821} + 0.7(37{,}158)^{0.1772}\right]$$

$$j_H = 65.6$$

$$h_o = j_H (k/D_e) Pr^{1/3} (\mu/\mu_w)^{0.14}$$

$$= 65.6(0.079/0.0825)\ (7.24)^{1/3}(1.0)$$

$$h_o = 122\ \text{Btu/h}\cdot\text{ft}^2\cdot{}^\circ\text{F}$$

(c) Calculate U_D.

$$U_D = \left[\frac{D_o}{h_i D_i} + \frac{D_o \ln(D_o/D_i)}{2k_{tube}} + \frac{1}{h_o} + R_D\right]^{-1}$$

$$= \left[\frac{1.0}{156 \times 0.834} + \frac{(1.0/12)\ \ln(1.0/0.839)}{2 \times 26} + \frac{1}{122} + 0.0056\right]^{-1}$$

$$U_D = 46\ \text{Btu/h}\cdot\text{ft}^2\cdot{}^\circ\text{F}$$

(d) Calculate tube length.

$$q = U_D n_t \pi D_o L F (\Delta T_{\ln})_{cf}$$

$$L = \frac{q}{U_D n_t \pi D_o F (\Delta T_{\ln})_{cf}} = \frac{3{,}717{,}000}{46 \times 124\pi(1.0/12) \times 0.97 \times 191.2} = 13.4\ \text{ft}$$

Therefore, take $L = 14$ ft.

(e) Tube-side pressure drop.

$$f = 0.4137\ Re^{-0.2585} = 0.4137\ (10{,}189)^{-0.2585} = 0.03807$$

$$G = \frac{\dot{m}(n_p/n_t)}{(\pi D_i^2/4)} = \frac{150{,}000(4/124)}{\left[\pi(0.0695)^2/4\right]} = 1{,}275{,}469\ \text{1bm/h}\cdot\text{ft}^2$$

$$\Delta P_f = \frac{f\ n_p L G^2}{7.50 \times 10^{12} D_i s \phi} = \frac{0.03807 \times 4 \times 14(1{,}275{,}469)^2}{7.50 \times 10^{12} \times 0.0695 \times 0.85 \times 1.0}$$

$$\Delta P_r = 1.334 \times 10^{-13}(2n_p - 1.5)G^2/s = 1.334 \times 10^{-13}(6.5)(1{,}275{,}469)^2/0.85$$

$$\Delta P_r = 1.66\text{psi}$$

Table 5.3 indicates that 4-in. nozzles are appropriate for a 19.25-in. shell. Assuming schedule 40 pipe is used for the nozzles,

$$Re_n = \frac{4\dot{m}}{\pi D_n \mu} = \frac{4 \times 150{,}000}{\pi(4.026/12) \times 8.7} = 65{,}432$$

$$G_n = \dot{m}/\left(\pi D_n^2/4\right) = 150{,}000/\left[\pi(4.026/12)^2/4\right] = 1{,}696{,}744\ \text{lbm/h}\ \cdot\text{ft}^2$$

Since the flow in the nozzles is turbulent, Equation (5.4) is applicable.

$$\Delta P_n = 2.0\ \times 10^{-13} N_s G_n^2/s = 2.0 \times 10^{-13} \times 1 \times (1{,}696{,}744)^2/0.85$$

$$\Delta P_n = 0.68\ \text{psi}$$

The total tube-side pressure drop is:

$$\Delta P_i = \Delta P_f + \Delta P_r + \Delta P_n = 7.83 + 1.66 + 0.68 = 10.17 \cong 10.2 \text{ psi}$$

(f) Shell-side pressure drop.
Since $B/d_s = 0.2$, the friction factor is given by:

$$f = 144 f_2 = 144\left(0.0016 + 5.8 \times 10^{-5} d_s\right) Re^{-0.157}$$

$$= 144\left(0.0016 + 5.8 \times 10^{-5} \times 19.25\right)(37,158)^{-0.157}$$

$$f = 0.07497$$

$n_B + 1 \cong L/B = (14 \times 12)/3.85 = 43.6 \Rightarrow 43$ (Rounded downward to keep $B \geq B_{\min}$)

$$\Delta P_f = \frac{f\, G^2 d_s (n_b + 1)}{7.50 \times 10^{12} d_e s \phi} = \frac{0.07497(436,893)^2 \times 19.25 \times 43}{7.50 \times 10^{12} \times 0.99 \times 0.785 \times 1.0}$$

$$\Delta P_f = 2.03 \text{ psi}$$

Since the flow rate of kerosene is much less than that of the crude oil, 3-in. nozzles should be adequate for the shell. Assuming schedule 40 pipe is used,

$$Re_n = \frac{4\dot{m}}{\pi D_n \mu} = \frac{4 \times 45,000}{\pi(3.068/12) \times 0.97} = 231,034 \text{ (turbulent)}$$

$$G_n = \frac{\dot{m}}{\left(\pi D_n^2\right)/4)} = \frac{45,000}{\left[\pi(3.068/12)^2/4\right]} = 876,545 \text{ lbm/h·ft}^2$$

$$\Delta P_n = 2.0 \times 10^{-13} N_s G_n^2 / s = 2.0 \times 10^{-13} \times 1 \times (876,545)^2/0.785$$

$$\Delta P_n = 0.20 \text{ psi}$$

Note: $\rho V_n^2 = 1210$ (lbm/ft^3) (ft/s)^2, so impingement protection for the tube bundle will not be required to prevent erosion.
The total shell-side pressure drop is:

$$\Delta P_o = \Delta P_f + \Delta P_n = 2.03 + 0.20 = 2.23 \cong 2.2 \text{ psi}$$

(g) Over-surface and over-design.

$$U_C = [1/U_D - R_D]^{-1} = [1/46 - 0.0056]^{-1} = 62 \text{ Btu/h} \cdot \text{ft}^2 \cdot {}^\circ\text{F}$$

$$A = n_t \pi D_o L = 124\pi \times (1.0/12) \times 14 = 454 \text{ ft}^2$$

$$U_{req} = \frac{q}{AF(\Delta T_{\ln})_{cf}} = \frac{3,717,000}{454 \times 0.97 \times 191.2} = 44 \text{ Btu/h} \cdot \text{ft}^2 \cdot {}^\circ\text{F}$$

$$\text{over-surface} = U_C/U_{req} - 1 = 62/44 - 1 = 41\%$$

$$\text{over-design} = U_D/U_{req} - 1 = 46/44 - 1 = 4.5\%$$

All design criteria are satisfied. The shell-side pressure drop is still quite low, but the shell-side heat-transfer coefficient (122 Btu/h · ft^2 · °F) does not differ greatly from the tube-side coefficient (156 Btu/h · ft^2 · °F). The reader can also verify that the tube-side fluid velocity is 6.7 ft/s, which is within the recommended range. Therefore, the design is acceptable.

Final design summary

Tube-side fluid: crude oil.

Shell-side fluid: kerosene.

Shell: Type AES, 19.25-in. ID

Tube bundle: 124 tubes, 1-in. OD, 14 BWG, 14-ft long, on 1.25-in. square pitch, arranged for four passes.

Heat-transfer area: 454 ft^2

Baffles: 20% cut segmental type with spacing approximately 3.85 in.

Sealing strips: one pair per ten tube rows.

Nozzles: 4-in. schedule 40 on tube side; 3-in. schedule 40 on shell side.

Materials: plain carbon steel throughout.

Example 5.2

350,000 lb/h of a light oil are to be cooled from 240°F to 150°F using cooling water with a range of 85°F to 120°F. A maximum pressure drop of 7 psi has been specified for each stream, and fouling factors of 0.003 h · ft^2 · °F/Btu for the oil and 0.001 h · ft^2 · °F/Btu for the water are required. Fluid properties are given in the table below. Design a shell-and-tube heat exchanger for this service.

Fluid Property	Oil at 195°F	Water at 102.5°F
C_P (Btu/lbm · °F)	0.55	1.0
k (Btu/h · ft · °F)	0.08	0.37
μ (cp)	0.68*	0.72
Specific gravity	0.80	0.99
Pr	11.31	4.707

*μ_{oil} (cp) = 0.03388 exp [1965.6/T(°R)]

Solution

(a) Make initial specifications.

(i) Fluid placement

According to Table 3.4, cooling water should be placed in the tubes even though the required fouling factors indicate that the oil has a greater fouling tendency. Also, this service (organic fluid versus water) is a good application for a finned-tube exchanger, which requires that the oil be placed in the shell.

(ii) Shell and head types

With oil in the shell, the exterior tube surfaces will require cleaning. Therefore, a floating-head type AES exchanger is selected. This configuration will also accommodate differential thermal expansion resulting from the large temperature difference between the two streams.

(iii) Tubing

Finned tubes having 19 fins per inch will be used. For water service, the design guidelines indicate $^3/_4$-in., 16 BWG tubes. A tube length of 16 ft is chosen.

The tubing dimensions in Table B.4 will be assumed for the purpose of this example. The last three digits of the catalog number (or part number) indicate the average wall thickness of the finned section in thousandths of an inch. Since a 16 BWG tube has a wall thickness of 0.065 in., the specified tubing corresponds to catalog number 60-195065.

(iv) Tube layout

Square pitch is specified to allow mechanical cleaning of the tube exterior surfaces. Following the design guidelines, for $^3/_4$-in.tubes a pitch of 1.0 in. is specified.

(v) Baffles

Segmental baffles with a 20% cut are specified, and the baffle spacing is set at 0.3 shell diameters.

(vi) Sealing strips

One pair of sealing strips per 10 tube rows is specified as required for the Simplified Delaware method.

(vii) Construction materials

Admiralty brass (71% Cu, 28% Zn, 1% Sn; k = 64 Btu/h · ft · °F) will be used for the tubes and navel brass for the tubesheets.[1] Plain carbon steel will be used for all other components, including the shell, heads, baffles, and tube-pass partitions. Although the heads and pass partitions will be exposed to the water, the corrosion potential is not considered sufficient to use alloys for these components. Brass tubesheets are specified for compatibility with the tubes in order to preclude electrolytic attack.

(b) Energy balances.

$$q = (\dot{m}C_P\Delta T)_{oil} = 350,000 \times 0.55 \times 90 = 17,325,000 \text{ Btu/h}$$

$$17,325,000 = (\dot{m}C_P\Delta T)_{water} = \dot{m}_{water} \times 1.0 \times 35$$

$$\dot{m}_{water} = 495,000 \text{ lb/h}$$

(c) LMTD.

$$(\Delta T_{\ln})_{cf} = \frac{120 - 65}{\ln(120/65)} = 89.7°\text{F}$$

(d) LMTD correction factor.

[1] At present, specification of copper and its alloys for process heat exchangers is comparatively infrequent. They formerly found widespread use as corrosion-resistant materials for water services, and are still commonly employed in HVAC (heating, ventilating, and air conditioning) applications. Depending on water chemistry, various grades of stainless steel, as well as titanium, are now preferred for many water services.

$$R = \frac{T_a - T_b}{t_b - t_a} = \frac{240 - 150}{120 - 85} = 2.57$$

$$P = \frac{t_b - t_a}{T_a - t_a} = \frac{120 - 85}{240 - 85} = 0.226$$

From Figure 3.14 or Equation (3.15), for a 1-2 exchanger $F \cong 0.93$.

(e) Estimate U_D.

Perusal of Table 3.5 indicates that the best matches are a kerosene-or-gas-oil/water exchanger and a low-viscosity-lube-oil/water exchanger, both with U_D between 25 and 50 Btu/h · ft² · °F. Therefore, assume $U_D = 40$ Btu/h · ft² · °F.

(f) Calculate heat-transfer area and number of tubes.

$$A = \frac{q}{U_D F(\Delta T_{\ln})_{cf}} = \frac{17,325,000}{40 \times 0.93 \times 89.7} = 5192 \text{ ft}^2$$

From Table B.4, the external surface area of a $^3/_4$-in. tube with 19 fins per inch is 0.507 ft² per foot of tube length. Therefore,

$$n_t = \frac{5192}{0.507 \times 16} = 640$$

(g) Number of tube passes.

$$Re = \frac{4\dot{m}(n_p/n_t)}{\pi D_i \mu}$$

$$D_i = 0.495 \text{ in.} = 0.04125 \text{ ft} \quad \text{(Table B.4)}$$

$$Re = \frac{4 \times 495,000\ (n_p/640)}{\pi \times 0.04125 \times 0.72 \times 2.419} = 13,707\ n_p$$

Thus, one or two passes will suffice. For two passes, the fluid velocity is:

$$V = \frac{\dot{m}(n_p/n_t)}{\rho \pi D_i^2/4} = \frac{(495,000/3600)\ (2/640)}{0.99 \times 62.43\pi\ (0.04125)^2/4} = 5.2 \text{ ft/s}$$

For one pass, $V = 2.6$ ft/s (too low) and for four passes, $V = 10.4$ ft/s (too high). Therefore, two passes will be used.

(h) Determine shell size and actual tube count.

From the tube-count table for $^3/_4$-in. tubes on 1-in. square pitch (Table C.3), the closest count is 624 tubes in a 31-in. shell. This completes the initial design of the unit. The rating calculations follow.

(i) Calculate the required overall coefficient.

$$U_{req} = \frac{q}{n_t A_{Tot} F(\Delta T_{\ln})_{cf}} = \frac{17,325,000}{624\ (0.507 \times 16) \times 0.93 \times 89.7} = 41 \text{ Btu/h}\cdot\text{ft}^2\cdot{}^\circ\text{F}$$

(j) Calculate h_i assuming $\phi_i = 1.0$.

$$Re = \frac{4\dot{m}(n_p/n_t)}{\pi D_i \mu} = \frac{4 \times 495,000(2/624)}{\pi \times 0.04125 \times 0.72 \times 2.419} = 28,117$$

$$h_i = (k/D_i) \times 0.023\ Re^{0.8}\ Pr^{1/3}(\mu/\mu_w)^{0.14}$$

$$= (0.37/0.04125) \times 0.023 \times (28,117)^{0.8}(4.707)^{1/3}(1.0)$$

$$h_i = 1253 \text{ Btu /h}\cdot\text{ft}^2\cdot{}^\circ\text{F}$$

(k) Calculate h_o assuming $\phi_o = 1.0$.

$$B = 0.3\ d_s = 0.3 \times 31 = 9.3 \text{ in.}$$

$$C' = 0.34 \text{ in.} \quad \text{(from Figure 3.17)}$$

$$a_s = \frac{d_s C' B}{144 P_T} = \frac{31 \times 0.34 \times 9.3}{144 \times 1.0} = 0.681 \text{ ft}^2$$

$$G = \dot{m}/a_s = 350,000/0.681 = 513,950 \text{ lbm/h}\cdot\text{ft}^2$$

$$d_e = 1.27 \text{ in.} \quad \text{(from Figure 3.17)}$$

$$D_e = 1.27/12 = 0.1058 \text{ ft}$$

$$Re = D_e G/\mu = 0.1058 \times 513,950/(0.68 \times 2.419) = 33,057$$

$$j_H = 0.5(1 + B/d_s)\left(0.08Re^{0.6821} + 0.7Re^{0.1772}\right)$$
$$= 0.5(1 + 0.3)\left[0.08(33,057)^{.6821} + 0.7(33,057)^{01772}\right]$$
$$j_H = 65.77$$
$$h_o = j_H(k/D_e)Pr^{1/3}(\mu/\mu_w)^{0.14}$$
$$= 65.77(0.08/0.1058)(11.31)^{1/3}(1.0)$$
$$h_o = 112 \text{ Btu/h·ft}^2\text{·°F}$$

(l) Calculate fin efficiency.

The fin efficiency is calculated from Equations (2.27) and (5.12) with $k = 64$ Btu/h·ft^2·°F for Admiralty brass. Fin dimensions are obtained from Table B.4.

τ = fin thickness = 0.011 in. = 0.0009167 ft

r_1 = root tube radius = 0.625/2 = 0.3125 in.

$r_2 = r_1$ + fin height = 0.3125 + 0.05 = 0.3625 in.

$r_{2c} = r_2 + \tau/2 = 0.3625 + 0.011/2 = 0.3680$ in.

$\psi = (r_{2c} - r_1)\,[1 + 0.35\ln(r_{2c}/r_1)] = (0.3680 - 0.3125)\,[1 + 0.35\ln(0.3680/0.3125)]$

$\psi = 0.058676$ in. = 0.0048897 ft

$m = (2h_o/k\tau)^{0.5} = (2 \times 112/64 \times 0.0009167)^{0.5} = 61.7903$ ft^{-1}

$m\psi = 61.7903 \times 0.0048897 = 0.302136$

$$\eta_f = \frac{\tanh(0.302136)}{0.302136} = 0.971$$

The weighted efficiency of the finned surface is given by Equation (2.31). The fin and prime surface areas per inch of tube length are first calculated to determine the area ratios:

$$A_{fins} = 2N_f\pi\left(r_{2c}^2 - r_1^2\right) = 19 \times 2\pi\left[(0.3680)^2 - (0.3125)^2\right] = 4.5087 \text{ in.}^2$$
$$A_{pnime} = 2\pi r_1\left(L - N_f\tau\right) = 2\pi \times 0.3125 \times (1.0 - 19 \times 0.011) = 1.5531 \text{ in.}^2$$
$$A_{fins}/A_{Tot} = 4.5087/(4.5087 + 1.5531) = 0.744$$
$$A_{prime}/A_{Tot} = 1 - 0.744 = 0.256$$
$$\eta_w = \left(A_{prime}/A_{Tot}\right) + \eta_f\left(A_{fins}/A_{Tot}\right) = 0.256 + 0.971 \times 0.744$$
$$\eta_w = 0.978$$

(m) Calculate wall temperatures.

The wall temperatures used to obtain viscosity correction factors are given by Equations (4.38) and (4.39). From Table B.4, the ratio of external to internal tube surface area is $A_{Tot}/A_i = 3.91$. Hence,

$$T_p = \frac{h_i t_{ave} + h_o\eta_w(A_{Tot}/A_i)T_{ave}}{h_i + h_o\eta_w(A_{Tot}/A_i)}$$
$$= \frac{1253 \times 102.5 + 112 \times 0.978 \times 3.91 \times 195}{1253 + 112 \times 0.978 \times 3.91}$$
$$T_p = 126\text{°F}$$

$$T_{wtd} = \frac{h_i\eta_w t_{ave} + [h_i(1 - \eta_w) + h_o\eta_w(A_{Tot}/A_i)]T_{ave}}{h_i + h_o\eta_w(A_{Tot}/A_i)}$$
$$= \frac{1253 \times 0.978 \times 102.5 + [1253 \times 0.022 + 112 \times 0.978 \times 3.91] \times 195}{1253 + 112 \times 0.978 \times 3.91}$$
$$T_{wtd} = 128\text{°F}$$

(n) Calculate viscosity correction factors and corrected heat-transfer coefficients. From Figure A.1, the viscosity of water at 126°F is approximately 0.58 cp. Hence,

$$\phi_i = (0.72/0.58)^{-0.14} = 1.031$$

The oil viscosity at 128°F = 588°R is:

$$\mu_{oil} = 0.03388 \exp(1965.6/588) = 0.96 \text{ cp}$$

Therefore,

$$\phi_o = (0.68/0.96)^{0.14} = 0.953$$

The corrected heat-transfer coefficients are:

$$h_i = 1253 \times 1.031 = 1292 \text{ Btu/h} \cdot \text{ft}^2 \cdot {}^\circ\text{F}$$

$$h_o = 112 \times 0.953 = 107 \text{ Btu/h} \cdot \text{ft}^2 \cdot {}^\circ\text{F}$$

(o) Calculate the clean overall coefficient.
The clean overall coefficient for an exchanger with finned tubes is given by Equation (4.27):

$$U_C = \left[\frac{A_{Tot}}{h_i A_i} + \frac{A_{Tot} \ln(Do/D_i)}{2\pi k_{tube} L} + \frac{1}{h_o \eta_w}\right]^{-1}$$

$$= \left[\frac{3.91}{1292} + \frac{8.112 \ \ln(0.625/0.495)}{2\pi \times 64 \times 16} + \frac{1}{107 \times 0.978}\right]^{-1}$$

$$U_C = 77.7 \ \text{Btu/h} \cdot \text{ft}^2 \cdot {}^\circ\text{F}$$

Since $U_C > U_{req}$, continue.

(p) Fouling allowance

$$R_D = R_{Di}(A_{Tot}/A_i) + R_{Do}/\eta_w = 0.001 \times 3.91 + 0.003/0.978$$

$$R_D = 0.00698 \text{ h} \cdot \text{ft}^2 \cdot {}^\circ\text{F/Btu}$$

(q) Calculate the design overall coefficient.

$$U_D = (1/U_C + R_D)^{-1} = (1/77.7 + 0.00698)^{-1} = 50.4 \text{ Btu/h} \cdot \text{ft}^2 \cdot {}^\circ\text{F}$$

Since $U_D > U_{req}$, the exchanger is thermally workable, but somewhat over-sized (over-design = 23%).

(r) Tube-side pressure drop

$$f = 0.4137 Re^{-0.2585} = 0.4137(28,117)^{-0.2585} = 0.02928$$

$$G = \frac{\dot{m}(n_p/n_t)}{(\pi D_i^2/4)} = \frac{495,000(2/624)}{[\pi(0.04125)^2/4]} = 1,187,170 \ \text{lbm/h} \cdot \text{ft}^2$$

$$\Delta P_f = \frac{f \, n_p L G^2}{7.50 \times 10^{12} D_i s \phi} = \frac{0.02928 \times 2 \times 16 \times (1,187,170)^2}{7.50 \times 10^{12} \times 0.04125 \times 0.99 \times 1.031}$$

$$\Delta P_f = 4.18 \text{ psi}$$

$$\Delta P_r = 1.334 \times 10^{-13}(2n_p - 1.5)G^2/s = 1.334 \times 10^{-13} \ (2.5) \ \ (1,187,170)^2/0.99$$

$$\Delta P_r = 0.475 \text{ psi}$$

For a 31-in. shell, Table 5.3 indicates that 8-in. nozzles are appropriate. For schedule 40 nozzles,

$$Re_n = \frac{4\dot{m}}{\pi D_n \mu} = \frac{4 \times 495,000}{\pi(7.981/12) \times 0.72 \times 2.419} = 544,090$$

$$G_n = \dot{m}/(\pi D_n^2/4) = 495,000/[\pi(7.98/12)^2/4] = 1,424,830 \ \text{lbm/h} \cdot \text{ft}^2$$

Since the nozzle flow is highly turbulent, Equation (5.4) is applicable:

$$\Delta P_n = 2.0 \times 10^{-13} N_s G_n^2/s = 2.0 \times 10^{-13} \times 1 \times (1,424.830)^2/0.99$$

$$\Delta P_n = 0.41 \ \text{psi}$$

The total tube-side pressure drop is:

$$\Delta P_i = \Delta P_f + \Delta P_r + \Delta P_n = 4.18 + 0.475 + 0.41 \cong 5.1 \text{ psi}$$

(s) Shell-side pressure drop

$$f_1 = (0.0076 + 0.000166 \, d_s) Re^{-0.125}$$

$$= (0.0076 + 0.000166 \times 31)(33,057)^{-0.125}$$

$$f_1 = 0.00347 \, \text{ft}^2/\text{in.}^2$$

$$f_2 = (0.0016 + 5.8 \times 10^{-5} d_s) Re^{-0.157}$$

$$= (0.0016 + 5.8 \times 10^{-5} \times 23.25)(33,057)^{-0.157}$$

$$f_2 = 0.000576 \, \text{ft}^2/\text{in.}^2$$

$$f = 144\{f_1 - 1.25(1 - B/d_s)(f_1 - f_2)\}$$
$$= 144\{0.00347 - 1.25\,(1 - 0.3)(0.00347 - 0.000576)\}$$
$$f = 0.135$$

$$n_b + 1 = L/B = (16 \times 12)/9.3 = 20.6 \Rightarrow 21$$

$$\Delta P_f = \frac{f\,G^2 d_s(n_b + 1)}{7.50 \times 10^{12} d_e s\phi} = \frac{0.135 \times (513,950)^2 \times 31 \times 21}{7.50 \times 10^{12} \times 1.27 \times 0.8 \times 0.953}$$

$$\Delta P_f = 3.20 \text{ psi}$$

Assuming 8-in. schedule 40 nozzles are also used for the shell, we first check the Reynolds number at the outlet conditions ($T = 150°F$) where the oil viscosity is highest.

$$\mu_{oil} = 0.03388 \exp\,(1965.6/610) = 0.85 \text{ cp}$$

$$Re_n = 4\dot{m}/\pi D_n \mu = \frac{4 \times 350,000}{\pi(7.981/12) \times 0.85 \times 2.419} = 325,872$$

Since the flow is turbulent, Equation (5.4) can be used to estimate the nozzle losses.

$$G_n = \dot{m}/\left(\pi D_n^2/4\right) = 350,000/\left[\pi(7.981/12)^2/4\right] = 1,007,456 \text{ lbm /h}\cdot\text{ft}^2$$

$$\Delta P_n = 2.0 \times 10^{-13} N_s G_n^2/s = 2.0 \times 10^{-13} \times 1 \times (1,007,456)^2/0.8$$

$$\Delta P_n = 0.25 \text{ psi}$$

Note: $\rho V_n^2 = 1,569$ (lbm/ft^2) (ft/s)^2, so impingement protection will be required, which will increase the pressure drop slightly. This could be avoided by using a larger diameter for the inlet nozzle.

The total shell-side pressure drop is:

$$\Delta P_o = \Delta P_f + \Delta P_n = 3.20 + 0.25 \cong 3.5 \text{ psi}$$

This completes the rating of the initial configuration of the exchanger. To summarize, all design criteria are satisfied, but the exchanger is larger than necessary.

The size of the unit can be reduced by decreasing the number of tubes and/or the tube length. In order to determine the extent of the modifications required, we first calculate the required area assuming that the overall coefficient remains unchanged.

$$A_{req} = \frac{q}{U_D F(\Delta T_{\ln})_{cf}} = \frac{17,325,000}{50.4 \times 0.93 \times 89.7} = 4121 \text{ ft}^2$$

Next, we consider reducing the tube length while keeping the number of tubes fixed at 624. The required tube length is:

$$L = \frac{4121 \text{ ft}^2}{624 \times 0.507 \text{ ft}^2/\text{ft}} = 13.0 \text{ ft}$$

Since this change will not affect the heat-transfer coefficients and will reduce the pressure drops, it is not necessary to repeat the rating calculations. The new pressure drops can be obtained by using the length ratio (13/16) as a scale factor for the friction losses. The minor losses will not change. For the tube side,

$$\Delta P_f = 4.18(13/16) = 3.40 \text{ psi}$$

$$\Delta P_i = \Delta P_f + \Delta P_r + \Delta P_n = 3.40 + 0.475 + 0.41 \cong 4.3 \text{ psi}$$

The calculation for the shell side is approximate because the requirement for an integral number of baffles is ignored:

$$\Delta P_f \cong 3.20(13/16) = 2.60 \text{ psi}$$

$$\Delta P_o = \Delta P_f + \Delta P_n \cong 2.60 + 0.25 \cong 2.9 \text{ psi}$$

The over-surface for the modified design is next computed.

$$A = n_t \pi D_o L = 624 \times 0.507 \times 13 = 4113 \text{ ft}^2$$

$$U_{req} = \frac{q}{AF(\Delta T\ln)_{cf}} = \frac{17,325,000}{4113 \times 0.93 \times 89.7} = 50.5 \text{ Btu/h}\cdot\text{ft}^2\cdot°\text{F}$$

$$\text{over-surface} = U_C/U_{req} - 1 = 77.7/50.5 - 1 \cong 54\%$$

The over-surface is fairly high, but as was the case with the finned-pipe exchanger of Example 4.2, this is simply a reflection of the high total fouling factor of 0.00698 $h \cdot ft^2 \cdot °F/Btu$ that was required for this exchanger.

The over-design for this unit is effectively zero since the actual tube length is equal to the required tube length. If an additional safety margin is desired, the tube length can be increased. A length of 14 ft, for example, provides an over-design of 7.7%, while both pressure drops remain well below the specified maximum.

Finally, the reader can verify that the tube-side fluid velocity is 5.3 ft/s, which is within the recommended range. The final design parameters are summarized below.

Design summary

Tube-side fluid: cooling water.

Shell-side fluid: oil.

Shell: Type AES, 31-in. ID

Tube bundle: 624 tubes, $^3/_4$-in. OD, 16 BWG, radial low-fin tubes, 19 fins per inch, 13-ft long, on 1-in. square pitch, arranged for two passes.

Heat-transfer area: 4113 ft^2

Baffles: 20% cut segmental type with spacing approximately 9.3 in.

Sealing strips: one pair per 10 tube rows.

Nozzles: 8-in. schedule 40 on both tube side and shell side.

Materials: Admiralty brass tubes, naval brass tubesheets, all other components of plain carbon steel.

Now we consider modifying the initial design by reducing the number of tubes. If the tube length and overall heat-transfer coefficient remain constant, the number of tubes required is:

$$n_t = \frac{4121 \text{ ft}^2}{(0.507 \times 16) \text{ ft}^2 \text{per tube}} = 508$$

From the tube-count table, the closest match with two tube passes is 532 tubes in a 29-in. shell. Reducing the number of tubes and the shell ID will cause both h_i and h_o to increase, so it may be possible to reduce the tube length as well. It will also cause the pressure drops to increase. To estimate the effect on tube-side pressure drop, note that from Equation (5.19), $\Delta P_f \sim n_t^{-1.74}$. Therefore, assuming no change in the tube length,

$$\Delta P_f \rightarrow 4.18(532/624)^{-1.74} = 5.52 \text{ psi}$$

Also, since G $\sim n_t^{-1}$,

$$\Delta P_r \rightarrow 0.475(532/624)^{-2} = 0.65 \text{ psi}$$

Hence,

$$\Delta P_i \rightarrow 5.52 + 0.65 + 0.41 = 6.6 \text{ psi}$$

For the shell side, if we neglect the effect of the changes on the friction factor, then Equation (5.28) shows that:

$$\Delta P_f \sim B^{-3} d_s^{-2}$$

Since $B = 0.3\ d_s$, the pressure drop is proportional to the –5 power of shell ID. Hence,

$$\Delta P_f \rightarrow 3.20(29/31)^{-5} = 4.47 \text{ psi}$$

Thus,

$$\Delta P_o \rightarrow 4.47 + 0.25 \cong 4.7 \text{ psi}$$

This configuration may provide a less expensive alternative to the one obtained above. To finalize the design, the overall heat-transfer coefficient must be recalculated, the required tube length determined, and the pressure drops checked, as was done for the second trial in Example 5.1. The calculations are left as an exercise for the reader.

5.9 Computer Software

Commercial software packages for designing shell-and-tube exchangers include ***Xist*** (Heat Transfer Research, Inc.), Aspen Shell & Tube Exchanger (Aspen Technology, Inc.), ProMax (Bryan Research & Engineering, Inc.), and HEXTRAN (Invensys Operations Management). Of these, ***Xist*** and HEXTRAN are covered in this book. HEXTRAN is discussed in this section, while ***Xist*** is considered in Chapter 7.

The shell-and-tube heat-exchanger module (STE) in HEXTRAN is very flexible. It can handle all of the TEMA shell types with either plain or radial low-fin tubes. Both single-phase and two-phase flows are accommodated on either side of the exchanger. Therefore, this module is used for condensers, vaporizers, and reboilers as well as single-phase exchangers. Both un-baffled and baffled exchangers are accepted, including the no-tubes-in window configuration.

The STE module can operate in either rating mode (TYPE=Old) or design mode (TYPE=New). In design mode, certain design parameters are automatically varied to meet a given performance specification (such as the heat duty or a stream outlet temperature) and pressure drop constraints. The following parameters can be varied automatically:

- Number of tubes
- Number of tube passes
- Tube length
- Shell ID
- Number of shells in series and parallel
- Baffle spacing
- Baffle cut

Baffle spacing and baffle cut cannot be varied independently; they must be varied as a pair. As previously noted, these two parameters are highly correlated in practice, and HEXTRAN takes this correlation into account.

For applications in which the shell-side fluid does not undergo a phase change, the full Delaware method is used to calculate the shell-side heat-transfer coefficient and pressure drop. There is also an option (DPSMETHOD = Stream) to calculate the pressure drop using a version of the stream analysis method. These methods will be discussed in more detail in subsequent chapters.

The STE module is not suitable for rod-baffle exchangers or for units, such as air-cooled exchangers, that employ radial high-fin tubes. HEXTRAN provides separate modules for these types of exchangers.

It is important to understand that the results generated by the STE module in design mode are not necessarily optimal. Therefore, it is incumbent on the user to scrutinize the results carefully and to make design modifications as necessary. The software does not eliminate the necessity of iteration in the design process; it just makes the process much faster, easier and less error prone.

Use of the STE module in HEXTRAN version 9.2 is illustrated by the following two examples.

Example 5.3

Use HEXTRAN to rate the final configuration obtained for the kerosene/crude-oil exchanger in Example 5.1, and compare the results with those obtained previously by hand.

Solution

The procedure for problem setup and data entry is similar to that discussed in Example 4.4. Under Units of Measure, the viscosity unit is changed from cp to lb/ft · h for convenience, and the flowsheet is then constructed in the usual manner. The two feed streams are defined as bulk property streams, and the physical properties are entered on the appropriate forms. Note that fluid density (49.008 lbm/ft^3 for kerosene and 53.066 lbm/ft^3 for crude oil), not specific gravity, must be entered. Since stream pressures were not given in Example 5.1 (they are not needed for the calculations, but must be specified in HEXTRAN), a value of 50 psia is arbitrarily assigned for each stream. Flow rates and temperatures of the feed streams are entered as given in Example 5.1.

The physical parameters of the heat exchanger are entered on the appropriate forms exactly as obtained from Example 5.1. In addition, on the Baffles form the baffle thickness is specified as 0.1875 in. (the default value) and the total tubesheet thickness as 4.0 in. (2.0 in. for each tubesheet), in accordance with the design guidelines given in Section 5.7. Note that the total thickness of both tubesheets must be entered, not the individual tubesheet thickness. Only the central baffle spacing (3.85 in.) is specified; no value is entered for the inlet or outlet spacing. One pair of sealing strips is also specified on this form. Finally, under Pressure Drop Options, the *Two-Phase Film/DP Method* is set to *HEXTRAN 5.0x Method,* and the *Shell-side DP Method* is set to *Bell.* (These settings are translated to TWOPHASE = Old and DPSMETHOD = Bell in the input file under UNIT OPERATIONS/STE/CALC.) With these settings, the program uses the Delaware method for both shell-side heat transfer and pressure drop calculations.

The input file generated by the HEXTRAN GUI is given below, followed by the Exchanger Data Sheet and Extended Data Sheet, which were extracted from the output file. Information obtained from the data sheets was used to prepare the following comparison between computer and hand calculations.

Item	Hand Calculation	HEXTRAN
Re_i	10,189	10,189
Re_o	37,158	45,148
h_i (Btu/h · ft^2 · °F)	156	156.2
h_o (Btu/h · ft^2 · °F)	122	191.2
U_D (Btu/h · ft^2 · °F)	46	53.3

(*Continued*)

—cont'd

Item	Hand Calculation	HEXTRAN
ΔP_i (psi)	10.2	10.06
ΔP_o (psi)	2.2	2.10
Over-design, %	4.5	21

The tube-side Reynolds number, heat-transfer coefficient, and pressure drop computed by HEXTRAN agree almost exactly with the hand calculations. There is also excellent agreement between the shell-side pressure drops. However, there are significant differences in the shell-side Reynolds numbers and heat-transfer coefficients calculated by the two methods. The difference in Reynolds numbers is due to differences in the way *Re* is defined in the two methods. As expected, the Simplified Delaware method gives a smaller value for the heat-transfer coefficient than does the full Delaware method. However, the difference is somewhat greater than expected, amounting to about 36% of the HEXTRAN value. This difference is reflected in the over-design values and in the kerosene outlet temperature computed by HEXTRAN, 236°F versus the target temperature of 250°F. Thus, according to HEXTRAN, the exchanger is somewhat over-sized. Also, notice that HEXTRAN adjusts the number of baffles and the end spacing to account for the thickness of the tubesheets. The baffle spacing is interpreted as baffle pitch since the tube length satisfies:

$$L = (n_b - 1)B + B_{in} + B_{out} + \text{total tubesheet thickness}$$

HEXTRAN Input File for Example 5.3

```
$ GENERATED FROM HEXTRAN KEYWORD EXPORTER
$
$      General Data Section
$
TITLE PROJECT=EX5-3, PROBLEM=KEROIL, SITE=
$
DIME English, AREA=FT2, CONDUCTIVITY=BTUH, DENSITY=LB/FT3, *
     ENERGY=BTU, FILM=BTUH, LIQVOLUME=FT3, POWER=HP, *
     PRESSURE=PSIA, SURFACE=DYNE, TIME=HR, TEMPERATURE=F, *
     UVALUE=BTUH, VAPVOLUME=FT3, VISCOSITY=LBFH, WT=LB, *
     XDENSITY=API, STDVAPOR=379.490
$
PRINT ALL, *
      RATE=M
$
CALC  PGEN=New, WATER=Saturated
$
$      Component Data Section
$
$      Thermodynamic Data Section
$
$Stream Data Section
$
STREAM DATA

$
PROP STRM=KEROSENE, NAME=KEROSENE, TEMP=390.00, PRES=50.000, *
        LIQUID(W)=45000.000, LCP(AVG)=0.59, Lcond(AVG)=0.079, *
        Lvis(AVG)=0.97, Lden(AVG)=49.008
$
 PROP STRM=4, NAME=4
$
 PROP STRM=OIL, NAME=OIL, TEMP=100.00, PRES=50.000, *
         LIQUID(W)=150000.000, LCP(AVG)=0.49, Lcond(AVG)=0.077, *
         Lvis(AVG)=8.7, Lden(AVG)=53.066
$
 PROP STRM=3, NAME=3
$
$ Calculation Type Section
```

```
$
SIMULATION
$
 TOLERANCE TTRIAL=0.01
$
 LIMITS AREA=200.00, 6000.00, SERIES=1, 10, PDAMP=0.00, *
      TTRIAL=30
$
 CALC  TWOPHASE=Old, DPSMETHOD=Bell, MINFT=0.80
$
 PRINT UNITS, ECONOMICS, STREAM, STANDARD, *
       EXTENDED, ZONES
$
ECONOMICS DAYS=350, EXCHANGERATE=1.00, CURRENCY=USDOLLAR
$
 UTCOST OIL=3.50, GAS=3.50, ELECTRICITY=0.10, *
        WATER=0.03, HPSTEAM=4.10, MPSTEAM=3.90, *
        LPSTEAM=3.60, REFRIGERANT=0.00, HEATINGMEDIUM=0.00
$
 HXCOST BSIZE=1000.00, BCOST=0.00, LINEAR=50.00, *
        EXPONENT=0.60, CONSTANT=0.00, UNIT
$
$     Unit Operations Data
$
UNIT OPERATIONS
$

STE UID=STE1
 TYPE Old, TEMA=AES, HOTSIDE=Shellside, ORIENTATION=Horizontal, *
      FLOW=Countercurrent, *
      UESTIMATE=50.00, USCALER=1.00

TUBE  FEED=OIL, PRODUCT=3, *
      LENGTH=14.00, OD=1.000, *
      BWG=14, NUMBER=124, PASS=4, PATTERN=90, *
      PITCH=1.2500, MATERIAL=1, *
      FOUL=0.003, LAYER=0, *
      DPSCALER=1.00
$
SHELL FEED=KEROSENE, PRODUCT=4, *
      ID=19.25, SERIES=1, PARALLEL=1, *
      SEALS=1, MATERIAL=1, *
      FOUL=0.002, LAYER=0, *
       DPSCALER=1.00
$
BAFF  Segmental=Single, *
      CUT=0.20, *
      SPACING=3.850, *
      THICKNESS=0.1875, SHEETS=4.000
$
 TNOZZ TYPE=Conventional, ID=4.026, 4.026, NUMB=1, 1
$
 SNOZZ TYPE=Conventional , ID=3.068, 3.068, NUMB=1, 1
$
 CALC TWOPHASE=Old, *
      DPSMETHOD=Bell, *
      MINFT=0.80
$
 PRINT STANDARD, *
       EXTENDED, *
       ZONES
$
 COST BSIZE=1000.00, BCOST=0.00, LINEAR=50.00, *
      CONSTANT=0.00, EXPONENT=0.60, Unit
$

$ End of keyword file...
```

HEXTRAN Output Data for Example 5.3

```
==============================================================================
                    SHELL AND TUBE EXCHANGER DATA SHEET
I----------------------------------------------------------------------------I
I EXCHANGER  NAME                               UNIT ID STE1                 I
I SIZE   19x  168  TYPE AES,   HORIZONTAL     CONNECTED 1 PARALLEL 1 SERIES I
I AREA/UNIT   444. FT2 (   444. FT2 REQUIRED) AREA/SHELL   444. FT2         I
I----------------------------------------------------------------------------I
I PERFORMANCE OF ONE UNIT          SHELL-SIDE                 TUBE-SIDE      I
I----------------------------------------------------------------------------I
I FEED STREAM NUMBER                KEROSENE                  OIL            I
I FEED STREAM NUMBER                KEROSENE                  OIL            I
I TOTAL FLUID         LB /HR           45000.                   150000.      I
I     VAPOR  (IN/OUT) LB /HR        0./          0.           0./        0. I
I     LIQUID          LB /HR    45000./      45000.      150000./   150000. I
I     STEAM           LB /HR        0./          0.           0./        0. I
I     WATER           LB /HR        0./          0.           0./        0. I
I     NON CONDENSIBLE LB /HR             0.                        0.        I
I TEMPERATURE (IN/OUT) DEG F     390.0 /   236.2           100.0 /  155.6    I
I PRESSURE    (IN/OUT) PSIA      50.00 /    47.90          50.00 /   39.94   I
I----------------------------------------------------------------------------I
I SP. GR., LIQ (60F /  60F H2O)  0.785 /   0.785           0.850 /  0.850    I
I          VAP (60F /  60F AIR)  0.000 /   0.000           0.000 /  0.000    I
I DENSITY, LIQUID      LB/FT3   49.008 /  49.008          53.066 / 53.066    I
I          VAPOR       LB/FT3    0.000 /   0.000           0.000 /  0.000    I
I VISCOSITY, LIQUID  LB/FT-HR    0.970 /   0.970           8.700 /  8.700    I
I            VAPOR   LB/FT-HR    0.000 /   0.000           0.000 /  0.000    I
I THRML COND,LIQ  BTU/HR-FT-F   0.0790 /  0.0790          0.0770 / 0.0770    I
I            VAP  BTU/HR-FT-F   0.0000 /  0.0000          0.0000 / 0.0000    I
I SPEC.HEAT,LIQUID BTU /LB F    0.5900 /  0.5900          0.4900 / 0.4900    I
I            VAPOR   BTU /LB F  0.0000 /  0.0000          0.0000 / 0.0000    I
I LATENT HEAT         BTU  /LB        0.00                      0.00         I
I VELOCITY             FT/SEC         1.81                      6.68         I
I DP/SHELL(DES/CALC)   PSI        0.00 /  2.10              0.00 / 10.06     I
I FOULING RESIST FT2-HR-F/BTU 0.00200 (0.00200 REQD)           0.00300       I
I----------------------------------------------------------------------------I
I TRANSFER RATE  BTU/HR-FT2-F SERVICE   53.32 (  53.32 REQD), CLEAN   76.01 I
I HEAT EXCHANGED MMBTU /HR    4.083,     MTD(CORRECTED) 172.6,     FT 0.954 I
I----------------------------------------------------------------------------I
I CONSTRUCTION OF ONE SHELL        SHELL-SIDE                TUBE-SIDE       I
I----------------------------------------------------------------------------I
I DESIGN PRESSURE/TEMP PSIA  /F    125./ 400.            125./ 400.          I
I NUMBER OF PASSES                     1                        4            I
I MATERIAL                         CARB STL                 CARB STL         I
I INLET  NOZZLE ID/NO       IN       3.1/ 1                 4.0/ 1           I
I OUTLET NOZZLE ID/NO       IN       3.1/ 1                 4.0/ 1           I
I----------------------------------------------------------------------------I
I TUBE: NUMBER   124, OD  1.000 IN , BWG     14        , LENGTH 14.0 FT      I
I       TYPE BARE,                 PITCH    1.2500 IN,  PATTERN 90 DEGREES  I
I SHELL:  ID    19.25 IN,             SEALING STRIPS  1 PAIRS                I
I BAFFLE: CUT  .200, SPACING(IN): IN    5.00, CENT   3.85, OUT   5.00,SING I
I RHO-V2: INLET NOZZLE  1210.3 LB/FT-SEC2                                    I
I TOTAL WEIGHT/SHELL,LB    3205.5 FULL OF WATER  0.738E+04 BUNDLE    2590.1 I
I----------------------------------------------------------------------------I
```

```
==============================================================================
                  SHELL AND TUBE EXTENDED DATA SHEET
I----------------------------------------------------------------------------I
I EXCHANGER  NAME                         UNIT ID STE1 I
I SIZE    19x 168   TYPE AES,   HORIZONTAL   CONNECTED 1 PARALLEL  1 SERIES  I
I AREA/UNIT   444. FT2 (    444. FT2 REQUIRED) I
I----------------------------------------------------------------------------I
I PERFORMANCE OF ONE UNIT          SHELL-SIDE                TUBE-SIDE       I
I----------------------------------------------------------------------------I
I FEED STREAM NUMBER                 KEROSENE                 OIL            I
I FEED STREAM NAME                  KEROSENE                 OIL            I
I WT FRACTION LIQUID (IN/OUT)     1.00 / 1.00               1.00 / 1.00     I
I REYNOLDS NUMBER                    45148.                   10189.         I
I PRANDTL NUMBER                     7.244                    55.364         I
I UOPK,LIQUID                    0.000 /    0.000          0.000 /    0.000  I
I      VAPOR                     0.000 /    0.000          0.000 /    0.000  I
I SURFACE TENSION    DYNES/CM    0.000 /    0.000          0.000 /    0.000  I
I FILM COEF(SCL) BTU/HR-FT2-F      191.2 (1.000)            156.2 (1.000)    I
I FOULING LAYER THICKNESS  IN        0.000                    0.000          I
I----------------------------------------------------------------------------I
I THERMAL RESISTANCE                                                         I
I UNITS:   (FT2-HR-F/BTU)      (PERCENT)   (ABSOLUTE)                        I
I SHELL FILM                     27.88       0.00523                         I
I TUBE  FILM                     40.93       0.00767                         I
I TUBE  METAL                     1.34       0.00025                         I
I TOTAL FOULING                  29.85       0.00560                         I
I ADJUSTMENT                      0.00       0.00000                         I
I----------------------------------------------------------------------------I
I PRESSURE DROP                     SHELL-SIDE                TUBE-SIDE      I
I UNITS: (PSIA  )              PERCENT)   (ABSOLUTE)   (PERCENT)   (ABSOLUTE)I
I WITHOUT NOZZLES                90.07         1.89       92.82          9.34 I
I INLET   NOZZLES                 6.20         0.13        4.48          0.45 I
I OUTLET  NOZZLES                 3.72         0.08        2.69          0.27 I
I TOTAL   /SHELL                               2.10                     10.06 I
I TOTAL   /UNIT                                2.10                     10.06 I
I DP SCALER                                    1.00                      1.00 I
I----------------------------------------------------------------------------I
I CONSTRUCTION OF ONE SHELL                                                  I
I----------------------------------------------------------------------------I
I TUBE:OVERALL LENGTH       14.0          FT EFFECTIVE LENGTH    13.67    FT I
I      TOTAL TUBESHEET THK   4.0          IN AREA RATIO (OUT/IN)  1.199      I
I      THERMAL COND.        30.0BTU/HR-FT-F DENSITY              490.80 LB/FT3I
I----------------------------------------------------------------------------I
I BAFFLE: THICKNESS          0.188        IN NUMBER              41          I
I----------------------------------------------------------------------------I
I BUNDLE: DIAMETER           17.0         IN TUBES IN CROSSFLOW  70          I
I         CROSSFLOW AREA     0.086       FT2 WINDOW AREA          0.142  FT2 I
I         TUBE-BFL LEAK AREA 0.033       FT2 SHELL-BFL LEAK AREA  0.022  FT2 I
I----------------------------------------------------------------------------I
```

Example 5.4

Use HEXTRAN to design a shell-and-tube heat exchanger for the kerosene/crude-oil service of Example 5.1, and compare the resulting unit with the one designed previously by hand.

Solution

The STE module in HEXTRAN is configured in design mode by right-clicking on the unit and selecting *Change Configuration* from the pop-up menu. Exchanger parameters are set as in Example 5.3, except for the following items that are left unspecified to be calculated by the program: number of tubes, tubesheet thickness, baffle spacing, number of sealing strips, and nozzle sizes. Design constraints are set as shown below, and the kerosene outlet temperature (250°F) is given as the design specification on the Specifications panel.

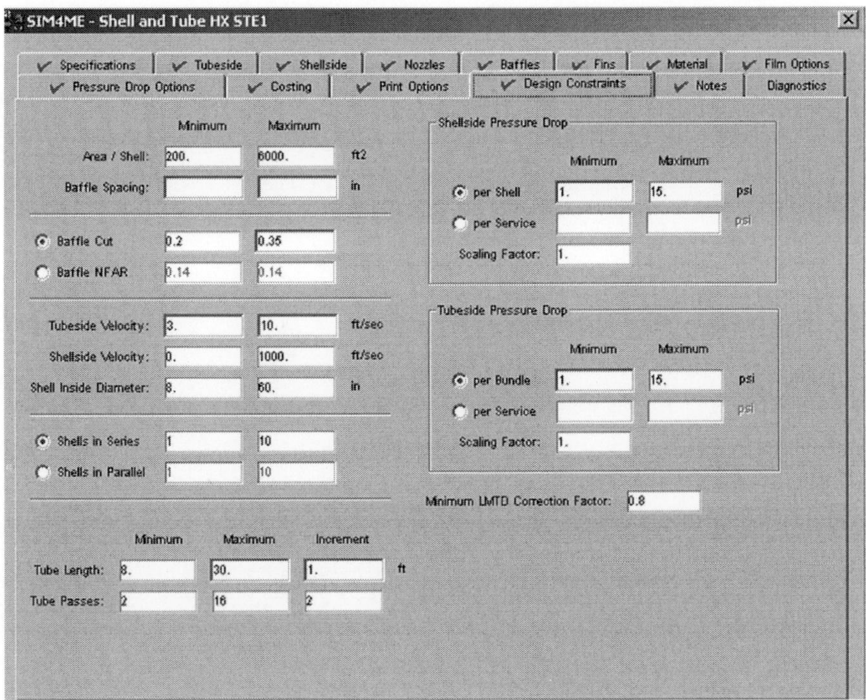

The HEXTRAN input file and Exchanger Data Sheets for this run (Run 1) are given below, and the results are compared with the hand calculations in the following table. In all cases, the tubes are 1-in. OD, 14 BWG, on 1.25-in. square pitch. Notice that the nozzle diameters are rounded to one decimal place in the HEXTRAN output. The exact nozzle sizes determined by the program are the same as used in the hand calculations, namely, 4-in. schedule 40 on the tube side and 3-in. schedule 40 on the shell side.

Item	Hand	Run 1	Run 2	Run 3
Shell ID (in.)	19.25	13.25	17.0	19.0
Number of tubes	124	47	79	99
Tube length (ft)	14	26	20	14
Number of tube passes	4	2	2	4
Baffle cut (%)	20	24.9	22.9	22.3
(Central) baffle spacing (in.)	3.85	6.14	6.00	6.18
Tube-side nozzle ID (in.)	4.026	4.0	4.0	4.0
Shell-side nozzle ID (in.)	3.068	3.1	3.1	3.1
Re_i	10,189	13,441	7997	12,762
ΔP_i (psi)	10.2	13.75	4.86	14.66
ΔP_o (psi)	2.2	3.55	1.74	1.04
Surface area (ft^2)	454	322	419	366

All design criteria are met, but with a length of 26 ft and a diameter of only 13.25 in., the configuration is awkward. In an attempt to obtain a more compact design, the maximum tube length was changed from 30 to 20 ft. The results for this run (Run 2) are given in the above table. As can be seen, the tube-side Reynolds number is in the transition region, which is undesirable.

A third run was made with the minimum number of tube passes set at four and the maximum tube length set at 20 ft. The Exchanger Data Sheet for this run is given below and the results are compared with those from the other runs in the table above. It can be seen that this run produced a configuration very similar to the one obtained by hand, albeit with less heat-transfer area due to the higher shell-side coefficient computed by HEXTRAN. Although the tube-side velocity is on the high side at 8.36 ft/s, it is acceptable with carbon steel tubes. Checking the appropriate tube-count table (Table C.5) shows that a 17.25 in. shell should be adequate for a bundle containing 99 tubes. Since HEXTRAN specifies a 19 in. shell for this bundle, it is apparent that it uses a different set of tube-count data. Notice that in the HEXTRAN designs the kerosene outlet temperature exactly matches the design specification of 250°F; thus, the over-design for these units is effectively zero.

HEXTRAN Input File for Example 5.4, Run 1

```
$
GENERATED FROM HEXTRAN KEYWORD EXPORTER $
$
$      General Data Section
$
TITLE PROJECT=Ex5-4, PROBLEM=KEROIL, SITE=
$
DIME  English, AREA=FT2, CONDUCTIVITY=BTUH, DENSITY=LB/FT3, *
      ENERGY=BTU, FILM=BTUH, LIQVOLUME=FT3, POWER=HP, *
      PRESSURE=PSIA, SURFACE=DYNE, TIME=HR, TEMPERATURE=F, *
      UVALUE=BTUH, VAPVOLUME=FT3, VISCOSITY=LBFH, WT=LB, *
      XDENSITY=API, STDVAPOR=379.490
$
PRINT ALL, *
      RATE=M
$
CALC  PGEN=New, WATER=Saturated
$
$      Component Data Section
$

$
$      Thermodynamic Data Section
$

$
$Stream Data Section
$
STREAM DATA

$
 PROP STRM=PS2, NAME=PS2
$
 PROP STRM=OIL, NAME=OIL, TEMP=100.00, PRES=50.000, *
         LIQUID(W)=150000.000, LCP(AVG)=0.49, Lcond(AVG)=0.077, *
         Lvis(AVG)=8.7, Lden(AVG)=53.066
$
 PROP STRM=PS4, NAME=PS4
$
 PROP STRM=KEROSENE, NAME=KEROSENE, TEMP=390.00, PRES=50.000, *
         LIQUID(W)=45000.000, LCP(AVG)=0.59, Lcond(AVG)=0.079, *
         Lvis(AVG)=0.97, Lden(AVG)=49.008
$
$ Calculation Type Section
$
SIMULATION
$
 TOLERANCE TTRIAL=0.01
$
 LIMITS AREA=200.00, 6000.00, SERIES=1, 10, PDAMP=0.00, *
        TTRIAL=30
$
CALC    TWOPHASE=New, DPSMETHOD=Stream, MINFT=0.80
$
PRINT   UNITS, ECONOMICS, STREAM, STANDARD, *
        EXTENDED, ZONES
```

```
$
ECONOMICS DAYS=350, EXCHANGERATE=1.00, CURRENCY=USDOLLAR
$
 UTCOST OIL=3.50, GAS=3.50, ELECTRICITY=0.10, *
        WATER=0.03, HPSTEAM=4.10, MPSTEAM=3.90, *
        LPSTEAM=3.60, REFRIGERANT=0.00, HEATINGMEDIUM=0.00
$
 HXCOST BSIZE=1000.00, BCOST=0.00, LINEAR=50.00, *
        EXPONENT=0.60, CONSTANT=0.00, UNIT
$
$     Unit Operations Data
$
UNIT OPERATIONS
$
STE UID=STE1
 TYPE  New, TEMA=AES, HOTSIDE=Shellside, ORIENTATION=Horizontal, *
       FLOW=Countercurrent, AREA=200.00, 6000.00, *
       UESTIMATE=50.00, USCALER=1.00

 TUBE  FEED=OIL, PRODUCT=PS4, *
       LENGTH=8.00, 30.00, 1.00, OD=1.000, *
       BWG=14, PASS=2, 16, 2, PATTERN=90, *
       PITCH=1.2500, MATERIAL=1, *
       FOUL=0.003, LAYER=0, *
       DPSCALER=1.00, DPSHELL=1.000, 15.000, VELOCITY=3.0, 10.0
$
SHELL  FEED=KEROSENE, PRODUCT=PS2, *
       ID=8.00, 60.00, SERIES=1, 10, *
       MATERIAL=1, *
       FOUL=0.002, LAYER=0, *
       DPSCALER=1.00, DPSHELL=1.000, 15.000, VELOCITY=0.0, 1000.0
$
BAFF   Segmental=Single, *
       CUT=0.20, 0.35, *
       THICKNESS=0.1875
$
 TNOZZ  NUMB=1, 1
$
 SNOZZ  NUMB=1, 1
$
 CALC  TWOPHASE=Old, *
       DPSMETHOD=Bell, *
       MINFT=0.80
$
 SPEC  Shell, Temp=250.000000
$
 PRINT STANDARD, *
       EXTENDED, *
       ZONES
$
COST   BSIZE=1000.00, BCOST=0.00, LINEAR=50.00, *
       CONSTANT=0.00, EXPONENT=0.60, Unit
$
$ End of keyword file...
```

HEXTRAN Output Data for Example 5.4, Run 1

```
==============================================================================
                      SHELL AND TUBE EXCHANGER DATA SHEET
I----------------------------------------------------------------------------I
I EXCHANGER  NAME                                  UNIT ID STE1              I
I SIZE   13x   312  TYPE AES,    HORIZONTAL        CONNECTED 1 PARALLEL 1 SERIES I
I AREA/UNIT    322. FT2 (    322. FT2 REQUIRED) AREA/SHELL    322. FT2       I
I----------------------------------------------------------------------------I
I PERFORMANCE OF ONE UNIT             SHELL-SIDE                TUBE-SIDE     I
I----------------------------------------------------------------------------I
I FEED STREAM NUMBER                   KEROSENE                  OIL          I
I FEED STREAM NAME                     KEROSENE                  OIL          I
I TOTAL FLUID             LB /HR         45000.                   150000.     I
I     VAPOR   (IN/OUT) LB /HR          0./          0.           0./        0. I
I     LIQUID           LB /HR      45000./     45000.       150000./  150000. I
I     STEAM            LB /HR          0./          0.           0./        0. I
I     WATER            LB /HR          0./          0.           0./        0. I
I     NON CONDENSIBLE LB /HR              0.                         0.       I
I TEMPERATURE (IN/OUT) DEG F       390.0 /  250.0           100.0 / 150.6     I
I PRESSURE     (IN/OUT) PSIA       50.00 /   46.45          50.00 /  36.25    I
I----------------------------------------------------------------------------I
I SP. GR., LIQ    (60F / 60F H2O)  0.786 /  0.786           0.851 /  0.851    I
I          VAP    (60F / 60F AIR)  0.000 /  0.000           0.000 /  0.000    I
I DENSITY,    LIQUID      LB/FT3  49.008 / 49.008          53.066 / 53.066    I
I             VAPOR       LB/FT3   0.000 /  0.000           0.000 /  0.000    I
I VISCOSITY, LIQUID      LB/FT-HR  0.970 /  0.970           8.700 /  8.700    I
I             VAPOR       LB/FT-HR  0.000 /  0.000           0.000 /  0.000    I
I THRML COND,LIQ  BTU/HR-FT-F    0.0790 /  0.0790          0.0770 /  0.0770   I
I             VAP  BTU/HR-FT-F    0.0000 /  0.0000          0.0000 /  0.0000   I
I SPEC.HEAT,LIQUID BTU /LB F      0.5900 /  0.5900          0.4900 /  0.4900   I
I             VAPOR  BTU /LB F    0.0000 /  0.0000          0.0000 /  0.0000   I
I LATENT HEAT          BTU /LB         0.00                     0.00          I
I VELOCITY              FT/SEC         2.35                     8.81          I
I DP/SHELL(DES/CALC)    PSI        15.00 / 3.55             15.00 / 13.75     I
I FOULING RESIST FT2-HR-F/BTU  0.00200 (0.00200 REQD)            0.00300      I
I----------------------------------------------------------------------------I
I TRANSFER RATE  BTU/HR-FT2-F  SERVICE    62.45 (  62.45 REQD), CLEAN   96.02 I
I HEAT EXCHANGED MMBTU /HR      3.717,      MTD(CORRECTED) 184.8,     FT 0.966 I
I----------------------------------------------------------------------------I
I CONSTRUCTION OF ONE SHELL            SHELL-SIDE                TUBE-SIDE    I
I----------------------------------------------------------------------------I
I DESIGN PRESSURE/TEMP PSIA /F       125./ 400.               125./ 400.      I
I NUMBER OF PASSES                        1                        2          I
I MATERIAL                             CARB STL                 CARB STL      I
I INLET  NOZZLE ID/NO        IN         3.1/ 1                  4.0/ 1        I
I OUTLET NOZZLE ID/NO        IN         3.1/ 1                  4.0/ 1        I
I----------------------------------------------------------------------------I
I TUBE: NUMBER     47, OD  1.000  IN , BWG   14         , LENGTH 26.0 FT      I
I        TYPE BARE,               PITCH   1.2500 IN,   PATTERN 90 DEGREES     I
I SHELL: ID    13.25 IN,              SEALING STRIPS  1 PAIRS                 I
I BAFFLE: CUT .249, SPACING(IN): IN      7.67, CENT   6.14, OUT      7.67,SING I
I RHO-V2: INLET NOZZLE  1209.7 LB/FT-SEC2                                     I
I TOTAL WEIGHT/SHELL,LB     2759.5 FULL OF WATER   0.611E+04 BUNDLE    1926.1 I
I----------------------------------------------------------------------------I
```

```
============================================================================
                    SHELL AND TUBE EXTENDED DATA SHEET
I--------------------------------------------------------------------------I
I EXCHANGER  NAME                            UNIT ID STE1                  I
I SIZE  13x  312    TYPE AES,   HORIZONTAL   CONNECTED 1 PARALLEL  1 SERIES I
I AREA/UNIT   322. FT2 (     322. FT2 REQUIRED)                            I
I--------------------------------------------------------------------------I
I PERFORMANCE OF ONE UNIT               SHELL-SIDE              TUBE-SIDE  I
I--------------------------------------------------------------------------I
I FEED STREAM NUMBER                     KEROSENE                OIL       I
I FEED STREAM NAME                      KEROSENE                 OIL       I
I WT FRACTION LIQUID (IN/OUT)         1.00 / 1.00             1.00 / 1.00  I
I REYNOLDS NUMBER                      44611.                  13441.      I
I PRANDTL NUMBER                       7.244                  55.364       I
I UOPK,LIQUID                       0.000 /    0.000          0.000 / 0.000  I
I      VAPOR                        0.000 /    0.000          0.000 / 0.000  I
I SURFACE TENSION    DYNES/CM       0.000 /    0.000          0.000 / 0.000  I
I FILM COEF(SCL) BTU/HR-FT2-F         249.2 (1.000)          195.0 (1.000)  I
I FOULING LAYER THICKNESS IN            0.000                  0.000       I
I--------------------------------------------------------------------------I
I THERMAL RESISTANCE                                                       I
I UNITS: (FT2-HR-F/BTU)           (PERCENT)   (ABSOLUTE)                   I
I SHELL FILM                         25.07      0.00401                    I
I TUBE FILM                          38.40      0.00615                    I
I TUBE METAL                          1.57      0.00025                    I
I TOTAL FOULING                      34.96      0.00560                    I
I ADJUSTMENT                          0.00      0.00000                    I
I--------------------------------------------------------------------------I
I PRESSURE DROP                        SHELL-SIDE              TUBE-SIDE   I
I UNITS: (PSIA )                  (PERCENT)   (ABSOLUTE)  (PERCENT)   (ABSOLUTE)I
I WITHOUT NOZZLES                    94.13         3.34      94.75       13.03 I
I INLET   NOZZLES                     3.67         0.13       3.28        0.45 I
I OUTLET  NOZZLES                     2.20         0.08       1.97        0.27 I
I TOTAL   /SHELL                                   3.55                  13.75 I
I TOTAL   /UNIT                                    3.55                  13.75 I
I DP SCALER                                        1.00                   1.00 I
I--------------------------------------------------------------------------I
I CONSTRUCTION OF ONE SHELL                                                I
I--------------------------------------------------------------------------I
I TUBE:OVERALL LENGTH       26.0        FT  EFFECTIVE LENGTH     25.82    FT I
I      TOTAL TUBESHEET THK  2.1         IN  AREA RATIO (OUT/IN)   1.199      I
I      THERMAL COND.        30.0BTU/HR-FT-F  DENSITY             490.80 LB/FT3I
I--------------------------------------------------------------------------I
I BAFFLE: THICKNESS           0.125       IN  NUMBER                 49      I
I--------------------------------------------------------------------------I
I BUNDLE: DIAMETER            11.2        IN TUBES IN CROSSFLOW     25       I
I         CROSSFLOW AREA      0.087     FT2 WINDOW AREA           0.128 FT2 I
I         TUBE-BFL LEAK AREA  0.012     FT2 SHELL-BFL LEAK AREA   0.010 FT2 I
I--------------------------------------------------------------------------I
```

HEXTRAN Output Data for Example 5.4, Run 3

```
==============================================================================
                     SHELL AND TUBE EXCHANGER DATA SHEET
I----------------------------------------------------------------------------I
I EXCHANGER  NAME                                  UNIT ID STE1              I
I SIZE   19x  168  TYPE AES,   HORIZONTAL          CONNECTED 1 PARALLEL 1 SERIES I
I AREA/UNIT   366. FT2 (   366. FT2 REQUIRED)      AREA/SHELL 366. FT2       I
I----------------------------------------------------------------------------I
I PERFORMANCE OF ONE UNIT             SHELL-SIDE             TUBE-SIDE       I
I----------------------------------------------------------------------------I
I FEED STREAM NUMBER                   KEROSENE                 OIL          I
I FEED STREAM NAME                     KEROSENE                 OIL          I
I TOTAL FLUID          LB /HR             45000.                  150000.    I
I     VAPOR  (IN/OUT) LB /HR          0./          0.          0./         0. I
I     LIQUID          LB /HR      45000./      45000.    150000./    150000. I
I     STEAM           LB /HR          0./          0.          0./         0. I
I     WATER           LB /HR          0./          0.          0./         0. I
I     NON CONDENSIBLE LB /HR               0.                        0.      I
I TEMPERATURE (IN/OUT) DEG F      390.0 /    250.0        100.0 /    150.6   I
I PRESSURE     (IN/OUT) PSIA      50.00 /     48.96       50.00 /     35.34  I
I----------------------------------------------------------------------------I
I SP. GR., LIQ  (60F / 60F H2O)   0.786 /   0.786         0.851 /    0.851   I
I          VAP  (60F / 60F AIR)   0.000 /   0.000         0.000 /    0.000   I
I DENSITY,   LIQUID    LB/FT3    49.008 /  49.008        53.066 /   53.066   I
I            VAPOR     LB/FT3     0.000 /   0.000         0.000 /    0.000   I
I VISCOSITY, LIQUID  LB/FT-HR     0.970 /   0.970         8.700 /    8.700   I
I            VAPOR   LB/FT-HR     0.000 /   0.000         0.000 /    0.000   I
I THRML COND,LIQ  BTU/HR-FT-F    0.0790 /  0.0790        0.0770 /   0.0770   I
I            VAP BTU/HR-FT-F     0.0000 /  0.0000        0.0000 /   0.0000   I
I SPEC.HEAT,LIQUID BTU /LB F     0.5900 /  0.5900        0.4900 /   0.4900   I
I            VAPOR  BTU /LB F    0.0000 /  0.0000        0.0000 /   0.0000   I
I LATENT HEAT         BTU /LB          0.00                    0.00          I
I VELOCITY            FT/SEC           1.40                    8.36          I
I DP/SHELL(DES/CALC)    PSI       15.00 / 1.04            15.00 / 14.66      I
I FOULING RESIST FT2-HR-F/BTU  0.00200 (0.00206 REQD)        0.00300         I
I----------------------------------------------------------------------------I
I TRANSFER RATE  BTU/HR-FT2-F  SERVICE   54.98 (  54.98 REQD), CLEAN   79.80 I
I HEAT EXCHANGED MMBTU /HR     3.717,      MTD(CORRECTED) 184.8,      FT 0.966 I
I----------------------------------------------------------------------------I
I CONSTRUCTION OF ONE SHELL            SHELL-SIDE              TUBE-SIDE     I
I----------------------------------------------------------------------------I
I DESIGN PRESSURE/TEMP PSIA  /F      125./  400.            125./  400.      I
I NUMBER OF PASSES                       1                       4           I
I MATERIAL                            CARB STL                CARB STL       I
I INLET NOZZLE  ID/NO       IN         3.1/ 1                  4.0/ 1        I
I OUTLET NOZZLE ID/NO       IN         3.1/ 1                  4.0/ 1        I
I----------------------------------------------------------------------------I
I TUBE: NUMBER    99, OD  1.000  IN , BWG    14          , LENGTH 14.0 FT    I
I       TYPE BARE,                  PITCH     1.2500 IN    PATTERN 90 DEGREES I
I SHELL:  ID    19.00 IN,              SEALING STRIPS     2 PAIRS            I
I BAFFLE: CUT  .223, SPACING(IN): IN 8.63, CENT         6.18, OUT   8.63,SING I
I RHO-V2: INLET NOZZLE  1209.7 LB/FT-SEC2                                    I
I TOTAL WEIGHT/SHELL,LB 3155.9 FULL OF WATER        0.690E+04 BUNDLE   2166.7 I
I----------------------------------------------------------------------------I
```

```
==============================================================================
                    SHELL AND TUBE EXTENDED DATA SHEET
I----------------------------------------------------------------------------I
I EXCHANGER  NAME                               UNIT ID STE1 I
I SIZE   19x 168    TYPE AES,      HORIZONTAL   CONNECTED 1 PARALLEL  1 SERIES I
I AREA/UNIT   366. FT2 (   366.   FT2 REQUIRED)                              I
I----------------------------------------------------------------------------I
I PERFORMANCE OF ONE UNIT               SHELL-SIDE             TUBE-SIDE      I
I----------------------------------------------------------------------------I
I FEED STREAM NUMBER                      KEROSENE             OIL            I
I FEED STREAM NAME                       KEROSENE              OIL            I
I WT FRACTION LIQUID (IN/OUT)          1.00 / 1.00           1.00 / 1.00      I
I REYNOLDS NUMBER                        28556.                12762.         I
I PRANDTL NUMBER                         7.244                55.364          I
I UOPK,LIQUID                        0.000 /    0.000       0.000 /    0.000 I
I      VAPOR                         0.000 /    0.000       0.000 /    0.000 I
I SURFACE TENSION    DYNES/CM        0.000 /    0.000       0.000 /    0.000 I
I FILM COEF(SCL) BTU/HR-FT2-F           170.4 (1.000)         187.1 (1.000)  I
I FOULING LAYER THICKNESS  IN             0.000                 0.000         I
I----------------------------------------------------------------------------I
I THERMAL RESISTANCE                                                          I
I UNITS: (FT2-HR-F/BTU)          (PERCENT)     (ABSOLUTE)                      I
I SHELL FILM                       32.38         0.00587                       I
I TUBE  FILM                       35.36         0.00641                       I
I TUBE  METAL                       1.39         0.00025                       I
I TOTAL FOULING                    30.88         0.00560                       I
I ADJUSTMENT                        0.00         0.00000                       I
I----------------------------------------------------------------------------I
I PRESSURE DROP                        SHELL-SIDE              TUBE-SIDE      I
I UNITS: (PSIA  )              (PERCENT)     (ABSOLUTE)   (PERCENT)    (ABSOLUTE)I
I WITHOUT NOZZLES                79.86          0.83       95.07         13.94 I
I INLET   NOZZLES                12.59          0.13        3.08          0.45 I
I OUTLET  NOZZLES                 7.55          0.08        1.85          0.27 I
I TOTAL   /SHELL                                1.04                     14.66 I
I TOTAL   /UNIT                                 1.04                     14.66 I
I DP SCALER                                     1.00                      1.00 I
I----------------------------------------------------------------------------I
I CONSTRUCTION OF ONE SHELL                                                   I
I----------------------------------------------------------------------------I
I TUBE:OVERALL LENGTH          14.0          FT  EFFECTIVE LENGTH     13.80   FT I
I      TOTAL TUBESHEET THK     2.4           IN  AREA RATIO (OUT/IN)    1.199    I
I      THERMAL COND.          30.0BTU/HR-FT-F  DENSITY              490.80 LB/FT3I
I----------------------------------------------------------------------------I
I BAFFLE: THICKNESS              0.188       IN  NUMBER                25        I
I----------------------------------------------------------------------------I
I BUNDLE: DIAMETER              16.8         IN  TUBES IN CROSSFLOW  52          I
I         CROSSFLOW AREA         0.135       FT2 WINDOW AREA          0.201  FT2 I
I         TUBE-BFL LEAK AREA     0.026       FT2 SHELL-BFL LEAK AREA  0.021  FT2 I
I----------------------------------------------------------------------------I
```

Example 5.5

Temperature profile is an important and often overlooked design consideration. This example demonstrates the unforeseen consequences of neglecting the temperature profile in design selection. The process conditions for a component cooling water heat exchanger (hot and cold fluids are water) are shown below. The function of this exchanger is to remove the decay heat from a nuclear reactor following shutdown.

Duty, MW	4.39
Hot water inlet temperature, °C	115
Hot water flow rate, kg/s	17
Cold water inlet temperature, °C	25
Cold water flow rate, kg/s	42

The duty shown is for design conditions and represents the minimum duty for the conditions shown. The outlet temperatures at design conditions are calculated based on these process conditions and a specific heat, C_P, of 4200 J/kg · K:

$$T_o = T_i - \frac{q}{\dot{m}_h C_P} = 115 - \frac{4.39 \times 10^6}{17 \times 4200} \cong 53.5°\text{C}$$

$$t_o = t_i + \frac{q}{\dot{m}_c C_P} = 25 + \frac{4.39 \times 10^6}{42 \times 4200} \cong 50°\text{C}$$

These exchangers are often TEMA E-shells with two tube passes. With this configuration, the LMTD correction factor is calculated based on Equations (3.10) to (3.15):

$$R = \frac{T_i - T_o}{t_o - t_i} = \frac{115 - 53.5}{50 - 25} = \frac{61.5}{25} = 2.46$$

$$P = \frac{t_o - t_i}{T_i - t_i} = \frac{25}{115 - 25} = 0.278$$

$$\alpha = \left(\frac{1 - RP}{1 - P}\right)^{1/N} = \left(\frac{1 - 2.46 \times 0.278}{1 - 0.278}\right)^{1/1} = 0.438$$

$$S = \frac{\alpha - 1}{\alpha - R} = \frac{0.438 - 1}{0.438 - 2.46} = 0.278$$

$$F = \frac{\sqrt{R^2 + 1}\ \ln\left(\frac{1 - S}{1 - RS}\right)}{(R - 1)\ln\left[\frac{2 - S\left(R + 1 - \sqrt{R^2 + 1}\right)}{2 - S\left(R + 1 + \sqrt{R^2 + 1}\right)}\right]}$$

$$= \frac{\sqrt{2.46^2 + 1}\ \ln\left(\frac{1 - 0.278}{1 - 2.46 \times 0.278}\right)}{(2.46 - 1)\ln\left(\frac{2 - 0.278\left(2.46 + 1 - \sqrt{2.46^2 + 1}\right)}{2 - 0.278\left(2.46 + 1 + \sqrt{2.46^2 + 1}\right)}\right)} = 0.843$$

As discussed in Section 3.6, this value is considered acceptable since it is greater than 0.8. In other words, an efficient temperature profile is expected. The design is completed based on the following parameters:

Shell ID, mm	508
Tube length, mm	7315
Tube count	320
Tube OD and schedule, mm	19.05, BWG 16
Tube pitch and layout, mm	23.8, triangular
Tube material	Type 304 stainless steel
Baffle number and type	14 segmental
Baffle cut, %	28.7
Cold water fouling factor, $m^2 \cdot K/W$	0.0004
Hot water fouling factor, $m^2 \cdot K/W$	0.0002

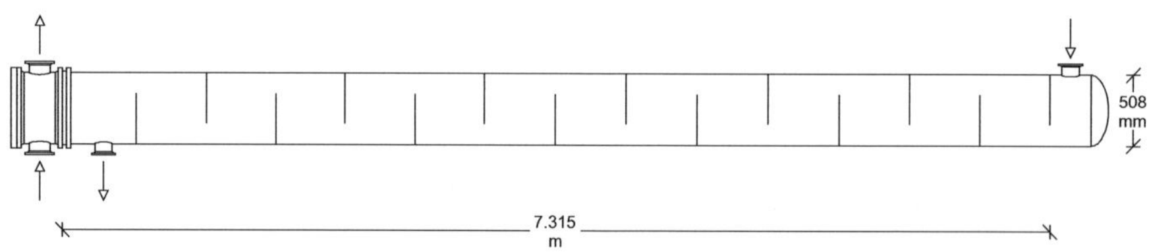

Using HTRI shell-and-tube software *Xist*, the thermal rating shows the following overall performance:

Overall heat-transfer coefficient, $W/m^2 \cdot K$	879
Mean temperature difference, °C	36.7
Over-design, %	4.84
Heat-transfer area, m^2	143
Average tube-side heat-transfer coefficient, $W/m^2 \cdot K$	6811
Average shell-side heat-transfer coefficient, $W/m^2 \cdot K$	6044

Xist shows the following temperature profiles, which are acceptable since there are no temperature pinches or reversals.

Temperature Profiles for Example 5.5: Design Conditions

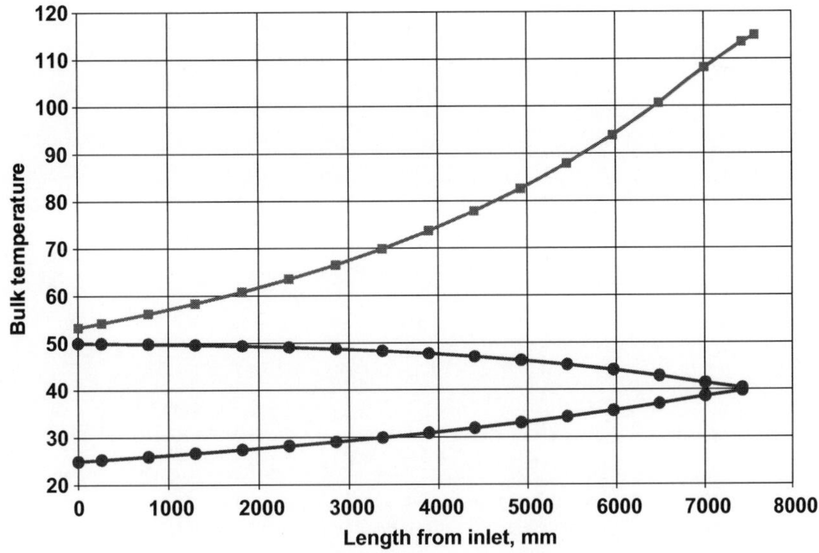

At design conditions, 60% of the overall thermal resistance is attributed to the fouling factors. However, the heat exchanger is relatively clean most of the time in order to meet appropriate regulatory commitments. The overall heat-transfer coefficient under clean conditions can be calculated as follows:

$$\frac{1}{U_C} = \frac{1}{U_D} - R_{Do} - (D_o/D_i)R_{Di} = \frac{1}{879} - 0.0002 - (19.05/15.75) \times 0.0004$$

$$U_C = 2203\ \mathrm{W/m^2 \cdot K}$$

The clean duty can be calculated using the ϵ-NTU method as described in Section 3.10:

$$C_{min} = 17(4200) = 71400\ \mathrm{W/K}$$

$$C_{max} = 42(4200) = 176400\ \mathrm{W/K}$$

$$r = 71400/176400 = 0.404$$

$$NTU = \frac{UA}{C_{min}} = \frac{2203(143)}{71400} = 4.41$$

$$\epsilon = \frac{2}{1 + r + (1+r^2)^{1/2}\dfrac{1+\exp\left[-NTU(1+r^2)^{1/2}\right]}{1-\exp\left[-NTU(1+r^2)^{1/2}\right]}}$$

$$\epsilon = \frac{2}{1 + 0.404 + (1+0.404^2)^{1/2}\dfrac{1+\exp\left(-4.41(1+0.404^2)^{1/2}\right)}{1-\exp\left(-4.41(1+0.404^2)^{1/2}\right)}} = 0.800$$

$$q = \epsilon C_{min}(T_i - t_i) = 0.8(71400)(115 - 25) = 5.14 \text{ MW}$$

The outlet temperatures under clean conditions are calculated using this duty:

$$T_o = T_i - \frac{q}{\dot{m}_h C_P} = 115 - \frac{5.14 \times 10^6}{17 \times 4200} = 43.0°\text{C}$$

$$t_o = t_i + \frac{q}{\dot{m}_c C_P} = 25 + \frac{5.14 \times 10^6}{42 \times 4200} = 54.1°\text{C}$$

To check the effectiveness of the temperature profile under clean conditions, the LMTD correction factor is again calculated:

$$R = \frac{T_i - T_o}{t_o - t_i} = \frac{115 - 43}{54.1 - 25} = \frac{72}{29.1} = 2.47$$

$$P = \frac{t_o - t_i}{T_i - t_i} = \frac{29.1}{115 - 25} = 0.323$$

$$\alpha = \left(\frac{1 - RP}{1 - P}\right)^{1/N} = \left(\frac{1 - 2.47 \times 0.323}{1 - 0.323}\right)^{1/1} = 0.299$$

$$S = \frac{\alpha - 1}{\alpha - R} = \frac{0.299 - 1}{0.299 - 2.47} = 0.323$$

$$F = \frac{\sqrt{R^2 + 1}\ \ln\left(\frac{1 - S}{1 - RS}\right)}{(R - 1)\ln\left[\frac{2 - S\left(R + 1 - \sqrt{R^2 + 1}\right)}{2 - S\left(R + 1 + \sqrt{R^2 + 1}\right)}\right]}$$

$$= \frac{\sqrt{2.47^2 + 1}\ \ln\left(\frac{1 - 0.323}{1 - 2.47 \times 0.323}\right)}{(2.47 - 1)\ln\left(\frac{2 - 0.323\left(2.47 + 1 - \sqrt{2.47^2 + 1}\right)}{2 - 0.323\left(2.47 + 1 + \sqrt{2.47^2 + 1}\right)}\right)} = 0.482$$

Since F << 0.8, the temperature profile is not very effective. Using *Xist* to perform the simulation under clean conditions, a temperature reversal is observed over approximately $^1/_4$ of the heat transfer area as shown below. With a temperature reversal, heat is transferred from the cold fluid to the hot fluid. This profile is not desirable, and re-design of this application is discussed in Chapter 7, Example 7.5. This type of unintentional temperature profile is fairly common in the process industries and reduces the overall process efficiency.

Temperature Profiles for Example 5.5: Clean Conditions

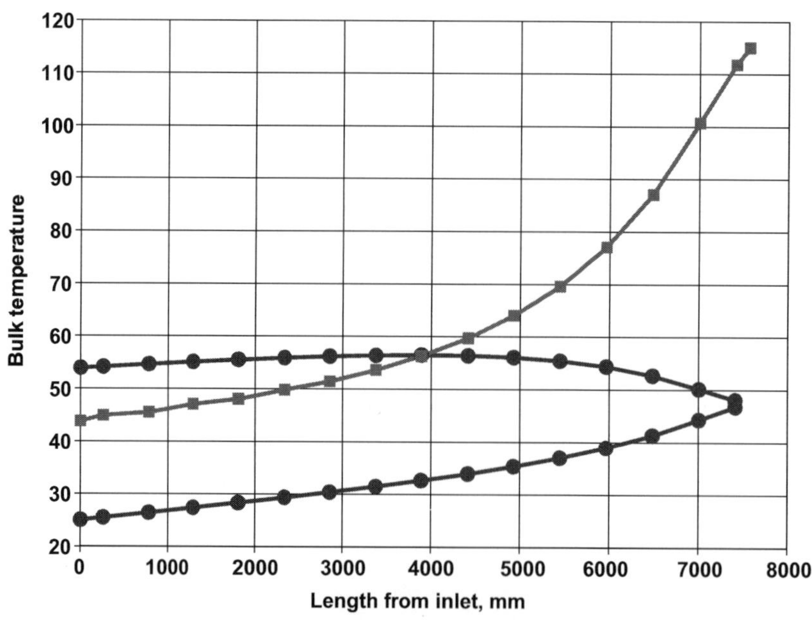

References

[1] Kern DQ, Kraus AD. Extended Surface Heat Transfer. New York: McGraw-Hill; 1972.
[2] Henry JAR. Headers, nozzles and turnarounds. In: Heat Exchanger Design Handbook, vol. 2. New York: Hemisphere Publishing Corp; 1988.
[3] Bell KJ, Mueller AC. Wolverine Engineering Data Book II. Wolverine Tube, Inc; 2001. www.wlv.com.
[4] Kern DQ. Process Heat Transfer. New York: McGraw-Hill; 1950.
[5] Kakac S, Liu H. Heat Exchangers: Selection, Rating and Thermal Design. Boca Raton, FL: CRC Press; 1998.
[6] HEXTRAN Keyword Manual. Lake Forest, CA: Invensys Systems, Inc; 2002.
[7] Perry RH, Chilton CH, editors. Chemical Engineers' Handbook. 5th ed. New York: McGraw-Hill; 1973.
[8] Perry RH, Green DW, editors. Perry's Chemical Engineers' Handbook. 7th ed. New York: McGraw-Hill; 1997.
[9] Taborek J. Shell-and-tube heat exchangers. In: Heat Exchanger Design Handbook, vol. 3. New York: Hemisphere Publishing Corp; 1988.
[10] Standards of the Tubular Exchanger Manufacturers Association. 8th ed. Tarrytown, NY: Tubular Exchanger Manufacturers Association, Inc; 1999.
[11] Mukherjee R. Effectively design shell-and-tube heat exchangers. Chem Eng Prog 1998;94(No. 2):21–37.

Appendix 5.A Hydraulic Equations in SI Units

The working equations for hydraulic calculations given in Section 5.3 are reformulated here in terms of SI units.

The pressure drop due to fluid friction in the tubes is calculated by the following equation:

$$\Delta P_f = \frac{f\, n_p L G^2}{2000 D_i s \phi} \tag{5.A.1}$$

where

ΔP_f = pressure drop (Pa)
f = Darcy friction factor (dimensionless)
n_p = number of tube passes
L = tube length (m)
G = mass flux (kg/s · m^2)
D_i = tube ID (m)
s = fluid specific gravity (dimensionless)
ϕ = viscosity correction factor (dimensionless)
= $(\mu/\mu_w)^{0.14}$ for turbulent flow
= $(\mu/\mu_w)^{0.25}$ for laminar flow

The above units for pressure drop and mass flux apply in the equations that follow as well. The minor losses on the tube side are estimated using the following equation:

$$\Delta P_r = 5.0 \times 10^{-4} \alpha_r G^2/s \tag{5.A.2}$$

where α_r is the number of velocity heads allocated for minor losses (Table 5.1). The pressure drop in the nozzles is estimated as follows:

$$\Delta P_n = 7.5 \times 10^{-4} N_s G_n^2/s \quad \text{(turbulent flow)} \tag{5.A.3}$$

$$\Delta P_n = 1.5 \times 10^{-3} N_s G_n^2/s \quad \text{(laminar flow, } Re_n \geq 100) \tag{5.A.4}$$

In these equations, N_s is the number of shells connected in series.

The shell-side pressure drop, excluding nozzle losses, is computed using the following equation:

$$\Delta P_f = \frac{f\, G^2 d_s (n_b + 1)}{2000 D_e s \phi} \tag{5.A.5}$$

where

ΔP_f = pressure drop (Pa)
f = shell-side friction factor (dimensionless)
G = mass flux = $\dot{m}/a_s$ (kg/s · m^2)
a_s = flow area across tube bundle (m^2)
= $d_s C' B / P_T$
d_s = shell ID (m)
C' = gap (m)
B = baffle spacing (m)
P_T = tube pitch (m); replaced by $P_T/\sqrt{2}$ for 45° tube layouts
n_b = number of baffles
D_e = equivalent diameter from Figure 3.17 (m)
s = fluid specific gravity (dimensionless)
ϕ = viscosity correction factor = $(\mu/\mu_w)^{0.14}$ (dimensionless)

Appendix 5.B Maximum Tube-Side Fluid Velocities

The data presented here are from Bell and Mueller [3]. The maximum velocities are based on prevention of tube wall erosion and are material specific. They are intended to serve as a supplement to the general guideline of $V_{max} = 8$ ft/s for liquids given in Section 5.7.4.

Maximum velocities for water are given in Table 5.B.1. For liquids other than water, multiply the values from the table by the factor $(\rho_{water}/\rho_{fluid})^{0.5}$.

TABLE 5.B.1 Maximum Recommended Velocities for Water in Heat-Exchanger Tubes

Tube Material	V_{max} (ft/s)
Plain carbon steel	10
Stainless steel	15
Aluminum	6
Copper	6
90–10 cupronickel	10
70–30 cupronickel	15
Titanium	>50

For gases flowing in plain carbon steel tubing, the following equation can be used to estimate the maximum velocity:

$$V_{max} = \frac{1800}{(PM)^{0.5}} \tag{5.B.1}$$

where

V_{max} = maximum velocity (ft/s)
P = gas pressure (psia)
M = molecular weight of gas

For tubing materials other than plain carbon steel, assume the maximum velocities are in the same ratio as given in Table 5.B.1 for water.

For example, to estimate the maximum velocity for air at 50 psia flowing in aluminum tubes, first calculate the velocity for plain carbon steel tubing using Equation 5.B.1:

$$(V_{max})_{cs} = \frac{1800}{(50 \times 29)^{0.5}} = 43.7 \text{ ft/s}$$

Then multiply by the ratio (6/10) from Table 5.B.1 to obtain the velocity for aluminum tubing:

$$V_{max} = 0.6 \times 47.3 = 28.4 \cong 28 \text{ ft/s}$$

Appendix 5.C Maximum Unsupported Tube Lengths

In order to prevent tube vibration and sagging, TEMA standards specify maximum unsupported tube lengths for two groups of materials. Material group A consists of steel, steel alloys, nickel, nickel-copper alloys, and nickel-chromium-steel alloys. Material group B consists of aluminum and its alloys, copper and its alloys, and titanium alloys at their upper temperature limit. For tube diameters between 19 mm and 51 mm, the standards are well-approximated by the following equations [9]:

$$\text{Group A}: \text{ maximum unsupported length(mm)} = 52\, D_o(\text{mm}) + 532 \tag{5.C.1}$$

$$\text{Group B}: \text{ maximum unsupported length(mm)} = 46\, D_o(\text{mm}) + 436 \tag{5.C.2}$$

TABLE 5.C.1 Maximum Unsupported Straight Tube Lengths in Inches (mm)

Tube OD	Material Group A	Material Group B
0.75 (19.1)	60 (1525)	52 (1315)
0.875 (22.2)	66 (1686)	57 (1457)
1.0 (25.4)	73 (1853)	63 (1604)
1.25 (31.8)	86 (2186)	75 (1899)
1.5 (38.1)	99 (2513)	86 (2189)
2.0 (50.8)	125 (3174)	109 (2773)

These equations apply to un-finned tubes. The standards for finned tubes are more complicated, but can be estimated by using the above equations with the tube OD replaced by the root-tube diameter. The standards also include temperature limits above which the unsupported length must be reduced [10]. For convenience, values computed from Equations (5.C.1) and (5.C.2) are set out in Table (5.C.1).

The baffle spacing is generally restricted to be no greater than half the tabled values because tubes in the baffle windows are supported by every other baffle. However, the inlet and outlet baffle spacings are often larger than the central baffle spacing. In this case, the central spacing must satisfy the following relation:

$$B \leq \text{tabled value} - \max(B_{in}, B_{out})$$

In practice, the actual unsupported tube length should be kept safely below (80% or less) the TEMA limit.

Appendix 5.D Comparison of Head Types for Shell-and-Tube Exchangers

TABLE 5.D.1 Comparison of Stationary Head Types

Head Type	Advantages	Disadvantages
A, L	(1) Tubesheet easily accessible by removing channel cover (2) Head can be removed if unrestricted access to tube-sheet is required	(1) Most expensive type except for D (2) Not well suited for high tube-side pressures; tube-side fluid can leak to environment through gasket at tubesheet (3) Type L rear head used only with fixed tubesheets
B, M	(1) Low cost (2) Removal of head provides unrestricted access to tubesheet (3) Absence of channel cover eliminates one external gasket where leakage to environment can occur	(1) Head must be disconnected from process piping and removed to access tubesheet (2) Not well suited for high tube-side pressures; tube-side fluid can leak to environment through gasket at tubesheet (3) Type M rear head used only with fixed tubesheets
C	(1) Low cost (2) Tubesheet easily accessed by removing channel cover (3) Suitable for high pressures; channel cover seal is the only external gasket	(1) Head and tubesheet materials must be compatible for welding (2) All tube-side maintenance must be done with channel in place (3) Used only with removable tube bundles
D	(1) Least prone to leakage (2) Best option when product of channel diameter and tube-side pressure exceeds about 86,000 in. · psia	(1) Not cost effective unless tube-side pressure is high
N	(1) Least expensive (2) Tubesheet easily accessed by removing channel cover (3) Suitable for high pressures; channel cover seal is the only external gasket	(1) Head, tubesheet, and shell materials must be compatible for welding (2) Used only with fixed tubesheets (3) All tube-side maintenance must be done with channel in place

TABLE 5.D.2 Comparison of Floating Head Types

Head Type	Advantages	Disadvantages
P	(1) No internal gaskets where leakage and fluid mixing can occur	(1) Shell-side fluid can leak through packing to environment (2) Shell-side T (<600°F) and P (<300 psia) are limited by packing
S	(1) Largest heat-transfer area per shell for an internal floating-head design (2) Leakage is contained within shell	(1) Floating head must be disassembled to remove tube bundle (2) Single-tube pass requires special construction (3) Leakage at internal gasket can result in mixing of fluids
T	(1) Bundle can be removed without removing rear shell cover or disassembling floating head (2) Leakage is contained within shell (3) Good option for kettle reboilers	(1) Smallest heat-transfer area per shell (2) Most expensive design (3) Single-tube pass requires special construction (4) Leakage at internal gasket can result in mixing of fluids
W	(1) Least expensive floating head design (2) No internal gaskets where leakage and fluid mixing can occur	(1) Both fluids can leak through packing to environment (2) Both tube-side and shell-side T (<375°F) and P (<150–300 psia depending on shell size) are limited by packing (3) Maximum of two tube passes

U-tubes provide a less expensive alternative to a floating head. In common with type P and W floating heads (and all stationary heads), U-tube bundles have no internal gaskets where leakage and fluid mixing can occur. The main disadvantages are:

(1) Cleaning interior tube surfaces is more difficult due to U-bends.
(2) Except for outermost tubes in bundle, individual tube replacement is not practical.
(3) Cannot be used if a single tube pass is required.

Notations

A	Heat-transfer surface area
A_{fins}	Surface area of fins
A_i	$\pi D_i L$ = Internal surface area of tube
A_{prime}	Prime surface area
A_{req}	Required heat-transfer surface area
A_{Tot}	$A_{prime} + A_{fins}$ = Total external surface area of a finned tube
a_s	Flow area across tube bundle
B	Baffle spacing
B_c	Baffle cut
B_{in}	Inlet baffle spacing
B_{out}	Outlet baffle spacing
BWG	Birmingham wire gage
b	Fin height
C_{max}	Larger of the values of $\dot{m}C_P$ for hot and cold streams
C_{min}	Smaller of the values of $\dot{m}C_P$ for hot and cold streams
C_P	Heat capacity at constant pressure
C'	Gap between tubes in bundle
D	Diameter
D_e	Equivalent diameter
D_i	Internal diameter of tube
D_n	Internal diameter of nozzle
D_o	External diameter of tube
D_r	External diameter of root tube
$\hat{D}_r$	Effective root tube diameter
d_e	Equivalent diameter expressed in inches
d_s	Internal diameter of shell
F	LMTD correction factor
f	Darcy friction factor
f_1	Friction factor from Figure 5.1 for $B/d_s = 1.0$
f_2	Friction factor from Figure 5.1 for $B/d_s = 0.2$
G	Mass flux
G_n	Mass flux in nozzle
h	Heat-transfer coefficient
h_i	Tube-side heat-transfer coefficient
h_o	Shell-side heat-transfer coefficient
j_H	$(h_o D_e/k)Pr^{1/3}\ (\mu/\mu_w)^{-0.14}$ = Modified Colburn factor for shell-side heat-transfer
k	Thermal conductivity
L	Tube length
M	Molecular weight
$\dot{m}$	Mass flow rate
$\dot{m}_c, \dot{m}_h$	Mass flow rate of cold and hot streams, respectively
$\dot{m}_t$	Total mass flow rate of tube-side fluid
N_s	Number of shells connected in series
NTU UA/C_{min}	= Number of transfer units
Nu	Nusselt number
n_b	Number of baffles

n_f Number of fins per unit length of tube
n_p Number of tube passes
n_t Number of tubes in tube bundle
$(n_t)_{req}$ Required number of tubes
P Pressure; parameter used to calculate LMTD correction factor
Pr Prandtl number
P_T Tube pitch
q Rate of heat-transfer
R Parameter used to calculate LMTD correction factor
R_D Total fouling factor
R_{Di} Fouling factor for tube-side fluid
R_{Do} Fouling factor for shell-side fluid
Re Reynolds number
$r\ C_{min}/C_{max}$ = Parameter used to calculate heat-exchanger effectiveness
r_1 External radius of root tube
r_2 Outer radius of annular fin
r_{2c} $r_2 + \tau/2$ = Corrected fin radius
S Parameter used to calculate LMTD correction factor
s Specific gravity
T Temperature
T_a, T_b Inlet and outlet temperatures, respectively, of shell-side fluid
T_{ave} Average temperature of shell-side fluid
T_i, T_o Inlet and outlet temperatures, respectively, of shell-side fluid in Example 5.5
T_p Average temperature of prime surface in a finned-tube heat exchanger
T_{wtd} Weighted average temperature of finned surface in a finned-tube heat exchanger
t_a, t_b Inlet and outlet temperatures, respectively, of tube-side fluid
t_{ave} Average temperature of tube-side fluid
t_i, t_o Inlet and outlet temperatures, respectively, of tube-side fluid in Example 5.5
U Overall heat-transfer coefficient
U_C Clean overall heat-transfer coefficient
U_D Design overall heat-transfer coefficient
U_{req} Required overall heat-transfer coefficient
V Fluid velocity
V_{max} Maximum fluid velocity based on erosion prevention
V_n Fluid velocity in nozzle

Greek Letters

α Parameter used to calculate LMTD correction factor
α_r Number of velocity heads allocated for tube-side minor pressure losses
β Parameter in Equation (5.15)
ΔP Pressure drop
ΔP_f Pressure drop due to fluid friction in straight sections of tubes or in shell
ΔP_i Total pressure drop for tube-side fluid
ΔP_n Pressure drop in nozzles
ΔP_o Total pressure drop for shell-side fluid
ΔP_r Tube-side pressure drop due to tube entrance, exit and return losses
ΔT Temperature difference
ΔT_{ln} Logarithmic mean temperature difference
$(\Delta T_{ln})_{cf}$ Logarithmic mean temperature difference for counter-current flow
ϵ Heat-exchanger effectiveness
η_f Fin efficiency
η_w Weighted efficiency of finned surface
μ Viscosity

μ_w	Fluid viscosity evaluated at average temperature of tube wall
ρ	Fluid density
τ	Fin thickness
ϕ	Viscosity correction factor
ϕ_i	Viscosity correction factor for tube-side fluid
ϕ_o	Viscosity correction factor for shell-side fluid
ψ	Effective height of annular fin

Problems

(5.1) Consider the following modified configuration for the kerosene/oil exchanger of Example 5.1:

$d_s = 17.25$ in.	tubes: 1-in., 14 BWG on 1.25-in. square pitch
$n_t = 104$	tube-side nozzles: 4-in. schedule 40
$n_p = 4$	shell-side nozzles: 3-in. schedule 40
$B/d_s = 0.2$	

As discussed in Example 5.1, the values of Re_i and h_i are the same for this configuration as for the initial configuration. Complete the design calculations for the modified configuration by calculating the following:

(a) h_o
(b) U_D
(c) The required tube length, rounded upward to the nearest foot.
(d) ΔP_i
(e) ΔP_o
(f) Over-surface and over-design
(g) Tube-side velocity

Ans. (a) 141 Btu/h · ft^2 · °F. (b) 51 Btu/h · ft^2 · °F. (c) 15 ft (d) 14.4 psi. (e) 3.4 psi. (f) 45% and 4.1%. (g) 7.96 ft/s.

(5.2) A fixed-tubesheet heat exchanger will use $^3/_4$ in. tubes, 20-ft long on 1 in. triangular pitch. It will require two shell-side passes and a total of four tube-side passes. A minimum of 1100 tubes will be required to handle the heat duty. If standard type E shells are used:

(a) What size shells will be required?
(b) What will be the approximate tube count for the exchanger?

Ans. (a) $d_s = 27$ in. (b) $n_t = 1148$.

(5.3) A shell-and-tube heat exchanger on the used equipment lot at a chemical plant has the following configuration.

Shell	Tube bundle
Type = AES	Number of tubes = 84
ID = 13.25 in.	Tube OD = 0.75 in.
Baffle type = segmental	BWG = 16
Baffle cut = 20%	Length = 20 ft
Baffle spacing = 5.3 in. (central)	Number of passes = 6
= 8.7 in. (inlet/outlet)	Pitch = 1 in. (square)
Number of baffles = 43	Sealing strips: 1 pair
Nozzles: 3-in. schedule 40	Nozzles: 3-in. schedule 40

It is contemplated to use this unit in a new service for which the shell-side fluid will have the following properties:

Shell-Side Fluid Property	Value
$\dot{m}$ (lb/h)	50,000
C_P (Btu/lbm · °F)	0.5

(Continued)

—cont'd

Shell-Side Fluid Property	Value
k (Btu/h · ft · °F)	0.25
μ (cp)	1.1
s	0.86

The tube-side fluid will be water with a flow rate of 30,000 lb/h and a range of 80°F to 120°F. A maximum pressure drop of 12 psi has been specified for each stream. Determine if the exchanger will be hydraulically suitable for the service by calculating:
(a) The tube-side pressure drop
(b) The shell-side pressure drop
(c) The tube-side velocity

Ans. (a) 11 psi. (b) 2.9 psi. (c) 4.6 ft/s.

(5.4) A shell-and-tube heat exchanger in a chemical process was designed to cool the shell-side fluid to a specified temperature. When clean, the heat exchanger performs adequately, but after two months of operation, the outlet temperature begins to rise above the design temperature, which causes inefficient operation of down-stream equipment. A process engineer has suggested that the problem can be solved by increasing the flow rate of the cold (tube-side) fluid, which is available in excess. She has stated that this will improve the performance of the exchanger by:

- Reducing fouling
- Increasing the heat-transfer coefficient
- Increasing the mean temperature difference in the unit

Comment on the engineer's proposed solution by addressing each of her three points as follows:

- Is the argument reasonable? Why or why not?
- Are there possible fallacies in her reasoning?

Finally, discuss any potential adverse consequences that might be associated with the proposed solution.

(5.5) The second design modification considered for the finned-tube heat exchanger of Example 5.2 was to use a 29-in. shell containing 532 tubes. Complete the design of this unit by following the procedure outlined at the end of Example 5.2.

(5.6) 60,000 lb/h of a petroleum fraction will be cooled from 250°F to 150°F using water with a range of 85°F to 120°F. For each stream, the inlet pressure will be 50 psia and a maximum pressure drop of 10 psi is specified. A fouling factor of 0.002 h · ft^2 · °F/Btu is required for each stream. The petroleum fraction is not corrosive and its properties may be assumed constant at the values given below.

Petroleum Fraction Property	Value
C_P (Btu/lbm · °F)	0.52
k (Btu/h · ft · °F)	0.074
μ (cp)	2.75
s	0.82
Pr	46.7

(a) Develop an initial configuration for a shell-and-tube heat exchanger with plain (un-finned) tubes to meet this service.
(b) Perform the rating calculations for the initial configuration developed in part (a).
(c) Based on the results of the rating calculations in part (b), specify what changes, if any, should be made to the configuration of the heat exchanger for the next trial.
(d) Use HEXTRAN or other available software to rate the initial configuration developed in part (a).

(5.7) Work problem 5.6 for a shell-and-tube exchanger with finned tubes.

(5.8) 90,000 lb/h of benzene will be heated from 75°F to 145°F by cooling a stream of ortho-toluidine from 230°F to 150°F. Inlet pressures are 45 psia for ortho-toluidine and 50 psia for benzene. A maximum pressure drop of 10 psi is specified for each stream. Properties of ortho-toluidine at 190°F are given below.

o-Toluidine Property	Value at 190°F
C_P (Btu/lbm · °F)	0.51
k (Btu/h · ft · °F)	0.0765
μ (cp)	0.91*

(*Continued*)

—cont'd

o-Toluidine Property	Value at 190°F
ρ (lbm/ft^3)	58.6
Pr	14.7

*$\log_{10}\mu(\text{cp}) = 1085.1\left\{\frac{1}{T(\text{K})} - \frac{1}{356.46}\right\}$.

(a) Develop an initial configuration for a shell-and-tube heat exchanger to meet this service.
(b) Perform the rating calculations for the initial configuration developed in part (a).
(c) Based on the results of the rating calculations in part (b), specify what changes, if any, should be made to the configuration of the heat exchanger for the next trial.
(d) Use HEXTRAN or other available software to rate the initial configuration developed in part (a).

(5.9) 85,000 lb/h of amyl acetate (n-pentyl acetate, $C_7H_{14}O_2$) will be cooled from 175°F to 125°F by heating a stream of amyl alcohol (1-pentanol) from 75°F to 120°F. Inlet pressures are 45 psia for the acetate and 50 psia for the alcohol. A maximum pressure drop of 7 psi is specified for each stream. Specific gravities are 0.81 for the alcohol and 0.83 for the acetate.
(a) Develop an initial configuration for a shell-and-tube heat exchanger to meet this service.
(b) Perform the rating calculations for the initial configuration developed in part (a).
(c) Based on the results of the rating calculations in part (b), specify what changes, if any, should be made to the configuration of the heat exchanger for the next trial.
(d) Obtain a final design for the heat exchanger.

(5.10) Develop an initial configuration for a shell-and-tube heat exchanger to meet the service of problem 5.9 if the temperature ranges are 175°F to 100°F for amyl acetate and 80°F to 150°F for amyl alcohol.

(5.11) 320,000 lb/h of oil (32°API, $K_w = 12.0$) will be cooled from 260°F to 130°F using treated water from a cooling tower with a range of 80°F to 120°F. For each streacm, the inlet pressure will be 50 psia and a maximum pressure drop of 12 psi is specified. A fouling factor of 0.002 h · ft^2 · °F/Btu is required for the oil stream, which is not corrosive. Properties of the oil are given below.

Oil Property	Value at 195°F
C_P (Btu/lbm · °F)	0.53
k (Btu/h · ft · °F)	0.068
μ (cp)	3.12*
ρ (lbm/ft^3)	50.5
Pr	58.8

*μ (cp) $=0.00488 \exp\left[4230.8/T\,(^\circ\text{R})\right]$.

(a) Develop an initial configuration for a shell-and-tube heat exchanger to meet this service. Use plain (un-finned) tubes.
(b) Perform the rating calculations for the initial configuration obtained in part (a).
(c) Based on the results of the rating calculations in part (b), specify what changes, if any, should be made to the configuration of the heat exchanger for the next trial.
(d) Obtain a final design for the heat exchanger.

(5.12) Work problem 5.11 for a shell-and-tube exchanger with finned tubes. Use 90-10 copper-nickel alloy tubes with 26 fins per inch.

(5.13) 150,000 lb/h of kerosene will be heated from 75°F to 120°F by cooling a gasoline stream from 160°F to 120°F. Inlet pressure will be 50 psia for each stream and maximum pressure drops of 7 psi for gasoline and 10 psi for kerosene are specified. Published fouling factors for oil refinery streams should be used for this application. Stream properties may be assumed constant at the values given below. Design a shell-and-tube heat exchanger for this service.

Property	Gasoline	Kerosene
C_P (Btu/lbm · °F)	0.55	0.48
k (Btu/h · ft · °F)	0.087	0.081
μ (cp)	0.45	1.5

(*Continued*)

—cont'd

Property	Gasoline	Kerosene
s	0.74	0.82
Pr	6.88	21.5

(5.14) Rate the initial configuration obtained for the acetate-alcohol heat exchanger in Problem 5.10.

(5.15) 60,000 lb/h of acetone will be cooled from 250°F to 100°F using 185,000 lb/h of 100% acetic acid, which is available at 90°F. Inlet pressures will be 100 psia for acetone and 50 psia for acetic acid. A maximum pressure drop of 10 psi is specified for each stream.
(a) Develop an initial configuration for a shell-and-tube heat exchanger to meet this service.
(b) Rate the initial configuration obtained in part (a).
(c) Use HEXTRAN or other available software to rate the configuration obtained in part (a).
Note: Particular care is required in selecting materials of construction for this application. Corrosion rate data, such as the charts and tables presented in Chapter 23 of the 5th edition of *Chemical Engineers' Handbook,* should be carefully analyzed as part of the decision-making process.

(5.16) For the service of Problem 5.6, use HEXTRAN or other available software to design a shell-and-tube heat exchanger using:
(a) Plain tubes
(b) Finned tubes

(5.17) Use HEXTRAN or other available software to design a shell-and-tube heat exchanger for the service of Problem 5.8.

(5.18) Use HEXTRAN or other available software to design a shell-and-tube heat exchanger for the service of Problem 5.9.

(5.19) Use HEXTRAN or other available software to design a shell-and-tube heat exchanger for the service of Problem 5.10.

(5.20) Use HEXTRAN or other available software to design a shell-and-tube heat exchanger for the service of Problem 5.11 using:
(a) Plain tubes
(b) 90-10 copper-nickel alloy tubes with 26 fins per inch.

(5.21) Use HEXTRAN or other available software to design a shell-and-tube heat exchanger for the service of Problem 5.13.

(5.22) Use HEXTRAN or other available software to design a shell-and-tube heat exchanger for the service of Problem 5.15.

(5.23) 250,000 lb/h of a hydrocarbon liquid (30°API, $K_w = 11.5$) are to be cooled from 225°F to 135°F using treated water from a cooling tower that is available at 80°F. The hydrocarbon is not corrosive, but a fouling factor of 0.0025 h · ft^2 °F/Btu is required for this stream. Inlet pressure will be 50 psia for each stream with a maximum pressure drop of 10 psi. Use HEXTRAN or other suitable software to design a shell-and-tube heat exchanger for this service using:
(a) Plain tubes
(b) Finned tubes

(5.24) 200,000 lb/h of lube oil (26°API, $K_w = 12.2$) will be cooled from 440°F to 350°F by heating a light gas oil (33°API, $K_w = 11.8$) from 310°F to 370°F. The inlet pressure of each stream will be 50 psia and maximum pressure drops of 15 psi for the lube oil and 10 psi for the gas oil are specified. Design a shell-and-tube heat exchanger for this service using HEXTRAN or other suitable software.

(5.25) A tube bundle consists of low-fin tubes containing 26 fins per inch laid out on a 1-in. triangular pitch. The fins have an average height of 0.049 in. and an average thickness of 0.013 in. The diameter of the root tube is 0.652 in. Calculate the gap and equivalent diameter to be used in the Simplified Delaware Method.

Ans. 0.313 and 0.908 in.

(5.26) A tube bundle consists of 1-in., 14 BWG, 26 fpi low-fin tubes on a 1.25-in. square pitch. Tube and fin dimensions are as given in Table B.5. Calculate the gap and equivalent diameter to be used in the Simplified Delaware Method.

Ans. 0.314 and 1.19 in.

(5.27) In Examples 5.1 and 5.2 over-surface and over-design were calculated using the equations:

$$\text{over-surface} = \frac{U_C}{U_{req}} - 1$$

$$\text{over-design} = \frac{U_D}{U_{req}} - 1$$

(5.28) Show that these equations follow from the definitions of over-surface and over-design given in Chapter 4.
Use HEXTRAN or other available software to rate the final design obtained for the finned-tube heat exchanger of Example 5.2. Compare the results of the computer calculations with those of the hand calculations in Example 5.2.

6 The Delaware Method

6.1 Introduction

The Delaware method for calculating the shell-side heat-transfer coefficient and pressure drop derives its name from a large industry sponsored study of shell-and-tube heat exchangers conducted at the University of Delaware, the final report of which was published in 1963 [1]. The method is often referred to as the Bell–Delaware method, after one of the principal investigators on the project, Kenneth J. Bell.

The Delaware method utilizes empirical correlations for the heat-transfer coefficient and friction factor in flow perpendicular to banks of tubes; these are referred to as ideal tube bank correlations. In baffled heat exchangers, this type of flow is approximated in the regions between the baffle tips. In the baffle windows, however, the flow is partly parallel to the tubes. Furthermore, only part of the shell-side fluid follows the main flow path through the exchanger due to the presence of leakage and bypass streams in the shell. These deviations from ideal tube bank conditions are accounted for by a set of empirical correction factors for heat transfer and pressure drop. The correction factors for leakage and bypass flows are correlated in terms of the flow areas for the leakage, bypass, and main cross-flow streams.

Slightly different versions of the Delaware method have been published in various handbooks and textbooks, e.g., [2–6]. The presentation in this chapter is based on the version given by Taborek [2]. Consideration is restricted to exchangers with standard type E shells, single-cut segmental baffles, and un-finned tubes. The extension of the method to other configurations is discussed in Refs. [2,5,6].

6.2 Ideal Tube Bank Correlations

The flow perpendicular to banks of tubes has been widely studied and a number of different correlations for heat transfer and pressure drop are available in the literature. Any of these correlations can, in principle, be used with the Delaware method. The correlation originally used in the development of the Delaware method has been recommended by Taborek [2], and is used herein. The correlation is in the form of three graphs (**Figures 6.1–6.3**), one for each of the commonly used tube bundle configurations.

The nomenclature for **Figures 6.1 to 6.3** is as follows:

f = Fanning friction factor, dimensionless

$j = \dfrac{hPr^{2/3}}{C_P G \phi}$, dimensionless

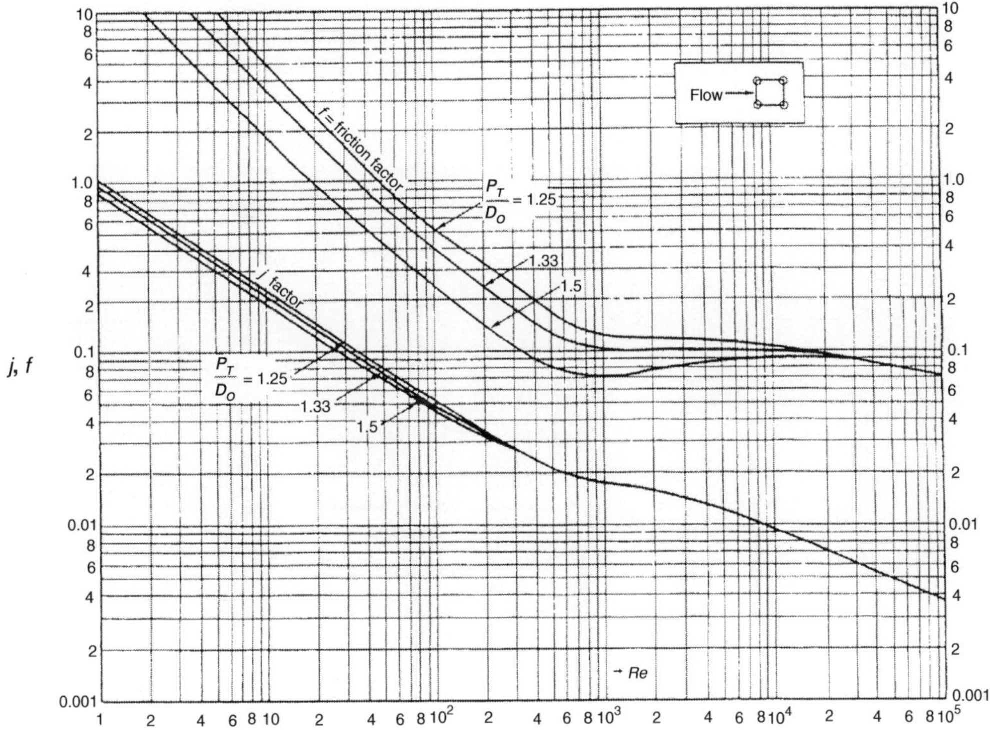

FIGURE 6.1 Ideal tube bank correlation for square pitch (Source: Ref. [2]).

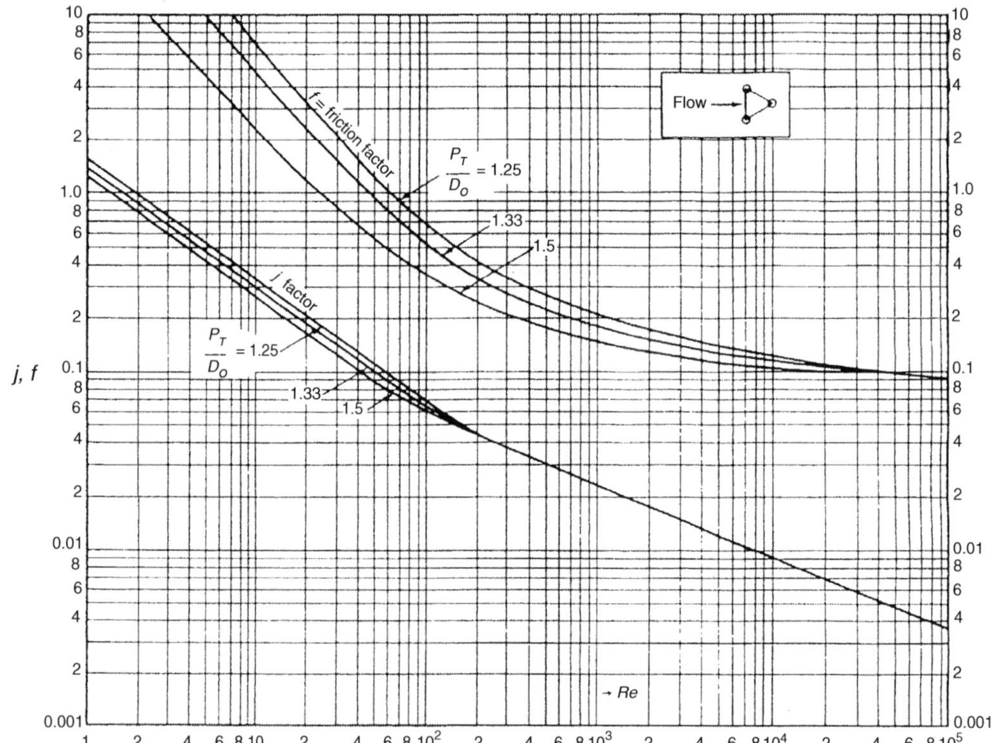

FIGURE 6.2 Ideal tube bank correlation for triangular pitch (Source: Ref. [2]).

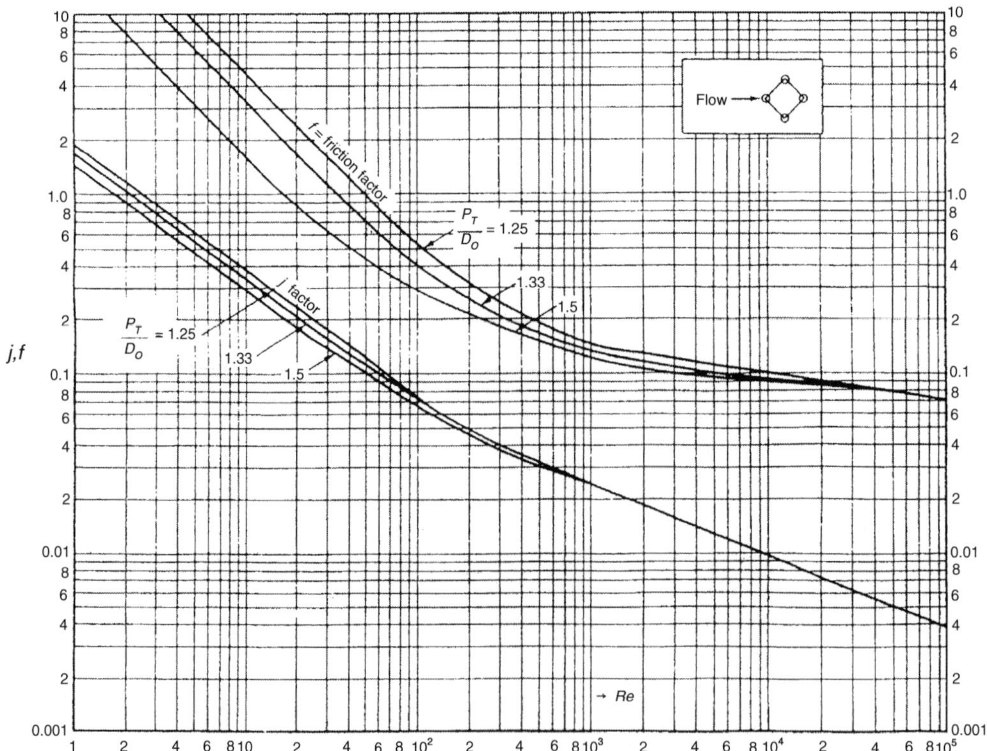

FIGURE 6.3 Ideal tube bank correlation for rotated square pitch (Source: Ref. [2]).

TABLE 6.1 Constants for use with Equations 6.1 to 6.4

Layout angle	Reynolds number	a_1	a_2	a_3	a_4	b_1	b_2	b_3	b_4
30°	10^5–10^4	0.321	−0.388	1.450	0.519	0.372	−0.123	7.00	0.500
	10^4–10^3	0.321	−0.388			0.486	−0.152		
	10^3–10^2	0.593	−0.477			4.570	−0.476		
	10^2–10	1.360	−0.657			45.100	−0.973		
	<10	1.400	−0.667			48.000	−1.000		
45°	10^5–10^4	0.370	−0.396	1.930	0.500	0.303	−0.126	6.59	0.520
	10^4–10^3	0.370	−0.396			0.333	−0.136		
	10^3–10^2	0.730	−0.500			3.500	−0.476		
	10^2–10	0.498	−0.656			26.200	−0.913		
	<10	1.550	−0.667			32.000	−1.000		
90°	10^5–10^4	0.370	−0.395	1.187	0.370	0.391	−0.148	6.30	0.378
	10^4–10^3	0.107	−0.266			0.0815	−0.220		
	10^3–10^2	0.408	−0.460			6.0900	−0.602		
	10^2–10	0.900	−0.631			32.1000	−0.963		
	<10	0.970	−0.667			35.0000	−1.000		

Source: Ref. [2].

h = heat-transfer coefficient, Btu/h · ft² · °F (W/m² · K)
C_P = heat capacity of shell-side fluid, Btu/lbm · °F (J/kg · K)
$G = \dot{m}_o/S_m$ = mass flux of shell-side fluid, lbm/h · ft² (kg/s · m²)
$\dot{m}_o$ = total mass flow rate of shell-side fluid, lbm/h (kg/s)
S_m = cross-flow area at shell centerline, ft² (m²)
Pr = Prandtl number, dimensionless
ϕ = viscosity correction factor, dimensionless
P_T = tube pitch, ft (m)
D_o = tube OD, ft (m)
$Re = D_oG/\mu$, dimensionless
μ = viscosity of shell-side fluid, lbm/ft · h (kg/m · s)

Note that the Reynolds number is based on the tube OD rather than the equivalent diameter used in the Simplified Delaware method. The cross-flow area, S_m, used to calculate the mass flux is approximated by the area, a_s, used in the Simplified Delaware method. More precise equations for calculating S_m are given in Section 6.5 below.

Approximate curve fits for **Figures 6.1 to 6.3** are as follows [2]:

$$j = a_1\left(\frac{1.33}{P_T/D_O}\right)^a (Re)^{a_2} \tag{6.1}$$

$$f = b_1\left(\frac{1.33}{P_T/D_o}\right)^b (Re)^{b_2} \tag{6.2}$$

where

$$a = \frac{a_3}{1 + 0.14\,(Re)^{a_4}} \tag{6.3}$$

$$b = \frac{b_3}{1 + 0.14\,(Re)^{b_4}} \tag{6.4}$$

The constants to be used in these equations are given in **Table 6.1**.

6.3 Shell-Side Heat-Transfer Coefficient

The heat-transfer coefficient for an ideal tube bank is designated h_{ideal}. This value is obtained from the appropriate graph, **Figures 6.1 to 6.3**, or from Equations (6.1) to (6.4). The shell-side heat-transfer coefficient, h_o, is obtained by multiplying h_{ideal} by a set of correction factors that account for the non-idealities in a baffled heat exchanger:

$$h_o = h_{ideal}(J_C J_L J_B J_R J_S) \tag{6.5}$$

where

J_C = correction factor for baffle window flow
J_L = correction factor for baffle leakage effects
J_B = correction factor for bundle bypass effects
J_R = laminar flow correction factor
J_S = correction factor for unequal baffle spacing

The factor J_C accounts for heat transfer in the baffle windows. It has a value of 1.0 for exchangers with no tubes in the windows. For other exchangers, it ranges from about 0.65 for very large baffle cuts to about 1.15 for small baffle cuts. For well-designed exchangers, the value of J_C is usually close to 1.0 [2].

The J_L correction factor accounts for both the tube-to-baffle and shell-to-baffle leakage streams. These streams flow from one baffle space to the next through the gaps between the tubes and the baffle and the gap between the shell and the baffle. The shell-to-baffle leakage stream is the more detrimental to heat transfer because it flows outside the tube bundle near the shell wall and does not contact any of the tubes. Hence, J_L decreases as the shell-to-baffle leakage fraction increases. The practical range of J_L is from about 0.2 to 1.0, with values of 0.7 to 0.8 being typical. For a well-designed exchanger, J_L should not be less than about 0.6 [2]. Smaller values indicate the need for design changes to decrease the size of the leakage streams, e.g., increasing the baffle spacing.

The bundle bypass stream flows around the periphery of the tube bundle from one baffle window to the next in the gap between the outermost tubes and the shell. It is accounted for by the factor J_B, which typically has values in the range of 0.7 to 0.9. Lower values indicate the need to add sealing strips, which force the bypass stream back into the tube bundle. The resulting improvement in heat transfer is reflected in an increase in the value of J_B with the number of pairs of sealing strips.

The baffle spacing in the inlet and outlet sections is often larger than in the remainder of the exchanger in order to accommodate the nozzles and, in the case of U-tube exchangers, the return bends. The resulting decrease in heat transfer rate is accounted for by the factor J_S, which is usually in the range of 0.85 to 1.0 [2]. If the baffle spacing is the same throughout the exchanger, J_S has a value of 1.0.

The factor J_R accounts for the decrease in the heat-transfer coefficient with downstream distance in laminar flow. The effect is analogous to the $(L)^{-1/3}$ dependence of the tube-side coefficient as expressed in the Seider–Tate equation for laminar flow. The range of J_R is from about 0.4 to 1.0, and it is equal to 1.0 for $Re \geq 100$.

For well-designed heat exchangers, the product of all the correction factors should not be less than about 0.5 [2]. A smaller value obtained during the design process indicates the need for appropriate design modifications. Correlations for calculating the correction factors are given in Section 6.6 below.

6.4 Shell-Side Pressure Drop

The nozzle-to-nozzle pressure drop in a type E shell is composed of three parts:

$$\Delta P_f = \Delta P_c + \Delta P_w + \Delta P_e \tag{6.6}$$

where

ΔP_c = pressure drop in all central baffle spaces
ΔP_w = pressure drop in all baffle windows
ΔP_e = pressure drop in the entrance and exit baffle spaces

The three regions of the shell corresponding to ΔP_c, ΔP_w, and ΔP_e are shown schematically in **Figure 6.4**.

6.4.1 Calculation of ΔP_c

Between the baffle tips the flow pattern is considered to be pure cross flow. Therefore, the pressure drop in one central baffle space is equal to the ideal tube bank pressure drop corrected for leakage and bypass effects. The ideal tube bank pressure drop is given by:

$$\Delta P_{ideal} = \frac{2 f_{ideal} N_c G^2}{g_c \rho \phi} \tag{6.7}$$

where

f_{ideal} = ideal tube bank friction factor, dimensionless
N_c = number of tube rows crossed between baffle tips, dimensionless
$G = \dot{m}_o / S_m$ = mass flux, lbm/h · ft^2 (kg/s · m^2)

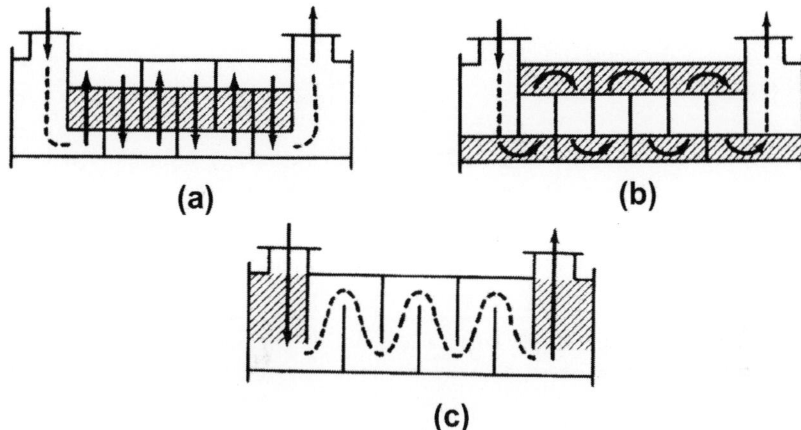

FIGURE 6.4 Flow regions considered for shell-side pressure drop. (a) Cross-flow region between baffle tips in the central baffle spaces, (b) window-flow region, and (c) cross-flow region for inlet and outlet baffle spaces (Source: Ref. [2]).

$g_c = 4.17 \times 10^8$ (lbm · ft/h^2)/lbf = 1.0 (kg · m/s^2)/N
ρ = fluid density, lbm/ft^3 (kg/m^3)
$\Delta P_{ideal} \propto$ lbf/ft^2 (Pa)

The number of tube rows crossed by the fluid between baffle tips is calculated as:

$$N_c = \frac{D_s(1 - 2B_c)}{P'_T} \tag{6.8}$$

where

D_s = shell ID, ft (m)
B_c = (fractional) baffle cut, dimensionless
$P'_T = P_T$ for square tube layout
$= P_T \cos\theta_{tp}$ for triangular ($\theta_{tp} = 30°$) and rotated square ($\theta_{tp} = 45°$) tube layouts

The numerator in Equation (6.8) is the length of the baffle over which cross flow occurs (refer to **Figure 6.4**(a)), while the denominator is the center-to-center distance between tube rows in the flow direction, i.e., the tube pitch projected onto the flow direction.

The pressure drop in all central baffle spaces is obtained by multiplying the ideal tube bank pressure drop by correction factors to account for the leakage and bypass streams, and then multiplying by the number of central baffle spaces. Thus,

$$\Delta P_c = (n_b - 1)\, \Delta P_{ideal} R_L R_B \tag{6.9}$$

where

n_b = number of baffles
R_L = leakage correction factor
R_B = bypass correction factor

The correction factors R_L and R_B are analogous to the factors J_L and J_B for heat transfer. The practical range of R_L is from about 0.1 to 1.0, with values of 0.4 to 0.6 being typical. For R_B the practical range is from about 0.3 to 1.0, with values of 0.4 to 0.7 being typical. As is the case for J_B, values of R_B increase with the number of pairs of sealing strips.

6.4.2 Calculation of ΔP_w

In the baffle windows the fluid undergoes a 180° change of direction (**Figure 6.4**(b)). The Delaware method allows two velocity heads for the associated pressure loss. There is an additional pressure loss in the windows due to fluid friction. The latter is handled by empirical correlations, one each for laminar ($Re < 100$) and turbulent ($Re \geq 100$) flow. (The Reynolds number here is the same one used in the ideal tube bank correlation.)

The ideal (uncorrected) pressure drop in one baffle window for turbulent flow is given by:

$$\Delta P_{w,ideal} = \frac{(2 + 0.6N_{cw})\dot{m}_o^2}{2g_c \rho S_m S_w} \tag{6.10}$$

where

N_{cw} = effective number of tube rows crossed in one baffle window
S_w = window-flow area, ft^2 (m^2)
$\Delta P_{w,ideal} \propto$ lbf/ft^2 (Pa)

Note that the total mass flow of shell-side fluid is used here, even though the leakage streams do not flow through the windows. Also, the flow area used in Equation (6.10) is the geometric mean of the cross-flow and window-flow areas, i.e., $(S_m\ S_w)^{1/2}$. The effective number of tube rows crossed in the window is given by:

$$N_{cw} = \frac{0.8 B_c D_s}{P_T'} \tag{6.11}$$

In this equation, $B_c D_s$ is the length of the baffle cut, while the denominator gives the center-to-center distance between tube rows in the cross-flow direction. Hence, without the factor of 0.8, Equation (6.11) would give the number of tube rows in the cross-flow direction in one baffle window. The empirical factor of 0.8 accounts for the fact that the flow in the baffle windows is partly across the tubes and partly parallel to the tubes.

For laminar flow,

$$\Delta P_{w,ideal} = \frac{26 \nu \dot{m}_o}{g_c \sqrt{S_m S_w}} \left[\frac{N_{cw}}{P_T - D_o} + \frac{B_c D_s}{D_w^2} \right] + \frac{\dot{m}_o^2}{g_c \rho S_m S_w} \tag{6.12}$$

where

ν = kinematic viscosity of shell-side fluid, ft^2/h (m^2/s)
D_w = equivalent diameter for window flow, ft (m)

The equivalent diameter for a baffle window is defined in the usual manner as four times the flow area divided by the wetted perimeter, except that the baffle edge is omitted from the wetted perimeter term:

$$D_w = \frac{4 S_w}{\pi D_o n_t \times 0.5 (1 - F_c) + D_s \theta_{ds}} \tag{6.13}$$

where

n_t = number of tubes in bundle
F_c = fraction of tubes in cross flow between baffle tips
θ_{ds} = baffle window angle (rad) defined in **Figure 6.5**

Note that $0.5(1-F_c)$ is the fraction of tubes in one baffle window, so that the first term in the denominator of Equation (6.13) represents the total wetted perimeter of all the tubes in the baffle window. See Section 6.5.2 below for calculation of F_c.

The pressure drop in all the baffle windows is obtained by multiplying $\Delta P_{w,ideal}$ by the number of baffle windows (equal to the number of baffles, n_b) and by the leakage correction factor, R_L. No correction is made for the bundle bypass flow because the bypass stream is considered to flow through the windows along with the main cross-flow stream. Thus,

$$\Delta P_w = n_b \Delta P_{w,ideal} R_L \tag{6.14}$$

6.4.3 Calculation of ΔP_e

The inlet and outlet baffle spaces differ in several respects from the central baffle spaces. As previously mentioned, the baffle spacing in one or both of the end zones may differ from the central baffle spacing in order to accommodate the shell-side nozzles or the return bends in U-tube exchangers. Also, the number of tube rows crossed in the end baffle spaces includes the rows in one baffle window, in addition to the tube rows between the baffle tips (refer to **Figure 6.4**(c)). Finally, the correction for leakage is not applicable to the inlet and outlet spaces because the leakage streams have not yet developed at the inlet and have already rejoined the main flow at the outlet.

The pressure drop in the entrance and exit baffle spaces is obtained by correcting the ideal pressure drop given by Equation (6.7) for the additional number of tube rows crossed, for the altered baffle spacing, and for the effects of the bundle bypass stream. The result is [2,3]:

$$\Delta P_e = 2 \Delta P_{ideal} \left(1 + \frac{N_{cw}}{N_c}\right) R_B R_S \tag{6.15}$$

where R_s is the correction factor for unequal baffle spacing. Notice that the effective number, rather than the actual number, of rows crossed in each window is used here, presumably to account for the distortion of the cross-flow pattern in the vicinity of the nozzles.

The practical range of R_s is from about 0.3 to 1.0. When the baffle spacing is uniform throughout the shell, the value of R_s is 1.0.

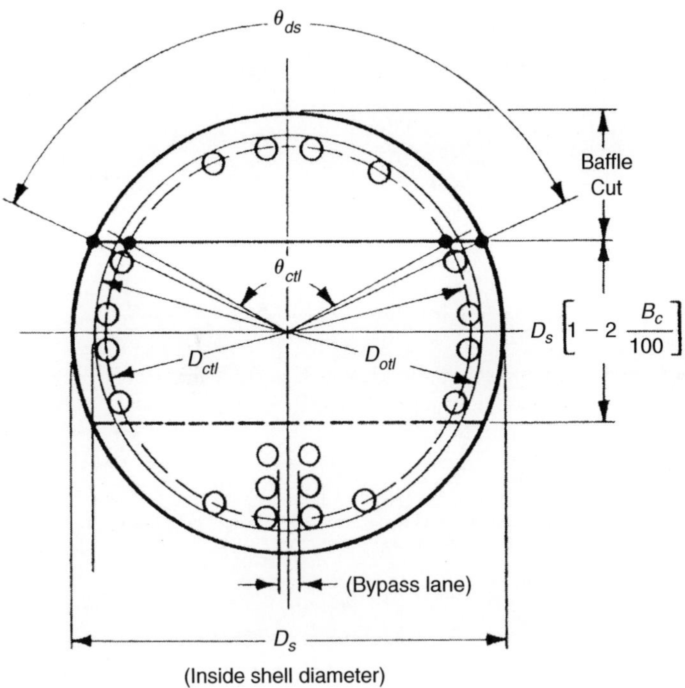

FIGURE 6.5 Segmental baffle geometry. B_c = baffle cut (%) (Source: Ref. [2]).

6.4.4 Summary

Equations (6.6), (6.9), (6.14), and (6.15) can be combined to give the following equation for shell-side pressure drop (excluding nozzle losses):

$$\Delta P_f = \left[(n_b - 1)\Delta P_{ideal} R_B + n_b \Delta P_{w,ideal}\right] R_L + 2\Delta P_{ideal}(1 + N_{cw}/N_c) R_B R_S \tag{6.16}$$

The ideal tube bank pressure drop, ΔP_{ideal}, is obtained from Equation (6.7). The uncorrected window pressure drop, $\Delta P_{w,ideal}$, is calculated using Equation (6.10) for turbulent flow ($Re \geq 100$) or Equation (6.13) for laminar flow ($Re < 100$).

Correlations for calculating the correction factors R_L, R_B, and R_S are given in Section 6.6 below.

6.5 The Flow Areas

6.5.1 The Cross-Flow Area

The cross-flow area, S_m, is the minimum flow area in one central baffle space at the center of the tube bundle. It is calculated by the following equation:

$$S_m = B\left[(D_s - D_{otl}) + \frac{(D_{otl} - D_o)}{(P_T)_{eff}}(P_T - D_o)\right] \tag{6.17}$$

where

B = central baffle spacing
D_{otl} = outer tube limit diameter
$(P_T)_{eff}$ = P_T for square and triangular layouts
= $P_T/\sqrt{2}$ for rotated square layout

The outer tube limit diameter is the diameter of the circle that circumscribes the tube bundle (see **Figure 6.5**). It is also referred to as the tube bundle diameter.

From **Figure 6.5**, it can be seen that the term ($D_s - D_{otl}$) in Equation (6.17) is equal to twice the clearance between the tube bundle and the shell. The corresponding flow area on each side of the bundle is a rectangle of width 0.5 ($D_s - D_{otl}$) and length equal to the baffle spacing, B. The sum of these two areas is B ($D_s - D_{otl}$), which is the first term in Equation (6.17).

The second term in Equation (6.17) represents the minimum flow area between tubes within the bundle. For square (90°) layouts the tubes are aligned in the flow direction (**Figure 6.6**), and the width, L_1, of the openings through which the fluid flows is

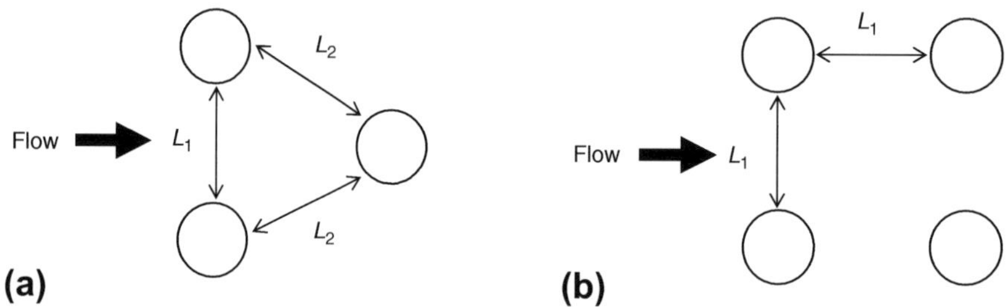

FIGURE 6.6 Aligned and staggered tube layouts. (a) Staggered and (b) inline square layout.

the gap, $C' = P_T - D_o$. The length of the openings is the baffle spacing, B, and the number of openings across the center of the bundle is $D_{ctl}/P_T = (D_{otl} - D_o)/P_T$, where D_{ctl} is the central tube limit diameter defined in **Figure 6.5**. Thus, the total area of these openings is given by the second term in Equation (6.17), with $(P_T)_{eff} = P_T$.

For triangular and rotated-square tube layouts, the tubes are staggered in the flow direction (**Figure 6.6**). The fluid first flows through an opening of width L_1 and then through two openings, each of width L_2. The minimum flow area through the bundle is determined by the smaller of L_1 and $2L_2$. For triangular layouts, $L_1 = L_2 = P_T - D_o$, and hence, $L_1 < 2L_2$. The minimum opening is therefore a rectangle of width $(P_T - D_o)$ and length B. The number of openings across the center of the bundle is again $(D_{otl} - D_o)/P_T$, which results in the same expression for the total area as for square layouts.

For rotated-square layouts, $L_1 = \sqrt{2}P_T - D_o$ and $L_2 = P_T - D_o$. With the dimensions used in standard tube layouts, $2L_2 < L_1$ in this case. Therefore, the minimum flow area per opening is $2B(P_T - D_o)$, and the number of openings across the center of the bundle is $(D_{otl} - D_o) / (\sqrt{2}P_T)$. Hence, the total flow area within the bundle is given by the second term in Equation (6.17) with the factor of $\sqrt{2}$ included in $(P_T)_{eff}$.

6.5.2 Tube-to-Baffle Leakage Area

The holes in the baffles through which the tubes pass are slightly larger in diameter than the tubes, allowing leakage of fluid through the gaps from one baffle space to the next. The tube-to-baffle leakage area is the total area of the gaps in one baffle. Denoting the tube-to-baffle clearance by δ_{tb}, the area of one gap is:

$$(S_{tb})_1 = \frac{\pi}{4}\left[(D_o + 2\delta_{tb})^2 - D_o^2\right] \cong \pi D_o\delta_{tb} \tag{6.18}$$

The approximation in Equation (6.18) is valid because $\delta_{tb} << D_o$.

Now the fraction of tubes that pass through one baffle is equal to the fraction, F_c, between the baffle tips plus the fraction, F_w, in one baffle window. These quantities are related by:

$$F_c + 2F_w = 1.0 \tag{6.19}$$

Therefore,

$$F_w = (1 - F_c)/2 \tag{6.20}$$

It follows that:

$$F_c + F_w = F_c + 0.5\,(1 - F_c) = 0.5\,(1 + F_c) \tag{6.21}$$

Hence, the total leakage area for one baffle is given by:

$$S_{tb} = 0.5\pi D_o\delta_{tb}n_t\,(1 + F_c) \tag{6.22}$$

The fraction of tubes in one baffle window is estimated as the fractional area of the circle of radius D_{ctl} that lies in the window (refer to **Figure 6.5**). The central tube limit diameter, D_{ctl}, is the diameter of the circle that passes through the centers of the outermost tubes in the bundle. It is obvious from **Figure 6.5** that:

$$D_{ctl} = D_{otl} - D_o \tag{6.23}$$

The fractional area in the window depends on the angle θ_{ctl} shown in **Figure 6.5**. The triangle formed by the baffle edge and two radii of length $\frac{1}{2}D_{ctl}$ is shown in **Figure 6.7**. Also shown is the right triangle formed by cutting the large triangle in half. The height of this triangle is half the length of the central portion of the baffle, i.e., $\frac{1}{2}D_s(1 - 2B_c)$. It can be seen from **Figure 6.7** that:

$$\cos\left(\frac{1}{2}\theta_{ctl}\right) = \frac{\frac{1}{2}D_s(1 - 2B_c)}{\frac{1}{2}D_{ctl}} = \frac{D_s(1 - 2B_c)}{D_{ctl}} \tag{6.24}$$

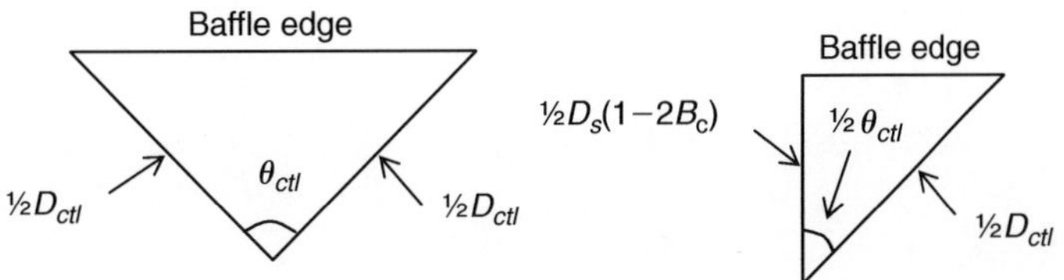

FIGURE 6.7 Baffle geometry used for calculating the fraction, F_w, of tubes in one baffle window; B_c = fractional baffle cut.

Solving for θ_{ctl} gives:

$$\theta_{ctl} = 2\cos^{-1}\left[\frac{D_s(1-2B_c)}{D_{ctl}}\right] \tag{6.25}$$

Now the length of the baffle edge in the small triangle of **Figure 6.7** is $\frac{1}{2}D_{ctl}\sin\left(\frac{1}{2}\theta_{ctl}\right)$, and this is equal to one-half the base of the large triangle. Thus, the area of the large triangle is:

$$A_{triangle} = \frac{1}{2}\text{base} \times \text{height}$$

$$A_{triangle} = \frac{1}{2}D_{ctl}\sin\left(\frac{1}{2}\theta_{ctl}\right)\left[\frac{1}{2}D_s(1-2B_c)\right]$$

Multiplying and dividing by D_{ctl} on the right side of this equation gives:

$$A_{triangle} = \frac{1}{4}D_{ctl}^2\sin\left(\frac{1}{2}\theta_{ctl}\right)\left[\frac{D_s\,(1-2B_c)}{D_{ctl}}\right] \tag{6.26}$$

Using Equation (6.24), Equation (6.26) can be written as:

$$A_{triangle} = \frac{1}{4}D_{ctl}^2\sin\left(\frac{1}{2}\theta_{ctl}\right)\cos\left(\frac{1}{2}\theta_{ctl}\right) \tag{6.27}$$

Using the double-angle formula,

$$\sin\theta_{ctl} = 2\sin\left(\frac{1}{2}\theta_{ctl}\right)\cos\left(\frac{1}{2}\theta_{ctl}\right) \tag{6.28}$$

Combining Equations (6.27) and (6.28) yields:

$$A_{triangle} = \frac{1}{8}D_{ctl}^2\sin\theta_{ctl} \tag{6.29}$$

The area of the circular sector subtended by angle θ_{ctl} is:

$$A_{sector} = \frac{\pi}{4}D_{ctl}^2\frac{\theta_{ctl}}{2\pi} = \frac{1}{8}D_{ctl}^2\theta_{ctl} \tag{6.30}$$

where θ_{ctl} is in radians. Now the area of the circle lying in the baffle window is:

$$A_{sector} - A_{triangle} = \frac{1}{8}D_{ctl}^2\,(\theta_{ctl} - \sin\theta_{ctl}) \tag{6.31}$$

Finally, the fractional area lying in the window is equated with F_w:

$$F_w = \frac{A_{sector} - A_{triangle}}{A_{circle}} \tag{6.32}$$

Substituting for the areas gives:

$$F_w = \frac{\frac{1}{8}D_{ctl}^2(\theta_{ctl} - \sin\theta_{ctl})}{(\pi/4)D_{ctl}^2}$$

or

$$F_w = \frac{1}{2\pi}\,(\theta_{ctl} - \sin\theta_{ctl}) \tag{6.33}$$

From Equation (6.19),

$$F_c = 1 - 2F_w \tag{6.34}$$

Combining Equations (6.33) and (6.34) yields:

$$F_c = 1 + \frac{1}{\pi}(\sin \theta_{ctl} - \theta_{ctl}) \tag{6.35}$$

In summary, the calculation of the tube-to-baffle leakage area involves the following steps:

- Calculate θ_{ctl} using Equation (6.25).
- Calculate F_c using Equation (6.35).
- Calculate S_{tb} using Equation (6.22).

6.5.3 Shell-to-Baffle Leakage Area

The shell-to-baffle leakage area, S_{sb}, is shown in **Figure 6.8**. Denoting the clearance between the baffle and shell by δ_{sb} and noting that $\delta_{sb} << D_s$, the leakage area can be written as follows:

$$S_{sb} = \pi\, D_s \delta_{sb}\left(\frac{2\pi - \theta_{ds}}{2\pi}\right) \tag{6.36}$$

or

$$S_{sb} = D_s \delta_{sb}(\pi - 0.5\theta_{ds}) \tag{6.37}$$

where the baffle window angle, θ_{ds} (**Figure 6.5**), is expressed in radians.

The baffle geometry in **Figure 6.7** remains valid if D_{ctl} and θ_{ctl} are replaced by D_s and θ_{ds}, respectively. It follows that:

$$\cos\left(\frac{1}{2}\theta_{ds}\right) = \frac{\frac{1}{2}D_s(1 - 2B_c)}{\frac{1}{2}D_s} = 1 - 2B_c \tag{6.38}$$

and

$$\theta_{ds} = 2\cos^{-1}(1 - 2B_c) \tag{6.39}$$

6.5.4 The Bundle Bypass Flow Area

The bundle bypass flow area, S_b, is the area between the outermost tubes and the shell at the shell centerline in one central baffle space. It is part of the cross-flow area, S_m, and in fact is represented by the first term in Equation (6.17). Thus,

$$S_b = B(D_s - D_{otl}) \tag{6.40}$$

6.5.5 The Window-Flow Area

The flow area in one baffle window, S_w, is the gross window area, S_{wg}, minus the area occupied by the tubes in the window. The gross window area is the open area between the shell and the baffle edge formed by the baffle cut (**Figure 6.9**). It is equal to the area of the circular sector subtended by angle θ_{ds} minus the area of the triangle formed by the baffle edge and two radii of length 0.5 D_s. The derivation is the same as that for Equation (6.31) with D_{ctl} and θ_{ctl} replaced by D_s and θ_{ds}, respectively. Thus,

$$S_{wg} = \frac{1}{8}D_s^2(\theta_{ds} - \sin\theta_{ds}) \tag{6.41}$$

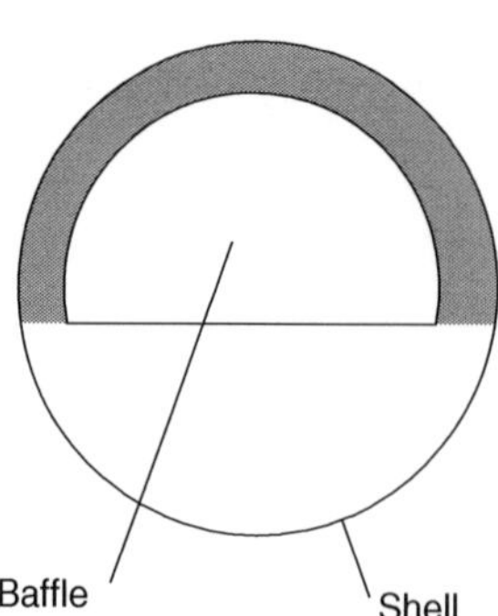

FIGURE 6.8 Shell-to-baffle leakage area. The shaded region between the baffle and shell is the leakage area, S_{sb}.

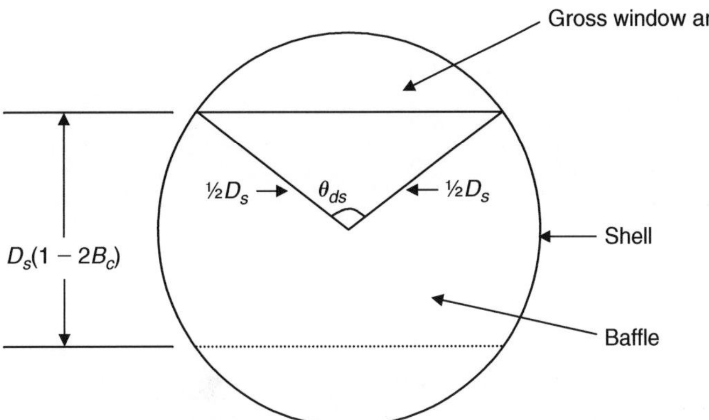

FIGURE 6.9 Gross window area; B_c = fractional baffle cut.

The area occupied by the tubes in the window is simply the cross-sectional area per tube times the number of tubes in the window:

$$A_{tubes} = n_t F_w \left(\pi D_o^2/4\right) \tag{6.42}$$

The window flow area is given by:

$$S_w = S_{wg} - A_{tubes} \tag{6.43}$$

Substituting from Equations (6.41) and (6.42) gives:

$$S_w = \frac{1}{8} D_s^2 (\theta_{ds} - \sin\,\theta_{ds}) - \frac{1}{4} n_t F_w \pi\, D_o^2 \tag{6.44}$$

6.6 Correlations for the Correction Factors

6.6.1 Correction Factor for Baffle Window Flow

The correction factor J_C expresses the effect on heat transfer of flow in the baffle windows. It depends on the number of tubes in the windows versus the number in cross flow between the baffle tips, which is determined by the baffle cut, B_c, and the diametral ratio D_s/D_{ctl}. The correction factor is correlated in terms of a single parameter, F_c, the fraction of tubes in cross flow between the baffle tips. For the practical range of baffle cuts (15–45%), the correlation is well represented by a linear relationship [2]:

$$J_C = 0.55 + 0.72 F_c \tag{6.45}$$

6.6.2 Correction Factors for Baffle Leakage

The correction factors J_L, and R_L express the effects of the tube-to-baffle and shell-to-baffle leakage streams on heat transfer and pressure drop, respectively. Both factors are correlated in terms of the following area ratios:

$$r_s = \frac{S_{sb}}{S_{sb} + S_{tb}} \tag{6.46}$$

$$r_l = \frac{S_{sb} + S_{tb}}{S_m} \tag{6.47}$$

The correlations are well approximated by the following equations [2]:

$$J_L = 0.44\,(1 - r_s) + [1 - 0.44\,(1 - r_s)]\exp\,(-2.2 r_l) \tag{6.48}$$

$$R_L = \exp\left[-1.33\,(1 + r_s)(r_l)^p\right] \tag{6.49}$$

where

$$p = 0.8 - 0.15\,(1 + r_s) \tag{6.50}$$

6.6.3 Correction Factors for Bundle Bypass Flow

The correction factors J_B and R_B express the effects of the bundle bypass flow on heat transfer and pressure drop, respectively. Both factors depend on the ratio, S_b/S_m, of bypass flow area to cross-flow area, and the number of pairs of sealing strips. The latter enters the correlation as the ratio, r_{ss}:

$$r_{ss} = N_{ss}/N_c \tag{6.51}$$

where

N_{ss} = number of pairs of sealing strips
N_c = number of tube rows crossed between baffle tips

The correlations can be expressed in terms of the following equations [2]:

$$\begin{aligned} J_B &= \exp[-C_J\,(S_b/S_m)\,(1-\sqrt[3]{2r_{ss}})] \quad \text{for } r_{ss} < 0.5 \\ J_B &= 1.0 \quad\quad\quad\quad\quad\quad\quad\quad\quad\quad\quad\quad \text{for } r_{ss} \geq 0.5 \end{aligned} \tag{6.52}$$

where

$C_J = 1.35$ for $Re < 100$
$= 1.25$ for $Re \geq 100$

$$\begin{aligned} R_B &= \exp\left[-C_R\,(S_b/S_m)(1-\sqrt[3]{2r_{ss}})\right] \quad \text{for } r_{ss} < 0.5 \\ R_B &= 1.0 \quad\quad\quad\quad\quad\quad\quad\quad\quad\quad\quad\quad \text{for } r_{ss} \geq 0.5 \end{aligned} \tag{6.53}$$

where

$C_R = 4.5$ for $Re < 100$
$= 3.7$ for $Re \geq 100$

6.6.4 Correction Factors for Unequal Baffle Spacing

The correction factors J_S and R_S depend on the inlet baffle spacing, B_{in}, the outlet baffle spacing, B_{out}, the central baffle spacing, B, and the number of baffles, n_b. They are calculated from the following equations [2]:

$$J_S = \frac{(n_b - 1) + (B_{in}/B)^{(1-n_1)} + (B_{out}/B)^{(1-n_1)}}{(n_b - 1) + (B_{in}/B) + (B_{out}/B)} \tag{6.54}$$

where

$n_1 = 0.6$ for $Re \geq 100$
$n_1 = \frac{1}{3}$ for $Re < 100$

$$R_S = 0.5\left[(B/B_{in})^{(2-n_2)} + (B/B_{out})^{(2-n_2)}\right] \tag{6.55}$$

where

$n_2 = 0.2$ for $Re \geq 100$
$n_2 = 1.0$ for $Re < 100$

Note that for $B_{in} = B_{out} = B$, Equations (6.54) and (6.55) give $J_S = R_S = 1.0$.

6.6.5 Laminar Flow Correction Factor

The correction factor J_R accounts for the fact that in laminar flow the heat-transfer coefficient decreases with downstream distance, which in the present context is interpreted as the number of tube rows crossed. The correlating parameter is the total number, N_{ct}, of tube rows crossed in the entire exchanger, i.e., in the inlet, outlet, and central baffle spaces, and in the baffle windows. For the present purpose, this parameter is approximated as follows [2]:

$$N_{ct} = (n_b + 1)\,(N_c + N_{cw}) \tag{6.56}$$

where N_c and N_{cw} are given by Equations (6.8) and (6.11), respectively.

The correlation for J_R is [2]:

$$J_R = (10/N_{ct})^{0.18} \quad \text{for } Re \leq 20$$
$$J_R = 1.0 \quad \text{for } Re \geq 100 \tag{6.57}$$

For $20 < Re < 100$, J_R is calculated by linear interpolation between the above values.

6.7 Estimation of Clearances

Determination of the flow areas (Section 6.5) requires the values of three clearances; namely, the tube-to-baffle clearance, δ_{tb}, the shell-to-baffle clearance, δ_{sb}, and the bundle-to-shell diametral clearance, $D_s - D_{otl}$. When rating an existing exchanger, these values may be available from the manufacturer's drawings and specifications. When the latter are unavailable, and for all design calculations, the clearances must be estimated. The estimation procedures given here are based on the recommendations of Taborek [2], which incorporate Tubular Exchanger Manufacturers Association (TEMA) standards and prevailing practice.

The TEMA specifications [7] for tube-to-baffle clearance are based on tube bundle assembly and tube vibration considerations, and depend on the tube size. For tube OD greater than 1.25 in., δ_{tb} is specified as 0.4 mm. For tubes with OD of 1.25 in. or less, the specified clearance depends on the longest unsupported tube length in the exchanger. If this length is less than 3.0 ft, then $\delta_{tb} = 0.4$ mm; otherwise, $\delta_{tb} = 0.2$ mm.

The longest unsupported length occurs for tubes that pass through the baffle windows, and is equal to twice the baffle spacing for exchangers with evenly spaced baffles. If the inlet and/or outlet baffle spaces are larger than the central baffle spaces, as is often the case, the longest unsupported length is the larger of the inlet and outlet spacings plus the central baffle spacing.

The shell-to-baffle clearance reflects manufacturing tolerances for both the shell and the baffles. The average clearance specified by TEMA [7] can be represented by the following linear function of shell diameter [2]:

$$\delta_{sb}(\text{mm}) = 0.8 + 0.002 D_s\,(\text{mm}) \tag{6.58}$$

To account for out-of-roundness tolerances in both the shell and the baffles, a safety factor of 0.75 mm is often added to the value of δ_{sb} calculated from Equation (6.58). Although this practice provides a conservative estimate for the heat-transfer coefficient, the larger clearance results in a smaller (non-conservative) value for the calculated pressure drop.

The shell-to-bundle clearance depends on shell diameter, head type, and in some cases, the design operating pressure. In order to accommodate the rear head, floating-head units require larger clearances than fixed-tubesheet and U-tube exchangers. Pull-through floating-head units require the largest clearances to provide room for the rear-head bolts, the size of which increases with the design operating pressure. Recommended values for the diametral shell-to-bundle clearance are given in **Figure 6.10**.

Application of the Delaware method is illustrated in the following example.

Example 6.1

Use the Delaware method to calculate the shell-side heat-transfer coefficient and pressure drop for the final configuration of the kerosene/crude oil exchanger of Example 5.1. Compare the results with those obtained previously using the Simplified Delaware method and HEXTRAN.

Solution

(a) The following data are obtained from Example 5.1.

Exchanger type = AES

Shell-side fluid = kerosene

Mass flow rate of kerosene = 45,000 lb/h

$D_s = 19.25$ in.

$n_t = 124$

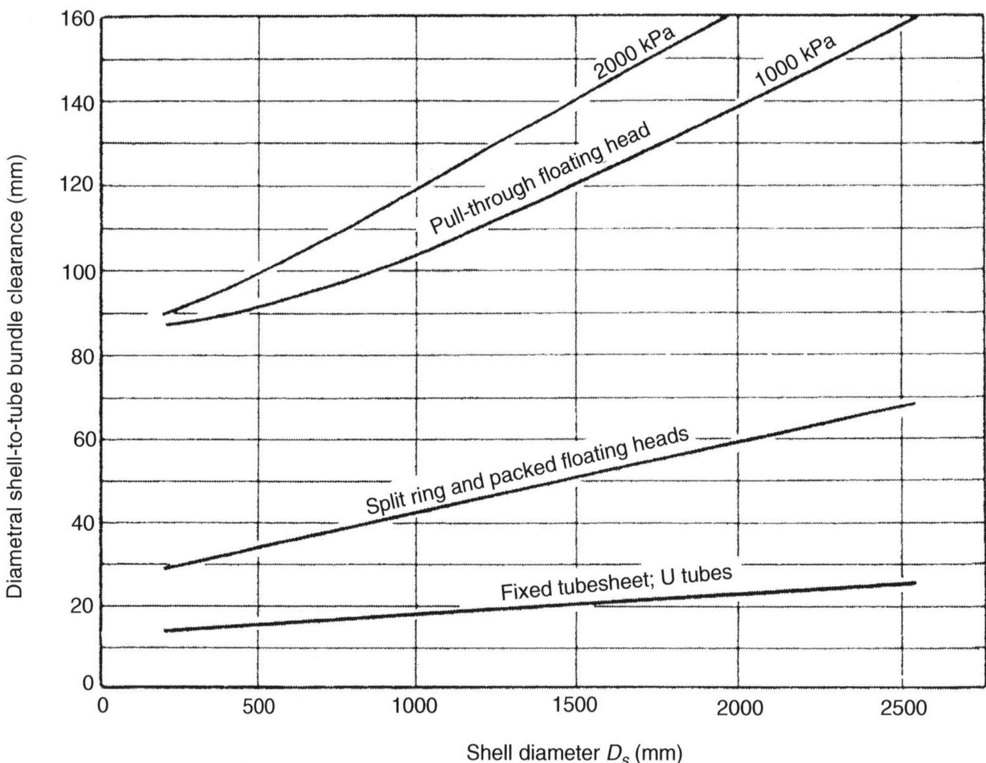

FIGURE 6.10 Diametral shell-to-bundle clearance, $D_s - D_{otl}$ (Source: Ref. [2]).

$D_o = 1.0$ in.
$L = 14$ ft
$P_T = 1.25$ in. (square)
$B = B_{in} = B_{out} = 3.85$ in.
$B_c = 0.2 =$ fractional baffle cut
$n_b = 42$
$N_{ss}/N_c = 0.1$

(b) Estimate leakage and bypass clearances.
Since the tube OD is less than 1.25 in., the tube-to-baffle clearance depends on the longest unsupported tube length, which is twice the baffle spacing, or 7.7 in. As this length is less than 3.0 ft, the tube-to-baffle clearance is taken as:

$$\delta_{tb} = 0.4 \text{ mm} = 0.015748 \text{ in.}$$

Equation (6.58) is used to estimate the shell-to-baffle clearance:

$$\delta_{sb} = 0.8 + 0.002\,D_s = 0.8 + 0.002 \times 19.25 \times 25.4$$

$$\delta_{sb} = 1.7779 \text{ mm}$$

Adding the safety factor (for heat transfer) of 0.75 mm gives:

$$\delta_{sb} = 2.5279 \text{ mm} = 0.09952 \text{ in.}$$

The shell-to-bundle diametral clearance is obtained from **Figure 6.10** with $D_s = 489$ mm. Reading the graph for split ring and packed floating heads gives:

$$D_s - D_{otl} \cong 34 \text{ mm} = 1.34 \text{ in.}$$

The outer tube limit and central tube limit diameters follow from this value:

$$D_{otl} = D_s - 1.34 = 19.25 - 1.34 = 17.91 \text{ in.}$$

$$D_{ctl} = D_{otl} - D_o = 17.91 - 1.0 = 16.91 \text{ in.}$$

(c) Calculate the flow areas.

(i) Cross-flow area

The cross-flow area is calculated from Equation (6.17):

$$S_m = B\left[(D_s - D_{otl}) + \frac{(D_{otl} - D_o)}{(P_T)_{eff}}(P_T - D_o)\right]$$

$$= 3.85\left[1.34 + \frac{(17.91 - 1.0)}{1.25}(1.25 - 1.0)\right]$$

$$S_m = 18.18 \text{ in.}^2 = 0.1262 \text{ ft}^2$$

(ii) Tube-to-baffle leakage area

Equation (6.25) is first used to calculate the value of angle θ_{ctl}:

$$\theta_{ctl} = 2\cos^{-1}\left[\frac{D_s(1 - 2B_c)}{D_{ctl}}\right] = 2\cos^{-1}\left[\frac{19.25\,(1 - 2 \times 0.2)}{16.91}\right]$$

$$\theta_{ctl} = 1.6378 \text{ rad}$$

Next, the fraction of tubes in cross flow is calculated from Equation (6.35):

$$F_c = 1 + \frac{1}{\pi}(\sin\theta_{ctl} - \theta_{ctl}) = 1 + \frac{1}{\pi}(\sin 1.6378 - 1.6378)$$

$$F_c \cong 0.7963$$

The tube-to-baffle leakage area is given by Equation (6.22):

$$S_{tb} = 0.5\pi\, D_o \delta_{tb} n_t\,(1 + F_c)$$

$$= 0.5\pi \times 1.0 \times 0.015748 \times 124\,(1 + 0.7963)$$

$$S_{tb} = 5.5099 \text{ in.}^2 = 0.03826 \text{ ft}^2$$

(iii) Shell-to-baffle leakage area

Equation (6.38) is first used to calculate the value of angle θ_{ds}:

$$\theta_{ds} = 2\cos^{-1}(1 - 2B_c) = 2\cos^{-1}(1 - 2 \times 0.2)$$

$$\theta_{ds} = 1.8546 \text{ rad}$$

The shell-to-baffle leakage area is given by Equation (6.37):

$$S_{sb} = D_s\delta_{sb}(\pi - 0.5\theta_{ds}) = 19.25 \times 0.09952\,(\pi - 0.5 \times 1.8546)$$

$$S_{sb} = 4.24205 \text{ in.}^2 = 0.02946 \text{ ft}^2$$

(iv) Shell-to-bundle bypass flow area

The bypass flow area is given by Equation (6.40):

$$S_b = B\,(D_s - D_{otl}) = 3.85 \times 1.34 = 5.1590 \text{ in.}^2 \cong 0.03583 \text{ ft}^2$$

(v) Window-flow area

The fraction of tubes in one baffle window is calculated from Equation (6.19):

$$F_w = 0.5\,(1 - F_c) = 0.5\,(1 - 0.7963) = 0.10185$$

The window-flow area is calculated using Equation (6.44):

$$S_w = \frac{1}{8}D_s^2(\theta_{ds} - \sin\theta_{ds}) - \frac{1}{4}n_t F_w \pi\, D_o^2$$

$$= \frac{1}{8}(19.25)^2\,(1.8546 - \sin 1.8546) - \frac{1}{4} \times 124 \times 0.10185 \times \pi\,(1.0)^2$$

$$S_w = 31.5192 \text{ in.}^2 = 0.2189 \text{ ft}^2$$

(d) Ideal tube bank correlations.

$$G = \frac{\dot{m}_o}{S_m} = \frac{45{,}000}{0.1262} = 356{,}577 \text{ lbm / h} \cdot \text{ft}^2$$

From Example 5.1, the viscosity of kerosene is 0.97 lbm/ft · h. Thus,

$$Re = D_o G/\mu = \frac{(1.0/12) \times 356{,}577}{0.97} = 30{,}634$$

Also,

$$P_T/D_o = 1.25/1.0 = 1.25$$

From **Figure 6.1**, $j \cong 0.0060$ and $f \cong 0.09$. (Equations (6.1) and (6.2) give $j = 0.0063$ and $f = 0.089$.) The ideal tube bank heat-transfer coefficient is given by:

$$h_{ideal} = jC_P G\phi \, Pr^{-2/3}$$

From Example 5.1, for kerosene:

$$C_P = 0.59 \text{ Btu/lbm} \cdot {}^\circ\text{F}$$

$$Pr = 7.24$$

$$\phi = 1.0 \text{ (constant fluid properties assumed)}$$

Therefore, using the j-value read from **Figure 6.1**:

$$h_{ideal} = 0.0060 \times 0.59 \times 356{,}577 \times 1.0\,(7.24)^{-2/3}$$

$$h_{ideal} = 337 \text{ Btu / h} \cdot \text{ft}^2 \cdot {}^\circ\text{F}$$

(e) Calculate the correction factors.

(i) Correction factor for baffle window flow

From Equation (6.45),

$$J_C = 0.55 + 0.72F_c = 0.55 + 0.72 \times 0.7963$$

$$J_C = 1.1233$$

(ii) Baffle leakage correction factors

The area ratios r_s and r_l are calculated from Equations (6.46) and 6.47):

$$r_s = \frac{S_{sb}}{S_{sb} + S_{tb}} = \frac{0.02946}{0.02946 + 0.03826} = 0.43503$$

$$r_l = \frac{S_{sb} + S_{tb}}{S_m} = \frac{0.02946 + 0.03826}{0.1262} = 0.53661$$

From Equation (6.50),

$$p = 0.8 - 0.15\,(1 + r_s) = 0.8 - 0.15\,(1 + 0.43503)$$

$$p = 0.58475$$

The leakage correction factors for heat transfer and pressure drop are calculated from Equations (6.48) and (6.49):

$$\begin{aligned} J_L &= 0.44\,(1 - r_s) + [1 - 0.44\,(1 - r_s)] \exp(-2.2 r_l) \\ &= 0.44\,(1 - 0.43503) + [1 - 0.44(1 - 0.43503)] \exp(-2.2 \times 0.53661) \\ J_L &= 0.4794 \end{aligned}$$

$$\begin{aligned} R_L &= \exp\left[-1.33\,(1 + r_s)\,(r_l)^p\right] \\ &= \exp\left[-1.33\,(1 + 0.43503)\,(0.53661)^{0.58475}\right] \\ R_L &= 0.2655 \end{aligned}$$

Note: The value of J_L is well below 0.6, indicating that the baffle spacing may be too small. In Example 5.1, the baffle spacing was set to the minimum because the shell-side pressure drop was very low. (Omitting the safety factor of 0.75 mm in δ_{sb} gives $J_L = 0.5410$.)

(iii) Bundle bypass correction factors

The correction factors for the bundle bypass flow are given by Equations (6.52) and (6.53) with $r_{ss}=N_{ss}/N_c = 0.1$:

$$J_B = \exp\left[-C_J(S_b/S_m)\left(1-\sqrt[3]{2r_{ss}}\right)\right]$$
$$= \exp\left[-1.25\,(0.03583/0.1262)\left(1-\sqrt[3]{0.2}\right)\right]$$
$$J_B = 0.8630$$
$$R_B = \exp\left[-C_R(S_b/S_m)\left(1-\sqrt[3]{2r_{ss}}\right)\right]$$
$$= \exp\left[-3.7\,(0.03583/0.1262)\left(1-\sqrt[3]{0.2}\right)\right]$$
$$R_B = 0.6465$$

(iv) Correction factors for unequal baffle spacing

Since $B_{in} = B_{out} = B$, both J_S and R_s are equal to 1.0.

(v) Laminar flow correction factor

Since $Re > 100$, $J_R = 1.0$.

(f) Calculate the shell-side heat-transfer coefficient.

The shell-side coefficient is given by Equation (6.5):

$$h_o = h_{ideal}\,(J_C\,J_L\,J_B\,J_R\,J_S)$$
$$= 337\,(1.1233 \times 0.4794 \times 0.8630 \times 1.0 \times 1.0)$$
$$= 337\,(0.4647)$$
$$h_o = 156.6\ \text{Btu/h}\cdot\text{ft}^2\cdot{}^\circ\text{F}$$

Note: The overall correction factor is outside the acceptable range (greater than 0.5). The leakage correction factor, J_L, is by far the most significant. (Omitting the safety factor in δ_{sb} gives an overall correction factor of 0.5244. The corresponding value of h_o is 176.7 Btu/h · ft^2 · °F.)

(g) Calculate the shell-side pressure drop.

(i) Number of tube rows crossed

The number of tube rows crossed between baffle tips is calculated using Equation (6.8):

$$N_c = \frac{D_s\,(1-2B_c)}{P'_T} = \frac{19.25\,(1-2\times 0.2)}{1.25}$$
$$N_c = 9.24$$

The effective number of tube rows crossed in one baffle window is given by Equation (6.11):

$$N_{cw} = \frac{0.8B_cD_s}{P'_T} = \frac{0.8\times 0.2\times 19.25}{1.25}$$
$$N_{cw} = 2.464$$

Neither N_c nor N_{cw} is rounded to the nearest integer.

(ii) Ideal tube bank pressure drop

From Example 5.1, the density of kerosene is 49 lbm/ft^3. The ideal tube bank pressure drop for one baffle space is given by Equation (6.7):

$$\Delta P_{ideal} = \frac{2f_{ideal}N_cG^2}{g_c\rho\phi} = \frac{2\times 0.09\times 9.24\times (356{,}577)^2}{4.17\times 10^8\times 49\times 1.0}$$
$$\Delta P_{ideal} = 10.35\ \text{lbf/ft}^2$$

(iii) Ideal window pressure drop

Since the flow is turbulent, the ideal pressure drop in one baffle window is calculated from Equation (6.10):

$$\Delta P_{w,ideal} = \frac{(2+0.6\,N_{cw})\,\dot{m}_o^2}{2g_c\rho\,S_mS_w} = \frac{(2+0.6\times 2.464)\,(45{,}000)^2}{2\times 4.17\times 10^8\times 49\times 0.1262\times 0.2189}$$
$$\Delta P_{w,ideal} = 6.24\ \text{lbf/ft}^2$$

(iv) Nozzle-to-nozzle pressure drop

Equation (6.16) gives the shell-side pressure drop between nozzles:

$$\Delta P_f = \left[(n_b - 1)\, \Delta P_{ideal} R_B + n_b \Delta P_{w,ideal} \right] R_L + 2\Delta P_{ideal} \left(1 + N_{cw}/N_c\right) R_B R_S$$

$$= [41 \times 10.35 \times 0.6465 + 42 \times 6.24] \times 0.2655 + 2 \times 10.35(1 + 2.464\ /\ 9.24) \times 0.6465 \times 1.0$$

$$\Delta P_f = 159.4\ \mathrm{lbf}\ /\ \mathrm{ft}^2 = 1.11\ \mathrm{psi}$$

(Omitting the safety factor in δ_{sb} gives $\Delta P_f = 1.31$ psi.)

(v) Pressure drop in nozzles

The pressure-drop calculation for the nozzles is the same as in the Simplified Delaware method. Hence, from Example 5.1 we obtain:

$$\Delta P_n = 0.20\ \mathrm{psi}$$

(vi) Total shell-side pressure drop

$$\Delta P_o = \Delta P_f + \Delta P_n = 1.11 + 0.20 = 1.31\ \mathrm{psi}$$

The results of the calculations are summarized in the following table, along with results obtained previously using the Simplified Delaware method and HEXTRAN.

Item	Simplified Delaware method	Delaware method	HEXTRAN
h_o (Btu / h · ft^2 · °F)	122	156.6 (176.7)*	191.2
ΔP_o (psi)	2.2	1.31 (1.51)*	2.10
Re_o	31,158	30,634	45,148
S_m (ft^2)	–	0.1262	0.086
S_{tb} (ft^2)	–	0.03826	0.033
S_{sb} (ft^2)	–	0.02946	0.022
S_w (ft^2)	–	0.2189	0.142
D_{otl} (inches)	–	17.91	17.0
F_c	–	0.7963	0.565
Tubes in cross flow	–	99	70

* Without δ_{sb} safety factor.

Comparing the Delaware method with the Simplified Delaware method, the difference in heat-transfer coefficients is about 22% using the δ_{sb} safety factor and about 30% without it. These values bracket the expected difference of 25% due to the safety factor built into the simplified method. Both methods are in agreement that the shell-side pressure drop is quite low, although the difference between the two methods is large on a percentage basis. The Simplified Delaware method gives a conservative value of pressure drop compared with the full method. The Reynolds numbers do not agree exactly because *Re* is defined differently in the two methods.

Comparing the Delaware and HEXTRAN results, there are significant differences despite the fact that HEXTRAN uses a version of the Delaware method. The two heat-transfer coefficients differ by about 20% if the δ_{sb} safety factor is used, but the difference is only about 8% when it is omitted. The discrepancy is in large part due to different ideal tube bank correlations. HEXTRAN uses the correlation presented by Gnielinski et al. [8] for the ideal tube bank heat-transfer coefficient. The difference in pressure drops is large on a percentage basis and may be partly due to the correlation used by HEXTRAN for the ideal tube bank friction factor. The correlation is not identified in the HEXTRAN documentation.

The difference in Reynolds numbers calculated by HEXTRAN and the Delaware method is due to the difference in cross-flow areas. HEXTRAN adjusts the cross-flow area for the presence of sealing strips by not including the bundle bypass flow area, which amounts to about 0.06 ft^2 with the clearance used by HEXTRAN. Without sealing strips it gives a cross-flow area of 0.145 ft^2 and a Reynolds number of 26,586, which differ from the values obtained by hand mainly due to the difference in clearances. HEXTRAN does not use this Reynolds number in calculating the heat-transfer coefficient because the correlation given by Gnielinski et al. is based on an entirely different definition of Reynolds number.

The difference in tube-to-baffle leakage areas is due to the different values of F_c. Back calculation using $S_{tb} = 0.033$ ft^2 and $F_c = 0.565$ gives $\delta_{tb} = 0.396 \cong 0.4$ mm, demonstrating that HEXTRAN uses the TEMA specification for tube-to-baffle clearance.

The difference in shell-to-baffle leakage areas is primarily due to the 0.75-mm safety factor for shell-to-baffle clearance used in the hand calculation. Without the safety factor, the hand calculation would have given a leakage area of 0.0207 ft^2, which is close to the HEXTRAN value. The difference in window areas is due to the different values of F_c. With fewer tubes in cross flow between baffle tips, HEXTRAN calculates more tubes in the window with correspondingly less free area. The window area of 0.142 ft^2 is consistent with $F_c = 0.565$.

The small fraction of tubes in cross flow is the most puzzling aspect of the HEXTRAN results. With only 70 tubes in cross flow, there are 54 tubes in the windows, i.e., 27 tubes in each window. However, for the specified exchanger with 1.0-in. tubes on 1.25-in. square pitch and 20% baffle cut, a single baffle window can accommodate only about 15 tubes. Therefore, the number (12–13 tubes) estimated in the hand calculation is reasonable, whereas the number (27 tubes) found by HEXTRAN is physically impossible. This observation casts doubt on the validity of the entire HEXTRAN calculation.

There are other inconsistencies in the results produced by HEXTRAN, one of which is shown in the table below. Significantly different values of the heat-transfer coefficient are obtained for the three possible combinations of TWOPHASE and DPSMETHOD. (For TWOPHASE = New, HEXTRAN automatically takes DPSMETHOD = Stream, regardless of the user-specified pressure-drop method.) Since this problem involves single-phase flow, neither the method used for two-phase flow calculations nor the method used to calculate pressure drop should have any effect on the calculated value of the heat-transfer coefficient.

TWOPHASE	DPSMETHOD	h_o (Btu/h · ft^2 · °F)
Old	Bell	191.2
Old	Stream	394.1
New	Stream	223.4

These observations emphasize the need to scrutinize results obtained from even well-established commercial software, and the importance of using standardized test problems for validation purposes. It is not unusual for errors to be introduced when a new version of a software product is developed. For this reason, testing and validation should be performed whenever a new version of an existing code is obtained.

References

[1] Bell KJ. Final report of the cooperative research program on shell-and-tube heat exchangers. University of Delaware Engineering Experiment Station Bulletin No. 5; 1963.
[2] Taborek J. Shell-and-tube heat exchangers. In: Heat Exchanger Design Handbook, Vol. 3. New York: Hemisphere Publishing Corp; 1988.
[3] Perry RH, Green DW, editors. Perry's Chemical Engineers' Handbook. 7th ed. New York: McGraw-Hill; 1997.
[4] Hewitt GF, Shires GL, Bott TR. Process Heat Transfer. Boca Raton, FL: CRC Press; 1994.
[5] Bell KJ, Mueller AC. Wolverine Engineering Data Book II. Wolverine Tube, Inc; 2001. www.wlv.com.
[6] Shah RK, Sekulic DP. Fundamentals of Heat Exchanger Design. Hoboken, NJ: Wiley; 2003.
[7] Standards of the Tubular Exchanger Manufacturers Association. 8th ed. Tarrytown, NY: Tubular Exchanger Manufacturers Association, Inc.; 1999.
[8] Gnielinski A, Zukauskas A, Skrinska A. Banks of plain and finned tubes. In: Heat Exchanger Design Handbook, Vol. 2. New York: Hemisphere Publishing Corp.; 1988.

Notations

A	Area
$a, a_1, a_2,$	Curve-fit parameters for **Figures 6.1 to 6.3**
B	Baffle spacing
B_c	Baffle cut
B_{in}	Inlet baffle spacing
B_{out}	Outlet baffle spacing
$b, b_1, b_2,$	Curve-fit parameters for **Figures 6.1 to 6.3**.
C_J	Parameter in correlation for J_B, Equation (6.52)
C_P	Heat capacity at constant pressure
C_R	Parameter in correlation for R_B, Equation (6.53)
D_{ctl}	Central tube limit diameter
D_o	Outside diameter of tube
D_{otl}	Outer tube limit diameter
D_s	Inside diameter of shell
D_w	Equivalent diameter for window flow
F_c	Fraction of tubes in cross flow between baffle tips
F_w	Fraction of tubes in one baffle window

f	Fanning friction factor
f_{ideal}	Fanning friction factor for ideal tube bank flow
G	$\dot{m}_o/S_m$ = Mass flux of shell-side fluid based on cross-flow area
g_c	Unit conversion factor = 4.17×10^8 lbm · ft/h² · lbf = 1.0 kg · m/s² · N
h	Heat-transfer coefficient
h_o	Shell-side heat-transfer coefficient
h_{ideal}	Ideal tube bank heat-transfer coefficient
J_B	Heat-transfer correction factor for bundle bypass effects
J_C	Heat-transfer correction factor for effect of baffle window flow
J_L	Heat-transfer correction factor for baffle leakage effects
J_R	Heat-transfer correction factor for laminar flow
J_S	Heat-transfer correction factor for unequal baffle spacing
j	$\dfrac{hPr^{2/3}}{C_P G \phi}$ = Colburn factor for heat transfer
L_1, L_2	Clearances between tubes in aligned and staggered tube layouts, **Figure 6.6**
$\dot{m}_o$	Mass flow rate of shell-side fluid
N_c	Number of tube rows crossed in flow between two baffle tips
N_{ct}	Total number of tube rows crossed in flow through shell
N_{cw}	Effective number of tube rows crossed in flow through one baffle window
N_{ss}	Number of pairs of sealing strips
n_b	Number of baffles
n_t	Number of tubes in bundle
n_1	Parameter in correlation for J_S, Equation (6.54)
n_2	Parameter in correlation for R_S, Equation (6.55)
P	Pressure
Pr	Prandtl number
P_T	Tube pitch
P'_T	Tube pitch parallel to flow direction
$(P_T)_{eff}$	P_T (for square and triangular layouts) or $P_T/\sqrt{2}$ (for rotated square layout)
p	$0.8 - 0.15(1 + r_s)$ = Parameter in correlation for R_L, Equation (6.50)
R_B	Pressure-drop correction factor for bundle bypass effects
Re	$D_o G/\mu$ = Shell-side Reynolds number
R_L	Pressure-drop correction factor for baffle leakage effects
R_S	Pressure-drop correction factor for unequal baffle spacing
r_l	Parameter in correlations for J_L and R_L, defined by Equation (6.47)
r_s	Parameter in correlations for J_L and R_L, defined by Equation (6.46)
r_{ss}	N_{ss}/N_c
S_b	Bundle bypass flow area
S_m	Cross-flow area
S_{sb}	Shell-to-baffle leakage area
S_{tb}	Tube-to-baffle leakage area
$(S_{tb})_1$	Tube-to-baffle leakage area for one tube
S_w	Window-flow area
S_{wg}	Gross window area

Greek Letters

ΔP_c	Total pressure drop in all central baffle spaces
ΔP_e	Sum of pressure drops in the entrance and exit baffle spaces
ΔP_f	Nozzle-to-nozzle shell-side pressure drop
ΔP_{ideal}	Ideal tube bank pressure drop
ΔP_n	Pressure drop in nozzles
ΔP_o	Total shell-side pressure drop
ΔP_w	Total pressure drop in all baffle windows
$\Delta P_{w,ideal}$	Pressure drop in one baffle window, uncorrected for baffle leakage effects

δ_{sb}	Shell-to-baffle clearance
δ_{tb}	Tube-to-baffle clearance
θ_{ctl}	Angle defined in **Figure 6.5** and given by Equation (6.25)
θ_{ds}	Baffle window angle defined in **Figure 6.5** and given by Equation (6.39)
θ_{tp}	Tube layout angle
μ	Viscosity
v	Kinematic viscosity
ρ	Fluid density
ϕ	Viscosity correction factor

Problems

(6.1) Use the data given below to calculate the shell-side heat-transfer coefficient and nozzle-to-nozzle pressure drop by the Delaware method. Neglect the corrections for unequal baffle spacing.

Shell-side fluid properties

$\dot{m} = 1.382$ kg/s	$\rho = 1200$ kg/m^3
$C_P = 2800$ J/kg · K	$Pr = 3.5$
$\mu = 5 \times 10^{-4}$ N · s /m^2	

Shell-side geometry

Cross-flow area: 0.035096 m^2
Tube-to-baffle leakage area: 0.006893 m^2
Shell-to-baffle leakage area: 0.003442 m^2
Bundle bypass flow area: 0.004088 m^2
Flow area in one baffle window: 0.0255 m^2
Fraction of tubes in cross flow: 0.66
Number of tube rows crossed between baffle tips: 12
Effective number of tube rows crossed in one window: 3.2
Number of baffles: 15
Tube OD: 0.0254 m
Tube pitch: 0.033782 m (square)
Sealing strips: none

Ans. $h_o = 450$ W/m^2 · K, $\Delta P_f = 40$ Pa

(6.2) Use the data given below to calculate the shell-side heat-transfer coefficient by the Delaware method. Neglect the correction for unequal baffle spacing.

Data

Shell-side Reynolds number: 20,695
Ideal tube bank heat-transfer coefficient: 376 Bth/h · ft^2 · °F
Cross-flow area: 0.37770 ft^2
Tube-to-baffle leakage area: 0.074194 ft^2
Shell-to-baffle leakage area: 0.037048 ft^2
Bundle bypass flow area: 0.044002 ft^2
Fraction of tubes in cross flow: 0.657
Sealing strips: None

Ans. 220 Bth/h · ft^2 · °F.

(6.3) Use the following data to calculate the shell-side heat-transfer coefficient and nozzle-to-nozzle pressure drop by the Delaware method. Neglect the corrections for unequal baffle spacing.

Shell-side fluid properties

$\dot{m} = 6.3$ kg/s	$\rho = 850$ kg/m^3
$C_P = 2500$ J/ kg · K	$Pr = 23.7$
$\mu = 0.0015$ N · s/m^2	

Shell-side geometry

Shell ID: 0.7874 m

Bundle diameter: 0.7239 m
Number of tubes: 368
Number of tubes between baffle tips: 232
Tube OD: 0.0254 m
Tube pitch: 0.0318 m
Tube layout: rotated square (45°)
Number of baffles: 29
Baffle cut: 20%
Central baffle spacing: 0.1575 m
Sealing strips: none
Tube-to-baffle leakage area: 0.0095 m^2
Shell-to-baffle leakage area: 0.0039 m^2

Ans. $\Delta P_f = 1900$ Pa.

(6.4) Consider again the heat-transfer service of Problem 5.3. In addition to the data given in that problem, the following information has been obtained from blueprints and manufacturer's specifications.
Number of tubes between baffle tips: 60
Tube-to-baffle clearance: $\delta_{tb} = 0.007874$ in.
Shell-to-baffle clearance: $\delta_{sb} = 0.07874$ in.
Shell-to-bundle diametral clearance: 0.6 in.
(a) What is the value of the outer-tube-limit diameter?
(b) Calculate the cross-flow area for the shell-side fluid.
(c) Calculate the tube-to-baffle leakage area.
(d) Calculate the shell-to-baffle leakage area.
(e) Calculate the bundle bypass flow area.
(f) Use the Delaware method to calculate the shell-side heat-transfer coefficient.
(g) What is the value of the correction factor, J_S, for unequal baffle spacing?
(h) Use the Delaware method to calculate the shell-side pressure drop.

Ans. (a) 12.65 in. (b) 0.13158 ft^2. (c) 0.00902 ft^2. (d) 0.01604 ft^2. (e) 0.02208 ft^2. (f) 435 Btu/h · ft^2 · °F. (g) 0.98135.

(6.5) Use the following data to calculate the shell-side heat-transfer coefficient and pressure drop by the Delaware method. Neglect the corrections for unequal baffle spacing.
Shell-side fluid Properties

$\dot{m} = 15$ kg/s	$\rho = 900$ kg/m^3
$C_P = 2500$ J/kg · K	$Pr = 4.0$
$\mu = 5 \times 10^{-4}$ N · s/m^3	

Shell-side geometry
Cross-flow area: 0.05 m^2
Tube-to-baffle leakage area: 0.01 m^2
Shell-to-baffle leakage area: 0.005 m^2
Bundle bypass flow area: 0.005 m^2
Flow area in one baffle window: 0.03 m^2
Fraction of tubes in cross flow: 0.5
Number of tube rows crossed between baffle tips: 10
Effective number of tube rows crossed in one window: 5
Number of baffles: 90
Tube OD: 0.0254 m
Tube pitch: 0.0381 m (triangular)
Sealing strips: one pair
Nozzle ID: 0.1282 m

(6.6) A shell-and-tube heat exchanger has the following specifications:

Shell ID = 0.635 m	Tube–baffle clearance: $\delta_{tb} = 0.2$ mm
Number of tubes = 454	Shell–baffle clearance: $\delta_{sb} = 2.75$ mm
Tube OD = 0.0191 m	Shell–bundle diametral clearance = 25 mm

(*Continued*)

—cont'd

Tube pitch = 0.0254 m	Sealing strips: none
Tube layout = Triangular (30°)	Shell-side nozzles: ID = 0.0779 m
Number of baffles = 29	
Baffle spacing = 0.254 m	
Baffle cut =25%	

The shell-side fluid properties are:

$\dot{m} = 7$ kg/s	$\rho = 800$ kg/m^3
$C_P = 2200$ J/kg · K	$Pr = 22.0$
$\mu = 0.001$ N · s/m^2	

Calculate the following:
(a) The outer tube limit and central tube limit diameters.
(b) The cross-flow area and bundle bypass flow area.
(c) The heat-transfer and pressure-drop correction factors for the bypass flow.
(d) The number of tube rows crossed between baffle tips and the effective number of tube rows crossed in one baffle window.
(e) The ideal cross-flow pressure drop in one baffle space.
(f) The actual (non-ideal) pressure drop in one baffle space.
(g) The shell-side heat-transfer coefficient.
(h) The total shell-side pressure drop.

Ans. (a) 0.610 and 0.5909 m. (b) 0.043577 and 0.00635 m^2. (c) 0.83 and 0.58. (d) $N_c = 14.4$, $N_{cw} = 5.77$. (e) 130 Pa. (f) 37 Pa. (g) 425 W/m^2 · K. (h) 4640 Pa.

(6.7) A computer program was used to rate a shell-and-tube heat exchanger having the following configuration:

Shell ID = 31 in.	Number of baffles = 29
Number of tubes = 368	Baffle spacing = 6.2 in.
Tube OD = 1.0 in.	Baffle cut = 20%
Tube pitch = 1.25 in. (square)	Sealing strips: none
Shell-side nozzles: 3-in. sch. 40	

The shell-side fluid properties are as follows:

$\dot{m} = 50,000$ lb/h	$\rho = 53$ lbm/ft^3
$C_P = 0.545$ Btu/lbm · °F	$Pr = 19.78$
$\mu = 1.5$ cp	

The output file generated by the computer program contained the following data:

Bundle diameter = 28.5 in.	Shell-baffle leakage area = 0.042 ft^2
Cross-flow area = 0.346 ft^2	Window area = 0.377 ft^2
Tube-baffle leakage area = 0.102 ft^2	Tubes in cross flow = 232

Based on the data in the output file:
(a) What value did the program calculate for the fraction of tubes in cross flow?
(b) What value did the program use for the diametral shell-to-bundle clearance?
(c) What value did the program use for the shell-to-baffle clearance, δ_{sb}?
(d) What flow area did the program use for the bundle bypass stream?
(e) Use the data from the program's output file to calculate the shell-side heat-transfer coefficient by the Delaware method. Neglect the correction for unequal baffle spacing.

(f) Use the data from the program's output file to calculate the shell-side pressure drop by the Delaware method. Neglect the correction for unequal baffle spacing.

(g) If the inlet/outlet baffle spacing is 12 in., calculate the correction factors for unequal baffle spacing.

Ans. (a) 0.63. (b) 2.5 in. (c) 0.0881 in. (d) 0.1076 ft^2. (e) 60 Btu/h · ft^2 · °F. (g) $J_S = 0.960$, $R_S = 0.305$.

(6.8) Consider the modified design for the kerosene/oil exchanger of Problem 5.1. For this configuration:

(a) Estimate values for the tube-to-baffle clearance, shell-to-baffle clearance, and shell-to-bundle diametral clearance.

(b) Use the values obtained in part (a) to calculate the shell-side heat-transfer coefficient by the Delaware method.

(6.9) For a shell-and-tube heat exchanger employing radial low-fin tubes, the following modifications are made in the Delaware method [2]:

(1) The shell-side Reynolds number is calculated as follows:

$$Re = D_{re}G/\mu$$

where

$D_{re} = D_r + 2n_f b\tau$ = equivalent root-tube diameter
D_r = root-tube diameter
n_f = number of fins per unit length
b = average fin height
τ = average fin thickness

(2) The cross-flow area is calculated as follows:

$$S_m = B\left\{D_s - D_{otl} + \frac{(D_{otl} - D_o)(P_T - D_{re})}{(P_T)_{eff}}\right\}$$

where $D_o = D_r + 2b$ = fin OD.

(3) In all other calculations, D_o is interpreted as the fin OD.

(4) The ideal tube bank friction factor obtained from **Figures 6.1 to 6.3** is multiplied by a correction factor of 1.4.

(5) If the shell-side Reynolds number is less than 1000, the Colburn *j*-factor obtained from **Figures 6.1 to 6.3** is multiplied by a correction factor, j_{cf}, given by the following equation:

$$j_{cf} = 0.566 + 0.00114Re - 1.01 \times 10^{-6}Re^2 + 2.99 \times 10^{-10}Re^3$$

Consider the final configuration of the oil/water heat exchanger obtained in Example 5.2. For this configuration:

(a) Estimate values for the tube-to-baffle clearance, shell-to-baffle clearance, and shell-to-bundle diametral clearance.

(b) Calculate the shell-side heat-transfer coefficient by the Delaware method.

(c) Calculate the shell-side pressure drop by the Delaware method.

(d) Compare the results obtained in parts (b) and (c) with the corresponding values calculated using the Simplified Delaware method in Example 5.2.

(6.10) Consider the modified design for the oil/water heat exchanger of Problem 5.5. For this configuration:

(a) Estimate values for the tube-to-baffle clearance, shell-to-baffle clearance, and shell-to-bundle diametral clearance.

(b) Use the Delaware method as modified for finned-tube exchangers (see Problem 6.9) to calculate the shell-side heat-transfer coefficient.

(6.11) A shell-and-tube heat exchanger has the following configuration:

Type: AEU	Number of baffles = 11
Shell ID = 39 in.	Baffle spacing = 15.6 in.
Number of tubes = 1336	Baffle cut = 35%
Tube OD = 0.75 in.	Sealing strips: none
Tube pitch = 15/16 in. (triangular)	

The shell-side fluid is a gas with the following properties:

$\dot{m} = 180{,}000$ lb/h	$\rho = 0.845$ lbm/ft^3
$C_P = 0.486$ Btu/lbm · °F	$Pr = 0.84$
$\mu = 0.0085$ *cp*	

(a) Estimate values for the tube-to-baffle clearance, shell-to-baffle clearance, and shell-to-bundle diametral clearance.

(b) Use the Delaware method to calculate the shell-side heat-transfer coefficient.

(c) Use the Delaware method to calculate the nozzle-to-nozzle pressure drop for the shell-side fluid.

7 The Stream Analysis Method

7.1 Introduction

In the Stream Analysis method a hydraulic model is used to calculate the pressure drop and flow rates of the cross flow, leakage and bypass streams in the shell of a shell-and-tube exchanger. These flow rates are then used to calculate the shell-side heat-transfer coefficient. This approach, although still highly empirical, is considered more fundamental than that of the Delaware method because it is based on sound hydraulic principles that properly account for interactions among the shell-side streams. The method was conceived by Tinker [1–3] and further developed by others, notably Palen and Taborek [4] at Heat Transfer Research, Inc. (HTRI). By computerizing the model and fitting it to HTRI's extensive experimental database on industrial-size exchangers, the latter workers produced a powerful tool for design and analysis of shell-and-tube exchangers.

Although the basic ideas and equations involved in the Stream Analysis method have been published, values of the many empirical parameters required for its implementation remain proprietary. Therefore, commercial software is required in order to use the method, with one exception. Wills and Johnston [5] published a simplified, but complete, version that can be used to calculate shell-side pressure drop. The HTRI version is presented below, followed by the Wills–Johnston version.

7.2 The Equivalent Hydraulic Network

The shell-side flow is represented by an equivalent hydraulic network as shown in **Figure 7.1**. The streams are defined as follows:

A = tube-to-baffle leakage
B = cross flow
C = bundle-to-shell bypass
E = shell-to-baffle leakage
F = tube-pass-partition bypass

These streams are the same as in the Delaware method with the exception of stream F, which was neglected in Chapter 6. Tube holes cannot be drilled in the tubesheets where they are intersected by the tube pass partition plates. When the resulting gaps in the tube bundle are aligned with the cross-flow direction, fluid can flow across each baffle compartment through the gaps, bypassing most of the heat-transfer surface. It is possible to fill the gaps using dummy tubes, seal rods, or sealing strips to eliminate or partially block stream F.

The valve symbols in **Figure 7.1** represent flow resistances. Note in particular that there is no resistance between the inlet node and central node in each baffle compartment. The essential features of this rather confusing model representation are as follows. Streams B, C, and F flow in parallel across each baffle compartment and then recombine to flow through the baffle windows. Leakage streams A and E, by contrast, flow in parallel from one baffle compartment to the next and do not flow through the windows. (Note that the depiction in **Figure 7.1** is strictly valid only for a single baffle and window; otherwise, the total flow out, which includes streams A and E, would flow through the next window.)

7.3 The Hydraulic Equations

The equations that describe the hydraulic network of **Figure 7.1** are presented in this section.

7.3.1 Stream Pressure Drops

The pressure drop of each stream in one baffle space is expressed in terms of a flow resistance (pressure loss) coefficient as follows:

$$\Delta P_j = \frac{K_j\,(\dot{m}_j/S_j)^2}{2\rho g_c \phi} \qquad j = A, B, C, E, F \tag{7.1}$$

where

$\dot{m}_j$ = mass flow rate of stream j
S_j = flow area for stream j
K_j = flow resistance coefficient for stream j
ρ = fluid density
g_c = unit conversion factor
$\phi = (\mu/\mu_w)^{0.14}$ = viscosity correction factor

The flow areas for streams A, B, C, and E are the same as in the Delaware method (Chapter 6). For stream F, the flow area is the baffle spacing multiplied by the width of the gap in the tube bundle created by the missing tubes.

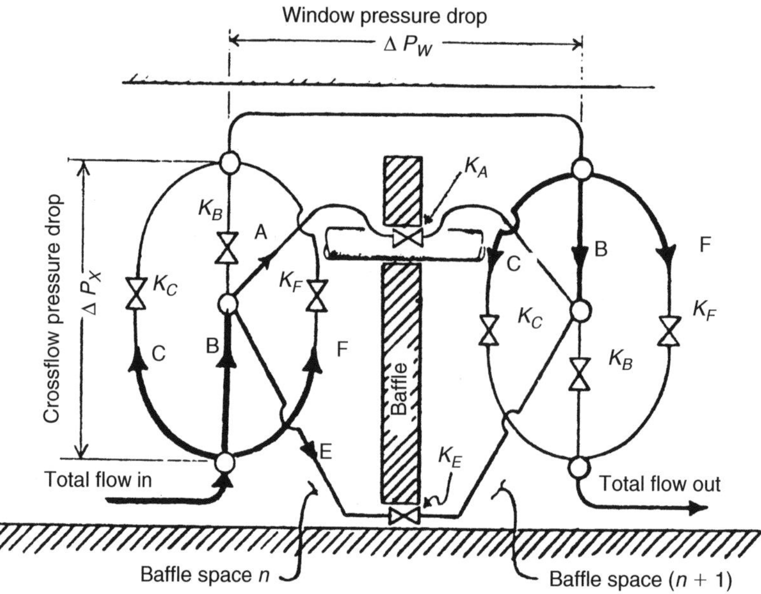

FIGURE 7.1 Equivalent hydraulic network for shell-side flow (Source: Ref. [4]).

7.3.2 Balanced Pressure Drop Requirements

The pressure difference between any two points in a hydraulic network must be the same for all paths connecting the points. This requirement leads to the following equations:

$$\Delta P_B = \Delta P_C = \Delta P_F \equiv \Delta P_x \tag{7.2}$$

$$\Delta P_A = \Delta P_E = \Delta P_x + \Delta P_w \tag{7.3}$$

In these equations ΔP_x denotes the common value of the pressure drop experienced by streams *B*, *C*, and *F* in one baffle space, and ΔP_w is the window pressure drop. Since the last equality in Equation (7.2) is just a definition of ΔP_x, Equations (7.2) and (7.3) constitute four independent relationships among the pressure drops.

7.3.3 Mass Conservation

Conservation of mass requires that the mass flow rates of the individual streams sum to the total mass flow rate of the shell-side fluid, which is fixed. Thus,

$$\dot{m}_o = \dot{m}_A + \dot{m}_B + \dot{m}_C + \dot{m}_E + \dot{m}_F \tag{7.4}$$

7.3.4 Correlations for Flow Resistance Coefficients

Each flow resistance coefficient, K_j, is a function of the Reynolds number for stream *j* and the flow geometry, i.e.,

$$K_j = K_j(Re_j, geometry) \quad j = A, B, C, E, F \tag{7.5}$$

The empirical correlations used to represent these relationships are discussed in some detail by Palen and Taborek [4]. However, complete details are publicly available only for the cross-flow stream, for which the ideal tube bank correlation is used. Equating ΔP_B with the ideal tube bank pressure drop, ΔP_{ideal}, from Equation (6.7) gives:

$$\frac{K_B(\dot{m}_B/S_m)^2}{2\rho g_c \phi} = \frac{2 f_{ideal} N_c G^2}{g_c \rho \phi} \tag{7.6}$$

Since in this context $G = G_B = \dot{m}_B/S_m$, where S_m is the cross-flow area given by Equation (6.17), it follows that:

$$K_B = 4 N_c f_{ideal} \tag{7.7}$$

where N_c is the number of tube rows crossed between baffle tips, and f_{ideal} is the ideal tube bank friction factor based on $Re_B = D_o G_B/\mu$.

7.3.5 Window Pressure Drop

The pressure drop in one baffle window is calculated from the following equation [4]:

$$\Delta P_w = 2\alpha\left(\frac{G_w^2}{2\rho g_c}\right) + \frac{4 f_w N_{cw} G_w^2 \beta}{2\rho g_c} \tag{7.8}$$

where

$G_w = \dfrac{\dot{m}_B + \dot{m}_C + \dot{m}_F}{S_w}$

S_w = flow area in window

N_{cw} = effective number of tube rows crossed in window

f_w = ideal tube bank friction factor based on $Re_w = D_o G_w/\mu$

α, β = empirical correction factors

The first term in Equation (7.8) is the pressure drop due to the 180° change of direction in the window. The correction factor, α, varies from 1.0 for small baffle spacing to zero at very large baffle spacing. Thus, the pressure drop due to the change of direction is equal to two velocity heads for small baffle spacing and approaches zero for very large baffle spacing.

The second term in Equation (7.8) represents the friction loss in the window, and is equal to the pressure drop in an ideal tube bank multiplied by the correction factor, β, that accounts for non-ideal flow conditions in the window. For laminar flow, β is close to unity, but in turbulent flow it is a function of the baffle-cut-to-baffle-spacing ratio, (B_c/B) [4].

7.3.6 Window Friction Factor

The friction factor for window flow is obtained from an empirical correlation of the form

$$f_w = f_w\,(Re_w, geometry) \tag{7.9}$$

This correlation is, again, the ideal tube bank correlation.

7.3.7 Computational Approach

The equivalent hydraulic network is described by a set of 17 equations, consisting of:

- Five stream pressure drop equations, Equation (7.1)
- Four balanced pressure drop equations, Equations (7.2) and (7.3)
- One mass conservation equation, Equation (7.4)
- Five correlations for the flow resistance coefficients, Equation (7.5)
- One equation for window pressure drop, Equation (7.8)
- One correlation for the window friction factor, Equation (7.9)

The equations are to be solved for the following 17 variables:

- Five flow rates, $\dot{m}_j\ (j = A, B, C, E, F)$
- Six pressure drops, ΔP_w and $\Delta P_j\ (j = A, B, C, E, F)$
- Five flow resistance coefficients, $K_j\ (j = A, B, C, E, F)$
- The window friction factor, f_w

The stream-flow rates and the pressure drops are the desired quantities, while the resistance coefficients and friction factor are obtained as a necessary part of the solution procedure. Since Equations (7.1), (7.5), (7.8), and (7.9) are nonlinear, an iterative procedure (described in Ref. [4]) is required to obtain the solution. However, the number of equations is small and the solution is easily obtained on a computer.

7.4 Shell-Side Pressure Drop

The total shell-side pressure drop is the sum of the pressure drops, ΔP_x and ΔP_w, for each central baffle space and baffle window, plus the pressure drops, ΔP_{in} and ΔP_{out}, for the inlet and outlet baffle spaces, plus the pressure drop, ΔP_n, in the nozzles. Thus,

$$\Delta P_o = (n_b - 1)\Delta P_x + n_b \Delta P_w + \Delta P_{in} + \Delta P_{out} + \Delta P_n \tag{7.10}$$

The values of ΔP_x and ΔP_w are obtained from the solution of the hydraulic equations as discussed above. Separate hydraulic calculations are required to obtain the values of ΔP_{in} and ΔP_{out} due to differences in baffle spacing and number of tube rows crossed in the end baffle spaces. In addition, the leakage streams join the cross-flow stream in the end baffle compartments, resulting in a higher pressure drop across these compartments. The pressure drop in the nozzles is calculated in the same manner as in the Delaware and Simplified Delaware methods.

7.5 Shell-Side Heat-Transfer Coefficient

The shell-side heat-transfer coefficient is calculated as a weighted average of the coefficients, h_x and h_w, for cross flow and window flow, respectively:

$$h_o = \gamma(F_c h_x + 2F_w h_w) \tag{7.11}$$

where

F_c = fraction of tubes in cross flow
F_w = fraction of tubes in one baffle window
γ = empirical correction factor

Calculation of F_c and F_w is discussed in Chapter 6. Both h_x and h_w are obtained from the ideal tube bank correlation using appropriate Reynolds numbers given below. The factor γ accounts for departures from ideal tube bank conditions. An additional correction factor, not shown in Equation (7.11), is used to account for differences between the central baffle spaces and the end spaces [4].

Each of the shell-side streams has a different degree of effectiveness for heat transfer, with the cross-flow stream (B) being the most effective and the shell-to-baffle leakage stream (E) being the least effective. The effectiveness of each stream for heat transfer in cross flow and in the windows is accounted for by means of empirically determined effectiveness factors, ϵ_{xj} and ϵ_{wj}. These factors are incorporated in the Reynolds numbers used to calculate h_x and h_w. The Reynolds numbers, designated $Re_{\epsilon x}$ and $Re_{\epsilon w}$, respectively, are computed as follows [4]:

$$Re_{\epsilon x} = \frac{D_o}{S_m \mu}(\epsilon_{xA} F_c \dot{m}_A + \epsilon_{xB} \dot{m}_B + \epsilon_{xC} \dot{m}_C + \epsilon_{xE} \dot{m}_E + \epsilon_{xF} \dot{m}_F) \tag{7.12}$$

$$Re_{\epsilon w} = \frac{D_o}{S_w \mu}(2\epsilon_{wA} F_w \dot{m}_A + \epsilon_{wB} \dot{m}_B + \epsilon_{wC} \dot{m}_C + \epsilon_{wE} \dot{m}_E + \epsilon_{wF} \dot{m}_F) \tag{7.13}$$

Equations (7.11) to (7.13) contain a total of eleven empirical parameters, and Equation (7.8) contains two additional parameters, α and β. Furthermore, the correlations for the flow resistance coefficients contain a number of constants that were back-calculated from the HTRI data bank. With this many adjustable parameters, it is not surprising that the HTRI model provides a good representation of the experimental data for shell-side heat transfer and pressure drop. Comparison of model predictions with experimental data and with predictions of other methods is given in Ref. [4].

7.6 Temperature Profile Distortion

Since each of the shell-side streams has a different effectiveness for heat transfer, and since mixing of these streams in the shell is incomplete, each stream exhibits a different temperature profile along the exchanger (see **Figure 7.2**). The cross-flow stream is the most effective for heat transfer, and therefore its temperature profile lies closest to that of the tube-side fluid, resulting in the lowest driving force for heat transfer. The least effective stream for heat transfer is the shell-to-baffle leakage stream, which also experiences the least amount of mixing with the other streams. Hence, it experiences the largest driving force for heat transfer. The net result of these different temperature profiles is that the mean temperature difference in the exchanger can be less than that calculated using

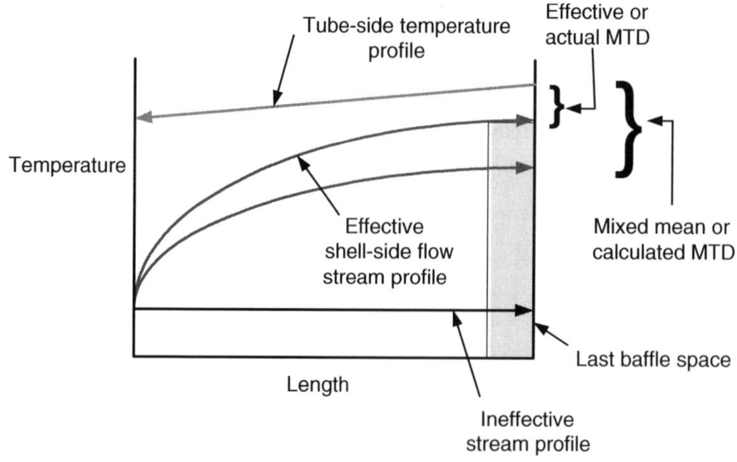

FIGURE 7.2 Effect of temperature profile distortion on mean temperature difference.

the logarithmic mean temperature difference (LMTD) correction factor, which does not take account of these effects. This phenomenon is referred to as temperature profile distortion.

An early method to account for this profile distortion is attributed to Whistler [6] and later Fisher and Parker [7]. They realized that thermal performance can be severely reduced if the change in shell-side temperature is much greater than the minimum temperature approach. A distortion function, δ, was proposed to reduce the mean temperature difference based on the traditional mixed mean shell-side temperature:

$$\Delta T_{mean} = \delta F(\Delta T_{\ln})_{cf} \tag{7.14}$$

where F is the LMTD correction factor and the value of δ ranges from 0 to 1.0. The overall rate of heat transfer in the exchanger is then calculated as:

$$q = UA\delta F(\Delta T_{\ln})_{cf} \tag{7.15}$$

These early analyses did not have the benefit of a complete stream analysis and the effects of bypassing were estimated based on parameters such as number of baffles and fluid viscosity. Palen and Taborek [4] proposed an improvement in the calculation of the distortion function by considering the Reynolds number of the B-stream and the E-stream flow fraction. An empirical correlation was developed for δ of the form [4]:

$$\delta = \delta\left(\frac{T_b - t_a}{T_b - T_a}, \frac{\dot{m}_E}{\dot{m}_o}, Re_B\right) \tag{7.16}$$

where

T_a = inlet temperature of shell-side fluid
T_b = outlet temperature of shell-side fluid
t_a = inlet temperature of tube-side fluid

For the reasons indicated above, the shell-to-baffle leakage stream (E) is the most important for temperature profile distortion effects, and the value of δ decreases as the flow fraction of this stream increases. The extent of mixing increases as the Reynolds number increases, and δ approaches 1.0 for large values of Re_B. The least amount of mixing occurs in laminar flow. At low Reynolds numbers, δ also decreases as the temperature difference ratio in Equation (7.16) decreases [4].

Fortunately, for properly designed exchangers with recommended shell-to-baffle clearances, δ should be close to 1.0 for turbulent flow and greater than 0.85 for laminar flow [4]. Nevertheless, values of δ ranged from about 0.4 to 1.0 for the HTRI test units, "none of which was of extreme design" [4]. A value of δ in the lower end of this range, if not accounted for in the design procedure, could result in severe under-design of the exchanger. Conditions favoring high leakage flow rates and low Reynolds numbers (hence, low δ) are high shell-side fluid viscosity and small baffle spacing. Thus, care must be taken in the design of viscous liquid coolers such as vacuum residue coolers in oil refineries [8]. The same is true for exchangers operating in laminar flow with a close temperature approach between streams. The effect of temperature profile distortion can be mitigated by using multiple shells in series because the shell-side fluid becomes well mixed as it exits one shell and enters the next.

The following example, due to Mukherjee [8], illustrates the effect of temperature profile distortion.

Example 7.1

In the design of a naphtha cooler for an oil refinery, the initial configuration of the exchanger was under surfaced by about 21%. The tube bundle consisted of 188 tubes, 0.75-in. OD, 20 ft long on 1.25-in. square pitch. Single segmental baffles were used with a cut of 21% and a spacing of 140 mm, corresponding to $B/D_s = 0.28$. The initial configuration was subsequently modified by progressively increasing the baffle spacing to a maximum of 210 mm while keeping all other design parameters fixed. Results of the calculations are shown in the table below [8]. Analyze and explain these results.

Parameter	Case 1	Case 2	Case 3	Case 4	Case 5
Baffle spacing (mm)	140	150	175	190	210
B/D_s	0.28	0.30	0.35	0.38	0.42
Stream-flow fraction					
A	0.189	0.173	0.163	0.154	0.143
B	0.463	0.489	0.506	0.521	0.539
C	0.109	0.113	0.116	0.118	0.121

(*Continued*)

—cont'd

Parameter	Case 1	Case 2	Case 3	Case 4	Case 5
E	0.239	0.225	0.215	0.207	0.196
F	0	0	0	0	0
h_o (Btu/h·ft^2·°F)	126	117	115	113	105
ΔP_o (psi)	0.48	0.41	0.38	0.37	0.33
U_D (Btu/h·ft^2·°F)	77.8	74.2	73.5	72.5	69.2
ΔT_{mean} (°F)	24.7	28.6	30.4	31.6	32.8
Under-design (%)	21.1	12.8	8.26	5.73	6.61

Solution

As expected, increasing the baffle spacing reduces both the shell-side pressure drop and heat-transfer coefficient. Nevertheless, the performance of the exchanger improves significantly up to a baffle spacing of 190 mm, beyond which it begins to deteriorate. Therefore, a baffle spacing of about 190 mm, corresponding to $B/D_s = 0.38$, is optimal for this configuration.

But how can the rate of heat transfer increase when the heat-transfer coefficient decreases? The answer is that the mean driving force for heat transfer, ΔT_{mean}, increases from 24.7°F to 31.6°F, an increase of nearly 28%, as the baffle spacing is increased from 140 mm to 190 mm. This effect can be explained in terms of temperature profile distortion. Note that as the baffle spacing is increased, the flow rates of the leakage streams (Streams A and E) decrease, while the flow rate of the cross-flow stream (Stream B) increases. As a result, there is less temperature profile distortion at the larger baffle spacings. Consequently, the temperature profile distortion factor, δ, and hence, ΔT_{mean}, increases as the baffle spacing is increased. For the conditions of this example, the increase in ΔT_{mean} outweighs the decrease in h_o for baffle spacing up to 190 mm.

Note that these results could not be predicted by the Delaware or Simplified Delaware methods. Only a version of the Stream Analysis method that accounts for temperature profile distortion is capable of generating such results. In fact, the HTRI software package was used to perform the calculations for this example [8].

7.7 Good Design Practice

HTRI's consortium meets on a regular basis to discuss good design practice. The design guidelines for stream analysis have evolved over the years as lessons have been learned and additional research performed. The relative distribution of flow streams has been shown to be a useful guide for shell-side design. **Tables 7.1 and 7.2** give suggested minimum B-stream and maximum bypass stream-flow fractions for TEMA designs.

The suggested flow fractions are helpful when optimizing performance because, if the B-stream flow fraction exceeds the values shown, the designer need not make additional changes to further increase the B-stream, and other details of the design can be attended to.

It is important to note that there is no consensus on the value of the shell-side Reynolds number that marks the threshold between turbulent and laminar flow. For the purpose of using **Tables 7.1 and 7.2**, however, Reynolds numbers greater than 800 can be considered turbulent and Reynolds numbers less than 800 can be considered laminar.

7.8 The Wills-Johnston Method

Wills and Johnston [5] published a simplified set of correlations for the flow resistance coefficients that permit a relatively simple solution of the hydraulic equations. Iteration is required only for triangular tube layouts, and even then the computational

TABLE 7.1 Typical Flow Fractions for Single-Phase Turbulent Applications

	Flow Fraction			
Shell Type	A	B^*	C	E
Fixed or U-tube	0.20	0.55	0.05	0.20
Split-ring 3 strips	0.15	0.45	0.20	0.20

*B-stream flow fractions greater than shown are acceptable.

TABLE 7.2 Typical Flow Fractions for Single-Phase Laminar Applications

	Flow Fraction			
Shell Type	*A*	*B**	*C*	*E*
Fixed or U-tube	0.05	0.50	0.05	0.40
Split ring 3 strips	0.05	0.30	0.25	0.40

**B*-stream flow fractions greater than shown are acceptable.

requirements are minimal. The method is available as an option (DPSMETHOD = Stream) for calculating shell-side pressure drop in the HEXTRAN software package. The basic method described here is applicable to exchangers with type E or J shells, single segmental baffles and single-phase flow in the shell.

7.8.1 Streams and Flow Areas

The streams are the same as in the HTRI method except that the *C* and *F* streams are combined to form a single bypass stream. Denoting this stream as *CF*, the streams are:

A = tube-to-baffle leakage
B = cross flow
CF = combined bypass flow
E = shell-to-baffle leakage

The cross-flow and window-flow areas are the same as in the Delaware method; namely, S_m and S_w, respectively. The bypass flow area is the sum of the flow areas for the *C* and *F* streams. Denoting this area as S_{bp}, we have:

$$S_{bp} = B(D_s - D_{otl} + N_p\delta_p) \tag{7.17}$$

where

B = central baffle spacing
D_{otl} = outer tube limit diameter
N_p = number of tube pass partitions aligned with the cross-flow direction
δ_p = pass partition clearance

The flow areas for the two leakage streams are intentionally over estimated by neglecting the effect of the baffle cut. This is done to improve the agreement between calculated and measured pressure drop. Denoting these modified flow areas by S_t (tube-to-baffle leakage) and S_s (shell-to-baffle leakage), the equations are:

$$S_t = n_t \pi D_o \delta_{tb} \tag{7.18}$$

$$S_s = \pi D_s\, \delta_{sb} \tag{7.19}$$

where

n_t = number of tubes in bundle
δ_{tb} = tube-to-baffle clearance
δ_{sb} = shell-to-baffle clearance

The ideal tube bank correlation used by Wills and Johnston involves an additional area that will be denoted as S_{BW}. It is calculated as follows:

$$S_{BW} = \frac{BD_{otl}^2(\pi - \theta_{otl} + \sin\theta_{otl})}{4D_s\,(1 - 2B_c)} - BN_p\delta_p \tag{7.20}$$

where

$$\theta_{otl} = 2\cos^{-1}\left[\frac{D_s\,(1 - 2B_c)}{D_{otl}}\right] \tag{7.21}$$

Here, B_c is the fractional baffle cut and θ_{otl} is expressed in radians.

7.8.2 Pressure Drops and Stream-Flow Rates

The pressure drop for each stream is written as:

$$\Delta P_j = \xi_j \dot{m}_j^2 \quad j = A, B, CF, E \tag{7.22}$$

Comparison with Equation (7.1) shows that:

$$\xi_j = \frac{K_j}{2\rho g_c S_j^2 \phi} \tag{7.23}$$

The viscosity correction factor, ϕ, is taken to be 1.0 in this method. The window pressure drop is written in the same format:

$$\Delta P_w = \xi_w \dot{m}_w^2 \tag{7.24}$$

Now the flow rate in the window is the sum of the flow rates of the cross-flow and bypass streams, i.e.,

$$\dot{m}_w = \dot{m}_B + \dot{m}_{CF} \tag{7.25}$$

From Equation (7.22), the flow rates on the right-hand side can be expressed as:

$$\dot{m}_B = (\Delta P_B/\xi_B)^{1/2} \tag{7.26}$$

$$\dot{m}_{CF} = (\Delta P_{CF}/\xi_{CF})^{1/2} \tag{7.27}$$

Referring to **Figure 7.1**, it is seen that:

$$\Delta P_B = \Delta P_{CF} = \Delta P_x \tag{7.28}$$

Upon substitution, Equation (7.25) becomes:

$$\dot{m}_w = \Delta P_x^{1/2}\left(\xi_B^{-1/2} + \xi_{CF}^{-1/2}\right) \tag{7.29}$$

Solving for ΔP_x gives:

$$\Delta P_x = \left(\xi_B^{-1/2} + \xi_{CF}^{-1/2}\right)^{-2} \dot{m}_w^2 = \xi_x \dot{m}_w^2 \tag{7.30}$$

where

$$\xi_x \equiv \left(\xi_B^{-1/2} + \xi_{CF}^{-1/2}\right)^{-2} \tag{7.31}$$

From the balanced pressure drop conditions we have:

$$\Delta P_A = \Delta P_E = \Delta P_x + \Delta P_w \tag{7.3}$$

Denote this common pressure drop by ΔP_y and substitute for ΔP_w and ΔP_x from Equations (7.24) and (7.30) to obtain:

$$\Delta P_y = (\xi_w + \xi_x)\dot{m}_w^2 = \xi_y \dot{m}_w^2 \tag{7.32}$$

where

$$\xi_y \equiv \xi_w + \xi_x \tag{7.33}$$

Now the total mass flow rate, $\dot{m}_o$, is given by:

$$\dot{m}_o = \dot{m}_A + \dot{m}_E + \dot{m}_w \tag{7.34}$$

Each flow rate on the right-hand side of this equation can be expressed in terms of ΔP_y to obtain:

$$\dot{m}_o = \Delta P_y^{1/2}\left(\xi_A^{-1/2} + \xi_E^{-1/2} + \xi_y^{-1/2}\right) \tag{7.35}$$

Solving for ΔP_y gives:

$$\Delta P_y = \left(\xi_A^{-1/2} + \xi_E^{-1/2} + \xi_y^{-1/2}\right)^{-2} \dot{m}_o^2 = \xi_o \dot{m}_o^2 \tag{7.36}$$

where

$$\xi_o \equiv \left(\xi_A^{-1/2} + \xi_E^{-1/2} + \xi_y^{-1/2}\right)^{-2} \tag{7.37}$$

Notice that Equation (7.36) gives the total pressure drop for one baffle space and one baffle window in terms of the flow resistances.

Next, Equations (7.36) and (7.22) are used to obtain the flow fractions of the two leakage streams. We have:

$$\Delta P_A = \Delta P_y$$

$$\xi_A \dot{m}_A^2 = \xi_o \dot{m}_o^2$$

Therefore,

$$\dot{m}_A/\dot{m}_o = (\xi_o/\xi_A)^{1/2} \tag{7.38}$$

Similarly,

$$\dot{m}_E/\dot{m}_o = (\xi_o/\xi_E)^{1/2} \tag{7.39}$$

In the same manner, the window-flow fraction is obtained from Equations (7.32) and (7.36):

$$\Delta P_y = \xi_y \dot{m}_w^2 = \xi_o \dot{m}_o^2$$

$$\dot{m}_w/\dot{m}_o = \left(\xi_o/\xi_y\right)^{1/2} \tag{7.40}$$

Finally, Equations (7.30) and (7.40) are used to obtain the flow fractions of the cross-flow and bypass streams. For the cross-flow stream we have:

$$\Delta P_B = \Delta P_x$$

$$\xi_B \dot{m}_B^2 = \xi_x \dot{m}_w^2$$

Substituting for $\dot{m}_w$ from Equation (7.40) gives:

$$\xi_B \dot{m}_B^2 = \xi_x \left(\xi_o/\xi_y\right) \dot{m}_o^2 \tag{7.41}$$

Therefore,

$$\frac{\dot{m}_B}{\dot{m}_o} = \left(\frac{\xi_x \xi_o}{\xi_B \xi_y}\right)^{1/2} \tag{7.42}$$

Similarly, for the bypass stream we have:

$$\Delta P_{CF} = \Delta P_x$$

$$\xi_{CF} \dot{m}_{CF}^2 = \xi_x \dot{m}_w^2 = \xi_x \left(\xi_o/\xi_y\right) \dot{m}_o^2$$

$$\frac{\dot{m}_{CF}}{\dot{m}_o} = \left(\frac{\xi_x \xi_o}{\xi_{CF} \xi_y}\right)^{1/2} \tag{7.43}$$

All the stream-flow rates have now been expressed explicitly as simple functions of the flow resistances alone. Therefore, once the resistances are known, the stream-flow rates and pressure drops can be easily computed.

7.8.3 Flow Resistances

The resistances of the two leakage streams, the bypass stream, and the window flow are assumed to be independent of the stream flow rates. Therefore, these resistances can be treated as constants for a given rating problem. Justification for these assumptions is given in Ref. [5]. Only the cross-flow resistance need be treated as a variable in the calculations. It is obtained from an ideal tube bank friction factor correlation and is a function of the cross-flow Reynolds number.

The Cross-Flow Resistance

The cross-flow resistance is given by the following equation [5]:

$$\xi_B = \frac{4aD_oD_sD_V(1-2B_c)(P_T-D_o)^{-3}\left(\dfrac{\dot{m}_B D_o}{\mu S_{BW}}\right)^{-b}}{2\rho g_c S_{BW}^2} \tag{7.44}$$

where

$a = 0.061$, $b = 0.088$ for square and rotated-square pitch
$a = 0.450$, $b = 0.267$ for triangular pitch

$$D_V = \left(\Omega_1 P_T^2 - D_o^2\right)/D_o \tag{7.45}$$

$\Omega_1 = 1.273$ for square and rotated-square pitch
$= 1.103$ for triangular pitch

Note that the Reynolds number in Equation (7.44) is based on S_{BW} rather than the cross-flow area, S_m, that is used in most ideal tube bank correlations. Also, notice that the Reynolds number dependence is very weak for square and rotated-square tube layouts. For this reason, iteration is not required to determine the value of ξ_B for these layouts. For these cases, $\dot{m}_B$ can be set to 0.5 $\dot{m}_o$ in Equation (7.44). For triangular pitch, iteration is required to determine the values of $\dot{m}_B$ and ξ_B using Equations (7.42) and (7.44). A starting value of $\dot{m}_B = 0.5\,\dot{m}_o$ is used in Equation (7.44) to calculate ξ_B, which is then used in equation (7.42) to calculate a new value of $\dot{m}_B$. Iteration is continued until successive values of $\dot{m}_B$ agree to within the desired accuracy. Convergence is usually rapid [5]; therefore, the use of a convergence acceleration method is not required.

The Bypass Flow Resistance

The bypass flow resistanceThe bypass flow resistance is computed as follows [5]:

$$\xi_{CF} = \frac{0.3164 D_s \left(\dfrac{1 - 2B_C}{\Omega_2 P_T}\right)\left(\dfrac{\dot{m}_o D_e}{\mu\, S_{bp}}\right)^{-0.025} + 2N_{ss}}{2\rho g_c S_{bp}^2} \tag{7.46}$$

where

N_{ss} = number of pairs of sealing strips
Ω_2 = 1.0 for square pitch
= 1.414 for rotated-square pitch
= 1.732 for triangular pitch
D_e = equivalent diameter for the bypass flow

The equivalent diameter is given by the following equation:

$$D_e = \frac{2S_{bp}}{D_s - D_{otl} + 2B + N_p(B + \delta_p)} \tag{7.47}$$

Due to the small exponent on the Reynolds number in Equation (7.46), little error is introduced in most cases by treating this term as a constant. The resulting simplified form of the equation is [9,10]:

$$\xi_{CF} = \frac{0.266 D_s \left(\dfrac{1 - 2B_c}{\Omega_2 P_T}\right) + 2N_{ss}}{2\rho g_c S_{bp}^2} \tag{7.48}$$

The Tube-To-Baffle Leakage Flow Resistance

The flow resistance for the tube-to-baffle leakage stream is given by the following equation [5], which is based on data for flow through eccentric orifices:

$$\xi_A = \frac{(2B_t/\delta_{tb})\left[0.0035 + 0.197\left(\dfrac{\dot{m}_o \delta_{tb}}{\mu S_t}\right)^{-0.42}\right] + 2.3(B_t/\delta_{tb})^{-0.177}}{2\rho g_c S_t^2} \tag{7.49}$$

where B_t is the baffle thickness. A simplified form of this equation can also be used [9,10]:

$$\xi_A = \frac{0.036 B_t/\delta_{tb} + 2.3(B_t/\delta_{tb})^{-0.177}}{2\rho g_c S_t^2} \tag{7.50}$$

The Shell-To-Baffle Leakage Flow Resistance

The flow resistance for the shell-to-baffle leakage stream is given by an equation analogous to (7.49) [5]:

$$\xi_E = \frac{(2B_t/\delta_{sb})\left[0.0035 + 0.197\left(\dfrac{\dot{m}_o \delta_{sb}}{\mu S_s}\right)^{-0.42}\right] + 2.3(B_t/\delta_{sb})^{-0.177}}{2\rho g_c S_s^2} \tag{7.51}$$

The simplified form analogous to Equation (7.50) is [9,10]:

$$\xi_E = \frac{0.036 B_t/\delta_{sb} + 2.3(B_t/\delta_{sb})^{-0.177}}{2\rho g_c S_s^2} \tag{7.52}$$

The Window-Flow Resistance

The window-flow resistance is given by the following simple expression that is independent of the flow rate [5]:

$$\xi_w = \frac{1.9 \exp\,(0.6856\; S_w/S_m)}{2\rho g_c S_w^2} \tag{7.53}$$

7.8.4 Inlet and Outlet Baffle Spaces

The pressure-drop calculation in the end baffle spaces is modified to account for differences in baffle spacing, number of tube rows crossed and flow distribution. The flow rate in the end spaces is taken as the arithmetic mean of the window and total flow rates, i.e.,

$$\dot{m}_e = 0.5\,(\dot{m}_o + \dot{m}_w) \tag{7.54}$$

where the subscript "e" denotes end baffle space, either inlet or outlet. The cross-flow resistance, ξ_e, in the end spaces is estimated by the following equation [5]:

$$\xi_e = 0.5\xi_x\,(B/B_e)^2\{1 + D_{otl}/[D_s(1 - 2B_c)]\} \tag{7.55}$$

where B_e represents either the inlet or outlet baffle spacing.

Half the pressure drop in the end baffle windows is allocated to the end baffle spaces. The flow resistance, ξ_{we}, in the end windows is calculated as follows [5]:

$$\xi_{we} = \frac{1.9 \exp\,[0.6856 S_w B/(S_m B_e)]}{2\rho g_c\; S_w^2} \tag{7.56}$$

The pressure drop in the inlet or outlet baffle space is then given by:

$$\Delta P_e = \xi_e \dot{m}_e^2 + 0.5\xi_{we}\dot{m}_w^2 \tag{7.57}$$

7.8.5 Total Shell-Side Pressure Drop

The total shell-side pressure drop is obtained by summing the pressure drops in all the central baffle spaces and windows, the pressure drops in the inlet and outlet baffle spaces, and the pressure drops in the nozzles. Thus,

$$\Delta P_o = (n_b - 1)\;\Delta P_y + \Delta P_{in} + \Delta P_{out} + \Delta P_n \tag{7.58}$$

Here, n_b is the number of baffles, and ΔP_{in} and ΔP_{out} are computed from Equation (7.57).

In order to improve the agreement between calculated and measured pressure drop at low Reynolds numbers, Wills and Johnston [5] modified Equation (7.58) by introducing an empirical correction factor. The cross-flow Reynolds number based on the cross-flow area, S_m, is first calculated:

$$Re_B = \frac{D_o \dot{m}_B}{\mu S_m} \tag{7.59}$$

For $Re_B < 1000$, the flow in some or all of the streams may not be fully turbulent, thus necessitating use of a correction factor. The correction factor, ψ, is computed as follows:

$$\begin{aligned} \psi &= 3.646 Re_B^{-0.1934} \qquad & Re_B < 1000 \\ &= 1.0 & Re_B \geq 1000 \end{aligned} \tag{7.60}$$

Equation (7.58) is then modified by multiplying the nozzle-to-nozzle pressure drop by ψ to obtain:

$$\Delta P_o = \psi\,\left[(n_b - 1)\;\Delta P_y + \Delta P_{in} + \Delta P_{out}\right] + \Delta P_n \tag{7.61}$$

Wills and Johnston do not recommend using their method for $Re_B < 300$. However, their data indicate that for $100 \leq Re_B < 300$, the worst errors are on the high (safe) side. Therefore, a reasonable lower limit for use of the method is $Re_B = 100$.

Example 7.2

Use the Wills-Johnston method to calculate the shell-side pressure drop for the final configuration of the kerosene/crude-oil exchanger of Example 5.1 and compare the results with those obtained previously using the Delaware and Simplified Delaware methods.

Solution

(a) The following data are obtained from Examples 5.1 and 6.1:

Exchanger type = *AES*	$B_c = 0.2$
Shell-side fluid = kerosene	$n_b = 42$

(Continued)

—cont'd

Kerosene viscosity = 0.97 lbm/ft • h	$N_{ss}/N_c = 0.1$
Kerosene density = 49 lbm/ft^3	$\delta_{tb} = 0.015748$ in.
$\dot{m}_o = 45{,}000$ lb/h	$\delta_{sb} = 0.09952$ in.
$D_s = 19.25$ in.	$D_{otl} = 17.91$ in.
$n_t = 124$	$D_{ctl} = 16.91$ in.
$n_p = 4$	$S_m = 0.1262$ ft^2
$D_o = 1.0$ in.	$S_w = 0.2189$ ft^2
$P_T = 1.25$ in. (square)	$S_b = 0.03583$ ft^2
$B = B_{in} = B_{out} = 3.85$ in.	$N_c = 9.24$

It is assumed that all tube-pass partitions are aligned normal to the cross-flow direction so that the flow rate of the F stream is zero. Also, the safety factor (for heat transfer) is included in the value of δ_{sb} from Example 6.1.

(b) Calculate the flow areas.

The bypass flow area is given by Equation (7.17) with $N_p = 0$ (since the tube pass partitions are normal to the cross-flow direction). Thus,

$$S_{bp} = B\,(D_s - D_{otl}) = S_b = 0.03583 \text{ ft}^2$$

The modified leakage flow areas are calculated from Equations (7.18) and (7.19):

$$S_t = n_t \pi D_o \delta_{tb} = 124\pi \times 1.0 \times 0.015748$$

$$S_t = 6.13475 \text{ in.}^2 = 0.04260 \text{ ft}^2$$

$$S_s = \pi D_s \delta_{sb} = \pi \times 19.25 \times 0.09952$$

$$S_s = 6.01854 \text{ in.}^2 = 0.04180 \text{ ft}^2$$

To find the area S_{BW}, angle θ_{otl} is first calculated using Equation (7.21):

$$\theta_{otl} = 2\cos^{-1}\left[\frac{D_s(1-2B_c)}{D_{otl}}\right]$$

$$= 2\cos^{-1}\left[\frac{19.25\,(1-2\times 0.2)}{17.91}\right]$$

$$\theta_{otl} = 1.73983 \text{ radians}$$

Equation (7.20) is then used to compute S_{BW}.

$$S_{BW} = \frac{BD_{otl}^2\,(\pi - \theta_{otl} + \sin\theta_{otl})}{4D_s(1-2B_c)} - BN_p\delta_p$$

$$= \frac{3.85(17.91)^2(\pi - 1.73983 + \sin 1.73983)}{4\times 19.25\,(1-2\times 0.2)} - 0$$

$$S_{BW} = 63.8198 \text{ in}^2 = 0.44319 \text{ ft}^2$$

The cross-flow area, S_m, and window-flow area, S_w, were obtained from Example 6.1, and therefore are not calculated here.

(c) Calculate the flow resistances.

(i) The cross-flow resistance is given by Equation (7.44) with $a = 0.061$ and $b = 0.088$ for square pitch. The value of D_V is first calculated from Equation (7.45):

$$D_V = \left(\Omega_1 P_T^2 - D_o^2\right)/D_o = \left[1.273(1.25)^2 - (1.0)^2\right]/1.0$$
$$D_V = 0.98906 \text{ in.}$$

Setting $\dot{m}_B$ to $0.5\dot{m}_o = 22{,}500$ lb/h in Equation (7.44), we have:

$$\xi_B = \frac{4aD_oD_sD_V\,(1-2B_c)\,(P_T-D_o)^{-3}\left(\dfrac{\dot{m}_B D_o}{\mu S_{BW}}\right)^{-b}}{2\rho g_c S_{BW}^2}$$

$$= \frac{4\times 0.061\times 1.0\times 19.25\times 0.98906\,(1-2\times 0.2)\,(0.25)^{-3}\left(\dfrac{22{,}500\times 1.0/12}{0.97\times 0.44319}\right)^{-0.088}}{2\times 49\times 32.174\,(0.44319)^2}$$

$$\xi_B = 0.13778 \text{ lbf}\cdot\text{s}^2/\text{lbm}^2\cdot\text{ft}^2$$

(ii) The bypass flow resistance is calculated using the simplified expression given by Equation (7.48). The number of pairs of sealing strips is first determined as follows:

$$N_{ss} = (N_{ss}/N_c)N_c = 0.1 \times 9.24 = 0.924 \rightarrow 1$$

Equation (7.48) is then used with $\Omega_2 = 1.0$ for square pitch:

$$\xi_{CF} = \frac{0.266 D_s \left(\dfrac{1-2B_c}{\Omega_2 P_T}\right) + 2N_{ss}}{2\rho g_c S_{bp}^2}$$

$$= \frac{0.266 \times 19.25 \left(\dfrac{1-2\times 0.2}{1.0\times 1.25}\right) + 2\times 1}{2\times 49\times 32.174\,(0.03583)^2}$$

$$\xi_{CF} = 1.10129 \text{ lbf}\cdot\text{s}^2/\text{lbm}^2\cdot\text{ft}^2$$

(iii) The tube-to-baffle leakage flow resistance is obtained from Equation (7.50). The baffle thickness, B_t, is 0.1875 in. from **Table 5.2**:

$$\xi_A = \frac{0.036 B_t/\delta_{tb} + 2.3\,(B_t/\delta_{tb})^{-0.177}}{2\rho g_c S_t^2}$$

$$= \frac{0.036\times 0.1875/0.015748 + 2.3\,(0.1875/0.015748)^{-0.177}}{2\times 49\times 32.174\,(0.04260)^2}$$

$$\xi_A = 0.33419 \text{ lbf}\cdot\text{s}^2/\text{lbm}^2\cdot\text{ft}^2$$

(iv) Equation (7.52) is used to calculate the shell-to-baffle leakage flow resistance:

$$\xi_E = \frac{0.036 B_t/\delta_{sb} + 2.3\,(B_t/\delta_{sb})^{-0.177}}{2\rho g_c S_s^2}$$

$$= \frac{0.036\times 0.1875/0.09952 + 2.3(0.1875/0.09952)^{-0.177}}{2\times 49\times 32.174\,(0.04180)^2}$$

$$\xi_E = 0.38552 \text{ lbf}\cdot\text{s}^2/\text{lbm}^2\cdot\text{ft}^2$$

(v) The window-flow resistance is computed from Equation (7.53):

$$\xi_w = \frac{1.9\exp\,(0.6856 S_w/S_m)}{2\rho\, g_c\, S_w^2}$$

$$= \frac{1.9\exp\,(0.6856\times 0.2189/0.1262)}{2\times 49\times 32.174\,(0.2189)^2}$$

$$\xi_w = 0.041304 \text{ lbf}\cdot\text{ s}^2/\text{lbm}^2\cdot\text{ft}^2$$

(vi) Equation (7.31) is used to calculate ξ_x:

$$\xi_x = \left(\xi_B^{-1/2} + \xi_{CF}^{-1/2}\right)^{-2} = \left[(0.13778)^{-1/2} + (1.10129)^{-1/2}\right]^{-2}$$

$$\xi_x = 0.075186 \text{ lbf}\cdot\text{s}^2/\text{lbm}^2\cdot\text{ ft}^2$$

(vii) Equation (7.33) is used to calculate ξ_y:

$$\xi_y = \xi_w + \xi_x = 0.041304 + 0.075186$$
$$\xi_y = 0.116490 \text{ lbf}\cdot\text{s}^2/\text{lbm}^2\cdot\text{ft}^2$$

(viii) Finally, Equation (7.37) is used to calculate ξ_o:

$$\xi_o = \left(\xi_A^{-1/2} + \xi_E^{-1/2} + \xi_y^{-1/2}\right)^{-2} = \left[(0.33419)^{-1/2} + (0.38552)^{-1/2} + (0.116490)^{-1/2}\right]^{-2}$$

$$\xi_o = 0.025434 \text{ lbf}\cdot\text{s}^2/\text{lbm}^2\cdot\text{ft}^2$$

(d) Calculate the flow fractions.
The flow resistances are now used to calculate the flow fractions, starting with the cross-flow stream. From Equation (7.42),

$$\dot{m}_B/\dot{m}_o = \left(\frac{\xi_x\xi_o}{\xi_B\xi_y}\right)^{1/2} = \left(\frac{0.075186\times 0.025434}{0.13778\times 0.116490}\right)^{1/2} = 0.345$$

Since the tube layout is square, iteration is not required, and this value is accepted as final. In a similar manner, Equation (7.43) is used to calculate the bypass flow fraction:

$$\dot{m}_{CF}/\dot{m}_o = \left(\frac{\xi_x\xi_o}{\xi_{CF}\xi_y}\right)^{1/2} = \left(\frac{0.075186\times 0.025434}{1.10129\times 0.116490}\right)^{1/2} = 0.122$$

The leakage flow fractions are obtained from Equations (7.38) and (7.39):

$$\dot{m}_A/\dot{m}_o = (\xi_o/\xi_A)^{1/2} = (0.025434/0.33419)^{1/2} = 0.276 \qquad \dot{m}_E/\dot{m}_o = (\xi_o/\xi_E)^{1/2} = (0.025434/0.38552)^{1/2} = 0.257$$

Note that the sum of the flow fractions is 1.000.
Finally, the window-flow fraction is given by Equation (7.40):

$$\dot{m}_w/\dot{m}_o = (\xi_o/\xi_y)^{1/2} = (0.025434/0.116490)^{1/2} = 0.467$$

Notice that the window-flow fraction equals the sum of the cross-flow and bypass-flow fractions.

(e) Calculate ΔP_y.
The total pressure drop in one baffle space and one baffle window is given by Equation (7.36):

$$\Delta P_y = \xi_o \dot{m}_o^2 = 0.025434\,(45,000/3600)^2$$
$$\Delta P_y = 3.974 \text{ lbf/ft}^2$$

Note: The factor of 3600 converts the flow rate to lbm/s.

(f) Calculations for the end baffle spaces.
The window-flow rate is obtained from the window-flow fraction:

$$\dot{m}_w = 0.467\,\dot{m}_o = 0.467\,(45,000/3600) = 5.8375 \text{ lbm/s}$$

The flow rate in the end baffle spaces is given by Equation (7.54):

$$\dot{m}_e = 0.5\,(\dot{m}_o + \dot{m}_w) = 0.5\,(45,000/3600 + 5.8375)$$
$$\dot{m}_e = 9.16875 \text{ lbm/s}$$

The flow resistance in the end spaces is calculated from Equation (7.55) with $B_e = B$:

$$\xi_e = 0.5\xi_x(B/B_e)^2\{1 + D_{otl}/[D_s(1 - 2B_c)]\}$$
$$= 0.5 \times 0.075186\,(1.0)^2\{1 + 17.91/[19.25(1 - 2 \times 0.2)]\}$$
$$\xi_e = 0.095887 \text{ lbf} \cdot \text{s}^2/\text{lbm}^2 \cdot \text{ft}^2$$

The flow resistance, ξ_{we}, in the end baffle windows is given by Equation (7.56). However, since $B_e = B$, this equation reduces to Equation (7.53) for ξ_w. Hence,

$$\xi_{we} = \xi_w = 0.041304 \text{ lbf} \cdot \text{s}^2/\text{lbm}^2 \cdot \text{ft}^2$$

Now the pressure drop in each end space is calculated from Equation (7.57):

$$\Delta P_e = \xi_e \dot{m}_e^2 + 0.5\xi_{we}\,\dot{m}_w^2$$
$$= 0.095887\,(9.16875)^2 + 0.5 \times 0.041304\,(5.8375)^2$$
$$\Delta P_e = 8.765 \text{ lbf/ft}^2$$

(g) Calculate shell-side pressure drop.
First, the cross-flow Reynolds number is calculated according to Equation (7.59):

$$Re_B = \frac{D_o \dot{m}_B}{\mu S_m} = \frac{(1.0/12)\,(0.345 \times 45,000)}{0.97 \times 0.1262} = 10,569$$

Since the Reynolds number is greater than 1000, the correction factor, ψ, is equal to 1.0 from Equation (7.60). The total shell-side pressure drop is then obtained from Equation (7.61).

$$\Delta P_o = \psi\,[(n_b - 1)\,\Delta P_y + \Delta P_{in} + \Delta P_{out}] + \Delta P_n$$
$$= 1.0[(42 - 1) \times 3.974 + 8.765 + 8.765] + \Delta P_n$$
$$\Delta P_o = 180.5 + \Delta P_n \text{ lbf/ft}^2 = 1.25 + \Delta P_n \text{ psi}$$

From Example 5.1, the nozzle pressure drop is 0.20 psi. Hence,

$$\Delta P_o = 1.25 + 0.20 = 1.45 \text{ psi}$$

The following table compares this result with those obtained previously by other methods:

Method	ΔP_o (psi)
Wills–Johnston	1.45
Delaware	1.31

(*Continued*)

—cont'd

Method	ΔP_o (psi)
Simplified Delaware	2.2
HEXTRAN (DPSMETHOD = Bell)	2.10
HEXTRAN (DPSMETHOD = Stream)	2.11

All methods are consistent in predicting a small pressure drop for the exchanger. The Wills-Johnston value is within about 10% of the value found by the Delaware method. However, it differs by a large percentage from the value given by HEXTRAN (DPSMETHOD = Stream), even though the latter uses the Wills–Johnston method. The anomalies discussed in Example 6.1 may be partly responsible for the discrepancy. Also, the Wills–Johnston value would be somewhat higher without the safety factor in δ_{sb}.

The Wills–Johnston method predicts that the two leakage streams account for more than 50% of the total shell-side flow. The large amount of leakage is consistent with the result obtained by the Delaware method in Example 6.1. Recall that the leakage correction factor, J_L, was found to be less that 0.6, indicating abnormally high flow fractions for the leakage streams. Note in particular that the *E* stream, which is the least effective for heat transfer, accounts for approximately 25% of the total flow.

7.9 Computer Software

The HTRI software package, *Xchanger Suite,* is a research-based thermal performance analysis tool which has been validated by thousands of industrial exchangers designed, manufactured and operated by member companies. *Xchanger Suite* consists of the following modules:

- *Xist* for shell and tube exchangers, including a detailed tube layout program and screening analysis for tube vibration
- *Xace* for air coolers and economizers
- *Xphe* for plate-and-frame heat exchangers
- *Xpfe* for brazed plate-fin heat exchangers
- *Xspe* for spiral plate heat exchangers
- *Xhpe* for hairpin heat exchangers
- *Xjpe* for jacketed pipe heat exchangers
- *Xfh* for fired heaters
- *Xvib* for finite element vibration analysis of a single tube

The software contains pure-component physical property data for a large number of compounds, and includes a third-party thermodynamics package (VMGThermo). Hence, *Xchanger Suite* can be used on a stand-alone basis. It also interfaces with CAPE-OPEN compliant process simulator software such as Aspen Plus and HYSYS. CAPE-OPEN standards describe rules and interfaces for Computer Aided Process Engineering (CAPE) software. In addition, *Xchanger Suite* can operate as an embedded native application in selected process simulators such as Honeywell's UniSim Design and Invensys' PRO/II. An educational version of *Xchanger Suite* is available to universities for a nominal fee. This version includes only three of the nine modules, *Xace, Xist* and *Xphe.*

Xist utilizes incrementation for all TEMA shell types. The heat exchanger is first divided into increments along the tube length. For baffled exchangers, one increment is used for each baffle space unless a larger number is specified by the user. For non-baffled exchangers, 20 increments are used, with the exception of X-shells (10 increments) and K-shells (5 increments). Next, the exchanger is divided into rows (top to bottom) and sections (side to side). The program uses enough rows and sections so that each three-dimensional subdivision of the exchanger involves only one type of flow (window flow or cross flow) and one tube pass. Thermal and hydraulic calculations are performed for each subdivision using local values of temperatures, pressures and physical properties, and the individual results are integrated to obtain overall results that are displayed in the output report files.

Xist can be run in any one of three modes as specified on the Input Summary form shown in **Figure 7.3**. In rating mode, the user specifies inlet stream conditions, the exchanger geometry, and either the duty or the outlet temperature of one stream. The program then uses energy balances to calculate missing outlet temperatures and, if not given, the duty. Based on these temperatures and the given geometry, the pressure drops and the achievable duty are computed along with the percentage over-design. The latter is based on the computed achievable duty versus the specified (required) duty; it is calculated as $(U_D / U_{req} - 1) \times 100$. The program is most frequently used in this mode.

Simulation mode is similar to rating mode except that neither the duty nor the outlet temperatures are specified. The program calculates the actual outlet temperatures, pressure drops, and duty based on the given exchanger geometry.

In design mode, *Xist* finds the smallest exchanger that meets a given performance specification and pressure drop constraints. Parameters that can be automatically varied are:

- Shell type and diameter
- Number of shells in series and parallel

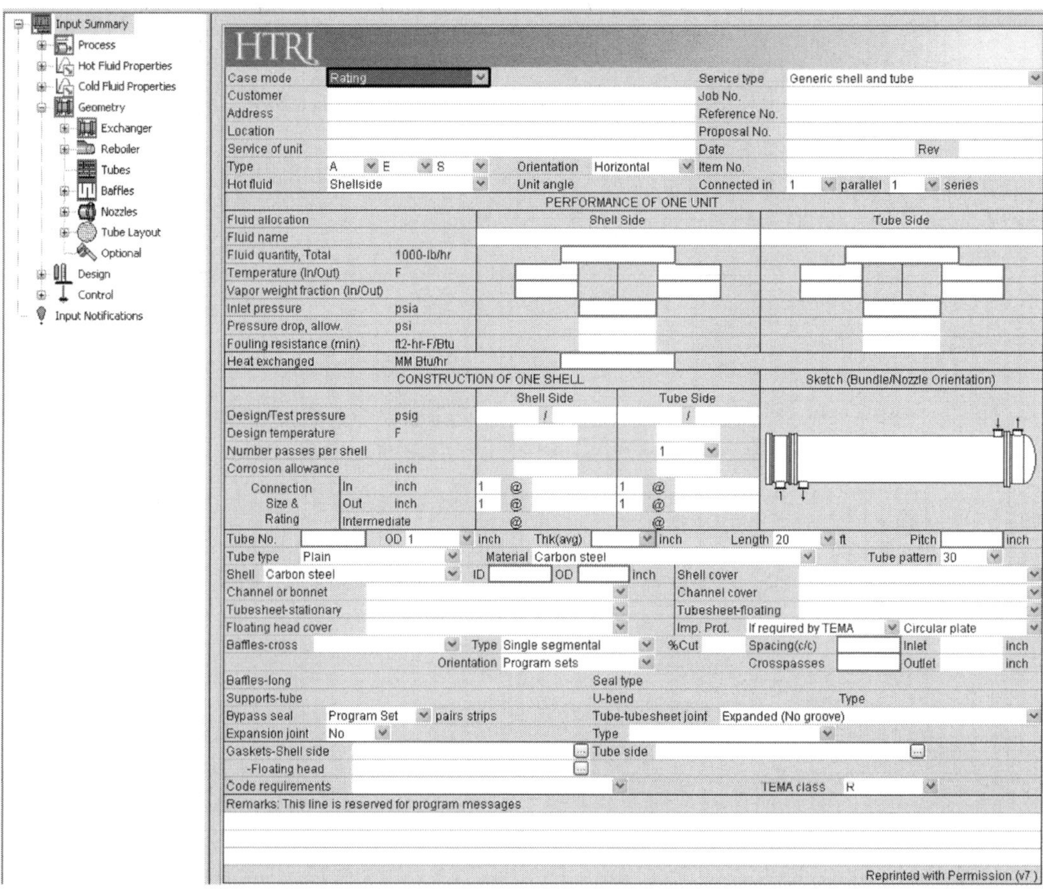

FIGURE 7.3 Input Summary screen in ***X*ist**, version 7.

- Number, diameter and length of tubes
- Number of tube passes
- Baffle type, cut, and spacing
- Tube pitch

The range over which each parameter is allowed to vary is user specified.

There are six major categories of input data for *Xist* as shown in the tree at the left of **Figure 7.3**. Key geometrical parameters and process conditions can be entered and/or modified on the Input Summary form. They can also be entered on the more detailed individual forms by expanding the tree at the left of the screen. Exchanger geometry can be specified in great detail using individual forms for the shell, tubes, baffles, nozzles, etc. However, many of the input parameters either have a default value or can be calculated by the program. Text boxes for required input items are outlined in red until a suitable value is entered.

In addition to geometry and process data, physical properties of the hot and cold fluids must be supplied. As in HEXTRAN, this can be accomplished by entering bulk property data for the stream or by specifying the composition of the stream and selecting a thermodynamic method to be used for calculating mixture properties.

The Design input category is used to specify which parameters are to be automatically varied in design mode, and the allowed ranges of these parameters. It is not used when the program is run in rating or simulation mode.

Under the Control input category, alternate computational methods can be chosen for certain calculations performed by the program, the number of increments per baffle can be specified, and default convergence tolerances can be over-ridden. For most applications, no input is required in this category.

The use of *Xist* is illustrated in the following examples.

Example 7.3

Use the *Xist* module of the HTRI *Xchanger Suite* to rate the final configuration of the kerosene/crude-oil exchanger of Example 5.1, and compare the results with those obtained previously using other methods.

Solution

Since stream pressures were not given in Example 5.1, the inlet pressure is arbitrarily set at 50 psia for each stream as in previous examples. Other input data are obtained from Example 5.1. Rating mode is selected for this problem on the Input Summary panel. The configuration of the heat exchanger is then specified using the appropriate forms, starting with the Geometry/ Exchanger panel shown below, where the shell ID is entered. Other items on this form, including exchanger type (AES), are default settings.

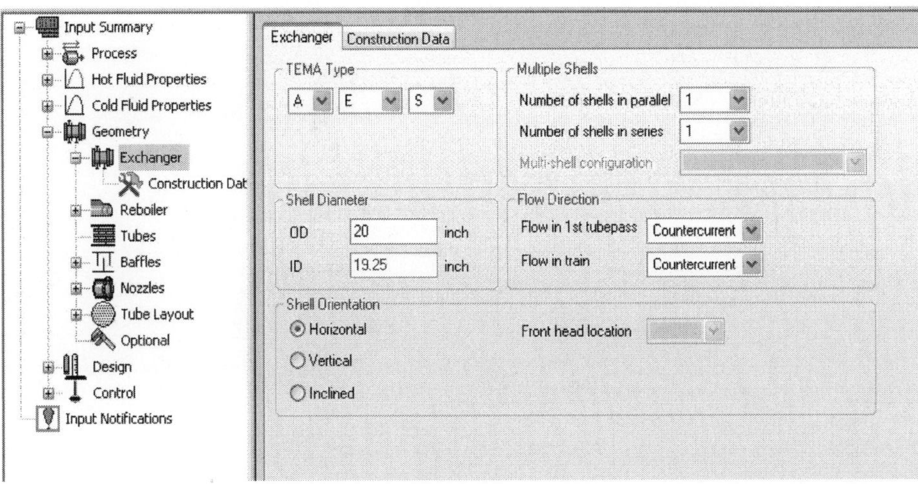

The total thickness of both tubesheets is entered on the Geometry/Exchanger/Construction Data panel as shown below. If this value is not supplied, the program calculates the tubesheet thickness based on Part UHX of the ASME Boiler and Pressure Vessel Code.

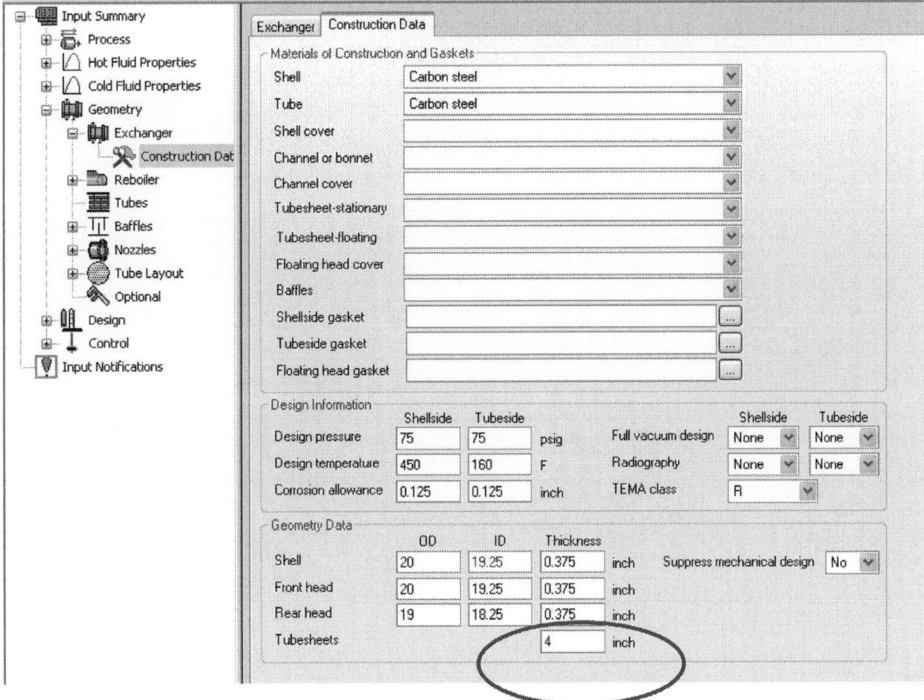

Next, tube-side data are entered on the Geometry/Tubes panel as shown below. Notice that the box for rigorous tube count is checked. This is the default setting and instructs the program to generate a detailed tube layout. The specified tube count of 124 is used in the calculations regardless of whether or not this box is checked.

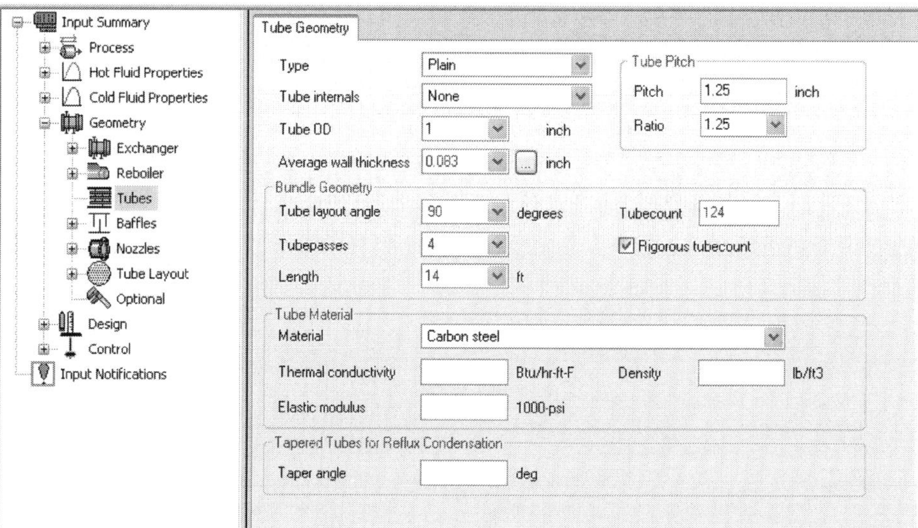

The baffle cut, central spacing, and thickness are entered on the Geometry/Baffles panel as shown. The number of baffles (42) is specified by setting the number of cross passes to 43. The latter is the number of passes that the shell-side fluid makes across the tube bundle, which is equal to the number of baffle spaces. The default setting for the *Adjust baffle cut* option is *Program set*. With this setting, the program will adjust the baffle cut for conformity with the detailed tube layout when a rigorous tube count is requested.

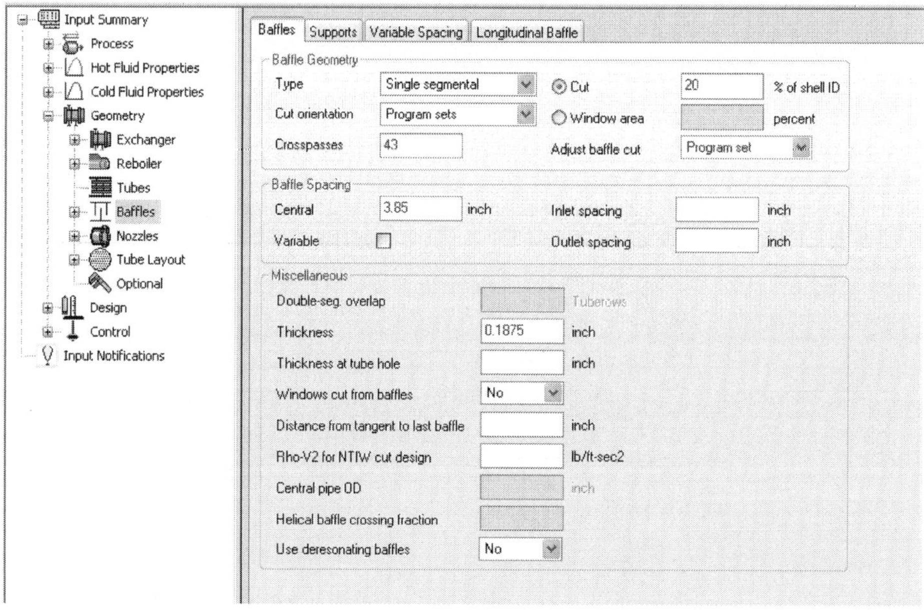

The Geometry/Tube Layout/Bundle Clearances panel is used to specify the number of pairs of sealing strips, which is one in this case. For this problem, no other entries are required on this form.

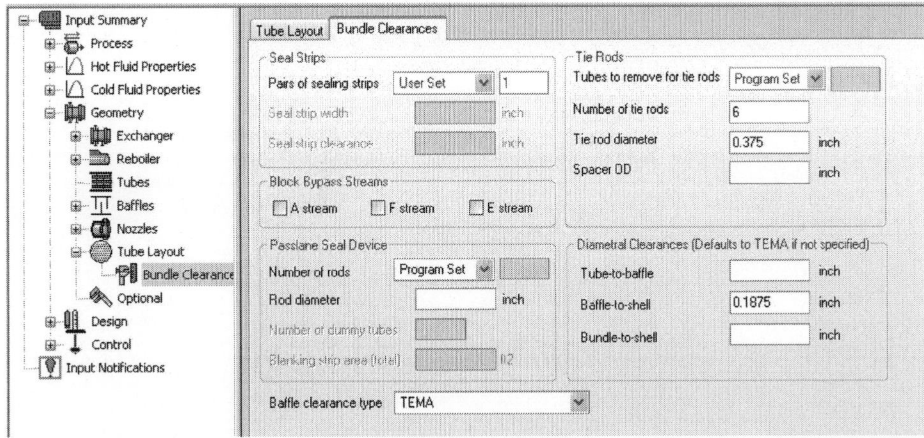

Nozzle data are entered next. The first entry on the Geometry/Nozzles panel is the piping standard to be used. With this entry and the schedule number, the program brings up a table of pipe sizes from which the nozzle OD and/or ID can be selected. The piping standard is used solely to provide this convenience to the user.

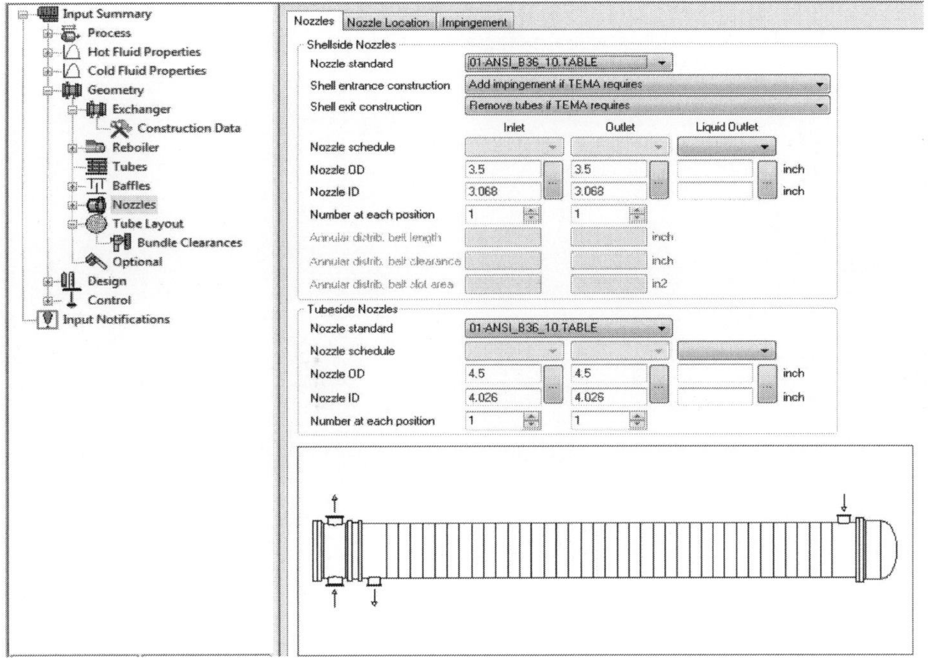

Stream data and fouling factors are entered on the Process panel as shown below. Notice that only one outlet temperature (or the exchanger duty) need be given.

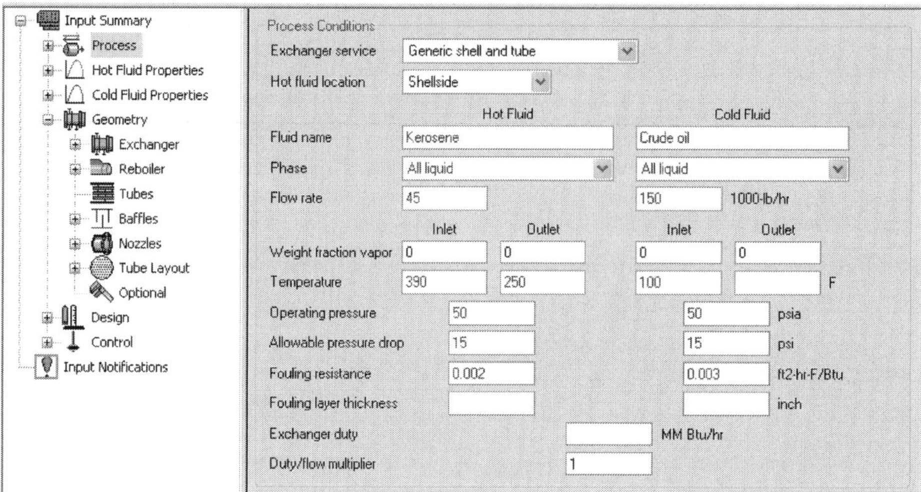

Physical properties of the hot fluid (kerosene) are entered next. Several forms are needed to accomplish this task, the first of which is used to specify the input option for the data. For bulk property data, as we have in this case, the *Program calculated* option is chosen. For a stream with defined composition, clicking on the *Property Generator* button accesses the thermodynamics interface. If the UniSim Design, HYSYS, PRO/II, PetroSim or ProMax process simulators or if PPDS or REFPROP property generators are installed on the computer running *Xist*, then the thermodynamic routines and physical property data banks of these programs can be used directly in *Xist*. Alternatively, the VMGThermo package can be selected.

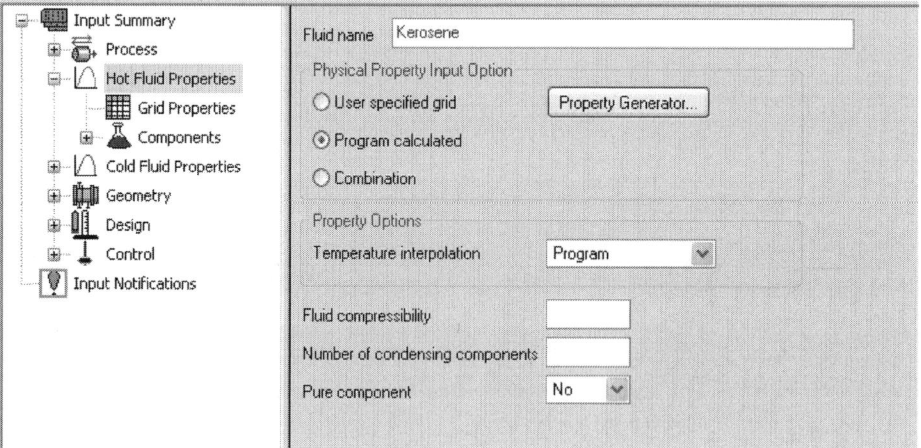

Proceed to the Components panel and choose *HTRI* as the Package to be used. Then select <User Defined> from the list of components and click the *Add* button. This opens a form where the properties can be entered at one or two temperatures by clicking on the *Liquid Properties* tab; no pressure dependence can be specified. An option is included on the *Constants* tab to generate hydrocarbon properties based on user-supplied values of API gravity and the Watson characterization factor (see Appendix E).

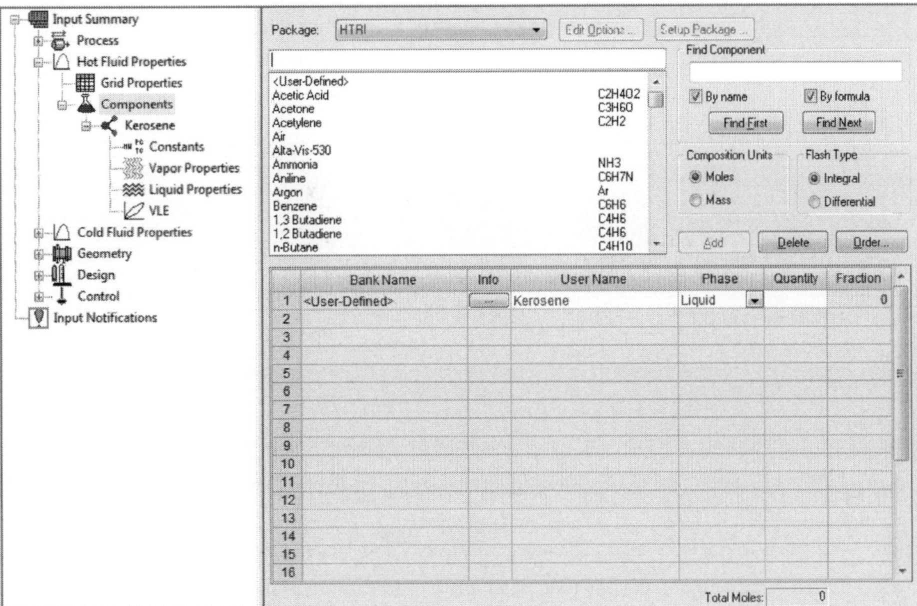

Values of thermal conductivity (0.079 Btu/h·ft·°F), density (49.008 lbm/ft^3), heat capacity (0.59 Btu/lbm·°F), and viscosity (0.4008cp) are entered on the Liquid Properties subpanel as shown below. The units of any property can be changed by clicking on the cell containing those units and selecting from a list of alternatives.

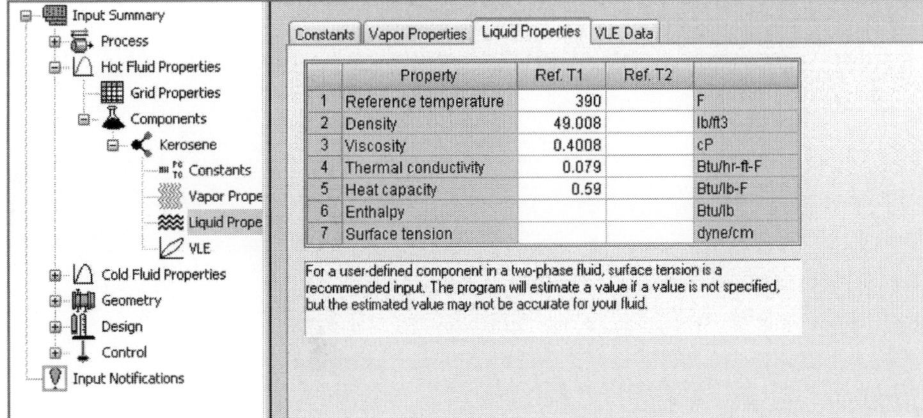

The above procedure is repeated using identical forms for the cold fluid (crude oil). Values of thermal conductivity (0.077 Btu/h·ft·°F), density (53.066 lbm/ft^3), heat capacity (0.49 Btu/lbm·°F), and viscosity (3.595 cp) are then entered on the Liquid Properties subpanel. This completes the data entry for the present case.

When all necessary data have been entered, the traffic signal icon on the toolbar shows a green light, and clicking on the icon then executes the program. Upon completion, the Output Summary is displayed on the screen. More detailed results, including the tube layout and setting plan graphics, are contained in a number of additional report files. Warning messages, if any, are contained in the Runtime Messages report file. The Output Summary for the present case is shown below.

Xist Output Summary for Example 7.3

	Xist Ver. 7.00 5/31/2013 12:09 SN: 14337-877388691						US Units
	Example 7.3						
	Rating - Horizontal Multipass Flow TEMA AES Shell With Single-Segmental Baffles						
1	**No Data Check Messages.**						
2	**See Runtime Message Report for Warning Messages.**						
3	**Process Conditions**		**Hot Shellside**			**Cold Tubeside**	
4	Fluid name			Kerosene			Crude oil
5	Flow rate	(1000-lb/hr)		45.000			150.00
6	Inlet/Outlet Y	(Wt. frac vap.)	0.0000	0.0000		0.0000	0.0000
7	Inlet/Outlet T	(Deg F)	390.00	250.00		100.00	150.57
8	Inlet P/Avg	(psia)	50.000	49.036		50.000	44.827
9	dP/Allow.	(psi)	1.927	15.000		10.346	15.000
10	Fouling	(ft2-hr-F/Btu)		0.00200			0.00300
11			**Exchanger Performance**				
12	Shell h	(Btu/ft2-hr-F)	251.39	Actual U		(Btu/ft2-hr-F)	65.42
13	Tube h	(Btu/ft2-hr-F)	219.33	Required U		(Btu/ft2-hr-F)	46.16
14	Hot regime	(--)	Sens. Liquid	Duty		(MM Btu/hr)	3.7170
15	Cold regime	(--)	Sens. Liquid	Eff. area		(ft2)	443.66
16	EMTD	(Deg F)	181.5	Overdesign		(%)	41.71
17	**Shell Geometry**			**Baffle Geometry**			
18	TEMA type	(--)	AES	Baffle type		(--)	Single-Seg.
19	Shell ID	(inch)	19.250	Baffle cut		(Pct Dia.)	21.13
20	Series	(--)	1	Baffle orientation		(--)	Perpend.
21	Parallel	(--)	1	Central spacing		(inch)	3.8500
22	Orientation	(deg)	0.00	Crosspasses		(--)	43
23	**Tube Geometry**			**Nozzles**			
24	Tube type	(--)	Plain	Shell inlet		(inch)	3.0680
25	Tube OD	(inch)	1.0000	Shell outlet		(inch)	3.0680
26	Length	(ft)	14.000	Inlet height		(inch)	1.6921
27	Pitch ratio	(--)	1.2500	Outlet height		(inch)	1.6921
28	Layout	(deg)	90	Tube inlet		(inch)	4.0260
29	Tubecount	(--)	124	Tube outlet		(inch)	4.0260
30	Tube Pass	(--)	4				
31	**Thermal Resistance, %**		**Velocities, ft/sec**			**Flow Fractions**	
32	Shell	26.02	Shellside		1.19	A	0.214
33	Tube	35.77	Tubeside		6.68	B	0.358
34	Fouling	36.62	Crossflow		1.93	C	0.166
35	Metal	1.59	Window		1.06	E	0.262
36						F	0.000

The Runtime Messages file contains a number of warning messages, two of which relate to the end baffle spacings. The values calculated by *Xist* for the inlet and outlet spacings are given in the Final Results file as 3.6515 in. and 2.4985 in., respectively, both of which are below the TEMA minimum of 3.85 in. These values satisfy the following equation for the tube length:

$$L = (n_b - 1)\,B + B_{in} + B_{out} + \text{total tubesheet thickness}$$

In Example 5.1, the tubesheet thickness was neglected in calculating the number of baffles, and hence the discrepancy in the end spacings. Notice that the baffle thickness is not included in the above equation, indicating that the program interprets baffle spacing as center-to-center distance, i.e., as the baffle pitch.

The tube layout generated by *Xist* contains 124 tubes laid out on a ribbon pattern as shown below. This layout has all tube-pass-partition lanes oriented normal to the cross-flow direction and, hence, there is no *F*-stream bypass. However, the tubes are not allocated equally among the passes; as a result, the flow in passes 1 and 4 is in the transition region. Notice that the program has adjusted the baffle cut to 21.13%, which places the baffle edge between adjacent tube rows.

The program also indicates that six tie rods of diameter 0.375 in. are required for the tube bundle. The purpose of tie rods is to hold the bundle assembly together. They are metal rods that are screwed into the tubesheet at the stationary (front) end, pass through holes in the baffles, and are fastened to the last baffle with lock nuts. Between the baffles, the rods are fitted with sleeves that act as spacers to hold the baffles securely in place. The minimum number of tie rods required by TEMA standards varies from 4 to 12 and the rod diameter varies from 3/8 to 5/8 in., depending on the shell diameter. Tie rods and spacers also help block the bundle bypass flow area, and extra sets may be installed for this purpose. However, they are not as effective as sealing strips in reducing the bypass flow. Tie rods and sealing strips can be added to the diagram using the tube layout editor.

Note: To start the tube layout editor, first go to the tube layout panel on the input tree and select the appropriate settings for *Tube pass layout* and *First tube pass location*. Then under *Drawings*, select *2D tube layout*. Finally, in the upper toolbar, click on the tube layout icon to open the editor.

Xist Tube Layout for Example 7.3

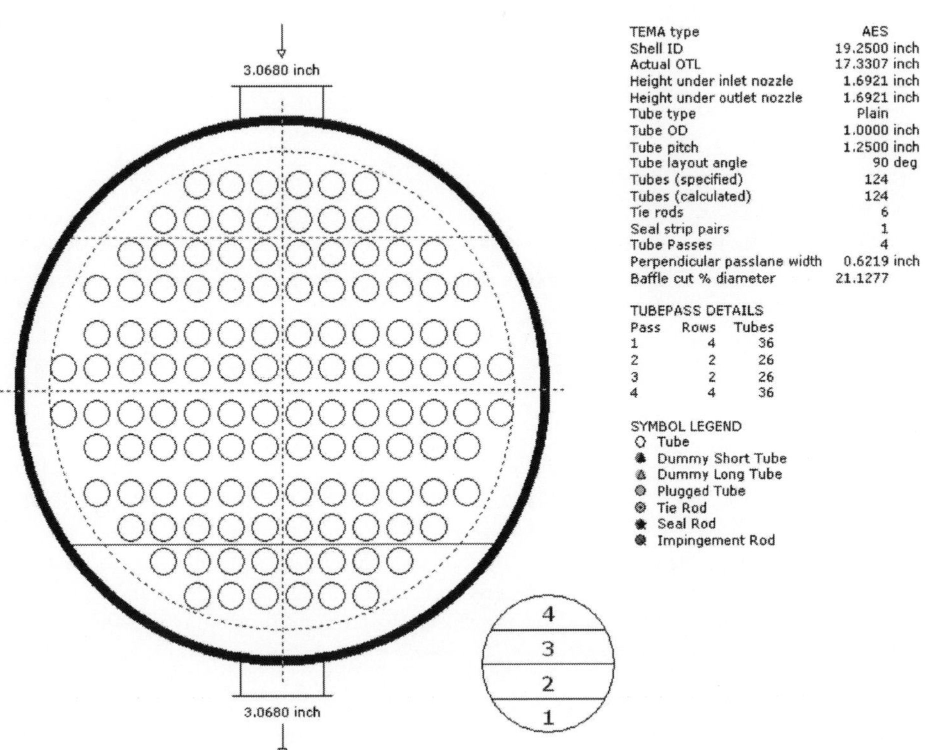

The results obtained with the HTRI method are compared with those from other methods in the two tables below. The greatest differences are in the heat-transfer coefficients; *Xist* gives significantly higher (less conservative) values for both h_i and h_o. As a result, the exchanger is over-designed by about 42% according to the HTRI results. The difference in the values of the tube-side coefficient is due to the fact that *Xist* uses a proprietary method developed by HTRI to calculate h_i, as opposed to the Seider–Tate correlation used by HEXTRAN and for the hand calculations. The differences in values of the shell-side coefficient obviously reflect the different computational methods employed.

Item	Simplified Delaware	Delaware	HEXTRAN	HTRI
h_i (Btu/h · ft^2 · °F)	156	–	156.2	219.3
h_o (Btu/h · ft^2 · °F)	122	156.6 (176.7)*	191.2	251.4
ΔP_i (psi)	10.2	–	10.06	10.35
ΔP_o (psi)	2.2	1.31 (1.51)*	2.11	1.93

* Without δ_{sb} safety factor.

Item	Wills–Johnston	HTRI
ΔP_o (psi)	1.45	1.93
Stream-flow fraction		
A	0.276	0.214
B	0.345	0.358
C	0.122	0.166
E	0.257	0.262
F	0	0

The flow fractions predicted by the Wills–Johnston method are in general agreement with the HTRI values, although the differences for the *A* stream (25%) and *C* stream (31%) are quite large. The two methods differ by about 28% in the shell-side pressure drop, which is consistent with the difference in *A*-stream flow fractions.

Example 7.4

Use *Xist* to obtain a final design for the kerosene/crude-oil exchanger of Example 5.1.

Solution

The design in Example 7.3 would probably work satisfactorily in service if the number of baffles were reduced to provide room for the tubesheets (which were neglected in the hand calculations) and sufficiently large end spaces to accommodate the shell-side nozzles. The length could also be reduced to eliminate the over-sizing. However, the design is not optimal to mitigate fouling. For refinery applications such as the kerosene-crude oil exchanger, fouling potential should be minimized. The *B*-stream flow fraction is less than the recommended minimum of 0.45 as specified in **Table 7.1**. Experience shows that high bypass-flow fractions contribute to shell-side fouling due to stagnant flow regions which form with a high degree of flow maldistribution. The large *E*-stream flow fraction is caused by the narrow baffle spacing, and while the shell-side heat-transfer coefficient is maximized at TEMA minimum spacing, this also maximizes the tube-wall temperature, which promotes crude oil fouling. The design can be improved by applying the following three specifications, which will mitigate fouling:

- Ensure that the *B*-stream flow fraction is greater than 0.45. This will mitigate shell-side fouling potential.
- Maintain the tube-side velocity between 5 and 8 ft/s. This will ensure that the wall temperature remains low and the wall shear stress high, both of which will mitigate crude oil fouling.
- Ensure that the over-design is not greater than 10%. Excessive margin often contributes to accelerated fouling due to lower fluid velocities. In the present case, with high fouling factors specified, little additional excess area is needed.

In order to meet these criteria, the baffle spacing should be increased. Increasing the baffle spacing lowers the *E*-stream flow fraction while increasing the *B*-stream fraction, and has the additional benefit of facilitating bundle fabrication without tube damage. With baffle spacing less than 4 in., damage to carbon steel tubes is difficult to avoid.

The above specifications were attained by adjusting the baffle spacing, tube length and tube layout in the *Xist* rating case for Example 7.3. Program defaults were selected to maximize design flexibility; parameters calculated by the program included number of tubes, number of cross passes, baffle thickness, and tubesheet thickness. Designs were obtained for both ribbon and H-banded tube layouts. A comparison of these two 4-pass arrangements is warranted since an H-banded layout often has a higher tube count but introduces an *F*-stream bypass. A quadrant layout is another four-pass arrangement that can be selected, but it is less popular in plant service because offset, or "hillside," channel nozzles are required, which complicate the piping layout.

Key attributes common to the designs for both tube layouts are summarized in the following table. Notice that the number of sealing strips, number of tie rods and the tie rod diameter have been increased over the values recommended by *Xist* (1 pair of seals, 6 tie rods of diameter 0.375 in.) in order to provide a more robust tube bundle assembly.

Item	Value	Comment
TEMA type	AES	A removable front-channel cover facilitates tube-side cleaning. A split-ring rear head permits bundle removal with straight tubes.
Tube OD and wall thickness	1-in., BWG 14	1-in. tube OD is consistent with API 660 guidelines; 0.75-in. OD is also commonly specified. BWG 14 has minimum wall thickness of 0.083 in. The average wall thickness can be up to 10% more than minimum for ferrous tubes.
Tube layout and pitch	90° square, 1.25 in. pitch	This provides for 0.25 in. cleaning lanes, which accommodate most high-pressure water cleaning equipment.
Tube material	Carbon steel	Carbon steel is commonly selected for low temperature crude oil preheat applications. Alloy tubes are sometimes selected if corrosion potential is high.
Shell ID	19.25 in.	
Tube passes	4	This provides for adequate, but not excessive, tube velocity within allowable pressure drop (15 psi) for a shell ID of 19.25 in.
Baffle type and cut orientation	Single segmental, horizontal	Facilitates mixing and reduces stratification when within allowable pressure drop (15 psi).
Tie rod number and diameter	8, 0.625 in.	This provides for uniform distribution of tie rods around bundle periphery without locating in high-velocity region directly under inlet nozzle.
Seal-strip pairs	2	This provides for symmetrical placement, which is common for smaller tube bundles.

Other key attributes of the final designs obtained with the ribbon and H-banded layouts are summarized in the following table. The tube-layout diagrams from *Xist* are also shown below; tie rods, sealing strips, and seal rods were added manually using the tube layout editor.

Item	Ribbon Layout	H-Banded Layout
Tube count	124	124
Tube length, ft	11.5	12
Heat transfer area, ft^2	368	384
Central baffle spacing, in.	8	9
Baffle cut, %	21.13	28.59
B-stream flow fraction	0.529	0.532
F-stream flow fraction	0	0.090
Tube-side velocity, ft/s		
Passes 1,4	5.75	7.96
Passes 2,3	7.96	5.75
Over-design, %	7.78	5.99

Xist Tube Layouts for Example 7.4

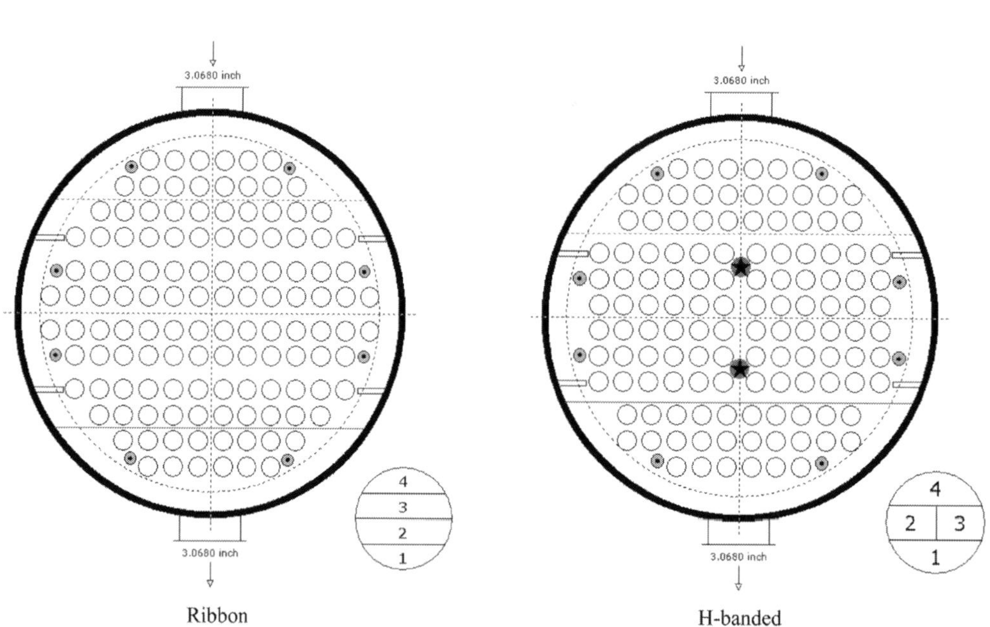

Ribbon

H-banded

For this application, the H-banded layout provides no advantage in tube count over the ribbon layout. Also, two seal rods (as specified by *Xist*) are employed in the H-banded layout to partially block the vertical pass-partition lane and reduce the *F*-stream flow fraction. Both layouts result in a tube-side Reynolds number (approximately 8800) that is in the transition region in two of the four passes. This is not a concern in the present context, however, because the HTRI correlations are based on a large amount of experimental data in the transition region and industrial experience has shown that they are reliable in this regime for liquids flowing in horizontal tubes.

The two designs are very similar, but the ribbon layout provides a slightly more compact unit and is therefore selected for this application. The *Xist* Output Summary for the final rating run with the ribbon layout is shown below, along with the exchanger drawing generated by the program. For the final runs the actual tube layouts were used as input; this is accomplished by clicking the box at the bottom of the tube layout panel on the input tree, or the identical box in the tube-layout editor.

Xist Output Summary for Example 7.4: Ribbon Tube Layout

#							
	Xist Ver. 7.00 6/1/2013 11:45 SN: 14337-877388691						US Units
	Example 7.4						
	Rating - Horizontal Multipass Flow TEMA AES Shell With Single-Segmental Baffles						
1	**No Data Check Messages.**						
2	**See Runtime Message Report for Warning Messages.**						
3	**Process Conditions**		**Hot Shellside**		**Cold Tubeside**		
4	Fluid name			Kerosene		Crude Oil	
5	Flow rate	(1000-lb/hr)		45.000		150.00	
6	Inlet/Outlet Y	(Wt. frac vap.)	0.0000	0.0000	0.0000	0.0000	
7	Inlet/Outlet T	(Deg F)	390.00	250.00	100.00	150.57	
8	Inlet P/Avg	(psia)	50.000	49.707	50.000	45.166	
9	dP/Allow.	(psi)	0.586	15.000	9.667	15.000	
10	Fouling	(ft2-hr-F/Btu)		0.00200		0.00300	
11			**Exchanger Performance**				
12	Shell h	(Btu/ft2-hr-F)	189.18	Actual U	(Btu/ft2-hr-F)	59.41	
13	Tube h	(Btu/ft2-hr-F)	210.18	Required U	(Btu/ft2-hr-F)	55.12	
14	Hot regime	(--)	Sens. Liquid	Duty	(MM Btu/hr)	3.7170	
15	Cold regime	(--)	Sens. Liquid	Eff. area	(ft2)	367.92	
16	EMTD	(Deg F)	183.3	Overdesign	(%)	7.78	
17	**Shell Geometry**			**Baffle Geometry**			
18	TEMA type	(--)	AES	Baffle type	(--)	Single-Seg.	
19	Shell ID	(inch)	19.250	Baffle cut	(Pct Dia.)	21.13	
20	Series	(--)	1	Baffle orientation	(--)	Perpend.	
21	Parallel	(--)	1	Central spacing	(inch)	8.0000	
22	Orientation	(deg)	0.00	Crosspasses	(--)	15	
23	**Tube Geometry**			**Nozzles**			
24	Tube type	(--)	Plain	Shell inlet	(inch)	3.0680	
25	Tube OD	(inch)	1.0000	Shell outlet	(inch)	3.0680	
26	Length	(ft)	11.500	Inlet height	(inch)	1.6921	
27	Pitch ratio	(--)	1.2500	Outlet height	(inch)	1.6921	
28	Layout	(deg)	90	Tube inlet	(inch)	4.0260	
29	Tubecount	(--)	124	Tube outlet	(inch)	4.0260	
30	Tube Pass	(--)	4				
31	**Thermal Resistance, %**		**Velocities, ft/sec**		**Flow Fractions**		
32	Shell	31.41	Shellside	0.80	A	0.141	
33	Tube	33.89	Tubeside	7.96	B	0.529	
34	Fouling	33.25	Crossflow	0.96	C	0.142	
35	Metal	1.45	Window	1.08	E	0.188	
36					F	0.000	

Xist Exchanger Drawing for Example 7.4

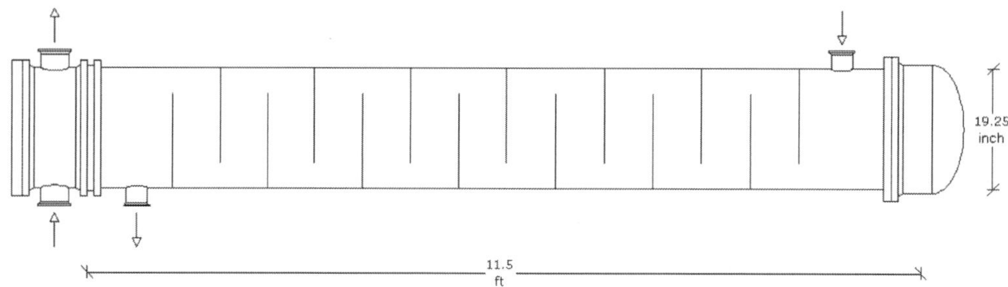

Note: The configuration of the tube-pass partitions in the H-banded tube layout is shown in the following diagram:

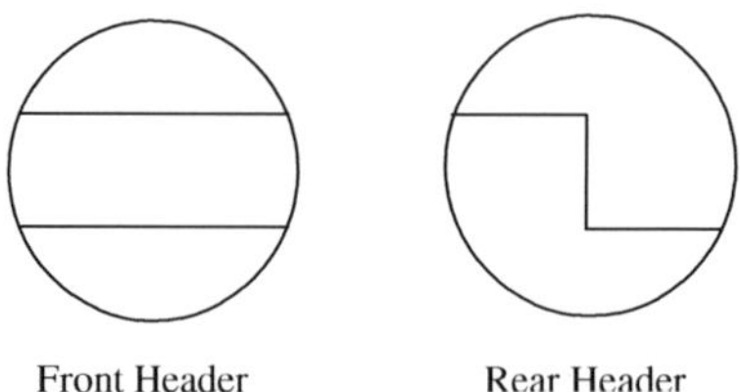

Example 7.5

Redesign the component cooling water heat exchanger of Example 5.5 so as to eliminate the temperature reversal and increase the duty under clean conditions. The redesign should consider the following additional constraints:

- Tube-side velocity should be greater than 1 m/s to minimize sedimentation fouling.
- The tube-side and shell-side pressure drops (31.5 and 10.9 kPa, respectively) should be reduced to lower the loading of the emergency diesel generator, which is often a limiting factor in nuclear power plant operation.

The key process specifications are as follows:

Duty, MW	4.39
Hot water inlet temperature, °C	115
Hot water outlet temperature, °C	53.5

(*Continued*)

—cont'd

Hot water flow rate, kg/s	17
Cold water inlet temperature, °C	25
Cold water outlet temperature, °C	50
Cold water flow rate, kg/s	42

Solution

The original design was a TEMA E-shell with two tube passes. Under clean conditions, this unit exhibits a temperature reversal over a significant portion of its length. In this region the hot stream temperature is less than the cold stream temperature and heat is transferred from the cold fluid back to the hot fluid. One change to consider is to switch the location of the shell-side inlet and outlet nozzles. In the original design the shell-side fluid enters at the rear end and exits at the front end of the exchanger (see Example 5.5). With this nozzle arrangement, the flow in the first tube pass is counter-current with the shell-side flow while the flow in the second pass is co-current. Switching the nozzles reverses the shell-side flow direction, putting the second tube pass in counter flow. *Xist* is run in simulation mode for this case to check the performance of the exchanger under clean conditions. The results confirm that the clean duty is increased from 5.05 to 5.12 MW, and no temperature reversal is observed in the temperature profiles shown below.

Temperature Profiles for Modified E-shell Design under Clean Conditions

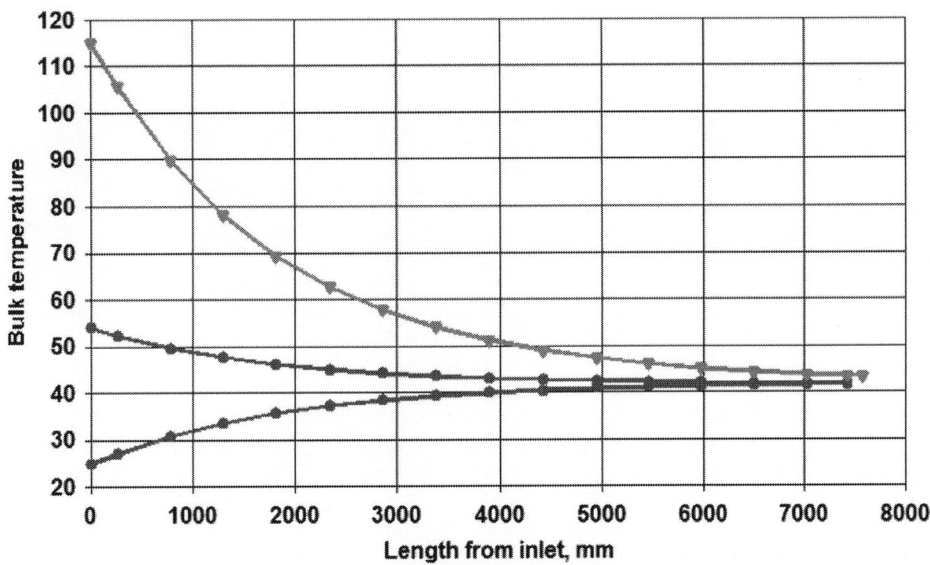

We next consider a more extensive redesign that employs an exchanger with two shell passes to improve the temperature profile and further increase the duty under clean conditions. With four passes and a quadrant tube layout, an F-shell with a welded longitudinal baffle provides a 2-4 exchanger with a removable bundle. The tube-side pressure drop can be reduced by decreasing the bundle length and increasing the tube OD from 19 to 25.4 mm. The shell-side pressure drop can be reduced by using a double-segmental baffle configuration. Optimizing the design is a trade-off of tube-side pressure drop and velocity. The Rating Data Sheet and tube layout below show a design with lower tube-side and shell-side pressure drops and velocities.

Xist Rating Data Sheet for Example 7.5: F-shell Design

HTRI

HEAT EXCHANGER RATING DATA SHEET

Page 1

SI Units

Service of Unit Water-Water			Item No. Exercise 1		
Type AFU	Orientation Horizontal		Connected In 1 Parallel 1 Series		
Surf/Unit (Gross/Eff) 137.02 / 135.52 m2	Shell/Unit 1		Surf/Shell (Gross/Eff) 137.02 / 135.52 m2		

PERFORMANCE OF ONE UNIT

Fluid Allocation		Shell Side		Tube Side	
Fluid Name		CC Water		Raw Water	
Fluid Quantity, Total	kg/s	17.000		42.000	
Vapor (In/Out)	wt%	0.00	0.00	0.00	0.00
Liquid	wt%	100.00	100.00	100.00	100.00
Temperature (In/Out)	C	115.00	53.58	25.00	50.00
Density	kg/m3	947.17	986.49	997.10	988.10
Viscosity	mN-s/m2	0.2429	0.5156	0.8901	0.5469
Specific Heat	kJ/kg-C	4.2376	4.1799	4.1816	4.1793
Thermal Conductivity	W/m-C	0.6827	0.6445	0.6076	0.6406
Critical Pressure	kPa				
Inlet Pressure	kPa	344.74		206.84	
Velocity	m/s		0.20		1.20
Pressure Drop, Allow/Calc	kPa	0.000	3.224	0.000	25.743
Average Film Coefficient	W/m2-K	4278.0		5620.8	
Fouling Resistance (min)	m2-K/W	0.000200		0.000400	

Heat Exchanged 4.3900 MegaWatts	MTD (Corrected) 41.4 C	Overdesign 6.12 %
Transfer Rate, Service 781.79 W/m2-K	Calculated 829.64 W/m2-K	Clean 1833.0 W/m2-K

CONSTRUCTION OF ONE SHELL

Sketch (Bundle/Nozzle Orientation)

		Shell Side	Tube Side
Design Pressure	kPaG	1000.0	1000.0
Design Temperature	C	150.01	100.01
No Passes per Shell		2	4
Flow Direction		Downward	Upward
Connections	In mm	1 @ 154.05	1 @ 154.05
Size &	Out mm	1 @ 154.05	1 @ 154.05
Rating	Liq. Out mm	@	1 @

Tube No. 370.00 OD 25.400 mm	Thk(Avg) 1.651 mm Length 4.400 m	Pitch 31.750 mm Tube pattern 60
Tube Type Plain	Material 304 Stainless steel (18 Cr, 8 Ni)	Pairs seal strips 2
Shell ID 750.00 mm	Kettle ID mm	Passlane Seal Rod No. 0
Cross Baffle Type Parallel Double-Seg.	%Cut (Diam) 27.27	Impingement Plate None
Spacing(c/c) 533.04 mm	Inlet 610.00 mm	No. of Crosspasses 9
Rho-V2-Inlet Nozzle 878.27 kg/m-s2	Shell Entrance 275.33 kg/m-s2	Shell Exit 264.35 kg/m-s2
	Bundle Entrance 80.42 kg/m-s2	Bundle Exit 77.22 kg/m-s2
Weight/Shell 3956.6 kg	Filled with Water 6174.0 kg	Bundle 2282.3 kg

Notes:	Thermal Resistance, %		Velocities, m/s		Flow Fractions	
	Shell	19.39	Shellside	0.20	A	0.004
	Tube	16.97	Tubeside	1.20	B	0.863
	Fouling	54.74	Crossflow	0.18	C	0.060
	Metal	8.90	Window	0.30	E	0.073
					F	0.000

Xist Tube Layout for Example 7.5: F-shell Design

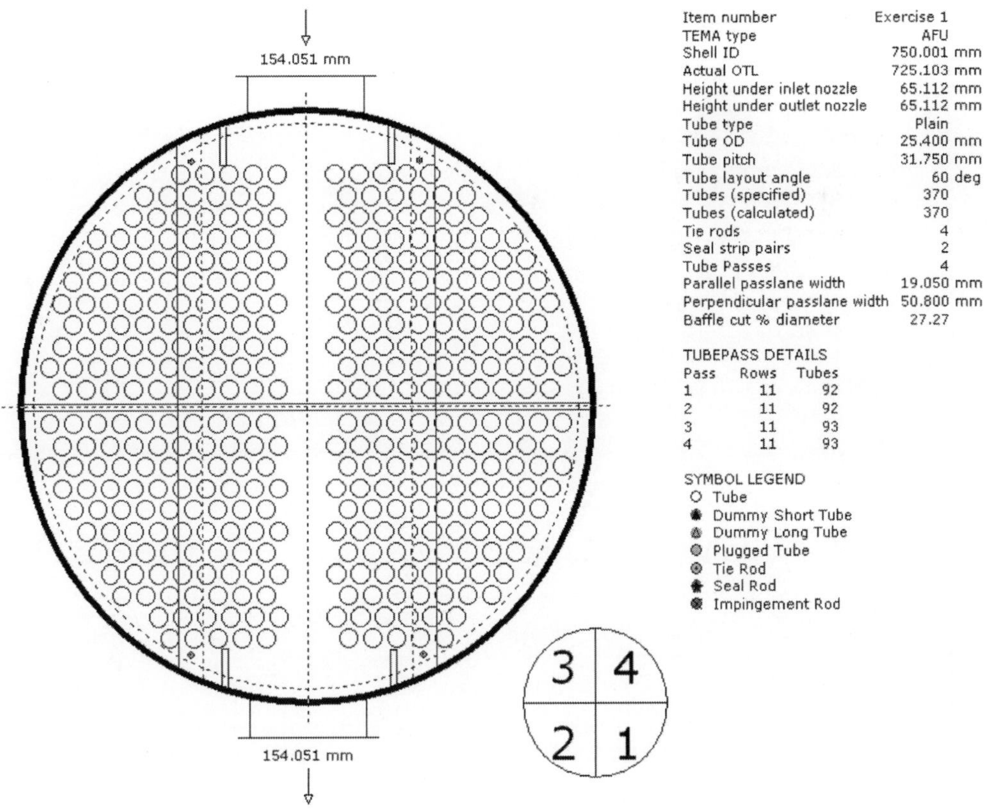

Xist is again run in simulation mode to check the performance of the exchanger under clean conditions. The Output Summary for this run is given below, from which the duty is seen to be 5.57 MW. This represents a 10% increase over the original design. Furthermore, the temperature profiles shown below do not exhibit a temperature reversal.

Xist Output Summary for Example 7.5: Simulation Run for F-shell Design

Xist Ver. 7.00 6/9/2013 8:17 SN: 14337-877388691					**SI Units**
Example 7.5					
Simulation - Horizontal Multipass Flow TEMA AFU Shell With Double-Segmental Baffles					
No Data Check Messages.					
See Runtime Message Report for Informative Messages.					
Process Conditions		**Hot Shellside**		**Cold Tubeside**	
Fluid name			CC Water		Raw Water
Flow rate	(kg/s)		17.000		42.000
Inlet/Outlet Y	(Wt. frac vap.)	0.0000	0.0000	0.0000	0.0000
Inlet/Outlet T	(Deg C)	115.00	37.00	25.00	56.71
Inlet P/Avg	(kPa)	344.74	343.11	206.84	194.01
dP/Allow.	(kPa)	3.250	0.000	25.662	0.000
Fouling	(m2-K/W)		0.000000		0.000000
Exchanger Performance					
Shell h	(W/m2-K)	3843.4	Actual U	(W/m2-K)	1760.0
Tube h	(W/m2-K)	5727.6	Required U	(W/m2-K)	1754.1
Hot regime	(--)	Sens. Liquid	Duty	(MegaWatts)	5.5691
Cold regime	(--)	Sens. Liquid	Eff. area	(m2)	135.52
EMTD	(Deg C)	23.4	Overdesign	(%)	0.34
Shell Geometry			**Baffle Geometry**		
TEMA type	(--)	AFU	Baffle type	(--)	Double-Seg.
Shell ID	(mm)	750.00	Baffle cut	(Pct Dia.)	27.27
Series	(--)	1	Baffle orientation	(--)	Parallel
Parallel	(--)	1	Central spacing	(mm)	533.04
Orientation	(deg)	0.00	Crosspasses	(--)	9
Tube Geometry			**Nozzles**		
Tube type	(--)	Plain	Shell inlet	(mm)	154.05
Tube OD	(mm)	25.40	Shell outlet	(mm)	154.05
Length	(m)	4.40	Inlet height	(mm)	65.11
Pitch ratio	(--)	1.2500	Outlet height	(mm)	65.11
Layout	(deg)	60	Tube inlet	(mm)	154.05
Tubecount	(--)	370	Tube outlet	(mm)	154.05
Tube Pass	(--)	4			
Thermal Resistance, %		**Velocities, m/s**		**Flow Fractions**	
Shell	45.79	Shellside	0.20	A	0.003
Tube	35.32	Tubeside	1.20	B	0.861
Fouling	0.00	Crossflow	0.18	C	0.061
Metal	18.89	Window	0.30	E	0.075
				F	0.000

Temperature Profiles for F-shell Design under Clean Conditions

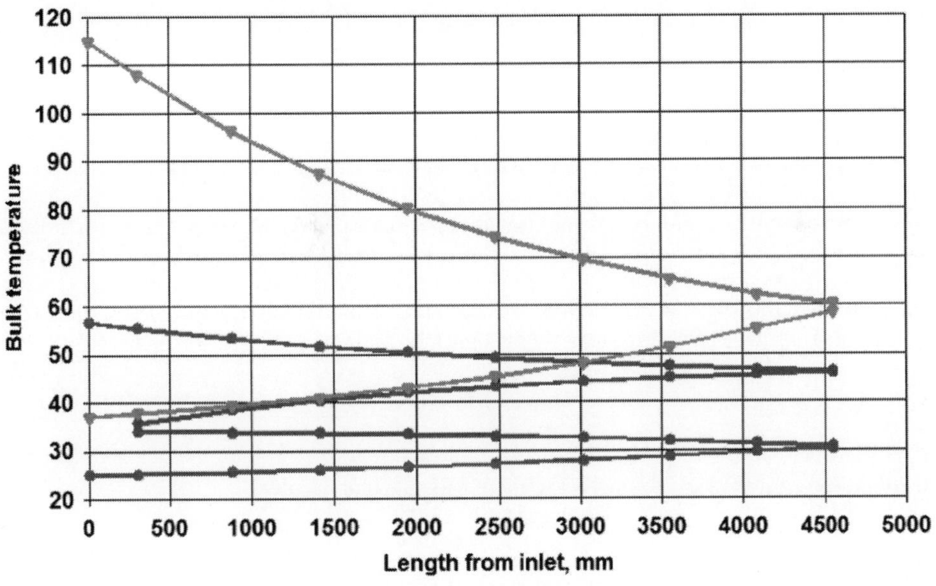

Example 7.6

300,000 lb/h of a liquid hydrocarbon are to be heated from 80 to 150°F by heat exchange with 176,000 lb/h of a petroleum fraction with a temperature range of 265 to 155°F. The inlet pressure of each stream will be 50 psia with a maximum pressure drop of 10 psi. Fouling factors of 0.0005 h· ft^2·°F/Btu for the liquid hydrocarbon and 0.001 h· ft^2·°F/Btu for the petroleum fraction are specified. Neither stream is corrosive to carbon steel. Average physical properties of the two streams are given in the following table. Use *Xist* to design a heat exchanger for this service.

Fluid Property	Liquid Hydrocarbon	Petroleum Fraction
C_P (Btu/lbm · °F)	0.483	0.524
k (Btu/h · ft · °F)	0.0724	0.0659
μ (cp)	2.06	0.475
Density (lbm/ft^3)	52.368	48.436

Solution

We begin by specifying the basic configuration of the exchanger.

(i) Fluid placement
Neither fluid is corrosive or highly fouling. The petroleum fraction is significantly less viscous (a tube-side criterion), but it will be placed in the shell due to its large ΔT of 110°F.

(ii) Shell and head types
A standard E-shell is suitable for this service. Since the tube-side fluid is clean, U-tubes are chosen to allow for differential thermal expansion (the two inlet stream temperatures differ by more than 100°F). A bonnet is selected for the front head because frequent cleaning of tube interiors is not anticipated with a clean tube-side fluid. Hence, a type BEU exchanger is specified.

(iii) Tubing
For a clean hydrocarbon service, 0.75-inch, 14 BWG tubes are selected based on the design guidelines in Chapter 5. Since the tube bundle will be removable for cleaning and maintenance, a maximum tube length of 20 ft is specified.

(iv) Tube layout
Although the shell-side fluid is relatively clean, there is sufficient fouling potential with this type of stream that cleaning may be required. Therefore, a square layout is specified with a pitch of 1.0 in. to provide cleaning lanes.

(v) Baffles
Single-segmental baffles are expected to be suitable for this service. Baffle cut and spacing will be determined during the design process.

(vi) Construction materials
Since neither fluid is corrosive, plain carbon steel is suitable for all components.

Next, *Xist* is run in design mode to obtain an estimate of the size required for the exchanger. Input data are as follows.

(a) Input Summary

Case mode: Design	Type: BEU

(b) Geometry/Tubes:

Tube OD: 0.75 in.	Pitch: 1.0 in.
Ave. wall thickness: 0.083 in.	Tube layout angle: 90

(c) Design/Geometry

Min. tube length: 8 ft	Step size: 2 ft
Max. tube length: 20 ft	

Note that the shortcut design method (the default) is specified here.

(d) Design/Constraints/Cold Fluid

Minimum Velocity: 3 ft/s	Maximum Velocity: 8 ft/s

(e) Process

	Hot Fluid	Cold Fluid
Fluid name	Petroleum Fraction	Liquid Hydrocarbon
Phase	All liquid	All liquid
Flow rate (1000 lb/h)	176	300
Inlet Temperature	265	80
Outlet Temperature	–	150

(*Continued*)

—cont'd

	Hot Fluid	Cold Fluid
Operating pressure (psia)	50	50
Allowable pressure drop (psi)	10	10
Fouling resistance (h·ft^2·°F/Btu)	0.001	0.0005

(f) Hot and Cold fluid properties
The bulk stream properties given in the problem statement are entered in the usual way. This completes the data entry.

The *Xist* Output Summary for the design run is given below, showing an exchanger with 612 tubes in a 30-in. shell. However, this design cannot be accepted as final due to the following problems:

Xist Output Summary for Example 7.6: Design Run

	Xist Ver. 7.00 6/7/2013 16:49 SN: 14337-877388691					US Units
	Example 7.6					
	Design - Horizontal Multipass Flow TEMA BEU Shell With Single-Segmental Baffles					
1	**No Data Check Messages.**					
2	**See Runtime Message Report for Warning Messages.**					
3	**Process Conditions**		**Hot Shellside**		**Cold Tubeside**	
4	Fluid name			Petroleum Fraction		Liquid Hydrocarbon
5	Flow rate	(1000-lb/hr)		176.00		300.00
6	Inlet/Outlet Y	(Wt. frac vap.)	0.0000	0.0000	0.0000	0.0000
7	Inlet/Outlet T	(Deg F)	265.00	155.02	80.00	150.00
8	Inlet P/Avg	(psia)	50.000	48.433	50.000	45.190
9	dP/Allow.	(psi)	3.133	10.000	9.620	10.000
10	Fouling	(ft2-hr-F/Btu)		0.00100		0.00050
11	**Exchanger Performance**					
12	Shell h	(Btu/ft2-hr-F)	249.92	Actual U	(Btu/ft2-hr-F)	88.95
13	Tube h	(Btu/ft2-hr-F)	239.62	Required U	(Btu/ft2-hr-F)	77.08
14	Hot regime	(--)	Sens. Liquid	Duty	(MM Btu/hr)	10.143
15	Cold regime	(--)	Sens. Liquid	Eff. area	(ft2)	1765.5
16	EMTD	(Deg F)	74.5	Overdesign	(%)	15.40
17	**Shell Geometry**			**Baffle Geometry**		
18	TEMA type	(--)	BEU	Baffle type	(--)	Single-Seg.
19	Shell ID	(inch)	30.000	Baffle cut	(Pct Dia.)	15.84
20	Series	(--)	1	Baffle orientation	(--)	Perpend.
21	Parallel	(--)	1	Central spacing	(inch)	6.0367
22	Orientation	(deg)	0.00	Crosspasses	(--)	27
23	**Tube Geometry**			**Nozzles**		
24	Tube type	(--)	Plain	Shell inlet	(inch)	6.0650
25	Tube OD	(inch)	0.7500	Shell outlet	(inch)	6.0650
26	Length	(ft)	14.000	Inlet height	(inch)	0.8765
27	Pitch ratio	(--)	1.3333	Outlet height	(inch)	0.8765
28	Layout	(deg)	90	Tube inlet	(inch)	6.0650
29	Tubecount	(--)	612	Tube outlet	(inch)	6.0650
30	Tube Pass	(--)	4			
31	**Thermal Resistance, %**		**Velocities, ft/sec**		**Flow Fractions**	
32	Shell	35.59	Shellside	1.32	A	0.373
33	Tube	47.67	Tubeside	5.59	B	0.316
34	Fouling	14.61	Crossflow	3.01	C	0.015
35	Metal	2.13	Window	2.79	E	0.182
36					F	0.113

- The cross-flow fraction (0.316) is below the recommended value of 0.55 for a U-tube exchanger with turbulent flow given in **Table 7.1**. (The shell-side Reynolds number is over 15,000 for this design.) Also, the *A*-stream flow fraction (0.373) is much higher than the recommended maximum value of 0.20.
- The baffle cut (15.84%) is outside the recommended range of 20% to 35% for applications involving single-phase fluids.
- *Xist* generates a warning message concerning tube vibration problems in the shell entrance region.

To correct these deficiencies the program is switched to rating mode and after entering the requisite input data, the following changes are made:

- A full support plate at the U-bend is specified on the Geometry/Baffles/Supports panel in order to mitigate tube vibration in the entrance region. This reduces the effective heat-transfer area somewhat because there is little flow (hence, little heat transfer) in the region behind the support plate.
- The central baffle spacing is increased (by trial) until the desired cross-flow fraction is attained at a spacing of 16.6 in. In this process the number of cross-passes is left unspecified, enabling the program to determine the number of baffles, along with inlet and outlet baffle spacings. The baffle cut is also adjusted in the range 20% to 25% in accordance with Figure 5.3. The final baffle cut is set by the program for compatibility with the tube layout. (For a square layout the baffle edge is situated in the nearest gap between adjacent tube rows.)
- The tube length is increased to 14.5 ft.

With these changes, the above problems are eliminated and the over-design is 3.0%, which is still feasible. No new warning messages are generated by *Xist*, but examination of the Rating Data Sheet for this case reveals an outlet baffle spacing of 6.18 in., which is just slightly larger than the nozzle ID. Although technically feasible, this is not a good configuration. The problem is also evident on the exchanger drawing, which is shown below.

Xist Exchanger Drawing Showing Poor Baffle Configuration

To fix this problem, the number of cross-passes is set to 10 and the baffle spacing is left unspecified, to be calculated by the

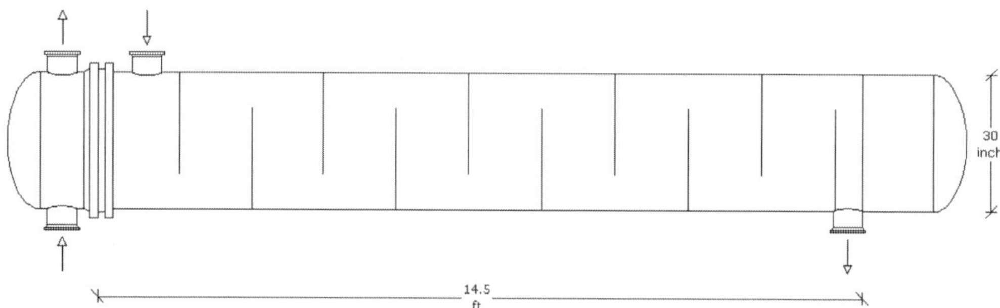

program. This reduces the number of baffles from 10 to 9 and produces a uniform baffle spacing of 17.25 in. The *B*-stream flow fraction is 0.554, but the over-design is reduced to 2.0%. The tube-side pressure drop (9.88 psi) is near the specified maximum, so the tube length cannot be increased significantly to provide additional margin. However, the *F*-stream fraction (0.204) is quite high, even though the program employs four seal rods to partially block the vertical pass-partition lane. This suggests that improved performance can be achieved by increasing the number of seal rods. Using eight rods and increasing the number of cross passes to 12 gives an over-design of about 6.4%. The *F*-stream fraction is reduced to 0.154 and the *B*-stream fraction is 0.556. The Output Summary for the final rating of this case is shown below.

Xist Output Summary for Example 7.6: Final Rating Run

	Xist Ver. 7.00 6/7/2013 13:17 SN: 14337-877388691					US Units
	Example 7.6					
	Rating - Horizontal Multipass Flow TEMA BEU Shell With Single-Segmental Baffles					
1	**No Data Check Messages.**					
2	**See Runtime Message Report for Warning Messages.**					
3	**Process Conditions**		**Hot Shellside**		**Cold Tubeside**	
4	Fluid name			Petroleum Fraction		Liquid Hydrocarbon
5	Flow rate	(1000-lb/hr)		176.00		300.00
6	Inlet/Outlet Y	(Wt. frac vap.)	0.0000	0.0000	0.0000	0.0000
7	Inlet/Outlet T	(Deg F)	265.00	155.02	80.00	150.00
8	Inlet P/Avg	(psia)	50.000	49.401	50.000	45.060
9	dP/Allow.	(psi)	1.197	10.000	9.880	10.000
10	Fouling	(ft2-hr-F/Btu)		0.00100		0.00050
11	**Exchanger Performance**					
12	Shell h	(Btu/ft2-hr-F)	200.51	Actual U	(Btu/ft2-hr-F)	81.78
13	Tube h	(Btu/ft2-hr-F)	239.62	Required U	(Btu/ft2-hr-F)	76.88
14	Hot regime	(--)	Sens. Liquid	Duty	(MM Btu/hr)	10.143
15	Cold regime	(--)	Sens. Liquid	Eff. area	(ft2)	1727.4
16	EMTD	(Deg F)	76.4	Overdesign	(%)	6.38
17	**Shell Geometry**			**Baffle Geometry**		
18	TEMA type	(--)	BEU	Baffle type	(--)	Single-Seg.
19	Shell ID	(inch)	30.00	Baffle cut	(Pct Dia.)	25.84
20	Series	(--)	1	Baffle orientation	(--)	Perpend.
21	Parallel	(--)	1	Central spacing	(inch)	14.27
22	Orientation	(deg)	0.00	Crosspasses	(--)	12
23	**Tube Geometry**			**Nozzles**		
24	Tube type	(--)	Plain	Shell inlet	(inch)	6.0650
25	Tube OD	(inch)	0.75	Shell outlet	(inch)	6.0650
26	Length	(ft)	14.50	Inlet height	(inch)	0.67
27	Pitch ratio	(--)	1.3333	Outlet height	(inch)	0.67
28	Layout	(deg)	90	Tube inlet	(inch)	6.0650
29	Tubecount	(--)	612	Tube outlet	(inch)	6.0650
30	Tube Pass	(--)	4			
31	**Thermal Resistance, %**		**Velocities, ft/sec**		**Flow Fractions**	
32	Shell	40.78	Shellside	0.92	A	0.164
33	Tube	43.83	Tubeside	5.59	B	0.556
34	Fouling	13.43	Crossflow	1.25	C	0.019
35	Metal	1.96	Window	1.57	E	0.107
36					F	0.154

Examining the temperature profiles reveals that the shell-side fluid is in counter flow with the fluid in the final tube pass. For this reason there is no temperature reversal in this unit under clean conditions, as can be verified by running *Xist* in simulation mode. Hence, no further design modifications are indicated.

The final design parameters for the exchanger are summarized in the following table. The exchanger drawing and tube layout from *Xist* are also shown below. Eight tie rods of diameter 0.625 in. are used to consolidate the tube bundle as in Example 7.4. Note that a quadrant tube layout is the only available option for this unit.

Design Summary for Example 7.6

Parameter	Value
Exchanger type	BEU
Shell size (in.)	30
Heat-transfer area (ft^2)	1727
Number of tubes	612
Tube OD (in.)	0.75
Straight tube length (ft)	14.5
Tube BWG	14
Tube passes	4
Tube pitch (in.)	1.0
Tube layout	Square
Number of baffles	11
Baffle cut (%)	25.84
Baffle thickness (in.)	0.3125
Central baffle spacing (in.)	14.27
Inlet baffle spacing (in.)	15.58
Outlet baffle spacing (in.)	14.27
Sealing strip pairs	0
Pass-lane seal rods	8
Tube-side nozzles	6-in. schedule 40
Shell-side nozzles	6-in. schedule 40
Full U-bend support	Included
Tube-side velocity (ft/s)	5.59
Re_i	10,295
ΔP_i (psi)	9.88
ΔP_o (psi)	1.20
Over-design (%)	6.38

Exchanger Drawing for Example 7.6: Final Design

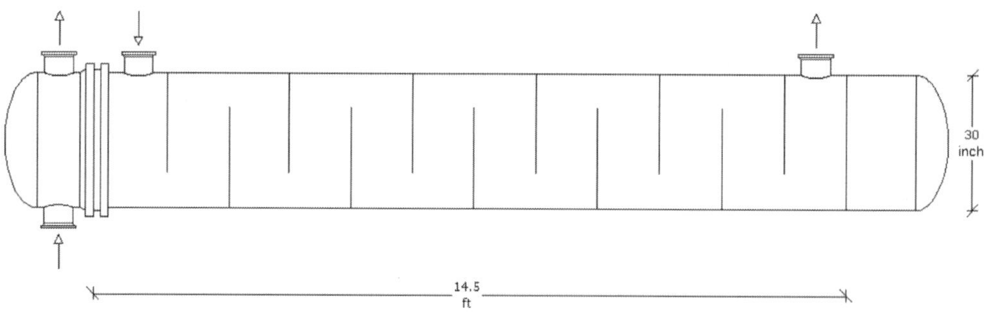

Tube Layout for Example 7.6

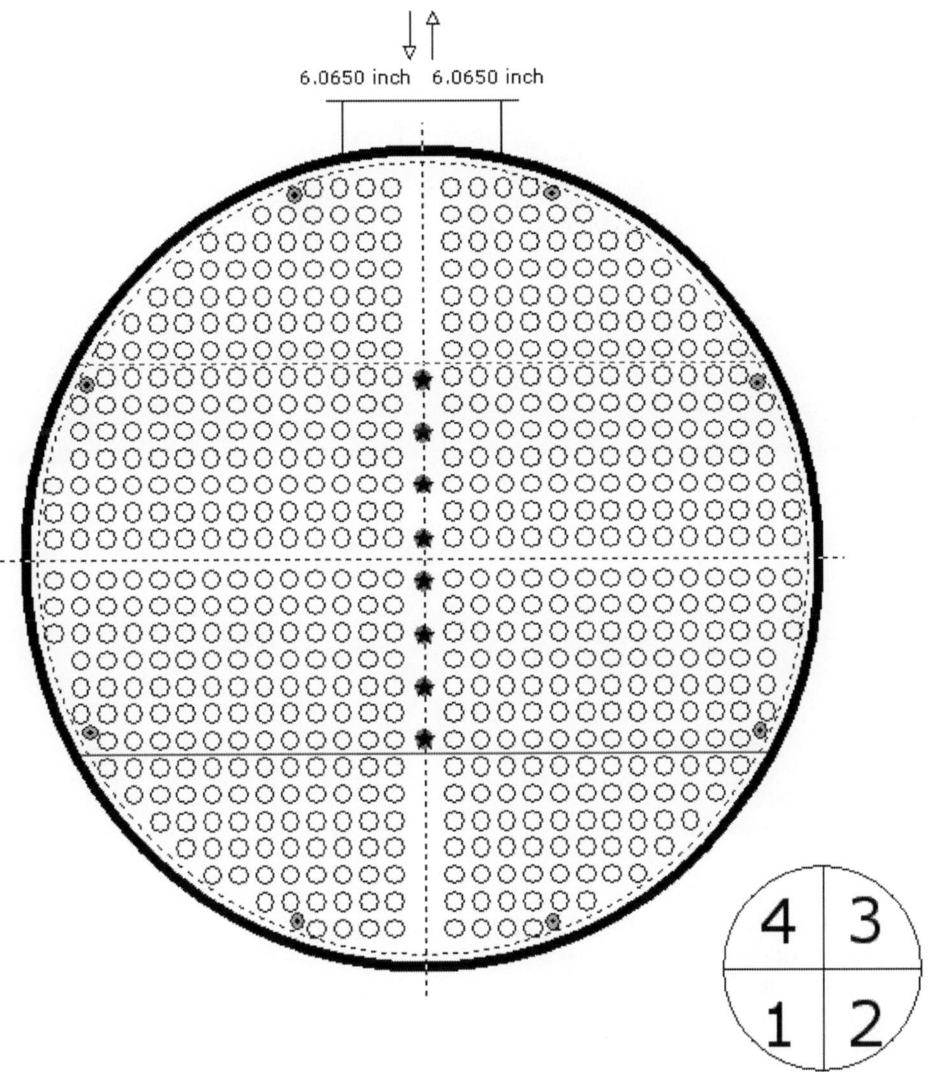

References

[1] Tinker T. Shell-side heat transfer characteristics of segmentally baffled shell-and-tube exchangers, ASME Paper No. 47-A-130. Am Soc Mech Eng 1947.

[2] Tinker T. Shell-Side characteristics of shell-and-tube heat exchangers, Parts I, II and III, Proc. General Discussion on Heat Transfer. London: Institute of Mechanical Engineers; 1951. 89–116.

[3] Tinker T. Shell-side characteristics of shell-and-tube heat exchangers-a simplified rating system for commercial heat exchangers. Trans Am Soc Mech Engrs 1958;80:36–52.

[4] Palen JW, Taborek J. Solution of shell side flow pressure drop and heat transfer by stream analysis method. Chem Eng Prog Symp Series 1969;65(No. 92):53–63.

[5] Wills MJN, Johnston D. A new and accurate hand calculation method for shell-side pressure drop and flow distribution, Proc. 22nd National Heat Transfer Conference, HTD Vol. 36. Am Soc Mech Eng 1984:67–79.

[6] Whistler AM. Effect of leakage around cross-baffles in a heat exchanger. Petroleum Refiner 1947;26(No. 10):114–8.

[7] Fisher J, Parker RO. New ideas on heat exchanger design. Hydrocarbon Proc 1969;48:147–54.

[8] Mukherjee R. Effectively design shell-and-tube heat exchangers. Chem Eng Prog 1998;94(No. 2):21–37.

[9] Wills MJN. A hand calculation method for shellside pressure drop and flow distribution in a shell-and-tube heat exchanger. Report AERE-R11136. Harwell, U.K: Atomic Energy Research Establishment; 1984.

[10] Hewitt GF, Shires GL, Bott TR. Process Heat Transfer. Boca Raton, FL: CRC Press; 1994.

Notations

A	Label for tube-to-baffle leakage stream
a	Parameter in Equation (7.44)
B	Label for cross-flow stream; (central) baffle spacing
B_c	Baffle cut
B_e	End (inlet or outlet) baffle spacing
B_{in}	Inlet baffle spacing
B_{out}	Outlet baffle spacing
B_t	Baffle thickness
b	Parameter in Equation (7.44)
C	Label for bundle-to-shell bypass stream
CF	Label for combined (bundle-to-shell and pass partition) bypass stream
D_{ctl}	Central tube limit diameter
D_e	Equivalent diameter for bypass flow
D_i	Internal diameter of tube
D_o	External diameter of tube
D_{otl}	Outer tube limit diameter
D_s	Internal diameter of shell
D_V	$(\Omega_1 P_T^2 - D_o^2)/\ D_o$
E	Label for shell-to-baffle leakage stream
F	Label for tube pass partition bypass stream; LMTD correction factor
F_c	Fraction of tubes in cross flow between baffle tips
F_w	Fraction of tubes in one baffle window
f_{ideal}	Ideal tube bank friction factor for cross flow
f_w	Ideal tube bank friction factor for window flow
G	Mass flux
G_B	Mass flux for cross flow
G_w	Mass flux for window flow
g_c	Unit conversion factor $= 32.174 \dfrac{\text{lbm}\cdot\text{ft/s}^2}{\text{lbf}} = 1.0 \dfrac{\text{kg}\cdot\text{m/s}^2}{N}$
h	Heat-transfer coefficient
h_i	Tube-side heat-transfer coefficient
h_o	Shell-side heat-transfer coefficient
h_x	Heat-transfer coefficient for cross flow
h_w	Heat-transfer coefficient for window flow
K_j	Flow resistance coefficient for stream j ($j = A, B, C, CF, E, F$)
L	Tube length
$\dot{m}$	Mass flow rate
$\dot{m}_e$	$0.5(\dot{m}_o + \dot{m}_w)$ = Effective mass flow rate in end baffle spaces
$\dot{m}_j$	Mass flow rate of stream j ($j = A, B, C, CF, E, F$)
$\dot{m}_o$	Total mass flow rate of shell-side fluid
N_c	Number of tube rows crossed between baffle tips
N_{cw}	Effective number of tube rows crossed in one baffle window
N_p	Number of tube pass partitions aligned in cross-flow direction
N_{ss}	Number of pairs of sealing strips
n_b	Number of baffles
n_p	Number of tube-side passes
n_t	Number of tubes in bundle
P	Pressure
P_T	Tube pitch
Re	Reynolds number
Re_i	Tube-side Reynolds number
Re_j	Reynolds number for stream j ($j = A, B, C, E, F$)
Re_w	Reynolds number for window flow

Re_{ew}	Effective Reynolds number for heat transfer in window flow
Re_{ex}	Effective Reynolds number for heat transfer in cross flow
S_{BW}	Flow area defined by Equation (7.20)
S_{bp}	Flow area for combined bypass stream
S_j	Flow area for stream j ($j = A, B, C, CF, E, F$)
S_m	Cross-flow area
S_s	Approximate flow area for shell-to-baffle leakage stream, defined by Equation (7.19)
S_t	Approximate flow area for tube-to-baffle leakage stream, defined by Equation (7.18)
S_w	Window-flow area
T	Temperature
T_a	Inlet temperature of shell-side fluid
T_b	Outlet temperature of shell-side fluid
t_a	Inlet temperature of tube-side fluid
U_D	Design overall heat-transfer coefficient
V_n	Fluid velocity in nozzle

Greek Letters

α	Empirical correction factor for window pressure drop, Equation (7.8)
β	Empirical correction factor for window pressure drop, Equation (7.8)
γ	Empirical correction factor for shell-side heat-transfer coefficient, Equation (7.11)
ΔP_e	Pressure drop in end (inlet or outlet) baffle space
ΔP_i	Tube-side pressure drop
ΔP_{in}	Pressure drop in inlet baffle space
ΔP_j	Pressure drop for stream j ($j = A, B, C, CF, E, F$)
ΔP_o	Total shell-side pressure drop
ΔP_{out}	Pressure drop in outlet baffle space
ΔP_w	Pressure drop in one baffle window
ΔP_x	Pressure drop in one central baffle space
ΔP_y	$\Delta P_x + \Delta P_w =$ Pressure drop across one baffle space and one baffle window
$(\Delta T_{\ln})_{cf}$	Logarithmic mean temperature difference for counter-current flow
ΔT_{mean}	Mean temperature difference in a heat exchanger
δ	Temperature profile distortion factor
δ_p	Tube-pass-partition clearance
δ_{sb}	Shell-to-baffle clearance
δ_{tb}	Tube-to-baffle clearance
ϵ_{wj}	Effectiveness of stream j ($j = A, B, C, E, F$) for heat transfer in baffle window
ϵ_{xj}	Effectiveness of stream j ($j = A, B, C, E, F$) for heat transfer in cross flow
θ_{otl}	Angle defined by Equation (7.21)
μ	Viscosity
μ_w	Viscosity of fluid at average tube wall temperature
ξ_e	Cross-flow resistance in end (inlet or outlet) baffle space
ξ_j	Flow resistance for stream j ($j = A, B, CF, E$)
ξ_o	Flow resistance defined by Equation (7.37)
ξ_w	Flow resistance for baffle window
ξ_{we}	Flow resistance in end (inlet or outlet) baffle window
ξ_x	Flow resistance defined by Equation (7.31)
ξ_y	$\xi_w + \xi_x$
ρ	Fluid density
ϕ	Viscosity correction factor
ψ	Correction factor for shell-side pressure drop, defined in Equation (7.60)
Ω_1	Parameter in correlation for cross-flow resistance, Equation (7.44)
Ω_2	Parameter in correlation for bypass flow resistance, Equation (7.46)

Problems

(7.1) The data given below pertain to the shell-side flow in a shell-and-tube heat exchanger. The tube bundle consists of 1-in. OD tubes laid out on 1.25-in. square pitch. The shell-side fluid has a flow rate of 1.38 kg/s and a density of 1,200 kg/m^3.

Stream	Flow Area (m^2)	Flow Resistance (kg^{-1} m^{-1})
Cross flow	0.035096	3.1
Bypass	0.007400	23.6
Tube-baffle leakage	0.006893	12.0
Shell-baffle leakage	0.003942	48.0
Window flow	0.025500	—

Use the Wills–Johnston method to calculate the following:

(a) The flow fraction of the cross-flow stream.
(b) The pressure drop across one baffle compartment. (This does not include the baffle window.)
(c) The pressure drop in one baffle window.
(d) The pressure drop in one end (inlet or outlet) baffle space if $B/B_e = 0.6$ and $\xi_e = 0.90$ kg^{-1} m^{-1}.
(e) The nozzle-to-nozzle shell side pressure drop if the exchanger contains 20 baffles and the flow regime for all shell-side streams is fully turbulent.

Ans. (a) 0.40. (b) 0.94 Pa. (c) 1.14 Pa. (d) 1.49 Pa. (e) 42.5 Pa.

(7.2) Rework Example 7.2 for the case in which one of the tube-pass partitions is aligned with the cross-flow direction, as in an H-banded tube layout. Assume that the pass partition clearance is equal to the tube OD.

(7.3) The data given below pertain to the shell-side flow in a shell-and-tube heat exchanger. The tube bundle consists of 0.75-in. OD tubes laid out on 1.0-in. square pitch. The flow rate of the shell-side fluid is 50,000 lb/h and its density is 55 lbm/ft^3.

Stream	Flow Area (ft^2)	Flow Resistance $\left(\frac{\text{lbf}\cdot\text{s}^2}{\text{lbm}^2\cdot\text{ft}^2}\right)$
Cross flow	0.180	0.30
Bypass	0.050	11.10
Tube-baffle leakage	0.055	4.20
Shell-baffle leakage	0.040	4.80
Window flow	0.275	—

Use the Wills–Johnston method to calculate the following:

(a) The flow fractions of the cross flow, leakage and bypass streams.
(b) The pressure drop across one baffle compartment.
(c) The pressure drop in one baffle window.
(d) The pressure drop in one end (inlet or outlet) baffle space if $B/B_e = 0.5$ and $\xi_e = 0.1\frac{\text{lbf}\cdot\text{s}^2}{\text{lbm}^2\cdot\text{ft}^2}$.
(e) The nozzle-to-nozzle shell-side pressure drop if the exchanger contains 37 baffles and the flow regime for all shell-side streams is fully turbulent.

Ans. (a) cross flow = 0.587, bypass = 0.096, tube-baffle leakage = 0.164, shell-baffle leakage = 0.153. (b) 19.9 lbf/ft^2. (c) 1.8 lbf/ft^2. (d) 14.2 lbf/ft^2. (e) 5.6 psi.

(7.4) Use the Wills–Johnston method to calculate the shell-side pressure drop for the heat exchanger of Problems 5.3 and 6.4. Assume that all tube pass partitions are oriented normal to the cross-flow direction, so that $N_p = 0$.

(7.5) Use the Wills–Johnston method to calculate the shell-side pressure drop for the heat exchanger of Problem 6.6. Assume that all tube pass partitions are oriented normal to the cross-flow direction, so that $N_p = 0$.

(7.6) Use the Wills–Johnston method to calculate the shell-side pressure drop for the modified configuration of the kerosene/oil exchanger considered previously in Problems 5.1 and 6.8. Assume that all tube pass partitions are oriented normal to the cross-flow direction, so that $N_p = 0$.

(7.7) Use the Wills–Johnston method to calculate the nozzle-to-nozzle shell-side pressure drop for the heat exchanger of Problem 6.11. Assume that the tube pass partition is oriented normal to the cross-flow direction.

(7.8) Use *Xist* to design a heat exchanger with two tube passes for the kerosene/oil service of Example 5.1.

(7.9) Use *Xist* to design a heat exchanger for the service of Problem 5.6 with:
(a) Plain tubes.
(b) Finned tubes.

(7.10) Design a heat exchanger for the service of Problem 5.8 using *Xist*.

(7.11) Design a heat exchanger for the service of Problem 5.9 using *Xist*.

(7.12) Use *Xist* to design a heat exchanger for the service of Problem 5.11 with:
(a) Plain tubes.
(b) Finned tubes.

(7.13) Design a heat exchanger for the service of Problem 5.13 using *Xist*.

(7.14) Use *Xist* to design a heat exchanger for the service of Problem 5.23 with:
(a) Plain tubes.
(b) Finned tubes.

(7.15) Solve Problem 5.5 using *Xist*.

(7.16) Design a heat exchanger for the service of Problem 5.10 using *Xist*.

(7.17) Design a heat exchanger for the service of Problem 5.15 using *Xist*.

(7.18) Design a heat exchanger for the service of Problem 5.24 using *Xist*.

(7.19) For the oil cooler of Example 5.2:
(a) Use *Xist* to rate the final design obtained by hand in Chapter 5. What is the over-design for the unit according to *Xist*?
(b) Run *Xist* in design mode to develop a *de novo* design for this unit.

(7.20) Repeat Example 7.2 without the δ_{sb} safety factor, i.e., using $\delta_{sb} = 1.7779$ mm $= 0.069996$ in.

(7.21) For the final configuration of the kerosene/oil exchanger of Example 5.1, use Equation (2.38) to calculate the tube-side heat-transfer coefficient. Compare your result with the value computed by *Xist* in Example 7.3.

(7.22) A shell-and-tube heat exchanger has an LMTD of 50°F and an LMTD correction factor of 0.9. When the exchanger is rated using *Xist*, the program output shows an effective mean temperature difference of 40.4°F. What is the explanation for the apparent discrepancy in mean temperature difference?

8 HEAT-Exchanger Networks

8.1 Introduction

The transport of energy in chemical plants and petroleum refineries is accomplished by means of heat-exchanger networks (HENs). Hence, the design of such networks is an important aspect of chemical process design. Each individual heat exchanger in the network can be designed by the methods discussed in previous chapters. But before this can be done, the overall configuration of the HEN must be determined. Decisions must be made as to how the process streams will be paired for heat exchange and the extent of the heat exchange between each pair.

The Pinch Design method provides a systematic approach for configuring HENs based on fundamental thermodynamic principles and simple heuristic rules. This chapter provides an introduction to the Pinch Design method.

8.2 An Example: TC3

Test Case Number 3, or TC3 for short, is a simple HEN problem devised by Linnhoff and Hindmarsh [1] to illustrate the concepts involved in the Pinch Design method. The problem has four process streams as shown in Table 8.1. The nomenclature used in this table is as follows:

$CP \equiv \dot{m}C_P$ = heat capacity flow rate
TS = supply temperature
TT = target temperature

In addition to the four process streams, there is a single hot utility available at a temperature above 150°C and a single cold utility available at a temperature below 20°C. The minimum approach, ΔT_{min}, for all heat exchangers is specified as 20°C. This is the smallest allowable temperature difference between hot and cold streams (including utilities) at either end of a co-current or counter-current exchanger. For purposes of this example, all exchangers are assumed to operate counter-currently.

The problem is to design a HEN to bring each process stream from its supply temperature to its target temperature. A trivial solution is obtained by supplying all heating and cooling requirements from utilities. The resulting network consists of four units (two heaters and two coolers) and consumes a total of 487.5 kW of hot utility and 420 kW of cold utility. This represents the limiting case of zero heat recovery.

A non-trivial solution can be obtained with a little more effort, e.g., by pairing each hot stream with the cold stream that most closely matches its duty. The resulting network is shown on a grid diagram in Figure 8.1. It consists of two process exchangers, two heaters and one cooler. The total hot utility consumption is 182.5 kW, cold utility consumption is 115 kW, and the heat recovery amounts to 305 kW (the total rate of heat exchange between process streams). Notice that in this network the heat exchange between Streams 2 and 3 is limited to 125 kW by the temperature difference at the hot end of the exchanger. A larger duty would result in a violation of ΔT_{min} = 20°C.

Other solutions to TC3 are clearly possible. The optimal solution is the one that minimizes the total cost, i.e., the capital cost of the heat exchange equipment and the operating cost of the system, including the cost of the utilities consumed.

8.3 Design Targets

Capital and operating costs are inconvenient to calculate, especially in the early stages of process design. Therefore, HEN design methods have traditionally used surrogates for these costs, namely, the number of units (process exchangers, heaters, and coolers) in the HEN and the amounts of hot and cold utilities consumed. Optimal networks are expected to have close to the minimum number of units because extra units require additional foundations, piping, fittings, and instrumentation that greatly increase the capital cost. Likewise, using more utilities than necessary increases the operating cost. Therefore, networks for which the number of units and the utility usage are both close to their respective minimum values should be close to optimal.

TABLE 8.1 Stream Data for TC3

Stream	Type	TS (°C)	TT (°C)	CP (kW/°C)	Duty (kW)
1	Hot	150	60	2.0	180
2	Hot	90	60	8.0	240
3	Cold	20	125	2.5	262.5
4	Cold	25	100	3.0	225

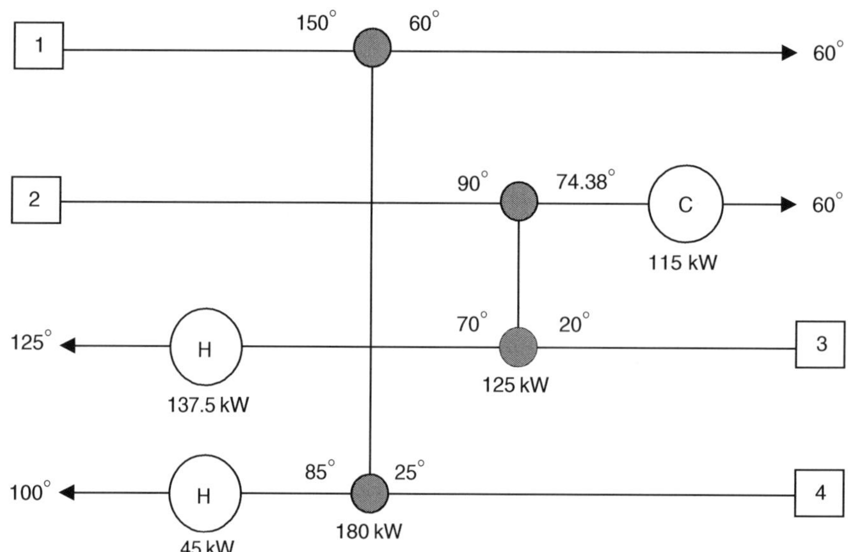

FIGURE 8.1 A non-trivial solution for TC3.

Design targets for minimum number of units and minimum-utility usage are easy to calculate and are independent of any specific network design. Hence, they can be computed at the outset and used as standards of comparison for all HEN designs that are subsequently developed. In a similar manner, the design targets can be used to assess the potential improvement that could be obtained by retrofitting an existing HEN.

The minimum number of units, U_{min}, for a given HEN problem can be estimated as follows [1]:

$$U_{min} = N - 1 \tag{8.1}$$

where N is the number of process streams and utilities. For example, in problem TC3 there are four process streams and two utilities, giving $N = 6$ and $U_{min} = 5$. Notice that this is not the absolute minimum because the trivial solution with no heat recovery requires only four units. Equation (8.1) also does not account for the situation in which hot and cold stream duties exactly match, so that a single exchanger brings both streams to their target temperatures. These cases are of little practical interest, however, and with these exceptions, Equation (8.1) usually gives the correct result.

Minimum-utility requirements for a given HEN problem can be calculated in two ways, using the problem table algorithm of Linnhoff and Flower [2] or using composite curves. These methods are described in the following sections.

8.4 The Problem Table

In problem TC3, the total duty of the hot streams is 420 kW while that of the cold streams is 487.5 kW. Thus, if all the heat available from the hot streams were transferred to the cold streams, the hot utility would have to supply the remaining 67.5 kW of heating and no cold utility would be needed. Thus, based on an energy balance (first law of thermodynamics), it appears that the minimum hot utility required for this problem is 67.5 kW and the minimum cold utility is zero. However, this result is incorrect because it ignores the constraint imposed by the second law of thermodynamics, i.e., heat exchange between hot and cold streams can occur only when the temperature of the hot stream exceeds that of the cold stream. In fact, since $\Delta T_{min} = 20°C$ for this problem, the temperature of the hot stream must exceed that of the cold stream by at least 20°C. (It is shown in the following section that it is possible to utilize all the heat available in the hot streams of TC3 if ΔT_{min} is reduced by a sufficient amount. This result does not hold in general, however.)

The problem table algorithm accounts for second law constraints and the effect of ΔT_{min} by dividing the temperature range into intervals, referred to as subnetworks (SNs), that are determined by the supply and target temperatures of each hot and cold stream. The heating and cooling requirements are then calculated for each SN, and any excess heat is transferred to the next lowest temperature interval. This procedure enforces the requirement that heat available at a given temperature level can be used only at lower levels. The problem table for TC3 is given below as Table 8.2.

The steps in constructing the problem table are as follows:

Step 1. *Construct the temperature scales:* Separate temperature scales differing by $\Delta T_{min} = 20°C$ are constructed for the hot and cold streams using the supply and target temperatures of the streams. For the cold streams, these temperatures are 20°C, 25°C, 100°C,

TABLE 8.2 The Problem Table for TC3

Subnetwork	Cold streams		T (°C)		Hot streams		Deficit	Heat flow		Adj. heat flow	
	(3)	(4)			(1)	(2)		In	Out	In	Out
			130	150							
SN1			125	145	↓	↓	−10	0	10	107.5	117.5
SN2	↑		100	120	↓	↓	12.5	10	−2.5	117.5	105
SN3	↑	↑	70	90	↓	↓	105	−2.5	−107.5	105	0
SN4	↑	↑	40	60	↓	↓	−135	−107.5	27.5	0	135
SN5	↑	↑	25	45			82.5	27.5	−55	135	52.5
SN6	↑		20	40			12.5	−55	−67.5	52.5	40

and 125°C. To this list are added the values obtained by subtracting ΔT_{min} from each hot stream terminal temperature, giving 130°C, 70°C, and 40°C. The resulting seven temperatures are placed in ascending order on the cold stream temperature grid. Adding ΔT_{min} to each of the seven temperatures then gives the hot stream temperature scale.

Step 2. *Add SNs and streams:* Each interval on the temperature scales corresponds to a SN. Thus, there are six SNs for TC3. For each stream, arrows are drawn from the supply temperature level to the target temperature level to indicate which streams occur in each SN.

Step 3. *Calculate deficits:* The energy deficit in each SN is the difference between the energy required to heat the cold streams and the energy available from cooling the hot streams. It is calculated as follows:

$$\text{deficit:} = \left[\sum_i (CP)_{cold,i} - \sum_i (CP)_{hot,i}\right]\Delta T \tag{8.2}$$

where ΔT is the magnitude of the temperature difference across the SN and the summations are over only those streams that exist in the SN. Thus, for SN1, only Stream 1 occurs and Equation (8.2) becomes:

$$\text{deficit} = -CP_1\Delta T = -2.0 \times 5 = -10 \text{ kW}$$

Note that a negative deficit represents a surplus of energy that can be used at lower temperature levels in the network. In SN2, Streams 1 and 3 occur. Therefore,

$$\text{deficit} = [CP_3 - CP_1]\Delta T = [2.5 - 2.0] \times 25 = 12.5 \text{ kW}$$

The remaining deficits are similarly computed.

Step 4. *Calculate heat flows:* For each SN, the output is the input minus the deficit. The energy output is transferred to the next lower temperature level and becomes the input for the next SN. To start the calculation, the input to SN1 is assumed to be zero. Thus, for SN1 the output is 10 kW, which becomes the input for SN2. Subtracting the deficit of 12.5 kW for SN2 gives an output of -2.5 kW, and so on.

Step 5. *Calculate adjusted heat flows:* Negative heat flows must be eliminated by addition of heat from the hot utility. Since there is only one hot utility in the problem and it is available above 150°C, the heat is added at the top of the energy cascade at SN1. The amount of energy that must be supplied corresponds to the largest negative heat flow, namely, 107.5 kW. Taking this value as the input to SN1, the remaining heat flows are computed as before to give the adjusted heat flows.

Now the minimum hot utility requirement for the network is the energy supplied as input to SN1, i.e., 107.5 kW. Likewise, the minimum cold utility requirement is the energy removed from SN6 at the bottom of the cascade, i.e., 40 kW.

The point of zero energy flow in the cascade is called the pinch. The problem table shows that the pinch occurs at a hot stream temperature of 90°C and a cold stream temperature of 70°C. The significance of the pinch is discussed in subsequent sections.

8.5 Composite Curves

Composite curves are essentially temperature–enthalpy diagrams for the combined hot streams and combined cold streams, respectively. Their construction is illustrated by reference to problem TC3.

TABLE 8.3 Enthalpy Calculation for Hot Composite Curve of TC3

Temperature (°C) Interval	ΔH (kW) = ($\Sigma CP_{hot,i}$) ΔT
60–90	$(2 + 8) \times 30 = 300$
90–150	$2 \times 60 = 120$

TABLE 8.4 Data Points for Hot Composite Curve of TC3

T (°C)	H (kW)
60	0
90	300
150	420

To construct the composite curve for the hot streams, we start at the lowest temperature level, 60°C, and calculate the change in enthalpy (actually enthalpy per unit time) for each temperature subinterval determined from the hot stream terminal temperatures. The calculations are shown in Table 8.3.

Now the reference temperature for enthalpy is taken to be the lowest temperature level, 60°C, for the hot streams, and the enthalpy at this temperature is set to zero. The ΔH values from Table 8.3 are then used to obtain the data points shown in Table 8.4. The points are plotted in Figure 8.2.

Exactly the same procedure is followed to construct the cold composite curve. The calculations are summarized in Table 8.5 and the data points to be plotted are given in Table 8.6. The cold composite curve is plotted in Figure 8.2 along with the hot composite curve.

It can be seen from Figure 8.2 that the point of closest approach between the hot and cold curves occurs at an enthalpy of 300 kW. The hot stream temperature at this point is 90°C and the cold stream temperature is between 70°C and 80°C. The exact cold stream temperature can be found from the equation for the line segment between 25°C and 100°C, which is easily determined using the data points in Table 8.5:

$$T = 22.727 + 0.18182\,H \tag{8.3}$$

Setting $H = 300$ in this equation gives $T = 77.273$°C. Thus, the minimum temperature difference between the hot and cold curves is:

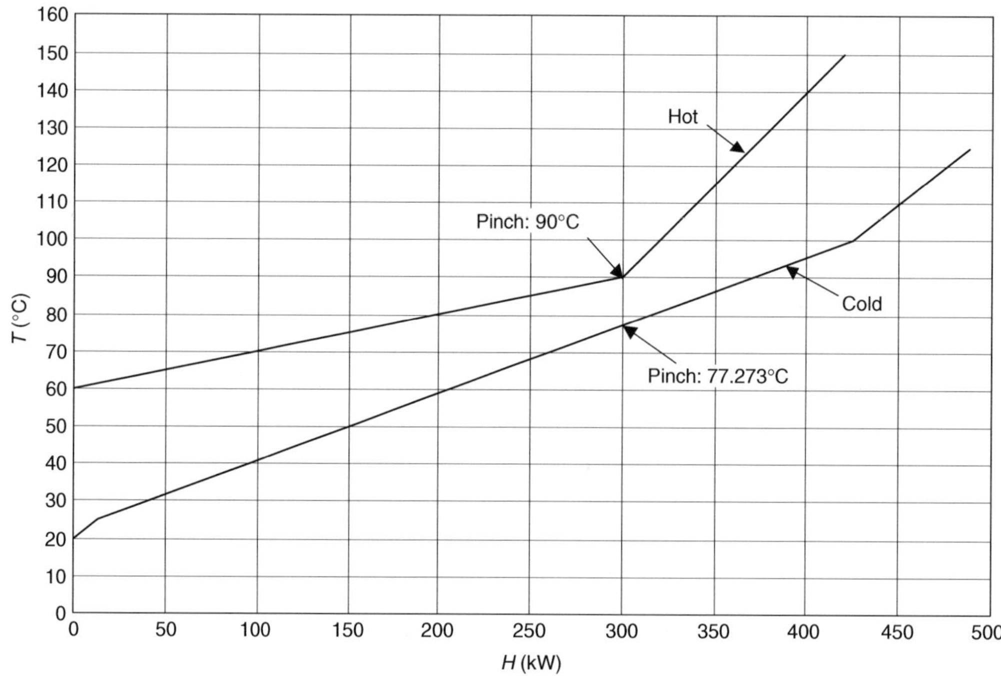

FIGURE 8.2 Composite curves for TC3.

TABLE 8.5 Enthalpy Calculation for Cold Composite Curve of TC3

Temperature (°C) Interval	ΔH (kW) $= (\Sigma CP_{cold,i})\ \Delta T$
20–25	$2.5 \times 5 = 12.5$
25–100	$(2.5 + 3) \times 75 = 412.5$
100–125	$2.5 \times 25 = 62.5$

TABLE 8.6 Data Points for Cold Composite Curve of TC3

T (°C)	H (kW)
20	0
25	12.5
100	425
125	487.5

$$\Delta T_{min} = 90 - 77.273 = 12.727°\text{C}$$

This is less than the minimum approach of 20°C for TC3. Since the reference point for enthalpy was set arbitrarily for each curve, either (or both) curve(s) can be shifted parallel to the H-axis. From Figure 8.2, it can be seen that shifting the cold curve to the right will increase ΔT_{min}. In order to determine the minimum heating and cooling requirements, the value of ΔT_{min} must be increased to 20°C. This will occur when the cold curve is shifted so that its temperature at $H = 300$ kW is 70°C. From Equation (8.3) we find that $H = 260$ kW when $T = 70$°C. Therefore, a shift of 40 kW will bring the two curves into proper alignment, as shown in Figure 8.3.

The region where the hot and cold curves overlap represents (potential) heat exchange between hot and cold process streams. In Figure 8.3 this region extends from $H = 40$ kW to $H = 420$ kW. The difference, 380 kW, represents the maximum possible heat exchange for the network (with $\Delta T_{min} = 20$°C). In the region from $H = 0$ to $H = 40$ kW, where only the hot curve exists, cold utility must be used to cool the hot streams. The duty of 40 kW represents the minimum cold utility requirement for the network. Similarly, in the region from $H = 420$ kW to $H = 527.5$ kW, hot utility must be used. The duty of 107.5 kW represents the minimum hot utility requirement for the network. The point of closest approach between the hot and cold curves is the pinch. Thus, Figure 8.3 gives the hot and cold stream temperatures at the pinch, 90°C and 70°C, respectively.

In Figure 8.2 it can be seen that the maximum possible amount of overlap between the hot and cold curves is attained, and as previously determined, $\Delta T_{min} = 12.727$°C. The maximum possible heat recovery (420 kW) occurs in this situation. Any further

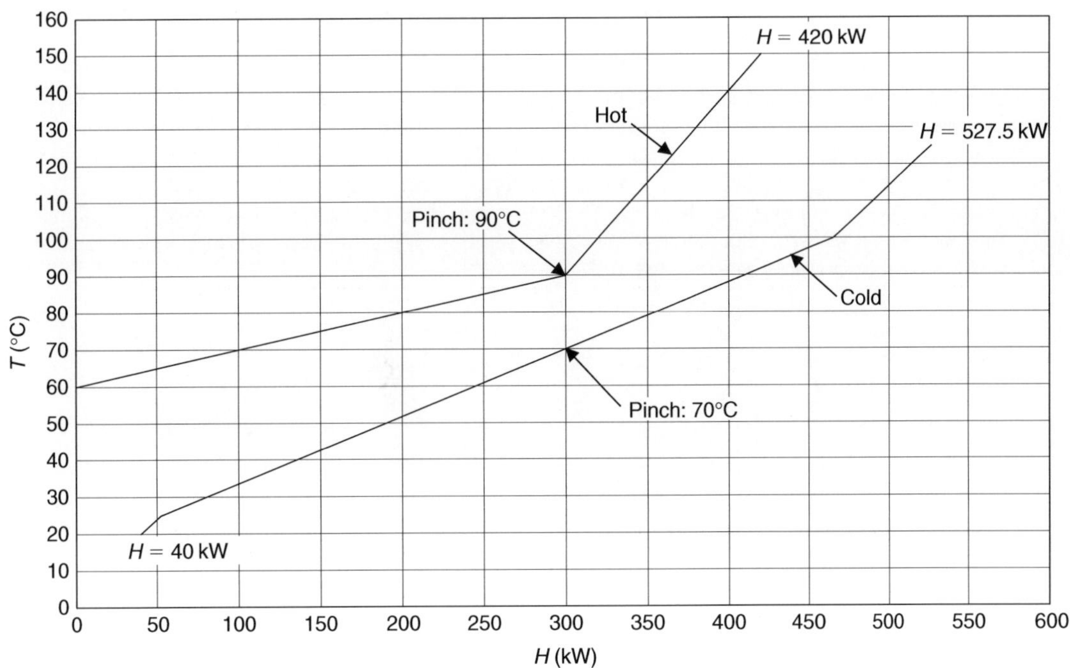

FIGURE 8.3 Shifted composite curves for TC3.

reduction in ΔT_{min} does not increase the amount of heat that is recoverable. At the other extreme, by shifting the cold curve to the right by 420 kW, the point of zero overlap between the hot and cold curves is reached. Here there is no heat recovery and all heating and cooling is provided by utilities. The value of ΔT_{min} at this point is 150°C – 20°C = 130°C. Thus, the practical range for ΔT_{min} is 12.727°C to 130°C. For any value in this range, the maximum heat exchange and minimum-utility requirements can be determined from the composite curves.

Although the above analysis provides useful information regarding the operating limits of any potential HEN, it ignores constraints imposed by utility temperatures. For example, suppose the temperature of the hot utility in TC3 is 200°C. Since the highest target temperature for the cold streams is 125°C, an allowable driving force of 75°C is required for heating. Therefore, a value of ΔT_{min} greater than 75°C is incompatible with the operating requirements of utility exchangers. In general, both hot and cold utility temperatures must be considered in determining the maximum allowable value for ΔT_{min}.

As ΔT_{min} is decreased from its maximum value, the extent of heat recovery increases while the amount and cost of utility usage decreases. At the same time, the surface area required for process heat exchangers increases, due to greater total heat duty and smaller average driving force for heat transfer. Thus, for any HEN problem there is an optimal value of ΔT_{min} that minimizes the total cost. The procedure for estimating this optimum is referred to as super targeting [3,4].

It should be noted that the method of composite curves is applicable to systems in which the stream enthalpies are nonlinear functions of temperature, including those involving condensation and boiling. Although the calculations are more complicated in these cases, there is no fundamental problem in applying the method. The problem table algorithm presented in the previous section can also be applied in these cases by using piecewise linear functions to approximate the nonlinear temperature–enthalpy relationships.

8.6 The Grand Composite Curve

The grand composite curve is a plot of adjusted heat flow versus the average of the hot and cold stream temperatures. The points to be plotted are obtained directly from the problem table, and for TC3 are listed in Table 8.7. The data are plotted in Figure 8.4.

The grand composite curve embodies the same basic information as the problem table. The intercept at 80°C on the temperature axis represents the pinch. The two end points of the curve give the minimum hot and cold utility requirements. The various line segments that constitute the graph display the magnitudes of heat sources and sinks and the temperature intervals over which they occur. Net heat is produced by the system over temperature intervals where the line segments slope downward and to the right. Conversely, net heat is consumed over intervals where the line segments slope downward and to the left.

When utilities are available at different temperature levels, as is normally the case in chemical and petroleum processing, these levels are indicated on the diagram to show their positions relative to heat sources and sinks in the process. The grand composite curve thus serves as an aid in properly matching utilities with process requirements. Another use is in process integration work, where it is used to determine the appropriate placement of unit operations within the overall network for effective energy utilization. Unit operations involving heat pumps, heat engines, distillation columns, evaporators, furnaces, etc., that can be represented in terms of heat sources and sinks, can be analyzed in this manner [3,4].

As an example, suppose there is a second hot utility in TC3 that is available at a temperature of 110°C. For simplicity, we will assume that the temperature change for this utility is negligible, so that it can be used to supply heat to cold streams at temperatures of 90°C or less. This temperature corresponds to 100°C on the graph because the temperatures used for the grand composite curve are averages of hot and cold stream temperatures. Using the procedure shown in Figure 8.5, it is found that 70 kW of hot utility could be supplied by the second utility (HU2), leaving 37.5 kW to be supplied by the hotter (and presumably more expensive) utility (HU1). The remaining cold stream duty of 10 kW above the pinch is supplied by heat exchange with the hot streams. This heat exchange is represented by the uppermost triangle in Figure 8.5.

TABLE 8.7 Data Points for Grand Composite Curve of TC3

T (°C)	H (kW)
140	107.5
135	117.5
110	105
80	0
50	135
35	52.5
30	40

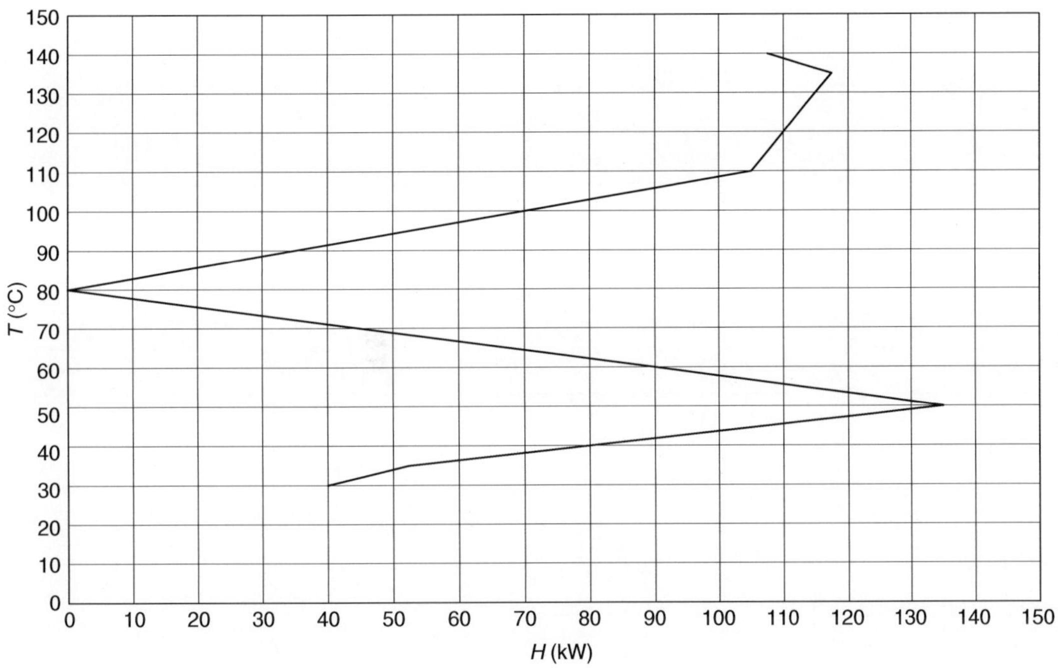

FIGURE 8.4 Grand composite curve for TC3.

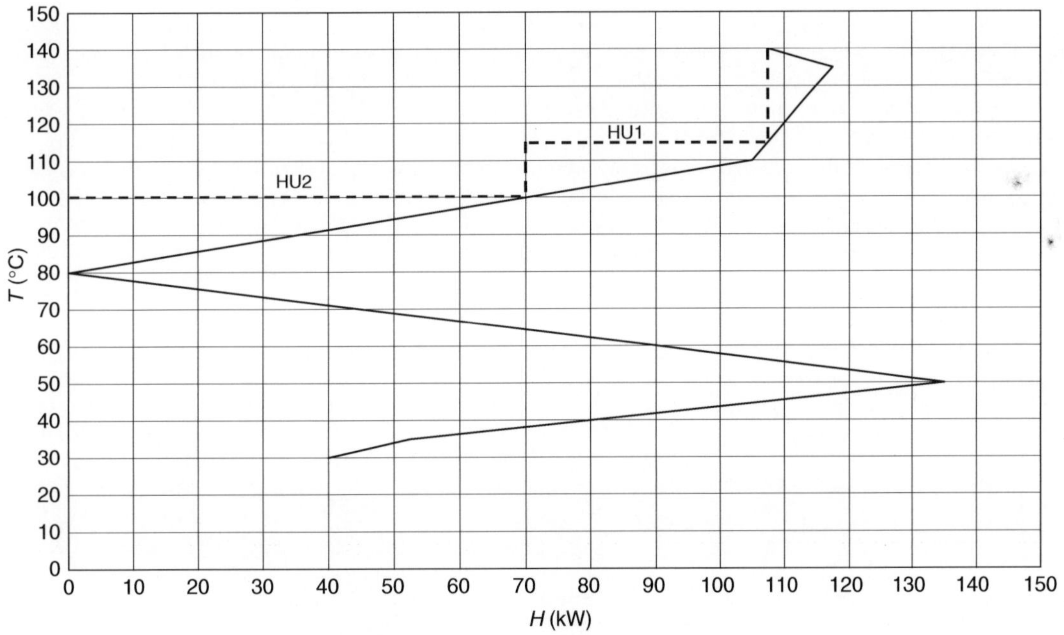

FIGURE 8.5 Using the grand composite curve to determine utility requirements when two hot utilities are available.

8.7 Significance of the Pinch

The pinch is the point of closest approach between the hot and cold composite curves. It is also the point of zero energy flow in a network with minimum-utility usage. The latter representation has some important consequences for HEN design.

Referring to the problem table for TC3 (Table 8.2), suppose we transfer some energy across the pinch, say 10 kW. To do this, we simply increase all the adjusted heat flows in the last two columns of the table by 10 kW. The hot utility requirement is now 117.5 kW and the cold utility requirement is 50 kW. Thus, we see that transferring energy across the pinch increases both hot and

TABLE 8.8 The Hot-End Problem for TC3

Stream	Type	*TS* (°C)	*TT* (°C)	*CP* (kW/°C)	Duty (kW)
1	Hot	150	90	2.0	120
3	Cold	70	125	2.5	137.5
4	Cold	70	100	3.0	90

TABLE 8.9 The Cold-End Problem for TC3

Stream	Type	*TS* (°C)	*TT* (°C)	*CP* (kW/°C)	Duty (kW)
1	Hot	90	60	2.0	60
2	Hot	90	60	8.0	240
3	Cold	20	70	2.5	125
4	Cold	25	70	3.0	135

cold utility requirements. It therefore tends to be inefficient and expensive, and should generally be avoided to the extent that is practical.

One way to transfer energy across the pinch is to put a heater below the pinch. In such a unit heat is transferred directly from the hot utility, which is hotter than the pinch temperature, to a cold stream that is colder than the pinch temperature. Thus, the heat flows from the hot utility temperature, across the pinch to the cold stream temperature. Similarly, placing a cooler above the pinch results in heat flow directly from the hot stream temperature above the pinch to the cold utility temperature below the pinch. It follows that heaters should be used only above the pinch, while coolers should be used only below the pinch.

Since any energy transfer across the pinch increases utility requirements, it is clear that for a minimum-utility design, there can be no energy flow across the pinch. This allows the HEN design problem to be decomposed into two separate problems, the hot-end problem above the pinch and the cold-end problem below the pinch. The two problems can be solved separately and then combined to obtain a solution for the entire network. The hot- and cold-end problems for TC3 are summarized in Tables 8.8 and 8.9, respectively. Note that stream 2 does not appear in the hot-end problem because its temperature range, from 90°C to 60°C, falls below the pinch temperature of 90°C for the hot streams.

8.8 Threshold Problems and Utility Pinches

Virtually all industrial HEN problems contain a pinch. However, it is possible to construct problems that do not exhibit a pinch. This can be done by setting ΔT_{min} below the value at which the maximum possible amount of heat recovery is attained. For TC3, this value is approximately 12.727°C, as previously shown. The problem table for TC3 with ΔT_{min} set at 10°C is given as Table 8.10. Although the composite curves still exhibit a point of closest approach, at 90°C on the hot side and 80°C on the cold side in this

TABLE 8.10 The Problem Table for TC3 with $\Delta T_{min} = 10°C$.

Subnetwork	Cold streams (3)	Cold streams (4)	*T* (°C)		Hot streams (1)	Hot streams (2)	Deficit	Heat flow In	Heat flow Out	Adj. heat flow In	Adj. heat flow Out
			140	150							
SN1			125	135			−30	0	30	67.5	97.5
SN2			100	110			12.5	30	17.5	97.5	85
SN3			80	90			70	17.5	− 52.5	85	15
SN4			50	60			−135	−52.5	82.5	15	150
SN5			25	35			137.5	82.5	−55	150	12.5
SN6			20	30			12.5	−55	−67.5	12.5	0

case, the energy flow is not equal to zero at this point. The energy flow becomes zero at the very bottom of the energy cascade, indicating that no cold utility is required. In effect, the entire HEN problem consists of a hot-end problem; there is no cold-end problem.

As long as ΔT_{min} for TC3 is less than 12.727°C, it places no restriction on the amount of heat that can be transferred between the hot and cold streams. Only when this threshold value is exceeded does the heat transfer become restricted. For this reason, problems for which ΔT_{min} is below the threshold value are referred to as threshold problems [1]. For a pinch to occur, it is necessary to have $\Delta T_{min} \geq \Delta T_{threshold}$.

It is also possible for HEN problems to contain more than a single pinch. This situation occurs when utilities are available at different temperature levels; e.g., high-, intermediate-, and low-pressure steam. Since there is usually an excess of low-pressure steam in chemical plants and petroleum refineries, it is economical to utilize this utility whenever possible. Higher-pressure steam is normally used only when the required steam temperature exceeds that of the lower pressure steam. This situation can result in additional pinches at the temperature levels of the available utilities, thereby allowing further decomposition of the HEN problem. For obvious reasons, these additional pinches are referred to as utility pinches.

8.9 Feasibility Criteria at the Pinch

The pinch represents the most temperature-constrained region of a HEN, and as such, it is the most critical region for HEN design. Due to the low driving force, exchangers operating near the pinch tend to be relatively large and expensive units. Identification of appropriate stream matches for exchangers operating at the pinch is a key aspect of the Pinch Design method. Exchangers that operate at the pinch have a temperature difference equal to ΔT_{min} on at least one end. Three feasibility criteria applied to the stream data at the pinch are used in determining appropriate matches. These criteria are discussed in the following subsections.

8.9.1 Number of Process Streams and Branches

For a minimum-utility design, coolers are not permitted above the pinch. Therefore, in the hot-end problem, each hot stream that intersects the pinch must be brought to the pinch temperature by heat exchange with a cold process stream. For this to occur, the following inequality must be satisfied in the hot-end problem:

$$NH \leq NC \quad \text{(hot end)} \tag{8.4}$$

where

NH = number of hot streams or branches at the pinch
NC = number of cold streams or branches at the pinch

A similar argument shows that for the cold-end problem, the inequality is reversed:

$$NH \geq NC \quad \text{(cold end)} \tag{8.5}$$

Stream splitting, i.e., dividing a stream into two or more branches, may be required in order to satisfy these inequalities.

8.9.2 The *CP* Inequality

For an exchanger operating at the pinch, $\Delta T = \Delta T_{min}$ at the pinch end of the exchanger. Therefore, the temperature difference between the two streams cannot decrease from the pinch end to the other end of the exchanger. This requirement leads to the following inequalities that must be satisfied for every stream match at the pinch:

$$CPH \leq CPC \quad \text{(hot end)} \tag{8.6}$$

$$CPH \geq CPC \quad \text{(cold end)} \tag{8.7}$$

where CPH and CPC are the heat capacity flow rates of the hot and cold streams, respectively. Violation of these inequalities will result in $\Delta T < \Delta T_{min}$ in the heat exchanger. Figure 8.6 shows a pinch match between Streams 1 and 4 in the cold-end problem for TC3. Note that inequality (8.7) is violated and as a result, the driving force in the exchanger decreases from ΔT_{min} = 20°C at the pinch end to 10°C at the opposite end.

Note that the *CP* inequalities hold only for exchangers operating at the pinch. Away from the pinch, temperature differences between hot and cold streams may be large enough to permit matches that violate these inequalities.

8.9.3 The *CP* Difference

For an individual exchanger operating at the pinch, the *CP* difference is defined as follows:

$$\begin{aligned} CP\text{ difference} &= CPC - CPH \quad \text{(hot end)} \\ &= CPH - CPC \quad \text{(cold end)} \end{aligned} \tag{8.8}$$

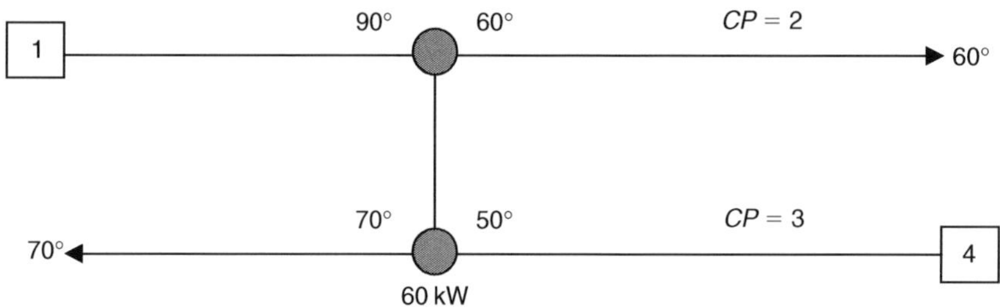

FIGURE 8.6 A pinch match that violates the CP inequality for the cold end of TC3.

Note that the *CP* difference must be non-negative according to the *CP* inequality criterion. The overall *CP* difference is defined as follows:

$$\begin{aligned}\text{overall } CP \text{ difference} &= \sum CPC - \sum CPH \quad \text{(hot end)} \\ &= \sum CPH - \sum CPC \quad \text{(cold end)}\end{aligned} \tag{8.9}$$

The summations in Equation (8.9) are taken over only those streams that intersect the pinch.

The third feasibility condition states that the *CP* difference for any exchanger operating at the pinch must not exceed the overall *CP* difference. The reason is that if one exchanger has a *CP* difference greater than the total, then another must have a negative *CP* difference that violates the *CP* inequality.

8.9.4 The *CP* Table

The *CP* table is a convenient means of identifying feasible pinch matches. The *CP* table for the hot end problem of TC3 is given as Table 8.11. The *CP* values for the hot and cold streams are listed separately in descending order, and the sums of these values are entered below each column. As a reminder, the feasibility criteria are given at the top of the table. Note that only those streams that intersect the pinch are included in the *CP* table.

It can be seen by inspection that the number of hot streams is less than the number of cold streams, and that the *CP* inequality is satisfied for each of the possible matches. In addition, the overall *CP* difference is immediately obtained from the *CP* totals as 3.5. Now the individual *CP* differences for the two possible matches are as follows:

Stream 1 and Stream 4: *CP* difference = 3 − 2 = 1
Stream 1 and Stream 3: *CP* difference = 2.5 − 2 = 0.5

Since both of these values are less than the overall *CP* difference, both satisfy all of the feasibility criteria.

8.10 Design Strategy

The pinch method attempts to develop HENs that meet the design targets, i.e., that have the minimum number of units and that consume the minimum amounts of utilities. Toward this end, the problem is decomposed at the pinch(es) into two or more subproblems that are solved independently. In each subproblem (e.g., the hot-end and cold-end problems), the design is begun at the pinch where the restrictions on stream matches are greatest. The feasibility criteria at the pinch are used to identify appropriate matches and the need for stream splitting.

When an appropriate stream match has been identified, the next step is to determine the duty that should be assigned to the corresponding exchanger. A simple rule called the tick-off heuristic [1] can be used for this purpose. It can be stated as follows:

Tick-off heuristic: In order to minimize the number of units, each exchanger should bring one stream to its target temperature (or exhaust a utility).

For process exchangers operating at the pinch, the duty can be taken as the lower of the two stream duties since the *CP* inequality will ensure an adequate driving force in the unit.

TABLE 8.11 The CP Table for the Hot-End Problem of TC3

$CPH \le CPC$		$NH \le NC$	
Stream number	Hot	Cold	Stream number
1	2.0	3.0	4
		2.5	3
Total	2.0	5.5	

The tick-off heuristic can sometimes result in excess utility usage that is incompatible with a minimum-utility design. This situation occurs when excessive temperature driving force is utilized in pinch exchangers, leaving too little driving force available for exchangers elsewhere in the network. In these cases, the duty on the offending exchanger(s) can be reduced (which may lead to more than the minimum number of units) or a different set of pinch matches can be tried [1]. Methods for analyzing these situations are given by Linnhoff and Ahmad [5,6] and in Section 8.16.

Once the pinch matches have been made, the remaining problem can be solved with much greater flexibility. The feasibility criteria that apply at the pinch need not be satisfied away from the pinch, leaving other factors, such as process constraints, plant layout, controllability, and engineering judgment, as paramount in determining stream matches. (Safety is an overriding consideration in all stream matches, including those at the pinch.)

A complete network design is obtained by joining the subproblem solutions at the pinches. The resulting network is usually too complex to be practical, however. Therefore, further analysis is required to simplify the network. Some energy flow across the pinch is generally necessary in order to obtain a practical network design.

The basic pinch design strategy is applied to obtain a minimum-utility design for TC3 in the following section. Network simplification is considered in subsequent sections.

8.11 Minimum-Utility Design for TC3

8.11.1 Hot-End Design

The hot-end problem for TC3 is summarized in Table 8.8. Before starting the design, consider the design targets for this problem. Since Stream 2 does not appear in the hot end, there are only three process streams. Also, there is only one utility (hot) because no cooling is permitted in the hot end. Therefore, according to Equation (8.1), the minimum number of units is:

$$U_{min} = (3 + 1) - 1 = 3$$

All the heating for the entire network must be done in the hot end because no heaters are allowed in the cold end. Therefore, the minimum hot utility requirement for the hot end is 107.5 kW, the minimum for the entire network. The minimum cold utility requirement is obviously zero, since no coolers are allowed in the hot end.

The *CP* table for the hot-end problem is given as Table 8.11. As previously shown, there are two feasible pinch matches, Stream 1 with Stream 3, and Stream 1 with Stream 4. We consider each of these in turn.

Case A: Match Stream 1 with Stream 3

From Table 8.8, the stream duties are 120 kW for Stream 1 and 137.5 kW for Stream 3. The lower value, 120 kW, is the duty assigned to the exchanger, bringing Stream 1 to its target temperature. The remaining duty on Stream 3 is 17.5 kW, which must be supplied by a heater since no hot streams remain to exchange heat with the cold streams. Likewise, a heater must be used to supply the entire duty of 90 kW for Stream 4. The resulting network is shown in Figure 8.7. Note that there are three units (a process exchanger and two heaters), the total hot utility consumption is 107.5 kW, and no cold utility is used. Comparing these values with the design targets for the hot-end problem, it is seen that this is a minimum-utility, minimum-number-of-units design.

Case B: Match Stream 1 with Stream 4

The stream duties in this case are 120 kW for Stream 1 and 90 kW for Stream 4. Assigning the lower value of 90 kW to the exchanger brings Stream 4 to its target temperature, while leaving a residual duty of 30 kW for Stream 1. This residual duty can be handled with

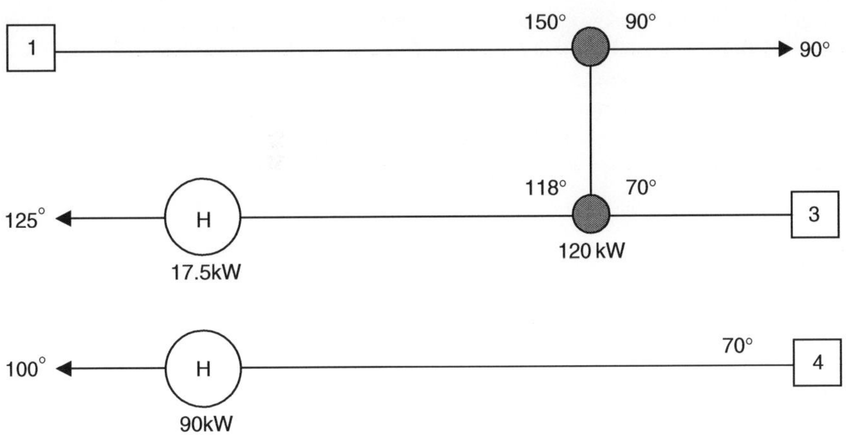

FIGURE 8.7 Hot-end design for TC3: Case A.

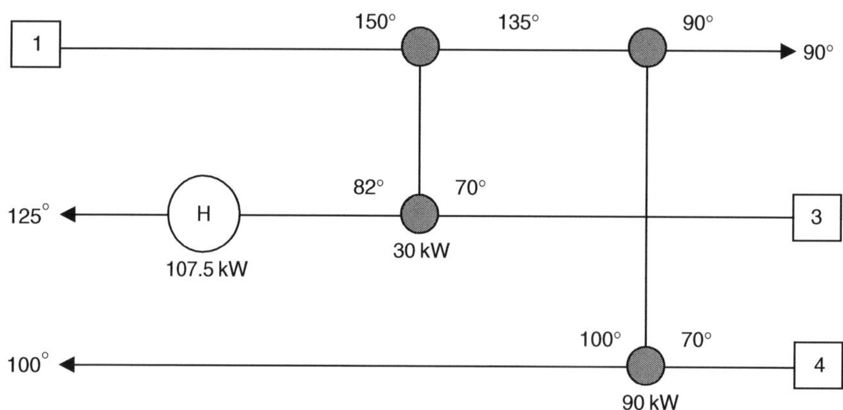

FIGURE 8.8 Hot-end design for TC3: Case B.

a second exchanger matching Stream 1 with Stream 3. The remaining duty of 107.5 kW for Stream 3 must be supplied by a heater. The network is shown in Figure 8.8. Note that it, too, is a minimum-utility, minimum-number-of-units design.

8.11.2 Cold-End Design

The cold-end problem for TC3 is summarized in Table 8.9. Considering first the design targets for this problem, there are four process streams and one (cold) utility. Therefore, Equation (8.1) gives:

$$U_{min} = (4 + 1) - 1 = 4$$

The minimum hot utility requirement is zero (no heaters allowed in cold end) and the minimum cold utility requirement is equal to the minimum for the entire network; namely, 40 kW.

The *CP* table for the cold end problem is given as Table 8.12. It is seen that hot Stream 1 cannot be matched with either cold stream due to violation of the *CP* inequality. Hot Stream 2 can be matched with either cold stream without violating the *CP* inequality, but the *CP* differences are 5 and 5.5, respectively, for these matches, while the overall *CP* difference is 4.5. Therefore, neither of these matches is feasible.

Since there are no feasible pinch matches, stream splitting is required. For instance, one of the cold streams can be split to allow a match with Stream 1. However, this will result in $NC > NH$, thereby requiring a hot stream split as well. Alternatively, Stream 2 can be split to allow a match with either cold stream. For example, an equal split giving $CP = 4$ in each branch permits matches with both cold streams. The exact split ratio need not be specified at this point; it will be determined when duties and stream temperatures are specified for the matches. In order to minimize the number of stream splits and the associated complexity of the HEN, it is clearly preferable to split Stream 2 rather than one of the cold streams.

From Table 8.9, the cold-end duties are 240 kW for Stream 2, 135 kW for Stream 4 and 125 kW for Stream 3. Therefore, in matching the two branches of Stream 2 with the cold streams, one cold stream can be brought to its target temperature along with Stream 2. There are two cases, depending on which cold stream is "ticked-off."

Case A: Tick-Off Stream 3

The match between Streams 2 and 3 is assigned a duty of 125 kW to satisfy the duty on Stream 3. The remaining duty on Stream 2 is 115 kW, which is less than the duty for Stream 4. Therefore, the match between Streams 2 and 4 is assigned a duty of 115 kW, leaving a duty of 20 kW on Stream 4. A match is now possible between Streams 1 and 4. Since the exchanger will not be operating at the pinch, the feasibility criteria are not applicable to this unit. The duty for Stream 1 is 60 kW, so the lower value of 20 kW is assigned, bringing Stream 4 to its target temperature and leaving a duty of 40 kW on Stream 1. Since both cold streams have now been ticked-off, a 40-kW cooler must be used to satisfy the remaining duty on Stream 1. The resulting network is shown in

TABLE 8.12 The CP Table for the Cold-End Problem of TC3

CPH ≥ *CPC*			*NH* ≥ *NC*
Stream Number	Hot	Cold	Stream Number
2	8.0	3.0	4
1	2.0	2.5	3
Total	10.0	5.5	

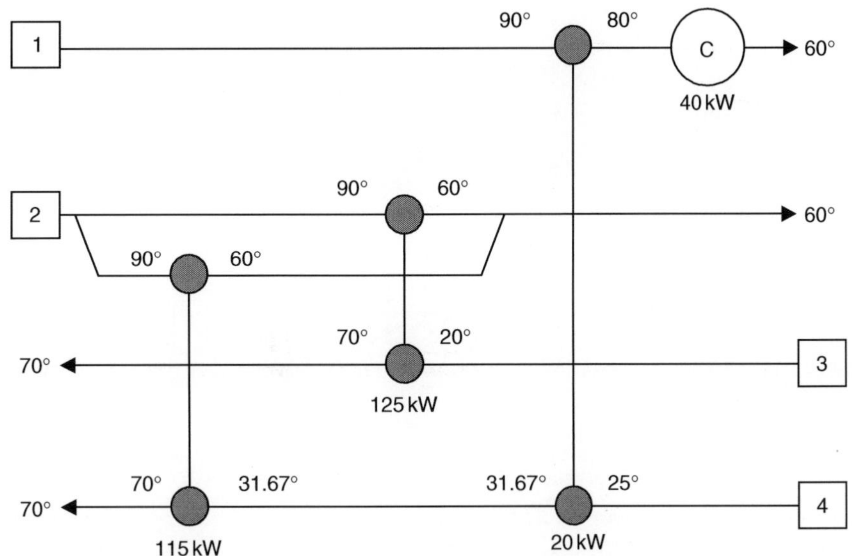

FIGURE 8.9 Cold-end design for TC3: Case A.

Figure 8.9. Comparison with the cold-end design targets shows that it is a minimum-utility, minimum-number-of-units design. If the outlet temperature of Stream 2 from each exchanger is set at 60°C as in Figure 8.9, then the *CP* values are 3.83 kW/°C for the branch matched with Stream 4 and 4.17 kW/°C for the branch matched with Stream 3.

Case B: Tick-Off Stream 4

In this case, the match between Streams 2 and 4 is assigned a duty of 135 kW to satisfy the duty on Stream 4. The remaining duty on Stream 2 is 105 kW, which is less than the duty of 125 kW on Stream 3. Therefore, the match between Streams 2 and 3 is assigned a duty of 105 kW, thereby ticking off Stream 2 and leaving a duty of 20 kW on Stream 3. A 20-kW match between Streams 1 and 3 brings Stream 3 to its target temperature, and a 40-kW cooler on Stream 1 completes the design. The resulting network is shown in Figure 8.10. Note that it, too, is a minimum-utility, minimum-number-of-units design. With the outlet temperature of Stream 2 set at 60°C for each exchanger, the *CP* values are 4.5 kW/°C for the branch matched with Stream 4 and 3.5 kW/°C for the branch matched with Stream 3.

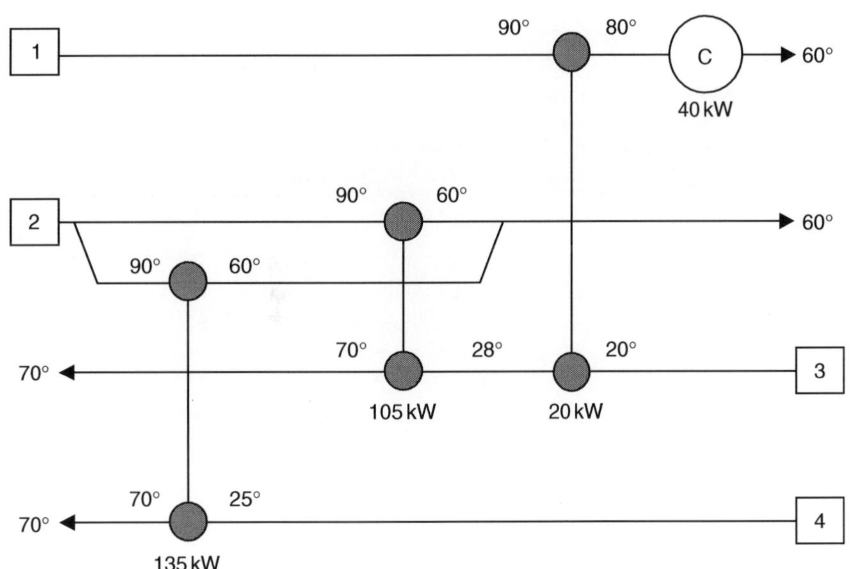

FIGURE 8.10 Cold-end design for TC3: Case B.

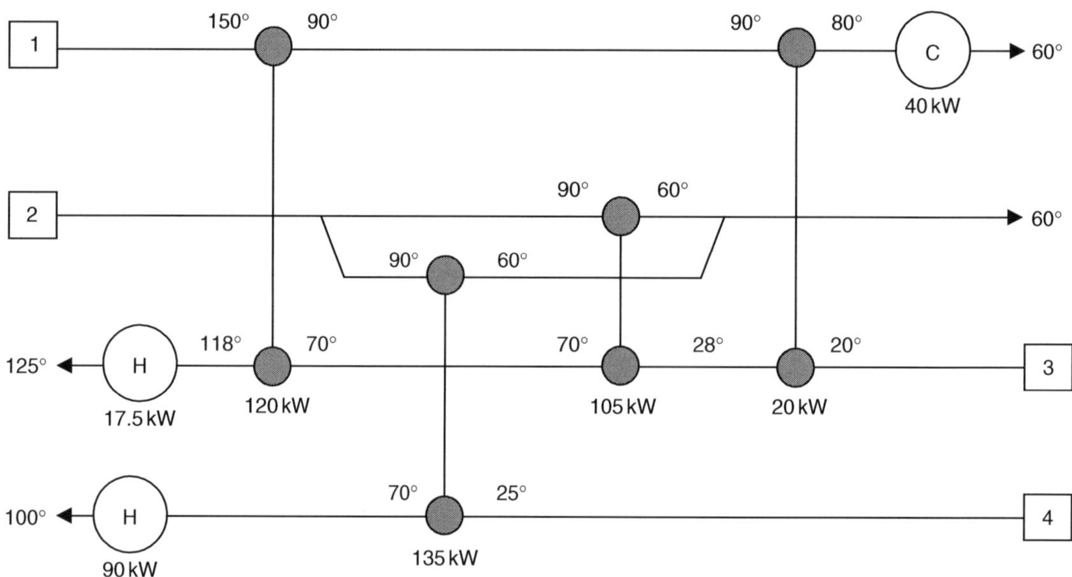

FIGURE 8.11 Network obtained by combining the hot A and cold B cases for TC3.

8.11.3 Complete Network Design

To obtain a complete network design, a hot-end design and a cold-end design are spliced together at the pinch. Each hot-end design can be combined with either cold-end design, leading to four possible networks for TC3. Figure 8.11 shows the network obtained by combining the hot A and cold B cases. It contains seven units and uses the minimum amounts of hot and cold utilities. Recall that the target for minimum number of units in the entire network was previously determined to be five. We now see that the actual minimum number of units consistent with minimum-utility usage is seven. The difference is due to the presence of the pinch. Equation (8.1) must be applied separately on either side of the pinch and the results summed to obtain a value consistent with a minimum-utility design.

The effect of the pinch on the number of units can be clearly seen in Figure 8.11. An unusual feature of this HEN is that Streams 1 and 3 are matched twice, with one exchanger operating above the pinch and one below the pinch. One of these units can obviously be eliminated, but doing so will result in energy flow across the pinch and increased utility usage.

8.12 Network Simplification

The basic pinch design strategy illustrated in the previous section generally results in networks that are too complex, i.e., have too many units, to be considered practical. The two exchangers involving Streams 1 and 3 in Figure 8.11 are an example of this complexity. Therefore, further analysis is required to simplify the design. An obvious way to simplify the HEN of Figure 8.11, for example, is to simply remove the small 20-kW exchanger operating between Streams 1 and 3. To satisfy the stream duties, the duty on the Stream 1 cooler can be increased to 60 kW and the duty on the Stream 3 heater can be increased to 37.5 kW. However, the simplification can be done in a more energy-efficient manner by using the concepts of heat load loop and heat load path.

8.12.1 Heat Load Loops

A heat load loop is a loop in the HEN around which duties can be shifted from one exchanger to another without affecting stream duties. Changing the duty on an exchanger may result in a violation of ΔT_{min}, however. The number of heat load loops is the difference between the actual number of units in the network and the minimum number of units calculated by applying Equation (8.1) to the network as a whole, i.e., ignoring the pinch. For the HEN of Figure 8.11, there are seven units compared with the minimum number of five. Hence, there are two loops as shown in Figures 8.12 and 8.13. The loop shown in Figure 8.12 is obvious and is formed by the two exchangers operating between Streams 1 and 3. The loop shown in Figure 8.13 is less obvious, passing through the heaters on Streams 3 and 4, which are not physically connected. Notice that both loops cross the pinch.

Now we can use the heat load loop in Figure 8.12 to simplify the network by eliminating the 20-kW exchanger. To do this, simply shift the 20-kW duty around the loop to the 120-kW exchanger, bringing its duty to 140 kW. The resulting network is shown in Figure 8.14. Notice that the temperature difference on the cold end of the 140-kW exchanger is now 18°C, a violation of ΔT_{min}. However, the utility consumption has not increased. Thus, by using the heat load loop to transfer the 20 kW across the pinch, a violation of ΔT_{min} has been incurred rather than increased utility usage.

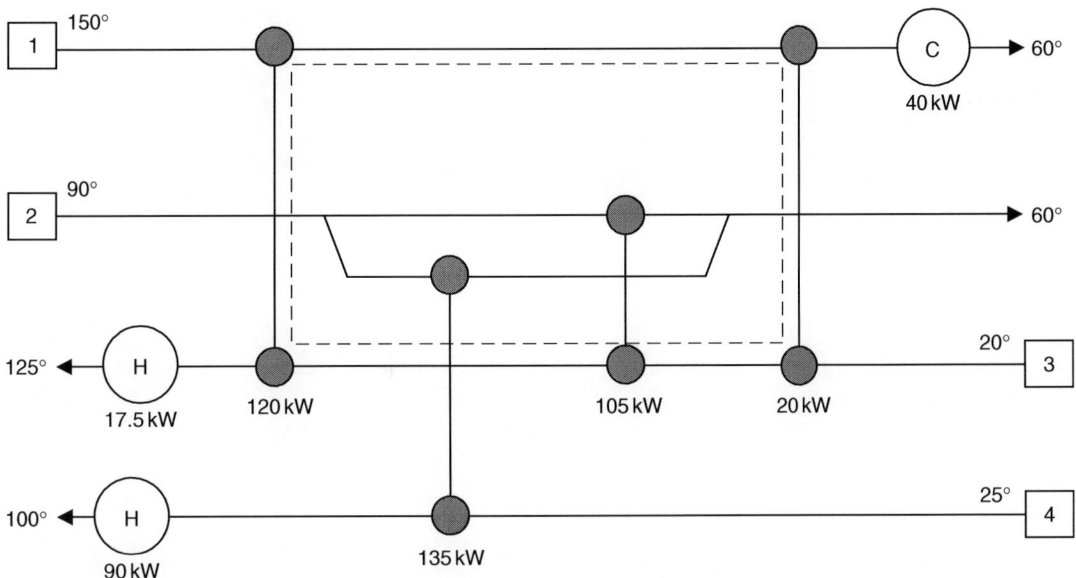

FIGURE 8.12 HEN for TC3 showing the heat load loop formed by the 20-kW and 120-kW exchangers.

In a similar manner, the loop shown in Figure 8.12 can be used to further simplify the network by eliminating the small 17.5-kW heater on Stream 3. The reader can verify that this results in a temperature difference of 11°C on the cold side of the 140-kW exchanger. This is a large violation of ΔT_{min}, and although it might nevertheless be acceptable in practice, it will not be considered further here. Instead, the 17.5-kW heater on Stream 3 will be used to illustrate the concept of a heat load path.

8.12.2 Heat Load Paths

A heat load path is a continuous connection in the network between a heater, one or more heat exchangers, and a cooler. Heat load can be shifted along a path by alternately adding and subtracting a duty from each successive unit. This procedure will not affect stream duties, but intermediate stream temperatures will change due to the change in exchanger duties. Therefore, a heat load path can be used to eliminate a ΔT_{min} violation incurred during network simplification.

Consider again the 140-kW exchanger in Figure 8.14. Although the 18°C driving force on the cold end would normally be considered entirely adequate in practice, suppose for the sake of argument that it is desired to restore this value to 20°C. This can be done by increasing the exit temperature of Stream 1 from 80°C to 82°C. Since the *CP* of Stream 1 is 2 kW/°C, the exchanger duty

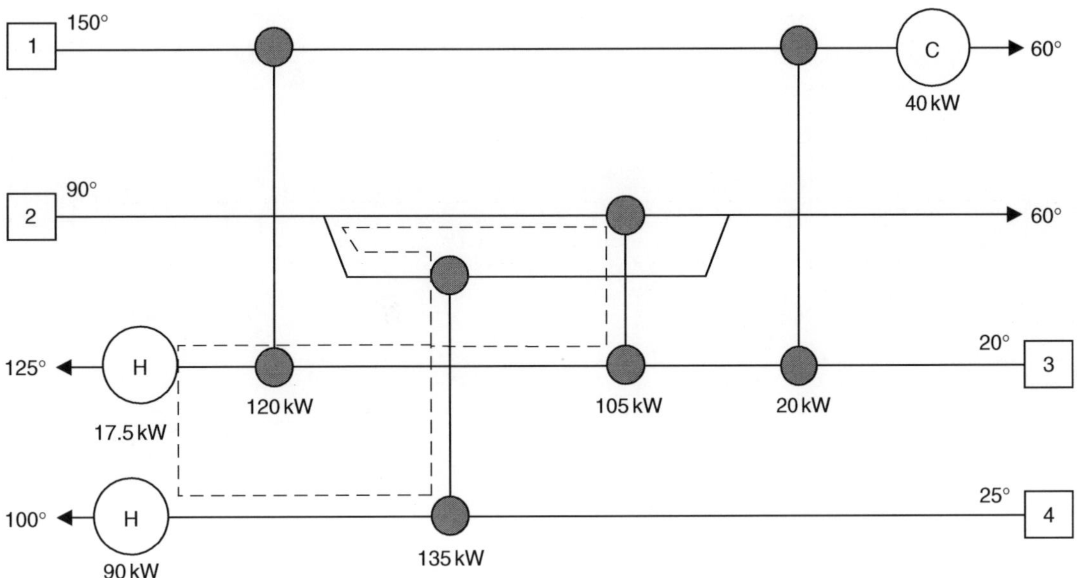

FIGURE 8.13 HEN for TC3 showing the second heat load loop.

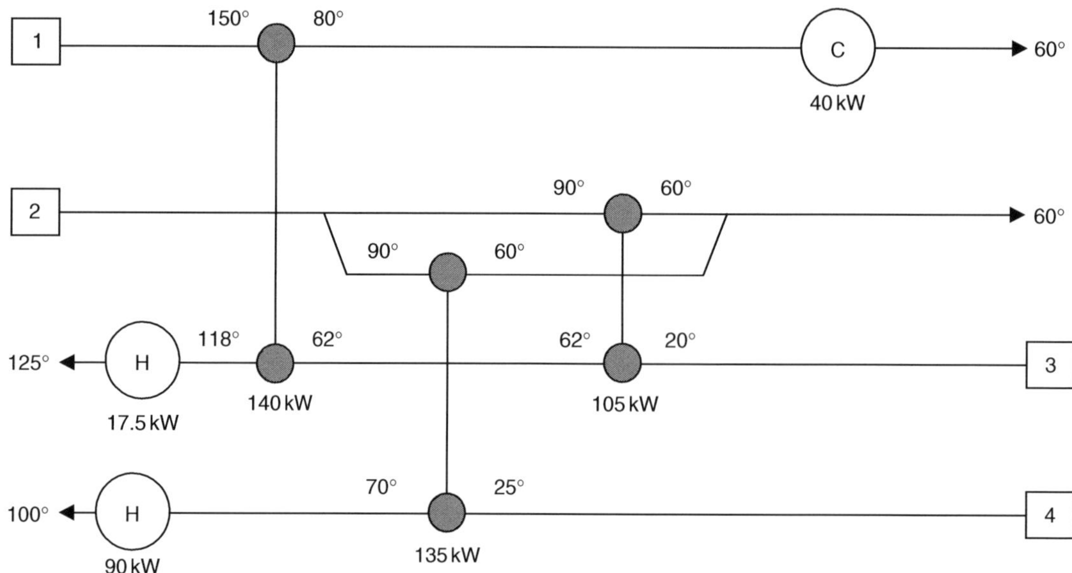

FIGURE 8.14 Modified HEN for TC3 after using heat load loop to eliminate the 20-kW exchanger.

must be reduced by 4 kW. The exchanger forms part of a heat load path from the heater on Stream 3 to the cooler on Stream 1. Therefore, the desired result can be obtained by adding 4 kW to the heater duty, subtracting the same amount from the exchanger (bringing its duty to 136 kW), and adding 4 kW to the cooler. The resulting network is shown in Figure 8.15.

The net result of the above modifications is that the 20-kW exchanger has been eliminated from the network without violating the ΔT_{min} on any remaining exchanger, and the hot and cold utility consumption have each increased by only 4 kW. By comparison, simply removing the 20-kW exchanger increases both utility loads by 20 kW. Furthermore, the network of Figure 8.15 compares favorably with the original design targets: 6 units versus 5, 111.5 kW hot utility versus 107.5 kW, and 44 kW cold utility versus 40 kW.

8.13 Number of Shells

Up to this point we have assumed that all exchangers in the network operate counter-currently. Most exchangers in the chemical process industries, however, are multi-pass shell-and-tube exchangers. Although the network of Figure 8.15 is valid regardless of the type of exchangers employed, two of the three process exchangers have large temperature crosses. A temperature cross exists in an

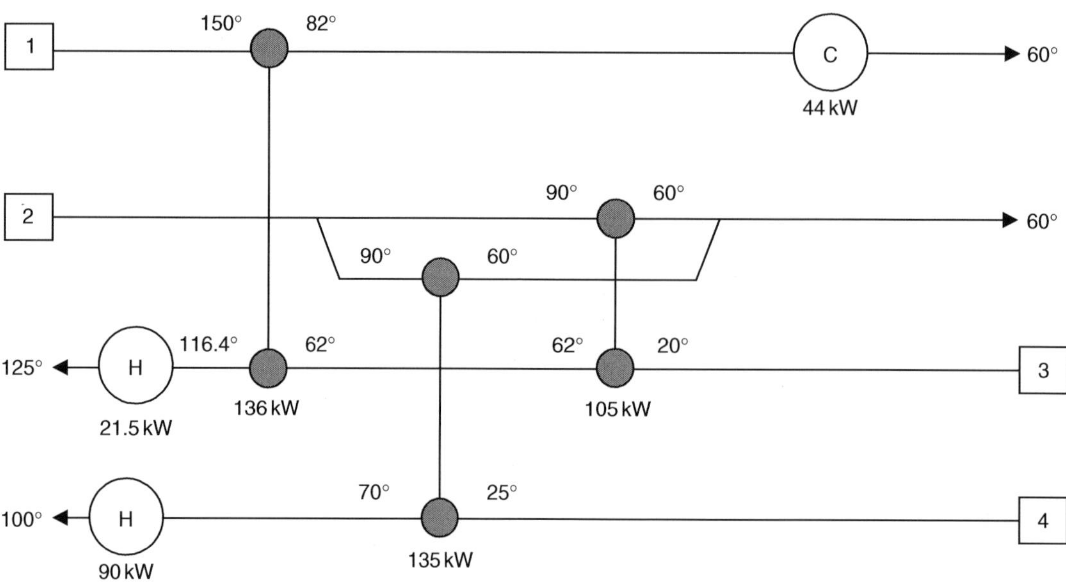

FIGURE 8.15 HEN for TC3 after using heat load path to restore ΔT_{min} to 20°C.

TABLE 8.13 Shells Required for the HEN of Figure 8.15

Exchanger (kW)	Temperature Cross (°C)	LMTD Correction Factor for 1-2 Exchanger	Shells Required
105	2	0.77	1–2
135	10	<0.5	2
136	34.4	<0.5	3

exchanger whenever the cold fluid outlet temperature exceeds the hot fluid outlet temperature, and the magnitude of the cross is the difference between these two temperatures. Due to the limitation imposed by the parallel pass in a 1-2 exchanger, only a small temperature cross can be supported. As the magnitude of the cross increases, the logarithmic mean temperature difference (LMTD) correction factor rapidly decreases, indicating the need for multiple shell passes. The most common way of achieving multiple shell passes is by connecting E shells in series, making these units relatively expensive compared with 1-2 or counter-flow exchangers that comprise a single shell.

Table 8.13 gives the number of shells required for each of the process exchangers in the HEN of Figure 8.15 in order to achieve a satisfactory LMTD correction factor. Although $F < 0.8$ for the 105-kW exchanger with one shell pass, the operating point is not on the steeply sloping portion of the *F*-curve, so a single shell pass could in fact be used. Alternatively, the small temperature cross on this exchanger can be eliminated by increasing the *CP* value of the Stream 2 branch allocated to this unit to 3.75 kW/°C. The outlet temperature of Stream 2 is then 62°C and the LMTD correction factor is 0.80 for one shell pass, which is definitely acceptable. The *CP* value for the Stream 2 branch allocated to the 135-kW exchanger becomes 4.25 kW/°C, resulting in an outlet temperature of 58.2°C, and a temperature cross of 11.8°C. Two shell passes are still adequate for this unit. Therefore, assuming that each of the utility exchangers in the HEN requires a single shell pass, a total of nine shells is required if the exchangers are multi-tube-pass shell-and-tube units. Although this appears to be a rather complicated solution to a very simple problem, large temperature crosses are the inevitable result of attempting to maximize heat exchange using a minimal number of exchangers.

The HEN of Figure 8.15 can be simplified by eliminating the temperature cross on the 136-kW exchanger. This is accomplished by shifting 38 kW of duty along the heat load path used previously. The resulting HEN, shown in Figure 8.16, has only seven shells, but uses 149.5 kW of hot utility and 82 kW of cold utility. It is not close to a minimum-utility design, but the trade-off of added utility consumption for capital cost reduction might be economically justifiable.

8.14 Targeting for Number of Shells

The composite curves can be used to estimate the minimum number of shells required for a HEN composed of shell-and-tube exchangers. Two methods are considered here, a very simple graphical procedure and a somewhat more accurate, albeit more computationally intensive, analytical procedure.

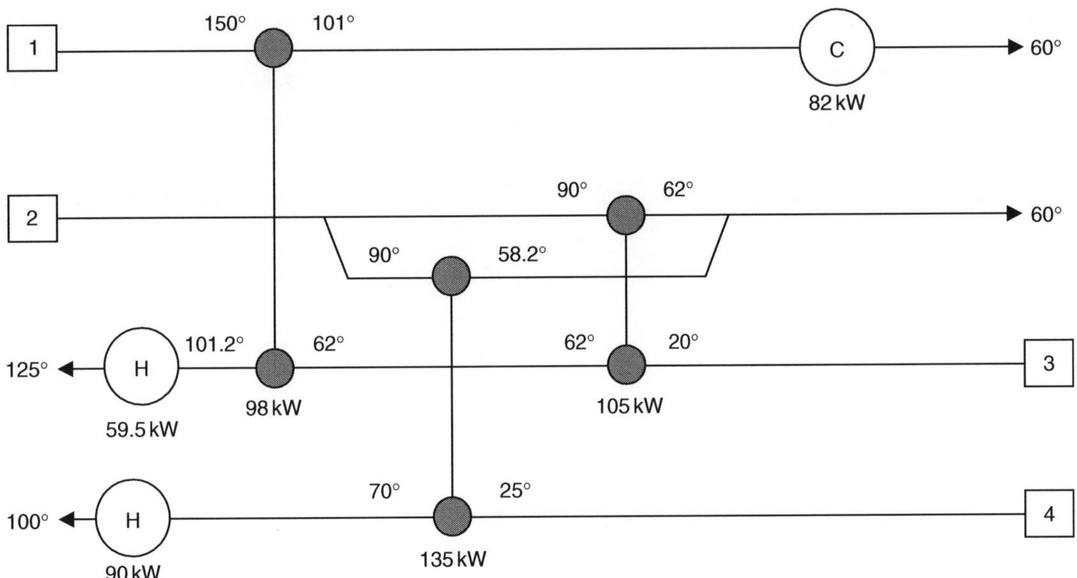

FIGURE 8.16 HEN for TC3 after modification to eliminate temperature crosses on two exchangers.

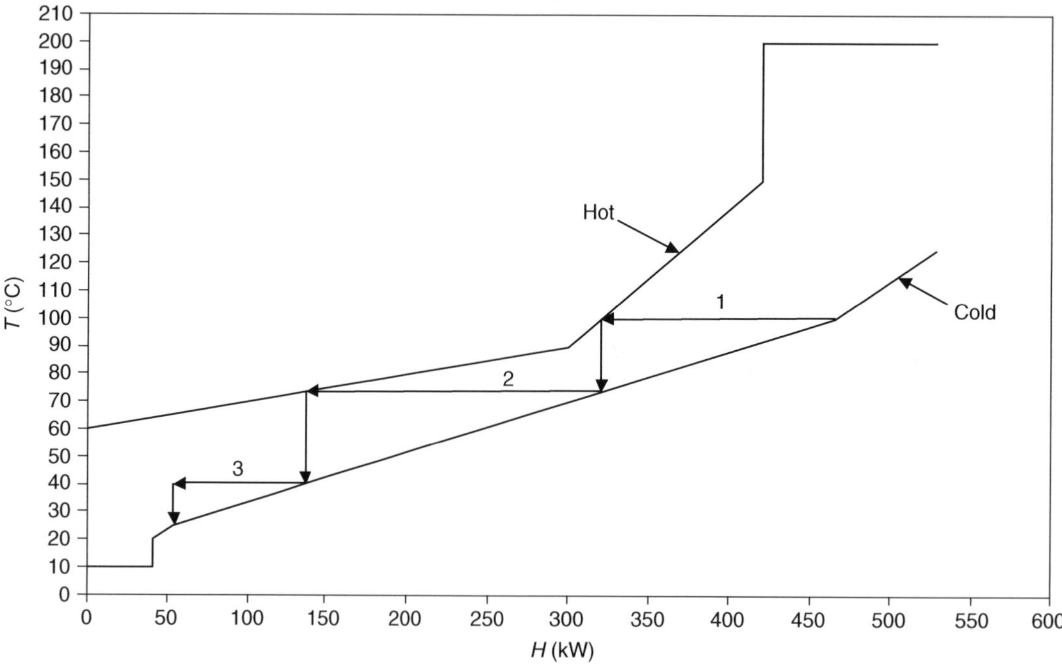

FIGURE 8.17 Estimating the minimum number of shells for cold Stream 4 of TC3.

8.14.1 Graphical Method

In this method the hot and cold composite curves are used to step off shells in the same manner that ideal stages are stepped off between the equilibrium curve and operating line for a distillation column on a McCabe–Thiele diagram. The number of shells is determined for each individual stream in the system, including the utilities. For this reason, the utilities are usually included in the composite curves as shown in Figure 8.17 for TC3. In this form, the composite curves are referred to as balanced. Since utility temperatures were not specified for TC3, values of 200°C and 10°C were assumed for the hot and cold utilities, respectively. For convenience, the temperature changes of the utility streams were assumed to be negligible.

Figure 8.17 shows the construction to estimate the number of shells for Stream 4 of TC3. Starting from the stream's target temperature (100°C) on the appropriate (cold) curve, triangles are stepped off between the hot and cold curves until the supply temperature (25°C) of the stream is reached. Trivedi et al. [7] have shown that each horizontal line segment between the two composite curves represents a single shell with no temperature cross. Thus, it is seen that three shells are required to cover the entire temperature interval for Stream 4.

Repeating the graphical procedure for each of the hot and cold process streams of TC3 yields the results shown in the first column of Table 8.14. The same procedure can by applied separately to the hot-end problem (above the pinch) and the cold-end problem (below the pinch) to obtain a result that is consistent with minimum energy usage. The values so obtained are also shown in Table 8.14.

The total number of shells required is the larger of the sums for the hot and cold streams, respectively, including utilities. Thus, for the entire network neglecting the pinch, seven shells are required. Four shells are required in the hot end and five in the cold end treated separately, giving a total of nine for the entire network when the pinch is accounted for. Trivedi et al. [7] point out that the graphical method can sometimes give values that are less than the minimum number of units calculated by Equation (8.1). In these cases, the estimated number of shells should be equated with U_{min}. For TC3, the minimum number of units is five for the entire network neglecting the pinch, three for the hot end and four for the cold end. Therefore, no adjustments are needed for the number-of-shells estimates.

TABLE 8.14 Estimated Number of Shells Required for the Streams of TC3

Stream	Entire Network	Hot End	Cold End
1	3	2	2
2	2	0	2
3	3	2	2
4	3	2	2
HU	1	1	0
CU	1	0	1

The actual number of shells required in the hot end of TC3 is four (Figures 8.7 and 8.8), which agrees with the estimated value. In the cold end, however, six shells are required (Figures 8.9 and 8.10) compared with the estimated value of five. For the entire network, ten shells are required for a strict minimum-utility design (Figure 8.11) and nine shells are needed for the simplified network of Figure 8.15, compared with the predicted number of nine shells for a minimum-utility design. Neglecting the pinch, the estimate of seven shells can be compared with the network of Figure 8.16, which has seven shells. It is seen that the graphical method gives reasonable estimates for the minimum number of shells, but it is not as accurate as Equation (8.1), the minimum-number-of-units estimator.

8.14.2 Analytical Method

The analytical method is based on a procedure developed by Ahmad et al. [8] for calculating the required number of shell passes in a shell-and-tube heat exchanger. The basic idea is that when a multi-pass exchanger is comprised of TEMA E shells connected in series, an acceptable value of the LMTD correction factor should be attained in each individual shell. Inspection of Figure 3.14 shows that for any value of the parameter R, there is a maximum value of the parameter P that is, in fact, attained asymptotically as $F \to -\infty$. This maximum value, P_{max} is given by [9]:

$$P_{max} = \frac{2}{R + 1 + \sqrt{R^2 + 1}} \tag{8.10}$$

Any exchanger must operate at some fraction, X_P, of P_{max} to achieve an acceptable value of F. The number of E shells in series required to attain $P = X_P P_{max}$ in each shell is given by the following equations [8]:

$$N_s = \ln[(1 - RP)/(1 - P)]/\ln W \quad (R \neq 1) \tag{8.11}$$

$$N_s = [P/(1 - P)]\left(1 - X_P + 1/\sqrt{2}\right)/X_P \quad (R = 1) \tag{8.12}$$

$$W = \left(R + 1 + \sqrt{R^2 + 1} - 2RX_P\right)/\left(R + 1 + \sqrt{R^2 + 1} - 2X_P\right) \tag{8.13}$$

Ahmad et al. [8] show that a reasonable value for X_P is 0.9. However, the somewhat more conservative value of 0.85 is recommended here. With $X_P = 0.9$, the smallest value of F for a 1-2 exchanger is approximately 0.75 at an R-value of about 1.0. Higher values of F are obtained at larger and smaller values of R where the slopes of the F-curves are steeper. With $X_P = 0.85$, the minimum value of F is approximately 0.8, again at an R-value of about 1.0. Thus, $X_P = 0.85$ is more in line with the standard procedure of designing for $F \geq 0.8$.

Equations (8.11) to (8.13) can be applied to any individual heat exchanger, e.g., the 136-kW exchanger of Figure 8.15. The hot fluid enters at 150°C and leaves at 82°C, while the cold fluid enters at 62°C and leaves at 116.4°C. Either fluid may be assumed to flow in the shell due to the symmetry of F with respect to fluid placement. Assuming the hot fluid is in the shell, Equations (3.11) and (3.12) give:

$$R = \frac{150 - 82}{116.4 - 62} = 1.25$$

$$P = \frac{116.4 - 62}{150 - 62} = 0.6182$$

From Equation (8.13) with $X_P = = 0.85$, we obtain:

$$W = \frac{1.25 + 1 + \sqrt{(1.25)^2 + 1} - 2 \times 1.25 \times 0.85}{1.25 + 1 + \sqrt{(1.25)^2 + 1} - 2 \times 0.85} = 0.8024$$

The required number of shells is obtained from Equation (8.11):

$$N_s = \frac{\ln[(1 - 1.25 \times 0.6182)/(1 - 0.6182)]}{\ln[0.8024]} = 2.3568 \rightarrow 3$$

The result is rounded to the next largest integer, giving three shells, in agreement with the value given previously in Table 8.13, which was based on the criterion $F \geq 0.8$.

In order to use Equations (8.11) to (8.13) to predict the number of shells required in a HEN, the composite curves for the HEN are divided into enthalpy intervals as shown in Figure 8.18 for TC3. For each interval, the terminal temperatures of the hot and cold curves are used as stream temperatures to calculate the number of shells required in that interval. The data for TC3 are shown in Table 8.15. Notice that the number of shells in each interval is not rounded to an integral value.

The number of shells required for each stream, including the utilities, is now obtained by adding the number of shells for each interval in which the stream occurs. This is done separately for intervals above and below the pinch. If the number of shells for any stream turns out to be greater than zero but less than one, it is reset to unity. The results for TC3 are given in Table 8.16.

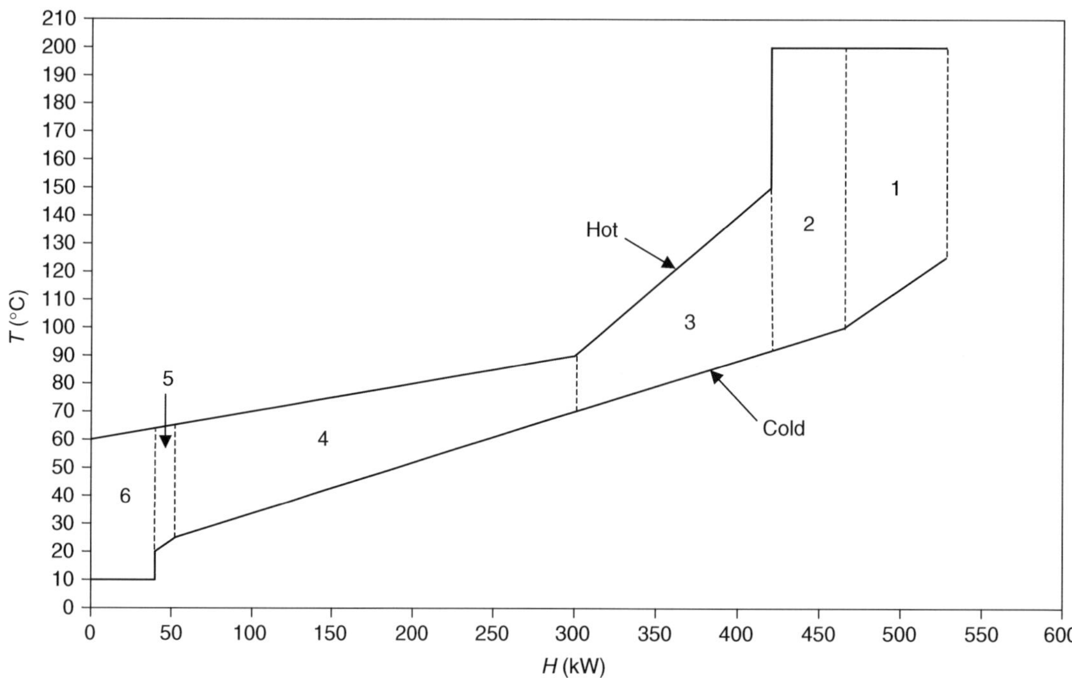

FIGURE 8.18 Enthalpy intervals for TC3.

Ahmad and Smith [10] have shown that the number of shells required for the network is given by the following equation:

$$(N_S)_{Network} = \sum_{Streams} (N_S)_{Stream\ j} - \sum_{Intervals} (N_S)_{Interval\ i} \tag{8.14}$$

The first three intervals for TC3 are above the pinch, and the sum of the shells for these intervals is 1.3659. From Table 8.16, the sum of the shells required for the streams above the pinch is 4.7531. Thus, the number of shells required for the network above the pinch is:

$$(N_s)_{above\ pinch} = 4.7531 - 1.3659 = 3.3872 \Rightarrow 4$$

TABLE 8.15 Enthalpy Interval Data for TC3

Interval	Streams	Hot Temperature (°C)	Cold Temperature (°C)	Number of Shells
1	HU, 3	200	100–125	0.1516
2	HU, 3, 4	200	91.82–100	0.0414
3	1, 3, 4	150–90	70–91.82	1.1729
4	1, 2, 3, 4	90–65.25	25–70	1.2227
5	1, 2, 3	65.25–64	20–25	0.0767
6	1, 2, CU	64–60	10	0.0406

TABLE 8.16 Shells Required for the Streams of TC3: Analytical Method

	Number of Shells	
Stream	Above Pinch	Below Pinch
HU	0.1930 → 1.0	0
1	1.1729	1.3400
2	0	1.3400
3	1.3659	1.2994
4	1.2140	1.2227
CU	0	0.0406 → 1.0
Total	4.7531	6.2021

At this point, the number of shells is rounded to the next largest integer, giving a total of four shells above the pinch. Below the pinch, the sum of the shells for intervals 4, 5, and 6 is 1.3400 and the sum of the shells required for the streams is 6.2021. Therefore,

$$(N_s)_{below\ pinch} = 6.2021 - 1.3400 = 4.8621 \Rightarrow 5$$

The number of shells required for the entire network is the sum of the values required above and below the pinch, or nine in this case. The analytical method can also be applied to the entire network disregarding the pinch, but due to the additivity of shells over the intervals, the same result is obtained except for rounding to integral values and (possibly) resetting of values to unity for individual streams. For TC3, the sum of shells for the streams above and below the pinch is 10.9552 and the sum of shells for all six intervals is 2.7059, giving:

$$(N_s)_{network} = 10.9552 - 2.7059 = 8.2493 \Rightarrow 9$$

In this case, exactly the same result is obtained when the pinch is ignored. Comparison with results obtained previously shows that both the graphical and analytical methods give the same number of shells when the pinch is accounted for. This is not always the case, however, and in general, the analytical method is considered to be the more reliable of the two [10].

8.15 Area Targets

In order to determine the heat-transfer area in a HEN, it is necessary to perform a detailed design of each unit in the network and then sum the areas of all the individual units. It is rather remarkable, therefore, that it is possible to estimate the minimum area required for a HEN before even starting the design process. The estimation procedure relies on estimates of heat-transfer coefficients for individual streams such as the compilation presented by Bell [11].

Suppose we have estimates, h_1 and h_2, for the film coefficients, including fouling allowances, of two streams. We can estimate the overall heat-transfer coefficient for a match between the two streams by neglecting the tube wall resistance and setting the area ratio, A_o/A_i, equal to unity. With these approximations, the equation for the design coefficient becomes:

$$U_D \cong (1/h_1 + 1/h_2)^{-1} \tag{8.15}$$

(Note that assuming $A_o/A_i \cong 1.0$ is reasonable for plain tubes, but not for finned tubes.)

Now suppose that in a particular enthalpy interval on the composite curves for the HEN, these two streams are the only streams. Assuming the streams are matched in a counter-current exchanger, the heat-transfer area required for the interval is given by:

$$A_{interval} = \frac{q}{U_D \Delta T_{ln}} = \frac{\Delta H}{\Delta T_{ln}}\left(\frac{1}{h_1} + \frac{1}{h_2}\right) \tag{8.16}$$

where ΔH is the width of the enthalpy interval and ΔT_{ln} is computed from the terminal temperatures of the hot and cold composite curves on the interval. When the interval contains more than two streams, Linnhoff and Ahmad [5] have shown that the area is given by:

$$A_{interval\ i} = \frac{1}{(\Delta T_{ln})_i} \sum_j \left(q_j/h_j\right)_i \tag{8.17}$$

where q_j is the duty for stream j in the ith interval and the summation extends over all streams that exist in the ith interval. The minimum area required for the network is obtained by summing the areas for all intervals.

When multi-pass shell-and-tube exchangers are used, Equation (8.17) is modified by introducing the LMTD correction factor:

$$A_{interval\ i} = \frac{1}{F_i(\Delta T_{ln})_i} \sum_j \left(q_j/h_j\right)_i \tag{8.18}$$

The LMTD correction factor should be computed based on the number of shells required for the interval. For example, Table 8.15 shows that 1.2227 shells are required for interval 4 of TC3. Rounding to the next largest integer gives two shells, so F for this interval would be computed for an exchanger with two E shells in series, i.e., for a 2-4 exchanger.

To illustrate the procedure, heat-transfer coefficients were (arbitrarily) assigned to the streams of TC3 as shown in Table 8.17. The stream duties in each of the six intervals are given in Table 8.18.

The calculations were performed both for counter-flow exchangers and multi-pass shell-and-tube exchangers, and are summarized in Table 8.19. The predicted minimum network area for counter-flow exchangers is 33.74 m^2 and the value for multi-pass exchangers is 35.46 m^2. These areas are rather small because the stream duties specified for TC3 are small. Note that $F = 1.0$ for intervals 1, 2, and 6 because the utility streams were assumed to be isothermal.

Details of the calculations for interval 4 are given here. From Table 8.15, the hot temperature range is 90°C to 65.25°C and the cold temperature range is 25°C to 70°C. Therefore,

$$(\Delta T_{ln})_{cf} = \frac{(65.25 - 25) - (90 - 70)}{\ln\left(\frac{65.25 - 25}{90 - 70}\right)} = 28.95°\text{C}$$

TABLE 8.17 Heat-transfer Coefficients Assigned to Streams of TC3

Stream	h (kW/m^2 · K)
1	1.1
2	0.6
3	0.8
4	1.0
HU	2.0
CU	1.5

TABLE 8.18 Stream Duties (kW) in Enthalpy Intervals of TC3

	Interval					
Stream	1	2	3	4	5	6
HU	62.5	45	–	–	–	–
1	–	–	120	49.5	2.5	8
2	–	–	–	198	10	32
3	62.5	20.45	54.55	112.5	12.5	–
4	–	24.55	65.45	135	–	–
CU	–	–	–	–	–	40

Assuming that the hot stream is in the shell, Equations (3.11) and (3.12) give

$$R = \frac{90 - 65.25}{70 - 25} = 0.55$$

$$P = \frac{70 - 25}{90 - 25} = 0.6923$$

Since two shells are required in this interval, F is calculated for an exchanger with two shell passes. From Equations (3.13) and (3.14),

$$\alpha = \left(\frac{1 - RP}{1 - P}\right)^{1/2} = \left(\frac{1 - 0.55 \times 0.6923}{1 - 0.6923}\right)^{1/2} = 1.41861$$

$$S = \frac{\alpha - 1}{\alpha - R} = \frac{1.41861 - 1}{1.41861 - 0.55} = 0.481933$$

Also,

$$\sqrt{R^2 + 1} = \sqrt{(0.55)^2 + 1} = 1.14127$$

Substitution into Equation (3.15) gives the LMTD correction factor:

$$F = \frac{\sqrt{R^2 + 1}\ \ln\left(\dfrac{1 - S}{1 - RS}\right)}{(R - 1)\ \ln\left[\dfrac{2 - S\left(R + 1 - \sqrt{R^2 + 1}\right)}{2 - S\left(R + 1 + \sqrt{R^2 + 1}\right)}\right]}$$

$$= \frac{1.14127\left(\dfrac{1 - 0.481933}{1 - 0.6923 \times 0.481933}\right)}{(0.55 - 1)\ \ln\left[\dfrac{2 - 0.481933(0.55 + 1 - 1.14127)}{2 - 0.481933(0.55 + 1 + 1.14127)}\right]}$$

$$F = 0.9416$$

For interval 4, Equation (8.18) becomes:

$$A_{interval\ 4} = \frac{1}{F_4(\Delta T_{\ln})_4}\left(\frac{q_1}{h_1} + \frac{q_2}{h_2} + \frac{q_3}{h_3} + \frac{q_4}{h_4}\right)_4$$

TABLE 8.19 Area Targeting Calculations for TC3

Interval	Streams	ΔT_{ln} (°C)	Shells	F	Counter-Flow Area (m^2)	Multi-Pass Area (m^2)
1	HU, 3	86.90	1	1.0	1.26	1.26
2	HU, 3, 4	104.04	1	1.0	0.70	0.70
3	1, 3, 4	35.76	2	0.9550	6.79	7.11
4	1, 2, 3, 4	28.95	2	0.9416	22.47	23.87
5	1, 2, 3	42.10	1	0.9994	0.82	0.82
6	1, 2, CU	51.97	1	1.0	1.70	1.7
Total					33.74	35.46

The stream duties for interval 4 are obtained from Table 8.18, and the heat-transfer coefficients from Table 8.17. Substituting gives:

$$A_{interval\ 4} = \frac{1}{0.9416 \times 28.95}\left(\frac{49.5}{1.1} + \frac{19.8}{0.6} + \frac{112.5}{0.8} + \frac{135}{1.0}\right)$$

$$A_{interval\ 4} = 23.87\ \text{m}^2$$

This is the area for multi-pass exchangers. Setting $F_4 = 1.0$ in the above calculation gives the area for counter-flow exchangers.

The target area can be compared with the area for the network of Figure 8.15, which was obtained by the Pinch Design method. The area of each unit in the network can be calculated using the film coefficients from Table 8.17 and Equation (8.15) for the overall coefficient. The calculations are summarized in Table 8.20. For counter-flow exchangers, the total network area of 36.53 m^2 is about 8% above the target value of 33.74 m^2. For multi-pass shell-and-tube exchangers, the estimated network area of 41.84 m^2 is about 18% above the target area of 35.46 m^2.

8.16 The Driving Force Plot

The minimum area target described in the preceding section is based on temperature differences for heat transfer that are obtained from the composite curves. Therefore, in order for a HEN design to approach the target area, the temperature differences in individual exchangers should mimic those of the composite curves. For exchangers operating at the pinch, the CP inequalities (8.6) and (8.7) ensure that the temperature difference increases with distance from the pinch end of the exchanger. This same general trend is exhibited by the composite curves, i.e., the difference between the hot and cold curves increases as one moves away from the pinch in either direction along the curves. In fact, the trends would be identical if the CP ratio of the two streams in a pinch exchanger exactly matched the CP ratio of the composite curves at the pinch. Thus, to approach minimum area in a HEN design, exchangers operating at the pinch should satisfy the following condition [5]:

$$\left(\frac{CPH}{CPC}\right)_{pinch\ exchanger} \cong \left(\frac{CP_{hot\ composite}}{CP_{cold\ composite}}\right)_{pinch} \tag{8.19}$$

For example, consider the hot-end problem for TC3, where Stream 1 can be matched with either Stream 3 or Stream 4. The CP ratios for these matches are as follows:

$$CP_1/CP_3 = 2/2.5 = 0.8$$

$$CP_1/CP_4 = 2/3 = 0.67$$

TABLE 8.20 Heat-transfer Area for the HEN of Figure 8.15

Unit	Streams	ΔT_{ln} (°C)	U_D (kW / $m^2 \cdot$ K)	F	Counter-Flow Area (m^2)	Multi-Pass Area (m^2)
44-kW cooler	1, CU	60.33	0.6346	1.0	1.15	1.15
21.5-kW heater	HU, 3	79.22	0.5714	1.0	0.47	0.47
90-kW heater	HU, 4	114.34	0.6667	1.0	1.18	1.18
136-kW exchanger	1, 3	26.21	0.4632	0.8906	11.20	12.58
135-kW exchanger	2, 4	26.80	0.3750	0.9157	13.43	14.67
105-kW exchanger	2, 3	33.64	0.3429	0.7718	9.10	11.79
Total					36.53	41.84

TABLE 8.21 Data for Constructing the Driving Force Plot for TC3

T_{cold} (°C)	T_{hot} (°C)	ΔT (°C)
10	60	50
10	64	54
20	64	44
25	65.25	40.25
70	90	20
91.82	150	58.18
91.82	200	108.18
100	200	100
125	200	75

Above the pinch, Stream 1 is the only hot stream, so the *CP* for the hot composite curve is 2 kW/°C. The *CP* of the cold composite curve is the sum of the *CP* values for Streams 3 and 4, or 5.5 kW/°C. Therefore, the *CP* ratio for the composite curves at the pinch is:

$$CP_{hot\ composite}/CP_{cold\ composite} = 2/5.5 \cong 0.36$$

Although neither stream match closely approximates this value, the match between Streams 1 and 4 is the better of the two.

The driving force plot provides a convenient means of examining the driving forces in individual exchangers *vis-à-vis* the composite curves for both pinch and non-pinch exchangers. It is a plot of the temperature difference between the hot and cold composite curves versus the temperature of either the hot or cold streams. For TC3, the data needed to construct the plot are extracted from Table 8.15 and shown in Table 8.21. The temperature differences are plotted against the cold stream temperatures in Figure 8.19.

The driving forces for the two hot-end pinch matches are also plotted in Figure 8.19, from which it can be seen that the match between Streams 1 and 4 conforms more closely to the driving force obtained from the composite curves. It is also clear from the graph that both matches significantly under-utilize the available driving force. Due to the simplicity of the problem, however, this under-utilization does not cause a problem in completing the hot-end design for TC3.

Two exchangers from the cold-end design for Case B (Figure 8.10) are also shown in Figure 8.19. The 135-kW pinch exchanger agrees closely with the driving force plot, but the 20-kW non-pinch exchanger significantly over-utilizes the available driving force. Since the 20-kW exchanger was the last process exchanger needed to complete the cold-end design, the over-utilization does no harm in this case.

The driving force plot can be used to guide the design process toward the minimum area target and to analyze designs that are grossly over target. One reason for failure to achieve the desired heat-transfer area is over-utilization of driving force due to use of the

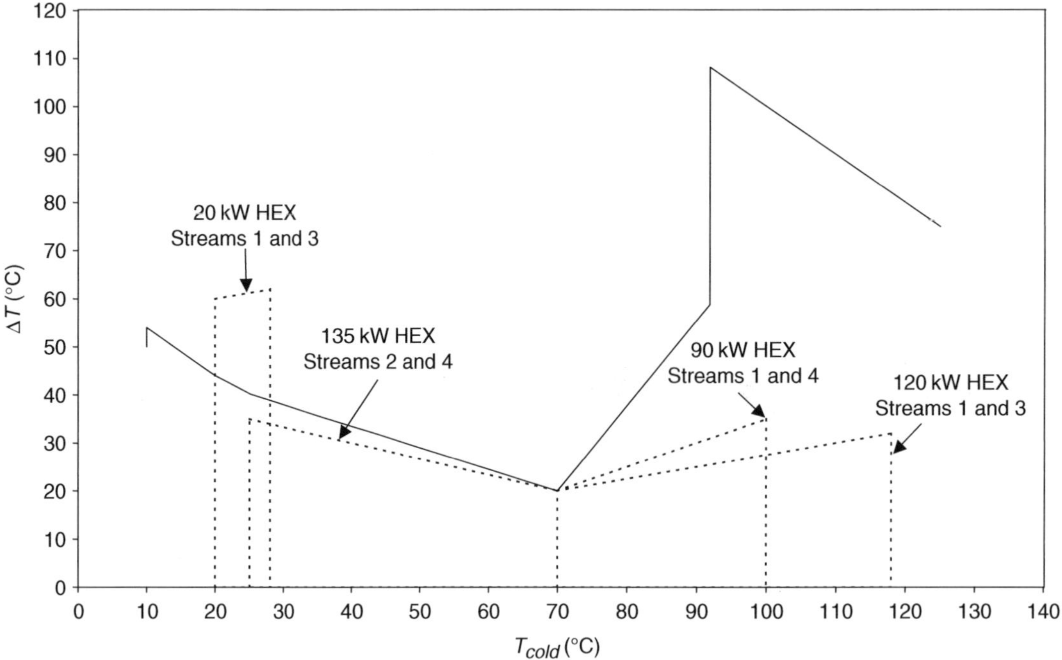

FIGURE 8.19 The driving force plot for TC3.

tick-off heuristic. In such cases it may be necessary to reduce the duty on one or more exchangers to obtain a better fit to the driving force plot. This procedure will generally result in more than the minimum number of units in the HEN. However, designs that achieve a good fit to the driving force plot while maintaining the number of units within 10% of the minimum are usually within 10% of the area target [5].

Although the driving force plot is intuitive and easy to use, the information that it provides is qualitative in nature. A quantitative assessment of the area penalty incurred by the improper use of driving force in a given match can be obtained by recalculating the area target after each successive exchanger is added to the evolving network design. While such a procedure is inconvenient for hand calculations, it is easily implemented in computer software.

8.17 Super Targeting

It has been shown in the preceding sections that for a given value of ΔT_{min} it is possible to estimate the minimum number of units, minimum-utility usage, minimum number of shells, and minimum heat-transfer area for a HEN. These values can be used to also estimate the minimum capital and operating costs of the HEN. All these estimates can be made prior to starting the actual design of the HEN.

Correlations for estimating the capital cost of heat exchangers are typically of the form:

$$\text{cost} = a + b(\text{area})^c \tag{8.20}$$

where a, b, and c are constants for a given type of heat exchanger, and the area to be used is the heat-transfer surface area of the exchanger. Ahmad et al. [6] have shown that the minimum capital cost of the HEN can be estimated by using the average area per shell in Equation (8.20) as follows:

$$\text{network cost} = aU_{\min} + bN_s(A_{\min}/N_s)^c \tag{8.21}$$

where U_{min}, N_s, and A_{min} are the target values for minimum number of units, minimum number of shells, and minimum heat-transfer area, respectively. Note that the fixed cost, a, is included only once per unit, regardless of the number of shells that comprise the unit. Ahmad et al. [6] and Hall et al. [12] also show how other factors such as materials of construction, pressure ratings, and different exchanger types can be included in the capital cost estimate. If counter-flow heat exchangers are assumed, the number of shells equals the number of units and N_s is replaced by U_{min} in Equation (8.21).

The minimum annual operating cost for the HEN can be estimated as the annual utility cost for a minimum-utility design. The other major factor involved in the operating cost is pumping cost, which depends on the pressure drop experienced by the streams in each heat exchanger. Methods for estimating these pressure drops are discussed in Refs. [13–15]. The heat exchanger design equations are used to relate stream pressure drop to heat-transfer area and stream heat-transfer coefficient. This is possible because both pressure drop and heat-transfer coefficient depend on the stream velocity. In this way, the estimated heat-transfer coefficients for the streams are used to estimate pressure drops for a given heat-transfer area. Although the accuracy of such estimates is questionable, neglecting pressure-drop effects can lead to serious errors in HEN design [3,4]. Hence, it is preferable to use both utility and pumping costs in estimating the annual operating cost for a HEN. With estimates of stream pressure drops, the capital cost for pumps and compressors can also be estimated and included in the total cost estimate for the HEN [15].

By using a capital cost recovery factor to annualize the capital cost, the operating and capital costs can be combined to obtain an estimate of the total annual cost of the HEN. These calculations can be performed for different values of ΔT_{min} within the range of feasible values determined from the composite curves. The optimum value of ΔT_{min} to use in designing the HEN is the one that minimizes the total annual cost. Figure 8.20 depicts the situation schematically and illustrates the effect of neglecting pumping costs on the optimum value of ΔT_{min} that is calculated.

8.18 Targeting by Linear Programming

The targeting procedures presented in preceding sections cannot be used when there are forbidden stream matches. Forbidden matches may arise from considerations such as safety, process operability, or plant layout. These additional constraints may result in higher utility requirements for the HEN. An alternative targeting procedure that can handle these constraints uses linear programming (LP).

A linear program is an optimization problem in which a linear function is to be maximized or minimized subject to a set of linear equality and/or inequality constraints. Here we will show that the problem table algorithm presented in Section 8.4 can be formulated as a linear program. To this end, let Q_j be the heat flow leaving the jth SN and let Q_0 be the amount of hot utility supplied to SN1. Then the energy balance for SN j can be written as:

$$Q_j = Q_{j-1} + \left\{\sum (CP)_{hot,i} - \sum (CP)_{cold,i}\right\}\Delta T_j \tag{8.22}$$

where the summations extend over the streams that exist in SN j. Using Equation (8.2), Equation (8.22) can be written as:

$$Q_j = Q_{j-1} - (\text{deficit})_j \tag{8.23}$$

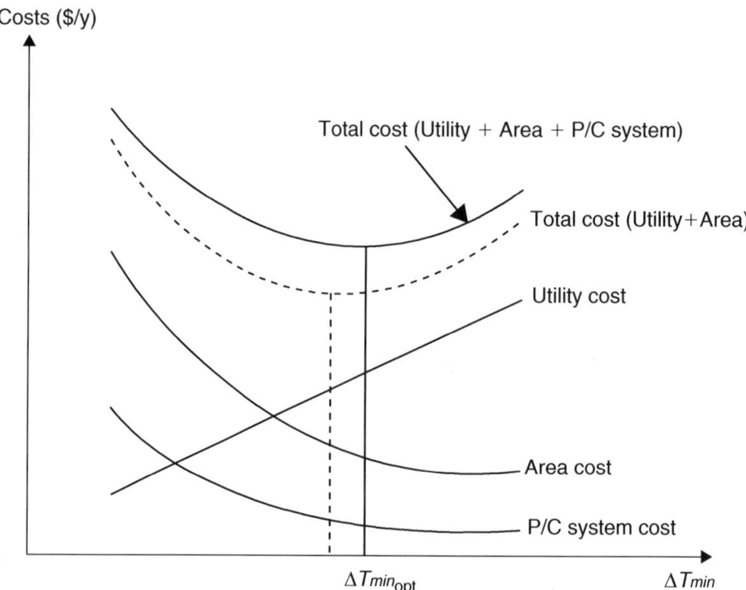

FIGURE 8.20 Determination of the optimal value of ΔT_{min} for HEN design. P/C system cost is the annualized capital and operating cost of pumps and compressors. Area cost is the annualized capital cost of heat exchangers (Source: Ref. [15]).

or

$$Q_j - Q_{j-1} + (\text{deficit})_j = 0 \tag{8.24}$$

Now for a problem with one hot and one cold utility and N SNs, Q_N is the heat flow to the cold utility, i.e., the cold utility requirement. Therefore, to find the minimum-utility requirements for the problem we need to minimize Q_0 and Q_N while satisfying the energy balance equation for each SN. In addition, all the heat flows must be non-negative. These conditions can be stated mathematically as follows:

$$\min f = Q_0 + Q_N \tag{8.25}$$

$$\text{subject to}: \; Q_j - Q_{j-1} + (\text{deficit})_j = 0 \; (j = 1, 2, ..., N)$$

$$Q_j \geq 0 \; (j = 0, 1, 2, ..., N)$$

Notice that both the objective function, f, and the constraints are linear in the variables, Q_j. Therefore, this is a linear program. The deficits for TC3 are given in Table 8.2. Substituting these values into (8.25) and setting $N = 6$ gives:

$$\min f = Q_0 + Q_6 \tag{8.26}$$

$$\text{subject to}: \; Q_1 - Q_0 - 10 = 0$$

$$Q_2 - Q_1 + 12.5 = 0$$

$$Q_3 - Q_2 + 105 = 0$$

$$Q_4 - Q_3 - 135 = 0$$

$$Q_5 - Q_4 + 82.5 = 0$$

$$Q_6 - Q_5 + 12.5 = 0$$

$$Q_0, Q_1, Q_2, Q_3, Q_4, Q_5, Q_6 \geq 0$$

This linear program can be solved using any of the widely available LP software packages, such as EXCEL Solver. The solution, given in Table 8.22, shows that the minimum hot and cold utility requirements are 107.5 and 40 kW, respectively, in agreement with the results obtained previously. Notice that the heat flow leaving SN3 is zero, indicating that the pinch occurs between SN3 and SN4, i.e., at 70°C for the cold streams and 90°C for the hot streams.

TABLE 8.22 Solution to Linear Program (8.26)

j	Q_j (kW)
0	107.5
1	117.5
2	105
3	0
4	135
5	52.5
6	40

When the problem contains more that a single hot or cold utility, the total utility cost is minimized rather than the total utility usage. The energy balance equations are also modified to allow introduction of utilities at the appropriate SNs. More substantive modifications are required to accommodate forbidden matches. In this case the problem is formulated as a mixed integer linear program (MILP), i.e., a linear program in which some variables take on only integer values. The MILP arises from the introduction of binary variables used to track the stream matches in the SNs. Each of these variables is equal to one if the two streams that it tracks are matched; otherwise, it is equal to zero. A good presentation of this topic can be found in the textbook by Biegler et al. [16]; the use of mathematical programming for the automatic synthesis of HENs is also covered in some detail. Further information can be found in Refs. [15,17–20].

8.19 Computer Software

Commercial software packages for HEN design include SuperTarget from KBC Advanced Technologies plc, HX-Net from Aspen Technology, Inc., and HEXTRAN by SimSci-Esscor (Invensys Operations Management). With the exception of HEXTRAN, these packages are interactive programs that allow the user to develop HEN designs on a grid diagram. They all have the capability to generate HEN designs automatically as well. HEXTRAN and HX-Net are considered in the following subsections.

8.19.1 HEXTRAN

Commercialized in 1980, HEXTRAN is the oldest of the HEN design packages and by far the most primitive. In Targeting mode it calculates minimum-utility targets and pinch temperatures, and constructs composite and grand composite curves. A variety of other targeting plots can also be generated. Although the program's graphing capability is minimal, all plots can be exported to EXCEL for manipulation by the user.

In Synthesis mode, HEXTRAN automatically generates HEN designs using a dual temperature approach method. The Heat Recovery Approach Temperature (HRAT) is the ΔT_{min} used for calculating utility and heat recovery targets. The Exchanger Minimum Approach Temperature (EMAT) is the ΔT_{min} imposed on individual heat exchangers. Setting EMAT$<$ HRAT allows some heat transfer across the pinch, which is required for practical HEN configurations. The algorithm identifies feasible stream matches in each SN and then combines the SNs to obtain an initial network configuration. Network simplification is accomplished by identifying and breaking heat load loops until a configuration is obtained that approaches the minimum-number-of-units target [21]. Use of the program for retrofit applications is discussed in Ref. [22].

The use of HEXTRAN for targeting and HEN design is considered in the following examples.

Example 8.1

Use HEXTRAN to perform targeting calculations for TC3 by proceeding as follows:

(a) Determine the range of ΔT_{min} for heat recovery, i.e., HRAT.

(b) Find the minimum-utility requirements, maximum amount of heat recovery, and the hot and cold stream temperatures at the pinch for a set of HRAT values covering the range found in part (a).

Solution

(a) When HEXTRAN is run in targeting mode with HRAT set to zero, it calculates the feasible range for HRAT. (The effect of utility temperatures is ignored in this calculation.) To begin, SI units are selected, the temperature unit is changed from Kelvin to Celsius and the unit of time is changed from hour to second. Next, the *Calculation Type* is set to *Targeting*, and under *Calculation Options* HRAT is set to zero (the default value).

The flowsheet for this problem consists simply of six unconnected streams (four process streams and two utility streams) as shown below:

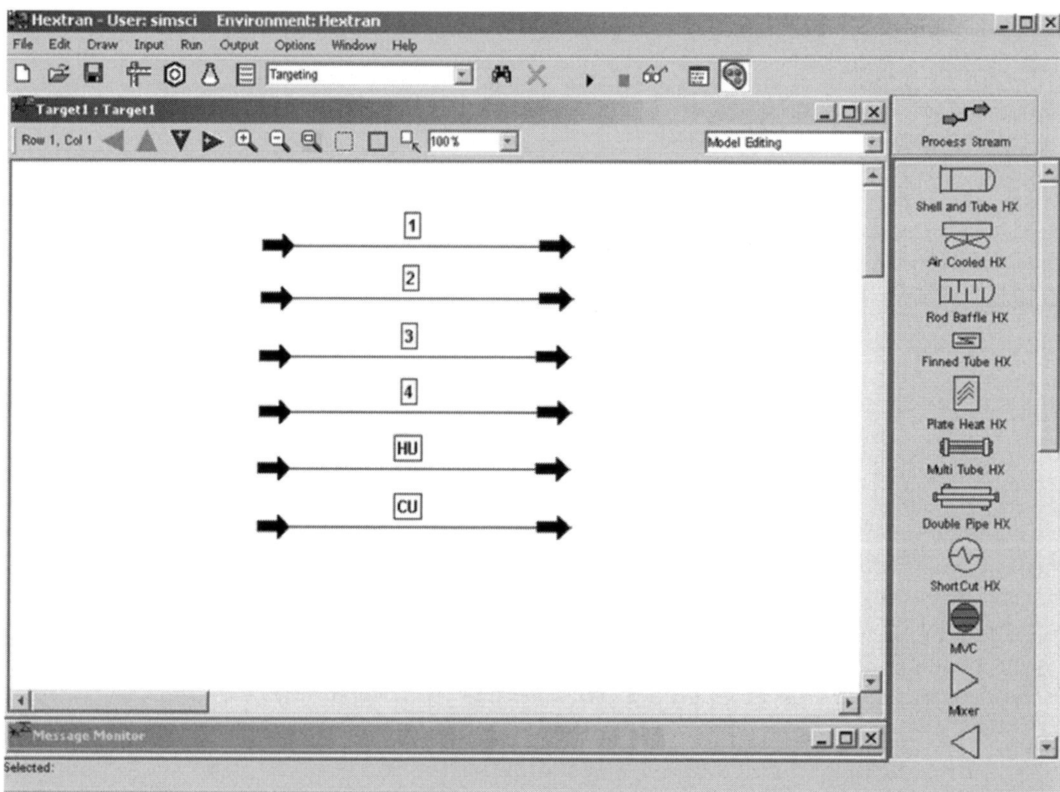

The stream type is changed to *Bulk Property* for each of the four process streams and to *Utility* for the two utility streams. The following data are entered for each process stream:

- Flow rate = 5 kg/s (arbitrary)
- Pressure = 500 kPa (arbitrary)
- Temperature = Supply temperature
- Outlet (exit) temperature = Target temperature
- Stream duty as an average value
- Heat-transfer coefficient ($W/m^2 \cdot K$) as an average value

The stream flow rate and pressure are required by HEXTRAN but are not needed for this problem, so the values were chosen arbitrarily. The unit for duty (MMkJ/s) is very large and only one or two decimal places are typically printed in the output file. Therefore, to avoid loss of significant figures, the stream duties are scaled by a factor of 10^5 (a factor of 10^6 causes the program to hang up during execution). For example, the duty for Stream 1 is entered as 18 MMkJ/s. This scaling does not affect the targeting calculations. The same results are obtained when different scale factors and/or units are used. The default value (100 $W/m^2 \cdot K$) for heat-transfer coefficients can be used if estimated values are not available or if area estimates are not required.

Required data for the utility streams are inlet and outlet temperatures and available duty. HEXTRAN does not accept an isothermal utility stream, so for the hot utility, inlet and outlet temperatures of 200°C and 199°C are specified. For the cold utility, 10°C and 10.1°C are used. The duty for each utility is set at 50 MMkJ/s (equivalent to 500 kW in the original problem).

The input file generated by the HEXTRAN GUI is given below, followed by the relevant data extracted from the output file. Under heat recovery limits, the range for HRAT is seen to be from 12.5°C to 130°C, corresponding to a threshold value of 12.5°C. In fact, the correct threshold value was found by hand calculation to be 12.727°C. The output data for the composite curves shows a minimum temperature difference of 12.7°C, which is correct to the given number of significant figures.

HEXTRAN Input File for Example 8.1, Part (a)

```
$ GENERATED FROM HEXTRAN KEYWORD EXPORTER
$
$       General Data Section
$
TITLE PROJECT=EX8-1, PROBLEM=TARGETING, SITE=
$
DIME  SI, AREA=M2, CONDUCTIVITY=WMK, DENSITY=KG/M3, *
      ENERGY=KJ, FILM=WMK, LIQVOLUME=M3, POWER=KW, *
      PRESSURE=KPA, SURFACE=NM, TIME=SEC, TEMPERATURE=C, *
      UVALUE=WMK, VAPVOLUME=M3, VISCOSITY=PAS, WT=KG, *
      XDENSITY=DENS, STDVAPOR=22.414
$
OUTD  SI, AREA=M2, CONDUCTIVITY=WMK, DENSITY=KG/M3, *
      ENERGY=KJ, FILM=WMK, LIQVOLUME=M3, POWER=KW, *
      PRESSURE=KPA, SURFACE=NM, TIME=SEC, TEMPERATURE=C, *
      UVALUE=WMK, VAPVOLUME=M3, VISCOSITY=PAS, WT=KG, *
      XDENSITY=DENS, STDVAPOR=22.414, ADD
$
PRINT ALL, *
      RATE=M
$
CALC  PGEN=New, WATER=Saturated
$
$       Component Data Section
$
$       Thermodynamic Data Section
$
$Stream Data Section
$
STREAM DATA

$
 PROP STRM=1, NAME=1, TEMP=150.00, PRES=500.000, *
      TOUT=60.00, LIQUID(W)=5.000, Duty(AVG)=18, *
         Film(AVG)=1100
$
 PROP STRM=2, NAME=2, TEMP=90.00, PRES=500.000, *
      TOUT=60.00, LIQUID(W)=5.000, Duty(AVG)=24, *
         Film(AVG)=600
$


 PROP STRM=3, NAME=3, TEMP=20.00, PRES=500.000, *
      TOUT=125.00, LIQUID(W)=5.000, Duty(AVG)=26.25, *
         Film(AVG)=800
$
 PROP STRM=4, NAME=4, TEMP=25.00, PRES=500.000, *
      TOUT=100.00, LIQUID(W)=5.000, Duty(AVG)=22.5, *
         Film(AVG)=1000
$
 UTILITY STRM=HU, TEMP=200.00, TOUT=199.00, FILM=2000.00, *
         DUTY=50.00, BSIZE=92.90, BCOST=0.00, LINEAR=538.20, *
         CONSTANT=0.00, EXPONENT=0.60
$
 UTILITY STRM=CU, TEMP=10.00, TOUT=10.10, FILM=1500.00, *
         DUTY=50.00, BSIZE=92.90, BCOST=0.00, LINEAR=538.20, *
         CONSTANT=0.00, EXPONENT=0.60
$
$ Calculation Type Section
$
TARGETING
$
 SPEC   HRAT=0.00
$
 PARAMETER FILM=100.00, ALPHA=1.00, EXPONENT=1.00, *
           DELTA=0.00
$
 PRINT  DUTY, COMPOSITE, GRAND, SUMMARY
```

```
$
ECONOMICS DAYS=350, EXCHANGERATE=1.00, CURRENCY=USDOLLAR, RATE=10.0, *
           LIFE=30
$
 HXCOST BSIZE=92.90, BCOST=0.00, LINEAR=538.20, *
        EXPONENT=0.60, CONSTANT=0.00
$
$ End of keyword file...
```

HEXTRAN Results for Example 8.1, Part (a)

```
===============================================================================
                            HEAT RECOVERY LIMITS

                                                MAXIMUM       MINIMUM

  HRAT                    - DEG C                 12.5         130.0
  QPROCESS DUTY           - MMKJ /SEC             42.0           0.0
  QHEATING DUTY           - MMKJ /SEC              6.8          48.8
  QCOOLING DUTY           - MMKJ /SEC              0.0          42.0
  THEATING TEMPERATURE - DEG C                    99.1          20.0
  TCOOLING TEMPERATURE - DEG C                    60.0         150.0
  MTD                     - DEG C                 24.0         130.0

===============================================================================

RUN NO.  1    HRAT =    0.0 DEG C   PROCESS EXCHANGE DUTY =    42.0000 MMKJ /SEC

------- TEMPERATURE(DEG C) --------           STREAMS
   HOT          COLD        APPROACH     ENTERING OR LEAVING
COMPOSITE    COMPOSITE    TEMPERATURE

  150.0         99.1          50.9          1
  146.0         97.6          48.4
  142.0         96.2          45.8
  138.0         94.7          43.3
  134.0         93.3          40.7
  130.0         91.8          38.2
  126.0         90.4          35.6
  122.0         88.9          33.1
  118.0         87.5          30.5
  114.0         86.0          28.0
  110.0         84.5          25.5
  106.0         83.1          22.9
  102.0         81.6          20.4
   98.0         80.2          17.8
   94.0         78.7          15.3
   90.0         77.3          12.7     MIN 2
   87.8         73.3          14.5
   85.6         69.3          16.3
   83.4         65.3          18.1
   81.2         61.3          19.9
   79.0         57.3          21.7
   76.8         53.3          23.5
   74.6         49.3          25.3
   72.4         45.3          27.1
   70.2         41.3          28.9
   68.0         37.3          30.7
   65.8         33.3          32.5
   63.6         29.3          34.3
   61.4         25.3          36.1
   61.2         25.0          36.2          4
   60.2         21.0          39.2
   60.0         20.0          40.0          1             2            3
```

```
================================================================================

                          SUMMARY OF CASES

   TEMPERATURE SUMMARY TABLE

    HRAT      QPROCESS     THEATING     TCOOLING   HOT APPROACH  CLD APPROACH  LMTD
    DEG C     MMKJ /SEC    DEG C        DEG C         DEG C          DEG C         DEG C

       0.0       42.00        99.1         60.0       90.0              77.3          24.0

   DUTY SUMMARY TABLE

    HRAT      QPROCESS     QHEATING     QCOOLING      QUTILITY     AREA      U-VALUE
    DEG C     MMKJ /SEC    MMKJ /SEC    MMKJ /SEC     MMKJ /SEC    M2      WATTS / M2-K

       0.0       42.00          6.75         0.00          6.75       4267042.  409.6

================================================================================
```

(b) A second run is made with the following set of HRAT values:

12.73	50.0
12.80	100.0
15.0	120.0
20.0	125.0
30.0	130.0

The summary of cases shown below was extracted from the output file and contains all the required information for this problem. Note that the areas and duties must be divided by the scale factor of 10^5 to obtain the correct values.

HEXTRAN Results for Example 8.1, Part (b)

```
================================================================================

                          SUMMARY OF CASES

   TEMPERATURE SUMMARY TABLE

    HRAT      QPROCESS     THEATING     TCOOLING    HOT APPROACH  CLD APPROACH  LMTD
    DEG C     MMKJ /SEC    DEG C        DEG C          DEG C          DEG C         DEG C

     130.0        0.00         0.0       150.0           0.0            0.0           0.0
     125.0        1.00         0.0       145.0         145.0           20.0         125.5
     120.0        1.80         0.0       141.0         141.0           20.0         122.1
     100.0        5.80        33.3       121.0         121.0           20.0         107.6
      50.0       21.50        61.8        80.5          90.0           40.0          60.7
      30.0       32.50        81.8        69.5          90.0           60.0          40.6
      20.0       38.00        91.8        64.0          90.0           70.0          31.1
      15.0       40.55        96.5        61.5          90.0           74.6          26.7
      12.8       41.65        98.5        60.4          90.0           76.6          24.7
      12.7       41.65        98.5        60.4          90.0           76.6          24.7
```

```
DUTY SUMMARY TABLE

 HRAT      QPROCESS     QHEATING     QCOOLING     QUTILITY     AREA      U-VALUE
 DEG C     MMKJ /SEC    MMKJ /SEC    MMKJ /SEC    MMKJ /SEC    M2      WATTS / M2-K

  130.0       0.00        48.75        42.00        90.75          0.       0.0
  125.0       1.00        47.75        41.00        88.75      17714.     449.9
  120.0       1.80        46.95        40.20        87.15      35069.     420.5
  100.0       5.80        42.95        36.20        79.15     109324.     493.2
   50.0      21.50        27.25        20.50        47.75     804602.     440.5
   30.0      32.50        16.25         9.50        25.75    1913414.     418.2
   20.0      38.00        10.75         4.00        14.75    2959870.     412.2
   15.0      40.55         8.20         1.45         9.65    3713650.     409.6
   12.8      41.65         7.10         0.35         7.45    4129781.     408.8
   12.7      41.65         7.10         0.35         7.45    4129781.     408.8

===============================================================================
```

Here again it can be seen that some of the results are surprisingly inaccurate. Two of the hot stream pinch temperatures are off by one degree (141 versus 140 and 121 versus 120). Likewise, the cold stream pinch temperatures are inaccurate for HRAT values of 15, 12.8, and 12.73. The hot and cold utility requirements for HRAT = 12.73 are 71 and 3.5 kW, respectively, whereas the correct values are approximately 67.5 kW and zero. Although these minor discrepancies may be of little practical significance, they nevertheless seem inexcusable considering that accurate results are easily obtained by hand for this problem. The error in the area estimates is more substantial. For HRAT = 20°C, the value of 29.6 m^2 is about 10% below the counter-flow area of 33.74 m^2 obtained by hand in Table 8.19.

Example 8.2

Use HEXTRAN to design a HEN for TC3.

Solution

For this problem, HEXTRAN is run in synthesis mode with HRAT = 20°C. To obtain a practical design, a value of EMAT must be selected that is less than HRAT. Two values of EMAT are tried, 17°C and 18°C. The HEXTRAN data from Example 8.1 are modified for this problem by changing the *Calculation Type* to *Synthesis* and entering the desired values of HRAT and EMAT under *Calculation Options/Specifications.* Note that both cases can be executed in a single HEXTRAN run. In addition, under *Calculation Options/Limits,* the maximum area per shell is set to 5×10^6 m^2, an absurdly large value, in order to accommodate the scale factor of 10^5 applied to the stream duties. The resulting input file generated by the GUI is shown below.

Data for the network generated with EMAT = 17°C were extracted from the output file and are shown below following the input file. These data and the corresponding results for EMAT = 18°C were used to construct the grid diagrams for the networks shown below following the output data. Notice that both networks meet the minimum-utility targets for ΔT_{min} = HRAT=20°C, but have small ΔT_{min} violations as allowed by the specified values of EMAT.

HEXTRAN Input File for Example 8.2

```
$ GENERATED FROM HEXTRAN KEYWORD EXPORTER
$
$       General Data Section
$
TITLE PROJECT=EX8-2, PROBLEM=SYNTHESIS, SITE=
$
DIME  SI, AREA=M2, CONDUCTIVITY=WMK, DENSITY=KG/M3, *
      ENERGY=KJ, FILM=WMK, LIQVOLUME=M3, POWER=KW, *
      PRESSURE=KPA, SURFACE=NM, TIME=SEC, TEMPERATURE=C, *
      UVALUE=WMK, VAPVOLUME=M3, VISCOSITY=PAS, WT=KG, *
      XDENSITY=DENS, STDVAPOR=22.414
$
OUTD  SI, AREA=M2, CONDUCTIVITY=WMK, DENSITY=KG/M3, *
      ENERGY=KJ, FILM=WMK, LIQVOLUME=M3, POWER=KW, *
      PRESSURE=KPA, SURFACE=NM, TIME=SEC, TEMPERATURE=C, *
      UVALUE=WMK, VAPVOLUME=M3, VISCOSITY=PAS, WT=KG, *
      XDENSITY=DENS, STDVAPOR=22.414, ADD
```

```
$
PRINT ALL, *
      RATE=M
$
CALC  PGEN=New, WATER=Saturated
$
$       Component Data Section
$
$       Thermodynamic Data Section
$
$Stream Data Section
$
STREAM DATA

$
 PROP STRM=1, NAME=1, TEMP=150.00, PRES=500.000, *
      TOUT=60.00, LIQUID(W)=5.000, Duty(AVG)=18, *
         Film(AVG)=1100
$
 PROP STRM=2, NAME=2, TEMP=90.00, PRES=500.000, *
      TOUT=60.00, LIQUID(W)=5.000, Duty(AVG)=24, *
         Film(AVG)=600
$
 PROP STRM=3, NAME=3, TEMP=20.00, PRES=500.000, *
      TOUT=125.00, LIQUID(W)=5.000, Duty(AVG)=26.25, *
         Film(AVG)=800
$
 PROP STRM=4, NAME=4, TEMP=25.00, PRES=500.000, *
      TOUT=100.00, LIQUID(W)=5.000, Duty(AVG)=22.5, *
         Film(AVG)=1000
$
 UTILITY STRM=HU, TEMP=200.00, TOUT=199.00, FILM=2000.00, *
         DUTY=50.00, BSIZE=92.90, BCOST=0.00, LINEAR=538.20, *
         CONSTANT=0.00, EXPONENT=0.60
$
 UTILITY STRM=CU, TEMP=10.00, TOUT=10.10, FILM=1500.00, *
         DUTY=50.00, BSIZE=92.90, BCOST=0.00, LINEAR=538.20, *
         CONSTANT=0.00, EXPONENT=0.60
$
$ Calculation Type Section
$
SYNTHESIS
$
 SPEC    HRAT=20.00, 20.00, EMAT=17.00, 18.00
$
 PARAMETER FILM=100.00
$
 PRINT   SPLIT=Short, UNSPLIT=Last
$
 PLOT    ALL
$
ECONOMICS DAYS=350,  EXCHANGERATE=1.00,  CURRENCY=USDOLLAR,  RATE=10.0, *
           LIFE=30
$
 HXCOST BSIZE=92.90, BCOST=0.00, LINEAR=538.20, *
        EXPONENT=0.60, CONSTANT=0.00
$
 LIMITS MAXP=10, MAXS=10, MAXAREA=5000000.000000, *
 MINFT=0.800000
$
$ End of keyword file...
```

HEXTRAN Results for Example 8.2 with EMAT = 17°C

NETWORK WITHOUT STREAM SPLITTING FOR RUN NO. 1

HRAT = 20.0 DEG C EMAT = 17.0 DEG C

HEAT EXCHANGER SUMMARY - CONNECTIVITY DATA

UNIT NUMBER	HOT SIDE STRM ID	HOT SIDE FROM/TO UNIT	COLD SIDE STRM ID	COLD SIDE FROM/TO UNIT
1	1	IN/ 3	4	2/OUT
2	2	IN/ 4	4	IN/ 1
3	1	1/OUT	3	4/501
4	2	2/500	3	IN/ 3
500	2	4/OUT	CU	
501	HU		3	3/OUT

HEAT EXCHANGER SUMMARY - TEMPERATURE DATA

UNIT NO.	HOT SIDE STRM ID	HOT SIDE TEMP(DEG C) IN / OUT	HOT SIDE FLOW RATIO	COLD SIDE STRM ID	COLD SIDE TEMP(DEG C) IN / OUT	COLD SIDE FLOW RATIO
1	1	150.0 / 109.5	1.000	4	73.0 / 100.0	1.000
2	2	90.0 / 72.0	1.000	4	25.0 / 73.0	1.000
3	1	109.5 / 60.0	1.000	3	42.4 / 82.0	1.000
4	2	72.0 / 65.0	1.000	3	20.0 / 42.4	1.000
500	2	65.0 / 60.0	1.000	CU	10.0 / 10.1	
501	HU	200.0 / 199.0		3	82.0 / 125.0	1.000

HEAT EXCHANGER SUMMARY - GENERAL DATA

UNIT NUMBER	DUTY MMKJ /SEC	AREA M2	U-VALUE WATTS / M2-K	FT	NUMBER OF SHELLS SERIES	PARALLEL
1	8.1000	404645.3	523.810	0.891	1	1
2	14.4000	1360883.7	375.000	0.957	2	1
3	9.9000	1200684.0	463.158	0.803	2	1
4	5.6000	453297.4	342.857	0.980	1	1
500	4.0000	178097.0	428.571	1.000	1	1
501	10.7500	199360.6	571.429	0.999	1	1

HEN for TC3 Generated By HEXTRAN with EMAT= 17°C

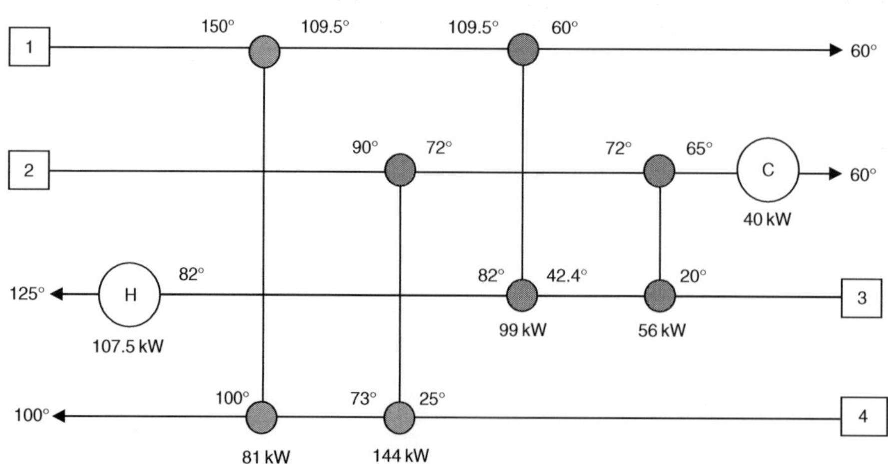

HEN for TC3 Generated By HEXTRAN with EMAT = 18°C

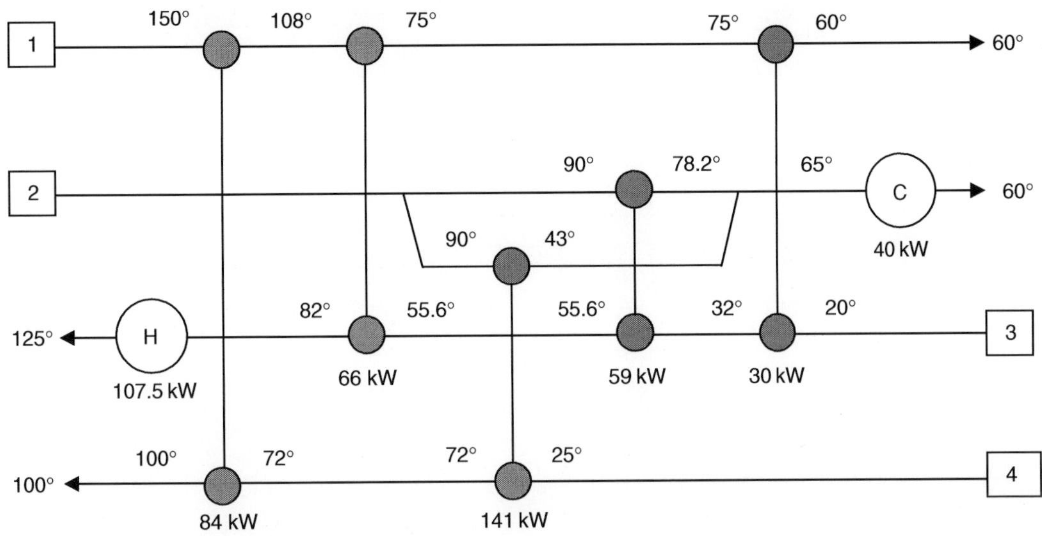

With EMAT = 17°C, the network contains six units and no stream splits. The 99-kW and 144-kW exchangers operate across the pinch and thus have ΔT_{min} violations. Only the 99-kW exchanger has a significant temperature cross, and it requires two shells. The counter-flow area estimate for this network is 34.46 m^2, which is about 1.4% over target. For multi-pass exchangers, the estimated area (calculated by hand) is 40.85 m^2, which is approximately 15% over target. With six units, seven shells, no stream splits, on-target utility usage, near minimum heat-transfer area, and only small ΔT_{min} violations, this network is clearly superior to any of those obtained by hand.

Notice that HEXTRAN uses two shells in series for the 144-kW exchanger. The reason is that the minimum value for the LMTD correction factor was set at the default value of 0.8, whereas the actual *F*-value for this unit is 0.789 with one shell pass. In fact, a single shell pass is adequate; checking the operating point on Figure 3.14 shows that it does not lie on the steeply sloping part of the graph. The higher *F*-value for this unit used by HEXTRAN results in a smaller network area. Re-running with the minimum *F*-value set at 0.78 results in a network area of 40.86 m^2, in agreement with the hand-calculated value given above.

With EMAT = 18°C, a network without stream splitting was obtained comprising eight units and nine shells. This network is not shown since it is clearly inferior to the one obtained with EMAT = 17°C. A second network was obtained with Stream 2 split into two branches as shown above. It contains seven units and ten shells. The network obtained by hand and shown in Figure 8.14 also meets the minimum-utility targets with ΔT_{min} = 18°C and Stream 2 split into two branches. However, the latter network contains only six units and nine shells, and is therefore simpler than the configuration generated by HEXTRAN.

8.19.2 HX-Net (Aspen Energy Analyzer)

HX-Net was developed by Hyprotech, Ltd. and is now marketed by Aspen Technology, Inc. under the name Aspen Energy Analyzer. As noted above, HX-Net is an interactive program that provides a grid diagram on which the user can develop a HEN design. The software calculates targets for utility consumption, heat-transfer area, and number of shells. It contains an MILP routine that is used for targeting when forbidden stream matches are specified by the user. Super targeting is done based on the trade-off between the capital cost of heat exchangers and utility costs; pressure-drop effects are neglected. The software also generates composite curves, the grand composite curve, and driving force plots.

HX-Net can generate HEN designs automatically, starting either from scratch or from a partially completed network developed by the user. The algorithm employs a two-stage procedure. In the first stage, an MILP is solved to minimize an annualized cost function comprised of exchanger capital cost and utility cost. The solution gives the optimal utility loads, stream matches and duties. Details of the model are given by Shethna et al. [23]. In the second stage, another MILP is solved to obtain a network configuration that conforms to the optimal stream matches and duties obtained in the first stage. Unfortunately, the automatic design feature is not available with an academic license.

HX-Net interfaces with both the HYSYS and Aspen Plus process simulators. It can, for example, automatically extract data from a HYSYS or Aspen Plus flowsheet and set up the corresponding HEN on an interactive grid diagram. Modifications made to the HEN can subsequently be implemented in the simulator flowsheet so that their effect on the overall process can be evaluated. The simulator flowsheet must be modified manually, however.

Example 8.3

Use HX-Net to perform targeting calculations for TC3. Use the following economic data:

Hot utility cost	\$150/kW · year
Cold utility cost	\$200/kW · year
Rate of return	10%
Plant life	5 years
Heat exchanger capital cost (\$) = 4000 + 7500(area)$^{0.8}$	

Solution

After starting HX-Net, select *Tools* on the toolbar, then *Preferences,* and under *Variables* set the units to *Energy Integration* – SI. Then add new units as follows:

Energy	kW
MC_P	kW/°C
Heat-transfer coefficient	kW/m^2 · °C

Next, open a new Heat Integration Case and enter the data for the four process streams. In the column labeled *Clean HTC,* enter the heat-transfer coefficients from Table 8.17, and enter zero for the fouling factors. This will give the correct values for the stream coefficients, which are automatically entered in the column labeled *Film HTC.* The columns labeled *Flowrate* and *Effective* C_P are left blank.

Now enter the data for the two utility streams, including unit costs. No flow rate, load, or heat capacity data are required.

Next, select the *Economics* tab and set the default parameters for capital cost as follows:

$a = 4000$
$b = 7500$
$c = 0.8$

Notice that the rate of return and plant life specified for this problem are the default values in HX-Net.

Clicking on the targeting icon opens the targets window shown below. The user can change the value of ΔT_{min} and new target values are automatically calculated. Notice that the target values for utility usage, number of units (MER stands for Maximum Energy Recovery), number of shells, and counter-flow area are in agreement with the hand calculations. However, the shell-and-tube target area of 43.5 m^2 is significantly higher than the value of 35.46 m^2 calculated by hand. HX-Net clearly overestimates this target because the optimal network generated by HEXTRAN in Example 8.2 has an area of 40.85 m^2, which is 6% *below* the HX-Net value. The "sufficient" designation for heating and cooling at the bottom of the targets form indicates that the temperature driving force between the utility streams and process streams is greater than or equal to the specified value of ΔT_{min}.

Example 8.3: Targets Window in HX-Net

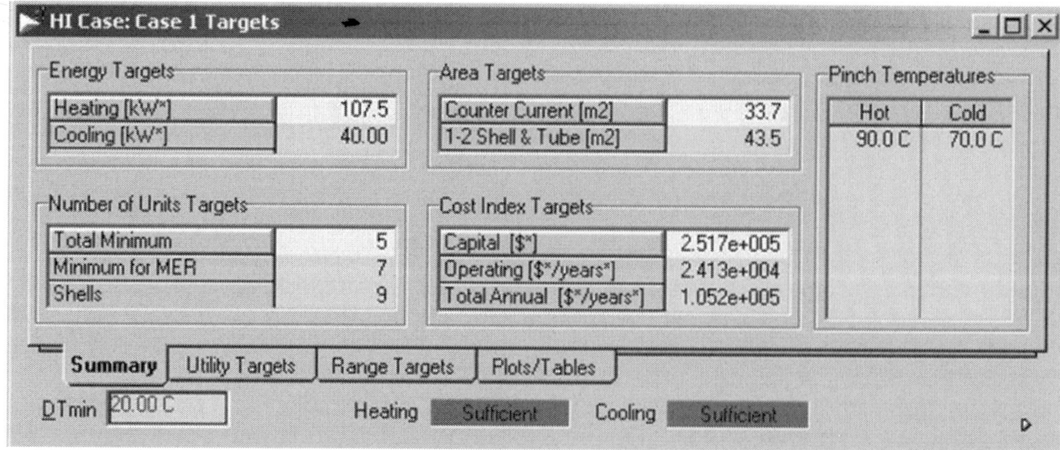

Super targeting is performed by clicking the *Range Targets* tab and entering the range of ΔT_{min} values desired and the increment to be used. The results shown in the graph below were obtained with an increment of 1.0°C. From the graph, the optimum value of ΔT_{min} for this problem is approximately 24.5°C.

Various types of composite curve and driving force plots are available under the *Plots/Tables* tab; several of these are shown below. By right-clicking on the plot area, the user can change the title, axis labels, and format of each graph as desired. Graphs can be printed on a stand-alone basis or included in a report generated by HX-Net, the contents of which are selected by the user.

Example 8.3: Super Targeting Results from HX-Net

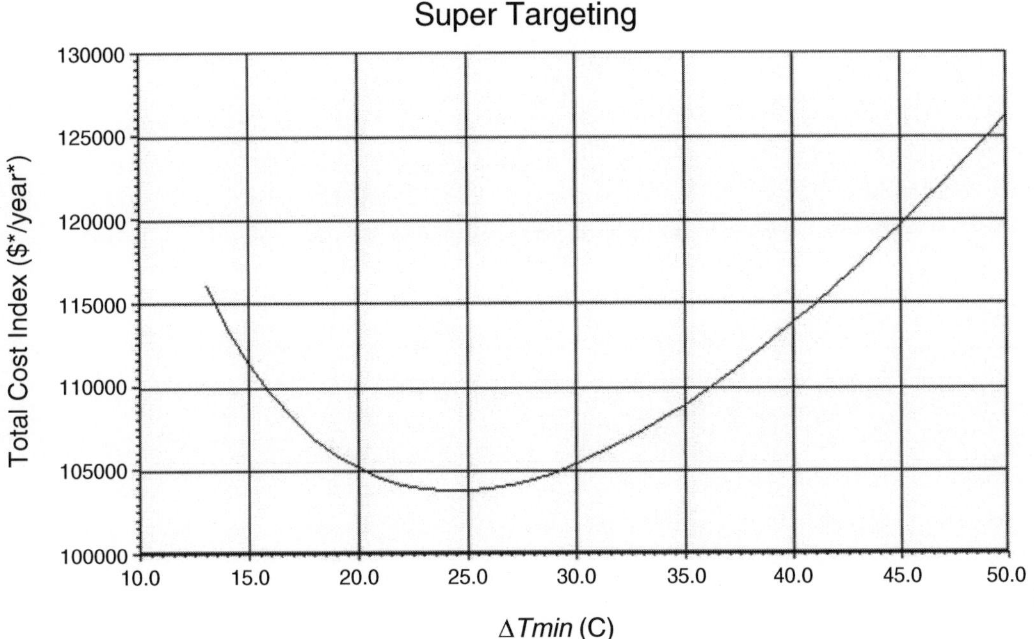

Example 8.3: Targeting Graphs Generated by HX-Net.

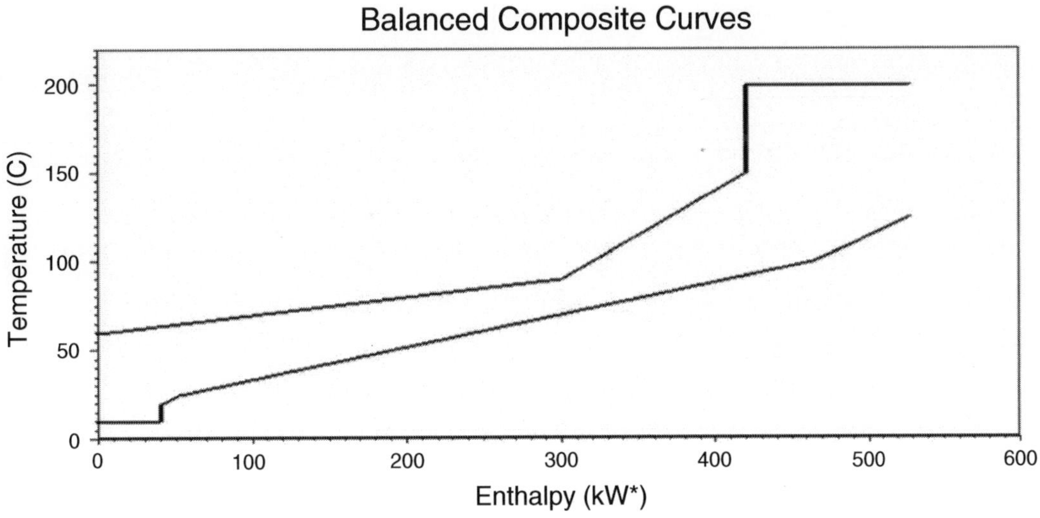

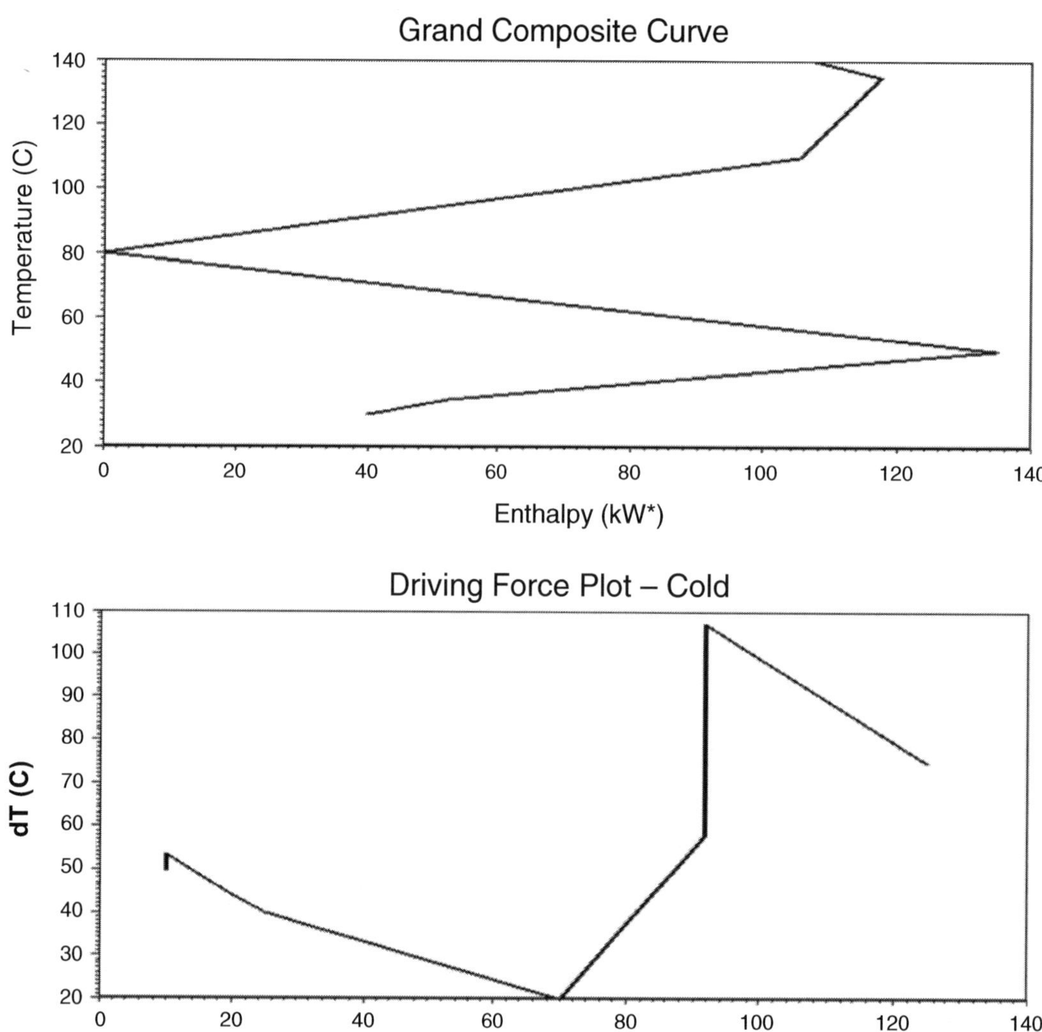

8.20 A Case Study: Gasoline Production from Bio-Ethanol

A process for converting bio-ethanol to gasoline was proposed by Whitcraft et al. [24] and Aldridge et al. [25]. It employs the same zeolite catalyst used in the Exxon-Mobil process for producing gasoline from methanol. The process (Figure 8.21) consists of two main units, a distillation column in which the ethanol concentration is increased from 10% to 60% by weight and a catalytic reactor in which ethanol is converted to hydrocarbons. The two units are thermally integrated to make the process highly energy efficient.

The feedstock for the process is a 10% by weight solution of ethanol in water produced in a fermentation unit. The feed is preheated to its saturation temperature in three stages before entering the distillation column. The first heat exchanger in the preheat train serves as the column condenser. The bottom product from the column, which contains about 0.1% ethanol by weight, exchanges heat with the feed in exchanger H-2 before proceeding to wastewater treatment and/or disposal. The overhead vapor product is compressed and heated prior to entering the reactor, which operates at 300°C and 8 atm. In the reactor, ethanol is dehydrated and converted to hydrocarbons using the ZSM-5 zeolite catalyst. The heat produced by the exothermic reaction is removed from the fixed-bed reactor by circulating a heat-transfer oil through the unit. The product stream from the reactor is partially condensed in exchanger H-4, which together with exchanger H-8 serves as the reboiler for the distillation column. The vapor and liquid fractions are separated in unit S-1, and the vapor fraction, which consists of light hydrocarbons, is compressed and recycled to the reactor feed. The liquid fraction, consisting of gasoline and water, is cooled by heat exchange with the distillation feed in unit H-3. After further cooling to 20°C in exchanger H-7, the stream is separated by decantation to produce the gasoline product and another wastewater stream. The hot heat-transfer oil from the reactor is used to preheat the reactor feed in exchanger H-5 and to supply heat to the reboiler for the distillation column in exchanger H-8.

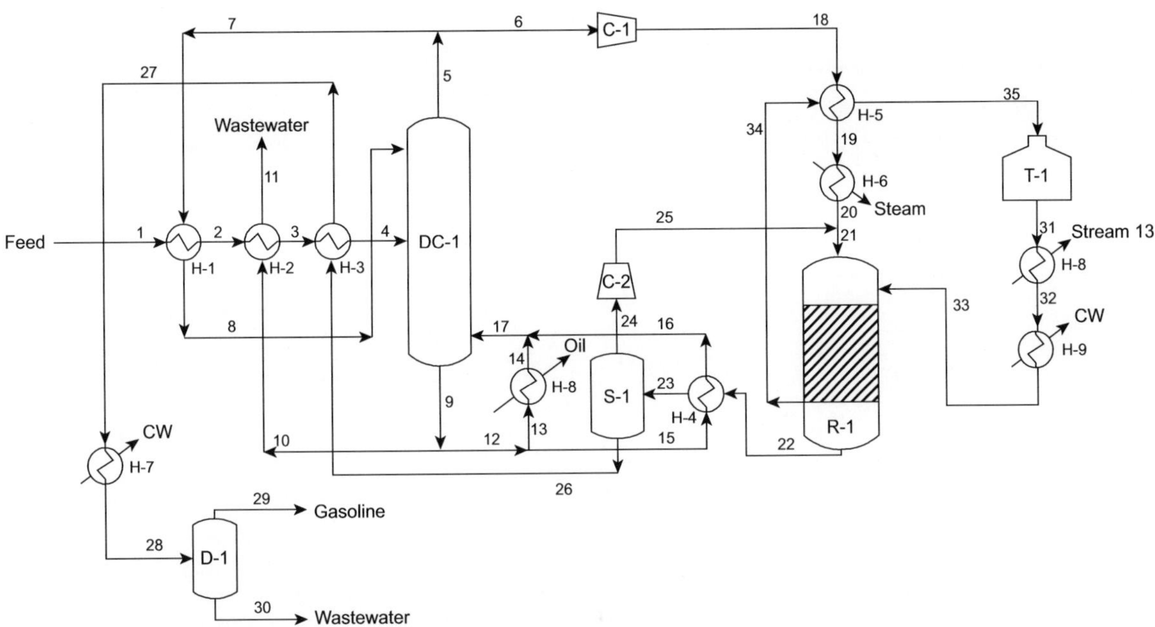

FIGURE 8.21 Flowsheet for production of gasoline from bio-ethanol. Redrawn from Ref. [26].

Stream data for the process are shown in Table 8.23 and are based primarily on information given in Ref. [25]. As in that reference, pressure changes across process units are generally neglected except for the compressors and the reactor. The temperatures of the streams from the distillation column (overhead, bottoms, reflux and boil-up) were calculated using the PRO/II process simulator. The UNIQUAC thermodynamic package was used for the simulation, and the reflux was assumed to enter the column as a saturated liquid. There is some uncertainty in the data for the heat-transfer oil since little information was provided for these streams in the original publications. The temperature of the oil leaving the reactor was taken as 290°C based on a 10°C approach for heat exchange in the reactor. (Although it is stated in [25] that the process, as given, is based on a 10°C approach, there is an inherent inconsistency here since an approach of 10°C in the reactor results in an approach of only 6°C in exchanger H-5, and vice versa.) The temperatures of the other oil streams are based on an assumed temperature range of 90°C and the duty for the oil, which was calculated from heat-of-reaction data given in [24].

Hot utility consists of steam to preheat the reactor feed and amounts to 33.4 kW [25]. Cold utility consists of cooling water used to remove excess heat from the heat-transfer oil and the liquid product stream, and amounts to 280 kW. This value was calculated based on data obtained from the original publications [24, 25].

Aldridge et al. [25] compared the energy requirements for the process with those for other methods of converting the ethanol solution from fermentation to a usable fuel. They considered only steam and compressor power (126 kW), neglecting pumping requirements and the potential need for chilled water in exchanger H-7. Their analysis showed that gasoline production compares very favorably with other methods in terms of energy requirements, while having the further advantage of producing a conventional fuel. Although this work was published in a reputable peer-reviewed journal, it contains an egregious error that invalidates the energy analysis. As shown in Figure 8.22, heat exchanger H-3 is thermodynamically impossible since the hot stream outlet temperature (74.2°C) is less than the cold stream inlet temperature (87°C). (This is not simply the result of a typographical error; the stream temperatures are consistent with stream enthalpies given in [25], and these enthalpies satisfy the energy balance around the exchanger.)

Khan and Riverol [26] performed a pinch analysis to obtain a new heat-exchanger network for the process. Unfortunately, their work was seriously flawed due to, among other problems, an inappropriate choice of hot and cold streams. Furthermore, they inexplicably retained the impossible heat exchanger in their HEN, thereby rendering it thermodynamically invalid.

The objective of this section is to use HEXTRAN to perform a pinch analysis and develop a thermodynamically valid HEN for the process. The first step is to identify the hot and cold streams. The two streams (22 and 34) leaving the reactor are clearly hot streams. Other streams that require cooling are Stream 7, which must be condensed to provide the liquid reflux for distillation; Stream 26, the liquid from the separator, which must be cooled prior to decantation; and Stream 10, the bottom product from the distillation column, which is essentially hot water at its boiling point. The cold streams are those that require heating. The process feed, Stream 1, must be heated to its bubble point prior to entering the distillation column. Likewise, Stream 18 must be heated in order to attain the required reactor feed temperature. Finally, Stream 12 must be vaporized to provide boil-up to the distillation column.

The stream data to be used for the pinch analysis are summarized in Table 8.24. The supply temperatures are the stream temperatures from Table 8.23. The target temperatures are also obtained from this table based on the final destination of each stream. For example, Stream 26 ultimately becomes Stream 28, which has a temperature of 20°C. Similarly, Stream 7, after condensation, becomes Stream 8, which has a temperature of 80.7°C. The target temperature of Stream 10 is somewhat arbitrary since it is a waste stream. The value of 35°C in Table 8.24 is the lowest practical temperature at which heat can be recovered in the

TABLE 8.23 Stream Data for Figure 8.21

Stream	Mass Flow Rate, kg/h (lbm/h)	Temperature, °C	Pressure, psia	Description
1	9,071.9 (20,000)	25	14.7	Raw feedstock
2	9,071.9 (20,000)	46.8	14.7	Heated feedstock
3	9,071.9 (20,000)	87	14.7	Heated feedstock
4	9,071.9 (20,000)	92	14.7	Distillation feed
5	1,866.6 (4,115)	89	14.7	Distillation overhead vapor
6	1,496.9 (3,300)	89	14.7	Distillate vapor
7	369.7 (815)	89	14.7	Distillation reflux (vapor)
8	369.7 (815)	80.7	14.7	Distillation reflux (liquid)
9	9,433.0 (20,796)	99.9	14.7	Distillation bottoms
10	7,575.1 (16,700)	99.9	14.7	Distillation bottom product
11	7,575.1 (16,700)	52.4	14.7	Wastewater from distillation column
12	1,857.9 (4,096)	99.9	14.7	Reboiler feed (total)
13	532.0 (1,172.9)	99.9	14.7	Reboiler feed (fractional)
14	532.0 (1,172.9)	100	14.7	Boil-up (fractional)
15	1,325.9 (2,923.1)	99.9	14.7	Reboiler feed (fractional)
16	1,325.9 (2,923.1)	100	14.7	Boil-up (fractional)
17	1,857.9 (4,096)	100	14.7	Boil-up (total)
18	1,496.9 (3,300)	238.7	125	Vapor from compressor C-1
19	1,496.9 (3,300)	284	125	Heated compressed vapor
20	1,496.9 (3,300)	321.7	125	Fresh reactor feed
21	1,673.8 (3,690)	300	117.6	Reactor feed (total)
22	1,673.8 (3,690)	300	110	Reactor product
23	1,673.8 (3,690)	110	105	Separator feed
24	176.9 (390)	110	105	Vapor from separator
25	176.9 (390)	117.2	117.6	Reactor recycle
26	1,496.9 (3,300)	110	105	Liquid from separator
27	1,496.9 (3,300)	74.2	105	Cooled separator liquid
28	1,496.9 (3,300)	20	105	Decanter feed
29	547.0 (1,206)	20	105	Gasoline product
30	949.8 (2,094)	20	105	Wastewater from decanter
31	–	283.8	14.7	Heat-transfer oil to reboiler
32	–	231.7	14.7	Heat-transfer oil from reboiler
33	–	200	14.7	Heat-transfer oil to reactor
34	–	290	14.7	Heat-transfer oil from reactor
35	–	283.8	14.7	Heat-transfer oil to holding tank

process, and is also a safe discharge temperature. A higher discharge temperature of 45°C would probably be acceptable, depending on the final destination for the stream. (Recycle back to the fermentation unit would be one possible option.) Hence, a target temperature of 45°C is considered as an alternative. A discharge temperature of 52.4°C was specified on the original flowsheet (Stream 11), but this is hot enough to pose a potential hazard.

The duties for Streams 7 and 10 were calculated using PRO/II. The duty for Stream 7 represents the rate of heat transfer required to convert the reflux from saturated vapor to saturated liquid at a constant pressure of one atmosphere. PRO/II was used to compute

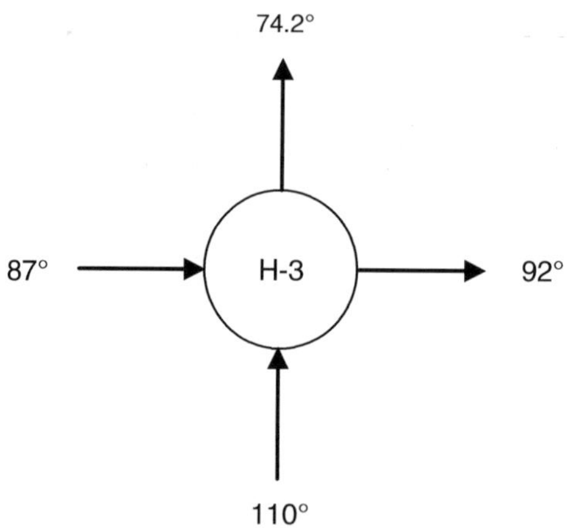

FIGURE 8.22 An impossible heat exchanger in the proposed process for conversion of bio-ethanol to gasoline.

TABLE 8.24 Stream Data for Pinch Analysis

Stream	Type	*TS* (°C)	*TT* (°C)	Duty (kW)
7	Hot	89	80.7	147.7
10	Hot	99.9	35 (45)	571.2 (483.6)
22	Hot	300	110	835.5
26	Hot	110	20	130.3
34	Hot	290	200	577.4
1	Cold	25	92	682.9
12	Cold	99.9	100	1170.7
18	Cold	238.7	321.7	73.3

this duty because the value obtained from [25] was found to be physically unrealistic. The duty for Stream 10 represents the rate of heat transfer for cooling the distillation bottom product, and the values from PRO/II are in close agreement with those obtained by extrapolating the enthalpy data in [25]. The remaining stream duties in Table 8.24 were calculated from enthalpy and heat recovery data given in [25], along with the aforementioned heat-of-reaction data from [24].

The hot utility is steam and it must be available at a temperature that exceeds the highest stream temperature (321.7°C) by at least ΔT_{min}. For this analysis, a steam supply temperature of 340°C is assumed with a range of 1°C. The cold utility is cooling water, and a single supply is assumed for simplicity. It must be available at a temperature that is at least ΔT_{min} degrees below the lowest stream temperature (20°C). A supply temperature of 5°C is assumed here with a range of 10°C. These utility temperatures are consistent with a minimum approach of 15°C or less. Previous work [25, 26] was based on a 10°C approach. Since the process as originally proposed is only at the conceptual design stage, there is little point in attempting to optimize ΔT_{min} via super-targeting calculations. Hence, values in the range 10 °C to 15°C are assumed to be appropriate for the present analysis.

HEXTRAN was first run in targeting mode with HRAT values of 10, 12, and 15°C specified. The eight process streams were specified as bulk property streams and values of temperature (*TS*), outlet temperature (*TT*) and average duty for each stream were entered from Table 8.24. The duties were scaled by an appropriate factor to avoid numerical problems as discussed in Section 8.19.1. Also, arbitrary values of flow rate and pressure were entered for each stream to satisfy HEXTRAN input requirements; they are irrelevant for the calculations. Separate runs were made for each of the two target temperatures for Stream 10.

Treating Stream 7 as a bulk property stream is equivalent to assuming a linear condensing curve, i.e., a linear variation of enthalpy with temperature. Although the ethanol-water solution is highly non-ideal, this approximation is quite adequate due to the relatively narrow condensing range of 8.3°C. Treating Stream 22 in this manner is not a good approximation, however, since it is a superheated vapor that undergoes cooling and partial condensation over a large temperature range. The approximation is used nevertheless in order to avoid the problem of reconciling HEXTRAN thermodynamic calculations for the partial-phase-change operation with the data from [25]. The validity of the results will be considered subsequently in light of this approximation.

Results of the targeting calculations are shown in Table 8.25. It can be seen that the results do not vary greatly over the specified range of HRAT. The target temperature for Stream 10 affects the cold utility requirement, but has no effect on heat recovery or hot utility usage. For comparison with the original process, a run was also made with HRAT = 10°C and the Stream 10 target temperature set at 52.4°C. The resulting cold utility target is 210.8 kW. The original process is ostensibly based on a minimum approach of 10°C and the hot utility usage is 33.4 kW; the cold utility usage is 280 kW; and the energy recovery amounts to 1,893 kW. These values are very close to the targets for HRAT = 15°C, and are thus somewhat over-target for an approach of 10°C. In fact, the cold utility is about 33% above target based on the actual Stream 10 discharge temperature of 52.4°C.

HEXTRAN was next run in synthesis mode using the same three values of HRAT and EMAT values of 8 and 10°C. All combinations produced similar results, with only minor variations. The network obtained with HRAT = 10°C and EMAT = 8°C is shown in Figure 8.23. It contains 9 units and has no split streams. The only temperature cross amounts to 0.3°C on the 45.3-kW exchanger, and this is not large enough to require multiple shell passes. The smallest temperature approach in the network is 10°C at the 45.3-kW exchanger, although the approach at the 790.2-kW unit is nearly identical (10.1°C). For a Stream 10 discharge temperature of 35°C, the duty on the Stream 10 cooler becomes 232.9 kW instead of 145.3 kW.

Notice that the order in which Stream 12 exchanges heat with Streams 22 and 34 is essentially arbitrary, since Stream 12 is for all practical purposes isothermal. However, the two exchangers involved here comprise the reboiler for the distillation column, and

TABLE 8.25 Targeting Results from HEXTRAN

HRAT (°C)	Heat Recovery (kW)	Hot Utility (kW)	Cold Utility (kW)
10	1,899	28.3	363.5* (275.9)**
12	1,897	30.0	365.2 (277.6)
15	1,894	32.7	367.9 (280.3)

*Stream 10 target temperature = 35°C
**Stream 10 target temperature = 45°C

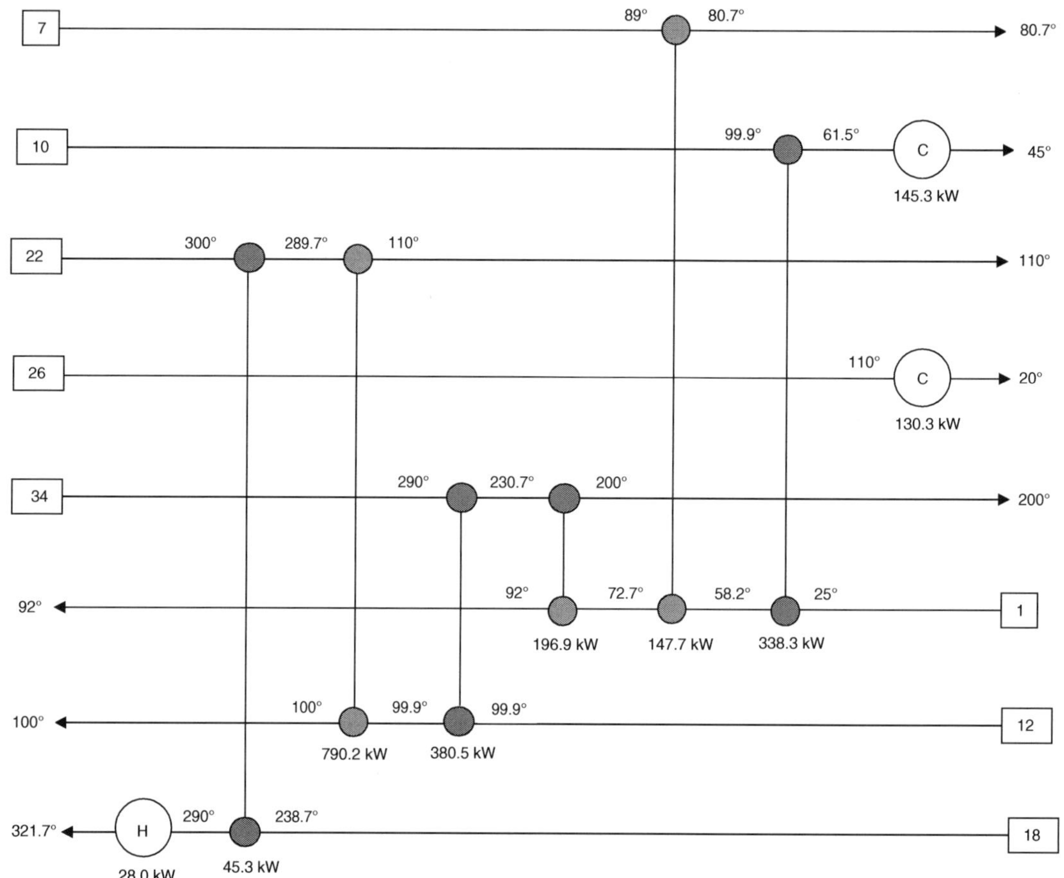

FIGURE 8.23 HEN generated by HEXTRAN with HRAT = 10, EMAT = 8 and Stream 10 target temperature = 45 °C.

would most likely be thermosyphon units (see Chapter 10). As such, they would actually operate in parallel, as shown in Figure 8.21, rather than in series. This requires splitting Stream 12 into two branches, which is unnecessary from the standpoint of the pinch analysis. In this case, however, the configuration is dictated by operational considerations.

Now the effect of treating Stream 22 as a bulk property stream can be seen by examining the terminal temperatures on the 45.3-kW exchanger. The small duty for this unit ensures that Stream 22 remains entirely in the vapor phase, so that this is a gas-to-gas exchanger. Furthermore, the mass flow rates of the two streams are approximately the same, differing by the amount of the small reactor recycle stream. Therefore, the temperature change experienced by each stream across the exchanger should be approximately the same, which is clearly not the case. The small temperature change for Stream 22 is the result of the assumed linear variation of enthalpy with temperature, which has the effect of spreading the latent heat of condensation over the entire temperature range. This makes the *CP* value too high in the superheated vapor range, which causes the temperature change across the exchanger to be too small.

In order to determine the correct temperature of Stream 22 exiting the 45.3-kW exchanger, the heat exchange was modeled using PRO/II. The composition of Stream 22 is given in Table 8.26. These data, along with the total molar flow rate of the stream

TABLE 8.26 Composition of Stream 22

Component	Mole %
Water	83.08
Ethylene	7.52
n-Pentane	2.45
n-Hexane	2.04
n-Heptane	0.95
n-Octane	2.15
Toluene	0.79
o-Xylene	1.02

TABLE 8.27 Comparison of Heat-exchanger Networks for Bio-ethanol Conversion Process

Item	Original HEN	New HEN
Number of units	9	9
Hot utility, kW	33.4	28.0
Cold utility, kW	280	276 (363)
Heat recovery, kW	1,893	1,899
Minimum approach, °C	6	10
Stream 10 discharge temp., °C	52.4	45 (35)
Thermodynamic feasibility	Infeasible	Feasible

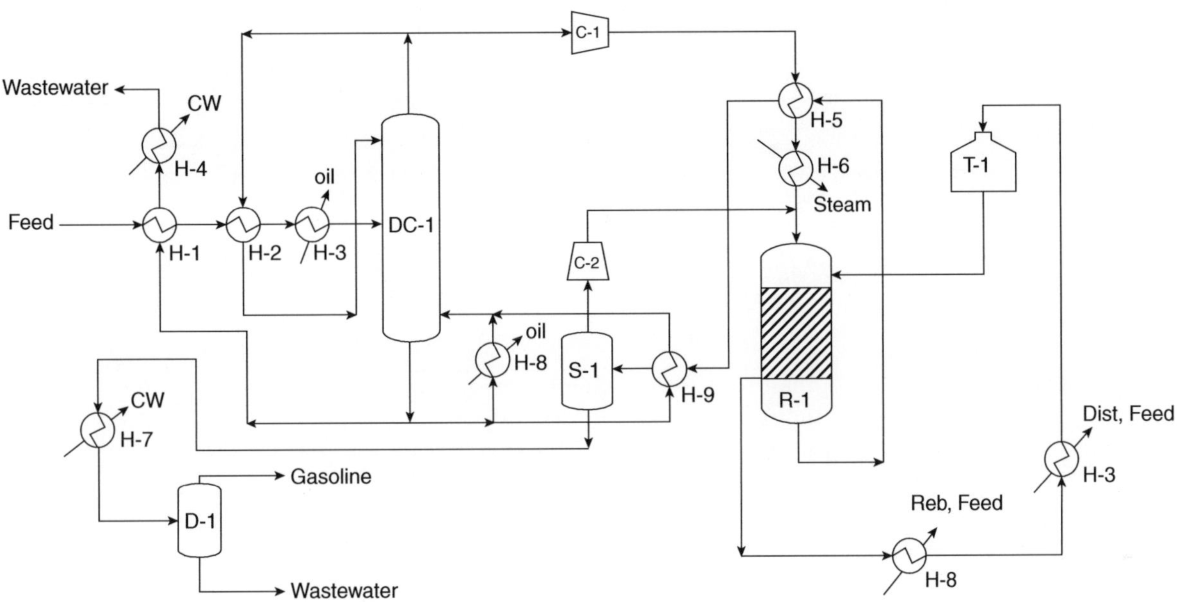

FIGURE 8.24 Revised flowsheet for production of gasoline from bio-ethanol.

(142.9 lb mole/h), were obtained from [25], which does not distinguish hydrocarbon isomers. The SRK-SIMSCI thermodynamic package was used for the simulation, which was also run with the Lee-Kessler-Plocker thermodynamic package as a check. Both methods gave the same exit temperature of approximately 257°C. (The calculation can also be done using the simple cooler module in HEXTRAN, with similar results.) This change in temperature does not compromise the validity of the HEN. However, it increases the temperature cross on the 45.3-kW exchanger to 33°C, which would require four shell passes. Fortunately, the small duty for this unit makes it likely that a hairpin exchanger will prove suitable for the service, thereby obviating the need for multiple shells. It should be noted that the original network contains an exchanger (H-2) with a temperature cross of comparable magnitude. To summarize, the only modification needed in the HEN of Figure 8.23 is to change the temperature of Stream 22 exiting the 45.3-kW exchanger from 289.7°C to 257°C.

The new HEN is compared with the original network in Table 8.27. The process flowsheet corresponding to the new HEN is shown in Figure 8.24. The two exchangers (H-8 and H-9) comprising the column reboiler are assumed to operate in parallel, as discussed above.

References

[1] Linnhoff B, Hindmarsh E. The pinch design method for heat exchanger networks. Chem Eng Sci 1983;38:745–63.

[2] Linnhoff B, Flower JR. Synthesis of heat exchanger networks. AIChE J 1978;24:633–54.

[3] Linnhoff B. Use pinch analysis to knock down capital costs and emissions. Chem Eng Prog 1994;18(No. 8):32–57.

[4] Linnhoff B. Pinch analysis – a state-of-the-art overview. Trans IChemE 1993;71(Part A):503–22.

[5] Linnhoff B, Ahmad S. Cost optimum heat exchanger networks. 1. Minimum energy and capital using simple models for capital cost. Comp Chem Eng 1990;14:729–50.

[6] Ahmad S, Linnhoff B, Smith R. Cost optimum heat exchanger networks. 2. Targets and design for detailed capital cost models. Comp Chem Eng 1990;14:751–67.

[7] Trivedi KK, Roach JR, O'Neill BK. Shell targeting in heat exchanger networks. AIChE J 1987;33:2087–90.
[8] Ahmad S, Linnhoff B, Smith R. Design of multipass heat exchangers: an alternative approach. J Heat Transfer 1988;110:304–9.
[9] Taborek J. Mean temperature difference. In: Heat Exchanger Design Handbook, vol. 3. New York: Hemisphere Publishing Corp.; 1988.
[10] Ahmad S, Smith R. Targets and design for minimum number of shells in heat exchanger networks. Chem Eng Res Des 1989;67:481–94.
[11] Bell KJ. Approximate sizing of shell-and-tube heat exchangers. In: Heat Exchanger Design Handbook, vol. 3. New York: Hemisphere Publishing Corp.; 1988.
[12] Hall SG, Ahmad S, Smith R. Capital cost targets for heat exchanger networks comprising mixed materials of construction, pressure ratings and exchanger types. Comp Chem Eng 1990;14:319–35.
[13] Polley GT, Panjeh Shaki MH, Jegede FO. Pressure drop considerations in the retrofit of heat exchanger networks. Trans IChemE 1990;68(Part A):211–20.
[14] Jegede FO, Polley GT. Optimum heat exchanger design. Trans IChemE 1992;70(Part A):133–41.
[15] Zhu XX, Nie XR. Pressure drop considerations for heat exchanger network grassroots design. Comp Chem Eng 2002;26:1661–76.
[16] Biegler LT, Grossman IE, Westerberg AW. Systematic Methods of Chemical Process Design. Upper Saddle River, NJ: Prentice Hall; 1997.
[17] Cerda J, Westerberg AW, Mason D, Linnhoff B. Minimum utility usage in heat exchanger network synthesis – a transportation problem. Chem Eng Sci 1983;38:373–87.
[18] Floudas CA, Ciric AR, Grossman IE. Automatic synthesis of optimum heat exchanger network configurations. AIChE J 1986;32:276–90.
[19] Yee TF, Grossman IE, Kravanja Z. Simultaneous optimization models for heat integration. I. Area and energy targeting and modeling of multistream exchangers. Comp Chem Eng 1990;14:1151–64.
[20] Yee TF, Grossman IE. Simultaneous optimization models for heat integration. II. Heat exchanger network synthesis. Comp Chem Eng 1990;14:1165–84.
[21] Challand TB, Colbert RW, Venkatesh CK. Computerized heat exchanger networks. Chem Eng Prog 1981;5(No. 7):65–71.
[22] Jones DA, Asuman NY, Tilton BE. Synthesis techniques for retrofitting heat recovery systems. Chem Eng Prog 1986;10(No. 7):28–33.
[23] Shethna HA, Jezowski JM, Castillo FJL. A new methodology for simultaneous optimization of capital and operating cost targets in heat exchanger network design. App Thermal Eng 2000;20:1577–87.
[24] Whitcraft DR, Veryklos XE, Mutharasan R. Recovery of ethanol from fermentation broths by catalytic conversion to gasoline. Ind Eng Chem Process Des Dev 1983;22:452–7.
[25] Aldridge GA, Veryklos XE, Mutharasan R. Recovery of ethanol from fermentation broths by catalytic conversion to gasoline. 2. Energy analysis. Ind Eng Chem Process Des Dev 1984;23:733–7.
[26] Khan S, Riverol C. Performance of a pinch analysis for the recovery of ethanol from fermentation. Chem Eng Technol 2007;30:1328–39.

Notations

A	Heat transfer surface area
A_i, A_o	Internal and external surface area, respectively, of a heat exchanger tube
A_{min}	Target for minimum heat-transfer area in a heat exchanger network
a	Constant in capital cost equation for heat exchangers, Equation (8.20)
b	Constant in capital cost equation for heat exchangers, Equation (8.20)
C_P	Heat capacity at constant pressure
CP	$\dot{m}C_P$ = Heat capacity flow rate
c	Constant in capital cost equation for heat exchangers, Equation (8.20)
$EMAT$	Minimum temperature approach for an individual heat exchanger in a network
F	LMTD correction factor
F_i	LMTD correction factor for enthalpy interval i
f	Objective function for linear program
H	Stream enthalpy per unit time
$HRAT$	Temperature approach for heat recovery
h	Heat-transfer coefficient with fouling allowance included
h_j	Heat-transfer coefficient, including fouling allowance, for stream j
$\dot{m}$	Mass flow rate
N	Number of process streams and utilities in a network
N_s	Number of E shells connected in series
NC	Number of cold streams or branches at the pinch
NH	Number of hot streams or branches at the pinch
P	Parameter used to calculate LMTD correction factor
P_{max}	Maximum value of P for a given value of R
Q_j	Heat flow leaving jth subnetwork
Q_N	Heat flow to cold utility
Q_0	Amount of hot utility supplied to first subnetwork
q	Rate of heat transfer
q_j	Rate of heat transfer in enthalpy interval j
R	Parameter used to calculate LMTD correction factor
S	Parameter used in calculating LMTD correction factor
T	Temperature
TS	Supply temperature

TT	Target temperature
U_D	Design overall heat-transfer coefficient
U_{min}	Minimum number of units required in a heat exchanger network
W	Parameter defined by Equation (8.13)
X_P	P/P_{max}

Greek Letters

α	Parameter used in calculating LMTD correction factor
ΔH	Enthalpy difference (per unit time)
ΔT	Temperature difference
$\Delta T_{\ln}$	Logarithmic mean temperature difference
$(\Delta T_{\ln})_{cf}$	LMTD for counter-current flow
ΔT_{min}	Minimum temperature approach

Problems

(8.1)

Stream	TS (°C)	TT (°C)	CP (kW/°C)	h (kW/m^2 · K)
1	220	150	2.0	1.1
2	240	60	3.0	0.6
3	50	190	2.5	0.8
4	100	210	4.0	1.0
HU	250	250	–	2.0
CU	20	25	–	1.5

A HEN will involve the four process streams and two utilities shown above. A minimum approach of 10°C will be used for design purposes. Set up the problem table for the network and use it to determine:
(a) The minimum hot and cold utility requirements.
(b) The hot and cold stream temperatures at the pinch.

Ans. (a) 135 kW (hot), 25 kW (cold). (b) 110°C and 100°C.

(8.2) For the system of Problem 8.1:
(a) Construct the hot and cold composite curves.
(b) From the graph, determine the value of ΔT_{min} that results in the maximum possible heat recovery, and the value (in kW) of the maximum possible heat recovery.
(c) What is the largest value of ΔT_{min} that can be specified for this system if utility temperatures are neglected?
(d) What is the largest value of ΔT_{min} that can be specified for this system if utility temperatures are accounted for?

Ans. (b) 1.67°C, 680 kW. (c) 190°C. (d) 40°C.

(8.3) For the system of Problem 8.1, use the Pinch Design method to develop a minimum-utility design for:
(a) The hot-end problem.
(b) The cold-end problem.
(c) The complete network.

(8.4) (a) For the HEN designed in Problem 8.3, what is the largest temperature cross in the network?
(b) If the exchanger having the largest temperature cross is a shell-and-tube unit, how many shells will it require?

Ans. (a) 87.5°C. (b) 6.

(8.5) For the system of Problem 8.1, the given heat-transfer coefficients include fouling allowances. Use the given data to solve the following problems.
(a) For the enthalpy interval from 680 to 735 kW, the temperature ranges for the hot and cold composite curves are:
Hot: 250°C
Cold: 181.54 to 190°C
Calculate the counter-flow area target for this interval.

(b) Repeat the calculation of part (a) for each of the remaining enthalpy intervals and determine the counter-flow area target for the entire network.

Ans. (a) 1.37 m^2.

(8.6) Use the analytical method to calculate the number-of-shells target for the system of Problem 8.1.

(8.7) Use HX-Net or other available software to perform targeting calculations for the system of Problem 8.1. Since the software will not accept an isothermal utility stream, use a value of 249°C for the target temperature of the hot utility. Compare the results with those obtained by hand in Problems 8.1, 8.2, 8.5, and 8.6.

(8.8) Use HEXTRAN or other suitable software to synthesize one or more HENs for the system of Problem 8.1. Since the software will not accept an isothermal utility stream, use a value of 249°C for the target temperature of the hot utility. Display each network that you obtain on a grid diagram and compare each of the following parameters with the corresponding target value found in Problem 8.7:

- Number of units
- Number of shells
- Hot and cold utility usage
- Counter-flow area for the network
- Multi-pass area for the network

(8.9)

Stream	*TS* (°F)	*TT* (°F)	*CP* (Btu/h · °F)	*h* (Btu/h · ft^2 · °F)
1	250	120	10,000	120
2	200	100	40,000	150
3	90	150	30,000	100
4	130	190	60,000	125
HU	300	299	–	750
CU	70	85	–	200

A HEN will involve the four process streams and two utilities shown above. A minimum approach of 10°F will be used for design purposes. Set up the problem table for the network and use it to determine:
(a) The minimum hot and cold utility requirements.
(b) The hot and cold stream temperatures at the pinch.

Ans. (a) 700,000 Btu/h (hot); 600,000 Btu/h (cold). (b) 140°F and 130°F.

(8.10) For the system of Problem 8.9:
(a) Construct the hot and cold composite curves.
(b) From the graph determine the value of ΔT_{min} that results in the maximum possible heat recovery, and the value of the maximum possible heat recovery.
(c) What is the largest value of ΔT_{min} that can be specified for this system? What are the minimum hot and cold utility requirements and maximum energy recovery corresponding to this value of ΔT_{min}? (Remember to account for the utility temperatures.)

Ans. (b) 0°F, 5.2×10^6 Btu/h. (c) 30°F, 1.7×10^6 Btu/h (hot), 1.6×10^6 Btu/h (cold), 3.7×10^6 Btu/h (MER).

(8.11) For the system of Problem 8.9, do the following:
(a) Set up the *CP* table for the hot-end problem and use it to identify the feasible pinch matches. Then develop a minimum-utility design for the hot-end problem.
(b) Repeat part (a) for the cold-end problem.
(c) Combine the hot and cold-end designs from parts (a) and (b) to obtain a minimum-utility design for the complete network.
(d) Determine the number of heat load loops that exist in the network. Then identify the loops on your grid diagram.
(e) Simplify the network of part (c) by eliminating the unit with the smallest duty using heat load loops and paths. For consistency, any unit whose duty falls below that of the unit eliminated should also be eliminated.

(8.12) For the final network obtained in Problem 8.11 (e), assume that all units are multi-pass shell-and-tube exchangers. How many type E shells will be required to realize this network?

(8.13) For the system of Problem 8.9, the given heat-transfer coefficients include fouling allowances. Use the given data to calculate the minimum counter-flow area target for the network.

Ans. 4,000 ft^2.

(8.14) Use the analytical method to calculate the number-of-shells target for the system of Problem 8.9.

Ans. 13.

(8.15) Use HX-Net or other available software to perform targeting calculations for the system of Problem 8.9. Compare the results with those obtained by hand in Problems 8.9, 8.10, 8.13, and 8.14.

(8.16) Use HEXTRAN or other suitable software to synthesize one or more HENs for the system of Problem 8.9. Display each network that you obtain on a grid diagram and compare each of the following parameters with the corresponding target value found in Problem 8.15.

- Number of units
- Number of shells
- Hot and cold utility usage
- Counter-flow area for the network
- Multi-pass area for the network

(8.17)

Stream	*TS* (°C)	*TT* (°C)	*CP* (kW/ °C)	Duty (kW)
1	280	60	3.0	660
2	180	20	4.5	720
3	20	160	4.0	560
4	120	260	6.0	840
HU	300	299	–	–
CU	–20	–15	–	–

A HEN will involve the four process streams and two utilities shown above. A minimum approach of 30°C will be used for design purposes. Set up the problem table for the network and use it to determine the hot and cold stream temperatures at the pinch and the minimum-utility targets for the system.

Ans. 150°C, 120°C, 475 kW (hot), 455 kW (cold).

(8.18) Construct the hot and cold composite curves for the system of Problem 8.17. Then use the composite curves to determine the following to within graphical accuracy.

(a) The range of values that can be specified for ΔT_{min} in this system, neglecting utility temperatures.
(b) The maximum amount of heat exchange (kW) between hot and cold streams that is theoretically possible.
(c) The minimum cold utility (kW) required for $\Delta T_{min} = 20°C$.
(d) The maximum amount of heat exchange (kW) that is possible for $\Delta T_{min} = 20°C$.
(e) The hot stream pinch temperature corresponding to a minimum hot utility requirement of 620 kW.

Ans. (a) 0–260°C. (b) 1150 kW. (c) 380 kW. (d) 1000 kW. (e) 169°C.

(8.19) (a) Use the Pinch Design method to find a minimum-utility design for the system of Problem 8.17.
(b) Modify the network obtained in part (a) by using a heat load path to eliminate the temperature cross on one of the heat exchangers.

(8.20)

Stream	*TS* (°C)	*TT* (°C)	*CP* (kW/°C)	*h* (kW/m^2 · K)
1	180	40	2.0	0.9
2	150	40	4.0	1.2
3	60	180	3.0	1.0
4	30	105	2.6	0.7
HU	220	219	–	2.5
CU	0	15	–	1.6

A HEN will involve the four process streams and two utilities shown above. A minimum approach of 20°C will be used for design purposes. Set up the problem table for the network and use it to determine the pinch temperatures and minimum-utility targets.

Ans. 150°C, 130°C, 90 kW (hot), 255 kW (cold).

(8.21) For the system of Problem 8.20, construct the hot and cold composite curves. Then use the composite curves to determine the following:

(a) The smallest value of ΔT_{min} that can be specified and the maximum amount of heat recovery corresponding to this value of ΔT_{min}.

(b) The largest value of ΔT_{min} that can be specified and the maximum amount of heat recovery corresponding to this value of ΔT_{min}. (Remember to account for the utility temperatures.)

(c) The minimum utility requirements for $\Delta T_{min} = 10°C$.

Ans. (a) 0°C, 525 kW. (b) 40°C, 405 kW.

(8.22) (a) Use the Pinch Design method to develop a minimum-utility design for the system of Problem 8.20.

(b) Simplify the network obtained in part (a) as appropriate using heat load loops and paths.

(8.23) For the final network obtained in Problem 8.22 (b):

(a) Assuming all units are counter-flow exchangers, use the heat-transfer coefficients given in Problem 8.20 to calculate the total heat-transfer area in the network.

(b) If all units are multi-pass shell-and-tube exchangers, how many type E shells will be required to realize the network?

(8.24) (a) For the system of Problem 8.20, construct the balanced composite curves for $\Delta T_{min} = 20°C$.

(b) Delineate the enthalpy intervals for the system.

(c) Use the heat-transfer coefficients given in Problem 8.20 to calculate the counter-flow area target for the network.

(d) Use the graphical method to calculate the minimum-number-of-shells target for the network.

(8.25) Use HX-Net or other available software to perform targeting calculations for the system of Problem 8.20. Compare the results with those obtained by hand in Problems 8.20, 8.21, and 8.24.

(8.26) Use HEXTRAN or other suitable software to synthesize one or more HENs for the system of Problem 8.20. Display each network obtained on a grid diagram and compare the network parameters with the targets calculated in Problem 8.25.

(8.27)

Stream	*TS* (°F)	*TT* (°F)	*CP* (Btu/h · °F)	*h* (Btu/h · ft^2 · °F)
1	260	160	30,000	350
2	250	130	15,000	250
3	180	240	40,000	300
4	120	235	20,000	320
HU	290	289	–	800
CU	60	75	–	400

A HEN will involve the four process streams and two utilities shown above. A minimum approach of 10°F will be used for design purposes. Set up the problem table and use it to determine the pinch temperatures and minimum-utility targets for the network.

Ans. 190°F, 180°F; 500,000 Btu/h (hot); 600,000 Btu/h (cold).

(8.28) Construct the hot and cold composite curves for the system of Problem 8.27. Use the composite curves to determine:

(a) The smallest value of ΔT_{min} that can be specified and the maximum amount of heat recovery corresponding to this value of ΔT_{min}.

(b) The largest value of ΔT_{min} that can be specified and the maximum amount of heat recovery corresponding to this value of ΔT_{min}. (Remember to consider utility temperatures.)

(c) The minimum utility requirements for $\Delta T_{min} = 20°F$.

Ans. (a) 0°F, 4.64×10^6 Btu/h. (b) 50°F, 3.04×10^6 Btu/h. (c) 950,000 Btu/h (hot), 1.05×10^6 Btu/h (cold).

(8.29) (a) Use the Pinch Design method to develop a minimum-utility design for the system of Problem 8.27.

(b) Simplify the network obtained in part (a) as appropriate using heat load loops and paths.

(8.30) For the final network obtained in Problem 8.29 (b):

(a) Assuming all units are counter-flow exchangers, use the heat-transfer coefficients given in Problem 8.27 to calculate the total heat-transfer area in the network.

(b) If all units are multi-pass shell-and-tube exchangers, how many type E shells will be required to realize the network?

(8.31) (a) For the system of Problem 8.27, construct the balanced composite curves for $\Delta T_{min} = 10°F$.

(b) Delineate the enthalpy intervals for the system.

(c) Use the heat-transfer coefficients given in Problem 8.27 to calculate the counter-flow area target for the network.

(d) Use the graphical method to calculate the minimum-number-of-shells target for the network.

(8.32) Use HX-Net or other available software to perform targeting calculations for the system of Problem 8.27. Compare the results with those obtained by hand in Problems 8.27, 8.28, and 8.31.

(8.33) Use HEXTRAN or other suitable software to synthesize one or more HENs for the system of Problem 8.27. Display each network obtained on a grid diagram and compare the network parameters with the targets calculated in Problem 8.32.

(8.34)

Stream	*TS* (°C)	*TT* (°C)	*CP* (kW/ °C)	*h* (kW/m^2 · K)
1	250	120	16.5	0.8
2	205	65	13.5	0.6
3	40	205	11.5	1.1
4	65	180	13.0	0.9
5	95	205	13.0	0.7
HU	280	279	–	3.0
CU	25	30	–	1.2

A HEN will involve the five process streams and two utilities shown above.

(a) Set up the problem table for $\Delta T_{min} = 10°C$ and use it to determine the minimum-utility targets for the network.

(b) Calculate the minimum-number-of-units target for the network.

(c) Construct the hot and cold composite curves for the system and use them to determine the smallest value of ΔT_{min} for which a pinch exists, i.e., find $\Delta T_{threshold}$.

(d) Set up the problem table for $\Delta T_{min} = 30°C$ and use it to find the pinch temperatures and minimum-utility targets.

Ans. (a) 787.5 kW(hot), 0 kW(cold). (b) 6. (c) 21.3°C. (d) 95°C, 65°C, 905 kW (hot), 117.5 kW (cold).

(8.35) Use HX-Net or other available software to perform targeting calculations for the system of Problem 8.34 and verify the results obtained by hand in the latter.

(8.36) For the system of Problem 8.34 with $\Delta T_{min} = 30°C$, use HEXTRAN or other suitable software to synthesize one or more HENs. Display the results on a grid diagram and compare the network parameters with the targets calculated in Problem 8.35.

(8.37) Repeat Example 8.3 for the case in which there is a second hot utility having the following properties:

TS = 110°C

TT = 109.5°C

Cost = \$50/kW · year

h = 2.0 kW/m^2 · K (fouling included)

All other data are the same as in Example 8.3. Use HX-Net to determine the following:

- Pinch temperatures.
- Minimum requirements for each of the three utilities.
- Minimum number of units.
- Minimum number of shells.
- Counter-flow area target.
- Multi-pass area target.
- Optimum value of ΔT_{min} (super targeting).
- A graph of the balanced composite curves.

(8.38) For example problem TC3 discussed in the text, do the following:

(a) Combine the hot B and cold A designs given in the text to obtain a minimum-utility design for the network. Identify the heat load loops on the grid diagram.

(b) Simplify the network from part (a) using heat load loops and paths.

(c) Compare the parameters for the final network from part (b) with those for TC3 calculated in the text.

(d) Consider the case in which there is a second hot utility available at 110°C. Assume that the temperature change for this utility is negligible and that it is the less expensive of the two hot utilities. Set up the problem table for this case and use it to determine pinch temperatures and minimum requirements for each of the three utilities.

(8.39)

Stream	*TS* (°F)	*TT* (°F)	*CP* (Btu/h · °F)	*h* (Btu/h · ft^2 · °F)
1	590	400	23,760	320
2	471	200	15,770	260
3	533	150	13,200	190
4	200	400	16,000	230
5	100	430	16,000	150
6	300	400	41,280	280
7	150	280	26,240	250
HU	500	499	–	1000
CU	75	115	–	450

A HEN will involve the seven process streams and two utilities shown above. The given heat-transfer coefficients include appropriate fouling allowances. A minimum approach of 20°F will be used for design purposes. Use HX-Net or other available software to perform targeting calculations for this system.

(8.40) Use HEXTRAN or other suitable software to synthesize a HEN for the system of Problem 8.39. Display the resulting network on a grid diagram and compare the network parameters with the target values found in Problem 8.39.

(8.41)

Stream	*TS* (°C)	*TT* (°C)	*CP* (kW/°C)	*h* (kW/m^2 · K)
1	327	40	100	1.6
2	220	160	160	0.7
3	220	60	60	1.2
4	160	45	400	1.0
5	100	300	100	0.8
6	35	164	70	0.6
7	85	138	350	1.5
8	60	170	60	1.0
9	140	300	200	0.9
HU1	350	349	–	4.3
HU2	210	209	–	4.3
CU1	–5	10	–	2.2
CU2	25	40	–	2.2

A HEN will involve the nine process streams and four utilities shown above. The given heat-transfer coefficients include appropriate fouling allowances. A minimum approach of 20°C will be used for design purposes. Use HX-Net or other available software to perform targeting calculations for this system.

9 Boiling Heat Transfer

9.1 Introduction

In reboilers and vaporizers, boiling usually takes place either on the exterior surface of submerged tubes or on the interior surface as the fluid flows through the tubes. In the former case the tubes are oriented horizontally, while in the latter they may be either horizontal or vertical, but the most common configuration is vertical with the fluid flowing upward through the tubes. In both situations, the rate of heat transfer is normally much higher than in ordinary forced convection, and different correlations are required to predict heat-transfer coefficients in boiling systems.

A boiling fluid consists of a two-phase mixture of vapor and liquid. When such a fluid flows through a tube, a number of distinct flow regimes can occur depending on the flow rate and the relative amounts of vapor and liquid present. Two-phase flow is thus more complex than single-phase flow, and special methods are needed to calculate the pressure drop in equipment handling boiling fluids.

Correlations used in the thermal and hydraulic analysis of reboilers and vaporizers are presented in this chapter. Equipment design is covered in Chapter 10.

9.2 Pool Boiling

Pool boiling refers to vaporization that takes place at a solid surface submerged in a quiescent liquid. When the temperature, T_s, of the solid surface exceeds the saturation temperature, T_{sat}, of the liquid, vapor bubbles form at nucleation sites on the surface, grow and subsequently detach from the surface. The driving force for heat transfer is $\Delta T_e = T_s - T_{sat}$, called the excess temperature. Liquid circulation in pool boiling occurs by natural convection and by the agitation resulting from bubble growth and detachment.

The boiling curve is a plot of surface heat flux versus excess temperature, and has the general features illustrated in Figure 9.1. Point A on the curve marks the onset of nucleate boiling (ONB). At lower excess temperatures, heat transfer occurs by natural

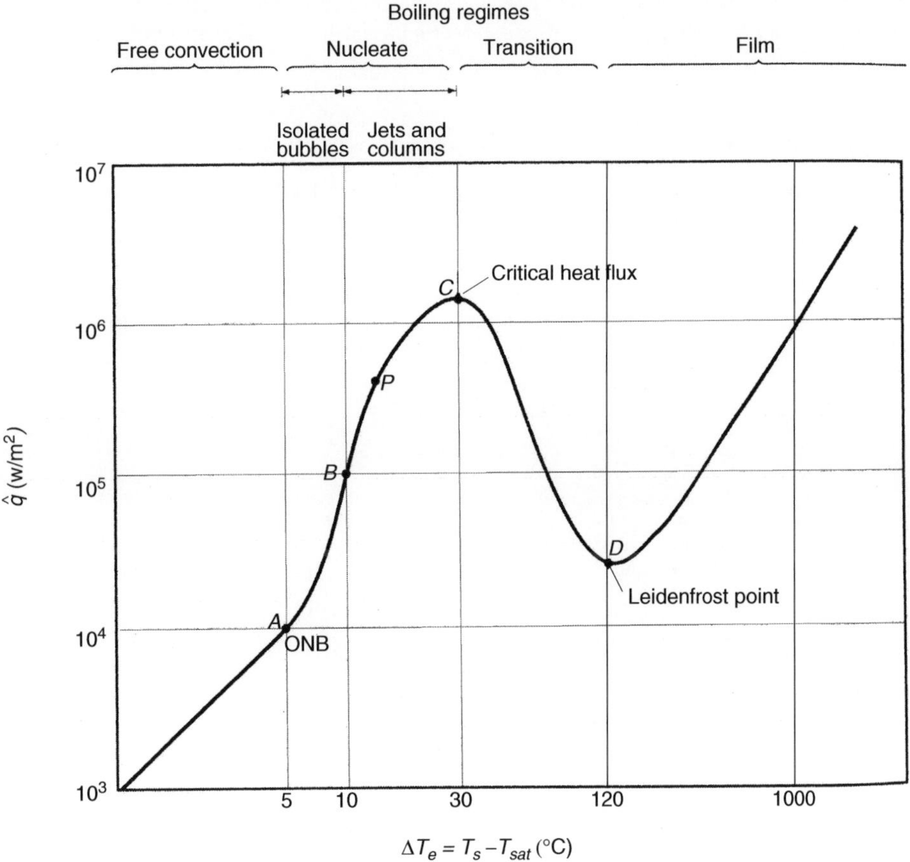

FIGURE 9.1 Pool boiling curve for water at one atmosphere pressure (Source: Ref. [1]).

convection alone. Nucleate boiling exists between points A and C on the curve. Two different boiling regimes can be distinguished in this region. Between points A and B, the boiling is characterized by the formation of isolated vapor bubbles at nucleation sites dispersed on the solid surface. Bubble growth and detachment result in significant fluid mixing near the solid surface that greatly increases the rate of heat transfer. In this regime, heat is transferred primarily from the solid surface directly to the liquid flowing across the surface.

As the heat flux increases beyond point B, the number of active nucleation sites and the rate of vapor formation become so great that bubble interference and coalescence occur. The vapor leaves the solid surface in jets or columns that subsequently merge and form large slugs of vapor. The high rate of vapor formation begins to inhibit the flow of liquid across the solid surface, causing the slope of the boiling curve to decrease. An inflection point occurs at point P; here, the heat-transfer coefficient reaches a maximum. The heat flux continues to increase between points P and C since the increase in temperature driving force more than compensates for the decreasing heat-transfer coefficient.

The heat flux attains a maximum, called the critical heat flux, at point C. At this point, the rate of vapor formation is so great that some parts of the surface are covered by a continuous vapor film. Since heat-transfer rates for gases are generally much lower than for liquids, the overall rate of heat transfer begins to decrease. Although the vapor film tends to be unstable, breaking up and reforming at any given point, the fraction of the solid surface covered by vapor continues to increase from point C to point D. This region, in which the heat flux decreases as ΔT_e increases, is referred to as the transition region.

The heat flux reaches a minimum at point D, the so-called Leidenfrost point, where the entire solid surface is covered by a vapor blanket. Beyond this point, heat is transferred from the solid surface across the vapor film to the liquid. Hence, this regime is called film boiling. As indicated in Figure 9.1, very high surface temperatures may be reached in film boiling, and consequently radiative heat transfer can be significant in this regime.

Most reboilers and vaporizers are designed to operate in the nucleate boiling regime. Although film boiling is sometimes employed, the much higher temperature driving force (corresponding to a much lower heat-transfer coefficient) in this regime generally makes it unattractive compared with nucleate boiling. The transition region, with its unusual characteristic of decreasing heat flux with increasing driving force, is always avoided in equipment design.

9.3 Correlations for Nucleate Boiling on Horizontal Tubes

Kettle, internal, and horizontal thermosyphon reboilers (see Section 10.2) all involve boiling on the external surfaces of tubes configured in horizontal bundles. Correlations for the heat-transfer coefficient and critical heat flux in this geometry are presented in the following subsections.

9.3.1 Heat-Transfer Coefficients for Pure Component Nucleate Boiling On a Single Tube

Nucleate pool boiling has been widely studied and many heat-transfer correlations have been proposed for this regime. The presentation here is limited to several methods that have been among the most widely used.

The Forster–Zuber Correlation

Published in 1955, the Forster–Zuber correlation is the oldest of the methods presented here. It is usually stated in the following form [2,3]:

$$h_{nb} = 0.00122 \frac{k_L^{0.79} C_{P,L}^{0.45} \rho_L^{0.49} g_c^{0.25} \Delta T_e^{0.24} \Delta P_{sat}^{0.75}}{\sigma^{0.5} \mu_L^{0.29} \lambda^{0.24} \rho_V^{0.24}} \tag{9.1}$$

where

h_{nb} = nucleate boiling heat-transfer coefficient, Btu/h · ft² · °F (W/m² · K)
k_L = liquid thermal conductivity, Btu/h · ft · °F (W/m · K)
$C_{P,L}$ = liquid heat capacity, Btu/lbm · °F (J/kg · K)
ρ_L = liquid density, lbm/ft³ (kg/m³)
μ_L = liquid viscosity, lbm/ft · h (kg/m · s)
σ = surface tension, lbf/ft (N/m)
ρ_V = vapor density, lbm/ft³ (kg/m³)
λ = latent heat of vaporization, Btu/lbm (J/kg)
g_c = unit conversion factor = 4.17×10^8 lbm · ft/lbf · h² (1.0 kg · m/N · s²)
$\Delta T_e = T_w - T_{sat}$, °F (K)
T_w = tube-wall temperature, °F (K)
T_{sat} = saturation temperature at system pressure, °F (K)
$\Delta P_{sat} = P_{sat}(T_w) - P_{sat}(T_{sat})$, lbf/ft² (Pa)
$P_{sat}(T)$ = vapor pressure of fluid at temperature T, lbf/ft² (Pa)

Any consistent set of units can be used with Equation (9.1), including the English and SI units shown above.

The Mostinski Correlation

The Mostinski correlation is based on the principle of corresponding states and can be expressed as follows:

$$h_{nb} = 0.00622 P_c^{0.69} \hat{q}^{0.7} F_P \tag{9.2a}$$

where

$h_{nb} \propto$ Btu/h · ft^2 · °F
P_c = fluid critical pressure, psia
$\hat{q}$ = heat flux, Btu/h · ft^2
F_P = pressure correction factor, dimensionless

In terms of SI units, the above equation becomes [4]:

$$h_{nb} = 0.00417 P_c^{0.69} \hat{q}^{0.7} F_P \tag{9.2b}$$

where

$h_{nb} \propto$ W/m^2 · K
$P_c \propto$ kPa
$\hat{q} \propto$ W/m^2

An alternative form of the Mostinski correlation that is explicit in the temperature difference can be obtained by making the substitution $\hat{q} = h_{nb}\Delta T_e$. The result corresponding to Equation (9.2a) is:

$$h_{nb} = 4.33 \times 10^{-8} P_c^{2.3} \Delta T_e^{2.333} F_P^{3.333} \tag{9.3a}$$

where

$h_{nb} \propto$ Btu/h · ft^2 · °F
$P_c \propto$ psia
$\Delta T_e \propto$ °F

In terms of SI units, the result is:

$$h_{nb} = 1.167 \times 10^{-8} P_c^{2.3} \Delta T_e^{2.333} F_P^{3.333} \tag{9.3b}$$

where

$h_{nb} \propto$ W/m^2 · K
$P_c \propto$ kPa
$\Delta T_e \propto$ K

The pressure correction factor given by Mostinski is:

$$F_P = 1.8 P_r^{0.17} + 4 P_r^{1.2} + 10 P_r^{10} \tag{9.4}$$

where

$P_r = P/P_c$ = reduced pressure

Various modifications of this relationship have been suggested by other workers, and Palen [4] recommends the following alternative expression for design purposes:

$$F_P = 2.1 P_r^{0.27} + \left[9 + \left(1 - P_r^2\right)^{-1}\right] P_r^2 \tag{9.5}$$

The Cooper Correlation

Cooper [5] developed the following correlation, which is of the same general type as the Mostinski correlation:

$$h_{nb} = 21\, \hat{q}^{0.67} P_r^{0.12} (-\log_{10} P_r)^{-0.55} M^{-0.5} \tag{9.6a}$$

where M is the molecular weight of the fluid and the other terms and units are the same as in Equation (9.2a). In terms of SI units, as in Equation (9.2b), the corresponding equation is:

$$h_{nb} = 55\, \hat{q}^{0.67} P_r^{0.12} (-\log_{10} P_r)^{-0.55} M^{-0.5} \tag{9.6b}$$

A surface roughness term in the exponent on P_r has been omitted in Equation (9.6) because it is usually dropped in practice.

The Stephan–Abdelsalam Correlation

Stephan and Abdelsalam [6] used regression analysis to correlate a large set of experimental data in terms of the following dimensionless groups of physical properties:

$$Z_1 = \frac{\hat{q} d_B}{k_L T_{sat}} \tag{9.7}$$

$$Z_2 = \frac{\alpha_L^2 \rho_L}{g_c \sigma d_B} \tag{9.8}$$

$$Z_3 = \frac{g_c \lambda d_B^2}{\alpha_L^2} \tag{9.9}$$

$$Z_4 = \rho_V / \rho_L \tag{9.10}$$

$$Z_5 = \frac{\rho_L - \rho_V}{\rho_L} \tag{9.11}$$

where

$$d_B = 0.0146\theta_c \left[\frac{2 g_c \sigma}{g(\rho_L - \rho_V)} \right]^{0.5} \tag{9.12}$$

d_B = theoretical diameter of bubbles leaving surface, ft (m)
θ_c = contact angle in degrees
g = gravitational acceleration, ft/h^2 (m/s^2)
$g_c = 4.17 \times 10^8$ lbm · ft/lbf · h^2 = 1.0 kg · m/N · s^2
α_L = liquid thermal diffusivity, ft^2/h (m^2/s)
$\hat{q} \propto$ Btu/h · ft^2 (W/m^2)
$k_L \propto$ Btu/h · ft · °F (W/m · K)
$T_{sat} \propto$ °R (K)
$\sigma \propto$ lbf/ft (N/m)
$\rho_L, \rho_v \propto$ lbm/ft^3 (kg/m^3)
$\lambda \propto$ ft · lbf/lbm (J/kg)

Note that in this correlation the latent heat has units of ft · lbf/lbm in the English system (1 Btu = 778 ft · lbf). The heat-transfer coefficient is given by the following equation:

$$\frac{h_{nb} d_B}{k_L} = 0.23 Z_1^{0.674} Z_2^{0.35} Z_3^{0.371} Z_4^{0.297} Z_5^{-1.73} \tag{9.13}$$

Fluids in the database were classified in four groups, each with a characteristic contact angle, as shown in Table 9.1. Note that the group labels are somewhat misleading, as alcohols are included in the hydrocarbon group and some of the light hydrocarbons are in the refrigerant and cryogenic groups. A contact angle of 35° can be assumed for organic compounds that do not fit in any of the listed groups.

Stephan and Abdelsalam [6] also gave separate correlations for each of the four fluid groups. These group-specific correlations provide somewhat better fits to the data within each group than does the general correlation, Equation (9.13).

The following example due to Hewitt et al. [2] illustrates the use of the above correlations.

TABLE 9.1 Fluid Groups and Contact Angles for Stephan-Abdelsalam Correlation

Fluid Group	Contact Angle (θ_c)
Water	45°
Hydrocarbons (including alcohols)	35°
Refrigerants (including CO_2, propane, *n*-butane)	35°
Cryogenic fluids (including methane, ethane)	1°

Example 9.1

Pure component nucleate boiling takes place on the surface of a 1-in. OD horizontal tube immersed in a saturated liquid organic compound. The tube-wall temperature is 453.7 K and the system pressure is 310.3 kPa. Fluid properties are given in the table below. Calculate the heat-transfer coefficient and wall heat flux using:

(a) The Forster–Zuber correlation.
(b) The Mostinski correlation.
(c) The Mostinski correlation with Palen's recommendation for F_P.
(d) The Cooper correlation.
(e) The Stephan–Abdelsalam correlation.

Fluid Property	Value
Vapor density (kg/m^3)	18.09
Liquid density (kg/m^3)	567
Liquid heat capacity (J/kg · K)	2730
Liquid viscosity (kg/m · s)	156×10^{-6}
Vapor viscosity (kg/m · s)	7.11×10^{-6}
Liquid thermal conductivity (W/m · K)	0.086
Surface tension (dyne/cm)	8.2
Latent heat of vaporization (J/kg)	272,000
Critical pressure (kPa)	2550
Saturation temperature (K) at 310.3 kPa	437.5
Vapor pressure (kPa) at 453.7 K	416.6
Molecular weight	110.37

Solution

(a) For the Forster–Zuber correlation, first calculate ΔT_e and ΔP_{sat}.

$$\Delta T_e = T_w - T_{sat} = 453.7 - 437.5 = 16.2 \text{ K}$$

$$\Delta P_{sat} = P_{sat}(T_w) - P_{sat}(T_{sat})$$

$$= 416.6 - 310.3$$

$$\Delta P_{sat} = 106.3 \text{ kPa} = 106{,}300 \text{ Pa}$$

All the physical properties appearing in the correlation are given in proper SI units except the surface tension, which is:

$$\sigma = 8.2 \text{ dyne/cm} = 8.2 \times 10^{-3} \text{ N/m}$$

Substituting the appropriate values into Equation (9.1) yields:

$$h_{nb} = 0.00122 \frac{k_L^{0.79} C_{P,L}^{0.45} \rho_L^{0.49} g_c^{0.25} \Delta T_e^{0.24} \Delta P_{sat}^{0.75}}{\sigma^{0.5} \mu_L^{0.29} \lambda^{0.24} \rho_V^{0.24}}$$

$$= \frac{0.00122(0.086)^{0.79}(2730)^{0.45}(567)^{0.49}(1.0)^{0.25}(16.2)^{0.24}(106,300)^{0.75}}{(8.2 \times 10^{-3})^{0.5}(156 \times 10^{-6})^{0.29}(272,000)^{0.24}(18.09)^{0.24}}$$

$$h_{nb} = 5512 \text{ W/m}^2 \cdot \text{K}$$

The heat flux at the pipe wall is given by:

$$\hat{q} = h_{nb}\Delta T_e = 5512 \times 16.2 = 89,294 \cong 89,300 \text{ W/m}^2$$

(b) For the Mostinski correlation, first calculate the pressure correction factor using Equation (9.4).

$$P_r = P/P_c = \frac{310.3}{2550} = 0.1217$$

$$F_P = 1.8P_r^{0.17} + 4P_r^{1.2} + 10P_r^{10}$$

$$F_P = 1.8(0.1217)^{0.17} + 4(0.1217)^{1.2} + 10(0.1217)^{10} = 1.5777$$

Since $\Delta T_e = 16.2$ K is known, use Equation (9.3b) to calculate the heat-transfer coefficient.

$$h_{nb} = 1.167 \times 10^{-8} P_c^{2.3}\,\Delta T_e^{2.333}\,F_P^{3.333}$$

$$= 1.167 \times 10^{-8}(2550)^{2.3}(16.2)^{2.333}(1.5777)^{3.333}$$

$$h_{nb} = 2421\ \text{W/m}^2\cdot\text{K}$$

The heat flux is calculated as before:

$$\hat{q} = h_{nb}\,\Delta T_e = 2421 \times 16.2 = 39,220\ \text{W/m}^2$$

(c) The calculation is the same as in part (b) except that Equation (9.5) is used to calculate the pressure correction factor.

$$F_P = 2.1P_r^{0.27} + \left[9 + \left(1 - P_r^2\right)^{-1}\right]P_r^2$$

$$= 2.1(0.1217)^{0.27} + \left\{9 + \left[1 - (0.1217)^2\right]^{-1}\right\}(0.1217)^2$$

$$F_P = 1.3375$$

$$h_{nb} = 1.167 \times 10^{-8}(2550)^{2.3}(16.2)^{2.333}(1.3375)^{3.333}$$

$$h_{nb} = 1396\ \text{W/m}^2\cdot\text{K}$$

$$\hat{q} = h_{nb}\Delta T_e = 1396 \times 16.2 = 22,615\ \text{W/m}^2$$

(d) To use the Cooper correlation, substitute $\hat{q} = h_{nb}\Delta T_e$ in Equation (9.6b) to obtain:

$$h_{nb} = 55(h_{nb}\Delta T_e)^{0.67}P_r^{0.12}(-\log_{10}P_r)^{-0.55}M^{-0.5}$$

Solving for h_{nb} gives:

$$h_{nb} = \left\{55\Delta T_e^{0.67}P_r^{0.12}(-\log_{10}P_r)^{-0.55}M^{-0.5}\right\}^{3.03}$$

$$= \left\{55(16.2)^{0.67}(0.1217)^{0.12}(-\log_{10}0.1217)^{-0.55}(110.37)^{-0.5}\right\}^{3.03}$$

$$h_{nb} = 23,214\ \text{W/m}^2\cdot\text{K}$$

The heat flux is then:

$$\hat{q} = h_{nb}\Delta T_e = 23,214 \times 16.2 \cong 376,070\ \text{W/m}^2$$

(e) For the Stephan–Abdelsalam correlation, the theoretical bubble diameter is calculated using Equation (9.12). A contact angle of 35° is assumed for a non-cryogenic organic compound.

$$d_B = 0.0146\theta_c\left[\frac{2g_c\sigma}{g(\rho_L - \rho_V)}\right]^{0.5} = 0.0146 \times 35\left[\frac{2 \times 1.0 \times 8.2 \times 10^{-3}}{9.81(567 - 18.09)}\right]^{0.5}$$

$$d_B = 8.918 \times 10^{-4}\text{m}$$

The five dimensionless parameters are calculated using Equations (9.7) to (9.11). Since the heat flux is unknown, it is retained in Z_1.

$$Z_1 = \frac{\hat{q}d_B}{k_L T_{sat}} = \frac{\hat{q} \times 8.918 \times 10^{-4}}{0.086 \times 437.5} = 2.370 \times 10^{-5}\hat{q}$$

$$Z_2 = \frac{\alpha_L^2\rho_L}{g_c\sigma d_B} = \frac{\left(5.556 \times 10^{-8}\right)^2 \times 567}{1.0 \times 8.2 \times 10^{-3} \times 8.918 \times 10^{-4}} = 2.393 \times 10^{-7}$$

$$Z_3 = \frac{g_c\lambda d_B^2}{\alpha_L^2} = \frac{1.0 \times 272,000\left(8.918 \times 10^{-4}\right)^2}{(5.556 \times 10^{-8})^2} = 7.008 \times 10^{13}$$

$$Z_4 = \rho_V/\rho_L = 18.09/567 = 0.031905$$

$$Z_5 = (\rho_L - \rho_V)/\rho_L = (567 - 18.09)/567 = 0.9681$$

Substituting these values into Equation (9.13) yields:

$$\frac{h_{nb}d_B}{k_L} = 0.23\, Z_1^{0.674}\, Z_2^{0.35}\, Z_3^{0.371} Z_4^{0.297} Z_5^{-1.73}$$

$$= 0.23\left(2.370 \times 10^{-5}\hat{q}\right)^{0.674}\left(2.393 \times 10^{-7}\right)^{0.35}\left(7.008 \times 10^{13}\right)^{0.371}(0.031905)^{0.297}(0.9681)^{-1.73}$$

$$\frac{h_{nb}d_B}{k_L} = 0.04402\hat{q}^{0.674}$$

Substituting $\hat{q} = h_{nb}\Delta T_e = h_{nb} \times 16.2$ gives:

$$h_{nb} = (k_L/d_B) \times 0.04402(16.2h_{nb})^{0.674}$$

$$= \left(0.086/8.918 \times 10^{-4}\right) \times 0.04402(16.2)^{0.674} h_{nb}^{0.674}$$

$$h_{nb} = 27.739\, h_{nb}^{0.674}$$

$$h_{nb} = 26,709\ \mathrm{W/m^2 \cdot K}$$

Finally, the heat flux is given by:

$$\hat{q} = h_{nb}\Delta T_e = 26,709 \times 16.2 = 432,686\ \mathrm{W/m^2}$$

The calculated values of the heat-transfer coefficient are summarized in the following table:

Correlation	h_{nb} ($\mathrm{W/m^2 \cdot K}$)
Forster–Zuber	5512
Mostinski	2421
Modified Mostinski	1396
Cooper	23,214
Stephan–Abdelsalam	26,709

The predictions of the various correlations differ by a huge amount, the ratio of largest to smallest value being nearly 20. One reason for the wide variation is that nucleate boiling is very sensitive to the precise condition of the surface on which boiling occurs. Although the factors that govern the nucleation process are reasonably well understood (7–10), it is not practical to specify the detailed surface characteristics required to rigorously model the phenomenon. Even if this could be done, it would be of dubious utility in equipment design because the surface characteristics change in an uncontrollable manner over time due to aging, corrosion, fouling, cleaning procedures, etc. As a result, the variability exhibited by the calculated values actually reflects the variability observed among the sets of experimental data upon which the various correlations are based [2]. Fortunately, tests with commercial tube bundles indicate that the effect of surface condition is much less pronounced in the operation of industrial equipment [11]. Nevertheless, significant conservatism in equipment design is warranted by the level of uncertainty in the fundamental boiling correlations.

9.3.2 Mixture Effects

It is a well-established fact, based on experimental results, that the rate of heat transfer for nucleate boiling of mixtures is lower than for pure substances. The reason is that the more volatile components accumulate preferentially in the vapor bubbles, leaving the surrounding liquid enriched in the less volatile (higher boiling) components. As a result, the temperature of the liquid in the immediate vicinity of the heating surface increases and the effective driving force for heat transfer is reduced. Furthermore, the concentration gradient generates a mass-transfer resistance in addition to the thermal resistance. In fact, the apparent heat-transfer coefficient for a mixture can be lower than for any of the pure components at equivalent conditions.

Schlünder [12] performed a fundamental analysis of the heat and mass transfer associated with mixture boiling and derived the following equation for the heat-transfer coefficient:

$$h_{nb} = h_{ideal}\left\{1 + (h_{ideal}/\hat{q})\left[1 - \exp\left(\frac{-\hat{q}}{\rho_L \lambda \beta}\right)\right]\sum_{i=1}^{n-1}\left(T_{sat,n} - T_{sat,i}\right)(y_i - x_i)\right\}^{-1} \tag{9.14}$$

where

n = number of components and the index of the highest boiling component

$T_{sat,i}$ = boiling point of pure component i at system pressure

y_i = mole fraction of component i in vapor

x_i = mole fraction of component i in liquid
$\beta = 0.0002$ m/s (SI units) = 2.36 ft/h (English units)
= approximate mass-transfer coefficient

The coefficient, h_{ideal}, is an average of the pure component values that is calculated as follows:

$$h_{ideal} = \left[\sum_{i=1}^{n} x_i/h_{nb,i}\right]^{-1} \tag{9.15}$$

where $h_{nb,i}$ is the heat-transfer coefficient for pure component i.

Note that in Equation (9.14) "n" represents both the number of components in the mixture and the index of the component with the highest boiling point. Also, the vapor and liquid phases are assumed to be in equilibrium. In general, thermodynamic calculations are required to determine the phase compositions. For complex mixtures such as petroleum fractions, these calculations are not feasible. Although it might be possible to use pseudo-components to represent these mixtures, the validity of this procedure in conjunction with Equation (9.14) has not been established.

Thome and Shakir [13] derived a modified version of Equation (9.14) in which the boiling range of the fluid replaces the term involving the phase compositions, making the method more convenient to use. Their result is as follows:

$$h_{nb} = h_{ideal}\left\{1 + (BR \cdot h_{ideal}/\hat{q})\left[1 - \exp\left(-\frac{\hat{q}}{\rho_L \lambda \beta}\right)\right]\right\}^{-1} \tag{9.16}$$

where

$BR = T_D - T_B$ = boiling range
T_D = dew-point temperature
T_B = bubble-point temperature
$\beta = 0.0003$ m/s (SI units) = 3.54 ft/h (English units)

Notice that the value of the mass-transfer coefficient, β, is different from that in Schlünder's equation. Thome and Shakir [13] also showed that h_{ideal} can be calculated with sufficient accuracy by using mixture properties in a pure component correlation for the heat-transfer coefficient, rather than using Equation (9.15). They used the Stephan–Abdelsalam correlation for this purpose. Since the boiling range of petroleum fractions can be readily determined, Equation (9.16) is directly applicable to these fluids.

Palen [4] presented a simple empirical method that involves a (multiplicative) mixture correction factor that is applied to the heat-transfer coefficient calculated from the Mostinski correlation. The correction factor, F_m, is given by the following equation:

$$F_m = \left(1 + 0.0176\,\hat{q}^{0.15}\,BR^{0.75}\right)^{-1} \tag{9.17a}$$

Here, $\hat{q} \propto$ Btu/h · ft^2 and $BR \propto$ °F. The corresponding equation in terms of SI units is:

$$F_m = \left(1 + 0.023\,\hat{q}^{0.15}\,BR^{0.75}\right)^{-1} \tag{9.17b}$$

where $\hat{q} \propto$ W/m^2 and $BR \propto$ K.

In conjunction with Equation (9.17), for $P_r > 0.2$ the pressure correction factor given by Equation (9.5) is replaced by the following:

$$F_P = 1.8P_r^{0.17} \tag{9.18}$$

According to Palen [4], this pressure correction is necessary to ensure a conservative estimate for the heat-transfer coefficient. For this reason, Equation (9.17) should not be used with any correlation other than the Mostinski correlation, Equation (9.3), for calculating the heat-transfer coefficient. Equations (9.14) and (9.16), on the other hand, can be used with any method for calculating pure component heat-transfer coefficients.

The following example is based on data from Ref. [12].

Example 9.2

Nucleate pool boiling takes place in a liquid consisting of 54.5 mole percent SF_6 (1) and 45.5 mole percent CCl_2F_2 (2) at a pressure of 23 bar with a heat flux of 10^4 W/m^2. The vapor in equilibrium with the liquid contains 71.1 mole percent SF_6. The liquid density of the mixture is 10 kmol/m^3. At 23 bar, the boiling point of SF_6 is 22°C and that of CCl_2F_2 is 80°C. Experimental values for the pure component heat-transfer coefficients at this pressure are 11,787 W/m^2 · K for SF_6 and 6,105 W/m^2 · K for CCl_2F_2. Additional data are given in the table below. Estimate the heat-transfer coefficient for the mixture using:

(a) Schlünder's method.
(b) The method of Thome and Shakir.
(c) Palen's method.

Property	SF_6	$CC1_2F_2$
Critical temperature (°C)	45.5	111.5
Critical pressure (bar)	37.6	41.2
Normal boiling point (°C)	-63.5	-29.8
Latent heat of vaporization at normal boiling point (J/mol)	16,790	20,207

Solution

(a) Since CCl_2F_2 has the higher boiling point, it is designated as component number 2 and SF_6 as component number 1. Values of all the parameters appearing in Equation (9.14) are given except for the latent heat of vaporization of the mixture. In order to estimate this value, the pure component latent heats must first be determined at the system pressure of 23 bar. The Watson correlation, which relates the latent heats at two different temperatures, is used for this purpose:

$$\lambda(T_2) = \lambda(T_1)\left[\frac{T_c - T_2}{T_c - T_1}\right]^{0.38}$$

In the present application, T_1 is taken as the normal boiling point and T_2 is taken as the boiling point at system pressure. Thus, for SF_6 we have:

$$\lambda_1(22°\text{C}) = 16,790\left[\frac{45.5 - 22}{45.5 - (-63.5)}\right]^{0.38}$$

$$\lambda_1 = 9372 \text{ J/mol}$$

Similarly, for CCl_2F_2 we have:

$$\lambda_2(80°\text{C}) = 20,207\left[\frac{111.5 - 80}{111.5 - (-29.8)}\right]^{0.38}$$

$$\lambda_2 = 11,424 \text{ J/mol}$$

For the mixture, the mole fraction weighted average of the pure component latent heats is used:

$$\lambda = x_1\lambda_1 + x_2\lambda_2 = 0.545 \times 9372 + 0.455 \times 11,424$$

$$\lambda \cong 10,300 \text{ J/mol}$$

Next, h_{ideal} is computed using Equation (9.15):

$$\begin{aligned} h_{ideal} &= \left(x_1/h_{nb,1} + x_2/h_{nb,2}\right)^{-1} \\ &= (0.545/11,787 + 0.455/6105)^{-1} \\ h_{ideal} &= 8280 \text{ W/m}^2\cdot\text{K} \end{aligned}$$

The heat-transfer coefficient for the mixture can now be obtained by substituting the data into Equation (9.14):

$$h_{nb} = h_{ideal}\left\{1 + (h_{ideal}/\hat{q})\left[1 - \exp\left(\frac{-\hat{q}}{\rho_L\lambda\beta}\right)\right]\sum_{i=1}^{n-1}(T_{sat,n} - T_{sat,i})(y_i - x_i)\right\}^{-1}$$

$$= 8280\left\{1 + (8280/10,000)\left[1 - \exp\left(\frac{-10,000}{10,000 \times 10,300 \times 0.0002}\right)\right] \times (80 - 22)(0.711 - 0.545)\right\}^{-1}$$

$$h_{nb} = 2036 \text{ W/m}^2\cdot\text{K}$$

The measured heat-transfer coefficient for this mixture is 2096 $\text{W/m}^2 \cdot \text{K}$ [12], which differs from the calculated value by about 3%.

(b) Since the dew-point and bubble-point temperatures for the mixture were not given in Ref. [12], the PRO/II (SimSci-Esscor) process simulator was used to obtain the following values at a pressure of 23 bars: $T_D = 326.5$ K and $T_B = 316.9$ K. The boiling range for the mixture is, therefore:

$$BR = T_D - T_B = 326.5 - 316.9 = 9.6 \text{ K}$$

The heat-transfer coefficient for the mixture is obtained from Equation (9.16) using the values of h_{ideal} and λ from part (a):

$$h_{nb} = h_{ideal}\left\{1 + (BR \cdot h_{ideal}/\hat{q})\left[1 - \exp\left(\frac{-\hat{q}}{\rho_L \lambda \beta}\right)\right]\right\}^{-1}$$

$$= 8280\left\{1 + \left(\frac{9.6 \times 8280}{10,000}\right)\left[1 - \exp\left(\frac{-10,000}{10,000 \times 10,300 \times 0.0003}\right)\right]\right\}^{-1}$$

$$h_{nb} = 2589\ \text{W/m}^2 \cdot \text{K}$$

This value is about 24% higher than the experimental value of 2096 W/m^2 · K.

(c) For a mixture, the pseudo-critical pressure, P_{pc}, can be used in place of the critical pressure in the Mostinski correlation. The pseudo-critical pressure is the mole fraction weighted average of the pure component critical pressures. Thus,

$$P_{pc} = x_1 P_{c1} + x_2 P_{c2} = 0.545 \times 37.6 + 0.435 \times 41.2$$

$$P_{pc} \cong 39.24\ \text{bar} = 3924\ \text{kPa}$$

The pseudo-reduced pressure is then:

$$P_{pr} = P/P_{pc} = 23/39.24 = 0.586$$

Since this value is greater than 0.2, Equation (9.18) is used to calculate the pressure correction factor with P_{pr} in place of P_r.

$$F_p = 1.8P_{pr}^{0.17} = 1.8(0.586)^{0.17} = 1.6437$$

The mixture correction factor is obtained from Equation (9.17b).

$$F_m = \left(1 + 0.023\hat{q}^{0.15} BR^{0.75}\right)^{-1}$$

$$= \left[1 + 0.023(10,000)^{0.15}(9.6)^{0.75}\right]^{-1}$$

$$F_m = 0.6669$$

Including this factor in Equation (9.2b) gives:

$$h_{nb} = 0.00417 P_c^{0.69} \hat{q}^{0.7} F_P F_m$$

$$= 0.00417(3924)^{0.69}(10,000)^{0.7} \times 1.6437 \times 0.6669$$

$$h_{nb} = 870\ \text{W/m}^2 \cdot \text{K}$$

This value is very conservative compared with the measured value of 2096 W/m^2 · K for this system. Note, however, that the mixture correction factor itself is not conservative, since $F_m \cdot h_{ideal} = 5522$ W/m^2 · K, which is much higher than the measured heat-transfer coefficient. This result emphasizes the point that Palen's mixture correction factor is intended for use only with the Mostinski correlation for calculating h_{nb}.

9.3.3 Convective Effects in Tube Bundles

Heat-transfer coefficients for boiling on tube bundles are generally higher than for boiling on single tubes under the same conditions. The enhancement in heat-transfer rate is due to the convective circulation that is set up within and around the tube bundle. The circulation is driven by the density difference between the liquid surrounding the bundle and the two-phase mixture within the bundle. In this respect it is similar to the circulation in a thermosyphon reboiler. The circulation rate can be calculated with a computer model and used to determine the convective contribution to the heat transfer.

For hand calculations, Palen [4] presented an approximate method for calculating convective effects. The average boiling heat-transfer coefficient, h_b, is expressed as follows:

$$h_b = h_{nb} F_b + h_{nc} \tag{9.19}$$

where h_{nc} is a heat-transfer coefficient for liquid-phase natural convection and F_b is a factor that accounts for the effect of the thermosyphon-type circulation in the tube bundle. The bundle convection factor is correlated in terms of bundle geometry by the following empirical equation [4]:

$$F_b = 1.0 + 0.1\left[\frac{0.785 D_b}{C_1 (P_T/D_o)^2 D_o} - 1.0\right]^{0.75} \tag{9.20}$$

where

D_b = bundle diameter (outer-tube-limit diameter)
D_o = tube OD
P_T = tube pitch
C_1 = 1.0 for square and rotated square layouts
= 0.866 for triangular layouts

The coefficient, h_{nc}, can be estimated using Equation (2.59) for free convection from a horizontal cylinder. However, liquid-phase natural convection makes a relatively small contribution to the total heat transfer except when the temperature difference, ΔT_e, is less than about 4°C. For larger temperature differences, therefore, Palen [4] suggests using a rough approximation for h_{nc} of 250 W/m$^2 \cdot$K (44 Btu/h$\cdot$ft$^2 \cdot$°F) for hydrocarbons and 1000 W/m$^2 \cdot$K (176 Btu/h$\cdot$ft$^2 \cdot$°F) for water and aqueous solutions.

9.3.4 Critical Heat Flux

At high heat fluxes, the vapor bubbles leave the heated surface in columns or jets. When the velocity of the vapor jets reaches a sufficiently large value, the jets become unstable and collapse. The resultant accumulation of vapor at the heated surface prevents an adequate supply of liquid from reaching the surface, and the heat flux begins to decrease. Zuber [14] performed a stability analysis to derive equations for the maximum vapor velocity and critical heat flux in nucleate pool boiling on a large flat upward-facing surface. The equation for critical heat flux is generally used in the following form:

$$\hat{q}_c = 0.149\lambda\rho_V^{0.5}[\sigma g g_c(\rho_L - \rho_V)]^{0.25} \tag{9.21}$$

The coefficient of 0.149 is based on experimental data; the theoretical value is $\pi/24 \cong 0.131$. Equation (9.21) is often used irrespective of the geometry of the heated surface.

Results for other geometries can be found in Refs. [9,15]. For boiling on a horizontal cylinder, the result is:

$$\hat{q}_c = K\lambda\rho_V^{0.5}[\sigma g g_c(\rho_L - \rho_V)]^{0.25} \tag{9.22}$$

where

$K = 0.118$ for $R^* \geq 1.17$
$= 0.123(R^*)^{-0.25}$ for $0.12 < R^* < 1.17$

$$R^* = R\left[\frac{g(\rho_L - \rho_V)}{g_c\sigma}\right]^{0.5} = \text{dimensionless radius}$$

R = radius of cylinder

Although Equations (9.21) and (9.22) do not contain an explicit pressure dependence, the critical heat flux is a strong function of pressure through its effect on λ, ρ_V, and σ.

A correlation for boiling on a single horizontal tube based on the principle of corresponding states and due to Mostinski is the following [4,7]:

$$\hat{q}_c = 803 P_c P_r^{0.35}(1 - P_r)^{0.9} \tag{9.23a}$$

where $\hat{q}_c \propto$ Btu/h $\cdot$ ft^2 and $P_c \propto$ psia.

The corresponding equation in terms of SI units is:

$$\hat{q}_c = 367 P_c P_r^{0.35}(1 - P_r)^{0.9} \tag{9.23b}$$

where $\hat{q}_c \propto$ W/m^2 and $P_c \propto$ kPa.

For tube bundles, Palen [4] presented the following correlation:

$$\hat{q}_{c,bundle} = \hat{q}_{c,tube}\phi_b \tag{9.24}$$

where

$\hat{q}_{c,bundle}$ = critical heat flux for tube bundle
$\hat{q}_{c,tube}$ = critical heat flux for a single tube
ϕ_b = bundle correction factor
= $3.1\psi_b$ for $\psi_b < 1.0/3.1 \cong 0.323$
= 1.0 otherwise
ψ_b = dimensionless bundle geometry parameter $= \dfrac{\pi D_b L}{A}$
D_b = bundle diameter
A = bundle surface area $= n_t\pi D_o L$ for plain tubes
D_o = tube OD
L = tube length

n_t = number of tubes in bundle

It is recommended that vapor release lanes be provided to aid vapor flow out of the bundle if $\phi_b < 0.1$ and the heat flux exceeds 50% of the critical heat flux for the bundle [4].

The above correlations apply to pure component nucleate boiling. The critical heat flux for a mixture may be higher or lower than for a pure component under similar conditions. Although a higher value is usually found, as would be expected from the effect of boiling point elevation and mass-transfer resistance in mixture boiling, the variation of surface tension with composition can cause the mixture critical heat flux to be lower than the pure component value in some cases. In any event, the mixture effect is generally neglected in design work since it cannot be reliably predicted. The above correlations are used with the mixture physical properties replacing the pure component properties. For the Mostinski correlation, the pseudo-critical and pseudo-reduced pressures can be used in place of the true mixture values.

Example 9.3

Estimate the critical heat flux for the conditions of Example 9.1 using the following methods:

(a) The Zuber equation.
(b) The Zuber-type equation for a horizontal cylinder.
(c) The Mostinski correlation.

Solution

(a) The following data are obtained from Example 9.1:

$\lambda = 272{,}000$ J/kg	$\rho_L = 567$ kg/m^3
$\sigma = 8.2 \times 10^{-3}$ N/m	$\rho_V = 18.09$ kg/m^3

Substituting these values into Equation (9.21) with $g = 9.81$ m/s^2 and $g_c = 1.0$ kg · m/N · s^2 gives:

$$\hat{q}_c = 0.149\lambda\, \rho_V^{0.5}[\sigma g g_c(\rho_L - \rho_V)]^{0.25}$$

$$= 0.149 \times 272{,}000(18.09)^{0.5}\left[8.2 \times 10^{-3} \times 9.81 \times 1.0(567 - 18.09)\right]^{0.25}$$

$$\hat{q}_c = 444{,}345 \cong 444{,}000 \text{ W/m}^2$$

(b) The first step is to calculate the dimensionless radius, R^*. For a 1-in. tube, $R = 0.0127$ m. Hence,

$$R^* = R\left[\frac{g(\rho_L - \rho_V)}{g_c\sigma}\right]^{0.5}$$

$$= 0.0127\left[\frac{9.81(567 - 18.09)}{1.0 \times 8.2 \times 10^{-3}}\right]^{0.5}$$

$$R^* = 10.3$$

Since $R^* > 1.17$, set $K = 0.118$ in Equation (9.22):

$$\hat{q}_c = 0.118\lambda\rho_V^{0.5}[\sigma g g_c(\rho_L - \rho_V)]^{0.25}$$

$$= 0.118 \times 272{,}000(18.09)^{0.5}\left[8.2 \times 10^{-3} \times 9.81 \times 1.0(567 - 18.09)\right]^{0.25}$$

$$\hat{q}_c = 351{,}898 \cong 352{,}000 \text{ W/m}^2$$

(c) From Example 9.1, $P = 310.3$ kPa and $P_c = 2550$ kPa. Therefore,

$$P_r = P/P_c = 0.1217$$

Using Equation (9.23b) gives:

$$\hat{q}_c = 367 P_c P_r^{0.35}(1 - P_r)^{0.9}$$

$$= 367 \times 2550(0.1217)^{0.35}(1 - 0.1217)^{0.9}$$

$$\hat{q}_c = 398{,}416 \cong 398{,}000 \text{ W/m}^2$$

The Zuber and Mostinski values differ by about 11%. The results from Equations (9.21) and (9.22) differ by about 23%, of which 10% is due to the effect of surface geometry. The other 13% is due to the empirical coefficient of 0.149 in Equation (9.21) as opposed to the theoretical coefficient in Equation (9.22).

Example 9.4

The nucleate boiling of Example 9.1 takes place on a tube bundle consisting of 520 1-in. OD tubes on a 1.25-in. square pitch. The bundle diameter is approximately 34 in. Estimate the critical heat flux and heat-transfer coefficient for this situation.

Solution

The critical heat flux for the tube bundle is given by Equation (9.24):

$$\hat{q}_{c,bundle} = \hat{q}_{c,tube}\phi_b$$

The critical heat flux for a single tube was calculated in Example 9.3, and any of the three results obtained there can be used here. Choosing the result calculated by the Mostinski correlation, we have:

$$\hat{q}_{c,tube} = 398,000 \text{ W/m}^2$$

To determine the bundle correction factor, first calculate the bundle geometry parameter, ψ_b, which for plain tubes is:

$$\psi_b = \frac{D_b}{n_t D_o} = \frac{34}{520 \times 1.0} = 0.0654$$

Since this value is less than 0.323, the bundle correction factor is:

$$\phi_b = 3.1\psi_b = 3.1 \times 0.0654 \cong 0.203$$

Therefore,

$$\hat{q}_{c,bundle} = 398,000 \times 0.203 \cong 81,000 \text{ W/m}^2$$

The heat-transfer coefficient for a tube bundle is given by Equation (9.19):

$$h_b = h_{nb}F_b + h_{nc}$$

We will use the value of h_{nb} calculated in Example 9.1 by the modified Mostinski correlation, i.e., $h_{nb} = 1396 \text{ W/m}^2 \cdot \text{K}$. Also, from Example 9.1, $\Delta T_e = 16.2$ K. Since this value is much greater than 4 K, a rough estimate is sufficient for the natural convection coefficient, h_{nc}. For an organic compound, Palen's recommendation for hydrocarbons is adequate, so we take $h_{nc} \cong 250 \text{ W/m}^2 \cdot \text{K}$. The value of F_b is calculated using Equation (9.20):

$$F_b = 1.0 + 0.1\left[\frac{0.785D_b}{C_1(P_T/D_o)^2 D_o} - 1.0\right]^{0.75}$$

$$= 1.0 + 0.1\left[\frac{0.785 \times 34}{1.0(1.25/1.0)^2 \times 1.0} - 1.0\right]^{0.75}$$

$$F_b \cong 1.803$$

Therefore,

$$h_b = 1396 \times 1.803 + 250 = 2767 \cong 2770 \text{ W/m}^2 \cdot \text{K}$$

9.4 Two-Phase Flow

9.4.1 Two-Phase Flow Regimes

When a vapor-liquid mixture flows through a circular tube, a number of different flow regimes can occur, depending on the vapor fraction, flow rate, and orientation of the tube. For vertical tubes, the following flow regimes are distinguished (Figure 9.2(a)):

- *Bubbly flow*: At low vapor fractions, vapor bubbles are dispersed in a continuous liquid phase.
- *Slug flow*: At moderate vapor fractions and relatively low flow rates, large bullet-shaped vapor bubbles flow through the tube separated by slugs of liquid in which smaller bubbles may be dispersed. A percolating coffee pot exemplifies this type of flow.
- *Churn flow*: At higher flow rates, the large vapor bubbles present in slug flow become unstable and break apart, resulting in an oscillatory, or churning, motion of the liquid upward and downward in the tube.

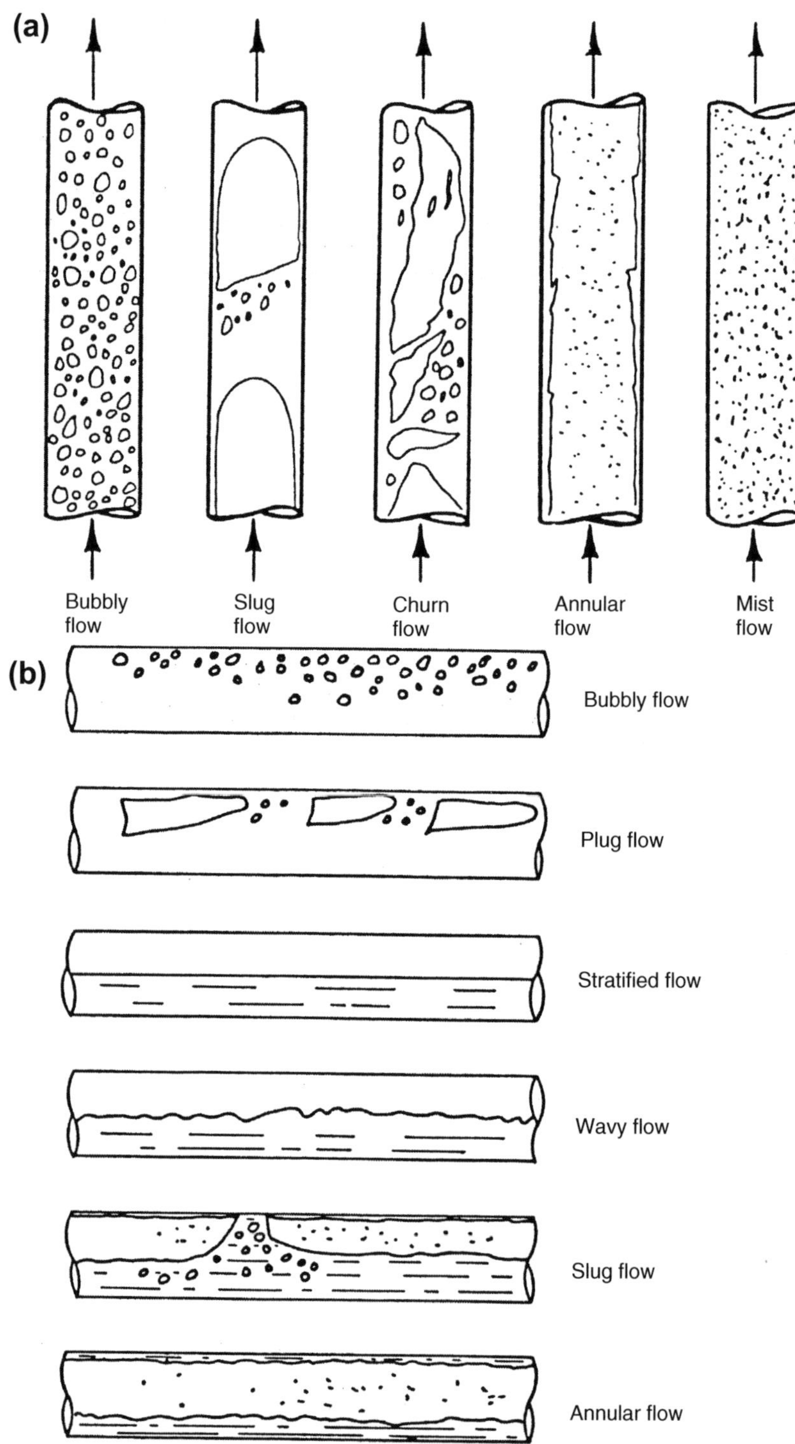

FIGURE 9.2 Two-phase flow regimes in circular tubes: (a) vertical tube and (b) horizontal tube (Source: Ret [8]).

- *Annular flow*: At high vapor fractions and high flow rates, the liquid flows as a film along the tube wall while the vapor flows at a higher velocity in the central region of the tube. Small liquid droplets are usually entrained in the vapor phase, and vapor bubbles may be dispersed in the liquid film as well. At sufficiently high liquid flow rates, the droplets coalesce to form large streaks or wisps of liquid entrained in the vapor phase. This condition, which is characteristic of flows with a high mass flux, is referred to as wispy annular flow.
- *Mist Flow*: At very high vapor fractions, the liquid phase exists entirely as droplets entrained in a continuous vapor phase.

In horizontal tubes the situation is somewhat different due to stratification of the flow resulting from the gravitational force. The following flow regimes are observed (Figure 9.2(b)):

- *Bubbly flow*: The flow pattern is similar to that in vertical tubes except the bubbles tend to concentrate in the upper part of the tube.
- *Plug flow*: The flow pattern is similar to slug flow in vertical tubes, but the bullet-shaped bubbles tend to flow closer to the top of the tube and occupy less of the tube cross-section.
- *Stratified flow*: The two phases are completely stratified, with the liquid flowing along the bottom of the tube and the vapor flowing along the top.
- *Wavy flow*: In stratified flow at higher vapor velocities, waves are formed on the surface of the liquid that characterize this flow regime.
- *Slug flow*: In this regime, intermittent slugs of liquid pass through the tube. The slugs occupy the entire tube cross-section and contain a large number of entrained vapor bubbles that impart a frothy character to the liquid.
- *Annular flow*: The flow pattern is similar to annular flow in vertical tubes except that the liquid film thickness is non-uniform, being greater on the bottom than the top of the tube.
- *Mist flow*: The flow pattern is the same as in a vertical tube and, hence, is not illustrated for the horizontal tube in Figure 9.2.

Flow pattern maps, which show the operating regions over which the various flow regimes exist, can be found in Refs. [2,8], among others. The latter also discusses modeling and empirical correlations for predicting the frictional pressure drop in each flow regime. However, most process equipment design is based on generalized pressure drop correlations that do not explicitly account for the two-phase flow regime. These correlations are presented in the following subsection.

9.4.2 Pressure Drop Correlations

Two general models of two-phase flow are used, the homogeneous flow model and the separated flow model. In the former, the two phases are assumed to have the same velocity. In the separated flow model, it is assumed that the two phases flow in separate zones, as in the annular and stratified regimes. In this case the two phases generally have different velocities, but can interact with each other. Under most conditions, the separated flow model provides a better representation of the pressure drop in pipe flow. Several widely used pressure drop correlations based on this model are presented below.

The Lockhart–Martinelli Correlation

In this method the two-phase pressure gradient is expressed as a multiple of the pressure gradient that would occur if the liquid phase flowed alone in the conduit. Thus,

$$\left(\frac{\Delta P_f}{L}\right)_{tp} = \phi_L^2 \left(\frac{\Delta P_f}{L}\right)_L \tag{9.25}$$

where

ϕ_L^2 = two-phase multiplier
$(\Delta P_f/L)_L$ = negative pressure gradient for liquid alone
$(\Delta P_f/L)_{tp}$ = negative two-phase pressure gradient

The negative pressure gradient is the pressure drop per unit length or, more precisely, the rate of decrease of pressure with distance in the flow direction. Equation (9.25) is formulated in terms of pressure gradients rather than pressure drops because the vapor fraction and other flow parameters may vary in the flow direction. In this case the pressure drop is obtained by integrating the two-phase pressure gradient along the flow path. When the vapor fraction and other flow parameters are essentially constant, either pressure gradients or pressure drops can be used in Equation (9.25). The two-phase multiplier is a function of the parameter, X, which is defined as follows:

$$X = \left[\frac{(\Delta P_f/L)_L}{(\Delta P_f/L)_V}\right]^{0.5} \tag{9.26}$$

where $(\Delta P_f/L)_V$ is the pressure gradient that would occur if the vapor phase flowed alone in the conduit. The relationship between ϕ_L^2 and X was given in graphical form by Lockhart and Martinelli [16], and subsequently expressed analytically by Chisholm [17] as follows:

$$\phi_L^2 = 1 + \frac{C}{X} + \frac{1}{X^2} \tag{9.27}$$

The constant, C, depends on whether the flow in each phase is laminar or turbulent, as shown in Table 9.2. Approximate critical Reynolds numbers for determining the flow condition of each phase, as suggested by Lockhart and Martinelli [16], are also listed. These Reynolds numbers are calculated as if each phase flowed alone in the conduit.

TABLE 9.2 Values of the Constant in Equation (9.27)

Liquid	Vapor	Notation	Re_L	Re_V	C
Turbulent	Turbulent	tt	>2000	>2000	20
Viscous (laminar)	Turbulent	vt	<1000	>2000	12
Turbulent	Viscous (laminar)	tv	>2000	<1000	10
Viscous (laminar)	viscous (laminar)	vv	<1000	<1000	5

In practice, if the liquid phase is turbulent, the vapor phase will usually be turbulent as well, and this is by far the most important case. Therefore, Equation (9.27) is restated for the turbulent-turbulent case as follows:

$$\phi_L^2 = 1 + \frac{20}{X_{tt}} + \frac{1}{X_{tt}^2} \tag{9.28}$$

The calculation of X can also be simplified if the friction factor has the same functional form for each phase, namely:

$$f = \text{const.}/Re^n \tag{9.29}$$

From Equation (4.5) or (5.1), the negative pressure gradient has the following functionality:

$$\Delta P_f/L \sim f\,G^2 s^{-1} \tag{9.30}$$

Combining Equations (9.29) and (9.30) yields:

$$\Delta P_f/L \sim G^{2-n}\mu^n s^{-1} \sim \dot{m}^{2-n}\mu^n s^{-1} \tag{9.31}$$

Taking square roots gives:

$$(\Delta P_f/L)^{0.5} \sim \dot{m}^{(2-n)/2}\mu^{n/2} s^{-0.5} \tag{9.32}$$

From the definition of X given by Equation (9.26), it then follows that:

$$X = (\dot{m}_L/\dot{m}_V)^{(2-n)/2}(\mu_L/\mu_V)^{n/2}(s_V/s_L)^{0.5} \tag{9.33}$$

Now the mass flow rates of the two phases are related to the vapor mass fraction, x, by:

$$\dot{m}_L = \dot{m}(1-x) \tag{9.34}$$

$$\dot{m}_V = \dot{m}x \tag{9.35}$$

Using these relations and replacing the specific gravity ratio by the density ratio, Equation (9.33) becomes:

$$X = \left(\frac{1-x}{x}\right)^{(2-n)/2}(\mu_L/\mu_V)^{n/2}(\rho_V/\rho_L)^{0.5} \tag{9.36}$$

For the turbulent-turbulent case, Lockhart and Martinelli [16] assumed the value $n = 0.2$ to obtain:

$$X_{tt} = \left(\frac{1-x}{x}\right)^{0.9}(\rho_V/\rho_L)^{0.5}(\mu_L/\mu_V)^{0.1} \tag{9.37}$$

Equation (9.37) is generally used in practice. Although the exponent, n, is closer to 0.25 for commercial pipe and tubing (cf. Equations (4.8) and (5.2)), the difference is insignificant compared with the inherent uncertainty associated with the Lockhart-Martinelli correlation. It will be noted from Equation (9.37) that X_{tt} is zero when the vapor fraction, x, equals 1.0. In this case, Equation (9.28) shows that the two-phase multiplier, ϕ_L^2, becomes infinite. For high vapor fractions, an alternate formulation in terms of the (negative) vapor-phase pressure gradient, $(\Delta P_f/L)_V$, can be used (see Problem 9.1). For process engineering work, however, the formulation in terms of the liquid-phase pressure gradient is used almost exclusively.

The Chisholm Correlation

Chisholm [18] expressed the two-phase pressure gradient as a multiple of the pressure gradient that would occur if the entire flow (vapor and liquid) had the properties of the liquid phase. The result is analogous to Equation (9.25):

$$\left(\frac{\Delta P_f}{L}\right)_{tp} = \phi_{LO}^2\left(\frac{\Delta P_f}{L}\right)_{LO} \tag{9.38}$$

where

ϕ_{LO}^2 = two-phase multiplier

$\left(\frac{\Delta P_f}{L}\right)_{LO}$ = negative pressure gradient for total flow as liquid

The two-phase multiplier is a function of the parameter, Y, defined as:

$$Y = \left[\frac{(\Delta P_f/L)_{VO}}{(\Delta P_f/L)_{LO}}\right]^{0.5} \tag{9.39}$$

where $(\Delta P_f/L)_{VO}$ is the negative pressure gradient that would occur if the entire flow had the properties of the vapor phase. Comparison with Equation (9.26) shows that Y is analogous to the reciprocal of the Lockhart–Martinelli parameter. Since the mass flux, G, is the same for the numerator and denominator, it follows from Equation (9.30) that Equation (9.39) can be written in the alternate form:

$$Y = [(f_{VO}/f_{LO})(s_L/s_V)]^{0.5} = [(f_{VO}/f_{LO})(\rho_L/\rho_V)]^{0.5} \tag{9.40}$$

where f_{VO} and f_{LO} are friction factors calculated with vapor and liquid properties, respectively. Assuming that the friction factor is the same function of Reynolds number in both phases as given by Equation (9.29), it follows that:

$$f_{VO}/f_{LO} = (\mu_V/\mu_L)^n \tag{9.41}$$

Substituting this relationship in Equation (9.40) yields:

$$Y = (\rho_L/\rho_V)^{0.5}(\mu_V/\mu_L)^{n/2} \tag{9.42}$$

The correlation for the two-phase multiplier is the following [18]:

$$\phi_{LO}^2 = 1 + (Y^2 - 1)\left\{B\,[x(1-x)]^{(2-n)/2} + x^{2-n}\right\} \tag{9.43}$$

where x is the vapor fraction. The parameter, B, depends on both Y and G. Chisholm [18] determined this relationship empirically using results from the literature. He then adjusted the correlation to obtain a conservative estimate of pressure drop for design purposes. As a result, his final correlation for B is an awkward discontinuous function that tends to over-predict the pressure drop in certain ranges of G and Y. Therefore, we will follow Hewitt's [2,19] recommendation and use Chisholm's unadjusted correlation for B, which is represented by the following piecewise continuous function (with G in units of $\text{lbm/h}\cdot\text{ft}^2$):

$$\begin{aligned} B &= 1500/G^{0.5} && (0 < Y < 9.5) \\ &= 14,250/(YG^{0.5}) && (9.5 < Y < 28) \\ &= 399,000/(Y^2G^{0.5}) && (Y > 28) \end{aligned} \tag{9.44}$$

For calculations in SI units, the following conversion can be used:

$$G(\text{lbm/h}\cdot\text{ft}^2) = 737.35\, G(\text{kg/s}\cdot\text{m}^2)$$

The Friedel Correlation

The two-phase pressure gradient is expressed in the same manner as the Chisholm method, Equation (9.38), with the following correlation for the two-phase multiplier [20]:

$$\phi_{LO}^2 = E + \frac{3.24FH}{Fr^{0.045}We^{0.035}} \tag{9.45}$$

where

$$E = (1-x)^2 + x^2(f_{VO}/f_{LO})(\rho_L/\rho_V) \tag{9.46}$$

$$F = x^{0.78}(1-x)^{0.24} \tag{9.47}$$

$$H = (\rho_L/\rho_V)^{0.91}(\mu_V/\mu_L)^{0.19}(1-\mu_V/\mu_L)^{0.7} \tag{9.48}$$

$$Fr = \frac{G^2}{gD_i\rho_{tp}^2} = \text{Froude number} \tag{9.49}$$

$$We = \frac{G^2D_i}{g_c\rho_{tp}\sigma} = \text{Weber number} \tag{9.50}$$

D_i = internal diameter of conduit

ρ_{tp} = two-phase density

For the purpose of this correlation, the two-phase density is calculated as follows:

$$\rho_{tp} = [x/\rho_V + (1-x)/\rho_L]^{-1} \tag{9.51}$$

Equation (9.46) can be restated in a more convenient form by using Equation (9.41) to eliminate the friction factors:

$$E = (1-x)^2 + x^2(\mu_V/\mu_L)^n(\rho_L/\rho_V) \tag{9.52}$$

The correlation as given here is valid for horizontal and vertical upward flow.

The Müller–Steinhagen and Heck Correlation

The correlation is given in Ref. [21], and can be reformulated in the Chisholm format of Equation (9.38) with the two-phase multiplier given by the following equation:

$$\phi_{LO}^2 = Y^2x^3 + \left[1 + 2x\left(Y^2 - 1\right)\right](1-x)^{1/3} \tag{9.53}$$

where x is the vapor mass fraction and Y is the Chisholm parameter. Equation (9.53) is an interpolation formula between all liquid flow ($x = 0$) and all vapor flow ($x = 1$). For $x = 0$, $\phi_{LO}^2 = 1$ and the negative two-phase pressure gradient reduces to $(\Delta P_f/L)_{LO}$. For $x = 1$, $\phi_{LO}^2 = Y$ and the negative two-phase pressure gradient reduces to $(\Delta P_f/L)_{VO}$.

The Friedel correlation has long been regarded as the most reliable general method for computing two-phase pressure losses. However, in two independent studies [22,23] the performance of the simpler Müller–Steinhagen and Heck (MSH) method was found to be slightly better than that of the Friedel method. It should be recognized that the uncertainty associated with all of these methods is much greater than for single-phase pressure drop calculations, and relatively large errors are possible in any given application.

Example 9.5

The return line from a thermosyphon reboiler consists of 10-in. schedule 40 pipe with an ID of 10.02 in. (0.835 ft). The total mass flow rate in the line is 300,000 lbm/h, of which 60,000 lbm/h (20%) is vapor and 240,000 lbm/h is liquid. Physical properties of the vapor and liquid fractions are given in the following table:

Property	Liquid	Vapor
μ (cp)	0.177	0.00885
ρ (lbm/ft^3)	38.94	0.4787
σ (dyne/cm)	11.4	–

Calculate the friction loss per unit length in the line using:

(a) The Lockhart–Martinelli correlation.
(b) The Chisholm correlation.
(c) The Friedel correlation.
(d) The MSH correlation.

Solution

(a) The friction loss for the liquid fraction flowing alone in the pipe is calculated first.

$$G_L = \frac{\dot{m}_L}{(\pi/4)D_i^2} = \frac{240,000}{(\pi/4)(0.835)^2} = 438,277 \text{ lbm/h}\cdot\text{ft}^2$$

$$Re_L = \frac{D_i G_L}{\mu_L} = \frac{0.835 \times 438,277}{0.177 \times 2.419} = 854,724$$

$$s_L = \rho_L/\rho_{water} = 38.94/62.43 = 0.6237$$

Equation (4.8) is used to calculate the friction factor:

$$f_L = 0.3673 Re_L^{-0.2314} = 0.3673\,(854,724)^{-0.2314} = 0.01557$$

The friction loss per unit length is given by:

$$\left(\frac{\Delta P_f}{L}\right)_L = \frac{f_L G_L^2}{7.50 \times 10^{12} D_i s_L \phi_L} = \frac{0.01557(438,277)^2}{7.5 \times 10^{12} \times 0.835 \times 0.6237 \times 1.0}$$

$$\left(\frac{\Delta P_f}{L}\right)_L = 0.000766 \text{ psi/ft}$$

Since the flow is turbulent, the Lockhart–Martinelli parameter is calculated using Equation (9.37).

$$X_{tt} \cong \left(\frac{1-x}{x}\right)^{0.9}\left(\frac{\rho_V}{\rho_L}\right)^{0.5}\left(\frac{\mu_L}{\mu_V}\right)^{0.1}$$

$$= (0.8/0.2)^{0.9}(0.4787/38.94)^{0.5}(0.177/0.00885)^{0.1}$$

$$X_{tt} \cong 0.521$$

For comparison, the reader can verify that using Equation (9.36) with $n = 0.2314$ to calculate X_{tt} gives a value of 0.534, which differs from the value calculated above by less than 3%. For turbulent flow, the two-phase multiplier is given by Equation (9.28):

$$\phi_L^2 = 1 + \frac{20}{X_{tt}} + \frac{1}{X_{tt}^2} = 1 + \frac{20}{0.521} + \frac{1}{(0.521)^2} = 43.07$$

The two-phase friction loss is then:

$$\left(\frac{\Delta P_f}{L}\right)_{tp} = \phi_L^2\left(\frac{\Delta P_f}{L}\right)_L = 43.07 \times 0.000766 = 0.033 \text{ psi/ft}$$

(b) For the Chisholm method, the friction loss is calculated assuming the total flow has the properties of the liquid. Thus,

$$G = \frac{\dot{m}}{(\pi/4)D_i^2} = \frac{300{,}000}{(\pi/4)(0.835)^2} = 547{,}846 \text{ lbm/h·ft}^2$$

$$Re_{LO} = \frac{D_i G}{\mu_L} = \frac{0.835(547{,}846)}{0.177 \times 2.419} = 1{,}068{,}405$$

$$f_{LO} = 0.3673 Re_{LO}^{-0.2314} = 0.3673(1{,}068{,}405)^{-0.2314} = 0.01479$$

$$\left(\frac{\Delta P_f}{L}\right)_{LO} = \frac{f_{LO}G^2}{7.50 \times 10^{12} D_i s_L \phi_L} = \frac{0.01479(547{,}846)^2}{7.5 \times 10^{12} \times 0.835 \times 0.6237 \times 1.0}$$

$$\left(\frac{\Delta P_f}{L}\right)_{LO} = 0.001136 \text{ psi/ft}$$

Since the flow is turbulent, the Chisholm parameter is calculated using Equation (9.42) with $n = 0.2314$.

$$Y = (\rho_L/\rho_V)^{0.5}(\mu_V/\mu_L)^{n/2} = (38.94/0.4787)^{0.5}(0.00885/0.177)^{0.1157}$$

$$Y \cong 6.38$$

Since $Y < 9.5$, Equation (9.44) gives:

$$B = 1500/G^{0.5} = 1500/(547{,}846)^{0.5} \cong 2.027$$

The two-phase multiplier is calculated using Equation (9.43):

$$\phi_{LO}^2 = 1 + (Y^2 - 1)\left\{B[x(1-x)]^{(2-n)/2} + x^{2-n}\right\}$$

$$= 1 + \left[(6.38)^2 - 1\right]\left\{2.027[0.2(1-0.2)]^{0.8843} + (0.2)^{1.7686}\right\}$$

$$\phi_{LO}^2 \cong 19.22$$

Finally, the two-phase friction loss is given by Equation (9.38):

$$\left(\frac{\Delta P_f}{L}\right)_{tp} = \phi_{LO}^2\left(\frac{\Delta P_f}{L}\right)_{LO} = 19.22 \times 0.001136 \cong 0.022 \text{ psi/ft}$$

(c) For the Friedel method, we need only recalculate ϕ_{LO}^2 according to Equation (9.45). The parameters E, F, H, Fr, and We are first calculated using Equations (9.47)–(9.52).

$$E = (1-x)^2 + x^2(\mu_V/\mu_L)^n(\rho_L/\rho_V)$$

$$= (0.8)^2 + (0.2)^2(0.00885/0.177)^{0.2314}(38.94/0.4787)$$

$$E = 2.2668$$

$$F = x^{0.78}(1-x)^{0.24} = (0.2)^{0.78}(0.8)^{0.24} = 0.2701$$

$$H = (\rho_L/\rho_V)^{0.91}(\mu_V/\mu_L)^{0.19}(1-\mu_V/\mu_L)^{0.7}$$

$$= (38.94/0.4787)^{0.91}(0.00885/0.177)^{0.19}(1-0.00885/0.177)^{0.7}$$

$$H = 29.896$$

$$\rho_{tp} = [x/\rho_V + (1-x)/\rho_L]^{-1} = [0.2/0.4787 + 0.8/38.94]^{-1}$$

$$\rho_{tp} = 2.2813 \text{ lbm/ft}^3$$

$$Fr = \frac{G^2}{gD_i\rho_{tp}^2} = \frac{(547,846)^2}{4.17\times 10^8 \times 0.835(2.2813)^2} = 165.63$$

The surface tension must be converted to English units for consistency.

$$\sigma = 11.4 \text{ dyne/cm} \times 6.8523\times 10^{-5}\left(\frac{\text{lbf /ft}}{\text{dyne/cm}}\right) = 7.812\times 10^{-4}\text{lbf /ft}$$

$$We = \frac{G^2D_i}{g_c\rho_{tp}\sigma} = \frac{(547,846)^2\times 0.835}{4.17\times 10^8\times 2.2813\times 7.812\times 10^{-4}}$$

$$We = 337,227$$

$$\phi_{LO}^2 = E + \frac{3.24FH}{Fr^{0.045}We^{0.035}}$$

$$= 2.2668 + \frac{3.24\times 0.2701\times 29.896}{(165.63)^{0.045}(337,227)^{0.035}}$$

$$\phi_{LO}^2 = 15.58$$

The two-phase friction loss is calculated using the value of $(\Delta P_f/L)_{LO}$ found in part (b):

$$\left(\frac{\Delta P_f}{L}\right)_{tp} = \phi_{LO}^2\left(\frac{\Delta P_f}{L}\right)_{LO} = 15.58\times 0.001136 \cong 0.018 \text{ psi/ft}$$

(d) For the MSH method, the two-phase multiplier is calculated using Equation (9.53).

$$\phi_{LO}^2 = Y^2x^3 + \left[1+2x(Y^2-1)\right](1-x)^{1/3}$$

$$= (6.38)^2(0.2)^3 + \left[1+2\times 0.2((6.38)^2-1)\right](1-0.2)^{1/3}$$

$$\phi_{LO}^2 \cong 16.00$$

The two-phase friction loss is again given by Equation (9.38):

$$\left(\frac{\Delta P_f}{L}\right)_{tp} = \phi_{LO}^2\left(\frac{\Delta P_f}{L}\right)_{LO} = 16.00\times 0.001136 \cong 0.018 \text{ psi/ft}$$

The predictions of the Friedel and MSH correlations are in very close agreement, while the Chisholm method predicts a value about 20% higher and the Lockhart-Martinelli prediction is about 80% higher.

9.4.3 Void Fraction and Two-Phase Density

Void Fraction

The void fraction, ϵ_V, is the fraction of the total volume that is occupied by the vapor phase. Since the two-phase density is the total mass (vapor and liquid) of fluid divided by the total volume of fluid, it follows that:

$$\rho_{tp} = \epsilon_V\rho_V + (1-\epsilon_V)\,\rho_L \tag{9.54}$$

The void fraction is also equal to the fractional area of the conduit cross-section occupied by the vapor. This alternate interpretation is helpful in relating the void fraction to the vapor weight fraction, or quality, x. If we denote the average vapor velocity by U_V, then the mass flux of vapor is:

$$\rho_V U_V = \dot{m}_V / A_V \tag{9.55}$$

where A_V is the cross-sectional area occupied by the vapor phase. Since $x = \dot{m}_V/\dot{m}$, it follows that:

$$x/\rho_V U_V = (\dot{m}_V/\dot{m})(A_V/\dot{m}_V) = A_V/\dot{m} \tag{9.56}$$

Similarly, for the liquid phase:

$$(1-x)/\rho_L U_L = A_L/\dot{m} \tag{9.57}$$

where A_L is the cross-sectional area occupied by the liquid. Using Equations (9.56) and (9.57) we obtain the void fraction as follows:

$$\frac{x/\rho_V U_V}{x/\rho_V U_V + (1-x)/\rho_L U_L} = \frac{A_V/\dot{m}}{A_V/\dot{m} + A_L/\dot{m}} = \frac{A_V}{A_V + A_L} = \epsilon_V \tag{9.58}$$

Rearranging this equation yields:

$$\epsilon_V = \frac{x}{x + SR(1-x)\rho_V/\rho_L} \tag{9.59}$$

where $SR \equiv U_V / U_L$ is the slip ratio. A flow model or empirical correlation is required to evaluate SR.

Homogeneous Flow Model

If homogeneous flow is assumed, both phases have the same velocity and $SR = 1$. Therefore,

$$\epsilon_V = \frac{x}{x + (1-x)\rho_V/\rho_L} \tag{9.60}$$

Substituting this result in Equation (9.54) and rearranging leads to the following expression for the two-phase density:

$$\rho_{tp} = [x/\rho_V + (1-x)/\rho_L]^{-1}$$

This equation was given previously in connection with the Friedel correlation as Equation (9.51).

The homogeneous model generally gives reliable results for the mist flow and bubbly flow regimes where the velocities of the two phases do not differ greatly. For other flow regimes, it tends to overestimate the void fraction, which causes the two-phase density to be underestimated. One of the following methods should be used in these cases.

Lockhart–Martinelli Correlation

In their classic paper, Lockhart and Martinelli [16] presented a single correlation for void fraction covering all four laminar–turbulent combinations. Their graphical correlation can be represented analytically by the following equation:

$$\epsilon_V = \frac{\phi_L - 1}{\phi_L} \tag{9.61}$$

where ϕ_L is calculated for the turbulent-turbulent case using Equation (9.28). Substituting the above expression for ϵ_V in Equation (9.54) yields the following equation for the two-phase density:

$$\rho_{tp} = \frac{\rho_L + (\phi_L - 1)\rho_V}{\phi_L} \tag{9.62}$$

The Chisholm Correlation

Chisholm [24] presented the following simple correlation for the slip ratio:

$$\begin{aligned} SR &= (\rho_L/\rho_{hom})^{0.5} \quad \text{for} \quad X > 1 \\ &= (\rho_L/\rho_V)^{0.25} \quad \text{for} \quad X \leq 1 \end{aligned} \tag{9.63}$$

where X is the Lockhart–Martinelli parameter and ρ_{hom} is the homogeneous two-phase density given by Equation (9.51). This correlation is valid for flow in both vertical and horizontal tubes over a wide range of conditions, including void fractions from zero to unity. Chisholm states that in a test of 14 methods it proved to be the best for density prediction.

The CISE Correlation

The CISE correlation was devised by Premoli et al. [25] so as to provide the correct asymptotic behavior for the slip ratio as various fluid properties and flow parameters approach their theoretical limits. The correlation is based on experimental data for upward flow in vertical channels. The equation for the slip ratio is:

$$SR = 1 + E_1\left[\left(\frac{y}{1+yE_2}\right) + yE_2\right]^{0.5} \tag{9.64}$$

where

$$y = \frac{(\epsilon_V)_{hom}}{1-(\epsilon_V)_{hom}} \tag{9.65}$$

$$(\epsilon_V)_{hom} = \text{homogeneous void fraction from Equation (9.60)}$$

$$E_1 = 1.578\ Re_{LO}^{-0.19}(\rho_L/\rho_V)^{0.22} \tag{9.66}$$

$$E_2 = 0.0273\ We_{LO}Re_{LO}^{-0.51}(\rho_L/\rho_V)^{-0.08} \tag{9.67}$$

As the subscripts indicate, both the Reynolds and Weber numbers in this correlation are computed using the total mass flux, G, and physical properties of the liquid phase.

In the article by Premoli et al. [25], the last term in square brackets in Equation (9.64) was incorrectly printed with a minus sign in place of the plus sign, and this typographical error has been repeated many times in the subsequent literature. The plus sign is required on theoretical grounds in order for the slip ratio to exhibit the correct asymptotic behavior, and to prevent the square root of negative numbers from occurring at low Reynolds numbers or high Weber numbers. It is also needed to reproduce the graphical results presented in Ref. [25].

Example 9.6

Calculate the void fraction and two-phase density for the reboiler return line of Example 9.5 using:

(a) The homogeneous flow model.
(b) The Lockhart–Martinelli correlation.
(c) The Chisholm correlation.
(d) The CISE correlation.

Solution

(a) For the homogeneous flow model, the void fraction is given by Equation (9.60):

$$\epsilon_V = \frac{x}{x+(1-x)\rho_V/\rho_L} = \frac{0.2}{0.2+0.8\ (0.4787/38.94)}$$

$$\epsilon_V = 0.9531$$

The two-phase density can be calculated using either Equation (9.54) or (9.51). The latter was used in Example 9.5 to obtain $\rho_{tp} = 2.2813$ lbm/ft^3. Using Equation (9.54) we obtain:

$$\rho_{tp} = \epsilon_V\rho_V + (1-\epsilon_V)\rho_L = 0.9531 \times 0.4787 + (1-0.9531)\times 38.94$$

$$\rho_{tp} = 2.2825 \cong 2.28\ \text{lbm/ft}^3$$

The difference between the two values of ρ_{tp} is due to round-off error in the value of ϵ_V, and is negligible in the present context.

(b) A value of $\phi_L^2 = 43.07$ was calculated in Example 9.5, giving $\phi_L = 6.5628$. The void fraction is computed from Equation (9.61):

$$\epsilon_V = (\phi_L - 1)/\phi_L = (6.5628-1)/6.5628 = 0.8476$$

The two-phase density is calculated using Equation (9.62):

$$\rho_{tp} = \frac{\rho_L + (\phi_L-1)\rho_V}{\phi_L} = \frac{38.94+(6.5628-1)\times 0.4787}{6.5628}$$

$$\rho_{tp} = 6.3392 \cong 6.34\ \text{lbm/ft}^3$$

(c) From Example 9.5, the Lockhart–Martinelli parameter is $X_{tt} = 0.521 < 1$. Therefore, Equation (9.63) gives:

$$SR = (\rho_L/\rho_V)^{0.25} = (38.94/0.4787)^{0.25} = 3.0032$$

Substitution into Equation (9.59) gives the void fraction:

$$\epsilon_V = \frac{x}{x + SR(1-x)\rho_V/\rho_L} = \frac{0.2}{0.2 + 3.0032 \times 0.8 \times 0.4787/38.94}$$

$$\epsilon_V = 0.8713$$

The two-phase density is obtained from Equation (9.54).

$$\rho_{tp} = \epsilon_V \rho_V + (1-\epsilon_V)\rho_L = 0.8713 \times 0.4787 + (1 - 0.8713) \times 38.94$$

$$\rho_{tp} = 5.4287 \cong 5.43 \text{ lbm/ft}^3$$

(d) The following values are obtained from Example 9.5:

$$D_i = 0.835 \text{ ft}$$

$$\sigma = 7.812 \times 10^{-4} \text{ lbf/ft}$$

$$G = 547,846 \text{ lbm/ h·ft}^2$$

$$Re_{LO} = 1,068,405$$

The Weber number is calculated as follows:

$$We_{LO} = \frac{G^2 D_i}{g_c \rho_L \sigma} = \frac{(547,846)^2 \times 0.835}{4.17 \times 10^8 \times 38.94 \times 7.812 \times 10^{-4}} = 19,756$$

The parameters in the CISE correlation are calculated using Equations (9.65) to (9.67). From part (a), the homogeneous void fraction is $(\epsilon_V)_{hom} = 0.9531$.

$$y = \frac{(\epsilon_V)_{hom}}{1 - (\epsilon_V)_{hom}} = \frac{0.9531}{1 - 0.9531} = 20.322$$

$$E_1 = 1.578\, Re_{LO}^{-0.19}(\rho_L/\rho_V)^{0.22} = 1.578\,(1,068,405)^{-0.19}(38.94/0.4787)^{0.22}$$

$$E_1 = 0.2973$$

$$E_2 = 0.0273\, We_{LO} Re_{LO}^{-0.51}(\rho_L/\rho_V)^{-0.08}$$

$$= 0.0273 \times 19,756(1,068,405)^{-0.51}(38.94/0.4787)^{-0.08}$$

$$E_2 = 0.3194$$

The slip ratio is calculated from Equation (9.64):

$$SR = 1 + E_1\left[\left(\frac{y}{1 + yE_2}\right) + yE_2\right]^{0.5}$$

$$= 1 + 0.2973\left[\frac{20.322}{1 + 20.322 \times 0.3194} + 20.322 \times 0.3194\right]^{0.5}$$

$$SR = 1.9013$$

The void fraction is calculated using Equation (9.59):

$$\epsilon_V = \frac{x}{x + SR(1-x)\rho_V/\rho_L} = \frac{0.2}{0.2 + 1.9013 \times 0.8 \times 0.4787/38.94}$$

$$\epsilon_V = 0.9145$$

Finally, the two-phase density is calculated using Equation (9.54).

$$\rho_{tp} = \epsilon_V \rho_V + (1-\epsilon_V)\rho_L = 0.9145 \times 0.4787 + (1 - 0.9145) \times 38.94$$

$$\rho_{tp} = 3.7671 \cong 3.77 \text{ lbm/ft}^3$$

Notice that while the void fractions predicted by the four methods are within about 12% of one another, the two-phase densities differ by large percentages.

9.4.4 Other Losses

Pressure losses due to valves, fittings, expansions, and contractions can be calculated using flow resistance coefficients that specify the number of velocity heads corresponding to the loss in each element of a pipeline. Methods for estimating these coefficients for two-phase flow can be found in Refs. [2,8,24,26].

A more expedient approach is commonly used in design work, namely, the equivalent length concept. The equivalent length of a piping element is the length of straight pipe that generates the same pressure loss as the given valve, fitting, etc. Tables of equivalent lengths used for single-phase flow are assumed to be valid for two-phase flow as well, and are used to determine the effective length of a pipeline in the usual manner by summing the equivalent lengths of all fittings, etc. and the lengths of all straight sections of pipe. The effective pipe length is then multiplied by the two-phase pressure gradient to obtain the total pressure loss. Although equivalent lengths for two-phase flow tend to be somewhat higher than for single-phase flow [26], the longstanding use of this method for industrial design work indicates that it is generally reliable, at least in the context of two-phase flow calculations, which, as previously noted, involve a significant level of uncertainty. Computational details are given in Chapter 10.

9.4.5 Recommendations

The MSH correlation is recommended for calculating two-phase pressure drop, while the Chisholm method is recommended for calculating two-phase density. These methods are easy to use, applicable over a wide range of conditions, and are among the more reliable of their respective genres.

9.5 Convective Boiling in Tubes

9.5.1 Boiling Regimes in a Vertical Tube

Figure 9.3 depicts the variation in flow regime and heat-transfer coefficient for convective boiling in a vertical tube. The fluid enters the tube as a subcooled liquid, and initially heat is transferred by convection alone. With the onset of boiling and the

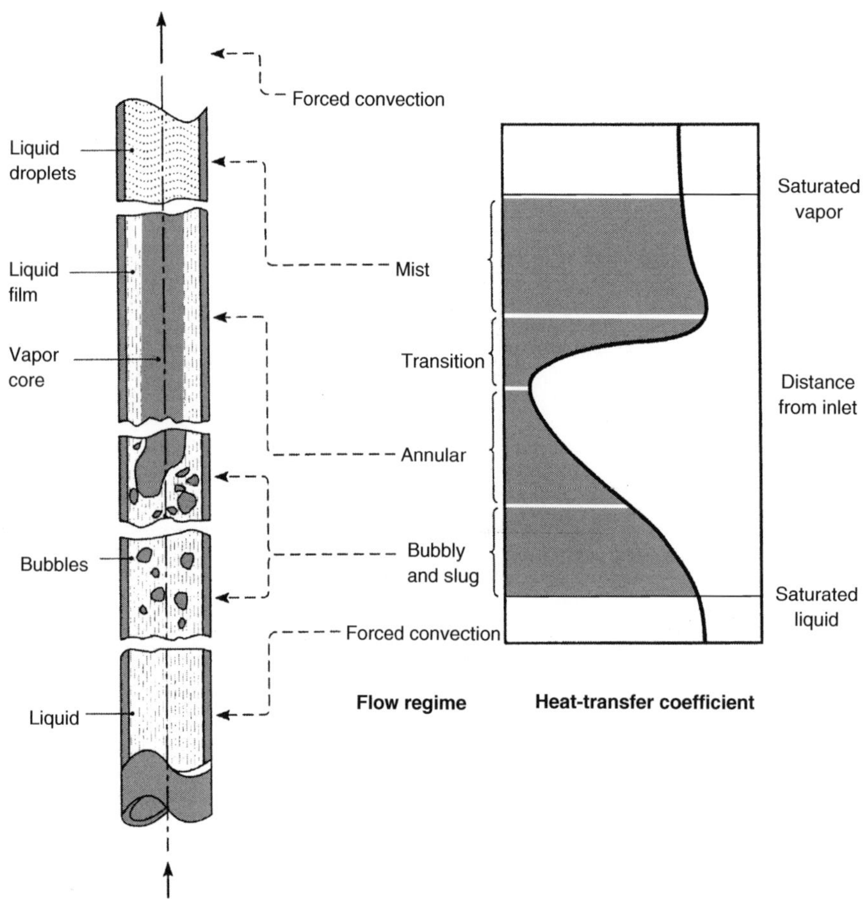

FIGURE 9.3 Convective boiling in a vertical tube (Source: Ref. [1]).

establishment of the bubbly flow regime, the heat-transfer coefficient starts to rise sharply as the nucleate boiling mechanism contributes to the heat transfer. As the vapor fraction increases, the flow regime changes from bubbly to slug flow, and then to annular flow (the churn flow regime is omitted here) while the heat-transfer coefficient continues to increase. Eventually, the amount of liquid is reduced to the point where dry spots begin to appear on the tube wall, and the heat-transfer coefficient begins to decrease. It continues to decrease until the tube wall is completely dry and all remaining liquid is in the form of droplets entrained in the vapor phase (mist flow regime). The heat-transfer coefficient remains relatively constant as the droplets are gradually vaporized and the vapor becomes superheated. In the final stages, heat is transferred solely by gas-phase convection.

In vertical thermosyphon reboilers, which employ the configuration shown in Figure 9.3, the vapor fraction is kept sufficiently low that the mist flow regime does not occur and the sharp drop in heat-transfer coefficient is avoided. Nevertheless, the rate of heat transfer can vary greatly over the length of the reboiler, and an incremental analysis is therefore required to accurately predict performance.

9.5.2 The Chen Correlation

The correlation developed by Chen [27] has been the most widely used method for calculating heat-transfer coefficients in convective boiling. Chen assumed that the convective and nucleate boiling contributions to heat transfer were additive. However, it is known that convection tends to suppress nucleate boiling. Chen attributed this effect to the steepening of the temperature gradient near the wall with increasing flow rate, which reduces the effective temperature difference between the tube wall and bubbles growing outward from the wall. He therefore introduced a suppression factor, S_{CH}, that is the ratio of the effective temperature difference for bubble growth to the overall temperature difference, $T_w - T_{sat}$.

Chen also expressed the convective heat-transfer coefficient in terms of the coefficient, h_L, for the liquid flowing alone in the tube. Thus,

$$h_{fc} = F(X_{tt})h_L \tag{9.68}$$

where

h_{fc} = two-phase convective heat-transfer coefficient
h_L = convective heat-transfer coefficient for liquid phase flowing alone
$F(X_{tt})$ = function of Lockhart–Martinelli parameter

Chen's graphs for S_{CH} and $F(X_{tt})$, which were based on experimental data, have been replaced by curve-fit equations to obtain the following form of the Chen correlation (2):

$$h_b = S_{CH}h_{nb} + F(X_{tt})h_L \tag{9.69}$$

where

h_b = convective boiling heat-transfer coefficient
h_{nb} = nucleate boiling heat-transfer coefficient

$$S_{CH} = \left(1 + 2.53 \times 10^{-6} Re^{1.17}\right)^{-1} \tag{9.70}$$

$$\begin{aligned} F(X_{tt}) &= 2.35\left(X_{tt}^{-1} + 0.213\right)^{0.736} \quad (X_{tt} < 10) \\ &= 1.0 \quad (X_{tt} \geq 10) \end{aligned} \tag{9.71}$$

$$Re = Re_L[F(X_{tt})]^{1.25} \tag{9.72}$$

The heat flux is given by:

$$\hat{q} = h_b(T_w - T_{sat}) \tag{9.73}$$

Although Chen developed his correlation for saturated boiling, it has been applied to subcooled boiling as well. The latter occurs when the bulk liquid temperature is below the saturation temperature, but the tube-wall temperature is high enough for vapor bubbles to form and grow. In this case, the suppression factor is computed by setting $Re = Re_L$ in Equation (9.70), and the heat flux is calculated as follows [2]:

$$\hat{q} = S_{CH}h_{nb}(T_w - T_{sat}) + h_L(T_w - T_b) \tag{9.74}$$

where T_b is the bulk liquid temperature.

The convective coefficient, h_L, can be calculated using any appropriate correlation for forced convection in pipes and ducts, such as the Seider–Tate equation (Chapter 2). However, Chen [27] used the Dittus–Boelter equation, a predecessor of the Seider–Tate equation for turbulent flow, as follows:

$$h_L = 0.023(k_L/D_i)Re_L^{0.8}Pr_L^{0.4} \tag{9.75}$$

Note that the Reynolds number in this equation is calculated using the flow rate of the liquid phase alone.

Chen used the Forster–Zuber correlation for the nucleate boiling heat-transfer coefficient. In order to obtain a conservative estimate for mixtures, however, Palen [4] suggested using the Mostinski correlation together with the mixture correction factor given by Equation (9.17).

Although the Chen correlation was developed using a limited amount of experimental data (for water, methanol, benzene, cyclohexane, pentane, and heptane, all in vertical tubes), it has a sound physical basis and reduces correctly in the limiting cases of zero flow rate ($S_{CH} = 1$), infinite flow rate ($S_{CH} \rightarrow 0$), and zero vapor fraction [$F(X_{tt}) = 1.0$], characteristics not always found in later correlations. It has also been widely used for many years, and is still the standard of comparison for new correlations.

9.5.3 The Gungor–Winterton correlation

The correlation of Gungor and Winterton [28] is similar in form to the Chen correlation in that the nucleate boiling and convection terms are additive. A nucleate boiling suppression factor is included, along with a convective enhancement factor, E_{GW}, to account for the effect of boiling on the convective heat transfer. The correlation is as follows:

$$h_b = S_{GW} h_{nb} + E_{GW} h_L \tag{9.76}$$

$$S_{GW} = \left(1 + 1.15 \times 10^{-6} E_{GW}^2 Re_L^{1.17}\right)^{-1} \tag{9.77}$$

$$E_{GW} = 1 + 24,000 Bo^{1.16} + 1.37 X_{tt}^{-0.86} \tag{9.78}$$

The boiling number, Bo, is defined by the following equation:

$$Bo = \frac{\hat{q}}{\lambda G} \tag{9.79}$$

For horizontal tubes and $Fr_{LO} < 0.05$, S_{GW} is to be multiplied by $Fr_{LO}^{0.5}$, and E_{GW} is to be multiplied by $Fr_{LO}^{(0.1-2Fr_{LO})}$. The nucleate boiling heat-transfer coefficient is calculated using the Cooper correlation, and the Dittus–Boelter equation is used for the forced convection coefficient.

For subcooled boiling, Equation (9.74) is used with S_{CH} replaced by S_{GW}. Liquid properties are evaluated at T_{sat} for saturated boiling and at T_b for subcooled boiling. The correlation was developed using a database consisting of approximately 3700 data points for water, ethylene glycol, and five refrigerants.

Due to the boiling number term in Equation (9.79), the enhancement factor does not reduce to unity when the vapor fraction is zero. Nevertheless, the correlation provides a relatively good representation of the data upon which it is based, particularly for refrigerants in the saturated boiling regime. The constants in Equations (9.77) and (9.78) were determined by a complex iterative regression procedure, and the nucleate and convective terms are strongly coupled through Equation (9.77). Therefore, it is not advisable to substitute another nucleate boiling correlation for the Cooper correlation in this method.

9.5.4 The Liu–Winterton Correlation

The correlation developed by Liu and Winterton [29] is described by the following equations:

$$h_b = \left[(S_{LW} h_{nb})^2 + (E_{LW} h_L)^2\right]^{1/2} \tag{9.80}$$

$$S_{LW} = \left[1 + 0.055 E_{LW}^{0.1} Re_L^{0.16}\right]^{-1} \tag{9.81}$$

$$E_{LW} = [1 + x Pr_L (\rho_L - \rho_V)/\rho_V]^{0.35} \tag{9.82}$$

For horizontal tubes, the low Froude number corrections are applied to S_{LW} and E_{LW} as in the Gungor–Winterton method. The Cooper and Dittus–Boelter correlations are used to calculate h_{nb} and h_L. Notice that the enhancement factor, E_{LW}, correctly reduces to unity when the vapor fraction, x, is zero.

Equation (9.80) is a special case of the general formula:

$$h_b = \left[(a h_{nb})^m + (b h_{fc})^m\right]^{1/m} \tag{9.83}$$

where $1 \leq m < \infty$. Formulas of this type provide a smooth transition between two limiting cases, in this instance between forced convection and nucleate boiling. The abruptness of the transition is governed by the size of the exponent, with larger values of m resulting in more abrupt transitions. This approach has been used to develop many very successful correlations for heat, mass, and momentum transport, as exemplified by the Churchill correlations presented in Chapter 2.

The Liu–Winterton correlation was developed using the same database as used for the Gungor–Winterton correlation. It provides a significantly better fit to the data than the latter correlation in both the saturated and subcooled boiling regimes.

9.5.5 Other Correlations

More complex correlations have been developed by Shah [30], Kandlikar [31], Steiner and Taborek [32], and Kattan et al. [33]. The Shah correlation appears to be no more reliable than the simpler methods given above. The Kandlikar correlation provides a good representation of test data, but it contains a fluid-specific parameter, values of which are given only for a small number of fluids. The Steiner-Taborek correlation takes the form of Equation (9.83) with $a = b = 1$ and $m = 3$. It was developed using a database containing over 13,000 data points for water, three alcohols, four hydrocarbons, four refrigerants, ammonia, nitrogen, and helium. All data are for convective boiling in vertical tubes. Although this method is among the best currently available in the open literature, the overall computational procedure is rather complicated and to some extent fluid specific. The correlation of Kattan et al. is based on data for five refrigerants flowing in horizontal tubes, and is restricted to vapor fractions above 0.15. Equation (9.83) is used with $a = b = 1$ and $m = 3$, and the Cooper correlation is used to calculate the nucleate boiling coefficient. Different models are used to calculate the heat-transfer coefficient in each of four flow regimes: annular, stratified, stratified-wavy, and annular flow with partial dryout at high vapor fractions.

All convective boiling correlations are developed and tested with experimental data for pure components, or in some cases, binary mixtures. Even for pure component boiling, they are not highly reliable when applied beyond the range of data used in their development. For the complex, often very non-ideal mixtures encountered in chemical and petroleum processing, their performance is still more problematic. Therefore, a relatively conservative approach to design, such as the use of Palen's method for nucleate boiling of mixtures in conjunction with Chen's method, seems warranted. Note, however, that while this procedure will provide a conservative estimate for the nucleate boiling coefficient, the end result may not be conservative if the convective contribution or the suppression factor is overestimated by Chen's method. For pure components, and refrigerants in particular, the Liu–Winterton correlation is likely to be more reliable than the Chen correlation.

Example 9.7

The fluid of Example 9.1 boils while flowing vertically upward through a 1-in., 14 BWG tube (ID = 2.12 cm) with a mass flux of 300 kg/m^2 · s. At a point in the tube where the wall temperature is 453.7 K, the pressure is 310.3 kPa and the quality is 0.2, calculate the heat-transfer coefficient using:

(a) Chen's method.
(b) Chen's method with the Mostinski correlation for nucleate boiling.
(c) The Gungor–Winterton correlation.
(d) The Liu-Winterton correlation.

Solution

(a) The first step is to calculate the heat-transfer coefficient for forced convection using Equation (9.75). The Reynolds number is based on the flow rate of the liquid phase alone, and fluid properties are obtained from Example 9.1.

$$Re_L = \frac{D_i G_L}{\mu_L} = \frac{0.0212 \times 0.8 \times 300}{156 \times 10^{-6}} = 32,615$$

$$Pr_L = \frac{C_{P,L}\mu_L}{k_L} = \frac{2730 \times 156 \times 10^{-6}}{0.086} = 4.9521$$

$$h_L = 0.023(k_L/D_i)\mathrm{Re}_L^{0.8}\mathrm{Pr}_L^{0.4}$$

$$= 0.023(0.086/0.0212)(32,615)^{0.8}(4.9521)^{0.4}$$

$$h_L = 722\ \mathrm{W/m^2 \cdot K}$$

Next, the Lockhart–Martinelli parameter is calculated using Equation (9.37).

$$X_{tt} = \left(\frac{1-x}{x}\right)^{0.9}(\rho_V/\rho_L)^{0.5}(\mu_L/\mu_V)^{0.1}$$

$$= (0.8/0.2)^{0.9}(18.09/567)^{0.5}(156 \times 10^{-6}/7.11 \times 10^{-6})^{0.1}$$

$$X_{tt} = 0.847$$

Equation (9.71) is used to calculate $F(X_{tt})$. Since $X_{tt} < 10$, we have:

$$F(X_{tt}) = 2.35\left(X_{tt}^{-1} + 0.213\right)^{0.736} = 2.35\left[(0.847)^{-1} + 0.213\right]^{0.736}$$

$$F(X_{tt}) = 3.00$$

Equations (9.72) and (9.70) are used to calculate the suppression factor.

$$Re = Re_L[F(X_{tt})]^{1.25} = 32,615(3.00)^{1.25} = 128,771$$

$$S_{CH} = \left(1 + 2.53 \times 10^{-6} Re^{1.17}\right)^{-1} = \left[1 + 2.53 \times 10^{-6}(128,771)^{1.17}\right]^{-1}$$

$$S_{CH} = 0.2935$$

In Example 9.1, the Forster–Zuber correlation was used to obtain $h_{nb} = 5{,}512\ \text{W/m}^2 \cdot \text{K}$, and this value is still valid here. Therefore, h_b can be found by substituting in Equation (9.69):

$$h_b = S_{CH}h_{nb} + F(X_{tt})h_L = 0.2935 \times 5512 + 3.00 \times 722$$

$$h_b = 3784\ \text{W/m}^2 \cdot \text{K}$$

(b) The values of h_L, $F(X_{tt})$, and S_{CH} are the same as in part (a). The Mostinski correlation in the form of Equation (9.3) was used in Example 9.1 to calculate the nucleate boiling heat-transfer coefficient. However, Equation (9.3) was derived from Equation (9.2) by substituting $\hat{q} = h_{nb}\Delta T_e$, which for convective boiling is replaced by $\hat{q} = h_b \Delta T_e$. Therefore, the calculation must be repeated using Equation (9.2b).

$$h_{nb} = 0.00417 P_c^{0.69} \hat{q}^{0.7} F_P$$

Using Palen's recommendation for the pressure correction factor, we obtain $F_P = 1.3375$ from Example 9.1, part (c). Also, $P_c = 2500$ kPa as given in Example 9.1. Thus,

$$h_{nb} = 0.00417(2500)^{0.69} \hat{q}^{0.7} \times 1.3375$$

$$h_{nb} = 1.2331 \hat{q}^{0.7}$$

Substituting in Equation (9.69) gives:

$$h_b = S_{CH}h_{nb} + F(X_{tt})h_L$$

$$= 0.2935 \times 1.2331 \hat{q}^{0.7} + 3.00 \times 722$$

$$h_b = 0.3619 \hat{q}^{0.7} + 2166$$

Since $\Delta T_e = 16.2$ K from Example 9.1,

$$\hat{q} = h_b \Delta T_e = \left(0.3619 \hat{q}^{0.7} + 2166\right) \times 16.2$$

$$\hat{q} = 5.8628 \hat{q}^{0.7} + 35,089.2$$

$$\hat{q} - 5.8628 \hat{q}^{0.7} - 35,089.2 = 0$$

Using the nonlinear equation solver on a TI calculator, the solution to this equation is found to be $\hat{q} = 45,826\ \text{W/m}^2$. The heat-transfer coefficient is obtained from this value as follows:

$$h_b = \frac{\hat{q}}{\Delta T_e} = \frac{45,826}{16.2} = 2829\ \text{W/m}^2 \cdot \text{K}$$

(c) The following results from part (a) apply here as well:

$$Re_L = 32,615$$

$$h_L = 722\ \text{W/m}^2 \cdot \text{k}$$

$$X_{tt} = 0.847$$

Also, from Example 9.1, the latent heat of vaporization is 272,000 J/kg.
The boiling number is given by Equation (9.79):

$$Bo = \frac{\hat{q}}{\lambda G} = \frac{\hat{q}}{272,000 \times 300} = 1.2255 \times 10^{-8} \hat{q}$$

The convective enhancement factor is computed by substituting in Equation (9.78).

$$E_{GW} = 1 + 24{,}000Bo^{1.16} + 1.37X_{tt}^{-0.86}$$

$$= 1 + 24{,}000\left(1.2255 \times 10^{-8}\hat{q}\right)^{1.16} + 1.37(0.847)^{-0.86}$$

$$= 1 + 1.5946 \times 10^{-5}\hat{q}^{1.16} + 1.5803$$

$$E_{GW} = 2.5803 + 1.5946 \times 10^{-5}\hat{q}^{1.16} \quad \text{(i)}$$

The suppression factor is computed using Equation (9.77).

$$S_{GW} = \left(1 + 1.15 \times 10^{-6}E_{GW}^2 Re_L^{1.17}\right)^{-1}$$

$$= \left(1 + 1.15 \times 10^{-6}E_{GW}^2 (32{,}615)^{1.17}\right)^{-1}$$

$$S_{GW} = \left(1 + 0.2195E_{GW}^2\right)^{-1} \quad \text{(ii)}$$

The Cooper nucleate boiling correlation in SI units is given by Equation (9.6b):

$$h_{nb} = 55\hat{q}^{0.67}P_r^{0.12}(-\log_{10}P_r)^{-0.55}M^{-0.5}$$

From Example 9.1, $P_r = 0.1217$ and the molecular weight is 110.37. Substitution gives:

$$h_{nb} = 55\hat{q}^{0.67}(0.1217)^{0.12}(-\log_{10}0.1217)^{-0.55}(110.37)^{-0.5}$$

$$h_{nb} = 4.2704\hat{q}^{0.67}$$

Substituting in Equation (9.76) gives the following result for h_b:

$$h_b = S_{GW}h_{nb} + E_{GW}h_L$$

$$h_b = 4.2704S_{GW}\hat{q}^{0.67} + 722E_{GW}$$

The equation for the heat flux is obtained as follows:

$$\hat{q} = h_b \Delta T_e = h_b \times 16.2$$

$$\hat{q} = 69.1805S_{GW}\hat{q}^{0.67} + 11{,}696.4E_{GW}$$

$$\hat{q} - 69.1805S_{GW}\hat{q}^{0.67} - 11{,}696.4E_{GW} = 0 \quad \text{(iii)}$$

Equation (iii) is to be solved in conjunction with Equations (i) and (ii) for the heat flux. However, these equations have no real solution, as the reader can verify using the nonlinear equation solver on a TI calculator or by graphing the left side of Equation (iii). The crux of the problem is the term involving the boiling number in the enhancement factor, which, as previously noted, is not physically realistic. Therefore, the Gungor–Winterton correlation cannot be used to solve this problem.

(d) The following results from part (a) are applicable:

$$Re_L = 32{,}615 \quad h_L = 722\ \text{W/m}^2\cdot\text{K} \quad Pr_L = 4.9521$$

The enhancement and suppression factors are calculated from Equations (9.82) and (9.81).

$$E_{LW} = [1 + xPr_L(\rho_L - \rho_V)/\rho_V]^{0.35}$$

$$= [1 + 0.2 \times 4.9521(567 - 18.09)/18.09]^{0.35}$$

$$E_{LW} = 3.3284$$

$$S_{LW} = \left[1 + 0.055E_{LW}^{0.1}Re_L^{0.16}\right]^{-1}$$

$$= \left[1 + 0.055(3.3284)^{0.1}(32{,}615)^{0.16}\right]^{-1}$$

$$S_{LW} = 0.7535$$

From part (c), the Cooper correlation gives:

$$h_{nb} = 4.2704\hat{q}^{0.67}$$

Substituting in Equation (9.80) for the convective boiling coefficient gives:

$$h_b = \left\{(S_{LW}h_{nb})^2 + (E_{LW}h_L)^2\right\}^{1/2}$$

$$= \left\{\left(0.7535 \times 4.2704\hat{q}^{0.67}\right)^2 + (3.3284 \times 722)^2\right\}^{1/2}$$

$$h_b = \left\{10.3539\hat{q}^{1.34} + 5,774,913\right\}^{1/2}$$

Setting $h_b = \hat{q}/\Delta T_e$ gives the following nonlinear equation for the heat flux:

$$\frac{\hat{q}}{16.2} - \left(10.3539\hat{q}^{1.34} + 5,774,913\right)^{0.5} = 0$$

Using the nonlinear equation solver on a TI calculator, the solution is found to be $\hat{q} = 172,788\ \text{W/m}^2$. Thus, the heat-transfer coefficient is:

$$h_b = \hat{q}/\Delta T_e = 172,788/16.2 = 10,666\ \text{W/m}^2\cdot\text{K}$$

This value is much higher than the one obtained with the Chen correlation, primarily due to the difference in the nucleate boiling terms. Recall that in Example 9.1 the Cooper correlation predicted a much higher value for the nucleate boiling heat-transfer coefficient than either the Forster–Zuber or Mostinski correlation.

9.5.6 Critical Heat Flux

At low vapor fractions, the mechanism responsible for critical heat flux in convective boiling is essentially the same as in pool boiling. At high vapor fractions, the phenomenon is caused by dryout of the tube wall due to depletion of the liquid phase. The transition between the two mechanisms is gradual, however, and there is no clear demarcation between them. In horizontal tubes, dryout occurs at lower heat fluxes than in vertical tubes under the same conditions because the liquid preferentially accumulates on the bottom of the tube, allowing the upper portion of the tube wall to dry more readily.

Vertical Tubes

The majority of the work on critical heat flux in convective boiling has been done with water due to the importance of the problem in relation to nuclear reactor cooling systems. Two methods that have been tested on a variety of fluids for flow in vertical tubes are presented here. The following simple correlation for vertical thermosyphon reboilers was given by Palen [4]:

$$\hat{q}_c = 16,070\left(D^2/L\right)^{0.35} P_c^{0.61} P_r^{0.25}(1 - P_r) \tag{9.84a}$$

where

$\hat{q}_c$ = critical heat flux, Btu/h · ft^2
D = tube ID, ft
L = tube length, ft
P_r = reduced pressure in tube
P_c = critical pressure of fluid, psia

The corresponding relationship in SI units is:

$$\hat{q}_c = 23,660\left(D^2/L\right)^{0.35} P_c^{0.61} P_r^{0.25}(1 - P_r) \tag{9.84b}$$

where

$\hat{q}_c \propto \text{W/m}^2$
$D, L \propto \text{m}$
$P_c \propto \text{kPa}$

The correlation of Katto and Ohno [34] is considerably more complicated and rather cumbersome to state. The critical heat flux is given by the following equation:

$$\hat{q}_c = \hat{q}_o(1 + \Gamma\Delta H_{in}/\lambda) \tag{9.85}$$

where

$\hat{q}_o$ = basic critical heat flux
Γ = inlet subcooling parameter

ΔH_{in} = inlet enthalpy of subcooling
λ = latent heat of vaporization

The following five equations are used in determining the value of $\hat{q}_o$:

$$\frac{\hat{q}_{oA}}{G\lambda} = C_2\left(\frac{g_c\sigma\rho_L}{G^2L}\right)^{0.043}(L/D)^{-1} \tag{9.86}$$

$$\frac{\hat{q}_{oB}}{G\lambda} = 0.10(\rho_V/\rho_L)^{0.133}\left(\frac{g_c\sigma\rho_L}{G^2L}\right)^{1/3}(1+0.0031L/D)^{-1} \tag{9.87}$$

$$\frac{\hat{q}_{oC}}{G\lambda} = 0.098(\rho_V/\rho_L)^{0.133}\left(\frac{g_c\sigma\rho_L}{G^2L}\right)^{0.433}(L/D)^{0.27}(1+0.0031L/D)^{-1} \tag{9.88}$$

$$\frac{\hat{q}_{oD}}{G\lambda} = 0.0384(\rho_V/\rho_L)^{0.6}\left(\frac{g_c\sigma\rho_L}{G^2L}\right)^{0.173}\left\{1+0.28\left(\frac{g_c\sigma\rho_L}{G^2L}\right)^{0.233}(L/D)\right\}^{-1} \tag{9.89}$$

$$\frac{\hat{q}_{oE}}{G\lambda} = 0.234(\rho_V/\rho_L)^{0.513}\left(\frac{g_c\sigma\rho_L}{G^2L}\right)^{0.433}(L/D)^{0.27}(1+0.0031L/D)^{-1} \tag{9.90}$$

In the above equations, L is the heated length of the tube and D is the inside diameter. The constant, C_2, in Equation (9.86) is given by:

$$\begin{aligned} C_2 &= 0.25 && (L/D < 50) \\ &= 0.25 + 0.0009(L/D - 50) && (50 \le L/D \le 150) \\ &= 0.34 && (L/D > 150) \end{aligned} \tag{9.91}$$

The following three equations are used in determining the value of Γ:

$$\Gamma_A = \frac{1.043}{4C_2(g_c\sigma\rho_L/G^2L)^{0.043}} \tag{9.92}$$

$$\Gamma_B = \frac{(5/6)(0.0124 + D/L)}{(\rho_V/\rho_L)^{0.133}(g_c\sigma\rho_L/G^2L)^{1/3}} \tag{9.93}$$

$$\Gamma_C = \frac{1.12\left\{1.52(g_c\sigma\rho_L/G^2L)^{0.233} + D/L\right\}}{(\rho_V/\rho_L)^{0.6}(g_c\sigma\rho_L/G^2L)^{0.173}} \tag{9.94}$$

The values of $\hat{q}_o$ and Γ are determined as follows, depending on the value of ρ_V / ρ_L.
For $\rho_V / \rho_L \le 0.15$:

$$\begin{aligned} \hat{q}_o &= \hat{q}_{oA} && (\hat{q}_{oA} \le \hat{q}_{oB}) \\ &= \text{Min}(\hat{q}_{oB}, \hat{q}_{oC}) && (\hat{q}_{oA} > \hat{q}_{oB}) \end{aligned} \tag{9.95}$$

$$\Gamma = \text{Max}(\Gamma_A, \Gamma_B) \tag{9.96}$$

For $\rho_V / \rho_L > 0.15$:

$$\begin{aligned} \hat{q}_o &= \hat{q}_{oA} && (\hat{q}_{oA} \le \hat{q}_{oE}) \\ &= \text{Max}(\hat{q}_{oD}, \hat{q}_{oE}) && (\hat{q}_{oA} > \hat{q}_{oE}) \end{aligned} \tag{9.97}$$

$$\begin{aligned} \Gamma &= \Gamma_A && (\Gamma_A \ge \Gamma_B) \\ &= \text{Min}(\Gamma_B, \Gamma_C) && (\Gamma_A < \Gamma_B) \end{aligned} \tag{9.98}$$

Horizontal Tubes

For flow in horizontal tubes, the dimensionless correlation of Merilo [35] is recommended by Hewitt et al. [2].

$$\frac{\hat{q}_c}{G\lambda} = 575\gamma_H^{-0.34}(L/D)^{-0.511}\left(\frac{\rho_L - \rho_V}{\rho_V}\right)^{1.27}(1+\Delta H_{in}/\lambda)^{1.64} \tag{9.99}$$

where

$$\gamma_H = \left(\frac{GD}{\mu_L}\right)\left(\frac{\mu_L^2}{g_c\sigma D\rho_L}\right)^{-1.58}\left[\frac{(\rho_L-\rho_V)gD^2}{g_c\sigma}\right]^{-1.05}\left(\frac{\mu_L}{\mu_V}\right)^{6.41} \quad (9.100)$$

The correlation is based on 605 data points for water and Freon-12 covering the ranges $5.3 \leq D \leq 19.1$ mm, $700 \leq G \leq 8100$ kg/s · m^2, $13 \leq \rho_L/\rho_V \leq 21$.

Example 9.8

The fluid of Example 9.1 boils while flowing through a vertical 1-in. 14 BWG tube (ID = 2.12 cm) with a mass flux of 300 kg/s · m^2. The tube length is 10 ft (3.048 m) and the inlet subcooling of the fluid is 23,260 J/kg. The tube is uniformly heated over its entire length, and the average internal pressure is 310 kPa. Estimate the critical heat flux using:

(a) The Palen correlation.
(b) The Katto-Ohno correlation.

Solution

(a) From Example 9.1, the critical pressure is 2550 kPa. Hence, the reduced pressure is:

$$P_r = P/P_c = 310/2550 \cong 0.12157$$

Equation (9.84b) is used to calculate the critical heat flux.

$$\hat{q}_c = 23,660\,(D^2/L)^{0.35}P_c^{0.61}P_r^{0.25}(1-P_r)$$
$$= 23,660\left[(0.0212)^2/3.048\right]^{0.35}(2550)^{0.61}(0.12157)^{0.25}(1-0.12157)$$
$$\hat{q}_c = 66,980\ \text{W/m}^2$$

Notice that this method does not take into account either the inlet subcooling or the flow rate of the fluid.

(b) From Example 9.1, the liquid and vapor densities are 567 kg/m^3 and 18.09 kg/m^3, respectively. Hence,

$$\rho_V/\rho_L = 18.09/567 = 0.031905$$

Since this value is less than 0.15, Equations (9.95) and (9.96) are used to evaluate $\hat{q}_o$ and Γ. We first calculate $\hat{q}_{oA}$ and $\hat{q}_{oB}$ using Equations (9.86) and (9.87). Since $L/D = 3.048/0.0212 = 143.8$, Equation (9.91) gives the value of C_2 as:

$$C_2 = 0.25 + 0.0009(143.8 - 50) = 0.334$$

From Example 9.1, $\sigma = 8.2 \times 10^{-3}$ N/m. Hence,

$$\frac{g_c\sigma\rho_L}{G^2L} = \frac{1.0\times 8.2\times 10^{-3}\times 567}{(300)^2\times 3.048} = 1.6949\times 10^{-5}$$

$$\hat{q}_{oA}/G\lambda = C_2\left(\frac{g_c\sigma\rho_L}{G^2L}\right)^{0.043}(L/D)^{-1}$$

$$= 0.334\left(1.6949\times 10^{-5}\right)^{0.043}(143.8)^{-1}$$

$$\hat{q}_{oA}/G\lambda = 1.4482\times 10^{-3}$$

$$\hat{q}_{oA} = 1.4482\times 10^{-3}\times 300\times 272,000 = 118,176\ \text{W/m}^2$$

$$\hat{q}_{oB}/G\lambda = 0.10\,(\rho_V/\rho_L)^{0.133}\left(\frac{g_c\sigma\rho_L}{G^2L}\right)^{1/3}(1+0.0031L/D)^{-1}$$

$$= 0.10(0.031905)^{0.133}\left(1.6949\times 10^{-5}\right)^{1/3}(1+0.0031\times 143.8)^{-1}$$

$$\hat{q}_{oB}/G\lambda = 1.1236\times 10^{-3}$$

$$\hat{q}_{oB} = 1.1236\times 10^{-3}\times 300\times 272,000 = 91,688\ \text{W/m}^2$$

Since $\hat{q}_{oA} > \hat{q}_{oB}$ we need to calculate $\hat{q}_{oC}$ from Equation (9.88).

$$\hat{q}_{oC}/G\lambda = 0.098(\rho_V/\rho_L)^{0.133}\left(\frac{g_c\sigma\rho_L}{G^2L}\right)^{0.433}(L/D)^{0.27}(1+0.0031L/D)^{-1}$$

$$= 0.098(0.031905)^{0.133}\left(1.6949\times10^{-5}\right)^{0.433}(143.8)^{0.27}(1+0.0031\times143.8)^{-1}$$

$$\hat{q}_{oC}/G\lambda = 1.4091\times10^{-3}$$

$$\hat{q}_{oC} = 1.4091\times10^{-3}\times300\times272,000 = 114,985\ \text{W/m}^2$$

From Equation (9.95), since $\hat{q}_{oA} > \hat{q}_{oB}$, we have:

$$\hat{q}_o = \text{Min}(\hat{q}_{oB}, \hat{q}_{oC}) = \text{Min}(91,688;\ 114,985) = 91,688\ \text{W/m}^2$$

The next step is to calculate Γ_A and Γ_B from Equations (9.92) and (9.93).

$$\Gamma_A = \frac{1.043}{4C_2(g_c\sigma\rho_L/G^2L)^{0.043}} = \frac{1.043}{4\times0.334(1.6949\times10^{-5})^{0.043}}$$

$$\Gamma_A = 1.2521$$

$$\Gamma_B = \frac{(5/6)(0.0124+D/L)}{(\rho_V/\rho_L)^{0.133}(g_c\sigma\rho_L/G^2L)^{1/3}}$$

$$= \frac{(5/6)(0.0124+0.0212/3.048)}{(0.031905)^{0.133}(1.6949\times10^{-5})^{1/3}}$$

$$\Gamma_B = 0.9929$$

From Equation (9.96) we have:

$$\Gamma = \text{Max}(\Gamma_A, \Gamma_B) = \text{Max}(1.2521, 0.9929) = 1.2521$$

The critical heat flux is given by Equation (9.85):

$$\hat{q}_c = \hat{q}_o(1+\Gamma\Delta H_{in}/\lambda) = 91,688(1+1.2521\times23,260/272,000)$$
$$\hat{q}_c \cong 101,500\ \text{W/m}^2$$

This value is about 50% higher than the result obtained using Palen's correlation.

Example 9.9

Calculate the critical heat flux for the conditions of Example 9.8 assuming that the tube is horizontal rather than vertical.

Solution

The factor γ_H is calculated from Equation (9.100) using fluid properties from Example 9.1.

$$\gamma_H = \left(\frac{GD}{\mu_L}\right)\left(\frac{\mu_L^2}{g_c\sigma D\rho_L}\right)^{-1.58}\left[\frac{(\rho_L-\rho_V)gD^2}{g_c\sigma}\right]^{-1.05}\left(\frac{\mu_L}{\mu_V}\right)^{6.41}$$

$$= \left(\frac{300\times0.0212}{156\times10^{-6}}\right)\left[\frac{(156\times10^{-6})^2}{1.0\times8.2\times10^{-3}\times0.0212\times567}\right]^{-1.58}$$

$$\times\left[\frac{(567-18.09)\times9.81(0.0212)^2}{1.0\times8.2\times10^{-3}}\right]^{-1.05}\left(\frac{156\times10^{-6}}{7.11\times10^{-6}}\right)^{6.41}$$

$$\gamma_H = 1.1326\times10^{21}$$

Equation (9.99) is now used to calculate the critical heat flux:

$$\frac{\hat{q}_c}{G\lambda} = 575\gamma_H^{-0.34}(L/D)^{-0.511}\left(\frac{\rho_L-\rho_V}{\rho_V}\right)^{1.27}(1+\Delta H_{in}/\lambda)^{1.64}$$

$$= 575\left(1.1326\times10^{21}\right)^{-0.34}(143.8)^{-0.511}\left(\frac{567-18.09}{18.09}\right)^{1.27}(1+23,260/272,000)^{1.64}$$

$$\frac{\hat{q}_c}{G\lambda} = 2.750 \times 10^{-4}$$

$$\hat{q}_c = 2.750 \times 10^{-4} \times 300 \times 272,000 \cong 22,440 \text{ W/m}^2$$

This result is much lower than the critical heat flux for vertical flow calculated in Example 9.8, demonstrating that the effect of tube orientation on critical heat flux can be very significant.

9.6 Film Boiling

As previously noted, most reboilers and vaporizers operate in the nucleate boiling regime. In some situations, however, process constraints or economics may make it impractical to match the temperature of the process stream and heating medium so as to obtain a temperature difference low enough for nucleate boiling. Film boiling may be a viable option for these situations provided that the higher tube-wall temperature required for film boiling will not result in fluid decomposition and/or heavy fouling. Methods for predicting heat-transfer coefficients in film boiling are presented in this section.

For saturated film boiling on the outside of a single horizontal tube, the semi-empirical equation of Bromley [36] has been widely used. It can be stated in dimensionless form as follows:

$$\frac{h_{fb}D_o}{k_V} = 0.62\left[\frac{g\rho_V(\rho_L-\rho_V)D_o^3(\lambda+0.76C_{P,V}\Delta T_e)}{k_V\mu_V\Delta T_e}\right]^{0.25} \quad (9.101)$$

Here, D_o is the tube OD and h_{fb} is the heat-transfer coefficient for film boiling. The coefficient of 0.76 in the bracketed term of Equation (9.101) is based on the analysis of Sadasivan and Lienhard [37]. In Equation (9.101), vapor properties are evaluated at the film temperature, $T_f = 0.5\ (T_w + T_{sat})$, and the liquid density is evaluated at T_{sat}.

Radiative heat transfer is often significant in film boiling. The convective and radiative effects are not simply additive, however, because the radiation acts to increase the thickness of the vapor film, which reduces the rate of convective heat transfer. A combined heat-transfer coefficient, h_t, for both convection and radiation can be calculated from the following equation [36]:

$$h_t^{4/3} = h_{fb}^{4/3} + h_r h_t^{1/3} \quad (9.102)$$

Here, h_r is the radiative heat-transfer coefficient calculated from Equation (2.73), which takes the following form:

$$h_r = \frac{\epsilon\sigma_{SB}(T_w^4 - T_{sat}^4)}{T_w - T_{sat}} \quad (9.103)$$

where

ϵ = emissivity of tube wall
σ_{SB} = Stefan–Boltzmann constant

If $h_r < h_{fb}$, Equation (9.102) can be approximated by the following explicit formula for h_t [36]:

$$h_t = h_{fb} + 0.75h_r \quad (9.104)$$

A more rigorous correlation for film boiling on a horizontal tube has been developed by Sakurai et al. [38, 39]. However, the method is rather complicated and will not be given here. The heat-transfer coefficient for a single tube provides a somewhat conservative approximation for film boiling on horizontal tube bundles [4].

Two types of convective dry wall boiling occur in tubes. At low vapor quality, the vapor film blanketing the tube wall is in contact with a continuous liquid phase that flows in the central region of the tube. This regime is analogous to pool film boiling, and is sometimes referred to as inverted annular flow. At high vapor quality, boiling takes place in the mist flow regime wherein the vapor phase is continuous over the entire tube cross-section, with the liquid phase in the form of entrained droplets. For the latter situation, the empirical correlation developed by Groeneveld is applicable [26]:

$$\frac{hD}{k_V} = 1.09 \times 10^{-3}\{Re_V[x + (1-x)\rho_V/\rho_L]\}^{0.989} Pr_V^{1.41} Y_G^{-1.15} \quad (9.105)$$

where

$Re_V = DGx/\mu_V$
$Pr_V = C_{P,V}\,\mu_V/k_V$
D = tube ID
x = vapor weight fraction

$$Y_G = 1 - 0.1[(\rho_L - \rho_V)/\rho_V]^{0.4}(1-x)^{0.4} \quad (9.106)$$

In this correlation, the Prandtl number is evaluated at the tube-wall temperature; other properties are evaluated at the fluid temperature.

For the inverted annular flow regime, Palen [4] presented the following equation that is intended to provide a conservative estimate for the heat-transfer coefficient:

$$h_{fb} = 59P_c^{0.5}\Delta T_e^{-0.33}P_r^{0.38}(1-P_r)^{0.22} \tag{9.107}$$

where

$h_{fb} \propto$ Btu/h · ft^2 · °F (W/m^2 · K)
$\Delta T_e \propto$ °F (K)
$P_c \propto$ psia (kPa)

Either English or SI units may be used with this equation, as indicated above. Since this equation is based on a pool film boiling correlation, it should be more conservative at higher flow rates.

The following equation for film boiling on a flat vertical surface is given in Ref. [26]:

$$h_{fb} = 0.056Re_V^{0.2}Pr_V^{1/3}\left[\frac{k_V^3 g\rho_V(\rho_L-\rho_V)}{\mu_V^2}\right]^{1/3} \tag{9.108}$$

This equation is based on turbulent flow in the vapor film and should be applicable to inverted annular flow in vertical tubes. However, no comparison with experimental data has been given in the literature.

Example 9.10

The pool boiling of Example 9.1 takes place with a tube-wall temperature of 537.5 K, at which film boiling occurs. Estimate the heat-transfer coefficient for this situation. In addition to the data given in Example 9.1, the heat capacity and thermal conductivity of the vapor are $C_{P,V} = 2360$ J/kg · K and $k_V = 0.011$ W/m · K.

Solution

For pool film boiling on a horizontal tube, the Bromley equation is applicable. The following data are obtained from Example 9.1:

$$\rho_L = 567 \text{ kg/m}^3 \qquad \mu_V = 7.11\times10^{-6} \text{ kg/m·s}$$
$$\rho_V = 18.09 \text{ kg/m}^3 \qquad D_o = 1 \text{ in.} = 0.0254 \text{ m}$$
$$\lambda = 272{,}000 \text{ J/kg} \qquad T_{Sat} = 437.5 \text{ K}$$

The temperature difference is:

$$\Delta T_e = T_w - T_{sat} = 537.5 - 437.5 = 100 \text{ K}$$

The heat-transfer coefficient for film boiling is obtained by substituting into Equation (9.101).

$$\frac{h_{fb}D_o}{k_V} = 0.62\left[\frac{g\rho_V(\rho_L-\rho_V)D_o^3(\lambda+0.76C_{P,V}\Delta T_e)}{k_V\mu_V\Delta T_e}\right]^{0.25}$$

$$= 0.62\left[\frac{9.81\times18.09(567-18.09)(0.0254)^3(272{,}000+0.76\times2360\times100)}{0.011\times7.11\times10^{-6}\times100}\right]^{0.25}$$

$$\frac{h_{fb}D_o}{k_V} = 341.57$$

$$h_{fb} = 341.57\times k_V/D_o = 341.57\times0.011/0.0254$$

$$h_{fb} = 148 \text{ W/m}^2\text{·K}$$

The radiative heat-transfer coefficient is estimated using Equation (9.103). Assuming an emissivity of 0.8 for the tube wall, we have:

$$h_r = \frac{\epsilon\sigma_{SB}(T_w^4 - T_{sat}^4)}{T_w - T_{sat}} = \frac{0.8\times5.67\times10^{-8}\left[(537.5)^4-(437.5)^4\right]}{537.5-437.5}$$

$$h_r = 21 \text{ W/m}^2\text{·K}$$

Since $h_r < h_{fb}$, Equation (9.104) is used to obtain the effective heat-transfer coefficient:

$$h_t = h_{fb} + 0.75h_r = 148 + 0.75 \times 21 \cong 164 \text{ W/m}^2\cdot\text{K}$$

This value is one to two orders of magnitude lower than the heat-transfer coefficient for nucleate boiling calculated in Example 9.1.

References

[1] Incropera FP, DeWitt DP. Introduction to Heat Transfer. 4th ed. New York: Wiley; 2002.
[2] Hewitt GF, Shires GL, Bott TR. Process Heat Transfer. Boca Raton, FL: CRC Press; 1994.
[3] Forster HK, Zuber N. Dynamics of vapor bubbles and boiling heat transfer. AIChE J 1955;1:531–5.
[4] Palen JW. Shell-and-tube reboilers. In: Heat Exchanger Design Handbook, vol. 3. New York: Hemisphere Publishing Corp; 1988.
[5] Cooper MG. Saturation nucleate pool boiling: a simple correlation. I Chem Eng Symp Ser 1984;86(No. 2):785–93.
[6] Stephan K, Abdelsalam M. Heat-transfer correlations for natural convection boiling. Int J Heat Mass Trans 1980;23:73–87.
[7] Bell KJ, Mueller AC. Wolverine Engineering Data Book II. Wolverine Tube, Inc; 2001. www.wlv.com.
[8] Tong LS, Tang YS. Boiling Heat Transfer and Two-Phase Flow. 2nd ed. Bristol, PA: Taylor & Francis; 1997.
[9] Heat Exchanger Design Handbook, vol. 2. New York: Hemisphere Publishing Corp; 1988.
[10] Kandlikar SG, Shoji M, Dhir VK, editors. Handbook of Phase Change: Boiling and Condensation. Philadelphia: Taylor and Francis; 1999.
[11] Palen JW, Yarden A, Taborek J. Characteristics of boiling outside large-scale horizontal multitube bundles. Chem Eng Prog Symp Ser 1972;68(No. 118):50–61.
[12] Schlünder EU. Heat transfer in nucleate boiling of mixtures. Int Chem Eng 1983;23:589–99.
[13] Thome JR, Shakir S. A new correlation for nucleate pool boiling of aqueous mixtures,. AIChE Symp Ser 1987;83(No. 257):46–51.
[14] Zuber N. On the stability of boiling heat transfer. Trans ASME 1958;80:711–20.
[15] Kreith F, Bohn MS. Principles of Heat Transfer. 6th ed. Pacific Grove, CA: Brooks/Cole; 2001.
[16] Lockhart RW, Martinelli RC. Proposed correlation of data for isothermal two-phase, two-component flow in pipes. Chem Eng Prog 1949;45(No. 1):39–48.
[17] Chisholm D. A theoretical basis for the Lockhart-Martinelli correlation for two-phase flow. Int J Heat Mass Trans 1967;10:1767–78.
[18] Chisholm D. Pressure gradients due to friction during the flow of evaporating two-phase mixtures in smooth tubes and channels. Int J Heat Mass Trans 1973;16:347–58.
[19] Hewitt GF. Fluid mechanics aspects of two-phase flow. In: Kandlikar SG, Shoji M, Dhir VK, editors. Handbook of Phase Change: Boiling and Condensation. Philadelphia: Taylor and Francis; 1999.
[20] Friedel L. Improved friction pressure drop correlations for horizontal and vertical two-phase pipe flow, Paper E2. Ispra, Italy: European Two-Phase Flow Group Meeting; 1979.
[21] Müller-Steinhagen H, Heck K. A simple friction pressure drop correlation for two-phase flow in pipes. Chem Eng Process 1986;20:297–308.
[22] Tribbe C, Müller-Steinhagen H. An evaluation of the performance of phenomenological models for predicting pressure gradient during gas-liquid flow in horizontal pipelines. Int J Multiphase Flow 2000;26:1019–36.
[23] Ould Didi MB, Kattan N, Thome JR. Prediction of two-phase pressure gradients of refrigerants in horizontal tubes. Int J Refrig 2002;25:935–47.
[24] Chisholm D. Gas-liquid flow in pipeline systems. In: Cheremisinoff NP, Gupta R, editors. Handbook of Fluids in Motion. Boston: Butterworth; 1983.
[25] Premoli A, Francesco D, Prina A. Una Correlazione Adimensionale per la Determinazione della Densita di Miscele Bifasiche. La Termotecnica 1971;25:17 26.
[26] Collier JG, Thome JR. Convective Boiling and Condensation. 3rd ed. Oxford: Clarendon Press; 1994.
[27] Chen J. C, Correlation for boiling heat transfer to saturated fluids in convective flow. I & EC Proc Des Dev 1966;5(No. 3):322–9.
[28] Gungor KE, Winterton RHS. A general correlation for flow boiling in tubes and annuli. Int J Heat Mass Trans 1986;29(No. 3):351–8.
[29] Liu Z, Winterton RHS. A general correlation for saturated and subcooled boiling in tubes and annuli, based on a nucleate pool boiling equation. Int J Heat Mass Trans 1991;34(No. 11):2759–66.
[30] Shah MM. Chart correlation for saturated boiling heat transfer equations and further study. ASHRAE Trans 1982;88:185–96.
[31] Kandlikar SG. A general correlation for two-phase flow boiling heat transfer coefficient inside horizontal and vertical tubes. J Heat Transfer 1990;112:219–28.
[32] Steiner D, Taborek J. Flow boiling heat transfer in vertical tubes correlated by an asymptotic model. Heat Transfer Eng 1992;13(No. 2):43–69.
[33] Kattan N, Thome JR, Favrat D. Flow boiling in horizontal tubes: Part 3. Development of a new heat transfer model based on flow pattern. J Heat Transfer 1998;120:156–65.
[34] Katto Y, Ohno H. An improved version of the generalized correlation of critical heat flux for the forced convective boiling in uniformly heated vertical tubes. Int J Heat Mass Trans 1984;27(No. 9):1641–8.
[35] Merilo M. Fluid-to-fluid modeling and correlation of flow boiling crisis in horizontal tubes. Int J Multiphase Flow 1979;5:313–25.
[36] Bromley LA. Heat transfer in stable film boiling. Chem Eng Prog 1950;46(No. 5):221–7.
[37] Sadasivan P, Lienhard JH. Sensible heat correction in laminar film boiling and condensation. J Heat Transfer 1987;109:545–7.
[38] Sakurai A, Shiotsu M, Hata K. A general correlation for pool film boiling heat transfer from a horizontal cylinder to subcooled liquid: Part 1. A theoretical pool film boiling heat transfer model including radiation contributions and its analytical solution. J Heat Transfer 1990;112:430–40.
[39] Sakurai A, Shiotsu M, Hata K. A general correlation for pool film boiling heat transfer from a horizontal cylinder to subcooled liquid: Part 2. Experimental data for various liquids and its correlation. J Heat Transfer 1990;112:441–50.

Notations

A	Total heat-transfer surface area in tube bundle
A_L	Cross-sectional area of conduit occupied by liquid phase
A_V	Cross-sectional area of conduit occupied by vapor phase
a	Constant in Equation (9.83)
B	Parameter in Chisholm correlation for two-phase pressure drop
Bo	Boiling number
BR	$T_D - T_B$ = Boiling range
b	Constant in Equation (9.83)
C	Constant in Equation (9.27)
$C_{P,L}$	Constant-pressure heat capacity of liquid

$C_{P,V}$	Constant-pressure heat capacity of vapor
C_1	Parameter in Equation (9.20)
C_2	Parameter in Equation (9.86), defined in Equation (9.91)
D	Diameter of tube
D_b	Diameter of tube bundle
D_i	Internal diameter of tube
D_o	External diameter of tube
d_B	Theoretical diameter of bubbles leaving solid surface in nucleate boiling
E	Parameter in Friedel correlation for two-phase pressure drop
E_{GW}	Convective enhancement factor in Gungor–Winterton correlation
E_{LW}	Convective enhancement factor in Liu–Winterton correlation
E_1, E_2	Parameters in CISE correlation for two-phase density
F	Parameter in Friedel correlation for two-phase pressure drop
F_b	Factor defined by Equation (9.20) that accounts for convective effects in boiling on tube bundles
F_m	Mixture correction factor for Mostinski correlation, defined by Equation (9.17)
F_P	Pressure correction factor in Mostinski correlation
Fr	Froude number
Fr_{LO}	Froude number based on total flow as liquid
$F(X_{tt})$	Convective enhancement factor in Chen correlation
f	(Darcy) friction factor
f_L	Friction factor for liquid phase
f_{LO}	Friction factor for total flow as liquid
f_{VO}	Friction factor for total flow as vapor
G	Mass flux
G_L	Mass flux for liquid phase
g	Gravitational acceleration
g_c	Unit conversion factor
H	Parameter in Friedel correlation
h	Heat-transfer coefficient
h_b	Convective boiling heat-transfer coefficient
h_{fb}	Heat-transfer coefficient for film boiling
h_{fc}	Forced convection heat-transfer coefficient for two-phase flow
h_{ideal}	Ideal mixture heat-transfer coefficient defined by Equation (9.15)
h_L	Heat-transfer coefficient for liquid phase flowing alone
h_{nb}	Nucleate boiling heat-transfer coefficient
$h_{nb,i}$	Nucleate boiling heat-transfer coefficient for pure component i
h_{nc}	Natural convection heat-transfer coefficient
h_r	Radiative heat-transfer coefficient
h_t	Combined heat-transfer coefficient for convection and radiation in film boiling
K	Parameter in Equation (9.22)
k_L	Thermal conductivity of liquid
k_V	Thermal conductivity of vapor
L	Length of conduit
M	Molecular weight
m	Exponent in Equation (9.83)
$\dot{m}$	Mass flow rate
$\dot{m}_L$	Mass flow rate of liquid phase
$\dot{m}_V$	Mass flow rate of vapor phase
n	Exponent in friction factor versus Reynolds number relationship
n_t	Number of tubes in tube bundle
P_c	Critical pressure
P_{pc}	Pseudo-critical pressure
P_{pr}	Pseudo-reduced pressure
P_r	Reduced pressure
Pr_L	Prandtl number for liquid phase

Pr_V	Prandtl number for vapor phase
P_{sat}	Vapor pressure
P_T	Tube pitch
$\hat{q}$	Heat flux
$\hat{q}_c$	Critical heat flux
$\hat{q}_{c,bundle}$	Critical heat flux for boiling on a tube bundle
$\hat{q}_{c,tube}$	Critical heat flux for boiling on a single tube
$\hat{q}_o$	Critical heat flux for convective boiling in vertical tubes with no inlet subcooling
$\hat{q}_{oA}, \hat{q}_{oB}, \hat{q}_{oC}, \hat{q}_{oD}, \hat{q}_{oE}$	Quantities used to determine the value of $\hat{q}_o$ in Katto-Ohno correlation
R	Tube radius
R^*	$R\left[\frac{g(\rho_L - \rho_V)}{g_c\sigma}\right]^{0.5}$ = Dimensionless radius
Re	Reynolds number
Re_L	Reynolds number for liquid phase
Re_V	Reynolds number for vapor phase
Re_{LO}	Reynolds number calculated for total flow as liquid
S_{CH}	Nucleate boiling suppression factor in Chen correlation
S_{GW}	Nucleate boiling suppression factor in Gungor–Winterton correlation
S_{LW}	Nucleate boiling suppression factor in Liu–Winterton correlation
SR	U_V/U_L = Slip ratio in two-phase flow
s	Specific gravity
s_L	Specific gravity of liquid phase
s_V	Specific gravity of vapor phase
T	Temperature
T_B	Bubble-point temperature
T_b	Temperature of bulk liquid phase
T_D	Dew-point temperature
T_f	Film temperature
T_s	Temperature of solid surface
T_{sat}	Saturation temperature at system pressure
$T_{sat,i}$	Saturation temperature of component i at system pressure
$T_{sat,n}$	Saturation temperature of highest boiling component at system pressure
T_w	Wall temperature
T_1, T_2	Arbitrary temperatures in Watson correlation
U_L	Velocity of liquid phase
U_V	Velocity of vapor phase
We	Weber number
We_{LO}	Weber number calculated for total flow as liquid
X	Lockhart–Martinelli parameter for two-phase flow
X_{tt}	Lockhart–Martinelli parameter for the case in which the flow in both phases is turbulent
x	Vapor mass fraction in a vapor-liquid mixture
x_i	Mole fraction of component i in liquid phase
Y	Chisholm parameter for two-phase flow
Y_G	Parameter in Groeneveld correlation, defined by Equation (9.106)
y	Parameter in CISE correlation, defined by Equation (9.65)
y_i	Mole fraction of component i in vapor phase
Z_1, Z_2, Z_3, Z_4, Z_5	Dimensionless groups used in the Stephan–Abdelsalam correlation

Greek Letters

α_L	Thermal diffusivity of liquid
β	Approximate mass-transfer coefficient used in nucleate boiling correlations for mixtures
Γ	Inlet sub-cooling parameter in Katto-Ohno correlation
$\Gamma_A, \Gamma_B, \Gamma_C$	Quantities used to determine the value of Γ in Katto–Ohno correlation
γ_H	Parameter in Merilo correlation, defined by Equation (9.100)
ΔH_{in}	Enthalpy of subcooling at inlet of tube

ΔP_f	Pressure loss due to fluid fraction
ΔP_{sat}	$P_{sat}(T_w)-P_{sat}(T_{sat})$
$(\Delta P_f/L)_L$	Frictional negative pressure gradient for liquid phase flowing alone
$(\Delta P_f/L)_{Lo}$	Frictional negative pressure gradient for total flow as liquid
$(\Delta P_f/L)_{tp}$	Frictional negative two-phase pressure gradient
$(\Delta P_f/L)_V$	Frictional negative pressure gradient for vapor phase flowing alone
$(\Delta P_f/L)_{VO}$	Frictional negative pressure gradient for total flow as vapor
ΔT_e	$T_s - T_{sat}$ = Excess temperature
ϵ	Emissivity of tube wall
ϵ_V	Void fraction
$(\epsilon_V)_{hom}$	Void fraction for homogeneous two-phase flow
θ_c	Contact angle
λ	Latent heat of vaporization
μ	Viscosity
μ_L	Viscosity of liquid
μ_V	Viscosity of vapor
ρ_{hom}	Two-phase density for homogeneous flow
ρ_L	Density of liquid
ρ_{tp}	Two-phase density
ρ_V	Density of vapor
σ	Surface tension
σ_{SB}	Stefan–Boltzmann constant
ϕ_b	Correction factor for critical heat flux in tube bundles
ϕ_L	Square root of two-phase multiplier applied to pressure gradient for liquid phase flowing alone
ϕ_{LO}	Square root of two-phase multiplier applied to pressure gradient for total flow as liquid
ψ_b	Dimensionless bundle geometry parameter

Problems

(9.1) (a) Show that the Chisholm, Friedel and MSH correlations exhibit the correct behavior at low and high vapor fractions, i.e., $\phi_{LO}^2 = 1$ for $x = 0$ and $\phi_{LO}^2 = Y^2$ for $x = 1$.

(b) Show that the two-phase multiplier, ϕ_L^2, in the Lockhart–Martinelli correlation has the following properties:

$$\phi_L^2 = 1 \quad \text{for x} = 0$$

$$\phi_L^2 \rightarrow \infty \quad \text{as } x \rightarrow 1$$

(c) The Lockhart–Martinelli correlation can be formulated in terms of the vapor-phase pressure gradient as follows:

$$\left(\frac{\Delta P_f}{L}\right)_{tp} = \phi_V^2 \left(\frac{\Delta P_f}{L}\right)_V$$

where

$$\phi_V^2 = 1 + CX + X^2$$

The parameter, X, and constant, C, are the same as specified in Equation (9.18) and Table 9.1, respectively. Show that the two-phase multiplier, ϕ_V^2, has the following properties:

$$\phi_V^2 = 1 \quad \text{for } x = 1$$

$$\phi_V^2 \rightarrow \infty \quad \text{as } x \rightarrow 0$$

(9.2) A stream with a flow rate of 500 lb/h and a quality of 50% flows through a 3/4-in., 14 BWG heat exchanger tube. Physical properties of the fluid are as follows:

$\rho_L = 49$ lbm/ft^3	$\mu_L = 0.73$ cp
$\rho_V = 0.123$ lbm/ft^3	$\mu_V = 0.0095$ cp

(a) Calculate the following Reynolds numbers: Re_{LO}, Re_L, Re_{VO}, Re_V.

(b) Calculate the negative pressure gradients $(\Delta P_f/L)_L$ and $(\Delta P_f/L)_{LO}$.

(c) Calculate the Lockhart–Martinelli parameter, X, and the Chisholm parameter, Y.
(d) Use the Lockhart–Martinelli correlation to calculate the two-phase multiplier, ϕ_L^2, and the negative two-phase pressure gradient.
(e) Use the MSH correlation to calculate the two-phase multiplier, ϕ_{LO}^2, and the negative two-phase pressure gradient.

Ans. (a) $Re_{LO} = 7408$; $Re_L = 3704$; $Re_{VO} = 569{,}231$; $Re_V = 284{,}616$.
(b) $(\Delta P_f/L)_L = 0.00312$ psi/ft; $(\Delta P_f/L)_{LO} = 0.0104$ psi/ft.
(c) $X = 0.07734$; $Y = 11.39$.
(d) $\phi_L^2 = 426.8$; $(\Delta P_f/L)_{tp} = 1.33$ psi/ft.
(e) $\phi_{LO}^2 = 119.2$; $(\Delta P_f/L)_{tp} = 1.24$ psi/ft.

(9.3) For the conditions specified in Problem 9.2, calculate the two-phase pressure gradient using:
(a) The Chisholm correlation.
(b) The Friedel correlation.
The surface tension of the fluid is 20 dynes/cm.

Ans. (a) $(\Delta P_f/L)_{tp} = 1.38$ psi/ft.

(9.4) For the conditions of Problems 9.2 and 9.3, calculate the void fraction and two-phase density using:
(a) The homogeneous flow model.
(b) The Lockhart–Martinelli correlation.
(c) The Chisholm correlation.
(d) The CISE correlation.

Ans. (a) $\epsilon_V = 0.9975$; $\rho_{tp} = 0.245$ lbm/ft^3. (b) $\epsilon_V = 0.9516$; $\rho_{tp} = 2.49$ lbm/ft^3.
(c) $\epsilon_V = 0.9889$; $\rho_{tp} = 0.666$ lbm/ft^3. (d) $\epsilon_V = 0.9789$; $\rho_{tp} = 1.15$ lbm/ft^3.

(9.5) Repeat Problem 9.2 for a flow rate of 45.87 lb/h.

Ans. (a) $Re_{LO} = 680$; $Re_L = 340$; $Re_{VO} = 52{,}221$; $Re_V = 26{,}111$.
(b) $(\Delta P_f/L)_L = 9.99 \times 10^{-5}$ psi/ft; $(\Delta P_f/L)_{LO} = 2.00 \times 10^{-4}$ psi/ft.
(c) $X = 0.1259$; $Y = 10.29$
(d) $\phi_L^2 = 159.4$; $(\Delta P_f/L)_{tp} = 0.0159$ psi/ft.
(e) $\phi_{LO}^2 = 97.28$; $(\Delta P_f/L)_{tp} = 0.0195$ psi/ft.

(9.6) Repeat Problem 9.4 for a flow rate of 45.87 lb/h.

Ans. (a) Same as Problem 9.4. (b) $\epsilon_V = 0.9208$; $\rho_{tp} = 3.99$ lbm/ft^3.
(c) Same as Problem 9.4. (d) $\epsilon_V = 0.9269$; $\rho_{tp} = 3.70$ lbm/ft^3.

(9.7) Nucleate boiling takes place on the surface of a 3/4-in. OD horizontal tube immersed in saturated liquid toluene. The tube-wall temperature is 300°F and the system pressure is 25 psia. Properties of toluene at these conditions are given in the table below. Calculate the heat-transfer coefficient and wall heat flux using:
(a) The Forster–Zuber correlation.
(b) The Mostinski correlation.
(c) The Cooper correlation.
(d) The Stephan–Abdelsalam correlation.

Ans. (a) 902 Btu/h · ft^2 · °F; 29, 950 Btu/h · ft^2. (b) 571 Btu/h · ft^2 · °F; 18,950 Btu/h · ft^2. (c) 2431 Btu/h · ft^2 · °F; 80,710 Btu/h · ft^2. (d)1594 Btu/h · ft^2 · °F; 52,920 Btu/h·ft^2.

Toluene Property	Value
ρ_V (lbm/ft^3)	0.310
ρ_L (lbm/ft^3)	47.3
$C_{P,V}$ (Btu/lbm · °F)	0.369
$C_{P,L}$ (Btu/lbm · °F)	0.491
μ_V (cp)	0.00932
μ_L (cp)	0.213
k_V (Btu/h · ft · °F)	0.010
k_L (Btu/h · ft · °F)	0.059
σ (dyne/cm)*	16.2
λ (Btu/lbm)	151.5

—cont'd

Toluene Property	Value
P_c (psia)	595.9
T_{sat} (°F) at 25 psia	266.8
Vapor pressure (psia) at 300° F	39.0
Molecular weight	92.14

* 1 dyne/cm = 6.8523×10^{-5} lbf/ft.

(9.8) Nucleate boiling takes place on the surface of a 3/4-in. OD horizontal tube immersed in saturated liquid refrigerant 134a (1,1,1,2-tetrafluoroethane). The tube-wall temperature is 300 K and the system pressure is 440 kPa. Properties of R-134a at these conditions are given in the table below. Calculate the heat-transfer coefficient and wall heat flux using:
(a) The Forster–Zuber correlation.
(b) The Mostinski correlation.
(c) The Cooper correlation.
(d) The Stephan–Abdelsalam correlation.

R-134a Property	Value
ρ_V (kg/m^3)	20.97
ρ_L (kg/m^3)	1253
$C_{P,V}$ (J/kg · K)	865
$C_{P,L}$ (J/kg · K)	1375
μ_V (kg/m· s)	1.185×10^{-5}
μ_L (kg/m· s)	2.41×10^{-4}
k_V (W/m · K)	0.0122
k_L (W/m · K)	0.087
σ (N/m)	0.00993
λ (J/kg)	192,800
P_c (kPa)	4064
T_{sat} (K) at 440 kPa	285
Vapor pressure (kPa) at 300 K	700
Molecular weight	102.03

(9.9) Calculate the critical heat flux for the conditions of Problem 9.7 using:
(a) The Zuber equation.
(b) The Zuber-type equation for a horizontal cylinder.
(c) The Mostinski correlation.

Ans. (a) 122,640 Btu/h · ft^2. (b) 97,127 Btu/h · ft^2. (c) 151,740 Btu/h · ft^2.

(9.10) Calculate the critical heat flux for the conditions of Problem 9.8 using:
(a) The Zuber equation.
(b) The Zuber-type equation for a horizontal cylinder.
(c) The Mostinski correlation.

(9.11) The nucleate boiling of Problem 9.7 takes place on a tube bundle consisting of 10,200 tubes having an OD of 3/4 in. and laid out on a 1.0-in. triangular pitch. The bundle diameter is 106.8 in. Calculate the critical heat flux and heat-transfer coefficient for this situation.

Ans. 6570 Btu/h · ft^2 and 2020 Btu/h · ft^2 · °F (based on Mostinski correlations).

(9.12) The nucleate boiling of Problem 9.8 takes place on a tube bundle consisting of 1270 tubes having an OD of 19 mm and laid out on a 23.8 mm triangular pitch. The bundle diameter is approximately 920 mm. Calculate the critical heat flux and heat-transfer coefficient for this situation.

(9.13) Nucleate boiling takes place on the surface of a 1.0-in. OD horizontal tube immersed in a saturated liquid consisting of 40 mole percent *n*-pentane and 60 mole percent toluene. The system pressure is 35 psia and the heat flux is 25,000 Btu/h · ft^2. The vapor in equilibrium with the liquid contains 87.3 mole percent *n*-pentane. At system pressure, the boiling point of *n*-pentane is 147.4°F while that of toluene is 291.6°F. The critical pressures are 488.6 psia for *n*-pentane and 595.9 psia for toluene. Mixture properties at system conditions are given in the table below.

(a) Use the mixture properties together with the Stephan–Abdelsalam correlation to calculate the ideal heat-transfer coefficient for the mixture.

(b) Use Schlünder's method with the ideal heat-transfer coefficient from part (a) to calculate the heat-transfer coefficient for the mixture.

(c) Repeat part (b) using the method of Thome and Shakir.

(d) Use Palen's method to calculate the mixture heat-transfer coefficient.

Ans. (a) 1127 Btu/h · ft^2 · °F. (b) 325 Btu/h · ft^2 · °F. (c) 349 Btu/h · ft^2 · °F. (d) 198 Btu/h · ft^2 · °F.

Mixture Property	Value
ρ_L (lbm/ft^3)	43.7
ρ_V (lbm/ft^3)	0.404
k_L (Btu/h · ft · °F)	0.0607
$C_{P,L}$ (Btu/lbm · °F)	0.498
μ_L (cp)	0.231
μ_V (cp)	0.00833
σ (dyne/cm)*	16.5
λ (Btu/lbm)	149.5
Bubble-point temperature (°F)	183.5
Dew-point temperature (°F)	258.3

* 1 dyne/cm = 6.8523 × 10^{-5} lbf/ft.

(9.14) Estimate the value of the critical heat flux for the conditions of Problem 9.13.

Ans. 159,350 Btu/h · ft^2 (from Mostinski correlation).

(9.15) The nucleate boiling of Problem 9.13 takes place on a horizontal tube bundle consisting of 390 tubes having an OD of 1.0 in. and laid out on a 1.25-in. square pitch. The bundle diameter is approximately 30 in. Calculate the heat-transfer coefficient and critical heat flux for this situation.

Ans. 1.h_b = 605 Btu/h · ft^2 · °F.

(9.16) The Gorenflo correlation (Gorenflo, D., Pool boiling, *VDI Heat Atlas*, VDI Verlag, Düsseldorf, 1993) is a fluid-specific method for calculating nucleate boiling heat-transfer coefficients that, when applicable, is reputed to be very reliable. It utilizes experimental heat-transfer coefficients obtained at the following reference conditions:

Reduced pressure =0.1

Heat flux = 20,000 W/m^2

Surface roughness = 0.4 μm

The heat-transfer coefficient is calculated as follows:

$$h_{nb} = h_{ref} F_P \left(\hat{q}/\hat{q}_{ref}\right)^m \left(\Xi/\Xi_{ref}\right)^{0.133}$$

where

h_{ref} = nucleate boiling heat-transfer coefficient at reference conditions

F_P = pressure correction factor

$\hat{q}_{ref}$ = 20,000 W/m^2 = 6,342 Btu/h · ft^2

Ξ = surface roughness

Ξ_{ref} = 0.4 μm = 1.31 × 10^{-6} ft

The pressure correction factor and exponent, m, are functions of reduced pressure that are calculated as follows:

$$F_P = 1.73P_r^{0.27} + \left(6.1 + \frac{0.68}{1 - P_r}\right)P_r^2 \qquad \text{(water)}$$

$$F_P = 1.2P_r^{0.27} + 2.5P_r + P_r/(1 - P_r) \quad \text{(other fluids)}$$

$$m = 0.9 - 0.3P_r^{0.15} \qquad \text{(water)}$$

$$m = 0.9 - 0.3P_r^{0.3} \qquad \text{(other fluids)}$$

If the surface roughness is unknown, it is set to 0.4 μm. Reference heat-transfer coefficients for a few selected fluids are given in the table below. Values for a number of other fluids are available on the web (Thome, J. R., *Engineering Data Book III*, Wolverine Tube, Inc., www.wlv.com, 2004).

Fluid	h_{ref} (W /m^2)
n-pentane	3400
n-heptane	3200
Toluene	2650
n-propanol	3800
R-134a	4500
Water	5600

Use the Gorenflo correlation to calculate:
(a) The heat-transfer coefficient and heat flux for the conditions of Problem 9.7.
(b) The heat-transfer coefficient and heat flux for the conditions of Problem 9.8.
(c) The ideal heat-transfer coefficient for the mixture of Problem 9.13.

(9.17) The Stephan–Abdelsalam correlation for the "hydrocarbon" group is as follows:

$$\frac{h_{nb}d_B}{k_L} = 0.0546\left(Z_1 Z_4^{0.5}\right)^{0.67} Z_3^{0.248} Z_5^{-4.33}$$

Use this correlation to calculate the nucleate boiling heat-transfer coefficient for the conditions of:
(a) Example 9.1.
(b) Problem 9.7.
(c) Problem 9.8.
(d) Problem 9.13.

Ans. (a) 21,000 W/m^2 · K.

(9.18) Toluene boils while flowing vertically upward through a 1-in. OD, 14 BWG tube with a mass flux of 250,000 lbm/h · ft^2. At a point in the tube where the wall temperature is 300°F, the pressure is 25 psia and the quality is 0.1, calculate the heat-transfer coefficient using:
(a) Chen's method.
(b) Chen's method with the Mostinski–Palen correlation for nucleate boiling.
(c) The Liu–Winterton correlation.
See Problem 9.7 for fluid property data.

Ans. (a) 690 Btu/h · ft^2 · °F (b) 554 Btu/h · ft^2 · °F. (c) 1377 Btu/h · ft^2 · °F.

(9.19) Refrigerant 134a boils while flowing vertically upward through a 3/4-in. OD, 14 BWG tube with a mass flux of 400 kg/s · m^2. At a point in the tube where the wall temperature is 300 K, the pressure is 440 kPa and the quality is 0.25, calculate the heat-transfer coefficient using:
(a) Chen's method.
(b) Chen's method with the Mostinski-Palen correlation for nucleate boiling.
(c) The Liu–Winterton correlation.
See Problem 9.8 for fluid property data.

(9.20) A mixture of n-pentane and toluene boils while flowing vertically upward through a 1-in. OD, 14 BWG tube with a mass flux of 300,000 lbm/h · ft^2. At a point in the tube where the conditions specified in Problem 9.13 exist and the quality is 0.15, calculate the heat-transfer coefficient using the Chen–Palen method.

Ans. 578 Btu/h · ft^2 · °F.

(9.21) For the conditions specified in Problem 9.18, the tube length is 8 ft and the enthalpy of subcooling at the tube inlet is 10 Btu/lbm. Calculate the critical heat flux using:
(a) Palen's method.
(b) The Katto–Ohno correlation.

Ans. (a) 25,660 Btu/h · ft^2. (b) 54,740 Btu/h · ft^2.

(9.22) For the conditions specified in Problem 9.19, the tube length is 10 ft and the liquid entering the tube is subcooled by 2 K. Calculate the critical heat flux using:
(a) Palen's method.
(b) The Katto–Ohno correlation.

(9.23) For the conditions specified in Problem 9.20, the tube length is 12 ft and the enthalpy of subcooling at the tube inlet is 5 Btu/lbm. Estimate the critical heat flux using:
(a) Palen's method.
(b) The Katto–Ohno correlation.

(9.24) Calculate the critical heat flux for the conditions of Problem 9.21 assuming that the tube is horizontal rather than vertical. Is the result consistent with the values calculated in Problem 9.21 for a vertical tube?

Ans. 62,000 Btu/h · ft^2.

(9.25) Calculate the critical heat flux for the conditions of Problem 9.22 assuming that the tube is horizontal rather than vertical. Is the result consistent with the values calculated for a vertical tube in Problem 9.22?

(9.26) Calculate the critical heat flux for the conditions of Problem 9.23 if the tube is horizontal rather than vertical. Is the result consistent with the values calculated for a vertical tube in Problem 9.23?

(9.27) The pool boiling of Problem 9.7 takes place with a tube-wall temperature of 400°F, at which film boiling occurs. Assuming an emissivity of 0.8 for the tube wall, estimate the heat-transfer coefficient for this situation.

Ans. 33 Btu/h · ft^2 · °F.

(9.28) The pool boiling of Problem 9.8 takes place with a tube-wall temperature of 380 K, at which film boiling occurs. Assuming an emissivity of 0.8 for the tube wall, calculate the heat-transfer coefficient for this situation.

(9.29) The convective boiling of Problem 9.18 takes place with a tube-wall temperature of 400°F. Inverted annular flow exists under these conditions. Estimate the heat-transfer coefficient for this situation using:
(a) Equation (9.107).
(b) Equation (9.108).

Ans. (a) 85 Btu/h · ft^2 · °F. (b) 114 Btu/h · ft^2 · °F.

(9.30) The convective boiling of Problem 9.19 takes place with a tube-wall temperature of 380 K. Inverted annular flow exists under these conditions. Estimate the heat-transfer coefficient for this situation using:
(a) Equation (9.107).
(b) Equation (9.108).

10 Reboilers

10.1 Introduction

A reboiler is a heat exchanger that is used to generate the vapor supplied to the bottom tray of a distillation column. The liquid from the bottom of the column is partially vaporized in the exchanger, which is usually of the shell-and-tube type. The heating medium is most often condensing steam, but commercial heat-transfer fluids and other process streams are also used. Boiling takes place either in the tubes or in the shell, depending on the type of reboiler. Exchangers that supply vapor for other unit operations are referred to as vaporizers, but are similar in most respects to reboilers.

Thermal and hydraulic analyses of reboilers are generally more complex than for single-phase exchangers. Some of the complicating factors are the following:

- Distillation bottom liquids are often mixtures having substantial boiling ranges. Hence, the physical properties of the liquid and vapor fractions can exhibit large variations throughout the reboiler. Thermodynamic calculations are required to determine the phase compositions and other properties within the reboiler.
- A zone or incremental analysis is generally required for rigorous calculations.
- Two-phase flow occurs in the boiling section of the reboiler and, in the case of thermosyphon units, in the return line to the distillation column.
- For recirculating thermosyphon reboilers, the circulation rate is determined by the hydraulics in both the reboiler and the piping connecting the distillation column and reboiler. Hence, the reboiler and connecting piping must be considered as a unit. The hydraulic circuit adds another iterative loop to the design procedure.

Even with simplifying assumptions, the complete design of a reboiler system can be a formidable task without the aid of computer software. For rigorous calculations, commercial software is a practical necessity.

10.2 Types of Reboilers

Reboilers are classified according to their orientation and the type of circulation employed. The most commonly used types are described below.

10.2.1 Kettle Reboilers

A kettle reboiler (Figure 10.1) consists of a horizontally mounted TEMA K-shell and a tube bundle comprised of either U-tubes or straight tubes (regular or finned) with a pull-through (type T) floating head. The tube bundle is unbaffled, so support plates are provided for tube support. Liquid is fed by gravity from the column sump and enters at the bottom of the shell through one or more nozzles. The liquid flows upward across the tube bundle, where boiling takes place on the exterior surface of the tubes. Vapor and liquid are separated in the space above the bundle, and the vapor flows overhead to the column, while the liquid flows over a weir and is drawn off as the bottom product. Low circulation rates, horizontal configuration, and all-vapor return flow make kettle reboilers relatively insensitive to system hydraulics. As a result, they tend to be reliable even at very low (vacuum) or high (near critical) pressures where thermosyphon reboilers are most prone to operational problems. Kettles can also operate efficiently with small temperature driving forces, as high heat fluxes can be obtained by increasing the tube pitch [1]. On the negative side, low circulation rates make kettles very susceptible to fouling, and the over-sized K-shell is relatively expensive.

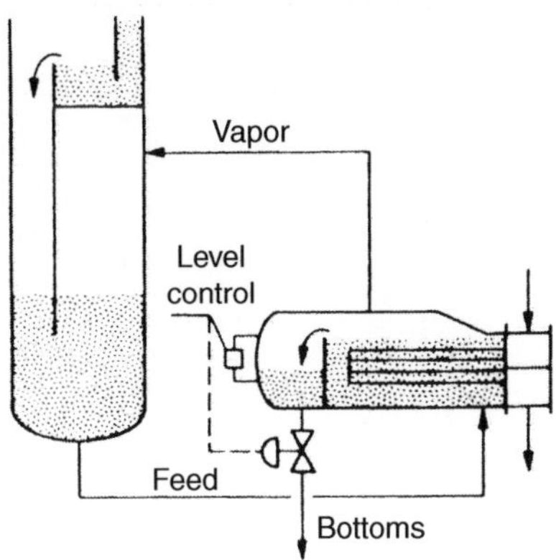

FIGURE 10.1 Typical configuration for a kettle reboiler (Source: Ref. [1]).

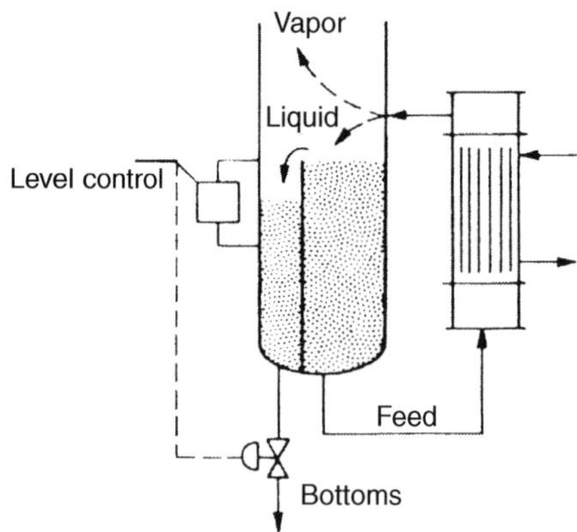

FIGURE 10.2 Typical configuration for a vertical thermosyphon reboiler (Source: Ref. [1]).

10.2.2 Vertical Thermosyphon Reboilers

A vertical thermosyphon reboiler (Figure 10.2) consists of a TEMA E-shell with a single-pass tube bundle. The boiling liquid usually flows through the tubes as shown, but shell-side boiling may be used in special situations, e.g., with a corrosive heating medium. A mixture of vapor and liquid is returned to the distillation column, where phase separation occurs. The driving force for the flow is the density difference between the liquid in the feed circuit and the two-phase mixture in the boiling region and return line. Except for vacuum services, the liquid in the column sump is usually maintained at a level close to that of the upper tubesheet in the reboiler to provide an adequate static head. For vacuum operations, the liquid level is typically maintained at 50% to 70% of the tube height to reduce the boiling point elevation of the liquid fed to the reboiler [2].

Vertical thermosyphon reboilers are usually attached directly to distillation columns, so the costs of support structures and piping are minimized, as is the required plot space. The TEMA E-shell is also relatively inexpensive. Another advantage is that the relatively high velocity attained in these units tends to minimize fouling. On the other hand, tube length is limited by the height of liquid in the column sump and the cost of raising the skirt height to increase the liquid level. This limitation tends to make these units relatively expensive for services with very large duties. The boiling point increase due to the large static head is another drawback for services with small temperature driving forces. Also, the vertical configuration makes maintenance more difficult, especially when the heating medium causes fouling on the outside of the tubes and/or the area near the unit is congested.

10.2.3 Horizontal Thermosyphon Reboilers

Horizontal thermosyphon reboilers (Figure 10.3) usually employ a TEMA G-, H-, or X-shell, although E- and J-shells are sometimes used. The tube bundle may be configured for a single pass as shown, or for multiple passes. In the latter case, either U-tubes or straight tubes (plain or finned) may be used. Liquid from the column is fed to the bottom of the shell and flows upward across the tube bundle. Boiling takes place on the exterior tube surface, and a mixture of vapor and liquid is returned to the column. As with vertical thermosyphons, the circulation is driven by the density difference between the liquid in the column sump and the two-phase mixture in the reboiler and return line.

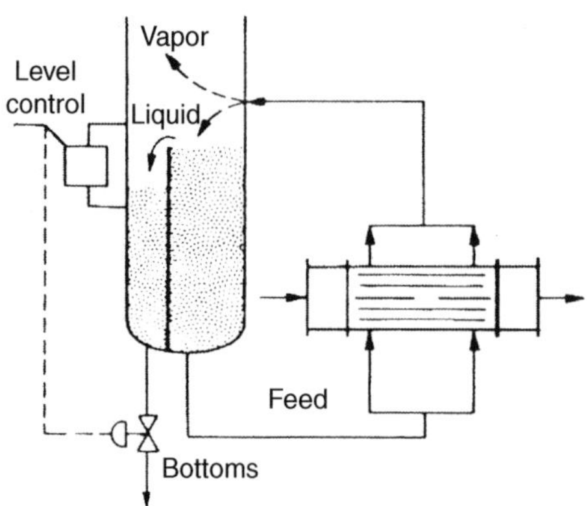

FIGURE 10.3 Typical configuration for a horizontal thermosyphon reboiler (Source: Ref. [1]).

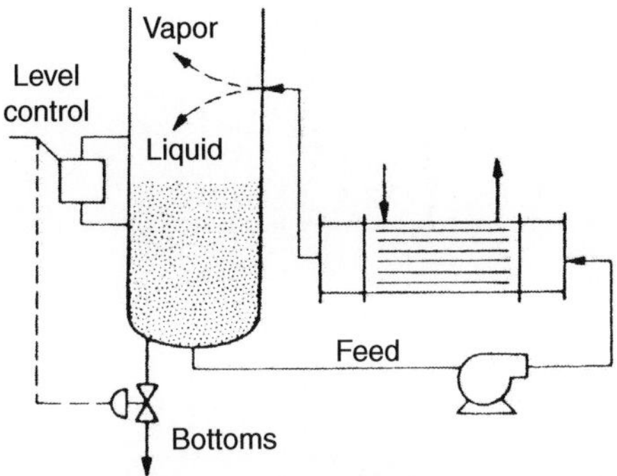

FIGURE 10.4 Typical configuration for a forced flow reboiler (Source: Ref. [1]).

The flow pattern in horizontal thermosyphon reboilers is similar to that in kettle reboilers, but the higher circulation rates and lower vaporization fractions in horizontal thermosyphons make them less susceptible to fouling. Due to the horizontal configuration and separate support structures, these units are not subject to restrictions on weight or tube length. As a result, they are generally better suited than vertical thermosyphons for services with very large duties. The horizontal configuration is also advantageous for handling liquids of moderately high viscosity, because a relatively small static head is required to overcome fluid friction and drive the flow. A rule of thumb is that a horizontal rather than a vertical thermosyphon should be considered if the feed viscosity exceeds 0.5 cp.

10.2.4 Forced Flow Reboilers

In a forced flow reboiler system (Figure 10.4) the circulation is driven by a pump rather than by gravity. The boiling liquid usually flows in the tubes, and the reboiler may be oriented either horizontally or vertically. A TEMA E-shell is usually used with a tube bundle configured for a single pass. These units are characterized by high tube-side velocities and very low vaporization fractions (usually less than 1% [1]) in order to mitigate fouling. The main use of forced flow reboilers is in services with severe fouling problems and/or highly viscous (greater than 25 cp) liquids for which kettle and thermosyphon reboilers are not well suited. Pumping costs render forced flow units uneconomical for routine services.

10.2.5 Internal Reboilers

An internal reboiler (Figure 10.5) consists of a tube bundle (usually U-tubes) that is inserted directly into the sump of the distillation column. Since no shell or connecting piping is required, it is the least expensive type of reboiler. However, the amount of heat-transfer area that can be accommodated is severely limited. Also, formation of froth and foam in the column sump can cause operational problems. As a result, this type of reboiler is infrequently used.

10.2.6 Recirculating Versus Once-Through Operation

Thermosyphon reboiler systems can be of either the recirculating type, as in Figures 10.2 and 10.3, or the once-through type shown in Figure 10.6. In the latter case, the liquid from the bottom tray is collected in a trap-out, from which it flows to the reboiler. The liquid fraction of the return flow collects in the column sump, from which it is drawn as the bottom product. Thus, the liquid passes through the reboiler only once, as with a kettle reboiler.

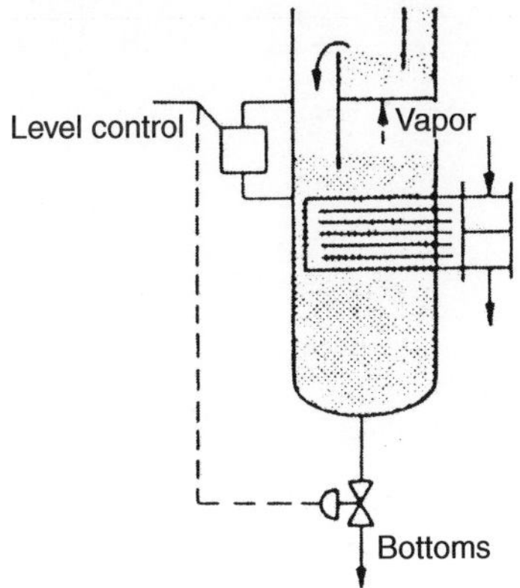

FIGURE 10.5 Typical configuration for an internal reboiler (Source: Ref. [1]).

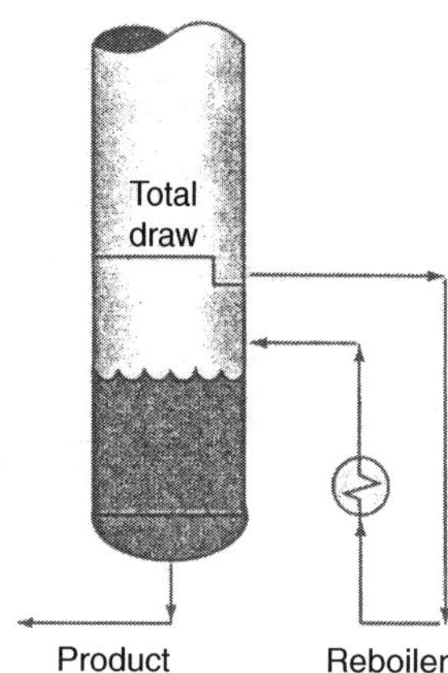

FIGURE 10.6 Typical configuration for a once-through thermosyphon reboiler system (Source: Ref. [2]).

Once-through operation requires smaller feed lines and generally provides a larger temperature driving force in the reboiler. For mixtures, the boiling point of the liquid fed to a recirculating reboiler is elevated due to the addition of the liquid returned from the reboiler, which is enriched in the higher boiling components. As a result, the mean temperature difference in the boiling zone of the exchanger is reduced. Recirculation can also result in increased fouling in some systems, e.g., when exposure to high temperatures results in chemical decomposition or polymerization.

For reliable design and operation, the vapor weight fraction in thermosyphon reboilers should be limited to about 25% to 30% for organic compounds and about 10% for water and aqueous solutions [1,2]. If these limits cannot be attained with once-through operation, then a recirculating system should be used.

10.2.7 Reboiler Selection

In some applications the choice of reboiler type is clear-cut. For example, severely fouling or very viscous liquids dictate a forced flow reboiler. Similarly, a dirty or corrosive heating medium together with a moderately fouling process stream favors a horizontal thermosyphon reboiler. In most applications, however, more than one type of reboiler will be suitable. In these situations the selection is usually based on considerations of economics, reliability, controllability, and experience with similar services. The guidelines presented by Palen [1] and reproduced in Table 10.1 provide useful information in this regard. Kister [3] also gives a good concise comparison of reboiler types and the applications in which each is preferred.

TABLE 10.1 Guidelines for Reboiler Selection

	Reboiler Type			
Process Conditions	Kettle or Internal	Horizontal Shell-Side Thermosyphon	Vertical Tube-Side Thermosyphon	Forced Flow
Operating pressure				
Moderate	E	G	B	E
Near critical	B-E	R	Rd	E
Deep vacuum	B	R	Rd	E
Design ΔT				
Moderate	E	G	B	E
Large	B	R	G-Rd	E
Small (mixture)	F	F	Rd	P
Very small (pure component)	B	F	P	P
Fouling				
Clean	G	G	G	E
Moderate	Rd	G	B	E
Heavy	P	Rd	B	G
Very heavy	P	P	Rd	B
Mixture boiling range				
Pure component	G	G	G	E
Narrow	G	G	B	E
Wide	F	G	B	E
Very wide, with viscous liquid	F-P	G-Rd	P	B

Category abbreviations: B: best; G: good operation; F: fair operation, but better choice is possible; Rd: risky unless carefully designed, but could be best choice in some cases; R: risky because of insufficient data; P: poor operation; and E: operable but unnecessarily expensive.
Source: Ref. [1]

Sloley [2] surveyed the use of vertical versus horizontal thermosyphon reboilers in the petroleum refining, petrochemical and chemical industries. Of the thermosyphons used in petroleum refining, 95% are horizontal units; in the petrochemical industry, 70% are vertical units; and in the chemical industry, nearly 100% are vertical units. He attributes this distribution to two factors, size and fouling tendency. For the relatively small, clean services typical of the chemical industry, vertical thermosyphons are favored, whereas the large and relatively dirty services common in petroleum refining dictate horizontal thermosyphons. Services in the petrochemical industry also tend to be relatively large, but to a lesser extent than in petroleum refining, and they are generally cleaner as well. Hence, the use of horizontal thermosyphons in petrochemical applications is less extensive compared with petroleum refining, but greater than in the chemical industry. The above analysis is somewhat contradictory with Table 10.1 because size permitting, a vertical thermosyphon is indicated for moderate to heavy fouling on the boiling side. The reason is that in a vertical unit the boiling fluid is on the tube side, which is relatively easy to clean, the vertical configuration notwithstanding.

Overall, however, the vertical thermosyphon is the most frequently used type of reboiler [3]. Size permitting, it will generally be the reboiler type of choice unless the service is such that one of the other types offers distinct advantages, as discussed above.

10.3 Design of Kettle Reboilers

10.3.1 Design Strategy

A schematic representation of the circulation in a kettle reboiler is shown in Figure 10.7. The circulation rate through the tube bundle is determined by a balance between the static head of liquid outside the bundle and the pressure drop across the bundle. A two-phase mixture exists in the bundle and the vapor fraction varies with position. Therefore, the bundle hydraulics are coupled with the heat transfer, and a computer model (such as that in the HTRI software package) is required to perform these calculations.

Since the circulation rate in a kettle reboiler is relatively low, the pressure drop in the unit is usually quite small. Therefore, a reasonable approximation is to neglect the pressure drop in the unit and size the bundle using the heat-transfer correlations given in Section 9.3. Since kettles utilize once-through operation, the feed rate is equal to the liquid flow rate from the bottom tray of the distillation column. Hence, the feed and return lines can be sized to accommodate the required liquid and vapor flows based on the available static head of liquid in the column sump. Because the flow in each line is single phase (liquid feed and vapor return), the hydraulic calculations are straightforward. Furthermore, the heat-transfer and hydraulic calculations are independent of one another, making the entire approximate design procedure relatively simple and suitable for hand calculations.

10.3.2 Mean Temperature Difference

In exchangers with boiling or condensing mixtures, the true mean temperature difference is not generally equal to $F(\Delta T_{ln})_{cf}$ because the stream enthalpy varies nonlinearly with temperature over the boiling or condensing range, violating an underlying premise of the F-factor method. Computer algorithms handle this situation by performing a zone analysis (incremental calculations) in which each zone or section of the exchanger is such that the stream enthalpy is nearly linear within the zone. For the approximate design

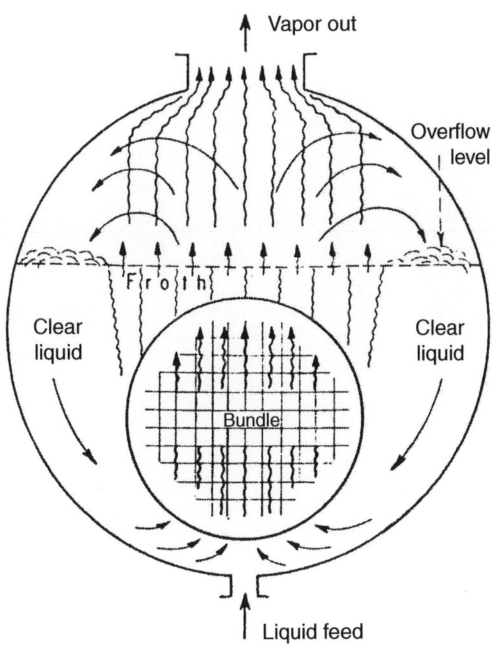

FIGURE 10.7 Schematic representation of the circulation in a kettle reboiler. Source: Ref. [4].

procedure outlined above, however, an effective mean temperature difference for the reboiler is required. For kettle reboilers, Palen [1] recommends using the logarithmic mean temperature difference (LMTD) based on the exit vapor temperature as an approximation for the mean temperature difference. That is, the LMTD is calculated assuming that the shell-side fluid temperature is constant and equal to the temperature of the vapor leaving the reboiler.

Palen's approximation is conservative for mixtures with a wide boiling range since the cold fluid temperature increases from the bottom to the top of the bundle. For pure components and mixtures with narrow boiling ranges (less than 5°C), the approximation may overestimate the true mean temperature difference. The temperature profile for pure and narrow-boiling-range fluids includes a temperature increase near the bottom of the bundle in the subcooled region, followed by a decreasing temperature region in the boiling zone as the saturation temperature decreases along with pressure. For accurate calculation of mean temperature difference in kettles, an incremental analysis is required as described above.

10.3.3 Fouling Factors

Since heat-transfer coefficients are generally high in reboilers, the specified fouling allowance can account for a substantial fraction of the total thermal resistance. Therefore, it is important to use realistic values for the fouling factors in order to avoid gross overdesign that could result in operational problems as well as needless expense. The recommendations of Palen and Small [5] are given in Table 10.2. TEMA fouling factors or those given in Table 3.3 may also be useful for some applications. As always, however, the best source for fouling factors is prior experience with the same or similar application.

Kettle reboilers are susceptible to buildup of heavy components (sometimes referred to as "high boilers") since the circulation forces, which tend to push the heavy components toward the liquid (bottom product) outlet nozzle, are relatively weak. As a result, the mean temperature difference and duty are reduced. This buildup of heavy components is often interpreted as fouling, and some designers might be tempted to increase the fouling factor to account for the needed margin. High fouling factors should be avoided in applications where heavy-component buildup is possible, and an alternative method should be used to specify design margin, as discussed in Section 3.9.

10.3.4 Number of Nozzles

In order to obtain a reasonably uniform flow distribution along the length of the tube bundle, an adequate number of feed and vapor return nozzles should be used. For a tube bundle of length L and diameter D_b, the number, N_n, of nozzle pairs (feed and return) is determined from the following empirical equation [1,6]:

$$N_n = \frac{L}{5\,D_b} \tag{10.1}$$

The calculated value is rounded upward to the next largest integer.

This rule-of-thumb applies when nozzles can be evenly distributed along the bundle length. With a tapered shell and a large kettle diameter, there may be very little horizontal length at the top of the kettle. There is little benefit to installing multiple nozzles in close proximity at the top of the kettle. When a single pair of nozzles is used, the inlet nozzle is usually placed at the front of the shell near the tubesheet. This location helps prevent the feed from bypassing the tube bundle and flowing directly over the weir into the overflow reservoir.

10.3.5 Shell Diameter

The diameter of the K-shell is chosen to provide adequate space above the surface of the boiling liquid for vapor–liquid disengagement. A rule of thumb is that the distance from the uppermost tube to the top of the shell should be at least 40% of the shell diameter. A somewhat more rigorous sizing procedure is based on the following empirical equation for the vapor loading [5,6]:

$$VL = 2290\ \rho_V \left(\frac{\sigma}{\rho_L - \rho_V}\right)^{0.5} \tag{10.2}$$

TABLE 10.2 Recommended Fouling Factors for Reboiler Design

Boiling-Side Stream	Fouling Factor (h·ft^2·°F/Btu)
C_1 – C_8 normal hydrocarbons	0–0.001
Heavier normal hydrocarbons	0.001–0.003
Diolefins and polymerizing hydrocarbons	0.003–0.005
Heating-Side Stream	
Condensing steam	0–0.0005
Condensing organic	0.0005–0.001
Organic liquid	0.0005–0.002

Source: Ref. [5]

where

VL = vapor loading (lbm/h · ft^3)
ρ_V, ρ_L = vapor and liquid densities (lbm/ft^3)
σ = surface tension (dyne/cm)

The vapor loading is the mass flow rate of vapor divided by the volume of the vapor space. The value given by Equation (10.2) is intended to provide a sufficiently low vapor velocity to allow gravitational settling of entrained liquid droplets. The dome segment area, SA, is calculated from the vapor loading as follows:

$$SA = \frac{\dot{m}_V}{L \times VL} \tag{10.3}$$

where

$\dot{m}_V$ = vapor mass flow rate (lbm/h)
L = length of tube bundle (ft)
$VL \propto$ lbm/h · ft^3
$SA \propto$ ft^2

Considering the K-shell cross-section shown in Figure 10.7, the dome segment area is the area of the circular segment that lies above the liquid surface. For known bundle diameter and dome segment area, the shell diameter can be determined (by trial and error) from the table of circular segment areas in Appendix 10.A. Since the liquid level is usually maintained slightly above the top row of tubes, the height of liquid in the shell is approximately equal to the bundle diameter plus the clearance between the bundle and the bottom of the shell. However, to account for the effect of foaming and froth formation, this height may be incremented by 3 in. to 5 in. for the purpose of calculating the shell diameter [6]. Splash baffles and demister pads can also be installed in the vapor outlet nozzles to further reduce entrainment.

Example 10.1

A kettle reboiler requires a dome segment area of 5.5 ft^2. The bundle diameter plus clearance is approximately 22.4 in. What shell diameter is required?

Solution

Adding 4 in. to the liquid height to account for foaming gives an effective liquid height of 26.4 in. = 2.2 ft. For the first trial, assume the effective liquid height is approximately 60% of the shell diameter. Then,

$$D_s = 2.2/0.6 = 3.67 \text{ ft}$$

Further, the ratio of sector height, h, to circle (shell) diameter is 40%, i.e.,

$$h/D = 1 - 0.6 = 0.4$$

From the table in Appendix 10.A with $h/D = 0.4$, the sector area factor is $A = 0.29337$. This value must be multiplied by the square of the shell diameter to obtain the actual segment area. Thus,

$$SA = 0.29337\,(3.67)^2 = 3.95 \text{ ft}^2$$

Since this is less than the required dome segment area, a larger shell diameter is needed. For the second trial, assume the effective liquid height is 55% of the shell diameter. Then,

$$D_s = \frac{2.2}{0.55} = 4.0 \text{ ft}$$

$$h/D = 1 - 0.55 = 0.45$$

$$A = 0.34278 \text{ (Appendix 10.A)}$$

$$SA = 0.34278(4.0)^2 = 5.48 \cong 5.5 \text{ ft}^2$$

Therefore, a shell diameter of approximately 4 ft is required.

10.3.6 Liquid Overflow Reservoir

With a kettle reboiler, surge capacity is provided by the liquid overflow reservoir in the kettle, as opposed to the column sump when a thermosyphon reboiler is used. The liquid holdup time in the overflow reservoir is usually significantly less than in the column sump due to the cost of extending the length of the K-shell, of which only the bottom portion is useable. The small size and limited

TABLE 10.3 Guidelines for Sizing Steam and Condensate Nozzles

		Nominal Nozzle Diameter (in.)	
Shell OD (in.)	Heat-Transfer Area (ft^2)	Steam	Condensate
16	130	4	1.5
20	215	4	2
24	330–450	6	3
30	525–1065	6–8	3–4
36	735–1520	8	4
42	1400–2180	8	4

Source: Ref. [9]

holdup time can make the liquid level in the reservoir difficult to control, and can lead to relatively large fluctuations in the bottom product flow rate. These fluctuations can adversely affect the operation of downstream units unless a separate surge vessel is provided downstream of the reboiler, or the bottom product flows to storage. These problems can be avoided by eliminating the overflow weir in the kettle [7]. However, a drawback of this strategy is that incomplete separation of reboiler feed and reboiled liquid results in the (partial) loss of one theoretical distillation stage.

10.3.7 Finned Tubing

Radial low-fin tubes and tubes with surface enhancements designed to improve nucleate boiling characteristics can be used in reboilers and vaporizers. They are particularly effective when the temperature driving force is small, and hence they are widely used in refrigeration systems. In addition to providing a large heat-transfer surface per unit volume, finned tubes can result in significantly higher boiling heat-transfer coefficients compared with plain tubes due to the convective effect of two-phase flow between the fins [1]. As the temperature driving force increases, the boiling-side resistance tends to become small compared with the thermal resistances of the tube wall and heating medium, and the advantage of finned tubes is substantially diminished. A detailed treatment of the performance benefits associated with enhanced boiling surfaces is beyond the scope of this book; however, a thorough discussion of various enhancements is provided by Thome [8].

10.3.8 Steam as Heating Medium

When condensing steam is used as a heating medium, it is common practice to use an approximate heat-transfer coefficient on the heating side for design purposes. Typically, a value of 1500 Btu/h · ft^2 · °F (8500 W/m^2 · K) is used. This value is referred to the external tube surface and includes a fouling allowance. Thus, for steam condensing inside plain tubes we have:

$$[(D_o/D_i)\,(1/h_i + R_{Di})]^{-1} \cong 1500\ \text{Btu/h}\cdot\text{ft}^2\cdot{}^\circ\text{F} \cong 8500\ \text{W/m}^2\cdot\text{K}$$

Some guidelines for sizing steam and condensate nozzles are presented in Table 10.3. The data are taken from Ref. [9] and are for vertical thermosyphon reboilers. However, they can be used as general guidelines for all types of reboilers of similar size.

10.3.9 Two-Phase Density Calculation

In order to calculate the static head in the reboiler, the density of the two-phase mixture in the boiling region must be determined. For cross flow over tube bundles, this calculation is usually made using either the homogeneous model, Equation (9.51), or one of the methods for separated flow in tubes, such as the Chisholm correlation, Equation (9.63). Experimental data indicate that neither approach is particularly accurate [10], but there is no entirely satisfactory alternative. The homogeneous model is somewhat easier to use, but the Chisholm correlation will generally give a more conservative (larger) result for the static head.

The following example illustrates the design procedure for kettle reboilers.

Example 10.2

96,000 lb/h of a distillation bottoms having the following composition will be partially vaporized in a reboiler:

Component	Mole %	Critical Pressure (psia)
Propane	15	616.3
i-butane	25	529.0
n-butane	60	551.1

The stream will enter the reboiler as a (nearly) saturated liquid at 250 psia. The dew-point temperature of the stream at 250 psia is 205.6°F. Saturated steam at a design pressure of 20 psia will be used as the heating medium. The reboiler is to supply 48,000 lb/h of vapor to the distillation column. The reboiler feed line will be approximately 23 ft long, while the vapor return line will have a total length of approximately 20 ft. The available elevation difference between the liquid level in the column sump and the reboiler inlet is 9 ft. Physical property data are given in the following table. Design a kettle reboiler for this service.

Property	Reboiler Feed	Liquid Overflow	Vapor Return
T (°F)	197.6	202.4	202.4
H (Btu/lbm)	106.7	109.9	216.4
C_P (Btu/lbm · °F)	0.805	0.811	0.576
k (Btu/h · ft · °F)	0.046	0.046	0.014
μ (cp)	0.074	0.074	0.0095
ρ (lbm/ft^3)	28.4	28.4	2.76
σ (dyne/cm)	3.64	3.59	–
Molecular weight	56.02	56.57	55.48

Solution

(a) Make initial specifications.

(i) Fluid placement

There is no choice here; the boiling fluid must be placed in the shell and the heating medium in the tubes.

(ii) Tubing

One-inch, 14 BWG, U-tubes with a length of 16 ft are specified. A tubing diameter of $^3/_4$ in. could also be used.

(iii) Shell and head types

A TEMA K-shell is chosen for a kettle reboiler, and a type B head is chosen since the tube-side fluid (steam) is clean. Thus, a BKU configuration is specified.

(iv) Tube layout

A square layout with a tube pitch of 1.25 in. is specified to permit mechanical cleaning of the external tube surfaces. Although this service should be quite clean, contaminants in distillation feed streams tend to concentrate in the bottoms, and kettle reboilers are very prone to fouling.

(v) Baffles and sealing strips

None are specified for a kettle reboiler. Support plates will be used to provide tube support and vibration suppression. Four plates are specified to give an unsupported tube length that is safely below the maximum of 73 in. listed in Table 5.C1.

(vi) Construction materials

Since neither stream is corrosive, plain carbon steel is specified for all components.

(b) Energy balance and steam flow rate.

The reboiler duty is obtained from an energy balance on the process stream (boiling fluid):

$$q = \dot{m}_V H_V + \dot{m}_L H_L - \dot{m}_F H_F$$

where the subscripts F, L, and V refer to the reboiler feed, liquid overflow, and vapor return streams, respectively. Substituting the appropriate enthalpies and flow rates gives:

$$q = 48{,}000 \times 216.4 + 48{,}000 \times 109.9 - 96{,}000 \times 106.7$$

$$q \cong 5.42 \times 10^6 \text{ Btu/h}$$

From Table A.8, the latent heat of condensation for steam at 20 psia is 960.1 Btu/lbm. Therefore, the steam flow rate will be:

$$\dot{m}_{steam} = q/\lambda_{steam} = 5.42 \times 10^6/960.1 = 5645 \text{ lbm/h}$$

(c) Mean temperature difference.

The effective mean temperature difference is computed as if the boiling-side temperature were constant at the vapor exit temperature, which in this case is 202.4°F. The temperature of the condensing steam is also constant at the saturation temperature, which is 228.0°F at 20 psia from Table A.8. Therefore, the effective mean temperature difference is:

$$\Delta T_m = 228.0 - 202.4 = 25.6°\text{F}$$

(d) Approximate overall coefficient.
Referring to Table 3.5, it is seen that for light hydrocarbons boiling on the shell side with condensing steam on the tube side, $200 \le U_d \le 300$ Btu/h · ft² · °F. Taking the mid-range value gives $U_D = 250$ Btu/h. · ft² · °F for preliminary design purposes.

(e) Heat-transfer area and number of tubes.

$$A = \frac{q}{U_D\,\Delta T_m} = \frac{5.42 \times 10^6}{250 \times 25.6} \cong 847\ \text{ft}^2$$

$$n_t = \frac{A}{\pi D_o L} = \frac{847}{\pi(1/12) \times 16} = 202$$

Note that n_t represents the number of straight sections of tubing in the bundle, i.e., the number of tube holes in the tubesheet. For U-tubes, this is twice the actual number of tubes. However, it corresponds to the value listed in the tube-count tables, and so will be referred to as the number of tubes.

(f) Number of tube passes.
For condensing steam, two passes are sufficient.

(g) Actual tube count and bundle diameter.
From Table C.5, the closest tube count is 212 tubes in a 23.25 in. shell. This shell size is the smaller diameter of the K-shell at the tubesheet. The bundle diameter will, of course, be somewhat smaller, but a value of 23 in. will be sufficiently accurate for design calculations.

(h) Required overall coefficient.
The required overall heat-transfer coefficient is calculated in the usual manner:

$$U_{req} = \frac{q}{n_t \pi D_o L \Delta T_m} = \frac{5.42 \times 10^6}{212 \times \pi \times (1/12) \times 16 \times 25.6} = 238\ \text{Btu/h} \cdot \text{ft}^2 \cdot {}^\circ\text{F}$$

(i) Inside coefficient, h_i.
For condensing steam we take:

$$[(D_o/D_i)\,(1/h_i + R_{Di})]^{-1} \cong 1500\ \text{Btu/h} \cdot \text{ft}^2 \cdot {}^\circ\text{F}$$

(j) Outside coefficient, $h_o = h_b$.
Palen's [1] method, which was presented in Chapter 9, will be used in order to ensure a safe (i.e., conservative) design. It is based on the Mostinski correlation for the nucleate boiling heat-transfer coefficient, to which correction factors are applied to account for mixture effects and convection in the tube bundle.

(i) Nucleate boiling coefficient, h_{nb}
The first step is to compute the pseudo-critical and pseudo-reduced pressures for the mixture, which will be used in place of the true values in the Mostinski correlation:

$$P_{pc} = \sum x_i\, P_{c,i} = 0.15 \times 616.3 + 0.25 \times 529.0 + 0.60 \times 551.1 = 555.4\ \text{psi}$$

$$P_{pr} = P/P_{pc} = 250/555.4 = 0.45$$

The Mostinski correlation is used as given in Equation (9.2a), along with the mixture correction factor as given by Equation (9.17a). Also, since $P_{pr} > 0.2$, Equation (9.18) is used to calculate the pressure correction factor. Thus,

$$h_{nb} = 0.00622\ P_c^{0.69}\ \hat{q}^{0.7}\ F_P\, F_m$$
$$F_P = 1.8\, P_r^{0.17} = 1.8(0.45)^{0.17} = 1.5715$$
$$F_m = \left(1 + 0.0176\, \hat{q}^{0.15}\, BR^{0.75}\right)^{-1}$$
$$BR = T_D - T_B = 205.6 - 197.6 = 8.0^\circ\text{F}$$

The heat flux is computed using the required duty and the heat-transfer area for the initial configuration:

$$\hat{q} = \frac{q}{n_t \pi\ D_o L} = \frac{5.42 \times 10^6}{212\pi(1/12) \times 16} = 6103\ \ \text{Btu/h} \cdot \text{ft}^2$$

$$F_m = \left[1 + 0.0176(6103)^{0.15}\,(8)^{0.75}\right]^{-1} = 0.7636$$

$$h_{nb} = 0.00622(555.4)^{0.69}(6103)^{0.7} \times\ 1.5715 \times 0.7636$$

$$h_{nb} = 261\ \text{Btu/h} \cdot \text{ft}^2 \cdot {}^\circ\text{F}$$

(ii) Bundle boiling coefficient, h_b
The boiling heat-transfer coefficient for the tube bundle is given by Equation (9.19):

$$h_b = h_{nb}\, F_b + h_{nc}$$

Although the tube wall temperature is unknown, with an overall temperature difference of 25.6°F, the heat transfer by natural convection should be small compared to the boiling component. Therefore, h_{nc} is roughly estimated as 44 Btu/h · ft^2 · °F. The bundle convection factor is computed using Equation (9.20) with $D_b \cong 23$ in.:

$$F_b = 1.0 + 0.1\left[\frac{0.785\ D_b}{C_1(P_T/D_o)^2 D_o} - 1.0\right]^{0.75}$$

$$= 1.0 + 0.1\left[\frac{0.785 \times 23}{1.0(1.25/1.0)^2 \times 1.0} - 1.0\right]^{0.75}$$

$$F_b = 1.5856$$

The outside coefficient is then:

$$h_o = h_b = 261 \times 1.5856 + 44 \cong 458 \text{ Btu/h} \cdot \text{ft}^2 \cdot {}^\circ\text{F}$$

(k) Overall coefficient.

$$U_D = \left[(1/h_i + R_{Di})\ (D_o/\ D_i) + \frac{D_o\ \ln\ (D_o/\ D_i)}{2\ k_{tube}} + 1/h_o + R_{Do}\right]^{-1}$$

Based on the values in Table 10.2, a boiling-side fouling allowance of 0.0005 h · ft^2 · °F/Btu is deemed appropriate for this service. For 1-in. 14 BWG tubes, $D_i = 0.834$ in. from Table B.1. Taking $k_{tube} \cong 26$ Btu/h · ft · °F for carbon steel, we obtain:

$$U_D = \left[(1/1500) + \frac{(1.0/12)\ \ln(1.0/0.834)}{2 \times 26} + 1.0/458 + 0.0005\right]^{-1}$$

$$U_D = 275 \text{ Btu/h} \cdot \text{ft}^2 \cdot {}^\circ\text{F}$$

(l) Check heat flux and iterate if necessary.

A new value of the heat flux can be obtained using the overall coefficient:

$$\hat{q} = U_D\,\Delta T_m = 275 \times 25.6 = 7040 \text{ Btu/h} \cdot \text{ft}^2$$

Since this value differs significantly from that used to calculate h_{nb}, steps (j) and (k) should be repeated until consistent values for $\hat{q}$ and U_D are obtained. Due to the uncertainty in both the heat-transfer coefficient and the mean temperature difference, exact convergence is not required. The following values are obtained after several more iterations:

$$h_b \cong 523 \text{ Btu/h} \cdot \text{ft}^2 \cdot {}^\circ\text{F}$$

$$U_D \cong 297 \text{ Btu/h} \cdot \text{ft}^2 \cdot {}^\circ\text{F}$$

$$\hat{q} \cong 7600 \text{ Btu/h} \cdot \text{ft}^2$$

The overall coefficient exceeds the required coefficient of 238 Btu/h · ft^2 · °F by a significant amount (over-design = 25%), indicating that the reboiler is over-sized.

(m) Critical heat flux.

The critical heat flux for nucleate boiling on a single tube is calculated using the Mostinski correlation, Equation (9.23a):

$$\hat{q}_c = 803\ P_c P_r^{0.35}(1 - P_r)^{0.9}$$

$$= 803 \times 555.4(0.45)^{0.35}(1 - 0.45)^{0.9}$$

$$\hat{q}_c = 196{,}912 \text{ Btu/h} \cdot \text{ft}^2$$

The critical heat flux for the bundle is obtained from Equation (9.24):

$$\hat{q}_{c,bundle} = \hat{q}_{c,tube}\ \phi_b = 196{,}912\ \phi_b$$

The bundle geometry parameter is given by:

$$\psi_b = \frac{D_b}{n_t\ D_o} = \frac{23}{212 \times 1.0} = 0.1085$$

Since this value is less than 0.323, the bundle correction factor is calculated as:

$$\phi_b = 3.1\ \psi_b = 3.1 \times 0.1085 = 0.3364$$

Therefore, the critical heat flux for the bundle is:

$$\hat{q}_{c,bundle} = 196{,}912 \times 0.3364 \cong 66{,}240 \text{ Btu/h} \cdot \text{ft}^2$$

Now the ratio of the actual heat flux to the critical heat flux is:

$$\hat{q}/\hat{q}_{c,bundle} = 7600/66,240 \cong 0.11$$

This ratio should not exceed 0.7 in order to provide an adequate safety margin for reliable operation of the reboiler. In the present case, this criterion is easily met.

Note: In addition, the process-side temperature difference, ΔT_e, must be in the nucleate boiling range. In operation, the value of ΔT_e may exceed the maximum value for nucleate boiling, particularly when the unit is clean. This situation can usually be rectified by adjusting the steam pressure. Maximum values of ΔT_e are tabulated for a number of substances in Ref. [11], and they provide guidance in specifying an appropriate design temperature for the heating medium in these and similar cases. For a given substance, the critical ΔT_e decreases markedly with increasing pressure. It is sometimes stated that the overall ΔT should not exceed about 90°F to 100°F in order to ensure nucleate boiling. However, this rule is not generally valid owing (in part) to the effect of pressure on the critical ΔT_e.

(n) Design modification.

The simplest way to modify the initial design in order to reduce the amount of heat-transfer area is to shorten the tubes. The required tube length is calculated as follows:

$$L_{req} = \frac{q}{n_t \pi D_o U_D \Delta T_m} = \frac{5.42 \times 10^6}{212\pi(1/12) \times 297 \times 25.6}$$

$$L_{req} = 12.8 \text{ ft}$$

Therefore, a tube length of 13 ft will be sufficient. (Note that a length of 12.8 ft is consistent with a heat flux of 7600 Btu/h · ft^2.) A second option is to reduce the number of tubes. From the tube-count table, the next smallest standard bundle (21.25 in.) contains 172 tubes. This modification will not be pursued here; it is left as an exercise for the reader to determine the suitability of this configuration.

(o) Number of nozzles.

Equation (10.1) gives the number of pairs of nozzles:

$$N_n = \frac{L}{5D_b} = \frac{13}{5(23/12)} = 1.36$$

Rounding upward to the next largest integer gives two pairs of inlet and outlet nozzles. They will be spaced approximately 4.4 ft apart. Alternatively, a single pair of nozzles could be used with the inlet nozzle located near the front of the shell.

(p) Shell diameter.

We first use Equation (10.2) to calculate the vapor loading:

$$VL = 2290\,\rho_V\left(\frac{\sigma}{\rho_L - \rho_V}\right)^{0.5} = 2290 \times 2.76\left(\frac{3.59}{28.4 - 2.76}\right)^{0.5}$$

$$VL = 2365 \text{ lbm/h} \cdot \text{ft}^3$$

The required dome segment area is then found using Equation (10.3):

$$SA = \frac{\dot{m}_V}{L \times VL} = \frac{48,000}{13 \times 2365} \cong 1.56 \text{ ft}^2$$

Next, the effective liquid height in the reboiler is estimated by adding 4 in. to the approximate bundle diameter (23 in.) to account for foaming, giving a value of 27 in. Assuming as a first approximation that the liquid height is 60% of the shell diameter, we obtain:

$$D_s = \frac{27}{0.6} = 45.0 \text{ in.} \cong 3.75 \text{ ft}$$

$$h/D = 1 - 0.6 = 0.4$$

The sector area factor is obtained from Appendix 10.A:

$$A = 0.29337$$

Multiplying this factor by the square of the diameter gives the segment area:

$$SA = 0.29337(3.75)^2 = 4.13 \text{ ft}^2$$

Since this is much greater than the required area, a smaller diameter is needed. Assuming (after several more trials) that the effective liquid height is 73% of the shell diameter, the next trial gives:

$$D_s = \frac{27}{0.73} = 36.99 \text{ in.} \cong 3.08 \text{ ft}$$

$$h/D = 1 - 0.73 = 0.27$$

$$A = 0.17109 \text{ (Appendix 10.A)}$$

$$SA = 0.17109(3.08)^2 = 1.62 \text{ ft}^2$$

This value is slightly larger than the required dome segment area, which is acceptable. Therefore, a shell diameter of about 37 in. will suffice.

(q) Liquid overflow reservoir.
The reservoir is sized to provide adequate holdup time for control purposes. We first calculate the volumetric flow rate of liquid over the weir:

$$\text{volumetric flow rate} = \frac{48,000 \text{ lbm/h}}{(28.4 \text{ lbm/ft}^3)(60 \text{ min/h})} = 28.17 \text{ ft}^3/\text{min}$$

Next, the cross-sectional area of the shell sector below the weir is calculated. The sector height is equal to the weir height, which is about 23 in. Therefore,

$$h/D = 23/37 = 0.62$$

$$1 - h/D = 0.38$$

The sector area factor corresponding to this value is 0.27386 from Appendix 10.A. Hence,

$$\text{sector area above weir} = 0.27386(37/12)^2 = 2.60 \text{ ft}^2$$

$$\text{sector area below weir} = \pi(37/12)^2/4 - 2.60 = 4.87 \text{ ft}^2$$

Now the shell length required is:

$$L_s = \frac{28.17 \text{ ft}^3/\text{min}}{4.87 \text{ ft}^2} \cong 5.8 \text{ ft/min of holdup}$$

Therefore, a reservoir length of 3 ft will provide a holdup time of approximately 30 s, which is adequate to control the liquid level using a standard cascaded level-to-flow control loop. With allowances for U-tube return bends and clearances, the overall length of the shell will then be about 17 ft. It is assumed that relatively large fluctuations in the bottom product flow rate are acceptable in this application.

(r) Feed and return lines.
The available liquid head between the reboiler inlet and the surface of the liquid in the column sump is 9 ft. The corresponding pressure difference is:

$$\Delta P_{available} = \rho_L(g/g_c)\,\Delta h_L = 28.4\,(1.0) \times 9$$

$$\Delta P_{available} = 255.6 \text{ lbf/ft}^2 = 1.775 \text{ psi}$$

This pressure difference must be sufficient to compensate for the friction losses in the feed line, vapor return line, and the reboiler itself; the static heads in the reboiler and return line; and the pressure loss due to acceleration of the fluid in the reboiler resulting from vapor formation. Of these losses, only the friction losses in the feed and return lines can be readily controlled, and these lines must be sized to meet the available pressure drop. We consider each of the pressure losses in turn.

(i) Static heads
The static head consists of two parts, namely, the two-phase region between the reboiler inlet and the surface of the boiling fluid, and the vapor region from the surface of the boiling fluid through the return line and back down to the liquid surface in the column sump. We estimate the two-phase head loss using the average vapor fraction in the boiling region, $x_{ave} = 0.25$. The average density is calculated using the homogeneous model, which is sufficiently accurate for the present purpose:

$$\rho_{ave} = \left[\frac{1 - x_{ave}}{\rho_L} + \frac{x_{ave}}{\rho_V}\right]^{-1} = \left[\frac{0.75}{28.4} + \frac{0.25}{2.76}\right]^{-1} \cong 8.55 \text{ lbm/ft}^3$$

The vertical distance between the reboiler inlet and the surface of the boiling fluid is approximately 23 in. The corresponding static pressure difference is:

$$\Delta P_{tp} = \frac{8.55 \times (23/12)}{144} = 0.114 \text{ psi}$$

The elevation difference between the boiling fluid surface in the reboiler and the liquid surface in the column sump is:

$$\Delta h = 9 - 23/12 = 7.08 \text{ ft}$$

The pressure difference corresponding to this head of vapor is:

$$\Delta P_V = \frac{2.76 \times 7.08}{144} \cong 0.136 \text{ psi}$$

The total pressure difference due to static heads is the sum of the above values:

$$\Delta P_{static} = 0.114 + 0.136 = 0.250 \text{ psi}$$

(ii) Friction and acceleration losses in reboiler
The friction loss is small due to the low circulation rate characteristic of kettle reboilers. The large vapor volume provided in the kettle results in a relatively low vapor velocity, and therefore the acceleration loss is also small. Hence, both these losses can be neglected. However, as a safety factor, an allowance of 0.2 psi will be made for the sum of these losses. (A range of 0.2–0.5 psi is typical for thermosyphon reboilers, so an allowance of 0.2 psi should be more than adequate for a kettle.)

(iii) Friction loss in feed lines
We begin by assuming the configuration shown below for the feed lines. The total length of the primary line between the column sump and the tee is approximately 23 ft as given in the problem statement. Each branch of the secondary line between the tee and the reboiler has a horizontal segment of length 2.2 ft and a vertical segment of length 1.0 ft.

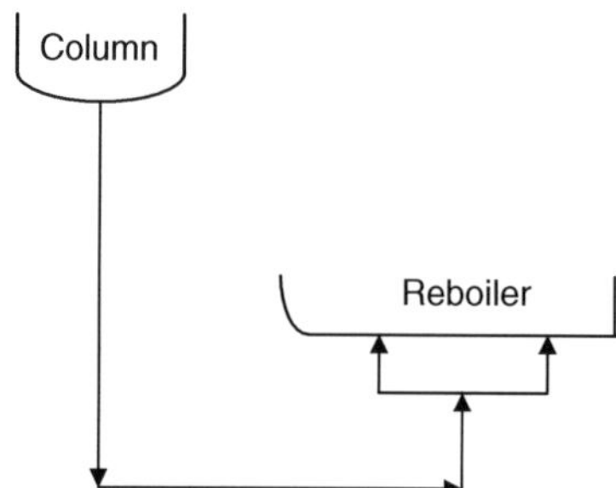

The pipe diameter is chosen to give a liquid velocity of about 5 ft/s. Thus, for the primary line:

$$D_i = \left(\frac{4\dot{m}}{\pi \rho V}\right)^{1/2} = \left[\frac{4(96,000/3600)}{\pi \times 28.4 \times 5}\right]^{1/2}$$

$$D_i = 0.49 \text{ ft} = 5.87 \text{ in.}$$

From Table B.2, a 6-in. schedule 40 pipe with an inside diameter of 6.065 in. is appropriate.

For the secondary line, the flow rate is halved. Therefore,

$$D_i = \left[\frac{4(48,000/3600)}{\pi \times 28.4 \times 5}\right]^{1/2} = 0.0346 \text{ ft} = 4.15 \text{ in.}$$

A 4-in. schedule 40 pipe with an inside diameter of 4.026 in. is the closest match. However, with 4-in. inlet nozzles, the value of ρV^2 will exceed the TEMA erosion prevention limit of 500 lbm/ft · s^2 for bubble-point liquids. Therefore, in order to avoid the need for impingement protection, 5-in. nozzles with inside diameter of 5.047 in. and matching piping will be used.

The pressure drop is computed using the equivalent pipe lengths for flow resistance of fittings given in Appendix D. The equivalent lengths for the two pipe sizes are tabulated below. Note that only one branch of the 5-in. pipe is used because the pressure drop is the same for each parallel branch.

Item	Equivalent Length of 6-in. Pipe (ft)	Equivalent Length of 5-in. Pipe (ft)
Straight pipe sections	23	3.2
90° elbows	20	8.5
Tee	30	–
6″ × 5″ reducer	–	4
Entrance loss	18	–
Exit loss	–	28
Total	91	43.7

The Reynolds number for the 6-in. pipe is:

$$Re = \frac{4\dot{m}}{\pi D_i \mu} = \frac{4 \times 96,000}{\pi(6.065/12) \times 0.074 \times 2.419} = 1.351 \times 10^6$$

The friction factor is calculated using Equation (4.8):

$$f = 0.3673\, Re^{-0.2314} = 0.3673\left(1.351 \times 10^6\right)^{-0.2314}$$

$$f = 0.014$$

The pressure drop is given by Equation (4.5) with the equivalent pipe length used in place of the actual length. The mass flux and specific gravity are computed first:

$$G = \dot{m}/A_{flow} = \frac{96,000}{(\pi/4)(6.065/12)^2} = 478,500 \text{ lbm/h} \cdot \text{ft}^2$$

$$s = \rho/\rho_{water} = 28.4/62.43 = 0.455$$

$$\Delta P_f = \frac{f\, LG^2}{7.50 \times 10^{12} D_i\, s\, \phi} = \frac{0.014 \times 91(478,500)^2}{7.50 \times 10^{12}(6.065/12) \times 0.455 \times 1.0}$$

$$\Delta P_f \cong 0.169 \text{ psi}$$

The calculations for the 5-in. pipe are similar:

$$Re = \frac{4 \times 48,000}{\pi(5.047/12) \times 0.074 \times 2.419} = 811,768$$

$$f = 0.3673(811,768)^{-0.2314} \cong 0.0158$$

$$G = \frac{48,000}{(\pi/4)(5.047/12)^2} = 345,499 \text{ lbm/h} \cdot \text{ft}^2$$

$$\Delta P_f = \frac{0.0158 \times 43.7(345,499)^2}{7.50 \times 10^{12}(5.047/12) \times 0.455 \times 1.0}$$

$$\Delta P_f \cong 0.0574 \text{ psi}$$

The total friction loss in the feed lines is therefore:

$$\Delta P_{feed} = 0.169 + 0.0574 \cong 0.226 \text{ psi}$$

(iv) Friction loss in return lines
A return line configuration similar to that of the feed line is assumed as shown below. The primary line has a total length of 20 ft as given in the problem statement. Each branch of the line connected to the reboiler has a vertical segment of length 1.0 ft and a horizontal segment of length 2.2 ft.

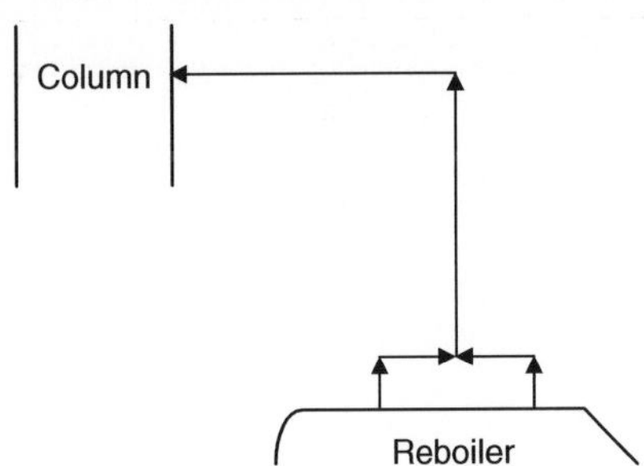

We begin by calculating the maximum recommended vapor velocity using Equation (5.B.1):

$$V_{max} = \frac{1800}{(P\,M)^{0.5}} = \frac{1800}{(250 \times 55.48)^{0.5}} = 15.3 \text{ ft/s}$$

The lines will be sized for a somewhat lower velocity of about 12 ft/s. For the main line, the required diameter is:

$$D_i = \left(\frac{4\dot{m}}{\pi\rho V}\right)^{1/2} = \left[\frac{4(48,000/3600)}{\pi \times 2.76 \times 12}\right]^{1/2}$$

$$D_i = 0.716 \text{ ft} = 8.59 \text{ in.}$$

From Table B.2, the closest match is an 8-in. schedule 40 pipe with an internal diameter of 7.981 in. For the split-flow section, we have:

$$D_i = \left[\frac{4(29,000/3600)}{\pi \times 2.76 \times 12}\right]^{1/2} = 0.506 \text{ ft} = 6.07 \text{ in.}$$

Six-inch schedule 40 pipe (ID = 6.065 in.) is appropriate for this section. Equivalent pipe lengths are summarized in the following table:

Item	Equivalent Length of 8-in. Pipe (ft)	Equivalent Length of 6-in. Pipe (ft)
Straight pipe sections	20	3.2
90° elbow	14	10
Tee	40	–
6″ × 8″ expander	–	7
Entrance loss	–	18
Exit loss	48	–
Total	122	38.2

The calculations for the 8-in. line are as follows:

$$Re = \frac{4\dot{m}}{\pi D_i \mu} = \frac{4 \times 48,000}{\pi(7.981/12) \times 0.0095 \times 2.419} = 3.999 \times 10^6$$

$$f = 0.3673\, Re^{-0.2314} = 0.3673\left(3.999 \times 10^6\right)^{-0.2314} \cong 0.0109$$

$$G = \dot{m}/A_{flow} = \frac{48,000}{(\pi/4)(7.981/12)^2} = 138,165 \text{ lbm/h} \cdot \text{ft}^2$$

$$s = \rho/\rho_{water} = 2.76/62.43 = 0.0442$$

$$\Delta P_f = \frac{f\,LG^2}{7.50 \times 10^{12} D_i\, s\,\phi} = \frac{0.0109 \times 122(138,165)^2}{7.50 \times 10^{12}(7.981/12) \times 0.0442 \times 1.0}$$

$$\Delta P_f \cong 0.115 \text{ psi}$$

The calculations for the 6-in. line are similar, but the flow rate is halved:

$$Re = \frac{4 \times 24,000}{\pi(6.065/12) \times 0.0095 \times 2.419} = 2.631 \times 10^6$$

$$f = 0.3673\left(2.631 \times 10^6\right)^{-0.2314} = 0.012$$

$$G = \frac{24,000}{(\pi/4)(6.065/12)^2} = 119,625 \text{ lbm/h} \cdot \text{ft}^2$$

$$\Delta P_f = \frac{0.012 \times 38.2(119,625)^2}{7.50 \times 10^{12}(6.065/12) \times 0.0442 \times 1.0}$$

$$\Delta P_f \cong 0.0392 \text{ psi}$$

The total friction loss in the return lines is thus:

$$\Delta P_{return} = 0.115 + 0.0392 \cong 0.154 \text{ psi}$$

(v) Total pressure loss

The total pressure loss is the sum of the individual losses calculated above:

$$\Delta P_{total} = \Delta P_{static} + \Delta P_{reboiler} + \Delta P_{feed} + \Delta P_{return}$$

$$= 0.191 + 0.2 + 0.226 + 0.154$$

$$\Delta P_{total} = 0.770 \text{ psi}$$

Since this value is less than the available pressure drop of 1.775 psi, the piping configuration is acceptable. However, a smaller feed line can be used. Replacing the 6-in. line with a 5-in. line gives a fluid velocity of 6.8 ft/s and a total pressure drop of 1.0 psi, both of which are still comfortably on the safe side. In actual operation, the liquid level in the column sump will self-adjust to satisfy the pressure balance.

(s) Tube-side pressure drop.

The pressure drop for condensing steam is usually small due to the low flow rate compared with sensible heating media. For completeness, however, the pressure drop is estimated here. For a condensing vapor, the two-phase pressure drop in the straight sections of tubing can be approximated by half the pressure drop calculated at the inlet conditions (saturated steam at 20 psia, vapor fraction = 1.0). The requisite physical properties of steam are obtained from Tables A.8 and A.9:

$$\rho = 1/20.087 = 0.0498 \text{ lbm/ft}^3$$

$$s = \rho/\rho_{water} = 0.0498/62.43 = 0.000797$$

$$\mu = 0.012 \text{ cp}$$

$$\dot{m} = 5645 \text{ lbm/h (from step (b))}$$

$$\dot{m}_{per\ tube} = \dot{m}(n_p/n_t) = 5645(2/212) = 53.25 \text{ lbm/h}$$

$$G = \frac{\dot{m}_{per\ tube}}{(\pi/4)D_i^2} = \frac{53.25}{(\pi/4)(0.834/12)^2} = 14,037 \text{ lbm/h} \cdot \text{ft}^2$$

$$Re = \frac{D_i\, G}{\mu} = \frac{(0.834/12) \times 14,037}{0.012 \times 2.419} = 33,608$$

The friction factor is calculated using Equation (5.2):

$$f = 0.4137\, Re^{-0.2585} = 0.4137(33,608)^{-0.2585}$$

$$f = 0.0280$$

The pressure drop is calculated by incorporating a factor of 1/2 on the right side of Equation (5.1):

$$\Delta P_f \cong \frac{1}{2}\left[\frac{f\, n_p L G^2}{7.50 \times 10^{12} D_i s \phi}\right] = \frac{1}{2}\left[\frac{0.0280 \times 2 \times 13(14,037)^2}{7.50 \times 10^{12}(0.834/12) \times 0.000797 \times 1.0}\right]$$

$$\Delta P_f \cong 0.173 \text{ psi}$$

To this degree of approximation, the pressure drop in the return bends can be neglected. However, the pressure drop in the nozzles will be calculated to check the nozzle sizing. Based on Table 10.3, 6-in. and 3-in. schedule 40 nozzles are selected for steam and condensate, respectively. For the steam nozzle we have:

$$G_n = \frac{\dot{m}}{(\pi/4)D_i^2} = \frac{5645}{(\pi/4)(6.065/12)^2} = 28,137 \text{ lbm/h} \cdot \text{ft}^2$$

$$Re_n = \frac{D_i G_n}{\mu} = \frac{(6.065/12) \times (28,137)}{0.012 \times 2.419} = 489,903$$

Since the flow is turbulent, allow 1 velocity head for the inlet nozzle loss. From Equation (4.11), we obtain:

$$\Delta P_{n,steam} = 1.334 \times 10^{-13}\, G_n^2/s = \frac{1.334 \times 10^{-13}(28,137)^2}{0.000797}$$

$$\Delta P_{n,steam} = 0.133 \text{ psi}$$

For the condensate at 20 psia, the physical properties are obtained from Tables A.8 and A.9.

$$\rho = 1/0.016834 = 59.40 \text{ lbm/ft}^3$$

$$s = \rho/\rho_{water} = 59.40/62.43 = 0.9515$$

$$\mu = 0.255 \text{ cp}$$

$$G_n = \frac{\dot{m}}{(\pi/4)D_i^2} = \frac{5645}{(\pi/4)(3.068/12)^2} = 109,958 \text{ lbm/h} \cdot \text{ft}^2$$

$$Re_n = \frac{D_i\, G_n}{\mu} = \frac{(3.068/12) \times 109,958}{0.255 \times 2.419} = 45,575$$

Since the flow is turbulent, allow 0.5 velocity head for the loss in the exit nozzle:

$$\Delta P_{n,condensate} = \frac{0.5 \times 1.334 \times 10^{-13}(109,958)^2}{0.9515} = 0.00085 \text{ psi}$$

The total tube-side pressure drop is estimated as:

$$\Delta P_i \cong \Delta P_f + \Delta P_{n,steam} + \Delta P_{n,condensate}$$

$$\Delta P_i = 0.173 + 0.133 + 0.00085 \cong 0.3 \text{ psi}$$

The pressure drop is small, as it should be for condensing steam. Therefore, the tubing and nozzle configurations are acceptable. The final design parameters are summarized below.

Design summary
Shell type: BKU
Port diameter/Shell ID: 23.25 in./37 in.
Shell length: approximately 17 ft
Length beyond weir: 3 ft
Weir height: approximately 23 in.
Tube bundle: 212 tubes (106 U-tubes), 1 in. OD, 14 BWG, 13 ft long on 1.25 in. square pitch
Baffles: none
Support plates: 3 (One less plate is used due to the reduced tube length.)
Shell-side nozzles: two 5-in. schedule 40 inlet, two 6-in. schedule 40 vapor outlet, one 4-in. schedule 40 liquid outlet
Tube-side nozzles: 6-in. schedule 40 inlet, 3-in. schedule 40 outlet
Feed lines: 5-in. schedule 40 from column to inlet tee, 5-in. schedule 40 from tee to reboiler
Return lines: 6-in. schedule 40 from reboiler to outlet tee, 8-in. schedule 40 from tee to column
Materials: plain carbon steel throughout

Note: Due to the elevated pressure on the shell side of the reboiler, it is particularly important in this application to check the required wall thickness of the shell-side nozzles. By running a mechanical design program, it was found that schedule 40 pipe is inadequate for these nozzles. Schedule 80 pipe is required for the vapor exit nozzles, while schedule 120 pipe is needed for the liquid exit and feed nozzles.

10.4 Design of Horizontal Thermosyphon Reboilers

10.4.1 Design Strategy

The boiling-side circulation in a horizontal thermosyphon reboiler is similar to that in a kettle reboiler, particularly when a cross-flow shell (X-shell) is used. With G- and H-shells, the horizontal baffle(s) impart additional axial flow components, so the overall flow pattern is more a mixture of cross flow and axial flow. The higher circulation rate typical of thermosyphons also results in a higher shell-side heat-transfer coefficient and pressure drop relative to kettles, as well as a higher mean temperature difference due to better mixing in the shell.

The above differences notwithstanding, an approximate computational scheme similar to that used for kettle reboilers can be applied to horizontal thermosyphon units. Notice from Equation (9.20) that the bundle convection factor, F_b, depends only on the bundle geometry and is independent of the circulation rate. Therefore, to this degree of approximation, the heat-transfer coefficient is independent of circulation rate, and the heat transfer and hydraulics are decoupled. Clearly, this approximation is conservative for thermosyphon units.

Due to the difficulty of calculating the two-phase pressure drop in a horizontal tube bundle with a flow area that varies with vertical position, it is not practical to calculate the pressure drop in a horizontal thermosyphon reboiler within the framework of an approximate method suitable for hand calculations. As an expedient alternative, an average value of 0.35 psi can be used to estimate the sum of the friction and acceleration losses in the reboiler.

To account for the higher mean temperature difference in a horizontal thermosyphon relative to a kettle reboiler, Palen [1] recommends using a co-current LMTD as a conservative approximation for the mean driving force. That is, the LMTD is calculated as if the shell-side and tube-side fluids were flowing co-currently.

With the heat transfer and hydraulics decoupled, the hydraulic calculations can, in principle, be performed in a manner similar to that used for the kettle reboiler in Example 10.2. In the thermosyphon case, however, the calculations are considerably more difficult. The fluid in the return line from the reboiler is a vapor-liquid mixture, so two-phase flow calculations are required. Also, in a recirculating unit the circulation rate is determined by a balance between the available static head of liquid in the column sump and the losses in the feed lines, return lines, and reboiler. Therefore, closure of the pressure balance must be attained to within reasonable accuracy. Furthermore, the pressure drop in the return lines depends on the vapor fraction, which in turn depends on the circulation rate. The upshot is that an iterative procedure is required to size the connecting lines and determine the circulation rate and vapor fraction.

More rigorous computational methods, suitable for computer implementation, are discussed in Refs. [12,13].

10.4.2 Design Guidelines

The recommendations for fouling factors and number of nozzles given in Section 10.3 for kettle reboilers are also applicable to horizontal thermosyphon reboilers, as are the guidelines given for steam as the heating medium. The clearance between the top of the tube bundle and the shell is much less than in kettle reboilers, since vapor-liquid disengagement is not required in a thermosyphon unit. One rule of thumb is to make the clearance cross-sectional area equal to approximately half the outlet nozzle flow area [14].

TEMA G- and H-shells are preferred for wide boiling mixtures because the horizontal baffles in these units help to reduce flashing of the lighter components. Flashing leaves the liquid enriched in the higher boiling components, which reduces the temperature driving force and, hence, the rate of heat transfer. The total length of the horizontal baffle(s) in these units is about two-thirds of the shell length.

In order to prevent unstable operation of the reboiler system, the velocity of the two-phase mixture in the return line should not exceed the following value [15]:

$$V_{max} = \left(4000/\rho_{tp}\right)^{0.5} \tag{10.4}$$

where

V_{max} = maximum velocity (ft/s)
ρ_{tp} = density of two-phase mixture (lbm/ft^3)

A complete design problem will not be worked here due to the lengthiness of the calculations. However, the following example illustrates the thermal analysis of a horizontal thermosyphon reboiler.

Example 10.3

A reboiler for a revamped distillation column in a refinery must supply 60,000 lb/h of vapor consisting of a petroleum fraction. The stream from the column sump will enter the reboiler as a (nearly) saturated liquid at 35 psia. The dew-point temperature of this stream is 321°F at 35 psia, and approximately 20% by weight will be vaporized in the reboiler. The properties of the reboiler feed and the vapor and liquid fractions of the return stream are given in the following table:

Property	Reboiler Feed	Liquid Return	Vapor Return
T (°F)	289	298.6	298.6
H (Btu/lbm)	136.6	142.1	265.9
C_P (Btu/lbm · °F)	0.601	0.606	0.494
k (Btu/h · ft · °F)	0.055	0.054	0.014
μ (cp)	0.179	0.177	0.00885
ρ (lbm/ft^3)	39.06	38.94	0.4787
σ (dyne/cm)	11.6	11.4	–
P_{pc} (psia)	406.5	–	–

Heat will be supplied by a Therminol® synthetic liquid organic heat-transfer fluid with a temperature range of 420°F to 380°F. The allowable pressure drop is 10 psi. Average properties of the Therminol® are given in the table below:

Property	Therminol® at $T_{ave} = 400°F$
C_P (Btu/lbm · °F)	0.534
k (Btu/h · ft · °F)	0.0613
μ (cp)	0.84
s	0.882
Pr	17.70

A used horizontal thermosyphon reboiler consisting of a 23.25-in. ID TEMA X-shell with 145 U-tubes (tube count of 290) is available at the plant site. The tubes are $^3/_4$ -in. OD, 14 BWG, 16 ft long on a 1.0-in. square pitch, and the bundle, which is configured for two passes, has a diameter of approximately 20 in. Tube-side nozzles consist of 6-in. schedule 40 pipe. Material of construction is plain carbon steel throughout. Will the reboiler be suitable for this service?

Solution

(a) Energy balances.
The energy balance for the boiling fluid is:

$$q = \dot{m}_V H_V + \dot{m}_L H_L - \dot{m}_F H_F$$

The feed rate to the reboiler is 60,000/0.20 = 300,000 lbm/h, and the liquid return rate is 300,000 – 60,000 = 240,000 lbm/h. Therefore,

$$q = 60,000 \times 265.9 + 240,000 \times 142.1 - 300,000 \times 136.6$$

$$q = 9,078,000 \text{ Btu/h}$$

The energy balance for the Therminol® is:

$$q = (\dot{m} C_P \Delta T)_{Th}$$

$$9,078,000 = \dot{m}_{Th} \times 0.534(420 - 380)$$

$$\dot{m}_{Th} = 425,000 \text{ lbm/h}$$

(b) Mean temperature difference.
The effective mean temperature difference is computed as if the flow were co-current:

$$\Delta T = 131°F \left\{ \begin{array}{ccc} 289°F & \rightarrow & 298.6°F \\ 420°F & \rightarrow & 380°F \end{array} \right\} \Delta T = 81.4°F$$

$$\Delta T_{mean} \cong (\Delta T_{ln})_{co-current} = \frac{131 - 81.4}{\ln(131/81.4)} = 104.2°F$$

(c) Heat-transfer area.

$$A = n_t \pi D_o L = 290 \times \pi \times (0.75/12) \times 16 = 911 \text{ ft}^2$$

(d) Required overall coefficient.

$$U_{req} = \frac{q}{A \Delta T_{mean}} = \frac{9,078,000}{911 \times 104.2} = 96 \text{ Btu/h} \cdot \text{ft}^2 \cdot °F$$

(e) Inside coefficient, h_i.

$$D_i = 0.584 \text{ in.} \quad \text{(Table B.1)}$$

$$G = \frac{\dot{m}(n_p/n_t)}{(\pi/4)D_i^2} = \frac{425,000(2/290)}{(\pi/4)(0.584/12)^2} = 1,575,679 \text{ lbm/h} \cdot \text{ft}^2$$

$$Re = D_i G/\mu = \frac{(0.584/12) \times 1,575,679}{0.84 \times 2.419} = 37,738$$

Since the flow is turbulent, Equation (4.1) is used to calculate h_i:

$$Nu = 0.023Re^{0.8}Pr^{1/3}(\mu/\mu_w)^{0.14} = 0.023(37,738)^{0.8}(17.70)^{1/3}(1.0)$$

$$Nu = 274.9$$

$$h_i = (k/D_i)Nu = \frac{0.0613 \times 274.9}{(0.584/12)} = 346 \text{ Btu/h} \cdot \text{ft}^2 \cdot {}^\circ\text{F}$$

(f) Outside coefficient, $h_o = h_b$.

(i) Nucleate boiling coefficient

The pseudo-reduced pressure is used in place of the reduced pressure:

$$P_{pr} = P/P_{pc} = 35/406.5 = 0.0861$$

Since this value is less than 0.2, Equation (9.5) is used to calculate the pressure correction factor in the Mostinski correlation:

$$F_P = 2.1P_r^{0.27} + \left[9 + \left(1 - P_r^2\right)^{-1}\right]P_r^2$$

$$= 2.1(0.0861)^{0.27} + \left\{9 + \left[1 - (0.0861)^2\right]^{-1}\right\}(0.0861)^2$$

$$F_P = 1.1573$$

The required duty and actual heat-transfer area in the exchanger are used to calculate the heat flux:

$$\hat{q} = q/A = 9,078,000/911 = 9965 \text{ Btu/h} \cdot \text{ft}^2$$

The boiling range is calculated from the given data and used to compute the mixture correction factor using Equation (9.17a):

$$BR = T_D - T_B = 321 - 289 = 32^\circ\text{F}$$

$$F_m = \left(1 + 0.0176\,\hat{q}^{0.15}BR^{0.75}\right)^{-1}$$

$$= \left[1 + 0.0176(9965)^{0.15}(32)^{0.75}\right]^{-1}$$

$$F_m = 0.5149$$

The nucleate boiling coefficient is obtained by substituting the above values into the Mostinski correlation, Equation (9.2a):

$$h_{nb} = 0.00622P_c^{0.69}\hat{q}^{0.7}F_PF_m$$

$$= 0.0062(406.5)^{0.690}(9965)^{0.7} \times 1.1573 \times 0.5149$$

$$h_{nb} = 147 \text{ Btu/h} \cdot \text{ft}^2 \cdot {}^\circ\text{F}$$

(ii) Bundle boiling coefficient, h_b

The correction factor for bundle convective effects is calculated using Equation (9.20):

$$F_b = 1.0 + 0.1\left[\frac{0.785D_b}{C_1(P_T/D_O)^2 \times D_O} - 1.0\right]^{0.75}$$

$$= 1.0 + 0.1\left[\frac{0.785 \times 20}{1.0(1.0/0.75)^2 \times 0.75} - 1.0\right]^{0.75}$$

$$F_b = 1.5947$$

A rough approximation of 44 Btu/h · ft^2 · °F is adequate for the natural convection coefficient, h_{nc}, because the temperature difference is large. The boiling coefficient for the bundle is given by Equation (9.19):

$$h_b = h_{nb}F_b + h_{nc} = 147 \times 1.5947 + 44$$

$$h_b = 278 \text{ Btu/h} \cdot \text{ft}^2 \cdot {}^\circ\text{F} = h_o$$

(g) Fouling factors.
Based on the guidelines in Table 10.2, the fouling factors are chosen as follows:

$$R_{Di} = 0.0005\ \text{h} \cdot \text{ft}^2 \cdot {}^\circ\text{F/Btu} \quad \text{(organic liquid heating medium)}$$

$$R_{Do} = 0.001\ \text{h} \cdot \text{ft}^2 \cdot {}^\circ\text{F/Btu} \quad \text{(heavier normal hydrocarbon)}$$

(h) Overall coefficient.

$$U_D = \left[\frac{D_o}{h_i D_i} + \frac{D_o \ln(D_o/D_i)}{2k_{tube}} + \frac{1}{h_o} + \frac{R_{Di} D_o}{D_i} + R_{Do}\right]^{-1}$$

$$= \left[\frac{0.75}{346 \times 0.584} + \frac{(0.584/12)\ln(0.75/0.584)}{2 \times 26} + \frac{1}{278} + \frac{0.0005 \times 0.75}{0.584} + 0.001\right]^{-1}$$

$$U_D \cong 109\ \text{Btu/h} \cdot \text{ft}^2 \cdot {}^\circ\text{F}$$

Since this is strictly a rating problem, there is no need to reconcile the values of $\hat{q}$, U_D, and ΔT_m. The fact that the computed value of U_D exceeds U_{req} means that the exchanger contains sufficient heat-transfer area to meet the required duty.

(i) Critical heat flux.
The critical heat flux for a single tube is calculated using Equation (9.23a):

$$\hat{q}_c = 803 P_c P_r^{0.35}(1 - P_r)^{0.9}$$

$$= 803 \times 406.5(0.0861)^{0.35}(1 - 0.0861)^{0.9}$$

$$\hat{q}_c = 127{,}596\ \text{Btu/h} \cdot \text{ft}^2$$

The bundle geometry factor is given by:

$$\psi_b = \frac{D_b}{n_t D_o} = \frac{20}{290 \times 0.75} = 0.09195$$

Since this value is less than 0.323, the bundle correction factor is:

$$\phi_b = 3.1\psi_b = 3.1 \times 0.09195 = 0.285$$

The critical heat flux for the bundle is given by Equation (9.24):

$$\hat{q}_{c,bundle} = \hat{q}_{c,tube}\,\phi_b = 127{,}596 \times 0.285$$

$$\hat{q}_{c,bundle} = 36{,}365\ \text{Btu/h} \cdot \text{ft}^2$$

The ratio of the actual heat flux to the critical heat flux is:

$$\hat{q}/\hat{q}_{c,bundle} = 9965/36{,}365 \cong 0.27$$

Since the ratio is less than 0.7 and $U_D > U_{req}$, the reboiler is thermally acceptable.

(j) Tube-side pressure drop.
(i) Friction loss
The calculation uses Equation (5.2) for the friction factor and Equation (5.1) for the pressure drop:

$$f = 0.4137\,Re^{-0.2585} = 0.4137(37{,}738)^{-0.2585} = 0.0271$$

$$\Delta P_f = \frac{f\,n_p L\,G^2}{7.50 \times 10^{12} D_i s \phi_i} = \frac{0.0271 \times 2 \times 16(1{,}575{,}679)^2}{7.5 \times 10^{12}(0.584/12) \times 0.882 \times 1.0}$$

$$\Delta P_f = 6.69\ \text{psi}$$

(ii) Minor losses
From Table 5.1, the number of velocity heads allocated for minor losses with turbulent flow in U-tubes is:

$$\alpha_r = 1.6 n_p - 1.5 = 1.6 \times 2 - 1.5 = 1.7$$

Substituting in Equation (5.3) yields:

$$\Delta P_r = 1.334 \times 10^{-13} \alpha_r G^2/s = 1.334 \times 10^{-13} \times 1.7(1{,}575{,}679)^2/0.882$$

$$\Delta P_r = 0.64\ \text{psi}$$

(iii) Nozzle losses

For 6-in. schedule 40 nozzles we have:

$$G_n = \frac{\dot{m}}{(\pi/4)D_i^2} = \frac{425,000}{(\pi/4)(6.065/12)^2} = 2,118,361 \text{ lbm/h} \cdot \text{ft}^2$$

$$Re_n = \frac{D_i G_n}{\mu} = \frac{(6.065/12) \times 2,118,361}{0.84 \times 2.419} = 526,907$$

Since the flow is turbulent, Equation (5.4) is used to estimate the pressure drop:

$$\Delta P_n = 2.0 \times 10^{-13} N_s G_n^2/s = 2.0 \times 10^{-13} \times 1(2,118,361)^2/0.882$$

$$\Delta P_n = 1.02 \text{ psi}$$

(iv) Total pressure drop

$$\Delta P_i = \Delta P_f + \Delta P_r + \Delta P_n = 6.69 + 0.64 + 1.02$$

$$\Delta P_i = 8.4 \text{ psi}$$

Since the pressure drop is within the specified limit of 10 psi, the reboiler is hydraulically acceptable.

In summary, the reboiler is thermally and hydraulically suitable for this service.

10.5 Design of Vertical Thermosyphon Reboilers

10.5.1 Introduction

The procedure developed by Fair [11] for design of vertical thermosyphon reboilers is presented in this section. This method has been widely used for industrial reboiler design, and it incorporates some simplifications that help make the design problem more amenable to hand calculation. Newer correlations for two-phase flow and convective boiling are used in place of those given by Fair [11], but the basic design strategy is the same.

Figure 10.8 shows the configuration of the reboiler system. Point *A* is at the surface of the liquid in the column sump, while Points *B* and *D* are at the inlet and outlet tubesheets, respectively. Boiling begins at point *C*; between points *B* and *C*, it is assumed that only sensible heat transfer occurs. The reason for the sensible heating zone is that the liquid generally enters the reboiler subcooled to some extent due to the static head in the column sump and heat losses in the inlet line.

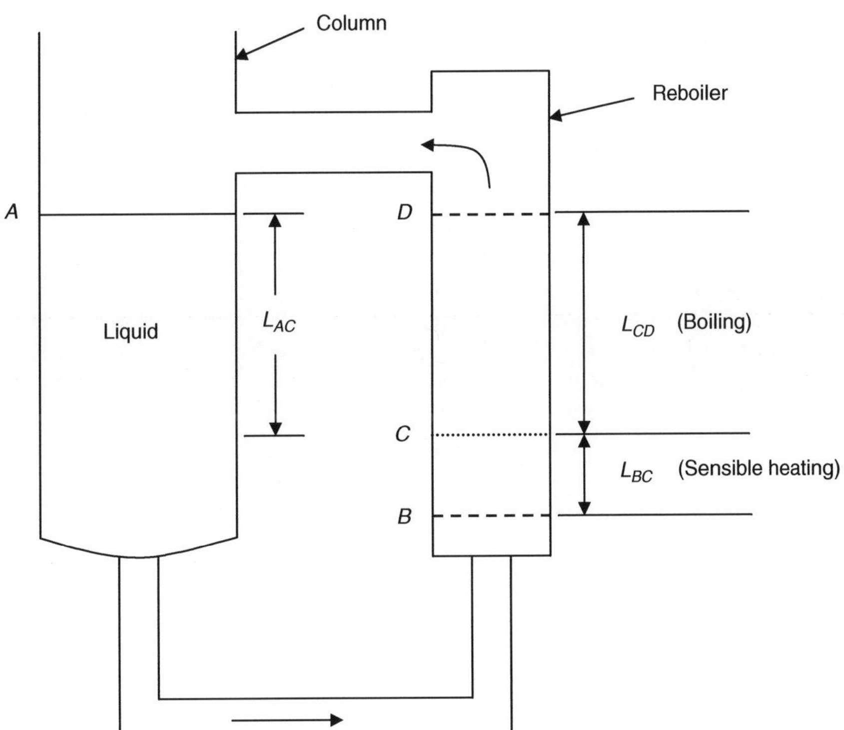

FIGURE 10.8 Configuration of vertical thermosyphon reboiler system.

10.5.2 Pressure Balance

With reference to Figure 10.8, the system pressure balance can be stated as follows:

$$(P_B - P_A) + (P_C - P_B) + (P_D - P_C) + (P_A - P_D) = 0 \tag{10.5}$$

The first pressure difference, $P_B - P_A$, consists of the static liquid head minus the friction loss in the inlet line. Expressing the pressure difference in units of psi and setting the viscosity correction factor to unity in Equation (4.5), we have:

$$P_B - P_A = \frac{\rho_L(g/g_c)\,(z_A - z_B)}{144} - \frac{f_{in}\,L_{in}\,G_{in}^2}{7.50 \times 10^{12} D_{in} s_L} \tag{10.6}$$

Here, z_A and z_B are the elevations at points A and B, respectively, and the subscript "*in*" refers to the inlet line to the reboiler. Also, L_{in} is an equivalent length that accounts for entrance, exit, and fitting losses.

A similar result holds for the second term, $P_C - P_B$, if the tube entrance loss is neglected:

$$P_C - P_B = \frac{\rho_L\,(g/g_c)\,L_{BC}}{144} - \frac{f_t\,L_{BC}\,G_t^2}{7.50 \times 10^{12} D_t s_L} \tag{10.7}$$

The subscript "*t*" in this equation refers to the reboiler tubes.

The pressure difference, $P_D - P_C$, across the boiling zone includes an acceleration loss term in addition to the static head and friction loss terms:

$$P_D - P_C = -\Delta P_{static,CD} - \Delta P_{f,CD} - \Delta P_{acc,CD} \tag{10.8}$$

The pressure difference due to the static head of fluid is obtained by integrating the two-phase density over the boiling zone, but the integral can be approximated using an appropriate average density:

$$\Delta P_{static,CD} = (g/144 g_c)\int_{Z_C}^{Z_D} \rho_{tp}\,dz \cong (g/144 g_c)\bar{\rho}_{tp} L_{CD} \tag{10.9}$$

Fair [11] recommends calculating the average density, $\bar{\rho}_{tp}$, at a vapor weight fraction equal to one-third the value at the reboiler exit.

The friction loss is obtained by integrating the two-phase pressure gradient over the boiling zone, but it, too, can be approximated, in this case using an average two-phase multiplier:

$$\Delta P_{f,CD} = \int_{Z_C}^{Z_D} \phi_{LO}^2 (\Delta P_f/L)_{LO} dz \cong \frac{f_t\,L_{CD}\,G_t^2\,\overline{\phi}_{LO}^2}{7.5 \times 10^{12} D_t s_L} \tag{10.10}$$

Fair [11] recommends calculating $\overline{\phi}_{LO}^2$ at a vapor weight fraction equal to two-thirds the value at the reboiler exit.

The pressure change due to acceleration of the fluid resulting from vapor formation is given by the following equation [11]:

$$\Delta P_{acc,CD} = \frac{G_t^2 \gamma}{144 g_c \rho_L} = \frac{G_t^2 \gamma}{144 g_c \rho_{water} s_L} = \frac{G_t^2 \gamma}{3.75 \times 10^{12} s_L} \tag{10.11}$$

where

$$\gamma = \frac{(1 - x_e)^2}{1 - \epsilon_{V,e}} + \frac{\rho_L x_e^2}{\rho_V \epsilon_{V,e}} - 1 \tag{10.12}$$

In this equation, x_e and $\epsilon_{V,e}$ are the vapor mass fraction and the void fraction at the reboiler exit.

The pressure difference, $P_A - P_D$, includes static head, friction, and acceleration effects. Since it is common practice (except for vacuum operation) to maintain the liquid level in the column sump near the elevation of the upper tubesheet in the reboiler, the static head effect is neglected. The effect of the velocity change from the reboiler tubes to the return line is accounted for explicitly. Other losses are lumped with the friction loss term by means of an equivalent length, L_{ex}. The result is as follows:

$$P_A - P_D = \frac{(G_t^2 - G_{ex}^2)\,(\gamma + 1)}{3.75 \times 10^{12} s_L} - \frac{f_{ex} L_{ex} G_{ex}^2 \phi_{LO,ex}^2}{7.50 \times 10^{12} D_{ex} s_L} \tag{10.13}$$

In this equation, the subscript "*ex*" designates conditions in the exit line from the reboiler.

Substituting for the four pressure differences in Equation (10.5) and combining terms leads to the following result:

$$\frac{\rho_L g L_{AC} - \bar{\rho}_{tp} g L_{CD}}{144 g_c} - \frac{G_{ex}^2(\gamma + 1) - G_t^2}{3.75 \times 10^{12} s_L} - \frac{f_{in} L_{in} G_{in}^2}{7.50 \times 10^{12} D_{in} s_L}$$

$$- \frac{f_t L_{BC} G_t^2}{7.50 \times 10^{12} D_t s_L} - \frac{f_t L_{CD} G_t^2 \overline{\phi}_{LO}^2}{7.50 \times 10^{12} D_t s_L} - \frac{f_{ex} L_{ex} G_{ex}^2 \phi_{LO,ex}^2}{7.50 \times 10^{12} D_{ex} s_L} = 0 \tag{10.14}$$

This equation provides a relationship between the circulation rate and the exit vapor fraction in the reboiler. It can be solved explicitly for the circulation rate if the dependence of the friction factors on flow rate is neglected. The solution is:

$$\dot{m}_i^2 = \frac{3.2 \times 10^{10} D_t^5 s_L (g/g_c)\,(\rho_L L_{AC} - \overline{\rho}_{tp} L_{CD})}{2D_t\left\{(\gamma+1)\left(\frac{D_t}{D_{ex}}\right)^4 - \frac{1}{n_t^2}\right\} + f_{in} L_{in}\left(\frac{D_t}{D_{in}}\right)^5 + \left(\frac{f_t}{n_t^2}\right)\left(L_{BC} + L_{CD}\overline{\phi}_{LO}^2\right) + f_{ex} L_{ex} \phi_{LO,ex}^2 \left(\frac{D_t}{D_{ex}}\right)^5} \tag{10.15}$$

where

$\dot{m}_i$ = tube-side mass flow rate (lbm/h)
n_t = number of tubes in reboiler
$\rho_L, \overline{\rho}_{tp} \propto$ lbm/ft^3
$L_{AC}, L_{CD}, L_{BC}, L_{in}, L_{ex} \propto$ ft
$D_t, D_{in}, D_{ex} \propto$ ft

For SI units, change the constant in Equation (10.15) from 3.2×10^{10} to 1,234. This will give $\dot{m}_i$ in kg/s when lengths and diameters are in m and densities are in kg/m^3. Note that the factor g/g_c equals 1.0 in English units and 9.81 in SI units.

Equation (10.15) can be solved iteratively to obtain the circulation rate and exit vapor fraction. For computer implementation, the integrals appearing in Equations (9.9) and (9.10) can be evaluated by numerical integration rather than using approximate average values of ρ_{tp} andϕ_{LO}^2

10.5.3 Sensible Heating Zone

In order to calculate the circulation rate using Equation (10.15), the length, L_{BC}, of the sensible heating zone must be determined. Fair's [11] method for estimating L_{BC} is described here. Boiling is assumed to begin when the liquid in the tubes becomes saturated; subcooled boiling is not considered, which is a conservative approach for design purposes.

In flowing from point B to point C, the fluid pressure decreases due to the elevation change and friction effects. At the same time, the fluid temperature increases due to heat transfer. A linear relationship between the temperature and pressure is assumed:

$$\frac{T_C - T_B}{P_C - P_B} = \frac{(\Delta T/L)}{(\Delta P/L)} \tag{10.16}$$

The saturation curve is linearized about point A to obtain:

$$\frac{T_{sat} - T_A}{P_{sat} - P_A} = (\Delta T/\Delta P)_{sat} \tag{10.17}$$

Now at point C, the fluid reaches saturation, so that $T_C = T_{sat}$ and $P_C = P_{sat}$. If heat losses in the reboiler feed line are neglected, then it also follows that $T_A = T_B$. With these equalities, Equations (10.16) and (10.17) can be combined to obtain the following expression for the pressure at point C:

$$\frac{P_B - P_C}{P_B - P_A} = \frac{(\Delta T/\Delta P)_{sat}}{(\Delta T/\Delta P)_{sat} - \frac{(\Delta T/L)}{(\Delta P/L)}} \tag{10.18}$$

If friction losses are neglected, then the pressure differences on the left side of this equation are proportional to elevation differences, i.e.,

$$(P_B - P_C)/(P_B - P_A) \cong L_{BC}/(z_A - z_B) \tag{10.19}$$

Furthermore, if the liquid level in the column sump is kept at approximately the upper tubesheet level, then $(z_A - z_B) \cong L_{BC} + L_{CD} =$ tube length. Equation (10.18) can then be written as:

$$\frac{L_{BC}}{L_{BC} + L_{CD}} \cong \frac{(\Delta T/\Delta P)_{sat}}{(\Delta T/\Delta P)_{sat} - \frac{(\Delta T/L)}{(\Delta P/L)}} \tag{10.20}$$

The left side of this equation is the fractional tube length required for sensible heating.

In order to evaluate $(\Delta T/\Delta P)_{sat}$, two points on the saturation curve are needed in the vicinity of (T_A, P_A). If the latter point is known from column design calculations, then only one additional point is needed at a temperature somewhat higher than T_A. For a pure component, this simply entails calculation of the vapor pressure at an appropriate temperature. For a mixture, a bubble-point pressure calculation is required.

The pressure gradient in the sensible heating zone is calculated as follows:

$$-(\Delta P/L) = \rho_L (g/g_c) + \Delta P_{f,BC}/L \tag{10.21}$$

The friction loss term in this equation can usually be neglected. Note that friction losses were neglected in deriving Equation (10.20).

The temperature gradient in the sensible heating zone is estimated as follows:

$$\Delta T/L = \frac{n_t \pi\, D_o U_D \Delta T_m}{\dot{m}_i C_{P,L}} \tag{10.22}$$

Here, U_D and ΔT_m are the overall coefficient and mean driving force, respectively, for the sensible heating zone.

10.5.4 Mist Flow Limit

The mist flow regime is avoided in reboiler design due to the large drop in heat-transfer coefficient that accompanies tube wall dryout. Fair [11] presented a simple empirical correlation for the onset of mist flow. Although the correlation was based on a very limited amount of data, it was later verified by Palen et al. [16] over a wide range of data for hydrocarbons, alcohols, water, and their mixtures. The correlation is as follows:

$$G_{t,mist} = 1.8 \times 10^6 X_{tt} \tag{10.23a}$$

where

$G_{t,mist}$ = tube-side mass flux at onset of mist flow (lbm/h · ft^2)
X_{tt} = Lockhart–Martinelli parameter, Equation (9.37)

In terms of SI units, the corresponding equation is:

$$G_{t,mist} = 2.44 \times 10^3 X_{tt} \tag{10.23b}$$

where $G_{t,mist} \propto$ kg/s · m^2. The tube-side mass flux should be kept safely below the value given by Equation (10.23). This will ensure that dryout does not occur, but it is still possible that the design heat flux may exceed the low-vapor-fraction critical heat flux. Hence, the critical heat flux should also be computed and compared with the design heat flux.

10.5.5 Flow Instabilities

Two-phase flow in pipes is subject to several types of instability that result from compressibility effects and the shape of the pressure-drop-versus-flow-rate relationship [14]. In thermosyphon reboilers, flow instability can result in "chugging" and "geysering," conditions that are characterized by periodic changes in the flow pattern. These conditions occur primarily in the slug flow and plug flow regimes when large slugs of liquid are alternately accelerated and decelerated. These instabilities can cause operational problems in the distillation column, and hence, must be prevented.

The flow in reboiler tubes tends to become more stable as the inlet pressure is reduced. Therefore, a valve or other flow restriction is often placed in the feed line to the reboiler to help stabilize the flow. The valve can also be used to compensate for discrepancies in the system pressure balance.

10.5.6 Size Limitations

As previously noted, vertical thermosyphon reboilers are subject to size limitations related to support and height considerations. General guidelines are that a maximum of three shells operating in parallel can be supported on a single distillation column, with a maximum total heat-transfer area of approximately 25,000 ft^2. Tube lengths are usually in the range of 8 ft to 20 ft, with values of 8ft to 16 ft being most common.

10.5.7 Design Strategy

The design procedure consists of three main steps: preliminary design, calculation of the circulation rate, and stepwise calculation of the rate of heat transfer and pressure drop in the reboiler tubes.

Preliminary Design

An initial configuration for the reboiler proper is obtained in the usual manner using an approximate overall heat-transfer coefficient along with an overall driving force to estimate the required surface area. The configuration of the feed and return lines must also be established. For recirculating units, the lines can be sized using an initial estimate for the recirculation rate (or equivalently, the exit vapor fraction).

Circulation Rate

The length of the sensible heating zone is first calculated using Equation (10.20). Then Equation (10.15) is solved iteratively to obtain the circulation rate and exit vapor fraction. The mass flux in the tubes should be checked against the value given by Equation (10.23) to ensure that the flow is not in or near the mist flow regime. If the calculated circulation rate and vapor fraction are not acceptable, the piping configuration is modified and the calculations repeated.

Stepwise Calculations

A zone analysis is performed by selecting an increment, Δx, of the vapor weight fraction. In each vapor-fraction interval, the arithmetic average vapor fraction is used to calculate the boiling heat-transfer coefficient, two-phase density, and friction loss. The overall heat-transfer coefficient and average driving force for the interval are used to calculate the tube length required to achieve the increment in vapor fraction.

The pressure drop for each interval is calculated by summing the static, friction, and acceleration losses. The acceleration loss for a given interval, k, is calculated using the following modification of Equation (10.11):

$$\Delta P_{acc,k} = \frac{G_t^2 \Delta \gamma_k}{3.75 \times 10^{12} s_L} \tag{10.24}$$

Here $\Delta \gamma_k = \Delta(\gamma_k + 1)$ is the change in γ from the beginning to the end of the kth interval.

For mixtures, thermodynamic (flash) calculations are required to determine the phase compositions and fluid temperature for each interval. These values are needed to obtain fluid physical properties, which in turn are needed for heat-transfer and pressure-drop calculations.

The calculations for each interval are iterative in nature. A value for the heat flux must be assumed to calculate the boiling heat-transfer coefficient, which is needed to calculate the tube length for the interval. From the tube length, a new value for the heat flux is obtained, thereby closing the iterative loop. The thermodynamic and pressure-drop calculations constitute another iterative sequence.

The sum of the pressure drops for all intervals provides an improved estimate for the pressure difference, $P_D - P_C$, and this value, when combined with the other terms in Equation (10.5), should satisfy the pressure balance. If there is a significant discrepancy, a new circulation rate is computed and the zone analysis is repeated.

Similarly, the sum of the tube lengths for all intervals, including the sensible heating zone, should equal or be slightly less than the actual tube length. If this is not the case, the reboiler configuration is modified and the calculations are repeated. Note that this will require the calculation of a new circulation rate. For an acceptable design, it is also necessary that the heat flux in each zone be less then the critical heat flux.

The accuracy of the stepwise calculations depends on the number of intervals used. A single interval, though generally not very accurate, is the most expedient option for hand calculations. In this case, the circulation rate is not adjusted (unless the reboiler configuration is modified) and only the heat-transfer calculations for the zone are performed.

The following example is a slightly modified version of a problem originally presented by Fair [11]. It involves some simplifying features, e.g., the boiling-side fluid is a pure component, the sizes of the feed and return lines are specified in the problem statement, and constant fluid properties are assumed.

Example 10.4

A reboiler is required to supply 15,000 lb/h of vapor to a distillation column that separates cyclohexane as the bottom product. The heating medium will be steam at a design pressure of 18 psia. The temperature and pressure below the bottom tray in the column are 182°F and 16 psia. Physical property data for cyclohexane at these conditions are given in the following table:

Property	Liquid	Vapor
ρ (lbm/ft^3)	45.0	0.200
μ (cp)	0.40	0.0086
C_P (Btu/lbm · °F)	0.45	–
k (Btu/h · ft · °F)	0.086	–
σ (lbf/ft)	0.00124	–
λ (Btu/lbm)	154	–
Pr	5.063	–

The vapor pressure of cyclohexane is given by the following equation [17], where $P_{sat} \propto$ torr and $T \propto K$:

$$P_{sat} = \exp\left[15.7527 - \frac{2766.63}{T - 50.50}\right]$$

The critical pressure of cyclohexane is 590.5 psia.

The feed line to the reboiler will consist of 100 equivalent feet of 6-in. schedule 40 pipe, and the return line will consist of 50 equivalent feet of 10-in. schedule 40 pipe. Design a recirculating vertical thermosyphon reboiler for this service.

Solution

(a) Make initial specifications.

(i) Fluid placement

Cyclohexane will flow in the tubes with steam in the shell.

(ii) Tubing

One-inch, 14 BWG tubes with a length of 8 ft are specified. Relatively short tubes are used in order to minimize the liquid height in the column sump.

(iii) Shell and head types

A TEMA E-shell is chosen for a vertical thermosyphon reboiler. Since condensing steam is a clean fluid, a fixed-tubesheet configuration can be used. Channel-type heads are selected for ease of tubesheet access. Thus, an AEL configuration is specified. A somewhat less expensive NEN configuration could also be used,

(iv) Tube layout

A triangular layout with a tube pitch of 1.25 in. is specified since mechanical cleaning of the external tube surfaces is not required.

(v) Baffles

Segmental baffles with a 35% cut and a spacing $B/D_s \cong 0.4$ are specified based on the recommendation for condensing vapors given in Figure 5.3.

(vi) Sealing strips

None are required for a fixed-tubesheet exchanger.

(vii) Construction materials

Since neither stream is corrosive, plain carbon steel is specified for all components.

(b) Energy balance and steam flow rate.

The reboiler duty is obtained from the vapor generation rate and the latent heat of vaporization for cyclohexane:

$$q = \dot{m}_V \lambda = 15{,}000 \times 154 = 2.31 \times 10^6 \text{ Btu/h}$$

From Table A.8, the latent heat of condensation for steam at 18 psia is 963.7 Btu/lbm. Therefore, the steam flow rate is:

$$\dot{m}_{steam} = q/\lambda_{steam} = 2.31 \times 10^6/963.7 = 2397 \quad \text{lbm/h}$$

(c) Mean temperature difference.

From Table A.8, the temperature of saturated steam at 18 psia is 222.4°F. Assuming that cyclohexane vaporizes at a constant temperature of 182°F, i.e., neglecting pressure effects in the reboiler system, we have:

$$\Delta T_m = 222.4 - 182 = 40.4°\text{F}$$

(d) Heat-transfer area and number of tubes.

Based on Table 3.5, an overall heat-transfer coefficient of 250 Btu/h · ft² · °F is assumed. The required area is then:

$$A = \frac{q}{U_D \Delta T_m} = \frac{2.31 \times 10^6}{250 \times 40.4} = 228.7 \text{ ft}^2$$

The corresponding number of tubes is:

$$n_t = \frac{A}{\pi D_o L} = \frac{228.7}{\pi(1/12) \times 8} \cong 109$$

(e) Number of tube passes and actual tube count.

A single tube pass is used for a vertical thermosyphon reboiler. From Table C.6, the closest tube count is 106 tubes in a 15.25-in. shell.

This completes the preliminary design of the reboiler system. Since the piping configuration was specified in the problem statement, sizing of the feed and return lines is not required here. The circulation rate is calculated in the steps that follow; only the final iteration is presented.

(f) Estimated circulation rate.

Assume an exit vapor fraction of 13.2%, i.e., $x_e = 0.132$. The corresponding circulation rate is:

$$\dot{m}_i = \frac{\dot{m}_V}{x_e} = \frac{15{,}000}{0.132} = 113{,}636 \quad \text{lbm/h}$$

(g) Friction factors.

The internal diameters for the tubes, inlet line, and exit line are obtained from Tables B.1 and B.2:

$$D_t = 0.834 \text{ in.} = 0.0695 \text{ ft}$$

$$D_{in} = 6.065 \text{ in.} = 0.5054 \text{ ft}$$

$$D_{ex} = 10.02 \text{ in.} = 0.835 \text{ ft}$$

The corresponding Reynolds numbers are computed next, based on all-liquid flow:

$$Re_t = \frac{4\,\dot{m}_i}{n_t \pi\, D_t \mu_L} = \frac{4 \times 113,636}{106 \times \pi \times 0.0695 \times 0.4 \times 2.419} = 20,297$$

$$Re_{in} = \frac{4\,\dot{m}_i}{\pi\, D_{in} \mu_L} = \frac{4 \times 113,636}{\pi \times 0.5054 \times 0.4 \times 2.419} = 295,866$$

$$Re_{LO,ex} = \frac{4\,\dot{m}_i}{\pi\, D_{ex} \mu_L} = \frac{4 \times 113,636}{\pi \times 0.835 \times 0.4 \times 2.419} = 179,079$$

Equations (4.8) and (5.2) are used to calculate the friction factors for the pipes and tubes, respectively:

$$f_t = 0.4137\, Re_t^{-0.2585} = 0.4137(20,297)^{-0.2585} = 0.0319$$

$$f_{in} = 0.3673\, Re_{in}^{-0.2314} = 0.3673(295,866)^{-0.2314} = 0.0199$$

$$f_{ex} = 0.3673\, Re_{LO,ex}^{-0.2314} = 0.3673(179,079)^{-0.2314} = 0.0224$$

(h) Sensible heating zone.

(i) Slope of saturation curve

Conditions in the column sump are first checked by calculating the vapor pressure of cyclohexane at the given temperature of 182°F = 356.7 K:

$$P_{sat} = \exp\left[15.7527 - \frac{2766.63}{356.7 - 50.50}\right] = 826.6 \text{ torr}$$

$$P_{sat} = 826.6 \text{ torr} \times \frac{14.696 \text{ psi/atm}}{760 \text{ torr/atm}} = 15.98 \text{ psia}$$

This value is in close agreement with the stated pressure of 16 psia below the bottom tray in the column. Next, the vapor pressure is calculated at a somewhat higher temperature, 192°F = 362.2 K:

$$P_{sat} = \exp\left[15.77527 - \frac{2766.63}{362.2 - 50.50}\right] = 969.5 \text{ torr}$$

$$P_{sat} = 969.5 \times 14.696/760 = 18.75 \text{ psia}$$

The required slope is obtained as follows:

$$(\Delta T/\Delta P)_{sat} = \frac{192 - 182}{18.75 - 15.98} = 3.61°\text{F/psi}$$

(ii) Pressure gradient

The pressure gradient in the sensible heating zone is estimated using Equation (10.21), neglecting the friction loss term:

$$-(\Delta P/L) = \frac{\rho_L (g/g_c)}{144} = \frac{45 \times 1.0}{144} = 0.3125 \text{ psi/ft}$$

(iii) Temperature gradient

To estimate the temperature gradient in the sensible heating zone, the heat-transfer coefficient for all-liquid flow in the tubes is calculated using the Seider–Tate equation:

$$h_{LO} = (k_L/D_t) \times 0.023 \mathrm{Re}_t^{0.8} Pr_L^{1/3} (\mu/\mu_w)^{0.14}$$

$$= (0.086/0.0695) \times 0.023(20,297)^{0.8}(5.063)^{1/3}(1.0)$$

$$h_{LO} = 136 \text{ Btu/h} \cdot \text{ft}^2 \cdot °\text{F}$$

The overall coefficient is calculated assuming a film coefficient (including fouling) of 1500 Btu/h · ft^2 · °F for steam and a fouling allowance for cyclohexane of 0.001 h · ft^2 · °F/Btu:

$$U_D = \left[(D_o/D_t)\left(\frac{1}{h_{LO}} + R_{Di}\right) + \frac{D_o \ln(D_o/D_t)}{2k_{tube}} + (1/h_o + R_{Do})\right]^{-1}$$

$$= \left[(1.0/0.834)\left(\frac{1}{136} + 0.001\right) + \frac{(1.0/12)\ln(1.0/0.834)}{2 \times 26} + \frac{1}{1500} \right]^{-1}$$

$$U_D \cong 91 \text{ Btu/h} \cdot \text{ft}^2 \cdot {}^\circ\text{F}$$

The temperature gradient is calculated using Equation (10.22) with a mean temperature difference of approximately 40°F:

$$\Delta T/L = \frac{n_t \pi D_o U_D \Delta T_m}{\dot{m}_i C_{P,L}} = \frac{106\pi(1.0/12) \times 91 \times 40}{113{,}636 \times 0.45} = 1.975^\circ\text{F/ft}$$

(iv) Length of sensible heating zone

The fractional length of the sensible heating zone is estimated using Equation (10.20):

$$\frac{L_{BC}}{L_{BC} + L_{CD}} = \frac{(\Delta T/\Delta P)_{sat}}{(\Delta T/\Delta P)_{sat} - \dfrac{(\Delta T/L)}{(\Delta P/L)}}$$

$$\frac{L_{BC}}{8} = \frac{3.61}{3.61 + \dfrac{1.975}{0.3125}} = 0.364$$

$$L_{BC} \cong 2.9 \text{ ft}$$

It follows that $L_{CD} \cong L_{AC} \cong 5.1$ ft. It is assumed that the liquid level in the column sump is maintained at approximately the elevation of the upper tubesheet in the reboiler.

(i) Average two-phase density.

The two-phase density is calculated at a vapor fraction of $x_e/3 = 0.044$. The Lockhart–Martinelli parameter is calculated using Equation (9.37):

$$X_{tt} = \left(\frac{1-x}{x}\right)^{0.9} (\rho_V/\rho_L)^{0.5} (\mu_L/\mu_V)^{0.1}$$

$$= \left(\frac{1-0.044}{0.044}\right)^{0.9} (0.2/45)^{0.5} (0.4/0.0086)^{0.1}$$

$$X_{tt} = 1.563$$

Since this value is greater than unity, the Chisholm correlation, Equation (9.63), gives the slip ratio as:

$$SR = (\rho_L/\rho_{\text{hom}})^{0.5}$$

The homogeneous density is given by Equation (9.51):

$$\rho_{\text{hom}} = [x/\rho_V + (1-x)/\rho_L]^{-1} = [0.044/0.2 + 0.956/45]^{-1}$$

$$\rho_{\text{hom}} = 4.1452 \text{ lbm/ft}^3$$

Substitution into the above equation gives the slip ratio:

$$SR = (45/4.1452)^{0.5} = 3.295$$

Next, the void fraction is computed using Equation (9.59).

$$\epsilon_V = \frac{x}{x + SR(1-x)\rho_V/\rho_L} = \frac{0.044}{0.044 + 3.295 \times 0.956 \times 0.2/45}$$

$$\epsilon_V = 0.7586$$

Finally, the two-phase density is computed from Equation (9.54):

$$\overline{\rho}_{tp} = \epsilon_V \rho_V + (1 - \epsilon_V)\rho_L = 0.7586 \times 0.2 + 0.2414 \times 45$$

$$\overline{\rho}_{tp} = 11.01 \text{ lbm/ft}^3$$

(j) Average two-phase multiplier.

The two-phase multiplier is calculated at a vapor fraction of $2x_e/3 = 0.088$. The Müller–Steinhagen and Heck (MSH) correlation, Equation (9.53), is used here:

$$\overline{\phi}_{LO}^2 = Y^2x^3 + \left[1 + 2x(Y^2 - 1)\right](1-x)^{1/3}$$

The Chisholm parameter, Y, is calculated using Equation (9.42) with $n = 0.2585$ for heat-exchanger tubes:

$$Y = (\rho_L/\rho_V)^{0.5}(\mu_V/\mu_L)^{n/2} = (45/0.2)^{0.5}(0.0086/0.4)^{0.2585/2}$$

$$Y = 9.13$$

Substituting in the MSH correlation gives:

$$\overline{\phi}_{LO}^2 = (9.13)^2(0.088)^3 + \{1 + 2 \times 0.088\,((9.13)^2 - 1)\}(0.912)^{1/3}$$

$$\overline{\phi}_{LO}^2 = 15.08$$

(k) Two-phase multiplier for exit line.
The above calculation is repeated with $x = x_e = 0.132$. For the exit pipe, however, the Chisholm parameter is calculated with $n = 0.2314$. Thus,

$$Y = (45/0.2)^{0.5}(0.0086/0.4)^{0.2314/2} = 9.62$$

$$\phi_{LO,ex}^2 = (9.62)^2(0.132)^3 + \{1 + 2 \times 0.132\,((9.62)^2 - 1)\}(0.868)^{1/3}$$

$$\phi_{LO,ex}^2 = 24.22$$

(l) Exit void fraction.
At $x = x_e = 0.132$, the Lockhart–Martinelli parameter is:

$$X_{tt} = \left(\frac{1 - 0.132}{0.132}\right)^{0.9}(0.2/45)^{0.5}(0.4/0.0086)^{0.1} = 0.533$$

Since this value is less than 1.0, the Chisholm correlation gives the slip ratio as:

$$SR = (\rho_L/\rho_V)^{0.25} = (45/0.2)^{0.25} = 3.873$$

From Equation (9.59), the void fraction is:

$$\epsilon_{V,e} = \frac{x_e}{x_e + SR(1 - x_e)\rho_V/\rho_L} = \frac{0.132}{0.132 + 3.873 \times 0.868 \times 0.2/45}$$

$$\epsilon_{V,e} = 0.8983$$

(m) Acceleration parameter.
The acceleration parameter, γ, is given by Equation (10.12):

$$\gamma = \frac{(1 - x_e)^2}{1 - \epsilon_{V,e}} + \frac{\rho_L x_e^2}{\rho_V \epsilon_{V,e}} - 1 = \frac{0.868}{0.1017} + \frac{45(0.132)^2}{0.2 \times 0.8983} - 1$$

$$\gamma = 10.77$$

(n) Circulation rate.
Equation (10.15) is used to obtain a new estimate of the circulation rate. Due to the complexity of the equation, the individual terms are computed separately, starting with the numerator:

$$\text{numerator} = 3.2 \times 10^{10} D_t^5 s_L (g/g_c)\left(\rho_L L_{AC} - \overline{\rho}_{tp} L_{CD}\right)$$

$$= 3.2 \times 10^{10}(0.0695)^5(45/62.43)\,(1.0)\,(45 \times 5.1 - 11.01 \times 5.1)$$

$$\text{numerator} = 6{,}483{,}575$$

Each of the four terms in the denominator is computed next:

$$\text{term 1} = 2D_t\left\{(\gamma + 1)\left(\frac{D_t}{D_{ex}}\right)^4 - \frac{1}{n_t^2}\right\} = 2 \times 0.06595\left\{11.77\left(\frac{0.0695}{0.835}\right)^4 - \frac{1}{(106)^2}\right\}$$

$\text{term 1} = 6.6150 \times 10^{-5}$

$\text{term 2} = f_{in}L_{in}(D_t/D_{in})^5 = 0.0199 \times 100(0.0695/0.5054)^5$

$\text{term 2} = 9.7859 \times 10^{-5}$

$\text{term 3} = (f_t/n_t^2)(L_{BC} + L_{CD}\overline{\phi}_{LO}^2) = \dfrac{0.0319}{(106)^2}(2.9 + 5.1 \times 15.08)$

$\text{term 3} = 2.2658 \times 10^{-4} = 22.658 \times 10^{-5}$

term 4 $= f_{ex}L_{ex}\phi^2_{LO,ex}(D_t/D_{ex})^5 = 0.0224 \times 50 \times 24.22(0.0695/0.835)^5$
term 4 $= 1.0836 \times 10^{-4} = 10.836 \times 10^{-5}$
Substituting the above values into Equation (10.15) gives:

$$\dot{m}_i^2 = \frac{6,483,575}{(6.6150 + 9.7859 + 22.658 + 10.836) \times 10^{-5}} = 1.2994 \times 10^{10}$$

$$\dot{m}_i = 113,991 \text{ lbm/h}$$

This value agrees with the assumed flow rate of 113,636 lbm/h to within about 0.3%, which is more than adequate for convergence. The average of the assumed and calculated values is taken as the final value, i.e.,

$$\dot{m}_i = \frac{113,991 + 113,636}{2} \cong 113,814 \text{ lbm/h}$$

(o) Mist flow limit.
The mass flux at the onset of mist flow is given by Equation (10.23a):

$$G_{t,mist} = 1.8 \times 10^6 X_{tt} = 1.8 \times 10^6 \times 0.533 = 959,400 \text{ lbm/h}\cdot\text{ft}^2$$

The actual mass flux in the tubes is:

$$G_t = \frac{\dot{m}_i}{n_t(\pi/4)D_t^2} = \frac{113,814}{106(\pi/4)(0.0695)^2} = 283,029 \text{ lbm/h}\cdot\text{ft}^2$$

The actual mass flux is far below the mist flow limit, as would be expected with a vapor fraction of only about 13%.

This completes the circulation rate calculation. The following steps deal with the zone analysis (stepwise calculations). To simplify matters, a single boiling zone is used. In this case, the pressure drop in the tubes is not recalculated and the circulation rate is not adjusted. Therefore, only heat-transfer calculations are involved in the zone analysis.

(p) Duty in boiling zone.
The cyclohexane temperature in the boiling zone is estimated based on the temperature gradient calculated above:

$$T_{cyhx} = 182 + (\Delta T/L)L_{BC} = 182 + 1.975 \times 2.9 \cong 187.7°\text{F}$$

Hence, the duty in the sensible heating zone is that required to raise the temperature of the liquid by 5.7°F:

$$q_{BC} = \dot{m}_i C_{P,L}\Delta T_{BC} = 113,814 \times 0.45 \times 5.7 = 291,933 \text{ Btu/h}$$

The duty for the boiling zone is the total duty minus the duty for the sensible heating zone. Thus,

$$q_{CD} = q - q_{BC} = 2.31 \times 10^6 - 291,933 \cong 2.018 \times 10^6 \text{ Btu/h}$$

(q) Boiling heat-transfer coefficient.
Since the boiling fluid is a pure component, the Liu–Winterton correlation, Equation (9.80), is used to calculate the heat-transfer coefficient. The average vapor weight fraction for the zone is used in the calculations, i.e., $x = 0.132/2 = 0.066$.

$$h_b = \left[(S_{LW}h_{nb})^2 + (E_{LW}h_L)^2\right]^{1/2}$$

(i) The enhancement factor
The convective enhancement factor, E_{LW}, is given by Equation (9.82):

$$E_{LW} = [1 + xPr_L(\rho_L - \rho_V)/\rho_V]^{0.35}$$

$$= [1 + 0.066 \times 5.063(45 - 0.2)/0.2]^{0.35}$$

$$E_{LW} = 4.550$$

(ii) The suppression factor
The nucleate boiling suppression factor, S_{LW}, is given by Equation (9.81):

$$S_{LW} = \left[1 + 0.055E_{LW}^{0.1}Re_L^{0.16}\right]^{-1}$$

The Reynolds number is calculated for the liquid phase flowing alone in the tubes:

$$Re_L = \frac{4\,(1-x)(\dot{m}_i/n_t)}{\pi\, D_t\mu_L} = \frac{4\,(1-0.066)\,(113,814/106)}{\pi \times 0.0695 \times 0.4 \times 2.419}$$

$$Re_L = 18,987$$

Substituting into the above equation for S_{LW} gives:

$$S_{LW} = \left[1 + 0.055(4.550)^{0.1}(18,987)^{0.16}\right]^{-1} = 0.7636$$

(iii) Convective heat-transfer coefficient

The Dittus–Boelter correlation, Equation (9.75), is used in conjunction with the Liu-Winterton correlation to calculate h_L.

$$h_L = 0.023(k_L/D_t)Re_L^{0.8}Pr_L^{0.4}$$

$$= 0.023\,(0.086/0.0695)\,(18,987)^{0.8}(5.063)^{0.4}$$

$$h_L = 144 \text{ Btu/h} \cdot \text{ft}^2 \cdot {}^\circ\text{F}$$

(iv) Nucleate boiling heat-transfer coefficient

The Cooper correlation in the form of Equation (9.6a) is used to calculate h_{nb}:

$$h_{nb} = 21\hat{q}^{0.67}P_r^{0.12}(-\log_{10}P_r)^{-0.55}M^{-0.5}$$

For cyclohexane, the molecular weight is $M = 84$. The pressure in the boiling zone is estimated as the vapor pressure of cyclohexane at 187.7°F = 359.8 K:

$$P_{sat} = \exp\left[15.7527 - \frac{2766.63}{359.8 - 50.50}\right] = 904.96 \text{ torr} = 17.5 \text{ psia}$$

The reduced pressure is then:

$$P_r = P/P_c = 17.5/590.5 = 0.0296$$

The heat flux is estimated using the total duty and total tube length, as follows:

$$\hat{q} \cong \frac{2.31 \times 10^6}{106\pi \times 0.0695 \times 8} = 12,476 \text{ Btu/h} \cdot \text{ft}^2$$

Substituting into the above equation for h_{nb} gives:

$$h_{nb} = 21(12,476)^{0.67}(0.0296)^{0.12}(-\log_{10}0.0296)^{-0.55}(84)^{-0.5}$$

$$h_{nb} = 660 \text{ Btu/h} \cdot \text{ft}^2 \cdot {}^\circ\text{F}$$

(v) Convective boiling coefficient

Substituting the results from the above steps into the Liu–Winterton correlation gives the following result for h_b:

$$h_b = \left[(0.7636 \times 660)^2 + (4.550 \times 144)^2\right]^{1/2} = 827 \text{ Btu/h} \cdot \text{ft}^2 \cdot {}^\circ\text{F}$$

(r) Overall coefficient.

Due to the higher velocity and greater agitation in the boiling zone, a fouling factor of 0.0005 h · ft² · °F/Btu is deemed appropriate for cyclohexane. A film coefficient, including fouling allowance, of 1500 Btu/h · ft² · °F is again assumed for steam. The overall heat-transfer coefficient for the boiling zone is then:

$$U_D = \left[\frac{D_o}{D_i}\left(\frac{1}{h_i} + R_{Di}\right) + \frac{D_o \ln(D_o/D_i)}{2k_{tube}} + \left(\frac{1}{h_o} + R_{Do}\right)\right]^{-1}$$

$$= \left[\frac{1.0}{0.834}\left(\frac{1}{827} + 0.0005\right) + \frac{(1.0/12)\ln(1.0/0.834)}{2 \times 26} + \frac{1}{1500}\right]^{-1}$$

$$U_D = 332.6 \text{ Btu/h} \cdot \text{ft}^2 \cdot {}^\circ\text{F}$$

(s) Check heat flux and iterate if necessary.

The mean temperature difference for the boiling zone is taken as:

$$\Delta T_m = T_{steam} - T_{cyhx} = 222.4 - 187.7 = 34.7^\circ\text{F}$$

Since the saturation temperature decreases with decreasing pressure, both the steam and cyclohexane temperatures will vary somewhat over the length of the boiling zone due to the pressure drops experienced by the two streams. These effects are neglected here. Thus, the heat flux is:

$$\hat{q} = U_D\Delta T_m = 332.6 \times 34.7 = 11,541 \text{ Btu/h} \cdot \text{ft}^2$$

This value is within 10% of the initial estimate of 12,476 Btu/h · ft². After a few more iterations, the following converged values are obtained:

$$h_{nb} = 622 \text{ Btu/h} \cdot \text{ft}^2 \cdot {}^\circ\text{F}$$

$$h_b = 809 \text{ Btu/h} \cdot \text{ft}^2 \cdot {}^\circ\text{F}$$

$$U_D = 329 \text{ Btu/h} \cdot \text{ft}^2 \cdot {}^\circ\text{F}$$

$$\hat{q} = 11,416 \text{ Btu/h} \cdot \text{ft}^2$$

(t) Tube length.
The tube length required for the boiling zone is calculated as follows:

$$L_{req} = \frac{q_{CD}}{n_t \pi D_o U_D \Delta T_m} = \frac{2.018 \times 10^6}{106\pi(1.0/12) \times 329 \times 34.7} \cong 6.4 \text{ ft}$$

This value is greater than the available length of 5.1 ft, indicating that the reboiler is somewhat under-sized.

(u) Critical heat flux.
For brevity, the critical heat flux is estimated using Palen's method as given by Equation (9.84a):

$$\hat{q}_c = 16,070 \left(D_t^2/L\right)^{0.35} P_c^{0.61} P_r^{0.25}(1 - P_r)$$

$$= 16,070\left[(0.0695)^2/8\right]^{0.35}(590.5)^{0.61}(0.0296)^{0.25}(1 - 0.0296)$$

$$\hat{q}_c \cong 23,690 \text{ Btu/h} \cdot \text{ft}^2$$

$$\hat{q}/\hat{q}_c = 11,416/23,690 \cong 0.48$$

Thus, the heat flux is safely below the critical value.

(v) Design modification.
Based on the above calculations, the only problem with the initial design is that the unit is under-sized. The under-surfacing is due to the presence of a significant sensible heating zone that was not considered in the preliminary design. Although a more rigorous analysis using more zones might yield a different result, it is assumed here that some modification of the initial design is required. Three possible design changes are the following:

(i) Increase the tube length from 8 to 10 ft. This change will increase the static head and, hence, the degree of subcooling at the reboiler entrance. It will thus tend to increase the length of the sensible heating zone.

(ii) Increase the number of tubes. From the tube-count table, the next largest unit is a 17.25-in. shell containing 147 tubes. This represents an increase of about 39% in heat-transfer area, whereas the initial design is under-surfaced by only about 20%.

(iii) Raise the steam temperature by 5 to 8°F, corresponding to a steam pressure of 20 to 21 psia. This change will reduce the tube length required in both the sensible heating and boiling zones. Of the three options considered here, this one appears to be the simplest and most cost effective.

Each of the above changes will affect the circulation rate; therefore, verification requires essentially complete recalculation for each case. Due to the lengthiness of the calculations, no further analysis is presented here.

10.6 Computer Software

10.6.1 HEXTRAN

The shell-and-tube module in HEXTRAN is used for reboilers and condensers, as well as for single-phase heat exchangers. For streams defined as compositional type, the software automatically detects phase changes and uses the appropriate computational methods. A zone analysis is always performed for operations involving a phase change. The HEXTRAN documentation states that Chen's method is used for boiling heat-transfer calculations, but little additional information is provided.

Connecting piping is not integrated with the heat-exchanger modules in HEXTRAN. A separate piping module exists that can be used to calculate pressure losses in the reboiler feed and return lines. Pipe fittings are handled by means of either flow resistance coefficients or equivalent lengths, and two-phase flow calculations are performed automatically. Pressure changes due to friction, acceleration, and elevation change are accounted for. However, the software does not automatically calculate the circulation rate for a thermosyphon reboiler, which is a significant drawback for design work.

The following two examples examine some of the attributes of HEXTRAN (version 9.2) with regard to reboiler applications.

Example 10.5

Use HEXTRAN to rate the kettle reboiler designed in Example 10.2, and compare the results with those obtained previously by hand.

Solution

Under Units of Measure, the English system of units is selected. Then, under Components and Thermodynamics, propane, *i*-butane, and *n*-butane are selected from the list of library components by double-clicking on each desired component. (Note that water is not required as a component for this problem.) The Peng–Robinson (PR) equation of state is selected as the principal thermodynamic method for the light hydrocarbon mixture. Thus, a New Method Slate called (arbitrarily) SET1 is defined on the Method tab and the options shown below are chosen from the pop-up lists obtained by right-clicking on the items in the thermodynamic data tree.

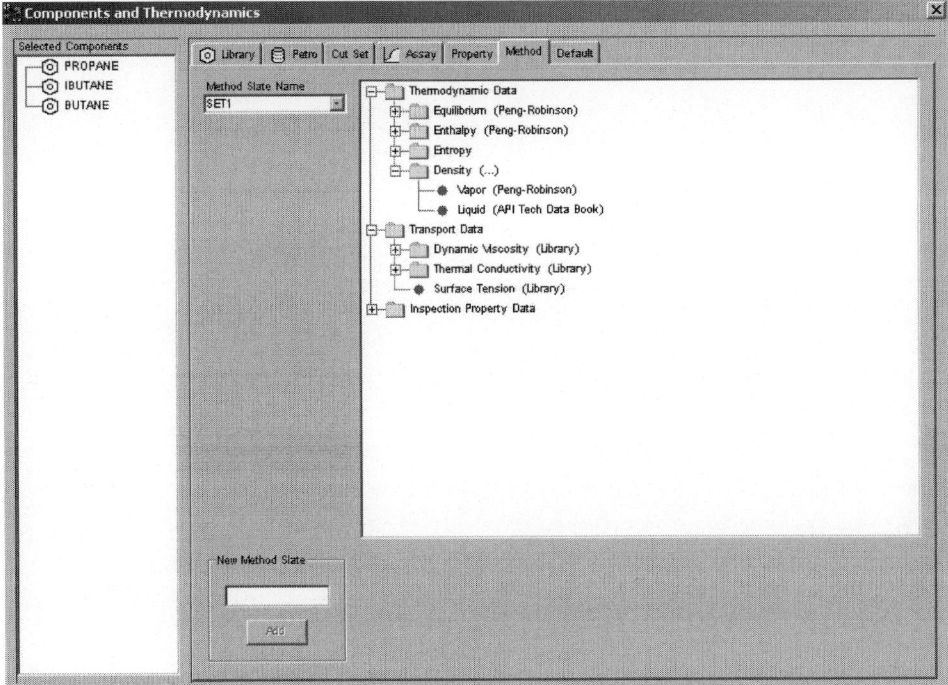

The API method for liquid density is chosen because it should be more reliable than the PR method for hydrocarbons. For transport properties, the Library method designates that property values are obtained from the program's pure-component databank. No methods are required for entropy or inspection property data in this problem.

After setting up the flowsheet, the tube-side feed stream is defined as a Water/Steam stream by right-clicking on the stream and selecting *Change Configuration* from the pop-up menu. Double-clicking on the stream brings up the Specifications form, where the pressure is set to 20 psia, and the flow rate is specified as 5645 lb/h of steam. Saturated steam tables will automatically be used by the program to obtain property values for this stream.

The shell-side feed stream is defined as a compositional stream, i.e., a stream having a defined composition, which is the default category. On the Specifications form its thermal condition is set by entering the pressure (250 psia) for the first specification and selecting *Bubble Point* for the second specification. The total stream flow rate (96,000 lb/h) is also entered. The stream composition is specified by entering the mole percent of each component in place of the component (molar) flow rates. When these values sum to 100, they are automatically interpreted as percentages by the program.

Data for the exchanger are obtained from Example 10.2 and entered on the appropriate forms, with the exception of the shell ID, which is not specified. The reason is that when the correct value of 23.25 in. is entered, the program gives an error message and fails to generate a solution, apparently due to a bug in the software. When the shell ID is not specified, the program calculates the diameter based on the tube data supplied. In the present case it calculates a diameter of 23 in., which is essentially the correct result. In addition to the data from Example 10.2, a fouling factor of 0.0005 h · ft^2 · °F/Btu is specified for steam. Fouling factors for both streams are entered on the Film Options form.

Finally, under Input/Calculation Options, the maximum number of iterations for the flowsheet is set to 100 because the default value of 30 proved to be insufficient for this problem.

The input file generated by the HEXTRAN GUI is given below, followed by a summary of results extracted from the HEXTRAN output file. From the latter it can be seen that the reboiler generates 48,571 lb/h of vapor, which is slightly more than the required rate of 48,000 lb/h. Thus, it appears that the unit is sized almost perfectly. In fact, however, the amount of vapor generated by the unit is limited by the amount of steam supplied, rather than by the available heat-transfer area. Referring to the zone analysis data given below, it is seen that all the steam condenses in the first five zones, leaving only condensate to be subcooled in the last zone. The area contained in the last zone is 117.2 ft^2, which is about 16% of the total surface area in the reboiler. If the steam flow rate is increased to 6850 lb/h, the subcooled condensate zone is eliminated and the amount of vapor generated increases to 58,349 lb/h. Taking the area of the first five zones (597 ft^2) as the required area gives an over-design for the unit of about 20%, which is a reasonable margin for this application.

The following table compares results from HEXTRAN with those obtained by hand in Example 10.2. As expected, the boiling heat-transfer coefficient calculated by hand is considerably more conservative than the value computed by HEXTRAN. However, the effective coefficient for steam used in Example 10.2 is actually much higher than the value computed by HEXTRAN. This result is due to the fouling factor used for steam in the present example, without which the effective steam coefficient for HEXTRAN would be about 1760 Btu/h · ft^2 · °F. The steam-side pressure drop found by HEXTRAN is comparable to the value estimated by hand. On the boiling side, the entire pressure drop calculated by HEXTRAN is due to the nozzles. The nozzle losses were not explicitly calculated in Example 10.2. They are obtained by using the appropriate equivalent lengths from Example 10.2; namely, the exit loss (28 ft of 5-in. pipe) for the inlet nozzles and the entrance loss (18 ft of 6-in. pipe) for the outlet nozzles. Finally, the mean temperature difference used in the hand calculations is quite close to the weighted average value from HEXTRAN.

Item	Hand	HEXTRAN
h_o (Btu/h · ft^2 · °F)	523	936[a]
$\{(D_o/D_i)(1/h_i + R_{Di})\}^{-1}$ (Btu/h · ft^2 · °F)	1,500 (assumed)	857[a]
U_D (Btu/h · ft^2 · °F)	297	335[a]
ΔP_i (psi)	0.3	0.43
ΔP_o (psi)[b]	0.2 (assumed)	–
ΔP_o (psi), nozzles only	0.05	0.05
ΔT_m (°F)	25.6	27. 1[a]

[a]Area-weighted average over first five zones; subcooled condensate zone not included.

[b]Friction and acceleration, excluding nozzle losses.

HEXTRAN Input File for Example 10.5

```
$ GENERATED FROM HEXTRAN KEYWORD EXPORTER
$
$      General Data Section
$
TITLE PROJECT=Example 10-5, PROBLEM=Kettle Reboiler, SITE=
$
DIME  English, AREA=FT2, CONDUCTIVITY=BTUH, DENSITY=LB/FT3, *
      ENERGY=BTU, FILM=BTUH, LIQVOLUME=FT3, POWER=HP, *
      PRESSURE=PSIA, SURFACE=DYNE, TIME=HR, TEMPERATURE=F, *
      UVALUE=BTUH, VAPVOLUME=FT3, VISCOSITY=CP, WT=LB, *
      XDENSITY=API, STDVAPOR=379.490
$
PRINT ALL, *
      RATE=M
$
CALC  PGEN=New, WATER=Saturated
$
$       Component Data Section
$
COMPONENT DATA
$
 LIBID     1, PROPANE /*
           2, IBUTANE /*
           3, BUTANE
$
$       Thermodynamic Data Section
$
THERMODYNAMIC DATA
$
 METHODS SET=SET1, KVALUE=PR, ENTHALPY(L)=PR, ENTHALPY(V)=PR, *
              DENSITY(L)=API, DENSITY(V)=PR, VISCOS(L)=LIBRARY, *
              VISCOS(V)=LIBRARY, CONDUCT(L)=LIBRARY, CONDUCT(V)=LIBRARY, *
              SURFACE=LIBRARY
$
 WATER DECANT=ON, SOLUBILITY = Simsci, PROP = Saturated
$
$Stream Data Section
$
STREAM DATA
$
 PROP STRM=PROD, NAME=PROD
```

```
    $
     PROP STRM=CONDENSATE, NAME=CONDENSATE
    $
     PROP STRM=STEAM, NAME=STEAM, PRES=20.000, STEAM=5645.000
    $
     PROP STRM=FEED, NAME=FEED, PRES=250.000, PHASE=L, *
             RATE(W)=96000.000, *
       COMP(M)=  1, 15 / *
            2, 25 /   3, 60
    $
    $ Calculation Type Section
    $
    SIMULATION
    $
     TOLERANCE TTRIAL=0.01
    $
     LIMITS AREA=200.00, 6000.00, SERIES=1, 10, PDAMP=0.00, *
            TTRIAL=100
    $
     CALC   TWOPHASE=New, DPSMETHOD=Stream, MINFT=0.80
    $
     PRINT UNITS, ECONOMICS, STREAM, STANDARD, *
           EXTENDED, ZONES


$
ECONOMICS DAYS=350, EXCHANGERATE=1.00, CURRENCY=USDOLLAR
$
 UTCOST OIL=3.50, GAS=3.50, ELECTRICITY=0.10, *
        WATER=0.03, HPSTEAM=4.10, MPSTEAM=3.90, *
        LPSTEAM=3.60, REFRIGERANT=0.00, HEATINGMEDIUM=0.00
$
 HXCOST BSIZE=1000.00, BCOST=0.00, LINEAR=50.00, *
        EXPONENT=0.60, CONSTANT=0.00, UNIT
$
$     Unit Operations Data
$
UNIT OPERATIONS
$
STE UID=KETTLE
 TYPE  Old, TEMA=BKU, HOTSIDE=Tubeside, ORIENTATION=Horizontal, *
       FLOW=Countercurrent, *
       UESTIMATE=50.00, USCALER=1.00

 TUBE  FEED=STEAM, PRODUCT=CONDENSATE, *
       LENGTH=13.00, OD=1.000, *
       BWG=14, NUMBER=212, PASS=2, PATTERN=90, *
       PITCH=1.2500, MATERIAL=1, *
       FOUL=0.0005, LAYER=0, *
       DPSCALER=1.00
$
SHELL  FEED=FEED, PRODUCT=PROD, *
       SERIES=1, PARALLEL=1, *
       MATERIAL=1, *
       FOUL=0.0005, LAYER=0, *
       DPSCALER=1.00
$
 BAFF  NONE
$
 TNOZZ TYPE=Conventional, ID=6.065, 3.068, NUMB=1, 1
$
 SNOZZ TYPE=Conventional , ID=5.047, 6.065, NUMB=2, 2
```

```
$
 LNOZZ ID=4.026, NUMB=1
$
 CALC  TWOPHASE=New, *
       DPSMETHOD=Stream, *
       MINFT=0.80
$
 PRINT STANDARD, *
       EXTENDED, *
       ZONES
$
 COST  BSIZE=1000.00, BCOST=0.00, LINEAR=50.00, *
       CONSTANT=0.00, EXPONENT=0.60, Unit
$
$ End of keyword file...
```

HEXTRAN Output Data for Example 10.5

```
==============================================================================
                     SHELL AND TUBE EXCHANGER DATA SHEET
I----------------------------------------------------------------------------I
I EXCHANGER  NAME                               UNIT ID KETTLE               I
I SIZE    23x  156   TYPE BKU,   HORIZONTAL     CONNECTED 1 PARALLEL 1 SERIES I
I AREA/UNIT   715.  FT2 (  714. FT2 REQUIRED)  AREA/SHELL   715. FT2         I
I----------------------------------------------------------------------------I
I PERFORMANCE OF ONE UNIT             SHELL-SIDE              TUBE-SIDE      I
I----------------------------------------------------------------------------I
I FEED STREAM NUMBER                  FEED                    STEAM          I
I FEED STREAM NAME                    FEED                    STEAM          I
I TOTAL FLUID          LB /HR            96000.                   5645.      I
I     VAPOR  (IN/OUT) LB /HR          0./      48571.            0./      0. I
I     LIQUID          LB /HR      96000./      47429.            0./      0. I
I     STEAM           LB /HR          0./          0.        5645./      0. I
I     WATER           LB /HR          0./          0.            0./   5645. I
I     NON CONDENSIBLE LB /HR                0.                     0.        I
I TEMPERATURE (IN/OUT) DEG F      197.6 /   202.4          228.3 /   217.2   I
I PRESSURE    (IN/OUT) PSIA      250.00 /   249.95         20.00 /   19.57   I
I----------------------------------------------------------------------------I
I SP. GR., LIQ  (60F / 60F H2O)   0.569 /   0.571          0.000 /   1.000   I
I          VAP  (60F / 60F AIR)   0.000 /   1.916          0.631 /   0.000   I
I DENSITY,  LIQUID    LB/FT3     28.406 /  28.369          0.000 /  59.738   I
I           VAPOR     LB/FT3      0.000 /   2.758          0.049 /   0.000   I
I VISCOSITY, LIQUID   CP          0.074 /   0.074          0.000 /   0.275   I
I           VAPOR     CP          0.000 /   0.009          0.012 /   0.000   I
I THRML COND,LIQ  BTU/HR-FT-F    0.0462 /  0.0459         0.0000 /  0.3942   I
I           VAP  BTU/HR-FT-F     0.0000 /  0.0141         0.0147 /  0.0000   I
I SPEC.HEAT,LIQUID BTU /LB F     0.8054 /  0.8106         0.0000 /  1.0080   I
I           VAPOR  BTU /LB F     0.0000 /  0.5763         0.5049 /  0.0000   I
I LATENT HEAT         BTU /LB       105.64                    0.00           I
I VELOCITY            FT/SEC          0.30                    0.13           I
I DP/SHELL(DES/CALC)   PSI         0.00 /  0.05             0.00 /   0.43    I
I FOULING RESIST FT2-HR-F/BTU  0.00050 (0.00050 REQD)          0.00050       I
I----------------------------------------------------------------------------I
I TRANSFER RATE  BTU/HR-FT2-F  SERVICE  282.91 ( 282.62 REQD), CLEAN  410.66 I
I HEAT EXCHANGED MMBTU /HR     5.479,      MTD(CORRECTED)  27.1,     FT 0.982 I
I----------------------------------------------------------------------------I
I CONSTRUCTION OF ONE SHELL            SHELL-SIDE              TUBE-SIDE     I
I----------------------------------------------------------------------------I
I DESIGN PRESSURE/TEMP PSIA  /F      325./  300.              75./  300.     I
I NUMBER OF PASSES                       1                        2          I
I MATERIAL                           CARB STL                 CARB STL       I
I INLET  NOZZLE ID/NO       IN         5.0/ 2                 6.1/ 1         I
I VAPOR  NOZZLE ID/NO       IN         6.1/ 2                 3.1/ 1         I
I INTERM NOZZLE ID/NO       IN         0.0/ 0                                I
I----------------------------------------------------------------------------I
I TUBE: NUMBER    212, OD  1.000  IN , BWG    14        , LENGTH 13.0 FT     I
I        TYPE BARE,               PITCH    1.2500 IN,   PATTERN 90 DEGREES   I
I SHELL:  ID    23.00 IN,           BUNDLE DIAMETER(DOTL)   22.50 IN         I
I RHO-V2: INLET NOZZLE  1297.0 LB/FT-SEC2                                    I
I TOTAL WEIGHT/SHELL,LB    6685.2 FULL OF WATER   0.138E+05 BUNDLE    4024.7 I
I----------------------------------------------------------------------------I
```

```
==============================================================================
                     SHELL AND TUBE EXTENDED DATA SHEET
I----------------------------------------------------------------------------I
I EXCHANGER  NAME                                UNIT ID KETTLE              I
I SIZE    23x 156   TYPE BKU,   HORIZONTAL      CONNECTED 1 PARALLEL  1 SERIES I
I AREA/UNIT   715. FT2 (   714. FT2 REQUIRED)                                I
I----------------------------------------------------------------------------I
I PERFORMANCE OF ONE UNIT             SHELL-SIDE               TUBE-SIDE     I
I----------------------------------------------------------------------------I
I FEED STREAM NUMBER                    FEED                    STEAM        I
I FEED STREAM NAME                     FEED                     STEAM        I
I WT FRACTION LIQUID (IN/OUT)        1.00 / 0.49               0.00 / 1.00   I
I REYNOLDS NUMBER                          0.                  13998.        I
I PRANDTL NUMBER                      0.000                    1.137         I
I UOPK,LIQUID                     13.722 /  13.681           0.000 /   0.000 I
I      VAPOR                       0.000 /  13.761           0.000 /   0.000 I
I SURFACE TENSION    DYNES/CM      3.637 /   3.586          55.448 /  56.997 I
I FILM COEF(SCL) BTU/HR-FT2-F         945.0 (1.000)          1066.0 (1.000)  I
I FOULING LAYER THICKNESS  IN           0.000                    0.000       I
I----------------------------------------------------------------------------I
I THERMAL RESISTANCE                                                         I
I UNITS:  (FT2-HR-F/BTU)      (PERCENT)   (ABSOLUTE)                         I
I SHELL FILM                       29.94      0.00106                        I
I TUBE FILM                        31.82      0.00112                        I
I TUBE METAL                        7.13      0.00025                        I
I TOTAL FOULING                    31.11      0.00110                        I
I ADJUSTMENT                        0.10      0.00000                        I
I----------------------------------------------------------------------------I
I PRESSURE DROP                        SHELL-SIDE               TUBE-SIDE    I
I UNITS: (PSIA  )              (PERCENT)    (ABSOLUTE)  (PERCENT)   (ABSOLUTE)I
I WITHOUT NOZZLES                   0.00          0.00     68.40          0.29 I
I INLET NOZZLES                    66.25          0.03     31.36          0.13 I
I OUTLET NOZZLES                   33.75          0.02      0.24          0.00 I
I TOTAL /SHELL                                    0.05                    0.43 I
I TOTAL /UNIT                                     0.05                    0.43 I
I DP SCALER                                       1.00                    1.00 I
I----------------------------------------------------------------------------I
I CONSTRUCTION OF ONE SHELL                                                  I
I----------------------------------------------------------------------------I
I TUBE:OVERALL LENGTH         13.0          FT  EFFECTIVE LENGTH    12.88   FT I
I      TOTAL TUBESHEET THK     1.4          IN  AREA RATIO (OUT/IN)  1.199     I
I      THERMAL COND.          30.0BTU/HR-FT-F  DENSITY             490.80 LB/FT3I
I----------------------------------------------------------------------------I
I BUNDLE: DIAMETER             22.5         IN  TUBES IN CROSSFLOW 212         I
I         CROSSFLOW AREA        5.201       FT2  WINDOW AREA          0.842 FT2  I
I         TUBE-BFL LEAK AREA   0.019        FT2  SHELL-BFL LEAK AREA  0.019 FT2  I
I----------------------------------------------------------------------------I
```

```
==============================================================================
                    ZONE ANALYSIS FOR EXCHANGER KETTLE

                      TEMPERATURE - PRESSURE SUMMARY

ZONE      TEMPERATURE IN/OUT DEG F           PRESSURE  IN/OUT  PSIA
          SHELL-SIDE        TUBE-SIDE       SHELL-SIDE      TUBE-SIDE

  1       201.0/ 202.4     228.3/ 228.0     250.0/ 249.9     20.0/ 19.9
  2       200.8/ 201.0     228.0/ 228.0     250.0/ 250.0     19.9/ 19.9
  3       199.5/ 200.8     228.0/ 227.7     250.0/ 250.0     19.9/ 19.8
  4       199.2/ 199.5     227.7/ 227.7     250.0/ 250.0     19.8/ 19.8
  5       197.7/ 199.2     227.7/ 227.5     250.0/ 250.0     19.8/ 19.7
  6       197.6/ 197.7     227.5/ 217.2     250.0/ 250.0     19.7/ 19.6

                  HEAT TRANSFER AND PRESSURE DROP SUMMARY

ZONE        HEAT TRANSFER         PRESSURE DROP (TOTAL)          FILM COEFF.
              MECHANISM                   PSIA                  BTU/HR-FT2-F
        SHELL-SIDE  TUBE-SIDE     SHELL-SIDE  TUBE-SIDE    SHELL-SIDE  TUBE-SIDE

  1  VAPORIZATION CONDENSATION      0.02        0.10         936.62     2093.49
  2  VAPORIZATION CONDENSATION      0.00        0.02         936.31     2027.16
  3  VAPORIZATION CONDENSATION      0.01        0.08         936.02     2294.64
  4  VAPORIZATION CONDENSATION      0.00        0.02         935.34     2528.96
  5  VAPORIZATION CONDENSATION      0.02        0.08         934.55     1897.63
  6  VAPORIZATION LIQ. SUBCOOL      0.00        0.13        4763.73       25.78
                                  --------    --------
          TOTAL PRESSURE DROP       0.05        0.43

                       HEAT TRANSFER SUMMARY (CONTD.)

   ZONE     ------ DUTY -------       U-VALUE        AREA       LMTD        FT
            MMBTU /HR    PERCENT   BTU/HR-FT2-F      FT2        DEG F

    1          1.81       33.0        334.22        208.2       26.4      0.982
    2          0.29        5.3        332.10         32.9       27.1      0.982
    3          1.52       27.7        339.84        163.8       27.7      0.982
    4          0.29        5.3        345.44         30.0       28.4      0.982
    5          1.52       27.7        327.49        162.0       29.1      0.982
    6          0.06        1.1         20.80        117.2       24.4      0.982
            ----------    -----                    -------
TOTAL          5.48      100.0                      714.1
WEIGHTED                              282.91                    27.6      0.982
OVERALL                                                         22.6      0.982
INSTALLED                                           714.9

 TOTAL DUTY = (WT. U-VALUE)(TOTAL AREA)(WT. LMTD)(OVL. FT)
  ZONE DUTY = (ZONE U-VALUE)(ZONE AREA)(ZONE LMTD)(OVL. FT)
```

Example 10.6

Use HEXTRAN to rate the horizontal thermosyphon reboiler of Example 10.3 and compare the results with those obtained previously by hand. Assay data (ASTM D86 distillation at atmospheric pressure) for the petroleum fraction fed to the reboiler are given in the following table. The feed stream has an average API gravity of 60°.

Volume Percent Distilled	Temperature (°F)
0	158.8[a]
10	210
30	240
50	260
70	275
90	290
100	309[b]

[a]Initial boiling point.
[b]End point.

Solution

For this problem, the tube-side feed (Therminol®) is defined as a bulk property stream and the values of C_P, k, μ and ρ (55.063 lbm/ft^3) given in Example 10.3 are entered as average liquid properties on the appropriate form. Note that the density, not the specific gravity, must be entered. Additional data required for this stream are the flow rate (425,000 lb/h), temperature (420°F) and pressure. Since the stream pressure was not specified in Example 10.3, a value of 40 psia is (arbitrarily) assumed.

The shell-side feed (petroleum fraction) is defined as an assay stream, and its flow rate (300,000 lb/h) and pressure (35 psia) are entered on the Specifications form. To complete the thermal specification of the stream, *Bubble Point* is selected from the list of available specifications. Next, under Components and Thermodynamics, the Assay tab is selected and a new assay name (A1) is entered. Clicking on the *Add* button activates the data entry tree that includes the listings *Distillation* and *Gravity*, as shown below. Clicking on each of these items in turn brings up the panels where the ASTM distillation data and average API gravity are entered. HEXTRAN uses the assay data to determine a set of pseudo components that represent the composition of the stream. The assay name, A1, is also entered on the Specifications form for the feed stream in order to link the assay with the stream to which it applies.

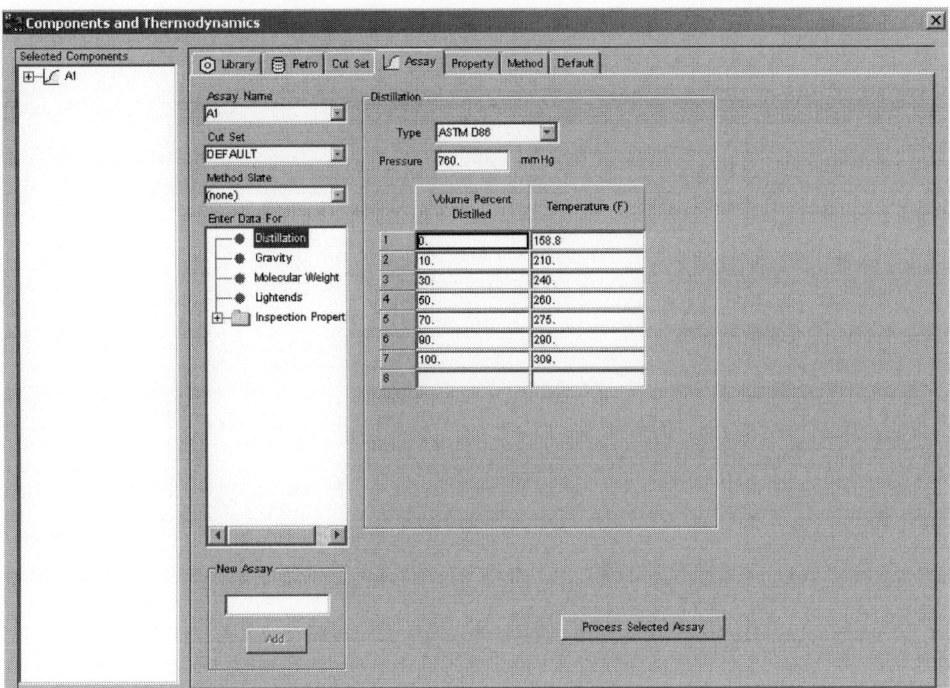

A set of thermodynamic procedures is also required for the assay stream. The PR EOS is selected as the method for equilibrium, enthalpy, and vapor density calculations; the API method is chosen for calculating liquid density. The petroleum method is selected for all transport properties (viscosity, thermal conductivity, and surface tension).

Data for the heat exchanger are entered as given in Example 10.3, including the number of tubes (290) and the shell ID (23.25 in). A type A front head and no baffles are assumed. Tubesheet thickness and shell-side nozzles are left unspecified; HEXTRAN will determine suitable values for these items, which were not specified in Example 10.3. Fouling factors from Example 10.3 are entered on the Film Options form.

The input file generated by the HEXTRAN GUI is given below, followed by a summary of results extracted from the HEXTRAN output file. It is seen that the reboiler generates 82,390 lb/h of vapor, about 37% more than the 60,000 lb/h required. The tube-side pressure drop is 9.34 psi, which is less than the maximum of 10 psi specified for the unit. HEXTRAN does not compute a critical heat flux, so this check must be done by hand. In the present case, the heat flux is approximately 13,000 Btu/h · ft^2, well below the critical value of 36,365 Btu/h · ft^2 calculated by hand in Example 10.3. (In actual operation, the heat flux would be about 37% lower.) Therefore, the reboiler is suitable for the service, in agreement with the result obtained in Example 10.3.

Results from HEXTRAN are compared with those calculated by hand in the following table. The shell-side (boiling) heat-transfer coefficient calculated by hand is very conservative compared with the value given by HEXTRAN, but the overall coefficients differ by less than 20%. The mean temperature difference used in the hand calculations is slightly higher than the value calculated by HEXTRAN. However, the heat flux ($U_D \Delta T_m$) calculated by hand is on the safe side, about 10% below the HEXTRAN value. Notice that virtually all of the shell-side pressure drop occurs in the nozzles. If two pairs of nozzles (6-in. inlet, 10-in. outlet) are assumed instead of the single pair used by HEXTRAN, the shell-side pressure drop is reduced to 0.33 psi.

Item	Hand	HEXTRAN
h_i (Btu/h · ft^2 ·°F)	346	346.2
h_o (Btu/h · ft^2 ·°F)	278	555[a]
U_D (Btu/h · ft^2 ·°F)	113	132[a]
ΔP_i (psi)	8.4	9.34
ΔP_o (psi)	–	1.39
ΔT_m (°F)	104.2	98.5[a]
$U_D \Delta T_m$ (Btu/h · ft^2)	11,775	13,002

[a]Area-weighted average from zone analysis.

HEXTRAN Input File for Example 10.6

```
$ GENERATED FROM HEXTRAN KEYWORD EXPORTER
$
$      General Data Section
$
TITLE PROJECT=Example 10-6, PROBLEM=Horizontal Thermosyphon Reboiler, SITE=
$
DIME  English, AREA=FT2, CONDUCTIVITY=BTUH, DENSITY=LB/FT3, *
      ENERGY=BTU, FILM=BTUH, LIQVOLUME=FT3, POWER=HP, *
      PRESSURE=PSIA, SURFACE=DYNE, TIME=HR, TEMPERATURE=F, *
      UVALUE=BTUH, VAPVOLUME=FT3, VISCOSITY=CP, WT=LB, *
      XDENSITY=API, STDVAPOR=379.490
$
PRINT ALL, *
      RATE=M
$
CALC  PGEN=New, WATER=Saturated
$
$   Component Data Section
$
COMPONENT DATA
$

$
 ASSAY    FIT= SPLINE, CHARACTERIZE=SIMSCI, MW= SIMSCI, *
          CONVERSION= API87, GRAVITY= WATSONK, TBPIP=1, TBPEP=98
$
 TBPCUTS 100.00, 800.00, 28 /*
         1200.00, 8 /*
         1600.00, 4
$
$      Thermodynamic Data Section
$
THERMODYNAMIC DATA
$
 METHODS SET=SET1, KVALUE=PR, ENTHALPY(L)=PR, ENTHALPY(V)=PR, *
                DENSITY(L)=API, DENSITY(V)=PR, VISCOS(L)=PETRO, *
                VISCOS(V)=PETRO, CONDUCT(L)=PETRO, CONDUCT(V)=PETRO, *
                SURFACE=PETRO
$
 WATER DECANT=ON, SOLUBILITY = Simsci, PROP = Saturated
$
$Stream Data Section
$
STREAM DATA
$
 PROP STRM=THERM_COLD, NAME=THERM_COLD
$
 PROP STRM=PROD, NAME=PROD
$
 PROP STRM=FEED, NAME=FEED, PRES=35.000, PHASE=L, *
         RATE(W)=300000.000, ASSAY=LV, BLEND
 D86    STRM=FEED, *
 DATA= 0.0, 158.80 / 10.0, 210.00 / 30.0, 240.00 / 50.0, 260.00 / *
        70.0, 275.00 / 90.0, 290.00 / 100.0, 309.00
 API    STRM=FEED, AVG=60.000
```

```
$
 PROP STRM=THERMINOL, NAME=THERMINOL, TEMP=420.00, PRES=40.000, *
        LIQUID(W)=425000.000, LCP(AVG)=0.534, Lcond(AVG)=0.0613, *
        Lvis(AVG)=0.84, Lden(AVG)=55.063
$
$ Calculation Type Section
$
SIMULATION


 $
  TOLERANCE TTRIAL=0.01
 $
  LIMITS AREA=200.00, 6000.00, SERIES=1, 10, PDAMP=0.00, *
         TTRIAL=30
 $
  CALC   TWOPHASE=New, DPSMETHOD=Stream, MINFT=0.80
 $
  PRINT  UNITS, ECONOMICS, STREAM, STANDARD, *
         EXTENDED, ZONES
 $
 ECONOMICS DAYS=350, EXCHANGERATE=1.00, CURRENCY=USDOLLAR
 $
  UTCOST OIL=3.50, GAS=3.50, ELECTRICITY=0.10, *
         WATER=0.03, HPSTEAM=4.10, MPSTEAM=3.90, *
         LPSTEAM=3.60, REFRIGERANT=0.00, HEATINGMEDIUM=0.00
 $
  HXCOST BSIZE=1000.00, BCOST=0.00, LINEAR=50.00, *
         EXPONENT=0.60, CONSTANT=0.00, UNIT
 $
 $     Unit Operations Data
 $
 UNIT OPERATIONS
 $
 STE UID=REBOILER
  TYPE  Old, TEMA=AXU, HOTSIDE=Tubeside, ORIENTATION=Horizontal, *
        FLOW=Countercurrent, *
        UESTIMATE=50.00, USCALER=1.00

  TUBE  FEED=THERMINOL, PRODUCT=THERM_COLD, *
        LENGTH=16.00, OD=0.750, *
        BWG=14, NUMBER=290, PASS=2, PATTERN=90, *
        PITCH=1.0000, MATERIAL=1, *
        FOUL=0.001, LAYER=0, *
        DPSCALER=1.00
 $
 SHELL  FEED=FEED, PRODUCT=PROD, *
        ID=23.25, SERIES=1, PARALLEL=1, *
        MATERIAL=1, *
        FOUL=0.0005, LAYER=0, *
        DPSCALER=1.00
 $
  BAFF  NONE
 $
  TNOZZ TYPE=Conventional, ID=6.065, 6.065, NUMB=1, 1
 $
  CALC  TWOPHASE=New, *
        DPSMETHOD=Stream, *
        MINFT=0.80
 $
  PRINT STANDARD, *
        EXTENDED, *
        ZONES
 $
  COST  BSIZE=1000.00, BCOST=0.00, LINEAR=50.00, *
        CONSTANT=0.00, EXPONENT=0.60, Unit
 $
 $ End of keyword file...
```

HEXTRAN Output Data for Example 10.6

```
================================================================================
                    SHELL AND TUBE EXCHANGER DATA SHEET
I------------------------------------------------------------------------------I
I EXCHANGER  NAME                               UNIT ID REBOILER               I
I SIZE    23x 192    TYPE AXU,    HORIZONTAL    CONNECTED 1 PARALLEL  1 SERIES I
I AREA/UNIT  904.   FT2 (    904. FT2 REQUIRED) AREA/SHELL    904. FT2         I
I------------------------------------------------------------------------------I
I PERFORMANCE OF ONE UNIT             SHELL-SIDE                TUBE-SIDE      I
I------------------------------------------------------------------------------I
I FEED STREAM NUMBER                    FEED                     THERMINOL     I
I FEED STREAM NAME                      FEED                     THERMINOL     I
I TOTAL FLUID          LB /HR            300000.                   425000.     I
I     VAPOR   (IN/OUT) LB /HR          0./      82390.           0./        0. I
I     LIQUID           LB /HR     300000./     217610.      425000./   425000. I
I     STEAM            LB /HR          0./          0.           0./        0. I
I     WATER            LB /HR          0./          0.           0./        0. I
I     NON CONDENSIBLE  LB /HR               0.                       0.        I
I TEMPERATURE (IN/OUT) DEG F       288.9 /    298.3          420.0 /  368.1    I
I PRESSURE     (IN/OUT) PSIA       35.00 /    33.61          40.00 /   30.66   I
I------------------------------------------------------------------------------I
I SP. GR., LIQ  (60F / 60F H2O)    0.739 /    0.742          0.883 /   0.883   I
I          VAP  (60F / 60F AIR)    0.000 /    3.577          0.000 /   0.000   I
I DENSITY,    LIQUID    LB/FT3    39.063 /  39.027          55.063 /  55.063   I
I             VAPOR     LB/FT3     0.000 /   0.463           0.000 /   0.000   I
I VISCOSITY, LIQUID     CP         0.179 /   0.179           0.840 /   0.840   I
I             VAPOR     CP         0.000 /   0.009           0.000 /   0.000   I
I THRML COND,LIQ  BTU/HR-FT-F     0.0547 /  0.0541          0.0613 /  0.0613   I
I             VAP  BTU/HR-FT-F    0.0000 /  0.0136          0.0000 /  0.0000   I
I SPEC.HEAT,LIQUID BTU /LB F      0.6013 /  0.6051          0.5340 /  0.5340   I
I             VAPOR BTU /LB F     0.0000 /  0.4936          0.0000 /  0.0000   I
I LATENT HEAT          BTU /LB        122.02                     0.00          I
I VELOCITY               FT/SEC         0.51                     7.95          I
I DP/SHELL(DES/CALC)  PSI            0.00 / 1.39             0.00 /  9.34      I
I FOULING RESIST FT2-HR-F/BTU  0.00050 (0.00050 REQD)          0.00100         I
I------------------------------------------------------------------------------I
I TRANSFER RATE  BTU/HR-FT2-F  SERVICE  132.17 ( 132.10 REQD), CLEAN  172.96 I
I HEAT EXCHANGED MMBTU /HR      11.772,      MTD(CORRECTED)  98.6,      FT 0.998 I
I------------------------------------------------------------------------------I
I CONSTRUCTION OF ONE SHELL           SHELL-SIDE                TUBE-SIDE      I
I------------------------------------------------------------------------------I
I DESIGN PRESSURE/TEMP PSIA  /F      100./  500.            100./  500.        I
I NUMBER OF PASSES                       1                       2             I
I MATERIAL                           CARB STL                CARB STL          I
I INLET  NOZZLE ID/NO        IN        6.1/ 1                  6.1/ 1          I
I OUTLET NOZZLE ID/NO        IN       10.0/ 1                  6.1/ 1          I
I------------------------------------------------------------------------------I
I TUBE: NUMBER   290, OD  0.750  IN , BWG    14        , LENGTH 16.0 FT        I
I       TYPE BARE,                PITCH    1.0000 IN,   PATTERN 90 DEGREES    I
I SHELL:  ID    23.25 IN,            SEALING STRIPS    0 PAIRS                 I
I RHO-V2: INLET NOZZLE  4416.7 LB/FT-SEC2                                      I
I TOTAL WEIGHT/SHELL,LB     3701.3 FULL OF WATER   0.113E+05 BUNDLE     5000.4 I
I------------------------------------------------------------------------------I
```

```
===============================================================================
                     SHELL AND TUBE EXTENDED DATA SHEET
I-----------------------------------------------------------------------------I
I EXCHANGER  NAME                               UNIT ID REBOILER              I
I SIZE    23x 192   TYPE AXU,   HORIZONTAL      CONNECTED 1 PARALLEL  1 SERIES I
I AREA/UNIT   904. FT2 (    904. FT2 REQUIRED)                                I
I-----------------------------------------------------------------------------I
I PERFORMANCE OF ONE UNIT             SHELL-SIDE               TUBE-SIDE      I
I-----------------------------------------------------------------------------I
I FEED STREAM NUMBER                    FEED                    THERMINOL     I
I FEED STREAM NAME                     FEED                     THERMINOL     I
I WT FRACTION LIQUID (IN/OUT)       1.00 / 0.73                 1.00 / 1.00   I
I REYNOLDS NUMBER                     13784.                    37732.        I
I PRANDTL NUMBER                      0.772                    17.705         I
I UOPK,LIQUID                     12.060 /  12.060             0.000 / 0.000  I
I      VAPOR                       0.000 /  12.060             0.000 / 0.000  I
I SURFACE TENSION    DYNES/CM     11.617 /  11.514             0.000 / 0.000  I
I FILM COEF(SCL) BTU/HR-FT2-F         552.0 (1.000)             346.2 (1.000) I
I FOULING LAYER THICKNESS  IN           0.000                       0.000     I
I-----------------------------------------------------------------------------I
I THERMAL RESISTANCE                                                          I
I UNITS: (FT2-HR-F/BTU)        (PERCENT)   (ABSOLUTE)                         I
I SHELL FILM                      23.94       0.00181                         I
I TUBE  FILM                      49.03       0.00371                         I
I TUBE  METAL                      3.44       0.00026                         I
I TOTAL FOULING                   23.58       0.00178                         I
I ADJUSTMENT                       0.06       0.00000                         I
I-----------------------------------------------------------------------------I
I PRESSURE DROP                        SHELL-SIDE               TUBE-SIDE     I
I UNITS: (PSIA )               (PERCENT)   (ABSOLUTE)   (PERCENT)  (ABSOLUTE)I
I WITHOUT NOZZLES                  0.01          0.00      88.38         8.26 I
I INLET   NOZZLES                 34.19          0.48       7.26         0.68 I
I OUTLET  NOZZLES                 65.79          0.92       4.36         0.41 I
I TOTAL   /SHELL                                 1.39                    9.34 I
I TOTAL   /UNIT                                  1.39                    9.34 I
I DP SCALER                                      1.00                    1.00 I
I-----------------------------------------------------------------------------I
I CONSTRUCTION OF ONE SHELL                                                   I
I-----------------------------------------------------------------------------I
I TUBE:OVERALL LENGTH        16.0          FT  EFFECTIVE LENGTH    15.88    FT I
I      TOTAL TUBESHEET THK    1.5          IN  AREA RATIO (OUT/IN)  1.284      I
I      THERMAL COND.         30.0BTU/HR-FT-F  DENSITY              490.80 LB/FT3I
I-----------------------------------------------------------------------------I
I BAFFLE: THICKNESS           0.500        IN  NUMBER                   1      I
I-----------------------------------------------------------------------------I
I BUNDLE: DIAMETER           22.7          IN  TUBES IN CROSSFLOW 290          I
I         CROSSFLOW AREA      8.003        FT2 WINDOW AREA            1.003 FT2 I
I         TUBE-BFL LEAK AREA 0.019         FT2 SHELL-BFL LEAK AREA  0.019  FT2 I
I-----------------------------------------------------------------------------I
```

```
================================================================================

                    ZONE ANALYSIS FOR EXCHANGER REBOILER

                     TEMPERATURE - PRESSURE SUMMARY

ZONE      TEMPERATURE IN/OUT DEG F               PRESSURE  IN/OUT PSIA
          SHELL-SIDE          TUBE-SIDE         SHELL-SIDE       TUBE-SIDE

  1       295.2/ 298.3      420.0/ 400.8        34.1/  33.6      40.0/  36.5
  2       292.1/ 295.2      400.8/ 383.6        34.5/  34.1      36.5/  33.5
  3       288.9/ 292.1      383.6/ 368.1        35.0/  34.5      33.5/  30.7

                     HEAT TRANSFER AND PRESSURE DROP SUMMARY

ZONE         HEAT TRANSFER           PRESSURE DROP (TOTAL)          FILM COEFF.
               MECHANISM                    PSIA                   BTU/HR-FT2-F
         SHELL-SIDE  TUBE-SIDE       SHELL-SIDE  TUBE-SIDE      SHELL-SIDE TUBE-SIDE

  1   VAPORIZATION LIQ. SUBCOOL        0.46        3.45         614.29     346.19
  2   VAPORIZATION LIQ. SUBCOOL        0.46        3.10         559.03     346.19
  3   VAPORIZATION LIQ. SUBCOOL        0.46        2.79         499.05     346.19
                                     --------    --------
           TOTAL PRESSURE DROP         1.39        9.34

                       HEAT TRANSFER SUMMARY (CONTD.)

     ZONE     ------ DUTY -------        U-VALUE          AREA       LMTD        FT
              MMBTU /HR    PERCENT    BTU/HR-FT2-F        FT2        DEG F

      1          4.35       37.0         135.46          283.5      113.5      0.998
      2          3.90       33.2         132.57          299.5       98.4      0.998
      3          3.52       29.9         128.89          320.7       85.2      0.998
             ----------    -----                        -------
TOTAL           11.77      100.0                         903.7
WEIGHTED                                 132.17                      98.7      0.998
OVERALL                                                              98.9      0.998
INSTALLED                                                904.2

 TOTAL DUTY = (WT. U-VALUE)(TOTAL AREA)(WT. LMTD)(OVL. FT)
  ZONE DUTY = (ZONE U-VALUE)(ZONE AREA)(ZONE LMTD)(OVL. FT)
```

10.6.2 HTRI

The *Xist* module of the HTRI *Xchanger Suite* is used for shell-and-tube reboilers. Although the HTRI technology is proprietary, some information has been published regarding the methodology used for kettle reboilers [18] and horizontal thermosyphon reboilers [13]. Additional information can be inferred from the detailed output files generated by the program.

For kettle reboilers, a recirculation model is used in which the internal circulation rate in the kettle is determined by a pressure balance. The internal circulation rate forms the basis for calculating the boiling heat-transfer coefficient, which is composed of nucleate boiling and convective terms, with correction factors for nucleate boiling suppression, convective enhancement, and mixture effects. As with single-phase exchangers, *Xist* performs incremental (stepwise) calculations using a three-dimensional grid. This feature allows local temperature gradients and heat-transfer coefficients to be computed and greatly improves the reliability of the method, especially for multi-component systems. Reliable simulation is possible with any type of heating medium, including multi-component condensing process streams [18].

For thermosyphon reboilers, the piping configuration is specified as input and the program calculates the circulation rate. Either a detailed or simplified piping configuration can be used. In the latter, only the total liquid head and the equivalent lengths of feed and return lines are entered. In the former, complete details of both lines are entered using equivalent lengths for pipe fittings. Either user-specified equivalent lengths or default values contained in the program can be used.

For all types of reboilers, the actual and critical heat fluxes are computed at each increment, along with the flow regime (bubble, slug, etc.) and the boiling mechanism (nucleate, film, etc.). For thermosyphons, a stability assessment is also performed to determine the potential for various types of flow instability in the hydraulic circuit.

The following examples illustrate the use of *Xist* for reboiler applications.

Example 10.7

Use *Xist* to rate the kettle reboiler designed in Example 10.2, and compare the results with those obtained previously by other methods.

Solution

Data from Example 10.2 are entered on the appropriate *Xist* input forms as indicated below. Parameters not listed are either left at their default settings or left unspecified to be calculated by the program.

(a) Input Summary.

Case mode: Rating	Service type: Kettle reboiler

(b) Geometry/Exchanger.

TEMA type: BKU	Shell ID: 23.25 in.

Note: The shell ID on this panel is actually the port diameter of the kettle.

(c) Geometry/Kettle Reboiler.

Kettle diameter: 37 in.	Reboiler pressure location: At top of bundle

Note: The kettle diameter on this panel is actually the shell ID.

(d) Geometry/Tubes.

Tube OD: 1 in.	Tube passes: 2
Average wall thickness: 0.083 in.	Straight tube length: 13 ft
Tube pitch: 1.25 in.	Tube count: 212
Tube layout angle: 90°	

(e) Geometry/Baffles/Supports.
Support plates/baffle space: User set: 3

(f) Geometry/Nozzles.

Shell side	Tube side
Inlet ID: 5.047 in.	Inlet ID: 6.065 in.
Number: 2	Number: 1
Outlet ID: 6.065 in.	Outlet ID: 3.068 in.
Number: 2	Number: 1
Liquid outlet ID: 4.026 in.	

(g) Control/Methods/Condensation
Pure component: Yes

(h) Process.

	Hot fluid	Cold fluid
Fluid name	Steam	Distillation Bottoms
Phase	Condensing	Boiling
Flow rate (1000 lb/h)	5.645	96
Inlet fraction vapor	1	0
Outlet fraction vapor	0	–
Operating pressure (psia)	20	250
Fouling resistance ($h \cdot ft^2 \cdot °F/Btu$)	0.0005	0.0005

(i) Hot fluid properties.
Physical property input option: Component by component
Heat release input method: Program calculated
Clicking on the *Property Generator* button opens the property generator as shown below. VMG Thermo is selected as the property package and Steam95 is selected from the list of thermodynamic methods for both the vapor and liquid phases. This method uses steam tables to obtain fluid properties.

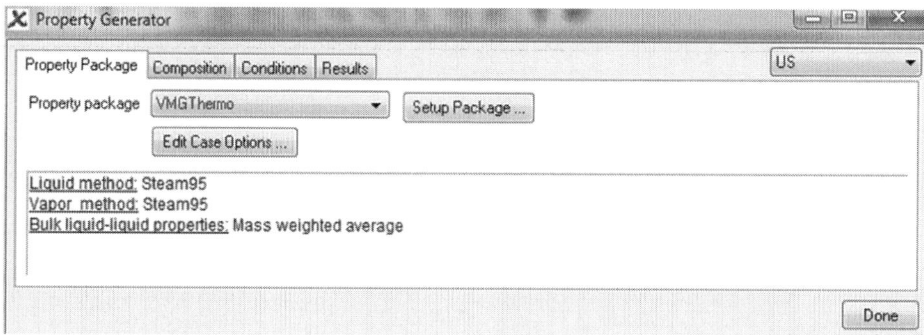

On the composition panel, water is selected from the list of components as shown below. Since it is the only component in the hot stream, its mole fraction is 1.0.

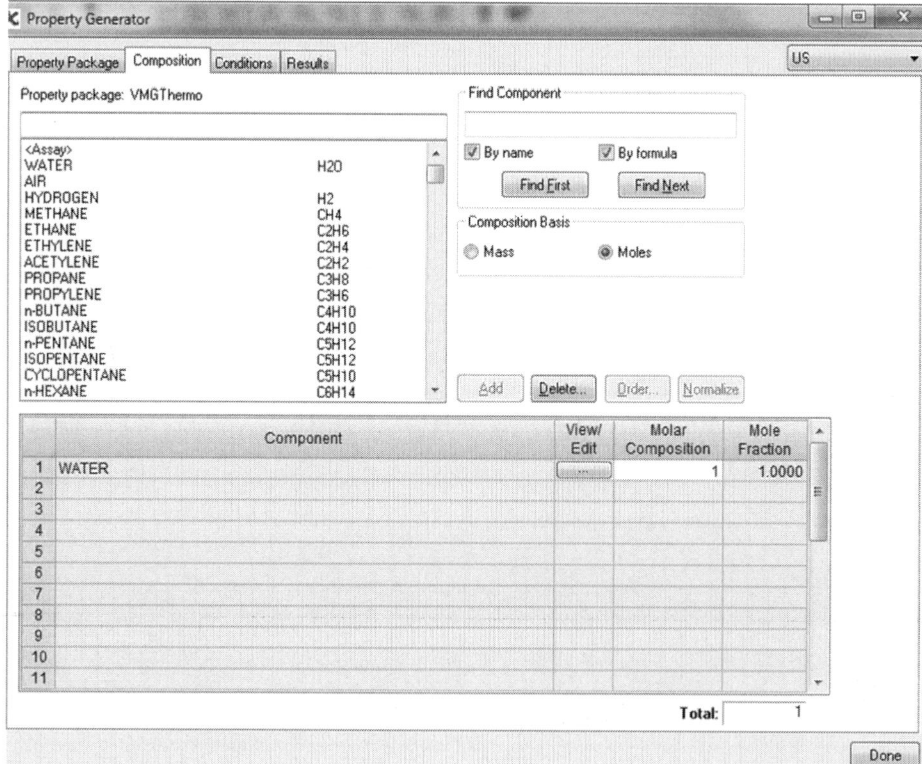

On the conditions panel, *property sets* is selected as the temperature point method. Two pressure levels, 20 and 19 psia, are specified and the temperature range for fluid properties is set as shown below. The number of points in this range at which properties are to be generated is set at 20.

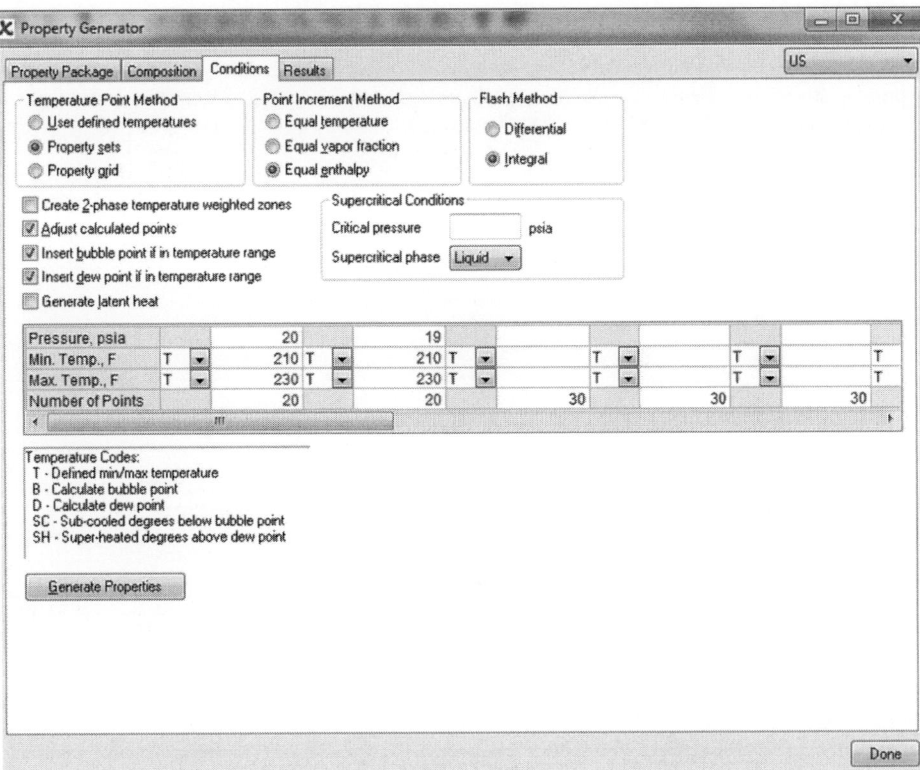

Clicking on the *Generate Properties* button produces the results shown below. The *Transfer* button is clicked to transfer the data to *Xist*. (Note that a maximum of 30 data points can be transferred.) Finally, clicking the *Done* button closes the property generator and returns control to the *Xist* input menu.

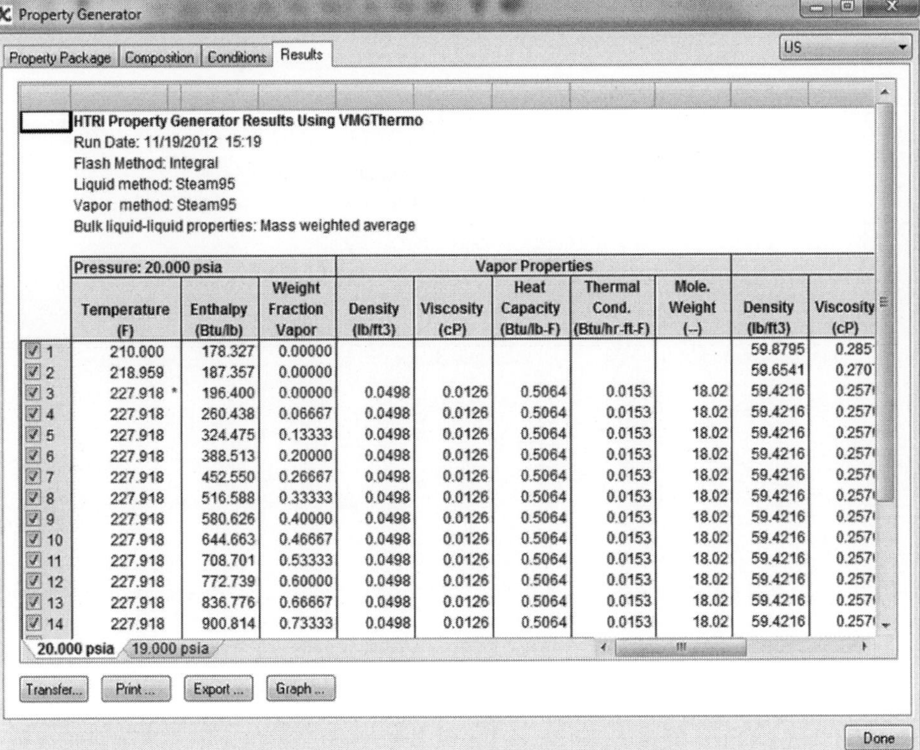
Property Generator

Property Package | Composition | Conditions | Results — US

HTRI Property Generator Results Using VMGThermo
Run Date: 11/19/2012 15:19
Flash Method: Integral
Liquid method: Steam95
Vapor method: Steam95
Bulk liquid-liquid properties: Mass weighted average

	Pressure: 20.000 psia			Vapor Properties						
	Temperature (F)	Enthalpy (Btu/lb)	Weight Fraction Vapor	Density (lb/ft3)	Viscosity (cP)	Heat Capacity (Btu/lb-F)	Thermal Cond. (Btu/hr-ft-F)	Mole. Weight (--)	Density (lb/ft3)	Viscosity (cP)
1	210.000	178.327	0.00000						59.8795	0.285
2	218.959	187.357	0.00000						59.6541	0.270
3	227.918 *	196.400	0.00000	0.0498	0.0126	0.5064	0.0153	18.02	59.4216	0.257
4	227.918	260.438	0.06667	0.0498	0.0126	0.5064	0.0153	18.02	59.4216	0.257
5	227.918	324.475	0.13333	0.0498	0.0126	0.5064	0.0153	18.02	59.4216	0.257
6	227.918	388.513	0.20000	0.0498	0.0126	0.5064	0.0153	18.02	59.4216	0.257
7	227.918	452.550	0.26667	0.0498	0.0126	0.5064	0.0153	18.02	59.4216	0.257
8	227.918	516.588	0.33333	0.0498	0.0126	0.5064	0.0153	18.02	59.4216	0.257
9	227.918	580.626	0.40000	0.0498	0.0126	0.5064	0.0153	18.02	59.4216	0.257
10	227.918	644.663	0.46667	0.0498	0.0126	0.5064	0.0153	18.02	59.4216	0.257
11	227.918	708.701	0.53333	0.0498	0.0126	0.5064	0.0153	18.02	59.4216	0.257
12	227.918	772.739	0.60000	0.0498	0.0126	0.5064	0.0153	18.02	59.4216	0.257
13	227.918	836.776	0.66667	0.0498	0.0126	0.5064	0.0153	18.02	59.4216	0.257
14	227.918	900.814	0.73333	0.0498	0.0126	0.5064	0.0153	18.02	59.4216	0.257

20.000 psia | 19.000 psia

Transfer... Print ... Export ... Graph ...

Done

(j) Cold fluid properties.

Properties of the hydrocarbon stream are generated in the same manner as for steam. In this case, the Advanced Peng–Robinson thermodynamic method is chosen for both the vapor and liquid phases. On the conditions panel, *property sets* is selected as the temperature point method. Two pressure levels, 250 and 255 psia, are specified (to allow for the static pressure effect in the reboiler) with a temperature range of 195°F to 225°F. The number of data points is again set at 20.

The *Xist* Output Summary for this case is given below, from which it is found that the over-design for the unit is about 30%. Thus, according to *Xist* the reboiler is over-sized.

Xist Output Summary for Example 10.7

	Xist Ver. 7.00 5/28/2013 16:05 SN: 14337-877388691					**US Units**
	Kettle Reboiler					
	Example 10.7					
	Rating - Horizontal Multipass Flow TEMA BKU Shell With No Baffles					
1	**No Data Check Messages.**					
2	**See Runtime Message Report for Warning Messages.**					
3	**Process Conditions**		**Cold Shellside**		**Hot Tubeside**	
4	Fluid name			Distillation Bottoms		Steam
5	Flow rate	(1000-lb/hr)		96.000		5.6450
6	Inlet/Outlet Y	(Wt. frac vap.)	0.0000	0.5007	1.0000	0.0000
7	Inlet/Outlet T	(Deg F)	197.56	202.39	227.92	226.95
8	Inlet P/Avg	(psia)	250.29	250.31	20.000	19.822
9	dP/Allow.	(psi)	0.126	0.000	0.355	0.000
10	Fouling	(ft2-hr-F/Btu)		0.00050		0.00050
11			**Exchanger Performance**			
12	Shell h	(Btu/ft2 hr F)	1014.3	Actual U	(Btu/ft2-hr-F)	367.63
13	Tube h	(Btu/ft2-hr-F)	2949.5	Required U	(Btu/ft2-hr-F)	282.40
14	Hot regime	(--)	Sens Liq	Duty	(MM Btu/hr)	5.4279
15	Cold regime	(--)	Flow	Eff. area	(ft2)	741.89
16	EMTD	(Deg F)	25.7	Overdesign	(%)	30.18
17	**Shell Geometry**			**Baffle Geometry**		
18	TEMA type	(--)	BKU	Baffle type	(--)	Support
19	Shell ID	(inch)	23.250	Baffle cut	(Pct Dia.)	
20	Series	(--)	1	Baffle orientation	(--)	
21	Parallel	(--)	1	Central spacing	(inch)	38.516
22	Orientation	(deg)	0.00	Crosspasses	(--)	1
23	**Tube Geometry**			**Nozzles**		
24	Tube type	(--)	Plain	Shell inlet	(inch)	5.0470
25	Tube OD	(inch)	1.0000	Shell outlet	(inch)	6.0650
26	Length	(ft)	13.000	Inlet height	(inch)	0.8750
27	Pitch ratio	(--)	1.2500	Outlet height	(inch)	0.3801
28	Layout	(deg)	90	Tube inlet	(inch)	6.0650
29	Tubecount	(--)	212	Tube outlet	(inch)	3.0680
30	Tube Pass	(--)	2			
31	**Thermal Resistance, %**		**Velocities, ft/sec**		**Flow Fractions**	
32	Shell	36.25	Shellside	1.05	A	0.000
33	Tube	14.95	Tubeside	43.41	B	1.000
34	Fouling	40.42	Crossflow	0.72	C	0.000
35	Metal	8.39	Window	0.00	E	0.000
36					F	0.000

The following table compares the results from *Xist* with those obtained in previous examples using other methods. It can be seen that both the boiling and condensing heat-transfer coefficients from *Xist* are significantly higher than the corresponding values from HEXTRAN. Furthermore, the boiling-side coefficient from *Xist* is nearly double the value computed using Palen's method in the hand calculation. This comparison again indicates that Palen's method leads to very conservative results in the present application, with consequent over-sizing of the reboiler. The combined friction and acceleration loss in the kettle computed by *Xist* is an order of magnitude less than the value assumed (as an upper bound) in the hand calculations. This result reinforces the point made in Example 10.2 that these losses can generally be neglected in calculating the pressure balance for the kettle reboiler system.

Item	Hand	HEXTRAN	*Xist*
h_o (Btu/h · ft^2 · °F)	523	936[a]	1014
$[(D_o/D_i)\ (1/h_i + R_{Di})]^{-1}$ (Btu/h · ft^2 · °F)	1500 (assumed)	857[a]	994
U_D (Btu/h · ft^2 · °F)	297	335[a]	368
ΔP_i (psi)	0.3	0.43	0.355
ΔP_o (psi), excluding nozzles	0.2 (assumed)	–	0.022
ΔP_o (psi), nozzles only	0.05	0.05	0.104
ΔT_m (°F)	25.6	27.1[a]	25.7
$U_D \Delta T_m$ (Btu/h · ft^2)	7603	9079[a]	9458
$(\hat{q}/\hat{q}_c)_{max}$	0.11	–	0.095

[a]Area-weighted average over first five zones; subcooled condensate zone not included.

The tube layout generated by *Xist* is shown below. It contains 216 tubes (108 U-tubes), which agrees well with the number (212 tubes) obtained from the tube-count table.

Xist Tube Layout for Kettle Reboiler

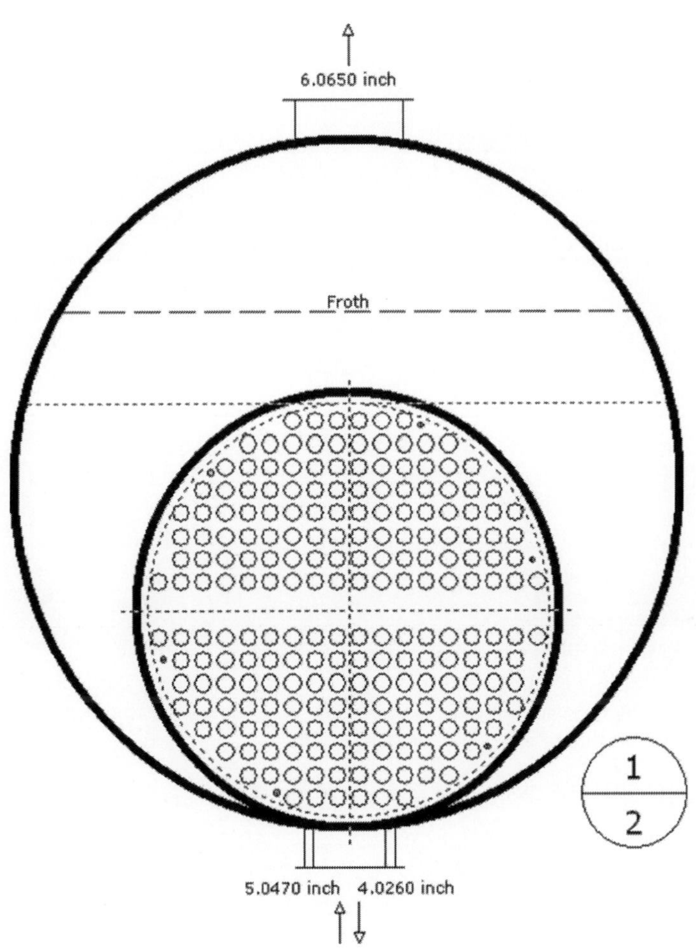

Example 10.8

Use *Xist* to rate the initial configuration for the vertical thermosyphon reboiler of Example 10.4 and compare the results with those obtained previously by hand.

Solution

Data from Example 10.4 are entered on the *Xist* input forms as indicated below. Parameters not listed are either left at their default settings or left unspecified to be calculated by the program.

(a) Input Summary.

Case mode: Rating	Service type: Thermosiphon reboiler
Hot fluid: Shell side	

(b) Geometry/Exchanger.

TEMA type: AEL	Shell Orientation: Vertical
Shell ID: 15.25 in.	

(c) Geometry/Tubes.

Tube OD: 1 in.	Tube passes: 1
Average wall thickness: 0.083 in.	Tube length: 8 ft
Tube pitch: 1.25 in.	Tube count: 106
Tube layout angle: 30°	

(d) Geometry/Baffles.

Baffle cut: 35%	Central baffle spacing: 6.1 in.

(e) Geometry/Tube Layout/Bundle Clearances.
Pairs of sealing strips: None

(f) Geometry/Nozzles.

Shell side	Tube side
Inlet ID: 4.026 in.	Inlet ID: 6.065 in.
Number: 1	Number: 1
Outlet ID: 2.067 in.	Outlet ID: 10.02 in.
Number: 1	Number: 1

(g) Geometry/Thermosyphon Reboiler.
Reboiler pressure location: At column bottom liquid surface

(h) Geometry/Thermosiphon Reboiler/Piping.
The detailed piping forms are used here to illustrate the procedure. They are invoked by checking the box for detailed piping on the main piping panel. The inlet piping panel is shown below:

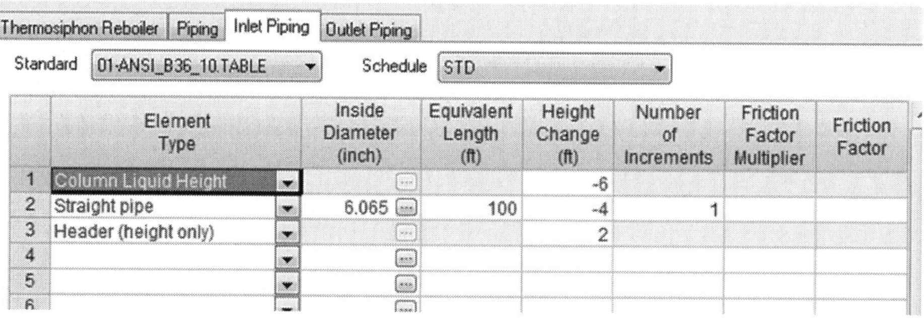

Thermosiphon Reboiler | Piping | Inlet Piping | Outlet Piping

Standard: 01-ANSI_B36_10.TABLE   Schedule: STD

	Element Type	Inside Diameter (inch)	Equivalent Length (ft)	Height Change (ft)	Number of Increments	Friction Factor Multiplier	Friction Factor
1	Column Liquid Height			-6			
2	Straight pipe	6.065	100	-4	1		
3	Header (height only)			2			
4							
5							
6							

The piping elements are selected from a list box that appears when a blank field in the first column is clicked. In this case there are only three elements because the straight pipe equivalent length is assumed to account for all entrance, exit, and fitting losses. Height changes are negative in the downward direction and positive in the upward direction. Height changes of individual elements are arbitrary here as long as they total to negative 8 ft. This puts the lower tubesheet a vertical distance of 8 ft below the liquid surface in the column sump. The outlet piping form is similar and is shown below.

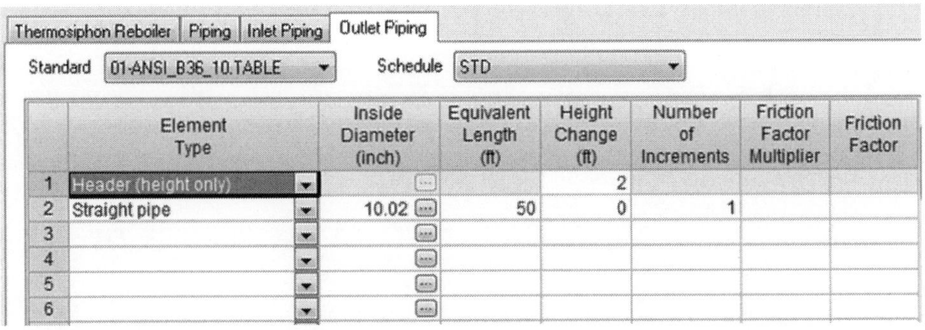

Thermosiphon Reboiler | Piping | Inlet Piping | Outlet Piping

Standard 01-ANSI_B36_10.TABLE   Schedule STD

	Element Type	Inside Diameter (inch)	Equivalent Length (ft)	Height Change (ft)	Number of Increments	Friction Factor Multiplier	Friction Factor
1	Header (height only)			2			
2	Straight pipe	10.02	50	0	1		
3							
4							
5							
6							

The outlet header extends from the upper tubesheet to the return pipe. Since the upper tubesheet is at the same elevation as the liquid surface in the column sump, the specified height change puts the return pipe a vertical distance of 2 ft above the surface of the liquid in the sump. The value assumed for this distance has a relatively small effect on the calculations. In practice, however, the bottom of the return line should be at least 6 in. above the highest liquid level expected in the column sump.

(i) Control/Methods/Condensation.
Pure component: Yes

(j) Process.

	Hot fluid	Cold fluid
Fluid name	Steam	Cyclohexane
Phase	Condensing	Boiling
Flow rate (1000 lb/h)	2.397	113.814
Inlet fraction vapor	1	0
Outlet fraction vapor	0	–
Operating pressure (psia)	18	16
Fouling resistance (h · ft^2 · °F/Btu)	0	0.0005

(k) Hot fluid properties.
VMGThermo and Steam95 are selected for the property package. On the conditions panel, *property sets* is selected as the temperature point method. Pressure levels of 20, 18, and 16 psia are specified with a temperature range of 200°F to 230°F, and the number of data points is set at 20.

(l) Cold fluid properties.
VMGThermo and the Advanced Peng–Robinson method are selected for the cyclohexane stream. Selecting *property sets* for the temperature point method, pressure levels of 20, 18, and 16 psia are specified, with a temperature range of 180°F to 220°F. The number of data points is again set at 20.

The *Xist* Output Summary for this case is shown below, from which the unit is seen to be under-designed by about 20%, in agreement with the hand calculation. Data from the Output Summary and detailed output files were used to prepare the results comparison shown in the table below. The heat-transfer coefficients and pressure drops calculated by *Xist* agree fairly well with the hand calculations despite the fact that the circulation rates computed by the two methods differ by a large amount.

Item	Hand	*Xist*
Circulation rate (lb/h)	113,814	70,919
h_i (Btu/h · ft^2 · °F)	565[a]	513
h_o (Btu/h · ft^2 · °F)	1500 (assumed)	1379
U_D (Btu/h · ft^2 · °F)	243[a]	257
ΔP_i (psi)[c]	0.862	0.863
ΔP_o (psi)	–	0.56
ΔT_m (°F)	34.7[b]	33.4
$(\hat{q}/\hat{q}_c)_{max}$	0.48	0.29
Under-design (%)	20	19.8

[a] Area-weighted average of values for sensible heating and boiling zones.

[b] Value for boiling zone.

[c] Friction and acceleration only; excluding entrance, exit, and static head losses.

Xist Output Summary for Example 10.8

Xist Ver. 7.00 5/28/2013 16:02 SN: 14337-877388691 **US Units**
Vertical Thermosiphon Reboiler
Example 10.8
Rating - Vertical Thermosiphon Reboiler TEMA AEL Shell With Single-Segmental Baffles

No Data Check Messages.
See Runtime Message Report for Warning Messages.

Process Conditions		**Hot Shellside**		**Cold Tubeside**	
Fluid name			Steam		Cyclohexane
Flow rate	(1000-lb/hr)		2.3970		70.919 *
Inlet/Outlet Y	(Wt. frac vap.)	1.0000	0.0000	0.0000	0.2146
Inlet/Outlet T	(Deg F)	222.36	220.66	182.53	183.15
Inlet P/Avg	(psia)	18.000	17.720	18.464	17.317
dP/Allow.	(psi)	0.560	0.000	2.294	0.000
Fouling	(ft2-hr-F/Btu)		0.00000		0.00050

Exchanger Performance					
Shell h	(Btu/ft2-hr-F)	1379.3	Actual U	(Btu/ft2-hr-F)	257.12
Tube h	(Btu/ft2-hr-F)	513.04	Required U	(Btu/ft2-hr-F)	320.61
Hot regime	(--)	Gravity	Duty	(MM Btu/hr)	2.3152
Cold regime	(--)	Conv	Eff. area	(ft2)	216.22
EMTD	(Deg F)	33.4	Overdesign	(%)	-19.80

Shell Geometry			**Baffle Geometry**		
TEMA type	(--)	AEL	Baffle type	(--)	Single-Seg.
Shell ID	(inch)	15.250	Baffle cut	(Pct Dia.)	35.8
Series	(--)	1	Baffle orientation	(--)	Parallel
Parallel	(--)	1	Central spacing	(inch)	6.1000
Orientation	(deg)	90.00	Crosspasses	(--)	13

Tube Geometry			**Nozzles**		
Tube type	(--)	Plain	Shell inlet	(inch)	4.0260
Tube OD	(inch)	1.0000	Shell outlet	(inch)	2.0670
Length	(ft)	8.000	Inlet height	(inch)	1.6250
Pitch ratio	(--)	1.2500	Outlet height	(inch)	0.2500
Layout	(deg)	30	Tube inlet	(inch)	6.0650
Tubecount	(--)	106	Tube outlet	(inch)	10.020
Tube Pass	(--)	1			

Thermal Resistance, %		**Velocities, ft/sec**		**Flow Fractions**	
Shell	18.64	Shellside	43.56	A	0.129
Tube	60.09	Tubeside	18.69	B	0.632
Fouling	15.41	Crossflow	51.76	C	0.113
Metal	5.85	Window	29.83	E	0.127
				F	0.000

The tube-side pressure drop was not explicitly calculated in Example 10.4, although all parameters needed for the calculation were evaluated. For completeness, the friction and acceleration losses are computed here.

$$\Delta P_{acc} = \frac{G_t^2 \gamma}{3.75 \times 10^{12} s_L} = \frac{(283,029)^2 \times 10.77}{3.75 \times 10^{12} \times 0.7208} = 0.3192 \text{ psi}$$

For the sensible heating zone, the friction loss is:

$$\Delta P_{f,BC} = \frac{f_t L_{BC} G_t^2}{7.50 \times 10^{12} D_t s_L} = \frac{0.0319 \times 2.9(283,029)^2}{7.50 \times 10^{12} \times 0.0695 \times 0.7208} = 0.0197 \text{ psi}$$

For the boiling zone, the friction loss is:

$$\Delta P_{f,CD} = \frac{f_t L_{CD} G_t^2 \overline{\phi}_{LO}^2}{7.50 \times 10^{12} D_t s_L} = \frac{0.0319 \times 5.1(283,029)^2 \times 15.08}{7.50 \times 10^{12} \times 0.0695 \times 0.7208} = 0.5231 \text{ psi}$$

The total friction loss is:

$$\Delta P_f = \Delta P_{f,BC} + \Delta P_{f,CD} = 0.0197 + 0.5231 = 0.5428 \text{ psi}$$

Therefore,

$$\Delta P_{acc} + \Delta P_f = 0.3192 + 0.5428 = 0.862 \text{ psi}$$

Example 10.9

Use *Xist* to obtain a final design for the vertical thermosyphon reboiler of Example 10.4.

Solution

Starting from the 15.25-in. unit rated in the previous example, the shell size is gradually increased until a suitable configuration is obtained. The following additional changes are made to the input data:

- The tube count is left unspecified so that it will be determined by the program based on the detailed tube layout.
- The central baffle spacing is adjusted to maintain B/d_s in the range 0.35 to 0.40, and the baffle cut is set accordingly at 35%.
- A fouling factor of 0.0005 h · ft^2 · °F/Btu is included for steam to provide an added safety margin.
- The steam pressure is increased to 20 psia as suggested in Example 10.4.
- The steam flow rate is increased to 2450 lb/h to ensure a vapor generation rate of at least 15,000 lb/h. With lower steam rates, the simulations may converge to solutions having vapor rates slightly below the required value.

With these settings, the smallest viable unit is found to be an 18-in. exchanger with an over-design of approximately 7.9%. The *Xist* Output Summary for this case is given below. Notice that the baffle cut has increased to 37.97%; *Xist* adjusts this value for conformity with the tube layout, placing the baffle edge at the center of the nearest tube row. Re-running the case using the tube layout as input yields essentially the same results.

Xist Output Summary for Example 10.9

Xist Ver. 7.00 5/28/2013 15:59 SN: 14337-877388691							US Units
Vertical Thermosiphon Reboiler							
Example 10.9							
Rating - Vertical Thermosiphon Reboiler TEMA AEL Shell With Single-Segmental Baffles							
No Data Check Messages.							
See Runtime Message Report for Warning Messages.							
Process Conditions		**Hot Shellside**			**Cold Tubeside**		
Fluid name			Steam				Cyclohexane
Flow rate	(1000-lb/hr)		2.4500				106.32 *
Inlet/Outlet Y	(Wt. frac vap.)	1.0000	0.0000		0.0000		0.1445
Inlet/Outlet T	(Deg F)	227.92	226.55		182.53		183.38
Inlet P/Avg	(psia)	20.000	19.753		18.385		17.308
dP/Allow.	(psi)	0.494	0.000		2.155		0.000
Fouling	(ft2-hr-F/Btu)		0.00050				0.00050
Exchanger Performance							
Shell h	(Btu/ft2-hr-F)	1326.7		Actual U	(Btu/ft2-hr-F)		197.77
Tube h	(Btu/ft2-hr-F)	403.05		Required U	(Btu/ft2-hr-F)		183.33
Hot regime	(--)	Gravity		Duty	(MM Btu/hr)		2.3568
Cold regime	(--)	Conv		Eff. area	(ft2)		322.23
EMTD	(Deg F)	39.9		Overdesign	(%)		7.88
Shell Geometry				**Baffle Geometry**			
TEMA type	(--)	AEL		Baffle type	(--)		Single-Seg.
Shell ID	(inch)	18.000		Baffle cut	(Pct Dia.)		37.97
Series	(--)	1		Baffle orientation	(--)		Parallel
Parallel	(--)	1		Central spacing	(inch)		6.3000
Orientation	(deg)	90.00		Crosspasses	(--)		13
Tube Geometry				**Nozzles**			
Tube type	(--)	Plain		Shell inlet	(inch)		4.0260
Tube OD	(inch)	1.0000		Shell outlet	(inch)		2.0670
Length	(ft)	8.000		Inlet height	(inch)		1.1250
Pitch ratio	(--)	1.2500		Outlet height	(inch)		0.3750
Layout	(deg)	30		Tube inlet	(inch)		6.0650
Tubecount	(--)	157		Tube outlet	(inch)		10.020
Tube Pass	(--)	1					
Thermal Resistance, %		**Velocities, ft/sec**			**Flow Fractions**		
Shell	14.91	Shellside	31.04		A		0.136
Tube	58.84	Tubeside	12.84		B		0.615
Fouling	21.75	Crossflow	40.26		C		0.072
Metal	4.51	Window	19.62		E		0.177
					F		0.000

The detailed output files from *Xist* were used to compile the design summary shown in the following table. Running a mechanical design program showed that 2-in. schedule 40 pipe is inadequate for the condensate nozzle; schedule 80 pipe is required for this nozzle. No other design modifications were indicated.

Design Summary for Example 10.9: Vertical Thermosyphon Reboiler

Item	Value
Steam design pressure (psia)	20
Exchanger type	AEL
Shell size (in.)	18
Surface area (ft^2)	322
Number of tubes	157
Tube OD (in.)	1.0
Tube length (ft)	8
Tube BWG	14
Tube passes	1
Tube pitch (in.)	1.25
Tube layout	Triangular
Number of baffles	12
Baffle cut (%)	37.97
Baffle thickness (in.)	0.1875
Central baffle spacing (in.)	6.3
Inlet baffle spacing (in.)	14.72
Outlet baffle spacing (in.)	10.05
Sealing strip pairs	0
Tube-side inlet nozzle	6-in. schedule 40
Tube-side outlet nozzle	10-in. schedule 40
Shell-side inlet nozzle	4-in. schedule 40
Shell-side outlet nozzle	2-in. schedule 40*
ΔP_i (psi)	2.16
ΔP_o (psi)	0.49
Circulation rate (lbm/h)	106,320
Exit vapor fraction	0.1445
Vapor generation rate (lbm/h)	15,363
Steam flow rate (lbm/h)	2450
$(\hat{q}/\hat{q}_c)_{max}$	0.20
Flow stability assessment	Stable
Two-phase flow regime(s)	Bubble, slug, annular
Boiling mechanism(s)	Nucleate, convective
Over-design (%)	7.88

*Schedule 80 required per mechanical design run.

The exchanger drawing and tube layout diagram from *Xist* are shown below. Tie rods were added manually using the tube layout editor. Notice that an impingement plate is provided at the steam inlet nozzle, as required for a saturated vapor. In addition, a shell-side vent should be included to purge any non-condensable gases that may enter with the steam. The vent nozzle, which is not shown on the exchanger drawing, should be located above and close to the condensate nozzle (see Section 11.2.3).

Xist **Exchanger Drawing for Example 10.9**

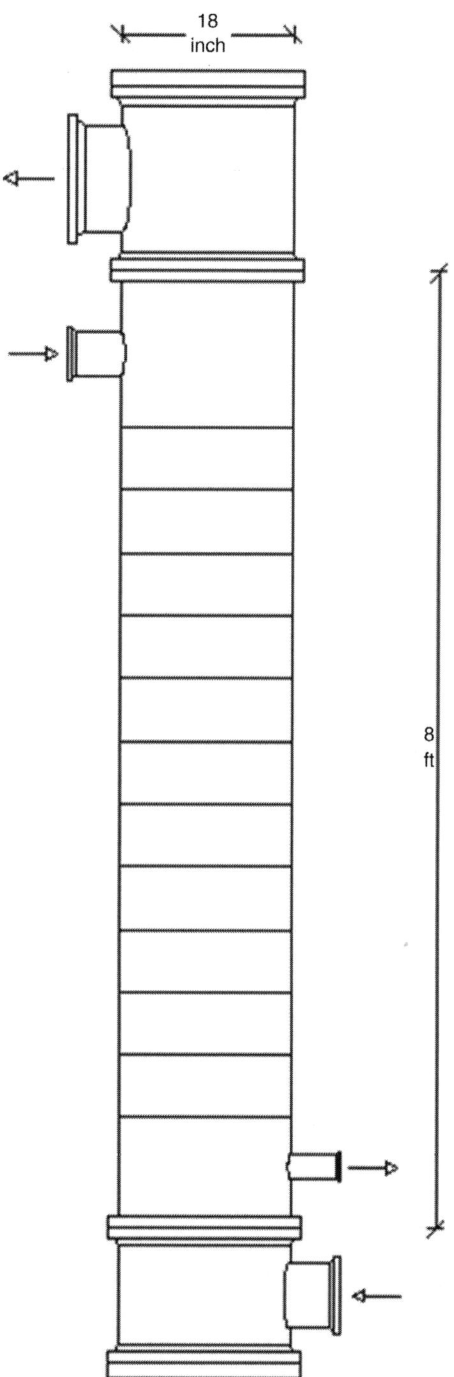

TEAM type	AEL	Total Tube Inlet Nozzles	1
Shell diameter	18 inch	Total Tube Outlet Nozzles	1
Tube length	8 ft	Total Shell Inlet Nozzles	1
Dry weight	2737 lb/shell	Total Shell Outlet Nozzles	1
Wet weight	4423 lb/shell		
Bundle weight	1108 lb/shell		

Xist Tube Layout for Example 10.9

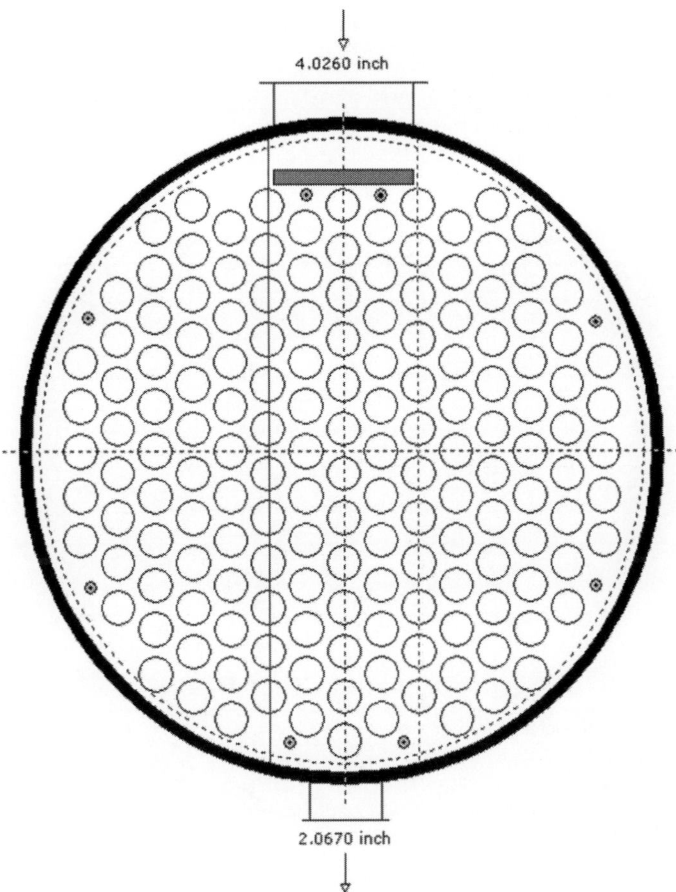

Example 10.10

This example illustrates a problem that can arise from application of fouling factors for cases with high heat flux. The *Xist* Output Summary shown below is a re-rating for the design conditions of an existing naphtha thermosyphon reboiler. The unit consists of a TEMA NEN shell with an ID of 47 in. containing 1153 tubes. The tubes are 1-in. OD, 13 BWG with a length of 12 ft. The heating medium is superheated steam. (The degree of superheating is quite high in this application, as indicated by the high steam inlet temperature. However, this condition is unrelated to the issues under consideration here.) As is common with re-ratings, the over-design is quite large because older design practices that predate the advent of modern design software tend to be very conservative.

The critical pressure of the naphtha stream is 457 psia, and the critical heat flux is calculated using Equation (9.84a):

$$\hat{q}_c = 16,070\left(D_t^2/L\right)^{0.35} P_c^{0.61} P_r^{0.25}(1 - P_r)$$

$$= 16,070\left[(0.0675)^2/12\right]^{0.35}(457)^{0.61}(45.6/457)^{0.25}(1 - 45.6/457)$$

$$\hat{q}_c = 21,651 \text{ Btu / h} \cdot \text{ft}^2$$

The average heat flux can be calculated from values of the overall coefficient and mean temperature difference obtained from the Output Summary, as follows:

$$\hat{q} = U_D \Delta T_m = 131.1 \times 148.3 = 19,442 \text{ Btu/ h} \cdot \text{ft}^2$$

Thus,

$$\hat{q}/\hat{q}_c = 19,442/21,651 = 0.90$$

While this ratio is less than 1, there is virtually no margin. (Recall that an important design guideline is to keep this ratio below 0.7.) The original designers expected a much lower design heat-flux ratio and they were inclined to increase heat flux conditions in order to attain the annular flow regime at the top of the tubes.

Xist Output Summary for Re-rating of an Existing Naphtha Reboiler

	Xist Ver. 7.00 5/29/2013 8:13 SN: 14337-877388691					**US Units**
	Naphtha Reboiler					
	Example 10.10					
	Rating - Vertical Thermosiphon Reboiler TEMA NEN Shell With Single-Segmental Baffles					
1	See Data Check Messages Report for Warning Messages.					
2	See Runtime Message Report for Warning Messages.					
3	**Process Conditions**		**Hot Shellside**		**Cold Tubeside**	
4	Fluid name			STEAM		NAPHTHA
5	Flow rate	(1000-lb/hr)		92.993		3565.3 *
6	Inlet/Outlet Y	(Wt. frac vap.)	1.0000	0.0000	0.0000	0.1556
7	Inlet/Outlet T	(Deg F)	726.62	482.47 *	339.02	347.40
8	Inlet P/Avg	(psia)	579.70	579.45	45.610	43.833
9	dP/Allow.	(psi)	0.508	0.000	3.554	0.000
10	Fouling	(ft2-hr-F/Btu)		0.00050		0.00200
11	**Exchanger Performance**					
12	Shell h	(Btu/ft2-hr-F)	573.40	Actual U	(Btu/ft2-hr-F)	131.04
13	Tube h	(Btu/ft2-hr-F)	491.74	Required U	(Btu/ft2-hr-F)	86.61
14	Hot regime	(--)	Gravity	Duty	(MM Btu/hr)	86.176
15	Cold regime	(--)	Conv	Eff. area	(ft2)	6691.1
16	EMTD	(Deg F)	148.3	Overdesign	(%)	51.30
17	**Shell Geometry**			**Baffle Geometry**		
18	TEMA type	(--)	NEN	Baffle type	(--)	Single-Seg.
19	Shell ID	(inch)	47.000	Baffle cut	(Pct Dia.)	26.97
20	Series	(--)	1	Baffle orientation	(--)	Perpend.
21	Parallel	(--)	2	Central spacing	(inch)	19.250
22	Orientation	(deg)	90.00	Crosspasses	(--)	6
23	**Tube Geometry**			**Nozzles**		
24	Tube type	(--)	Plain	Shell inlet	(inch)	8.0000
25	Tube OD	(inch)	1.0000	Shell outlet	(inch)	4.0000
26	Length	(ft)	12.000	Inlet height	(inch)	3.0145
27	Pitch ratio	(--)	1.2500	Outlet height	(inch)	3.5145
28	Layout	(deg)	30	Tube inlet	(inch)	18.000
29	Tubecount	(--)	1153	Tube outlet	(inch)	30.000
30	Tube Pass	(--)	1			
31	**Thermal Resistance, %**		**Velocities, ft/sec**		**Flow Fractions**	
32	Shell	22.85	Shellside	3.49	A	0.143
33	Tube	32.90	Tubeside	12.35	B	0.628
34	Fouling	38.91	Crossflow	4.49	C	0.062
35	Metal	5.34	Window	4.33	E	0.168
36					F	0.000

Under clean conditions the heat flux is higher, so the critical heat flux could potentially be attained. To consider this possibility in more detail, the clean heat-transfer coefficient is calculated assuming no change in the boiling regime. From the Output Summary, the fouling resistance accounts for 39% of the total thermal resistance. Hence, the clean coefficient is given by:

$$U_C = \left[\frac{1-0.39}{U_D}\right]^{-1} = 215 \text{ Btu} / \text{h} \cdot \text{ft}^2 \cdot {}^\circ\text{F}$$

Assuming the same mean temperature difference, the heat flux is then:

$$\hat{q} = 215 \times 148.3 = 31,885 \text{ Btu} / \text{h} \cdot \text{ft}^2$$

This value exceeds the estimated critical heat flux of 21,651 Btu / h · ft^2 by a substantial margin ($\hat{q}/\hat{q}_c \cong 1.5$). Hence, the uncertainty in the calculated heat fluxes notwithstanding, it is likely that the critical heat flux will be reached and that dry-wall heat transfer will occur when the reboiler is clean.

Based on proprietary methods, *Xist* performs calculations for a variety of potential dry-wall conditions; these include film boiling, mist flow, Boure instability, and transition flow. In the present case, under clean conditions *Xist* predicts transition boiling in the lower sections of the tubes followed by film boiling in the upper sections. Transition boiling is a regime between wet-wall nucleate boiling and complete dry-wall film boiling (see Figure 9.1). In this regime, dry-wall heat transfer is mixed with wet-wall heat transfer, which can result in unstable operation. Furthermore, with partial and complete dry-wall operation the wall temperature is elevated, which accelerates fouling in the case of naphtha. In summary, the application of a large fouling factor promotes fouling by elevating wall temperature.

The recommended solution to this problem is to reduce the steam pressure to eliminate dry-wall operation under clean conditions. Reducing the steam pressure from the original value of 579.7 psia to 250 psia reduces the saturation temperature from approximately 482°F to 401°F. This lowering of the hot-side temperature is sufficient to eliminate dry-wall operation; *Xist* predicts convective (wet-wall) boiling over the entire tube length for this case. The *Xist* Output Summary for this run with a steam pressure of

250 psia and zero fouling resistance is shown below. It can be seen that the lower steam pressure provides an over-design of 10% under clean conditions, which is an adequate safety margin. As fouling progresses between cleanings, the steam pressure is gradually increased as necessary to maintain proper operation of the reboiler.

Xist Output Summary for Naphtha Reboiler Using 250 psia Steam

	Xist Ver. 7.00 5/29/2013 8:23 SN: 14337-877388691					US Units
	Naphtha Reboiler					
	Example 10.10					
	Rating - Vertical Thermosiphon Reboiler TEMA NEN Shell With Single-Segmental Baffles					
1	See Data Check Messages Report for Warning Messages.					
2	See Runtime Message Report for Warning Messages.					
3	**Process Conditions**		**Hot Shellside**		**Cold Tubeside**	
4	Fluid name			STEAM		NAPHTHA
5	Flow rate	(1000-lb/hr)		83.424		3567.1 *
6	Inlet/Outlet Y	(Wt. frac vap.)	1.0000	0.0000	0.0000	0.1556
7	Inlet/Outlet T	(Deg F)	726.62	400.68 *	339.02	347.38
8	Inlet P/Avg	(psia)	250.00	249.57	45.607	43.827
9	dP/Allow.	(psi)	0.868	0.000	3.561	0.000
10	Fouling	(ft2-hr-F/Btu)		0.00000		0.00000
11			**Exchanger Performance**			
12	Shell h	(Btu/ft2-hr-F)	694.60	Actual U	(Btu/ft2-hr-F)	228.21
13	Tube h	(Btu/ft2-hr-F)	487.06	Required U	(Btu/ft2-hr-F)	206.97
14	Hot regime	(--)	Gravity	Duty	(MM Btu/hr)	86.184
15	Cold regime	(--)	Conv	Eff. area	(ft2)	6691.1
16	EMTD	(Deg F)	62.1	Overdesign	(%)	10.26
17	**Shell Geometry**			**Baffle Geometry**		
18	TEMA type	(--)	NEN	Baffle type	(--)	Single-Seg.
19	Shell ID	(inch)	47.000	Baffle cut	(Pct Dia.)	26.97
20	Series	(--)	1	Baffle orientation	(--)	Perpend.
21	Parallel	(--)	2	Central spacing	(inch)	19.250
22	Orientation	(deg)	90.00	Crosspasses	(--)	6
23	**Tube Geometry**			**Nozzles**		
24	Tube type	(--)	Plain	Shell inlet	(inch)	8.0000
25	Tube OD	(inch)	1.0000	Shell outlet	(inch)	4.0000
26	Length	(ft)	12.000	Inlet height	(inch)	3.0145
27	Pitch ratio	(--)	1.2500	Outlet height	(inch)	3.5145
28	Layout	(deg)	30	Tube inlet	(inch)	18.000
29	Tubecount	(--)	1153	Tube outlet	(inch)	30.000
30	Tube Pass	(--)	1			
31	**Thermal Resistance, %**		**Velocities, ft/sec**		**Flow Fractions**	
32	Shell	32.85	Shellside	6.98	A	0.142
33	Tube	57.84	Tubeside	11.80	B	0.628
34	Fouling	0.00	Crossflow	8.85	C	0.062
35	Metal	9.30	Window	8.53	E	0.168
36					F	0.000

References

[1] Palen JW. Shell-and-tube reboilers. In: Heat Exchanger Design Handbook, Vol. 3. New York: Hemisphere Publishing Corp.; 1988.
[2] Sloley AW. Properly design thermosyphon reboilers. Chem Eng Prog 1997;93(No. 3):52–64.
[3] Kister HZ. Distillation Operation. New York: McGraw-Hill; 1990.
[4] Fair JR. Vaporizer and reboiler design: Part 1. Chem Eng 1963;70(No. 14):119–24.
[5] Palen JW, Small WM. A new way to design kettle and internal reboilers. Hydrocarbon Proc 1964;43(No. 11):199–208.
[6] Bell KJ, Mueller AJ. Wolverine Engineering Data Book II. Wolverine Tube, Inc, www.wlv.com; 2001.
[7] Shinskey FG. Distillation Control. New York: McGraw-Hill; 1977.
[8] Thome JR. Engineering Data Book III. Wolverine Tube, Inc, www.wlv.com; 2004.
[9] Lee DC, Dorsey JW, Moore GZ, Mayfield FD. Design data for thermosyphon reboilers. Chem Eng Prog 1956;52(No. 4):,160–164.
[10] Dowlati R, Kawaji M. Two-phase flow and boiling heat transfer in tube bundles, Chapter 12. In: Kandlikar SG, Shoji M, Dhir VK, editors. Handbook of Phase Change: Boiling and Condensation. Philadelphia, PA: Taylor and Francis; 1999.
[11] Fair JR. What you need to design thermosyphon reboilers. Pet Refiner 1960;39(No. 2):105–23.
[12] Fair JR, Klip A. Thermal design of horizontal reboilers. Chem Eng Prog 1983;79(No. 3):86–96.
[13] Yilmaz SB. Horizontal shellside thermosyphon reboilers. Chem Eng Prog 1987;83(No. 11):64–70.
[14] Hewitt GF, Shires GL, Bott TR. Process Heat Transfer. Boca Raton, FL: CRC Press; 1994.
[15] Collins GK. Horizontal thermosyphon reboiler design. Chem Eng 1976;83(No. 15):149–52.
[16] Palen JW, Shih CC, Taborek J. Mist flow in thermosyphon reboilers. Chem Eng Prog 1982;78(No. 7):59–61.
[17] Poling BE, Prausnitz JM, O'Connell JP. The Properties of Gases and Liquids. 5th ed. New York: McGraw-Hill; 2000.
[18] Palen JW, Johnson DL. Evolution of kettle reboiler design methods and present status, Paper No. 13i. Houston: AIChE National Meeting; March 14–18, 1999.
[19] Perry RH, Chilton CH, editors. Chemical Engineers' Handbook. 5th ed. New York: McGraw-Hill; 1973.

Appendix 10.A Areas of Circular Segments

h/D	A	h/D	A	h/D	A	h/D	A	h/D	A	h/D	A	h/D	A	h/D	A	h/D	A	h/D	A
		0.050	0.01468	0.100	0.04087	0.150	0.07387	0.200	0.11182	0.250	0.15355	0.300	0.19817	0.350	0.24498	0.400	0.29337	0.450	0.34278
0.002	0.00012	0.052	0.01556	0.102	0.04208	0.152	0.07531	0.202	0.11343	0.252	0.15528	0.302	0.20000	0.352	0.24689	0.402	0.29533	0.452	0.34477
0.004	0.00034	0.054	0.01646	0.104	0.04330	0.154	0.07675	0.204	0.11504	0.254	0.15702	0.304	0.20184	0.354	0.24880	0.404	0.29729	0.454	0.34676
0.006	0.00062	0.056	0.01737	0.106	0.04452	0.156	0.07819	0.206	0.11665	0.256	0.15876	0.306	0.20368	0.356	0.25071	0.406	0.29926	0.456	0.34876
0.008	0.00095	0.058	0.01830	0.108	0.04576	0.158	0.07965	0.208	0.11827	0.258	0.16051	0.308	0.20553	0.358	0.25263	0.408	0.30122	0.458	0.35075
0.010	0.00133	0.060	0.01924	0.110	0.04701	0.160	0.08111	0.210	0.11990	0.260	0.16226	0.310	0.20738	0.360	0.25455	0.410	0.30319	0.460	0.35274
0.012	0.00175	0.062	0.02020	0.112	0.04826	0.162	0.08258	0.212	0.12153	0.262	0.16402	0.312	0.20923	0.362	0.25647	0.412	0.30516	0.462	0.35474
0.014	0.00220	0.064	0.02117	0.114	0.04953	0.164	0.08406	0.214	0.12317	0.264	0.16578	0.314	0.21108	0.364	0.25839	0.414	0.30712	0.464	0.35673
0.016	0.00268	0.066	0.02215	0.116	0.05080	0.166	0.08554	0.216	0.12481	0.266	0.16755	0.316	0.21294	0.366	0.26032	0.416	0.30910	0.466	0.35873
0.018	0.00320	0.068	0.02315	0.118	0.05209	0.168	0.08704	0.218	0.12646	0.268	0.16932	0.318	0.21480	0.368	0.26225	0.418	0.31107	0.468	0.36072
0.020	0.00375	0.070	0.02417	0.120	0.05338	0.170	0.08854	0.220	0.12811	0.270	0.17109	0.320	0.21667	0.370	0.26418	0.420	0.31304	0.470	0.36272
0.022	0.00432	0.072	0.02520	0.122	0.05469	0.172	0.09004	0.222	0.12977	0.272	0.17287	0.322	0.21853	0.372	0.26611	0.422	0.31502	0.472	0.36471
0.024	0.00492	0.074	0.02624	0.124	0.05600	0.174	0.09155	0.224	0.13144	0.274	0.17465	0.324	0.22040	0.374	0.26805	0.424	0.31699	0.474	0.36671
0.026	0.00555	0.076	0.02729	0.126	0.05733	0.176	0.09307	0.226	0.13311	0.276	0.17644	0.326	0.22228	0.376	0.26998	0.426	0.31897	0.476	0.36871
0.028	0.00619	0.078	0.02836	0.128	0.05866	0.178	0.09460	0.228	0.13478	0.278	0.17823	0.328	0.22415	0.378	0.27192	0.428	0.32095	0.478	0.37071
0.030	0.00687	0.080	0.02943	0.130	0.06000	0.180	0.09613	0.230	0.13646	0.280	0.18002	0.330	0.22603	0.380	0.27386	0.430	0.32293	0.480	0.37270
0.032	0.00756	0.082	0.03053	0.132	0.06135	0.182	0.09767	0.232	0.13815	0.282	0.18182	0.332	0.22792	0.382	0.27580	0.432	0.32491	0.482	0.37470
0.034	0.00827	0.084	0.03163	0.134	0.06271	0.184	0.09922	0.234	0.13984	0.284	0.18362	0.334	0.22980	0.384	0.27775	0.434	0.32689	0.484	0.37670
0.036	0.00901	0.086	0.03275	0.136	0.06407	0.186	0.10077	0.236	0.14154	0.286	0.18542	0.336	0.23169	0.386	0.27969	0.436	0.32887	0.486	0.37870
0.038	0.00976	0.088	0.03387	0.138	0.06545	0.188	0.10233	0.238	0.14324	0.288	0.18723	0.338	0.23358	0.388	0.28164	0.438	0.33086	0.488	0.38070
0.040	0.01054	0.090	0.03501	0.140	0.06683	0.190	0.10390	0.240	0.14494	0.290	0.18905	0.340	0.23547	0.390	0.28359	0.440	0.33284	0.490	0.38270
0.042	0.01133	0.092	0.03616	0.142	0.06822	0.192	0.10547	0.242	0.14666	0.292	0.19086	0.342	0.23737	0.392	0.28554	0.442	0.33483	0.492	0.38470
0.044	0.01214	0.094	0.03732	0.144	0.06963	0.194	0.10705	0.244	0.14837	0.294	0.19268	0.344	0.23927	0.394	0.28750	0.444	0.33682	0.494	0.38670
0.046	0.01297	0.096	0.03850	0.146	0.07103	0.196	0.10864	0.246	0.15009	0.296	0.19451	0.346	0.24117	0.396	0.28945	0.446	0.33880	0.496	0.38870
0.048	0.01382	0.098	0.03968	0.148	0.07245	0.198	0.11023	0.248	0.15182	0.298	0.19634	0.348	0.24307	0.398	0.29141	0.448	0.34079	0.498	0.39070
0.050	0.01468	0.100	0.04087	0.150	0.07387	0.200	0.11182	0.250	0.15355	0.300	0.19817	0.350	0.24498	0.400	0.29337	0.450	0.34278	0.500	0.39270

h: height; *D*: diameter; and *A*: area.
Rules for using table: (1) Divide height of segment by the diameter; multiply the area in the table corresponding to the quotient, height/diameter, by the diameter squared. When segment exceeds a semicircle, its area is: area of circle minus the area of a segment whose height is the circle diameter minus the height of the given segment. (2) To find the diameter when given the chord and the segment height: the diameter = $[(^1/_2 \text{ chord})^2/\text{height}] + \text{height}$.
Source: Ref. [19]

Notations

A	Heat-transfer surface area; circular sector area factor
A_{flow}	Flow area
B	Baffle spacing
BR	Boiling range
C_P	Heat capacity at constant pressure
$C_{P,L}$	Heat capacity of liquid
C_1	Parameter in Equation (9.20)
D	Diameter
D_b	Diameter of tube bundle
D_{ex}	Internal diameter of reboiler exit line
D_i	Internal diameter of tube
D_{in}	Internal diameter of reboiler inlet line
D_o	External diameter of tube
D_s	Internal diameter of shell
D_t	Internal diameter of tube in vertical thermosyphon reboiler
E_{LW}	Convective enhancement factor in Liu-Winterton correlation
F	LMTD correction factor
F_b	Factor defined by Equation (9.20) that accounts for convective effects in boiling on tube bundles
F_m	Mixture correction factor for Mostinski correlation
F_P	Pressure correction factor for Mostinski correlation
f	Darcy friction factor
f_{ex}	Darcy friction factor for reboiler exit line
f_{in}	Darcy friction factor for reboiler inlet line
f_t	Darcy friction factor for flow in vertical thermosyphon reboiler tubes
G	Mass flux
G_{ex}	Mass flux in reboiler exit line
G_{in}	Mass flux in reboiler inlet line
G_n	Mass flux in nozzle
G_t	Mass flux in vertical thermosyphon reboiler tubes
$G_{t,mist}$	Tube-side mass flux at onset of mist flow
g	Gravitational acceleration
g_c	Unit conversion factor
H	Specific enthalpy
H_F	Specific enthalpy of reboiler feed stream
H_L	Specific enthalpy of liquid
H_V	Specific enthalpy of vapor
h	Height of circular sector
h_b	Convective boiling heat-transfer coefficient
h_i	Tube-side heat-transfer coefficient
h_{LO}	Heat-transfer coefficient for total flow as liquid
h_{nb}	Nucleate boiling heat-transfer coefficient
h_{nc}	Natural convection heat-transfer coefficient
h_o	Shell-side heat-transfer coefficient
k	Thermal conductivity
k_L	Thermal conductivity of liquid
k_{tube}	Thermal conductivity of tube wall
L	Tube length
L_{BC}	Length of sensible heating zone in vertical thermosyphon reboiler
L_{CD}	Length of boiling zone in vertical thermosyphon reboiler
L_{ex}	Equivalent length of reboiler exit line
L_{in}	Equivalent length of reboiler inlet line
L_{req}	Required tube length
L_s	Shell length required for liquid overflow reservoir in kettle reboiler

M	Molecular weight
$\dot{m}$	Mass flow rate
$\dot{m}_F$	Mass flow rate of reboiler feed stream
$\dot{m}_i$	Mass flow rate of tube-side fluid
$\dot{m}_L$	Mass flow rate of liquid
$\dot{m}_{steam}$	Mass flow rate of steam
$\dot{m}_{Th}$	Mass flow rate of Therminol® heat-transfer fluid
$\dot{m}_V$	Mass flow rate of vapor
N_n	Number of pairs (inlet/outlet) of nozzles
Nu	Nusselt number
n_p	Number of tube passes
n_t	Number of tubes in bundle
P_A, P_B, P_C, P_D	Pressure at point *A B, C, D* in vertical thermosyphon reboiler system (Figure 10.8)
P_c	Critical pressure
$P_{c,i}$	Critical pressure of *i*th component in mixture
P_{pc}	Pseudo-critical pressure
P_{pr}	Pseudo-reduced pressure
P_r	Reduced pressure
Pr_L	Prandtl number of liquid
P_{sat}	Saturation pressure
P_T	Tube pitch
q	Rate of heat transfer
q_{BC}	Rate of heat transfer in sensible heating zone of vertical thermosyphon reboiler
q_{CD}	Rate of heat transfer in boiling zone of vertical thermosyphon reboiler
$\hat{q}$	Heat flux
$\hat{q}_c$	Critical heat flux
$\hat{q}_{c,bundle}$	Critical heat flux for boiling on tube bundle
$\hat{q}_{c,tube}$	Critical heat flux for boiling on a single tube
R_{Di}	Fouling factor for tube-side fluid
R_{Do}	Fouling factor for shell-side fluid
Re	Reynolds number
Re_i	Reynolds number for tube-side fluid
Re_{in}	Reynolds number for flow in reboiler inlet line
Re_L	Reynolds number for liquid phase flowing alone
$Re_{LO,ex}$	Reynolds number for total flow as liquid in reboiler exit line
Re_n	Reynolds number for flow in nozzle
SA	Dome segment area in kettle reboiler
S_{LW}	Nucleate boiling suppression factor in Liu–Winterton correlation
SR	Slip ratio
s	Specific gravity
s_L	Specific gravity of liquid
T	Temperature
T_B	Bubble-point temperature; temperature at inlet tubesheet (Figure 10.8)
T_C	Temperature at end of sensible heating zone (Figure 10.8)
T_{cyhx}	Temperature of cyclohexane
T_D	Dew-point temperature
T_{sat}	Saturation temperature
U_C	Clean overall heat-transfer coefficient
U_D	Overall heat-transfer coefficient for design
U_{req}	Required overall heat-transfer coefficient
V	Fluid velocity
VL	Vapor loading
V_{max}	Maximum fluid velocity
X_{tt}	Lockhart-Martinelli parameter
x_{ave}	Average value of vapor mass fraction

x_e	Vapor mass fraction at reboiler exit
Y	Chisholm parameter
z	Distance in vertical (upward) direction
z_A, z_B, z_C, z_D	Elevation at point *A B, C, D* in vertical thermosyphon reboiler system (Figure 10.8)

Greek Letters

α_r	Number of velocity heads allocated for minor losses on tube side
γ	Acceleration parameter defined by Equation (10.12)
Δh	Elevation difference between liquid surface in column sump and surface of boiling liquid in kettle reboiler
Δh_L	Available liquid head in kettle reboiler system
ΔP_{acc}	Pressure loss due to fluid acceleration
$\Delta P_{acc,k}$	Pressure loss due to fluid acceleration in *k*th interval of vapor weight fraction
ΔP_f	Pressure loss due to fluid friction in straight sections of tubes
ΔP_{feed}	Total frictional pressure loss in reboiler feed lines
ΔP_i	Total pressure drop for tube-side fluid
ΔP_n	Pressure loss in nozzles
ΔP_o	Total pressure drop for shell-side fluid
ΔP_r	Pressure drop due to minor losses on tube side
$\Delta P_{reboiler}$	Shell-side pressure drop due to friction and acceleration in kettle reboiler
ΔP_{return}	Total frictional pressure loss in return lines from reboiler
ΔP_{static}	Total pressure difference due to static heads in kettle reboiler system
ΔP_{total}	Total pressure loss in kettle reboiler system due to friction, acceleration and static heads
ΔP_{tp}	Pressure difference due to static head of boiling fluid in kettle reboiler
ΔP_V	Pressure difference due to static head of vapor in kettle reboiler
$(\Delta P_f/L)_{LO}$	Frictional pressure gradient for total flow as liquid
ΔT	Temperature difference
ΔT_{BC}	Temperature difference across sensible heating zone in vertical thermosyphon reboiler
$\Delta T_{\ln}$	Logarithmic mean temperature difference
$(\Delta T_{\ln})_{cf}$	Logarithmic mean temperature difference for counter-current flow
$(\Delta T_{\ln})_{co\text{-}current}$	Logarithmic mean temperature difference for co-current flow
ΔT_m	Mean temperature difference
$(\Delta T/\Delta P)_{sat}$	Slope of saturation curve
$\Delta\gamma_k$	Change in γ for *kth* vapor-weight-fraction interval
ϵ_V	Void fraction
$\epsilon_{V,e}$	Void fraction at reboiler exit
λ	Latent heat of vaporization or condensation
λ_{steam}	Latent heat of condensation of steam
μ	Viscosity
μ_w	Fluid viscosity at average tube wall temperature
ρ	Density
ρ_{ave}	Estimated average density of boiling fluid in kettle reboiler
ρ_{hom}	Homogeneous two-phase density
ρ_L	Density of liquid
$\overline{\rho}_{tp}$	Average two-phase density in boiling zone of vertical thermosyphon reboiler
ρ_V	Density of vapor
ρ_{water}	Density of water
σ	Surface tension
ϕ_b	Correction factor for critical heat flux in tube bundles
ϕ_i	Viscosity correction factor for tube-side fluid
ϕ_{LO}^2	Two-phase multiplier applied to pressure gradient for total flow as liquid
$\overline{\phi}_{LO}^2$	Average two-phase multiplier for boiling zone of vertical thermosyphon reboiler
$\phi_{LO,ex}^2$	Two-phase multiplier in reboiler exit line
ψ_b	Dimensionless bundle geometry parameter

Problems

(10.1) A kettle reboiler is being designed to generate 75,000 lb/h of vapor having a density of 0.40 lbm/ft^3. The liquid leaving the reboiler has a density of 41.3 lbm/ft^3 and a surface tension of 16 dyne/cm. The length of the tube bundle is 15 ft and the diameter plus clearance is 32 in.

(a) Calculate the vapor loading and dome segment area.
(b) Calculate the diameter required for the enlarged section of the K-shell.
(c) How many pairs of shell-side nozzles should be used?

Ans. (a) 572.9 lbm/h · ft^3 and 8.73 ft^2. (b) 63 in. (c) 2.

(10.2) The reboiler of Problem 10.1 is being designed for 65% vaporization. The feed to the reboiler has a density of 41.2 lbm/ft^3 and a viscosity of 0.25 cp. Assuming schedule 40 pipe is used:

(a) What size inlet nozzles are required to meet TEMA specifications without using impingement plates?
(b) The primary feed line from the column sump to the reboiler will contain 35 linear feet of pipe, two 90° elbows and a tee. The secondary lines (from the tee to the inlet nozzles) will each contain 4 linear feet of pipe, one 90° elbow and (if necessary) a reducer. The secondary lines will be sized to match the inlet nozzles. Size the primary line to give a fluid velocity of about 5 ft/s.
(c) Calculate the friction loss in the feed lines.

Ans. (a) 5-in. (b) All lines 5-in. (c) 0.58 psi.

(10.3) The horizontal thermosyphon reboiler of Example 10.3 contains two shell-side exit nozzles. The return lines from the exit nozzles meet at a tee, from which the combined stream flows back to the distillation column. Each section of line between exit nozzle and tee contains 8 linear feet of 8-in. schedule 40 pipe and one 90° elbow. Between the tee and the column there is an 8×10 expander, 50 linear feet of 10-in. schedule 40 pipe and one 90° elbow. Calculate the total friction loss in the return lines.

(10.4) For the reboiler of Example 10.3 and Problem 10.3, the vertical distance between the reboiler exit and the point at which the center of the return line enters the distillation column is 8 ft. Calculate the pressure drop in the return line due to the static head.

Ans. 0.30 psi.

(10.5) Considering the large uncertainty associated with convective boiling correlations, it might be deemed prudent for design purposes to include a safety factor, F_{sf}, in the Liu–Winterton correlation as follows:

$$h_b = F_{sf}\left[(S_{LW}h_{nb})^2 + (E_{LW}h_L)^2\right]^{0.5}$$

In Example 10.4, repeat steps (q)-v through (t) using a safety margin of 20% ($F_{sf} = 0.8$) with the Liu-Winterton correlation.

Ans. $L_{req} = 7.3$ ft.

(10.6) In Example 10.4, repeat steps (q) through (t) using the Chen correlation in place of the Liu–Winterton correlation.

(10.7) In Example 10.4, repeat step (u) using the Katto-Ohno correlation to calculate the critical heat flux. Compare the resulting value of $\hat{q}/\hat{q}_c$ with the value obtained in Example 10.8 using *Xist.*

(10.8) For the vertical thermosyphon reboiler of Example 10.4, suppose the tube length is increased from 8 ft to 12 ft and the surface of the liquid in the column sump is adjusted to remain at the level of the upper tubesheet.

(a) Assuming an exit vapor fraction of 0.132 corresponding to a circulation rate of 113,636 lbm/h, calculate a new circulation rate using Equation (10.15).
(b) Continue the iterations begun in part (a) to obtain a converged value for the circulation rate.
(c) Use the result obtained in part (b) to calculate the tube length required in the boiling zone and compare this value with the available tube length.

Ans. (a) 126,435 lbm/h.

(10.9) A kettle reboiler is required to supply 55,000 lb/h of hydrocarbon vapor to a distillation column. 80,000 lb/h of liquid at 360°F and 150 psia will be fed to the reboiler, and the duty is 6.2 × 10^6 Btu/h. Heat will be supplied by steam at a design pressure of 275 psia. An existing carbon steel kettle containing 390 tubes is available at the plant site. The tubes are 1-in. OD, 14 BWG, 12 ft long on 1.25-in. square pitch, and the bundle diameter is 30 in. Will this unit be suitable for the service?

Data for boiling-side fluid
Bubble point at 150 psia: 360°F
Dew point at 150 psia: 380°F
Vapor exit temperature: 370°F
Pseudo-critical pressure: 470 psia

(10.10) A reboiler must supply 15,000 kg/h of vapor to a distillation column at an operating pressure of 250 kPa. The reboiler duty is 5.2×10^6 kJ/h and the flow rate of the bottom product, which consists of an aromatic petroleum fraction, is specified to be 6000 kg/h. Heat will be supplied by a liquid organic heat-transfer fluid flowing on the tube side with a range of 220 – 190°C. A carbon steel kettle reboiler containing 510 tubes is available at the plant site. The tubes are 25.4-mm OD, 14 BWG, 4.57 m long on a 31.75-mm square pitch, and the bundle diameter is 863 mm. In this unit the organic heat-transfer fluid will provide a tube-side coefficient of 1100 $W/m^2 \cdot K$ with an acceptable pressure drop. Will the reboiler be suitable for this service?

Data for boiling-side fluid
Bubble point at 250 kPa: 165°C
Dew point at 250 kPa: 190°C
Vapor exit temperature: 182°C
Pseudo-critical pressure: 2200 kPa

(10.11) A reboiler must supply 80,000 lb/h of vapor to a distillation column at an operating pressure of 30 psia. The column bottoms, consisting of an aromatic petroleum fraction, will enter the reboiler as a (nearly) saturated liquid and the vapor fraction at the reboiler exit will be 0.2. Heat will be supplied by steam at a design pressure of 235 psia. A used horizontal thermosyphon reboiler consisting of an X-shell containing 756 carbon steel tubes is available at the plant site. The tubes are 1-in. OD, 14 BWG, 16 ft long on 1.25-in. square pitch, and the bundle diameter is 40.4 in. Will the unit be suitable for this service?

Data for boiling-side fluid
Bubble point at 30 psia: 335°F
Dew point at 30 psia: 370°F
Saturation temperature at 30 psia and 0.2 vapor fraction: 344°F
Enthalpy of liquid at 335°F: 245 Btu/lbm
Enthalpy of liquid at 344°F: 250 Btu/lbm
Enthalpy of vapor at 344°F: 385 Btu/lbm
Pseudo-critical pressure: 320 psia

(10.12) 105,000 lb/h of a distillation bottoms having the following composition will be partially vaporized in a kettle reboiler.

Component	Mole %	Critical Pressure (psia)
Toluene	84	595.9
m-Xylene	12	513.6
o-Xylene	4	541.4

The stream will enter the reboiler as a (nearly) saturated liquid at 35 psia. The dew-point temperature of the stream at 35 psia is 304.3°F. Saturated steam at a design pressure of 115 psia will be used as the heating medium. The reboiler must supply 75,000 lb/h of vapor to the distillation column. Physical property data are given in the following table. Design a kettle reboiler for this service.

Property	Reboiler Feed	Liquid Overflow	Vapor Return
T (°F)	298.6	302.1	302.1
H (Btu/lbm)	117.6	119.6	265.1
C_P (Btu/lbm · °F)	0.510	0.512	0.390
k (Btu/h · ft · °F)	0.057	0.057	0.011
μ (cp)	0.192	0.191	0.00965
ρ (lbm/ft^3)	46.5	46.4	0.429
σ (dyne/cm)	14.6	14.5	–
Molecular weight	94.39	95.42	93.98

(10.13) The feed line for the reboiler of Problem 10.12 will contain approximately 30 linear feet of pipe while the vapor return line will require about 25 linear feet of pipe. The available elevation difference between the liquid level in the column sump and the reboiler inlet is 8 ft. Size the feed and return lines for the unit.

(10.14) 100,000 lb/h of a distillation bottoms having the following composition will be partially vaporized in a kettle reboiler.

Component	Mole %	Critical Pressure (psia)
Cumene	60	465.4
m-diisopropylbenzene	20	355.3
p-diisopropylbenzene	20	355.3

The stream will enter the reboiler as a (nearly) saturated liquid at 60 psia. The dew-point temperature of the stream at 60 psia is 480.3°F. Saturated steam at a design pressure of 760 psia will be used as the heating medium. The reboiler must supply 60,000 lb/h of vapor to the distillation column. Physical property data are given in the following table. Design a kettle reboiler for this service.

Property	Reboiler Feed	Liquid Overflow	Vapor Return
T (°F)	455.3	471.4	471.4
H (Btu/lbm)	213.9	225.3	330.1
C_P (Btu/lbm · °F)	0.621	0.632	0.516
k (Btu/h · ft · °F)	0.0481	0.0481	0.0158
μ (cp)	0.153	0.150	0.010
ρ (lbm/ft^3)	41.8	41.5	0.905
σ (dyne/cm)	8.66	8.20	–
Molecular weight	137.0	143.5	133.8

(10.15) For the reboiler of Problem 10.14, the feed line will contain approximately 27 linear feet of pipe while the vapor return line will require about 24 linear feet of pipe. The available elevation difference between the liquid level in the column sump and the reboiler inlet is 7.5 ft. Size the feed and return lines for the unit.

(10.16) Use *Xist*, HEXTRAN or other available software to design a kettle reboiler for the service of Problem 10.12.

(10.17) Use *Xist*, HEXTRAN or other available software to design a kettle reboiler for the service of Problem 10.14.

(10.18) A distillation column bottoms having the composition specified in Problem 10.12 will be fed to a horizontal thermosyphon reboiler operating at 35 psia. The reboiler must supply 240,000 lb/h of vapor to the column. The lengths of feed and return lines, as well as the liquid level in the column sump, are as specified in Problem 10.13. Use *Xist* or other suitable software to design a reboiler system for this service.

(10.19) A distillation column bottoms having the composition specified in Problem 10.14 will be fed to a horizontal thermosyphon reboiler operating at 60 psia. The reboiler must supply 180,000 lb/h of vapor to the column. The length of feed and return lines, as well as the liquid level in the column sump, are as specified in Problem 10.15. Use *Xist* or other suitable software to design a reboiler system for this service.

(10.20) A distillation column bottoms has an average API gravity of 48° and the following assay (ASTM D86 distillation at atmospheric pressure).

Volume % Distilled	Temperature (°F)
0	100
10	153
20	190
30	224
40	257
50	284

—cont'd

Volume % Distilled	Temperature (°F)
60	311
70	329
80	361
90	397
100	423

This stream will be fed to a horizontal thermosyphon reboiler operating at a pressure of 25 psia. At this pressure, the bubble- and dew-point temperatures of the feed are 218.2°F and 353.6°F, respectively. The reboiler must supply 200,000 lb/h of vapor to the distillation column. Saturated steam at a design pressure of 70 psia will be used as the heating medium, and approximately 20% by weight of the feed will be vaporized in the reboiler. Physical properties of the feed and return streams are given in the following table. Design a reboiler for this service.

Property	Reboiler Feed	Liquid Return	Vapor Return
T (°F)	218.2	254.5	254.5
H (Btu/lbm)	89.1	107.3	247.2
C_P (Btu/lbm · °F)	0.516	0.533	0.437
k (Btu/h · ft · °F)	0.061	0.058	0.013
μ (cp)	0.250	0.245	0.0092
ρ (lbm/ft^3)	44.5	44.2	0.287
σ (dyne/cm)	16.9	16.2	–
P_{pc} (psia)	466.4	–	–

(10.21) Use HEXTRAN or other available software to design a horizontal thermosyphon reboiler for the service of Problem 10.20.

(10.22) For the service of Problem 10.20, the reboiler feed and return lines will each contain approximately 35 linear feed of pipe, and the available elevation difference between the liquid level in the column sump and the reboiler inlet will be 9.0 ft. Use *Xist* or other available software to design a horizontal thermosyphon reboiler system for this service. The size and configuration of the feed and return lines, along with the circulation rate, are to be determined in the design process.

(10.23) Use *Xist* or other suitable software to design a vertical thermosyphon reboiler for the service of Problem 10.12. Assume that the liquid level in the column sump will be maintained at approximately the elevation of the upper tubesheet in the reboiler. Also assume that the reboiler feed line will consist of 100 equivalent feet of pipe, while the return line will comprise 50 equivalent feet of pipe. Pipe diameters and circulation rate are to be determined in the design process.

(10.24) Use *Xist* or other suitable software to design a vertical thermosyphon reboiler for the service of Problem 10.14. The assumptions specified in Problem 10.23 are applicable here as well.

(10.25) A reboiler is required to supply 30,000 lb/h of vapor to a distillation column at an operating pressure of 23 psia. The reboiler feed will have the following composition:

Component	Mole %
Ethanol	1
Isopropanol	2
1-Propanol	57
2-Methyl-l-propanol	16
1-Butanol	24

Saturated steam at a design pressure of 55 psia will be used as the heating medium. Use *Xist* or other suitable software to design a vertical thermosyphon reboiler for this service. The assumptions stated in Problem 10.23 are also applicable to this problem. At operating pressure, the bubble point of the reboiler feed is 238°F and the dew point is 244°F. The specific enthalpy of the bubble-point liquid is 139.5 Btu/lbm and that of the dew-point vapor is 408.5 Btu/lbm.

(10.26) Rework Problem 10.14 for the case in which the heating medium is hot oil (30° API, $K_w = 12.0$) with a range of 600–500°F. Properties of the oil are given in the following table. Assume that the oil is available at a pressure of 50 psia.

Oil property	Value at 500°F	Value at 600°F
C_P (Btu/lbm · °F)	0.69	0.75
k (Btu/h · ft · °F)	0.049	0.044
μ (cp)	0.49	0.31
ρ (lbm/ft^3)	43.2	40.4

(10.27) Rework Problem 10.17 for the case in which the heating medium is hot oil as specified in Problem 10.26.

(10.28) Rework problem 10.19 for the case in which the heating medium is hot oil as specified in Problem 10.26.

(10.29) For the kettle reboiler of Example 10.2, a possible design modification (see step (n) of the solution) is to use a 21.25-in. bundle containing 172 tubes. Determine the suitability of this configuration.

(10.30) In Example 10.4 one of the suggested design modifications was to increase the tube length. Use *Xist* to implement this modification and obtain a final design for the reboiler.

(10.31) Treating the vapor and liquid phases separately, show that the term $G_t^2\gamma/\rho_L$ in Equation (10.11) represents the difference in total momentum flux (mass flow rate × velocity/cross – sectional area) across the boiling zone in a vertical thermosyphon reboiler tube.

11 Condensers

11.1 Introduction

Condensers are used in a variety of operations in chemical and petroleum processing, including distillation, refrigeration, and power generation. Virtually every distillation column employs either a partial or total condenser to liquefy some, or all, of the overhead vapor stream, thereby providing reflux for the column and (often) a liquid product stream. In refrigeration operations, condensers are used to liquefy the high-pressure refrigerant vapor leaving the compressor. Heat exchangers referred to as surface condensers are used to condense the exhaust from steam turbines that generate in-house power for plant operations.

Condensation is the reverse of boiling, and the condensing curve is the same as the boiling curve. Thus, many of the computational difficulties encountered in the analysis of reboilers are present in condensers as well. For wide-boiling mixtures, in particular, the nonlinearity of the condensing curve and the variation of liquid and vapor properties over the condensing range mean that a zone or incremental analysis is required for rigorous calculations. Mass-transfer effects may also be significant in the condensation of mixtures, as they are in boiling of mixtures.

On the other hand, condensing and boiling differ in important respects. In particular, condensation is more amenable to fundamental analysis, and useful heat-transfer correlations can be derived from first principles.

11.2 Condenser Geometries and Configurations

Most condensers used in the chemical process industries are shell-and-tube exchangers or air-cooled exchangers. In the latter, the condensing vapor flows inside a bank of finned tubes and ambient air blown across the tubes by fans serves as the coolant. Other types of equipment, such as double-pipe exchangers, plate-and-frame exchangers, and direct contact condensers are less frequently used. In direct contact condensing, the coolant is sprayed directly into the condensing vapor. This method is sometimes used for intermediate heat removal from distillation and absorption columns by means of pumparounds. While direct contacting provides a high rate of heat transfer with low pressure drop, it is obviously limited to applications in which mixing of coolant and condensate is permissible.

Surface condensers are tubular exchangers, but their construction differs somewhat from shell-and-tube equipment used to condense process vapors. They are most often designed for vacuum operation on the shell (steam) side, and hence must handle a large volumetric flow rate of vapor with very low pressure drop. Smaller units may have circular cross-flow shells similar to X-shells, but larger units usually employ a box-type shell.

In the remainder of this chapter, consideration is restricted to shell-and-tube condensers used for condensation of process vapors. These units are classified according to orientation (horizontal versus vertical) and placement of condensing vapor (shell side versus tube side).

11.2.1 Horizontal Shell-Side Condenser

Most large condensers in the process industries are oriented horizontally in order to minimize the cost of support structures and facilitate maintenance operations. The condensing vapor is often an organic compound or mixture and the coolant is most often water, a combination that favors the use of finned tubes. Therefore, the condensing vapor is most frequently placed in the shell (the finned side) and the cooling water, which is usually more prone to fouling, flows in the tubes.

The baffled E-shell condenser (Figure 11.1) is widely used and the least expensive type of horizontal condenser. It may have a floating head, as shown, or fixed tubesheets. The baffles are usually cut vertically for side-to-side flow and notched at the bottom to facilitate drainage of the condensate. An extra nozzle at the top of the shell near the rear head is used to vent non-condensable gases. A mechanism is required in all condensers to prevent accumulation of non-condensables that would otherwise impair the performance of the unit (see Section 11.2.6). Impingement protection is also required since the vapor generally enters the shell at or near its dew point.

If shell-side pressure drop is a problem, double-cut segmental baffles or rod baffles can be used. Alternatively, a J-shell (Figure 11.2) or X-shell (Figure 11.3) can be used. An X-shell provides very low pressure drop and, hence, is often used for vacuum services. Venting can be a problem with these units, however, since the vent(s) must be located where non-condensables collect, which is also where condensate tends to collect, as indicated in Figure 11.3.

11.2.2 Horizontal Tube-Side Condenser

This configuration, illustrated in Figure 11.4, is sometimes used to condense high-pressure or corrosive vapors. It also occurs in kettle and horizontal thermosyphon reboilers when the heating medium is steam or a condensing process stream. Multiple tube-side passes (usually two) can be used as well as the single pass shown in Figure 11.4. Two-phase flow regimes are an important consideration in design of these units since flow instabilities can cause operational problems [1].

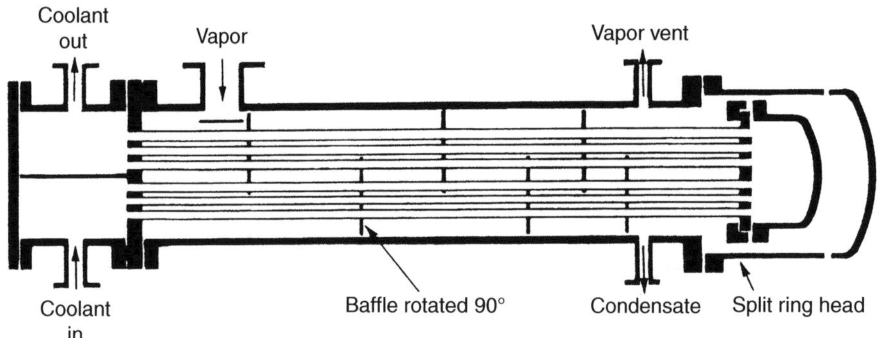

FIGURE 11.1 Horizontal shell-side condenser, type AES (Source: Ref. [1]).

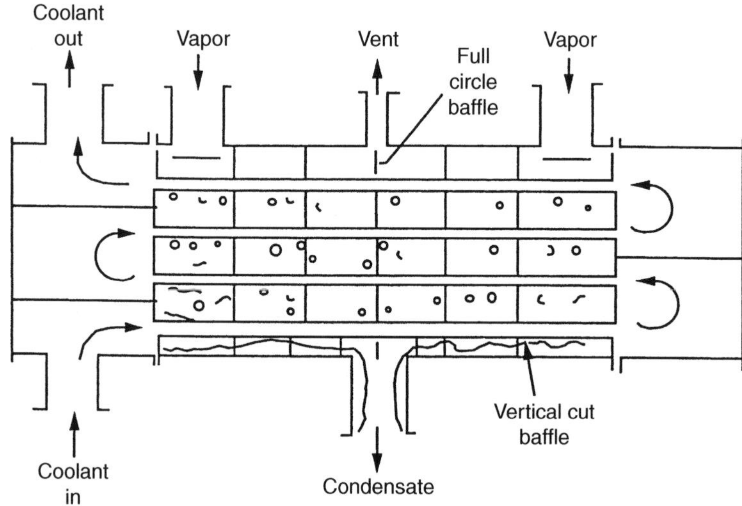

FIGURE 11.2 Horizontal shell-side condenser with split flow (type J) shell (Source: Ref. [2]).

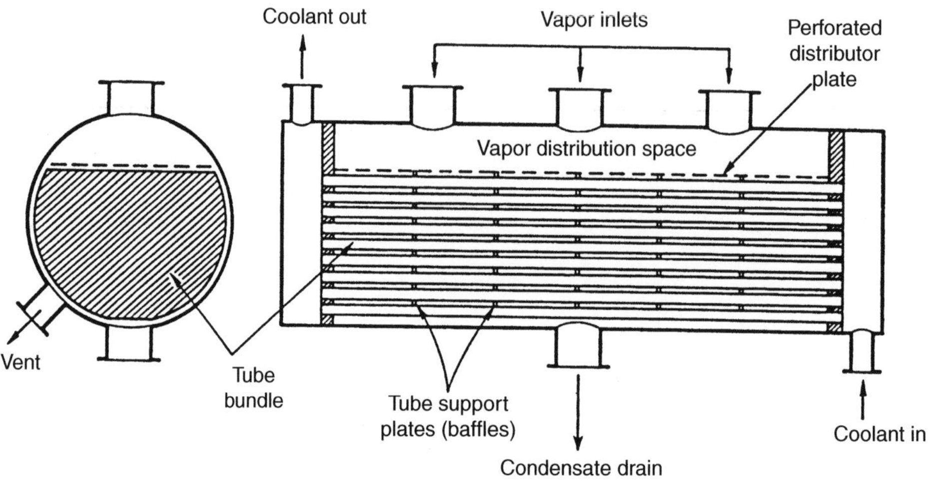

FIGURE 11.3 Horizontal shell-side condenser with cross-flow (type X) shell (Source: Ref. [3]).

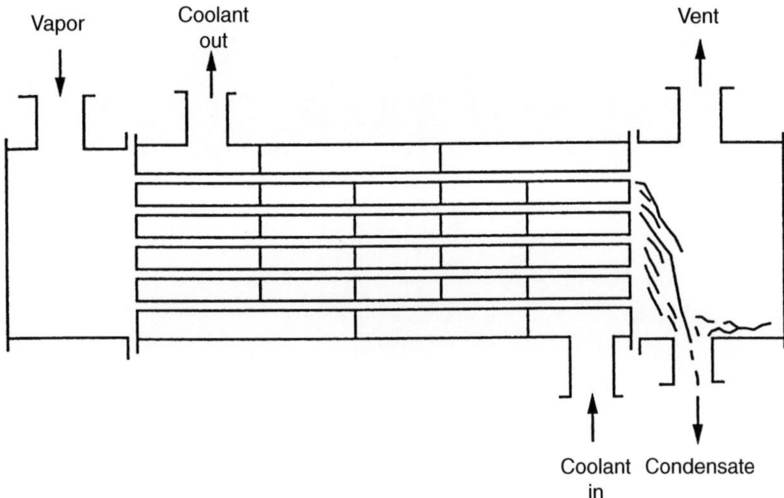

FIGURE 11.4 Horizontal tube-side condenser (Source: Ref. [2]).

11.2.3 Vertical Shell-Side Condenser

This configuration, illustrated in Figure 11.5, is most commonly encountered in vertical thermosyphon reboilers when the heating medium is steam or a condensing process stream. Vapor enters at the top of the shell and flows downward along with the condensate, which is removed at the bottom. The vent is placed near the bottom of the shell where non-condensables collect. The tube bundle may be either baffled or unbaffled.

11.2.4 Vertical Tube-Side Downflow Condenser

This configuration (Figure 11.6) is often used in the chemical industry [4]. It consists of an E-shell with either a floating head or fixed tubesheets. The lower head is over-sized to accommodate the condensate and a vent for non-condensables. The upper tubesheet is also provided with a vent to prevent any non-condensable gas, such as air, that may enter with the coolant from accumulating in the space between the tubesheet and the upper coolant nozzle.

The condensate flows down the tubes in the form of an annular film of liquid, thereby maintaining good contact with both the cooling surface and the remaining vapor. Hence, this configuration tends to promote the condensation of light components from wide-boiling mixtures. A disadvantage is that the coolant, which is often more prone to fouling, is on the shell side.

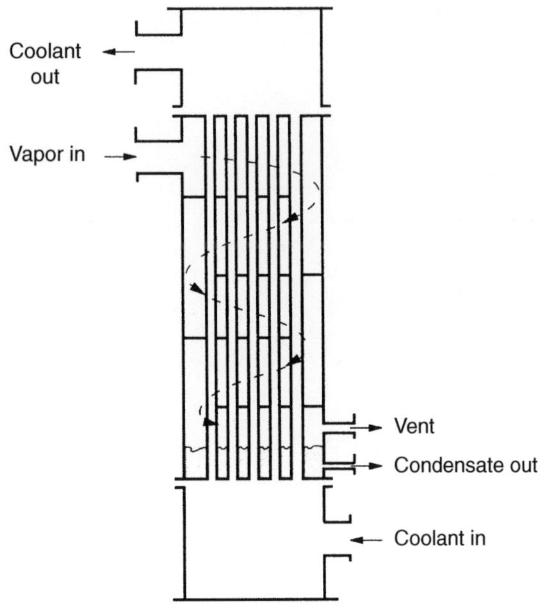

FIGURE 11.5 Vertical shell-side condenser (Source: Ref. [2]).

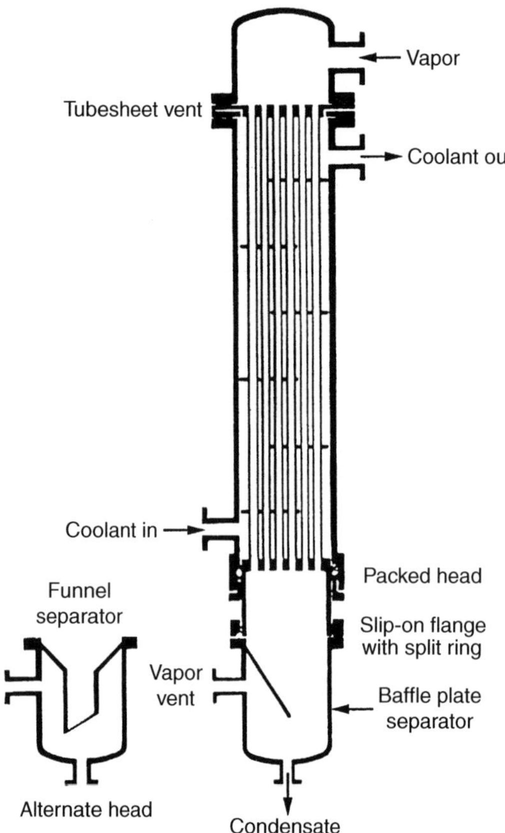

FIGURE 11.6 Vertical tube-side downflow condenser (Source: Ref. [1]).

11.2.5 Reflux Condenser

A reflux condenser, also called a vent condenser or knockback condenser, is a vertical tube-side condenser in which the vapor flows upward, as indicated in Figure 11.7. These units are typically used when relatively small amounts of light components are to be separated from a vapor mixture. The heavier components condense and flow downward along the tube walls, while the light components remain in the vapor phase and exit through the vent in the upper header. In distillation applications they are most often used as internal condensers [4], where the condensate drains back into the top of the distillation column to supply the reflux, or as secondary condensers attached to accumulators (Figure 11.8). These units have excellent venting characteristics, but the vapor velocity must be kept low to prevent excessive entrainment of condensate and the possibility of flooding (see Appendix 11.B). The fluid placement (coolant in shell) entails the same disadvantage as the tube-side downflow condenser.

11.2.6 Venting, Draining, and Subcooling

Venting of non-condensable gases and draining of condensate are essential considerations in condenser design. Industry experience indicates that buildups of non-condensables and/or condensate are common causes of performance deficiencies. Successful venting and draining involve equipment external to the condenser itself, and therefore these considerations are often overlooked by designers. A related consideration is condensate subcooling, which may be specified for a number of reasons, including ease of pumping and product storage, and the requirements of downstream processing units. There are a variety of different design approaches to address these issues, and some judgment is needed to account for non-ideal process conditions.

Self-Venting Nozzles

A self-venting line, in general, is one that runs part-full of liquid such that any vapor that is entrained with the liquid can disengage and flow separately through the line. In a condenser, a self-venting condensate nozzle allows for drainage via gravity with weir-type flow into the nozzle. That is, the condensate flows over the lip of the nozzle and down around the periphery of the pipe, while a vapor core occupies the center of the nozzle. An old rule of thumb for sizing nozzles is that the Froude number, *Fr*, should be less than 0.3 to ensure self-venting operation [5]. Thus,

$$Fr \equiv \frac{V_C}{\sqrt{gD_n}} < 0.3 \tag{11.1}$$

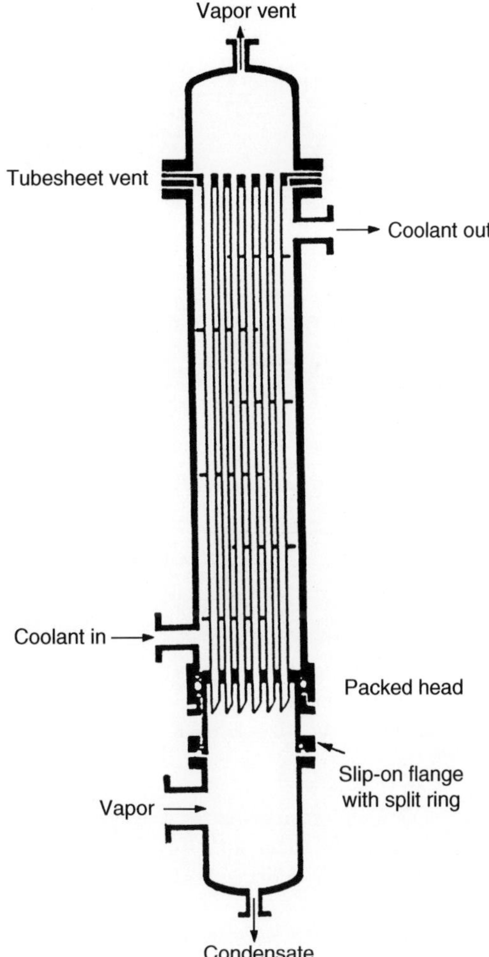

FIGURE 11.7 Reflux condenser (Source: Ref. [1]).

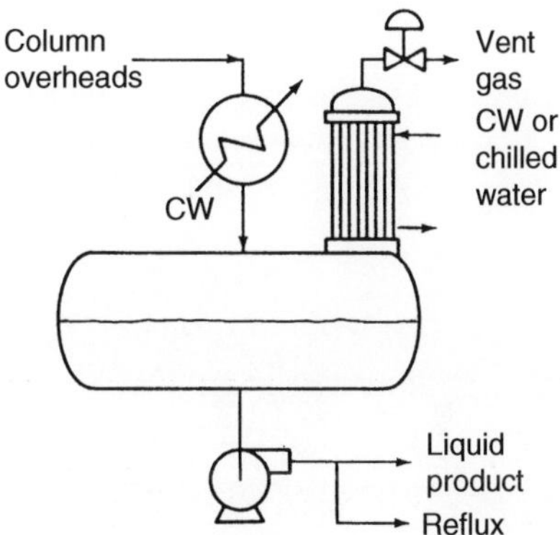

FIGURE 11.8 A reflux condenser used as a secondary condenser on an accumulator (Source: Ref. [4]).

where

V_C = superficial condensate velocity
D_n = nozzle ID
g = gravitational acceleration

Solving for the nozzle diameter yields:

$$D_n > \frac{V_C^2}{0.09g} \tag{11.2}$$

It is convenient to express this relationship in terms of the condensate flow rate rather than the condensate velocity. The superficial velocity is equal to the condensate volumetric flow rate, $\dot{v}_L$, divided by the cross-sectional area of the nozzle:

$$V_C = \frac{\dot{v}_L}{(\pi D_n^2/4)} \tag{11.3}$$

Substituting for V_C and again solving for D_n yields:

$$D_n > \left(\frac{16}{0.09\pi^2 g}\right)^{0.2} \dot{v}_L^{0.4} \tag{11.4}$$

Setting $g = 32.174$ ft/s^2 gives:

$$D_n(\text{ft}) > 0.89 \left[\dot{v}_L(\text{ft}^3/\text{s})\right]^{0.4} \tag{11.5a}$$

This important working relationship is more commonly presented with the nozzle diameter in inches and the condensate flow rate in gallons per minute. Making the appropriate unit conversions gives:

$$D_n(\text{in.}) > 0.93 \left[\dot{v}_L(\text{gpm})\right]^{0.4} \tag{11.5b}$$

A graphical correlation based on this relationship has been widely used in practice and can be found in the book by Kister [4]. Finally, taking $g = 9.8067$ m/s^2, Inequality (11.4) yields:

$$D_n(\text{m}) > 1.13 \left[\dot{v}_L(\text{m}^3/\text{s})\right]^{0.4} \tag{11.5c}$$

Venting Non-Condensables

Non-condensable gases are present in all vapor streams, and these components dominate the vapor volume above the condensate at the outlet of a total condenser. Therefore, it is generally not practical to remove all non-condensables; the objective is rather to ensure that the condenser design does not promote their buildup. Non-condensables tend to accumulate in low velocity regions of the condenser where the tube-wall temperatures are lowest. These regions of low velocity occur near the condenser outlet where little condensable vapor remains. Hence, one of the most effective ways to remove non-condensables is via the condensate nozzle(s). For this reason, all condensers should be provided with self-venting nozzles. A vent in addition to a self-venting condensate nozzle is often not needed for condensers operating at pressures above atmospheric. When needed (such as for vacuum conditions), a continuously operating vent should be installed near the outlet and near the tubes with the coldest wall temperatures. A typical horizontal shell-side condenser with fully equipped vent system is shown in Figure 11.9.

The purpose of the start-up vents, which are common to all process vessels, is to purge air from the shell during start-ups. These vents are closed during normal operation of the unit.

The purpose of the equalizing line in Figure 11.9 is to mitigate potential pressure fluctuations that could adversely affect the operation of the condenser. These can be caused, for example, by an under-sized condensate nozzle. Such under-sizing can result in condensate waves, or "sloshing," temporarily covering the outlet nozzle and creating a liquid static head with its associated pressure difference.

In contrast with Figure 11.9, the vents in Figures 11.1 and 11.2 are not well-situated for removal of non-condensables. Although they are at the correct axial positions (same as the condensate nozzles), their location at the top of the shell (the farthest point from the condensate nozzle) renders them largely ineffective. In fact, these vents can actually be counter-productive. For example, with

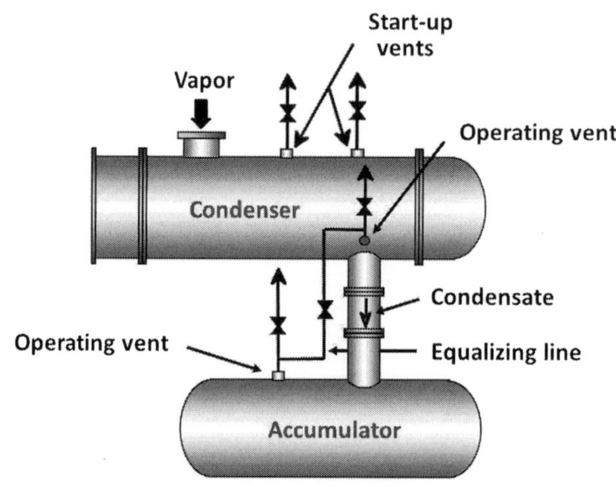

FIGURE 11.9 Vent system for a horizontal shell-side condenser (Source: HTRI).

a large degree of tube removal under the inlet nozzle and with inadequate bypass-flow blockage, a large amount of condensable vapor can escape through the vent in either the E-shell or J-shell condenser. Furthermore, the lack of a properly situated vent on these units may result in the buildup of non-condensables, thereby impairing heat transfer. Vacuum condensers with over-sized venting systems are susceptible to this mode of deficient operation. For the above reasons, the vents in Figures 11.1 and 11.2 are often used only during process start-ups, and are sometimes even welded shut to ensure that condensable vapor is not inadvertently vented by operating personnel.

Flooding and Condensate Subcooling

Flooded condensers are often specified to subcool condensate or to control the duty by adjusting the effective heat-transfer area. One of the most common methods to establish a flooded level is with a loop seal; a typical configuration is shown in Figure 11.10. This arrangement is common when steam is used to heat a process fluid stream in the tubes, as with the water heater depicted in the diagram. An equalizing line is an important detail needed to vent non-condensables and ensure pressure variations are mitigated. The steam trap shown can only be used for applications with operating pressures greater than atmospheric.

Other methods, such as dam baffles, can be used to flood the heat-transfer surface. A dam baffle (Figure 11.11) is a segmental baffle wherein the flooded tubes have no window. The baffle is cut vertically, but the cut goes only as far as the tube rows near the bottom that are to be flooded, thereby excluding those tubes from the window. Due to leakage flows (around the baffle edge and between the tubes and the baffle) the liquid level varies along the length of the bundle and as the condensing rate varies. In some instances, an active level control system is used, but that approach is the most expensive since it requires electric power and/or an instrument air system.

Designers often ask the question: "How much subcooling can you accomplish in a condenser?" For vertical shells, there is really no limit. For horizontal shell-side condensers, 40°F is a conservative estimate for design purposes. Many successful designs have operated with larger amounts of subcooling, but there is a risk for "flash hammering" or "knocking" due to rapid collapse of vapor in contact with highly subcooled condensate.

11.3 Condensation on a Vertical Surface: Nusselt Theory

11.3.1 Condensation on a Plane Wall

The basic heat-transfer correlations for film condensation were first derived by Nusselt [6]. Consider the situation depicted in Figure 11.12, in which a pure-component saturated vapor condenses on a vertical wall, forming a thin film of condensate that flows downward due to gravity. The following assumptions are made:

(i) The flow in the condensate film is laminar.
(ii) The temperature profile across the condensate film is linear. This assumption is reasonable for a very thin film.
(iii) The shear stress at the vapor–liquid interface is negligible.
(iv) The fluid velocity in the film is small so that the inertial terms (terms of second degree in velocity) in the Navier–Stokes equations are negligible.
(v) Only latent heat is transferred, i.e., subcooling of the condensate is neglected.

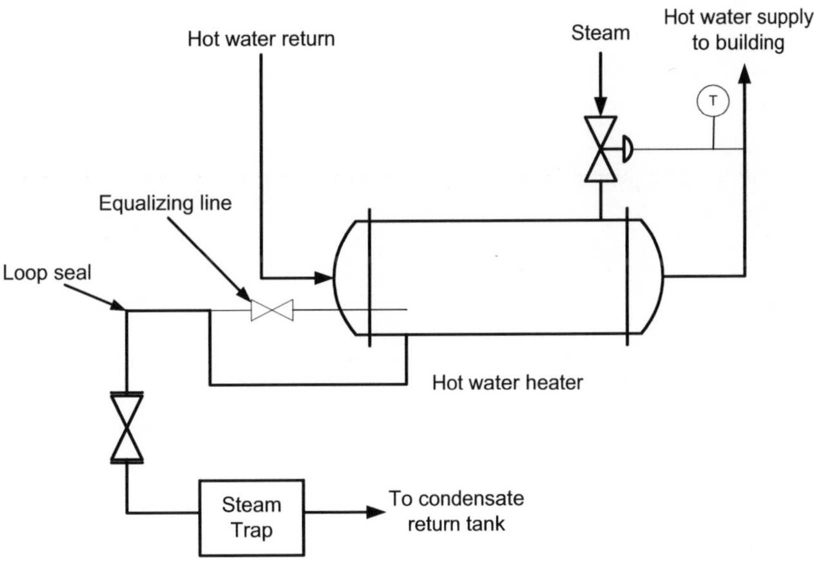

FIGURE 11.10 Typical loop seal configuration for a flooded condenser (Source: HTRI).

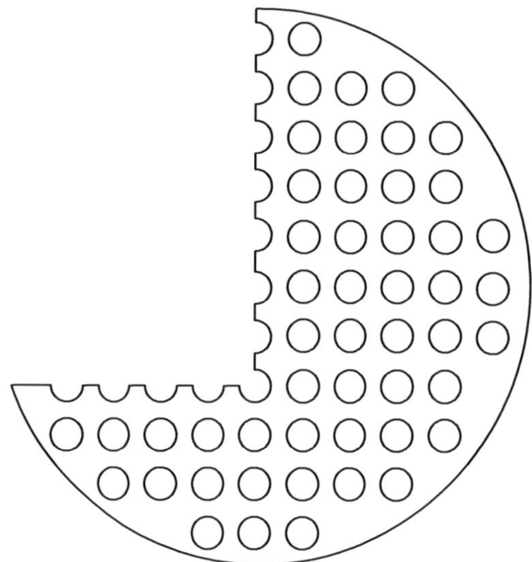

FIGURE 11.11 Illustration of a dam baffle.

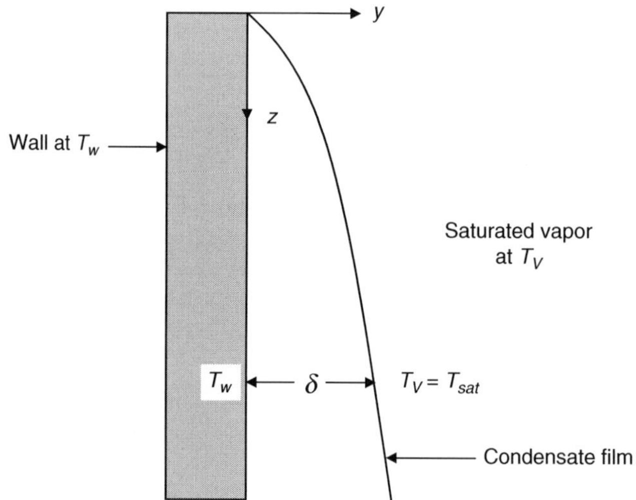

FIGURE 11.12 Film condensation on a vertical wall.

(vi) Fluid properties and the wall temperature, T_w, are constant.
(vii) The wall is flat (no curvature).
(viii) The system is at steady state.

The heat flux, $\hat{q}_y$, at the wall is given by Fourier's law:

$$\hat{q}_y = -k_L \frac{dT}{dy}\bigg|_{y=0} \tag{11.6}$$

It follows from assumption (ii) that the temperature gradient in the condensate film is independent of y and equal to:

$$\frac{dT}{dy} = \frac{\Delta T}{\Delta y} = \frac{T_V - T_w}{\delta} \tag{11.7}$$

Therefore,

$$\hat{q}_y = -\frac{k_L(T_V - T_w)}{\delta} \tag{11.8}$$

Letting $\hat{q} = |\hat{q}_y|$ and introducing the local heat-transfer coefficient, h_z, Equation (11.8) can be expressed as follows:

$$\hat{q} = h_z(T_V - T_w) = \frac{k_L(T_V - T_w)}{\delta} \tag{11.9}$$

Therefore,

$$h_z = k_L/\delta \tag{11.10}$$

Thus, the problem of determining the heat-transfer coefficient is equivalent to computing the film thickness, δ, which is a function of vertical position, z. To this end, the heat flux is written in terms of the condensation rate per unit area, $\hat{W}$, and the latent heat of vaporization, λ. By assumption (v):

$$\hat{q} = \hat{W}\lambda \tag{11.11}$$

Combining Equations (11.9) and (11.11) gives:

$$\hat{W} = \frac{k_L(T_V - T_w)}{\lambda\delta} \tag{11.12}$$

Next, an equation for the average velocity, $\bar{u}$, of the condensate is needed. The velocity in the film varies from zero at the wall to a maximum at the vapor–liquid interface. The velocity profile can be found by solving the Navier–Stokes equations using assumption (iv) and boundary conditions of zero velocity at the wall ($y = 0$) and zero shear stress at the vapor–liquid interface ($y = \delta$). This is a standard problem in fluid mechanics, and the solution for the average velocity is [7]:

$$\bar{u} = \frac{(\rho_L - \rho_V)g\delta^2}{3\mu_L} \tag{11.13}$$

A condensate mass balance is now made on a differential control volume of length Δz, as shown in Figure 11.13. The width of the wall is denoted by w and the balance is written for steady-state conditions (assumption (viii)).

$$\{\text{rate of mass entering}\} = \{\text{rate of mass leaving}\} \tag{11.14}$$

$$(\rho_L\bar{u}\delta w)\,|_z + \hat{W}\,w\,\Delta z = (\rho_L\bar{u}\delta w)\,|_{z+\Delta z} \tag{11.15}$$

$$\frac{(\rho_L\bar{u}\delta)|_{z+\Delta z} - (\rho_L\bar{u}\delta)|_z}{\Delta z} = \hat{W} \tag{11.16}$$

Taking the limit as $\Delta z \to 0$ gives:

$$\frac{d}{dz}(\rho_L\bar{u}\delta) = \hat{W} \tag{11.17}$$

Substituting for $\bar{u}$ and $\hat{W}$ using Equations (11.12) and (11.13) yields:

$$\frac{d}{dz}\left[\frac{\rho_L(\rho_L - \rho_V)g\delta^3}{3\mu_L}\right] = \frac{k_L(T_V - T_w)}{\lambda\delta} \tag{11.18}$$

Since ρ_L, ρ_V, and μ_L are constant by assumption (iv), they can be taken outside the derivative. Differentiating δ^3 then gives:

$$\left[\frac{\rho_L(\rho_L - \rho_V)g}{3\mu_L}\right] \times 3\delta^2\frac{d\delta}{dz} = \frac{k_L(T_V - T_w)}{\lambda\delta} \tag{11.19}$$

$$\delta^3\frac{d\delta}{dz} = \frac{\mu_L k_L(T_V - T_w)}{\rho_L(\rho_L - \rho_V)g\lambda} \tag{11.20}$$

Now all the terms on the right side of Equation (11.20) are constant. Hence, separating the variables and integrating gives:

$$\frac{\delta^4}{4} = \left[\frac{\mu_L k_L(T_V - T_w)}{\rho_L(\rho_L - \rho_V)g\lambda}\right]z + \text{constant} \tag{11.21}$$

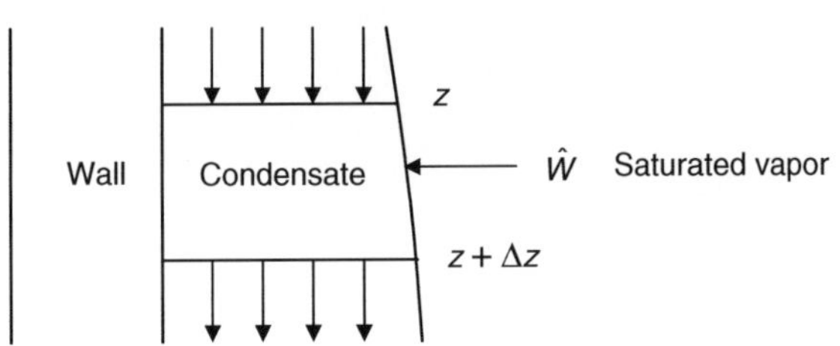

FIGURE 11.13 Control volume for condensate mass balance.

Applying the boundary condition $\delta = 0$ at $z = 0$, the constant of integration is zero and the equation for the film thickness becomes:

$$\delta = \left[\frac{4\mu_L k_L (T_V - T_w) z}{\rho_L(\rho_L - \rho_V) g \lambda}\right]^{1/4} \tag{11.22}$$

Substituting the above expression for the film thickness into Equation (11.10) gives the desired result for the local heat-transfer coefficient:

$$h_z = \frac{k}{\delta} = \left[\frac{k_L^3 \rho_L(\rho_L - \rho_V) g \lambda}{4\mu_L (T_V - T_w)}\right]^{1/4} z^{-1/4} \tag{11.23}$$

Notice that the heat-transfer coefficient decreases with distance down the wall because the thermal resistance is proportional to the film thickness, which increases with distance due to condensate accumulation.

The average heat-transfer coefficient, h, for the condensate film can now be obtained by integrating the local coefficient over the length, L, of the wall:

$$h = \frac{1}{L}\int_0^L h_z dz$$

$$= \frac{1}{L}\left[\frac{k_L^3 \rho_L(\rho_L - \rho_V) g \lambda}{4\mu_L (T_V - T_w)}\right]^{1/4} \int_0^L z^{-1/4} dz$$

$$= \frac{1}{L}\left[\frac{k_L^3 \rho_L(\rho_L - \rho_V) g \lambda}{4\mu_L (T_V - T_w)}\right]^{1/4} (4/3) L^{3/4}$$

$$h = 0.943\left[\frac{k_L^3 \rho_L(\rho_L - \rho_V) g \lambda}{\mu_L (T_V - T_w) L}\right]^{1/4} \tag{11.24}$$

The total rate of heat transfer across the condensate film is given by:

$$q = hwL(T_V - T_w) \tag{11.25}$$

Equation (11.24) can be reformulated in terms of the condensation rate, W, rather than the temperature difference across the film. This alternate formulation is often more convenient for computational purposes. Since the rate of heat transfer is λW, it follows from Equation (11.25) that

$$L(T_V - T_w) = \frac{q}{hw} = \frac{\lambda W}{hw} \tag{11.26}$$

Substituting this result for $L(T_V - T_w)$ in Equation (11.24) gives:

$$h = 0.943\left[\frac{k_L^3 \rho_L(\rho_L - \rho_V) g h w}{\mu_L W}\right]^{1/4} \tag{11.27}$$

A Reynolds number can be introduced at this point as follows:

$$Re = \frac{D_e G}{\mu_L} = \frac{4W}{w\mu_L} = \frac{4\Gamma}{\mu_L} \tag{11.28}$$

where

$$D_e = \frac{4 \times \textit{Flow Area}}{\textit{Wetted Perimeter}} = \frac{4A_f}{w} \tag{11.29}$$

$$G = W/A_f$$

and

A_f = flow area for condensate
$\Gamma = W/w$ = condensation rate per unit width of surface

In terms of Re, Equation (11.27) becomes:

$$h = 0.943\left[\frac{4k_L^3 \rho_L(\rho_L - \rho_V) g h}{\mu_L^2 Re}\right]^{1/4} \tag{11.30}$$

Solving for h yields:

$$h = 1.47\left[\frac{k_L^3\rho_L(\rho_L - \rho_V)g}{\mu_L^2 Re}\right]^{1/3} \quad (11.31)$$

11.3.2 Condensation on Vertical Tubes

For tube diameters of practical interest, the effect of wall curvature on the condensate film thickness is negligible. Hence, Equation (11.24) can be used for condensation on either the internal or external surfaces of vertical heat-exchanger tubes. The alternate form, Equation (11.31) is also applicable with the appropriate modification in equivalent diameter. For a bank of n_t tubes, the wetted perimeter is $n_t\pi D$, where D is either the inner or outer tube diameter, depending on whether condensation occurs inside or outside the tubes. Hence, Equations (11.28) and (11.29) are replaced by:

$$D_e = \frac{4A_f}{n_t\pi D} \quad (11.32)$$

$$Re = \frac{D_e G}{\mu_L} = \frac{4\ W}{n_t\pi D\mu_L} = \frac{4\Gamma}{\mu_L} \quad (11.33)$$

where now

$$\Gamma = \frac{W}{n_t\pi D} \quad (11.34)$$

11.4 Condensation on Horizontal Tubes

The Nusselt analysis for condensation on the external surface of a horizontal tube is similar to that for a vertical surface. The result corresponding to Equation (11.24) is:

$$h = 0.728\left[\frac{k_L^3\rho_L(\rho_L - \rho_V)g\lambda}{\mu_L(T_V - T_w)D_o}\right]^{1/4} \quad (11.35)$$

The coefficient in this equation is often given as 0.725, which Nusselt [6] obtained by numerical integration over the tube periphery. The more accurate value of 0.728 was obtained later by performing the integration analytically [8], but for practical purposes the difference is insignificant.

The alternate form corresponding to Equation (11.31) is:

$$h = 1.52\left[\frac{k_L^3\rho_L(\rho_L - \rho_V)g}{\mu_L^2 Re}\right]^{1/3} \quad (11.36)$$

where

$$Re = \frac{4\Gamma}{\mu_L}$$

$$\Gamma = \frac{W}{n_t L} \quad (11.37)$$

n_t = number of tubes
L = tube length

Equations (11.35) and (11.36) apply to a single tube or a single row of tubes. When tubes are stacked vertically, condensate drainage from tubes above causes an increase in the loading and film thickness on the lower tube rows. Nusselt [6] analyzed the situation for N_r tube rows stacked vertically by assuming that condensate flows downward in a continuous uniform sheet from tube to tube, and that the film remains laminar on all tubes. He found that the average heat-transfer coefficient, h_{Nr}, for the array of N_r tube rows is:

$$h_{Nr} = h_1 N_r^{-1/4} \quad (11.38)$$

Here, h_1 is the heat-transfer coefficient for a single tube row calculated from Equation (11.35) or (11.36). In reality, condensate tends to drain from tubes in droplets and rivulets that disturb the film on the tubes below and promote turbulence. As a result,

Equation (11.38) generally underestimates the heat-transfer coefficient. Based on experience with industrial condensers, Kern [9,10] proposed the following relationship:

$$h_{Nr} = h_1 N_r^{-1/6} \tag{11.39}$$

Butterworth [11] showed that this equation agrees well with experimental data for water and a refrigerant condensing on a tube bank with $N_r = 20$.

The circular tube bundles used in shell-and-tube condensers consist of a number of vertical stacks of tubes of different heights. For this situation, Kern [9,10] proposed calculating the average heat-transfer coefficient using Equation (11.36) with a modified condensate loading term, Γ^*, to account for the effects of condensate drainage. The result is:

$$h = 1.52\left[\frac{k_L^3 \rho_L(\rho_L - \rho_V)g}{4\mu_L \Gamma^*}\right]^{1/3} \tag{11.40}$$

where

$$\Gamma^* = \frac{W}{L n_t^{2/3}} \tag{11.41}$$

In effect, the number of tubes is replaced by $n_t^{2/3}$ in Nusselt's equation. For a single stack of $N_r = n_t$ tubes, Equations (11.40) and (11.41) are equivalent to Equation (11.39). To see this, substitute $N_r^{2/3}$ for N_r in the Nusselt relationship, Equation (11.38).

Example 11.1

A stream consisting of 5000 lb/h of saturated *n*-propyl alcohol (1-propanol) vapor at 207°F and approximately atmospheric pressure will be condensed using a tube bundle containing 3769 tubes arranged for a single pass. The tubes are 0.75 in. OD, 14 BWG, with a length of 12 ft. Physical properties of the condensate are as follows:

$k_L = 0.095$ Btu/h · ft · °F
$\rho_L = 49$ lbm/ft^3
$\mu_L = 0.5$ cp

Estimate the condensing-side heat-transfer coefficient for the following cases:

(a) The tube bundle is vertical and condensation occurs inside the tubes.
(b) The tube bundle is horizontal and condensation occurs outside the tubes.

Solution

(a) Since the tubes are vertical and the condensation rate is given, Equation (11.31) will be used to calculate h. The condensate loading per tube is first calculated using Equation (11.34) with $D = D_i = 0.584$ in. from Table B.l.

$$\Gamma = \frac{W}{n_t \pi D} = \frac{5000}{3769\pi(0.584/12)} = 8.677 \text{ lbm/ft}\cdot\text{h}$$

The Reynolds number is given by Equation (11.33):

$$Re = \frac{4\Gamma}{\mu_L} = \frac{4 \times 8.677}{0.5 \times 2.419} = 28.70$$

In Equation (11.31) the vapor density is neglected in comparison with the liquid density since the pressure is low (1 atm). Thus,

$$h = 1.47\left[\frac{k_L^3 \rho_L^2 g}{\mu_L^2 Re}\right]^{1/3} = 1.47\left[\frac{(0.095)^3(49)^2(4.17 \times 10^8)}{(0.5 \times 2.419)^2(28.70)}\right]^{1/3}$$

$$h = 402 \text{ Btu/h}\cdot\text{ft}^2\cdot{}^\circ\text{F}$$

(b) The calculation for the horizontal tube bundle utilizes Equations (11.40) and (11.41). The vapor density is again neglected.

$$\Gamma^* = \frac{W}{L n_t^{2/3}} = \frac{5000}{12(3769)^{2/3}} = 1.720 \text{ lbm/ft}\cdot\text{h}$$

$$h = 1.52\left[\frac{k_L^3 \rho_L^2 g}{4\mu_L \Gamma^*}\right]^{1/3} = 1.52\left[\frac{(0.095)^3(49)^2(4.17 \times 10^8)}{4(0.5 \times 2.419)(1.720)}\right]^{1/3}$$

$$h = 713 \text{ Btu/h}\cdot\text{ft}^2\cdot{}^\circ\text{F}$$

The horizontal configuration provides a substantial advantage under these conditions, which were chosen to ensure laminar flow of condensate in the vertical tube bundle. While the horizontal configuration is generally advantageous in this respect, in practice other factors such as interfacial shear and turbulence in the vertical condensate film may act to mitigate the effect of condenser orientation.

11.5 Modifications of Nusselt Theory

The basic Nusselt theory of film condensation has been modified by a number of workers, including Nusselt himself, in order to relax some of the assumptions made in the basic theory. These modifications are considered in the following subsections.

11.5.1 Variable Fluid Properties

The fluid properties, k_L, ρ_L, and μ_L, are functions of temperature, which varies across the condensate film. Of the three properties, viscosity is the most temperature sensitive. In practice, the condensate properties are evaluated at a weighted average film temperature, T_f, defined by:

$$T_f = \beta T_w + (1 - \beta) T_{sat} \tag{11.42}$$

The value of the weight factor, β, recommended in the literature ranges from 0.5 to 0.75. The latter value will be used herein, i.e.,

$$T_f = 0.75\, T_w + 0.25\, T_{sat} \tag{11.43}$$

11.5.2 Inclined Surfaces

For condensation on an inclined surface making an angle, θ, with the vertical, where $0° \le \theta \le 45°$, Equations (11.27) and (11.31) can be used by replacing g with $g \cos \theta$. For condensation outside an inclined tube making an angle, θ', with the horizontal, Equations (11.35) and (11.36) can be used if g is replaced with $g \cos \theta'$, provided $L/D > 1.8 \tan \theta'$ [11].

11.5.3 Turbulence in Condensate Film

For condensation on vertical surfaces having lengths typical of industrial heat-transfer equipment, the flow in the condensate film may become turbulent at some distance from the top of the surface. The portion of the surface experiencing turbulent conditions becomes significant at film Reynolds numbers exceeding about 1600. However, ripples and waves that form and propagate along the surface of the condensate film begin to affect the heat transfer at a much lower Reynolds number of about 30. Thus, the following three-flow regimes are recognized:

- Laminar wave free ($Re \le 30$)
- Laminar wavy ($30 < Re \le 1600$)
- Turbulent ($Re > 1600$)

The Nusselt relations, Equations (11.24) and (11.31), are valid for the strictly laminar regime ($Re \le 30$). For the laminar wavy regime, the average heat-transfer coefficient for the entire film, including the laminar wave-free portion, can be obtained from the following semi-empirical correlation [11]:

$$h = \frac{Re\left[k_L^3 \rho_L (\rho_L - \rho_V) g / \mu_L^2\right]^{1/3}}{1.08 Re^{1.22} - 5.2} \tag{11.44}$$

A similar correlation is available [11] for the turbulent regime ($Re > 1600$):

$$h = \frac{Re\left[k_L^3 \rho_L (\rho_L - \rho_V) g / \mu_L^2\right]^{1/3}}{8750 + 58 Pr_L^{-0.5} (Re^{0.75} - 253)} \tag{11.45}$$

In this equation h is again an average value for the entire film, including the laminar portions, and Pr_L is the Prandtl number for the condensate. Equation (11.45) is valid for $Pr_L \le 10$, which includes most cases for which a turbulent condensate film is likely to occur. For higher Prandtl numbers, it will tend to overestimate h. Therefore, if $Pr_L > 10$, Equation (11.45) should be used with $Pr_L = 10$ [11].

More complex correlations for the wavy and turbulent regimes have been developed (see, e.g., Ref. [12]). However, the correlations given above are easy to use and are sufficiently accurate for most purposes.

For condensation on horizontal tubes, the Nusselt relations, Equations (11.35) and (11.36), are valid for Reynolds numbers up to 3200, beyond which the condensate film becomes turbulent. This value is unlikely to be exceeded for a single tube or tube row, but higher Reynolds numbers may occur on tubes in a vertical stack due to cumulative condensate loading. Nevertheless, the usual practice is to use Equation (11.40) for horizontal tube bundles regardless of the Reynolds number.

11.5.4 Superheated Vapor

When the vapor is superheated, condensation can still take place if the wall temperature is below the saturation temperature. In this case, heat is transferred from the bulk vapor at T_V to the vapor-liquid interface at T_{sat}, and then through the condensate film to the wall at T_w. The total amount of heat transferred consists of the latent heat of condensation plus the sensible heat to cool the vapor from T_V to T_{sat}. Thus,

$$q = W\lambda + WC_{P,V}(T_V - T_{sat}) = W\lambda' \tag{11.46}$$

where

$$\lambda' \equiv \lambda \left[1 + \frac{C_{P,V}(T_V - T_{sat})}{\lambda}\right] \tag{11.47}$$

Repeating the Nusselt analysis of Section 11.3.1 with λ' replacing λ leads to the following expression for the average heat-transfer coefficient:

$$h = 0.943 \left[\frac{k_L^3 \rho_L(\rho_L - \rho_V) g \lambda'}{\mu_L (T_{sat} - T_w) L}\right]^{1/4} \tag{11.48}$$

Comparing this result with Equation (11.24), it follows that:

$$\frac{h}{h_{Nu}} = (\lambda'/\lambda)^{1/4} = \left[1 + \frac{C_{P,V}(T_V - T_{sat})}{\lambda}\right]^{1/4} \tag{11.49}$$

where h_{Nu} is the heat-transfer coefficient given by the basic Nusselt theory, Equation (11.24). The total rate of heat transfer is given by:

$$q = hwL(T_{sat} - T_w) \tag{11.50}$$

Equation (11.49) also holds for condensation on horizontal tubes. Because the sensible heat is usually small compared with the latent heat, the effect of vapor superheat on the heat-transfer coefficient is usually small. Equation (11.49) does not account for the heat-transfer resistance in the vapor phase, which is generally small compared with the resistance of the condensate film.

11.5.5 Condensate Subcooling

The temperature in the condensate film drops from T_{sat} at the vapor–liquid interface to T_w at the wall. Therefore, the average condensate temperature, T_L, is less than T_{sat}, and hence the condensate leaving the surface is subcooled. Accounting for subcooling, the rate of heat transfer is:

$$q = W\lambda + WC_{P,L}(T_{sat} - T_L) \tag{11.51}$$

The effect of subcooling on the heat-transfer coefficient was first considered by Nusselt [6], and subsequently by a number of other researchers. Sadisivan and Lienhard [13] used a boundary-layer analysis, which accounted for both subcooling and inertial effects, to derive the following equation:

$$\frac{h}{h_{Nu}} = \left[1 + \left(0.683 - 0.228 Pr_L^{-1}\right)\epsilon\right]^{1/4} \tag{11.52}$$

where

$$\epsilon = \frac{C_{P,L}(T_{sat} - T_w)}{\lambda}$$

$$Pr_L = \frac{C_{P,L}\mu_L}{k_L}$$

Equation (11.52) is valid for $Pr_L > 0.6$, and thus includes virtually all condensates except liquid metals, which have very small Prandtl numbers. It predicts an effect of subcooling that is similar to the effect of vapor superheat, i.e., a (usually) small increase in the heat-transfer coefficient over the value given by the basic Nusselt theory.

Chen [14] also used boundary-layer theory to study the condensation process. He replaced the boundary condition of zero shear stress at the vapor–liquid interface with that of zero vapor velocity far from the interface. His analysis thus included the effect of vapor drag on the condensate as well as subcooling and inertial effects. The following equation closely approximates the boundary-layer solutions for both a vertical plane wall and a single horizontal tube [14]:

$$\frac{h}{h_{Nu}} = \left(\frac{1 + 0.68\epsilon + 0.02\epsilon^2 Pr_L^{-1}}{1 + 0.85\epsilon Pr_L^{-1} - 0.15\epsilon^2 Pr_L^{-1}}\right)^{1/4} \tag{11.53}$$

The range of validity for this equation is as follows:

$$\epsilon \leq 2$$

$$\epsilon \leq 20\ Pr_L$$

$$Pr_L \geq 1 \text{ or } Pr_L \leq 0.05$$

Most condensates have Prandtl numbers of 1.0 or higher, and for these fluids Equations (11.52) and (11.53) both predict a relatively small effect on the heat-transfer coefficient (less than 14% for $\epsilon \leq 1.0$). In contrast to Equation (11.52), however, Equation (11.53) predicts that h is slightly less than h_{Nu} for Prandtl numbers near unity.

Most liquid metals have Prandtl numbers below 0.05, and for these fluids Equation (11.53) predicts that h can be much less than h_{Nu}. This prediction is in qualitative agreement with experimental data [14].

Example 11.2

A stream consisting of 5000 lb/h of saturated *n*-propyl alcohol (1-propanol) vapor at 207°F will be condensed using a tube bundle containing 109 tubes arranged for one pass. The tubes are 0.75 in. OD, 14 BWG, with a length of 12 ft. Estimate the condensing-side heat-transfer coefficient for the following cases:

(a) The tube bundle is vertical and condensation occurs inside the tubes.
(b) The tube bundle is horizontal and condensation occurs outside the tubes.

Solution

(a) The physical properties given in Example 11.1 will be used as a first approximation. Values of k_L and ρ_L will be assumed constant, but the variation of μ_L with temperature will be accounted for. As in Example 11.1, the density of the vapor will be neglected compared with that of the liquid. The first step is to compute the Reynolds number:

$$\Gamma = \frac{W}{n_t \pi D_i} = \frac{5000}{109\pi(0.584/12)} = 300.0\ \text{lbm/ft}\cdot\text{h}$$

$$Re = \frac{4\Gamma}{\mu_L} = \frac{4 \times 300.0}{0.5 \times 2.419} = 992.1$$

The flow regime is wavy laminar, so Equation (11.44) is used to calculate the heat-transfer coefficient:

$$h = \frac{Re\left[k_L^3 \rho_L^2 g/\mu_L^2\right]^{1/3}}{1.08 Re^{1.22} - 5.2}$$

$$= \frac{992.1\left[(0.095)^3 (49)^2 (4.17 \times 10^8)/(0.5 \times 2.419)^2\right]^{1/3}}{1.08(992.1)^{1.22} - 5.2}$$

$$h = 170\ \text{Btu/h}\cdot\text{ft}^2\cdot{}^\circ\text{F}$$

Next, the temperature drop across the condensate film is calculated. From Table A.17, the latent heat of vaporization is:

$$\lambda = 164.36\ \text{cal/g} \times 1.8 \frac{\text{Btu/lbm}}{\text{cal/g}} = 295.85\ \text{Btu/lbm}$$

Neglecting the sensible heat of subcooling, the duty is:

$$q = W\lambda = 5000 \times 295.85 = 1{,}479{,}250\ \text{Btu/h}$$

$$q = h n_t \pi D_i L \Delta T$$

$$\Delta T = \frac{q}{h n_t \pi D_i L} = \frac{1{,}479{,}250}{170 \times 109\pi(0.584/12) \times 12} = 43.5^\circ\text{F}$$

Therefore, the wall temperature is:

$$T_w = 207 - 43.5 = 163.5^\circ\text{F}$$

The weighted average film temperature is calculated from Equation (11.43):

$$T_f = 0.75\ T_w + 0.25\ T_{sat} = 0.75 \times 163.5 + 0.25 \times 207$$

$$T_f = 174.4\ ^\circ\text{F}$$

From Figure A.1, the condensate viscosity at this temperature is 0.7 cp. A second iteration is made using this value of μ_L:

$$Re = \frac{4 \times 300.0}{0.7 \times 2.419} = 708.7 \Rightarrow \text{wavy laminar flow}$$

$$h = \frac{708.7\left[(0.095)^3(49)^2(4.17 \times 10^8)/(0.7 \times 2.419)^2\right]^{1/3}}{1.08(708.7)^{1.22} - 5.2} \cong 146\ \text{Btu/h}\cdot\text{ft}^2\cdot{}^\circ\text{F}$$

$$\Delta T = \frac{1,479,250}{146 \times 109\pi(0.584/12) \times 12} = 50.7^\circ\text{F}$$

$$T_w = 207 - 50.7 = 156.3^\circ\text{F}$$

$$T_f = 0.75 \times 156.3 + 0.25 \times 207 \cong 169^\circ\text{F}$$

At this value of T_f, the condensate viscosity is 0.73 cp. Results of the third iteration are given below:

$$Re = 679.6$$

$$h \cong 144\ \text{Btu/h}\cdot\text{ft}^2\cdot{}^\circ\text{F}$$

$$\Delta T = 51.4^\circ\text{F}$$

$$T_w = 155.6^\circ\text{F}$$

$$T_f = 168.5^\circ\text{F}$$

Since the new value of T_f is close to the previous value, no further iterations are required.

To check the effect of subcooling on h, the condensate Prandtl number is computed next. The heat capacity of liquid n-propyl alcohol at $T_f \cong 169^\circ\text{F}$ is found from Figure A.3: $C_{P,L} = 0.72\ \text{Btu/lbm}\cdot{}^\circ\text{F}$. Thus,

$$Pr_L = \frac{C_{P,L}\mu_L}{k_L} = \frac{0.72(0.73 \times 2.419)}{0.095} = 13.4$$

$$\epsilon = \frac{C_{P,L}(T_{sat} - T_w)}{\lambda} = \frac{0.72(207 - 155.6)}{295.85} = 0.125$$

Using Equation (11.52) we obtain:

$$h/h_{Nu} = \left[1 + \left(0.683 - 0.228 Pr_L^{-1}\right)\epsilon\right]^{1/4}$$

$$= [1 + (0.683 - 0.228/13.4) \times 0.125]^{1/4}$$

$$h/h_{Nu} = 1.020$$

For comparison, the calculation is also done using Equation (11.53):

$$h/h_{Nu} = \left(\frac{1 + 0.68\epsilon + 0.02\epsilon^2 Pr_L^{-1}}{1 + 0.85\epsilon Pr_L^{-1} - 0.15\epsilon^2 Pr_L^{-1}}\right)^{1/4}$$

$$= \left\{\frac{1 + 0.68 \times 0.125 + 0.02(0.125)^2/13.4}{1 + 0.85 \times 0.125/13.4 - 0.15(0.125)^2/13.4}\right\}^{1/4}$$

$$h/h_{Nu} = 1.019$$

Thus, both methods predict an increase in h of about 2% due to condensate subcooling. Including this effect gives $h \cong 147$ Btu/h · ft² · °F. This correction is not large enough to warrant further iteration.

(b) The physical properties from Example 11.1 are again used to initialize the calculations and the vapor density is neglected. The heat-transfer coefficient is computed using Equations (11.40) and (11.41):

$$\Gamma^* = \frac{W}{Ln_t^{2/3}} = \frac{5000}{12(109)^{2/3}} = 18.26\ \text{lbm/ft}\cdot\text{h}$$

$$h = 1.52\left[\frac{k_L^3\rho_L^2 g}{4\mu_L\Gamma^*}\right]^{1/3} = 1.52\left[\frac{(0.095)^3(49)^2(4.17\times10^8)}{4(0.5\times2.419)\times18.26}\right]^{1/3}$$

$$h = 324 \text{ Btu/h}\cdot\text{ft}^2\cdot{}^\circ\text{F}$$

Next, the temperature difference across the condensate film is calculated:

$$\Delta T = T_{sat} - T_w = \frac{q}{hn_t\pi D_O L} = \frac{1{,}479{,}250}{324\times109\pi(0.75/12)\times12}$$

$$\Delta T = 17.8^\circ\text{F}$$

$$T_w = 207 - 17.8 = 189.2^\circ\text{F}$$

$$T_f = 0.75T_w + 0.25T_{sat} = 0.75\times189.2 + 0.25\times207 = 193.7^\circ\text{F}$$

From Figure A.1, the condensate viscosity at T_f is 0.58 cp. The above calculations are repeated using this value of viscosity.

$$h = 1.52\left[\frac{(0.095)^3(49)^2(4.17\times10^8)}{4(0.58\times2.419)\times18.26}\right]^{1/3} \cong 309 \text{ Btu/h}\cdot\text{ft}^2\cdot{}^\circ\text{F}$$

$$\Delta T = \frac{1{,}479{,}250}{309\times109\pi(0.75/12)\times12} = 18.6^\circ\text{F}$$

$$T_w = 207 - 18.6 = 188.4^\circ\text{F}$$

$$T_f = 0.75\times188.4 + 0.25\times207 = 193^\circ\text{F}$$

Since the new value of T_f is close to the previous value, the calculations have converged.

Using Equation (11.52) to estimate the effect of condensate subcooling for this case yields $h/h_{Nu} = 1.0076$. Including this minor correction gives $h = 311$ Btu/h · ft^2 · °F. Clearly, the effect of condensate subcooling is negligible in this case.

11.5.6 Interfacial Shear

In the basic Nusselt theory interfacial shear is neglected and the condensate is assumed to drain from the surface under the influence of gravity alone. By contrast, at very high vapor velocities the effect of interfacial shear is predominant and gravitational effects can be neglected. Methods for computing the heat-transfer coefficient under the latter conditions are presented in the following subsections. At intermediate velocities, the effects of both gravity and interfacial shear may be significant. In this situation, the following formula is often used to represent the combined effects of gravity and interfacial shear:

$$h = \left(h_{sh}^2 + h_{gr}^2\right)^{1/2} \tag{11.54}$$

In this equation h_{sh} and h_{gr} are the heat-transfer coefficients for shear-controlled and gravity-controlled film condensation, respectively. A more conservative approach that may be preferable for design purposes is to use either h_{gr} or h_{sh}, whichever is larger.

Condensation in Vertical Tubes with Vapor Upflow

Low vapor velocities are generally employed with this configuration in order to prevent flooding (the condition wherein condensate is forced out the top of the tubes rather than draining freely from the bottom). As a result, interfacial shear effects can usually be neglected in this type of condenser.

Condensation in Vertical Tubes with Vapor Downflow

In this configuration the interfacial shear stress tends to accelerate the flow of condensate down the tube wall and decrease the critical Reynolds number for the onset of turbulence. As a result, the heat-transfer coefficient tends to be enhanced.

For shear-controlled condensation, the correlation of Boyko and Kruzhilin [15] provides a simple method to estimate the average heat-transfer coefficient. The local coefficient is given by the following correlation:

$$h = h_{LO}[1 + x(\rho_L - \rho_V)/\rho_V]^{0.5} \tag{11.55}$$

where

x = vapor weight fraction

h_{LO} = heat-transfer coefficient for total flow as liquid

The value of h_{LO} can be computed using a correlation for single-phase heat transfer, e.g., the Seider–Tate correlation. Turbulent flow is assumed to exist at all positions along the length of the tube, including the region in a total condenser where condensation is complete and the flow is all liquid. The average coefficient for the entire tube is the arithmetic average of the coefficients computed at inlet and outlet conditions [15]. This average coefficient can be equated with h_{sh} in Equation (11.54).

The basis for the Boyko–Kruzhilin method is the analogy between heat transfer and fluid flow (momentum transfer). Many similar correlations have been published, some of which are discussed in Section 11.6.

Condensation Outside Horizontal Tubes

In horizontal shell-side condensers employing a J- or X-shell, the vapor flow is primarily perpendicular to the tubes, as it is between baffle tips of units employing an E-shell. At high vapor velocities, this cross flow causes condensate to be blown off the tubes, resulting in a very complex flow pattern in the tube bundle [2]. Despite this complexity, McNaught [16] developed the following simple correlation for shear-controlled condensation in tube bundles:

$$h/h_L = 1.26X_{tt}^{-0.78} \tag{11.56}$$

where

X_{tt} = Lockhart–Martinelli parameter, Equation (9.37)
h_L = heat-transfer coefficient for the liquid phase flowing alone through the bundle

The value of h_L can be computed by any of the methods discussed in Chapters 5, 6, and 7. Since X_{tt} depends on vapor quality, Equation (11.56) provides a local heat-transfer coefficient suitable for use with a zone analysis. It can also be used to assess the significance of interfacial shear at any position in a condenser where the quality is known or specified.

Equation (11.56) is based on experimental data for condensation of steam in downflow over tube bundles laid out on square and triangular patterns. In conjunction with Equation (11.54), it correlated 90% of the data to within ±25% [16].

Example 11.3

Evaluate the significance of interfacial shear for the conditions of Example 11.2 and the following specifications:

(a) For the vertical condenser, the vapor flows downward through the tubes.
(b) For the horizontal shell-side condenser, the unit consists of an E-shell with an ID of 12 in. The condenser has 20% cut segmental baffles with a spacing of 4.8 in., and a triangular tube layout with a pitch of 1.0 in.

Solution

(a) The following data are obtained from Example 11.2, part (a):

$T_V = T_{sat} = 207°F$	$\mu_L = 0.73$ cp
$T_f \cong 169°F$	$Pr_L = 13.4$
$T_w \cong 156°F$	$D_i = 0.584$ in.
$P_{sat} \cong 1$ atm $= 14.7$ psia	$L = 12$ ft
$k_L = 0.095$ Btu/h · ft · °F	$n_t = 109$
$\rho_L = 49$ lbm/ft^3	

Additional data needed for the calculations are:
Molecular weight of propyl alcohol = 60
$\mu_w = 0.85$ cp at $T_w = 156°F$ from Figure A.1

$$\rho_V = \frac{PM}{\tilde{R}T} = \frac{14.7 \times 60}{10.73(207 + 460)} = 0.123 \text{ lbm/ft}^3$$

The shear-controlled heat-transfer coefficient is given by Equation (11.55), which involves the coefficient, h_{LO}, for the total flow (5000 lb/h) as liquid. The Reynolds number is calculated first:

$$Re_{LO} = \frac{4\dot{m}/n_t}{\pi D_i \mu_L} = \frac{4 \times 5000/109}{\pi(0.584/12) \times 0.73 \times 2.419} = 679.6$$

Since the total condensate flow is laminar, the Boyko–Kruzhilin correlation is not strictly applicable. However, a conservative estimate for the shear-controlled coefficient can be obtained by using Equation (11.55) along with a laminar-flow correlation to calculate h_{LO}. (See Problem 11.2 for an alternative approach.) Hence, Equation (2.36) is used as follows:

$$Nu_{LO} = 1.86(Re_{LO}Pr_L D_i/L)^{1/3}(\mu_L/\mu_w)^{0.14}$$

$$= 1.86[679.6 \times 13.4(0.584/12)/12]^{1/3}(0.73/0.85)^{0.14}$$

$$Nu_{LO} = 6.06$$

$$h_{LO} = 6.06 \times k_L/D_i = 6.06 \times 0.095/(0.584/12) = 11.8 \text{ Btu/h·ft}^2\cdot°\text{F}$$

At the condenser inlet, the vapor fraction is $x_{in} = 1.0$. Hence, from Equation (11.55):

$$h_{in} = h_{LO}[1 + x_{in}(\rho_L - \rho_V)/\rho_V]^{0.5}$$

$$= 11.8[1 + 1.0(49 - 0.123)/0.123]^{0.5}$$

$$h_{in} = 236 \text{ Btu/h·ft}^2\cdot°\text{F}$$

At the condenser outlet the vapor fraction is zero, and therefore:

$$h_{out} = h_{LO} = 11.8 \text{ Btu/h·ft}^2\cdot°\text{F}$$

The average heat-transfer coefficient for shear-controlled condensation is the arithmetic average of the inlet and outlet values:

$$h_{sh} = 0.5(h_{in} + h_{out}) = 0.5(236 + 11.8) = 124 \text{ Btu/h·ft}^2\cdot°\text{F}$$

This result is the same order of magnitude as the average coefficient for gravity-controlled condensation (147 Btu/h · ft^2 · °F) calculated in Example 11.2, part (a). Therefore, both gravity and interfacial shear effects are significant, and Equation (11.54) can be used to estimate the resultant heat-transfer coefficient:

$$h = \left(h_{sh}^2 + h_{gr}^2\right)^{1/2} = \left[(124)^2 + (147)^2\right]^{1/2} = 192 \text{ Btu/h·ft}^2\cdot°\text{F}$$

This value is about 30% higher than the result based on gravity-controlled condensation alone. Comparison with other methods (see Problems 11.1 and 11.2 for two examples) indicates that the calculated value of h is, in fact, somewhat conservative.

(b) The following data are obtained from Example 11.2, part (b):

$T_f = 193°\text{F}$	$T_w \cong 188°\text{F}$	$\mu_L = 0.58$ cp

Additional data needed for the calculations are:

$C_{P,L} = 0.75$ Btu/lbm · °F	from Figure A.3
$\mu_V = 0.0095$ cp	from Figure A.2
$\mu_w = 0.60$ cp at $T_w = 188°\text{F}$	from Figure A.1

$$Pr_L = \frac{C_{P,L}\mu_L}{k_L} = \frac{0.75 \times 0.58 \times 2.419}{0.095} \cong 11.1$$

Other data are the same as in part (a).

The shear-controlled heat-transfer coefficient is given by Equation (11.56), which involves the coefficient, h_L, for the liquid phase flowing alone. The Simplified Delaware method is used here to calculate h_L. A vapor weight fraction of 0.9 is selected, giving a liquid flow rate of:

$$\dot{m}_L = (1 - x)\dot{m} = 0.1 \times 5000 = 500 \text{ lbm/h}$$

The flow area through the tube bundle is:

$$a_s = \frac{d_s C' B}{144 P_T} = \frac{12 \times 0.25 \times 4.8}{144 \times 1.0} = 0.10 \text{ ft}^2$$

The mass flux is:

$$G_L = \dot{m}_L/a_s = 500/0.10 = 5000 \text{ lbm/h·ft}^2$$

From Figure 3.17, the equivalent diameter is $D_e = 0.99/12 = 0.0825$ ft. Thus, the Reynolds number is:

$$Re_L = \frac{D_e G_L}{\mu_L} = \frac{0.0825 \times 5000}{0.58 \times 2.419} = 294.0$$

Next, the Colburn *j*-factor is calculated using Equation (3.21) with $B/d_s = 4.8/12 = 0.4$.

$$j_H = 0.5(1 + B/d_s)\left(0.08Re_L^{0.6821} + 0.7Re_L^{0.1772}\right)$$

$$= 0.5(1 + 0.4)\left\{0.08(294)^{0.6821} + 0.7(294)^{0.1772}\right\}$$

$$j_H = 4.04$$

Equation (3.20) is used to calculate h_L:

$$h_L = j_H(k_L/D_e)Pr_L^{1/3}(\mu_L/\mu_w)^{0.14}$$

$$= 4.04(0.095/0.0825)(11.1)^{1/3}(0.58/0.60)^{0.14}$$

$$h_L = 10.3 \text{ Btu/h·ft}^2\text{·°F}$$

Next, the Lockhart–Martinelli parameter is computed from Equation (9.37):

$$X_{tt} = \left(\frac{1-x}{x}\right)^{0.9}(\rho_V/\rho_L)^{0.5}(\mu_L/\mu_V)^{0.1}$$

$$= \left(\frac{1-0.9}{0.9}\right)^{0.9}(0.123/49)^{0.5}(0.58/0.0095)^{0.1}$$

$$X_{tt} = 0.0105$$

Substituting the values of h_L and X_{tt} in Equation (11.56) gives the local coefficient for shear-controlled condensation:

$$h_{sh} = 1.26X_{tt}^{-0.78}h_L = 1.26(0.0105)^{-0.78} \times 10.3$$

$$h_{sh} = 454 \text{ Btu/h·ft}^2\text{·°F}$$

This value is about 44% higher than the value of 315 Btu/h · ft² · °F found for the average coefficient based on gravity-controlled condensation in Example 11.2. Thus, interfacial shear effects are significant at this point in the condenser.

Repeating the above calculations for vapor weight fractions of 0.5 and 0.1 yields the following results:

$x = 0.5$	$x = 0.1$
$\dot{m}_L = 2500$ lbm/h	$\dot{m}_L = 4500$ lbm/h
$G_L = 25{,}000$ lbm/h · ft²	$G_L = 45{,}000$ lbm/h · ft²
$Re_L = 1470$	$Re_L = 2646$
$j_H = 9.89$	$j_H = 14.1$
$h_L = 25.3$ Btu/h · ft² · °F	$h_L = 36.0$ Btu/h · ft² · °F
$X_{tt} = 0.076$	$X_{tt} = 0.546$
$h_{sh} = 238$ Btu/h · ft² · °F	$h_{sh} = 73$ Btu/h · ft² · °F

Interfacial shear effects are negligible at $x = 0.1$, but they are clearly significant over a large portion of the condensing range.

11.6 Condensation Inside Horizontal Tubes

11.6.1 Flow Regimes

The analysis of condensation in horizontal tubes is complicated by the variety of two-phase flow regimes that can exist in this geometry. The flow pattern typically changes as condensation proceeds due to the increase in the amount of liquid present and the decrease in vapor velocity. Initially, the condensate forms an annular film around the tube periphery while the vapor flows in the core of the tube. Liquid droplets are normally entrained in the vapor due to the high shear forces, with the amount of entrainment decreasing as the vapor velocity falls. This flow regime is referred to as annular or mist-annular, depending on the amount of entrainment. In this regime the condensation is shear controlled.

As condensation proceeds and vapor shear decreases, the condensate begins to accumulate at the bottom of the tube due to gravity, causing an increasing asymmetry in the condensate film. At high flow rates, the flow pattern eventually changes to slug flow, followed by plug flow. In the final stages of condensation, the vapor flows as bubbles that are elongated in the flow direction and entrained in a continuous liquid phase. At lower flow rates, annular flow gradually changes to a stratified regime in which most of the condensate flows along the bottom portion of the tube with vapor above. The condensate finally drains from the tube under gravity. As noted by Butterworth [11], however, if the condensate discharges into a liquid-filled header, it will not drain freely, but will accumulate in the tube and be forced out under pressure as intermittent slugs of liquid. In the stratified flow regime, the condensation is gravity controlled.

Breber et al. [17] developed a simplified classification scheme for the flow regimes in horizontal tube-side condensation, as illustrated in Figure 11.14. The prevailing flow regime is determined by two parameters, the Lockhart–Martinelli parameter, X, and a dimensionless vapor mass flux, j^*, defined as follows:

$$j^* = \frac{xG}{[D_i g \rho_V (\rho_L - \rho_V)]^{0.5}} \tag{11.57}$$

In this equation, G is the total mass flux of vapor and liquid based on the entire tube cross section. The Lockhart–Martinelli parameter is correlated with the liquid fraction of the flow, while j^* is directly related to the ratio of shear force to gravitational force [17].

Quantitative criteria for determining the flow regime are given in Table 11.1. In addition to the four zones shown in Figure 11.14, the transition regions between zones I and II and between zones II and III are of practical importance. According to Breber et al. [17], the bubble regime, zone IV, occurs only at high reduced pressures.

11.6.2 Stratified Flow

For stratifying flow conditions, the condensation is gravity controlled and a Nusselt-type analysis can be used to derive an equation for the heat-transfer coefficient. As shown in Figure 11.15, the condensate forms a thin film on the upper portion of the tube wall and drains into the stratified layer that covers the bottom portion of the tube. The heat transfer across the stratified layer is usually negligible compared with that across the condensate film. Thus, the average heat-transfer coefficient over the entire circumference of the tube is given by a modified version of Equation (11.35):

$$h = \Omega \left[\frac{k_L^3 \rho_L (\rho_L - \rho_V) g \lambda}{\mu_L (T_V - T_w) D_i} \right]^{1/4} \tag{11.58}$$

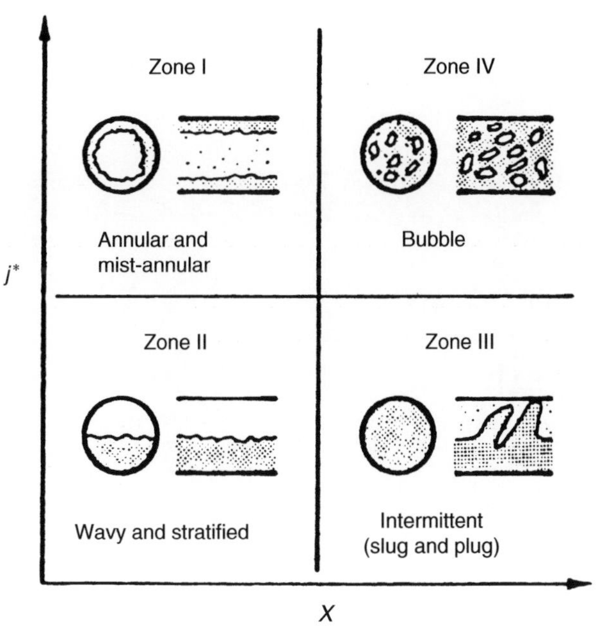

FIGURE 11.14 Simplified flow regime diagram for condensation in a horizontal tube (Source: Ref. [17]).

TABLE 11.1 Flow Regime Criteria for Condensation in Horizontal Tubes

Zone	Criterion
I	$j^* > 1.5$ and $X < 1.0$
II	$j^* < 0.5$ and $X < 1.0$
III	$j^* < 0.5$ and $X > 1.5$
IV	$j^* > 1.5$ and $X > 1.5$
Transition (I, II)	$0.5 \leq j^* \leq 1.5$ and $X < 1.0$
Transition (II, III)	$j^* < 0.5$ and $1.0 \leq X \leq 1.5$

It is assumed here that the vapor is saturated and the effect of condensate subcooling is neglected. The coefficient, Ω, depends on the angle, ϕ, shown in Figure 11.15. It is also related to the void fraction in the tube by the following simple formula [11]:

$$\Omega = 0.728\epsilon_V^{0.75} \tag{11.59}$$

Methods for calculating the void fraction are given in Chapter 9. An alternative that provides a somewhat conservative approximation is to use a constant value of $\Omega = 0.56$ [17].

11.6.3 Annular Flow

For the shear-controlled annular flow regime, Breber et al. [17] recommend the following correlation based on the analogy between heat transfer and fluid flow:

$$h = h_L\left(\phi_L^2\right)^{0.45} = h_{LO}\left(\phi_{LO}^2\right)^{0.45} \tag{11.60}$$

In this equation ϕ_L^2 and ϕ_{LO}^2 are two-phase multipliers for the pressure gradient; they can be calculated by the methods given in Chapter 9. The exponent of 0.45 in Equation (11.60) was chosen to provide a conservative estimate of h for design work [17]. An exponent of 0.5 actually gives a better fit to experimental data.

The Boyko–Kruzhilin correlation, Equation (11.55), is also applicable in the annular flow regime. A number of other correlations have been developed for this regime; see, e.g., Refs. [18-23]. Among these, the empirical correlation of Shah [23] is notable for the variety of fluids (water, methanol, ethanol, benzene, toluene, trichloroethylene, and several refrigerants) studied and the number (21) of independent data sets analyzed. The correlation had a mean deviation of less than 17% for 474 data points, although for some data sets it was in excess of 25%. The correlation can be stated as follows:

$$h = h_{LO}\left\{(1-x)^{0.8} + 3.8x^{0.76}(1-x)^{0.04}P_r^{-0.38}\right\} \tag{11.61}$$

where P_r is the reduced pressure. Shah used the Dittus–Boelter correlation, Equation (9.75), to calculate h_{LO} with all fluid properties evaluated at the saturation temperature. The Dittus–Boelter correlation is valid for turbulent flow, and values of Re_{LO} ranged from 100 to 63,000 in the data sets analyzed by Shah. Although Shah reported satisfactory results using this procedure, he recommended using the correlation only for $Re_{LO} \geq 350$. A conservative approach would be to use

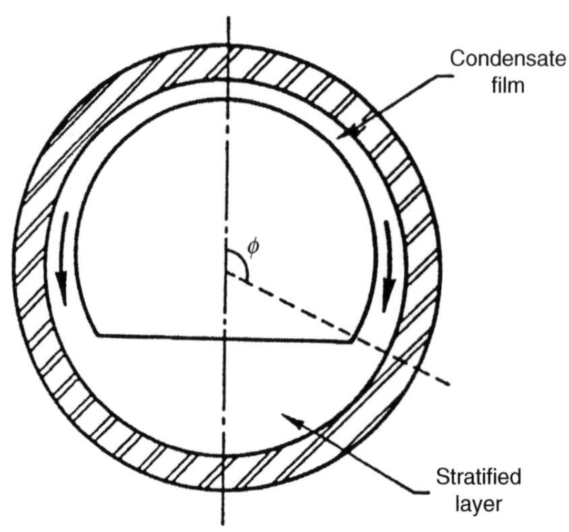

FIGURE 11.15 Illustration of condensate cross-sectional profile in the stratified flow regime (Source: Ref. [3]).

a laminar-flow correlation to calculate h_{LO} for $Re_{LO} < 350$. Equation (11.61) can be used for both vertical and horizontal tubes and qualities from zero to 0.999. Notice that the correlation breaks down as the quality approaches 1.0, since it predicts $h = 0$ for $x = 1.0$.

11.6.4 Other Flow Regimes

Specific correlations for the flow regimes of zones III and IV are not available, but Breber et al. [17] recommend using Equation (11.60) as an approximation for these cases.

For the transition region between zones I and II, the following linear interpolation formula is recommended [17]:

$$h = h_I + (j^* - 1.5)(h_I - h_{II}) \tag{11.62}$$

where h_I and h_{II} are the heat-transfer coefficients for zones I and II, respectively. A similar procedure can be used for the transition region between zones II and III:

$$h = h_{II} + 2(1 - X)(h_{II} - h_{III}) \tag{11.63}$$

Example 11.4

The tube bundle of Example 11.2 will be used to condense 50,000 lb/h of saturated *n*-propyl alcohol vapor at 207°F. The bundle will be oriented horizontally and condensation will take place inside the tubes. Estimate the condensing-side heat-transfer coefficient at a position in the condenser where the quality is 0.06 and the tube wall temperature is 156°F.

Solution

The first step is to determine the flow regime by computing j^* and X. The total mass flux, which is needed to compute j^* is:

$$G = \frac{\dot{m}}{n_t(\pi/4)D_i^2} = \frac{50{,}000}{109(\pi/4)(0.584/12)^2} = 246{,}598 \text{ lbm/h}\cdot\text{ft}^2$$

From Equation (11.57):

$$j^* = \frac{xG}{[D_i g\rho_V(\rho_L - \rho_V)]^{0.5}} = \frac{0.06 \times 246{,}598}{[(0.584/12) \times 4.17 \times 10^8 \times 0.123\,(49 - 0.123)]^{0.5}}$$

$$j^* = 1.34$$

To compute the Lockhart–Martinelli parameter, the Reynolds numbers of the liquid and vapor fractions flowing alone are calculated. From Examples 11.2 and 11.3, $\mu_L = 0.73$ cp and $\mu_V = 0.0095$ cp.

$$Re_L = D_i G_L/\mu_L = (0.584/12)(0.94 \times 246{,}598)/(0.73 \times 2.419)$$

$$Re_L = 6388$$

$$Re_V = D_i G_V/\mu_V = (0.584/12)(0.06 \times 246{,}598)/(0.0095 \times 2.419)$$

$$Re_V = 31{,}334$$

According to the Lockhart–Martinelli criteria given in Table 9.2, the flow is turbulent in both phases. Therefore, $X = X_{tt}$ is given by Equation (9.37).

$$X_{tt} = \left(\frac{1-x}{x}\right)^{0.9}\left(\frac{\rho_V}{\rho_L}\right)^{0.5}\left(\frac{\mu_L}{\mu_V}\right)^{0.1}$$

$$= \left(\frac{1-0.06}{0.06}\right)^{0.9}\left(\frac{0.123}{49}\right)^{0.5}\left(\frac{0.73}{0.0095}\right)^{0.1}$$

$$X_{tt} = 0.92$$

For the computed values of j^* and X, Table 11.1 shows that the flow is in the transition region between zones I and II. Therefore, heat-transfer coefficients must be calculated for both stratified and annular flow regimes. For the stratified flow calculation, the void fraction is required. Using the Chisholm correlation, Equation (9.63), the slip ratio is:

$$SR = (\rho_L/\rho_V)^{0.25} = (49/0.123)^{0.25} = 4.468$$

The void fraction is obtained from Equation (9.59):

$$\epsilon_V = \frac{x}{x + SR(1-x)\rho_V/\rho_L} = \frac{0.06}{0.06 + 4.468 \times 0.94 \times 0.123/49}$$

$$\epsilon_V = 0.85$$

The heat-transfer coefficient for zone II is given by the following equation obtained by combining Equations (11.58) and (11.59):

$$h_{II} = 0.728\epsilon_V^{0.75}\left[\frac{k_L^3 \rho_L(\rho_L - \rho_V)g\lambda}{\mu_L(T_V - T_w)D_i}\right]^{1/4}$$

$$= 0.728(0.85)^{0.75}\left[\frac{(0.095)^3 \times 49(49 - 0.123) \times 4.17 \times 10^8 \times 295.85}{0.73 \times 2.419(207 - 156)(0.584/12)}\right]^{1/4}$$

$$h_{II} = 316 \text{ Btu/h·ft}^2\cdot{}^\circ\text{F}$$

Equation (11.60) will be used for the zone I calculation. To calculate h_{LO}, the appropriate Reynolds number is:

$$Re_{LO} = D_i G/\mu_L = (0.584/12) \times 246{,}598/(0.73 \times 2.419)$$
$$Re_{LO} = 6796$$

Since this value is in the transition region for heat transfer in tubular flow, the Hausen correlation, Equation (2.37) is used. Needed physical property data are obtained from Example 11.3:

$$Nu = 0.116\left[Re^{2/3} - 125\right]Pr^{1/3}(\mu/\mu_w)^{0.14}\{1 + (\mathrm{D_i/L})^{2/3}\}$$

$$= 0.116\left\{(6796)^{2/3} - 125\right\}(13.4)^{1/3}(0.73/0.85)^{0.14}\left\{1 + [0.584/(12 \times 12)]^{2/3}\right\}$$

$$Nu = 64.66$$

$$h_{LO} = Nu\,(k_L/D_i) = 64.66 \times 0.095/(0.584/12)$$

$$h_{LO} \cong 126 \text{ Btu/h·ft}^2\cdot{}^\circ\text{F}$$

The Müller–Steinhagen and Heck (MSH) correlation will be used to calculate the two-phase multiplier, ϕ^2_{LO} . First, the Chisholm parameter is calculated from Equation (9.42) with $n = 0.2585$ for turbulent flow in heat-exchanger tubes:

$$Y = (\rho_L/\rho_V)^{0.5}(\mu_V/\mu_L)^{n/2} = (49/0.123)^{0.5}(0.0095/0.73)^{0.1293}$$

$$Y = 11.39$$

The two-phase multiplier is given by Equation (9.35):

$$\phi^2_{LO} = Y^2x^3 + \left[1 + 2x(Y^2 - 1)\right](1-x)^{1/3}$$

$$= (11.39)^2(0.06)^3 + \left\{1 + 2 \times 0.06\left((11.39)^2 - 1\right)\right\}(0.94)^{1/3}$$

$$\phi^2_{LO} = 16.14$$

The heat-transfer coefficient for zone I is obtained from Equation (11.60):

$$h_I = h_{LO}\left(\phi^2_{LO}\right)^{0.45} = 126\,(16.14)^{0.45} = 440 \text{ Btu/h·ft}^2\cdot{}^\circ\text{F}$$

Finally, the heat-transfer coefficient for the transition region is estimated by Equation (11.62):

$$h = h_I + (j^* - 1.5)(h_I - h_{II})$$

$$= 440 + (1.34 - 1.5)(440 - 316)$$

$$h = 420 \text{ Btu/h·ft}^2\cdot{}^\circ\text{F}$$

Example 11.5

Compare the correlations of Breber et al., Boyko and Kruzhilin, and Shah for the conditions of Example 11.4. The critical pressure of *n*-propyl alcohol is 51.0 atm.

Solution

All three correlations give the heat-transfer coefficient as the product of h_{LO} and an effective two-phase multiplier for heat transfer. For the correlation of Breber et al., the multiplier is:

$$\left(\phi_{LO}^2\right)^{0.45} = (16.14)^{0.45} = 3.50$$

However, this correlation includes a built-in safety factor for design work that the other two correlations lack. Hence, using an exponent of 0.5 in this method provides a fairer comparison:

$$\left(\phi_{LO}^2\right)^{0.5} = (16.14)^{0.5} = 4.02$$

The safety factor in this case amounts to roughly 15%.

From Equation (11.55), the multiplier in the Boyko–Kruzhilin correlation is:

$$[1 + x(\rho_L - \rho_V)/\rho_V]^{0.5} = [1 + 0.06(49 - 0.123)/0.123]^{0.5} = 4.98$$

From Equation (11.61), the multiplier for the Shah correlation is:

$$(1-x)^{0.8} + 3.8x^{0.76}(1-x)^{0.04}p_r^{-0.38} = (0.94)^{0.8} + 3.8(0.06)^{0.76}(0.94)^{0.04}(1/51.0)^{-0.38} = 2.94$$

Using the value of $h_{LO} = 126$ Btu/h · ft^2 · °F from Example 11.4, the two-phase heat-transfer coefficients for the Breber et al. and Boyko–Kruzhilin methods are as shown in the table below. The Shah correlation includes very specific instructions for calculating h_{LO}: use the Dittus–Boelter correlation with all fluid properties evaluated at the saturation temperature. At $T_{sat} = 207$°F, we have $\mu_L = 0.5$ cp and $C_{P,L} = 0.77$ Btu/lbm · °F. Using these values gives:

$$Re_{LO} = D_i G/\mu_L = (0.584/12) \times 246,598/(0.5 \times 2.419) = 9922$$

$$Pr_L = C_{P,L}\mu_L/k_L = 0.77 \times 0.5 \times 2.419/0.095 = 9.80$$

$$h_{LO} = 0.023\, Re_{LO}^{0.8} Pr_L^{0.4}(k_L/D_i) = 0.023(9922)^{0.8}(9.80)^{0.4} \times 0.095/(0.584/12)$$

$$h_{LO} = 176 \text{ Btu/h} \cdot \text{ft}^2 \cdot {}^\circ\text{F}$$

$$h = 176 \times 2.94 = 517 \text{ Btu/h} \cdot \text{ft}^2 \cdot {}^\circ\text{F}$$

The results are summarized in the following table.

Method	h (Btu/h · ft^2 · °F)
Breber et al., exponent = 0.45	440
Breber et al., exponent = 0.50	506
Boyko–Kruzhilin	627
Shah	517

With no safety factor, the predictions of the three methods agree to within 25%. The close agreement between results from the Shah and Breber et al. methods is probably fortuitous, as the uncertainties associated with all methods of this type are substantial. For design work, a safety factor of 15% to 25% appears warranted.

11.7 Condensation on Finned Tubes

Horizontal shell-side condensers equipped with radial low-fin tubes are commonly used to condense refrigerants and other organic fluids. The surface tension of these condensates is typically less than about 35 dyne/cm. With fluids having elevated surface tension, the condensate may bridge the gap between fins, resulting in poor condensate drainage, and heat-transfer coefficients that are lower than for plain tubes. Hence, fin spacing is an important consideration for these condensates. For example, with condensing steam ($\sigma_{water} \cong 60$ dyne/cm at 100°C and atmospheric pressure) a fin density no greater than 16 fpi should be used [24].

Beatty and Katz [25] studied the condensation of propane, *n*-butane, *n*-pentane, sulfur dioxide, methyl chloride, and Freon-22 on single horizontal finned tubes, and developed a modified Nusselt correlation that fit the experimental data to within ±10%. The correlation is based on an equivalent diameter that allows the heat transfer from both fins and prime surface to be represented by a single average heat-transfer coefficient. The correlation can be stated as follows:

$$h = 0.689\left[\frac{k_L^3\rho_L(\rho_L-\rho_V)g\lambda}{\mu_L\Delta T_f D_e}\right]^{1/4} \tag{11.64}$$

Here ΔT_f is the temperature difference across the condensate film and the equivalent diameter, D_e, is defined by the following equation:

$$D_e^{-0.25} = \frac{1.30\ \eta_f A_{fins}E^{-0.25} + A_{prime}D_r^{-0.25}}{\eta_w A_{Tot}} \tag{11.65}$$

where

η_f = fin efficiency
η_w= weighted efficiency of finned surface
A_{fins} = Area of all fins
$A_{Tot} = A_{fins} + A_{prime}$
D_r = root-tube diameter
$E = \pi(r_2^2 - r_1^2)/(2r_2)$
r_2 = fin radius
$r_1 = D_r/2$ = root-tube radius

The rate of heat transfer across the condensate film is given by:

$$q = W\lambda = hn_t\eta_w A_{Tot}\Delta T_f \tag{11.66}$$

Rearranging this equation gives:

$$\lambda/\Delta T_f = hn_t\eta_w A_{Tot}/W = h\eta_w A_{Tot}/(\Gamma L) \tag{11.67}$$

Here, $\Gamma - W/n_tL$ is the condensate loading per tube. Substituting this result in Equation (11.64) gives:

$$h = 0.689\left[\frac{k_L^3\rho_L(\rho_L-\rho_V)g\eta_w(A_{Tot}/L)}{\mu_L D_e\Gamma}\right]^{1/4} h^{1/4}$$

Solving for *h*, we obtain:

$$h = 0.609\left[\frac{k_L^3\rho_L(\rho_L-\rho_V)g\eta_w(A_{Tot}/L)}{\mu_L D_e\Gamma}\right]^{1/3} \tag{11.68}$$

This equation is valid for a single row of tubes. For a horizontal tube bundle, Γ is replaced by the effective loading, $\Gamma^* = W/Ln_t^{2/3}$.

Condensing coefficients for finned tubes tend to be substantially higher than for plain tubes, provided there is good condensate drainage from the finned surface. However, the Beatty-Katz correlation does not account for the effect of surface tension on condensate drainage. As a result, it may over-predict the heat-transfer coefficient at very small fin spacing if the surface tension of the condensate is relatively high. Correlations that include the effects of surface tension are discussed by Kraus et al. [24], who also present a method for estimating the minimum fin spacing that is compatible with good drainage for a given condensate.

11.8 Pressure Drop

The pressure change in a condensing fluid is comprised of three parts, namely, the static head, the momentum change, and the friction loss. Since the fluid velocity decreases from inlet to outlet in a condenser, the momentum change results in a pressure gain rather than a loss. This effect is generally small, and with the exception of vacuum operations, can be safely neglected.

For condensation inside vertical and horizontal tubes, the friction loss can be calculated using the methods given in Chapter 9 for two-phase flow. Since the quality usually changes greatly from inlet to outlet, an incremental or zone analysis is required for rigorous calculations. If desuperheating or subcooling zones are present, they must be treated separately as single-phase flow regimes. In calculating the friction loss, the effect of mass transfer due to condensation is neglected, although it can be significant in some circumstances [26]. For vertical units, calculation of the static head effect involves integration of the two-phase density over the length of the condensing zone. The procedure is essentially the same as that used in the analysis of vertical thermosyphon reboilers in Chapter 10. The contributions from desuperheating and subcooling zones must be added. For horizontal tube-side condensers, the static head is essentially determined by the height of liquid in the outlet header, and the corresponding pressure difference can usually be neglected.

TABLE 11.2 Values of Parameters in Chisholm Correlation for Calculating Friction Losses in Shell-side Condensers

Flow Geometry	Flow Regimes	B	n
Cross flow, vertical	Spray, bubbly	1.0	0.37
Cross flow, horizontal	Spray, bubbly	0.75	0.46
Cross-flow, horizontal	Stratified, stratified spray	0.25	0.46
Baffle window, vertical cut*	All	$2/(Y+1)$	0
Baffle window, horizontal cut**	All	ρ_{hom} / ρ_L	0

*Side-to-side flow pattern

**Up-and-down flow pattern

A method for calculating the friction loss in shell-side condensers based on the Chisholm correlation is presented in Refs. [1,2]. The Chisholm correlation for the two-phase pressure gradient multiplier was presented in Chapter 9 and is repeated here for convenience.

$$\phi_{LO}^2 = 1 + (Y^2 - 1)\left\{B[x(1-x)]^{(2-n)/2} + x^{2-n}\right\} \tag{9.43}$$

Values of the parameters B and n to be used for flow across tube banks and in baffle windows are given in Table 11.2. The homogeneous two-phase density, ρ_{hom}, appearing in this table is calculated according to Equation (9.51).

The flow regime can be determined using flow pattern maps given in Ref. [2]. However, the accuracy achievable with this procedure is limited, and hence, a conservative alternative is suggested here. For horizontal cross flow, use the values for the spray and bubbly flow regimes when vapor shear is significant and use the values for the stratified flow regimes when it is not. As discussed under "Condensation outside horizontal tubes," the significance of vapor shear can be determined using Equation (11.56). For vertical cross-flow characteristic of X-shell condensers, use the values for spray and bubbly flow in all situations, as no alternative is available. Note that for baffled E- and J-shell condensers, the two-phase multipliers must be applied individually in the cross-flow zones between baffle tips and in the baffle windows. Therefore, the method must be used in conjunction with either the Stream Analysis method or the Delaware method for single-phase pressure drop, both of which calculate the individual pressure drops for cross-flow and window zones. Combined with an incremental analysis, such a procedure obviously requires computer implementation.

An approximate method suitable for hand calculations is based on the pressure drop, $(\Delta P_f)_{VO}$, calculated for the total flow as vapor at inlet conditions:

$$\Delta P_f = \overline{\phi}_{VO}^2 (\Delta P_f)_{VO} \tag{11.69}$$

For shell-side condensation with a saturated vapor feed, Bell and Mueller [27] present a graph for the average two-phase multiplier, $\overline{\phi}_{VO}^2$, as a function of the exit vapor fraction, x_e. The following equation was obtained by regression analysis using values read from the graph:

$$\overline{\phi}_{VO}^2 = 0.33 + 0.22x_e + 0.61x_e^2 \quad (0 \le x_e \le 0.95) \tag{11.70}$$

Although based on experimental data, this correlation involves a number of assumptions [27], including constant condensation rate throughout the tube bundle. Note that for a total condenser, $x_e = 0$ and Equation (11.70) gives $\overline{\phi}_{VO}^2 = 0.33$. Kern and Kraus [28] noted that if the vapor velocity varies linearly from inlet to outlet, then the two-phase multiplier should be equal to 1/3 for a total condenser, which is consistent with Equation (11.70). However, they also noted that in actuality the multiplier tends to be slightly higher, and recommended a value of 0.5 as a conservative approximation for total condensers. If a more conservative estimate is desired for a partial condenser, the larger of 0.5 and the value given by Equation (11.70) can be used.

For tube-side condensation of saturated vapors, the following equation can be used to estimate the average two-phase multiplier [29]:

$$\overline{\phi}_{VO}^2 = 0.5(1 + u_{V,out}/u_{V,in}) \tag{11.71}$$

where $u_{V,in}$ and $u_{V,out}$ are the vapor velocities at the condenser inlet and outlet, respectively. For a total condenser, Equation (11.71) reduces to $\overline{\phi}_{VO}^2 = 0.5$. In fact, this method was used to estimate the pressure drop for condensing steam in Example 10.2.

11.9 Mean Temperature Difference

When one stream in a heat exchanger is isothermal, the LMTD correction factor is equal to 1.0 regardless of the flow pattern. This situation is closely approximated in the condensation of a saturated pure-component vapor because the pressure drop is generally small on the condensing side. In this case, the mean temperature difference in the condenser is simply the LMTD.

When a significant desuperheating or subcooling zone exists in the condenser, or when a multi-component vapor is condensed, the temperature of the condensing stream usually varies substantially from inlet to outlet. Furthermore, the stream enthalpy varies nonlinearly with temperature in these situations, so the mean temperature difference is not equal to $F(\Delta T_{ln})_{cf}$ as in a single-phase heat exchanger. (In a single-phase exchanger with constant stream heat capacities, the specific enthalpy, H, of each stream is linear in temperature since it satisfies $\Delta H = C_P \Delta T$. This is one of the fundamental assumptions upon which the methodology for single-phase exchangers is based.) The upshot is that a zone analysis is generally required for these cases, and with multiple coolant passes this is an iterative procedure suitable for computer implementation but not easily adapted to hand calculation.

In certain types of application, simplifying assumptions can be made to avoid a zone analysis and facilitate hand calculations. Two of these are the following:

- For shell-side condensation of a pure-component vapor in a horizontal E-shell unit, it can be assumed that all the heat is transferred at the saturation temperature in the shell. Provided that a condensate film exists over the entire desuperheating section, the temperature at the vapor-liquid interface will be T_{sat}, the same as in the condensing section. Dry-wall desuperheating is possible if the tube wall temperature exceeds T_{sat}. However, as discussed by Rubin [30], a dry desuperheating zone is unlikely to exist in a standard E-shell condenser due to mixing of the inlet vapor with condensate in the shell. Therefore, the above assumption is generally acceptable for this type of unit. It is reasonable for other types of condensers as well, provided that conditions permitting a dry desuperheating zone do not exist.
- For narrow-boiling mixtures, the condensing curve (a plot of stream temperature versus the amount of heat removed from the stream, equivalent to a temperature-enthalpy plot) may be approximately linear. Figure 11.16 shows the condensing curve for a mixture of 30 mole percent *n*-butane and 70 mole percent *n*-pentane at a pressure of 75 psia. Although some curvature is present, a straight line provides a reasonable approximation to the graph. If such a mixture is condensed without substantial desuperheating or subcooling zones, the LMTD correction factor can be used to estimate the mean temperature difference, as in a single-phase heat exchanger. For E-shell units, the method given in Chapter 3 is used to calculate the LMTD correction factor. Procedures for J- and X-shell units are given in Appendix 11.A.

For situations in which a zone analysis is unavoidable, Gulley [31] presented an approximate method for calculating a weighted mean temperature difference in multi-pass condensers that does not require iteration, and hence, is amenable to hand calculation. Even with this simplification, calculations for multi-component, multi-zone condensers can be very laborious, and will not be considered further here.

When using the above approximation for condensing mixtures, Bell and Mueller [27] recommend the following procedure to account for the thermal resistance due to the sensible heat transfer involved in cooling the vapor over the condensing range. The overall coefficient, U_D, is replaced with a modified design coefficient,U'_D, defined by:

$$U'_D = \left[U_D^{-1} + (q_{sen}/q_{Tot})h_V^{-1}\right]^{-1} \tag{11.72}$$

where

q_{sen} = sensible heat duty for vapor cooling
q_{Tot} = total duty
h_V = heat-transfer coefficient for vapor calculated using the average vapor flow rate

The sensible heat duty is estimated using the average heat capacity, $\overline{C}_{P,V}$, and average flow rate of the vapor:

$$q_{sen} = 0.5\overline{C}_{P,V}\left(\dot{m}_{V,in} + \dot{m}_{V,out}\right)\left(T_{V,in} - T_{V,out}\right) \tag{11.73}$$

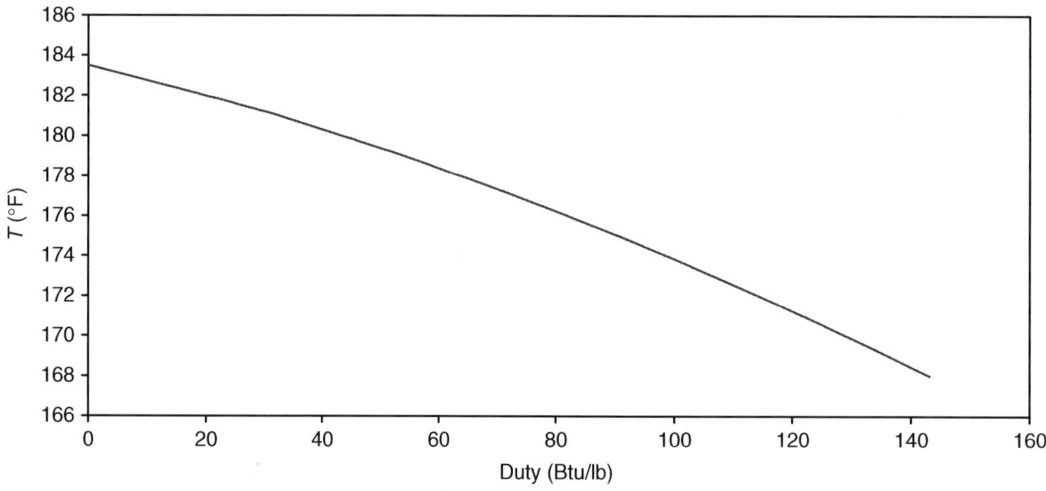

FIGURE 11.16 Condensing curve for mixture of butane and pentane.

Example 11.6

180,000 lb/h of a saturated vapor consisting of 30 mole percent *n*-butane and 70 mole percent *n*-pentane are to be condensed at a pressure of 75 psia. The condensing range at feed conditions is from 183.5°F to 168°F, and the enthalpy difference between saturated vapor and saturated liquid is 143 Btu/lbm. Cooling water is available at 85°F and 60 psia. Tubing material is specified to be 90/10 copper–nickel alloy ($k = 30$ Btu/h · ft · °F), and a fouling factor of 0.001 h · ft^2 · °F/Btu is recommended for the cooling water. Maximum pressure drops of 5 psi for the hydrocarbon stream and 10 psi for the coolant are specified. Physical properties of the feed and condensate are given in the table below. The viscosity of the condensate can be estimated using the following equation:

$$\mu_L(\text{cp}) = 0.00941 \exp[1668/T(°\text{R})]$$

The other condensate properties may be assumed constant at the tabled values. Design a condenser for this service using plain (un-finned) tubes.

Property	Feed	Condensate
C_P (Btu/lbm · °F)	0.486	0.61
k (Btu/h · ft · °F)	0.0119	0.057
μ (cp)	0.0085	–
ρ (lbm/ft^3)	0.845	35.5
Pr	0.84	–

Solution

(a) Make initial specifications.

(i) Fluid placement

A horizontal shell-side condenser will be used. Therefore, the condensing vapor is placed in the shell and cooling water in the tubes.

(ii) Shell and head types

An E-shell unit is specified since it is generally the most economical type of condenser. The difference between the inlet temperatures of the two streams is nearly 100°F at design conditions, and could be significantly higher if the temperature of the cooling water decreases during cold weather. Therefore, U-tubes are specified to allow for differential thermal expansion. Thus, an AEU exchanger is specified.

(iii) Tubing

For water service, 3/4-in., 16-BWG tubes are selected with a length of 16 ft.

(iv) Tube layout

Since the shell-side fluid is clean, triangular pitch is specified. A tube pitch of either 15/16 or 1.0 in. can be used. The smaller value is selected because it provides more heat-transfer surface for a given shell size. (See tube-count tables in Appendix C.)

(v) Baffles

Segmental baffles with a spacing of 0.4 times the shell diameter and a cut of 35% are appropriate for a condensing vapor (see Figure 5.3).

(vi) Sealing strips

The bundle-to-shell clearance in U-tube exchangers is generally small. However, impingement protection is required for a condenser, and some tubes will have to be omitted from the bundle to provide adequate flow area above the impingement plate for the entering vapor. Depending on the size of the resulting gap in the tube bundle, sealing strips may be needed to block the bundle bypass flow.

(vii) Construction materials

The 90/10 cupro–nickel alloy specified for the tubes will provide corrosion resistance and allow a maximum water velocity of 10 ft/s (Table 5.B1). For compatibility, this material is also specified for the tubesheets. Plain carbon steel is adequate for the shell, heads, and all other components.

(b) Energy balances.

The duty is calculated using the given enthalpy difference between saturated vapor and saturated liquid. Condensate subcooling is neglected, as is the effect of pressure drop on saturation temperature:

$$q = \dot{m}\Delta H = 180{,}000 \times 143 = 25{\cdot}74 \times 10^6 \text{Btu/h}$$

Neither the outlet temperature nor the flow rate of the cooling water is specified in this problem. The outlet temperature is frequently limited to 110°F to 125°F in order to minimize fouling due to deposition of minerals contained in the water. Economic considerations are also involved, since a higher outlet temperature reduces the amount of cooling water used (which tends to lower the operating cost), while also lowering the mean temperature difference in the exchanger (which tends to increase the capital cost). For the purpose of this example, an outlet temperature of 120°F will be used, giving an average water temperature of 102.5°F. The properties of water at this temperature are as follows:

Property	Water at 102.5 °F
C_P (Btu/lbm · °F)	1.0
k (Btu/h · ft · °F)	0.37
μ (cp)	0.72
Specific gravity	0.99
Pr	4.707

The cooling water flow rate is obtained from the energy balance:

$$q = 25.74 \times 10^6 = (\dot{m}C_P\Delta T)_{water} = \dot{m}_{water} \times 1.0 \times 35$$

$$\dot{m}_{water} = 735{,}429 \text{ lbm/h}$$

(c) Mean temperature difference.
Since the condensing curve for this system (Figure 11.16) is approximately linear, the mean temperature difference is estimated as follows:

$$\Delta T_m \cong F(\Delta T_{\ln})_{cf}$$

$$(\Delta T_{\ln})_{cf} = \frac{83 - 63.5}{\ln(83/63.5)} = 72.8°\text{F}$$

$$R = \frac{T_a - T_b}{t_b - t_a} = \frac{183.5 - 168}{120 - 85} = 0.443$$

$$P = \frac{t_b - t_a}{T_a - t_a} = \frac{120 - 85}{183.5 - 85} = 0.355$$

From Figure 3.14, $F \cong 0.98$. Therefore,

$$\Delta T_m \cong 0.98 \times 72.8 = 71.3°\text{F}$$

(d) Approximate overall heat-transfer coefficient.
From Table 3.5, for a condenser with low-boiling hydrocarbons on the shell side and cooling water on the tube side, $80 \leq U_D \leq 200$ Btu/h · ft² · °F. Taking the mid-range value gives $U_D = 140$ Btu/h · ft² · °F.

(e) Heat-transfer area and number of tubes

$$A = \frac{q}{U_D \Delta T_m} = \frac{25.74 \times 10^6}{140 \times 71.3} = 2579 \text{ ft}^2$$

$$n_t = \frac{A}{\pi D_o L} = \frac{2579}{\pi(0.75/12) \times 16} = 821$$

(f) Number of tube passes.
$D_i = 0.620$ in. $= 0.0517$ ft (Table B.1)

$$Re = \frac{4\dot{m}(n_p/n_t)}{\pi D_i \mu} = \frac{4 \times 735{,}429(n_p/821)}{\pi \times 0.0517 \times 0.72 \times 2.419} = 12{,}666\, n_p$$

Therefore, two tube passes will suffice to give fully turbulent flow in the tubes. Checking the fluid velocity,

$$V = \frac{\dot{m}(n_p/n_t)}{\rho \pi D_i^2/4} = \frac{(735{,}429/3600)(2/821)}{0.99 \times 62.43 \times \pi(0.0517)^2/4} = 3.84 \text{ ft/s}$$

The velocity is acceptable, and therefore two passes will be used. (Note that four passes could also be used, giving a velocity of about 7.7 ft/s.)

(g) Shell size and tube count.
From Table C.2, the closest count is 846 tubes in a 31-in. shell.

(h) Required overall coefficient.

$$U_{req} = \frac{q}{n_t \pi D_o L \Delta T_m} = \frac{25.74 \times 10^6}{846 \times \pi \times (0.75/12) \times 16 \times 71.3} = 136 \text{ Btu/h·ft}^2\text{·°F}$$

(i) Calculate h_i assuming $\phi_i = 1.0$.

$$Re = \frac{4\dot{m}(n_p/n_t)}{\pi D_i \mu} = \frac{4 \times 735{,}429(2/846)}{\pi \times 0.0517 \times 0.72 \times 2.419} = 24{,}584$$

$$h_i = (k/D_i) \times 0.023\, Re^{0.8} Pr^{1/3} (\mu/\mu_w)^{0.14}$$

$$= (0.37/0.0517) \times 0.023(24{,}584)^{0.8}(4.707)^{1/3} \times 1.0$$

$$h_i = 898 \text{ Btu/h·ft}^2\text{·°F}$$

(j) Calculate h_o.
The basic Nusselt theory will be used to obtain a conservative estimate for the condensing heat-transfer coefficient. A tube wall temperature of 125°F is assumed to start the calculation. The average film temperature is calculated using the average vapor temperature of 175.75°F:

$$T_f = 0.75T_w + 0.25T_V = 0.75 \times 125 + 0.25 \times 175.75 = 137.7\text{°F}$$

The condensate viscosity at this temperature is:

$$\mu_L = 0.00941 \exp\left[1668/(137.7 + 460) = 0.153 \text{ cp}\right]$$

The modified condensate loading is calculated from Equation (11.41):

$$\Gamma^* = \frac{W}{L n_t^{2/3}} = \frac{180{,}000}{16(846)^{2/3}} = 125.77 \text{ lbm/h·ft}$$

The heat-transfer coefficient is given by Equation (11.40):

$$h_o = 1.52\left[\frac{k_L^3 \rho_L(\rho_L - \rho_V)g}{4\mu_L \Gamma^*}\right]^{1/3} = 1.52\left[\frac{(0.057)^3 \times 35.5(35.5 - 0.845) \times 4.17 \times 10^8}{4 \times 0.153 \times 2.419 \times 125.77}\right]^{1/3}$$

$$h_o = 121 \text{ Btu/h·ft}^2\text{·°F}$$

The shell-side coefficient is less than U_{req}, indicating that the condenser is severely under-sized. Therefore, no further calculations will be made with the initial configuration.
The low value of h_o suggests that the value of U_D is likely to be less than 100 Btu/h · ft^2 · °F. The number of tubes needed to give U_{req} a value of, say, 90 Btu/h · ft^2 · °F is:

$$(n_t)_{req} = 846(136/90) = 1278$$

With this many tubes, four tube passes will be required to keep the water velocity above 3 ft/s. Referring to the tube-count table, a 39-in. shell containing 1336 tubes is selected for the next trial.

Second Trial

(a) Required overall coefficient.

$$U_{req} = \frac{q}{n_t \pi D_o L \Delta T_m} = \frac{25.74 \times 10^6}{1336 \times \pi(0.75/12) \times 16 \times 71.3} = 86 \text{ Btu/h·ft}^2\text{·°F}$$

(b) Calculate h_i assuming $\phi_i = 1.0$.

$$Re = \frac{4\dot{m}(n_p/n_t)}{\pi D_i \mu} = \frac{4 \times 735{,}429(4/1336)}{\pi \times 0.0517 \times 0.72 \times 2.419} = 31{,}135$$

$$h_i = (k/D_i) \times 0.023 Re^{0.8} Pr^{1/3} (\mu/\mu_w)^{0.14}$$

$$= (0.37/0.0517) \times 0.023(31{,}135)^{0.8}(4.707)^{1/3} \times 1.0$$

$$h_i = 1085 \text{ Btu/h}\cdot\text{ft}^2\cdot{}^\circ\text{F}$$

(c) Calculate h_o.

As before, a wall temperature of 125°F is assumed, giving an average film temperature of 137.7°F, at which $\mu_L = 0.153$ cp.

$$\Gamma^* = \frac{W}{Ln_t^{2/3}} = \frac{180{,}000}{16(1336)^{2/3}} = 92.74 \text{ lbm/h}\cdot\text{ft}$$

$$h_o = 1.52\left[\frac{(0.057)^3 \times 35.5(35.5 - 0.845) \times 4.17 \times 10^8}{4 \times 0.153 \times 2.419 \times 92.74}\right]^{1/3}$$

$$h_o = 134 \text{ Btu/h}\cdot\text{ft}^2\cdot{}^\circ\text{F}$$

(d) Calculate T_w and T_f.

$$T_w = \frac{h_i t_{ave} + h_o(D_o/D_i)T_{ave}}{h_i + h_o(D_o/D_i)} = \frac{1085 \times 102.5 + 134(0.75/0.62) \times 175.75}{1085 + 134(0.75/0.62)}$$

$$T_w = 112^\circ\text{F}$$

$$T_f = 0.75 \times 112 + 0.25 \times 175.75 = 128^\circ\text{F}$$

(e) Recalculate h_o.

The value of μ_L at 128°F is:

$$\mu_L = 0.00941 \exp[1668/(128 + 460)] = 0.161 \text{ cp}$$

The corresponding value of h_o is:

$$h_o = 1.52\left[\frac{(0.057)^3 \times 35.5(35.5 - 0.845) \times 4.17 \times 10^8}{4 \times 0.161 \times 2.419 \times 92.74}\right]^{1/3} = 132 \text{ Btu/h}\cdot\text{ft}^2\cdot{}^\circ\text{F}$$

Since this value is close to the previous one, no further iterations are required.

(f) Viscosity correction factor for cooling water.

Since the tube wall temperature is close to the bulk water temperature, the viscosity correction factor is neglected.

(g) Overall coefficient.

The recommended fouling factor for the cooling water was given in the problem statement as 0.001 h · ft^2 · °F/Btu. Since the condensing vapor is a very clean stream, a fouling factor of 0.0005 h · ft^2 · °F/Btu is appropriate. Thus,

$$U_D = \left[\frac{D_o}{h_i D_i} + \frac{D_o \ln(D_o/D_i)}{2\,k_{tube}} + \frac{1}{h_o} + \frac{R_{Di} \times D_o}{D_i} + R_{Do}\right]^{-1}$$

$$= \left[\frac{0.75}{1085 \times 0.62} + \frac{(0.75/12)\ln(0.75/0.62)}{2 \times 30} + \frac{1}{132} + \frac{0.001 \times 0.72}{0.62} + 0.0005\right]^{-1}$$

$$U_D = 94 \text{ Btu/h}\cdot\text{ft}^2\cdot{}^\circ\text{F}$$

(h) Correction for sensible heat transfer.

The effect of interfacial shear was neglected in calculating h_o, which will more than offset the effect of sensible heat transfer. Hence, this step could be omitted, but is included here to illustrate the procedure. The rate of sensible heat transfer is estimated using Equation (11.73). For convenience, the heat capacity at vapor inlet conditions is used as an approximation for $\overline{C}_{P,V}$:

$$q_{sen} = 0.5\overline{C}_{P,V}\left(\dot{m}_{V,in} + \dot{m}_{V,out}\right)\left(T_{V,in} - T_{V,out}\right)$$

$$= 0.5 \times 0.486 \times 180{,}000(183.5 - 168)$$

$$q_{sen} = 677{,}970 \text{ Btu/h}$$

$$q_{sen}/q_{Tot} = 677{,}970/25.74 \times 10^6 \cong 0.026$$

The Simplified Delaware method will be used to calculate h_V. Since the baffle cut is greater than 20%, this method will tend to overestimate h_V. However, the safety factor built into the method will help to compensate for this error:

$$B = 0.4\,d_s = 0.4 \times 39 = 15.6 \text{ in.}$$

$$C' = P_T - D_o = 15/16 - 0.75 = 0.1875 \text{ in.}$$

$$a_s = \frac{d_s C' B}{144 P_T} = \frac{39 \times 0.1875 \times 15.6}{144 \times 15/16} = 0.845 \text{ ft}^2$$

$$D_e = 0.55/12 = 0.04583 \text{ ft} \quad \text{(from Figure 3.17)}$$

The average vapor flow rate of 90,000 lb/h is used to calculate h_V. Physical properties of the vapor are taken at inlet conditions for convenience:

$$G = \dot{m}/a_s = 90{,}000/0.845 = 106{,}509 \text{ lbm/h}\cdot\text{ft}^2$$

$$Re = \frac{D_e G}{\mu_V} = \frac{0.04583 \times 106{,}509}{0.0085 \times 2.419} = 237{,}400$$

$$j_H = 0.5(1 + B/d_s)\left(0.08 Re^{0.6821} + 0.7 Re^{0.1772}\right)$$

$$= 0.5(1 + 0.4)\left\{0.08(237{,}400)^{0.6821} + 0.7(237{,}400)^{0.1772}\right\}$$

$$j_H = 264.3$$

$$h_V = j_H (k/D_e) Pr^{1/3} (\mu/\mu_w)^{0.14}$$

$$= 264.3(0.0119/0.04583)(0.84)^{1/3}(1.0)$$

$$h_V = 65 \text{ Btu/h}\cdot\text{ft}^2\cdot{}^\circ\text{F}$$

The modified overall coefficient is given by Equation (11.72):

$$U'_D = \left[U_D^{-1} + (q_{sen}/q_{Tot})\, h_V^{-1}\right]^{-1}$$

$$= \left[(94)^{-1} + 0.026(65)^{-1}\right]^{-1}$$

$$U'_D = 91 \text{ Btu/h}\cdot\text{ft}^2\cdot{}^\circ\text{F}$$

Since $U'_D > U_{req}$, the condenser is thermally acceptable

(i) Tube-side pressure drop.

$$f = 0.4137 Re^{-0.2585} = 0.4137(31{,}135)^{-0.2585} = 0.0285$$

$$G = \frac{\dot{m}(n_p/n_t)}{A_f} = \frac{735{,}429(4/1336)}{(\pi/4)(0.0517)^2} = 1{,}048{,}874 \text{ lbm/h}\cdot\text{ft}^2$$

$$\Delta P_f = \frac{f\, n_p L G^2}{7.50 \times 10^{12} D_i s \phi} = \frac{0.0285 \times 4 \times 16(1{,}048{,}874)^2}{7.50 \times 10^{12} \times 0.0517 \times 0.99 \times 1.0}$$

$$\Delta P_f = 5.23 \text{ psi}$$

$$\Delta P_r = 1.334 \times 10^{-13} \alpha_r G^2/s$$

From Table 5.1, for turbulent flow in U-tubes:

$$\alpha_r = 1.6 n_p - 1.5 = 1.6 \times 4 - 1.5 = 4.9$$

Hence,

$$\Delta P_r = 1.334 \times 10^{-13} \times 4.9(1{,}048{,}874)^2/0.99 = 0.75 \text{ psi}$$

Table 5.3 indicates that 10-in. nozzles are appropriate for this unit. Assuming that schedule 40 pipe is used, the Reynolds number for the nozzles is:

$$Re_n = \frac{4\dot{m}}{\pi D_i \mu} = \frac{4 \times 735{,}429}{\pi(10.02/12) \times 0.72 \times 2.419} = 643{,}867$$

$$G_n = \frac{\dot{m}}{(\pi/4)D_i^2} = \frac{735{,}429}{(\pi/4)(10.02/12)^2} = 1{,}343{,}006\ \text{lbm/h}\cdot\text{ft}^2$$

Since the flow is turbulent, Equation (5.4) is used to estimate the pressure loss in the nozzles:

$$\Delta P_n = 2.0\times10^{-13}N_sG_n^2/s = 2.0\times10^{-13}\times1(1{,}343{,}006)^2/0.99$$

$$\Delta P_n = 0.36\ \text{psi}$$

The total tube-side pressure drop is:

$$\Delta P_i = \Delta P_f + \Delta P_r + \Delta P_n = 5.23 + 0.73 + 0.36 \cong 6.3\ \text{psi}$$

(j) Shell-side pressure drop.
Equation (11.69) will be used to estimate the pressure drop, and the Simplified Delaware method will be used to calculate $(\Delta P_f)_{VO}$. Due to the relatively large baffle cut, the Simplified Delaware method will tend to overestimate the actual pressure drop. (The full Delaware method could be used to obtain a more accurate estimate of the pressure drop, but the additional computational effort is not justified in the present context.) From step (h) we have:

$$B = 15.6\ \text{in.}$$

$$a_s = 0.845\ \text{ft}^2$$

$$D_e = 0.04583\ \text{ft}$$

$$G = \dot{m}_V/a_s = 180{,}000/0.845 = 213{,}018\ \text{lbm/h}\cdot\text{ft}^2$$

$$Re = \frac{D_eG}{\mu_V} = \frac{0.04583\times213{,}018}{0.0085\times2.419} = 474{,}801$$

The friction factor is calculated using Equations (5.7) and (5.9). Note that since the shell diameter exceeds 23.25 in., d_s is set to 23.25 in Equation (5.9):

$$f_1 = (0.0076 + 0.000166\,d_s)Re^{-0.125}$$

$$= (0.0076 + 0.000166\times39)(474{,}801)^{-0.125}$$

$$f_1 = 0.00275\ \text{ft}^2/\text{in.}^2$$

$$f_2 = \left(0.0016 + 5.8\times10^{-5}\,d_s\right)Re^{-0.157}$$

$$= \left(0.0016 + 5.8\times10^{-5}\times23.25\right)(474{,}801)^{-0.157}$$

$$f_2 = 0.000379\ \text{ft}^2/\text{in.}^2$$

$$f = 144\{f_1 - 1.25(1 - B/d_s)(f_1 - f_2)\}$$

$$= 144\{0.00275 - 1.25(1 - 0.4)(0.00275 - 0.000379)\}$$

$$f = 0.140$$

The number of baffle spaces is estimated as:

$$n_b + 1 = L/B = \frac{16\times12}{15.6} \cong 12.3 \Rightarrow 12$$

The pressure drop for the total flow as vapor is calculated from Equation (5.6):

$$(\Delta P_f)_{VO} = \frac{f\,G^2d_s(n_b+1)}{7.50\times10^{12}D_es\phi} = \frac{0.140(213{,}018)^2(39/12)(12)}{7.50\times10^{12}\times0.04583(0.845/62.43)(1.0)}$$

$$(\Delta P_f)_{VO} = 53.3\ \text{psi}$$

From Equation (11.70), $\overline{\phi}_{VO}^2 = 0.33$ for a total condenser. Therefore,

$$\Delta P_f = \overline{\phi}_{VO}^2(\Delta P_f)_{VO} = 0.33\times53.3 = 17.6\ \text{psi}$$

The shell-side pressure drop greatly exceeds the allowable value of 5 psi. Therefore, we consider increasing the baffle spacing. From Appendix 5.C, the maximum unsupported tube length for 3/4-in. copper alloy tubes is 52 in. Since the tubes in the baffle windows are supported by every other baffle, the maximum allowable baffle spacing is 26 in., which is less than the shell diameter. So in this case, the baffle spacing is limited by tube support considerations. When adjusted for an integral number of baffles, the maximum spacing is decreased to 24 in., and this is too small to reduce the pressure drop to the required level.

Next we consider using a split-flow (type J) shell. The inlet vapor stream is divided into two parts that are fed to opposite ends of the shell and flow toward the center. Since both the flow rate and length of the flow path are halved, the shell-side pressure drop is reduced by a factor of approximately eight compared with an E-shell of the same size. Therefore, for the third trial an AJU exchanger with a 39-in. shell, 1336 tubes and four tube passes is specified. The baffle spacing is decreased somewhat to provide better tube support with seven baffle spaces in each half of the shell. Thus,

$$B = \frac{16 \times 12}{14} = 13.7 \text{ in.}$$

The middle baffle is a full circle baffle, as shown in Figure 11.2. These changes have no effect on the calculations for h_i, ΔP_i, and h_o. However, F, h_V, and ΔP_o must be recalculated.

Third trial

(a) LMTD correction factor.
From the graph in Appendix 11.A for a J shell with an even number of tube passes, $F \cong 0.98$, which is essentially the same as the value for the E shell. Therefore, the required overall coefficient remains unchanged at 86 Btu/h · ft^2 · °F.

(b) Correction for sensible heat transfer.

$$a_s = \frac{d_s C' B}{144 P_T} = \frac{39 \times 0.1875 \times 13.7}{144 \times 15/16} = 0.742 \text{ ft}^2$$

Half the average vapor flow rate of 90,000 lb/h is used to calculate h_V. Thus,

$$G = \dot{m}/a_s = \frac{0.5 \times 90{,}000}{0.742} = 60{,}647 \text{ lbm/h} \cdot \text{ft}^2$$

$$Re = \frac{D_e G}{\mu_V} = \frac{0.04583 \times 60{,}647}{0.0085 \times 2.419} = 135{,}177$$

$$B/d_s = 13.7/39 = 0.351$$

$$j_H = 0.5(1 + B/d_s)\left(0.08\, Re^{0.6821} + 0.7\, Re^{0.1772}\right)$$

$$= 0.5(1 + 0.351)\left\{0.08(135{,}177)^{0.6821} + 0.7(135{,}177)^{0.1772}\right\}$$

$$j_H = 174.6$$

$$h_V = j_H(k/D_e)Pr^{1/3}(\mu/\mu_w)^{0.14}$$

$$h_V = 174.6(0.0119/0.04583)(0.84)^{1/3}(1.0)$$

$$h_V \cong 43 \text{ Btu/h} \cdot \text{ft}^2 \cdot °\text{F}$$

The modified overall coefficient then becomes:

$$U'_D = \left[U_D^{-1} + (q_{sen}/q_{Tot})h_V^{-1}\right]^{-1} = \left[(94)^{-1} + 0.026(43)^{-1}\right]$$

$$U'_D = 89 \text{ Btu/h} \cdot \text{ft}^2 \cdot °\text{F}$$

Since $U'_D > U_{req}$, the unit is thermally acceptable.

(c) Shell-side pressure drop.
Half the total mass flow rate of vapor entering the condenser is used to compute $(\Delta P_f)_{VO}$ for the J shell. Thus,

$$G = 0.5\dot{m}_V/a_s = 0.5 \times 180{,}000/0.742 = 121{,}294 \text{ lbm/h} \cdot \text{ft}^2$$

$$Re = \frac{D_e G}{\mu_V} = \frac{0.04583 \times 121{,}294}{0.0085 \times 2.419} = 270{,}355$$

$$f_1 = (0.0076 + 0.000166 d_s) Re^{-0.125}$$

$$= (0.0076 + 0.000166 \times 39)(270,355)^{-0.125}$$

$$f_1 = 0.00295 \text{ ft}^2/\text{in.}^2$$

$$f_2 = \left(0.0016 + 5.8 \times 10^{-5}\, d_s\right) Re^{-0.157}$$

$$= \left(0.0016 + 5.8 \times 10^{-5} \times 23.25\right)(270,355)^{-0.157}$$

$$f_2 = 0.00041 \text{ ft}^2/\text{in.}^2$$

$$f = 144\{f_1 - 1.25(1 - B/d_s)(f_1 - f_2)\}$$

$$= 144\{0.00295 - 1.25(1 - 0.351)(0.00295 - 0.000414)\}$$

$$f = 0.1285$$

$$(\Delta P_f)_{VO} = \frac{f\, G^2 d_s (n_b + 1)}{7.50 \times 10^{12} D_e s \phi} = \frac{0.1285(121{,}294)^2(39/12)(7)}{7.50 \times 10^{12} \times 0.04583(0.845/62.43)(1.0)}$$

$$(\Delta P_f)_{VO} = 9.24 \text{ psi}$$

Notice that $(n_b + 1) = 7$ here, rather than 14, because the flow traverses only half the length of the J-shell.

$$\Delta P_f = \overline{\phi}_{VO}^2 (\Delta P_f)_{VO} = 0.33 \times 9.24 = 3.05 \text{ psi}$$

Assuming 10-in. schedule 40 nozzles for the vapor and allowing one velocity head for the inlet loss gives:

$$G_{n,in} = \frac{90{,}000}{(\pi/4)\,(10.02/12)^2} = 164{,}354 \text{ lbm/h} \cdot \text{ft}^2$$

$$\Delta P_{n,in} = 1.334 \times 10^{-13} G_{n,in}^2 / s_V = \frac{1.334 \times 10^{-13}(164{,}354)^2}{(0.845/62.43)}$$

$$\Delta P_{n,in} = 0.266 \text{ psi}$$

The condensate nozzle will be sized for self-venting operation. The volumetric flow rate is:

$$\dot{v}_L = \dot{m}/\rho = (180{,}000/3600)/35.5 = 1.408 \text{ ft}^3/\text{s}$$

The minimum nozzle size is obtained from Inequality (11.5a):

$$D_n > 0.89\, \dot{v}_L^{0.4} = 0.89\,(1.408)^{0.4} = 1.02 \text{ ft} = 12.24 \text{ in.}$$

Assuming schedule 40 pipe is used, the smallest size that satisfies this condition is 14-in. pipe (from Table B.2).

The pressure drop across the self-venting condensate nozzle will be very small and can be neglected. Hence, the total shell-side pressure drop is:

$$\Delta P_o = \Delta P_f + \Delta P_{n,in} = 3.05 + 0.266 \cong 3.3 \text{ psi}$$

All design criteria are satisfied and the over-design is approximately 3.5%. Hence, the unit is acceptable as configured.

Final design summary

Tube-side fluid: cooling water
Shell-side fluid: condensing hydrocarbons
Shell: Type AJU, 39 in. ID, oriented horizontally
Number of tubes: 1336
Tube size: 0.75 in. OD, 16BWG, 16-ft long
Tube layout: 15/16-in. triangular pitch
Tube passes: 4
Baffles: 35% cut segmental type with vertical cut; middle baffle is full circle
Baffle spacing: approximately 13.7 in.
Number of baffles: 13

Sealing strips: as required based on detailed tube layout
Tube-side nozzles: 10-in. schedule 40 inlet and outlet
Shell-side nozzles: two 10-in. schedule 40 inlet (top), one 14-in. schedule 40 outlet (bottom)
Materials: Tubes and tubesheets, 90/10 copper-nickel alloy; all other components, carbon steel

Example 11.7

Design a finned-tube condenser for the service of Example 11.6.

Solution

(a) Make initial specifications.
The following changes are made to the initial specifications of Example 11.6.
(i) Shell and head types
Based on the results of Example 11.6 and the fact that a smaller shell will result from the use of finned tubes, difficulty in meeting the shell-side pressure drop specification while providing adequate tube support is anticipated. Therefore, a cross-flow shell is selected and an AXU configuration is specified.
(ii) Tubing
Referring to Table B.5, 3/4-in., 16 BWG, 26 fpi tubing is selected; the corresponding part number is 265065. A tube length of 16 ft and a triangular layout with tube pitch of 15/16 in. are also specified.
(iii) Baffles
Tube support plates are used in a cross-flow exchanger rather than standard baffles. A sufficient number of plates must be provided for adequate tube support and suppression of tube vibration. Considering the potential for tube vibration problems in this application, a plate spacing of 24 in. is a reasonable initial estimate, but this figure has no effect on the thermal or hydraulic calculations.

(b) Energy balances.
From Example 11.6 we have:

$$q = 25.74 \times 10^6 \text{ Btu/h}$$

$$\dot{m}_{water} = 735{,}729 \text{ lbm/h}$$

(c) Mean temperature difference.
The following data are obtained from Example 11.6:

$$(\Delta T_{\ln})_{cf} = 72.8°\text{F}$$

$$R = 0.443 \quad P = 0.355$$

The LMTD correction factor chart for an X-shell in Appendix 11.A is for a single tube pass; no chart is available for an X-shell with an even number of tube passes. However, with the given values of R and P, F will not differ greatly from the value of approximately 0.98 for a single tube pass. Therefore,

$$\Delta T_m \cong 0.98 \times 72.8 = 71.3°\text{F}$$

(d) Approximate overall heat-transfer coefficient.
Based on the results of Example 11.6, the following values are estimated for the film coefficients:

$$h_i \cong 1000 \text{ Btu/h}\cdot\text{ft}^2\cdot°\text{F} \quad h_o \cong 250 \text{ Btu/h}\cdot\text{ft}^2\cdot°\text{F}$$

The value of h_o for finned tubes is expected to be significantly higher than the value of 132 Btu/h · ft^2 · °F calculated for plain tubes, while the value of h_i should be similar to that for plain tubes (1085 Btu/h · ft^2 · °F). The following data are obtained from Table B.5 for #265065 finned tubes:

$$A_{Tot}/L = 0.596\text{ft}^2/\text{ft} = \text{external surface area per unit length}$$

$$A_{Tot}/A_i = 4.35$$

$$D_r = 0.652 \text{ in.} = 0.0543 \text{ ft} = \text{root-tube diameter}$$

$$D_i = 0.522 \text{ in.} = 0.0435 \text{ ft}$$

Using Equation (4.25) and assuming $\eta_w \cong 1.0$ we obtain:

$$U_D = \left[\frac{A_{Tot}}{h_i A_i} + \frac{R_{Di} A_{Tot}}{A_i} + \frac{A_{Tot} \ln(D_r/D_i)}{2\pi k_{tube} L} + \frac{1}{h_o \eta_w} + \frac{R_{Do}}{\eta_w}\right]^{-1}$$

$$= \left[\frac{4.35}{1000} + 0.001 \times 4.35 + \frac{0.596 \ln(0.652/0.522)}{2\pi \times 30} + \frac{1}{250 \times 1.0} + \frac{0.0005}{1.0}\right]^{-1}$$

$$U_D = 72 \text{ Btu/h·ft}^2\text{·°F}$$

(e) Heat-transfer area and number of tubes.

$$A = \frac{q}{U_D \Delta T_m} = \frac{25.74 \times 10^6}{72 \times 71.3} = 5014 \text{ ft}^2$$

$$n_t = \frac{A}{(A_{Tot}/L) \times L} = \frac{5014}{0.596 \times 16} = 526$$

(f) Number of tube passes.
Assuming two tube passes, the velocity is:

$$V = \frac{\dot{m}(n_P/n_t)}{\rho\pi(D_i^2)} = \frac{(735{,}429/3600)(2/526)}{0.99 \times 62.43 \times \pi(0.0435)^2/4} = 8.5 \; ft/s$$

Since the velocity is in the acceptable range for the tubing material, two tube passes will be used.

(g) Shell size and tube count.
From Table C.2, for a U-tube exchanger with 2 tube passes, the closest tube count is 534 tubes in a 25-in. shell.

(h) Required overall coefficient.

$$U_{req} = \frac{q}{n_t(A_{Tot}/L)\Delta T_m} = \frac{25.74 \times 10^6}{534 \times 0.596 \times 16 \times 71.3} = 71 \text{ Btu/h·ft}^2\text{·°F}$$

(i) Calculate h_i assuming $\phi_i = 1.0$.

$$Re = \frac{4\dot{m}(n_p/n_t)}{\pi D_i \mu} = \frac{4 \times 735{,}429(2/534)}{\pi \times 0.0435 \times 0.72 \times 2.419} = 46{,}289$$

$$h_i = (k/D_i) \times 0.023 Re^{0.8} Pr^{1/3} (\mu/\mu_w)^{0.14}$$

$$= (0.37/0.0435) \times 0.023(46{,}289)^{0.8}(4.707)^{1/3} \times 1.0$$

$$h_i = 1770 \text{ Btu/h·ft}^2\text{·°F}$$

(j) Calculate h_o.
For finned tubes, the equivalent diameter defined by Equation (11.65) is required:

$$D_e^{-0.25} = \frac{1.30\eta_f A_{fins} E^{-0.25} + A_{prime} D_r^{-0.25}}{\eta_w A_{Tot}}$$

The fin efficiency for low-fin tubes is usually quite high unless the material of construction has a relatively low thermal conductivity. Therefore, η_f and η_w can be set to unity as a first approximation. The remaining parameters are calculated from the tube and fin dimensions:

$$E = \pi(r_2^2 - r_1^2)/2r_2$$

$$r_1 = D_r/2 = 0.652/2 = 0.326 \text{ in.}$$

$$r_2 = 0.75/2 = 0.375 \text{ in.}$$

$$E = \pi\{(0.375)^2 - (0.326)^2\}/(2 \times 0.375)$$

$$E = 0.14388 \text{ in} = 0.0120 \text{ ft}$$

For convenience, the fin and prime surface areas are calculated per inch of tube length:

$$A_{fins} = 2N_f\pi(r_2^2 - r_1^2) = 2 \times 26\pi\{(0.375)^2 - (0.326)^2\} = 5.611 \text{ in.}^2$$

$$\tau = \text{fin thickness} = 0.013 \text{ in. (Table B.5)}$$

$$A_{prime} = 2\pi r_1(L - N_f\tau) = 2\pi \times 0.326(1.0 - 26 \times 0.013) = 1.356 \text{ in.}^2$$

$$A_{fins}/A_{Tot} = 5.611/(5.611 + 1.356) = 0.805$$

$$A_{prime}/A_{Tot} = 1 - A_{fins}/A_{Tot} = 0.195$$

Substituting into the above equation for D_e gives:

$$D_e^{-0.25} = \frac{1.30 \times 1.0 \times 0.805(0.0120)^{-0.25} + 0.195(0.0543)^{-0.25}}{1.0} = 3.5658 \text{ ft}^{-0.25}$$

$$D_e = (3.5658)^{-4} = 0.0062 \text{ ft}$$

Next, the modified condensate loading is computed:

$$\Gamma^* = \frac{W}{L(n_t)^{2/3}} = \frac{180{,}000}{16(534)^{2/3}} = 170.92 \text{ lbm/ft}\cdot\text{h}$$

From Example 11.6, a tube wall temperature of 112°F is assumed, which gives $T_f = 128$°F and $\mu_L = 0.161$ cp. The shell-side heat-transfer coefficient is calculated using Equation (11.68) with Γ replaced by Γ^*:

$$h_o = 0.609\left[\frac{k_L^3\rho_L(\rho_L - \rho_V)g\eta_w(A_{Tot}/L)}{\mu_L D_e \Gamma^*}\right]^{1/3}$$

$$= 0.609\left[\frac{(0.057)^3 \times 35.5(35.5 - 0.845) \times 4.17 \times 10^8 \times 1.0 \times 0.596}{0.161 \times 2.419 \times 0.0062 \times 170.92}\right]^{1/3}$$

$$h_o = 314 \text{ Btu/h}\cdot\text{ft}^2\cdot°\text{F}$$

(k) Calculate T_p, T_{wtd} and T_f.
For $\eta_w = 1.0$, Equations (4.38) and (4.39) give:

$$T_p = T_{wtd} = \frac{h_i t_{ave} + h_o(A_{Tot}/A_i)T_{ave}}{h_i + h_o(A_{Tot}/A_i)} = \frac{1770 \times 102.5 + 314 \times 4.35 \times 175.75}{1770 + 314 \times 4.35}$$

$$T_p = T_{wtd} = 134.4°\text{F}$$

$$T_f = 0.75T_{wtd} + 0.25T_V = 0.75 \times 134.4 + 0.25 \times 175.75 \cong 145°\text{F}$$

At this value of T_f, $\mu_L = 0.148$ cp. Using this value to recalculate h_o and the wall temperatures gives:

$$h_o = 323 \text{ Btu/h}\cdot\text{ft}^2\cdot°\text{F}$$

$$T_p = T_{wtd} = 134.9°\text{F}$$

$$T_f = 145.1°\text{F}$$

Since the two values of T_f are essentially the same, no further iteration is required.

(l) Calculate fin efficiency.
Equation (2.27) is used to calculate the fin efficiency.

$$r_{2c} = r_2 + \tau/2 = 0.375 + 0.013/2 = 0.3815 \text{ in.}$$

$$\tau = 0.013/12 = 0.001083 \text{ ft}$$

$$\psi = (r_{2c} - r_1)\left[1 + 0.35\ln(r_{2c}/r_1)\right] = (0.3815 - 0.326)\left[1 + 0.35 \ln(0.3815/0.326)\right]$$

$$\psi = 0.5855 \text{ in.} = 0.004879 \text{ ft}$$

$$m = (2h_o/k\tau)^{0.5} = (2 \times 323/30 \times 0.001083)^{0.5} = 141.0 \text{ ft}^{-1}$$

$$m\psi = 141.0 \times 0.004879 = 0.6879$$

$$\eta_f = \tanh(m\psi)/(m\psi) = \tanh(0.6879)/0.6879 = 0.867$$

The weighted efficiency of the finned surface is computed using Equation (2.31):

$$\eta_w = (A_{prime}/A_{Tot}) + \eta_f(A_{fins}/A_{Tot}) = 0.195 + 0.867 \times 0.805 \cong 0.893$$

(m) Recalculate h_o and fin efficiency.
Repeating steps (j) and (k) with $\eta_f = 0.867$ and $\eta_w = 0.893$ yields the following results:

$D_e = 0.0065$ ft	$T_p = 132.2°F$	$T_f = 146.6°F$
$h_o = 311$ Btu/h · ft^2 ·°F	$T_{wtd} = 136.9°F$	

Repeating step (l) with the new value of h_o gives:

$\eta_f = 0.871$	$\eta_w = 0.896$

Since the new values of efficiency are essentially the same as the previous set, no further iteration is required and the above results are accepted as final.

(n) Viscosity correction factor for cooling water.
At 132°F, the viscosity of water is 0.52 cp from Figure A.1. Hence,

$$h_i = 1770(0.72/0.52)^{0.14} = 1853 \text{ Btu/h·ft}^2\text{·°F}$$

(o) Overall heat-transfer coefficient.

$$U_D = \left[\frac{A_{Tot}}{h_i A_i} + \frac{R_{Di} A_{Tot}}{A_i} + \frac{(A_{Tot}/L)\ln(D_r/D_i)}{2\pi k_{tube}} + \frac{1}{h_o \eta_w} + \frac{R_{Do}}{\eta_w}\right]^{-1}$$

$$= \left[\frac{4.35}{1852} + 0.001 \times 4.35 + \frac{0.596 \ln(0.652/0.522)}{2\pi \times 30} + \frac{1}{311 \times 0.896} + \frac{0.0005}{0.896}\right]^{-1}$$

$$U_D = 86.6 \text{ Btu/h· ft}^2\text{·°F}$$

(p) Correction for sensible heat transfer.
The effect of sensible heat transfer is neglected here since the effects of interfacial shear and condensate subcooling have also been neglected, which will compensate for this factor. Therefore, since $U_D > U_{req}$, the condenser is thermally acceptable, but somewhat over-sized (over-design = 22%).

(q) Tube-side pressure drop.

$$f = 0.4137\, Re^{-0.2585} = 0.4137(46,289)^{-0.2585} = 0.0257$$

$$G = \frac{\dot{m}(n_p/n_t)}{A_f} = \frac{735,429(2/534)}{(\pi/4)(0.0435)^2} = 1,853,366 \text{ lbm/h·ft}^2$$

$$\Delta P_f = \frac{f\, n_p L G^2}{7.50 \times 10^{12} D_i s \phi} = \frac{0.0257 \times 2 \times 16(1,853,366)^2}{7.50 \times 10^{12} \times 0.0435 \times 0.99 \times 1.0}$$

$$\Delta P_f = 8.40 \text{ psi}$$

$$\Delta P_r = 1.334 \times 10^{-13}(1.6 n_p - 1.5)G^2/s$$

$$= 1.334 \times 10^{-13}(1.6 \times 2 - 1.5)(1,853,366)^2/0.99$$

$$\Delta P_r = 0.79 \text{ psi}$$

Assuming 10-in. schedule 40 nozzles are used, the nozzle losses will be the same as calculated in Example 11.6, namely:

$$\Delta P_n = 0.36 \text{ psi}$$

The total tube-side pressure drop is:

$$\Delta P_i = \Delta P_f + \Delta P_r + \Delta P_n = 8.4 + 0.79 + 0.36$$

$$\Delta P_i \cong 9.6 \text{ psi}$$

(r) Shell-side pressure drop.
The ideal tube bank correlations in Chapter 6 can be used to calculate the pressure drop in a cross-flow shell. The cross-flow area, S_m, is approximated by a_s with the baffle spacing equal to the length of the shell. For finned tubes, an effective clearance is used that accounts for the area occupied by the fins:

$$C'_{eff} = P_T - D_{re}$$

where
$D_{re} = D_r + 2n_f b\tau$ = equivalent root-tube diameter
n_f = number of fins per unit length

Note that BC'_{eff} is the flow area between two adjacent tubes in one baffle space. Substituting the values of the parameters in the present problem gives:

$$D_{re} = 0.652 + 2 \times 26 \times 0.049 \times 0.013 = 0.6851 \text{ in.}$$

$$C'_{eff} = 15/16 - 0.6851 = 0.2524 \text{ in.}$$

$$a_s = \frac{d_s C'_{eff} B}{144 P_T} = \frac{25 \times 0.2524 \times (16 \times 12)}{144 \times 15/16} = 8.974 \text{ ft}^2$$

$$G = \dot{m}_V / a_s = 180{,}000/8.974 = 20{,}058 \text{ lbm/h} \cdot \text{ft}^2$$

In calculating the Reynolds number for finned tubes, the equivalent root-tube diameter is used in place of D_o [32]:

$$Re_V = \frac{D_{re} G}{\mu_V} = \frac{(0.6851/12) \times 20{,}058}{0.0085 \times 2.419} = 55{,}694$$

The friction factor for plain tubes is obtained from Figure 6.2: $f_{ideal} = 0.10$. The friction factor for finned tubes is estimated as 1.4 times the value for plain tubes [32]. Thus,

$$f'_{ideal} = 1.4 f_{ideal} = 1.4 \times 0.10 = 0.14$$

The number of tube rows crossed is estimated using Equation (6.8) with the baffle cut taken as zero.

$$N_c = \frac{d_s(1 - 2B_c)}{P_T \cos \theta_{tp}} = \frac{25}{(15/16) \cos 30°} = 30.8$$

The pressure drop for all-vapor flow is calculated using Equation (6.7). The effect of the bundle bypass flow is neglected here:

$$(\Delta P_f)_{VO} = \frac{2 f'_{ideal} N_c G^2}{g_c \rho_V \phi} = \frac{2 \times 0.14 \times 30.8(20{,}058)^2}{4.17 \times 10^8 \times 0.845 \times 1.0}$$

$$(\Delta P_f)_{VO} = 9.85 \text{ lbf/ft}^2 = 0.068 \text{ psi}$$

The two-phase friction loss is calculated using an average two-phase multiplier of 0.33 for a total condenser:

$$\Delta P_f = \overline{\phi}^2_{VO} (\Delta P_f)_{VO} = 0.33 \times 0.068 = 0.022 \text{ psi}$$

As would be expected, the pressure drop in the X-shell is very small. Assuming two 10-in. schedule 40 inlet nozzles are used, the nozzle losses will be the same as calculated in Example 11.6, namely:

$$\Delta P_{n,in} = 0.266 \text{ psi}$$

Two condensate nozzles are specified, corresponding to the two inlet nozzles. The condensate nozzles are sized for self-venting operation. Since each nozzle handles half the condensate, the volumetric flow rate is:

$$\dot{v}_L = \dot{m}/\rho = (90{,}000/3600)/35.5 = 0.704 \text{ ft}^3/\text{s}$$

Using Inequality (11.5a), we obtain:

$$D_n > 0.89\, \dot{v}_L^{0.4} = 0.89\,(0.704)^{0.4} = 0.773 \text{ ft} = 9.28 \text{ in.}$$

From Table B.2, for schedule 40 pipe the smallest size that satisfies this condition is 10 in. Therefore, two 10-in. outlet nozzles are required.

The pressure drop across the self-venting condensate nozzles will be very small and can be neglected. Hence, the total shell-side pressure drop is:

$$\Delta P_o = \Delta P_f + \Delta P_{n,in} = 0.022 + 0.266 \cong 0.29 \text{ psi}$$

All design criteria are satisfied; however, the condenser is somewhat over-sized. The number of tubes cannot be reduced because the tube-side pressure drop is close to the maximum. Therefore, we consider using shorter tubes. The required tube length is:

$$L_{req} = \frac{q}{(A_{Tot}/L)\, n_t U_D \Delta T_m} = \frac{25.74 \times 10^6}{0.596 \times 534 \times 86.6 \times 71.3}$$

$$L_{req} = 13.1 \text{ ft}$$

Hence, the tube length is reduced to 14 ft, which gives an over-design of about 7%. This change will decrease the tube-side pressure drop and increase the shell-side pressure drop slightly. Although these changes will not affect the viability of the design, the new pressure drops are calculated here for completeness. For the tube side, we have:

$$\Delta P_f = 8.40(14/16) = 7.35 \text{ psi}$$

$$\Delta P_i = \Delta P_f + \Delta P_r + \Delta P_n = 7.35 + 0.79 + 0.36$$

$$\Delta P_i = 8.5 \text{ psi}$$

For the shell side:

$$a_s = \frac{25 \times 0.2524(14 \times 12)}{144(15/16)} = 7.85 \text{ ft}^2$$

$$G = 180{,}000/7.85 = 22{,}930 \text{ lbm/h}\cdot\text{ft}^2$$

$$Re = \frac{(0.6851/12) \times 22{,}930}{0.0085 \times 2.419} = 63{,}668$$

$$f_{ideal} \cong 0.095 \text{ (Figure 6.2)}$$

$$f'_{ideal} = 1.4 f_{ideal} = 1.4 \times 0.095 = 0.133$$

$$\left(\Delta P_f\right)_{VO} = \frac{2 \times 0.133 \times 30.8(22{,}930)^2}{4.17 \times 10^8 \times 0.845 \times 1.0}$$

$$\left(\Delta P_f\right)_{VO} = 12.2 \text{ lbf/ft}^2 = 0.085 \text{ psi}$$

$$\Delta P_f = 0.33 \times 0.085 = 0.028 \text{ psi}$$

$$\Delta P_o = 0.028 + 0.266 \cong 0.29 \text{ psi}$$

The final design parameters are summarized below.

Final design summary
Tube-side fluid: cooling water
Shell-side fluid: condensing hydrocarbon
Shell: type AXU, 25-in. ID, oriented horizontally
Number of tubes: 534
Tube size: 0.75 in., 16 BWG, radial low-fin tubes, 14 ft long
Fins: 26 fpi, 0.049 in. high, 0.013 in. thick
Tube layout: 15/16-in. triangular pitch
Tube passes: 2
Baffles/support plates: 6 support plates
Sealing strips: as needed based on tube layout
Tube-side nozzles: 10-in. schedule 40 inlet and outlet
Shell-side nozzles: two 10-in. schedule 40 inlet (top), two 10-in. schedule 40 outlet (bottom)
Materials: tubes and tubesheets, 90/10 copper-nickel alloy; all other components, carbon steel

11.10 Multi-Component Condensation

11.10.1 The General Problem

Analysis of the condensation process for a general multi-component mixture entails a much greater level of complexity compared to pure-component condensation. Among the factors responsible for the added complexity are the following:

- As previously noted, multi-component condensation is always non-isothermal, and the condensing range can be large (greater than 100°F). Thus, there are sensible heat effects in both the vapor and liquid phases. Sensible heat transfer in the vapor phase can have a significant effect on the condensation process due to the low heat-transfer coefficients that are typical for gases.
- The compositions of both phases vary from condenser inlet to outlet because the less volatile components condense preferentially. As a result, the physical properties of both phases can vary significantly over the length of the condenser.

- As discussed in the previous section, the condensing curve may be highly nonlinear, invalidating the use of the LMTD correction factor for calculating the mean temperature difference.
- Thermodynamic (equilibrium flash) calculations are required to obtain the condensing curve and determine the phase compositions. Equilibrium ratios (K-values) and enthalpies are needed for this purpose, and the mixture may be highly non-ideal.
- Equilibrium exists at the vapor-liquid interface, not between the bulk phases. Hence, the thermodynamic calculations should be performed at the interfacial temperature, which is unknown.
- Since the interfacial composition differs from the bulk phase compositions, there are mass-transfer as well as heat-transfer resistances in both the vapor and liquid phases. Therefore, mass-transfer coefficients are needed in addition to heat-transfer coefficients in order to model the process. Furthermore, the heat and mass-transfer effects are coupled, and the equations describing the transport processes are complex. Thus, there are computational difficulties involved, as well as a lack of data for mass-transfer coefficients.

A rigorous formulation of the general multi-component condensation problem has been presented by Taylor et al. [33]. Due to the inherent complexity of the model, however, it has not been widely used for equipment design. An approximate method developed by Bell and Ghaly [34] has formed the basis for the condenser algorithms used in most commercial software packages. This method is described in the following subsection.

11.10.2 The Bell–Ghaly method

The Bell–Ghaly method neglects the mass-transfer resistances, but attempts to compensate by overestimating the thermal resistances, primarily the vapor-phase resistance. The following approximations are made [34]:

- Vapor and liquid phases are assumed to be in equilibrium at the temperature, T_V, of the bulk vapor phase, rather than at the interfacial temperature. Thus, the thermodynamic calculations are performed at T_V.
- The liquid and vapor properties are assumed to be those of the equilibrium phases at temperature T_V.
- The heat-transfer coefficient for sensible heat transfer in the vapor phase is calculated assuming that the vapor flows alone through the condenser. In effect, a two-phase multiplier of unity is assumed for vapor-phase heat transfer, which tends to significantly overestimate the corresponding thermal resistance.
- The entire heat load (latent and sensible heats) is assumed to be transferred through the entire thickness of the condensate film. This assumption produces a slight overestimation of the thermal resistance in the liquid phase.

A differential analysis is employed. The rate of heat transfer from the bulk vapor phase to the vapor-liquid interface in a differential condenser element of area dA is:

$$dq_V = h_V dA(T_V - T_{sat}) \tag{11.74}$$

where

dq_V = rate of sensible heat transfer in element of area dA due to vapor cooling
h_V = vapor-phase heat-transfer coefficient

The total rate of heat transfer in the differential element is given by:

$$dq = U_D dA(T_{sat} - T_c) \tag{11.75}$$

where

dq = total rate of heat transfer in element of area dA
U_D = overall coefficient between the vapor–liquid interface and the coolant, including fouling allowances
T_c = coolant temperature

Equation (11.74) can be solved for the interfacial temperature to yield:

$$T_{sat} = T_V - (1/h_V)\frac{dq_V}{dA} \tag{11.76}$$

Substituting this expression for T_{sat} in Equation (11.75) gives:

$$dq = U_D dA\left(T_V - T_c - (1/h_V)\frac{dq_V}{dA}\right)$$

$$dq = U_D dA(T_V - T_c) - \frac{U_D dq_V}{h_V} \tag{11.77}$$

Now let $\Lambda \equiv dq_V/dq$ and substitute in Equation (11.77) to obtain:

$$dq = U_D dA(T_V - T_c) - (U_D \Lambda/h_V)dq$$

$$dq(1 + U_D\Lambda/h_V) = U_D dA(T_V - T_c) \tag{11.78}$$

Solving Equation (11.78) for dA yields:

$$dA = \frac{(1 + U_D\Lambda/h_V)}{U_D(T_V - T_c)} dq \tag{11.79}$$

The total heat-transfer area is obtained by integrating Equation (11.79) from $q = 0$ to $q = q_{Tot}$, where q_{Tot} is the total duty:

$$A = \int_0^A dA = \int_0^{q_{Tot}} \frac{(1 + U_D\Lambda/h_V)}{U_D(T_V - T_c)} dq \tag{11.80}$$

In practice, the design integral is evaluated numerically by performing an incremental analysis. For counter-current flow of vapor and coolant, the computational procedure is straightforward, while for multi-pass condensers it is somewhat more involved. Procedures for both cases are given in Ref. [34]. Due to the complexity of the overall procedure, which includes the thermodynamic calculations, implementation is practical only with the use of commercial software.

Bell and Ghaly [34] report that tests of the method using the HTRI data bank showed that it gives results for the heat-transfer area that range from correct to about 100% high. They also state that proprietary modifications of the method made by HTRI greatly improve the accuracy.

Following is a highly simplified example designed to illustrate the basic computational procedure.

Example 11.8

100,000 lb/h of saturated vapor consisting of 30 mole percent *n*-butane and 70 mole percent *n*-pentane are to be condensed at a pressure of 75 psia. A single-pass horizontal shell-and-tube condenser will be used with the condensing vapor in the shell. Cooling water with a range of 85-120°F will flow in the tubes. The flow area across the tube bundle is $a_s = 0.572$ ft^2, the equivalent diameter is $D_e = 0.06083$ ft and the baffle spacing is $B/d_s = 0.45$. The bundle contains 107 ft^2 of external surface area per foot of length. The overall heat-transfer coefficient between the vapor–liquid interface and coolant is $U_D = 120$ Btu/h · ft^2 · °F. Use the Bell–Ghaly method to calculate the required surface area and length of the condenser.

Solution

For convenience, the condensing range (183.5–168°F) is divided into three intervals: 168°F to 173°F, 173°F to 178°F and 178°F to 183.5°F. Equilibrium flash calculations are then performed at each of the above four temperatures using a flowsheet simulator (PRO/II by SimSci-Esscor). The results are summarized in the following table. (Note that physical properties of the condensate are not required because U_D is given in this problem.)

T_V (°F)	$\dot{m}_V$ (lbm/h)	Duty (Btu/h)	$C_{P,V}$ (Btu/lbm · °F)	k_V (Btu/h · ft · °F)	μ_V (cp)
183.5	100,000	0	0.486	0.01194	0.00853
178	54,590	6,373,860	0.483	0.01185	0.00851
173	24,850	10,661,170	0.480	0.01177	0.00850
168	0	14,314,980	0.477	0.01168	0.00848

The flow rate of the cooling water is obtained from the overall energy balance:

$$\dot{m}_c = \frac{q}{C_{P,c}\Delta T_c} = \frac{14{,}314{,}980}{1.0\,(120 - 85)} = 408{,}999 \text{ lbm/h}$$

Next, the temperature profile for the cooling water is determined using energy balances over the individual temperature intervals. Counter-current flow and constant heat capacity of the cooling water are assumed. For the interval from 168°F to 173°F, we have:

$$\Delta q = 14{,}314{,}980 - 10{,}661{,}170 = 3{,}653{,}810 \text{ Btu/h}$$

$$\Delta T_c = \frac{\Delta q}{\dot{m}_c C_{P,c}} = \frac{3{,}653{,}810}{408{,}999 \times 1.0} = 8.93°\text{F}$$

$$T_{c,out} = 85 + 8.93 = 93.93°\text{F}$$

Similarly, for the interval from 173°F to 178°F we obtain:

$$\Delta q = 10{,}661{,}170 - 6{,}373{,}860 = 4{,}287{,}310 \text{ Btu/h}$$

$$\Delta T_c = \frac{4{,}287{,}310}{408{,}999 \times 1.0} = 10.48°\text{F}$$

$$T_{c,out} = 93.93 + 10.48 = 104.41°\text{F}$$

The temperature profiles of the vapor and cooling water are summarized in the following table.

T_V (°F)	T_c (°F)
183.5	120
178	104.41
173	93.93
168	85

The calculations for the temperature interval from 168°F to 173°F are performed next. The rate of sensible heat transfer to cool the vapor in this interval is calculated using the average vapor flow rate and heat capacity for the interval. The averages are:

$$\overline{C}_{P,V} = 0.5(0.477 + 0.480) = 0.4785\ \text{Btu/lbm}\cdot°\text{F}$$

$$(\dot{m}_V)_{ave} = 0.5(0 + 24{,}850) = 12{,}425\ \text{lbm/h}$$

Hence, the rate of sensible heat transfer is:

$$\Delta q_V = (\dot{m}_V)_{ave}\overline{C}_{P,V}\Delta T_V = 12{,}425 \times 0.4785 \times 5 = 29{,}727\ \text{Btu/h}$$

The value of Λ for the interval is:

$$\Lambda = \Delta q_V/\Delta q = 29{,}727/3{,}653{,}810 = 0.0081$$

The Reynolds number for the vapor is computed next using the average vapor flow rate and average viscosity for the interval.

$$\overline{\mu}_V = 0.5(0.00848 + 0.00850) = 0.00849\ \text{cp}$$

$$Re_V = \frac{D_e\{(\dot{m}_V)_{ave}/a_s\}}{\overline{\mu}_v} = \frac{0.06083\ \{12{,}425/0.572\}}{0.00849 \times 2.419} = 64{,}339$$

Equations (3.20) and (3.21) are used to calculate the heat-transfer coefficient for the vapor phase:

$$j_H = 0.5(1 + B/d_s)\left(0.08\ Re^{0.6821} + 0.7Re^{0.1772}\right)$$

$$= 0.5(1 + 0.45)\left\{0.08(64{,}339)^{0.6821} + 0.7(64{,}339)^{0.1772}\right\}$$

$$j_H = 114.09$$

$$\overline{k}_V = 0.5(0.01168 + 0.01177) = 0.011725\ \text{Btu/h}\cdot\text{ft}\cdot°\text{F}$$

$$Pr_V = \frac{\overline{C}_{P,V}\overline{\mu}_V}{\overline{k}_V} = \frac{0.4785 \times 0.00849 \times 2.419}{0.011725} = 0.8381$$

$$h_V = \frac{j_H\overline{k}_V Pr_V^{1/3}\phi}{D_e} = \frac{114.09 \times 0.011725(0.8381)^{1/3} \times 1.0}{0.06083}$$

$$h_V = 20.7\ \text{Btu/h}\cdot\text{ft}^2\cdot°\text{F}$$

The heat-transfer area required for the temperature interval is obtained from the finite difference form of Equation (11.79) using the average vapor and coolant temperatures:

$$\Delta A = \left\{\frac{1 + U_D\Lambda/h_V}{U_D(T_{V,ave} - T_{c,ave})}\right\}\Delta q$$

$$\Delta A = \left\{\frac{1 + 120 \times 0.0081/20.7}{120(170.5 - 89.465)}\right\} \times 3{,}653{,}810$$

$$\Delta A = 393.4\ \text{ft}^2$$

The above calculations are repeated for the other two temperature intervals to obtain the results shown in the following table.

Parameter	Temperature Interval		
	168–173°F	173–178°F	178–183.5°F
$(\dot{m}_V)_{ave}$ (lbm/h)	12,425	39,720	77,295
Δq_V (Btu/h)	29,727	95,626	205,972
Δq (Btu/h)	3,653,810	4,287,310	6,373,860
Λ	0.0081	0.0223	0.0323
Re_V	64,339	205,315	398,839
j_H	114.09	248.23	388.46
h_V (Btu/h · ft^2 · °F)	20.7	45.5	71.7
$T_{V,ave}$ (°F)	170.5	175.5	180.75
$T_{c,ave}$ (°F)	89.465	99.17	112.205
ΔA (ft^2)	393.4	495.6	816.8

The total heat-transfer area required in the exchanger is the sum of the incremental areas:

$$A = 393.4 + 495.6 + 816.8 \cong 1706 \text{ ft}^2$$

The required length of the exchanger is:

$$L = \frac{A}{(A/L)} = \frac{1706}{107} = 15.9 \cong 16 \text{ ft}$$

11.11 Computer Software

Condenser design can be accomplished using any of the commercial software packages mentioned in Section 5.9. HEXTRAN and HTRI *Xchanger Suite* are considered here.

The shell-and-tube module (STE) in HEXTRAN is used for segmentally baffled or unbaffled condensers. A separate module is available for rod-baffle units, but it supports only single-phase operation on the shell side. The software automatically detects the phase change for a condensing stream and uses the appropriate computational methods. A zone analysis is always performed for applications involving a phase change.

The *Xist* module of HTRI *Xchanger Suite* is used for all types of shell-and-tube condensers. The phase change (condensation in this case) is specified on the Process input panel. *Xist* performs a three-dimensional incremental analysis for condensers as well as other types of exchangers. A number of options for calculating the condensing heat-transfer coefficient are available on the Control/Methods panel. The default (Proration method) is a proprietary method developed by HTRI and is recommended for all cases. Among the alternatives, the Literature method is the Bell-Ghaly method as described in Section 11.10.2. The Rose–Briggs method provides improved accuracy for condensation on low-finned tubes; it is based on a method from the open literature [24]. A special method (REFLUX) is also provided for reflux condensers.

The following examples illustrate the use of *Xist* and HEXTRAN for condenser calculations.

Example 11.9

Use HEXTRAN and *Xist* to rate the final design for the C_4–C_5 condenser obtained in Example 11.6, and compare the results with those from the hand calculations.

Solution

The problem setup in HEXTRAN is done in the usual manner using the data from Example 11.6. The cooling water stream is defined as a Water/Steam stream to invoke the SimSci databank for the thermodynamic data and transport properties of water. The hydrocarbon stream is defined as a compositional stream and the Peng–Robinson equation of state is selected as the principal thermodynamic method. The thermodynamic state of the stream (saturated vapor) is fixed by entering the stream pressure (75 psia) as the first specification and selecting *Dew Point* for the second specification. The API method is chosen for liquid density of the hydrocarbon stream, and the Library method is selected for all transport properties to designate that the property values are to be obtained from the program's databank.

A J-shell with two inlet nozzles and one outlet nozzle is designated as type J2 in HEXTRAN. Hence, a type AJ2U exchanger is specified. A single shell-side inlet nozzle is specified because HEXTRAN automatically uses half the total flow rate to calculate the nozzle pressure drop. If two inlet nozzles are specified, the pressure drop is incorrectly calculated. The default settings for calculation options (TWOPHASE = New and DPSMETHOD = Stream) are used. Zero pairs of sealing strips are specified. On the specification form for the unit the shell-side product liquid fraction is set to 1.0, which forces the condensate to exit the condenser as a saturated liquid, i.e., with no subcooling. This specification is made to facilitate comparison of results with the other methods.

Xist designates a J-shell with two inlet nozzles and one outlet nozzle as a J21 shell. Hence, a type AJ21U exchanger is specified on the Geometry/Exchanger panel. On the Geometry/Tube Layout/ Bundle Clearances panel, the option for pairs of sealing strips is set to *None*. On the Geometry/Nozzles panel, the shell-side inlet nozzles are specified to be on the top of the shell with the outlet nozzle on the opposite side. Circular impingement plates are specified on the Geometry/Nozzles/Impingement panel. On the process panel, for the hot fluid the weight fraction of vapor at the inlet is set to 1.0 and the value at the outlet is set to 0.0. Neither the inlet nor the outlet temperature of the stream is specified. Together, these settings imply that the stream enters the unit as a saturated vapor and leaves as a saturated liquid. VMGThermo is used for both fluids. For cooling water, the Steam95 property package is selected and pressure levels of 50 and 60 psia are specified. A temperature range of 80°F to 160°F is used with 20 data points. For the hydrocarbon stream, the Advanced Peng–Robinson method is chosen with pressure levels of 70 and 75 psia. A temperature range of 85°F to 185°F is used with 20 data points. The default method (HTRI Proration) is used for the condensing calculations.

Results summaries obtained from the two programs are given below, along with the HEXTRAN input file. Data from the output files were used to construct the following table comparing the computer solutions with the hand calculations of Example 11.6. Perusal of this table reveals that the hand calculations for both shell-side heat transfer and pressure drop are very conservative compared with the computer solutions. The basic Nusselt theory was used to calculate h_o in Example 11.6, and this method tends to significantly underestimate the heat-transfer coefficient in baffled condensers. Vapor shear is very significant in this unit; the *Runtime Monitor* report file from *Xist* shows that the condensation is initially shear controlled and gradually transitions to the gravity-controlled regime toward the outlet. Thus, the results for shell-side heat transfer are as expected. The difference in ΔP_o between the hand and computer calculations is primarily due to the difference between the Delaware and Stream Analysis methods in the calculation of the vapor-phase pressure drop, $(\Delta P_f)_{VO}$. The Delaware method yields a value for this problem that is roughly three times the value given by the Stream Analysis method, and this difference is reflected in the Simplified Delaware method that was used in the hand calculations.

Item	Hand	HEXTRAN	*Xist*
h_i (Btu/h · ft^2 · °F)	1085	1134	1262
h_o (Btu/h · ft^2 · °F)	132	269	208
U_D (Btu/h · ft^2 · °F)	89*	149	131
ΔP_i (psi)	6.3	7.27	6.79
ΔP_o (psi)	3.3	0.58	1.42
ΔT_m (°F)	71.3	72.0	70.8
$U_D \Delta T_m$ (Btu/h · ft^2 · °F)	6346	10,728	9275
Over-design (%)	3.5	71.6	58.6

*Corrected for sensible heat transfer

Although there are significant differences between the two computer solutions, they both display large values for the over-design, indicating that the condenser is over-sized by a substantial amount. (The over-design for HEXTRAN is calculated from the value of U_D (148.75), denoted as "service" transfer rate in the output file, and U_{req} (86.67), denoted as "REQD" transfer rate.) Furthermore, the tube vibration analysis performed by *Xist* predicts flow-induced vibration problems in the unit, which could result in damage to the tubes. This information is contained in the *Runtime Messages* report file.

Xist Output Summary for Example 11.9

Xist Ver. 7.00 5/3/2013 8:27 SN: 14337-877388691 **US Units**
Example 11.9

Rating - Horizontal Multipass Flow TEMA AJ21U Shell With Single-Segmental Baffles

No Data Check Messages.
See Runtime Message Report for Warning Messages.

Process Conditions		Hot Shellside		Cold Tubeside	
Fluid name			C4-C5		Water
Flow rate	(1000-lb/hr)		180.00		735.43
Inlet/Outlet Y	(Wt. frac vap.)	1.0000	0.0000	0.0000	0.0000
Inlet/Outlet T	(Deg F)	183.32	166.42	85.00	120.30
Inlet P/Avg	(psia)	75.000	74.291	60.000	56.605
dP/Allow.	(psi)	1.417	5.000	6.789	10.000
Fouling	(ft2-hr-F/Btu)		0.00050		0.00100

Exchanger Performance					
Shell h	(Btu/ft2-hr-F)	207.66	Actual U	(Btu/ft2-hr-F)	130.94
Tube h	(Btu/ft2-hr-F)	1262.0	Required U	(Btu/ft2-hr-F)	82.58
Hot regime	(--)	Transition	Duty	(MM Btu/hr)	25.919
Cold regime	(--)	Sens. Liquid	Eff. area	(ft2)	4430.1
EMTD	(Deg F)	70.8	Overdesign	(%)	58.57

Shell Geometry			Baffle Geometry		
TEMA type	(--)	AJ21U	Baffle type	(--)	Single-Seg.
Shell ID	(inch)	39.000	Baffle cut	(Pct Dia.)	35
Series	(--)	1	Baffle orientation	(--)	Parallel
Parallel	(--)	1	Central spacing	(inch)	13.700
Orientation	(deg)	0.00	Crosspasses	(--)	14

Tube Geometry			Nozzles		
Tube type	(--)	Plain	Shell inlet	(inch)	10.020
Tube OD	(inch)	0.7500	Shell outlet	(inch)	13.124
Length	(ft)	16.000	Inlet height	(inch)	1.8501
Pitch ratio	(--)	1.2500	Outlet height	(inch)	0.2500
Layout	(deg)	30	Tube inlet	(inch)	10.020
Tubecount	(--)	1336	Tube outlet	(inch)	10.020
Tube Pass	(--)	4			

Thermal Resistance, %		Velocities, ft/sec		Flow Fractions	
Shell	63.06	Shellside	9.92	A	0.258
Tube	12.55	Tubeside	4.71	B	0.497
Fouling	22.39	Crossflow	16.92	C	0.039
Metal	2.01	Window	9.87	E	0.102
				F	0.104

HEXTRAN Input File for Example 11.9

```
$ GENERATED FROM HEXTRAN KEYWORD EXPORTER
$
$      General Data Section
$
TITLE PROJECT=Example 11.9, PROBLEM=LHC Condenser, SITE=
$
DIME  English, AREA=FT2, CONDUCTIVITY=BTUH, DENSITY=LB/FT3, *
      ENERGY=BTU, FILM=BTUH, LIQVOLUME=FT3, POWER=HP, *
      PRESSURE=PSIA, SURFACE=DYNE, TIME=HR, TEMPERATURE=F, *
      UVALUE=BTUH, VAPVOLUME=FT3, VISCOSITY=CP, WT=LB, *
      XDENSITY=API, STDVAPOR=379.490
$
PRINT ALL, *
      RATE=M
$
CALC  PGEN=New, WATER=Saturated
$
$      Component Data Section
$
COMPONENT DATA
$
 LIBID     1, BUTANE /*
           2, PENTANE
$
$
$      Thermodynamic Data Section
$
THERMODYNAMIC DATA
$
 METHODS SET=SET1, KVALUE=PR, ENTHALPY(L)=PR, ENTHALPY(V)=PR,  *
              DENSITY(L)=API, DENSITY(V)=PR, VISCOS(L)=LIBRARY,  *
              VISCOS(V)=LIBRARY, CONDUCT(L)=LIBRARY, CONDUCT(V)=LIBRARY,   *
              SURFACE=LIBRARY
$
 WATER DECANT=ON, SOLUBILITY = Simsci, PROP = Saturated
$
$Stream Data Section
$
STREAM DATA
$
 PROP STRM=WATER_IN, NAME=WATER_IN, TEMP=85.00, PRES=60.000, *
         WATER=735429.000
$
 PROP STRM=WATER_OUT, NAME=WATER_OUT
$
 PROP STRM=VAPOR, NAME=VAPOR, PRES=75.000, PHASE=V, *
         SET=SET1, RATE(W)=180000.000, *
   COMP(M)=  1, 0.3 / *
        2, 0.7, NORMALIZE
$
 PROP STRM=CONDENSATE, NAME=CONDENSATE
$
$ Calculation Type Section
$
SIMULATION
$
 TOLERANCE TTRIAL=0.01
$
LIMITS AREA=200.00, 6000.00, SERIES=1, 10, PDAMP=0.00, *
        TTRIAL=100
```

```
$
 CALC   TWOPHASE=New, DPSMETHOD=Stream, MINFT=0.80
$
 PRINT  UNITS, ECONOMICS, STREAM, STANDARD, *
        EXTENDED, ZONES
$
ECONOMICS DAYS=350, EXCHANGERATE=1.00, CURRENCY=USDOLLAR
$
 UTCOST OIL=3.50, GAS=3.50, ELECTRICITY=0.10, *
        WATER=0.03, HPSTEAM=4.10, MPSTEAM=3.90, *
        LPSTEAM=3.60, REFRIGERANT=0.00, HEATINGMEDIUM=0.00
$
 HXCOST BSIZE=1000.00, BCOST=0.00, LINEAR=50.00, *
        EXPONENT=0.60, CONSTANT=0.00, UNIT
$
$      Unit Operations Data
$
UNIT OPERATIONS
$
STE UID=CONDENSER
 TYPE  Old, TEMA=AJ2U, HOTSIDE=Shellside, ORIENTATION=Horizontal, *
       FLOW=Countercurrent, *
       UESTIMATE=50.00, USCALER=1.00
 TUBE  FEED=WATER_IN, PRODUCT=WATER_OUT, *
       LENGTH=16.00, OD=0.750, *
       BWG=16, NUMBER=1336, PASS=4, PATTERN=30, *
       PITCH=0.9380, MATERIAL=32, *
       FOUL=0.001, LAYER=0, *
       DPSCALER=1.00
$
SHELL  FEED=VAPOR, PRODUCT=CONDENSATE, *
       ID=39.00, SERIES=1, PARALLEL=1, *
       MATERIAL=1, *
       FOUL=0.0005, LAYER=0, *
       DPSCALER=1.00
$
 BAFF  Segmental=Single, *
       CUT=0.35, *
       SPACING=13.700, *
       THICKNESS=0.1875
$
 TNOZZ TYPE=Conventional, ID=10.020, 10.020,  NUMB=1, 1
$
 SNOZZ TYPE=Conventional , ID=10.020, 13.124,  NUMB=1, 1
$
 CALC  TWOPHASE=New, *
       DPSMETHOD=Stream, *
       MINFT=0.80
$
 SPEC  Shell, Lfrac=1.000000
$
 PRINT STANDARD, *
       EXTENDED, *
       ZONES
$
 COST  BSIZE=1000.00, BCOST=0.00, LINEAR=50.00, *
       CONSTANT=0.00, EXPONENT=0.60, Unit
$
$ End of keyword file...
```

HEXTRAN Output Data for Example 11.9

```
==============================================================================
                    SHELL AND TUBE EXCHANGER DATA SHEET
I----------------------------------------------------------------------------I
I EXCHANGER  NAME                           UNIT ID CONDENSER                I
I SIZE    39x  192  TYPE AJ2U,  HORIZONTAL     CONNECTED 1 PARALLEL  1 SERIES I
I AREA/UNIT  4144. FT2 (  2414. FT2 REQUIRED) AREA/SHELL  4144. FT2          I
I----------------------------------------------------------------------------I
I PERFORMANCE OF ONE UNIT            SHELL-SIDE                TUBE-SIDE      I
I----------------------------------------------------------------------------I
I FEED STREAM NUMBER                   VAPOR                   WATER_IN       I
I FEED STREAM NAME                     VAPOR                   WATER_IN       I
I TOTAL FLUID          LB /HR            180000.                  735429.     I
I     VAPOR  (IN/OUT) LB /HR      180000./         0.          0./         0. I
I     LIQUID          LB /HR           0./    180000.          0./         0. I
I     STEAM           LB /HR           0./         0.          0./         0. I
I     WATER           LB /HR           0./         0.     735429./    735429. I
I     NON CONDENSIBLE LB /HR                0.                          0.    I
I TEMPERATURE (IN/OUT) DEG F       183.5 /   167.5          85.0 /  120.2     I
I PRESSURE    (IN/OUT) PSIA        75.00 /    74.42         60.00 /   52.73   I
I----------------------------------------------------------------------------I
I SP. GR., LIQ  (60F / 60F H2O)    0.000 /   0.618          1.000 /   1.000   I
I          VAP  (60F / 60F AIR)    2.346 /   0.000          0.000 /   0.000   I
I DENSITY,    LIQUID    LB/FT3     0.000 /  34.262         62.081 /  61.605   I
I             VAPOR     LB/FT3     0.845 /   0.000          0.000 /   0.000   I
I VISCOSITY, LIQUID     CP         0.000 /   0.134          0.810 /   0.559   I
I             VAPOR     CP         0.009 /   0.000          0.000 /   0.000   I
I THRML COND,LIQ  BTU/HR-FT-F     0.0000 /  0.0539         0.3540 /  0.3696   I
I             VAP  BTU/HR-FT-F    0.0119 /  0.0000         0.0000 /  0.0000   I
I SPEC.HEAT,LIQUID BTU /LB F      0.0000 /  0.6373         0.9982 /  0.9986   I
I             VAPOR  BTU /LB F    0.4859 /  0.0000         0.0000 /  0.0000   I
I LATENT HEAT          BTU /LB         10.00                      0.00        I
I VELOCITY              FT/SEC          1.35                      4.72        I
I DP/SHELL(DES/CALC)    PSI         0.00 /  0.58            0.00 /  7.27      I
I FOULING RESIST FT2-HR-F/BTU  0.00050 (0.00532 REQD)           0.00100       I
I----------------------------------------------------------------------------I
I TRANSFER RATE  BTU/HR-FT2-F  SERVICE  148.75 (  86.67 REQD), CLEAN  199.49 I
I HEAT EXCHANGED MMBTU /HR    25.834,      MTD(CORRECTED)  71.9,     FT 0.982 I
I----------------------------------------------------------------------------I
I CONSTRUCTION OF ONE SHELL          SHELL-SIDE                TUBE-SIDE      I
I----------------------------------------------------------------------------I
I DESIGN PRESSURE/TEMP PSIA  /F     150./  200.            125./  200.       I
I NUMBER OF PASSES                        1                        4          I
I MATERIAL                            CARB STL                CUNI9010        I
I INLET  NOZZLE ID/NO       IN         10.0/ 2                 10.0/ 1        I
I OUTLET NOZZLE ID/NO       IN         13.1/ 1                 10.0/ 1        I
I----------------------------------------------------------------------------I
I TUBE: NUMBER  1336, OD  0.750   IN , BWG    16        , LENGTH 16.0 FT     I
I       TYPE BARE,                 PITCH    0.9380 IN,   PATTERN 30 DEGREES  I
I SHELL:  ID   39.00 IN,             SEALING STRIPS   0 PAIRS                I
I BAFFLE: CUT  .350, SPACING(IN): IN    19.99, CENT   13.70, OUT    19.99,SING I
I RHO-V2: INLET NOZZLE  2467.2 LB/FT-SEC2                                     I
I TOTAL WEIGHT/SHELL,LB     8928.0 FULL OF WATER   0.392E+05 BUNDLE   23250.1 I
I----------------------------------------------------------------------------I
```

```
==============================================================================
                    SHELL AND TUBE EXTENDED DATA SHEET
I----------------------------------------------------------------------------I
I EXCHANGER  NAME                                UNIT ID CONDENSER           I
I SIZE    39x  192  TYPE AJ2U,  HORIZONTAL       CONNECTED 1 PARALLEL  1 SERIES I
I AREA/UNIT  4144. FT2 (  2414. FT2 REQUIRED)                                I
I----------------------------------------------------------------------------I
I PERFORMANCE OF ONE UNIT            SHELL-SIDE                TUBE-SIDE     I
I----------------------------------------------------------------------------I
I FEED STREAM NUMBER                   VAPOR                    WATER_IN     I
I FEED STREAM NAME                     VAPOR                    WATER_IN     I
I WT FRACTION LIQUID (IN/OUT)        0.00 / 1.00              1.00 / 1.00    I
I REYNOLDS NUMBER                     161738.                  33337.        I
I PRANDTL NUMBER                       0.839                    4.547        I
I UOPK,LIQUID                       0.000 /  13.155          0.000 /   0.000 I
I      VAPOR                       13.155 /   0.000          0.000 /   0.000 I
I SURFACE TENSION    DYNES/CM       8.744 /   9.026         71.057 /  68.044 I
I FILM COEF(SCL) BTU/HR-FT2-F          269.0 (1.000)        1133.5 (1.000)   I
I FOULING LAYER THICKNESS  IN          0.000                    0.000        I
I----------------------------------------------------------------------------I
I THERMAL RESISTANCE                                                         I
I UNITS:  (FT2-HR-F/BTU)       (PERCENT)   (ABSOLUTE)                        I
I SHELL FILM                      55.29      0.00372                         I
I TUBE  FILM                      15.87      0.00107                         I
I TUBE  METAL                      3.40      0.00023                         I
I TOTAL FOULING                   25.43      0.00171                         I
I ADJUSTMENT                      71.63      0.00482                         I
I----------------------------------------------------------------------------I
I PRESSURE DROP                      SHELL-SIDE               TUBE-SIDE      I
I UNITS: (PSIA  )              (PERCENT)   (ABSOLUTE)  (PERCENT)  (ABSOLUTE) I
I WITHOUT NOZZLES                 53.52          0.31      94.67        6.88 I
I INLET   NOZZLES                 45.56          0.27       3.32        0.24 I
I OUTLET  NOZZLES                  0.92          0.01       2.01        0.15 I
I TOTAL   /SHELL                                 0.58                   7.27 I
I TOTAL   /UNIT                                  0.58                   7.27 I
I DP SCALER                                      1.00                   1.00 I
I----------------------------------------------------------------------------I
I CONSTRUCTION OF ONE SHELL                                                  I
I----------------------------------------------------------------------------I
I TUBE:OVERALL LENGTH         16.0         FT  EFFECTIVE LENGTH    15.80   FT I
I      TOTAL TUBESHEET THK    2.4          IN  AREA RATIO (OUT/IN)  1.210     I
I      THERMAL COND.         26.0BTU/HR-FT-F   DENSITY             559.00 LB/FT3I
I----------------------------------------------------------------------------I
I BAFFLE: THICKNESS            0.250       IN  NUMBER              10         I
I----------------------------------------------------------------------------I
I BUNDLE: DIAMETER            38.4         IN  TUBES IN CROSSFLOW 466         I
I         CROSSFLOW AREA       0.777      FT2  WINDOW AREA          1.254  FT2 I
I         TUBE-BFL LEAK AREA   0.230      FT2  SHELL-BFL LEAK AREA  0.044  FT2 I
I----------------------------------------------------------------------------I
```

ZONE ANALYSIS FOR EXCHANGER CONDENSER

TEMPERATURE - PRESSURE SUMMARY

ZONE	TEMPERATURE IN/OUT DEG F SHELL-SIDE	TEMPERATURE IN/OUT DEG F TUBE-SIDE	PRESSURE IN/OUT PSIA SHELL-SIDE	PRESSURE IN/OUT PSIA TUBE-SIDE
1	183.5/ 178.2	105.4/ 120.2	75.0/ 74.8	55.8/ 52.7
2	178.2/ 172.8	94.4/ 105.4	74.8/ 74.6	58.1/ 55.8
3	172.8/ 167.5	85.0/ 94.4	74.6/ 74.4	60.0/ 58.1

HEAT TRANSFER AND PRESSURE DROP SUMMARY

ZONE	HEAT TRANSFER MECHANISM SHELL-SIDE	HEAT TRANSFER MECHANISM TUBE-SIDE	PRESSURE DROP (TOTAL) PSIA SHELL-SIDE	PRESSURE DROP (TOTAL) PSIA TUBE-SIDE	FILM COEFF. BTU/HR-FT2-F SHELL-SIDE	FILM COEFF. BTU/HR-FT2-F TUBE-SIDE
1	CONDENSATION	LIQ. HEATING	0.19	3.06	395.35	1200.61
2	CONDENSATION	LIQ. HEATING	0.19	2.27	260.83	1115.27
3	CONDENSATION	LIQ. HEATING	0.19	1.94	172.70	1045.69
		TOTAL PRESSURE DROP	0.58	7.27		

HEAT TRANSFER SUMMARY (CONTD.)

ZONE	DUTY MMBTU /HR	DUTY PERCENT	U-VALUE BTU/HR-FT2-F	AREA FT2	LMTD DEG F	FT
1	10.88	42.1	182.63	893.3	67.9	0.982
2	8.06	31.2	145.84	745.4	75.6	0.982
3	6.89	26.7	112.54	775.7	80.4	0.982
TOTAL	25.83	100.0		2414.5		
WEIGHTED			148.75		73.3	0.982
OVERALL					72.5	0.982
INSTALLED				4143.9		

TOTAL DUTY = (WT. U-VALUE)(TOTAL AREA)(WT. LMTD)(OVL. FT)

ZONE DUTY = (ZONE U-VALUE)(ZONE AREA)(ZONE LMTD)(OVL. FT)

Example 11.10

Use *Xist* to obtain a final design for the C_4–C_5 condenser of Example 11.9.

Solution

Using *Xist* to rate the 39-in. J-shell condenser in the previous example, the following two problems were identified:

- The condenser is over-sized.
- The vibration analysis predicts tube vibration problems.

The tubes in the vicinity of the inlet nozzles are the most susceptible to vibration problems. Hence, the following changes are made to mitigate vibration in the inlet regions:

- A full support plate is specified at the U-bend to keep velocities negligible in the U-bend region. This specification is made on the Geometry/Baffles/Supports panel.
- The circular impingement plates are replaced with impingement rods to reduce the bundle and shell entrance velocities. The rod diameter is set at 0.75 in. to match the tube size, and the rod pitch is set at 1.125 in. (1.5 times the rod diameter). These changes are made on the Geometry/Nozzles/Impingement panel.
- A vibration support is installed under the impingement rods in each inlet region. These support plates encompass only the first few tube rows below the rods. This specification is made on the Geometry/Baffles/Supports panel.

The over-sizing is addressed by reducing the shell ID until an acceptably small over-design is achieved. In this process, the number of tubes is left unspecified, allowing *Xist* to determine the tube count from the tube layout. As the shell size is reduced, it is also necessary to reduce the number of tube passes from four to two in order to keep the tube-side pressure drop within the prescribed limits. In this manner, a 35-in. or 34-in. shell is found to be marginally acceptable with a small positive or small negative over-design, respectively. In either case, an acceptable margin can be achieved by increasing the tube length. For the purpose of this example, the 34-in. unit is selected and the design changes are summarized as follows:

- The shell ID is reduced from 39 in. to 34 in.
- The number of tube passes is reduced from 4 to 2.
- The tube length is increased from 16 ft to 18 ft.
- The baffle spacing is rounded to 14 in., giving $B/d_s = 0.41$.
- The baffle cut is automatically adjusted by *Xist* to 35.67% for compatibility with the tube layout.

The Output Summary for this case is shown below, from which it is seen that the over-design is about 10%. In addition, the baffle cut and baffle spacing fall within the recommended region for condensing vapors in Figure 5.3. The maximum unsupported tube length is 36.7 in., which is safely within the TEMA limit from Table 5.C.1 (material group B).

Despite the measures taken to minimize tube vibration, *Xist* still flags a potential vibration problem in the inlet regions. However, *Xist* performs a screening analysis only; a more precise analysis can be performed with *Xvib*, which is a software tool to perform finite element analysis of a single tube. Using *Xvib* to analyze the tubes in the vicinity of the nozzles confirms that vibration levels are satisfactory in the inlet regions and, therefore, should not be a cause of tube damage. Hence, the design is considered to be acceptable. However, acoustic vibrations are still a potential problem in this unit, and deresonating baffles may be needed for noise control.

Xist Output Summary for Example 11.10: Design 1 (J-shell Condenser)

	Xist Ver. 7.00 5/11/2013 16:36 SN: 14337-877388691					US Units
	Example 11.10					
	Rating - Horizontal Multipass Flow TEMA AJ21U Shell With Single-Segmental Baffles					
1	**No Data Check Messages.**					
2	**See Runtime Message Report for Warning Messages.**					
3	**Process Conditions**		**Hot Shellside**		**Cold Tubeside**	
4	Fluid name			C4-C5		Water
5	Flow rate	(1000-lb/hr)		180.00		735.43
6	Inlet/Outlet Y	(Wt. frac vap.)	1.0000	0.0000	0.0000	0.0000
7	Inlet/Outlet T	(Deg F)	183.32	166.80	85.00	120.23
8	Inlet P/Avg	(psia)	75.000	74.469	60.000	58.823
9	dP/Allow.	(psi)	1.062	5.000	2.355	10.000
10	Fouling	(ft2-hr-F/Btu)		0.00050		0.00100
11	**Exchanger Performance**					
12	Shell h	(Btu/ft2-hr-F)	203.41	Actual U	(Btu/ft2-hr-F)	124.99
13	Tube h	(Btu/ft2-hr-F)	990.25	Required U	(Btu/ft2-hr-F)	113.57
14	Hot regime	(--)	Transition	Duty	(MM Btu/hr)	25.876
15	Cold regime	(--)	Sens. Liquid	Eff. area	(ft2)	3202.3
16	EMTD	(Deg F)	71.1	Overdesign	(%)	10.06
17	**Shell Geometry**			**Baffle Geometry**		
18	TEMA type	(--)	AJ21U	Baffle type	(--)	Single-Seg.
19	Shell ID	(inch)	34.000	Baffle cut	(Pct Dia.)	35.67
20	Series	(--)	1	Baffle orientation	(--)	Parallel
21	Parallel	(--)	1	Central spacing	(inch)	14.000
22	Orientation	(deg)	0.00	Crosspasses	(--)	14
23	**Tube Geometry**			**Nozzles**		
24	Tube type	(--)	Plain	Shell inlet	(inch)	10.020
25	Tube OD	(inch)	0.7500	Shell outlet	(inch)	13.124
26	Length	(ft)	18.000	Inlet height	(inch)	2.0568
27	Pitch ratio	(--)	1.2500	Outlet height	(inch)	3.7812
28	Layout	(deg)	30	Tube inlet	(inch)	10.020
29	Tubecount	(--)	914	Tube outlet	(inch)	10.020
30	Tube Pass	(--)	2			
31	**Thermal Resistance, %**		**Velocities, ft/sec**		**Flow Fractions**	
32	Shell	61.45	Shellside	9.89	A	0.170
33	Tube	15.27	Tubeside	3.44	B	0.393
34	Fouling	21.37	Crossflow	15.77	C	0.107
35	Metal	1.91	Window	10.88	E	0.088
36					F	0.242

Due to the challenges of a J-shell design with respect to vibration, an X-shell design should be considered for this application. Vibration is mitigated in the X-shell condenser by implementing the following features:

- Eight support plates spaced uniformly along the tube bundle. Support plates are specified on the Geometry/Baffles/Supports panel.
- A full support plate at the U-bend, as in the J-shell design.
- Three 10-in. inlet nozzles to better distribute the vapor along the length of the tube bundle and to reduce the inlet velocity.

The following additional specifications are made:

- Three 8-in. schedule 40 outlet nozzles. These nozzles will be self-venting.
- Circular impingement plates, which are selected on the Geometry/Nozzles/Impingement panel.
- The number of sealing strip pairs is set to be chosen by the program on the Geometry/Tube Layout/Clearances panel.

Starting with a 35-in. shell and a maximum tube length of 20 ft, an acceptable design is readily found consisting of a 33-in. shell containing 878 tubes. The over-design is 10% and *Xist* generates no warning messages concerning potential vibration problems. Using the tube-layout editor, 8 tubes (4 U-tubes) are replaced by tie rods, leaving a bundle with 870 tubes. Re-running the program with this tube bundle as input produces the Output Summary shown below.

Xist Output Summary for Example 11.10: Design 2 (X-shell Condenser)

Xist Ver. 7.00 5/12/2013 17:53 SN: 14337-877388691 — **US Units**

Example 11.10

Rating - Horizontal Multipass Flow TEMA AXU Shell With No Baffles

1	See Data Check Messages Report for Warning Messages.					
2	See Runtime Message Report for Warning Messages.					
3	**Process Conditions**		**Hot Shellside**		**Cold Tubeside**	
4	Fluid name			C4-C5		Water
5	Flow rate	(1000-lb/hr)		180.00		735.43
6	Inlet/Outlet Y	(Wt. frac vap.)	1.0000	0.0000	0.0000	0.0000
7	Inlet/Outlet T	(Deg F)	183.32	167.64	85.00	120.10
8	Inlet P/Avg	(psia)	75.000	74.874	60.000	58.630
9	dP/Allow.	(psi)	0.253	5.000	2.739	10.000
10	Fouling	(ft2-hr-F/Btu)		0.00050		0.00100
11			**Exchanger Performance**			
12	Shell h	(Btu/ft2-hr-F)	175.74	Actual U	(Btu/ft2-hr-F)	114.53
13	Tube h	(Btu/ft2-hr-F)	1026.7	Required U	(Btu/ft2-hr-F)	104.45
14	Hot regime	(--)	Gravity	Duty	(MM Btu/hr)	25.780
15	Cold regime	(--)	Sens. Liquid	Eff. area	(ft2)	3355.1
16	EMTD	(Deg F)	72.8	Overdesign	(%)	9.65
17	**Shell Geometry**			**Baffle Geometry**		
18	TEMA type	(--)	AXU	Baffle type	(--)	Support
19	Shell ID	(inch)	33.000	Baffle cut	(Pct Dia.)	
20	Series	(--)	1	Baffle orientation	(--)	
21	Parallel	(--)	1	Central spacing	(inch)	26.465
22	Orientation	(deg)	0.00	Crosspasses	(--)	1
23	**Tube Geometry**			**Nozzles**		
24	Tube type	(--)	Plain	Shell inlet	(inch)	10.020
25	Tube OD	(inch)	0.7500	Shell outlet	(inch)	7.9810
26	Length	(ft)	20.000	Inlet height	(inch)	3.1334
27	Pitch ratio	(--)	1.2500	Outlet height	(inch)	3.6334
28	Layout	(deg)	30	Tube inlet	(inch)	10.020
29	Tubecount	(--)	870	Tube outlet	(inch)	10.020
30	Tube Pass	(--)	2			
31	**Thermal Resistance, %**		**Velocities, ft/sec**		**Flow Fractions**	
32	Shell	65.17	Shellside	2.83	A	0.000
33	Tube	13.49	Tubeside	3.61	B	0.869
34	Fouling	19.58	Crossflow	2.46	C	0.131
35	Metal	1.75	Window	0.00	E	0.000
36					F	0.000

Design details for both the J-shell and X-shell units are summarized in the table below. Tube layouts and exchanger drawings from *Xist* are also shown. Notice that the horizontal pass-partition lane, in combination with the side-to-side flow pattern, results in an *F*-stream bypass flow in the J-shell condenser. Four seal rods with diameters of 0.75 in. (as specified by *Xist*) are employed to partially block the bypass-flow area.

Design Summaries for Example 11.10

Item	Design 1	Design 2
Exchanger type	AJU	AXU
Shell size (in.)	34	33
Surface area (ft^2)	3202	3355
Number of tubes	914	870
Tube OD (in.)	0.75	0.75
Tube length (ft)	18	20
Tube BWG	16	16
Tube passes	2	2
Tube pitch (in.)	15/16	15/16
Tube layout	Triangular	Triangular
Number of baffles	13	–
Number of support plates	–	8
Baffle cut (%)	35.67	–
Baffle thickness (in.)	0.3125	–
Central baffle spacing (in.)	14.0	–
Front end baffle spacing (in.)	22.7	–
Max. unsupported length (in.)	36.7	26.5
Sealing strip pairs	0	5
Tube-side nozzles	10-in. schedule 40	10-in. schedule 40
Shell-side inlet nozzles	10-in. schedule 40 (2)	10-in. schedule 40 (3)
Shell-side outlet nozzles(s)	14-in. schedule 40 (1)	8-in. schedule 40 (3)
Full U-bend support	yes	yes
Tube-side velocity (ft/s)	3.4	3.6
$(Re_i)_{ave}$	25,140	25,891
ΔP_i (psi)	2.36	2.74
ΔP_o (psi)	1.06	0.25
Over-design (%)	10.1	9.7

Exchanger Drawing and Tube Layout for Design 1 (J-shell Condenser)

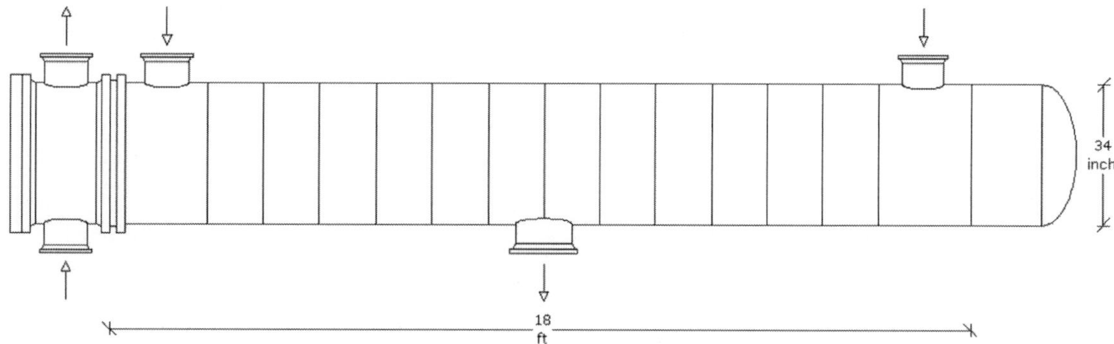

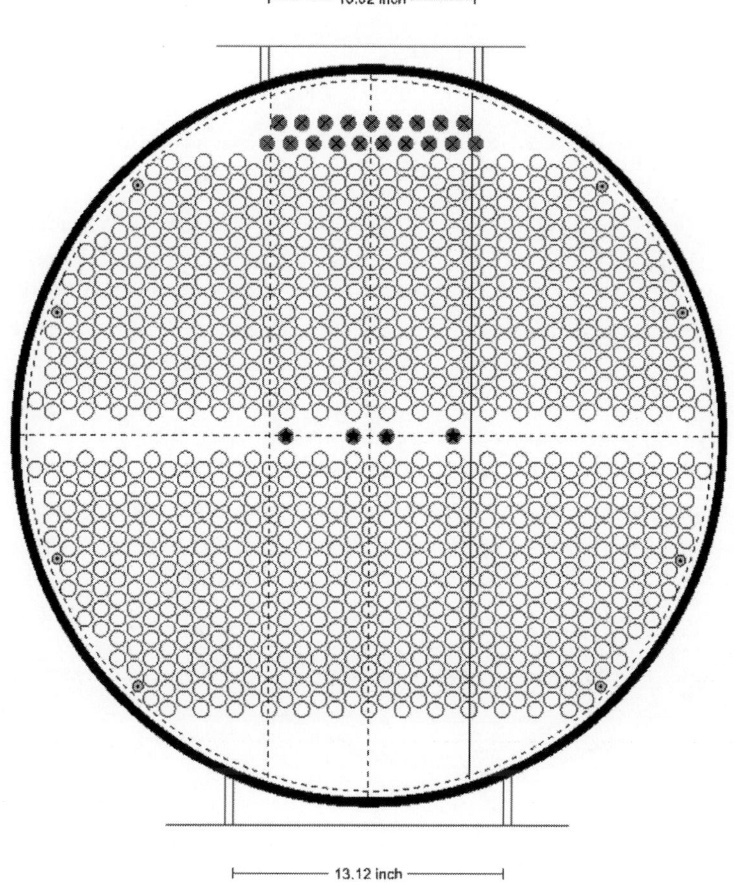

Exchanger Drawing and Tube Layout for Design 2 (X-shell Condenser)

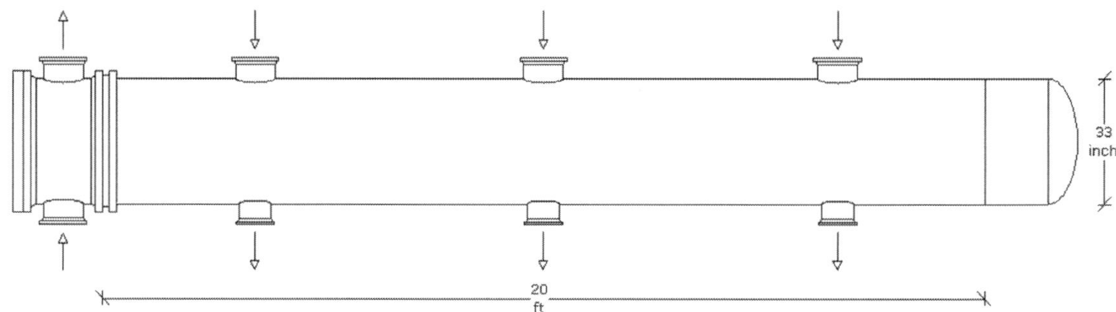

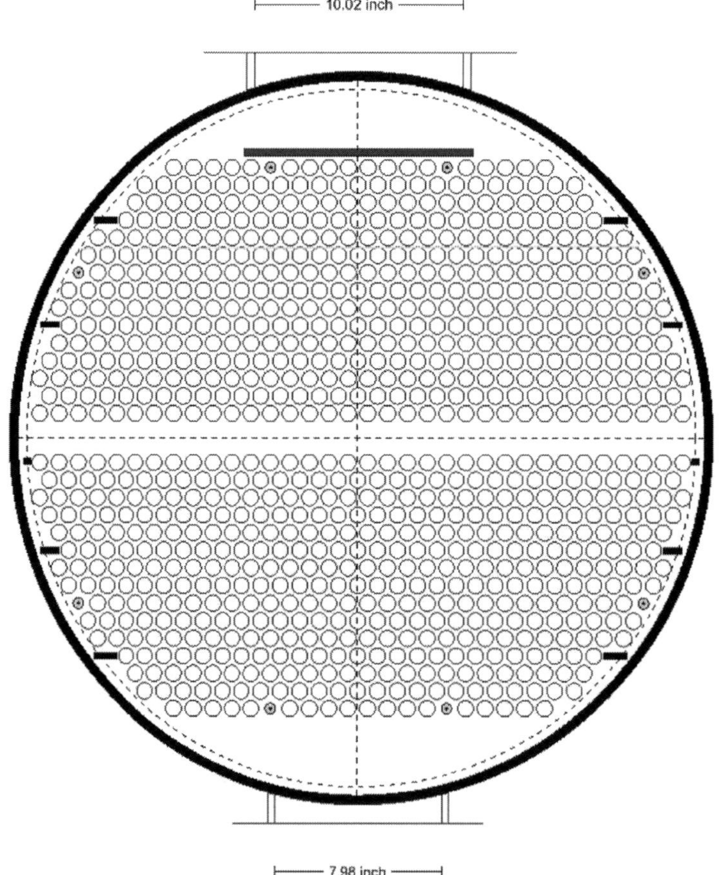

Example 11.11

100,000 kg/hr of propane are condensed on the shell-side of a horizontal TEMA E shell (type AEU). The inlet vapor temperature is 70.79°C and the outlet condensate temperature is 48.9°C. There is about 20°C of desuperheating and less than 1°C of subcooling. As with most industrial propane, small amounts of other hydrocarbons (mostly ethane and butane) are mixed with the propane so that the boiling range is 1.5°C. Cooling water is supplied at an inlet temperature of 30°C and the outlet temperature is 35°C. A fouling factor of 0.000176 $m^2 \cdot K/W$ is provided for each stream. Use *Xist* to compare design attributes of:

(a) Single and double segmental baffles.

(b) Plain tubes, low-finned tubes and double-enhanced tubes.

Solution

The overall attributes of a design with single segmental baffles are shown in the tube layout and exchanger setting plan given below. This design will work satisfactorily, but the baffles will cause a buildup of condensate. This buildup may interfere with the desired weir-type drainage in the outlet nozzle, and some tube rows may become submerged. Designers often cut the bottom of the baffles to facilitate drainage.

Tube Layout for Example 11.11: Design with Single Segmental Baffles

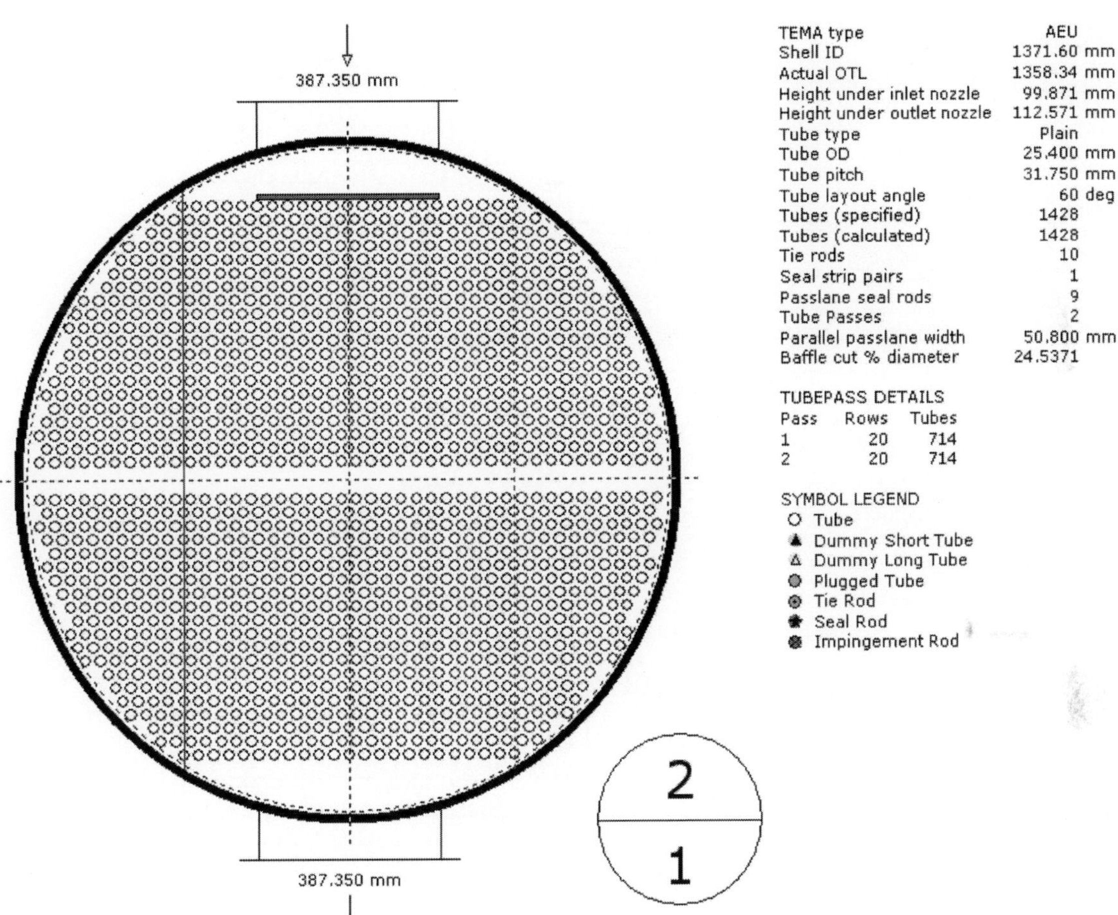

Setting Plan for Example 11.11: Design with Single Segmental Baffles

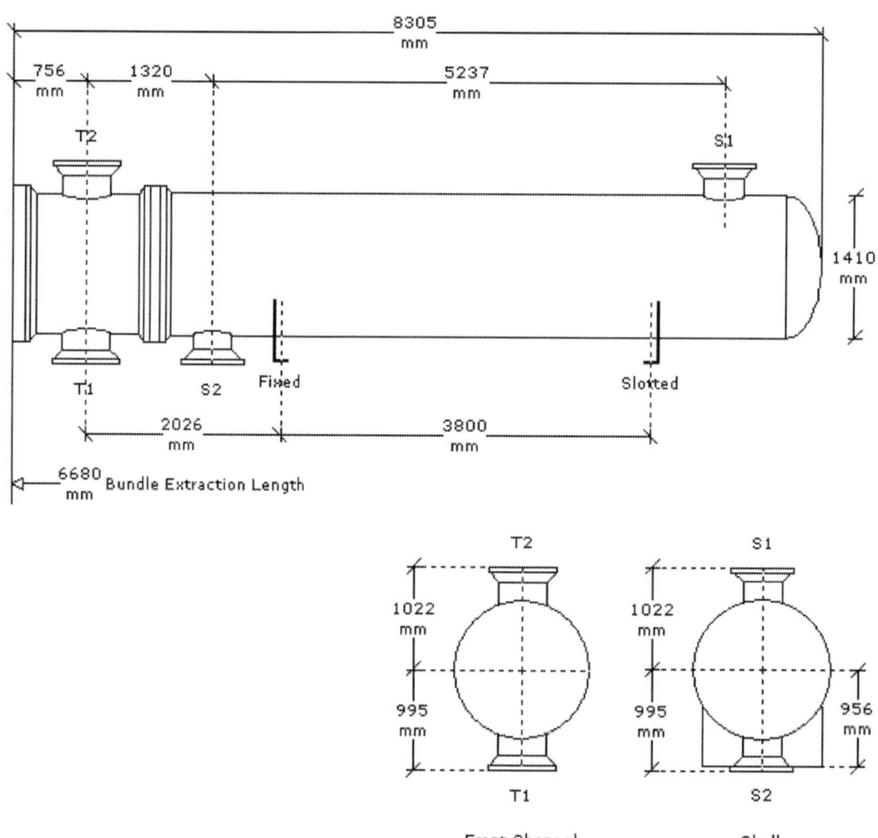

	Nozzles	OD, mm	Rating	Design	Shell	Tube	Weight	kg	Company	None		
S1	Inlet	406.4		Pres (kPaG)	1861.58	551.581	Bundle	8552	Customer		Ref	
S2	Outlet	406.4		Temp (C)	98.89	65.56	Dry	16380	Item			
T1	Inlet	501.65		Passes	1	2	Wet	30430	Service			
T2	Outlet	501.65		Thick (mm)	19.05	1.245			TEMA	AEU		Setting Plan
									Date	5/13/2013	By	
									Diagram		Rev	

Another approach is to use double segmental baffles. The tube layout for a double segmental design with two tube rows of overlap at the baffle cuts is shown below. As can be seen, only the center baffle blocks condensate drainage and designers often cut the bottom of the center baffles. The shell-side pressure drop decreases from 12.52 kPa to 3.49 kPa. In some situations this pressure change can significantly increase the mean temperature difference, but in this case the improvement is negligible since the inlet pressure is high. The Rating Data Sheet generated by *Xist* for this design is also shown below:

Tube Layout for Example 11.11: Design with Double Segmental Baffles

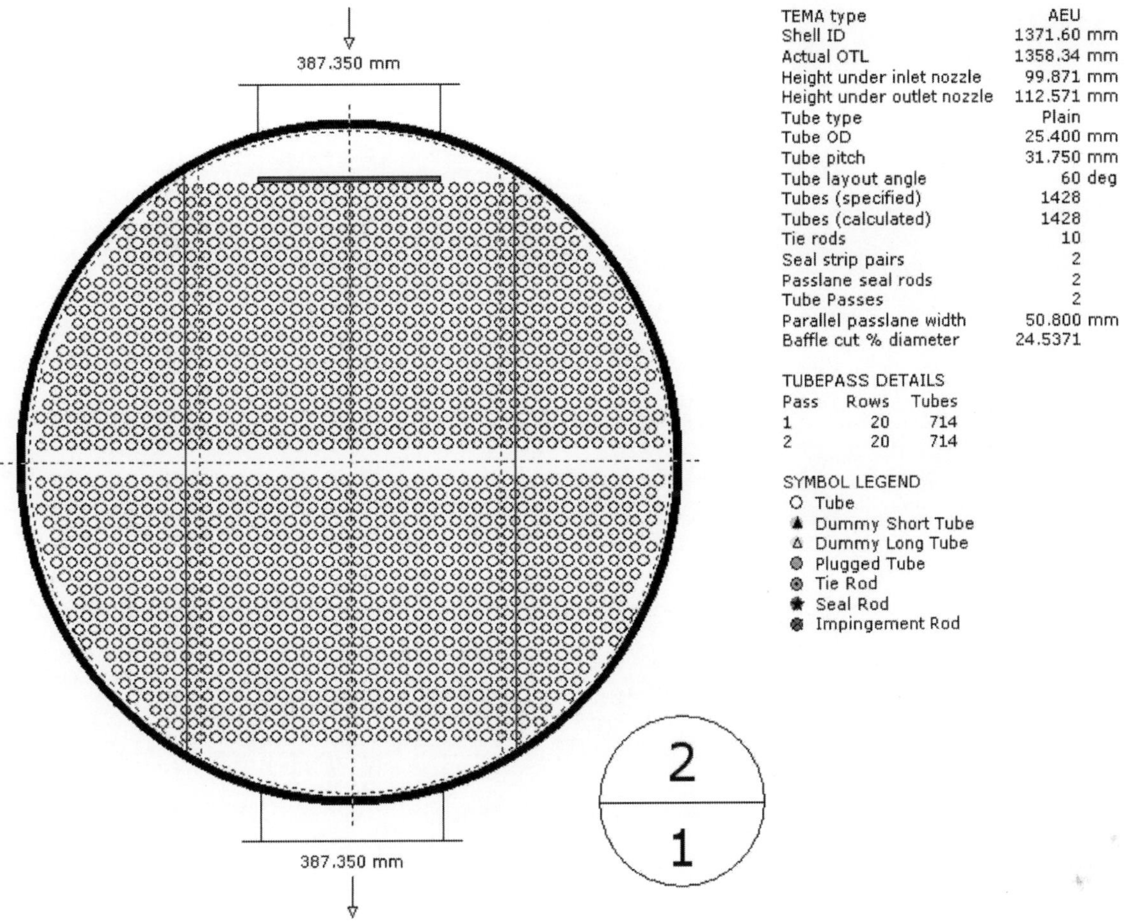

Rating Data Sheet for Example 11.11: Design with Double Segmental Baffles

HTRI	HEAT EXCHANGER RATING DATA SHEET					Page 1 **SI Units**
Service of Unit			Item No.			
Type AEU	Orientation Horizontal		Connected In	1 Parallel	1 Series	
Surf/Unit (Gross/Eff) 694.63 / 681.79 m2	Shell/Unit 1		Surf/Shell (Gross/Eff)	694.63 / 681.79 m2		
	PERFORMANCE OF ONE UNIT					
Fluid Allocation		Shell Side		Tube Side		
Fluid Name		Propane		water		
Fluid Quantity, Total	kg/s	27.778		441.63		
Vapor (In/Out)	wt%	100.00	0.00	0.00	0.00	
Liquid	wt%	0.00	100.00	100.00	100.00	
Temperature (In/Out)	C	70.79	48.90	30.01	35.01	
Density	kg/m3	34.679	451.48	995.86	994.25	
Viscosity	mN-s/m2	0.0107	0.0739	0.7971	0.7192	
Specific Heat	kJ/kg-C	2.1818	3.3522	4.1787	4.1777	
Thermal Conductivity	W/m-C	0.0256	0.0798	0.6153	0.6223	
Critical Pressure	kPa					
Inlet Pressure	kPa	1751.3		580.00		
Velocity	m/s		1.04		1.51	
Pressure Drop, Allow/Calc	kPa	68.948	3.490	80.000	25.266	
Average Film Coefficient	W/m2-K	1356.3		6445.6		
Fouling Resistance (min)	m2-K/W	0.000176		0.000176		
Heat Exchanged	9.2316 MegaWatts	MTD (Corrected)	18.4 C	Overdesign	4.45 %	
Transfer Rate, Service	734.78 W/m2-K	Calculated	767.50 W/m2-K	Clean	1073.5 W/m2-K	
	CONSTRUCTION OF ONE SHELL			Sketch (Bundle/Nozzle Orientation)		
		Shell Side	Tube Side			
Design Pressure	kPaG	1861.6	551.58			
Design Temperature	C	98.89	65.56			
No Passes per Shell		1	2			
Flow Direction		Downward	Upward			
Connections	In mm	1 @ 387.35	1 @ 488.95			
Size &	Out mm	1 @ 387.35	1 @ 488.95			
Rating	Liq. Out mm	@	1 @			
Tube No. 1428.0	OD 25.400 mm	Thk(Avg) 1.245 mm	Length 6.096 m	Pitch 31.750 mm	Tube pattern 60	
Tube Type Plain		Material Carbon steel		Pairs seal strips	2	
Shell ID 1371.6 mm		Kettle ID mm		Passlane Seal Rod No.	2	
Cross Baffle Type Parallel	Double-Seg.		%Cut (Diam) 24.54	Impingement Plate	Circular plate	
Spacing(c/c) 609.60 mm		Inlet 762.00 mm		No. of Crosspasses	9	
Rho-V2-Inlet Nozzle 1602.2 kg/m-s2		Shell Entrance	2037.7 kg/m-s2	Shell Exit	82.98 kg/m-s2	
		Bundle Entrance	2326.6 kg/m-s2	Bundle Exit	74.32 kg/m-s2	
Weight/Shell 16148 kg		Filled with Water	30229 kg	Bundle	8320.4 kg	
Notes:		Thermal Resistance, %		Velocities, m/s		Flow Fractions
		Shell	56.59	Shellside 1.04	A	0.020
		Tube	13.20	Tubeside 1.51	B	0.746
		Fouling	28.50	Crossflow 0.99	C	0.015
		Metal	1.71	Window 1.14	E	0.073
					F	0.145

From the Rating Data Sheet, the thermal resistance on the shell-side dominates (56.59% of the total resistance) and low-finned tubes are a suitable enhancement for condensation with low-surface-tension fluids. For this application with U-tubes, a tube supplier that can provide finning on the straight tube portion only would be needed to avoid wall-thinning under the fins. The tube velocity should remain greater than 1 m/s to reduce sedimentation fouling but less than 2.5 m/s to minimize the potential for erosion and avoid excessive pressure drop. Since fewer low-finned tubes are needed than plain tubes, lower cooling water flow rates can be applied, which reduces pumping costs but also lowers the EMTD. With enhanced surfaces, lower fouling margins should be considered to prevent fouling margin from becoming the dominant thermal resistance. In this case we consider zero fouling allowance on the shell side (since propane condensation is a clean service) and half the original fouling resistance on the tube side (since the velocity is sufficient to minimize sedimentation fouling).

Using the above guidelines, condenser designs were obtained with low-finned tubes and double-enhanced tubes in place of plain tubes. In the table below the results are compared with those for plain tubes obtained in part (a). Carbon steel tubes with a straight tube length of 6.096 m and double segmental baffles are employed in all three cases. Low-finned tubes provide a significant advantage over plain tubes with respect to shell size and associated tube count, which is reduced by more than 50%.

Compared with low-finned tubes, the double enhanced tubes provide a relatively small advantage, which is not surprising since the tube-side thermal resistance was not very high to begin with. (For the low-finned design, the shell-side and tube-side resistances are nearly equal.) Low-finned tubes, and possibly double-enhanced tubes, can provide a cost-effective alternative to plain tubes, particularly for process upgrades when duty increases are desired.

Parameter	Plain Tube*	Low-Finned Tube**	Double-Enhanced Tube***
Shell ID, mm	1372	1000	950
Tube count	1428	696	608
Coolant flow rate, kg/s	442	183	183
Heat-transfer area, m^2	682	1035	792
EMTD, °C	18.4	14.1	14.1

*1-in. OD carbon steel tubes

**1-in. OD plain tube end, 1024 fins/meter, 1.24 mm fin height

***Wieland GEWA-KS tubes with 1-in. OD plain ends, low fins on the outside (748 fins/meter, 1.5 mm fin height) and helical fins on the inside surface

References

[1] Mueller AC. Condensers. In: Heat Exchanger Design Handbook, vol. 3. New York: Hemisphere Publishing Corp.; 1988.
[2] Hewitt GF, Shires GL, Bott TR. Process Heat Transfer. Boca Raton, FL: CRC Press; 1994.
[3] Kakac S, Liu H. Heat Exchangers: Selection, Rating and Thermal Design. Boca Raton, FL: CRC Press; 1998.
[4] Kister HZ. Distillation Operation. New York: McGraw-Hill; 1990.
[5] Simpson LL. Sizing piping for process plants. Chem Eng 1968;75(No. 13):192.
[6] Nusselt W. Die Obertlachenkondensation des Wasserdamtes, parts I and II. Z Ver Deut Ing 1916;60. 541–546 and 569–575.
[7] Bird RB, Stewart WE, Lightfoot EN. Transport Phenomena. New York: Wiley; 1960.
[8] Rose J. Laminar film condensation of pure vapors. In: Kandlikar SG, Shoji M, Dhir VK, editors. Handbook of Phase Change: Boiling and Condensation. Philadelphia: Taylor and Francis; 1999.
[9] Kern DQ. Process Heat Transfer. New York: McGraw-Hill; 1950.
[10] Kern DQ. Mathematical development of tube loading in horizontal condensers. AIChE J 1958;4:157–60.
[11] Butterworth D. Film condensation of pure vapor. In: Heat Exchanger Design Handbook, vol. 2. New York: Hemisphere Publishing Corp.; 1988.
[12] Uehara H. Transition and turbulent film condensation. In: Kandlikar SG, Shoji M, Dhir VK, editors. Handbook of Phase Change: Boiling and Condensation. Philadelphia: Taylor and Francis; 1999.
[13] Sadisivan P, Lienhard JH. Sensible heat correction in laminar film boiling and condensation. J Heat Transfer 1987;109:545–7.
[14] Chen MM. An analytical study of laminar film condensation: Part 1-flat plates; Part 2-single and multiple horizontal tubes. J Heat Transfer 1961;83:48–60.
[15] Boyko LD, Kruzhilin GN. Heat transfer and hydraulic resistance during condensation of steam in a horizontal tube and in a bundle of tubes. Int J Heat Mass Transfer 1967;10:361–73.
[16] McNaught JM. Two-phase forced convection heat transfer during condensation on horizontal tube bundles. In: Proc. Seventh Int. Heat Transfer Conf, 5. New York: Hemisphere Publishing Corp.; 1982. pp. 125–31.
[17] Breber G, Palen JW, Taborek J. Prediction of horizontal tubeside condensation of pure components using flow regime criteria. J Heat Transfer 1980;102:471–6.
[18] Akers WW, Deans HA, Crosser OK. Condensation heat transfer within horizontal tubes. Chem Eng Prog Symposium Series 1959;55(No. 29):171–6.
[19] Akers WW, Rosson HF. Condensation inside a horizontal tube. Chem Eng Prog Symposium Series 1960;56(No. 30):145–9.
[20] Soliman M, Schuster JR, Berenson PJ. A general heat transfer correlation for annular flow condensation,. J Heat Transfer 1968;90:267–79.
[21] Traviss DP, Rhosenow WM, Baron AB. Forced convection condensation inside tubes: a heat transfer equation for condenser design. ASHRAE Trans 1972;79:157–65.
[22] Cavallini A, Zecchin R. A dimensionless correlation for heat transfer in forced convection condensation. Proc 5th Int Heat Transfer Conf Tokyo 1974:309–13.
[23] Shah MM. A general correlation for heat transfer during film condensation inside pipes. Int J Heat Mass Transfer 1979;22:547–56.
[24] Kraus AD, Aziz A, Welty J. Extended Surface Heat Transfer. New York: Wiley; 2001.
[25] Beatty KO, Katz DL. Condensation of vapors on outside of finned tubes. Chem Eng Prog 1948;44(No. 1):55–70.
[26] Collier JG, Thome JR. Convective Boiling and Condensation. 3rd ed. Oxford: Clarendon Press; 1994.
[27] Bell KJ, Mueller AJ. Wolverine Engineering Data Book II. Wolverine Tube, Inc; 2001. www.wlv.com.
[28] Kern DQ, Kraus AD. Extended Surface Heat Transfer. New York: McGraw-Hill; 1972.
[29] Minton PE. Heat Exchanger Design. In: McKetta JJ, editor. Heat Transfer Design Methods. New York: Marcel Dekker; 1991.
[30] Rubin FL. Multizone condensers: desuperheating, condensing, subcooling. Heat Transfer Eng 1981;3(No. 1):49–58.
[31] Gulley DL. How to calculate weighted MTD's. Hydrocarbon Processing 1966;45(No. 6):116–22.
[32] Taborek J. Shell-and-tube heat exchangers. In: Heat Exchanger Design Handbook, vol. 3. New York: Hemisphere Publishing Corp.; 1988.
[33] Taylor R, Krishnamurthy R, Furno JS, Krishna R. Condensation of vapor mixtures: 1. nonequilibrium models and design procedures; 2. comparison with experiment. Ind Eng Chem Process Des Dev 1986;25:83–101.
[34] Bell KJ, Ghaly MA. An approximate generalized design method for multicomponent / partial condensers. AIChE Symposium Series 1972;69(No. 131):72–9.
[35] Taborek J. Mean temperature difference. In: Heat Exchanger Design Handbook, vol. 1. New York: Hemisphere Publishing Corp.; 1988.
[36] Diehl JE, Koppany CR. Flooding velocity correlation for gas-liquid counterflow in vertical tubes. Chem Eng Prog Symposium Series 1969;65(No. 92):77–83.

Appendix 11.A LMTD Correction Factors for TEMA J- and X-Shells

The following notation is used for the charts:

A = heat-transfer surface area in exchanger

$C_1 = (\dot{m}\ C_P)_1$ = heat capacity flow rate of shell-side fluid
$C_2 = (\dot{m}\ C_P)_2$ = heat capacity flow rate of tube-side fluid
$\Delta T_m = F(\Delta T_{ln})_{cf}$ = mean temperature difference in exchanger
U = overall heat-transfer coefficient

All other symbols are explicitly defined on the charts themselves.

When the outlet temperatures of both streams are known, the lower charts can be used to obtain the LMTD correction factor in the usual manner. When the outlet temperatures are unknown but the product UA is known, the upper charts can be used to obtain θ, from which the mean temperature difference, ΔT_m, can be found. The exchanger duty is then obtained as $q = UA\Delta T_m$, and the outlet temperatures are computed using the energy balances on the two streams.

The F-factor for J-shells is the same if the flow direction of the shell-side fluid is reversed so that there are two entrance nozzles and one exit nozzle, as in a condenser. However, the F-factor for J-shells is not symmetric with respect to fluid placement; switching the fluids between tubes and shell will change the value of F.

The equations from which the charts were constructed are given by Taborek [35]. The equations for J-shells are readily solvable on a scientific calculator. For an X-shell, the equations are rather intractable, and mathematical software such as MATHCAD is recommended in this case. The following definitions are used in the equations:

$$\Psi = \frac{R-1}{\ln[(1-P)/(1-PR)]} \quad (R \neq 1) \tag{11.A.1}$$

$$\Psi = (1-P)/P \quad (R = 1) \tag{11.A.2}$$

$$\Phi = \exp\,(1/F\Psi) \tag{11.A.3}$$

For these configurations, the equations are solved explicitly for P, but are implicit in F.

(1) J-Shell with single tube pass

$$P = 1 - \left(\frac{2R-1}{2R+1}\right)\left[\frac{2R+\Phi^{-(R+0.5)}}{2R-\Phi^{-(R+0.5)}}\right] \quad (R \neq 0.5) \tag{11.A.4}$$

$$P = 1 - \left(\frac{1+\Phi^{-1}}{2+\ln\Phi}\right) \quad (R = 0.5) \tag{11.A.5}$$

(2) J-shell with even number of tube passes

$$P = \left(\frac{R\Phi^R}{\Phi^R - 1} + \frac{\Phi}{\Phi - 1} - \frac{1}{\ln\Phi}\right)^{-1} \tag{11.A.6}$$

(3) X-shell with single tube pass

The equation for this configuration represents the situation in which the streams flow perpendicular to one another and both streams flow through a large number of channels with no mixing between channels. This situation is referred to as unmixed–unmixed cross flow.

$$P = \frac{1}{R\ln\Phi}\sum_{n=0}^{\chi}\left\{\left[1 - \Phi^{-1}\sum_{m=0}^{n}\frac{(\ln\Phi)^m}{m!}\right]\left[1 - \Phi^{-R}\sum_{m=0}^{n}\frac{(R\ln\Phi)^m}{m!}\right]\right\} \tag{11.A.7}$$

The graph for this case was constructed with the number of channels, χ, set to 10 [35].

To calculate the F-factor for given values of P and R, the applicable equation is solved for Φ, from which F is easily obtained by taking logarithms.

Example 11.A.1

Calculate the LMTD correction factor for the following cases:

(a) A J-shell with one tube pass, $R = 0.5$ and $P = 0.75$.
(b) A J-shell with two tube passes, $R = 1.0$ and $P = 0.5$.

Solution

(a) Equation (11.A.5) is applicable for this case. Setting $P = 0.75$ gives the following equation for Φ:

$$0.75 = 1 - \left(\frac{1+\Phi^{-1}}{2+\ln\Phi}\right)$$

Using the nonlinear equation solver on a TI-80 series calculator, the solution is found to be $\Phi = 10.728$, and $\ln \Phi = 2.37286$. Next, Ψ is calculated using Equation (11.A.1):

$$\Psi = \frac{0.5 - 1.0}{\ln[(1 - 0.75)/(1 - 0.75 \times 0.5)]} = 0.54568$$

Finally, Equation (11.A.3) is solved for F:

$$F = (\Psi \ln \Phi)^{-1} = (0.54568 \times 2.37286)^{-1} = 0.7723$$

(b) Equation (11.A.6) is applicable for this case. Setting $R = 1.0$ and $P = 0.5$ gives the following equation for Φ:

$$0.5 = \left(\frac{\Phi}{\Phi - 1} + \frac{\Phi}{\Phi - 1} - \frac{1}{\ln \Phi}\right)^{-1}$$

The solution is found using a TI calculator: $\Phi = 3.51286$, from which $\ln \Phi = 1.25643$. Since $R = 1.0$, the value of Ψ is obtained from Equation (11.A.2):

$$\Psi = (1 - P)/P = (1 - 0.5)/0.5 = 1.0$$

The LMTD correction factor is found by solving Equation (11.A.3) as before:

$$F = (\Psi \ln \Phi)^{-1} = (1.0 \times 1.25643)^{-1} = 0.7959$$

The reader can verify that the calculated values of F are in agreement with the graphs.

Appendix 11.B. Flooding in Reflux Condensers

As noted in the text, the vapor velocity in a reflux condenser must be kept low enough to prevent excessive liquid entrainment and flooding of the tubes. The flooding correlation of Diehl and Koppany [36] can be used to estimate the vapor velocity at which flooding begins. This correlation, which is based on experimental data that cover a wide range of operating conditions, is as follows:

$$\begin{aligned} V_f &= F_1 F_2 (\sigma/\rho_V)^{0.5} \quad \text{for } F_1 F_2 (\sigma/\rho_V)^{0.5} \geq 1.0 \\ &= 1.0 \quad \text{otherwise} \end{aligned} \tag{11.B.1}$$

$$\begin{aligned} F_1 &= (80\, D_i/\sigma)^{0.4} \quad \text{for } 80\, D_i/\sigma < 1.0 \\ &= 1.0 \quad \text{otherwise} \end{aligned} \tag{11.B.2}$$

$$F_2 = (G_V/G_L)^{0.25} \tag{11.B.3}$$

where

V_f = vapor superficial velocity at incipient flooding, ft/s
D_i = tube internal diameter, in.
ρ_V = vapor density, lbm/ft^3
σ = liquid surface tension, dyne/cm

The flooding velocity can be increased by tapering the ends of the tubes to facilitate condensate drainage. For a tapering angle of 70° with the horizontal, the flooding velocity can be estimated by the following equation that was obtained by curve fitting the graphical correlation in Ref. [36].

$$V_{f,70}/V_f = 0.692 + 0.058\,\sigma - 0.00127\,\sigma^2 + 9.34 \times 10^{-6}\,\sigma^3 \tag{11.B.4}$$

where

$V_{f,70}$ = flooding velocity for tapered tube
V_f = flooding velocity for un-tapered tube from Equation (11.B.1)
σ = surface tension of liquid, dyne/cm

Equation (11.B.4) is valid for systems with liquid surface tension between 6 and 60 dyne/cm. The effect of tapering is negligible for surface tension below about 6 dyne/cm. Based on the data reported by Diehl and Koppany [36], the tapering angle should be in the range of 60° to 75°. Over this range, the effect of tapering angle on flooding velocity is insignificant.

A reflux condenser should be designed to operate at a vapor velocity that is safely below the flooding velocity. Thus, an operating velocity in the range of 50% to 70% of the flooding velocity should be used for design purposes. The amount of liquid entrainment decreases with decreasing vapor velocity. At 50% of the flooding velocity, the entrainment ratio (mass of entrained liquid per unit mass of vapor) is approximately 0.005 [36].

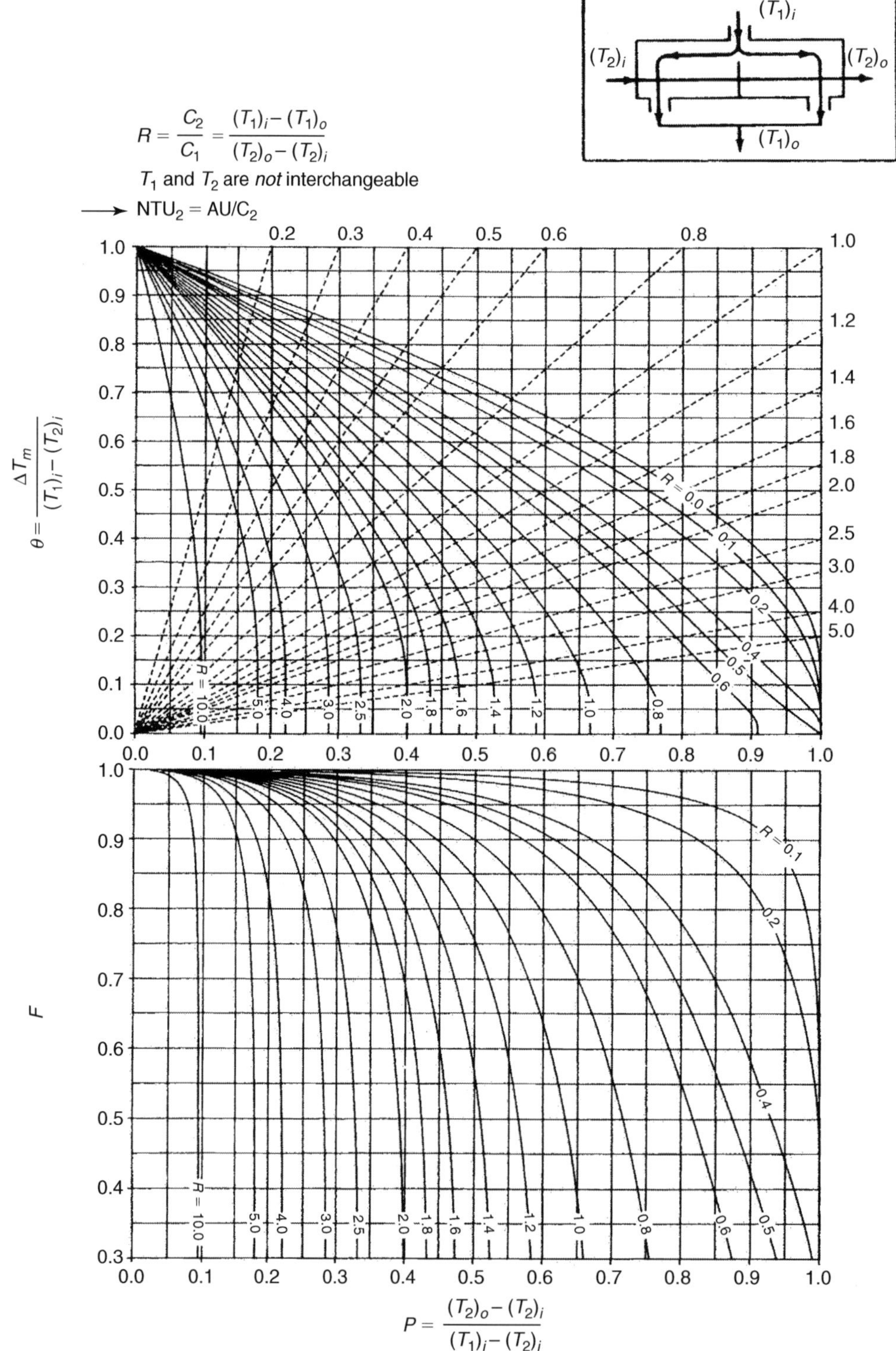

FIGURE 11.A.1 Mean temperature difference relationships for a TEMA J-shell exchanger with one tube pass (Source: Ref. [35]).

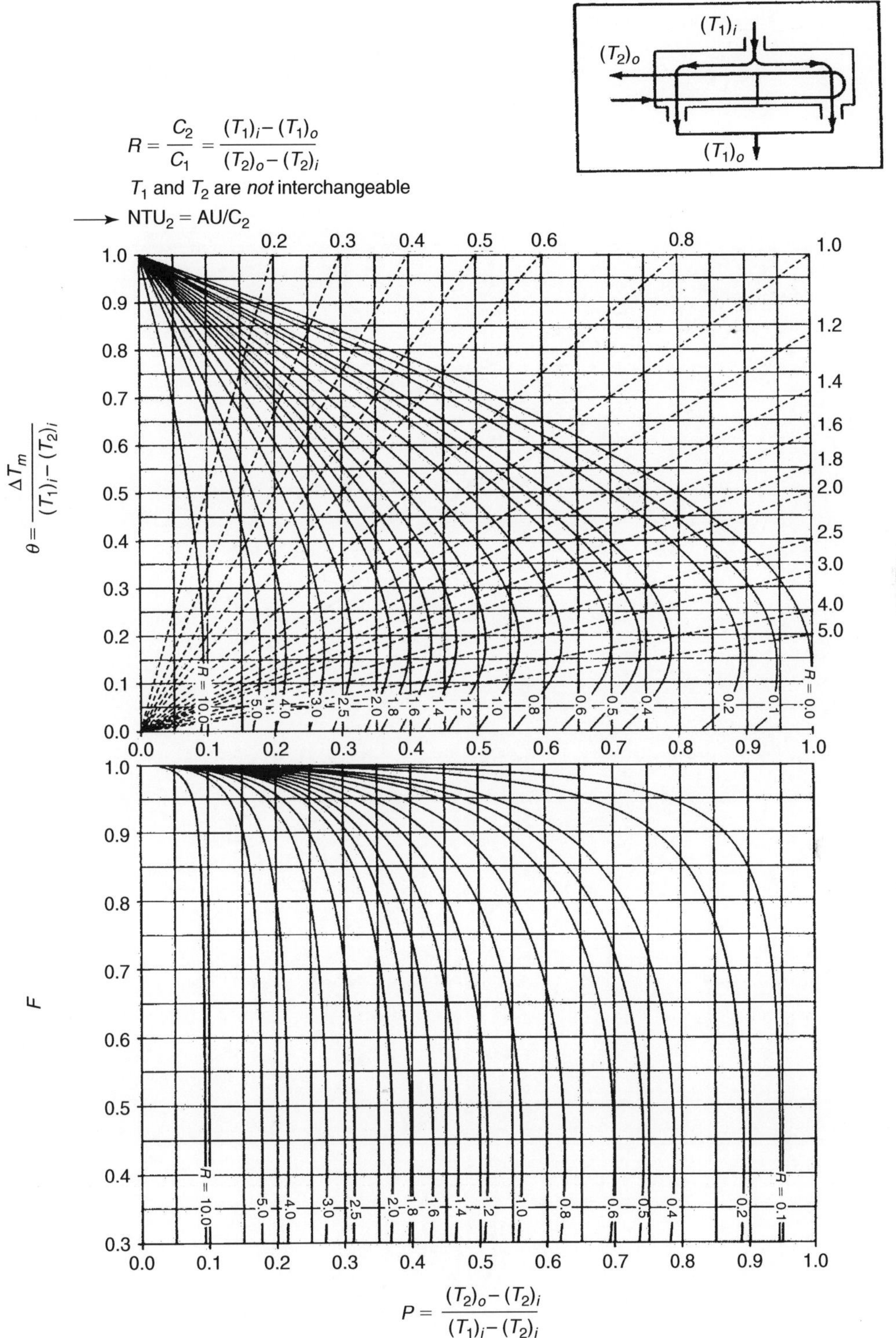

FIGURE 11.A.2 Mean temperature difference relationships for a TEMA J-shell exchanger with an even number of tube passes (Source: Ref. [35]).

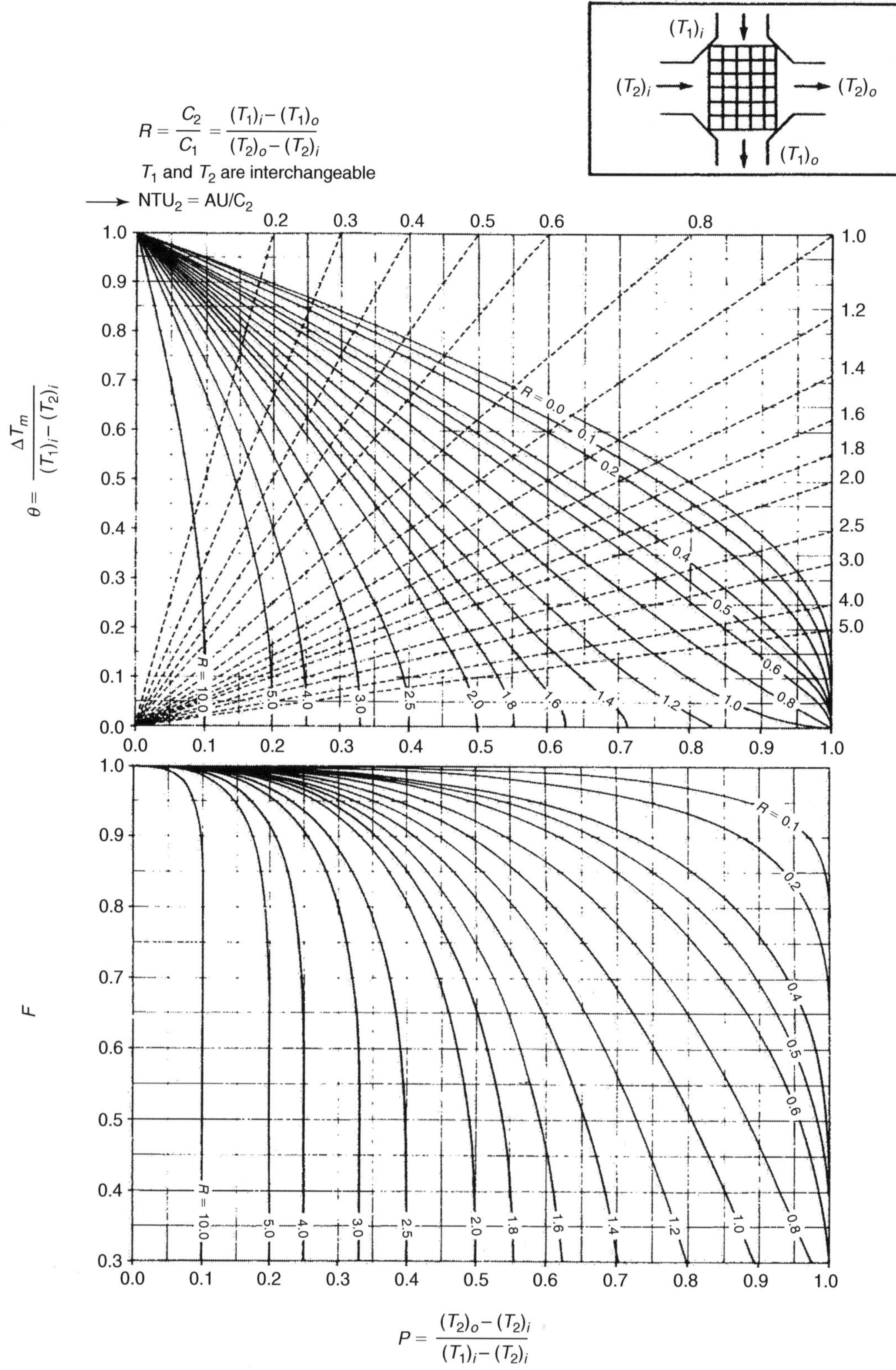

FIGURE 11.A.3 Mean temperature difference relationships for a TEMA X-shell exchanger with one tube pass (Source: Ref. [35]).

Notations

A	Heat-transfer surface area
A_f	Flow Area
A_{fins}	Surface area of all fins
A_i	Internal surface area of tube
A_{prime}	Prime surface area
A_{Tot}	Total area of fins and prime surface
a_s	Flow area across-tube bundle
B	Baffle spacing; parameter in Chisholm correlation for two-phase pressure drop
B_c	Baffle cut
B_{in}	Inlet baffle spacing
B_{out}	Outlet baffle spacing
$C_{P,c}$	Heat capacity of coolant
$C_{P,L}$	Heat capacity of condensate
$C_{P,V}$	Heat capacity of vapor
$\overline{C}_{P,V}$	Average heat capacity of vapor
C'	Tube gap
C'_{eff}	Effective gap for finned-tube bundle
D	Diameter
D_e	Equivalent diameter
D_i	Internal diameter of tube
D_n	Internal diameter of nozzle
D_o	External diameter of tube
D_r	Root-tube diameter
D_{re}	Equivalent root-tube diameter
d_s	Internal diameter of shell
E	Parameter in Beatty–Katz correlation
F	LMTD correction factor
F_1, F_2	Parameters in flooding correlation for reflux condensers
f	Darcy friction factor
f_{ideal}	Fanning friction factor for ideal flow through bank of plain tubes
f'_{ideal}	Fanning friction factor for ideal flow through bank of finned tubes
f_1, f_2	Friction factors from Figure 5.1 for B/d_s of 1.0 and 0.2, respectively
G	Mass flux
G_L	Mass flux of condensate
G_n	Mass flux in nozzle
$G_{n,in}$, $G_{n,out}$	Mass flux in inlet and outlet nozzle, respectively
G_V	Mass flux of vapor
g	Gravitational acceleration
g_c	Unit conversion factor
H	Specific enthalpy
h	Heat-transfer coefficient
h_{gr}	Heat-transfer coefficient for gravity-controlled condensation
h_i	Tube-side heat-transfer coefficient
h_{in}	Heat-transfer coefficient at condenser inlet
h_L	Heat-transfer coefficient for liquid phase flowing alone
h_{LO}	Heat-transfer coefficient for total flow as liquid
h_{Nr}	Heat-transfer coefficient for condensation on a vertical stack of N_r tube rows
h_{Nu}	Heat-transfer coefficient given by basic Nusselt theory
h_o	Shell-side heat-transfer coefficient
h_{out}	Heat-transfer coefficient at condenser outlet
h_{sh}	Heat-transfer coefficient for shear-controlled condensation
h_V	Heat-transfer coefficient for vapor phase
h_z	Local heat-transfer coefficient for condensation on a vertical surface

h_1	Heat-transfer coefficient for condensation on single tube row
h_I, h_{II}, h_{III}	Heat-transfer coefficient for condensation in horizontal tube with flow regime corresponding to zones I, II, and III, respectively
j_H	Modified Colburn factor for shell-side heat transfer
j^*	Dimensionless vapor mass flux defined by Equation (11.57)
k	Thermal conductivity
k_L	Thermal conductivity of condensate
k_{tube}	Thermal conductivity of tube wall
$\bar{k}_V$	Average thermal conductivity of vapor
L	Length
M	Molecular weight
$\dot{m}$	Mass flow rate
$\dot{m}_c$	Mass flow rate of coolant
$\dot{m}_L$	Mass flow rate of liquid phase
$\dot{m}_V$	Mass flow rate of vapor phase
$\dot{m}_{V,in}$	Mass flow rate of vapor entering condenser
$\dot{m}_{V,out}$	Mass flow rate of vapor leaving condenser
N_c	Number of tube rows crossed in flow between two baffle tips
N_f	Number of fins
N_r	Number of tube rows in vertical direction
Nu	Nusselt number
Nu_{LO}	Nusselt number for total flow as liquid
n	Exponent in friction factor Versus Reynolds number relationship
n_b	Number of baffles
n_f	Number of fins per unit length of tube
n_p	Number of tube passes
n_t	Number of tubes in bundle
$(n_t)_{req}$	Required number of tubes
P	Pressure; parameter used to calculate LMTD correction factor
P_r	Reduced pressure
Pr	Prandtl number
Pr_L	Prandtl number of condensate
Pr_V	Prandtl number of vapor
P_{sat}	Saturation pressure
P_T	Tube pitch
q	Rate of heat transfer
q_{sen}	Rate of sensible heat transfer
q_{Tot}	Total condenser duty
$\hat{q}$	Heat flux
$\hat{q}_y$	Heat flux in y-direction
R	Parameter used to calculate LMTD correction factor
$\tilde{R}$	Universal gas constant
R_{Di},R_{Do}	Fouling factors for tube-side and shell-side, respectively
Re	Reynolds number
Re_L	Reynolds number for liquid phase flowing alone
Re_{LO}	Reynolds number for total flow as liquid
Re_V	Reynolds number for vapor
r_1	Root-tube radius
r_2	Fin radius
r_{2c}	Corrected fin radius
SR	Slip ratio
s	Specific gravity
T	Temperature
T_a, T_b	Inlet and outlet temperatures of shell-side fluid
T_{ave}	Average bulk temperature of shell-side fluid

T_c	Coolant temperature
$T_{c,ave}$	Average coolant temperature in a condenser zone
$T_{c,out}$	Temperature of coolant leaving a condenser zone
T_f	Film temperature
T_L	Condensate temperature
T_p	Average temperature of prime surface
T_{sat}	Saturation temperature
T_V	Temperature of vapor
$T_{V,in}$, $T_{V,out}$	Temperature of vapor at condenser inlet and outlet, respectively
T_w	Wall temperature
T_{wtd}	Weighted average temperature of finned surface
t_a, t_b	Inlet and outlet temperatures of tube-side fluid
t_{ave}	Average bulk temperature of tube-side fluid
U	Overall heat-transfer coefficient
U_D	Design overall heat-transfer coefficient
U'_D	Design overall heat-transfer coefficient modified to account for thermal resistance due to vapor cooling
U_{req}	Required overall heat-transfer coefficient
$u_{V,in}$, $u_{V,out}$	Vapor velocity at condenser inlet and outlet, respectively
V	Fluid velocity
V_C	Superficial velocity of condensate in outlet nozzle
V_f	Flooding velocity in reflux condenser
$V_{f,70}$	Flooding velocity in reflux condenser having tubes with 70% taper
$\dot{v}_L$	Volumetric flow rate of condensate
W	Condensation rate
$\hat{W}$	Condensation rate per unit area
w	Width of vertical wall
X	Lockhart–Martinelli parameter
X_{tt}	Lockhart–Martinelli parameter for turbulent flow in both phases
x	Vapor weight fraction
x_e	Vapor weight fraction at condenser exit
x_{in}	Vapor weight fraction at condenser inlet
Y	Chisholm parameter
y	Coordinate in direction normal to condensing surface
z	Coordinate in vertically downward direction

Greek Letters

α_r	Number of velocity heads allocated for tube-side minor pressure losses
β	Weight factor used to calculate film temperature
Γ	Condensation rate per unit width of surface
Γ^*	Modified condensate loading defined by Equation (11.41)
ΔH	Specific enthalpy difference
ΔP_f	Frictional pressure drop in straight sections of tubes or in shell
$(\Delta P_f)_{VO}$	Frictional pressure drop for total flow as vapor
ΔP_i	Total tube-side pressure drop
ΔP_n	Pressure drop in nozzle
$\Delta P_{n,in}$, $\Delta P_{n,out}$	Pressure losses in inlet and outlet nozzles, respectively
ΔP_o	Total shell-side pressure drop
ΔP_r	Tube-side pressure drop due to entrance, exit, and return losses
Δq	Duty in a condenser zone
Δq_V	Duty due to vapor cooling in a condenser zone
ΔT	Temperature difference
ΔT_c	Coolant temperature difference
ΔT_f	Temperature difference across condensate film

ΔT_m	Mean temperature difference
$(\Delta T_{ln})_{cf}$	Logarithmic mean temperature difference for counter-current flow
Δy, Δz	Difference in y- or z-coordinate value
δ	Thickness of condensate film
ϵ	$C_{P,L}(T_{sat} - T_w)/\lambda$
ϵ_V	Void fraction
η_f	Fin efficiency
η_w	Weighted efficiency of finned surface
θ	Angle between an inclined surface and the vertical direction
θ'	Angle between an inclined tube and the horizontal direction
θ_{tp}	Tube layout angle
Λ	dq_V / dq
λ	Latent heat of condensation
λ'	$\lambda + C_{P,V}(T_V - T_{sat})$
μ_L	Viscosity of condensate
μ_V	Viscosity of vapor
$\overline{\mu}_V$	Average viscosity of vapor
μ_w	Viscosity of fluid at average wall temperature
ρ_L	Density of condensate
ρ_V	Density of vapor
ρ_{hom}	Homogenous two-phase density
σ	Surface tension
τ	Fin thickness
Φ	Parameter used to calculate LMTD correction factor for J- and X-shell exchangers
ϕ	Viscosity correction factor; angle defined in Figure 11.15
ϕ_L^2	Two-phase multiplier for pressure gradient based on liquid phase flowing alone
ϕ_{LO}^2	Two-phase multiplier for pressure gradient based on total flow as liquid
$\overline{\phi}_{VO}^2$	Average two-phase pressure-drop multiplier in Equation (11.69)
χ	Number of flow channels in unmixed-unmixed cross-flow configuration
Ψ	Parameter used to calculate LMTD correction factor for J- and X-shell exchangers
ψ	Parameter in equation for efficiency of annular fin
Ω	Parameter in correlation for condensation in stratified flow regime, Equation (11.58)

Other Symbols

$\vert_z$	Evaluated at z

Problems

(11.1) Use Shah's method to estimate the condensing-side heat-transfer coefficient for the condenser of Example 11.3 (a). Do the calculation for three positions in the condenser where the quality is, respectively, 75%, 50%, and 25%. Fluid properties can be obtained from Example 11.5.

Ans. 50% quality: $h = 288$ Btu/h · ft^2 · °F 25% quality: $h = 185$ Btu/h · ft^2 · °F.

(11.2) The average heat-transfer coefficient for a condensing vapor flowing downward in a vertical tube can be estimated using the correlation of Carpenter and Colburn (Carpenter, E. F. and A. P. Colburn, *Proc. of General Discussion on Heat Transfer, I* Mech E/ASME, 1951, 20–26):

$$\frac{h\mu_L}{k_L\rho_L^{0.5}} = 0.065\left(\frac{C_{P,L}\mu_L}{k_L}\right)^{0.5}\tau_i^{0.5}$$

$$\tau_i = \frac{f_V\overline{G}_V^2}{8\rho_V} = \text{average interfacial shear stress}$$

$$\overline{G}_V = \left(\frac{G_{V,in}^2 + G_{V,in}G_{V,out} + G_{V,out}^2}{3}\right)^{0.5}$$

$$f_V = 0.4137 Re_V^{-0.2585} = 0.4137 \left(\frac{D_i \overline{G}_V}{\mu_V}\right)^{-0.2585}$$

Note that for a total condenser, $\overline{G}_V = 0.58\ G_{V,in}$, where $G_{V,in}$ is the inlet mass flux of vapor. The Carpenter–Colburn correlation is strictly valid for shear-controlled condensation; it may under-predict the heat-transfer coefficient when this condition is not satisfied. Use the Carpenter–Colburn method to estimate the average heat-transfer coefficient for the condenser of Example 11.3(a).

Ans. $h = 217$ Btu/h · ft^2 · °F.

(11.3) The condenser of Examples 11.2 and 11.3 is oriented horizontally and the propyl alcohol condenses inside the tubes. At a position in the condenser where the quality is 50%, use the method of Breber et al. to determine the flow regime. Note that the Lockhart–Martinelli parameter for this situation was computed in Problem 9.5.

Ans. Transition region between zones I and II.

(11.4) Akers and Rosson [19] developed a correlation for convective condensation in horizontal tubes that is valid for both laminar and turbulent condensate films. It utilizes a vapor-phase Reynolds number, Re_V^*, defined as follows:

$$Re_V^* = \frac{D_i G_V (\rho_L/\rho_V)^{0.5}}{\mu_L}$$

The correlation consists of the following three equations, the choice of which depends on the values of Re_V^* and Re_L.
For $Re_L \leq 5000$ and $1000 \leq Re_V^* \leq 20{,}000$:

$$Nu = 13.8 Pr_L^{1/3} (Re_V^*)^{0.2} \left(\frac{\lambda}{C_{P,L}\Delta T}\right)^{1/6}$$

For $Re_L \leq 5000$ and $20{,}000 < Re_V^* \leq 100{,}000$:

$$Nu = 0.1 Pr_L^{1/3} (Re_v^*)^{2/3} \left(\frac{\lambda}{C_{P,L}\Delta T}\right)^{1/6}$$

For $Re_L > 5000$ and $Re_V^* \geq 20{,}000$:

$$Nu = 0.026 Pr_L^{1/3} (Re_V^* + Re_L)^{0.8}$$

In these equations, $\Delta T = T_{sat} - T_w$ and $Nu = hD_i / k_L$.
(a) For the conditions specified in Problem 11.3, assume $\Delta T = 51$°F and calculate h using the Akers–Rosson correlation.
(b) Using the value of h obtained in part (a), calculate a new value of ΔT. Iterate to obtain converged values of h and ΔT.

Ans. (a) $h = 526$ Btu/h · ft^2 · °F.

(11.5) 20,000 lb/h of saturated cyclohexane vapor will be condensed at 182°F and 16 psia using a tube bundle containing 147 tubes arranged for a single pass. The tubes are 1.0-in. OD, 14 BWG with a length of 20 ft. Physical properties of cyclohexane at these conditions are given in Example 10.4, and all properties may be assumed constant at these values. For the purpose of this problem, the effects of condensate subcooling and interfacial shear are to be neglected. Calculate the condensing-side heat-transfer coefficient for the following cases:
(a) The tube bundle is vertical and condensation occurs inside the tubes.
(b) The tube bundle is horizontal and condensation occurs outside the tubes.

Ans. (a) 185 Btu/h · ft^2 · °F. (b) 238 Btu/h · ft^2 · °F.

(11.6) Repeat Problem 11.5 taking into account the variation of condensate viscosity with temperature. Use the viscosity data in Figure A.1 for this purpose. Assume all other physical properties remain constant and neglect the effects of condensate subcooling and interfacial shear.

Ans. (a) 204 Btu/h · ft^2 · °F. (b) 249 Btu/h · ft^2 · °F.

(11.7) Estimate the effect of condensate subcooling on the heat-transfer coefficients computed in Problem 11.6.

Ans. (a) $h / h_{Nu} = 1.01$. (b) $h / h_{Nu} = 1.007$.

(11.8) Use the Boyko–Kruzhilin correlation to estimate the effect of interfacial shear in the vertical condenser of problem 11.6. Assume that the vapor flows downward through the tubes.

Ans. $h_{sh} = 148$ Btu/h · ft^2 · °F; $h = h_i = 252$ Btu/h · ft^2 · °F.

(11.9) Use the McNaught correlation to evaluate the significance of interfacial shear in the horizontal condenser of Problem 11.5. The tube bundle resides in an E-shell with an ID of 17.25 in. The tubes are laid out on 1.25-in. triangular pitch. Baffles are

20% cut segmental type with a spacing of 8.6 in. Make the calculation for a point in the condenser where the vapor weight fraction is 0.5. Use the physical property data given in Example 10.4 and neglect the viscosity correction factor.

Ans. $h_{sh} = 310$ Btu/h · ft² · °F.

(11.10) 50,000 lb/h of saturated acetone vapor will be condensed at 80°C and 31 psia using a tube bundle containing 316 tubes arranged for a single pass. The tubes are 1.0-in. OD, 16 BWG with a length of 25 ft. The molecular weight of acetone is 58.08 and the liquid specific gravity is 0.79. For the purpose of this problem, assume constant physical properties and neglect the effects of condensate subcooling and interfacial shear. Calculate the condensing-side heat-transfer coefficient for the following cases:
(a) The tube bundle is vertical and condensation occurs inside the tubes.
(b) The tube bundle is horizontal and condensation occurs outside the tubes.

(11.11) Repeat problem 11.10 taking into account the variation of condensate viscosity with temperature. Assume all other physical properties remain constant and neglect the effects of condensate subcooling and interfacial shear.

(11.12) Estimate the effect of condensate subcooling on the heat-transfer coefficients computed in Problem 11.11.

(11.13) Use the Boyko–Kruzhilin correlation to estimate the effect of interfacial shear in the vertical condenser of Problem 11.10. Assume that the vapor flows downward through the tubes.

(11.14) Use the McNaught correlation to evaluate the effect of interfacial shear in the horizontal condenser of Problem 11.10. The tube bundle resides in an E-shell with an ID of 25 in. and the tubes are laid out on 1.25-in. triangular pitch. Baffles are 35% cut segmental type with a spacing of 10 in. Make the calculation for a point in the condenser where the quality is 25%.

(11.15) Use the Shah correlation to calculate the condensing-side heat-transfer coefficient at a quality of 50% for:
(a) The vertical cyclohexane condenser of Problems 11.5 to 11.8.
(b) The vertical acetone condenser of Problems 11.10 to 11.13.
The critical pressure of cyclohexane is 40.4 atm, and that of acetone is 47.0 atm.

(11.16) Use the Carpenter–Colburn correlation given in problem 11.2 to estimate the average condensing-side heat-transfer coefficient for:
(a) The vertical cyclohexane condenser of Problems 11.5 to 11.8.
(b) The vertical acetone condenser of Problems 11.10 to 11.13.
Assume that the vapor flows downward through the tubes in each case.

(11.17) The Shah correlation, Equation (11.61), can be integrated over the length of the condenser to obtain an average heat-transfer coefficient. If the quality varies linearly with distance along the condenser from 100% at the inlet to zero at the outlet, the result is:

$$h = h_{LO}\left(0.55 + 2.09P_r^{-0.38}\right)$$

Values obtained from this equation differ by only about 5% from those calculated by Equation (11.61) for a quality of 50%. Use the above equation to estimate the average heat-transfer coefficient for:
(a) The vertical propanol condenser of Examples 11.3 to 11.5.
(b) The vertical cyclohexane condenser of Problems 11.5 to 11.8.
(c) The vertical acetone condenser of Problems 11.9 to 11.13.
Critical pressures of cyclohexane and acetone are given in Problem 11.15. Make a table comparing your results with the average heat-transfer coefficients calculated using the Carpenter–Colburn correlation in Problems 11.2 and 11.16, and with the values obtained using Equation (11.61) for a quality of 50% in Problems 11.1 and 11.15.

(11.18) The condenser of Problem 11.5 is oriented horizontally and the cyclohexane condenses inside the tubes. At a position where the quality is 50%:
(a) Use the method of Breber et al. to determine the flow regime.
(b) Estimate the value of the condensing-side heat-transfer coefficient.

(11.19) The condenser of Problem 11.10 is oriented horizontally and the acetone condenses inside the tubes. At a position where the quality is 25%:
(a) Use the method of Breber et al. to determine the flow regime.
(b) Estimate the value of the condensing-side heat-transfer coefficient.

(11.20) Use the Simplified Delaware method in conjunction with Equations (11.69) and (11.70) to estimate the nozzle-to nozzle shell-side pressure drop for the horizontal condenser of Problems 11.5 and 11.9. Consider the following cases:
(a) Complete condensation.
(b) Partial condensation with an exit quality of 30%.

Ans. (a) 13 psi. (b) 17.8 psi.

(11.21) Use Equations (11.69) and (11.71) to estimate the nozzle-to-nozzle tube-side pressure drop for the horizontal tube-side condenser of problem 11.18. Assume complete condensation.

Ans. 0.2 psi.

(11.22) Use the Simplified Delaware method in conjunction with Equations (11.69) and (11.70) to estimate the nozzle-to-nozzle shell-side pressure drop for the horizontal condenser of Problems 11.10 and 11.14. Consider the following cases:
(a) Complete condensation.
(b) Partial condensation with an exit quality of 50%.

(11.23) Use Equations (11.69) and (11.71) to estimate the nozzle-to-nozzle tube-side pressure drop for the horizontal tube-side condenser of Problem 11.19. Assume complete condensation.

(11.24) The horizontal tube bundle of problem 11.5 contains 26-fpi finned tubes, part number 267065 (Table B.5), made of 90-10 copper–nickel alloy. Assuming constant physical properties, calculate the condensing-side heat-transfer coefficient, the fin efficiency and the weighted efficiency of the finned surface. Neglect the effects of condensate subcooling and interfacial shear.

Ans. $h = 676$ Btu/h · ft^2 · °F; $\eta_f = 0.761$; $\eta_w = 0.808$.

(11.25) The horizontal tube bundle of Problem 11.10 contains 19-fpi finned tubes, Catalog Number 60-197065 (Table B.4), made of plain carbon steel. Assuming constant physical properties, calculate the condensing-side heat-transfer coefficient, the fin efficiency, and the weighted efficiency of the finned surface. Neglect the effects of condensate subcooling and interfacial shear.

(11.26) 50,000 lb/h of saturated acetone vapor will be completely condensed at 80°C and 31 psia. A used heat exchanger consisting of a 33-in. ID E-shell containing 730 tubes arranged for a single pass is available at the plant site. The tubes are 3/4-in. OD, 14 BWG with a length of 20 ft. The tube-side nozzles are made of 8-in. schedule 40 pipe. Material of construction is plain carbon steel throughout. The unit will be oriented vertically and the condensing vapor will flow downward inside the tubes. A coolant will flow through the shell with a range of 80°F to 140°F, providing a shell-side heat-transfer coefficient of 400 Btu/h · ft^2 · °F. A fouling factor of 0.001 h · ft^2 · °F/Btu is required for the coolant, and a maximum pressure drop of 5 psi is specified for the acetone stream. Condensate will drain from the condenser by gravity. Determine the unit's thermal and hydraulic suitability for this service. Neglect the effects of condensate subcooling and interfacial shear in the analysis. The specific gravity of liquid acetone is 0.79.

(11.27) A used heat exchanger consisting of a 29-in. ID J-shell containing 416 tubes is available for the service of Problem 11.26. The tubes are 1-in. OD, 16 BWG laid out on 1.25-in. triangular pitch with a length of 20 ft. There are eight baffles on each side of the central (full circle) baffle with a spacing of 13.3 in. On the shell side, the two inlet nozzles consist of 6-in. schedule 40 pipe and the single outlet nozzle is made of 8-in. schedule 40 pipe. Material of construction is plain carbon steel throughout. Acetone will flow in the shell and a coolant with a range of 80-140°F will flow through the tubes, providing an inside heat-transfer coefficient of 500 Btu/h · ft^2 · °F. Other specifications are the same as given in Problem 11.26. Determine the thermal and hydraulic suitability of the unit for this service. Neglect the effects of condensate subcooling and interfacial shear in the analysis.

(11.28) 70,000 lb/h of saturated cyclohexane vapor will be completely condensed at 196°F and 20 psia. A used carbon steel heat exchanger consisting of a 35-in. ID E-shell containing 645 tubes arranged for a single pass is available for this service. The tubes are 1-in. OD, 16 BWG with a length of 16 ft. Tube-side nozzles are 10-in. schedule 40 pipe. The unit will be oriented vertically and the condensing vapor will flow downward inside the tubes. A cold process liquid will flow in the shell with a range of 100°F to 175°F, providing a shell-side heat-transfer coefficient of 300 Btu/h · ft^2 · °F. A fouling factor of 0.001 h · ft^2 · °F/Btu is required for the coolant, and a maximum pressure drop of 4 psi is specified for the cyclohexane stream. Gravity drainage of condensate will be employed. Determine the thermal and hydraulic suitability of the unit for this service. Neglect the effects of condensate subcooling and interfacial shear in the analysis. At inlet conditions the density of cyclohexane vapor is 0.25 lbm/ft^3, the viscosity is 0.0087 cp and the latent heat of condensation is 149 Btu/lbm. Liquid viscosity data for cyclohexane are available in Figure A.1. Values of other physical properties for cyclohexane given in Example 10.4 may be used for this problem.

(11.29) A used heat exchanger consisting of a 31-in. ID J-shell containing 774 tubes is available for the service of Problem 11.28. The tubes are 3/4-in. OD, 14 BWG on 1.0-in. triangular pitch with a length of 18 ft. There are eight baffles on each side of the central (full circle) baffle with a spacing of 12 in. On the shell side, the two inlet nozzles consist of 6-in. schedule 40 pipe and the single outlet nozzle is 10-inch schedule 40 pipe. The tubes and tubesheets are made of 90-10 copper–nickel alloy; all other components are plain carbon steel. Cyclohexane will flow in the shell and a coolant with a range of 100°F to 175°F will flow through the tubes, providing a tube-side heat-transfer coefficient of 400 Btu/h · ft^2 · °F. Other specifications are the same as in Problem (11.28) Determine the thermal and hydraulic suitability of the unit for this service. Neglect the effects of condensate subcooling and interfacial shear in the analysis.

(11.30) Consider a horizontal shell-side condenser in which the coolant makes two passes through the exchanger. Assume the coolant and vapor enter at the same end of the condenser and denote the coolant temperature in the first and second passes by T'_c and T''_c, respectively. The temperature profiles along the length of the condenser are as shown below.

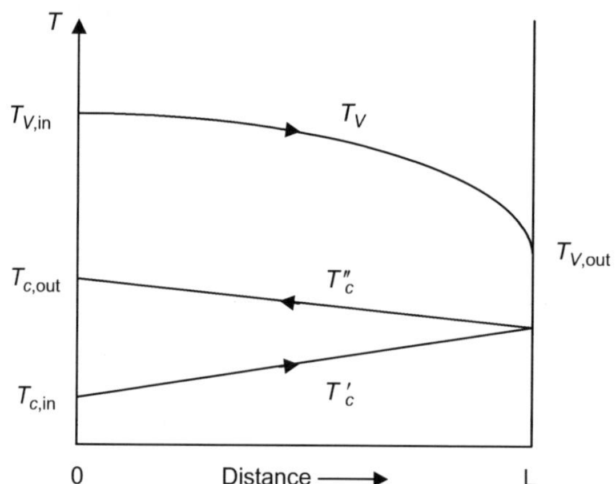

If the surface area is equally divided between the two passes and the overall coefficient is assumed to be the same for both passes, the rate of heat transfer in a differential condenser element can be written as:

$$dq = \frac{1}{2}U_D dA(T_{sat} - T'_c) + \frac{1}{2}U_D dA(T_{sat} - T''_c)$$

or

$$dq = \frac{1}{2}U_D dA(2T_{sat} - T'_c - T''_c) \tag{i}$$

(a) Using Equation (i) in place of Equation (11.75), show that the relationship corresponding to Equation (11.80) is:

$$A = 2\int_0^{q_{Tot}} \frac{(1 + U_D\Lambda/h_V)}{U_D(2T_V - T'_c - T''_c)} dq \tag{ii}$$

(b) Differential energy balances for the coolant are as follows:

$$dq = \dot{m}_c C_{P,c}(dT'_c - dT''_c)$$

or

$$dT'_c - dT''_c = \frac{dq}{\dot{m}_c C_{P,c}} \tag{iii}$$

$$\dot{m}_c C_{P,c} dT'_c = \frac{1}{2}U_D dA(T_V - T'_c) \tag{iv}$$

$$\dot{m}_c C_{P,c}(-dT''_c) = \frac{1}{2}U_D dA(T_V - T''_c) \tag{v}$$

Dividing Equation (iv) by Equation (v) gives:

$$\frac{dT'_c}{-dT''_c} = \frac{T_V - T'_c}{T_V - T''_c} \tag{vi}$$

Show that Equations (iii) and (vi) can be solved for dT'_c and dT''_c to obtain the following results:

$$dT'_c = \left(\frac{dq}{\dot{m}_c C_{P,c}}\right)\left(\frac{T_V - T'_c}{2T_V - T'_c - T''_c}\right) \tag{vii}$$

$$dT''_c = -\left(\frac{dq}{\dot{m}_c C_{P,c}}\right)\left(\frac{T_V - T''_c}{2T_V - T'_c - T''_c}\right) \tag{viii}$$

In practice, Equations (vii) and (viii) are applied to a condenser element of finite size to obtain $\Delta T'_c$ and $\Delta T''_c$. The values of T_V, T'_c and T''_c on the right-hand side are taken at the beginning of the element. The temperature profiles in the two coolant passes are easily determined in this manner starting from the inlet end of the condenser where $T_V = T_{V,in}$, $T'_c = T_{c,in}$ and $T''_c = T_{c,out}$.

(c) Assume that the coolant in Example 11.8 makes two passes through the condenser. Using the same three temperature intervals, calculate the temperature profiles in the two coolant passes and determine the temperature in the return header.

Ans. At the return header, $T'_c = T''_c = 105.17^\circ$F.

(d) Use the temperature profiles obtained in part (c) together with the data from Example 11.8 to compute the required surface area for the condenser according to Equation (ii).

Ans. $A = 1742$ ft^2.

(11.31)

T_V (°F)	Duty (Btu/h)
300	0
270	6.5×10^6
240	12.0×10^6
210	15.5×10^6

A condenser is to be designed based on the above condensing curve. Water with a range of 75°F to 125°F will be used as the coolant. Determine the temperature profile for the cooling water for the following cases:
(a) A single coolant pass and counter-current flow.
(b) Two coolant passes with coolant and vapor entering at the same end of the condenser. (Refer to Problem 11.30.)

(11.32) Use any available software to design a condenser for the service of Problem 11.26. For the coolant, use water that is available from a cooling tower at 85°F. Assume an inlet pressure of 50 psia and a maximum allowable pressure drop of 10 psi for the coolant.

(11.33) Use any available software to design a condenser for the service of problem 11.28. For the coolant, use water that is available from a cooling tower at 85°F. Assume an inlet pressure of 50 psia and a maximum allowable pressure drop of 10 psi for the coolant.

(11.34) A stream with a flow rate of 150,000 lb/h having the following composition is to be condensed.

Component	Mole percent
Propane	15
i-Butane	25
n-Butane	60

The stream will enter the condenser as a saturated vapor at 150 psia and is to leave as a subcooled liquid with a minimum 10°F of subcooling. A maximum pressure drop of 2 psi is specified. Cooling will be supplied by water from a cooling tower entering at 80°F and 50 psia, with a maximum allowable pressure drop of 10 psi. Use any available software to design a condenser for this service.

(11.35) A petroleum fraction has an average API gravity of 40° and the following assay (ASTM D85 distillation at atmospheric pressure):

Volume % distilled	T (°F)
0	112
10	157
20	201
30	230
40	262
50	291

(Continued)

—cont'd

Volume % distilled	T (°F)
60	315
70	338
80	355
90	376
100	390

90,000 lb/h of this material will be condensed using cooling water available at 80°F. The petroleum fraction will enter the condenser as a saturated vapor at 20 psia, and is to be completely condensed with a pressure drop of 3 psi or less. Maximum allowable pressure drop for the cooling water is 10 psi. Use HEXTRAN or other appropriate software to design a condenser for this service.

(11.36) Using any available software package, do the following.
  (a) Rate the final configuration for the finned-tube condenser of Example 11.7 and compare the results with those from the hand calculations in the text.
  (b) Modify the condenser configuration as appropriate and obtain a final design for the unit.

(11.37) A stream with a flowrate of 65,000 lb/h having the following composition is to be condensed.

Component	Mole percent
Ethanol	32
Isopropanol	10
1-Propanol	23
2-Methyl-1-Propanol	19
1-Butanol	16

The stream will enter the condenser as a saturated vapor at 18 psia, and a maximum pressure drop of 2 psia is specified. The temperature of the condensate leaving the unit should not exceed 180°F. Cooling will be supplied by water from a cooling tower entering at 90°F and 40 psia, with a maximum allowable pressure drop of 10 psi. Use any available software to design a condenser for this service.

(11.38) 40,000 lb/h of refrigerant 134a (1,1,1,2-tetrafluoroethane) will enter a condenser as a superheated vapor at 180°F and 235 psia. The condensate is to leave the unit as a subcooled liquid with a minimum of 10°F of subcooling. A maximum pressure drop of 3 psia is specified. The coolant will be water from a cooling tower entering at 80°F and 50 psia, with a maximum allowable pressure drop of 10 psi. Use any available software package to design a condenser for this service. Develop one or more designs for each of the following condenser types and select the best option for the service. Give the rationale for your selection.
  (a) Vertical tube-side downflow condenser.
  (b) Horizontal shell-side condenser.
  (c) Horizontal tube-side condenser.

(11.39) 230,000 lb/h of propane will enter a condenser as a superheated vapor at 180°F and 210 psia. The condensate is to leave the condenser as a subcooled liquid at a temperature of 95°F or less, and a maximum pressure drop of 5 psi is specified. The coolant will be water from a cooling tower entering at 80°F and 40 psia, with a maximum allowable pressure drop of 10 psi. Use any available software to design a condenser for this service.

(11.40) For the service of Example 11.6, use *Xist* to design a rod-baffle condenser with:
  (a) Plain tubes.
  (b) Finned tubes.

12 Air-Cooled Heat Exchangers

12.1 Introduction

Air-cooled heat exchangers are second only to shell-and-tube exchangers in frequency of occurrence in chemical and petroleum processing operations. These units are used to cool and/or condense process streams with ambient air as the cooling medium rather than water. Cooling with air is often economically advantageous, e.g., in arid or semi-arid locations, in areas where the available water requires extensive treatment to reduce fouling, or when additional investment would otherwise be required to expand a plant's existing cooling-water supply. Regulations governing water use and discharge of effluent streams to the environment also tend to favor air cooling. Although the capital cost of an air-cooled exchanger is generally higher, the operating cost is usually significantly lower compared with a water-cooled exchanger. Hence, high energy cost relative to capital cost favors air cooling. Air cooling also eliminates the fouling and corrosion problems associated with cooling water, and there is no possibility of leakage and mixing of water with the process fluid. Thus, maintenance costs are generally lower for air-cooled exchangers.

12.2 Equipment Description

12.2.1 Overall Configuration

In an air-cooled heat exchanger, the hot process fluid flows through a bank of finned tubes, and ambient air is blown across the tubes by one or more axial-flow fans. For applications involving only sensible heat transfer, the tubes are oriented horizontally as shown in Figures 12.1 and 12.2. For condensers, an A-frame configuration (Figure 12.3) is often used, with the condensing vapor flowing downward through the tubes, which are oriented at an angle of 60° with the horizontal.

In units employing horizontal tubes, the fan may be located either below (forced draft) or above (induced draft) the tube bank. In either case, the air flows upward across the tubes. The fan drive assembly in an induced-draft unit may be mounted below the tube bundle (either on the ground as shown in Figure 12.2 or suspended from the framework), or it may be mounted above the fan. With the former arrangement, the drive assembly is easily accessible for inspection and maintenance, and it is not exposed to the heated air leaving the unit. However, the drive shaft passes through the tube bundle, which entails the omission of some tubes and a lower drive efficiency.

The forced-draft configuration provides the simplest and most convenient fan arrangement. With all blower components located below the tube bundle, they are easily accessible for maintenance and are not exposed to the heated air leaving the unit. However, these exchangers are susceptible to hot air recirculation due to the wind. Induced-draft operation gives more uniform air flow over the tube bundle than in forced-draft operation, and the higher discharge elevation reduces the potential for hot air to be recirculated back to the intake of the unit or other nearby units. Hot air recirculation reduces the capacity of the heat exchanger, thereby requiring a higher air flow rate and/or more heat-transfer surface. The induced-draft configuration also provides some protection from the elements for the tube bundle, which helps to stabilize the operation of the unit when sudden changes in ambient conditions occur.

For a given mass flow rate of air, induced-draft operation in principle entails greater power consumption than forced-draft operation due to the higher volumetric flow rate of the heated air that is handled by the induced-draft fan. In practice, however, this potential disadvantage tends to be offset by the more uniform flow distribution and lower potential for hot gas recirculation

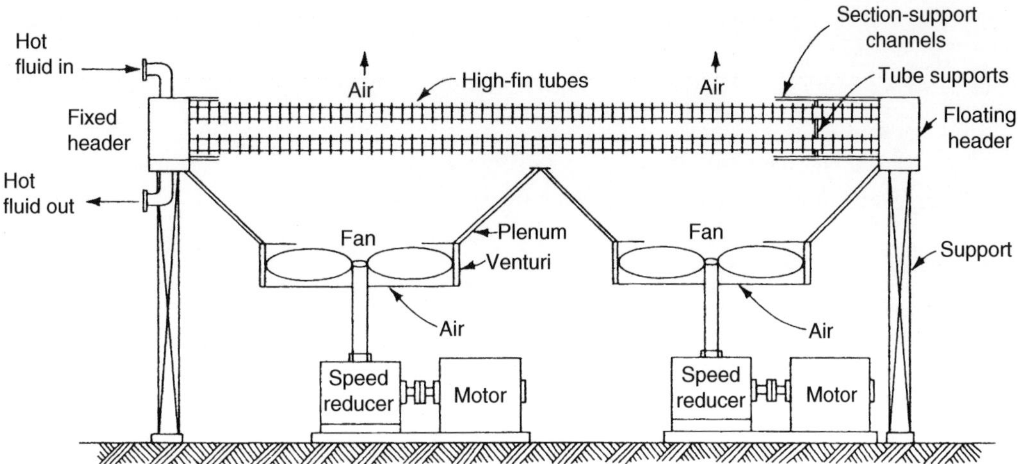

FIGURE 12.1 Configuration of a forced-draft air-cooled heat exchanger (Source: Ref. [1]).

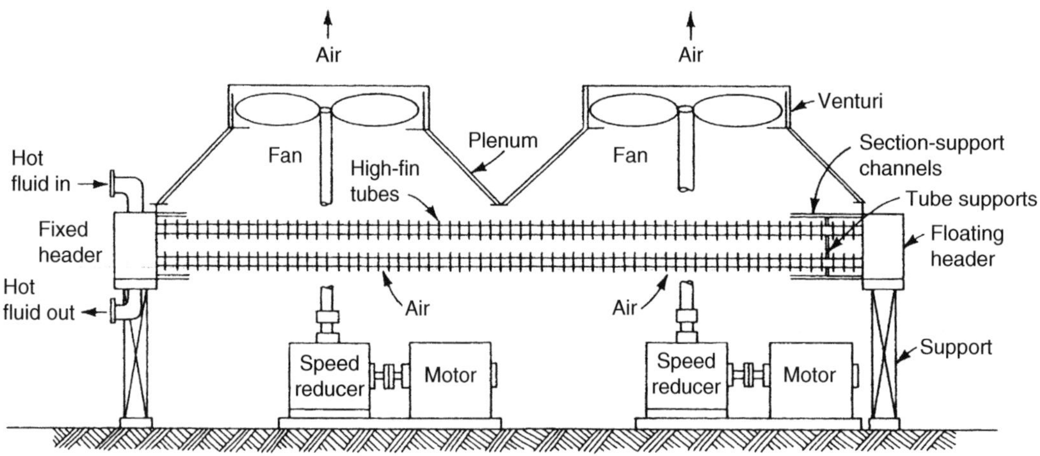

FIGURE 12.2 Configuration of an induced-draft air-cooled heat exchanger (Source: Ref. [1]).

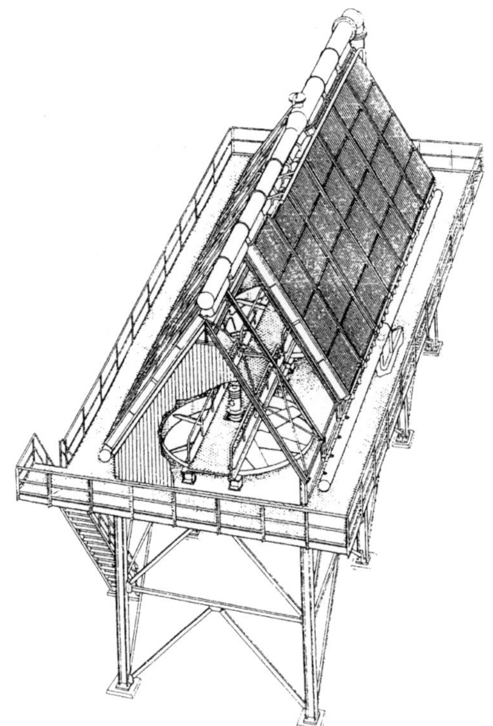

FIGURE 12.3 Configuration of an A-frame air-cooled condenser (Source: Spiro-Gills, Ltd. Originally published in Ref. [2]).

obtained with induced-draft operation. As a result, induced-draft units typically do not require significantly more power than forced-draft units, and in some cases may actually require less power [3].

12.2.2 High-Fin Tubing

Finned tubes are almost always used in air-cooled exchangers to compensate for the low air-side heat-transfer coefficient. Radial (annular) fins arranged in a helical pattern along the tube are used. The fin height is significantly larger than that of the low-fin tubes used in shell-and-tube exchangers. Hence, this type of tubing is referred to as high-fin tubing.

Various types of high-fin tubing are available, including:

- Integrally finned
- Bimetallic
- Tension-wound fin

- Embedded fin
- Brazed fin

Integrally finned (K-fin) tubing is made by extruding the fins from the tube metal. It is generally made from copper or aluminum alloys that are relatively soft and easily worked. Since the fins are integral with the root tube, perfect thermal contact is ensured under any operating conditions.

Bimetallic (E-fin) tubes consist of an inner tube, or liner, and an outer tube, or sleeve. The inner tube may be made from any tubing material and has the same dimensions as standard heat-exchanger tubing. The outer tube is integrally finned and is usually made of aluminum alloy. The sleeve thickness beneath and between the fins is usually 0.04 in. to 0.05 in. Since the contact between the two tubes is not perfect, there is a contact resistance at the interface between the tubes. Although this resistance is negligible at low temperatures, it can amount to 10% to 25% of the total thermal resistance in operations involving high tube-side fluid temperatures [4]. Hence, this type of tubing is not recommended for tube-side temperatures above 600°F.

Tension-wound finned tubing is widely used due to its relatively low cost. The fins are formed by winding a strip of the fin material around the tube under tension. The metal strip may be either straight (edge-wound or I-fin) or bent in the shape of the letter L (L-footed or L-fin) as shown in Figure 12.4. The latter configuration provides more contact area between the fin strip and tube surface and also helps protect the tube wall from atmospheric corrosion. Better adhesion and corrosion protection can be achieved by overlapping the "feet" of the Ls (LL-fin). The strip metal is subjected to controlled deformation under tension to provide good contact between the strip and the tube wall. Collars at both ends of the tube hold the fin strip in place and maintain the tension. Nevertheless, since the fins are held in place solely by the tension in the metal strip, they can be loosened by operation at high temperatures or by temperature cycling. Therefore, this type of tubing is used for continuous services with tube-side temperatures below 400°F (below 250°F in the case of I-fin).

Embedded-fin (G-fin) tubing is made by winding a strip of the fin metal into a helical groove machined in the surface of the tube and then securing the strip in place by backfilling the groove with the tube metal (peening). This type of tubing is much more robust than tension-wound tubing and is widely used for this reason. It is applicable for tube-side temperatures up to 750°F and in services involving cyclic operation. Shoulder-grooved-fin tubing is a type of embedded-fin tubing that combines the characteristics of L- and G-fin tubing; in this case the foot of the L-shaped finning contains a section that fits into the embedding groove. To accommodate the groove in the surface, an additional wall thickness of 1 BWG is used for embedded-fin tubing. Thus, for hydrocarbon liquids, 13 BWG tubes are normally used rather than 14 BWG tubes.

Brazed-fin tubing is made by first winding a strip of the fin material around the tube under tension. The fin and tube metal are then bonded together by brazing. This metallurgical bond minimizes contact resistance and allows operation under more severe

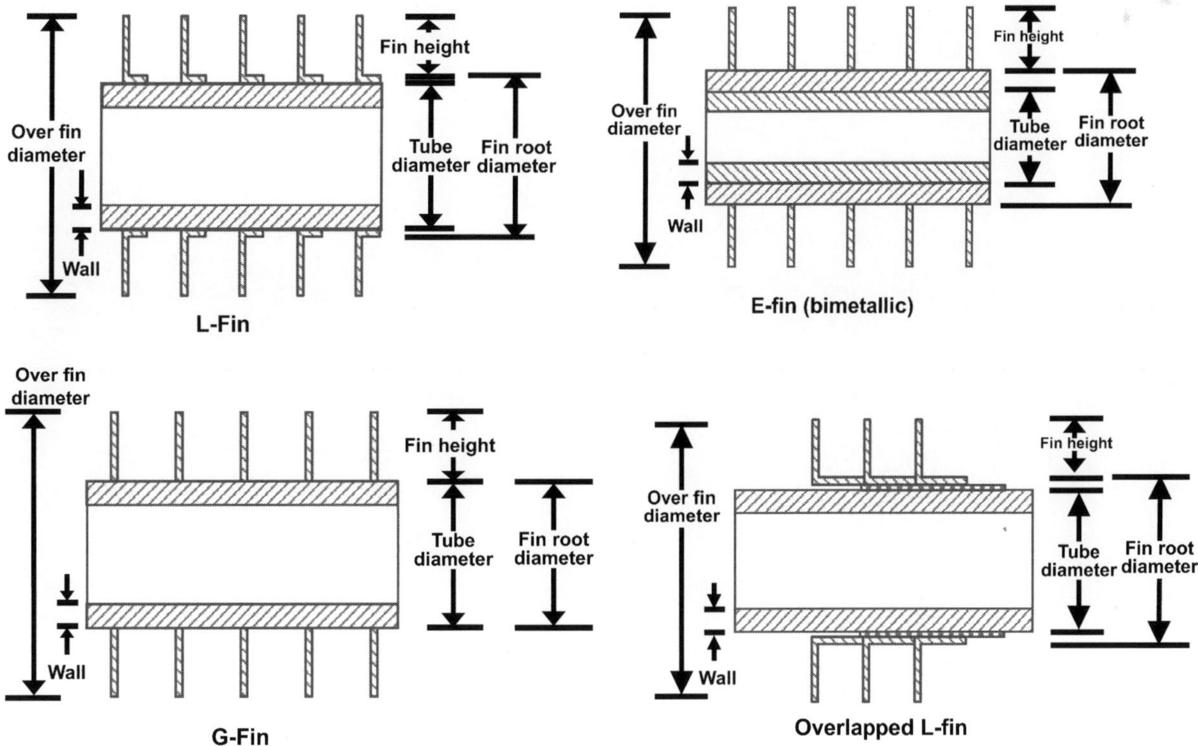

FIGURE 12.4 High-fin tubing (Source: *Xace* Help File).

conditions than is possible with standard tension-wound or embedded tubing. Tube-side temperatures up to 1000°F are permissible with copper fins and up to 1500°F with stainless steel fins.

In addition to loosening of the fins, corrosion at the base of the fins may cause the performance of an air-cooled heat exchanger to deteriorate over time. Tension-wound finned tubes are the most susceptible to corrosion since moisture can penetrate between the fin material and tube wall even with overlapping (LL) fins. Embedded-fin tubing is less susceptible to fin-root corrosion, while bimetallic tubing is very corrosion resistant.

High-fin tubing is made in a variety of sizes and configurations using both tubes and pipes as root stock. Diameters range from 0.5 in. to 8 in. with fin heights of 0.25 in. to 1.5 in. The number of fins per inch varies from less than 2 to 12, and the average fin thickness typically ranges from 0.012 in. to about 0.035 in., although tubing with thicker fins is available from some manufacturers. However, 1-in. OD tubing with a fin height of 0.5 in. or 0.625 in. is by far the most widely used in air-cooled heat exchangers. The tube layout is usually triangular with a clearance of 0.125 in. to 0.375 in. between fin tips. Increasing the clearance reduces air-side pressure drop but increases the size of the tube bundle. The tubes are arranged in shallow rectangular bundles with the number of tube rows usually between three and six. A small number of tube rows are used in order to keep the air-side pressure drop low. The axial-flow fans used in air-cooled heat exchangers are capable of developing a static pressure on the order of 1 in. of water or more.

Typical values of parameters for the most common tubing configurations used in air-cooled heat exchangers are given in Table 12.1.

12.2.3 Tube Bundle Construction

Tube bundles are rectangular in shape and usually 6 ft to 18 ft wide. Since tube bundles are factory assembled and shipped to the plant site, bundle width can impact transportation costs. (In the United States the maximum load width for interstate truck transport is 14 ft in most states.) The tubes are either welded to or rolled into long rectangular tubesheets that are welded to box-type headers. Both front and rear headers are commonly equipped with screwed plugs that are aligned with the tube holes as illustrated in Figure 12.5. The plugs can be removed to provide access to the tubes for cleaning and other maintenance.

Headers are also available with flanged end plates that can be removed to provide unencumbered access to the tubesheets. In addition to being more expensive, this type of header is prone to leakage because the long rectangular gasket is difficult to seat properly. It is normally used only when frequent tube-side maintenance is required and the operating pressure is relatively low. Box headers become impractical at pressures above about 4000 psia due to the large wall thickness required. For these high-pressure applications, the tubes are welded directly into a section of pipe of appropriate schedule number that serves as the header.

For multi-pass operation, the headers are equipped with pass partition plates as shown in Figure 12.5. The tubes are partitioned so that the process fluid flows from the top tube row to lower tube rows. The downward flow of process fluid in combination with

TABLE 12.1 Characteristics of Typical High-Fin Tube Arrays

Root tube OD (in.)	1.0	1.0
Fin height (in.)	0.500	0.625
Fin OD (in.)	2.00	2.25
Average fin thickness (in.)		
Tension wound or embedded	0.012–0.014	0.012–0.014
Bimetallic or integral*	0.015–0.025	0.015–0.025
Fins per inch	9	10
Tube layout angle	30°	30°
Tube pitch (in.)	2.25	2.50
A_{Tot}/L	3.80	5.58
A_{Tot}/A_o	14.5	21.4
A_{Tot}/A_i		
13 BWG	17.9	26.3
14 BWG	17.4	25.6
15 BWG	17.0	24.9
16 BWG	16.7	24.5
External surface area per unit bundle face area		
Three tube rows	60.6	80.4
Four tube rows	80.8	107.2
Five tube rows	101.0	134.0
Six tube rows	121.2	160.8

*Dimensions in this table are approximate for these types of tubing. The root tube OD of integrally finned tubing may be somewhat greater or less than 1 in. The root tube OD is greater for bimetallic tubes due to the sleeve thickness.
Source: Ref. [5]

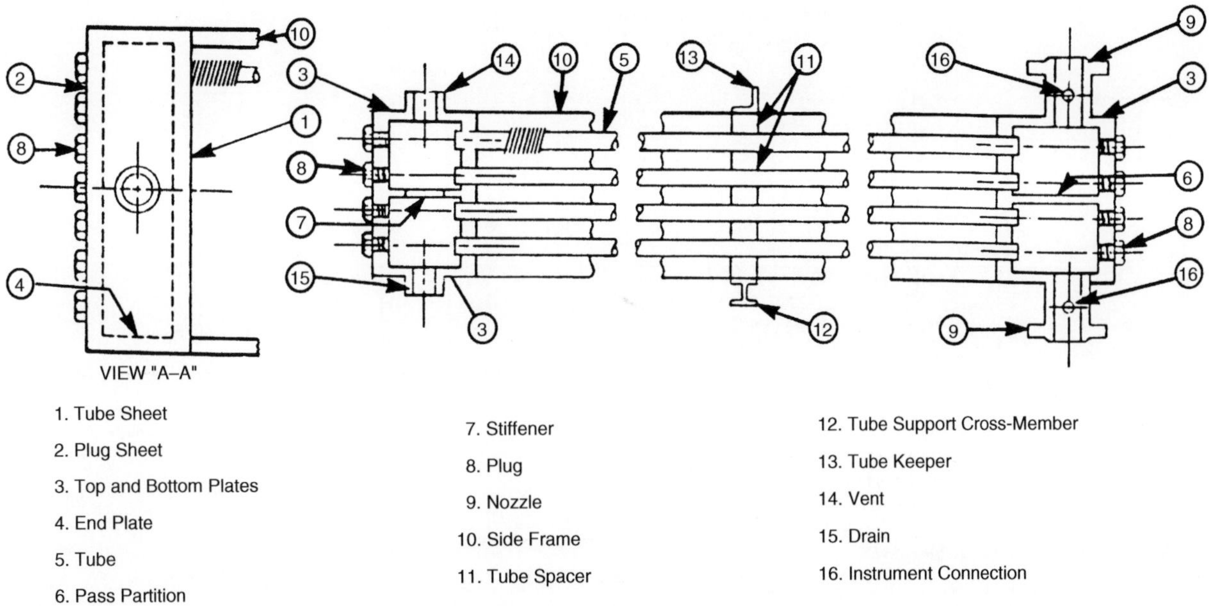

FIGURE 12.5 Typical construction of a tube bundle with plug-type box headers (Source: Ref [3]).

the upward flow of air gives an overall flow pattern that is a combination of counter-current and cross flow, and maximizes the mean temperature difference in the heat exchanger.

Tube supports and spacers are provided to hold the tubes securely in place and to dampen tube vibration. The bundle is held together and given structural integrity by side members that are bolted or welded to the headers, tube supports, and the framework that supports the unit.

12.2.4 Fans and Drivers

Axial-flow fans with four or six blades and diameters of 6 ft to 18 ft are typically employed in air-cooled heat exchangers, although larger and smaller fans are occasionally used as well. Plastic fan blades are used for air temperatures up to 175°F; metal (usually aluminum) blades are required for higher air temperatures.

Electric motors are most frequently used as fan drivers, generally 50 hp (37 kW) or less. Speed reduction is usually accomplished using high-torque-drive (HTD) belts or reduction gear boxes. Hydraulic variable-speed drives may also be used. HTD belts can be used with motor sizes up to 50 hp [6] and are the most widely employed method of speed reduction.

Variable-pitch fans are commonly used to provide process-side temperature control in air-cooled exchangers. The blade pitch is automatically adjusted to provide the required air flow to maintain the desired outlet temperature of the process fluid. This is accomplished using a temperature controller and a pneumatically operated blade adjustment mechanism. Reducing air flow also reduces power consumption when the ambient temperature is low. Similar results can be achieved using variable-speed drives.

The fans are situated in bays, which are self-contained sections of an air-cooled heat exchanger. A bay consists of one or more tube bundles, the fans and drive assemblies that supply air to the bundles, and the associated framework and support structures. Except in unusual circumstances, multiple tube bundles are placed side by side in the bay. Bays are usually designed for one to three fans, with two-fan bays being most common.

Fan bays can be preassembled and shipped to the plant site provided they are small enough to meet transportation requirements. Otherwise, they must be assembled in the field, which adds to the cost of the heat exchanger. An air-cooled heat exchanger consists of one or more fan bays, with multiple bays operating in parallel. Some typical configurations are illustrated in Figure 12.6.

Additional equipment associated with each fan includes the fan casing (also called the fan ring or shroud) and the plenum. The casing forms a cylindrical enclosure around the fan blades. It is often tapered on the intake end to reduce the pressure loss. A bell-shaped inlet is the most effective for this purpose, but conical and other types of inlets are also used. The plenum is the structure that connects the fan with the tube bundle. In forced-draft operation, the plenum serves to distribute the air delivered by the fan across the face of the tube bundle. With induced-draft operation, the plenum delivers the air from the top of the tube bundle to the fan intake. Box-type plenums are used most frequently in forced-draft units, whereas tapered plenums are the norm for induced-draft operation.

Induced-draft fans are sometimes equipped with diffusers in order to reduce power consumption. A diffuser is essentially a short stack with an expanding cross-section that serves to lower the velocity of the exhaust air. Although there is some friction loss in the diffuser, the net result is an increase in the static pressure of the exhaust air and a concomitant reduction in the power consumed by the fan.

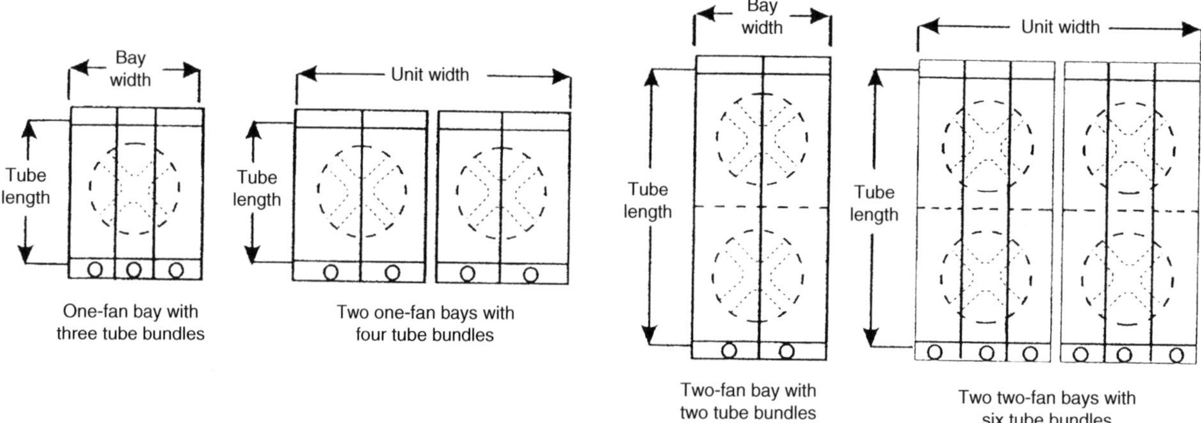

FIGURE 12.6 Some typical configurations of fan bays in air-cooled heat exchangers (Source: Ref. [5]).

12.2.5 Equipment for Cold Climates

Air-cooled heat exchangers are designed to operate over a wide range of environmental conditions, including ambient temperatures from −60°F to 130°F. Special design features are employed for operation in cold climates in order to prevent freezing of the process fluid. If the wall temperature of a tube carrying a hydrocarbon stream reaches the pour point of the hydrocarbon, the liquid will congeal around the wall, thereby reducing the flow area and increasing the tube-side pressure drop. If water is present in the process stream, ice can form around the tube wall with similar effect. Likewise, methane hydrates can form on the tube walls of natural gas coolers.

The standard method for preventing freezing is to intentionally recirculate some of the warm air leaving the unit in order to raise the temperature of the intake air. This can be accomplished in a number of ways, depending on whether forced-draft or induced-draft operation is employed and the severity of the winter climate [7]. Figure 12.7 shows a typical configuration for a forced-draft unit with external recirculation, which provides the most reliable freeze protection. The unit is completely contained in an enclosure equipped with adjustable louvers to control both exhaust and intake air rates. The manual louvers are adjusted seasonally while the automatic louvers are adjusted continuously via pneumatic mechanisms directed by a temperature controller that maintains the temperature of the air entering the tube bundle at an appropriate level. A recirculation chamber projects beyond the front and rear headers, providing ducts where cold ambient air mixes with warm recirculated air. The flow rate of recirculated air is controlled by

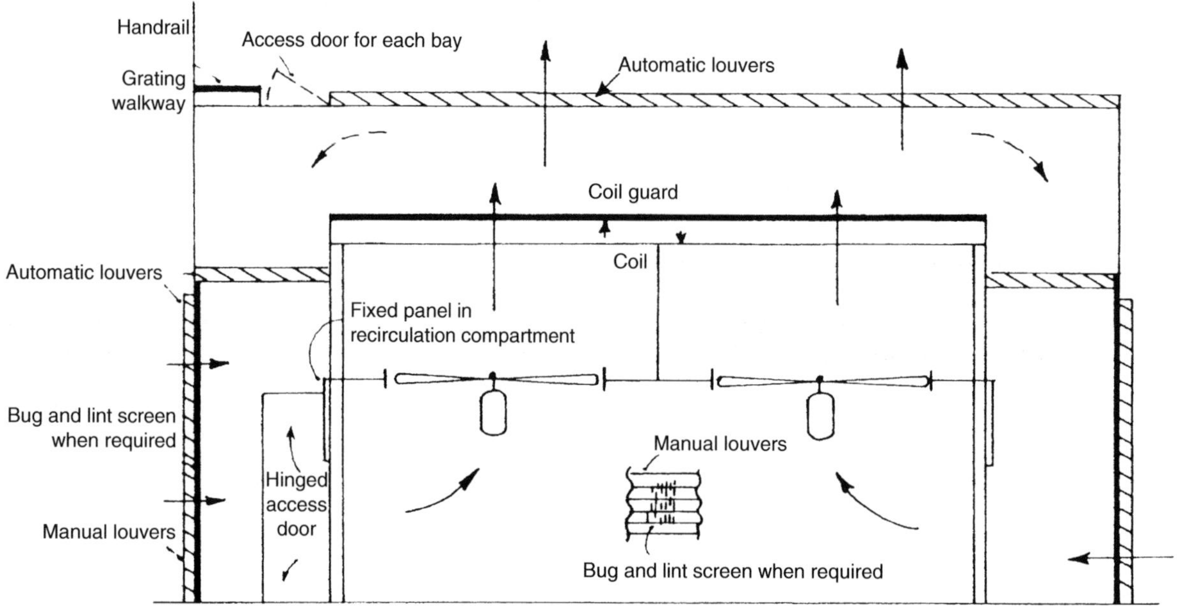

FIGURE 12.7 Configuration of a forced-draft air-cooled heat exchanger designed for external recirculation of warm air during cold weather. The coil in the diagram is the tube bundle (Source: Ref. [5]).

internal louvers in the ducts that open as the external intake louvers close. Either variable-pitch or variable-speed fans are used in these units. The pitch or speed is automatically adjusted by a second temperature controller that maintains the outlet temperature of the process fluid at the desired temperature.

A row of steam tubes is installed below the tube bundle to warm the air stream during startups and shutdowns in cold weather. These tubes are typically the same type and size as those in the tube bundle, but with a pitch equal to twice that of the bundle. The steam tubes are commonly referred to as a steam coil.

12.3 Air-Side Heat-Transfer Coefficient

The flow of air over banks of finned tubes has been extensively studied and numerous correlations for this geometry are available in the open literature. Among these, the correlation of Briggs and Young [8] has been widely used:

$$Nu = 0.134\, Re^{0.681}\, Pr^{1/3}(\ell/b)^{0.2}(\ell/\tau)^{0.1134} \tag{12.1}$$

where

$Nu = h_o D_r/k$
$Re = D_r V_{max} \rho/\mu$
V_{max} = maximum air velocity in tube bank
ℓ = fin spacing
b = fin height
τ = fin thickness
k = thermal conductivity of air
ρ = density of air
μ = viscosity of air
h_o = air-side heat-transfer coefficient

The correlation is based on experimental data for tube banks containing six rows of tubes laid out on equilateral triangular pitch. The data covered the following ranges of parameters:

$1000 \le Re \le 18{,}000$
0.438 in. $\le D_r \le$ 1.61 in.
0.056 in. $\le b \le$ 0.6525 in.
0.013 in. $\le \tau \le$ 0.0795 in.
0.035 in. $\le \ell \le$ 0.117 in.
0.96 in. $\le P_T \le$ 4.37 in.

The fin spacing is related to the number, n_f, of fins per unit length by the following equation:

$$\ell = 1/n_f - \tau \tag{12.2}$$

The maximum air velocity in the tube bank is related to the face velocity (the average air velocity approaching the first row of tubes) by the following equation:

$$V_{max}/V_{face} = A_{face}/A_{min} \tag{12.3}$$

where A_{min} is the minimum flow area in the tube bank and A_{face} is the face area. For equilateral triangular pitch, the minimum flow area is the open area between two adjacent tubes. The gap between adjacent tubes is the tube pitch minus the root diameter, giving a gross gap area of $(P_T - D_r)L$, where L is the tube length. The area occupied by the fins on both tubes is approximately $2n_f L b\tau$, giving:

$$A_{min} = (P_T - D_r)L - 2\eta_f L b\tau \tag{12.4}$$

The air that flows through this gap approaches the tube bank over the rectangle of length L and width P_T, extending from the center of one tube to the center of the adjacent tube. Thus, the face area corresponding to two adjacent tubes is simply $P_T L$. Substituting this value and A_{min} from Equation (12.4) into Equation (12.3) gives:

$$V_{max} = \frac{P_T V_{face}}{P_T - D_r - 2n_f b\tau} \tag{12.5}$$

Based on a study conducted at HTRI, the following correlation is recommended by Ganguli et al. [9]:

$$Nu = 0.38\, Re^{0.6} Pr^{1/3}(A_{Tot}/A_o)^{-0.15} \tag{12.6}$$

where

A_{Tot} = total external surface area of finned tube
$A_o = \pi D_r L$ = total external surface area of root tube

Equation (12.6) is valid for tube banks with three or more rows of tubes on triangular pitch and is based on data covering the following parameter ranges [10]:

$1800 \leq Re \leq 10^5$
$0.44 \text{ in.} \leq D_r \leq 2.0 \text{ in.}$
$0.23 \text{ in.} \leq b \leq 0.75 \text{ in.}$
$0.01 \text{ in.} \leq \tau \leq 0.022 \text{ in.}$
$1.08 \text{ in.} \leq P_T \leq 3.88 \text{ in.}$
$1 \leq A_{Tot}/A_O \leq 50$
$7 \leq$ fins per inch ≤ 11

12.4 Air-Side Pressure Drop

The pressure drop for flow across a bank of high-finned tubes is given by the following equation:

$$\Delta P_f = \frac{2f N_r G^2}{g_c \rho \phi} \tag{12.7}$$

where

f = Fanning friction factor
N_r = number of tube rows
$G = \rho V_{max}$
g_c = unit conversion factor
ϕ = viscosity correction factor

This equation is essentially the same as Equation (6.7). In air-cooled heat exchangers, the viscosity correction on the air side is negligible and, hence, ϕ can be set to unity. Furthermore, when English units are used, the pressure drop in these exchangers is usually expressed in inches of water. It is convenient for design work to incorporate the unit conversion factors into the constant in Equation (12.7) to obtain:

$$\Delta P_f = \frac{9.22 \times 10^{-10} f N_r G^2}{\rho} \tag{12.8a}$$

where

$\Delta P_f \propto$ in.H_2O
$G \propto$ lbm/h $\cdot$ ft^2
$\rho \propto$ lbm/ft^3

The corresponding relationship in SI units is:

$$\Delta P_f = \frac{2f N_r G^2}{\rho} \tag{12.8b}$$

where

$\Delta P_f \propto$ Pa
$G \propto$ kg/s $\cdot$ m^2
$\rho \propto$ kg/m^3

For the flow of air across tube banks with equilateral triangular pitch, the friction factor correlation of Robinson and Briggs [11] has been widely used:

$$f = 18.93\, Re^{-0.316} (P_T/D_r)^{-0.927} \tag{12.9}$$

The correlation is based on experimental data for tube banks containing six rows of tubes with the following ranges of parameters:

$2000 \leq Re \leq 50{,}000$
$0.734 \text{ in.} \leq D_r \leq 1.61 \text{ in.}$
$0.4135 \text{ in.} \leq b \leq 0.588 \text{ in.}$
$0.0158 \text{ in.} \leq \tau \leq 0.0235 \text{ in.}$
$0.0729 \text{ in.} \leq \ell \leq 0.1086 \text{ in.}$
$1.687 \text{ in.} \leq P_T \leq 4.500 \text{ in.}$

An alternative correlation was developed by Ganguli et al. [9]:

$$f = \left\{1 + \frac{2e^{-(a/4)}}{1+a}\right\} \left\{0.021 + \frac{27.2}{Re_{eff}} + \frac{0.29}{Re_{eff}^{0.2}}\right\} \tag{12.10}$$

where

$a = (P_T - D_f)/D_r$
$Re_{eff} = Re(\ell/b)$
$D_f = D_r + 2b = \text{fin OD}$

The parameter ranges covered by the data upon which this correlation is based were not given by Ganguli et al., but the database included that of Robinson and Briggs [11]. Hence, the parameter ranges exceed those given above for the Robinson-Briggs correlation.

In addition to the tube bank, other sources of friction loss on the air side include the following:

- The support structure and enclosure (including louvers, screens, and/or fencing if present)
- Fan casings and fan supports
- The plenums
- The steam coil (if present)
- Screens used as fan guards or hail guards for the tube bundle (if present)
- Other obstructions in the air flow path, such as the drive assemblies and walkways
- Diffusers (if present)

Although the friction loss due to each of these factors is usually small compared with the pressure loss in the tube bank, in aggregate the losses can amount to between 10% and 40% of the bundle pressure drop. Procedures for estimating these losses can be found in Ref. [10]; they will not be given here.

12.5 Overall Heat-Transfer Coefficient

Equation (4.26) is used to calculate the overall design heat-transfer coefficient for low-fin tubing. This equation is modified for high-fin tubing by adding a term to account for the contact resistance between the fin and the tube wall. The result is:

$$U_D = \left[\left(\frac{1}{h_i} + R_{Di}\right)\frac{A_{Tot}}{A_i} + \frac{A_{Tot}\ln(D_r/D_i)}{2\pi k_{tube} L} + \frac{R_{con} A_{Tot}}{A_{con}} + \frac{1}{\eta_w h_o} + \frac{R_{Do}}{\eta_w}\right]^{-1} \quad (12.11)$$

where

R_{con} = contact resistance between fin and tube wall
A_{con} = contact area between fin and tube wall

Equation (12.11) is applicable to all types of high-fin tubing with the exception of bimetallic (E-fin) tubes. (Note that for integrally finned tubes, the contact resistance is zero.) For E-fin tubing, the thermal resistance of the outer tube, or sleeve, must also be accounted for. In this case, the contact resistance is between the inner tube and the sleeve, so that $A_{con} = \pi D_o L$ = external surface area of inner tube. Thus,

$$U_D = \left[\left(\frac{1}{h_i} + R_{Di}\right)\frac{A_{Tot}}{A_i} + \frac{A_{Tot}\ln(D_o/D_i)}{2\pi k_{tube} L} + \frac{A_{Tot}\ln(D_{o,sl}/D_{i,sl})}{2\pi k_{sl} L} + \frac{R_{con}A_{Tot}}{\pi D_o L} + \frac{1}{\eta_w h_o} + \frac{R_{Do}}{\eta_w}\right]^{-1} \quad (12.12)$$

where

D_o = external diameter of inner tube
$D_{i,sl}$ = inner sleeve diameter
$D_{o,sl}$ = outer sleeve diameter
k_{sl} = thermal conductivity of sleeve

It is common practice to neglect contact resistance unless data are available from the tubing manufacturer. If available, an upper bound for the contact resistance can be used to provide a conservative estimate for the overall coefficient.

Approximate values of the overall heat-transfer coefficient suitable for preliminary design calculations are given in Table 12.2. These values are based on 1-in. OD tubes with 10 fins per inch and a fin height of 0.625 in. (*Note:* Overall heat-transfer coefficients are sometimes quoted on the basis of bare tube surface area. Such values are about 20 times larger than those given here, which are based on total external surface area.)

12.6 Fan and Motor Sizing

Fan performance is characterized in terms of the fan static pressure. The total pressure in a flowing fluid is defined as the sum of the static pressure and the dynamic (or velocity) pressure, the latter comprising the kinetic energy term in Bernoulli's equation. Thus,

$$P_{Total} = P + \frac{\alpha \rho V^2}{2 g_c} \quad (12.13)$$

TABLE 12.2 Typical Values of Overall Heat-Transfer Coefficient in Air-cooled Heat Exchangers

Service	U_D (Btu/h · ft^2 · °F)
Liquid Coolers	
Engine jacket water	6.1–7.3
Process water	5.7–6.8
Ethylene glycol (50%) - water	4.4–4.9
Light hydrocarbons	4.2–5.7
Light gas oil	3.3–4.2
Light naphtha	4.2
Hydroformer and platformer liquids	4.0
Residuum	0.5–1.4
Tar	0.2–0.5
Gas coolers	
Air or flue gas, 50 psig ($\Delta P = 1$ psi)	0.5
Air or flue gas, 100 psig ($\Delta P = 2$ psi)	0.9
Air or flue gas, 100 psig ($\Delta P = 3$ psi)	1.4
Hydrocarbon gases, 15–50 psig ($\Delta P = 1$ psi)	1.4–1.9
Hydrocarbon gases, 50–250 psig ($\Delta P = 3$ psi)	2.3–2.8
Hydrocarbon gases, 250–1500 psig ($\Delta P = 5$ psi)	3.3–4.2
Ammonia reactor stream	4.2–5.2
Condensers	
Light hydrocarbons	4.5–5.0
Light gasoline	4.5
Light naphtha	3.8–4.7
Heavy naphtha	3.3–4.2
Reactor effluent (platformers, hydroformers, reformers)	3.8–4.7
Ammonia	5.0–5.9
Amine reactivator	4.7–5.7
Freon 12	3.5–4.2
Pure steam (0–20 psig)	6.3–9.4
Steam with non-condensables	3.3

Source: Ref. [5] and Hudson Products Corporation, Sugarland, TX, www.hudsonproducts.com

where

α = kinetic energy correction factor
ρ = fluid density
V = mass-average fluid velocity

The kinetic energy correction factor depends on the velocity profile and is equal to unity for a uniform profile, which is usually a reasonable approximation for turbulent flow. The fan static pressure, *FSP*, is defined as follows:

$$FSP = (\Delta P_{Total})_{fan} - \frac{\alpha_{fr}\rho_{fr}V_{fr}^2}{2g_c} \tag{12.14}$$

where

$(\Delta P_{Total})_{fan}$ = change in total pressure between fan inlet and outlet

The subscript "fr" denotes conditions for the air leaving the fan ring. Note that *FSP* is the rise in static pressure across the fan only if the air velocity at the fan inlet is zero.

Assuming that the ambient air velocity is zero and neglecting the static pressure difference in the ambient air at inlet and outlet elevations, a pressure balance around an air-cooled heat exchanger yields the following result:

$$FSP = \sum_j \Delta P_j = \text{sum of all air-side losses in the unit} \tag{12.15}$$

Therefore, the fan static pressure can be determined by calculating (or estimating) the individual losses in the unit.

The power that must be supplied to the fan (the brake power) is given by the following equation:

$$\dot{W}_{fan} = \frac{(\Delta P_{Total})_{fan}\dot{v}_{fan}}{\eta_{fan}} \tag{12.16a}$$

where

$\dot{v}_{fan}$ = volumetric flow rate of air through fan
η_{fan} = (total) fan efficiency

With the pressure loss expressed in Pascals and the volumetric flow rate in cubic meters per second, Equation (12.16a) gives the brake power in Watts. For use with English units, it is convenient to include a unit conversion factor in the equation as follows:

$$\dot{W}_{fan} = \frac{(\Delta P_{Total})_{fan}\dot{v}_{fan}}{6342\eta_{fan}} \tag{12.16b}$$

where

$(\Delta P_{Total})_{fan} \propto$ in. H_2O
$\dot{v}_{fan} \propto$ acfm (actual cubic feet per minute)
$\dot{W}_{fan} \propto$ hp (horsepower)

Note: Two fan efficiencies are in common use, the total efficiency used here and the static efficiency. The latter is used in Equation (12.16) with the fan static pressure replacing the total pressure change in the numerator. The static efficiency is always lower than the total efficiency.

The power that must be supplied by the motor is:

$$\dot{W}_{motor} = \frac{\dot{W}_{fan}}{\eta_{sr}} \tag{12.17}$$

where η_{sr} is the efficiency of the speed reducer. Finally, the power drawn by the motor is given by:

$$\dot{W}_{used} = \frac{\dot{W}_{motor}}{\eta_{motor}} \tag{12.18}$$

where η_{motor} is the motor efficiency. For estimation purposes, reasonable values for the efficiencies are 70% to 75% for the fan and 95% each for the motor and speed reducer.

Fan selection is accomplished by means of fan performance curves and tables that are supplied by fan manufacturers. For each fan model, these graphs and/or tables present the relationships among air volumetric flow rate, fan static pressure, fan speed, brake power, and fan efficiency. The data are for air at standard conditions of 1 atm, 70°F, and 50% relative humidity, for which the density is 0.075 lbm/ft^3. Hence, corrections must be made for differences in air density between actual operating conditions and standard conditions. Another important aspect of fan selection is fan noise, which depends on blade-tip speed, number of blades, and blade design. To simplify matters, most fan manufacturers offer fan-selection software that is free upon request or in some cases can be downloaded directly from the manufacturer's website. These programs will select the best fan (or fans) for a specified service from the company's product line. A choice of power consumption, fan noise, or cost is usually offered for the selection criterion.

Another consideration in fan selection is the need to achieve a good distribution of air flow across the face of the tube bundle. The fan diameter should be such that the area covered by the fans is at least 40% of the total bundle face area. In addition, the fan diameter must be at least 6 in. less than the total width of all tube bundles in the fan bay.

Motors are rated according to their output power, which is calculated by Equation (12.17). The calculated value must be rounded upward to a standard motor size (see Appendix 12.B). Motors are frequently oversized to provide operational flexibility and allowance for contingencies.

12.7 Mean Temperature Difference

The LMTD correction factor for an air-cooled heat exchanger depends on the number of tube rows, the number of tube passes, the pass arrangement, and whether the tube-side fluid is mixed (in a header) or unmixed (in U-tubes) between passes. Charts for a number of industrially significant configurations are given by Taborek [12], and the most important of these are reproduced in Appendix 12.A. The charts can be grouped into three categories as follows:

(1) One tube pass and three (Figure 12.A.1), four (Figure 12.A.2), or more tube rows. With more than four tube rows, the *F*-factor is nearly the same as for unmixed-unmixed cross flow. Hence, the chart for an X-shell exchanger given in Appendix 11.A can be used for these cases.
(2) Multiple tube passes with one pass per tube row. Charts are given for three rows and three passes (Figure 12.A.3) and four rows and four passes (Figure 12.A.4). With more than four passes, the flow pattern approaches true counter flow [12], and the *F*-factor should be close to unity for most practical configurations. Figure 12.A.4 provides a conservative (lower bound) estimate of *F* for these cases.
(3) Multiple tube passes with multiple tube rows per pass. Of the many possible arrangements of this type, a chart is available only for the case of four tube rows and two passes with two rows per pass. The tube-side fluid is mixed (in a return header) between passes as shown in Figure 12.A.5. This chart also provides a conservative estimate of *F* for two other common configurations, namely, six tube rows with two passes (three tube rows per pass) and six tube rows with three passes (two tube rows per pass), both cases involving mixing of the tube-side fluid between passes.

12.8 Design Guidelines

12.8.1 Tubing

Tubing selection should be based on the tube-side fluid temperature and the potential for corrosion of the external tube surface as discussed in Section 12.2.2. Furthermore, it is recommended to choose one of the tubing configurations given in Table 12.1.

12.8.2 Air-Flow Distribution

In order to obtain an even distribution of air flow across the tube bundle, the fan area should be at least 40% of the bundle face area as previously noted. In addition, for two-fan bays, the ratio of tube length to bundle width should be in the range of 3 to 3.5. It is also desirable to have a minimum of four tube rows.

Note: In chemical plants and petroleum refineries, air-cooled heat exchangers are often mounted on pipe racks in order to conserve plot space. In this situation, the configuration of the unit may be dictated by the width of the pipe rack.

12.8.3 Design Air Temperature

An air-cooled heat exchanger must be designed to operate at summertime conditions. However, using the highest annual ambient air temperature to size the unit generally produces a very conservative and overly expensive design. Therefore, the usual practice is to use an air temperature corresponding to the 97th or 98th percentile, i.e., a temperature that is exceeded only 2% to 3% of the time. The appropriate design temperature can be determined from meteorological data for the plant site.

12.8.4 Outlet Air Temperature

For induced-draft operation, the outlet air temperature should be limited to about 220°F in order to prevent damage to fan blades and bearings. These parts may nevertheless be exposed to high temperatures in the event of fan failure. Therefore, forced-draft operation should be considered if the tube-side fluid temperature is greater than 350°F.

12.8.5 Air Velocity

The air velocity based on bundle face area and air at standard conditions is usually between 400 and 800 ft/min, with a value of 500 to 700 ft/min being typical for units with four to six tube rows. A value in this range will usually provide a reasonable balance between air-side heat transfer and pressure drop.

12.8.6 Construction Standards

Most air-cooled heat exchangers for industrial applications, in petroleum refineries and elsewhere, are manufactured in accordance with API Standard 661, *Air-cooled Heat Exchangers for General Refinery Services*, published by the American Petroleum Institute (www.api.org). Similar to the TEMA standards for shell-and-tube exchangers, API 661 provides specifications for the design, fabrication and testing of air-cooled heat exchangers. The recommendations given in this chapter are consistent with this standard, which should be consulted for further details, particularly in regard to structural and mechanical aspects of design. One additional item of note relates to differential thermal expansion. If the tube-side fluid temperatures entering one pass and leaving the next pass differ by more than 200°F, a split header is required. Such a header consists of two separate headers, one above the other, each containing the tubes from one of the two passes in question. This arrangement allows the tubes in each of the two passes to expand independently of one another, thereby preventing damage due to thermal stresses.

12.9 Design Strategy

The basic design procedure for air-cooled heat exchangers is similar to that for shell-and-tube exchangers. An initial configuration for the unit is obtained using an approximate overall heat-transfer coefficient together with the design guidelines given above. Rating calculations are then performed and the initial design is modified as necessary until an acceptable configuration is arrived at.

An important preliminary step in the design process is the selection of the outlet air temperature. This parameter has a major effect on exchanger economics. Increasing the outlet air temperature reduces the amount of air required, which reduces the fan power and, hence, the operating cost. However, it also reduces the air-side heat-transfer coefficient and the mean temperature difference in the exchanger, which increases the size of the unit and, hence, the capital investment. The same situation exists with water-cooled heat exchangers, but the feasible range of outlet temperatures tends to be significantly greater for air-cooled exchangers. Thus, optimization with respect to outlet air temperature (or equivalently, air flow rate) is an important aspect of air-cooled heat-exchanger design. In this chapter the primary concern is obtaining a workable (and reasonable) design rather than an optimal design. However, the importance of optimization in this context should not be overlooked.

Example 12.1

A liquid hydrocarbon stream with a flow rate of 250,000 lb/h is to be cooled from 250°F to 150°F in an air-cooled heat exchanger. The unit will be mounted at grade and there are no space limitations at the site. The design ambient air temperature is 95°F and the site elevation is 250 ft above mean sea level. An outlet air temperature of 150°F is specified for the purpose of this example. Average properties of the hydrocarbon and air are given in the table below. A fouling factor of 0.001 h · ft^2 · °F/Btu is required for the hydrocarbon, which is not corrosive, and a maximum pressure drop of 20 psi is specified for this stream. Inlet pressure will be 50 psia. The maximum allowable air-side pressure drop is 0.5 in. H_2O. Design an air-cooled heat exchanger for this service:

Property	Hydrocarbon at 200°F	Air at 122.5°F*
C_P (Btu/lbm · °F)	0.55	0.241
k (Btu/h · ft · °F)	0.082	0.0161
μ (lbm/ft · h)	1.21	0.0467
ρ (lbm/ft^3)	49.94	0.0685
Pr	8.12	0.70

*Data are for $T = 120°F$ from Table A5.

Solution

(a) Make initial specifications.

(i) Tubing type

G-fin tubing with carbon steel tubes and aluminum fins is specified based on its excellent durability. It is assumed that the environment at the plant site is not highly corrosive; otherwise, bimetallic tubing would be a better choice.

(ii) Tube size and layout

One inch OD, 13 BWG tubes with 10 fins per inch, fin height of 0.625 in. and average fin thickness of 0.013 in. are specified with reference to Table 12.1. The tube layout is triangular (30°) with a tube pitch of 2.5 in.

(iii) Draft type

Since the process fluid temperature is below 350°F, an induced-draft unit will be used. For simplicity, diffusers are not specified and it is assumed that winterization of the unit is unnecessary.

(iv) Headers

The pressure is low and based on the specified tube-side fouling factor, frequent cleaning is not anticipated. Therefore, plug-type headers will be used.

(b) Energy balances.

$$q = (\dot{m}C_P\Delta T)_{HC} = 250,000 \times 0.55 \times 100 = 13,750,000 \text{ Btu/h}$$

For the specified outlet air temperature of 150°F, the required mass flow rate of air is:

$$\dot{m}_{air} = \frac{q}{(C_P\Delta T)_{air}} = \frac{13,750,000}{0.241 \times 55} = 1,037,344 \text{ lbm/h}$$

(c) LMTD.

$$(\Delta T_{\ln})_{cf} = \frac{100 - 55}{\ln(100/55)} = 75.27°\text{F}$$

(d) LMTD correction factor.

This factor depends on the number of tube rows and tube passes, which have not yet been established. Therefore, in order to estimate the required heat-transfer surface area, $F = 0.9$ is assumed.

(e) Estimate U_D.

Based on Table 12.2, a value of 4.5 Btu/h · ft^2 · °F is assumed, which is in the expected range for light hydrocarbon liquid coolers.

(f) Calculate heat-transfer area.

$$A = \frac{q}{U_D F(\Delta T_{\ln})_{cf}} = \frac{13,750,000}{4.5 \times 0.9 \times 75.27} = 45,105 \text{ ft}^2$$

(g) Number of tube rows, tube length, and number of tubes.
The bundle face area required for a given (standard) face velocity is:

$$A_{face} = \frac{\dot{m}_{air}}{\rho_{std} V_{face}}$$

Assuming a (standard) face velocity of 600 ft/min from the design guidelines gives:

$$A_{face} = \frac{1,037,344/60}{0.075 \times 600} = 384.2 \ \text{ft}^2$$

Thus, the ratio of heat-transfer surface area to bundle face area is:

$$A/A_{face} = 45,105/384.2 = 117.4$$

From Table 12.1, the closest ratio is 107.2 for four tube rows. Using this value, the required face area is:

$$A_{face} = 45,105/107.2 = 420.8 \ \text{ft}^2$$

Based on the design guidelines, a tube length, L, of three times the bundle width, W, is assumed, giving:

$$A_{face} = 420.8 = WL = 3\ W^2$$

Thus,

$$W = 11.84 \ \text{ft}$$

$$L = 3 \times 11.84 \cong 36 \ \text{ft}$$

The number of tubes is found using the value of $A_{Tot}/L = 5.58$ from Table 12.1.

$$n_t = \frac{A}{(A_{Tot}/L) \times L} = \frac{45,105}{5.58 \times 36} = 224.5$$

Taking the closest integer divisible by four gives 224 tubes with 56 tubes per row. The corresponding bundle width is the tube pitch times the number of tubes per row. Allowing 2 in. for side clearances gives:

$$W = 2.5 \times 56 + 2 = 142 \ \text{in.} = 11.83 \ \text{ft}$$

The actual bundle face area and (standard) face velocity are:

$$A_{face} = WL = 11.83 \times 36 \cong 426 \ \text{ft}^2$$

$$V_{face,std} = \frac{\dot{m}_{air}}{\rho_{std} A_{face}} = \frac{1,037,344/60}{0.075 \times 426} = 541 \ \text{ft/min}$$

Note: It is assumed that the fan drives will be located above the tube bundle. Therefore, no allowance is made for blade-shaft lanes in the tube bundle.

(h) Number of tube passes.
The tube-side fluid velocity is:

$$V = \frac{\dot{m}_i(n_p/n_t)}{\rho \pi D_i^2/4} = \frac{(250,000/3600)(n_p/224)}{49.94 \times \pi \times (0.810/12)^2/4} = 1.73 n_p \ \text{ft/s}$$

Two, three, or four passes will give a velocity in the range of 3 to 8 ft/s. Since the tube-side pressure drop allowance is fairly generous, four passes are chosen for the first trial in order to maximize the heat-transfer coefficient and minimize fouling. Checking the Reynolds number:

$$Re = \frac{4\dot{m}_i(n_p/n_t)}{\pi D_i \mu} = \frac{4 \times 250,000 \times (4/224)}{\pi(0.81/12) \times 1.21} = 69,594$$

The flow is fully turbulent and, hence, the configuration is satisfactory. This completes the preliminary design calculations.

(i) LMTD correction factor.
The correct value of the LMTD correction factor can now be determined using Figure 12.A5:

$$R = \frac{250 - 150}{150 - 95} \cong 1.82$$

$$P = \frac{150 - 95}{250 - 95} \cong 0.35$$

$$F \cong 0.99 \ \text{from chart}$$

(j) Calculate required overall coefficient.

$$U_{req} = \frac{q}{AF(\Delta T_{\ln})_{cf}} = \frac{13,750,000}{(224 \times 5.58 \times 36) \times 0.99 \times 75.27} = 4.10 \text{ Btu/h}\cdot\text{ft}^2\cdot{}^\circ\text{F}$$

(k) Calculate h_i.

$$D_i = 0.81/12 = 0.0675 \text{ ft}$$

$$Re = 69,594 \text{ from step (h)}$$

$$h_i = (k/D_i) \times 0.023\, Re^{0.8} Pr^{1/3} (\mu/\mu_w)^{0.14}$$

$$= (0.082/0.0675) \times 0.023(69,594)^{0.8}(8.12)^{1/3}(1.0)$$

$$h_i = 420 \text{ Btu/h}\cdot\text{ft}^2\cdot{}^\circ\text{F}$$

(l) Calculate h_o.
The maximum air velocity in the tube bundle is calculated using Equation (12.5). The face velocity is first converted from standard conditions to conditions at the average air temperature. The fin thickness is taken as 0.013 in.

$$V_{face,ave} = V_{face,std}(\rho_{std}/\rho_{ave}) = 541(0.075/0.0685) = 592 \text{ ft/min}$$

$$V_{max} = \frac{P_T V_{face,\,ave}}{P_T - D_r - 2n_f b\tau} = \frac{2.5 \times 592}{2.5 - 1.0 - 2 \times 10 \times 0.625 \times 0.013}$$

$$V_{max} = 1106.5 \text{ ft/min} = 66,390 \text{ ft/h}$$

$$Re = \frac{D_r V_{max} \rho}{\mu} = \frac{(1.0/12) \times 66,390 \times 0.0685}{0.0467} = 8115$$

Note: The air density should be corrected for the average atmospheric pressure at the elevation of the plant site (see Appendix 12.C). In the present case, the site elevation (250 ft) is such that the pressure does not differ significantly from one atmosphere and, hence, no correction is needed.
Equation (12.6) is used to calculate the air-side heat-transfer coefficient with $A_{Tot}/A_o = 21.4$ from Table 12.1:

$$Nu = 0.38 Re^{0.6} Pr^{1/3} (A_{Tot}/A_o)^{-0.15} = 0.38(8115)^{0.6}(0.7)^{1/3}(21.4)^{-0.15}$$

$$Nu = 47.22$$

$$h_o = (k/D_r)Nu = \frac{0.0161 \times 47.22}{(1.0/12)} \cong 9.12 \text{ Btu/h}\cdot\text{ft}^2\cdot{}^\circ\text{F}$$

(m) Calculate fin efficiency.
Equations (2.27) and (5.12) are used to calculate the fin efficiency. From Table A.1, the thermal conductivity of aluminum is:

$$k \cong 238 \times 0.57782 = 137.5 \text{ Btu/h}\cdot\text{ft}\cdot{}^\circ\text{F}$$

This value is slightly optimistic because the aluminum alloys used for finning have somewhat lower thermal conductivities than the pure metal. However, the difference is not large enough to significantly affect the results.

$$r_1 = \text{root tube radius} = 0.5 \text{ in.}$$

$$r_2 = r_1 + \text{fin height} = 0.5 + 0.625 = 1.125 \text{ in.}$$

$$r_{2c} = r_2 + \tau/2 = 1.125 + 0.013/2 = 1.1315 \text{ in.}$$

$$\psi = (r_{2c} - r_1)[1 + 0.35 \ln(r_{2c}/r_1)]$$

$$= (1.1315 - 0.5)[1 + 0.35 \ln(1.1315/0.5)]$$

$$\psi = 0.812 \text{ in.} = 0.0677 \text{ ft.}$$

$$m = (2h_o/k\tau)^{0.5} = \left(\frac{2 \times 9.12}{137.5 \times (0.013/12)}\right)^{0.5} = 11.07 \text{ ft}^{-1}$$

$$m\psi = 11.07 \times 0.0677 = 0.7494$$

$$\eta_f = \frac{\tanh(m\psi)}{m\psi} = \frac{\tanh(0.7494)}{0.7494} = 0.8471$$

The extended and prime surface areas per inch of tube length are estimated as follows:

$$A_{fins} = 2N_f\pi\left(r_{2c}^2 - r_1^2\right) = 2 \times 10\pi\{(1.1315)^2 - (0.5)^2\} = 64.735 \text{ in.}^2$$

$$A_{prime} = 2\pi r_1\left(L - N_f\tau\right) = 2\pi \times 0.5(1.0 - 10 \times 0.013) = 2.733 \text{ in.}^2$$

$$A_{fins}/A_{Tot} = \frac{64.735}{64.735 + 2.733} \cong 0.96$$

$$A_{prime}/A_{Tot} = 1 - 0.96 = 0.04$$

The weighted efficiency of the finned surface is given by Equation (2.31):

$$\eta_w = \left(A_{prime}/A_{Tot}\right) + \eta_f\left(A_{fins}/A_{Tot}\right) = 0.04 + 0.8471 \times 0.96 \cong 0.853$$

(n) Wall temperatures and viscosity correction factors.
The wall temperatures, T_p and T_{wtd}, used to obtain viscosity correction factors are given by Equations (4.38) and (4.39). However, no viscosity correction is required for the air-side heat-transfer coefficient, so only T_p is needed for the tube-side correction. In the present case, no viscosity data were given for the tube-side fluid. Therefore, ϕ_i is assumed to be 1.0 and the wall temperature is not calculated.

(o) Calculate the clean overall coefficient.
The clean overall coefficient is given by Equation (12.12) with the fouling factors omitted. The contact resistance is neglected and $A_{Tot}/A_i = 26.3$ is obtained from Table 12.1.

$$U_C = \left[\frac{(A_{Tot}/A_i)}{h_i} + \frac{(A_{Tot}/L)\ln(D_r/D_i)}{2\pi\, k_{tube}} + \frac{1}{\eta_w h_o}\right]^{-1}$$

$$= \left[\frac{26.3}{420} + \frac{5.58\ln(1.0/0.81)}{2\pi \times 26} + \frac{1}{0.853 \times 9.12}\right]^{-1}$$

$$U_C = 5.04 \text{ Btu/h}\cdot\text{ft}^2\cdot{}^\circ\text{F}$$

Since $U_C > U_{req}$, continue.

(p) Fouling allowance.
The tube-side fouling factor was specified in the problem statement as 0.001 Btu/h · ft^2 · °F. Except in unusual circumstances, air-side fouling is minimal and, therefore, R_{Do} is taken as zero. Thus, the total fouling allowance is:

$$R_D = R_{Di}\left(A_{Tot}/A_i\right) + R_{Do}/\eta_w = 0.001 \times 26.3 + 0 = 0.0263 \text{ h}\cdot\text{ft}^2\cdot{}^\circ\text{F/Btu}$$

(q) Calculate the design overall coefficient.

$$U_D = (1/U_C + R_D)^{-1} = (1/5.04 + 0.0263)^{-1} = 4.45 \text{ Btu/h}\cdot\text{ft}^2\cdot{}^\circ\text{F}$$

Since $U_D > U_{req} = 4.10$ Btu/h · ft^2 · °F, the heat exchanger is thermally workable.

(r) Over-surface and over-design.

$$\text{Over-surface} = U_C/U_{req} - 1 = 5.04/4.10 - 1 \cong 23\%$$

$$\text{Over-design} = U_D/U_{req} - 1 = 4.45/4.10 - 1 \cong 8.5\%$$

Both values are reasonable and, hence, the unit is thermally acceptable.

(s) Tube-side pressure drop.
Equations (5.1) to (5.4) are used to calculate the tube-side pressure drop:

$$f = 0.4137Re^{-0.2585} = 0.4137(69{,}594)^{-0.2585} = 0.02317$$

$$G = \frac{\dot{m}\left(n_p/n_t\right)}{\left(\pi D_i^2/4\right)} = \frac{250{,}000(4/224)}{\left[\pi(0.0675)^2/4\right]} = 1{,}247{,}540 \text{ lbm/h}\cdot\text{ft}^2$$

$$s = 49.94/62.43 = 0.80$$

$$\Delta P_f = \frac{f\, n_p L G^2}{7.50 \times 10^{12} D_i s\phi} = \frac{0.02317 \times 4 \times 36\,(1{,}247{,}540)^2}{7.50 \times 10^{12} \times 0.0675 \times 0.80 \times 1.0} = 12.82 \text{ psi}$$

$$\Delta P_r = 1.334 \times 10^{-13}\left(2n_p - 1.5\right)G^2/s = 1.334 \times 10^{-13}(6.5)(1{,}247{,}540)^2/0.80$$

$$\Delta P_r = 1.69 \text{ psi}$$

Assuming 5-in. schedule 40 nozzles are used, the flow area per nozzle from Table B.1 is 0.1390 ft^2. Hence,

$$G_n = 250{,}000/0.1390 = 1{,}798{,}561 \text{ lbm/h}\cdot\text{ft}^2$$

$$Re_n = \frac{D_i G_n}{\mu} = \frac{(5.047/12) \times 1{,}798{,}561}{1.21} = 625{,}161$$

$$\Delta P_n = 2.0 \times 10^{-13} G_n^2/s = 2.0 \times 10^{-13}(1{,}798{,}561)^2/0.80 = 0.81\text{psi}$$

$$\Delta P_i = \Delta P_f + \Delta P_r + \Delta P_n = 12.82 + 1.69 + 0.81 = 15.3 \text{ psi}$$

(t) Air-side pressure drop.
Equation (12.10) will be used to calculate the friction factor. The parameters a and Re_{eff} are computed first.

$$D_f = D_r + 2b = 1.0 + 2 \times 0.625 = 2.25 \text{ in.}$$

$$a = (P_T - D_f)/D_r = (2.5 - 2.25)/1.0 = 0.25$$

The fin spacing is obtained from Equation (12.2):

$$\ell = 1/n_f - \tau = 1/10 - 0.013 = 0.087 \text{ in.}$$

$$Re_{eff} = Re(\ell/b) = 8115(0.087/0.625) = 1130$$

$$f = \left\{1 + \frac{2e^{-(a/4)}}{1+a}\right\}\left\{0.021 + \frac{27.2}{Re_{eff}} + \frac{0.29}{Re_{eff}^{0.2}}\right\}$$

$$= \left\{1 + \frac{2e^{-(0.25/4)}}{1+0.25}\right\}\left\{0.021 + \frac{27.2}{1130} + \frac{0.29}{(1130)^{0.2}}\right\}$$

$$f = 0.291$$

The pressure drop across the tube bundle is calculated using Equation (12.8a). The air mass flux is computed first:

$$G = \rho V_{max} = 0.0685 \times 66{,}390 = 4548 \text{ lbm/h}\cdot\text{ft}^2$$

$$\Delta P_f = \frac{9.22 \times 10^{-10} f N_r G^2}{\rho} = \frac{9.22 \times 10^{-10} \times 0.291 \times 4(4548)^2}{0.0685}$$

$$\Delta P_f \cong 0.324 \text{ in. } H_2O$$

This unit will not require louvers, steam coils, fan guards, or hail guards. Enclosure losses will be small, and other losses will be due primarily to the fan casings, plenums, and obstructions such as fan supports and walkways. Therefore, a relatively small allowance of 10% is made for other air-side losses, giving:

$$\Delta P_o \cong 1.1\Delta P_f = 1.1 \times 0.324 \cong 0.36 \text{ in. } H_2O$$

Both tube-side and air-side pressure drops are below the specified maximum values. Therefore, the unit is hydraulically acceptable.

(u) Fan sizing.
The fans should cover at least 40% of the bundle face area. Assuming a two-fan bay, this condition gives the following relation for the fan diameter:

$$2\left(\pi D_{fan}^2/4\right) \geq 0.4A_{face} = 0.4 \times 426 = 170.4 \text{ ft}^2$$

$$D_{fan} \geq 10.4 \text{ ft}$$

Therefore, two fans with diameters of 10.5–11.0 ft are required.

The fan static pressure is given by Equation (12.15). Thus, $FSP = 0.36$ in. H_2O. The induced-draft fans handle air at the outlet temperature of 150°F, for which the density is 0.065 lbm/ft^3 from Table A.5. Since there are two fans, the volumetric flow rate per fan is:

$$\dot{v}_{fan} = \frac{0.5\dot{m}_{air}}{\rho_{air}} = \frac{0.5(1{,}037{,}344/60)}{0.065} = 132{,}993 \text{ acfm}$$

Thus, each fan must deliver about 133,000 acfm at a static pressure of 0.36 in. H_2O.

(v) Motor sizing.
From Equation (12.14), the total pressure change across the fans is:

$$(\Delta P_{Total})_{fan} = FSP + \frac{\alpha_{fr}\,\rho_{fr}\,V_{fr}^2}{2g_c}$$

For a fan diameter of 10.5 ft, the air velocity in the fan ring is (neglecting the clearance between the blades and the housing):

$$V_{fr} = \frac{\dot{v}_{fan}}{\pi D_{fr}^2/4} = \frac{132{,}993}{\pi(10.5)^2/4} = 1536 \ \text{ft/min} = 25.6 \ \text{ft/s}$$

Thus, the velocity pressure is:

$$\frac{\alpha_{fr}\rho_{fr}V_{fr}^2}{2g_c} \cong \frac{1.0 \times 0.065(25.6)^2}{2 \times 32.174} = 0.662 \ \text{lbf/ft}^2$$

$$\frac{\alpha_{fr}\rho_{fr}V_{fr}^2}{2g_c} = 0.662 \ \text{lbf/ft}^2 \times 0.1922 \ \frac{\text{in.H}_2\text{O}}{\text{lbf/ft}^2} = 0.127 \ \text{in.H}_2\text{O}$$

The total pressure difference across the fan is:

$$(\Delta P_{Total})_{fan} = 0.36 + 0.127 \cong 0.49 \text{ in. H}_2\text{O}$$

The brake power is calculated using Equation (12.16b); a fan efficiency of 70% is assumed:

$$\dot{W}_{fan} = \frac{(\Delta P_{Total})_{fan}\dot{v}_{fan}}{6342\eta_{fan}} = \frac{0.49 \times 132{,}993}{6342 \times 0.7} = 14.7 \text{ hp}$$

The power supplied by the motor is given by Equation (12.17); an efficiency of 95% is assumed for the speed reducer:

$$\dot{W}_{motor} = \frac{\dot{W}_{fan}}{\eta_{sr}} = \frac{14.7}{0.95} = 15.5 \ \text{hp}$$

The next largest standard motor size is 20 hp (Appendix 12.B). This motor size will provide an adequate allowance for operational flexibility and contingencies. Therefore, each fan will be equipped with a 20 hp motor. This result is preliminary pending actual fan selection. Final values for fan efficiency and motor size will be based on the fan manufacturer's data. A fan that is well matched with the service may have a total efficiency as high as 80% to 85%. In that case, 15 hp motors might be sufficient.

The main design parameters for the unit are summarized below.

Design Summary

Number of fan bays: 1
Number of tube bundles per bay: 1
Number of fans per bay: 2
Bundle width and length: 11.8 ft × 37 ft (including headers)
Number of tube rows: 4
Number of tube passes: 4
Number of tubes: 224
Tubing type: G-fin
Tube size: 1 in. OD, 13 BWG, 36 ft long
Tube layout: Equilateral triangular with 2.5-in. pitch
Fins: 10 fpi, 0.625 in. high, 0.013 in. thick
Heat-transfer surface area: 45,000 ft^2
Draft type: Induced draft
Fan diameter: 10.5 ft
Motor size: 20 hp
Tube-side nozzles: 5-in. schedule 40
Headers: Plug-type box headers
Materials: Carbon steel tubes, aluminum fins, carbon steel headers, tubesheets, pass partitions, and nozzles

12.10 Computer Software

12.10.1 HEXTRAN

The air-cooled heat-exchanger module (ACE) in HEXTRAN is configured in a similar manner to the shell-and-tube exchanger module. It can operate in either rating mode (TYPE=Old) or design mode (TYPE=New). In design mode the following parameters can be varied automatically between user-specified limits to meet a given performance specification (usually tube-side outlet temperature or duty) and pressure drop constraints:

- Area per tube bundle
- Number of tube passes
- Number of tube rows
- Tube length
- Width of tube bundle
- Number of fan bays in parallel

HEXTRAN does not have a special thermodynamic package for air. Air is treated as a pure component, and methods must be selected for computing thermodynamic and transport properties. For the latter, the pure component data bank should be used by selecting the LIBRARY method. The choice of thermodynamic method is not critical. The SRKS method, which is a modified Soave-Redlich-Kwong cubic equation of state developed by SimSci-Esscor, is suggested here, but any equation-of-state method other than IDEAL can be used with similar results.

Two different tube pitches must be entered to specify the tube layout in HEXTRAN. The transverse pitch is the center-to-center distance between adjacent tubes in the same row. The longitudinal pitch is the center-to-center distance between adjacent tube rows. For an equilateral triangular layout, this is the height of an equilateral triangle with the length of a side equal to the transverse pitch, i.e.,

$$\text{longitudinal pitch} = 0.5 \times (\text{transverse pitch}) \times \tan 60^\circ \tag{12.19}$$

In general, however, both tube pitches can be independently specified.

Unlike the HTRI program considered in the following subsection, HEXTRAN has no provision for entering data needed to calculate air-side pressure losses other than the pressure drop across the tube bundle. Also, the only orientation allowed for the tube bundle is horizontal. Therefore, an A-frame condenser cannot be simulated.

HEXTRAN version 9.2 is used in the following example.

Example 12.2

Use HEXTRAN to rate the air-cooled heat exchanger designed by hand in Example 12.1.

Solution

The English system of units is selected and for convenience, the unit of viscosity is changed from cp to lb/ft · h. Under Components and Thermodynamics, air is selected from the list of components and a New Method Slate called SET1 is defined on the Method form. The SRKS equation of state is specified as the thermodynamic method and LIBRARY is specified for transport properties.

The flowsheet is constructed in the usual manner. The hydrocarbon feed is defined as a bulk property stream and the flow rate, temperature, pressure, and physical properties are entered on the appropriate forms. The inlet air is defined (by default) as a compositional stream and the composition (100% air), flow rate, temperature, and pressure (14.7 psia) are entered on the appropriate form.

Data for the air-cooled heat exchanger are obtained from Example 12.1 and entered on the appropriate panels as follows: Items not listed are either left at the default settings or left unspecified, in which case they are calculated by the program.

(a) Tube side

Tube length: 36 ft	Pattern: Staggered
Transverse pitch: 2.5 in.	Outside diameter: 1 in.
Longitudinal pitch: 2.165 in.	BWG: 13

(b) Air side

Number of tubes/bundle: 224	Flow direction: Counter current
Number of passes/bundle: 4	Hotside: Tube side
Number of rows/bundle: 4	

(c) Tube side nozzles
The inside diameter (5.047 in.) of inlet and outlet nozzles is entered on this form. The number (1) of each type of nozzle is the default setting.
(d) Fins

Number of fins/length: 10/in.	Thickness: 0.013 in.
Height above root: 0.625 in.	Area/length: 5.58 ft^2/ft

The area/length entry is optional and will be calculated by the program if not given.
(e) Material
The default materials of construction, carbon steel for the tubes and aluminum alloy 1060-H14 for the fins, are used.
(f) Film options
The fouling factors, 0.001 h · ft^2 · °F/Btu for the tube side and zero for the air side, are entered here.
(g) Fan

Draft type: Induced	Efficiency: 57%
Number of fans/bay: 2	Fan diameter: 10.5 ft

The efficiency entered here is the product of the fan static efficiency (estimated as 60%) and the speed reducer, or drive, efficiency (95%). HEXTRAN uses this value to calculate the power supplied by the fan motors. Although it is not stated in the program documentation, the static efficiency should be used here because HEXTRAN does not include the velocity pressure in the calculation of fan power.
(h) Multi-bundle
Number of bundles in series/bay: 1
Number of bundles in parallel/bay: 1
Number of bays in parallel: 1
These are the default settings in the program.

Finally, under Global Options, the Water Decant Option switch is unchecked (OFF). Water decanting is not relevant to this problem, but if the switch is left in the default (ON) position, an error results because the SRKS method does not support this option.

The input file generated by the HEXTRAN GUI is given below, followed by a summary of results in the form of the Exchanger Data Sheet and Extended Data Sheet. The data sheets were extracted from the HEXTRAN output file and used to prepare the following comparison between computer and hand calculations:

Item	Hand	HEXTRAN
Re_i	69,594	69,596
Re_o	8115	8080
h_i (Btu/h · ft^2 · °F)	420	420.1
h_o (Btu/h · ft^2 · °F)	9.12	7.2
U_D (Btu/h · ft^2 · °F)	4.45	4.05
ΔP_i (psi)	15.3	15.7
ΔP_o (in. H_2O)	0.36	0.40
$\dot{W}_{motor}$ (hp)	15.5[a]	14.8[b]
Over-design (%)	8.5	0

[a]Based on fan total efficiency of 70%
[b]Based on fan static efficiency of 60%

Overall, the two sets of results are in reasonably good agreement, but there are significant differences in the air-side heat-transfer coefficient (26%) and pressure drop (10%). Notice that the outlet temperature of the hydrocarbon stream calculated by HEXTRAN is 150°F. Thus, in agreement with the hand calculations, the heat exchanger is workable, but the over-design is essentially zero according to HEXTRAN.

HEXTRAN Input File for Example 12.2

```
$ GENERATED FROM HEXTRAN KEYWORD EXPORTER
$
$      General Data Section
$
TITLE PROJECT=Ex12-2, PROBLEM=HC-Cooler, SITE=
$
DIME  English, AREA=FT2, CONDUCTIVITY=BTUH, DENSITY=LB/FT3, *
      ENERGY=BTU, FILM=BTUH, LIQVOLUME=FT3, POWER=HP, *
      PRESSURE=PSIA, SURFACE=DYNE, TIME=HR, TEMPERATURE=F, *
      UVALUE=BTUH, VAPVOLUME=FT3, VISCOSITY=LBFH, WT=LB, *
      XDENSITY=API, STDVAPOR=379.490
$
PRINT ALL, *
      RATE=M
$
CALC  PGEN=New, WATER=Saturated
$
$      Component Data Section
$
COMPONENT DATA
$
 LIBID 1, AIR
$

$
$     Thermodynamic Data Section
$
THERMODYNAMIC DATA
$
 METHODS SET=SET1, KVALUE=SRKS, ENTHALPY(L)=SRKS, ENTHALPY(V)=SRKS, *
               ENTROPY(L)=SRKS, ENTROPY(V)=SRKS, DENSITY(L)=API, *
               DENSITY(V)=SRKS, VISCOS(L)=LIBRARY, VISCOS(V)=LIBRARY, *
               CONDUCT(L)=LIBRARY, CONDUCT(V)=LIBRARY, SURFACE=LIBRARY
$
 WATER DECANT=OFF


$
$Stream Data Section
$
STREAM DATA

$
 PROP STRM=HC, NAME=HC, TEMP=250.00, PRES=50.000, *
        LIQUID(W)=250000.000, LCP(AVG)=0.55, Lcond(AVG)=0.082, *
        Lvis(AVG)=1.21, Lden(AVG)=49.94
$
 PROP STRM=HCOUT, NAME=HCOUT
$
 PROP STRM=AIR, NAME=AIR, TEMP=95.00, PRES=14.700, *
        SET=SET1, RATE(W)=1037344.000, *
   COMP(M)= 1, 100, *
        NORMALIZE
$
 PROP STRM=AIROUT, NAME=AIROUT
$
$ Calculation Type Section
$
SIMULATION
$
 TOLERANCE TTRIAL=0.01
$
LIMITS AREA=200.00, 6000.00, SERIES=1, 10, PDAMP=0.00, *
       TTRIAL=30
```

```
$
 CALC   TWOPHASE=New, DPSMETHOD=Stream, MINFT=0.80
$
 PRINT  UNITS, ECONOMICS, STREAM, STANDARD, *
        EXTENDED, ZONES
$
ECONOMICS DAYS=350, EXCHANGERATE=1.00, CURRENCY=USDOLLAR
$
 UTCOST OIL=3.50, GAS=3.50, ELECTRICITY=0.10, *
        WATER=0.03, HPSTEAM=4.10, MPSTEAM=3.90, *
        LPSTEAM=3.60, REFRIGERANT=0.00, HEATINGMEDIUM=0.00
$
 HXCOST BSIZE=1000.00, BCOST=0.00, LINEAR=50.00, *
        EXPONENT=0.60, CONSTANT=0.00, UNIT
$
$       Unit Operations Data
$
UNIT OPERATIONS
$
ACE UID=ACE1
 TYPE  Old, HOTSIDE=Tubeside, *
       FLOW=Countercurrent, *
       UESTIMATE=5.00, USCALER=1.00
 TUBE  FEED=HC, PRODUCT=HCOUT, *
       LENGTH=36.00, *
       OD=1.000, BWG=13, *
       NUMBER=224, PASS=4, ROWS=4, PATTERN=Staggered, *
       PARA=1, SERIES=1, *
       TPITCH=2.500, LPITCH=2.165, MATERIAL=1, *
       FOUL=0.001, LAYER=0, *
       DPSCALER=1.00
$
 FINS  NUMBER=10.00, AREA=5.580, HEIGHT=0.625, *
       THICKNESS=0.013, BOND=0.000, *
       MATERIAL=20
$
 AIRS  FEED=AIR, PRODUCT=AIROUT, *
       PARALLEL=1, *
       FOUL=0, LAYER=0, *
       DPSCALER=1.00
$
 FAN   DRAFT=Induced, DIAM=10.50, NUMBER=2, EFFI=57.00
$
TNOZZ  ID=5.047, 5.047 NUMBER=1, 1
$
 CALC  TWOPHASE=New, *
       MINFT=0.80
$
 PRINT STANDARD, *
       EXTENDED, *
       ZONES
$
 COST  BSIZE=1000.00, BCOST=0.00, LINEAR=50.00, *
       CONSTANT=0.00, EXPONENT=0.60, Unit
$

$ End of keyword file...
```

HEXTRAN Output Data for Example 12.2

```
==============================================================================
                    AIR-COOLED EXCHANGER DATA SHEET
I----------------------------------------------------------------------------I
I EXCHANGER  NAME                        UNIT ID ACE1                        I
I SIZE                                   TYPE INDUCED       NO. OF BAYS  1   I
I AREA/UNIT-FINNED  44997. FT2 ( 45106. FT2 REQUIRED)      -BARE    2111. FT2 I
I HEAT EXCHANGED MMBTU /HR    13.758,      MTD(CORRECTED)  75.4,      FT 1.000 I
I TRANSFER RATE    FINNED-SERVICE   4.05, BARE-SERVICE  86.27,   CLEAN   4.53 I
I  BTU/HR-FT2-F          (REQUIRED   4.06)     (REQUIRED  86.48)             I
I----------------------------------------------------------------------------I
I PERFORMANCE OF ONE UNIT              AIR-SIDE              TUBE-SIDE       I
I----------------------------------------------------------------------------I
I FEED STREAM NUMBER                    AIR                     HC           I
I FEED STREAM NAME                     AIR                     HC           I
I TOTAL FLUID          LB/HR          1037344.                250000.        I
I     VAPOR  (IN/OUT)  LB/HR     1037344./  1037344.         0./         0. I
I     LIQUID           LB/HR          0./         0.     250000./  250000. I
I     STEAM            LB/HR          0./         0.         0./         0. I
I     WATER            LB/HR          0./         0.         0./         0. I
I     NON CONDENSIBLE  LB/HR              0.                     0.         I
I TEMPERATURE (IN/OUT) DEG F       95.0 /  149.9           250.0 /  150.1     I
I PRESSURE    (IN/OUT) PSIA        14.7 /      14.7          50.0 /     34.3  I
I FOULING RESIST FT2-HR-F/BTU          0.00000 (-.00060 REQD)        0.00100 I
I----------------------------------------------------------------------------I
I SP. GR., LIQ  (60F / 60F H2O)    0.801 /     0.801 I AIR QTY/UNIT          I
I          VAP  (60F / 60F AIR)    0.000 /     0.000 I  STD FT3/MIN   230521. I
I DENSITY,    LIQUID    LB/FT3    49.940 /   49.940 I AIR QTY/FAN            I
I             VAPOR     LB/FT3     0.000 /     0.000 I  ACT FT3/MIN   132916. I
I VISCOSITY, LIQUID  LB/FT-HR      1.210 /     1.210 I STATIC DP             I
I             VAPOR   LB/FT-HR     0.000 /     0.000 I        IN H2O     0.40 I
I THRML COND,LIQ  BTU/HR-FT-F    0.0820 /   0.0820 I FACE VELOCITY          I
I             VAP  BTU/HR-FT-F    0.0000 /   0.0000 I        FT/SEC       9.5 I
I SPEC.HEAT, LIQUID BTU /LB F    0.5500 /   0.5500 I                        I
I             VAPOR  BTU /LB F    0.0000 /   0.0000 I                        I
I LATENT HEAT         BTU /LB         0.00        I                        I
I PRESSURE DROP (CALC) PSI            15.70        I                        I
I----------------------------------------------------------------------------I
I                          CONSTRUCTION OF ONE BAY                           I
I----------------------------------------------------------------------------I
I BUNDLE                 I HEADER                    I TUBE                  I
I----------------------------------------------------------------------------I
I SIZE 11.8 FT X  36.0 FT I PASSES/BUNDLE     4       I MATERIAL   CARB STL I
I BUNDLES IN PARALLEL   1 I NOZZLES                  I OD          1.000 IN I
I         IN SERIES     1 I NO./SIZE INLET  1 /  5.0 IN I THICKNESS 0.095 IN I
I ROWS                  4 I NO./SIZE OUTLT  1 /  5.0 IN I NUMBER/BNDL     224 I
I-----------------------I---------------------------I LENGTH       36.0 FT I
I FAN                   I FIN                       I PITCH-TRAN  2.50IN I
I-----------------------I---------------------------I        -LONG  2.17IN I
I NUMBER/BAY            2 I MATERIAL          A1060H14 I LAYOUT      STAGGER I
I POWER/FAN      14.8 HP I OD   2.25 IN,THICK 0.013 IN I                     I
I DIAMETER       10.5 FT I NUMBER/IN          10.0   I                     I
I EFFICIENCY    57.0 PCNT I EFFICIENCY       87.8 PCNT I                     I
I                        I TYPE          TRANSVERSE I                     I
I----------------------------------------------------------------------------I
```

```
==============================================================================
                    AIR COOLED EXTENDED DATA SHEET
I----------------------------------------------------------------------------I
I EXCHANGER  NAME                             UNIT ID ACE1                   I
I AREA/UNIT 44997. FT2 ( 45106. FT2 REQUIRED)                                I
I----------------------------------------------------------------------------I
I PERFORMANCE OF ONE UNIT               AIR-SIDE                TUBE-SIDE    I
I----------------------------------------------------------------------------I
I FEED STREAM NUMBER                    AIR                     HC           I
I FEED STREAM NAME                     AIR                      HC           I
I WT FRACTION LIQUID (IN/OUT)         0.00 / 0.00              1.00 / 1.00   I
I REYNOLDS NUMBER                        8080.                  69596.       I
I PRANDTL NUMBER                        0.698                   8.116        I
I UOPK,LIQUID                        0.000 /    0.000          0.000 /    0.000 I
I        VAPOR                       5.981 /    5.981          0.000 /    0.000 I
I SURFACE TENSION    DYNES/CM        0.000 /    0.000          0.000 /    0.000 I
I FILM COEF(SCL) BTU/HR-FT2-F              7.2 (1.000)           420.1 (1.000)  I
I FOULING LAYER THICKNESS   IN             0.000                   0.000     I
I----------------------------------------------------------------------------I
I THERMAL RESISTANCE                                                         I
I UNITS:   (FT2-HR-F/BTU)        (PERCENT)     (ABSOLUTE)                    I
I SHELL FILM                        63.88         0.15783                    I
I TUBE   FILM                       25.35         0.06263                    I
I TUBE   METAL                       0.12         0.00029                    I
I TOTAL FOULING                     10.65         0.02631                    I
I ADJUSTMENT                        -0.24        -0.00060                    I
I----------------------------------------------------------------------------I
I PRESSURE DROP                         AIR-SIDE               TUBE-SIDE     I
I UNITS: (PSIA  )                (PERCENT)    (ABSOLUTE)   (PERCENT)    (ABSOLUTE)I
I WITHOUT NOZZLES                 100.00          0.01       94.50          14.84 I
I INLET    NOZZLES                  0.00          0.00        3.44           0.54 I
I OUTLET   NOZZLES                  0.00          0.00        2.06           0.32 I
I TOTAL    /BUNDLE                                0.01                      15.70 I
I TOTAL    /UNIT                                  0.01                      15.70 I
I DP SCALER                                       1.00                       1.00 I
I----------------------------------------------------------------------------I
```

12.10.2 HTRI

The *Xace* module of *Xchanger Suite* is used for design and rating of air-cooled heat exchangers. The program is similar in structure and format to the *Xist* module discussed in previous chapters. As in *Xist*, the computational method in *Xace* is fully incremental, and the same proprietary correlations for tube-side heat transfer and pressure drop are used in both programs. Proprietary correlations are also used for air-side calculations. The methods for handling tube-side fluid properties are identical in the two programs; the properties of air are generated automatically by *Xace*.

Xace can operate in rating, simulation, or design mode. The rating and simulation modes are the same as in *Xist*. Two options are available for design mode, namely, classic design and grid design. In classic design, the only parameters that can be varied are bundle width, air face velocity and number of tube passes, and the parameter ranges are controlled by the program. In grid design, transverse and longitudinal tube pitch, tube length, tube diameter, and number of tube rows can also be varied. Furthermore, the parameter ranges are user-specified.

Xace accounts for air-side pressure losses from a number of factors in addition to the tube bundle, namely, fan rings, plenums, steam coil, louvers, fan guard, hail screen, and fan area blockage. The user must supply values for the percent open area in the fan guard and hail screen, and the percentage of the fan area that is blocked by obstructions if these items are to be included in the pressure drop calculation. The type of plenum (box or tapered) and fan ring inlet (straight, flanged, 15° cone, 30° cone, bell) must always be specified.

Fin geometry can be specified by the user on the appropriate input form. However, *Xace* has a built-in databank of finned tubing that is available from selected manufacturers, and the user has the option of selecting the fin geometry from this databank.

Like *Xist*, *Xace* generates a tube layout that can be modified by the user. Also, for tube-side condensing, the inclination of the tube bundle can be specified in the range of 1–89°. A unique feature in *Xace* is an interface with software from several fan manufacturers. If the user selects one of the available manufacturers, a list of fan models that will meet the requirements of the unit is printed in the output file. The list includes pertinent information such as fan size, efficiency, speed, and power.

Example 12.3

Use *Xace* to rate the air-cooled heat exchanger designed by hand in Example 12.1 and compare the results with those obtained previously by other methods.

Solution

Xace is run in rating mode (the default option) for this problem. Data obtained from Example 12.1 are entered on the appropriate input forms as indicated below. Items not listed are either left at their default values or left blank to be computed by the program.

(a) Geometry/Unit
Fan arrangement: Induced
Number of bays in parallel per unit: 1 (default)
Number of bundles in parallel per bay: 1 (default)
Number of tube passes per bundle: 4
The number (1) and ID (5.047 in.) of the tube-side inlet and outlet nozzles are also given on this form.

(b) Geometry/Fans
Number of fans per bay: 2 (default)
Fan diameter: 10.5 ft
Total combined fan and drive efficiency: 66.5%
Fan ring type: 15°cone
The combined fan and drive efficiency of 66.5% is equivalent to the efficiencies assumed in Example 12.1; namely, total fan efficiency of 70% and drive efficiency of 95%. The type of fan ring entrance was not specified in Example 12.1. A 15° conical inlet is chosen because it gives a pressure loss in the mid-range of the values for the available options. The straight entrance has by far the greatest loss, while the bell-shaped entrance has the least.

(c) Geometry/Optional
The only entry made on this form is the selection of a tapered plenum, which is standard for an induced-draft fan.

(d) Geometry/Bundle
Number of tube rows/tube passes: 4/4
Number of tubes in each odd/even numbered row: 56/56
Tube layout: Staggered (default)
Tube form: Straight (default)
Tube length: 36 ft

(e) Geometry/Tube Types/Tube Type 1/ Geometry

Tube type: High Fin	Wall thickness: 0.095 in.
Tube material: Carbon steel (Default)	Transverse pitch: 2.5 in.
Tube OD: 1 in.	Longitudinal pitch: 2.165 in.

(f) Geometry/Tube Types/Tube Type 1/High Fin

Fin type: Circular fin (Default)	Fin base thickness: 0.013 in.
Fin density: 10 fin/in.	Material: Aluminum 1060-H14
Fin height: 0.625 in.	

Note: The fins are tapered from base to tip, but only the average thickness affects the results. If no value is entered for the fin tip thickness, then the value entered for the base thickness is interpreted as the average thickness.

(g) Process

	Tube-Side Fluid (Hot)	Air-Side Fluid
Fluid name	Hydrocarbon	–
Phase/air-side flow-rate units	All liquid	Mass flow rate
Flow rate (1000 lb/h)	250	1037.344
Inlet temperature (°F)	250	95
Outlet temperature (°F)	150	–
Inlet pressure/altitude of unit	50 psia	250 ft
Fouling resistance (h · ft^2·°F/Btu)	0.001	0

If the altitude of the unit above mean sea level is given here, the program accounts for the variation of atmospheric pressure with altitude in determining the air density.

(h) Hot Fluid Properties

Bulk properties of the hydrocarbon stream are entered by first choosing *Program calculated* as the Physical Property Input Option. On the Components panel, *HTRI* is selected as the Package to be used and <User Defined> is selected from the list of components. Then on the Liquid Properties subpanel the density (49.94 lbm/ft^3), viscosity (0.5 cp = 1.21 lbm/ft·h), thermal conductivity (0.082 Btu/h · ft · °F) and heat capacity (0.55 Btu/lbm · °F) are entered at a single reference temperature of 200°F.

This completes the data entry. No input data are required for the properties of air, which are automatically generated by the program.

The Output Summary for this case is given below along with the tube layout and exchanger drawings produced by *Xace*. The output data were used to construct the results comparison shown in the following table. Note that the air-side heat-transfer coefficient of 7.81 Btu/h· ft^2·°F given in the output file is the effective outside coefficient, i.e., $\eta_w h_o$. The weighted efficiency is not available in the *Xace* output files, but the fin efficiency computed by the program is given as 79.7% in the Output Summary. Using the percentages of fin (96%) and prime (4%) surface area calculated in Example 12.1 gives $\eta_w = 0.805$, from which $h_o = 9.70$ Btu/h · ft^2 · °F:

Item	Hand	HEXTRAN	*Xace*
Re_i	69,594	69,596	69,623
Re_o	8115	8080	8381
h_i (Btu/h · ft^2 · °F)	420	420	490
h_o (Btu/h · ft^2 · °F)	9.12	7.2	9.70
U_D (Btu/h · ft^2 · °F)	4.45	4.05	4.67
ΔP_i (psi)	15.3	15.7	16.3
ΔP_o (in. H_2O)	0.36	0.40	0.32
$\dot{W}_{motor}$ (hp)	15.5[a]	14.8[b]	14.2[a]
Over-design (%)	8.5	0	12.9

[a]Based on 70% fan total efficiency and 95% drive efficiency, or combined 66.5% fan and drive efficiency

[b]Based on 60% fan static efficiency and 95% drive efficiency, or equivalent combined efficiency of 57%

Both tube-side and air-side heat-transfer coefficients from HEXTRAN are significantly more conservative than those from *Xace*. By contrast, values of the air-side coefficient calculated by hand and by *Xace* agree quite well, the difference being about 6%. The tube-side results are consistent with those from examples in previous chapters.

The overall heat-transfer coefficient calculated by hand differs from the *Xace* value by less than 5%. However, *Xace* calculated a higher tube-side heat-transfer coefficient and lower fin efficiency. The difference in fin efficiency is attributed to the correction applied by *Xace* that accounts for a non-uniform heat-transfer coefficient (see Section 2.3).

The air-side pressure drop computed by *Xace* is the lowest (least conservative) of the three values compared here. Note, however, that no allowance for fan area blockage was included in the *Xace* calculation for this example.

Xace Output Summary for Example 12.3

Xace Ver. 7.00 5/2/2013 10:22 SN: 14337-877388691 **US Units**

Example 12.3
Rating-Horizontal air-cooled heat exchanger induced draft countercurrent to crossflow

#	Item	Units	Value 1	Value 2	Item	Units	Value 1	Value 2
1	**No Data Check Messages.**							
2	**See Runtime Message Report for Warning Messages.**							
3	**Process Conditions**		**Outside**		**Tubeside**			
4	Fluid name				Hydrocarbon			
5	Fluid condition			Sens. Gas				Sens. Liquid
6	Total flow rate	(1000-lb/hr)		1037.340				250.000
7	Weight fraction vapor, In/Out		1.0000	1.0000			0.0000	0.0000
8	Temperature, In/Out	(Deg F)	95.00	150.09			250.00	150.00
9	Skin temperature, Min/Max	(Deg F)	130.30	205.14			137.97	221.22
10	Pressure, Inlet/Outlet	(psia)	14.565	14.553			50.000	33.680
11	Pressure drop, Total/Allow	(inH2O) | (psi)	0.320	0.000			16.320	0.000
12	Midpoint velocity	(ft/sec)		18.94				6.94
13	- In/Out	(ft/sec)					6.94	6.94
14	Heat transfer safety factor	(--)		1.0000				1.0000
15	Fouling	(ft2-hr-F/Btu)		0.00000				0.00100
16	**Exchanger Performance**							
17	Outside film coef	(Btu/ft2-hr-F)		7.81	Actual U	(Btu/ft2-hr-F)		4.665
18	Tubeside film coef	(Btu/ft2-hr-F)		489.86	Required U	(Btu/ft2-hr-F)		4.133
19	Clean coef	(Btu/ft2-hr-F)		5.323	Area	(ft2)		44629
20	Hot regime			Sens. Liquid	Overdesign	(%)		12.86
21	Cold regime			Sens. Gas	**Tube Geometry**			
22	EMTD	(Deg F)		74.5	Tube type			High-finned
23	Duty	(MM Btu/hr)		13.750	Tube OD	(inch)		1.0000
24	**Unit Geometry**				Tube ID	(inch)		0.8100
25	Bays in parallel per unit			1	Length	(ft)		36.000
26	Bundles parallel per bay			1	Area ratio(out/in)	(--)		26.512
27	Extended area	(ft2)		44629	Layout			Staggered
28	Bare area	(ft2)		2078.2	Trans pitch	(inch)		2.5000
29	Bundle width	(ft)		11.813	Long pitch	(inch)		2.1650
30	**Nozzle**		**Inlet**	**Outlet**	Number of passes	(--)		4
31	Number	(--)	1	1	Number of rows	(--)		4
32	Diameter	(inch)	5.0470	5.0470	Tubecount	(--)		224
33	Velocity	(ft/sec)	10.01	10.01	Tubecount Odd/Even	(--)		56 / 56
34	R-V-SQ	(lb/ft-sec2)	5003.1	5003.1	Material			Carbon steel
35	Pressure drop	(psi)	0.594	0.378	**Fin Geometry**			
36	**Fan Geometry**				Type			Circular
37	No/bay	(--)		2	Fins/length	(fin/inch)		10.0
38	Fan ring type			15 deg	Fin root	(inch)		1.0000
39	Diameter	(ft)		10.500	Height	(inch)		0.6250
40	Ratio, Fan/bundle face area	(--)		0.41	Base thickness	(inch)		0.0130
41	Driver power	(hp)		14.24	Over fin	(inch)		2.2500
42	Tip clearance	(inch)		0.6250	Efficiency	(%)		79.7
43	Efficiency	(%)		66.500	Area ratio (fin/bare)	(--)		21.475
44	**Airside Velocities**		**Actual**	**Standard**	Material			Aluminum 1060 - H14
45	Face	(ft/min)	573.62	542.08	**Thermal Resistance, %**			
46	Maximum	(ft/sec)	18.38	17.37	Air			59.71
47	Flow	(1000 ft3/min)	243.93	230.52	Tube			25.25
48	Velocity pressure	(inH2O)	0.129		Fouling			12.37
49	Bundle pressure drop	(inH2O)	0.303		Metal			2.68
50	Bundle flow fraction	(--)	1.000		Bond			0.00
51	Bundle	94.78	**Airside Pressure Drop, %**			Louvers		0.00
52	Ground clearance	0.00	Fan guard	0.00		Hail screen		0.00
53	Fan ring	5.22	Fan area blockage	0.00		Steam coil		0.00

Xace Exchanger Drawings for Example 12.3

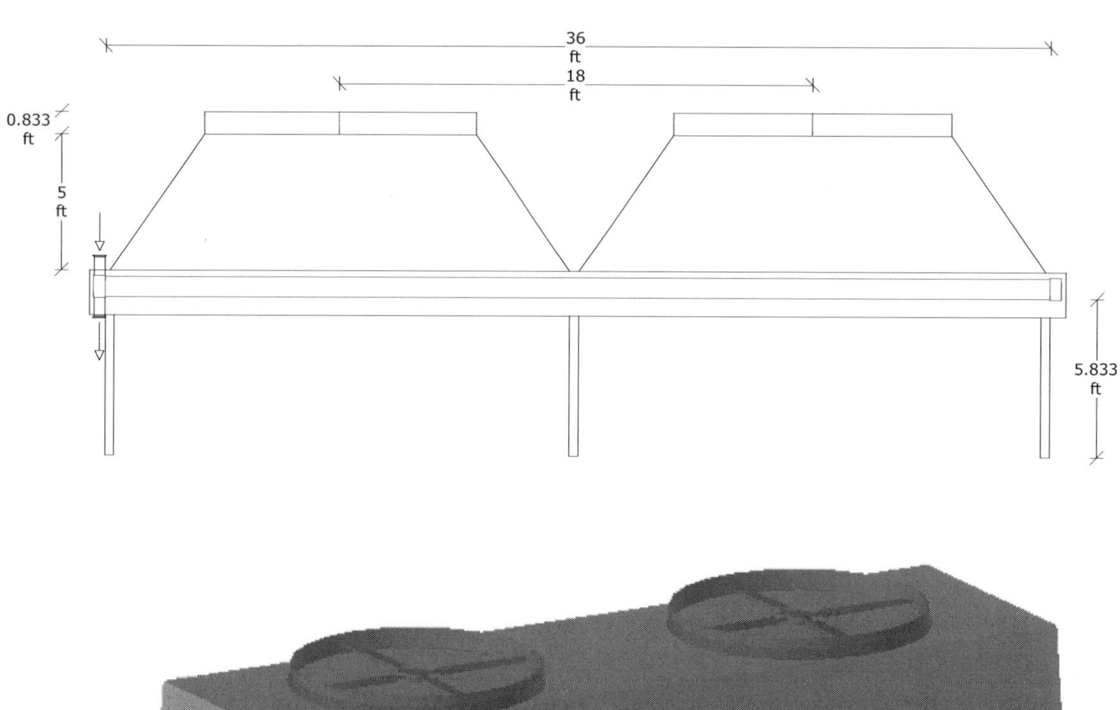

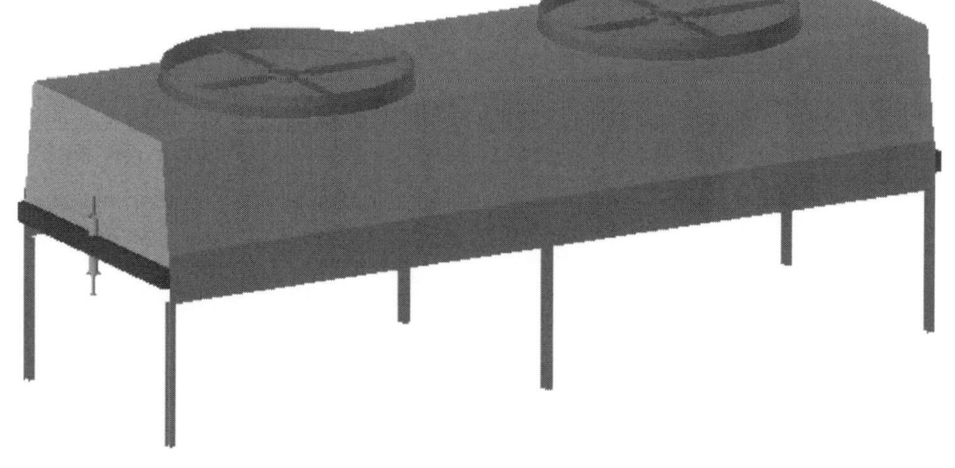

Xace Tube Layout for Example 12.3

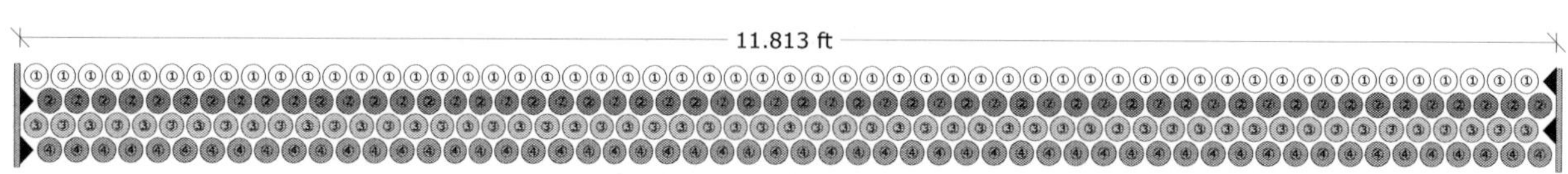

Name	Type	Outer Diameter (inch)	Wall Thickness (inch)	Transverse Pitch (inch)	Longitudinal Pitch (inch)	Fin Height (inch)
1 TubeType1	High-finned	1.0000	0.0950	2.5000	2.1650	0.6250

Row	Number of Tubes	Tube Type	Wall Clearance (inch)	Row	Number of Tubes	Tube Type	Wall Clearance (inch)
1	56	TubeType1	0.3750	3	56	TubeType1	0.3750
2	56	TubeType1	1.6250	4	56	TubeType1	1.6250

Bundle Information

Bundle width 11.813 ft
Number of tube rows 4
Number of tubes 224
Minimum wall clearance
Left 0.3750 inch
Right 0.3750 inch
Number of tubes per pass
○ Tubepass # 1: 56
● Tubepass # 2: 56
● Tubepass # 3: 56
● Tubepass # 4: 56

Air-cooled propane condensers are quite common in the petrochemical industry. Design of an air-cooled unit for the service of Example 11.11 is considered in the following example.

Example 12.4

100,000 kg/h (220,460 lb/h) of industrial propane at 1751.3 kPa absolute (254 psia) and 70.79°C (159.42°F) are to be condensed. The composition of the stream on a molar basis is 97% propane, 2% ethane, 0.5% *n*-butane and 0.5% *i*-butane. A fouling factor of 0.000176 $m^2 \cdot K/W$ (0.001 $h \cdot ft^2 \cdot °F/Btu$) and a maximum pressure drop of 17.24 kPa (2.5 psi) are specified for the propane stream. The design ambient air temperature is 95°F. Use *Xace* to design an air-cooled condenser for this service and compare the result with the water-cooled condenser of Example 11.11.

Solution

To obtain an initial configuration for the unit, *Xace* is first run in classic design mode with input parameters based on the design guidelines given in this chapter.

(a) Geometry/Unit

Tube orientation: Horizontal (Default)	Fan type: Forced (Default)

This is the most commonly used configuration for air coolers.

(b)

Number of fans per bay: 2 (3)	Fan diameter: 13 ft
Combined efficiency: 65% (Default)	Fan ring type: Bell

Either two or three fans are used per bay, depending on the tube length. The fan diameter is based on a maximum bundle width of 14 ft.

(c) Geometry/Optional
Free area in hail screen: 80%
A hail screen is specified to protect the tube bundle in a forced-draft unit. It is assumed that a steam coil will not be required.

(d) Geometry/Bundle

Number of tube rows: 4	Tube length: 42 (48, 54, 60) ft

Four runs are made with tube lengths ranging from 42 to 60 ft. Two fans per bay are used for the two shorter lengths, while three-fan bays are specified for the longer lengths. Four tube rows constitute a reasonable choice, but other values (3, 5, or 6) could also be tried. If the number of rows is not specified, *Xace* generates tube bundles with an unusually large number of rows for this problem.

(e) Geometry/Tube Types/Tube Type 1/ Geometry

Tube type: High Fin	Wall thickness: 0.095 in.(13 BWG)
Tube material: Carbon steel (Default)	Transverse pitch: 2.5 in.
Tube OD: 1 in.	

(f) Geometry/Tube Types/Tube Type 1/High Fin

Fin type: Circular fin (Default)	Fin base thickness: 0.012 in.
Fin density: 10 fin/in.	Fin height: 0.625 in.
Material: Aluminum 1100-annealed (Default)	

(g) Process

	Tube-Side Fluid (Hot)	Air-Side Fluid
Fluid name	Propane	–
Phase/air-side flow-rate units	Condensing	Mass flow rate
Flow rate (1000 lb/h)	220.46	–
Inlet temperature (°F)	159.42	95
Outlet temperature (°F)	–	–
Inlet vapor fraction	1	–
Outlet vapor fraction	0	–
Inlet pressure/altitude of unit	254 psia	–
Fouling resistance (h · ft^2·°F/Btu)	0.001	0

Note that neither the flow rate nor the outlet temperature of the air is specified here.

(h) Hot fluid properties

Since propane is a defined composition stream, *User specified grid* is chosen as the physical property input option and VMGThermo is selected as the property package. The Advanced Peng–Robinson method is used to calculate physical properties. After selecting the components on the Composition tab, *Property sets* is chosen as the temperature point method on the Conditions tab. Four pressure levels are specified (254, 253, 252, and 251 psia) to cover the operating pressure range. At each level, a temperature range of 95°F to 160°F is specified to cover the entire operating range, and the number of points is set at 20.

The *Xace* Output Summary for the run with 60 ft tubes is shown below, from which it is seen that five fan bays are used with a total of 15 fans. The designs for 48 and 54 ft tubes have six bays, while seven bays are used with 42 ft tubes. Although any of these designs could be taken as a starting point, the result for 60 ft tubes is chosen here because it has the fewest number of bays.

Xace Output Summary for Example 12.4: Design Run with 60 ft Tubes

Xace Ver. 7.00 5/1/2013 17:43 SN: 14337-877388691 **US Units**

Design-Horizontal air-cooled heat exchanger forced draft countercurrent to crossflow

1	**No Data Check Messages.**					
2	**See Runtime Message Report for Warning Messages.**					
3	**Process Conditions**		**Outside**		**Tubeside**	
4	Fluid name				Propane	
5	Fluid condition			Sens. Gas		Cond. Vapor
6	Total flow rate	(1000-lb/hr)		7560.000		220.460
7	Weight fraction vapor, In/Out		1.0000	1.0000	1.0000	0.0000
8	Temperature, In/Out	(Deg F)	95.00	112.33	159.42	120.04
9	Skin temperature, Min/Max	(Deg F)	104.99	123.30	106.84	126.11
10	Pressure, Inlet/Outlet	(psia)	14.697	14.691	254.00	251.89
11	Pressure drop, Total/Allow	(inH2O) | (psi)	0.170	0.000	2.111	5.000
12	Midpoint velocity	(ft/sec)		13.38		3.16
13	- In/Out	(ft/sec)			11.88	0.91
14	Heat transfer safety factor	(--)		1.0000		1.0000
15	Fouling	(ft2-hr-F/Btu)		0.00000		0.00100

16			**Exchanger Performance**				
17	Outside film coef	(Btu/ft2-hr-F)		6.88	Actual U	(Btu/ft2-hr-F)	3.900
18	Tubeside film coef	(Btu/ft2-hr-F)		335.42	Required U	(Btu/ft2-hr-F)	3.809
19	Clean coef	(Btu/ft2-hr-F)		4.350	Area	(ft2)	441379
20	Hot regime			Cond. Vapor	Overdesign	(%)	2.39
21	Cold regime			Sens. Gas	**Tube Geometry**		
22	EMTD	(Deg F)		18.7	Tube type		High-finned
23	Duty	(MM Btu/hr)		31.499	Tube OD	(inch)	1.0000
24	**Unit Geometry**				Tube ID	(inch)	0.8100
25	Bays in parallel per unit			5	Length	(ft)	60.000
26	Bundles parallel per bay			1	Area ratio(out/in)	(--)	26.497
27	Extended area	(ft2)		441379	Layout		Staggered
28	Bare area	(ft2)		20565	Trans pitch	(inch)	2.5000
29	Bundle width	(ft)		14.000	Long pitch	(inch)	2.1650
30	**Nozzle**		**Inlet**	**Outlet**	Number of passes	(--)	2
31	Number	(--)	1	1	Number of rows	(--)	4
32	Diameter	(inch)	5.0470	3.0680	Tubecount	(--)	266
33	Velocity	(ft/sec)	40.70	8.47	Tubecount Odd/Even	(--)	67 / 66
34	R-V-SQ	(lb/ft-sec2)	3588.0	2019.7	Material		Carbon steel
35	Pressure drop	(psi)	0.426	0.153	**Fin Geometry**		
36	**Fan Geometry**				Type		Circular
37	No/bay	(--)		3	Fins/length	(fin/inch)	10.0
38	Fan ring type			Bell	Fin root	(inch)	1.0000
39	Diameter	(ft)		13.000	Height	(inch)	0.6250
40	Ratio, Fan/bundle face area	(--)		0.47	Base thickness	(inch)	0.0120
41	Driver power	(hp)		6.14	Over fin	(inch)	2.2500
42	Tip clearance	(inch)		0.7500	Efficiency	(%)	80.9
43	Efficiency	(%)		65.000	Area ratio (fin/bare)	(--)	21.462
44	**Airside Velocities**		**Actual**	**Standard**	Material		Aluminum 1100-annealed
45	Face	(ft/min)	419.46	400.00	**Thermal Resistance, %**		
46	Maximum	(ft/sec)	13.19	12.58	Air		56.69
47	Flow	(1000 ft3/min)	1761.7	1680.0	Tube		30.81
48	Velocity pressure	(inH2O)	0.047		Fouling		10.33
49	Bundle pressure drop	(inH2O)	0.166		Metal		2.16
50	Bundle flow fraction	(--)	1.000		Bond		0.00
51	Bundle	97.86	**Airside Pressure Drop, %**		Louvers		0.00
52	Ground clearance	0.00	Fan guard	0.00	Hail screen		0.77
53	Fan ring	1.37	Fan area blockage	0.00	Steam coil		0.00

Switching to rating mode, the required additional input data are entered from the results of the design run: number of bays (5), number of tube passes (2), tubes in odd/even rows (67/66), bundle width (14 ft), and tube-side nozzles (6 in. schedule 40 inlet, 3 in. schedule 40 outlet). The mass flow rate of air is then adjusted to give an over-design of about 5%; this is done to provide a consistent basis for comparing alternate designs. The result is listed as Design 1 in the design summary table shown below.

We next attempt to obtain a more compact design by reducing the number of bays from five to four. The inlet nozzle size for this case is increased to 8 in. in order to accommodate the higher flow rate per bundle. (Alternatively, two smaller inlet nozzles could be used for each bundle.) With these changes the tube-side pressure drop is found to be about 2.7 psi, which is slightly higher than the specified maximum of 2.5 psi. A possible remedy is to increase the number of tubes by decreasing the gap between tubes. The minimum gap between fin tips recommended by equipment manufacturers is 1/16 in., which in this case corresponds to a transverse pitch of 2.3125 in. This change increases the number of tubes per bundle from 266 to 286 (72/71 tubes in odd/even rows) and reduces the pressure drop to 2.37 psi, which is acceptable. The number of fans can also be reduced from 12 to 6 by using two bundles per bay. The Output Summary for this case, which is designated Design 2, is given below.

Xace Output Summary for Example 12.4: Rating Run for Design 2

	Xace Ver. 7.00 5/1/2013 15:21 SN: 14337-877388691						US Units
	Rating-Horizontal air-cooled heat exchanger forced draft countercurrent to crossflow						
1	**No Data Check Messages.**						
2	See Runtime Message Report for Warning Messages.						
3	**Process Conditions**		**Outside**		**Tubeside**		
4	Fluid name				Propane		
5	Fluid condition			Sens. Gas			Cond. Vapor
6	Total flow rate	(1000-lb/hr)		8650.000			220.460
7	Weight fraction vapor, In/Out		1.0000	1.0000	1.0000		0.0000
8	Temperature, In/Out	(Deg F)	95.00	110.15	159.42		119.95
9	Skin temperature, Min/Max	(Deg F)	104.45	121.20	106.57		124.57
10	Pressure, Inlet/Outlet	(psia)	14.697	14.683	254.00		251.63
11	Pressure drop, Total/Allow	(inH2O) | (psi)	0.381	0.000	2.374		5.000
12	Midpoint velocity	(ft/sec)		20.63			3.56
13	- In/Out	(ft/sec)			13.81		1.06
14	Heat transfer safety factor	(--)		1.0000			1.0000
15	Fouling	(ft2-hr-F/Btu)		0.00000			0.00100
16			**Exchanger Performance**				
17	Outside film coef	(Btu/ft2-hr-F)		8.17	Actual U	(Btu/ft2-hr-F)	4.333
18	Tubeside film coef	(Btu/ft2-hr-F)		346.63	Required U	(Btu/ft2-hr-F)	4.125
19	Clean coef	(Btu/ft2-hr-F)		4.895	Area	(ft2)	379653
20	Hot regime			Cond. Vapor	Overdesign	(%)	5.04
21	Cold regime			Sens. Gas	**Tube Geometry**		
22	EMTD	(Deg F)		20.1	Tube type		High-finned
23	Duty	(MM Btu/hr)		31.514	Tube OD	(inch)	1.0000
24	**Unit Geometry**				Tube ID	(inch)	0.8100
25	Bays in parallel per unit			2	Length	(ft)	60.000
26	Bundles parallel per bay			2	Area ratio(out/in)	(--)	26.497
27	Extended area	(ft2)		379653	Layout		Staggered
28	Bare area	(ft2)		17689	Trans pitch	(inch)	2.3125
29	Bundle width	(ft)		14.000	Long pitch	(inch)	2.0026
30	**Nozzle**		**Inlet**	**Outlet**	Number of passes	(--)	2
31	Number	(--)	1	1	Number of rows	(--)	4
32	Diameter	(inch)	7.9810	3.0680	Tubecount	(--)	286
33	Velocity	(ft/sec)	20.34	10.58	Tubecount Odd/Even	(--)	72 / 71
34	R-V-SQ	(lb/ft-sec2)	896.55	3154.7	Material		Carbon steel
35	Pressure drop	(psi)	0.106	0.238	**Fin Geometry**		
36	**Fan Geometry**				Type		Circular
37	No/bay	(--)		3	Fins/length	(fin/inch)	10.0
38	Fan ring type			Bell	Fin root	(inch)	1.0000
39	Diameter	(ft)		18.000	Height	(inch)	0.6250
40	Ratio, Fan/bundle face area	(--)		0.45	Base thickness	(inch)	0.0120
41	Driver power	(hp)		39.39	Over fin	(inch)	2.2500
42	Tip clearance	(inch)		0.7500	Efficiency	(%)	77.3
43	Efficiency	(%)		65.000	Area ratio (fin/bare)	(--)	21.462
44	**Airside Velocities**		**Actual**	**Standard**	Material		Aluminum 1100-annealed
45	Face	(ft/min)	599.93	572.09	**Thermal Resistance, %**		
46	Maximum	(ft/sec)	20.39	19.44	Air		53.00
47	Flow	(1000 ft3/min)	2015.8	1922.2	Tube		33.12
48	Velocity pressure	(inH2O)	0.104		Fouling		11.48
49	Bundle pressure drop	(inH2O)	0.373		Metal		2.40
50	Bundle flow fraction	(--)	1.000		Bond		0.00
51	Bundle	97.94	**Airside Pressure Drop, %**		Louvers		0.00
52	Ground clearance	0.00	Fan guard	0.00	Hail screen		0.70
53	Fan ring	1.36	Fan area blockage	0.00	Steam coil		0.00

The size of the exchanger can be further reduced at the expense of additional fan power by reducing the tube length. With a length of 54 ft, for example, the total power requirement increases from 236 to 371 hp; this case is designated Design 3. The number of fans could be halved in this case as well by utilizing bays with two tube bundles. Further reducing the tube length to 50 ft increases the total fan power to nearly 550 hp. It seems unlikely that this case would be economical unless plot space considerations restricted the size of the unit. However, economic data are needed to select an optimal design.

The following table compares key attributes of air-cooled and water-cooled units for this service.

Parameter	Air-Cooled Condenser (Design 3)	Water-Cooled Condenser (Plain Tube, Double Segmental Baffles)
Heat-transfer area (ft^2)	341,620	7339
Number of tubes	1144	1428
Tube length (ft)	54	20
Effective mean temperature difference (°F)	21.2	33.2
Propane ΔP (psi)	2.19	0.51
Propane heat-transfer coefficient (Btu/h·ft²·°F)	346	239

Clearly, the air-cooled condenser requires much more heat-transfer area due to the low air-side heat-transfer coefficient. Also, notice that the cross-flow arrangement in the air cooler results in a lower mean temperature difference. While the operating cost of an air cooler tends to be lower than that of a water-cooled unit, the operating cost is still quite significant. These units have sizable power requirements (about 100–370 hp for the designs in this example), not to mention the maintenance of all the motors and drive trains.

Design Summaries for Example 12.4

Item	Design 1	Design 2	Design 3
Fan bays	5	2	4
Bundles per bay	1	2	1
Tubes per bundle	266	286	286
Total number of tubes	1330	1144	1144
Tube length, ft	60	60	54
Transverse pitch, in.	2.5	2.3125	2.3125
Number of fans	15	6	12
Fan diameter, ft	13	18	13
Driver power, hp	6.6	39.4	30.9
Total fan power, hp	99	236	371
Air flow rate, 1000 lb/h	7700	8650	9770
Heat-transfer area, ft^2	441,379	379,653	341,620
Tube-side ΔP, psi	1.88	2.37	2.19
Air-side ΔP, in. H_2O	0.177	0.381	0.551
Tube-side inlet nozzles	6 in. sch. 40	8 in. sch. 40	8 in. sch 40
Tube-side outlet nozzles	3 in. sch. 40	3 in. sch. 40	3 in. sch. 40
Over-design, %	5.1	5.0	5.0

All designs have 14 ft wide tube bundles with 4 tube rows and 2 passes. The carbon steel tubes are 1 in. OD, 13 BWG with embedded (G-fin) type 1100 annealed aluminum fins. Fins are 0.625 in. high with a density of 10 fpi. All cases employ plug-type box headers and forced-draft operation. Nozzle specifications are tentative pending mechanical design calculations.

Fan Bay Layout for Design 1

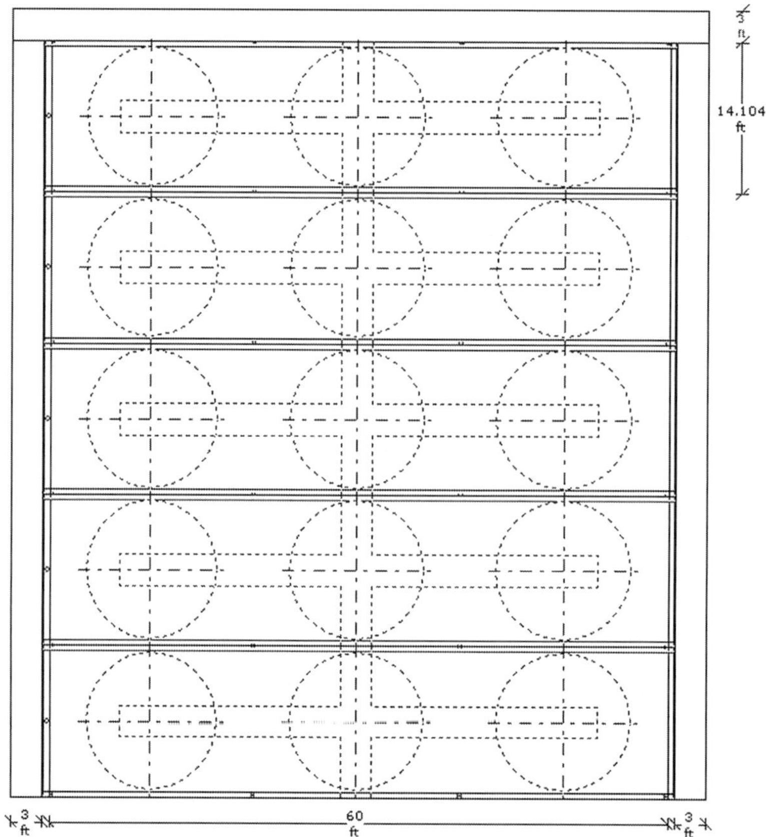

Fan Bay Layout for Design 2

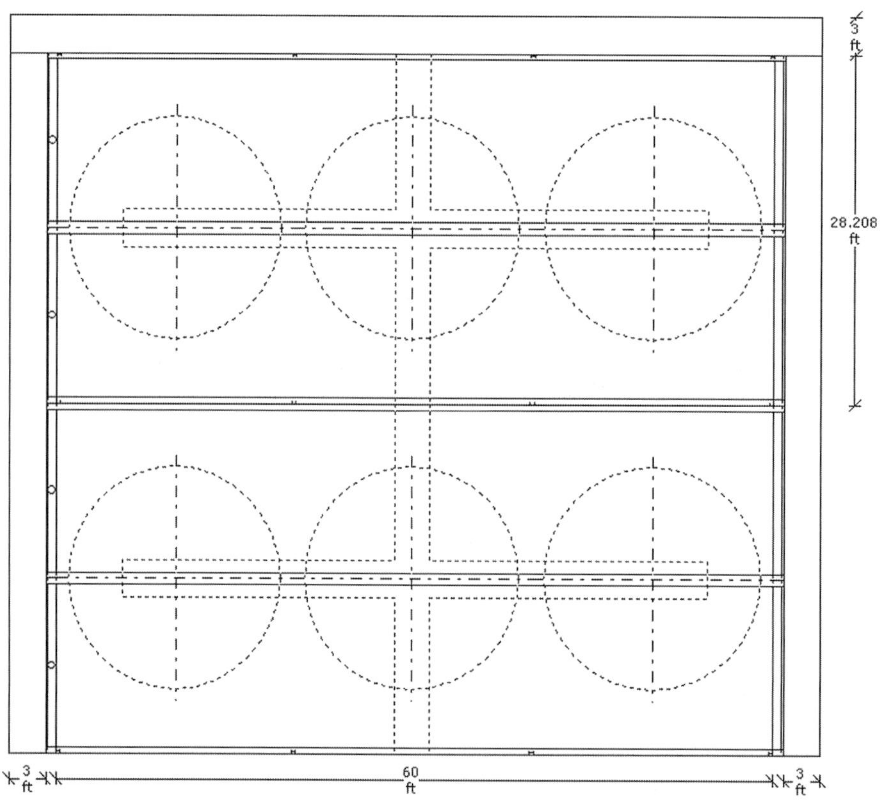

Tube Bundle Layout for Designs 2 and 3

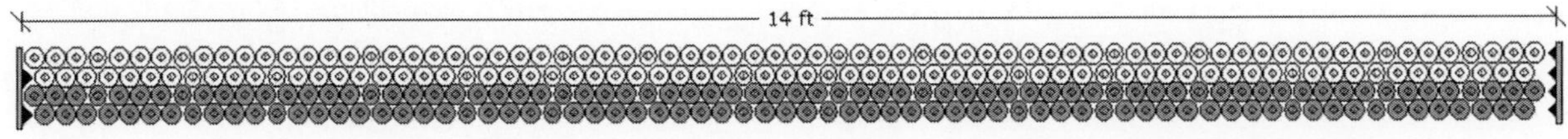

	Name	Type	Outer Diameter (inch)	Wall Thickness (inch)	Transverse Pitch (inch)	Longitudinal Pitch (inch)	Fin Height (inch)
1	TubeType1	High-finned	1.0000	0.0950	2.3125	2.0026	0.6250

Row	Number of Tubes	Tube Type	Wall Clearance (inch)	Row	Number of Tubes	Tube Type	Wall Clearance (inch)
1	72	TubeType1	0.3750	3	72	TubeType1	0.3750
2	71	TubeType1	1.5313	4	71	TubeType1	1.5313

Bundle Information

Bundle width	14.000 ft
Number of tube rows	4
Number of tubes	286
Minimum wall clearance	
Left	0.3750 inch
Right	1.1875 inch
Number of tubes per pass	
○ Tubepass # 1:	143
● Tubepass # 2:	143

Exchanger Drawing for Design 2 (1 of 2 Bays)

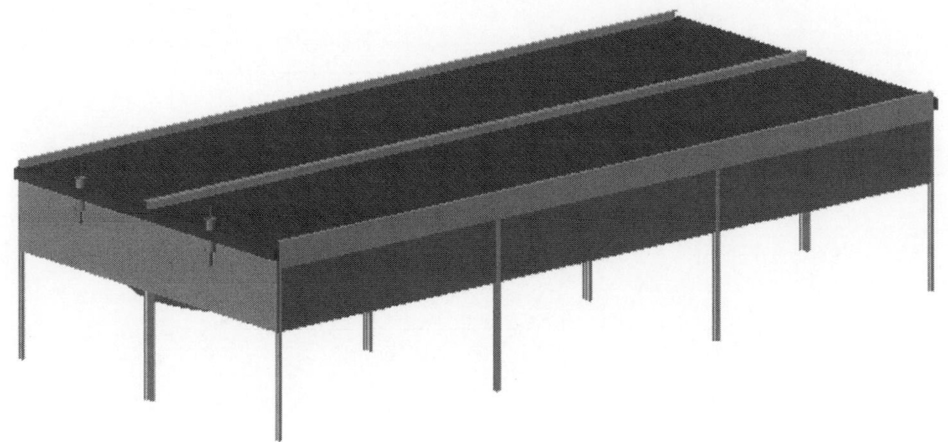

References

[1] Kraus AD, Aziz A, Welty J. Extended Surface Heat Transfer. New York: Wiley; 2001.

[2] Hewitt GE, Shires GL, Bott TR. Process Heat Transfer. Boca Raton, FL: CRC Press; 1994.

[3] Minton PE. Heat exchanger design. In: McKetta JJ, editor. Heat Transfer Design Methods. New York: Marcel Dekker; 1991.

[4] Bell KJ, Mueller AC. Wolverine Engineering Data Book II. Wolverine Tube, Inc; 2001. www.wlv.com.

[5] Anonymous. Engineering Data Book. 11th ed. Tulsa, OK: Gas Processors Suppliers Association; 1998.

[6] Mukherjee R. Effectively design air-cooled heat exchangers. Chem. Eng. Prog. 1987;93(No. 2):26–47.

[7] Shipes KV. Air-cooled exchangers in cold climates. Chem. Eng. Prog 1974;70(No. 7):53–8.

[8] Briggs DE, Young EH. Convection heat transfer and pressure drop of air flowing across triangular pitch banks of finned tubes. Chem. Eng. Prog. Symp. Ser 1963;59(No. 41):1–10.

[9] Ganguli A, Tung SS, Taborek J. Parametric study of air-cooled heat exchanger finned tube geometry. AIChE Symp. Ser 1985;81(No. 245):122–8.

[10] Kröger DG. Air-Cooled Heat Exchangers and Cooling Towers. New York: distributed by Begell House, Inc; 1998.

[11] Robinson KK, Briggs DE. Pressure drop of air flowing across triangular pitch banks of finned tubes. Chem. Eng Prog. Symp. Ser 1965;62(No. 64):177–84.

[12] Taborek J. Mean temperature difference. In: Heat Exchanger Design Handbook, Vol. 1. New York: Hemisphere Publishing Corp.; 1988.

[13] Roetzel W, Neubert J. Calculation of mean temperature difference in air-cooled cross-flow heat exchangers. J. Heat Transfer 1979;101:511–3.

Appendix 12.A LMTD Correction Factors for Air-Cooled Heat Exchangers

The following notation is used for the charts:

A = heat-transfer surface area in exchanger

$C_1 = (\dot{m}C_P)_1$ = heat capacity flow rate of tube-side fluid

$C_2 = (\dot{m}C_P)_2$ = heat capacity flow rate of cross-flow stream (air)

$\Delta T_m = F(\Delta T_{\ln})_{cf}$ = mean temperature difference in exchanger

U = overall heat-transfer coefficient

All other symbols are explicitly defined on the charts themselves.

When the outlet temperatures of both streams are known, the lower charts can be used to obtain the LMTD correction factor in the usual manner. When the outlet temperatures are unknown but the product UA is known, the upper charts can be used to obtain θ, from which the mean temperature difference, ΔT_m, can be found. The exchanger duty is then obtained as $q = UA\Delta T_m$, and the outlet temperatures are computed using the energy balances on the two streams.

The equations from which the charts were constructed are given by Taborek [12]. Unfortunately, some of the equations contain errors; therefore, the original sources listed in Ref. [12] should be consulted if the equations are to be used. An alternative computational method is given in Ref. [13] and is reproduced in Ref. [10]. Although approximate, this method is well suited for computer calculations. Neither the exact nor the approximate method is convenient for hand calculations. Therefore, only the charts are presented here.

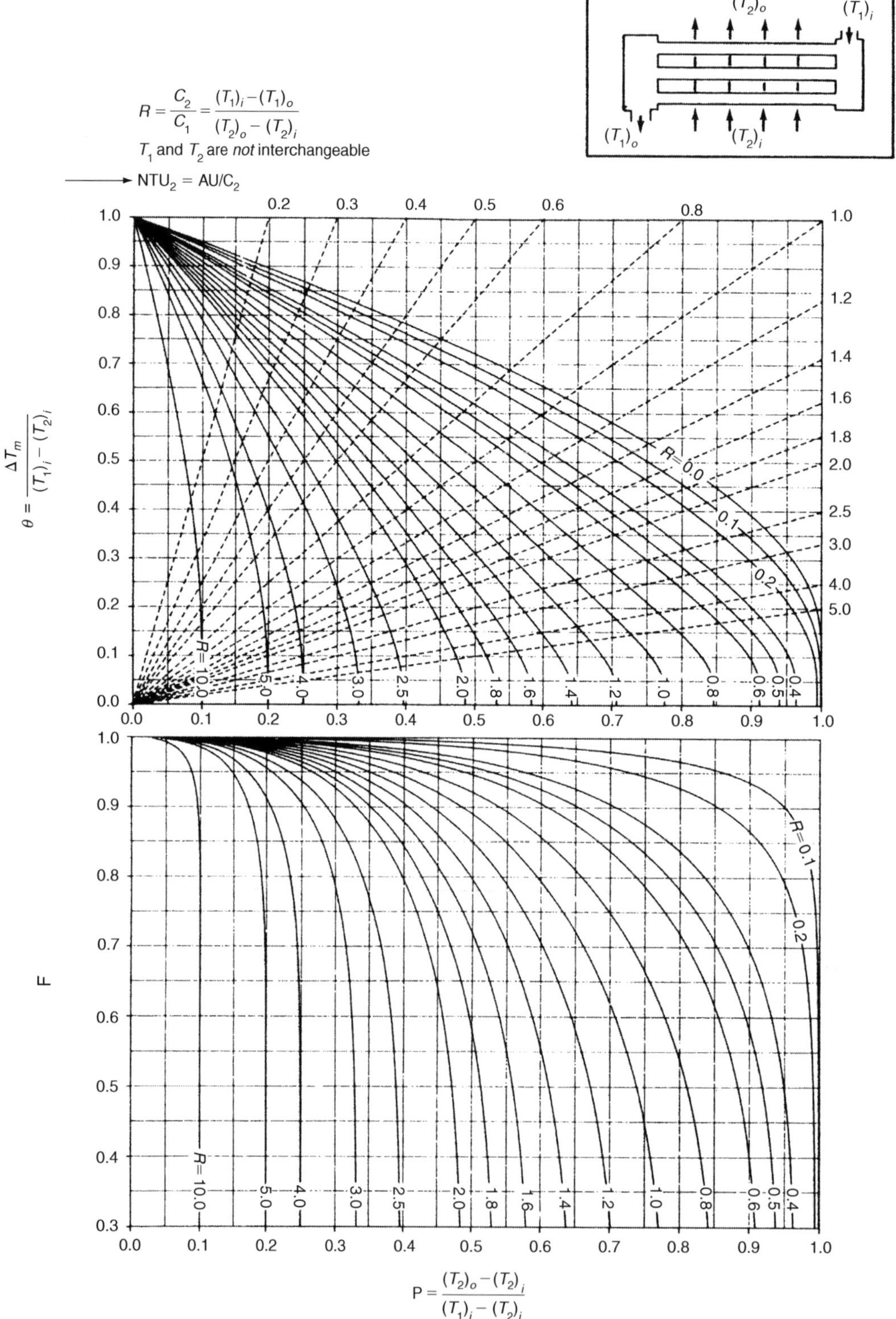

FIGURE 12.A.1 Mean temperature difference relationships for cross flow: three tube rows and one tube pass (Source: Ref [12]).

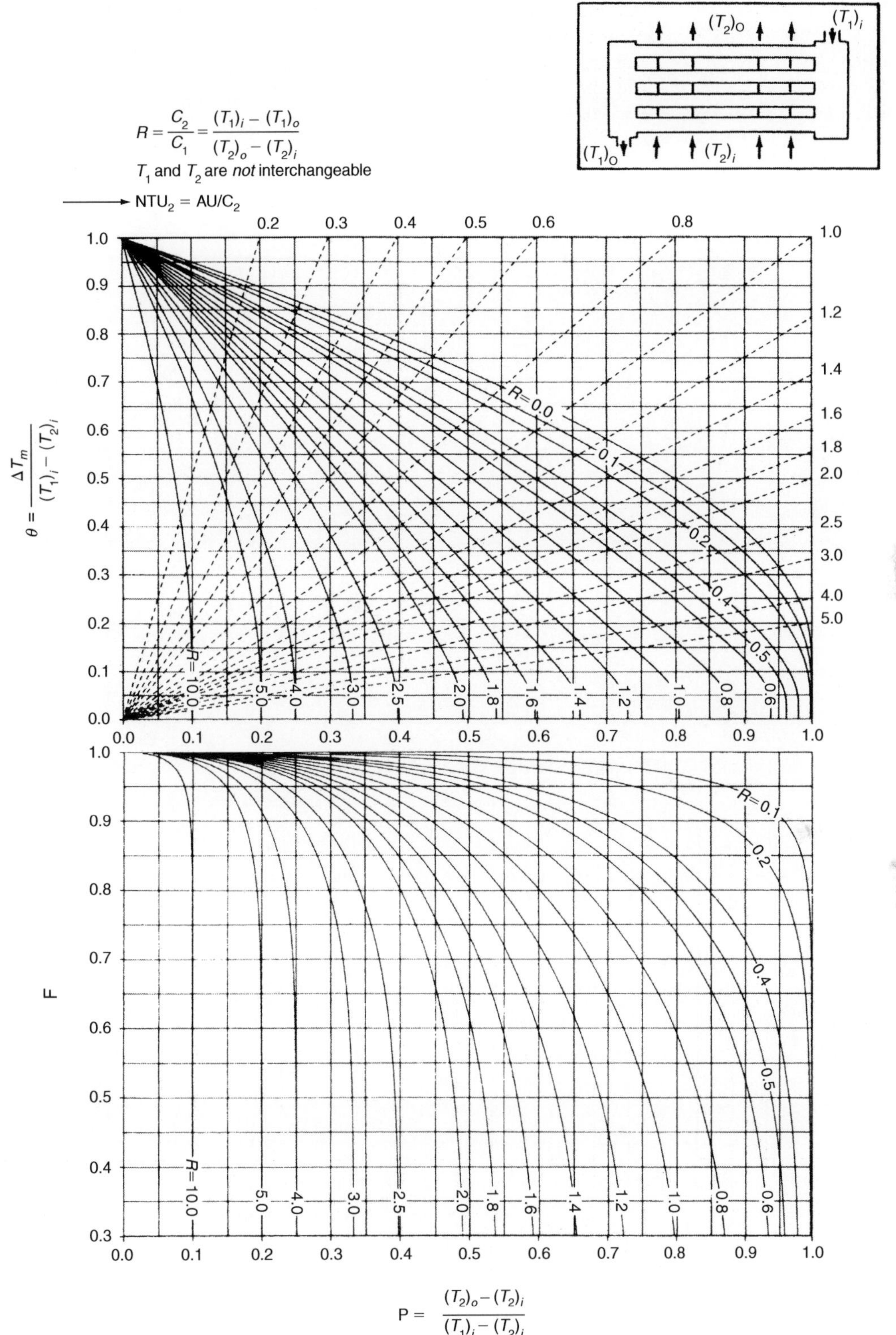

FIGURE 12.A.2 Mean temperature difference relationships for cross flow: four tube rows and one tube pass (Source: Ref. [12]).

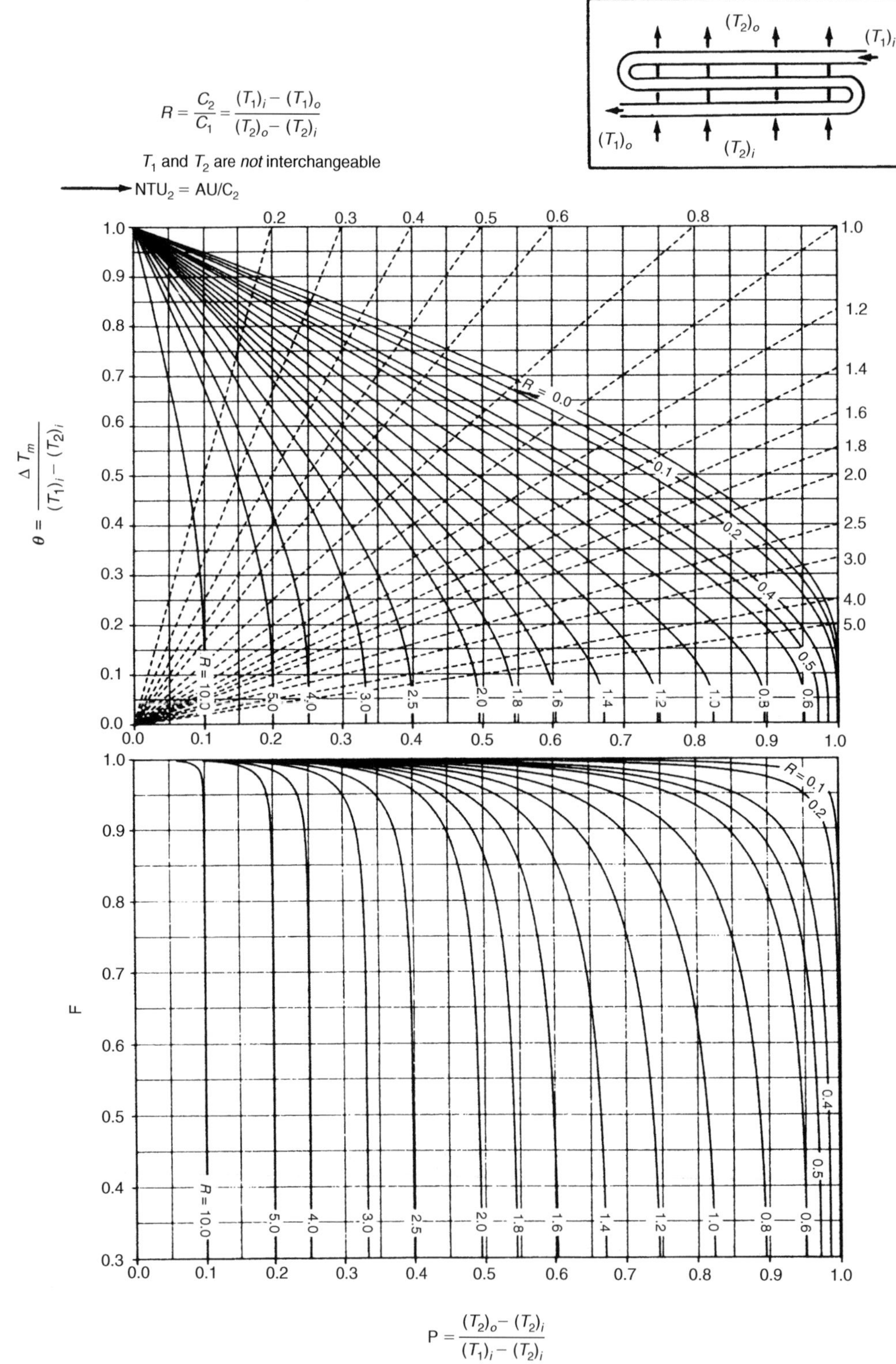

FIGURE 12.A.3 Mean temperature difference relationships for cross flow: three tube rows and three tube passes (Source: Ref. [12]).

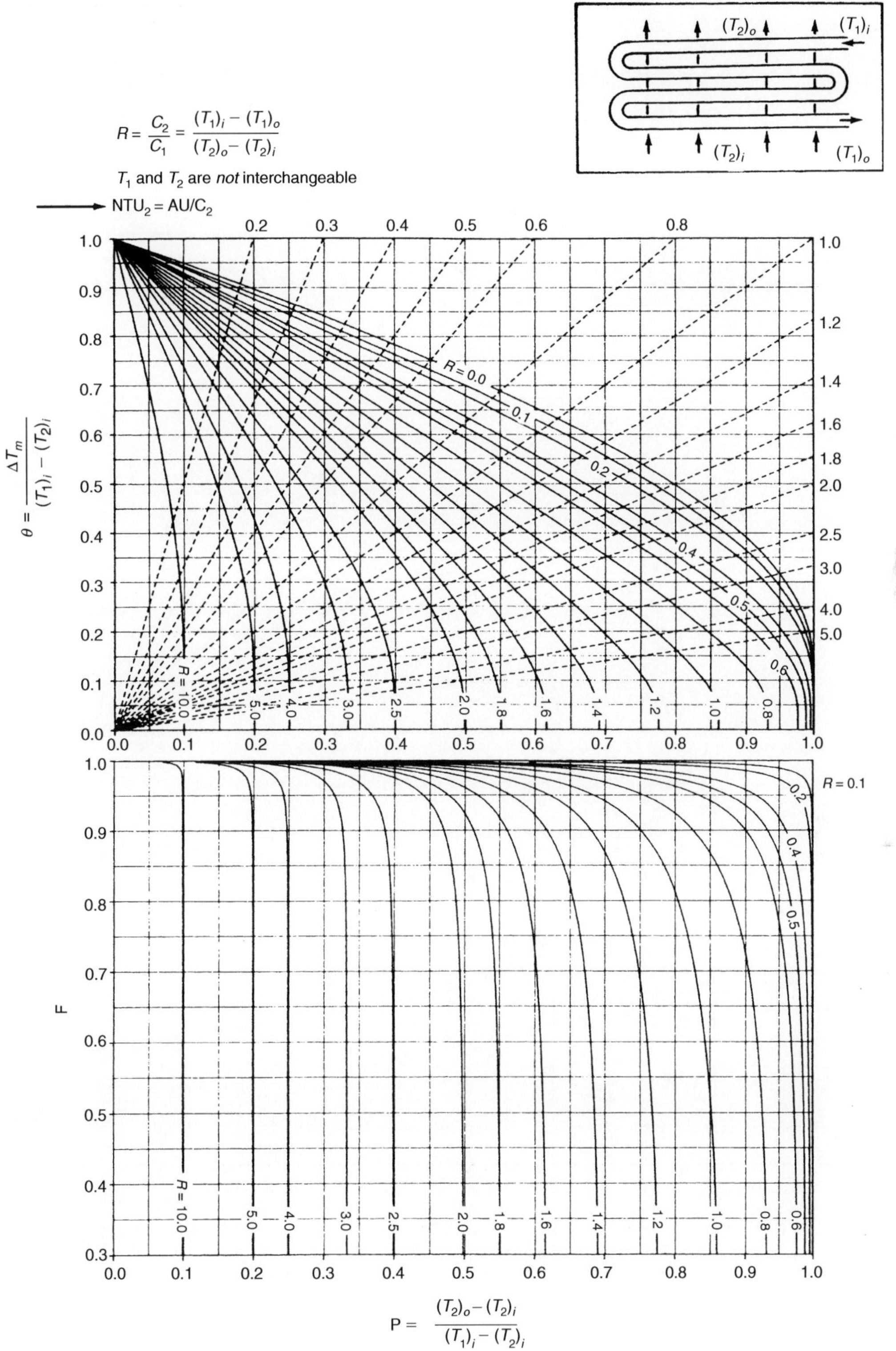

FIGURE 12.A.4 Mean temperature difference relationships for cross flow: four tube rows and four tube passes (Source: Ref. [12]).

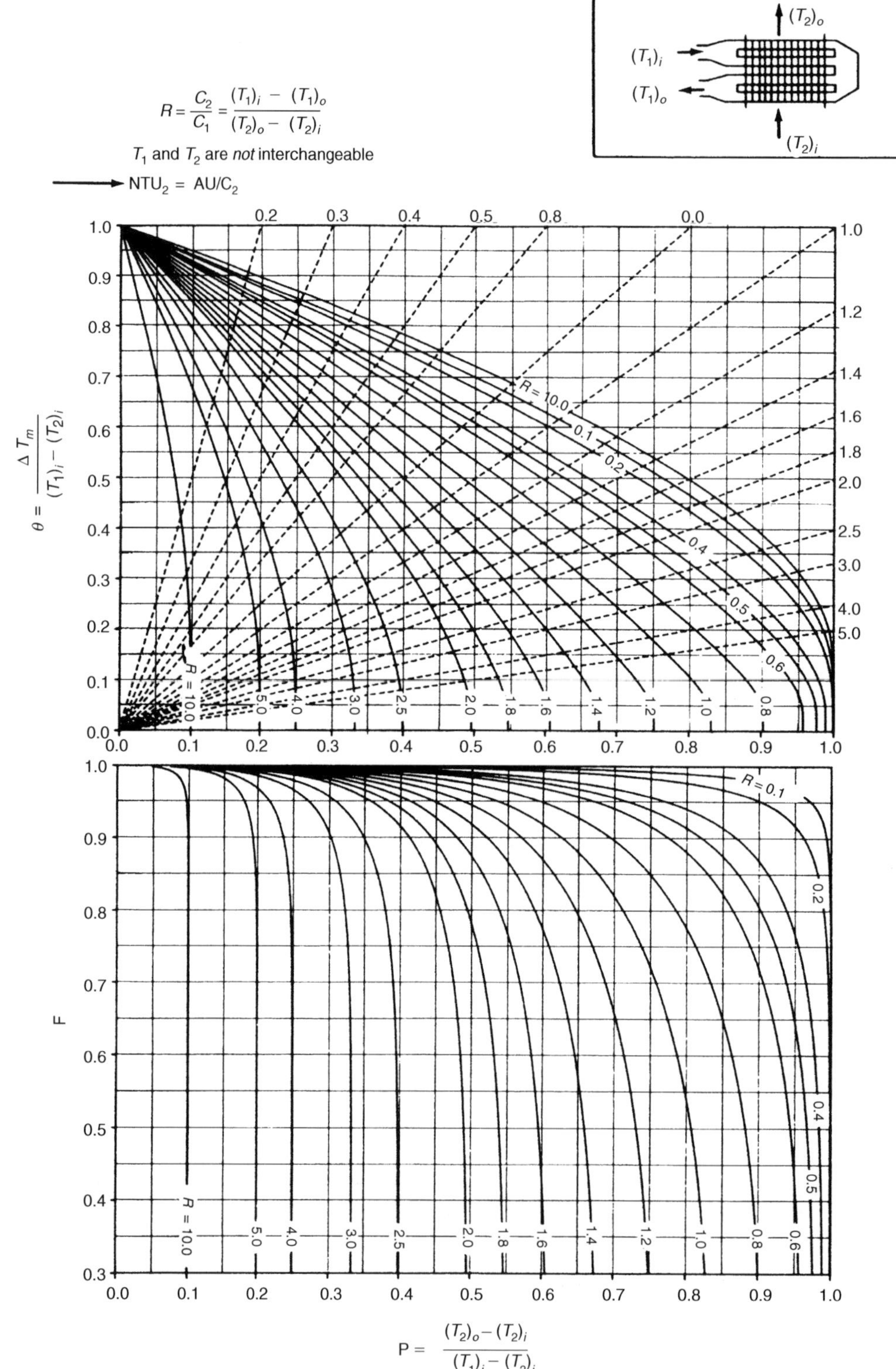

FIGURE 12.A.5 Mean temperature difference relationships for cross flow: four tube rows, two tube passes, two rows per pass, tube-side fluid mixed at return header (Source: Ref. [12]).

Appendix 12.B Standard US Motor Sizes

Motor size (hp)	Motor size (hp)
0.5	30
0.75	40
1.0	50
1.5	60
2	75
3	100
5	150
7.5	200
10	250
15	300
20	400
25	500

Appendix 12.C Correction of Air Density for Elevation

The variation of pressure with altitude in the atmosphere depends on the lapse rate, which is the rate of decrease of temperature with altitude. For an isothermal atmosphere (zero lapse rate), the pressure (and, hence, the density) decreases exponentially with altitude according to the following equation:

$$\frac{\rho}{\rho_0} = \frac{P}{P_0} = \exp\left[\frac{-M(g/g_c)z}{\tilde{R}T}\right] \tag{12.C.1}$$

where

M = molecular weight of air $\cong$ 29.0

$\tilde{R}$ = gas constant $= 1545 \dfrac{\text{ft}\cdot\text{lbf}}{\text{lb mole}\cdot{}^\circ\text{R}} = 8314 \dfrac{J}{\text{kmol}\cdot\text{K}}$

T = absolute temperature, °R (K)

z = elevation above mean sea level, ft (m)

P_0, ρ_0 = pressure and density of air at temperature T and mean sea level

P, ρ = pressure and density of air at temperature T and elevation z

Assuming P_0 is essentially one atmosphere, Equation (12.C.1) gives the air density at any elevation in terms of the density, ρ_0, at one atmosphere.

Example 12.C.1

Estimate the density of dry air at 100°F and an elevation of 3500 ft.

Solution

Equation (12.C.1) is used with $T = 100°\text{F} \cong 560°\text{R}$ and $z = 3500$ ft.

$$\frac{\rho}{\rho_0} = \exp\left[\frac{-M(g/g_c)z}{\tilde{R}T}\right] = \exp\left[\frac{-29(1.0)\times 3500}{1545\times 560}\right] = 0.8893$$

From Table A.5, the density of air at 100°F and 1 atm is $\rho_0 = 0.0709\ \text{lbm/ft}^3$. Hence,

$$\rho = 0.0709 \times 0.8893 \cong 0.0631\ \text{lbm/ft}^3$$

Notations

A	Heat-transfer surface area
A_{con}	Contact area between fin and tube wall
A_{face}	Tube bundle face area
A_{fins}	Surface area of fins
A_i	$\pi D_i L$ = Internal surface area of tube
A_{min}	Minimum flow area in tube bank
A_o	$\pi D_r L$ = External surface area of root tube
A_{prime}	Prime surface area
A_{Tot}	Total external surface area of finned tube
a	$(P_T - D_f)/D_r$ = Parameter in Equation (12.10)
BWG	Birmingham wire gage
b	Fin height
C_1, C_2	Heat capacity flow rate of fluid 1 and fluid 2
C_P	Heat capacity at constant pressure
D_f	Outer fin diameter
D_{fan}	Fan diameter
D_i	Internal diameter of tube
$D_{i,sl}$	Internal diameter of sleeve on bimetallic tube
D_o	External diameter of inner tube in bimetallic tube
$D_{o,sl}$	External diameter of sleeve on bimetallic tube
D_r	External diameter of root tube
F	LMTD correction factor
FSP	Fan static pressure
f	Friction factor
G	Mass flux
G_n	Mass flux in nozzle
g	Gravitational acceleration
g_c	Unit conversion factor
h_i	Tube-side heat-transfer coefficient
h_o	Air-side heat-transfer coefficient
k	Thermal conductivity
k_{sl}	Thermal conductivity of sleeve on bimetallic tube
k_{tube}	Thermal conductivity of tube wall
L	Tube length
ℓ	Fin spacing
M	Molecular weight of air
m	$(2h_o/k\tau)^{0.5}$ = Fin parameter
$\dot{m}$	Mass flow rate
$\dot{m}_{air}$	Mass flow rate of air
$\dot{m}_i$	Mass flow rate of tube-side fluid
N_f	Number of fins
N_r	Number of tube rows
Nu	Nusselt number
n_f	Number of fins per unit length
n_p	Number of tube passes
n_t	Number of tubes
P	Pressure; parameter used to calculated LMTD correction factor
P_0	Atmospheric pressure at mean sea level
Pr	Prandtl number
P_T	Tube pitch
P_{Total}	Sum of static and velocity pressures
q	Rate of heat transfer
R	Parameter used to calculate LMTD correction factor

$\tilde{R}$	Universal gas constant
R_{con}	Contact resistance
R_D	Total fouling allowance
R_{Di}	Tube-side fouling factor
R_{Do}	Air-side fouling factor
Re	Reynolds number
Re_{ff}	$Re(\ell/b)$ = Effective Reynolds number used in Equation (12.10)
Re_n	Nozzle Reynolds number
r_1	Inner radius of fin
r_2	Outer radius of fin
r_{2c}	$r_2 + \tau/2$ = Corrected fin radius
s	Specific gravity
T	Temperature
U	Overall heat-transfer coefficient
U_C	Clean overall heat-transfer coefficient
U_D	Design overall heat-transfer coefficient
U_{req}	Required overall heat-transfer coefficient
V	Fluid velocity
V_{face}	Air face velocity
$V_{face,\ ave}$	Air face velocity based on average air temperature
$V_{face,\ std}$	Air face velocity based on standard conditions (70°F, 1 atm)
V_{fr}	Velocity of air leaving fan ring
V_{max}	Maximum velocity of air in tube bundle
W	Width of tube bundle
$\dot{W}_{fan}$	Fan brake power
$\dot{W}_{motor}$	Power delivered by motor
$\dot{W}_{used}$	Power used by motor

Greek Letters

α	Kinetic energy correction factor
α_{fr}	Kinetic energy correction factor for air leaving fan ring
ΔP_f	Pressure drop due to fluid friction in straight sections of tubes or in flow across tube bundle
ΔP_i	Total pressure drop for tube-side fluid
ΔP_n	Pressure loss in tube-side nozzles
ΔP_o	Total air-side pressure loss
ΔP_r	Tube-side pressure drop due to tube entrance, exit, and return losses
$(\Delta P_{Total})_{fan}$	Change in total pressure between fan entrance and exit
ΔT	Temperature difference
ΔT_m	Mean temperature difference in heat exchanger
$(\Delta T_{ln})_{cf}$	Logarithmic mean temperature difference for counter-current flow
η_f	Fin efficiency
η_{fan}	Total fan efficiency
η_{motor}	Motor efficiency
η_{sr}	Efficiency of speed reducer
η_w	Weighted efficiency of finned surface
μ	Viscosity
μ_w	Fluid viscosity evaluated at average temperature of tube wall
ρ	Fluid density
ρ_{air}	Density of air
ρ_{ave}	Density of air at average air temperature
ρ_{fr}	Density of air leaving fan ring
ρ_0	Density of ambient air at mean sea level
ρ_{std}	Density of air at standard conditions (70°F, 1 atm, 50% relative humidity)

τ	Fin thickness
ϕ	Viscosity correction factor
ψ	Effective height of annular fin

Problems

(12.1) For Example 12.1:

(a) Repeat step (k) using Equation (2.38) to calculate the tube-side heat-transfer coefficient.

(b) Repeat step (1) using Equation (12.1) to calculate the air-side heat-transfer coefficient.

(c) Repeat step (t) using Equation (12.9) to calculate the air-side friction factor.

Compare the results of the above calculations with those obtained in Examples 12.1 through 12.3.

(12.2) Design an air-cooled heat exchanger for the service of Problem 5.6. Use the properties of dry air at one atmosphere pressure and assume an ambient air temperature of 100°F for design purposes. The air-side pressure loss should not exceed 0.5 in. H_2O. The unit will be mounted at grade and there are no space limitations at the plant site.

(12.3) Design an air-cooled heat exchanger for the service of Problem 5.11. Use the properties of dry air at one atmosphere pressure and assume an ambient air temperature of 95°F for design purposes. The air-side pressure loss should not exceed 0.5 in. H_2O. Assume that the unit will be mounted at grade and that there are no space limitations at the site.

(12.4) Rate the heat exchanger designed in Problem 12.3 using:

(a) HEXTRAN.

(b) *Xace*.

(12.5) Design an air-cooled heat exchanger for the service of Problem 5.23. The unit will be situated on the US Gulf Coast (a relatively corrosive environment). Design ambient air temperature is 95°F and the maximum air-side pressure loss is 0.5 in. H_2O. Assume that the unit will be mounted at grade and that there are no space limitations at the site. Properties of the hydrocarbon stream are given in Problem 7.20.

(12.6) Rate the heat exchanger designed in Problem 12.5 using:

(a) HEXTRAN.

(b) *Xace*.

(12.7) A stream consisting of 275,000 lb/h of a hydrocarbon liquid at 200°F is to be cooled to 130°F in an air-cooled heat exchanger. Four induced-draft fan bays are available on the used equipment lot. Each bay contains a single tube bundle having four rows of tubes with 45 tubes per row arranged for two passes. The root tubes are 1-in. OD, 14 BWG and are made of carbon steel. The tubes contain nine aluminum fins per inch with a height of 0.5 in. and an average thickness of 0.012 in. The tubes are laid out on 2.25-in. equilateral triangular pitch. Two of the bays each contain 24 ft long tubes and two 7.5 ft diameter fans equipped with 10 hp motors. The other two bays each contain 16 ft long tubes and two 6 ft diameter fans equipped with 7.5 hp motors. Design specifications include maximum pressure drops of 10 psi on the tube side and 0.5 in. H_2O on the air side.

Average properties of the hydrocarbon are as follows:

Hydrocarbon Property	Value at 165° F
C_P (Btu/lbm · °F)	0.53
k (Btu/h · ft · °F)	0.080
μ (cp)	0.91
ρ (lbm/ft^3)	48.5

Use HEXTRAN or other suitable software to determine if the existing heat exchangers are suitable for this service and if so, how they should be arranged. For the calculations, use dry air at one atmosphere pressure and an ambient air temperature of 95°F.

(12.8) (a) For the service of Example 12.1, run each of the following computer programs in design mode.

(i) HEXTRAN

(ii) *Xace*

(b) Based on the results obtained in part (a), develop a final design for the heat exchanger using each of the computer programs and compare the results with the design obtained by hand in Example 12.1.

(12.9) Use any available software to solve Problem 12.2.

(12.10) Use any available software to solve Problem 12.3.

(12.11) Use any available software to solve Problem 12.5.

(12.12) Design an air-cooled condenser for the service of Problem 11.26 using *Xace*. The elevation at the plant site is 1500 ft and ambient air at a temperature of 98°F with a relative humidity of 40% should be used for design purposes. The air-side pressure loss should not exceed 0.5 in. H_2O.

(12.13) Design an air-cooled condenser for the service of Problem 11.28 using *Xace*. Design conditions are the same as in Problem 12.12.

(12.14) Design an air-cooled condenser for the service of Problem 11.37 using *Xace*. Design conditions are as specified in Problem 12.12.

(12.15) Design an air-cooled condenser for the service of Problem 11.38 using *Xace*. The elevation at the plant site is 2500 ft and ambient air at a temperature of 95°F with a relative humidity of 40% should be used for design purposes. The air-side pressure loss should not exceed 0.5 in. H_2O.

Appendix A Thermophysical Properties of Materials

TABLE A.1 Properties of Metallic Elements

Element[a]	Thermal Conductivity k (W/m · K)[b]							Properties at 293 K (20 °C)				
	200 K −73 °C	273 K 0 °C	400 K 127 °C	600 K 327 °C	800 K 527 °C	1000 K 727 °C	1200 K 927 °C	ρ (kg/m^3)	c (J/kg · K)	k (W/m · K)	$\alpha \times 10^6$ (m^2/s)	Melting Temperature (K)
Aluminum	237	236	240	232	220			2702	896	236	97.5	933
Antimony	30.2	25.5	21.2	18.2	16.8			6684	208	24.6	17.7	904
Beryllium	301	218	161	126	107	89	73	1850	1750	205	63.3	1550
Bismuth[c]	9.7	8.2						9780	124	7.9	6.51	545
Boron[c]	52.5	31.7	18.7	11.3	8.1	6.3	5.2	2500	1047	28.6	10.9	2573
Cadmium[c]	99.3	97.5	94.7					8650	231	97	48.5	594
Cesium	36.8	36.1						1873	230	36	83.6	302
Chromium	111	94.8	87.3	80.5	71.3	65.3	62.4	7160	440	91.4	29.0	2118
Cobalt[c]	122	104	84.8					8862	389	100	29.0	1765
Copper	413	401	392	383	371	357	342	8933	383	399	116.6	1356
Germanium	96.8	66.7	43.2	27.3	19.8	17.4	17.4	5360		61.6		1211
Gold	327	318	312	304	292	278	262	19300	129	316	126.9	1336
Hafnium	24.4	23.3	22.3	21.3	20.8	20.7	20.9	13280		23.1		2495
Indium	89.7	83.7	74.5					7300		82.2		430
Iridium	153	148	144	138	132	126	120	22500	134	147	48.8	2716
Iron	94	83.5	69.4	54.7	43.3	32.6	28.2	7870	452	81.1	22.8	1810
Lead	36.6	35.5	33.8	31.2				11340	129	35.3	24.1	601
Lithium	88.1	79.2	72.1					534	3391	77.4	42.7	454
Magnesium	159	157	153	149	146			1740	1017	156	88.2	923
Manganese	7.17	7.68						7290	486	7.78	2.2	1517
Mercury[c]	28.9							13546				234
Molybdenum	143	139	134	126	118	112	105	10240	251	138	53.7	2883
Nickel	106	94	80.1	65.5	67.4	71.8	76.1	8900	446	91	22.9	1726
Niobium	52.6	53.3	55.2	58.2	61.3	64.4	67.5	8570	270	53.6	23.2	2741
Palladium	75.5	75.5	75.5	75.5	75.5	75.5		12020	247	75.5	25.4	1825
Platinum	72.4	71.5	71.6	73.0	75.5	78.6	82.6	21450	133	71.4	25.0	2042
Potassium	104	104	52					860	741	103	161.6	337
Rhenium	51	48.6	46.1	44.2	44.1	44.6	45.7	21100	137	48.1	16.6	3453
Rhodium	154	151	146	136	127	121	115	12450	248	150	48.6	2233
Rubidium	58.9	58.3						1530	348	58.2	109.3	312
Silicon	264	168	98.9	61.9	42.2	31.2	25.7	2330	703	153	93.4	1685
Silver	403	428	420	405	389	374	358	10500	234	427	173.8	1234
Sodium	138	135						971	1206	133	113.6	371
Tantalum	57.5	57.4	57.8	58.6	59.4	60.2	61	16600	138	57.5	25.1	3269
Tin[c]	73.3	68.2	62.2					5750	227	67.0	51.3	505
Titanium[c]	24.5	22.4	20.4	19.4	19.7	20.7	22	4500	611	22.0	8.0	1953
Tungsten[c]	197	182	162	139	128	121	115	19300	134	179	69.2	3653
Uranium[c]	25.1	27	29.6	34	38.8	43.9	49	19070	113	27.4	12.7	1407
Vanadium	31.5	31.3	32.1	34.2	36.3	38.6	41.2	6100	502	31.4	10.3	2192
Zinc	123	122	116	105				7140	385	121	44.0	693
Zirconium[c]	25.2	23.2	21.6	20.7	21.6	23.7	25.7	6570	272	22.8	12.8	2125

[a]Purity for all elements exceeds 99%.

[b]The expected percent errors in the thermal conductivity values are approximately within ±5% of the true values near room temperature and within about ±10% at other temperatures.

[c]For crystalline materials, the values are given for the polycrystalline materials.

Source: Refs. [1–4].

TABLE A.2 Properties of Alloys

		Properties at 293 K (20 °C)			
Metal	Composition (%)	ρ (kg/m^3)	c (J/kg · K)	k (W/m · K)	$\alpha \times 10^5$ (m^2/s)
Aluminum					
Duralumin	94-96 Al, 3-5 Cu, trace Mg	2787	833	164	6.676
Silumin	87 Al, 13 Si	2659	871	164	7.099
Copper					
Aluminum bronze	95 Cu, 5 Al	8666	410	83	2.330
Bronze	75 Cu, 25 Sn	8666	343	26	0.859
Red brass	85 Cu, 9 Sn, 5 Zn	8714	385	61	1.804
Brass	70 Cu, 30 Zn	8522	385	111	3.412
German silver	62 Cu, 15 Ni, 22 Zn	8618	394	24.9	0.733
Constantan	60 Cu, 40 Ni	8922	410	22.7	0.612
Iron					
Cast iron	~4C	7272	420	52	1.702
Wrought iron		7849	460	59	1.626
Steel					
Carbon steel	1 C	7801	473	43	1.172
	1.5 C	7753	486	36	0.970
Chrome steel	1 Cr	7865	460	61	1.665
	5 Cr	7833	460	40	1.110
	10 Cr	7785	460	31	0.867
Chrome-nickel steel	15 Cr, 10 Ni	7865	460	19	0.526
	20 Cr, 15 Ni	7833	460	15.1	0.415
Nickel steel	10 Ni	7945	460	26	0.720
	20 Ni	7993	460	19	0.526
	40 Ni	8169	460	10	0.279
	60 Ni	8378	460	19	0.493
Nickel-chrome steel	80 Ni, 15 Cr	8522	460	17	0.444
	60 Ni, 15 Cr	8266	460	12.8	0.333
	40 Ni, 15 Cr	8073	460	11.6	0.305
	20 Ni, 15 Cr	7865	460	14.0	0.390
Manganese steel	1 Mn	7865	460	50	1.388
	5 Mn	7849	460	22	0.637
Silicon steel	1 Si	7769	460	42	1.164
	5 Si	7417	460	19	0.555
Stainless steel	Type 304	7817	461	14.4	0.387
	Type 347	7817	461	14.3	0.387
Tungsten steel	1W	7913	448	66	1.858
	5 W	8073	435	54	1.525

Source: Refs. [1] and [2].

TABLE A.3 Properties of Insulations and Building Materials

Substance	Temperature (°C)	k (W/m · °C)	ρ (kg/m^3)	c (kJ/kg· °C)	$\alpha \times 10^7$ (m^2/s)
Structural and heat-resistant materials					
Asphalt	20–55	0.74–0.76			
Brick					
Building brick, common	20	0.69	1600	0.84	5.2
Face		1.32	2000		
Carborundum brick	600	18.5			
	1400	11.1			
Chrome brick	200	2.32	3000	0.84	9.2
	550	2.47			9.8
	900	1.99			7.9
Diatomaceous earth, molded and fired	200	0.24			
	870	0.31			

TABLE A.3 Properties of Insulations and Building Materials—cont'd

Substance	Temperature (°C)	k (W/m · °C)	ρ (kg/m^3)	c (kJ/kg· °C)	$\alpha \times 10^7$ (m^2/s)
Fireclay brick, burnt 2426°F	500	1.04	2000	0.96	5.4
	800	1.07			
	1100	1.09			
Burnt 2642°F	500	1.28	2300	0.96	5.8
	800	1.37			
	1100	1.40			
Missouri	200	1.00	2600	0.96	4.0
	600	1.47			
	1400	1.77			
Magnesite	200	3.81		1.13	
	650	2.77			
	1200	1.90			
Cement, Portland		0.29	1500		
Mortar	23	1.16			
Concrete, cinder	23	0.76			
Stone, 1-2-4 mix	20	1.37	1900–2300	0.88	8.2–6.8
Glass, window	20	0.78 (avg)	2700	0.84	3.4
Borosilicate	30–75	1.09	2200		
Plaster, gypsum	20	0.48	1440	0.84	4.0
Metal lath	20	0.47			
Wood lath	20	0.28			
Stone					
Granite		1.73–3.98	2640	0.82	8–18
Limestone	100–300	1.26–1.33	2500	0.90	5.6–5.9
Marble		2.07–2.94	2500–2700	0.80	10–13.6
Sandstone	40	1.83	2160–2300	0.71	11.2–11.9
Wood (across the grain)					
Balsa, 8.8 lb/ft^3	30	0.055	140		
Cypress	30	0.097	460		
Fir	23	0.11	420	2.72	0.96
Maple or oak	30	0.166	540	2.4	1.28
Yellow pine	23	0.147	640	2.8	0.82
White pine	30	0.112	430		
Insulating materials					
Asbestos					
Loosely packed	-45	0.149			
	0	0.154	470–570	0.816	3.3–4
	100	0.161			
Asbestos-cement boards	20	0.74			
Sheets	51	0.166			
Felt, 40 laminations/in.	38	0.057			
	150	0.069			
	260	0.083			
Felt, 20 laminations/in.	38	0.078			
	150	0.095			
	260	0.112			
Corrugated, 4 plies/in.	38	0.087			
	93	0.100			
	150	0.119			
Asbestos cement	–	2.08			
Balsam wool, 2.2 lb/ft^3	32	0.04	35		
Cardboard, corrugated	–	0.064			
Celotex	32	0.048			
Corkboard, 10 lb/ft^3	30	0.043	160		
Cork, regranulated	32	0.045	45–120	1.88	2–5.3
Ground	32	0.043	150		
Diatomaceous earth (Sil-o-cel)	0	0.061	320		
Felt, hair	30	0.036	130–200		
Wool	30	0.052	330		
Fiber, insulating board	20	0.048	240		

(Continued)

TABLE A.3 Properties of Insulations and Building Materials—cont'd

Substance	Temperature (°C)	k (W/m · °C)	ρ (kg/m^3)	c (kJ/kg· °C)	$\alpha \times 10^7$ (m^2/s)
Glass wool, 1.5 lb/ft^3	23	0.038	24	0.7	22.6
Insulex, dry	32	0.064			
Kapok	30	0.035			
Magnesia, 85%	38	0.067	270		
	93	0.071			
	150	0.074			
	204	0.080			
Rock wool, 10 lb/ft^3	32	0.040	160		
Loosely packed	150	0.067	64		
	260	0.087			
Sawdust	23	0.059			
Silica aerogel	32	0.024	140		
Wood shavings	23	0.059			

Source: Refs. [5] and [6].

TABLE A.4 Properties of Dry Air at Atmospheric Pressure: SI Units

T (K)	T (°C)	ρ (kg/m^3)	$\beta \times 10^3$ (1/K)	C_P (J/kg · K)	k (W/m · K)	$\alpha \times 10^6$ (m^2/s)	$\mu \times 10^6$ (N · s/m^2)	$\nu \times 10^6$ (m^2/s)	Pr	$g\beta/\nu^2 \times 10^{-8}$ (1/K · m^3)
273	0	1.252	3.66	1011	0.0237	19.2	17.456	13.9	0.71	1.85
293	20	1.164	3.41	1012	0.0251	22.0	18.240	15.7	0.71	1.36
313	40	1.092	3.19	1014	0.0265	24.8	19.123	17.6	0.71	1.01
333	60	1.025	3.00	1017	0.0279	27.6	19.907	19.4	0.71	0.782
353	80	0.968	2.83	1019	0.0293	30.6	20.790	21.5	0.71	0.600
373	100	0.916	2.68	1022	0.0307	33.6	21.673	23.6	0.71	0.472
473	200	0.723	2.11	1035	0.0370	49.7	25.693	35.5	0.71	0.164
573	300	0.596	1.75	1047	0.0429	68.9	39.322	49.2	0.71	0.0709
673	400	0.508	1.49	1059	0.0485	89.4	32.754	64.6	0.72	0.0350
773	500	0.442	1.29	1076	0.0540	113.2	35.794	81.0	0.72	0.0193
1273	1000	0.268	0.79	1139	0.0762	240	48.445	181	0.74	0.00236

Source: Refs. [1] and [3].

TABLE A.5 Properties of Dry Air at Atmospheric Pressure: English Units

T (°F)	C_P (Btu/lbm · °F)	k (Btu/h · ft · F)	μ (lbm/ft · h)	ρ (lbm/ft^3)	Pr
0	0.241	0.0131	0.0387	0.0864	0.71
20	0.241	0.0136	0.0402	0.0828	0.71
40	0.241	0.0141	0.0416	0.0795	0.71
60	0.241	0.0147	0.0428	0.0764	0.70
80	0.241	0.0152	0.0440	0.0735	0.70
100	0.241	0.0156	0.0455	0.0709	0.70
120	0.241	0.0161	0.0467	0.0685	0.70
140	0.242	0.0166	0.0479	0.0662	0.70
160	0.242	0.0171	0.0491	0.0640	0.69
180	0.242	0.0176	0.0503	0.0620	0.69
200	0.242	0.0180	0.0515	0.0601	0.69
250	0.243	0.0192	0.0544	0.0559	0.69
300	0.244	0.0203	0.0573	0.0522	0.69
350	0.245	0.0214	0.0600	0.0490	0.69
400	0.246	0.0225	0.0627	0.0461	0.69
450	0.247	0.0236	0.0653	0.0436	0.68

TABLE A.5 Properties of Dry Air at Atmospheric Pressure: English Units—cont'd

T (°F)	C_P (Btu/lbm · °F)	k (Btu/h · ft · F)	μ (lbm/ft · h)	ρ (lbm/ft^3)	Pr
500	0.248	0.0246	0.0677	0.0413	0.68
600	0.251	0.0266	0.0723	0.0374	0.68
700	0.254	0.0285	0.0767	0.0342	0.68
800	0.257	0.0303	0.0810	0.0315	0.69
900	0.260	0.0320	0.0849	0.0292	0.69
1000	0.263	0.0337	0.0885	0.0272	0.69

Data generated using the PRO/II process simulator.

TABLE A.6 Properties of Steam at Atmospheric Pressure

T										
(K)	(°C)	ρ (kg/m^3)	β × 10^3 (1/K)	C_P (J/kg · K)	k (W/m · K)	α × 10^4 (m^2/s)	μ × 10^6 (N · s/m^2)	ν × 10^6 (m^2/s)	Pr	$g\beta/\nu^2$ ×10^{-6} (1/K · m^3)
373	100	0.5977		2034	0.0249	0.204	12.10	20.2	0.987	
380	107	0.5863		2060	0.0246	0.204	12.71	21.6	1.060	
400	127	0.5542	2.50	2014	0.0261	0.234	13.44	24.2	1.040	41.86
450	177	0.4902	2.22	1980	0.0299	0.307	15.25	31.1	1.010	22.51
500	227	0.4405	2.00	1985	0.0339	0.387	17.04	38.6	0.996	13.16
550	277	0.4005	1.82	1997	0.0379	0.475	18.84	47.0	0.991	8.08
600	327	0.3652	1.67	2026	0.0422	0.573	20.67	56.6	0.986	5.11
650	377	0.3380	1.54	2056	0.0464	0.666	22.47	66.4	0.995	3.43
700	427	0.3140	1.43	2085	0.0505	0.772	24.26	77.2	1.000	2.35
750	477	0.2931	1.33	2119	0.0549	0.883	26.04	88.8	1.005	1.65
800	527	0.2739	1.25	2152	0.0592	1.001	27.86	102.0	1.010	1.18
850	577	0.2579	1.18	2186	0.0637	1.130	29.69	115.2	1.019	0.872

Source: Refs. [1] and [2].

TABLE A.7 Properties of Liquid Water at Saturation Pressure: SI Units

T										
(K)	(°C)	ρ (kg/m^3)	β × 10^4 (1/K)	C_P (J/kg · K)	k (W/m · K)	α × 10^6 (m^2/s)	μ × 10^6 (N · s/m^2)	ν × 10^6 (m^2/s)	Pr	$g\beta/\nu^2$ × 10^{-9} (1/K · m^3)
273	0	999.9	-0.7	4226	0.558	0.131	1794	1.789	13.7	–
278	5	1000	–	4206	0.568	0.135	1535	1.535	11.4	–
283	10	999.7	0.95	4195	0.577	0.137	1296	1.300	9.5	0.551
288	15	999.1	–	4187	0.585	0.141	1136	1.146	8.1	–
293	20	998.2	2.1	4182	0.597	0.143	993	1.006	7.0	2.035
298	25	997.1	–	4178	0.606	0.146	880.6	0.884	6.1	–
303	30	995.7	3.0	4176	0.615	0.149	792.4	0.805	5.4	4.540
308	35	994.1	–	4175	0.624	0.150	719.8	0.725	4.8	–
313	40	992.2	3.9	4175	0.633	0.151	658.0	0.658	4.3	8.833
318	45	990.2	–	4176	0.640	0.155	605.1	0.611	3.9	–
323	50	988.1	4.6	4178	0.647	0.157	555.1	0.556	3.55	14.59
348	75	974.9	–	4190	0.671	0.164	376.6	0.366	2.23	–
373	100	958.4	7.5	4211	0.682	0.169	277.5	0.294	1.75	85.09
393	120	943.5	8.5	4232	0.685	0.171	235.4	0.244	1.43	140.0
413	140	926.3	9.7	4257	0.684	0.172	201.0	0.212	1.23	211.7
433	160	907.6	10.8	4285	0.680	0.173	171.6	0.191	1.10	290.3
453	180	886.6	12.1	4396	0.673	0.172	152.0	0.173	1.01	396.5
473	200	862.8	13.5	4501	0.665	0.170	139.3	0.160	0.95	517.2
493	220	837.0	15.2	4605	0.652	0.167	124.5	0.149	0.90	671.4
513	240	809.0	17.2	4731	0.634	0.162	113.8	0.141	0.86	848.5
533	260	779.0	20.0	4982	0.613	0.156	104.9	0.135	0.86	1076
553	280	750.0	23.8	5234	0.588	0.147	98.07	0.131	0.89	1360
573	300	712.5	29.5	5694	0.564	0.132	92.18	0.128	0.98	1766

Source: Refs. [1] and [3].

TABLE A.8 Properties of Saturated Steam and Water: English Units

Absolute pressure (psi)	Absolute pressure (in. Hg)	Vacuum (in. Hg)	Temperature (°F)	Heat of the liquid (Btu/lb)	Latent heat of evaporation (Btu/lb)	Total heat of steam (Btu/lb)	Specific volume Water (ft³/lb)	Specific volume Steam (ft³/lb)
0.08859	0.02	29.90	32.018	0.0003	1075.5	1075.5	0.016022	3302.4
0.10	0.20	29.72	35.023	3.026	1073.8	1076.8	0.016020	2945.5
0.15	0.31	29.61	45.453	13.498	1067.9	1081.4	0.016020	2004.7
0.20	0.41	29.51	53.160	21.217	1053.5	1084.7	0.016025	1526.3
0.25	0.51	29.41	59.323	27.382	1060.1	1087.4	0.016032	1235.5
0.30	0.61	29.31	64.484	32.541	1057.1	1089.7	0.016040	1039.7
0.35	0.71	29.21	68.939	36.992	1054.6	1091.6	0.016048	898.6
0.40	0.81	29.11	72.869	40.917	1052.4	1093.3	0.016056	792.1
0.45	0.92	29.00	76.387	44.430	1050.5	1094.9	0.016063	708.8
0.50	1.02	28.90	79.586	47.623	1048.6	1096.3	0.016071	641.5
0.60	1.22	28.70	85.218	53.245	1045.5	1098.7	0.016085	540.1
0.70	1.43	28.49	90.09	58.10	1042.7	1100.8	0.016099	466.94
0.80	1.63	28.29	94.38	62.39	1040.3	1102.6	0.016112	411.69
0.90	1.83	28.09	98.24	66.24	1038.1	1104.3	0.016124	368.43
1.0	2.04	27.88	101.74	69.73	1036.1	1105.8	0.016136	333.60
1.2	2.44	27.48	107.91	75.90	1032.6	1108.5	0.016158	280.96
1.4	2.85	27.07	113.26	81.23	1029.5	1110.7	0.016178	243.02
1.6	3.26	26.66	117.98	85.95	1026.8	1112.7	0.016196	214.33
1.8	3.66	26.26	122.22	90.18	1024.3	1114.5	0.016213	191.85
2.0	4.07	25.85	126.07	94.03	1022.1	1116.2	0.016230	173.76
2.2	4.48	25.44	129.61	97.57	1020.1	1117.6	0.016245	158.87
2.4	4.89	25.03	132.88	100.84	1018.2	1119.0	0.016260	146.40
2.6	5.29	24.63	135.93	103.88	1016.4	1120.3	0.016274	135.80
2.8	5.70	24.22	138.78	106.73	1014.7	1121.5	0.016287	126.67
3.0	6.11	23.81	141.47	109.42	1013.2	1122.6	0.016300	118.73
3.5	7.13	22.79	147.56	115.51	1009.6	1125.1	0.016331	102.74
4.0	8.14	21.78	152.96	120.92	1006.4	1127.3	0.016358	90.64
4.5	9.16	20.76	157.82	125.77	1003.5	1129.3	0.016384	83.03
5.0	10.18	19.74	162.24	130.20	1000.9	1131.1	0.016407	73.532
5.5	11.20	18.72	166.29	134.26	998.5	1132.7	0.016430	67.249
6.0	12.22	17.70	170.05	138.03	996.2	1134.2	0.016451	61.984
6.5	13.23	16.69	173.56	141.54	994.1	1135.6	0.016472	57.506
7.0	14.25	15.67	176.84	144.83	992.1	1136.9	0.016491	53.650
7.5	15.27	14.65	179.93	147.93	990.2	1138.2	0.016510	50.294
8.0	16.29	13.63	182.86	150.87	988.5	1139.3	0.016527	47.345
8.5	17.31	12.61	185.63	153.65	986.8	1140.4	0.016545	44.733
9.0	18.32	11.60	188.27	156.30	985.1	1141.4	0.016561	42.402
9.5	19.34	10.58	190.80	158.84	983.6	1142.4	0.016577	40.310
10.0	20.36	9.56	193.21	161.26	982.1	1143.3	0.016592	38.420
11.0	22.40	7.52	197.75	165.82	979.3	1145.1	0.016622	35.142
12.0	24.43	5.49	201.96	170.05	976.6	1146.7	0.016650	32.394
13.0	26.47	3.45	205.88	174.00	974.2	1148.2	0.016676	30.057
14.0	28.50	1.42	209.56	177.71	971.9	1149.6	0.016702	28.043

Pressure (psi) Absolute	Pressure (psi) Gage	Temperature (°F)	Heat of the liquid (Btu/lb)	Latent heat of evaporation (Btu/lb)	Total heat of steam (Btu/lb)	Specific volume Water (ft³/lb)	Specific volume Steam (ft³/lb)
14.696	0.0	212.00	180.17	970.3	1150.3	0.016719	26.799
15.0	0.3	213.03	181.21	969.7	1150.9	0.016726	26.290
16.0	1.3	216.32	184.52	967.6	1152.1	0.016749	24.750
17.0	2.3	219.44	187.66	965.6	1153.2	0.016771	23.385
18.0	3.3	222.41	190.66	963.7	1154.3	0.016793	22.168
19.0	4.3	225.24	193.52	961.8	1155.3	0.016814	21.074
20.0	5.3	227.96	196.27	960.1	1156.3	0.016834	20.087
21.0	6.3	230.57	198.90	958.4	1157.3	0.016854	19.190
22.0	7.3	233.07	201.44	956.7	1158.1	0.016873	18.373
23.0	8.3	235.49	203.88	955.1	1159.0	0.016891	17.624
24.0	9.3	237.82	206.24	953.6	1159.8	0.016909	16.936
25.0	10.3	240.07	208.52	952.1	1160.6	0.016927	16.301

TABLE A.8 Properties of Saturated Steam and Water: English Units—cont'd

Pressure (psi)						Specific volume	
Absolute	Gage	Temperature (°F)	Heat of the liquid (Btu/lb)	Latent heat of evaporation (Btu/lb)	Total heat of steam (Btu/lb)	Water (ft^3/lb)	Steam (ft^3/lb)
26.0	11.3	242.25	210.7	950.6	1161.4	0.016944	15.7138
27.0	12.3	244.36	212.9	949.2	1162.1	0.016961	15.1684
28.0	13.3	246.41	214.9	947.9	1162.8	0.016977	14.6607
29.0	14.3	248.40	217.0	946.5	1163.5	0.016993	14.1869
30.0	15.3	250.34	218.9	945.2	1164.1	0.017009	13.7436
31.0	16.3	252.22	220.8	943.9	1164.8	0.017024	13.3280
32.0	17.3	254.05	222.7	942.7	1165.4	0.017039	12.9376
33.0	18.3	255.84	224.5	941.5	1166.0	0.017054	12.5700
34.0	19.3	257.58	226.3	940.3	1166.6	0.017069	12.2234
35.0	20.3	259.29	228.0	939.1	1167.1	0.017083	11.8959
36.0	21.3	260.95	229.7	938.0	1167.7	0.017097	11.5860
37.0	22.3	262.58	231.4	936.9	1168.2	0.017111	11.2923
38.0	23.3	264.17	233.0	935.8	1168.8	0.017124	11.0136
39.0	24.3	265.72	234.6	934.7	1169.3	0.017138	10.7487
40.0	25.3	267.25	236.1	933.6	1169.8	0.017151	10.4965
41.0	26.3	268.74	237.7	932.6	1170.2	0.017164	10.2563
42.0	27.3	270.21	239.2	931.5	1170.7	0.017177	10.0272
43.0	28.3	271.65	240.6	930.5	1171.2	0.017189	9.8083
44.0	29.3	273.06	242.1	929.5	1171.6	0.017202	9.5991
45.0	30.3	274.44	243.5	928.6	1172.0	0.017214	9.3988
46.0	31.3	275.80	244.9	927.6	1172.5	0.017226	9.2070
47.0	32.3	277.14	246.2	926.6	1172.9	0.017238	9.0231
48.0	33.3	278.45	247.6	925.7	1173.3	0.017250	8.8465
49.0	34.3	279.74	248.9	924.8	1173.7	0.017262	8.6770
50.0	35.3	281.02	250.2	923.9	1174.1	0.017274	8.5140
51.0	36.3	282.27	251.5	923.0	1174.5	0.017285	8.3571
52.0	37.3	283.50	252.8	922.1	1174.9	0.017296	8.2061
53.0	38.3	284.71	254.0	921.2	1175.2	0.017307	8.0606
54.0	39.3	285.90	255.2	920.4	1175.6	0.017319	7.9203
55.0	40.3	287.08	256.4	919.5	1175.9	0.017329	7.7850
56.0	41.3	288.24	257.6	918.7	1176.3	0.017340	7.6543
57.0	42.3	289.38	258.8	917.8	1176.6	0.017351	7.5280
58.0	43.3	290.50	259.9	917.0	1177.0	0.017362	7.4059
59.0	44.3	291.62	261.1	916.2	1177.3	0.017372	7.2879
60.0	45.3	292.71	262.2	915.4	1177.6	0.017383	7.1736
61.0	46.3	293.79	263.3	914.6	1177.9	0.017393	7.0630
62.0	47.3	294.86	264.4	913.8	1178.2	0.017403	6.9558
63.0	48.3	295.91	265.5	913.0	1178.6	0.017413	6.8519
64.0	49.3	296.95	266.6	912.3	1178.9	0.017423	6.7511
65.0	50.3	297.98	267.6	911.5	1179.1	0.017433	6.6533
66.0	51.3	298.99	268.7	910.8	1179.4	0.017443	6.5584
67.0	52.3	299.99	269.7	910.0	1179.7	0.017453	6.4662
68.0	53.3	300.99	270.7	909.3	1180.0	0.017463	6.3767
69.0	54.3	301.96	271.7	908.5	1180.3	0.017472	6.2896
70.0	55.3	302.93	272.7	907.8	1180.6	0.017482	6.2050
71.0	56.3	303.89	273.7	907.1	1180.8	0.017491	6.1226
72.0	57.3	304.83	274.7	906.4	1181.1	0.017501	6.0425
73.0	58.3	305.77	275.7	905.7	1181.4	0.017510	5.9645
74.0	59.3	306.69	276.6	905.0	1181.6	0.017519	5.8885
75.0	60.3	307.61	277.6	904.3	1181.9	0.017529	5.8144
76.0	61.3	308.51	278.5	903.6	1182.1	0.017538	5.7423
77.0	62.3	309.41	279.4	902.9	1182.4	0.017547	5.6720
78.0	63.3	310.29	280.3	902.3	1182.6	0.017556	5.6034
79.0	64.3	311.17	281.3	901.6	1182.8	0.017565	5.5364
80.0	65.3	312.04	282.1	900.9	1183.1	0.017573	5.4711
81.0	66.3	312.90	283.0	900.3	1183.3	0.017582	5.4074
82.0	67.3	313.75	283.9	899.6	1183.5	0.017591	5.3451
83.0	68.3	314.60	284.8	899.0	1183.8	0.017600	5.2843
84.0	69.3	315.43	285.7	898.3	1184.0	0.017608	5.2249
85.0	70.3	316.26	286.5	897.7	1184.2	0.017617	5.1669

(Continued)

TABLE A.8 Properties of Saturated Steam and Water: English Units—cont'd

Pressure (psi)						Specific volume	
Absolute	Gage	Temperature (°F)	Heat of the liquid (Btu/lb)	Latent heat of evaporation (Btu/lb)	Total heat of steam (Btu/lb)	Water (ft^3/lb)	Steam (ft^3/lb)
86.0	71.3	317.08	287.4	897.0	1184.4	0.017625	5.1101
87.0	72.3	317.89	288.2	896.4	1184.6	0.017634	5.0546
88.0	73.3	318.69	289.0	895.8	1184.8	0.017642	5.0004
89.0	74.3	319.49	289.9	895.2	1185.0	0.017651	4.9473
90.0	75.3	320.28	290.7	894.6	1185.3	0.017659	4.8953
91.0	76.3	321.06	291.5	893.9	1185.5	0.017667	4.8445
92.0	77.3	321.84	292.3	893.3	1185.7	0.017675	4.7947
93.0	78.3	322.61	293.1	892.7	1185.9	0.017684	4.7459
94.0	79.3	323.37	293.9	892.1	1186.0	0.017692	4.6982
95.0	80.3	324.13	294.7	891.5	1186.2	0.017700	4.6514
96.0	81.3	324.88	295.5	891.0	1186.4	0.017708	4.6055
97.0	82.3	325.63	296.3	890.4	1186.6	0.017716	4.5606
98.0	83.3	326.36	297.0	889.8	1186.8	0.017724	4.5166
99.0	84.3	327.10	297.8	889.2	1187.0	0.017732	4.4734
100.0	85.3	327.82	298.5	888.6	1187.2	0.017740	4.4310
101.0	86.3	328.54	299.3	888.1	1187.3	0.01775	4.3895
102.0	87.3	329.26	300.0	887.5	1187.5	0.01776	4.3487
103.0	88.3	329.97	300.8	886.9	1187.7	0.01776	4.3087
104.0	89.3	330.67	301.5	886.4	1187.9	0.01777	4.2695
105.0	90.3	331.37	302.2	885.8	1188.0	0.01778	4.2309
106.0	91.3	332.06	303.0	885.2	1188.2	0.01779	4.1931
107.0	92.3	332.75	303.7	884.7	1188.4	0.01779	4.1560
108.0	93.3	333.44	304.4	884.1	1188.5	0.01780	4.1195
109.0	94.3	334.11	305.1	883.6	1188.7	0.01781	4.0837
110.0	95.3	334.79	305.8	883.1	1188.9	0.01782	4.0484
111.0	96.3	335.46	306.5	882.5	1189.0	0.01782	4.0138
112.0	97.3	336.12	307.2	882.0	1189.2	0.01783	3.9798
113.0	98.3	336.78	307.9	881.4	1189.3	0.01784	3.9464
114.0	99.3	337.43	308.6	880.9	1189.5	0.01785	3.9136
115.0	100.3	338.08	309.3	880.4	1189.6	0.01785	3.8813
116.0	101.3	338.73	309.9	879.9	1189.8	0.01786	3.8495
117.0	102.3	339.37	310.6	879.3	1189.9	0.01787	3.8183
118.0	103.3	340.01	311.3	878.8	1190.1	0.01787	3.7875
119.0	104.3	340.64	311.9	878.3	1190.2	0.01788	3.7573
120.0	105.3	341.27	312.6	877.8	1190.4	0.01789	3.7275
121.0	106.3	341.89	313.2	877.3	1190.5	0.01790	3.6983
122.0	107.3	342.51	313.9	876.8	1190.7	0.01790	3.6695
123.0	108.3	343.13	314.5	876.3	1190.8	0.01791	3.6411
124.0	109.3	343.74	315.2	875.8	1190.9	0.01792	3.6132
125.0	110.3	344.35	315.8	875.3	1191.1	0.01792	3.5857
126.0	111.3	344.95	316.4	874.8	1191.2	0.01793	3.5586
127.0	112.3	345.55	317.1	874.3	1191.3	0.01794	3.5320
128.0	113.3	346.15	317.7	873.8	1191.5	0.01794	3.5057
129.0	114.3	346.74	318.3	873.3	1191.6	0.01795	3.4799
130.0	115.3	347.33	319.0	872.8	1191.7	0.01796	3.4544
131.0	116.3	347.92	319.6	872.3	1191.9	0.01797	3.4293
132.0	117.3	348.50	320.2	871.8	1192.0	0.01797	3.4046
133.0	118.3	349.08	320.8	871.3	1192.1	0.01798	3.3802
134.0	119.3	349.65	321.4	870.8	1192.2	0.01799	3.3562
135.0	120.3	350.23	322.0	870.4	1192.4	0.01799	3.3325
136.0	121.3	350.79	322.6	869.9	1192.5	0.01800	3.3091
137.0	122.3	351.36	323.2	869.4	1192.6	0.01801	3.2861
138.0	123.3	351.92	323.8	868.9	1192.7	0.01801	3.2634
139.0	124.3	352.48	324.4	868.5	1192.8	0.01802	3.2411
140.0	125.3	353.04	325.0	868.0	1193.0	0.01803	3.2190
141.0	126.3	353.59	325.5	867.5	1193.1	0.01803	3.1972
142.0	127.3	354.14	326.1	867.1	1193.2	0.01804	3.1757
143.0	128.3	354.69	326.7	866.6	1193.3	0.01805	3.1546
144.0	129.3	355.23	327.3	866.2	1193.4	0.01805	3.1337

TABLE A.8 Properties of Saturated Steam and Water: English Units—cont'd

Pressure (psi)						Specific volume	
Absolute	Gage	Temperature (°F)	Heat of the liquid (Btu/lb)	Latent heat of evaporation (Btu/lb)	Total heat of steam (Btu/lb)	Water (ft^3/lb)	Steam (ft^3/lb)
145.0	130.3	355.77	327.8	865.7	1193.5	0.01806	3.1130
146.0	131.3	356.31	328.4	865.2	1193.6	0.01806	3.0927
147.0	132.3	356.84	329.0	864.8	1193.8	0.01807	3.0726
148.0	133.3	357.38	329.5	864.3	1193.9	0.01808	3.0528
149.0	134.3	357.91	330.1	863.9	1194.0	0.01808	3.0332
150.0	135.3	358.43	330.6	863.4	1194.1	0.01809	3.0139
152.0	137.3	359.48	331.8	862.5	1194.3	0.01810	2.9760
154.0	139.3	360.51	332.8	861.6	1194.5	0.01812	2.9391
156.0	141.3	361.53	333.9	860.8	1194.7	0.01813	2.9031
158.0	143.3	362.55	335.0	859.9	1194.9	0.01814	2.8679
160.0	145.3	363.55	336.1	859.0	1195.1	0.01815	2.8336
162.0	147.3	364.54	337.1	858.2	1195.3	0.01817	2.8001
164.0	149.3	365.53	338.2	857.3	1195.5	0.01818	2.7674
166.0	151.3	366.50	339.2	856.5	1195.7	0.01819	2.7355
168.0	153.3	367.47	340.2	855.6	1195.8	0.01820	2.7043
170.0	155.3	368.42	341.2	854.8	1196.0	0.01821	2.6738
172.0	157.3	369.37	342.2	853.9	1196.2	0.01823	2.6440
174.0	159.3	370.31	343.2	853.1	1196.4	0.01824	2.6149
176.0	161.3	371.24	344.2	852.3	1196.5	0.01825	2.5864
178.0	163.3	372.16	345.2	851.5	1196.7	0.01826	2.5585
180.0	165.3	373.08	346.2	850.7	1196.9	0.01827	2.5312
182.0	167.3	373.98	347.2	849.9	1197.0	0.01828	2.5045
184.0	169.3	374.88	348.1	849.1	1197.2	0.01830	2.4783
186.0	171.3	375.77	349.1	848.3	1197.3	0.01831	2.4527
188.0	173.3	376.65	350.0	847.5	1197.5	0.01832	2.4276
190.0	175.3	377.53	350.9	846.7	1197.6	0.01833	2.4030
192.0	177.3	378.40	351.9	845.9	1197.8	0.01834	2.3790
194.0	179.3	379.26	352.8	845.1	1197.9	0.01835	2.3554
196.0	181.3	380.12	353.7	844.4	1198.1	0.01836	2.3322
198.0	183.3	380.96	354.6	843.6	1198.2	0.01838	2.3095
200.0	185.3	381.80	355.5	842.8	1198.3	0.01839	2.28728
205.0	190.3	383.88	357.7	840.9	1198.7	0.01841	2.23349
210.0	195.3	385.91	359.9	839.1	1199.0	0.01844	2.18217
215.0	200.3	387.91	362.1	837.2	1199.3	0.01847	2.13315
220.0	205.3	389.88	364.2	835.4	1199.6	0.01850	2.08629
225.0	210.3	391.80	366.2	833.6	1199.9	0.01852	2.04143
230.0	215.3	393.70	368.3	831.8	1200.1	0.01855	1.99846
235.0	220.3	395.56	370.3	830.1	1200.4	0.01857	1.95725
240.0	225.3	397.39	372.3	828.4	1200.6	0.01860	1.91769
245.0	230.3	399.19	374.2	826.6	1200.9	0.01863	1.87970
250.0	235.3	400.97	376.1	825.0	1201.1	0.01865	1.84317
255.0	240.3	402.72	378.0	823.3	1201.3	0.01868	1.80802
260.0	245.3	404.44	379.9	821.6	1201.5	0.01870	1.77418
265.0	250.3	406.13	381.7	820.0	1201.7	0.01873	1.74157
270.0	255.3	407.80	383.6	818.3	1201.9	0.01875	1.71013
275.0	260.3	409.45	385.4	816.7	1202.1	0.01878	1.67978
280.0	265.3	411.07	387.1	815.1	1202.3	0.01880	1.65049
285.0	270.3	412.67	388.9	813.6	1202.4	0.01882	1.62218
290.0	275.3	414.25	390.6	812.0	1202.6	0.01885	1.59482
295.0	280.3	415.81	392.3	810.4	1202.7	0.01887	1.56835
300.0	285.3	417.35	394.0	808.9	1202.9	0.01889	1.54274
320.0	305.3	423.31	400.5	802.9	1203.4	0.01899	1.44801
340.0	325.3	428.99	406.8	797.0	1203.8	0.01908	1.36405
360.0	345.3	434.41	412.8	791.3	1204.1	0.01917	1.28910
380.0	365.3	439.61	418.6	785.8	1204.4	0.01925	1.22177
400.0	385.3	444.60	424.2	780.4	1204.6	0.01934	1.16095
420.0	405.3	449.40	429.6	775.2	1204.7	0.01942	1.10573
440.0	425.3	454.03	434.8	770.0	1204.8	0.01950	1.05535

(Continued)

TABLE A.8 Properties of Saturated Steam and Water: English Units—cont'd

Pressure (psi)						Specific volume	
Absolute	Gage	Temperature (°F)	Heat of the liquid (Btu/lb)	Latent heat of evaporation (Btu/lb)	Total heat of steam (Btu/lb)	Water (ft^3/lb)	Steam (ft^3/lb)
460.0	445.3	458.50	439.8	765.0	1204.8	0.01959	1.00921
480.0	465.3	462.82	444.7	760.0	1204.8	0.01967	0.96677
500.0	485.3	467.01	449.5	755.1	1204.7	0.01975	0.92762
520.0	505.3	471.07	454.2	750.4	1204.5	0.01982	0.89137
540.0	525.3	475.01	458.7	745.7	1204.4	0.01990	0.85771
560.0	545.3	478.84	463.1	741.0	1204.2	0.01998	0.82637
580.0	565.3	482.57	467.5	736.5	1203.9	0.02006	0.79712
600.0	585.3	486.20	471.7	732.0	1203.7	0.02013	0.76975
620.0	605.3	489.74	475.8	727.5	1203.4	0.02021	0.74408
640.0	625.3	493.19	479.9	723.1	1203.0	0.02028	0.71995
660.0	645.3	496.57	483.9	718.8	1202.7	0.02036	0.69724
680.0	665.3	499.86	487.8	714.5	1202.3	0.02043	0.67581
700.0	685.3	503.08	491.6	710.2	1201.8	0.02050	0.65556
720.0	705.3	506.23	495.4	706.0	1201.4	0.02058	0.63639
740.0	725.3	509.32	499.1	701.9	1200.9	0.02065	0.61822
760.0	745.3	512.34	502.7	697.7	1200.4	0.02072	0.60097
780.0	765.3	515.30	506.3	693.6	1199.9	0.02080	0.58457
800.0	785.3	518.21	509.8	689.6	1199.4	0.02087	0.56896
820.0	805.3	521.06	513.3	685.5	1198.8	0.02094	0.55408
840.0	825.3	523.86	516.7	681.5	1198.2	0.02101	0.53988
860.0	845.3	526.60	520.1	677.6	1197.7	0.02109	0.52631
880.0	865.3	529.30	523.4	673.6	1197.0	0.02116	0.51333
900.0	885.3	531.95	526.7	669.7	1196.4	0.02123	0.50091
920.0	905.3	534.56	530.0	665.8	1195.7	0.02130	0.48901
940.0	925.3	537.13	533.2	661.9	1195.1	0.02137	0.47759
960.0	945.3	539.65	536.3	658.0	1194.4	0.02145	0.46662
980.0	965.3	542.14	539.5	654.2	1193.7	0.02152	0.45609
1000.0	985.3	544.58	542.6	650.4	1192.9	0.02159	0.44596
1050.0	1035.3	550.53	550.1	640.9	1191.0	0.02177	0.42224
1100.0	1085.3	556.28	557.5	631.5	1189.1	0.02195	0.40058
1150.0	1135.3	561.82	564.8	622.2	1187.0	0.02214	0.38073
1200.0	1185.3	567.19	571.9	613.0	1184.8	0.02232	0.36245
1250.0	1235.3	572.38	578.8	603.8	1182.6	0.02250	0.34556
1300.0	1285.3	577.42	585.6	594.6	1180.2	0.02269	0.32991
1350.0	1335.3	582.32	592.2	585.6	1177.8	0.02288	0.31536
1400.0	1385.3	587.07	598.8	567.5	1175.3	0.02307	0.30178
1450.0	1435.3	591.70	605.3	567.6	1172.9	0.02327	0.28909
1500.0	1485.3	596.20	611.7	558.4	1170.1	0.02346	0.27719
1600.0	1585.3	604.87	624.2	540.3	1164.5	0.02387	0.25545
1700.0	1685.3	613.13	636.5	522.2	1158.6	0.02428	0.23607
1800.0	1785.3	621.02	648.5	503.8	1152.3	0.02472	0.21861
1900.0	1885.3	628.56	660.4	485.2	1145.6	0.02517	0.20278
2000.0	1985.3	635.80	672.1	466.2	1138.3	0.02565	0.18831
2100.0	2085.3	642.76	683.8	446.7	1130.5	0.02615	0.17501
2200.0	2185.3	649.45	695.5	426.7	1122.2	0.02669	0.16272
2300.0	2285.3	655.89	707.2	406.0	1113.2	0.02727	0.15133
2400.0	2385.3	662.11	719.0	384.8	1103.7	0.02790	0.14076
2500.0	2485.3	668.11	731.7	361.6	1093.3	0.02859	0.13068
2600.0	2585.3	673.91	744.5	337.6	1082.0	0.02938	0.12110
2700.0	2685.3	679.53	757.3	312.3	1069.7	0.03029	0.11194
2800.0	2785.3	684.96	770.7	285.1	1055.8	0.03134	0.10305
2900.0	2885.3	690.22	785.1	254.7	1039.8	0.03262	0.09420
3000.0	2985.3	695.33	801.8	218.4	1020.3	0.03428	0.08500
3100.0	3085.3	700.28	824.0	169.3	993.3	0.03681	0.07452
3200.0	3185.3	705.08	875.5	56.1	931.6	0.04472	0.05663
3208.2	3193.5	705.47	906.0	0.0	906.0	0.05078	0.05078

Source: Ref. [7].

TABLE A.9 Viscosities of Steam and Water

Temperature (°F)	1 psia	2 psia	5 psia	10 psia	20 psia	50 psia	100 psia	200 psia	500 psia	1000 psia	2000 psia	5000 psia	7500 psia	10000 psia	12000 psia
	Viscosity of steam and water in centipoise														
Saturated water	0.667	0.524	0.388	0.313	0.255	0.197	0.164	0.138	0.111	0.094	0.078	...	...	...	...
Saturated steam	0.010	0.010	0.011	0.012	0.012	0.013	0.014	0.015	0.017	0.019	0.023	...	...	...	...
1500°	0.041	0.041	0.041	0.041	0.041	0.041	0.041	0.041	0.042	0.042	0.042	0.044	0.046	0.048	0.050
1450	0.040	0.040	0.040	0.040	0.040	0.040	0.040	0.040	0.040	0.041	0.041	0.043	0.045	0.047	0.049
1400	0.039	0.039	0.039	0.039	0.039	0.039	0.039	0.039	0.039	0.040	0.040	0.042	0.044	0.047	0.049
1350	0.038	0.038	0.038	0.038	0.038	0.038	0.038	0.038	0.038	0.038	0.039	0.041	0.044	0.046	0.049
1300	0.037	0.037	0.037	0.037	0.037	0.037	0.037	0.037	0.037	0.037	0.038	0.040	0.043	0.045	0.048
1250	0.035	0.035	0.035	0.035	0.035	0.035	0.035	0.036	0.036	0.036	0.037	0.039	0.042	0.045	0.048
1200	0.034	0.034	0.034	0.034	0.034	0.034	0.034	0.034	0.035	0.035	0.036	0.038	0.041	0.045	0.048
1150	0.034	0.034	0.034	0.034	0.034	0.034	0.034	0.034	0.034	0.034	0.034	0.037	0.041	0.045	0.049
1100	0.032	0.032	0.032	0.032	0.032	0.032	0.032	0.032	0.033	0.033	0.034	0.037	0.040	0.045	0.050
1050	0.031	0.031	0.031	0.031	0.031	0.031	0.031	0.031	0.032	0.032	0.033	0.036	0.040	0.047	0.052
1000	0.030	0.030	0.030	0.030	0.030	0.030	0.030	0.030	0.030	0.031	0.032	0.035	0.041	0.049	0.055
950	0.029	0.029	0.029	0.029	0.029	0.029	0.029	0.029	0.029	0.030	0.031	0.035	0.042	0.052	0.059
900	0.028	0.028	0.028	0.028	0.028	0.028	0.028	0.028	0.028	0.028	0.029	0.035	0.045	0.057	0.064
850	0.026	0.026	0.026	0.026	0.026	0.026	0.027	0.027	0.027	0.027	0.028	0.035	0.052	0.064	0.070
800	0.025	0.025	0.025	0.025	0.025	0.025	0.025	0.025	0.026	0.026	0.027	0.040	0.062	0.071	0.075
750	0.024	0.024	0.024	0.024	0.024	0.024	0.024	0.024	0.025	0.025	0.026	0.057	0.071	0.078	0.081
700	0.023	0.023	0.023	0.023	0.023	0.023	0.023	0.023	0.023	0.024	0.026*	0.071	0.079	0.085	0.086
650	0.022	0.022	0.022	0.022	0.022	0.022	0.022	0.022	0.023	0.023	0.023	0.082	0.088	0.092	0.096
600	0.021	0.021	0.021	0.021	0.021	0.021	0.021	0.021	0.021	0.021	0.087	0.091	0.096	0.101	0.104
550	0.020	0.020	0.020	0.020	0.020	0.020	0.020	0.020	0.020	0.019	0.095	0.101	0.105	0.109	0.113
500	0.019	0.019	0.019	0.019	0.019	0.019	0.019	0.018	0.018	0.103	0.105	0.111	0.114	0.119	0.122
450	0.018	0.018	0.018	0.018	0.017	0.017	0.017	0.017	0.115	0.116	0.118	0.123	0.127	0.131	0.135
400	0.016	0.016	0.016	0.016	0.016	0.016	0.016	0.016	0.131	0.132	0.134	0.138	0.143	0.147	0.150
350	0.015	0.015	0.015	0.015	0.015	0.015	0.015	0.152	0.153	0.154	0.155	0.160	0.164	0.168	0.171
300	0.014	0.014	0.014	0.014	0.014	0.014	0.182	0.183	0.183	0.184	0.185	0.190	0.194	0.198	0.201
250	0.013	0.013	0.013	0.013	0.013	0.228	0.228	0.228	0.228	0.229	0.231	0.235	0.238	0.242	0.245
200	0.012	0.012	0.012	0.012	0.300	0.300	0.300	0.300	0.301	0.301	0.303	0.306	0.310	0.313	0.316
150	0.011	0.011	0.427	0.427	0.427	0.427	0.427	0.427	0.427	0.428	0.429	0.431	0.434	0.437	0.439
100	0.680	0.680	0.680	0.680	0.680	0.680	0.680	0.680	0.680	0.680	0.680	0.681	0.682	0.683	0.683
50	1.299	1.299	1.299	1.299	1.299	1.299	1.299	1.299	1.299	1.298	1.296	1.289	1.284	1.279	1.275
32	1.753	1.753	1.753	1.753	1.753	1.753	1.753	1.752	1.751	1.749	1.745	1.733	1.723	1.713	1.705

Values directly below underscored viscosities are for water.
*Critical point.
Source: Ref. [8].

TABLE A.10 Properties of Carbon Dioxide at One Atmosphere

T (K)	*T* (°C)	ρ (kg/m^3)	$\beta \times 10^3$ (1/K)	C_P (J/kg · K)	*k* (W/m · K)	$\alpha \times 10^4$ (m^2/s)	$\mu \times 10^6$ (N · s/m^2)	$\nu \times 10^6$ (m^2/s)	*Pr*	$g\beta/\nu^2 \times 10^{-6}$ (1/K · m^3)
220	−53	2.4733	–	783	0.010805	0.0558	11.105	4.490	0.818	–
250	−23	2.1657	–	804	0.012884	0.0740	12.590	5.813	0.793	–
300	27	1.7973	3.33	871	0.016572	0.1059	14.958	8.321	0.770	472
350	77	1.5362	2.86	900	0.02047	0.1481	17.205	11.19	0.755	224
400	127	1.3424	2.50	942	0.02461	0.1946	19.32	14.39	0.738	118
450	177	1.1918	2.22	980	0.02897	0.2480	21.34	17.90	0.721	67.9
500	227	1.0732	2.00	1013	0.03352	0.3083	23.26	21.67	0.702	41.8
550	277	0.9739	1.82	1047	0.03821	0.3747	25.08	25.74	0.685	26.9
600	327	0.8938	1.67	1076	0.04311	0.4483	26.83	30.02	0.668	18.2

Source: Refs. [1] and [2].

TABLE A.11 Properties of Ammonia Vapor at Atmospheric Pressure

T (k)	ρ (kg/m^3)	C_P (J/kg · K)	$\mu \times 10^7$ (N · s/m^2)	$\nu \times 10^6$ (m^2/s)	$k \times 10^3$ (W/m · K)	$\alpha \times 10^6$ (m^2/s)	*Pr*
300	0.6894	2158	101.5	14.7	24.7	16.6	0.887
320	0.6448	2170	109	16.9	27.2	19.4	0.870
340	0.6059	2192	116.5	19.2	29.3	22.1	0.872
360	0.5716	2221	124	21.7	31.6	24.9	0.872
380	0.5410	2254	131	24.2	34.0	27.9	0.869
400	0.5136	2287	138	26.9	37.0	31.5	0.853
420	0.4888	2322	145	29.7	40.4	35.6	0.833
440	0.4664	2357	152.5	32.7	43.5	39.6	0.826
460	0.4460	2393	159	35.7	46.3	43.4	0.822
480	0.4273	2430	166.5	39.0	49.2	47.4	0.822
500	0.4101	2467	173	42.2	52.5	51.9	0.813
520	0.3942	2504	180	45.7	54.5	55.2	0.827
540	0.3795	2540	186.5	49.1	57.5	59.7	0.824
560	0.3708	2577	193	52.0	60.6	63.4	0.827
580	0.3533	2613	199.5	56.5	63.8	69.1	0.817

Source: Refs. [9] and [10].

TABLE A.12 Properties of Saturated Liquid Freon 12 ($C\,Cl_2\,F_2$)

T (K)	*T* (°C)	ρ (kg/m^3)	$\beta \times 10^3$ (1/K)	C_P (J/kg · K)	*k* (W/m · K)	$\alpha \times 10^8$ (m^2/s)	$\mu \times 10^4$ (N · s/m^2)	$\nu \times 10^6$ (m^2/s)	*Pr*	$g\beta/\nu^2 \times 10^{-10}$ (1/K · m^3)
223	−50	1547	2.63	875.0	0.067	5.01	4.796	0.310	6.2	26.84
233	−40	1519		884.7	0.069	5.14	4.238	0.279	5.4	
243	−30	1490		895.6	0.069	5.26	3.770	0.253	4.8	
253	−20	1461		907.3	0.071	5.39	3.433	0.235	4.4	
263	−10	1429		920.3	0.073	5.50	3.158	0.221	4.0	
273	0	1397	3.10	934.5	0.073	5.57	2.990	0.214	3.8	6.68
283	10	1364		949.6	0.073	5.60	2.769	0.203	3.6	
293	20	1330		965.9	0.073	5.60	2.633	0.198	3.5	
303	30	1295		983.5	0.071	5.60	2.512	0.194	3.5	
313	40	1257		1001.9	0.069	5.55	2.401	0.191	3.5	
323	50	1216		1021.6	0.067	5.45	2.310	0.190	3.5	

Source: Refs. [1] and [2].

TABLE A.13 Properties of Selected Organic Liquids at 20°C

Liquid	Chemical formula	ρ (kg/m^3)	$\beta \times 10^4$ (1/K)	C_P (J/kg · K)	k (W/m · K)	$\alpha \times 10^9$ (m^2/s)	$\mu \times 10^4$ (N · s/m^2)	$\nu \times 10^6$ (m^2/s)	Pr	$g\beta/\nu^2 \times 10^{-8}$ (1/K · m^3)
Acetic acid	$C_2H_4O_2$	1049	10.7	2031	0.193	90.6	-	-	-	-
Acetone	C_3H_6O	791	14.3	2160	0.180	105.4	3.31	0.418	3.97	802.6
Aniline	C_6H_7N	1022	8.5	2064	0.172	81.5	44.3	4.34	53.16	4.43
Benzene	C_6H_6	879	10.6	1738	0.154	100.8	6.5	0.739	7.34	190.3
n-Butyl alcohol	$C_4H_{10}O$	810	8.1	2366	0.167	87.1	29.5	3.64	41.79	5.99
Chloroform	$CHCl_3$	1489	12.8	967	0.129	89.6	5.8	0.390	4.35	825.3
Ethyl acetate	$C_4H_8O_2$	900	13.8	2010	0.137	75.7	4.49	0.499	6.59	543.5
Ethyl alcohol	C_2H_6O	790	11.0	2470	0.182	93.3	12.0	1.52	16.29	46.7
Ethylene glycol	$C_2H_6O_2$	1115	-	2382	0.258	97.1	199	17.8	183.7	-
Glycerine	$C_3H_8O_3$	1260	5.0	2428	0.285	93.2	14,800	1175	12,609	0.0000355
n-Heptane	C_7H_{16}	684	12.4	2219	0.140	92.2	4.09	0.598	6.48	340.1
n-Hexane	C_6H_{14}	660	13.5	1884	0.137	110.2	3.20	0.485	4.40	562.8
Isobutyl alcohol	$C_4H_{10}O$	804	9.4	2303	0.134	72.4	39.5	4.91	67.89	3.82
Methyl alcohol	CH_4O	792	11.9	2470	0.212	108.4	5.84	0.737	6.8	214.9
n-Octane	C_8H_{18}	720	11.4	2177	0.147	93.8	5.4	0.750	8.00	198.8
n-Pentane	C_5H_{12}	626	16.0	2179	0.136	99.8	2.29	0.366	3.67	1171
Toluene	C_7H_8	866	10.8	1675	0.151	104.1	5.86	0.677	6.5	231.1
Turpentine	$C_{10}H_{16}$	855	9.7	1800	0.128	83.2	14.87	1.74	20.91	31.4

Source: Refs. [1] and [3].

TABLE A.14 Properties of Selected Saturated Liquids

T (K)	ρ (kg/m^3)	C_P (kJ/kg · K)	$\mu \times 10^2$ (N · s/m^2)	$\nu \times 10^6$ (m^2/s)	$k \times 10^3$ (W/m · K)	$\alpha \times 10^7$ (m^2/s)	Pr	$\beta \times 10^3$ (K^{-1})
Engine oil (unused)								
273	899.1	1.796	385	4280	147	0.910	47,000	0.70
280	895.3	1.827	217	2430	144	0.880	27,500	0.70
290	890.0	1.868	99.9	1120	145	0.872	12,900	0.70
300	884.1	1.909	48.6	550	145	0.859	6400	0.70
310	877.9	1.951	25.3	288	145	0.847	3400	0.70
320	871.8	1.993	14.1	161	143	0.823	1965	0.70
330	865.8	2.035	8.36	96.6	141	0.800	1205	0.70
340	859.9	2.076	5.31	61.7	139	0.779	793	0.70
350	853.9	2.118	3.56	41.7	138	0.763	546	0.70
360	847.8	2.161	2.52	29.7	138	0.753	395	0.70
370	841.8	2.206	1.86	22.0	137	0.738	300	0.70
380	836.0	2.250	1.41	16.9	136	0.723	233	0.70
390	830.6	2.294	1.10	13.3	135	0.709	187	0.70
400	825.1	2.337	0.874	10.6	134	0.695	152	0.70
410	818.9	2.381	0.698	8.52	133	0.682	125	0.70
420	812.1	2.427	0.564	6.94	133	0.675	103	0.70
430	806.5	2.471	0.470	5.83	132	0.662	88	0.70
Ethylene glycol [$C_2H_4(OH)_2$]								
273	1130.8	2.294	6.51	57.6	242	0.933	617	0.65
280	1125.8	2.323	4.20	37.3	244	0.933	400	0.65
290	1118.8	2.368	2.47	22.1	248	0.936	236	0.65
300	1114.4	2.415	1.57	14.1	252	0.939	151	0.65
310	1103.7	2.460	1.07	9.65	255	0.939	103	0.65
320	1096.2	2.505	0.757	6.91	258	0.940	73.5	0.65
330	1089.5	2.549	0.561	5.15	260	0.936	55.0	0.65
340	1083.8	2.592	0.431	3.98	261	0.929	42.8	0.65
350	1079.0	2.637	0.342	3.17	261	0.917	34.6	0.65
360	1074.0	2.682	0.278	2.59	261	0.906	28.6	0.65
370	1066.7	2.728	0.228	2.14	262	0.900	23.7	0.65
373	1058.5	2.742	0.215	2.03	263	0.906	22.4	0.65
Glycerin [$C_3H_5(OH)_3$]								
273	1276.0	2.261	1060	8310	282	0.977	85,000	0.47
280	1271.9	2.298	534	4200	284	0.972	43,200	0.47
290	1265.8	2.367	185	1460	286	0.955	15,300	0.48
300	1259.9	2.427	79.9	634	286	0.935	6780	0.48
310	1253.9	2.490	35.2	281	286	0.916	3060	0.49
320	1247.2	2.564	21.0	168	287	0.897	1870	0.50

Source: Refs. [2] and [9].

TABLE A.15 Thermal Conductivities of Liquids

Liquid	T (°F)	k (Btu/h · ft · °F)	Liquid	T (°F)	k (Btu/h · ft · °F)
Acetic acid 100%	68	0.099	Ethyl alcohol 40%	68	0.224
Acetic acid 50%	68	0.20	Ethyl alcohol 20%	68	0.281
Acetone	86	0.102	Ethyl alcohol 100%	122	0.087
	167	0.095	Ethyl benzene	86	0.086
Allyl alcohol	77–86	0.104		140	0.082
Ammonia	5–86	0.29	Ethyl bromide	68	0.070
Ammonia, aqueous 26%	68	0.261	Ethyl ether	86	0.080
	140	0.29		167	0.078
Amyl acetate	50	0.083	Ethyl iodide	104	0.064
Amyl alcohol (n-)	86	0.094		167	0.063
	212	0.089	Ethylene glycol	32	0.153
Amyl alcohol (iso-)	86	0.088	Gasoline	86	0.078
	167	0.087	Glycerol 100%	68	0.164

TABLE A.15 Thermal Conductivities of Liquids—cont'd

Liquid	T (°F)	k (Btu/h · ft · °F)	Liquid	T (°F)	k (Btu/h · ft · °F)
Aniline	32–68	0.100	Glycerol 80%	68	0.189
Benzene	86	0.092	Glycerol 60%	68	0.220
	140	0.087	Glycerol 40%	68	0.259
Bromobenzene	86	0.074	Glycerol 20%	68	0.278
	212	0.070	Glycerol 100%	212	0.164
Butyl acetate (n-)	77–86	0.085	Heptane (n-)	86	0.081
Butyl alcohol (n-)	86	0.097		140	0.079
	167	0.095	Hexane (n-)	86	0.080
				140	0.078
Butyl alcohol (iso-)	50	0.091	Heptyl alcohol (n-)	86	0.094
Calcium chloride brine				167	0.091
30%	86	0.32	Hexyl alcohol (n-)	86	0.093
15%	86	0.34		167	0.090
Carbon disulfide	86	0.093	Kerosene	68	0.086
	167	0.088		167	0.081
Carbon tetrachloride	32	0.107	Mercury	82	4.83
	154	0.094	Methyl alcohol 100%	68	0.124
Chlorobenzene	50	0.083	Methyl alcohol 80%	68	0.154
Chloroform	86	0.080	Methyl alcohol 60%	68	0.190
Cymene (para-)	86	0.078	Methyl alcohol 40%	68	0.234
	140	0.079	Methyl alcohol 20%	68	0.284
Decane (n-)	86	0.085	Methyl alcohol 100%	122	0.114
	140	0.083	Methyl chloride	5	0.111
				86	0.089
Dichlorodifluoromethane	20	0.057	Nitrobenzene	86	0.095
	60	0.053		212	0.038
	100	0.048	Nitromethane	86	0.125
	140	0.043		140	0.120
	180	0.038			
Dichloroethane	122	0.082	Nonane (n-)	86	0.084
Dichloromethane	5	0.111		140	0.082
	86	0.096	Octane (n-)	86	0.083
Ethyl acetate	68	0.101		140	0.081
Ethyl alcohol 100%	68	0.105	Sulfuric acid 90%	86	0.21
Ethyl alcohol 80%	68	0.137	Sulfuric acid 60%	86	0.25
Ethyl alcohol 60%	68	0.176	Sulfuric acid 30%	86	0.30
Oils	86	0.079	Sulfur dioxide	5	0.128
Oils, castor	68	0.104		86	0.111
	212	0.100	Toluene	86	0.086
Oils, Olive	68	0.097		167	0.084
	212	0.095	β-Trichloroethane	122	0.077
Paraldehyde	86	0.084	Trichloroethylene	122	0.080
	212	0.078	Turpentine	59	0.074
Pentane (n-)	86	0.078	Vaseline	59	0.106
	167	0.074	Water	32	0.343
Perchloroethylene	122	0.092		100	0.363
Petroleum ether	86	0.075		200	0.393
	167	0.073		300	0.395
Propyl alcohol (n-)	86	0.099		420	0.376
	167	0.095		620	0.275
Propyl alcohol (iso-)	86	0.091	Xylene (ortho-)	68	0.090
	140	0.090	Xylene (meta-)	68	0.090
Sodium	212	49			
	410	46			
Sodium chloride brine					
25.0%	86	0.33			
12.5%	86	0.34			

Source: Ref. [11].

TABLE A.16 Thermal Conductivities of Tubing Materials

Material	k (Btu/h · ft · °F)	Material	k (Btu/h · ft · °F)
Carbon steel	24–30	Inconel 800	6.7–8
304 Stainless steel	8.6–12	Inconel 825	7.2
309 Stainless steel	29	Hastelloy B	6.1–9
310 Stainless steel	7.3–11	Hastelloy C	5.9–10
316 and 317 Stainless steel	7.7–12	Alloy 904L	7.5–9
321 and 347 Stainless steel	8–12	Alloy 28	6.5–9
25Cr–12Ni Steel	6.5–10	Cr–Mo Alloy XM–27	11.3
22Cr–5Ni–3Mo Steel	9.5	Alloy 20CB	7.6
3.5Ni Steel	23.5	Copper	225
Carbon–0.5Mo Steel	25	90–10 Cu–Ni	30
1.0 & 1.25Cr–0.5Mo Steel	21.5	70–30 Cu–Ni	18
2.25Cr–1.0Mo Steel	21	Admirality brass	64–75
5Cr–0.5Mo Steel	16.9–19	Naval brass	71–74
12Cr & 13Cr Steel	15.3	Muntz metal (60Cu–40Zn)	71
15Cr Steel	14.4	Aluminum bronze	71
17Cr Steel	13	Al–Ni Bronze	72
Nickel alloy 200	38.5	Aluminum alloy 3003	102–106
Nickel alloy 400	12.6–15	Aluminum alloy 6061	96–102
Inconel 600	9	Titanium	11.5–12.7
Inconel 625	7.5–9	Zirconium	12

This table lists typical values of thermal conductivity that can be used to estimate the thermal resistance of tube and pipe walls. These values may not be appropriate for operation at very high or very low temperatures.

TABLE A.17 Latent Heats of Vaporization of Organic Compounds

	Formula	Temperature (°C)	λ (cal./g)
Hydrocarbon compounds			
Paraffins			
Methane	CH_4	−161.6	121.87
Ethane	C_2H_6	−88.9	116.87
Propane	C_3H_8	25	81.76
		−42.1	101.76
n-Butane	C_4H_{10}	25	86.63
		−0.50	92.09
2-Methylpropane (isobutane)	C_4H_{10}	25	78.63
		−11.72	87.56
n-Pentane	C_5H_{12}	25	87.54
		36.08	85.38
2-Methylbutane (isopentane)	C_5H_{12}	25	81.47
		27.86	80.97
2,2-Dimethylpropane (neopentane)	C_5H_{12}	25	72.15
		9.45	75.37
n-Hexane	C_6H_{14}	25	87.50
		68.74	80.48
2-Methylpentane	C_6H_{14}	25	82.83
		60.27	76.89
3-Methylpentane	C_6H_{14}	25	83.96
		63.28	78.42
2,2-Dimethylbutane	C_6H_{14}	25	76.79
		49.74	73.75
2,3-Dimethylbutane	C_6H_{14}	25	80.77
		57.99	76.53
n-Heptane	C_7H_{16}	25	87.18
		98.43	76.45
2-Methylhexane	C_7H_{16}	25	83.02
		90.05	73.4
3-Methylhexane	C_7H_{16}	25	83.68
		91.95	74.1
3-Ethylpentane	C_7H_{16}	25	84.02
		93.47	74.3

TABLE A.17 Latent Heats of Vaporization of Organic Compounds—cont'd

	Formula	Temperature (°C)	λ (cal./g)
2,2-Dimethylpentane	C_7H_{16}	25	77.36
		79.20	69.7
2,3-Dimethylpentane	C_7H_{16}	25	81.68
		89.79	72.9
2,4-Dimethylpentane	C_7H_{16}	25	78.44
		80.51	70.9
3,3-Dimethylpentane	C_7H_{16}	25	78.76
		86.06	70.6
2,2,3-Trimethylbutane	C_7H_{16}	25	76.42
		80.88	69.3
n-Octane	C_8H_{18}	25	86.80
		125.66	73.19
2-Methylheptane	C_8H_{18}	25	83.02
		117.64	70.3
3-Methylheptane	C_8H_{18}	25	83.35
		118.92	71.3
4-Methylheptane	C_8H_{18}	25	83.01
		117.71	70.91
3-Ethylhexane	C_8H_{18}	25	82.95
		118.53	71.7
2,2-Dimethylhexane	C_8H_{18}	25	78.02
		106.84	67.7
2,3-Dimethylhexane	C_8H_{18}	25	81.17
		115.60	70.2
2,4-Dimethylhexane	C_8H_{18}	25	79.02
		109.43	68.5
2,5-Dimethylhexane	C_8H_{18}	25	79.21
		109.10	68.6
3,3-Dimethylhexane	C_8H_{18}	25	78.54
		111.97	68.5
3,4-Dimethylhexane	C_8H_{18}	25	81.55
		117.72	70.2
2-Methyl-3-ethylpentane	C_8H_{18}	25	80.60
		115.65	69.7
3-Methyl-3-ethylpentane	C_8H_{18}	25	79.49
		118.26	69.3
2,2,3-Trimethylpentane	C_8H_{18}	25	77.24
		109.84	67.3
2,2,4-Trimethylpentane	C_8H_{18}	25	73.50
		99.24	64.87
2,3,3-Trimethylpentane	C_8H_{18}	25	77.87
		114.76	68.1
2,3,4-Trimethylpentane	C_8H_{18}	25	78.90
		113.47	68.37
2,2,3,3-Tetramethylbutane	C_8H_{18}	106.30	66.2
Alkyl benzenes			
Benzene	C_6H_6	25	103.57
		80.10	94.14
Methylbenzene (toluene)	C_7H_8	25	98.55
		110.62	86.8
Ethylbenzene	C_8H_{10}	25	95.11
		136.19	81.0
1,2-Dimethylbenzene (*o*-xylene)	C_8H_{10}	25	97.79
		144.42	82.9
1,3-Dimethylbenzene (*m*-xylene)	C_8H_{10}	25	96.03
		139.10	82.0
1,4-Dimethylbenzene (*p*-xylene)	C_8H_{10}	25	95.40
		138.35	81.2
n-Propylbenzene	C_9H_{12}	25	91.93
		159.22	76.0
Isopropylbenzene	C_9H_{12}	25	89.77
		152.40	74.6

(*Continued*)

TABLE A.17 Latent Heats of Vaporization of Organic Compounds—cont'd

	Formula	Temperature (°C)	λ (cal./g)
l-Methyl-2-ethylbenzene	C_9H_{12}	25	94.9
		165.15	77.3
l-Methyl-3-ethylbenzene	C_9H_{12}	25	93.3
		161.30	76.6
l-Methyl-4-ethylbenzene	C_9H_{12}	25	92.7
		162.05	76.4
1,2,3-Trimethylbenzene	C_9H_{12}	25	97.56
		176.15	79.6
1,2,4-Trimethylbenzene (pseudocumene)	C_9H_{12}	25	95.33
		169.25	78.0
1,3,5-Trimethylbenzene (mesitylene)	C_9H_{12}	25	94.40
		164.70	77.6
Alkyl cyclopentanes			
Cyclopentane	C_5H_{10}	25	97.1
		49.26	93.1
Methylcyclopentane	C_6H_{12}	25	89.83
		71.81	83.2
Ethylcyclopentane	C_7H_{14}	25	88.6
		103.45	78.3
1,1-Dimethylcyclopentane	C_7H_{14}	25	82.5
		87.5	74.6
cis-1,2-Dimethylcyclopentane	C_7H_{14}	25	86.4
		99.3	77.0
trans-1,2-Dimethylcyclopentane	C_7H_{14}	25	83.9
		91.9	75.5
trans-1,3-Dimethylcyclopentane	C_7H_{14}	25	83.6
		90.8	75.3
Alkyl cyclohexanes			
Cyclohexane	C_6H_{12}	25	93.81
		80.74	85.6
Methylcyclohexane	C_7H_{14}	25	86.07
		100.94	76.9
Ethylcyclohexane	C_8H_{16}	25	86.21
		131.79	73.7
1,1-Dimethylcyclohexane	C_8H_{16}	25	80.9
		119.50	70.7
cis-1,2-Dimethylcyclohexane	C_8H_{16}	25	84.59
		129.73	72.9
trans-1,2-Dimethylcyclohexane	C_8H_{16}	25	81.70
		123.42	71.1
cis-1,3-Dimethylcyclohexane	C_8H_{16}	25	83.49
		124.45	72.1
trans-1,3-Dimethylcyclohexane	C_8H_{16}	25	81.42
		120.09	70.9
cis-1,4-Dimethylcyclohexane	C_8H_{16}	25	83.13
		124.32	71.9
trans-1,4-Dimethylcyclohexane	C_8H_{16}	25	80.67
		119.35	70.4
Monoolefins			
Ethene (ethylene)	C_2H_4	−103.71	115.39
Propene (propylene)	C_3H_6	−47.70	104.62
1-Butene	C_4H_8	25	86.8
		−6.25	93.36
cis-2-Butene	C_4H_8	25	94.5
		3.72	99.46
trans-2-Butene	C_4H_8	25	91.8
		0.88	96.94
2-Methylpropene (isobutene)	C_4H_8	25	87.7
		−6.90	94.22
Non-hydrocarbon compounds			
Acetal	$C_6H_{14}O_2$	102.9	66.18
Acetaldehyde	C_2H_4O	21	136.17

TABLE A.17 Latent Heats of Vaporization of Organic Compounds—cont'd

	Formula	Temperature (°C)	λ (cal./g)
Acetic acid	$C_2H_4O_2$	118.3	96.75
		140	94.37
		220	81.23
		321	0
Acetic anhydride	$C_4H_6O_3$	137	92.2
Acetone	C_3H_6O	0	134.74
		20	131.87
		40	128.05
		60	123.51
		80	118.26
		100	112.76
		235	0
Acetonitrile	C_2H_3N	80	173.68
Acetophenone	C_8H_8O	203.7	77.16
Acetyl chloride	C_3H_3ClO	51	78.84
Air	-	-	51.0
Allyl alcohol	C_3H_6O	96	163.41
Amyl alcohol (*n*-)	$C_5H_{11}OH$	131	120.17
Amyl alcohol (*t*-)	$C_5H_{11}OH$	102	105.83
Amyl amine (*n*-)	$C_5H_{13}N$	95	98.67
Amyl bromide (*n*-)	$C_5H_{11}Br$	129	48.26
Amyl ether (*n*-)	$C_{10}H_{22}O$	170	69.52
Amyl iodide (*n*-)	$C_5H_{11}I$	155	47.54
Amyl methyl ketone (*n*-)	$C_7H_{14}O$	149.2	82.66
Amylene	C_5H_{10}	12.5	75.01
Anethole (*p*-)	$C_{10}H_{12}O$	232	71.43
Aniline	C_6H_7N	183	103.68
Benzaldehyde	C_7H_6O	179	86.48
Benzonitrile	C_7H_5N	189	87.68
Benzyl alcohol	C_7H_8O	204.3	112.28
Butyl acetate (*n*-)	$C_6H_{12}O_2$	124	73.82
Butyl alcohol (*n*-)	$C_4H_{10}O$	116.8	141.26
Butyl alcohol (*s*-)	$C_4H_{10}O$	98.1	134.38
Butyl alcohol (*t*-)	$C_4H_{10}O$	83	130.44
Butyl formate	$C_4H_{10}O_2$	105.1	86.74
Butyl methyl ketone (*n*-)	$C_6H_{12}O$	127	82.42
Butyl propionate (*n*-)	$C_7H_{14}O_2$	144.9	71.74
Butyric acid (*n*-)	$C_4H_8O_2$	163.5	113.96
Butyronitrile (*n*-)	C_4H_7N	117.4	114.91
Bromobenzene	C_6H_5Br	155.9	57.60
Capronitrile	$C_6H_{11}N$	156	88.15
Carbon disulfide	CS_2	0	89.35
		46.25	84.09
		100	75.49
		140	67.37
Carbon tetrachloride	CCl_4	0	52.06
		76.75	46.42
		200	32.73
Carvacrol	$C_{10}H_{14}O$	237	68.09
Chloral	C_2HCl_3O	-	53.99
Chloral hydrate	$C_2H_3Cl_3O_2$	96	131.87
Chlorobenzene	C_6H_5Cl	130.6	77.59
Chloroethyl alcohol (2-)	C_2H_5ClO	126.5	122.94
Chloroethyl acetate (*β*-)	$C_4H_7ClO_2$	141.5	80.75
Chloroform	$CHCl_3$	0	64.74
		40	60.92
		61.5	59.01
		100	55.19
		260	0
Chlorotoluene (*o*-)	C_7H_7Cl	158.1	72.63
Chlorotoluene (*p*-)	C_7H_7Cl	160.4	73.13
Cresol (*m*-)	C_7H_8O	202	100.58

(Continued)

TABLE A.17 Latent Heats of Vaporization of Organic Compounds—cont'd

	Formula	Temperature (°C)	λ (cal./g)
Cyanogen	$(CN)_2$	0	102.97
Cyanogen chloride	CNCl	13	134.98
Cyclohexanol	$C_6H_{12}O$	161.1	108.22
Cycohexyl chloride	$C_6H_{11}Cl$	142.0	74.78
Dichloroacetic acid	$C_2H_2Cl_2O_2$	194.4	77.16
Dichlorodifluormethane	CCl_2F_2	−29.8	40.40
Diethylamine	$C_4H_{11}N$	58	91.02
Diethyl carbonate	$C_5H_{10}O_3$	126	73.10
Diethyl ketone	$C_5H_{10}O$	101	90.78
Diethyl oxalate	$C_6H_{10}O_4$	185	67.61
Di-isobutylamine	$C_8H_{19}N$	134	65.70
Dimethyl aniline	$C_8H_{11}N$	193	80.75
Dimethyl carbonate	$C_3H_6O_3$	90	88.15
Dipropyl ketone	$C_7H_{14}O$	143.5	75.73
Dipropylamine (*n*-)	$C_6H_{15}N$	108	75.73
Ethyl acetate	$C_4H_8O_2$	0.0	102.01
Ethyl alcohol	C_2H_6O	78.3	204.26
Ethylamine	C_2H_7N	15	145.97
Ethyl benzoate	$C_9H_{10}O_2$	213	64.50
Ethyl bromide	C_2H_5Br	38.4	59.92
Ethyl butyrate (*n*-)	$C_6H_{12}O_2$	118.9	74.68
Ethyl caprylate	$C_{10}H_{20}O_2$	207	60.44
Ethyl chloride	C_2H_5Cl	4.7	92.93
		15.0	92.45
		20.0	92.22
		25.0	91.98
Ethylene bromide	$C_2H_4Br_2$	130.8	46.23
Ethylene chloride	$C_2H_4Cl_2$	0	85.29
		82.3	77.33
Ethylene glycol	$C_2H_6O_2$	197	191.12
Ethylene oxide	C_2H_4O	13	138.56
Ethyl ether	$C_4H_{10}O$	34.6	83.85
Ethyl formate	$C_3H_6O_2$	53.3	97.18
Ethyl iodide	C_2H_5I	71.2	45.61
Ethylidine chloride	$C_2H_4Cl_2$	0.0	76.69
		60	67.13
Ethyl isobutyl ether	$C_6H_{14}O$	79.0	74.78
Ethyl isobutyrate	$C_6H_{12}O_2$	109.2	72.05
Ethyl isovalerate	$C_7H_{14}O_2$	144	67.85
Ethyl methyl ketone	C_4H_8O	78.2	105.93
Ethyl methyl ketoxime	C_4H_9NO	182	115.87
Ethyl nonylate	$C_{11}H_{22}O_2$	227	58.05
Ethyl propionate	$C_5H_{10}O_2$	97.6	80.08
Ethyl propyl ether	$C_5H_{12}O$	60	82.66
Ethyl valerate (*n*-)	$C_7H_{14}O_2$	98	77.16
Formic acid	CH_2O_2	101	119.93
Furane	C_4H_4O	31	95.32
Furfural	$C_5H_4O_2$	160.5	107.51
Heptyl alcohol (*n*-)	$C_7H_{16}O$	176	104.88
Hexylmethyl ketone	$C_8H_{16}O$	173	74.06
Hydrogen cyanide	HCN	20	210.23
Isoamyl acetate	$C_7H_{14}O_2$	143.6	69.04
Isoamyl alcohol	$C_5H_{12}O$	130.2	119.78
Isoamyl butyrate (*n*-)	$C_9H_{18}O_2$	169	61.88
Isoamyl formate	$C_6H_{12}O_2$	123	73.58
Isoamyl isobutyrate	$C_9H_{18}O_2$	168	57.57
Isoamyl propionate	$C_8H_{16}O_2$	161	65.22
Isoamyl valerate (*n*-)	$C_{10}H_{20}O_2$	187	56.14
Isobutyl acetate	$C_6H_{12}O_2$	115.3	73.75
Isobutyl alcohol	$C_4H_{10}O$	106.9	138.08
Isobutyl butyrate (*n*-)	$C_8H_{16}O_2$	157	64.50
Isobutyl formate	$C_5H_{10}O_2$	97	78.50

TABLE A.17 Latent Heats of Vaporization of Organic Compounds—cont'd

	Formula	Temperature (°C)	λ (cal./g)
Isobutyl isovalerate	$C_9H_{18}O_2$	169	60.44
Isobutyl isobutyrate	$C_8H_{16}O_2$	148	63.31
Isobutyl propionate	$C_7H_{14}O_2$	137	65.94
Isobutyl valerate (*n*-)	$C_9H_{18}O_2$	169	57.81
Isobutyric acid	$C_4H_8O_2$	154	111.57
Isopropyl alcohol	C_3H_8O	82.3	159.35
Isopropyl methyl ketone	$C_5H_{10}O$	92	89.83
Isovaleric acid	$C_5H_{10}O_2$	176.3	101.05
Limonene	$C_{10}H_{16}$	165	69.52
Mesityl oxide	$C_6H_{10}O$	128	85.77
Methyl acetate	$C_3H_6O_2$	0.0	113.96
		56.3	98.09
Methylal	$C_3H_8O_2$	42	89.83
Methyl alcohol	CH_4O	0	284.29
		64.7	262.79
		100	241.29
		160	193.51
		200	148.12
		220	109.89
		240	0
Methyl amyl ketone (*n*-)	$C_7H_{14}O$	149.2	82.66
Methyl aniline	C_7H_9N	194	95.56
Methyl butyl ketone (*n*-)	$C_6H_{12}O_2$	127	82.42
Methyl butyrate (*n*-)	$C_5H_{10}O_2$	102.6	79.79
Methyl chloride	CH_3Cl	−23.8	102.25
		15.0	96.04
		20.0	95.32
		25.0	94.60
Methyl ethyl ketone	C_4H_8O	78.2	105.93
Methyl ethyl ketoxime	C_4H_9NO	182	115.87
Methyl formate	$C_2H_4O_2$	31.3	112.35
Methyl hexyl ketone	$C_8H_{16}O$	173	74.06
Methyl iodide	CH_3I	42	45.87
Methyl isobutyrate	$C_5H_{10}O_2$	91.1	78.12
Methyl isopropyl ketone	$C_5H_{10}O$	92	89.83
Methyl isovalerate	$C_6H_{12}O_2$	116	72.39
Methyl phenyl ether	C_7H_8O	153	81.46
Methyl propionate	$C_4H_8O_2$	79.0	87.56
Methyl valerate (*n*-)	$C_6H_{12}O_2$	116	70.00
Naphthalene	$C_{10}H_8$	218	75.49
Nitrobenzene	$C_6H_5NO_2$	210	79.08
Nitromethane	CH_3NO_2	99.9	134.98
Octyl alcohol (*n*-)	$C_8H_{18}O$	196	97.47
Octyl alcohol (dl-) (sec-)	$C_8H_{18}O$	180	94.37
Phenyl methyl ether	C_7H_8O	153	81.46
Picoline (α-)	C_6H_7N	129	90.78
Piperidine	$C_5H_{11}N$	106	89.35
Propionic acid	$C_3H_6O_2$	139.3	98.81
Propionitrile	C_3H_5N	97	134.26
Propyl acetate (*n*-)	$C_5H_{10}O_2$	100.4	80.27
Propyl alcohol (*n*-)	C_3H_8O	97.2	164.36
Propyl butyrate (*n*-)	$C_7H_{14}O_2$	143.6	68.33
Propyl formate (*n*-)	$C_4H_8O_2$	80.0	88.13
Propyl isobutyrate (*n*-)	$C_7H_{14}O_2$	134	63.79
Propyl isovalerate (*n*-)	$C_8H_{16}O_2$	156	64.50
Propyl propionate (*n*-)	$C_6H_{12}O_2$	120.6	73.15
Pyridine	C_5H_5N	114.1	107.36
Salicylaldehyde	$C_7H_6O_2$	196	74.78
Tetrachloroethane (1,1,2,2-)	$C_2H_2Cl_4$	145	55.07
Tetrachloroethylene	C_2Cl_4	120.7	50.05
Toluidine (*o*-)	C_7H_9N	198	95.08
Trichloroethylene	C_2HCl_3	85.7	57.24
Valeronitrile (*n*-)	C_5H_9N	129	96.28

Source: Ref. [11].

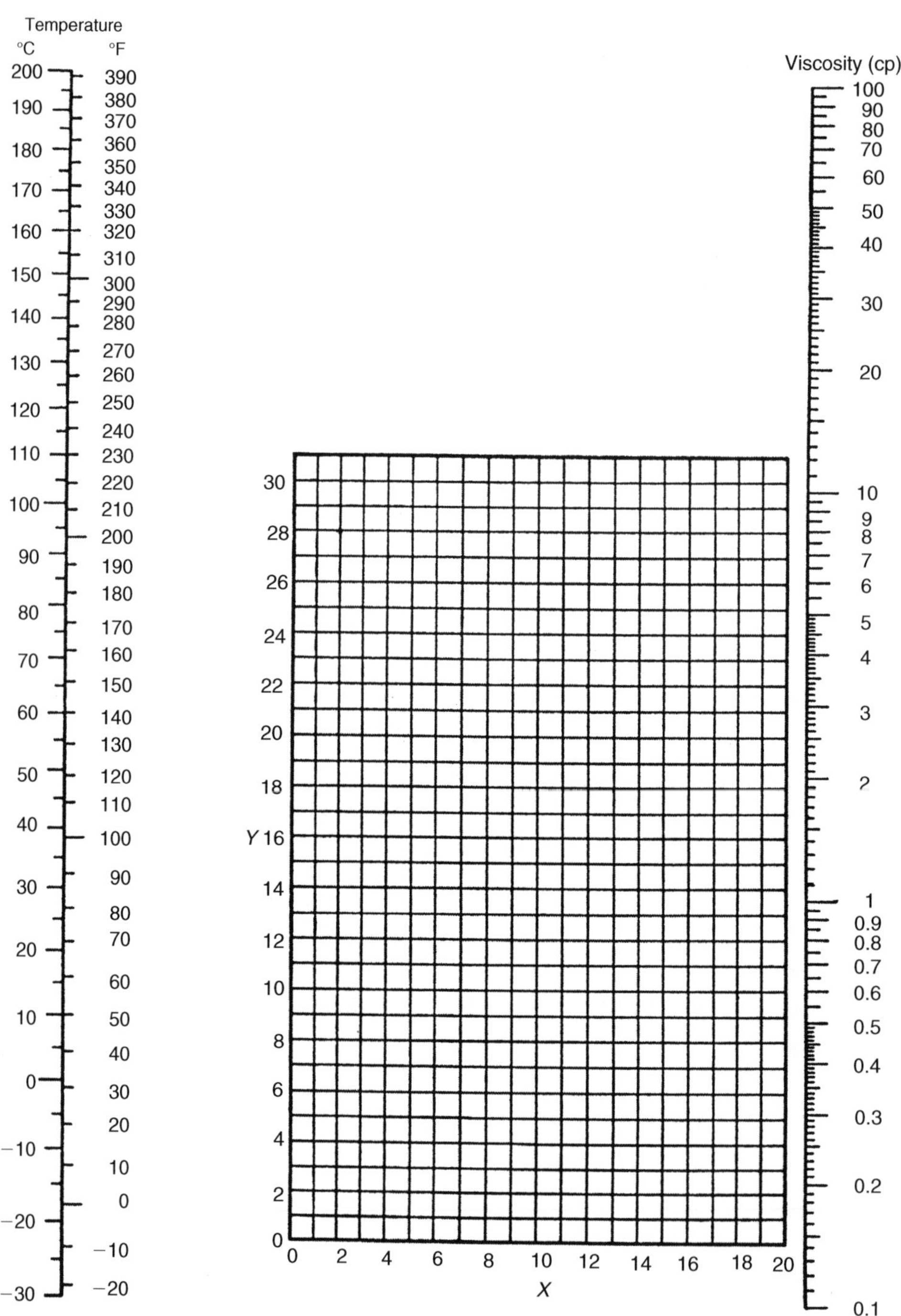

FIGURE A.1 Viscosities of liquids at atmospheric pressure (Source: Ref. [11]).

TABLE A.18 Coordinates to be Used with Figure A.1

Liquid	X	Y	Liquid	X	Y
Acetaldehyde	15.2	4.8	Ethyl alcohol, 95%	9.8	14.3
Acetic acid, 100%	12.1	14.2	Ethyl alcohol, 40%	6.5	16.6
Acetic acid, 70%	9.5	17.0	Ethyl benzene	13.2	11.5
Acetic anhydride	12.7	12.8	Ethyl bromide	14.5	8.1
Acetone, 100%	14.5	7.2	2-Ethyl butyl acrylate	11.2	14.0
Acetone, 35%	7.9	15.0	Ethyl chloride	14.8	6.0
Acetonitrile	14.4	7.4	Ethyl ether	14.5	5.3
Acrylic acid	12.3	13.9	Ethyl formate	14.2	8.4
Allyl alcohol	10.2	14.3	2-Ethyl hexyl acrylate	9.0	15.0
Allyl bromide	14.4	9.6	Ethyl iodide	14.7	10.3
Allyl iodide	14.0	11.7	Ethyl propionate	13.2	9.9
Ammonia, 100%	12.6	2.0	Ethyl propyl ether	14.0	7.0
Ammonia, 26%	10.1	13.9	Ethyl sulfide	13.8	8.9
Amyl acetate	11.8	12.5	Ethylene bromide	11.9	15.7
Amyl alcohol	7.5	18.4	Ethylene chloride	12.7	12.2
Aniline	8.1	18.7	Ethylene glycol	6.0	23.6
Anisole	12.3	13.5	Ethylidene chloride	14.1	8.7
Arsenic trichloride	13.9	14.5	Fluorobenzene	13.7	10.4
Benzene	12.5	10.9	Formic acid	10.7	15.8
Brine, $CaCl_2$, 25%	6.6	15.9	Freon-11	14.4	9.0
Brine, NaCl, 25%	10.2	16.6	Freon-12	16.8	15.6
Bromine	14.2	13.2	Freon-21	15.7	7.5
Bromotoluene	20.0	15.9	Freon-22	17.2	4.7
Butyl acetate	12.3	11.0	Freon-113	12.5	11.4
Butyl acrylate	11.5	12.6	Glycerol, 100%	2.0	30.0
Butyl alcohol	8.6	17.2	Glycerol, 50%	6.9	19.6
Butyric acid	12.1	15.3	Heptane	14.1	8.4
Carbon dioxide	11.6	0.3	Hexane	14.7	7.0
Carbon disulfide	16.1	7.5	Hydrochloric acid, 31.5%	13.0	16.6
Carbon tetrachloride	12.7	13.1	Iodobenzene	12.8	15.9
Chlorobenzene	12.3	12.4	Isobutyl alcohol	7.1	18.0
Chloroform	14.4	10.2	Isobutyric acid	12.2	14.4
Chlorosulfonic acid	11.2	18.1	Isopropyl alcohol	8.2	16.0
Chlorotoluene, ortho	13.0	13.3	Isopropyl bromide	14.1	9.2
Chlorotoluene, meta	13.3	12.5	Isopropyl chloride	13.9	7.1
Chlorotoluene, para	13.3	12.5	Isopropyl iodide	13.7	11.2
Cresol, meta	2.5	20.8	Kerosene	10.2	16.9
Cyclohexanol	2.9	24.3	Linseed oil, raw	7.5	27.2
Cyclohexane	9.8	12.9	Mercury	18.4	16.4
Dibromomethane	12.7	15.8	Methanol, 100%	12.4	10.5
Dichloroethane	13.2	12.2	Methanol, 90%	12.3	11.8
Dichloromethane	14.6	8.9	Methanol, 40%	7.8	15.5
Diethyl ketone	13.5	9.2	Methyl acetate	14.2	8.2
Diethyl oxalate	11.0	16.4	Methyl acrylate	13.0	9.5
Diethylene glycol	5.0	24.7	Methyl *i*-butyrate	12.3	9.7
Diphenyl	12.0	18.3	Methyl *n*-butyrate	13.2	10.3
Dipropyl ether	13.2	8.6	Methyl chloride	15.0	3.8
Dipropyl oxalate	10.3	17.7	Methyl ethyl ketone	13.9	8.6
Ethyl acetate	13.7	9.1	Methyl formate	14.2	7.5
Ethyl acrylate	12.7	10.4	Methyl iodide	14.3	9.3
Ethyl alcohol, 100%	10.5	13.8	Methyl propionate	13.5	9.0
Methyl propyl ketone	14.3	9.5	Sodium hydroxide, 50%	3.2	25.8
Methyl sulfide	15.3	6.4	Stannic chloride	13.5	12.8
Naphthalene	7.9	18.1	Succinonitrile	10.1	20.8
Nitric acid, 95%	12.8	13.8	Sulfur dioxide	15.2	7.1
Nitric acid, 60%	10.8	17.0	Sulfuric acid, 110%	7.2	27.4
Nitrobenzene	10.6	16.2	Sulfuric acid, 100%	8.0	25.1
Nitrogen dioxide	12.9	8.6	Sulfuric acid, 98%	7.0	24.8
Nitrotoluene	11.0	17.0	Sulfuric acid, 60%	10.2	21.3
Octane	13.7	10.0	Sulfuryl chloride	15.2	12.4

(*Continued*)

TABLE A.18 Coordinates to be Used with Figure A.1—cont'd

Liquid	*X*	*Y*	Liquid	*X*	*Y*
Octyl alcohol	6.6	21.1	Tetrachloroethane	11.9	15.7
Pentachloroethane	10.9	17.3	Thiophene	13.2	11.0
Pentane	14.9	5.2	Titanium tetrachloride	14.4	12.3
Phenol	6.9	20.8	Toluene	13.7	10.4
Phosphorus tribromide	13.8	16.7	Trichloroethylene	14.8	10.5
Phosphorus trichloride	16.2	10.9	Triethylene glycol	4.7	24.8
Propionic acid	12.8	13.8	Turpentine	11.5	14.9
Propyl acetate	13.1	10.3	Vinyl acetate	14.0	8.8
Propyl alcohol	9.1	16.5	Vinyl toluene	13.4	12.0
Propyl bromide	14.5	9.6	Water	10.2	13.0
Propyl chloride	14.4	7.5	Xylene, ortho	13.5	12.1
Propyl formate	13.1	9.7	Xylene, meta	13.9	10.6
Propyl iodide	14.1	11.6	Xylene, para	13.9	10.9
Sodium	16.4	13.9			

Source: Ref. [11].

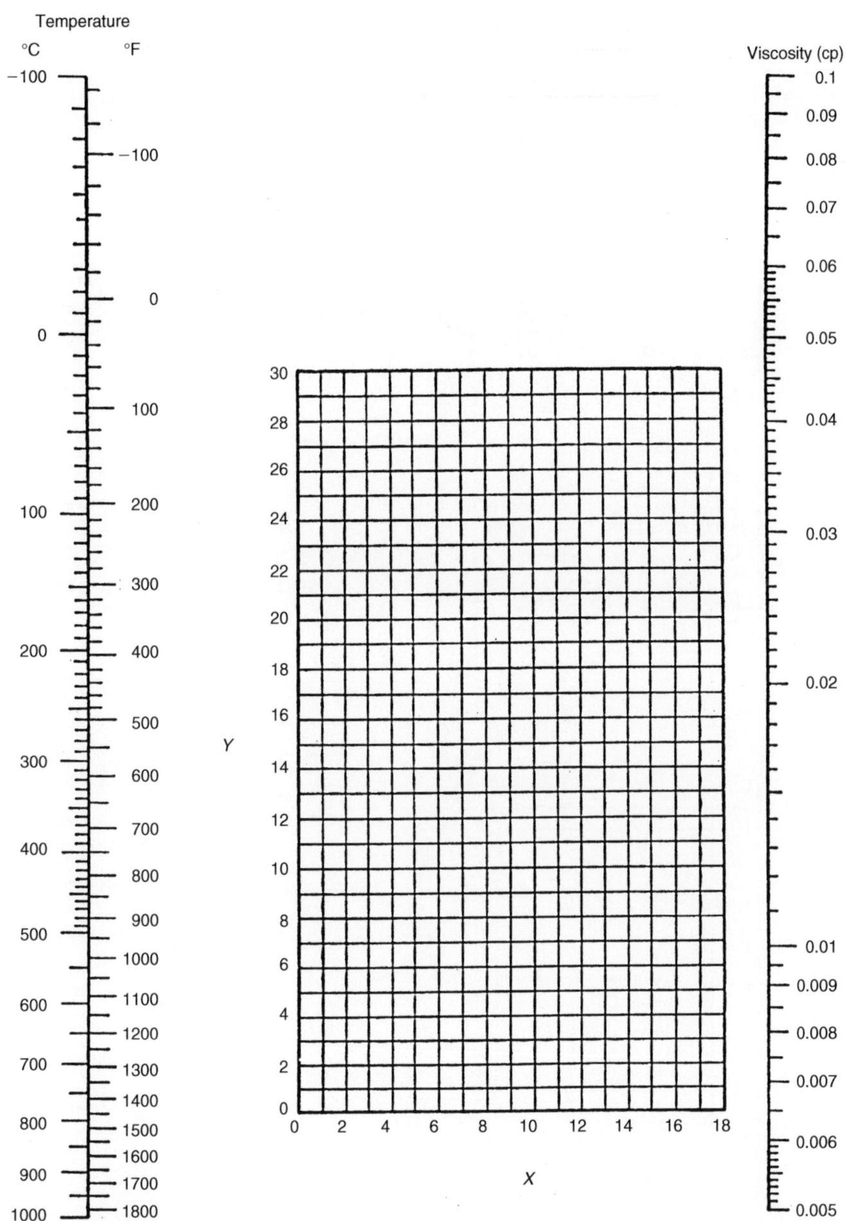

No.	Gas	X	Y	No.	Gas	X	Y	No.	Gas	X	Y	No.	Gas	X	Y
1	Acetic acid	7.7	14.3	15	Chloroform	8.9	15.7	29	Freon-113	11.3	14.0	43	Nitric oxide	10.9	20.5
2	Acetone	8.9	13.0	16	Cyanogen	9.2	15.2	30	Helium	10.9	20.5	44	Nitrogen	10.6	20.0
3	Acetylene	9.8	14.9	17	Cyclohexane	9.2	12.0	31	Hexane	8.6	11.8	45	Nitrosyl chloride	8.0	17.6
4	Air	11.0	20.0	18	Ethane	9.1	14.5	32	Hydrogen	11.2	12.4	46	Nitrous oxide	8.8	19.0
5	Ammonia	8.4	16.0	19	Ethyl acetate	8.5	13.2	33	$3H_2 + IN_2$	11.2	17.2	47	Oxygen	11.0	21.3
6	Argon	10.5	22.4	20	Ethyl alcohol	9.2	14.2	34	Hydrogen bromide	8.8	20.9	48	Pentane	7.0	12.8
7	Benzene	8.5	13.2	21	Ethyl chloride	8.5	15.6	35	Hydrogen chloride	8.8	18.7	49	Propane	9.7	12.9
8	Bromine	8.9	19.2	22	Ethyl ether	8.9	13.0	36	Hydrogen cyanide	9.8	14.9	50	Propyl alcohol	8.4	13.4
9	Butene	9.2	13.7	23	Ethylene	9.5	15.1	37	Hydrogen iodide	9.0	21.3	51	Propylene	9.0	13.8
10	Butylene	8.9	13.0	24	Fluorine	7.3	23.8	38	Hydrogen sulfide	8.6	18.0	52	Sulfur dioxide	9.6	17.0
11	Carbon dioxide	9.5	18.7	25	Freon-11	10.6	15.1	39	Iodine	9.0	18.4	53	Toluene	8.6	12.4
12	Carbon disulfide	8.0	16.0	26	Freon-12	11.1	16.0	40	Mercury	5.3	22.9	54	2,3,3-Trimethylbutane	9.5	10.5
13	Carbon monoxide	11.0	20.0	27	Freon-21	10.8	15.3	41	Methane	9.9	15.6	55	Water	8.0	16.0
14	Chlorine	9.0	18.4	28	Freon-22	10.1	17.0	42	Methyl alcohol	8.5	15.6	56	Xenon	9.3	23.0

FIGURE A.2 Viscosities of gases at atmospheric pressure (Source: Ref. [11]).

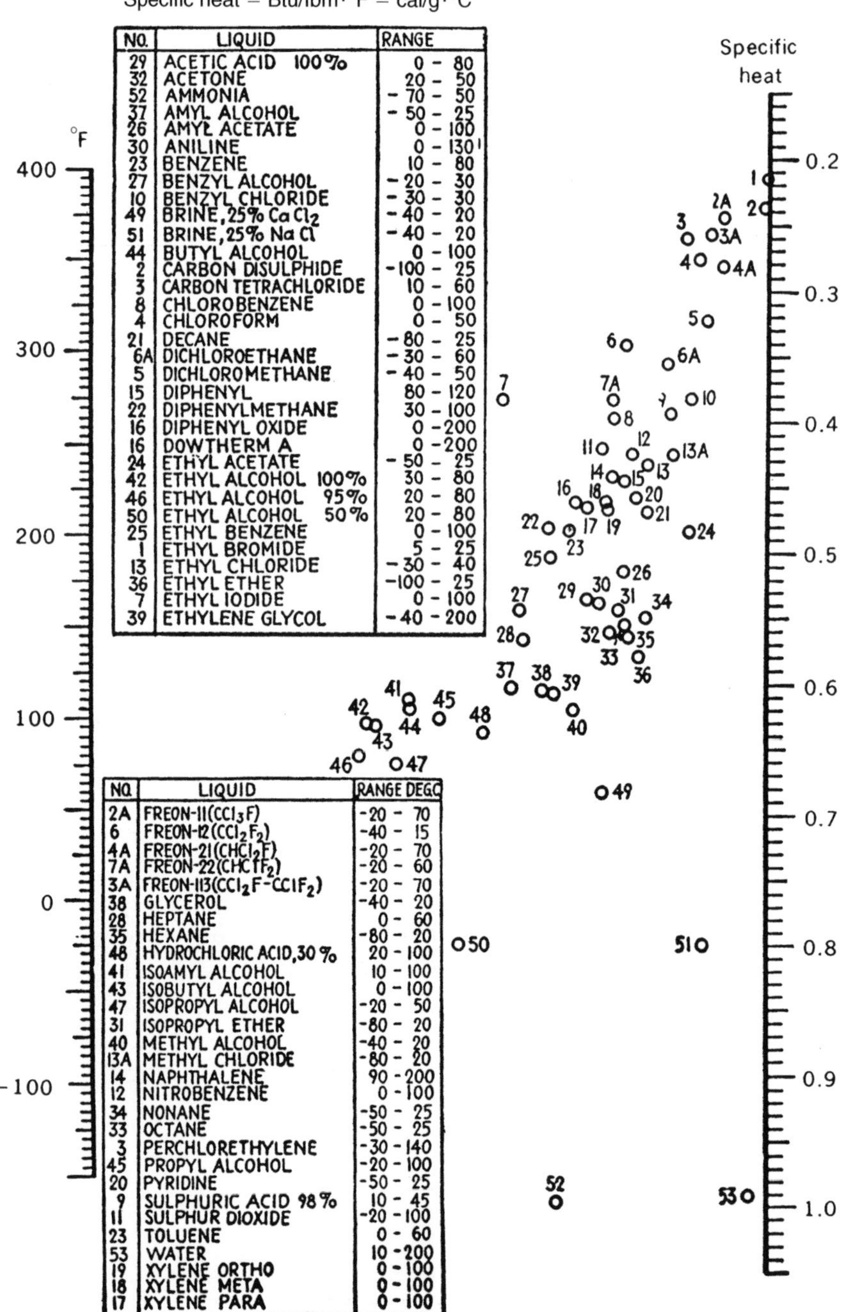

NO.	LIQUID	RANGE
29	ACETIC ACID 100%	0 - 80
32	ACETONE	20 - 50
52	AMMONIA	- 70 - 50
37	AMYL ALCOHOL	- 50 - 25
26	AMYL ACETATE	0 - 100
30	ANILINE	0 - 130
23	BENZENE	10 - 80
27	BENZYL ALCOHOL	- 20 - 30
10	BENZYL CHLORIDE	- 30 - 30
49	BRINE, 25% $CaCl_2$	- 40 - 20
51	BRINE, 25% NaCl	- 40 - 20
44	BUTYL ALCOHOL	0 - 100
2	CARBON DISULPHIDE	-100 - 25
3	CARBON TETRACHLORIDE	10 - 60
8	CHLOROBENZENE	0 - 100
4	CHLOROFORM	0 - 50
21	DECANE	- 80 - 25
6A	DICHLOROETHANE	- 30 - 60
5	DICHLOROMETHANE	- 40 - 50
15	DIPHENYL	80 - 120
22	DIPHENYLMETHANE	30 - 100
16	DIPHENYL OXIDE	0 - 200
16	DOWTHERM A	0 - 200
24	ETHYL ACETATE	- 50 - 25
42	ETHYL ALCOHOL 100%	30 - 80
46	ETHYL ALCOHOL 95%	20 - 80
50	ETHYL ALCOHOL 50%	20 - 80
25	ETHYL BENZENE	0 - 100
1	ETHYL BROMIDE	5 - 25
13	ETHYL CHLORIDE	- 30 - 40
36	ETHYL ETHER	-100 - 25
7	ETHYL IODIDE	0 - 100
39	ETHYLENE GLYCOL	- 40 - 200

NO.	LIQUID	RANGE DEG C
2A	FREON-11(CCl_3F)	-20 - 70
6	FREON-12(CCl_2F_2)	-40 - 15
4A	FREON-21($CHCl_2F$)	-20 - 70
7A	FREON-22($CHClF_2$)	-20 - 60
3A	FREON-113($CCl_2F-CClF_2$)	-20 - 70
38	GLYCEROL	-40 - 20
28	HEPTANE	0 - 60
35	HEXANE	-80 - 20
48	HYDROCHLORIC ACID, 30%	20 - 100
41	ISOAMYL ALCOHOL	10 - 100
43	ISOBUTYL ALCOHOL	0 - 100
47	ISOPROPYL ALCOHOL	-20 - 50
31	ISOPROPYL ETHER	-80 - 20
40	METHYL ALCOHOL	-40 - 20
13A	METHYL CHLORIDE	-80 - 20
14	NAPHTHALENE	90 - 200
12	NITROBENZENE	0 - 100
34	NONANE	-50 - 25
33	OCTANE	-50 - 25
3	PERCHLORETHYLENE	-30 - 140
45	PROPYL ALCOHOL	-20 - 100
20	PYRIDINE	-50 - 25
9	SULPHURIC ACID 98%	10 - 45
11	SULPHUR DIOXIDE	-20 - 100
23	TOLUENE	0 - 60
53	WATER	10 - 200
19	XYLENE ORTHO	0 - 100
18	XYLENE META	0 - 100
17	XYLENE PARA	0 - 100

FIGURE A.3 Specific heats of liquids (Source: Ref. [12]).

Appendix B: Dimensions of Pipe and Tubing

TABLE B.1 Dimensions of Heat Exchanger and Condenser Tubing (a) The Birmingham Wire Gage Scale

BWG	Thickness (in.)
7	0.180
8	0.165
10	0.134
11	0.120
12	0.109
13	0.095
14	0.083
15	0.072
16	0.065
17	0.058
18	0.049
20	0.035
22	0.028
24	0.022
26	0.018

TABLE B.1 Dimensions of Heat Exchanger and Condenser Tubing (b) Tubing Dimensions

Tube OD (in.)	BWG	Tube ID[a] (in.)	Internal Area[b] ($in.^2$)	External Surface Per Foot Length[c] (ft^2/ft)	OD/ID
1/2	16	0.370	0.1075	0.1309	1.351
	18	0.402	0.1269	0.1309	1.244
	20	0.430	0.1452	0.1309	1.163
	22	0.444	0.1548	0.1309	1.126
5/8	12	0.407	0.1301	0.1636	1.536
	13	0.435	0.1486	0.1636	1.437
	14	0.459	0.1655	0.1636	1.362
	15	0.481	0.1817	0.1636	1.299
	16	0.495	0.1924	0.1636	1.263
	17	0.509	0.2035	0.1636	1.228
	18	0.527	0.2181	0.1636	1.186
	19	0.541	0.2299	0.1636	1.155
	20	0.555	0.2419	0.1636	1.126
3/4	10	0.482	0.1825	0.1963	1.556
	11	0.510	0.2043	0.1963	1.471
	12	0.532	0.2223	0.1963	1.410
	13	0.560	0.2463	0.1963	1.339
	14	0.584	0.2679	0.1963	1.284
	15	0.606	0.2884	0.1963	1.238
	16	0.620	0.3019	0.1963	1.210
	17	0.634	0.3157	0.1963	1.183
	18	0.652	0.3339	0.1963	1.150
	20	0.680	0.3632	0.1963	1.103

(*Continued*)

TABLE B.1 Dimensions of Heat Exchanger and Condenser Tubing (b) Tubing Dimensions—cont'd

Tube OD (in.)	BWG	Tube ID[a] (in.)	Internal Area[b] ($in.^2$)	External Surface Per Foot Length[c] (ft^2/ft)	OD/ID
7/8	10	0.607	0.2894	0.2291	1.442
	11	0.635	0.3167	0.2291	1.378
	12	0.657	0.3390	0.2291	1.332
	13	0.685	0.3685	0.2291	1.277
	14	0.709	0.3948	0.2291	1.234
	15	0.731	0.4197	0.2291	1.197
	16	0.745	0.4359	0.2291	1.174
	17	0.759	0.4525	0.2291	1.153
	18	0.777	0.4742	0.2291	1.126
	20	0.805	0.5090	0.2291	1.087
1.0	8	0.670	0.3526	0.2618	1.493
	10	0.732	0.4208	0.2618	1.366
	11	0.760	0.4536	0.2618	1.316
	12	0.782	0.4803	0.2618	1.279
	13	0.810	0.5153	0.2618	1.235
	14	0.834	0.5463	0.2618	1.199
	15	0.856	0.5755	0.2618	1.168
	16	0.870	0.5945	0.2618	1.149
	18	0.902	0.6390	0.2618	1.109
	20	0.930	0.6793	0.2618	1.075
1.25	7	0.890	0.6221	0.3272	1.404
	8	0.920	0.6648	0.3272	1.359
	10	0.982	0.7574	0.3272	1.273
	11	1.010	0.8012	0.3272	1.238
	12	1.032	0.8365	0.3272	1.211
	13	1.060	0.8825	0.3272	1.179
	14	1.084	0.9229	0.3272	1.153
	16	1.120	0.9852	0.3272	1.116
	18	1.152	1.0423	0.3272	1.085
	20	1.180	1.0936	0.3272	1.059
1.5	10	1.232	1.1921	0.3927	1.218
	12	1.282	1.2908	0.3927	1.170
	14	1.334	1.3977	0.3927	1.124
	16	1.370	1.4741	0.3927	1.095
2.0	11	1.760	2.4328	0.5236	1.136
	12	1.782	2.4941	0.5236	1.122
	13	1.810	2.5730	0.5236	1.105
	14	1.834	2.6417	0.5236	1.091

[a] ID = OD − 2 × wall thickness from part (a) of table
[b] Internal area = $(\pi/4)(ID)^2$
[c] External surface per foot length = π (OD/12)

TABLE B.2 Properties of Steel Pipe

Nominal Pipe Size (in.)	Outside Diameter (in.)	Schedule No.	Wall Thickness (in.)	Inside Diameter (in.)	Cross-Sectional Area		Circumference (ft) or Surface (ft^2/ft of length)		Capacity at 1 ft/s Velocity		Weight of Plain-End Pipe (lb/ft)
					Metal ($in.^2$)	Flow (ft^2)	Outside	Inside	US gal/min	lb/h Water	
$\frac{1}{8}$	0.405	10S	0.049	0.307	0.055	0.00051	0.106	0.0804	0.231	115.5	0.19
		40ST, 40S	0.068	0.269	0.072	0.00040	0.106	0.0705	0.179	89.5	0.24
		80XS, 80S	0.095	0.215	0.093	0.00025	0.106	0.0563	0.113	56.5	0.31
$\frac{1}{4}$	0.540	10S	0.065	0.410	0.097	0.00092	0.141	0.107	0.412	206.5	0.33
		40ST, 40S	0.088	0.364	0.125	0.00072	0.141	0.095	0.323	161.5	0.42
		80XS, 80S	0.119	0.302	0.157	0.00050	0.141	0.079	0.224	112.0	0.54
$\frac{3}{8}$	0.675	10S	0.065	0.545	0.125	0.00162	0.177	0.143	0.727	363.5	0.42
		40ST, 40S	0.091	0.493	0.167	0.00133	0.177	0.129	0.596	298.0	0.57
		80XS, 80S	0.126	0.423	0.217	0.00098	0.177	0.111	0.440	220.0	0.74
$\frac{1}{2}$	0.840	5S	0.065	0.710	0.158	0.00275	0.220	0.186	1.234	617.0	0.54
		10S	0.083	0.674	0.197	0.00248	0.220	0.176	1.112	556.0	0.67
		40ST, 40S	0.109	0.622	0.250	0.00211	0.220	0.163	0.945	472.0	0.85
		80XS, 80S	0.147	0.546	0.320	0.00163	0.220	0.143	0.730	365.0	1.09
		160	0.188	0.464	0.385	0.00117	0.220	0.122	0.527	263.5	1.31
		XX	0.294	0.252	0.504	0.00035	0.220	0.066	0.155	77.5	1.71
$\frac{3}{4}$	1.050	5S	0.065	0.920	0.201	0.00461	0.275	0.241	2.072	1036.0	0.69
		10S	0.083	0.884	0.252	0.00426	0.275	0.231	1.903	951.5	0.86
		40ST, 40S	0.113	0.824	0.333	0.00371	0.275	0.216	1.665	832.5	1.13
		80XS, 80S	0.154	0.742	0.433	0.00300	0.275	0.194	1.345	672.5	1.47
		160	0.219	0.612	0.572	0.00204	0.275	0.160	0.917	458.5	1.94
		XX	0.308	0.434	0.718	0.00103	0.275	0.114	0.461	230.5	2.44
1	1.315	5S	0.065	1.185	0.255	0.00768	0.344	0.310	3.449	1725	0.87
		10S	0.109	1.097	0.413	0.00656	0.344	0.287	2.946	1473	1.40
		40ST, 40S	0.133	1.049	0.494	0.00600	0.344	0.275	2.690	1345	1.68
		80XS, 80S	0.179	0.957	0.639	0.00499	0.344	0.250	2.240	1120	2.17
		160	0.250	0.815	0.836	0.00362	0.344	0.213	1.625	812.5	2.84
		XX	0.358	0.599	1.076	0.00196	0.344	0.157	0.878	439.0	3.66
$1\frac{1}{4}$	1.660	5S	0.065	1.530	0.326	0.01277	0.435	0.401	5.73	2865	1.11
		10S	0.109	1.442	0.531	0.01134	0.435	0.378	5.09	2545	1.81
		40ST, 40S	0.140	1.380	0.668	0.01040	0.435	0.361	4.57	2285	2.27
		80XS, 80S	0.191	1.278	0.881	0.00891	0.435	0.335	3.99	1995	3.00
		160	0.250	1.160	1.107	0.00734	0.435	0.304	3.29	1645	3.76
		XX	0.382	0.896	1.534	0.00438	0.435	0.235	1.97	985	5.21
$1\frac{1}{2}$	1.900	5S	0.065	1.770	0.375	0.01709	0.497	0.463	7.67	3835	1.28
		10S	0.109	1.682	0.614	0.01543	0.497	0.440	6.94	3465	2.09
		40ST, 40S	0.145	1.610	0.800	0.01414	0.497	0.421	6.34	3170	2.72
		80XS, 80S	0.200	1.500	1.069	0.01225	0.497	0.393	5.49	2745	3.63
		160	0.281	1.338	1.429	0.00976	0.497	0.350	4.38	2190	4.86
		XX	0.400	1.100	1.885	0.00660	0.497	0.288	2.96	1480	6.41

(*Continued*)

TABLE B.2 Properties of Steel Pipe—cont'd

Nominal Pipe Size (in.)	Outside Diameter (in.)	Schedule No.	Wall Thickness (in.)	Inside Diameter (in.)	Cross-Sectional Area		Circumference (ft) or Surface (ft^2/ft of length)		Capacity at 1 ft/s Velocity		Weight of Plain-End Pipe (lb/ft)
					Metal (in.^{2})	Flow (ft^2)	Outside	Inside	US gal/min	lb/h Water	
2	2.375	5S	0.065	2.245	0.472	0.02749	0.622	0.588	12.34	6170	1.61
		10S	0.109	2.157	0.776	0.02538	0.622	0.565	11.39	5695	2.64
		40ST, 40S	0.154	2.067	1.075	0.02330	0.622	0.541	10.45	5225	3.65
		80ST, 80S	0.218	1.939	1.477	0.02050	0.622	0.508	9.20	4600	5.02
		160	0.344	1.687	2.195	0.01552	0.622	0.436	6.97	3485	7.46
		XX	0.436	1.503	2.656	0.01232	0.622	0.393	5.53	2765	9.03
$2\frac{1}{2}$	2.875	5S	0.083	2.709	0.728	0.04003	0.753	0.709	17.97	8985	2.48
		10S	0.120	2.635	1.039	0.03787	0.753	0.690	17.00	8500	3.53
		40ST, 40S	0.203	2.469	1.704	0.03322	0.753	0.647	14.92	7460	5.79
		80XS, 80S	0.276	2.323	2.254	0.02942	0.753	0.608	13.20	6600	7.66
		160	0.375	2.125	2.945	0.02463	0.753	0.556	11.07	5535	10.01
		XX	0.552	1.771	4.028	0.01711	0.753	0.464	7.68	3840	13.70
3	3.500	5S	0.083	3.334	0.891	0.06063	0.916	0.873	27.21	13,605	3.03
		10S	0.120	3.260	1.274	0.05796	0.916	0.853	26.02	13,010	4.33
		40ST, 40S	0.216	3.068	2.228	0.05130	0.916	0.803	23.00	11,500	7.58
		80XS, 80S	0.300	2.900	3.016	0.04587	0.916	0.759	20.55	10,275	10.25
		160	0.438	2.624	4.213	0.03755	0.916	0.687	16.86	8430	14.31
		XX	0.600	2.300	5.466	0.02885	0.916	0.602	12.95	6475	18.58
$3\frac{1}{2}$	4.0	5S	0.083	3.834	1.021	0.08017	1.047	1.004	35.98	17,990	3.48
		10S	0.120	3.760	1.463	0.07711	1.047	0.984	34.61	17,305	4.97
		40ST, 40S	0.226	3.548	2.680	0.06870	1.047	0.929	30.80	15,400	9.11
		80XS, 80S	0.318	3.364	3.678	0.06170	1.047	0.881	27.70	13,850	12.51
4	4.5	5S	0.083	4.334	1.152	0.10245	1.178	1.135	46.0	23,000	3.92
		10S	0.120	4.260	1.651	0.09898	1.178	1.115	44.4	22,200	5.61
		40ST, 40S	0.237	4.026	3.17	0.08840	1.178	1.054	39.6	19,800	10.79
		80XS, 80S	0.337	3.826	4.41	0.07986	1.178	1.002	35.8	17,900	14.98
		120	0.438	3.624	5.58	0.07170	1.178	0.949	32.2	16,100	18.98
		160	0.531	3.438	6.62	0.06647	1.178	0.900	28.9	14,450	22.52
		XX	0.674	3.152	8.10	0.05419	1.178	0.825	24.3	12,150	27.54
5	5.563	5S	0.109	5.345	1.87	0.1558	1.456	1.399	69.9	34,950	6.36
		10S	0.134	5.295	2.29	0.1529	1.456	1.386	68.6	34,300	7.77
		40ST, 40S	0.258	5.047	4.30	0.1390	1.456	1.321	62.3	31,150	14.62
		80XS, 80S	0.375	4.813	6.11	0.1263	1.456	1.260	57.7	28,850	20.78
		120	0.500	4.563	7.95	0.1136	1.456	1.195	51.0	25,500	27.04
		160	0.625	4.313	9.70	0.1015	1.456	1.129	45.5	22,750	32.96
		XX	0.750	4.063	11.34	0.0900	1.456	1.064	40.4	20,200	38.55
6	6.625	5S	0.109	6.407	2.23	0.2239	1.734	1.677	100.5	50,250	7.60
		10S	0.134	6.357	2.73	0.2204	1.734	1.664	98.9	49,450	9.29
		40ST, 40S	0.280	6.065	5.58	0.2006	1.734	1.588	90.0	45,000	18.97

		80XS, 80S	0.432	5.761	8.40	0.1810	1.734	1.508	81.1	40,550	28.57
		120	0.562	5.501	10.70	0.1650	1.734	1.440	73.9	36,950	36.42
		160	0.719	5.187	13.34	0.1467	1.734	1.358	65.9	32,950	45.34
		XX	0.864	4.897	15.64	0.1308	1.734	1.282	58.7	29,350	53.16
8	8.625	5S	0.109	8.407	2.915	0.3855	2.258	2.201	173.0	86,500	9.93
		10S	0.148	8.329	3.941	0.3784	2.258	2.180	169.8	84,900	13.40
		20	0.250	8.125	6.578	0.3601	2.258	2.127	161.5	80,750	22.36
		30	0.277	8.071	7.265	0.3553	2.258	2.113	159.4	79,700	24.70
		40ST, 40S	0.322	7.981	8.399	0.3474	2.258	2.089	155.7	77,850	28.55
		60	0.406	7.813	10.48	0.3329	2.258	2.045	149.4	74,700	35.66
		80XS, 80S	0.500	7.625	12.76	0.3171	2.258	1.996	142.3	71,150	43.39
		100	0.594	7.437	14.99	0.3017	2.258	1.947	135.4	67,700	50.93
		120	0.719	7.187	17.86	0.2817	2.258	1.882	126.4	63,200	60.69
		140	0.812	7.001	19.93	0.2673	2.258	1.833	120.0	60,000	67.79
		XX	0.875	6.875	21.30	0.2578	2.258	1.800	115.7	57,850	72.42
		160	0.906	6.813	21.97	0.2532	2.258	1.784	113.5	56,750	74.71
10	10.75	5S	0.134	10.842	4.47	0.5993	2.814	2.744	269.0	134,500	15.19
		10S	0.165	10.420	5.49	0.5922	2.814	2.728	265.8	132,900	18.65
		20	0.250	10.250	8.25	0.5731	2.814	2.685	257.0	128,500	28.04
		30	0.307	10.136	10.07	0.5603	2.814	2.655	252.0	126,000	34.24
		40ST, 40S	0.365	10.020	11.91	0.5475	2.814	2.620	246.0	123,000	40.48
		80S, 60XS	0.500	9.750	16.10	0.5185	2.814	2.550	233.0	116,500	54.74
		80	0.594	9.562	18.95	0.4987	2.814	2.503	223.4	111,700	64.40
		100	0.719	9.312	22.66	0.4729	2.814	2.438	212.3	106,150	77.00
		120	0.844	9.062	26.27	0.4479	2.814	2.372	201.0	100,500	89.27
		140, XX	1.000	8.750	30.63	0.4176	2.814	2.291	188.0	94,000	104.13
		160	1.125	8.500	34.02	0.3941	2.814	2.225	177.0	88,500	115.65
12	12.75	5S	0.156	12.438	6.17	0.8438	3.338	3.26	378.7	189,350	20.98
		10S	0.180	12.390	7.11	0.8373	3.338	3.24	375.8	187,900	24.17
		20	0.250	12.250	9.82	0.8185	3.338	3.21	367.0	183,500	33.38
		30	0.330	12.090	12.88	0.7972	3.338	3.17	358.0	179,000	43.77
		ST, 40S	0.375	12.000	14.58	0.7854	3.338	3.14	352.5	176,250	49.56
		40	0.406	11.938	15.74	0.7773	3.338	3.13	349.0	174,500	53.56
		XS, 80S	0.500	11.750	19.24	0.7530	3.338	3.08	338.0	169,000	65.42
		60	0.562	11.626	21.52	0.7372	3.338	3.04	331.0	165,500	73.22
		80	0.688	11.374	26.07	0.7056	3.338	2.98	316.7	158,350	88.57
		100	0.844	11.062	31.57	0.6674	3.338	2.90	299.6	149,800	107.29
		120, XX	1.000	10.750	36.91	0.6303	3.338	2.81	283.0	141,500	125.49
		140	1.125	10.500	41.09	0.6013	3.338	2.75	270.0	135,000	139.68
		160	1.312	10.126	47.14	0.5592	3.338	2.65	251.0	125,500	160.33
14	14	5S	0.156	13.688	6.78	1.0219	3.665	3.58	459	229,500	23.07
		10S	0.188	13.624	8.16	1.0125	3.665	3.57	454	227,000	27.73
		10	0.250	13.500	10.80	0.9940	3.665	3.53	446	223,000	36.71
		20	0.312	13.376	13.42	0.9750	3.665	3.50	438	219,000	45.68
		30, ST	0.375	13.250	16.05	0.9575	3.665	3.47	430	215,000	54.57
		40	0.438	13.124	18.66	0.9397	3.665	3.44	422	211,000	63.37
		XS	0.500	13.000	21.21	0.9218	3.665	3.40	414	207,000	72.09

(Continued)

TABLE B.2 Properties of Steel Pipe—cont'd

Nominal Pipe Size (in.)	Outside Diameter (in.)	Schedule No.	Wall Thickness (in.)	Inside Diameter (in.)	Cross-Sectional Area		Circumference (ft) or Surface (ft^2/ft of length)		Capacity at 1 ft/s Velocity		Weight of Plain-End Pipe (lb/ft)
					Metal ($in.^2$)	Flow (ft^2)	Outside	Inside	US gal/min	lb/h Water	
		60	0.594	12.812	25.02	0.8957	3.665	3.35	402	201,000	85.01
		80	0.750	12.500	31.22	0.8522	3.665	3.27	382	191,000	106.13
		100	0.938	12.124	38.49	0.8017	3.665	3.17	360	180,000	130.79
		120	1.094	11.812	44.36	0.7610	3.665	3.09	342	171,000	150.76
		140	1.250	11.500	50.07	0.7213	3.665	3.01	324	162,000	170.22
		160	1.406	11.188	55.63	0.6827	3.665	2.93	306	153,000	189.15
16	16	5S	0.165	15.670	8.21	1.3393	4.189	4.10	601	300,500	27.90
		10S	0.188	15.624	9.34	1.3314	4.189	4.09	598	299,000	31.75
		10	0.250	15.500	12.37	1.3104	4.189	4.06	587	293,500	42.05
		20	0.312	15.376	15.38	1.2985	4.189	4.03	578	289,000	52.36
		30, ST	0.375	15.250	18.41	1.2680	4.189	3.99	568	284,000	62.58
		40, XS	0.500	15.000	24.35	1.2272	4.189	3.93	550	275,000	82.77
		60	0.656	14.688	31.62	1.1766	4.189	3.85	528	264,000	107.54
		80	0.844	14.312	40.19	1.1171	4.189	3.75	501	250,500	136.58
		100	1.031	13.938	48.48	1.0596	4.189	3.65	474	237,000	164.86
		120	1.219	13.562	56.61	1.0032	4.189	3.55	450	225,000	192.40
		140	1.438	13.124	65.79	0.9394	4.189	3.44	422	211,000	223.57
		160	1.594	12.812	72.14	0.8953	4.189	3.35	402	201,000	245.22
18	18	5S	0.165	17.670	9.25	1.7029	4.712	4.63	764	382,000	31.43
		10S	0.188	17.624	10.52	1.6941	4.712	4.61	760	379,400	35.76
		10	0.250	17.500	13.94	1.6703	4.712	4.58	750	375,000	47.39
		20	0.312	17.376	17.34	1.6468	4.712	4.55	739	369,500	59.03
		ST	0.375	17.250	20.76	1.6230	4.712	4.52	728	364,000	70.59
		30	0.438	17.124	24.16	1.5993	4.712	4.48	718	359,000	82.06
		XS	0.500	17.000	27.49	1.5763	4.712	4.45	707	353,500	93.45
		40	0.562	16.876	30.79	1.5533	4.712	4.42	697	348,500	104.76
		60	0.750	16.500	40.64	1.4849	4.712	4.32	666	333,000	138.17
		80	0.938	16.124	50.28	1.4180	4.712	4.22	636	318,000	170.84
		100	1.156	15.688	61.17	1.3423	4.712	4.11	602	301,000	208.00
		120	1.375	15.250	71.82	1.2684	4.712	3.99	569	284,500	244.14

		140	1.562	14.876	80.66	1.2070	4.712	3.89	540	270,000	274.30
		160	1.781	14.438	90.75	1.1370	4.712	3.78	510	255,000	308.55
20	20	5S	0.188	19.624	11.70	2.1004	5.236	5.14	943	471,500	39.78
		10S	0.218	19.564	13.55	2.0878	5.236	5.12	937	467,500	46.06
		10	0.250	19.500	15.51	2.0740	5.236	5.11	930	465,000	52.73
		20, ST	0.375	19.250	23.12	2.0211	5.236	5.04	902	451,000	78.60
		30, XS	0.500	19.000	30.63	1.9689	5.236	4.97	883	441,500	104.13
		40	0.594	18.812	36.21	1.9302	5.236	4.92	866	433,000	123.06
		60	0.812	18.376	48.95	1.8417	5.236	4.81	826	413,000	166.50
		80	1.031	17.938	61.44	1.7550	5.236	4.70	787	393,500	208.92
		100	1.281	17.438	75.33	1.6585	5.236	4.57	744	372,000	256.15
		120	1.500	17.000	87.18	1.5763	5.236	4.45	707	353,500	296.37
		140	1.750	16.500	100.3	1.4849	5.236	4.32	665	332,500	341.10
		160	1.969	16.062	111.5	1.4071	5.236	4.21	632	316,000	379.14
24	24	5S	0.218	23.564	16.29	3.0285	6.283	6.17	1359	679,500	55.37
		10,10S	0.250	23.500	18.65	3.012	6.283	6.15	1350	675,000	63.41
		20, ST	0.375	23.250	27.83	2.948	6.283	6.09	1325	662,500	94.62
		XS	0.500	23.000	36.90	2.885	6.283	6.02	1295	642,500	125.49
		30	0.562	22.876	41.39	2.854	6.283	5.99	1281	640,500	140.80
		40	0.688	22.624	50.39	2.792	6.283	5.92	1253	626,500	171.17
		60	0.969	22.062	70.11	2.655	6.283	5.78	1192	596,000	238.29
		80	1.219	21.562	87.24	2.536	6.283	5.64	1138	569,000	296.53
		100	1.531	20.938	108.1	2.391	6.283	5.48	1073	536,500	367.45
		120	1.812	20.376	126.3	2.264	6.283	5.33	1016	508,000	429.50
		140	2.062	19.876	142.1	2.155	6.283	5.20	965	482,500	483.24
		160	2.344	19.312	159.5	2.034	6.283	5.06	913	456,500	542.09
30	30	5S	0.250	29.500	23.37	4.746	7.854	7.72	2130	1,065,000	79.43
		10,10S	0.312	29.376	29.10	4.707	7.854	7.69	2110	1,055,000	98.93
		ST	0.375	29.250	34.90	4.666	7.854	7.66	2094	1,048,000	118.65
		20, XS	0.500	29.000	46.34	4.587	7.854	7.59	2055	1,027,500	157.53
		30	0.625	28.750	57.68	4.508	7.854	7.53	2020	1,010,000	196.08

Source: Reprinted from ASME B36.10-1959 and B36.19-1965, by permission of the American Society of Mechanical Engineers;

TABLE B.3 Dimensions of Selected Radial Low-Fin Tubing (Type S/T Trufin©): 16 fpi Tubing

Catalog Number	Nominal Outside Diameter (in. (mm))	Average Plain End Wall Thickness (in. (mm))	Minimum wall Thickness Under Fin (in. (mm))	Finned Section Nominal Root Diameter (in. (mm))	Nominal Inside Diameter (in. (mm))	Actual Outside Area (ft^2/ft (m^2/m))	Area Ratio Actual Outside/ Actual Inside	Approximate Weight Per Unit Length (lb/ft (kg/m))
60-163049	1/2 (12.70)	0.065 (1.65)	0.044 (1.118)	0.375 (9.53)	0.277 (7.04)	0.261 (0.080)	3.60	0.288 (0.428)
60-164065	5/8 (15.88)	0.085 (2.16)	0.058 (1.473)	0.500 (12.70)	0.370 (9.40)	0.340 (0.104)	3.51	0.471 (0.701)
60-165065	3/4 (19.05)	0.085 (2.16)	0.058 (1.473)	0.625 (15.89)	0.495 (12.57)	0.418 (0.127)	3.23	0.478 (0.711)
60-165083	3/4 (19.05)	0.095 (2.41)	0.074 (1.880)	0.625 (15.89)	0.459 (11.66)	0.418 (0.127)	3.48	0.659 (0.981)
60-166065	7/8 (22.23)	0.085 (2.16)	0.058 (1.473)	0.750 (19.05)	0.620 (15.75)	0.496 (0.151)	3.06	0.681 (1.014)
60-166083	7/8 (22.23)	0.095 (2.41)	0.074 (1.880)	0.750 (19.05)	0.584 (14.83)	0.496 (0.151)	3.24	0.791 (1.176)
60-167083	1 (25.40)	0.095 (2.41)	0.074 (1.880)	0.875 (22.23)	0.709 (18.01)	0.574 (0.175)	3.09	0.899 (1.338)

For S/T Trufin® 16 fpi, the average fin height is 0.053 in. (1.35 mm) and the average fin width is 0.010 in. (0.254 mm).
Source: Wolverine Tube, Inc., www.wlv.com

TABLE B.4 Dimensions of Selected Radial Low-Fin Tubing (Type S/T Trufin®): 19 fpi Tubing

Catalog Number	Nominal Outside Diameter (in. (mm))	Average Plain End Wall Thickness (in. (mm))	Minimum Wall Thickness Under Fin (in. (mm))	Finned Section Nominal Root Diameter (in. (mm))	Nominal Inside Diameter (in. (mm))	Actual Outside Area (ft^2/ft (m^2/m))	Area Ratio Actual Outside/ Actual Inside	Approximate Weight Per Unit Length (lb/ft (kg/m))
60-193042	1/2 (12.70)	0.060 (1.52)	0.037 (0.940)	0.375 (9.53)	0.291 (7.39)	0.319 (0.097)	4.19	0.239 (0.355)
60-193049	1/2 (12.70)	0.065 (1.65)	0.044 (1.118)	0.375 (9.53)	0.277 (7.04)	0.319 (0.097)	4.40	0.284 (0.423)
60-193058	1/2 (12.70)	0.075 (1.91)	0.049 (1.245)	0.375 (9.53)	0.259 (6.58)	0.319 (0.097)	4.71	0.286 (0.425)
60-194049	5/8 (15.88)	0.065 (1.65)	0.044 (1.118)	0.500 (12.70)	0.402 (10.21)	0.414 (0.126)	3.94	0.362 (0.538)
60-194058	5/8 (15.88)	0.075 (1.91)	0.049 (1.245)	0.500 (12.70)	0.384 (9.75)	0.414 (0.126)	4.12	0.396 (0.589)
60-194065	5/8 (15.88)	0.085 (2.16)	0.058 (1.473)	0.500 (12.70)	0.370 (9.40)	0.414 (0.126)	4.27	0.456 (0.679)
60-194072	5/8 (15.88)	0.090 (2.29)	0.065 (1.651)	0.500 (12.70)	0.356 (9.04)	0.414 (0.126)	4.44	0.447 (0.665)
60-195049	3/4 (19.05)	0.065 (1.65)	0.044 (1.118)	0.625 (15.88)	0.527 (13.39)	0.507 (0.155)	3.67	0.447 (0.666)
60-195058	3/4 (19.05)	0.075 (1.91)	0.049 (1.245)	0.625 (15.88)	0.509 (12.93)	0.507 (0.155)	3.80	0.485 (0.722)
60-195065	3/4 (19.05)	0.085 (2.16)	0.058 (1.473)	0.625 (15.88)	0.495 (12.57)	0.507 (0.155)	3.91	0.568 (0.845)
60-195072	3/4 (19.05)	0.090 (2.29)	0.065 (1.651)	0.625 (15.88)	0.481 (12.22)	0.507 (0.155)	4.03	0.592 (0.881)
60-195083	3/4 (19.05)	0.095 (2.41)	0.074 (1.880)	0.625 (15.88)	0.459 (11.66)	0.507 (0.155)	4.22	0.618 (0.920)
60-196058	7/8 (22.23)	0.075 (1.91)	0.049 (1.245)	0.750 (19.05)	0.634 (16.10)	0.588 (0.179)	3.54	0.595 (0.885)
60-196065	7/8 (22.23)	0.085 (2.16)	0.058 (1.473)	0.750 (19.05)	0.620 (15.75)	0.588 (0.179)	3.62	0.675 (1.004)
60-196072	7/8 (22.23)	0.090 (2.29)	0.065 (1.651)	0.750 (19.05)	0.606 (15.39)	0.588 (0.179)	3.71	0.689 (1.024)
60-196083	7/8 (22.23)	0.095 (2.41)	0.074 (1.880)	0.750 (19.05)	0.584 (14.83)	0.588 (0.179)	3.85	0.762 (1.134)
60-196095	7/8 (22.23)	0.110 (2.79)	0.084 (2.134)	0.750 (19.05)	0.560 (14.22)	0.588 (0.179)	4.01	0.835 (1.243)
60-197058	1 (25.40)	0.075 (1.91)	0.049 (1.245)	0.875 (22.23)	0.759 (19.28)	0.695 (0.212)	3.50	0.697 (1.037)
60-197065	1 (25.40)	0.085 (2.16)	0.058 (1.473)	0.875 (22.23)	0.745 (18.92)	0.695 (0.212)	3.56	0.757 (1.126)
60-197072	1 (25.40)	0.090 (2.29)	0.065 (1.651)	0.875 (22.23)	0.731 (18.57)	0.695 (0.212)	3.63	0.813 (1.210)
60-197083	1 (25.40)	0.095 (2.41)	0.074 (1.880)	0.875 (22.23)	0.709 (18.01)	0.695 (0.212)	3.74	0.873 (1.299)
60-197095	1 (25.40)	0.110 (2.79)	0.084 (2.134)	0.875 (22.23)	0.685 (17.40)	0.695 (0.212)	3.88	0.988 (1.471)
60-197109	1 (25.40)	0.125 (3.18)	0.097 (2.464)	0.875 (22.23)	0.657 (16.69)	0.695 (0.212)	4.04	1.135 (1.688)

For S/T Trufin® 19 fpi, the minimum fin height is 0.050 in. (1.27 mm) and the average fin width is 0.011 in. (0.279 mm).

Source: Wolverine Tube, Inc., www.wlv.com

TABLE B.5 Dimensions of Selected Radial Low-Fin Tubing (Type 3 Fine-Fin®): 26 fpi Tubing

Part Number	Plain Section Average Diameter (in. (mm))	Plain Section Average Wall (in. (mm))	Wall Under Fin, Average (in. (mm))	Wall Under Fin, Minimum (in. (mm))	Nominal Root Diameter (in. (mm))	Fin Section ID (in. (mm))	Outside Area, A_o (ft²/ft (m²/m))	Inside Area, A_i (ft²/ft (m²/m))	Area Ratio A_o/A_i	ID Cross Sectional Area (in.² (cm²))
264028	0.625 (15.875)	0.049 (1.245)	0.028 (0.711)	0.025 (0.635)	0.527 (13.386)	0.471 (11.963)	0.492 (0.150)	0.123 (0.037)	4.000	0.174 (1.124)
264035	0.625 (15.875)	0.058 (1.473)	0.035 (0.889)	0.031 (0.787)	0.527 (13.386)	0.457 (11.608)	0.492 (0.150)	0.120 (0.037)	4.100	0.164 (1.058)
264042	0.625 (15.875)	0.065 (1.651)	0.042 (1.067)	0.037 (0.940)	0.527 (13.386)	0.443 (11.252)	0.492 (0.150)	0.116 (0.035)	4.241	0.154 (0.994)
264049	0.625 (15.875)	0.072 (1.829)	0.049 (1.245)	0.044 (1.118)	0.527 (13.386)	0.429 (10.897)	0.492 (0.150)	0.112 (0.034)	4.393	0.145 (0.933)
264065	0.625 (15.875)	0.083 (2.108)	0.065 (1.651)	0.058 (1.473)	0.527 (13.386)	0.397 (10.084)	0.492 (0.150)	0.104 (0.032)	4.731	0.124 (0.799)
265028	0.750 (19.050)	0.049 (1.245)	0.028 (0.711)	0.025 (0.635)	0.652 (16.561)	0.596 (15.138)	0.596 (0.182)	0.156 (0.048)	3.821	0.279 (1.800)
265035	0.750 (19.050)	0.058 (1.473)	0.035 (0.889)	0.031 (0.787)	0.652 (16.561)	0.582 (14.783)	0.596 (0.182)	0.152 (0.046)	3.921	0.266 (1.716)
265042	0.750 (19.050)	0.065 (1.651)	0.042 (1.067)	0.037 (0.940)	0.652 (16.561)	0.568 (14.427)	0.596 (0.182)	0.149 (0.045)	4.000	0.253 (1.635)
265049	0.750 (19.050)	0.072 (1.829)	0.049 (1.245)	0.044 (1.118)	0.652 (16.561)	0.554 (14.072)	0.596 (0.182)	0.145 (0.044)	4.110	0.241 (1.555)
265065	0.750 (19.050)	0.083 (2.108)	0.065 (1.651)	0.058 (1.473)	0.652 (16.561)	0.522 (13.259)	0.596 (0.182)	0.137 (0.042)	4.350	0.214 (1.381)
265083	0.750 (19.050)	0.109 (2.769)	0.083 (2.108)	0.074 (1.880)	0.652 (16.561)	0.486 (12.344)	0.596 (0.182)	0.127 (0.039)	4.693	0.186 (1.197)
266028	0.875 (22.225)	0.049 (1.245)	0.028 (0.711)	0.025 (0.635)	0.777 (19.736)	0.721 (18.313)	0.704 (0.215)	0.189 (0.058)	3.725	0.408 (2.634)
266035	0.875 (22.225)	0.058 (1.473)	0.035 (0.889)	0.031 (0.787)	0.777 (19.736)	0.707 (17.958)	0.704 (0.215)	0.185 (0.056)	3.805	0.393 (2.533)
266042	0.875 (22.225)	0.065 (1.651)	0.042 (1.067)	0.037 (0.940)	0.777 (19.736)	0.693 (17.602)	0.704 (0.215)	0.181 (0.055)	3.890	0.377 (2.433)
266049	0.875 (22.225)	0.072 (1.829)	0.049 (1.245)	0.044 (1.118)	0.777 (19.736)	0.679 (17.247)	0.704 (0.215)	0.178 (0.054)	3.955	0.362 (2.336)
266065	0.875 (22.225)	0.083 (2.108)	0.065 (1.651)	0.058 (1.473)	0.777 (19.736)	0.647 (16.434)	0.704 (0.215)	0.169 (0.052)	4.166	0.329 (2.121)
266083	0.875 (22.225)	0.109 (2.769)	0.083 (2.108)	0.074 (1.880)	0.777 (19.736)	0.611 (15.519)	0.704 (0.215)	0.160 (0.049)	4.400	0.293 (1.892)
267035	1.000 (25.400)	0.058 (1.473)	0.035 (0.889)	0.031 (0.787)	0.902 (22.911)	0.832 (21.133)	0.811 (0.247)	0.218 (0.066)	3.720	0.544 (3.508)
267042	1.000 (25.400)	0.065 (1.651)	0.042 (1.067)	0.037 (0.940)	0.902 (22.911)	0.818 (20.777)	0.811 (0.247)	0.214 (0.065)	3.790	0.526 (3.391)
267049	1.000 (25.400)	0.072 (1.829)	0.049 (1.245)	0.044 (1.118)	0.902 (22.911)	0.804 (20.422)	0.811 (0.247)	0.210 (0.064)	3.862	0.508 (3.275)
267065	1.000 (25.400)	0.083 (2.108)	0.065 (1.651)	0.058 (1.473)	0.902 (22.911)	0.772 (19.609)	0.811 (0.247)	0.202 (0.062)	4.015	0.468 (3.020)
267083	1.000 (25.400)	0.109 (2.769)	0.083 (2.108)	0.074 (1.880)	0.902 (22.911)	0.736 (18.694)	0.811 (0.247)	0.193 (0.059)	4.202	0.425 (2.745)

Type 3 26 FPI Fine-Fin® Materials: Ferritic Stainless Steel, Carbon Steel, Copper-Nickel Alloys, Monel. For Type 3 26 FPI Fine-Fin®, the average fin height is 0.049 in. (1.245 mm) and the average fin thickness is 0.013 in. (0.330 mm).
Source: High Performance Tube, Inc., www.highperformancetube.com

Appendix C: Tube-Count Tables

Source for this Appendix: Nooter Corporation, St. Louis, MO. Originally published in Ref. [11].

TABLE C.1 Tube Counts for 5/8-in. OD Tubes on 13/16-in. Square Pitch

	TEMA P or S				TEMA U		
	Number of Passes				Number of Passes		
Shell ID (in.)	1	2	4	6	2	4	6
8	55	48	34	24	52	40	32
10	88	78	62	56	90	80	74
12	140	138	112	100	140	128	108
13.25	178	172	146	136	180	164	148
15.25	245	232	208	192	246	232	216
17.25	320	308	274	260	330	312	292
19.25	405	392	352	336	420	388	368
21.25	502	484	442	424	510	488	460
23.25	610	584	536	508	626	596	562
25	700	676	618	600	728	692	644
27	843	812	742	716	856	816	780
29	970	942	868	840	998	956	920
31	1127	1096	1014	984	1148	1108	1060
33	1288	1250	1172	1148	1318	1268	1222
35	1479	1438	1330	1308	1492	1436	1388
37	1647	1604	1520	1480	1684	1620	1568
39	1840	1794	1700	1664	1882	1816	1754
42	2157	2112	2004	1968	2196	2136	2068
45	2511	2458	2326	2288	2530	2464	2402
48	2865	2808	2686	2656	2908	2832	2764
54	3656	3600	3462	3404	3712	3624	3556
60	4538	4472	4310	4256	4608	4508	4426

TABLE C.2 Tube Counts for 3/4-in. OD Tubes on 15/16-in. Triangular Pitch

	TEMA L or M				TEMA P or S				TEMA U		
	Number of Passes				Number of Passes				Number of Passes		
Shell ID (in.)	1	2	4	6	1	2	4	6	2	4	6
8	64	48	34	24	34	32	16	18	32	24	24
10	85	72	52	50	60	62	52	44	64	52	52
12	122	114	94	96	109	98	78	68	98	88	78
13.25	151	142	124	112	126	120	106	100	126	116	108
15.25	204	192	166	168	183	168	146	136	180	160	148
17.25	264	254	228	220	237	228	202	192	238	224	204
19.25	332	326	290	280	297	286	258	248	298	280	262
21.25	417	396	364	348	372	356	324	316	370	352	334
23.25	495	478	430	420	450	430	392	376	456	428	408
25	579	554	512	488	518	498	456	444	534	500	474
27	676	648	602	584	618	602	548	532	628	600	570
29	785	762	704	688	729	708	650	624	736	696	668
31	909	878	814	792	843	812	744	732	846	812	780
33	1035	1002	944	920	962	934	868	840	978	928	904

(Continued)

TABLE C.2 Tube Counts for 3/4-in. OD Tubes on 15/16-in. Triangular Pitch—cont'd

	TEMA L or M				TEMA P or S				TEMA U		
	Number of Passes				Number of Passes				Number of Passes		
Shell ID (in.)	1	2	4	6	1	2	4	6	2	4	6
35	1164	1132	1062	1036	1090	1064	990	972	1100	1060	1008
37	1304	1270	1200	1168	1233	1196	1132	1100	1238	1200	1152
39	1460	1422	1338	1320	1365	1346	1266	1244	1390	1336	1290
42	1703	1664	1578	1552	1611	1580	1498	1464	1632	1568	1524
45	1960	1918	1830	1800	1875	1834	1736	1708	1882	1820	1770
48	2242	2196	2106	2060	2132	2100	1998	1964	2152	2092	2044
54	2861	2804	2682	2660	2730	2684	2574	2536	2748	2680	2628
60	3527	3476	3360	3300	3395	3346	3228	3196	3420	3340	3286
66	4292	4228	4088	4044							
72	5116	5044	4902	4868							
78	6034	5964	5786	5740							
84	7005	6934	6766	6680							
90	8093	7998	7832	7708							
96	9203	9114	8896	8844							
108	11696	11618	11336	11268							
120	14459	14378	14080	13984							

TABLE C.3 Tube Counts for 3/4-in. OD Tubes on 1-in. Square Pitch

	TEMA P or S				TEMA U		
	Number of Passes				Number of Passes		
Shell ID (in.)	1	2	4	6	2	4	6
8	28	26	16	12	28	24	12
10	52	48	44	24	52	44	32
12	80	76	66	56	78	72	70
$13^1/_4$	104	90	70	80	96	92	90
$15^1/_4$	136	128	128	114	136	132	120
$17^1/_4$	181	174	154	160	176	176	160
$19^1/_4$	222	220	204	198	224	224	224
$21^1/_4$	289	272	262	260	284	280	274
$23^1/_4$	345	332	310	308	348	336	328
25	398	386	366	344	408	392	378
27	477	456	432	424	480	468	460
29	554	532	510	496	562	548	530
31	637	624	588	576	648	636	620
33	730	712	682	668	748	728	718
35	828	812	780	760	848	820	816
37	937	918	882	872	952	932	918
39	1048	1028	996	972	1056	1044	1020
42	1224	1200	1170	1140	1244	1224	1212
45	1421	1394	1350	1336	1436	1408	1398
48	1628	1598	1548	1536	1640	1628	1602
54	2096	2048	2010	1992	2108	2084	2068
60	2585	2552	2512	2476	2614	2584	2558

TABLE C.4 Tube Counts for 3/4-in. OD Tubes on 1-in. Triangular Pitch

	TEMA L or M				TEMA P or S				TEMA U		
	Number of Passes				Number of Passes				Number of Passes		
Shell ID (in.)	1	2	4	6	1	2	4	6	2	4	6
8	42	40	26	24	31	26	16	12	32	24	24
10	73	66	52	44	56	48	42	40	52	48	40
12	109	102	88	80	88	78	62	68	84	76	74
$13^1/_4$	136	128	112	102	121	106	94	88	110	100	98
$15^1/_4$	183	172	146	148	159	148	132	132	152	140	136
$17^1/_4$	237	228	208	192	208	198	182	180	206	188	182
$19^1/_4$	295	282	258	248	258	250	228	220	266	248	234
$21^1/_4$	361	346	318	320	320	314	290	276	330	316	296
$23^1/_4$	438	416	382	372	400	384	352	336	400	384	356
25	507	486	448	440	450	442	400	392	472	440	424
27	592	574	536	516	543	530	488	468	554	528	502
29	692	668	632	604	645	618	574	556	648	616	588
31	796	774	732	708	741	716	666	648	744	716	688
33	909	886	836	812	843	826	760	740	852	816	788
35	1023	1002	942	920	950	930	878	856	974	932	908
37	1155	1124	1058	1032	1070	1052	992	968	1092	1056	1008
39	1277	1254	1194	1164	1209	1184	1122	1096	1224	1180	1146
42	1503	1466	1404	1372	1409	1378	1314	1296	1434	1388	1350
45	1726	1690	1622	1588	1635	1608	1536	1504	1652	1604	1560
48	1964	1936	1870	1828	1887	1842	1768	1740	1894	1844	1794
54	2519	2466	2380	2352	2399	2366	2270	2244	2426	2368	2326
60	3095	3058	2954	2928	2981	2940	2832	2800	3006	2944	2884
66	3769	3722	3618	3576							
72	4502	4448	4324	4280							
78	5309	5252	5126	5068							
84	6162	6108	5964	5900							
90	7103	7040	6898	6800							
96	8093	8026	7848	7796							
108	10260	10206	9992	9940							
120	12731	12648	12450	12336							

TABLE C.5 Tube Counts for 1-in. OD Tubes on 1.25-in. Square Pitch

	TEMA P or S				TEMA U		
	Number of Passes				Number of Passes		
Shell ID (in.)	1	2	4	6	2	4	6
8	17	12	8	12	14	8	6
10	30	30	16	18	30	24	12
12	52	48	42	24	44	40	32
$13^1/_4$	61	56	52	50	60	48	44
$15^1/_4$	85	78	62	64	80	72	74
$17^1/_4$	108	108	104	96	104	100	100
$19^1/_4$	144	136	130	114	132	132	120
$21^1/_4$	173	166	154	156	172	168	148
$23^1/_4$	217	208	194	192	212	204	198
25	252	240	230	212	244	240	230
27	296	280	270	260	290	284	274
29	345	336	310	314	340	336	328
31	402	390	366	368	400	384	372
33	461	452	432	420	456	444	440
35	520	514	494	484	518	504	502
37	588	572	562	548	584	576	566
39	661	640	624	620	664	644	640
42	776	756	738	724	764	748	750
45	900	882	862	844	902	880	862
48	1029	1016	984	972	1028	1008	1004
54	1310	1296	1268	1256	1320	1296	1284
60	1641	1624	1598	1576	1634	1616	1614

TABLE C.6 Tube Counts for 1-in. OD Tubes on 1.25-in. Triangular Pitch

	TEMA L or M				TEMA P or S				TEMA U		
	Number of Passes				Number of Passes				Number of Passes		
Shell ID (in.)	1	2	4	6	1	2	4	6	2	4	6
8	27	26	8	12	18	14	8	12	14	12	6
10	42	40	34	24	33	28	16	18	28	24	24
12	64	66	52	44	51	48	42	44	52	40	40
$13^1/_4$	81	74	62	56	73	68	52	44	64	56	52
$15^1/_4$	106	106	88	92	93	90	78	76	90	80	78
$17^1/_4$	147	134	124	114	126	122	112	102	122	112	102
$19^1/_4$	183	176	150	152	159	152	132	136	152	140	136
$21^1/_4$	226	220	204	186	202	192	182	172	196	180	176
$23^1/_4$	268	262	236	228	249	238	216	212	242	224	216
25	316	302	274	272	291	278	250	240	286	264	246
27	375	360	336	324	345	330	298	288	340	320	300
29	430	416	390	380	400	388	356	348	400	380	352
31	495	482	452	448	459	450	414	400	456	436	414
33	579	554	520	504	526	514	484	464	526	504	486
35	645	622	586	576	596	584	548	536	596	572	548
37	729	712	662	648	672	668	626	608	668	636	614
39	808	792	744	732	756	736	704	692	748	728	700
42	947	918	874	868	890	878	834	808	890	856	830
45	1095	1068	1022	1000	1035	1008	966	948	1028	992	972
48	1241	1220	1176	1148	1181	1162	1118	1092	1180	1136	1100
54	1577	1572	1510	1480	1520	1492	1436	1416	1508	1468	1442
60	1964	1940	1882	1832	1884	1858	1800	1764	1886	1840	1794
66	2390	2362	2282	2260							
72	2861	2828	2746	2708							
78	3368	3324	3236	3216							
84	3920	3882	3784	3736							
90	4499	4456	4370	4328							
96	5144	5104	4986	4936							
108	6546	6494	6360	6300							
120	8117	8038	7870	7812							

TABLE C.7 Tube Counts for 1.25-in. OD Tubes on $1^9/_{16}$-in. Square Pitch

	TEMA P or S				TEMA U		
	Number of Passes				Number of Passes		
Shell ID (in.)	1	2	4	6	2	4	6
8	12	12	4	0	4	4	6
10	21	12	8	12	12	8	12
12	29	28	16	18	26	20	12
$13^1/_4$	38	34	34	24	36	28	15
$15^1/_4$	52	48	44	48	44	44	32
$17^1/_4$	70	66	56	50	60	60	56
$19^1/_4$	85	84	70	80	82	76	79
$21^1/_4$	108	108	100	96	100	100	100
$23^1/_4$	136	128	128	114	128	120	120
25	154	154	142	136	154	148	130
27	184	180	158	172	176	172	160
29	217	212	204	198	212	204	198
31	252	248	234	236	242	240	234
33	289	276	270	264	280	280	274
35	329	316	310	304	324	312	308
37	372	368	354	340	358	352	350
39	420	402	402	392	408	400	392
42	485	476	468	464	480	476	464
45	565	554	546	544	558	548	550
48	653	636	628	620	644	628	632
54	837	820	812	804	824	808	808
60	1036	1028	1012	1008	1028	1016	1008

TABLE C.8 Tube Counts for 1.25-in. OD Tubes on $1^{9}/_{16}$-in. Triangular Pitch

	TEMA L or M				TEMA P or S				TEMA U		
	Number of Passes				Number of Passes				Number of Passes		
Shell ID (in.)	1	2	4	6	1	2	4	6	2	4	6
8	15	10	8	12	13	10	4	0	6	4	6
10	27	22	16	12	18	20	8	12	14	12	12
12	38	36	26	24	33	26	26	18	28	20	18
$13^{1}/_{4}$	55	44	42	40	38	44	34	24	34	28	30
$15^{1}/_{4}$	66	64	52	50	57	58	48	44	52	48	40
$17^{1}/_{4}$	88	82	78	68	81	72	62	68	72	68	64
$19^{1}/_{4}$	117	106	98	96	100	94	86	80	90	84	78
$21^{1}/_{4}$	136	134	124	108	126	120	116	102	118	112	102
$23^{1}/_{4}$	170	164	146	148	159	146	132	132	148	132	120
25	198	188	166	168	183	172	150	148	172	160	152
27	237	228	208	192	208	206	190	180	200	188	180
29	268	266	242	236	249	238	224	220	242	228	216
31	312	304	284	276	291	282	262	256	282	264	250
33	357	346	322	324	333	326	298	296	326	308	292
35	417	396	372	364	372	368	344	336	362	344	336
37	446	446	422	408	425	412	394	384	416	396	384
39	506	490	472	464	478	468	442	432	472	444	428
42	592	584	552	544	558	546	520	512	554	524	510
45	680	676	646	632	646	634	606	596	636	624	592
48	788	774	736	732	748	732	704	696	736	708	692
54	1003	980	952	928	962	952	912	892	946	916	890
60	1237	1228	1188	1152	1194	1182	1144	1116	1176	1148	1116
66	1520	1496	1448	1424							
72	1814	1786	1736	1724							
78	2141	2116	2068	2044							
84	2507	2470	2392	2372							
90	2861	2840	2764	2744							
96	3275	3246	3158	3156							
108	4172	4136	4046	4020							
120	5164	5128	5038	5000							

Appendix D: Equivalent Lengths of Pipe Fittings

Source for this appendix: Ref. [13].

TABLE D.1 Equivalent Lengths (ft) of Elbows, Tees, and Bends

	90° Elbows*		90° Bends*		Tees	
Nominal Pipe Size (in.)	Short Radius, $R = 1\,D$	Long Radius, $R = 1.5\,D$	$R = 5\,D$	$R = 10\,D$	Flow-Through Branch	Flow-Through
$1^1/_2$	4.5	3	2.5	4	8	3
2	5.25	3.5	3	5	11	3.5
$2^1/_2$	6	4	3.5	6	13	4
3	7.5	5	4	7.5	16	5
4	10.5	7	5.5	10	20	7
6	15	10	8.5	15	30	10
8	21	14	11	20	40	14
10	24	16	14	25	50	16
12	32	21	16	30	60	21
14	33	22	19	33	65	22
16	39	26	21	38	75	26
18	44	29	24	42	86	29
20	48	32	27	50	100	32
24	57	38	32	60	120	38

For 180° returns, double the tabulated values.
*For 45° elbows and bends, estimate 50% of tabulated values.

TABLE D.2 Equivalent Lengths (ft) for Valves

		Globe* Fully Open Bevel or Plug Seat			Check			Three-Way Cock[2]		
Nominal Pipe Size (in.)	Gate Fully Open	90°	60°	45°	Swing	Ball	Straight-Through Cock[1]	Straight-Through Flows	Flow-Through Branch	Butterfly Fully Open
$1^1/_2$	1.75	46	23	18	17	20	2.5	6	20	6
2	2.25	60	30	24	22	25	3.5	7.5	24	8
$2^1/_2$	2.75	70	38	30	27	30	4	9	30	10
3	3.5	90	45	38	35	38	5	12	36	12
4	4.5	120	60	48	45	50	6.5	15	48	15
6	6.5	175	88	72	65	75	10	22	70	23
8	9	230	120	95	90	100	13	30	95	27
10	12	280	150	130	120	130	16	38	120	35
12	14	320	170	145	140	150	19			40
14	15	380	190	160	150	170	20			45
16	17	420	220	180	170	190	22			50
18	18	480	250	205	180	210	24			58
20	20	530	290	240	200	240	27			64
24	32	630	330	270	250	290	33			78

*For partially closed globe valves, multiply tabulated values by 3 for three-quarters open, by 12 for one-half open, and by 70 for one-quarter open.
[1]With port area open. Port area = pipe area.
[2]Port area equals 80% of pipe area.

TABLE D.3 Equivalent Lengths (ft) of Inlets and Outlets

Resistance Coefficient	$K = 1.0$	$K = 0.78$	$K = 0.5$	$K = 0.23$
Nominal Pipe Size (in.)	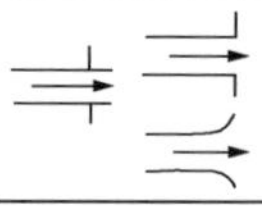	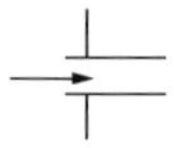	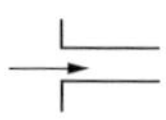	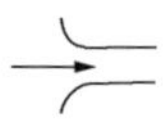
$^1/_2$	2	1.5	1	0.5
$^3/_4$	3	2.5	1.5	0.75
1	4	3	2	1
$1^1/_2$	7	5.5	3.5	1.75
2	9	7	4.5	2.25
3	15	12	7.5	3.75
4	20	16	10	5
6	36	29	18	9
8	48	38	24	12
10	62	49	31	15
12	78	60	39	19
14	88	70	44	22
16	100	78	50	25
18	120	95	60	30
20	136	107	68	34
24	170	135	85	42

TABLE D.4 Equivalent Lengths (ft) of Expanders and Reducers

Nominal Sizes (in.)			
D_1	D_2	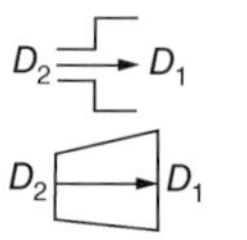 $D_2 \rightarrow D_1$	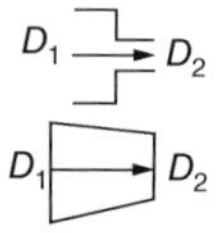 $D_1 \rightarrow D_2$
$^3/_4$	$^1/_2$	0.6	0.5
1	$^1/_2$	1.2	0.7
	$^3/_4$	0.6	0.6
$1^1/_2$	$^3/_4$	1.6	1.0
	1	1.2	0.9
2	1	2.2	1.3
	$1^1/_2$	1.3	1.3
3	$1^1/_2$	3.8	2.4
	2	2.7	2.3
4	2	5	3.2
	3	3	3
6	3	8	5
	4	4	4
8	4	12	7
	6	7	7
10	4	15	8
	6	14	9.5
	8	6	6
12	6	19	12
	8	14	12
	10	6.5	6.5
14	6	22	14
	8	22	14
	10	15	13
	12	6	6
16	8	27	17
	10	23	17
	12	15	15
	14	7	7
18	10	30	19
	12	23	19
	14	15	15
	16	4	4
20	12	30	23
	14	21	23
	16	13	13
	18	5	5
24	18	25	25
	20	12	12

Note: Add these equivalent lengths to the equivalent length of the smaller pipe and its components.

Appendix E: Properties of Petroleum Streams

Crude oils and petroleum fractions are complex mixtures containing large numbers of chemical species, primarily hydrocarbons. Methods have been devised for correlating the physical properties of these mixtures in terms of readily measurable parameters. Flowsheet simulators such as HEXTRAN have built-in correlations that are used to estimate the physical properties needed for simulation purposes. Assay data in the form of ASTM boiling curves can be used as input for petroleum streams. However, physical properties can also be estimated with just two pieces of information, namely, the standard density and the average normal boiling point of the liquid. This information is usually supplied in the form of two parameters commonly used in petroleum correlations, the API gravity and the Watson characterization factor (also called the UOP characterization factor).

The API gravity is defined as follows:

$$^{\circ}API = \frac{141.5}{s} - 131.5 \tag{E.1}$$

where s is the liquid specific gravity of the material at 60°F referenced to water at 60°F. The inverse relation is:

$$s = \frac{141.5}{^{\circ}API + 131.5} \tag{E.2}$$

The API gravity ranges from less than zero for heavy residual oils to 340° for methane (liquid specific gravity of 0.30 at 60°F). A range of 10 to 70° is typical of many petroleum liquids. Note that a specific gravity of 1.0 corresponds to 10° API. Fluids that are denser than water have lower API gravities, while a value above 10° API indicates that the liquid hydrocarbon is less dense than water.

The Watson (or UOP) characterization factor, K_w, is defined as follows:

$$K_w = \frac{T_B^{1/3}}{s} \tag{E.3}$$

where

T_b = average normal boiling point (°R)
s = liquid specific gravity at 60°F

Values of K_w typically fall in the range 10 to 13, although lower and higher values also occur. The value of the characterization factor is an indication of the aromatic content of the hydrocarbon mixture, with lower values generally corresponding to greater aromatic content. Benzene, for example, has a characterization factor of 10.0, while hexane has a value of 12.8.

Example E. 1

A 40° API hydrocarbon mixture has a Watson characterization factor of 12.0. Use HEXTRAN to estimate the liquid and vapor properties of the mixture at 350°F and 500°F and a pressure of 50 psia.

Solution

Two runs are made, one each for liquid and vapor properties. A flowsheet is constructed consisting of a feed stream, a heater, and a product stream. The feed is defined as a bulk property stream, and values of the API gravity and characterization factor are entered on the Specifications form, along with the stream temperature (350°F) and pressure (50 psia). The flow rate is (arbitrarily) specified as 1000 lb/h of liquid (run 1) or vapor (run 2). The outlet temperature of the heater is specified to be 500°F and the pressure drop is set to zero. After running the simulations, the following values are obtained from the output files.

	Liquid		Vapor	
Property	350°F	500°F	350°F	500°F
ρ (lbm/ft^3)	43.674	39.227	1.909	1.278
C_P (Btu/lbm · °F)	0.630	0.715	0.523	0.599
k (Btu/h · ft · °F)	0.0561	0.0471	0.0114	0.0160
μ (cp)	0.416	0.226	0.00283	0.00399
σ (dyne/cm)	16.98	10.79	–	–

References

[1] Kreith F, Bohn MS. Principles of Heat Transfer. 6th ed. Pacific Grove, CA: Brooks/Cole; 2001.
[2] Eckert ERG, Drake RM. Analysis of Heat and Mass Transfer. New York: McGraw-Hill; 1972.
[3] Raznjevic K. Handbook of Thermodynamic Tables and Charts. New York: Hemisphere Publishing Corp.; 1976.
[4] Touloukian YS, editor. Thermophysical Properties of High Temperature Solid Materials, vol. 1: Elements. New York: The MacMillan Company; 1967.
[5] Holman JP. Heat Transfer. 7th ed. New York: McGraw-Hill; 1990.
[6] Brown AI, Marco SM. Introduction to Heat Transfer. 3rd ed. New York: McGraw-Hill; 1958.
[7] Anonymous. ASME Steam Tables. New York: American Society of Mechanical Engineers; 1967.
[8] Anonymous. Flow of Fluids Through Valves, Fittings and Pipe, Technical Paper 410. New York: Crane Company; 1988.
[9] Incropera FP, DeWitt DP. Introduction to Heat Transfer. 4th ed. New York: Wiley; 2002.
[10] Vargaftik NB. Tables of Thermophysical Properties of Liquids and Gases. 2nd ed. New York: Hemisphere Publishing Corp.; 1975.
[11] Perry RH, Chilton CH, editors. Chemical Engineers' Handbook. 5th ed. New York: McGraw-Hill; 1973.
[12] McCabe WL, Smith JC, Harriott P. Unit Operations of Chemical Engineering. 4th ed. New York: McGraw-Hill; 1985.
[13] Kern R. How to compute pipe size. Chem Eng 1975;82(No. 1):115–20.

INDEX

Note: Page numbers followed by "f" denote figures; "t" tables.

E

F

G

H

I

T

U

V

W

X

Z

Printed and bound by CPI Group (UK) Ltd, Croydon, CR0 4YY
14/05/2025
01871118-0001